9(0)

$99.95 per copy, (In United States).
Price subject to change without prior notice.

0175

RSMeans

Residential Cost Data

24th Annual Edition

D1541065

2005

RSMeans
Construction Publishers & Consultants
63 Smiths Lane
Kingston, MA 02364-0800
(781) 585-7880

Copyright©2004 by Reed Construction Data, Inc.
All rights reserved.

Printed in the United States of America
ISSN 0738-1239
ISBN 0-87629-758-0

Reed Construction Data's RSMeans product line, its authors, editors and engineers, apply diligence and judgement in locating and using reliable sources for the information published. **However, RSMeans makes no express or implied warranty or guarantee in connection with the content of the information contained herein, including the accuracy, correctness, value, sufficiency, or completeness of the data, methods and other information contained herein. RSMeans makes no express or implied warranty of merchantability or fitness for a particular purpose.** RSMeans shall have no liability to any customer or third party for any loss, expense, or damage including consequential, incidental, special or punitive damage, including losts profits or lost revenue, caused directly or indirectly by any error or omission, or arising out of, or in connection with, the information contained herein.

No part of this publication may be reproduced, stored in a retrieval system, or transmitted in any form or by any means without prior written permission of Reed Construction Data.

9780876297582

Senior Editor
Robert W. Mewis, CCC

Contributing Editors
Barbara Balboni
Robert A. Bastoni
John H. Chiang, PE
Robert J. Kuchta
Robert C. McNichols
Melville J. Mossman, PE
John J. Moylan
Jeannene D. Murphy
Stephen C. Plotner
Michael J. Regan
Eugene R. Spencer
Marshall J. Stetson
Phillip R. Waier, PE

Senior Engineering Operations Manager
John H. Ferguson, PE

Senior Vice President & General Manager
John Ware

Product Manager
Jane Crudden

Vice President of Sales
John M. Shea

Production Manager
Michael Kokernak

Production Coordinator
Marion E. Schofield

Technical Support
Thomas J. Dion
Jonathan Forgit
Mary Lou Geary
Gary L. Hoitt
Alice McSharry
Paula Reale-Camelio
Robin Richardson
Kathryn S. Rodriguez
Sheryl A. Rose

Book & Cover Design
Norman R. Forgit

Editorial Advisory Board
James E. Armstrong, CPE, CEM
Program Manager,
Energy Conservation
NSTAR

William R. Barry, CCC
Cost Consultant

Robert F. Cox, PhD
Assistant Professor
ME Rinker Sr. School of Bldg. Constr.
University of Florida

Roy F. Gilley, AIA
Principal
Gilley-Hinkel Architects

Kenneth K. Humphreys, PhD, PE, CCE
Secretary-Treasurer
International Cost
Engineering Council

Patricia L. Jackson, PE
President
Jackson A&E Assoc., Inc.

Martin F. Joyce
Executive Vice President
Bond Brothers, Inc.

This book is recyclable.

This book is printed on recycled stock.

Reed Construction Data

First Printing

Foreword

RSMeans is a product line of Reed Construction Data, Inc., a leading provider of construction information, products, and services in North America and globally. Reed Construction Data's project information products include more than 100 regional editions, national construction data, sales leads, and local plan rooms in major business centers. Reed Construction Data's PlansDirect provides surveys, plans and specifications. The First Source suite of products consists of *First Source for Products,* SPEC-DATA™, MANU-SPEC™, CADBlocks, Manufacturer Catalogs and First Source Exchange (www.firstsourceexchange.com) for the selection of nationally available building products. Reed Construction Data also publishes ProFile, a database of more than 20,000 U.S. architectural firms. RSMeans provides construction cost data, training, and consulting services in print, CD-ROM and online. Reed Construction Data, headquartered in Atlanta, is owned by Reed Business Information (www.reedconstructiondata.com), a leading provider of critical information and marketing solutions to business professionals in the media, manufacturing, electronics, construction and retail industries. Its market-leading properties include more than 135 business-to-business publications, over 125 Webzines and Web portals, as well as online services, custom publishing, directories, research and direct-marketing lists. Reed Business Information is a member of the Reed Elsevier plc group (NYSE: RUK and ENL)—a world-leading publisher and information provider operating in the science and medical, legal, education and business-to-business industry sectors.

Our Mission

Since 1942, RSMeans has been actively engaged in construction cost publishing and consulting throughout North America.

Today, over 60 years after RSMeans began, our primary objective remains the same: to provide you, the construction and facilities professional, with the most current and comprehensive construction cost data possible.

Whether you are a contractor, an owner, an architect, an engineer, a facilities manager, or anyone else who needs a fast and reliable construction cost estimate, you'll find this publication to be a highly useful and necessary tool.

Today, with the constant flow of new construction methods and materials, it's difficult to find the time to look at and evaluate all the different construction cost possibilities. In addition, because labor and material costs keep changing, last year's cost information is not a reliable basis for today's estimate or budget.

That's why so many construction professionals turn to RSMeans. We keep track of the costs for you, along with a wide range of other key information, from city cost indexes . . . to productivity rates . . . to crew composition . . . to contractor's overhead and profit rates.

RSMeans performs these functions by collecting data from all facets of the industry, and organizing it in a format that is instantly accessible to you. From the preliminary budget to the detailed unit price estimate, you'll find the data in this book useful for all phases of construction cost determination.

The Staff, the Organization, and Our Services

When you purchase one of RSMeans' publications, you are in effect hiring the services of a full-time staff of construction and engineering professionals.

Our thoroughly experienced and highly qualified staff works daily at collecting, analyzing, and disseminating comprehensive cost information for your needs. These staff members have years of practical construction experience and engineering training prior to joining the firm. As a result, you can count on them not only for the cost figures, but also for additional background reference information that will help you create a realistic estimate.

The RSMeans organization is always prepared to help you solve construction problems through its five major divisions: Construction and Cost Data Publishing, Electronic Products and Services, Consulting Services, Insurance Services, and Educational Services.

Besides a full array of construction cost estimating books, Means also publishes a number of other reference works for the construction industry. Subjects include construction estimating and project and business management; special topics such as HVAC, roofing, plumbing, and hazardous waste remediation; and a library of facility management references.

In addition, you can access all of our construction cost data through your computer with RSMeans *CostWorks 2005* CD-ROM, an electronic tool that offers over 50,000 lines of Means detailed construction cost data, along with assembly and whole building cost data. You can also access Means cost information from our Web site at www.rsmeans.com

What's more, you can increase your knowledge and improve your construction estimating and management performance with a Means Construction Seminar or In-House Training Program. These two-day seminar programs offer unparalleled opportunities for everyone in your organization to get updated on a wide variety of construction-related issues.

RSMeans also is a worldwide provider of construction cost management and analysis services for commercial and government owners and of claims and valuation services for insurers.

In short, RSMeans can provide you with the tools and expertise for constructing accurate and dependable construction estimates and budgets in a variety of ways.

Robert Snow Means Established a Tradition of Quality That Continues Today

Robert Snow Means spent years building RSMeans, making certain he always delivered a quality product.

Today, at RSMeans, we do more than talk about the quality of our data and the usefulness of our books. We stand behind all of our data, from historical cost indexes to construction materials and techniques to current costs.

If you have any questions about our products or services, please call us toll-free at 1-800-334-3509. Our customer service representatives will be happy to assist you or visit our Web site at www.rsmeans.com

Table of Contents

Foreword	ii
How the Book Is Built: An Overview	iv
How To Use the Book: The Details	v
Square Foot Cost Section	1
Assemblies Cost Section	97
Unit Price Section	283
Reference Section	533
Reference Numbers	534
Crew Listings	562
Location Factors	588
Abbreviations	594
Index	599
Other Means Publications and Reference Works	Yellow Pages
Installing Contractor's Overhead & Profit	Inside Back Cover

RESIDENTIAL
SQUARE FOOT COSTS

UNIT PRICES
GENERAL REQUIREMENTS 1
SITE CONSTRUCTION 2
CONCRETE 3
MASONRY 4
METALS 5
WOOD & PLASTICS 6
THERMAL & MOISTURE PROTECTION 7
DOORS & WINDOWS 8
FINISHES 9
SPECIALTIES 10
EQUIPMENT 11
FURNISHINGS 12
SPECIAL CONSTRUCTION 13
CONVEYING SYSTEMS 14
MECHANICAL 15
ELECTRICAL 16

ASSEMBLIES
SITE WORK 1
FOUNDATIONS 2
FRAMING 3
EXTERIOR WALLS 4
ROOFING 5
INTERIORS 6
SPECIALTIES 7
MECHANICAL 8
ELECTRICAL 9

REFERENCE INFORMATION
REFERENCE NUMBERS
CREWS
LOCATION FACTORS
INDEX

Related Means Products and Services

Robert W. Mewis, Senior Editor of this cost data book, suggests the following Means products and services as **companion** information resources to *Residential Cost Data*:

Construction Cost Data Books

Open Shop Building Construction Cost Data 2004
Light Commercial Cost Data 2004

Reference Books

Cyberplaces 2nd Ed.: The Internet Guide for Architects, Engineers, Contractors & Facility Managers
Builder's Essentials: Framing and Rough Carpentry
Interior Home Improvement Costs
Exterior Home Improvement Costs
Illustrated Construction Dictionary, Condensed, 2nd Ed.
Builder's Essentials: How to Survive and Prosper in Construction
Plan Reading and Material Takeoff
Building Spec Homes Profitably

Means on the Internet

Visit Means at **http://www.rsmeans.com.** The site contains useful ***interactive*** cost and reference material. Request or download ***FREE*** estimating software demos. Visit our bookstore for convenient ordering, plus learn more about new publications and companion products.

Means Data on CD-ROM

Get the information found in Means traditional cost books on Means new, easy to use CD-ROM, Means *CostWorks 2005*. Means *CostWorks* users can now enhance their Costlist with the Means *CostWorks Estimator*. For more information see the special full color brochure inserted in this book.

Means Consulting Services

Means Consulting Group provides a number of construction cost-related services including: Cost Estimates, Customized Databases, Benchmark Studies, Estimating Audits, Feasibility Studies, Litigation Support, Staff Assessments and Customized Training.

Means Data for Computer Estimating

Construction costs for software applications.
Over 25 unit price and assemblies cost databases are available through a number of leading estimating and facilities management software providers (listed below). For more information see the "yellow" pages at the back of this publication.

- 3D International
- 4Clicks-Solutions, LLC
- Aepco, Inc.
- Applied Flow Technology
- ArenaSoft Estimating
- ARES Corporation
- Barchan Associates
- BSD – Building Systems Design, Inc.
- CMS – Computerized Micro Solutions
- Corecon Technologies, Inc.
- CorVet Systems
- Discover Software
- Estimating Systems, Inc.
- Entech Engineering
- FAMIS Software, Inc.
- ISES Corporation
- Magellan K-12
- Maximus Asset Solutions
- MC^2 – Management Computer Controls
- Neddam Software Technologies
- Prime Time
- Quest Solutions, Inc.
- RIB Software (Americas), Inc.
- Shaw Beneco Enterprises, Inc.
- Schwab Technology Solutions, Inc.
- Timberline Software Corp.
- TMA Systems, LLC
- US Cost, Inc.
- Vanderweil Facility Advisors
- Vertigraph, Inc.

How the Book Is Built: An Overview

A Powerful Construction Tool

You have in your hands one of the most powerful construction tools available today. A successful project is built on the foundation of an accurate and dependable estimate. This book will enable you to construct just such an estimate.

For the casual user the book is designed to be:

- quickly and easily understood so you can get right to your estimate
- filled with valuable information so you can understand the necessary factors that go into the cost estimate

For the regular user, the book is designed to be:

- a handy desk reference that can be quickly referred to for key costs
- a comprehensive, fully reliable source of current construction costs and productivity rates, so you'll be prepared to estimate any project
- a source book for preliminary project cost, product selections, and alternate materials and methods

To meet all of these requirements we have organized the book into the following clearly defined sections.

Square Foot Cost Section

This section lists Square Foot costs for typical residential construction projects. The organizational format used divides the projects into basic building classes. These classes are defined at the beginning of the section. The individual projects are further divided into ten common components of construction. An outline of a typical page layout, an explanation of Square Foot prices, and a Table of Contents are located at the beginning of the section.

Assemblies Cost Section

This section uses an "Assemblies" (sometimes referred to as "systems") format grouping all the functional elements of a building into 9 construction divisions.

At the top of each "Assembly" cost table is an illustration, a brief description, and the design criteria used to develop the cost. Each of the components and its contributing cost to the system is shown.

Material: These cost figures include a standard 10% markup for "handling". They are national average material costs as of January of the current year and include delivery to the job site.

Installation: The installation costs include labor and equipment, plus a markup for the installing contractor's overhead and profit.

For a complete breakdown and explanation of a typical "Assemblies" page, see "How To Use Assemblies Cost Tables" at the beginning of the Assembly Section.

Unit Price Section

All cost data has been divided into the 16 divisions according to the MasterFormat system of classification and numbering as developed by the Construction Specifications Institute (CSI) and Construction Specifications Canada (CSC). For a listing of these divisions and an outline of their subdivisions, see the Unit Price Section Table of Contents.

Estimating tips are included at the beginning of each division.

Reference Section

This section includes information on Reference Numbers, Crew Listings, Location Factors, and a listing of Abbreviations. It is visually identified by a vertical gray bar on the edge of pages.

Reference Numbers: At the beginning of selected major classifications in the Unit Price Section are "reference numbers" shown in bold squares. These numbers refer you to related information in the Reference Section.

In this section, you'll find reference tables, explanations, and estimating information that support how we develop the unit price data. Also included are alternate pricing methods, technical data, and estimating procedures, along with information on design and economy in construction. You'll also find helpful tips on what to expect and what to avoid when estimating and constructing your project.

It is recommended that you refer to the Reference Section if a "reference number" appears within the section you are estimating.

Crew Listings: This section lists all the crews referenced in the book. For the purposes of this book, a crew is composed of more than one trade classification and/or the addition of power equipment to any trade classification. Power equipment is included in the cost of the crew. Costs are shown both with bare labor rates and with the installing contractor's overhead and profit added. For each, the total crew cost per eight-hour day and the composite cost per labor-hour are listed.

Location Factors: Costs vary depending upon regional economy. You can adjust the "national average" costs in this book to over 930 major cities throughout the U.S. and Canada by using the data in this section.

Abbreviations: A listing of the abbreviations used throughout this book, along with the terms they represent, is included.

Index

A comprehensive listing of all terms and subjects in this book to help you find what you need quickly when you are not sure where it falls in MasterFormat.

The Scope of This Book

This book is designed to be as comprehensive and as easy to use as possible. To that end we have made certain assumptions and limited its scope in three key ways:

1. We have established material prices based on a "national average."
2. We have computed labor costs based on a 7 major region average of open shop wage rates.
3. We have targeted the data for projects of a certain size range.

Project Size

This book is intended for use by those involved primarily in Residential construction costing less than $750,000. This includes the construction of homes, row houses, townhouses, condominiums and apartments.

With reasonable exercise of judgment the figures can be used for any building work. For other types of projects, such as repair and remodeling or commercial buildings, consult the appropriate MEANS publication for more information.

Carlinville Public Library
P.O. Box 17
Carlinville, Illinois 62626

How to Use the Book: The Details

What's Behind the Numbers? The Development of Cost Data

The staff at RSMeans continuously monitors developments in the construction industry in order to ensure reliable, thorough and up-to-date cost information.

While *overall* construction costs may vary relative to general economic conditions, price fluctuations within the industry are dependent upon many factors. Individual price variations may, in fact, be opposite to overall economic trends. Therefore, costs are continually monitored and complete updates are published yearly. Also, new items are frequently added in response to changes in materials and methods.

Costs—$ (U.S.)

All costs represent U.S. national averages and are given in U.S. dollars. The Means Location Factors can be used to adjust costs to a particular location. The Location Factors for Canada can be used to adjust U.S. national averages to local costs in Canadian dollars.

Material Costs

The RSMeans staff contacts manufacturers, dealers, distributors, and contractors all across the U.S. and Canada to determine national average material costs. If you have access to current material costs for your specific location, you may wish to make adjustments to reflect differences from the national average. Included within material costs are fasteners for a normal installation. RSMeans engineers use manufacturers' recommendations, written specifications and/or standard construction practice for size and spacing of fasteners. Adjustments to material costs may be required for your specific application or location. Material costs do not include sales tax.

Labor Costs

Labor costs are based on the average of open shop wages from across the U.S. for the current year. Rates along with overhead and profit markups are listed on the inside back cover of this book.

- If wage rates in your area vary from those used in this book, or if rate increases are expected within a given year, labor costs should be adjusted accordingly.

Labor costs reflect productivity based on actual working conditions. These figures include time spent during a normal workday on tasks other than actual installation, such as material receiving and handling, mobilization at site, site movement, breaks, and cleanup.

Productivity data is developed over an extended period so as not to be influenced by abnormal variations and reflects a typical average.

Equipment Costs

Equipment costs include not only rental, but also operating costs for equipment under normal use. The operating costs include parts and labor for routine servicing such as repair and replacement of pumps, filters and worn lines. Normal operating expendables such as fuel, lubricants, tires and electricity (where applicable) are also included. Extaordinary operating expendables with highly variable wear patterns such as diamond bits and blades are excluded. These costs are included under materials. Equipment rental rates are obtained from industry sources throughout North America—contractors, suppliers, dealers, manufacturers, and distributors.

Crew Equipment Cost/Day— The power equipment required for each crew is included in the crew cost. The daily cost for crew equipment is based on dividing the weekly bare rental rate by 5 (number of working days per week), and then adding the hourly operating cost times 8 (hours per day). This "Crew Equipment Cost/Day" is listed in Subdivision 01590.

Mobilization/Demobilization—The cost to move construction equipment from an equipment yard or rental company to the job site and back again is not included in equipment costs. Mobilization (to the site) and demobilization (from the site) costs can be found in Section 02305-250. If a piece of equipment is already at the job site, it is not appropriate to utilize mob/demob costs again in an estimate.

Factors Affecting Costs

Costs can vary depending upon a number of variables. Here's how we have handled the main factors affecting costs.

Quality— The prices for materials and the workmanship upon which productivity is based represent sound construction work. They are also in line with U.S. government specifications.

Overtime— We have made no allowance for overtime. If you anticipate premium time or work beyond normal working hours, be sure to make an appropriate adjustment to your labor costs.

Productivity— The productivity, daily output, and labor-hour figures for each line item are based on working an eight-hour day in daylight hours in moderate temperatures. For work that extends beyond normal work hours or is performed under adverse conditions, productivity may decrease. (See the section in "How To Use the Unit Price Pages" for more on productivity.)

Size of Project— The size, scope of work, and type of construction project will have a significant impact on cost. Economies of scale can reduce costs for large projects. Unit costs can often run higher for small projects. Costs in this book are intended for the size and type of project as previously described in "How the Book Is Built: An Overview." Costs for projects of a significantly different size or type should be adjusted accordingly.

Location— Material prices in this book are for metropolitan areas. However, in dense urban areas, traffic and site storage limitations may increase costs. Beyond a 20-mile radius of large cities, extra trucking or transportation charges may also increase the material costs slightly. On the other hand, lower wage rates may be in effect. Be sure to consider both these factors when preparing an estimate, particularly if the job site is located in a central city or remote rural location.

In addition, highly specialized subcontract items may require travel and per diem expenses for mechanics.

Other factors—

- season of year
- contractor management
- weather conditions
- local union restrictions
- building code requirements
- availability of:
  - adequate energy
  - skilled labor
  - building materials
- owner's special requirements/restrictions
- safety requirements
- environmental considerations

General Conditions— The "Square Foot" and "Assemblies" sections of this book use costs that include the installing contractor's overhead and profit (O&P). The Unit Price Section presents cost data in two ways: Bare Costs and Total Cost including O&P (Overhead and Profit). General Conditions, when applicable, should also be added to the Total Cost including O&P. The costs for General Conditions are listed in Division 1 of the Unit Price Section and the Reference Section of this book. General Conditions for the *Installing Contractor* may range from 0% to 10% of the Total Cost including O&P. For the *General* or *Prime Contractor*, costs for General Conditions may range from 5% to 15% of the Total Cost including O&P, with a figure of 10% as the most typical allowance.

Overhead & Profit— Total Cost including O&P for the *Installing Contractor* is shown in the last column on both the Unit Price and the Assemblies pages of this book. This figure is the sum of the bare material cost plus 10% for profit, the base labor cost plus overhead and profit, and the bare equipment cost plus 10% for profit. Details for the calculation of Overhead and Profit on labor are shown on the inside back cover and in the Reference Section of this book. (See the "How to Use the Unit Price Pages" for an example of this calculation.)

Unpredictable Factors— General business conditions influence "in-place" costs of all items. Substitute materials and construction methods may have to be employed. These may affect the installed cost and/or life cycle costs. Such factors may be difficult to evaluate and cannot necessarily be predicted on the basis of the job's location in a particular section of the country. Thus, where these factors apply, you may find significant, but unavoidable cost variations for which you will have to apply a measure of judgment to your estimate.

Rounding of Costs

In general, all unit prices in excess of $5.00 have been rounded to make them easier to use and still maintain adequate precision of the results. The rounding rules we have chosen are in the following table.

Prices from ...	Rounded to the nearest ...
$.01 to $5.00	$.01
$5.01 to $20.00	$.05
$20.01 to $100.00	$.50
$100.01 to $300.00	$1.00
$300.01 to $1,000.00	$5.00
$1,000.01 to $10,000.00	$25.00
$10,000.01 to $50,000.00	$100.00
$50,000.01 and above	$500.00

Final Checklist

Estimating can be a straightforward process provided you remember the basics. Here's a checklist of some of the items you should remember to do before completing your estimate.

Did you remember to . . .

- factor in the Location Factor for your locale
- take into consideration which items have been marked up and by how much
- mark up the entire estimate sufficiently for your purposes
- read the background information on techniques and technical matters that could impact your project time span and cost
- include all components of your project in the final estimate
- double check your figures to be sure of your accuracy
- call RSMeans if you have any questions about your estimate or the data you've found in our publications

Remember, RSMeans stands behind its publications. If you have any questions about your estimate . . . about the costs you've used from our books . . . or even about the technical aspects of the job that may affect your estimate, feel free to call the RSMeans editors at 1-800-334-3509.

Square Foot Cost Section

Table of Contents

	Page
Introduction	
General	3
How to Use	4
Building Classes	8
Configurations	9
Building Types	10
Garage Types	12
Building Components	13
Exterior Wall Construction	14
Residential Cost Estimate Worksheets	
Instructions	15
Model Residence Example	17
Economy	21
Illustrations	23
1 Story	24
1-1/2 Story	26
2 Story	28
Bi-Level	30
Tri-Level	32
Wings & Ells	34
Average	35
Illustrations	37
1 Story	38
1-1/2 Story	40
2 Story	42
2-1/2 Story	44
3 Story	46
Bi-Level	48
Tri-Level	50
Solid Wall (log home) 1 Story	52
Solid Wall (log home) 2 Story	54
Wings & Ells	56
Custom	57
Illustrations	59
1 Story	60
1-1/2 Story	62
2 Story	64
2-1/2 Story	66
3 Story	68
Bi-Level	70
Tri-Level	72
Wings & Ells	74
Luxury	75
Illustrations	77
1 Story	78
1-1/2 Story	80
2 Story	82
2-1/2 Story	84
3 Story	86
Bi-Level	88
Tri-Level	90
Wings & Ells	92
Residential Modifications	
Kitchen Cabinets	93
Kitchen Counter Tops	93
Vanity Bases	94
Solid Surface Vanity Tops	94
Fireplaces & Chimneys	94
Windows & Skylights	94
Dormers	94
Appliances	95
Breezeway	95
Porches	95
Finished Attic	95
Alarm System	95
Sauna, Prefabricated	95
Garages	96
Swimming Pools	96
Wood & Coal Stoves	96
Sidewalks	96
Fencing	96
Carport	96

Introduction to Square Foot Cost Section

The Square Foot Cost Section of this manual contains costs per square foot for four classes of construction in seven building types. Costs are listed for various exterior wall materials which are typical of the class and building type. There are cost tables for Wings and Ells with modification tables to adjust the base cost of each class of building. Non-standard items can easily be added to the standard structures.

Cost estimating for a residence is a three-step process:

(1) Identification
(2) Listing dimensions
(3) Calculations

Guidelines and a sample cost estimating procedure are shown on the following pages.

Identification

To properly identify a residential building, the class of construction, type, and exterior wall material must be determined. Page 8 has drawings and guidelines for determining the class of construction. There are also detailed specifications and additional drawings at the beginning of each set of tables to further aid in proper building class and type identification.

Sketches for eight types of residential buildings and their configurations are shown on pages 10 and 11. Definitions of living area are next to each sketch. Sketches and definitions of garage types are on page 12.

Living Area

Base cost tables are prepared as costs per square foot of living area. The living area of a residence is that area which is suitable and normally designed for full time living. It does not include basement recreation rooms or finished attics, although these areas are often considered full time living areas by the owners.

Living area is calculated from the exterior dimensions without the need to adjust for exterior wall thickness. When calculating the living area of a 1-1/2 story, two story, three story or tri-level residence, overhangs and other differences in size and shape between floors must be considered.

Only the floor area with a ceiling height of six feet or more in a 1-1/2 story residence is considered living area. In bi-levels and tri-levels, the areas that are below grade are considered living area, even when these areas may not be completely finished.

Base Tables and Modifications

Base cost tables show the base cost per square foot without a basement, with one full bath and one full kitchen for economy and average homes and an additional half bath for custom and luxury models. Adjustments for finished and unfinished basements are part of the base cost tables. Adjustments for multi family residences, additional bathrooms, townhouses, alternative roofs, and air conditioning and heating systems are listed in Modifications, Adjustments and Alternatives tables below the base cost tables.

Costs for other modifications, adjustments and alternatives, including garages, breezeways and site improvements, are on pages 93 to 96.

Listing of Dimensions

To use this section of the manual only the dimensions used to calculate the horizontal area of the building and additions, modifications, adjustments and alternatives are needed. The dimensions, normally the length and width, can come from drawings or field measurements. For ease in calculation, consider measuring in tenths of feet, i.e., 9 ft. 6 in. = 9.5 ft., 9 ft. 4 in. = 9.3 ft.

In all cases, make a sketch of the building. Any protrusions or other variations in shape should be noted on the sketch with dimensions.

Calculations

The calculations portion of the estimate is a two-step activity:

(1) The selection of appropriate costs from the tables
(2) Computations

Selection of Appropriate Costs

To select the appropriate cost from the base tables, the following information is needed:

(1) Class of construction
- (a) Economy
- (b) Average
- (c) Custom
- (d) Luxury

(2) Type of residence
- (a) One story
- (b) 1-1/2 story
- (c) 2 story
- (d) 3 story
- (e) Bi-level
- (f) Tri-level

(3) Occupancy
- (a) One family
- (b) Two family
- (c) Three family

(4) Building configuration
- (a) Detached
- (b) Town/Row house
- (c) Semi-detached

(5) Exterior wall construction
- (a) Wood frame
- (b) Brick veneer
- (c) Solid masonry

(6) Living areas

Modifications, adjustments and alternatives are classified by class, type and size.

Computations

The computation process should take the following sequence:

(1) Multiply the base cost by the area
(2) Add or subtract the modifications, adjustments, alternatives
(3) Apply the location modifier

When selecting costs, interpolate or use the cost that most nearly matches the structure under study. This applies to size, exterior wall construction and class.

How to Use the Residential Square Foot Cost Pages

The following is a detailed explanation of a sample entry in the Residential Square Foot Cost Section. Each bold number below corresponds to the item being described on the facing page with the appropriate component of the sample entry following in parenthesis.

Prices listed are costs that include overhead and profit of the installing contractor. Total model costs include an additional mark-up for General Contractors' overhead and profit, and fees specific to class of construction.

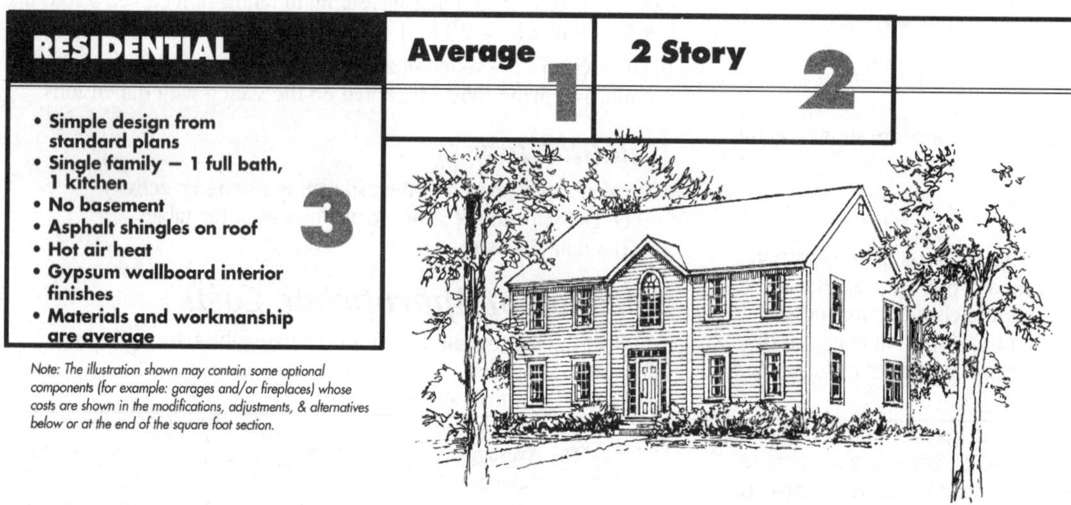

RESIDENTIAL
Average 1
2 Story 2
• Simple design from standard plans
• Single family — 1 full bath, 1 kitchen
• No basement
• Asphalt shingles on roof
• Hot air heat
• Gypsum wallboard interior finishes
• Materials and workmanship are average
3

Note: The illustration shown may contain some optional components (for example: garages and/or fireplaces) whose costs are shown in the modifications, adjustments, & alternatives below or at the end of the square foot section.

Base cost per square foot of living area 4 5 6 7 8

Exterior Wall	1000	1200	1400	1600	1800	Living Area 2000	2200	2600	3000	3400	3800
Wood Siding - Wood Frame	99.15	90.00	85.80	82.95	80.00	**76.75**	[illegible]	70.65	66.55	64.85	63.25
Brick Veneer - Wood Frame	106.65	97.00	92.25	89.15	85.80	82.40	[illegible]	75.45	71.00	69.05	67.30
Stucco on Wood Frame	98.40	89.35	85.20	82.35	79.40	76.20	74.15	70.15	66.15	64.45	62.90
Solid Masonry	118.35	107.80	102.45	98.85	94.90	91.15	88.30	83.05	78.00	75.70	73.60
Finished Basement, Add	16.05	15.85	15.35	15.00	14.55	14.35	[illegible]	13.55	13.20	12.95	12.75
Unfinished Basement, Add	6.40	6.00	5.65	5.40	5.15	5.05	4.85	4.50	4.35	4.15	4.05

Modifications

Add to the total cost 9

Upgrade Kitchen Cabinets	$ + 3157
Solid Surface Countertops (Included)	
Full Bath - including plumbing, wall and floor finishes	+ 4261
Half Bath - including plumbing, wall and floor finishes	+ 2660
One Car Attached Garage	+ 9490
One Car Detached Garage	+ 12,424
Fireplace & Chimney	+ 4670

Adjustments

For multi family - add to total cost

Additional Kitchen	$ + 5105
Additional Bath	+ 4261
Additional Entry & Exit	+ 1276
Separate Heating	+ 1248
Separate Electric	+ 1575

For Townhouse/Rowhouse - Multiply cost per square foot by 10

Inner Unit	.90
End Unit	.95

Alternatives

Add to or deduct from the cost per square foot of living area 11

Cedar Shake Roof	+ 1.17
Clay Tile Roof	+ 2.04
Slate Roof	+ 3.89
Upgrade Walls to Skim Coat Plaster	+ .42
Upgrade Ceilings to Textured Finish	+ .40
Air Conditioning, in Heating Ductwork	+ 1.70
In Separate Ductwork	+ 4.35
Heating Systems, Hot Water	+ 1.39
Heat Pump	+ 2.15
Electric Heat	– .51
Not Heated	– 2.46

Additional upgrades or components

12

Kitchen Cabinets & Countertops	Page 93
Bathroom Vanities	94
Fireplaces & Chimneys	94
Windows, Skylights & Dormers	94
Appliances	95
Breezeways & Porches	95
Finished Attic	95
Garages	96
Site Improvements	96
Wings & Ells	56

1 *Class of Construction (Average)*

The class of construction depends upon the design and specifications of the plan. The four classes are economy, average, custom and luxury.

2 *Type of Residence (2 Story)*

The building type describes the number of stories or levels in the model. The seven building types are 1 story, 1-1/2 story, 2 story, 2-1/2 story, 3 story, bi-level and tri-level.

3 *Specification Highlights (Hot Air Heat)*

These specifications include information concerning the components of the model, including the number of baths, roofing types, HVAC systems, and materials and workmanship. If the components listed are not appropriate, modifications can be made by consulting the information shown lower on the page or the Assemblies Section.

4 *Exterior Wall System (Wood Siding - Wood Frame)*

This section includes the types of exterior wall systems and the structural frame used. The exterior wall systems shown are typical of the class of construction and the building type shown.

5 *Living Areas (2000 SF)*

The living area is that area of the residence which is suitable and normally designed for full time living. It does not include basement recreation rooms or finished attics. Living area is calculated from the exterior dimensions without the need to adjust for exterior wall thickness. When calculating the living area of a 1-1/2 story, 2 story, 3 story or tri-level residence, overhangs and other differences in size and shape between floors must be considered. Only the floor area with a ceiling height of six feet or more in a 1-1/2 story residence is considered living area. In bi-levels and tri-levels, the areas that are below grade are considered living area, even when these areas may not be completely finished. A range of various living areas for the residential model are shown to aid in selection of values from the matrix.

6 *Base Costs per Square Foot of Living Area ($76.75)*

Base cost tables show the cost per square foot of living area without a basement, with one full bath and one full kitchen for economy and average homes and an additional half bath for custom and luxury models. When selecting costs, interpolate or use the cost that most nearly matches the residence under consideration for size, exterior wall system and class of construction. Prices listed are costs that include overhead and profit of the installing contractor, a general contractor markup and an allowance for plans that vary by class of construction. For additional information on contractor overhead and architectural fees, see the Reference Section.

7 *Basement Types (Finished)*

The two types of basements are finished or unfinished. The specifications and components for both are shown on the Building Classes page in the Introduction to this section.

8 *Additional Costs for Basements ($14.35)*

These values indicate the additional cost per square foot of living area for either a finished or an unfinished basement.

9 *Modifications and Adjustments (Upgrade Kitchen Cabinets $3157)*

Modifications and Adjustments are costs added to or subtracted from the total cost of the residence. The total cost of the residence is equal to the cost per square foot of living area times the living area. Typical modifications and adjustments include kitchens, baths, garages and fireplaces.

10 *Multiplier for Townhouse/ Rowhouse (Inner Unit .90)*

The multipliers shown adjust the base costs per square foot of living area for the common wall condition encountered in townhouses or rowhouses.

11 *Alternatives (Skim Coat Plaster $.42)*

Alternatives are costs added to or subtracted from the base cost per square foot of living area. Typical alternatives include variations in kitchens, baths, roofing, air conditioning and heating systems.

12 *Additional Upgrades or Components (Wings & Ells page 56)*

Costs for additional upgrades or components, including wings or ells, breezeways, porches, finished attics and site improvements, are shown in other locations in the Residential Square Foot Cost Section.

How to Use the Square Foot Cost Pages *(Continued)*

Average 2 Story
Living Area - 2000 S.F.
Perimeter - 135 L.F.

		Labor-Hours	Cost Per Square Foot Of Living Area		
			Mat.	**Labor**	**Total**
1 Site Work	Site preparation for slab; 4' deep trench excavation for foundation wall.	.034		.61	.61
2 Foundation	Continuous reinforced concrete footing 8" deep x 18" wide; dampproofed and insulated reinforced concrete foundation wall, 8" thick, 4' deep, 4" concrete slab on 4" crushed stone base and polyethylene vapor barrier, trowel finish.	.066	2.33	3.13	5.46
3 Framing	Exterior walls - 2" x 4" wood studs, 16" O.C.; 1/2" plywood sheathing; 2" x 6" rafters 16" O.C. with 1/2" plywood sheathing, 4 in 12 pitch; 2" x 6" ceiling joists 16" O.C.; 2" x 8" floor joists 16" O.C. with 5/8" plywood subfloor; 1/2" plywood subfloor on 1" x 2" wood sleepers 16" O.C.	.131	6.21	6.88	13.09
4 Exterior Walls	Beveled wood siding and building paper on insulated wood frame walls; 6" attic insulation; double hung windows; 3 flush solid core wood exterior doors with storms.	.111	8.82	4.61	13.43
5 Roofing	25 year asphalt shingles; #15 felt building paper; aluminum gutters, downspouts, drip edge and flashings.	.024	.48	.94	1.42
6 Interiors	Walls and ceilings, 1/2" taped and finished gypsum wallboard, primed and painted with 2 coats; painted baseboard and trim, finished hardwood floor 40%, carpet with 1/2" underlayment 40%, vinyl tile with 1/2" underlayment 15%, ceramic tile with 1/2" underlayment 5%; hollow core and louvered doors.	.232	11.38	11.41	22.79
7 Specialties	Average grade kitchen cabinets - 14 L.F. wall and base with solid surface counter top and kitchen sink; 40 gallon electric water heater.	.021	1.48	.68	2.16
8 Mechanical	1 lavatory, white, wall hung; 1 water closet, white; 1 bathtub with shower; enameled steel, white; gas fired warm air heat.	.060	2.27	2.22	4.49
9 Electrical	200 Amp. service; romex wiring; incandescent lighting fixtures, switches, receptacles.	.039	.89	1.29	2.18
10 Overhead	Contractor's overhead and profit and plans.		5.74	5.38	11.12
		Total	39.60	37.15	**76.75**

1 *Specifications*

The parameters for an example dwelling from the previous pages are listed here. Included are the square foot dimensions of the proposed building. Living Area takes into account the number of floors and other factors needed to define a building's total square footage. Perimeter dimensions are defined in terms of linear feet.

2 *Building Type*

This is a sketch of a cross section view through the dwelling. It is shown to help define the living area for the building type. For more information, see the Building Types pages in the Introduction.

3 *Components (4 Exterior Walls)*

This page contains the ten components needed to develop the complete square foot cost of the typical dwelling specified. All components are defined with a description of the materials and/or task involved. Use cost figures from each component to estimate the cost per square foot of that section of the project. The components listed on this page are typical of all sizes of residences from the facing page. Specific quantities of components required would vary with the size of the dwelling and the exterior wall system.

4 *Labor-Hours (.111)*

Use this column to determine the number of labor-hours needed to perform a task. This figure will give the builder labor-hours per square foot of building. The total labor-hours per component is determined by multiplying the living area times the labor-hours listed on that line.

5 *Materials (8.82)*

This column gives the amount needed to develop the cost of materials. The figures given here are not bare costs. Ten percent has been added to bare material cost for profit.

6 *Installation (4.61)*

Installation includes labor and equipment costs. The labor rates included here incorporate the total overhead and profit costs for the installing contractor. The average mark-up used to create these figures is 72.0% over and above bare labor costs. The equipment rates include 10% for profit.

7 *Line Totals (13.43)*

The extreme right-hand column lists the sum of two figures. Use this total to determine the sum of material cost plus installation cost. The result is a convenient total cost for each of the ten components.

8 *Overhead*

The costs in components 1 through 9 include overhead and profit for the installing contractor. Item 10 is overhead and profit for the general contractor. This is typically a percentage mark-up of all other costs. The amount depends on size and type of dwelling, building class and economic conditions. An allowance for plans or design has been included where appropriate.

9 *Bottom Line Total (76.75)*

This figure is the complete square foot cost for the construction project and equals the sum of total material and total labor costs. To determine total project cost, multiply the bottom line total times the living area.

Building Classes

Economy Class

An economy class residence is usually built from stock plans. The materials and workmanship are sufficient to satisfy building codes. Low construction cost is more important than distinctive features. The overall shape of the foundation and structure is seldom other than square or rectangular.

An unfinished basement includes 7′ high 8″ thick foundation wall composed of either concrete block or cast-in-place concrete.

Included in the finished basement cost are inexpensive paneling or drywall as the interior finish on the foundation walls, a low cost sponge backed carpeting adhered to the concrete floor, a drywall ceiling, and overhead lighting.

Custom Class

A custom class residence is usually built from plans and specifications with enough features to give the building a distinction of design. Materials and workmanship are generally above average with obvious attention given to construction details. Construction normally exceeds building code requirements.

An unfinished basement includes a 7′-6″ high 10″ thick cast-in-place concrete foundation wall or a 7′-6″ high 12″ thick concrete block foundation wall.

A finished basement includes painted drywall on insulated 2″ x 4″ wood furring as the interior finish to the concrete walls, a suspended ceiling, carpeting adhered to the concrete floor, overhead lighting and heating.

Average Class

An average class residence is a simple design and built from standard plans. Materials and workmanship are average, but often exceed minimum building codes. There are frequently special features that give the residence some distinctive characteristics.

An unfinished basement includes 7′-6″ high 8″ thick foundation wall composed of either cast-in-place concrete or concrete block.

Included in the finished basement are plywood paneling or drywall on furring that is fastened to the foundation walls, sponge backed carpeting adhered to the concrete floor, a suspended ceiling, overhead lighting and heating.

Luxury Class

A luxury class residence is built from an architect's plan for a specific owner. It is unique both in design and workmanship. There are many special features, and construction usually exceeds all building codes. It is obvious that primary attention is placed on the owner's comfort and pleasure. Construction is supervised by an architect.

An unfinished basement includes 8′ high 12″ thick foundation wall that is composed of cast-in-place concrete or concrete block.

A finished basement includes painted drywall on 2″ x 4″ wood furring as the interior finish, suspended ceiling, tackless carpet on wood subfloor with sleepers, overhead lighting and heating.

Configurations

Detached House

This category of residence is a free-standing separate building with or without an attached garage. It has four complete walls.

Semi-Detached House

This category of residence has two living units side-by-side. The common wall is a fireproof wall. Semi-detached residences can be treated as a row house with two end units. Semi-detached residences can be any of the building types.

Town/Row House

This category of residence has a number of attached units made up of inner units and end units. The units are joined by common walls. The inner units have only two exterior walls. The common walls are fireproof. The end units have three walls and a common wall. Town houses/row houses can be any of the building types.

Building Types

One Story

This is an example of a one-story dwelling. The living area of this type of residence is confined to the ground floor. The headroom in the attic is usually too low for use as a living area.

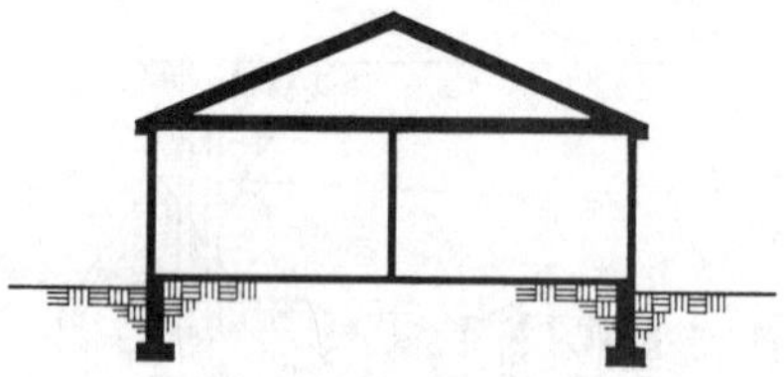

One-and-a-half Story

The living area in the upper level of this type of residence is 50% to 90% of the ground floor. This is made possible by a combination of this design's high-peaked roof and/or dormers. Only the upper level area with a ceiling height of 6 feet or more is considered living area. The living area of this residence is the sum of the ground floor area plus the area on the second level with a ceiling height of 6 feet or more.

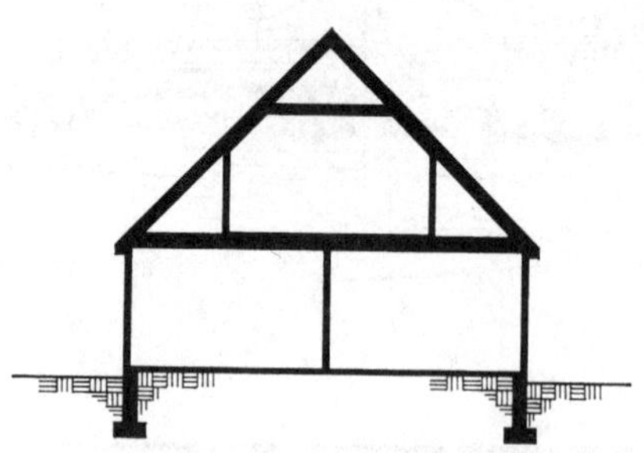

One Story with Finished Attic

The main living area in this type of residence is the ground floor. The upper level or attic area has sufficient headroom for comfortable use as a living area. This is made possible by a high peaked roof. The living area in the attic is less than 50% of the ground floor. The living area of this type of residence is the ground floor area only. The finished attic is considered an adjustment.

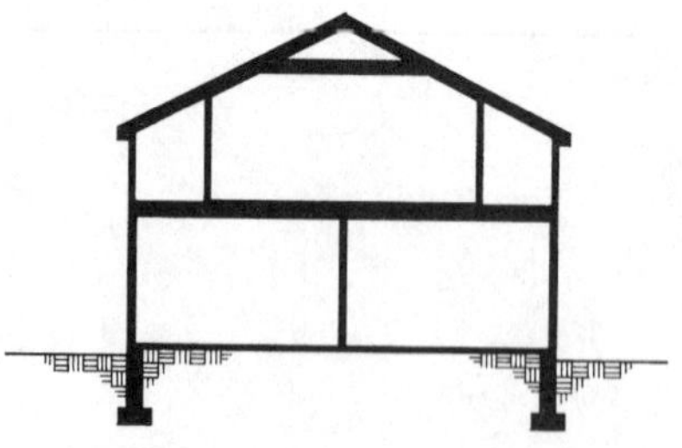

Two Story

This type of residence has a second floor or upper level area which is equal or nearly equal to the ground floor area. The upper level of this type of residence can range from 90% to 110% of the ground floor area, depending on setbacks or overhangs. The living area is the sum of the ground floor area and the upper level floor area.

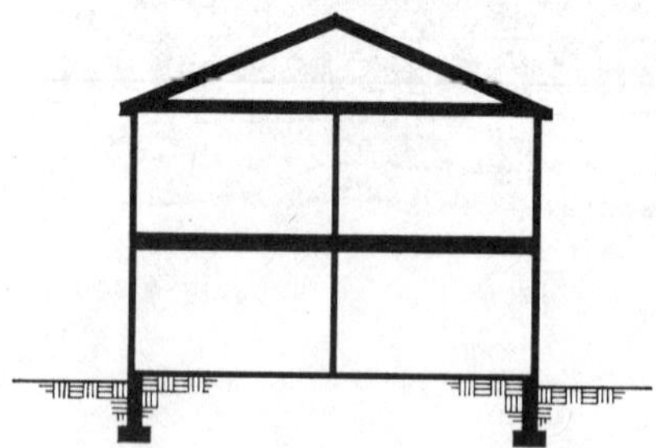

Two-and-one-half Story

This type of residence has two levels of equal or nearly equal area and a third level which has a living area that is 50% to 90% of the ground floor. This is made possible by a high peaked roof, extended wall heights and/or dormers. Only the upper level area with a ceiling height of 6 feet or more is considered living area. The living area of this residence is the sum of the ground floor area, the second floor area and the area on the third level with a ceiling height of 6 feet or more.

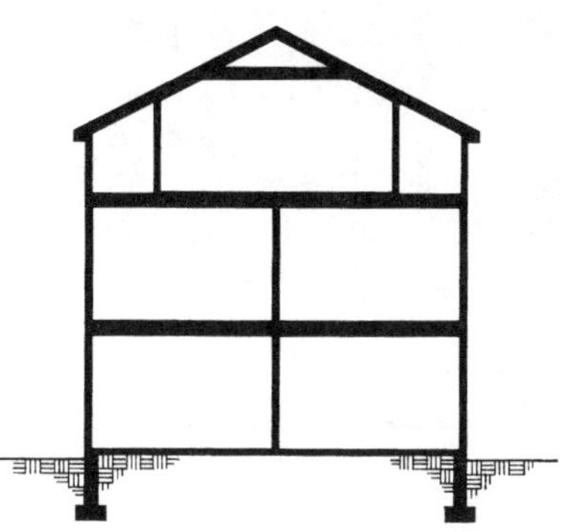

Bi-level

This type of residence has two living areas, one above the other. One area is about 4 feet below grade and the second is about 4 feet above grade. Both areas are equal in size. The lower level in this type of residence is originally designed and built to serve as a living area and not as a basement. Both levels have full ceiling heights. The living area is the sum of the lower level area and the upper level area.

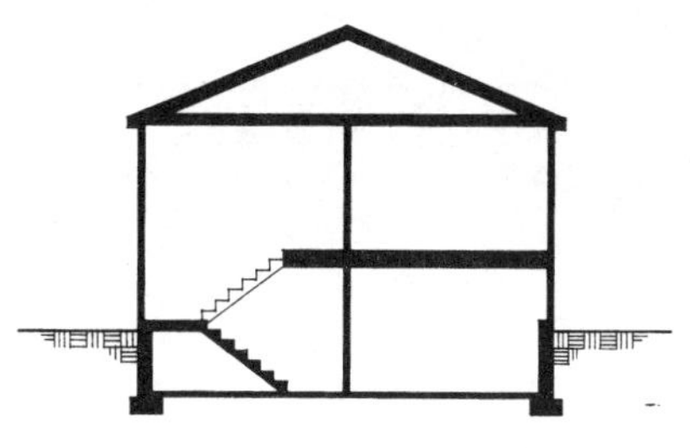

Three Story

This type of residence has three levels which are equal or nearly equal. As in the 2 story residence, the second and third floor areas may vary slightly depending on setbacks or overhangs. The living area is the sum of the ground floor area and the two upper level floor areas.

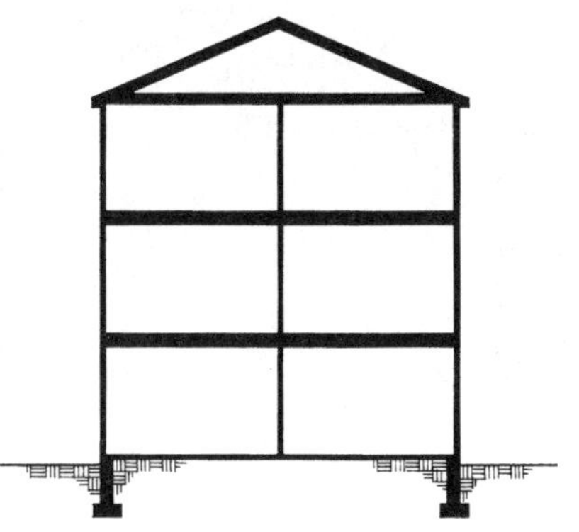

Tri-level

This type of residence has three levels of living area, one at grade level, one about 4 feet below grade and one about 4 feet above grade. All levels are originally designed to serve as living areas. All levels have full ceiling heights. The living area is a sum of the areas of each of the three levels.

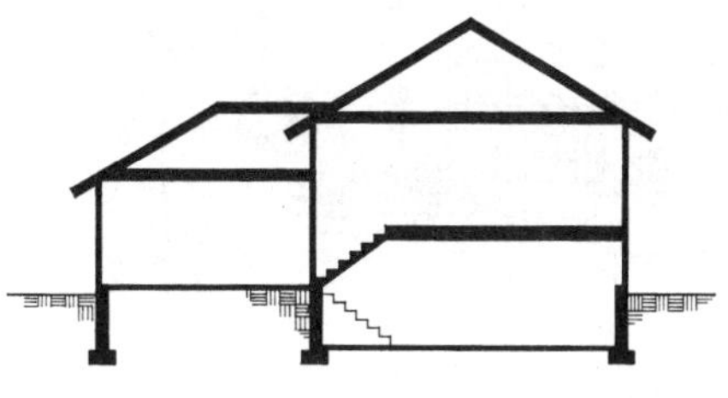

Garage Types

Attached Garage

Shares a common wall with the dwelling. Access is typically through a door between dwelling and garage.

Basement Garage

Constructed under the roof of the dwelling but below the living area.

Built-In Garage

Constructed under the second floor living space and above basement level of dwelling. Reduces gross square feet of living area.

Detached Garage

Constructed apart from the main dwelling. Shares no common area or wall with the dwelling.

Building Components

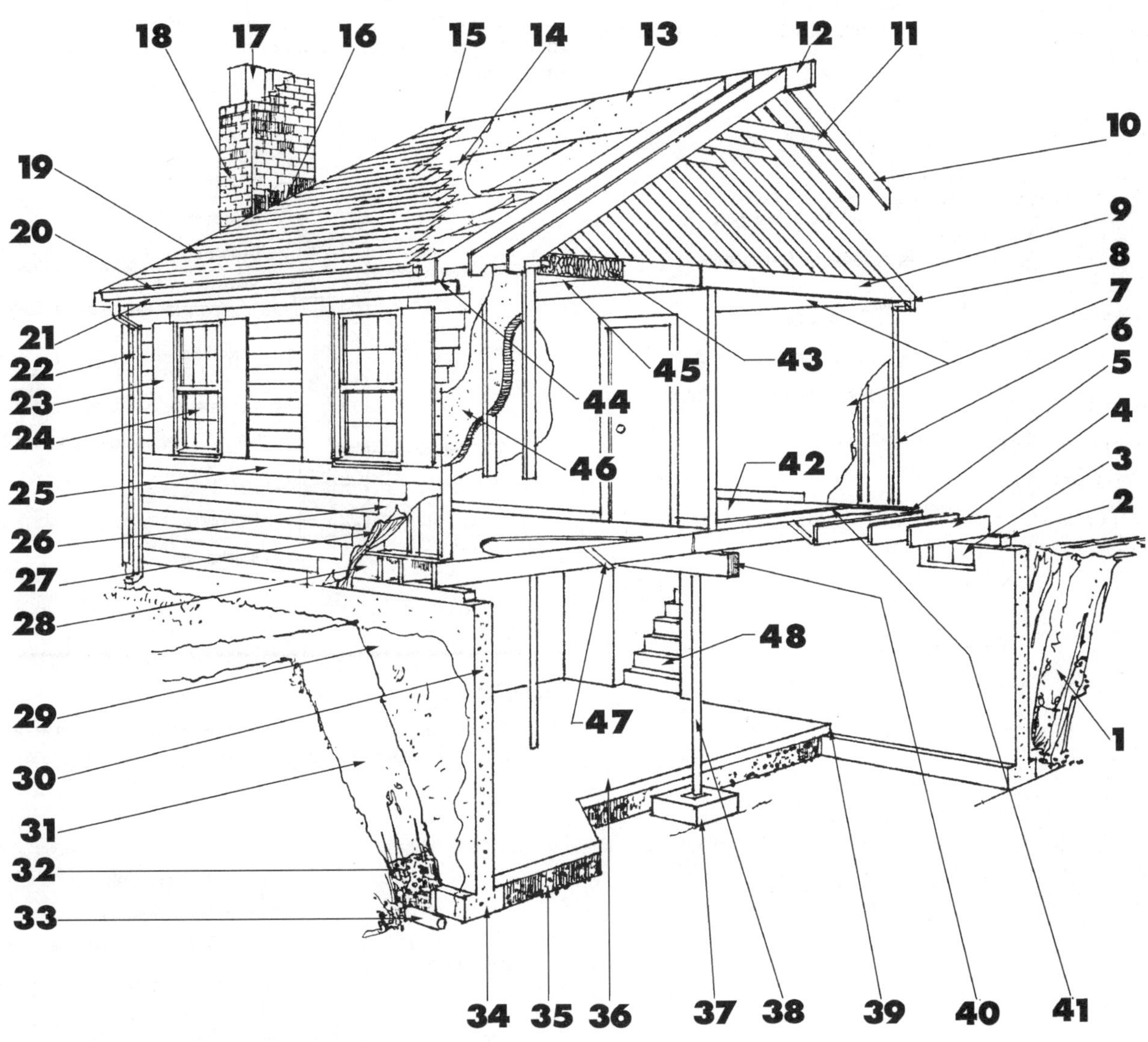

1. Excavation
2. Sill Plate
3. Basement Window
4. Floor Joist
5. Shoe Plate
6. Studs
7. Drywall
8. Plate
9. Ceiling Joists
10. Rafters
11. Collar Ties
12. Ridge Rafter
13. Roof Sheathing
14. Roof Felt
15. Roof Shingles
16. Flashing
17. Flue Lining
18. Chimney
19. Roof Shingles
20. Gutter
21. Fascia
22. Downspout
23. Shutter
24. Window
25. Wall Shingles
26. Building Paper
27. Wall Sheathing
28. Fire Stop
29. Dampproofing
30. Foundation Wall
31. Backfill
32. Drainage Stone
33. Drainage Tile
34. Wall Footing
35. Gravel
36. Concrete Slab
37. Column Footing
38. Pipe Column
39. Expansion Joint
40. Girder
41. Sub-floor
42. Finish Floor
43. Attic Insulation
44. Soffit
45. Ceiling Strapping
46. Wall Insulation
47. Cross Bridging
48. Bulkhead Stairs

Exterior Wall Construction

Typical Frame Construction

Typical wood frame construction consists of wood studs with insulation between them. A typical exterior surface is made up of sheathing, building paper and exterior siding consisting of wood, vinyl, aluminum or stucco over the wood sheathing.

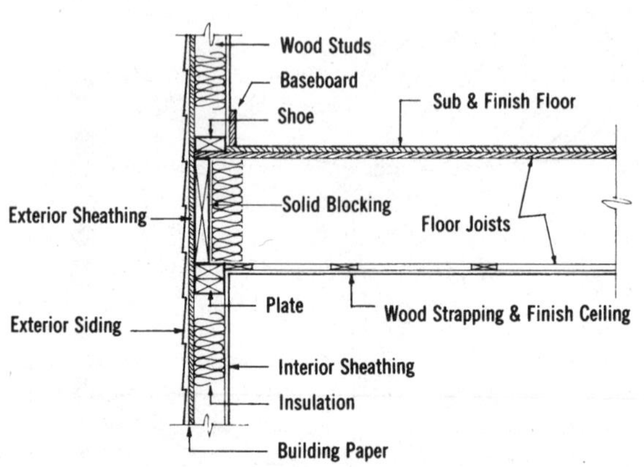

Brick Veneer

Typical brick veneer construction consists of wood studs with insulation between them. A typical exterior surface is sheathing, building paper and an exterior of brick tied to the sheathing, with metal strips.

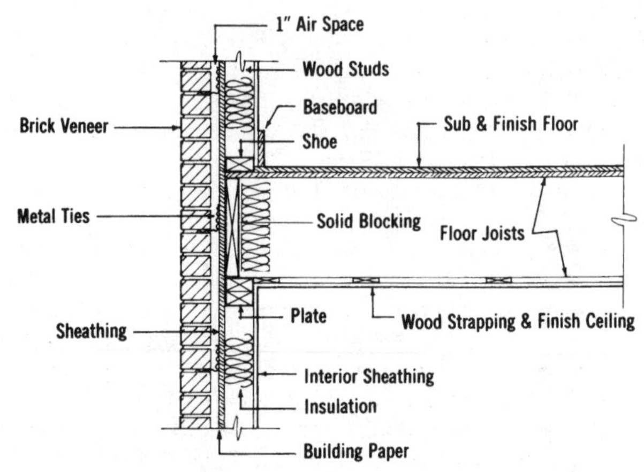

Stone

Typical solid masonry construction consists of a stone or block wall covered on the exterior with brick, stone or other masonry.

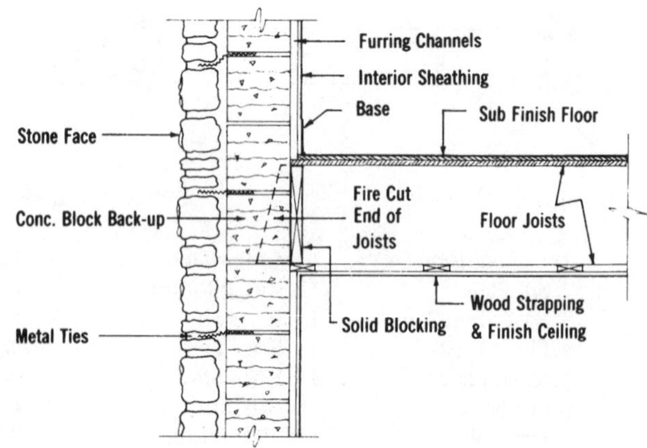

Residential Cost Estimate Worksheets

Worksheet Instructions

The residential cost estimate worksheet can be used as an outline for developing a residential construction or replacement cost. It is also useful for insurance appraisals. The design of the worksheet helps eliminate errors and omissions. To use the worksheet, follow the example below.

1. Fill out the owner's name, residence address, the estimator or appraiser's name, some type of project identifying number or code, and the date.
2. Determine from the plans, specifications, owner's description, photographs or any other means possible the class of construction. The models in this book use economy, average, custom and luxury as classes. Fill in the appropriate box.
3. Fill in the appropriate box for the residence type, configuration, occupancy, and exterior wall. If you require clarification, the pages preceding this worksheet describe each of these.
4. Next, the living area of the residence must be established. The heated or air conditioned space of the residence, not including the basement, should be measured. It is easiest to break the structure up into separate components as shown in the example. The main house (A), a one-and-one-half story wing (B), and a one story wing (C). The breezeway (D), garage (E) and open covered porch (F) will be treated differently. Data entry blocks for the living area are included on the worksheet for your use. Keep each level of each component separate, and fill out the blocks as shown.
5. By using the information on the worksheet, find the model, wing or ell in the following square foot cost pages that best matches the class, type, exterior finish and size of the residence being estimated. Use the *modifications, adjustments, and alternatives* to determine the adjusted cost per square foot of living area for each component.
6. For each component, multiply the cost per square foot by the living area square footage. If the residence is a townhouse/rowhouse, a multiplier should be applied based upon the configuration.
7. The second page of the residential cost estimate worksheet has space for the additional components of a house. The cost for additional bathrooms, finished attic space, breezeways, porches, fireplaces, appliance or cabinet upgrades, and garages should be added on this page. The information for each of these components is found with the model being used, or in the *modifications, adjustments, and alternatives* pages.
8. Add the total from page one of the estimate worksheet and the items listed on page two. The sum is the adjusted total building cost.
9. Depending on the use of the final estimated cost, one of the remaining two boxes should be filled out. Any additional items or exclusions should be added or subtracted at this time. The data contained in this book is a national average. Construction costs are different throughout the country. To allow for this difference, a location factor based upon the first three digits of the residence's zip code must be applied. The location factor is a multiplier that increases or decreases the adjusted total building cost. Find the appropriate location factor and calculate the local cost. If depreciation is a concern, a dollar figure should be subtracted at this point.
10. No residence will match a model exactly. Many differences will be found. At this level of estimating, a variation of plus or minus 10% should be expected.

Adjustments Instructions

No residence matches a model exactly in shape, material, or specifications.

The common differences are:

1. Two or more exterior wall systems
    a. Partial basement
    b. Partly finished basement
2. Specifications or features that are between two classes
3. Crawl space instead of a basement

EXAMPLES

Below are quick examples. See pages 17-19 for complete examples of cost adjustments for these differences:

1. Residence "A" is an average one-story structure with 1,600 S.F. of living area and no basement. Three walls are wood siding on wood frame, and the fourth wall is brick veneer on wood frame. The brick veneer wall is 35% of the exterior wall area.

    Use page 38 to calculate the Base Cost per S.F. of Living Area.

Wood Siding for 1,600 S.F.	=	$78.50 per S.F.
Brick Veneer for 1,600 S.F.	=	$92.15 per S.F.

    .65 ($78.50) + .35 ($92.15) = $83.28 per S.F. of Living Area.

2. a. Residence "B" is the same as Residence "A"; However, it has an unfinished basement under 50% of the building. To adjust the $83.28 per S.F. of living area for this partial basement, use page 38.

    $83.28 + .50 ($8.20) = $87.38 per S.F. of Living Area.

    b. Residence "C" is the same as Residence "A"; However, it has a full basement under the entire building. 640 S.F. or 40% of the basement area is finished.

    Using Page 38:

    $83.28 + .40 ($22.95) + .60 ($8.20) = $97.38 per S.F. of Living Area.

3. When specifications or features of a building are between classes, estimate the percent deviation, and use two tables to calculate the cost per S.F.

    A two-story residence with wood siding and 1,800 S.F. of livir area has features 30% better than Average, but 70% less than Custom.

    From pages 42 and 64:

Custom 1,800 S.F. Base Cost	=	$99.35 per S.F.
Average 1,800 S.F. Base Cost	=	$80.00 per S.F.
DIFFERENCE	=	$19.35 per S.F.

    Cost is $80.00 + .30 ($19.35) = $85.81 per S.F. of Living Area.

4. To add the cost of a crawl space, use the cost of an unfinished basement as a maximum. For specific costs of components to be added or deducted, such as vapor barrier, underdrain, and floor, see the "Assemblies" section, pages 101 to 281.

Model Residence Example

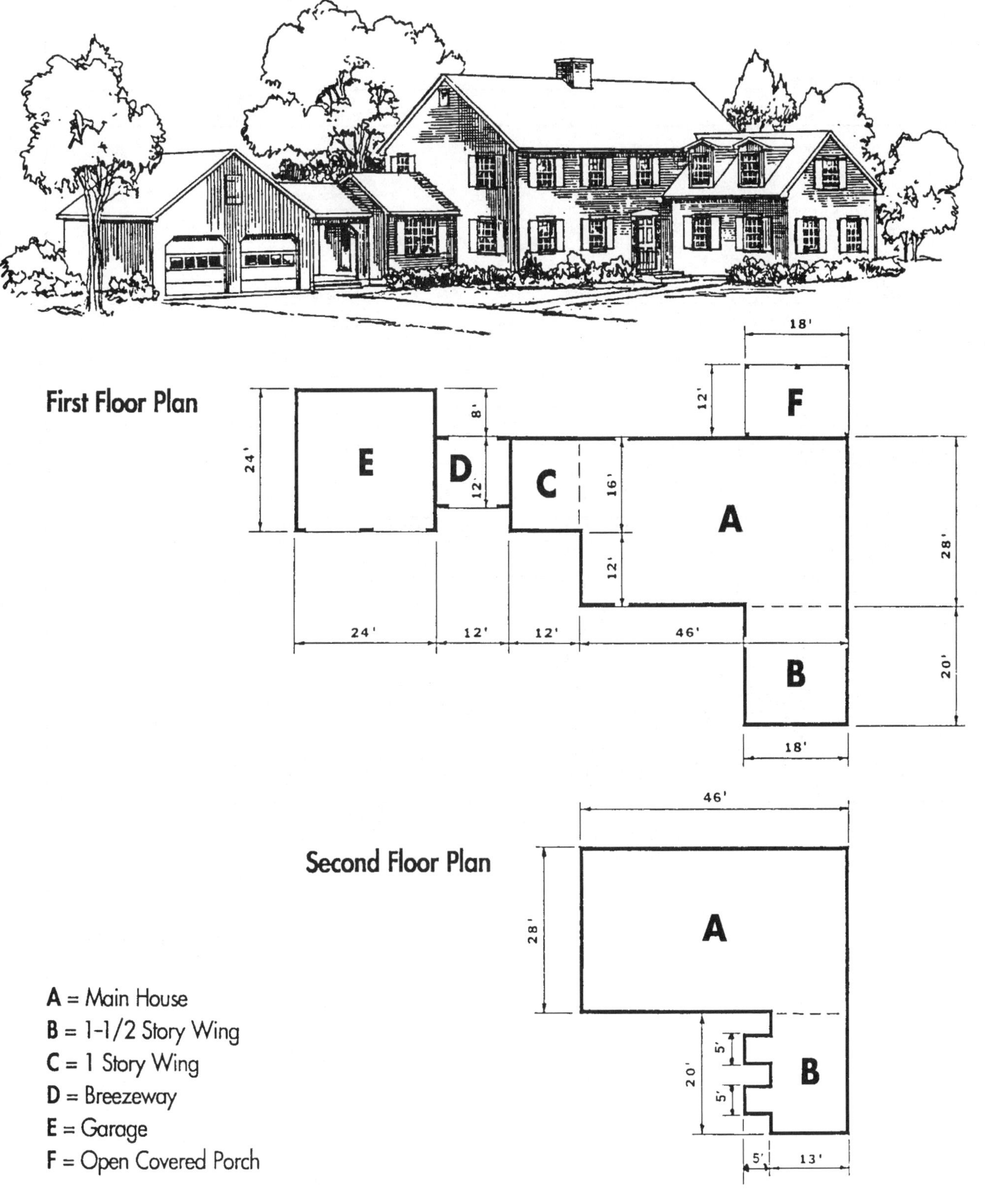

RESIDENTIAL
COST ESTIMATE

OWNERS NAME:	**Albert Westenberg**	APPRAISER:	**Nicole Wojtowicz**
RESIDENCE ADDRESS:	**300 Sygiel Road**	PROJECT:	**# 55**
CITY, STATE, ZIP CODE:	**Three Rivers, MA 01080**	DATE:	**Jan. 2, 2005**

CLASS OF CONSTRUCTION	RESIDENCE TYPE	CONFIGURATION	EXTERIOR WALL SYSTEM
☐ ECONOMY	☐ 1 STORY	☑ DETACHED	☑ WOOD SIDING - WOOD FRAME
☑ AVERAGE	☐ 11/2 STORY	☐ TOWN/ROW HOUSE	☐ BRICK VENEER - WOOD FRAME
☐ CUSTOM	☑ 2 STORY	☐ SEMI-DETACHED	☐ STUCCO ON WOOD FRAME
☐ LUXURY	☐ 2 1/2 STORY		☐ PAINTED CONCRETE BLOCK
	☐ 3 STORY	OCCUPANCY	☐ SOLID MASONRY (AVERAGE & CUSTOM)
	☐ BI-LEVEL	☑ ONE FAMILY	☐ STONE VENEER - WOOD FRAME
	☐ TRI-LEVEL	☐ TWO FAMILY	☐ SOLID BRICK (LUXURY)
		☐ THREE FAMILY	☐ SOLID STONE (LUXURY)
		☐ OTHER	

* LIVING AREA (Main Building)			* LIVING AREA (~~Wing~~ or Ell) (B)			* LIVING AREA (Wing or ~~Ell~~) (C)		
First Level	1288	S.F.	First Level	360	S.F.	First Level	192	S.F.
Second level	1288	S.F.	Second level	310	S.F.	Second level		S.F.
Third Level		S.F.	Third Level		S.F.	Third Level		S.F.
Total	2576	S.F.	Total	670	S.F.	Total	192	S.F.

* Basement Area is not part of living area.

MAIN BUILDING	COSTS PER S.F. LIVING AREA
Cost per Square Foot of Living Area, from Page 42 #	$ 70.65
Basement Addition: ____ % Finished, 100 % Unfinished	+ 4.50
Roof Cover Adjustment: Cedar Shake Type, Page 42 (Add or Deduct)	( ) 1.17
Central Air Conditioning: ☐ Separate Ducts ☑ Heating Ducts, Page 42	+ 1.70
Heating System Adjustment: ____ Type, Page ____ (Add or Deduct)	( ) —
Main Building: Adjusted Cost per S.F. of Living Area	$ 78.02

MAIN BUILDING TOTAL COST: $ 78.02 /S.F. (Cost per S.F. Living Area) x 2,576 S.F. (Living Area) x — (Town/Row House Multiplier (Use 1 for Detached)) = $ 200,980 (TOTAL COST)

~~WING~~ OR ELL (B) 1 - 1/2 STORY	COSTS PER S.F. LIVING AREA
Cost per Square Foot of Living Area, from Page 56	$ 68.15
Basement Addition: 100 % Finished, ____ % Unfinished	+ 20.65
Roof Cover Adjustment: ____ Type, Page ____ (Add or Deduct)	( ) –
Central Air Conditioning: ☐ Separate Ducts ☑ Heating Ducts, Page 30 #	+ 1.7
Heating System Adjustment: ____ Type, Page ____ (Add or Deduct)	( ) —
~~Wing~~ or Ell (B): Adjusted Cost per S.F. of Living Area	$ 90.50

WING OR ELL (B) TOTAL COST: $ 90.50 /S.F. (Cost per S.F. Living Area) x 670 S.F. (Living Area) = $ 60,635 (TOTAL COST)

WING OR ~~ELL~~ (C) 1 STORY	COSTS PER S.F. LIVING AREA
Cost per Square Foot of Living Area, from Page 56 (WOOD SIDING)	$ 99.50
Basement Addition: ____ % Finished, ____ % Unfinished	+ —
Roof Cover Adjustment: ____ Type, Page (Add or Deduct)	( ) —
Central Air Conditioning: ☐ Separate Ducts ☐ Heating Ducts, Page ____	+ —
Heating System Adjustment: ____ Type, Page ____ (Add or Deduct)	( ) —
Wing or Ell (C) Adjusted Cost per S.F. of Living Area	$ 99.50

WING OR ELL (C) TOTAL COST: $ 99.50 /S.F. (Cost per S.F. Living Area) x 192 S.F. (Living Area) = $ 19,184 (TOTAL COST)

TOTAL THIS PAGE: 280,719

Page 1 of 2

RESIDENTIAL
COST ESTIMATE

	QUANTITY		UNIT COST	
Total Page 1				$ 280,719
Additional Bathrooms: 2 Full, 1 Half 2 @ 4261 1 @ 2660				+ 11,182
Finished Attic: N/A Ft. x ____ Ft.		S.F.		
Breezeway: ☑ Open ☐ closed 12 Ft. x 12 Ft.	144	S.F.	18.20	+ 2,621
Covered Porch: ☑ Open ☐ Enclosed 18 Ft. x 12 Ft.	216	S.F.	27.00	+ 5,832
Fireplace: ☑ Interior Chimney ☐ Exterior Chimney ☑ No. of Flues (2) ☑ Additional Fireplaces 1 - 2nd Story				+ 7880
Appliances:				+ —
Kitchen Cabinets Adjustments: (±)				—
☑ Garage ☐ Carport: 2 Car(s) Description Wood, Attached (±)				16,130
Miscellaneous:				+

ADJUSTED TOTAL BUILDING COST $ 324,364

REPLACEMENT COST		
ADJUSTED TOTAL BUILDING COST	$	324,364
Site Improvements		
(A) Paving & Sidewalks	$	
(B) Landscaping	$	
(C) Fences	$	
(D) Swimming Pools	$	
(E) Miscellaneous	$	
TOTAL	$	324,364
Location Factor	x	1.027
Location Replacement Cost	$	333,122
Depreciation	-$	33,312
LOCAL DEPRECIATED COST	$	299,810

INSURANCE COST		
ADJUSTED TOTAL BUILDING COST	$	
Insurance Exclusions		
(A) Footings, sitework, Underground Piping	-$	
(B) Architects Fees	-$	
Total Building Cost Less Exclusion	$	
Location Factor	x	
LOCAL INSURABLE REPLACEMENT COST	$	

SKETCH AND ADDITIONAL CALCULATIONS

Page 2 of 2

Economy Class

1 Story

1-1/2 Story

2 Story

Bi-Level

Tri-Level

- **Mass produced from stock plans**
- **Single family — 1 full bath, 1 kitchen**
- **No basement**
- **Asphalt shingles on roof**
- **Hot air heat**
- **Gypsum wallboard interior finishes**
- **Materials and workmanship are sufficient to meet codes**

Note: The illustration shown may contain some optional components (for example: garages and/or fireplaces) whose costs are shown in the modifications, adjustments, & alternatives below or at the end of the square foot section.

©Home Planners, Inc.

Base cost per square foot of living area

	Living Area										
Exterior Wall	**600**	**800**	**1000**	**1200**	**1400**	**1600**	**1800**	**2000**	**2400**	**2800**	**3200**
Wood Siding - Wood Frame	88.95	80.90	74.70	69.65	65.20	62.40	61.00	59.15	55.25	52.40	50.50
Brick Veneer - Wood Frame	95.20	86.50	79.80	74.20	69.35	66.25	64.80	62.65	58.50	55.40	53.30
Stucco on Wood Frame	85.80	78.05	72.15	**67.35**	63.05	60.40	59.10	57.40	53.60	50.90	49.10
Painted Concrete Block	89.10	81.05	74.85	69.80	65.25	62.50	61.10	59.30	55.30	52.50	50.55
Finished Basement, Add	22.70	21.40	20.45	19.60	18.90	18.45	18.20	17.80	17.30	16.95	16.55
Unfinished Basement, Add	10.30	9.20	8.45	7.80	7.25	6.90	6.70	6.40	6.00	5.70	5.40

Modifications

Add to the total cost

Upgrade Kitchen Cabinets	$ + 697
Solid Surface Countertops	+ 684
Full Bath - including plumbing, wall and floor finishes	+ 3395
Half Bath - including plumbing, wall and floor finishes	+ 2120
One Car Attached Garage	+ 8870
One Car Detached Garage	+ 11,433
Fireplace & Chimney	+ 4185

Adjustments

For multi family - add to total cost

Additional Kitchen	$ + 2556
Additional Bath	+ 3395
Additional Entry & Exit	+ 1276
Separate Heating	+ 1248
Separate Electric	+ 842

For Townhouse/Rowhouse - Multiply cost per square foot by

Inner Unit	.95
End Unit	.97

Alternatives

Add to or deduct from the cost per square foot of living area

Composition Roll Roofing	– .63
Cedar Shake Roof	+ 2.65
Upgrade Walls and Ceilings to Skim Coat Plaster	+ .67
Upgrade Ceilings to Textured Finish	+ .40
Air Conditioning, in Heating Ductwork	+ 2.75
In Separate Ductwork	+ 5.30
Heating Systems, Hot Water	+ 1.46
Heat Pump	+ 1.77
Electric Heat	– 1.24
Not Heated	– 3.16

Additional upgrades or components

Kitchen Cabinets & Countertops	Page 93
Bathroom Vanities	94
Fireplaces & Chimneys	94
Windows, Skylights & Dormers	94
Appliances	95
Breezeways & Porches	95
Finished Attic	95
Garages	96
Site Improvements	96
Wings & Ells	34

Economy 1 Story

Living Area 1200 S.F.
Perimeter 146 L.F.

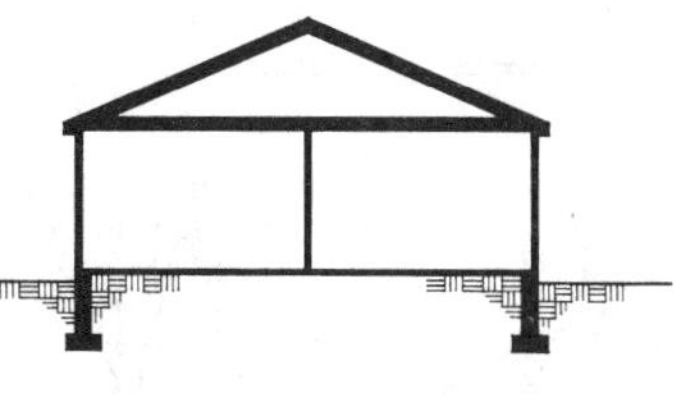

SQUARE FOOT COSTS

			Labor-Hours	Cost Per Square Foot Of Living Area		
				Mat.	Labor	Total
1	**Site Work**	Site preparation for slab; 4′ deep trench excavation for foundation wall.	.060		.98	.98
2	**Foundation**	Continuous reinforced concrete footing, 8″ deep x 18″ wide; dampproofed and insulated 8″ thick reinforced concrete block foundation wall, 4′ deep; 4″ concrete slab on 4″ crushed stone base and polyethylene vapor barrier, trowel finish.	.131	4.25	5.59	9.84
3	**Framing**	Exterior walls - 2″ x 4″ wood studs, 16″ O.C.; 1/2″ insulation board sheathing; wood truss roof frame, 24″ O.C. with 1/2″ plywood sheathing, 4 in 12 pitch.	.098	4.43	4.78	9.21
4	**Exterior Walls**	Metal lath reinforced stucco exterior on insulated wood frame walls; 6″ attic insulation; sliding sash wood windows; 2 flush solid core wood exterior doors with storms.	.110	4.90	5.04	9.94
5	**Roofing**	20 year asphalt shingles; #15 felt building paper; aluminum gutters, downspouts, drip edge and flashings.	.047	.93	1.82	2.75
6	**Interiors**	Walls and ceilings, 1/2″ taped and finished gypsum wallboard, primed and painted with 2 coats; painted baseboard and trim; rubber backed carpeting 80%, asphalt tile 20%; hollow core wood interior doors.	.243	6.97	9.49	16.46
7	**Specialties**	Economy grade kitchen cabinets - 6 L.F. wall and base with plastic laminate counter top and kitchen sink; 30 gallon electric water heater.	.004	1.37	.70	2.07
8	**Mechanical**	1 lavatory, white, wall hung; 1 water closet, white; 1 bathtub, enameled steel, white; gas fired warm air heat.	.086	2.89	2.60	5.49
9	**Electrical**	100 Amp. service; romex wiring; incandescent lighting fixtures, switches, receptacles.	.036	.68	1.14	1.82
10	**Overhead**	Contractor's overhead and profit		3.98	4.81	8.79
			Total	30.40	36.95	**67.35**

- **Mass produced from stock plans**
- **Single family — 1 full bath, 1 kitchen**
- **No basement**
- **Asphalt shingles on roof**
- **Hot air heat**
- **Gypsum wallboard interior finishes**
- **Materials and workmanship are sufficient to meet codes**

Note: The illustration shown may contain some optional components (for example: garages and/or fireplaces) whose costs are shown in the modifications, adjustments, & alternatives below or at the end of the square foot section.

Base cost per square foot of living area

Exterior Wall	Living Area 600	800	1000	1200	1400	1600	1800	2000	2400	2800	3200
Wood Siding - Wood Frame	102.85	85.90	76.80	72.55	69.55	65.00	62.70	60.40	55.50	53.65	51.70
Brick Veneer - Wood Frame	111.60	92.20	82.60	78.00	74.75	69.75	67.20	64.70	59.30	57.25	54.95
Stucco on Wood Frame	98.45	82.65	73.75	69.75	66.85	62.60	60.45	58.25	53.60	51.85	50.00
Painted Concrete Block	103.05	86.05	76.90	72.70	69.70	65.10	62.85	60.50	55.60	53.75	51.80
Finished Basement, Add	17.50	14.85	14.20	13.70	13.35	12.80	12.50	12.25	11.70	11.45	11.15
Unfinished Basement, Add	9.00	6.90	6.40	6.00	5.70	5.30	5.05	4.85	4.45	4.25	4.00

Modifications

Add to the total cost

Upgrade Kitchen Cabinets	$ + 697
Solid Surface Countertops	+ 684
Full Bath - including plumbing, wall and floor finishes	+ 3395
Half Bath - including plumbing, wall and floor finishes	+ 2120
One Car Attached Garage	+ 8870
One Car Detached Garage	+ 11,433
Fireplace & Chimney	+ 4185

Adjustments

For multi family - add to total cost

Additional Kitchen	$ + 2556
Additional Bath	+ 3395
Additional Entry & Exit	+ 1276
Separate Heating	+ 1248
Separate Electric	+ 842

For Townhouse/Rowhouse - Multiply cost per square foot by

Inner Unit	.95
End Unit	.97

Alternatives

Add to or deduct from the cost per square foot of living area

Composition Roll Roofing	– .4
Cedar Shake Roof	+ 1.9
Upgrade Walls and Ceilings to Skim Coat Plaster	+ .6
Upgrade Ceilings to Textured Finish	+ .4
Air Conditioning, in Heating Ductwork	+ 2.0
In Separate Ductwork	+ 4.6
Heating Systems, Hot Water	+ 1.4
Heat Pump	+ 1.9
Electric Heat	– .9
Not Heated	– 2.9

Additional upgrades or components

Kitchen Cabinets & Countertops	Page 9
Bathroom Vanities	9
Fireplaces & Chimneys	9
Windows, Skylights & Dormers	9
Appliances	9
Breezeways & Porches	9
Finished Attic	9
Garages	9
Site Improvements	9
Wings & Ells	3

Economy 1-1/2 Story

Living Area 1600 S.F.
Perimeter 135 L.F.

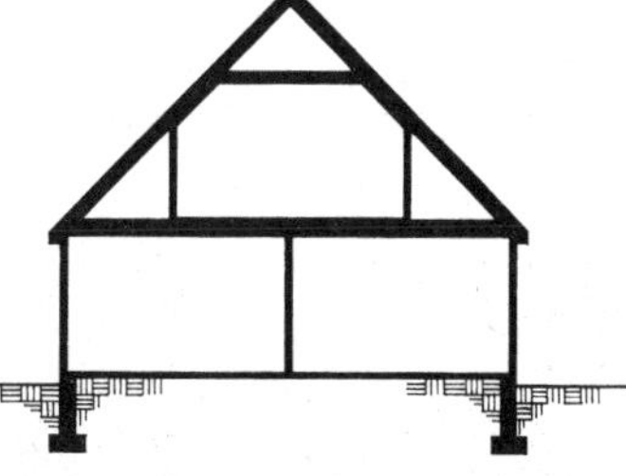

		Labor-Hours	Cost Per Square Foot Of Living Area		
			Mat.	Labor	Total
1 Site Work	Site preparation for slab; 4′ deep trench excavation for foundation wall.	.041		.73	.73
2 Foundation	Continuous reinforced concrete footing, 8″ deep x 18″ wide; dampproofed and insulated 8″ thick reinforced concrete block foundation wall, 4′ deep; 4″ concrete slab on 4″ crushed stone base and polyethylene vapor barrier, trowel finish.	.073	2.84	3.80	6.64
3 Framing	Exterior walls - 2″ x 4″ wood studs, 16″ O.C.; 1/2″ insulation board sheathing; 2″ x 6″ rafters, 16″ O.C. with 1/2″ plywood sheathing, 8 in 12 pitch; 2″ x 8″ floor joists 16″ O.C. with bridging and 5/8″ plywood subfloor.	.090	4.59	5.37	9.96
4 Exterior Walls	Beveled wood siding and building paper on insulated wood frame walls; 6″ attic insulation; double hung windows; 2 flush solid core wood exterior doors with storms.	.077	7.34	4.18	11.52
5 Roofing	20 year asphalt shingles; #15 felt building paper; aluminum gutters, downspouts, drip edge and flashings.	.029	.58	1.14	1.72
6 Interiors	Walls and ceilings, 1/2″ taped and finished gypsum wallboard, primed and painted with 2 coats; painted baseboard and trim; rubber backed carpeting 80%, asphalt tile 20%; hollow core wood interior doors.	.204	7.82	10.17	17.99
7 Specialties	Economy grade kitchen cabinets - 6 L.F. wall and base with plastic laminate counter top and kitchen sink; 30 gallon electric water heater.	.020	1.03	.53	1.56
8 Mechanical	1 lavatory, white, wall hung; 1 water closet, white; 1 bathtub, enameled steel, white; gas fired warm air heat.	.079	2.41	2.34	4.75
9 Electrical	100 Amp. service; romex wiring; incandescent lighting fixtures, switches, receptacles.	.033	.62	1.03	1.65
10 Overhead	Contractor's overhead and profit.		4.07	4.41	8.48
		Total	31.30	33.70	**65.00**

- **Mass produced from stock plans**
- **Single family — 1 full bath, 1 kitchen**
- **No basement**
- **Asphalt shingles on roof**
- **Hot air heat**
- **Gypsum wallboard interior finishes**
- **Materials and workmanship are sufficient to meet codes**

Note: The illustration shown may contain some optional components (for example: garages and/or fireplaces) whose costs are shown in the modifications, adjustments, & alternatives below or at the end of the square foot section.

Base cost per square foot of living area

Exterior Wall	Living Area 1000	1200	1400	1600	1800	2000	2200	2600	3000	3400	3800
Wood Siding - Wood Frame	81.65	73.70	70.25	67.80	65.35	62.55	60.70	57.25	53.65	52.20	50.85
Brick Veneer - Wood Frame	88.40	79.95	76.05	73.40	70.65	67.55	65.45	61.55	57.65	56.00	54.50
Stucco on Wood Frame	78.20	70.55	67.25	65.00	62.70	59.90	58.25	55.00	51.60	50.25	49.00
Painted Concrete Block	81.85	73.85	70.35	67.95	65.55	62.60	60.85	57.35	53.70	52.25	50.95
Finished Basement, Add	11.90	11.35	11.00	10.75	10.40	10.25	10.05	9.65	9.35	9.20	9.00
Unfinished Basement, Add	5.55	5.15	4.85	4.65	4.35	4.25	4.10	3.75	3.55	3.40	3.30

Modifications

Add to the total cost

Upgrade Kitchen Cabinets	$ + 697
Solid Surface Countertops	+ 684
Full Bath - including plumbing, wall and floor finishes	+ 3395
Half Bath - including plumbing, wall and floor finishes	+ 2120
One Car Attached Garage	+ 8870
One Car Detached Garage	+ 11,433
Fireplace & Chimney	+ 4670

Adjustments

For multi family - add to total cost

Additional Kitchen	$ + 2556
Additional Bath	+ 3395
Additional Entry & Exit	+ 1276
Separate Heating	+ 1248
Separate Electric	+ 842

For Townhouse/Rowhouse - Multiply cost per square foot by

Inner Unit	.93
End Unit	.96

Alternatives

Add to or deduct from the cost per square foot of living area

Composition Roll Roofing	– .32
Cedar Shake Roof	+ 1.33
Upgrade Walls and Ceilings to Skim Coat Plaster	+ .69
Upgrade Ceilings to Textured Finish	+ .40
Air Conditioning, in Heating Ductwork	+ 1.65
In Separate Ductwork	+ 4.25
Heating Systems, Hot Water	+ 1.36
Heat Pump	+ 2.07
Electric Heat	– .81
Not Heated	– 2.75

Additional upgrades or components

Kitchen Cabinets & Countertops	Page 93
Bathroom Vanities	94
Fireplaces & Chimneys	94
Windows, Skylights & Dormers	94
Appliances	95
Breezeways & Porches	95
Finished Attic	95
Garages	96
Site Improvements	96
Wings & Ells	34

Important: See the Reference Section for Location Factors (to adjust for your city) and Estimating Forms

Economy 2 Story

Living Area 2000 S.F.
Perimeter 135 L.F.

		Labor-Hours	Cost Per Square Foot Of Living Area		
			Mat.	Labor	Total
1 Site Work	Site preparation for slab; 4′ deep trench excavation for foundation wall.	.034		.59	.59
2 Foundation	Continuous reinforced concrete footing, 8″ deep x 18″ wide; dampproofed and insulated 8″ thick reinforced concrete block foundation wall, 4′ deep; 4″ concrete slab on 4″ crushed stone base and polyethylene vapor barrier, trowel finish.	.069	2.27	3.04	5.31
3 Framing	Exterior walls - 2″ x 4″ wood studs, 16″ O.C.; 1/2″ insulation board sheathing; wood truss roof frame, 24″ O.C. with 1/2″ plywood sheathing, 4 in 12 pitch; 2″ x 8″ floor joists 16″ O.C. with bridging and 5/8″ plywood subfloor.	.112	4.70	5.63	10.33
4 Exterior Walls	Beveled wood siding and building paper on insulated wood frame walls; 6″ attic insulation; double hung windows; 2 flush solid core wood exterior doors with storms.	.107	7.50	4.26	11.76
5 Roofing	20 year asphalt shingles; #15 felt building paper; aluminum gutters, downspouts, drip edge and flashings.	.024	.47	.91	1.38
6 Interiors	Walls and ceilings, 1/2″ taped and finished gypsum wallboard, primed and painted with 2 coats; painted baseboard and trim; rubber backed carpeting 80%, asphalt tile 20%; hollow core wood interior doors.	.219	7.73	10.17	17.90
7 Specialties	Economy grade kitchen cabinets - 6 L.F. wall and base with plastic laminate counter top and kitchen sink; 30 gallon electric water heater.	.017	.83	.43	1.26
8 Mechanical	1 lavatory, white, wall hung; 1 water closet, white; 1 bathtub, enameled steel, white; gas fired warm air heat.	.061	2.12	2.17	4.29
9 Electrical	100 Amp. service; romex wiring; incandescent lighting fixtures; switches, receptacles.	.030	.58	.96	1.54
10 Overhead	Contractor's overhead and profit		3.95	4.24	8.19
		Total	30.15	32.40	**62.55**

RESIDENTIAL | Economy | Bi-Level

- **Mass produced from stock plans**
- **Single family – 1 full bath, 1 kitchen**
- **No basement**
- **Asphalt shingles on roof**
- **Hot air heat**
- **Gypsum wallboard interior finishes**
- **Materials and workmanship are sufficient to meet codes**

Note: The illustration shown may contain some optional components (for example: garages and/or fireplaces) whose costs are shown in the modifications, adjustments, & alternatives below or at the end of the square foot section.

Base cost per square foot of living area

Exterior Wall	Living Area 1000	1200	1400	1600	1800	2000	2200	2600	3000	3400	3800
Wood Siding - Wood Frame	76.10	68.55	65.40	63.20	61.00	58.30	56.75	53.60	50.35	49.00	47.85
Brick Veneer - Wood Frame	81.15	73.30	69.80	67.40	64.95	62.10	60.30	56.85	53.35	51.90	50.55
Stucco on Wood Frame	73.45	66.20	63.15	61.10	59.05	56.35	54.90	51.95	48.80	47.60	46.45
Painted Concrete Block	76.20	68.70	65.50	63.30	61.10	58.45	56.85	53.70	50.45	49.10	47.95
Finished Basement, Add	11.90	11.35	11.00	10.75	10.40	10.25	10.05	9.65	9.35	9.20	9.00
Unfinished Basement, Add	5.55	5.15	4.85	4.65	4.35	4.25	4.10	3.75	3.55	3.40	3.30

Modifications

Add to the total cost

Upgrade Kitchen Cabinets	$ + 697
Solid Surface Countertops	+ 684
Full Bath - including plumbing, wall and floor finishes	+ 3395
Half Bath - including plumbing, wall and floor finishes	+ 2120
One Car Attached Garage	+ 8870
One Car Detached Garage	+ 11,433
Fireplace & Chimney	+ 4185

Adjustments

For multi family - add to total cost

Additional Kitchen	$ + 2556
Additional Bath	+ 3395
Additional Entry & Exit	+ 1276
Separate Heating	+ 1248
Separate Electric	+ 842

For Townhouse/Rowhouse - Multiply cost per square foot by

Inner Unit	.94
End Unit	.97

Alternatives

Add to or deduct from the cost per square foot of living area

Composition Roll Roofing	– .32
Cedar Shake Roof	+ 2.65
Upgrade Walls and Ceilings to Skim Coat Plaster	+ .65
Upgrade Ceilings to Textured Finish	+ .40
Air Conditioning, in Heating Ductwork	+ 1.65
In Separate Ductwork	+ 4.25
Heating Systems, Hot Water	+ 1.36
Heat Pump	+ 2.07
Electric Heat	– .81
Not Heated	– 2.75

Additional upgrades or components

Kitchen Cabinets & Countertops	Page 93
Bathroom Vanities	94
Fireplaces & Chimneys	94
Windows, Skylights & Dormers	94
Appliances	95
Breezeways & Porches	95
Finished Attic	95
Garages	96
Site Improvements	96
Wings & Ells	34

Economy Bi-Level

Living Area 2000 S.F.
Perimeter 135 L.F.

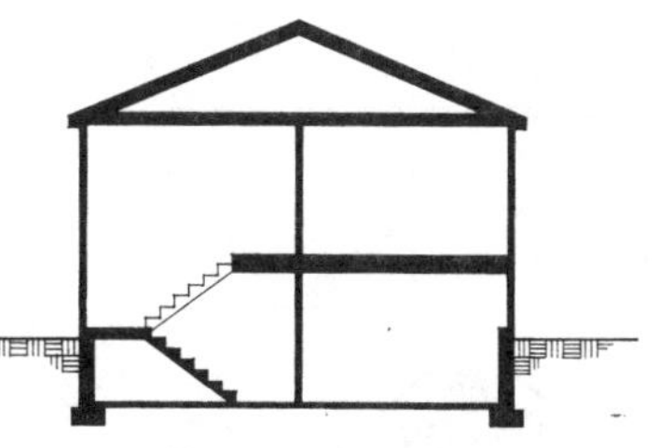

SQUARE FOOT COSTS

		Labor-Hours	Cost Per Square Foot Of Living Area		
			Mat.	Labor	Total
1 Site Work	Excavation for lower level, 4' deep. Site preparation for slab.	.029		.59	.59
2 Foundation	Continuous reinforced concrete footing, 8" deep x 18" wide; dampproofed and insulated 8" thick reinforced concrete block foundation wall, 4' deep; 4" concrete slab on 4" crushed stone base and polyethylene vapor barrier, trowel finish.	.069	2.27	3.04	5.31
3 Framing	Exterior walls - 2" x 4" wood studs, 16" O.C.; 1/2" insulation board sheathing; wood truss roof frame, 24" O.C. with 1/2" plywood sheathing, 4 in 12 pitch; 2" x 8" floor joists 16" O.C. with bridging and 5/8" plywood subfloor.	.107	4.41	5.29	9.70
4 Exterior Walls	Beveled wood siding and building paper on insulated wood frame walls; 6" attic insulation; double hung windows; 2 flush solid core wood exterior doors with storms.	.089	5.88	3.33	9.21
5 Roofing	20 year asphalt shingles; #15 felt building paper; aluminum gutters, downspouts, drip edge and flashings.	.024	.47	.91	1.38
6 Interiors	Walls and ceilings, 1/2" taped and finished gypsum wallboard, primed and painted with 2 coats; painted baseboard and trim, rubber backed carpeting 80%, asphalt tile 20%; hollow core wood interior doors.	.213	7.57	9.85	17.42
7 Specialties	Economy grade kitchen cabinets - 6 L.F. wall and base with plastic laminate counter top and kitchen sink; 30 gallon electric water heater.	.018	.83	.43	1.26
8 Mechanical	1 lavatory, white, wall hung; 1 water closet, white; 1 bathtub, enameled steel, white; gas fired warm air heat.	.061	2.12	2.17	4.29
9 Electrical	100 Amp. service; romex wiring; incandescent lighting fixtures; switches, receptacles.	.030	.58	.96	1.54
10 Overhead	Contractor's overhead and profit.		3.62	3.98	7.60
		Total	27.75	30.55	**58.30**

- **Mass produced from stock plans**
- **Single family – 1 full bath, 1 kitchen**
- **No basement**
- **Asphalt shingles on roof**
- **Hot air heat**
- **Gypsum wallboard interior finishes**
- **Materials and workmanship are sufficient to meet codes**

Note: The illustration shown may contain some optional components (for example: garages and/or fireplaces) whose costs are shown in the modifications, adjustments, & alternatives below or at the end of the square foot section.

©Design Basics, Inc.

Base cost per square foot of living area

Exterior Wall	Living Area 1200	1500	1800	2000	2200	2400	2800	3200	3600	4000	4400
Wood Siding - Wood Frame	70.00	64.40	60.15	58.50	55.95	54.00	52.45	50.35	47.80	46.95	45.00
Brick Veneer - Wood Frame	74.65	68.60	63.95	62.15	59.40	57.25	55.60	53.25	50.50	49.55	47.45
Stucco on Wood Frame	67.60	62.25	58.15	56.65	54.20	52.35	50.85	48.85	46.45	45.60	43.80
Solid Masonry	70.15	64.50	60.25	58.60	56.05	54.10	52.50	50.40	47.90	47.00	45.10
Finished Basement, Add*	14.30	13.65	13.10	12.80	12.60	12.30	12.15	11.80	11.55	11.45	11.25
Unfinished Basement, Add*	6.15	5.65	5.15	5.00	4.80	4.55	4.45	4.20	4.00	3.90	3.75

*Basement under middle level only.

Modifications

Add to the total cost

Upgrade Kitchen Cabinets	$ + 697
Solid Surface Countertops	+ 684
Full Bath - including plumbing, wall and floor finishes	+ 3395
Half Bath - including plumbing, wall and floor finishes	+ 2120
One Car Attached Garage	+ 8870
One Car Detached Garage	+ 11,433
Fireplace & Chimney	+ 4185

Adjustments

For multi family - add to total cost

Additional Kitchen	$ + 2556
Additional Bath	+ 3395
Additional Entry & Exit	+ 1276
Separate Heating	+ 1248
Separate Electric	+ 842

For Townhouse/Rowhouse - Multiply cost per square foot by

Inner Unit	.93
End Unit	.96

Alternatives

Add to or deduct from the cost per square foot of living area

Composition Roll Roofing	– .44
Cedar Shake Roof	+ 1.91
Upgrade Walls and Ceilings to Skim Coat Plaster	+ .60
Upgrade Ceilings to Textured Finish	+ .40
Air Conditioning, in Heating Ductwork	+ 1.40
In Separate Ductwork	+ 3.97
Heating Systems, Hot Water	+ 1.30
Heat Pump	+ 2.15
Electric Heat	– .72
Not Heated	– 2.66

Additional upgrades or components

Kitchen Cabinets & Countertops	Page 93
Bathroom Vanities	94
Fireplaces & Chimneys	94
Windows, Skylights & Dormers	94
Appliances	95
Breezeways & Porches	95
Finished Attic	95
Garages	96
Site Improvements	96
Wings & Ells	34

Important: See the Reference Section for Location Factors (to adjust for your city) and Estimating Forms

Economy Tri-Level

Living Area 2400 S.F.
Perimeter 163 L.F.

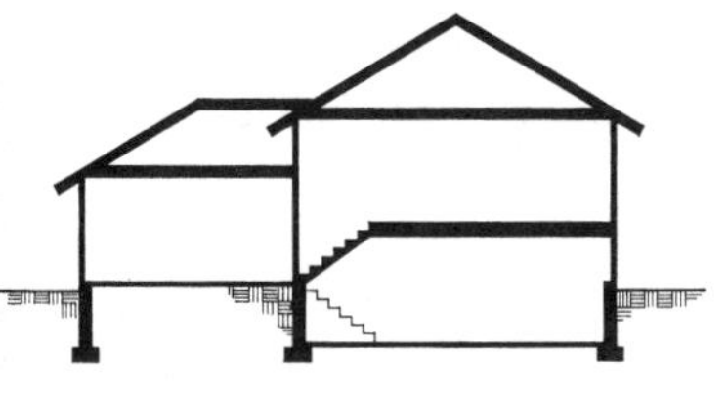

		Labor-Hours	Cost Per Square Foot Of Living Area		
			Mat.	Labor	Total
1 Site Work	Site preparation for slab; 4′ deep trench excavation for foundation wall, excavation for lower level, 4′ deep.	.027		.49	.49
2 Foundation	Continuous reinforced concrete footing, 8″ deep x 18″ wide; dampproofed and insulated 8″ thick reinforced concrete block foundation wall, 4′ deep; 4″ concrete slab on 4″ crushed stone base and polyethylene vapor barrier, trowel finish.	.071	2.59	3.31	5.90
3 Framing	Exterior walls - 2″ x 4″ wood studs, 16″ O.C.; 1/2″ insulation board sheathing; wood truss roof frame, 24″ O.C. with 1/2″ plywood sheathing, 4 in 12 pitch; 2″ x 8″ floor joists 16″ O.C. with bridging and 5/8″ plywood subfloor.	.094	4.14	4.76	8.90
4 Exterior Walls	Beveled wood siding and building paper on insulated wood frame walls; 6″ attic insulation; double hung windows; 2 flush solid core wood exterior doors with storms.	.081	5.09	2.90	7.99
5 Roofing	20 year asphalt shingles; #15 felt building paper; aluminum gutters, downspouts, drip edge and flashings.	.032	.62	1.21	1.83
6 Interiors	Walls and ceilings, 1/2″ taped and finished gypsum wallboard, primed and painted with 2 coats; painted baseboard and trim, rubber backed carpeting 80%, asphalt tile 20%; hollow core wood interior doors.	.177	6.60	8.73	15.33
7 Specialties	Economy grade kitchen cabinets - 6 L.F. wall and base with plastic laminate counter top and kitchen sink; 30 gallon electric water heater.	.014	.69	.36	1.05
8 Mechanical	1 lavatory, white, wall hung; 1 water closet, white; 1 bathtub, enameled steel, white; gas fired warm air heat.	.057	1.93	2.06	3.99
9 Electrical	100 Amp. service; romex wiring; incandescent lighting fixtures, switches, receptacles.	.029	.55	.92	1.47
10 Overhead	Contractor's overhead and profit		3.34	3.71	7.05
		Total	25.55	28.45	**54.00**

1 Story — Base cost per square foot of living area

Exterior Wall	Living Area 50	100	200	300	400	500	600	700
Wood Siding - Wood Frame	124.55	95.20	82.45	69.10	65.00	62.45	60.80	61.35
Brick Veneer - Wood Frame	140.30	106.40	91.75	75.35	70.60	67.70	65.80	66.15
Stucco on Wood Frame	116.55	89.45	77.65	65.95	62.15	59.75	58.25	58.90
Painted Concrete Block	124.95	95.50	82.70	69.25	65.10	62.60	60.90	61.45
Finished Basement, Add	34.50	28.20	25.55	21.15	20.30	19.80	19.45	19.20
Unfinished Basement, Add	19.10	14.30	12.30	8.90	8.25	7.80	7.60	7.35

1-1/2 Story — Base cost per square foot of living area

Exterior Wall	Living Area 100	200	300	400	500	600	700	800
Wood Siding - Wood Frame	97.35	78.20	66.50	59.80	56.25	54.55	52.40	51.75
Brick Veneer - Wood Frame	111.40	89.45	75.90	67.10	63.00	60.95	58.45	57.70
Stucco on Wood Frame	90.25	72.45	61.70	56.05	52.80	51.30	49.35	48.70
Painted Concrete Block	97.70	78.45	66.75	59.95	56.40	54.75	52.55	51.90
Finished Basement, Add	23.05	20.40	18.70	16.75	16.20	15.85	15.55	15.50
Unfinished Basement, Add	11.80	9.80	8.45	7.00	6.60	6.30	6.10	6.00

2 Story — Base cost per square foot of living area

Exterior Wall	Living Area 100	200	400	600	800	1000	1200	1400
Wood Siding - Wood Frame	99.35	73.95	62.80	52.20	48.65	46.40	44.95	45.70
Brick Veneer - Wood Frame	115.10	85.20	72.20	58.45	54.25	51.65	49.95	50.50
Stucco on Wood Frame	91.35	68.25	58.00	49.00	45.80	43.75	42.40	43.25
Painted Concrete Block	99.75	74.25	63.05	52.35	48.75	46.60	45.10	45.80
Finished Basement, Add	17.25	14.10	12.75	10.60	10.20	9.90	9.75	9.60
Unfinished Basement, Add	9.55	7.15	6.10	4.45	4.15	3.90	3.80	3.70

Base costs do not include bathroom or kitchen facilities. Use Modifications/Adjustments/Alternatives on pages 93-96 where appropriate.

Average Class

1 Story

1-1/2 Story

2 Story

2-1/2 Story

Bi-Level

Tri-Level

- **Simple design from standard plans**
- **Single family – 1 full bath, 1 kitchen**
- **No basement**
- **Asphalt shingles on roof**
- **Hot air heat**
- **Gypsum wallboard interior finishes**
- **Materials and workmanship are average**

Note: The illustration shown may contain some optional components (for example: garages and/or fireplaces) whose costs are shown in the modifications, adjustments, & alternatives below or at the end of the square foot section.

Base cost per square foot of living area

Exterior Wall	Living Area 600	800	1000	1200	1400	1600	1800	2000	2400	2800	3200
Wood Siding - Wood Frame	110.45	100.55	93.00	87.00	81.75	**78.50**	76.70	74.50	69.95	66.70	64.45
Brick Veneer - Wood Frame	126.70	116.05	107.95	101.35	95.65	92.15	90.20	87.70	82.85	79.35	76.85
Stucco on Wood Frame	119.10	109.25	101.80	95.80	90.60	87.40	85.60	83.45	78.90	75.70	73.45
Solid Masonry	137.55	125.70	116.70	109.30	102.85	98.85	96.70	93.90	88.45	84.50	81.65
Finished Basement, Add	27.75	26.85	25.60	24.50	23.55	22.95	22.60	22.10	21.45	20.95	20.40
Unfinished Basement, Add	11.65	10.60	9.80	9.15	8.55	8.20	7.95	7.70	7.25	7.00	6.70

Modifications

Add to the total cost

Upgrade Kitchen Cabinets	$ + 3157
Solid Surface Countertops (Included)	
Full Bath - including plumbing, wall and floor finishes	+ 4261
Half Bath - including plumbing, wall and floor finishes	+ 2660
One Car Attached Garage	+ 9490
One Car Detached Garage	+ 12,424
Fireplace & Chimney	+ 4185

Adjustments

For multi family - add to total cost

Additional Kitchen	$ + 5105
Additional Bath	+ 4261
Additional Entry & Exit	+ 1276
Separate Heating	+ 1248
Separate Electric	+ 1575

For Townhouse/Rowhouse - Multiply cost per square foot by

Inner Unit	.92
End Unit	.96

Alternatives

Add to or deduct from the cost per square foot of living area

Cedar Shake Roof	+ 2.34
Clay Tile Roof	+ 4.08
Slate Roof	+ 7.77
Upgrade Walls to Skim Coat Plaster	+ .36
Upgrade Ceilings to Textured Finish	+ .40
Air Conditioning, in Heating Ductwork	+ 2.85
In Separate Ductwork	+ 5.50
Heating Systems, Hot Water	+ 1.50
Heat Pump	+ 1.84
Electric Heat	– .68
Not Heated	– 2.63

Additional upgrades or components

Kitchen Cabinets & Countertops	Page 93
Bathroom Vanities	94
Fireplaces & Chimneys	94
Windows, Skylights & Dormers	94
Appliances	95
Breezeways & Porches	95
Finished Attic	95
Garages	96
Site Improvements	96
Wings & Ells	56

Average 1 Story

Living Area 1600 S.F.
Perimeter 163 L.F.

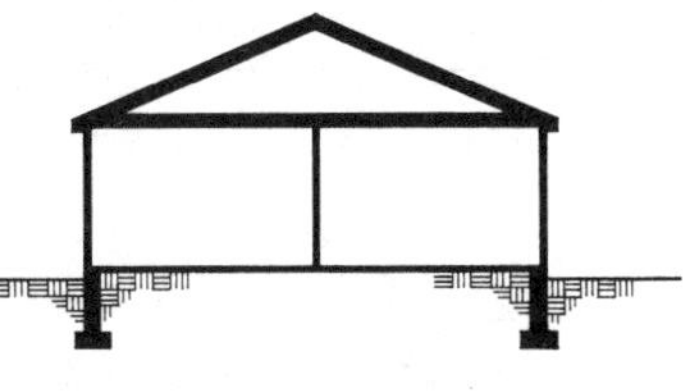

		Labor-Hours	Cost Per Square Foot Of Living Area		
			Mat.	Labor	Total
1 Site Work	Site preparation for slab; 4′ deep trench excavation for foundation wall.	.048		.75	.75
2 Foundation	Continuous reinforced concrete footing 8″ deep x 18″ wide; dampproofed and insulated reinforced concrete foundation wall, 8″ thick, 4′ deep; 4″ concrete slab on 4″ crushed stone base and polyethylene vapor barrier, trowel finish.	.113	3.99	5.11	9.10
3 Framing	Exterior walls - 2″ x 4″ wood studs, 16″ O.C.; 1/2″ plywood sheathing; 2″ x 6″ rafters 16″ O.C. with 1/2″ plywood sheathing, 4 in 12 pitch; 2″ x 6″ ceiling joists 16″ O.C.; 1/2″ plywood subfloor on 1″ x 2″ wood sleepers 16″ O.C.	.136	5.84	7.15	12.99
4 Exterior Walls	Beveled wood siding and building paper on insulated wood frame walls; 6″ attic insulation; double hung windows; 3 flush solid core wood exterior doors with storms.	.098	7.45	3.92	11.37
5 Roofing	25 year asphalt shingles; #15 felt building paper; aluminum gutters, downspouts, drip edge and flashings.	.047	.96	1.87	2.83
6 Interiors	Walls and ceilings, 1/2″ taped and finished gypsum wallboard, primed and painted with 2 coats; painted baseboard and trim, finished hardwood floor 40%, carpet with 1/2″ underlayment 40%, vinyl tile with 1/2″ underlayment 15%, ceramic tile with 1/2″ underlayment 5%; hollow core and louvered doors.	.251	9.91	10.08	19.99
7 Specialties	Average grade kitchen cabinets - 14 L.F. wall and base with solid surface counter top and kitchen sink; 40 gallon electric water heater.	.009	1.87	.84	2.71
8 Mechanical	1 lavatory, white, wall hung; 1 water closet, white; 1 bathtub with shower, enameled steel, white; gas fired warm air heat.	.098	2.59	2.40	4.99
9 Electrical	200 Amp. service; romex wiring; incandescent lighting fixtures, switches, receptacles.	.041	.98	1.39	2.37
10 Overhead	Contractor's overhead and profit and plans.		5.71	5.69	11.40
		Total	39.30	39.20	**78.50**

- **Simple design from standard plans**
- **Single family — 1 full bath, 1 kitchen**
- **No basement**
- **Asphalt shingles on roof**
- **Hot air heat**
- **Gypsum wallboard interior finishes**
- **Materials and workmanship are average**

Note: The illustration shown may contain some optional components (for example: garages and/or fireplaces) whose costs are shown in the modifications, adjustments, & alternatives below or at the end of the square foot section.

©By Designer

Base cost per square foot of living area

Exterior Wall	Living Area 600	800	1000	1200	1400	1600	1800	2000	2400	2800	3200
Wood Siding - Wood Frame	124.60	104.85	94.15	89.10	85.55	80.20	77.60	74.80	69.25	67.10	64.75
Brick Veneer - Wood Frame	134.30	111.90	100.65	95.25	91.35	85.55	82.65	79.55	73.45	71.10	68.40
Stucco on Wood Frame	123.70	104.20	93.50	88.55	84.95	79.75	77.10	74.40	68.85	66.75	64.35
Solid Masonry	149.45	122.90	110.80	104.75	100.35	93.75	90.45	87.00	80.00	77.30	74.15
Finished Basement, Add	22.65	19.80	18.95	18.30	17.85	17.15	16.75	16.40	15.65	15.35	14.95
Unfinished Basement, Add	10.00	7.95	7.35	6.95	6.65	6.20	6.00	5.75	5.35	5.15	4.90

Modifications

Add to the total cost

Upgrade Kitchen Cabinets	$ + 3157
Solid Surface Countertops (Included)	
Full Bath - including plumbing, wall and floor finishes	+ 4261
Half Bath - including plumbing, wall and floor finishes	+ 2660
One Car Attached Garage	+ 9490
One Car Detached Garage	+ 12,424
Fireplace & Chimney	+ 4185

Adjustments

For multi family - add to total cost

Additional Kitchen	$ + 5105
Additional Bath	+ 4261
Additional Entry & Exit	+ 1276
Separate Heating	+ 1248
Separate Electric	+ 1575

For Townhouse/Rowhouse - Multiply cost per square foot by

Inner Unit	.92
End Unit	.96

Alternatives

Add to or deduct from the cost per square foot of living area

Cedar Shake Roof	+ 1.69
Clay Tile Roof	+ 2.95
Slate Roof	+ 5.61
Upgrade Walls to Skim Coat Plaster	+ .41
Upgrade Ceilings to Textured Finish	+ .40
Air Conditioning, in Heating Ductwork	+ 2.15
In Separate Ductwork	+ 4.80
Heating Systems, Hot Water	+ 1.42
Heat Pump	+ 2.04
Electric Heat	- .60
Not Heated	- 2.54

Additional upgrades or components

Kitchen Cabinets & Countertops	Page 93
Bathroom Vanities	94
Fireplaces & Chimneys	94
Windows, Skylights & Dormers	94
Appliances	95
Breezeways & Porches	95
Finished Attic	95
Garages	96
Site Improvements	96
Wings & Ells	56

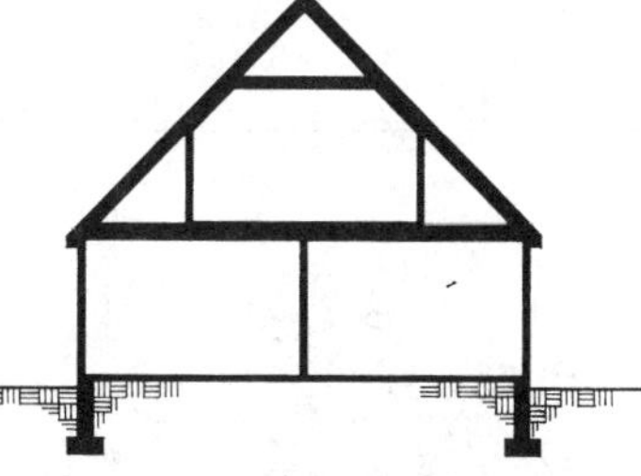

Average 1-1/2 Story

Living Area 1800 S.F.
Perimeter 144 L.F.

		Labor-Hours	Cost Per Square Foot Of Living Area		
			Mat.	Labor	Total
1 Site Work	Site preparation for slab; 4′ deep trench excavation for foundation wall.	.037		.67	.67
2 Foundation	Continuous reinforced concrete footing 8″ deep x 18″ wide; dampproofed and insulated reinforced concrete foundation wall, 8″ thick, 4′ deep; 4″ concrete slab on 4″ crushed stone base and polyethylene vapor barrier, trowel finish.	.073	2.82	3.75	6.57
3 Framing	Exterior walls - 2″ x 4″ wood studs, 16″ O.C.; 1/2″ plywood sheathing; 2″ x 6″ rafters 16″ O.C. with 1/2″ plywood sheathing, 8 in 12 pitch; 2″ x 8″ floor joists 16″ O.C. with 5/8″ plywood subfloor; 1/2″ plywood subfloor on 1″ x 2″ wood sleepers 16″ O.C.	.098	5.99	6.66	12.65
4 Exterior Walls	Beveled wood siding and building paper on insulated wood frame walls; 6″ attic insulation; double hung windows; 3 flush solid core wood exterior doors with storms.	.078	8.09	4.24	12.33
5 Roofing	25 year asphalt shingles; #15 felt building paper; aluminum gutters, downspouts, drip edge and flashings.	.029	.60	1.17	1.77
6 Interiors	Walls and ceilings, 1/2″ taped and finished gypsum wallboard, primed and painted with 2 coats; painted baseboard and trim, finished hardwood floor 40%, carpet with 1/2″ underlayment 40%, vinyl tile with 1/2″ underlayment 15%, ceramic tile with 1/2″ underlayment 5%; hollow core and louvered doors.	.225	11.54	11.43	22.97
7 Specialties	Average grade kitchen cabinets - 14 L.F. wall and base with solid surface counter top and kitchen sink; 40 gallon electric water heater.	.022	1.65	.75	2.40
8 Mechanical	1 lavatory, white, wall hung; 1 water closet, white; 1 bathtub with shower, enameled steel, white; gas fired warm air heat.	.049	2.41	2.29	4.70
9 Electrical	200 Amp. service; romex wiring; incandescent lighting fixtures, switches, receptacles.	.039	.93	1.34	2.27
10 Overhead	Contractor's overhead and profit and plans.		5.77	5.50	11.27
		Total	39.80	37.80	**77.60**

- **Simple design from standard plans**
- **Single family — 1 full bath, 1 kitchen**
- **No basement**
- **Asphalt shingles on roof**
- **Hot air heat**
- **Gypsum wallboard interior finishes**
- **Materials and workmanship are average**

Note: The illustration shown may contain some optional components (for example: garages and/or fireplaces) whose costs are shown in the modifications, adjustments, & alternatives below or at the end of the square foot section.

Base cost per square foot of living area

Exterior Wall	Living Area 1000	1200	1400	1600	1800	2000	2200	2600	3000	3400	3800
Wood Siding - Wood Frame	99.15	90.00	85.80	82.95	80.00	76.75	74.65	70.65	66.55	64.85	63.25
Brick Veneer - Wood Frame	106.65	97.00	92.25	89.15	85.80	82.40	80.00	75.45	71.00	69.05	67.30
Stucco on Wood Frame	98.40	89.35	85.20	82.35	79.40	76.20	74.15	70.15	66.15	64.45	62.90
Solid Masonry	118.35	107.80	102.45	98.85	94.90	91.15	88.30	83.05	78.00	75.70	73.60
Finished Basement, Add	16.05	15.85	15.35	15.00	14.55	14.35	14.10	13.55	13.20	12.95	12.75
Unfinished Basement, Add	6.40	6.00	5.65	5.40	5.15	5.05	4.85	4.50	4.35	4.15	4.05

Modifications

Add to the total cost

Upgrade Kitchen Cabinets	$ + 3157
Solid Surface Countertops (Included)	
Full Bath - including plumbing, wall and floor finishes	+ 4261
Half Bath - including plumbing, wall and floor finishes	+ 2660
One Car Attached Garage	+ 9490
One Car Detached Garage	+ 12,424
Fireplace & Chimney	+ 4670

Adjustments

For multi family - add to total cost

Additional Kitchen	$ + 5105
Additional Bath	+ 4261
Additional Entry & Exit	+ 1276
Separate Heating	+ 1248
Separate Electric	+ 1575

For Townhouse/Rowhouse - Multiply cost per square foot by

Inner Unit	.90
End Unit	.95

Alternatives

Add to or deduct from the cost per square foot of living area

Cedar Shake Roof	+ 1.17
Clay Tile Roof	+ 2.04
Slate Roof	+ 3.89
Upgrade Walls to Skim Coat Plaster	+ .42
Upgrade Ceilings to Textured Finish	+ .40
Air Conditioning, in Heating Ductwork	+ 1.70
In Separate Ductwork	+ 4.35
Heating Systems, Hot Water	+ 1.39
Heat Pump	+ 2.15
Electric Heat	– .51
Not Heated	– 2.46

Additional upgrades or components

Kitchen Cabinets & Countertops	Page 93
Bathroom Vanities	94
Fireplaces & Chimneys	94
Windows, Skylights & Dormers	94
Appliances	95
Breezeways & Porches	95
Finished Attic	95
Garages	96
Site Improvements	96
Wings & Ells	56

Average 2 Story

Living Area 2000 S.F.
Perimeter 135 L.F.

		Labor-Hours	Cost Per Square Foot Of Living Area		
			Mat.	Labor	Total
1 Site Work	Site preparation for slab; 4′ deep trench excavation for foundation wall.	.034		.61	.61
2 Foundation	Continuous reinforced concrete footing 8″ deep x 18″ wide; dampproofed and insulated reinforced concrete foundation wall, 8″ thick, 4′ deep, 4″ concrete slab on 4″ crushed stone base and polyethylene vapor barrier, trowel finish.	.066	2.33	3.13	5.46
3 Framing	Exterior walls - 2″ x 4″ wood studs, 16″ O.C.; 1/2″ plywood sheathing; 2″ x 6″ rafters 16″ O.C. with 1/2″ plywood sheathing, 4 in 12 pitch; 2″ x 6″ ceiling joists 16″ O.C.; 2″ x 8″ floor joists 16″ O.C. with 5/8″ plywood subfloor; 1/2″ plywood subfloor on 1″ x 2″ wood sleepers 16″ O.C.	.131	6.21	6.88	13.09
4 Exterior Walls	Beveled wood siding and building paper on insulated wood frame walls; 6″ attic insulation; double hung windows; 3 flush solid core wood exterior doors with storms.	.111	8.82	4.61	13.43
5 Roofing	25 year asphalt shingles; #15 felt building paper; aluminum gutters, downspouts, drip edge and flashings.	.024	.48	.94	1.42
6 Interiors	Walls and ceilings, 1/2″ taped and finished gypsum wallboard, primed and painted with 2 coats; painted baseboard and trim, finished hardwood floor 40%, carpet with 1/2″ underlayment 40%, vinyl tile with 1/2″ underlayment 15%, ceramic tile with 1/2″ underlayment 5%; hollow core and louvered doors.	.232	11.38	11.41	22.79
7 Specialties	Average grade kitchen cabinets - 14 L.F. wall and base with solid surface counter top and kitchen sink; 40 gallon electric water heater.	.021	1.48	.68	2.16
8 Mechanical	1 lavatory, white, wall hung; 1 water closet, white; 1 bathtub with shower; enameled steel, white; gas fired warm air heat.	.060	2.27	2.22	4.49
9 Electrical	200 Amp. service; romex wiring; incandescent lighting fixtures, switches, receptacles.	.039	.89	1.29	2.18
10 Overhead	Contractor's overhead and profit and plans.		5.74	5.38	11.12
		Total	39.60	37.15	**76.75**

RESIDENTIAL | Average | 2-1/2 Story

- **Simple design from standard plans**
- **Single family – 1 full bath, 1 kitchen**
- **No basement**
- **Asphalt shingles on roof**
- **Hot air heat**
- **Gypsum wallboard interior finishes**
- **Materials and workmanship are average**

Note: The illustration shown may contain some optional components (for example: garages and/or fireplaces) whose costs are shown in the modifications, adjustments, & alternatives below or at the end of the square foot section.

Base cost per square foot of living area

Exterior Wall	Living Area 1200	1400	1600	1800	2000	2400	2800	3200	3600	4000	4400
Wood Siding - Wood Frame	99.05	93.15	85.20	83.65	80.80	76.20	72.45	**68.80**	66.90	63.50	62.35
Brick Veneer - Wood Frame	107.05	100.40	91.85	90.30	87.05	81.85	77.85	73.70	71.60	67.80	66.55
Stucco on Wood Frame	98.25	92.45	84.55	83.00	80.20	75.70	71.95	68.30	66.50	63.15	62.00
Solid Masonry	119.50	111.65	102.25	100.70	96.80	90.60	86.30	81.30	78.80	74.55	73.05
Finished Basement, Add	13.55	13.30	12.80	12.75	12.45	11.90	11.70	11.30	11.10	10.85	10.70
Unfinished Basement, Add	5.35	4.90	4.60	4.55	4.35	4.05	3.85	3.60	3.50	3.35	3.30

Modifications

Add to the total cost

Upgrade Kitchen Cabinets	$ + 3157
Solid Surface Countertops (Included)	
Full Bath - including plumbing, wall and floor finishes	+ 4261
Half Bath - including plumbing, wall and floor finishes	+ 2660
One Car Attached Garage	+ 9490
One Car Detached Garage	+ 12,424
Fireplace & Chimney	+ 5305

Adjustments

For multi family - add to total cost

Additional Kitchen	$ + 5105
Additional Bath	+ 4261
Additional Entry & Exit	+ 1276
Separate Heating	+ 1248
Separate Electric	+ 1575

For Townhouse/Rowhouse - Multiply cost per square foot by

Inner Unit	.90
End Unit	.95

Alternatives

Add to or deduct from the cost per square foot of living area

Cedar Shake Roof	+ 1.02
Clay Tile Roof	+ 1.77
Slate Roof	+ 3.37
Upgrade Walls to Skim Coat Plaster	+ .40
Upgrade Ceilings to Textured Finish	+ .40
Air Conditioning, in Heating Ductwork	+ 1.55
In Separate Ductwork	+ 4.20
Heating Systems, Hot Water	+ 1.25
Heat Pump	+ 2.20
Electric Heat	– .94
Not Heated	– 2.87

Additional upgrades or components

Kitchen Cabinets & Countertops	Page 93
Bathroom Vanities	94
Fireplaces & Chimneys	94
Windows, Skylights & Dormers	94
Appliances	95
Breezeways & Porches	95
Finished Attic	95
Garages	96
Site Improvements	96
Wings & Ells	56

Important: See the Reference Section for Location Factors (to adjust for your city) and Estimating Forms

Average 2-1/2 Story

Living Area 3200 S.F.
Perimeter 150 L.F.

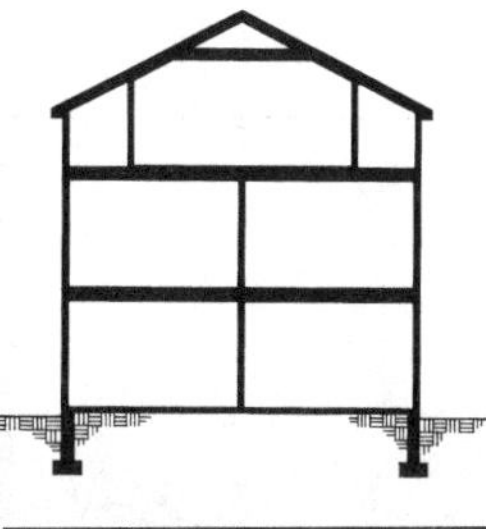

		Labor-Hours	Cost Per Square Foot Of Living Area		
			Mat.	Labor	Total
1 Site Work	Site preparation for slab; 4' deep trench excavation for foundation wall.	.046		.38	.38
2 Foundation	Continuous reinforced concrete footing 8" deep x 18" wide, dampproofed and insulated reinforced concrete foundation wall, 8" thick, 4' deep; 4" concrete slab on 4" crushed stone base and polyethylene vapor barrier, trowel finish.	.061	1.69	2.22	3.91
3 Framing	Exterior walls - 2" x 4" wood studs, 16" O.C.; 1/2" plywood sheathing; 2" x 6" rafters 16" O.C. with 1/2" plywood sheathing, 4 in 12 pitch; 2" x 6" ceiling joists 16" O.C.; 2" x 8" floor joists 16" O.C. with 5/8" plywood subfloor; 1/2" plywood subfloor on 1" x 2" wood sleepers 16" O.C.	.127	6.12	6.74	12.86
4 Exterior Walls	Beveled wood siding and building paper on insulated wood frame walls; 6" attic insulation; double hung windows; 3 flush solid core wood exterior doors with storms.	.136	7.42	3.87	11.29
5 Roofing	25 year asphalt shingles; #15 felt building paper; aluminum gutters, downspouts, drip edge and flashings.	.018	.37	.72	1.09
6 Interiors	Walls and ceilings, 1/2" taped and finished gypsum wallboard, primed and painted with 2 coats; painted baseboard and trim, finished hardwood floor 40%, carpet with 1/2" underlayment 40%, vinyl tile with 1/2" underlayment 15%, ceramic tile with 1/2" underlayment 5%; hollow core and louvered doors.	.286	11.23	11.03	22.26
7 Specialties	Average grade kitchen cabinets - 14 L.F. wall and base with solid surface counter top and kitchen sink; 40 gallon electric water heater.	.030	.91	.43	1.34
8 Mechanical	1 lavatory, white, wall hung; 1 water closet, white; 1 bathtub with shower, enameled steel, white; gas fired warm air heat.	.072	1.81	1.97	3.78
9 Electrical	200 Amp. service; romex wiring; incandescent lighting fixtures, switches, receptacles.	.046	.76	1.13	1.89
10 Overhead	Contractor's overhead and profit and plans.		5.14	4.86	10.00
	Total		35.45	33.35	**68.80**

RESIDENTIAL | Average | 3 Story

- **Simple design from standard plans**
- **Single family — 1 full bath, 1 kitchen**
- **No basement**
- **Asphalt shingles on roof**
- **Hot air heat**
- **Gypsum wallboard interior finishes**
- **Materials and workmanship are average**

Note: The illustration shown may contain some optional components (for example: garages and/or fireplaces) whose costs are shown in the modifications, adjustments, & alternatives below or at the end of the square foot section.

Base cost per square foot of living area

Exterior Wall	Living Area 1500	1800	2100	2500	3000	3500	4000	4500	5000	5500	6000
Wood Siding - Wood Frame	92.00	83.65	80.05	77.25	71.80	69.50	66.20	62.60	61.50	60.20	58.85
Brick Veneer - Wood Frame	99.50	90.60	86.55	83.45	77.45	74.80	71.05	67.05	65.85	64.35	62.80
Stucco on Wood Frame	91.30	83.00	79.45	76.65	71.25	69.00	65.70	62.15	61.10	59.75	58.45
Solid Masonry	111.20	101.45	96.70	93.05	86.25	83.15	78.60	74.05	72.55	70.80	68.90
Finished Basement, Add	11.70	11.60	11.25	11.00	10.55	10.35	10.00	9.80	9.65	9.55	9.40
Unfinished Basement, Add	4.30	4.00	3.80	3.70	3.45	3.25	3.10	2.95	2.90	2.80	2.70

Modifications

Add to the total cost

Upgrade Kitchen Cabinets	$ + 3157
Solid Surface Countertops (Included)	
Full Bath - including plumbing, wall and floor finishes	+ 4261
Half Bath - including plumbing, wall and floor finishes	+ 2660
One Car Attached Garage	+ 9490
One Car Detached Garage	+ 12,424
Fireplace & Chimney	+ 5305

Adjustments

For multi family - add to total cost

Additional Kitchen	$ + 5105
Additional Bath	+ 4261
Additional Entry & Exit	+ 1276
Separate Heating	+ 1248
Separate Electric	+ 1575

For Townhouse/Rowhouse - Multiply cost per square foot by

Inner Unit	.88
End Unit	.94

Alternatives

Add to or deduct from the cost per square foot of living area

Cedar Shake Roof	+ .7
Clay Tile Roof	+ 1.3
Slate Roof	+ 2.5
Upgrade Walls to Skim Coat Plaster	+ .4
Upgrade Ceilings to Textured Finish	+ .4
Air Conditioning, in Heating Ductwork	+ 1.5
In Separate Ductwork	+ 4.2
Heating Systems, Hot Water	+ 1.2
Heat Pump	+ 2.2
Electric Heat	– .7
Not Heated	– 2.6

Additional upgrades or components

Kitchen Cabinets & Countertops	Page 9
Bathroom Vanities	9
Fireplaces & Chimneys	9
Windows, Skylights & Dormers	9
Appliances	9
Breezeways & Porches	9
Finished Attic	9
Garages	9
Site Improvements	9
Wings & Ells	5

Important: See the Reference Section for Location Factors (to adjust for your city) and Estimating Forms

Average 3 Story

Living Area 3000 S.F.
Perimeter 135 L.F.

		Labor-Hours	Cost Per Square Foot Of Living Area		
			Mat.	Labor	Total
1 Site Work	Site preparation for slab; 4′ deep trench excavation for foundation wall.	.038		.40	.40
2 Foundation	Continuous reinforced concrete footing 8″ deep x 18″ wide, dampproofed and insulated reinforced concrete foundation wall, 8″ thick, 4′ deep; 4″ concrete slab on 4″ crushed stone base and polyethylene vapor barrier, trowel finish.	.053	1.56	2.09	3.65
3 Framing	Exterior walls - 2″ x 4″ wood studs, 16″ O.C.; 1/2″ plywood sheathing; 2″ x 6″ rafters 16″ O.C. with 1/2″ plywood sheathing, 4 in 12 pitch; 2″ x 6″ ceiling joists 16″ O.C.; 2″ x 8″ floor joists 16″ O.C. with 5/8″ plywood subfloor; 1/2″ plywood subfloor on 1″ x 2″ wood sleepers 16″ O.C.	.128	6.33	6.95	13.28
4 Exterior Walls	Horizontal beveled wood siding; building paper; 3-1/2″ batt insulation; wood double hung windows; 3 flush solid core wood exterior doors; storms and screens.	.139	8.43	4.43	12.86
5 Roofing	25 year asphalt shingles; #15 felt building paper; aluminum gutters, downspouts, drip edge and flashings.	.014	.32	.63	.95
6 Interiors	Walls and ceilings, 1/2″ taped and finished gypsum wallboard, primed and painted with 2 coats; painted baseboard and trim, finished hardwood floor 40%, carpet with 1/2″ underlayment 40%, vinyl tile with 1/2″ underlayment 15%, ceramic tile with 1/2″ underlayment 5%; hollow core and louvered doors.	.280	11.61	11.45	23.06
7 Specialties	Average grade kitchen cabinets - 14 L.F. wall and base with solid surface counter top and kitchen sink; 40 gallon electric water heater.	.025	.98	.45	1.43
8 Mechanical	1 lavatory, white, wall hung; 1 water closet, white; 1 bathtub with shower, enameled steel, white; gas fired warm air heat.	.065	1.85	2.00	3.85
9 Electrical	200 Amp. service; romex wiring; incandescent lighting fixtures, switches, receptacles.	.042	.77	1.15	1.92
10 Overhead	Contractor's overhead and profit and plans.		5.40	5.00	10.40
		Total	37.25	34.55	**71.80**

RESIDENTIAL | Average | Bi-Level

- **Simple design from standard plans**
- **Single family — 1 full bath, 1 kitchen**
- **No basement**
- **Asphalt shingles on roof**
- **Hot air heat**
- **Gypsum wallboard interior finishes**
- **Materials and workmanship are average**

Note: The illustration shown may contain some optional components (for example: garages and/or fireplaces) whose costs are shown in the modifications, adjustments, & alternatives below or at the end of the square foot section.

Base cost per square foot of living area

	Living Area										
Exterior Wall	**1000**	**1200**	**1400**	**1600**	**1800**	**2000**	**2200**	**2600**	**3000**	**3400**	**3800**
Wood Siding - Wood Frame	92.80	84.15	80.30	77.70	75.05	**71.95**	70.10	66.50	62.80	61.20	59.80
Brick Veneer - Wood Frame	98.45	89.35	85.20	82.35	79.45	76.20	74.15	70.15	66.15	64.40	62.85
Stucco on Wood Frame	92.30	83.70	79.80	77.25	74.65	71.55	69.75	66.15	62.50	60.90	59.50
Solid Masonry	107.20	97.50	92.80	89.60	86.30	82.80	80.40	75.80	71.40	69.35	67.55
Finished Basement, Add	16.05	15.85	15.35	15.00	14.55	14.35	14.10	13.55	13.20	12.95	12.75
Unfinished Basement, Add	6.40	6.00	5.65	5.40	5.15	5.05	4.85	4.50	4.35	4.15	4.05

Modifications

Add to the total cost

Upgrade Kitchen Cabinets	$ + 3157
Solid Surface Countertops (Included)	
Full Bath - including plumbing, wall and floor finishes	+ 4261
Half Bath - including plumbing, wall and floor finishes	+ 2660
One Car Attached Garage	+ 9490
One Car Detached Garage	+ 12,424
Fireplace & Chimney	+ 4185

Adjustments

For multi family - add to total cost

Additional Kitchen	$ + 5105
Additional Bath	+ 4261
Additional Entry & Exit	+ 1276
Separate Heating	+ 1248
Separate Electric	+ 1575

For Townhouse/Rowhouse - Multiply cost per square foot by

Inner Unit	.91
End Unit	.96

Alternatives

Add to or deduct from the cost per square foot of living area

Cedar Shake Roof	+ 2.3
Clay Tile Roof	+ 4.0
Slate Roof	+ 3.8
Upgrade Walls to Skim Coat Plaster	+ .3
Upgrade Ceilings to Textured Finish	+ .4
Air Conditioning, in Heating Ductwork	+ 1.7
In Separate Ductwork	+ 4.3
Heating Systems, Hot Water	+ 1.3
Heat Pump	+ 2.1
Electric Heat	– .5
Not Heated	– 2.4

Additional upgrades or components

Kitchen Cabinets & Countertops	Page 9
Bathroom Vanities	9
Fireplaces & Chimneys	9
Windows, Skylights & Dormers	9
Appliances	9
Breezeways & Porches	9
Finished Attic	9
Garages	9
Site Improvements	9
Wings & Ells	5

Important: See the Reference Section for Location Factors (to adjust for your city) and Estimating Form

Average Bi-Level

Living Area 2000 S.F.
Perimeter 135 L.F.

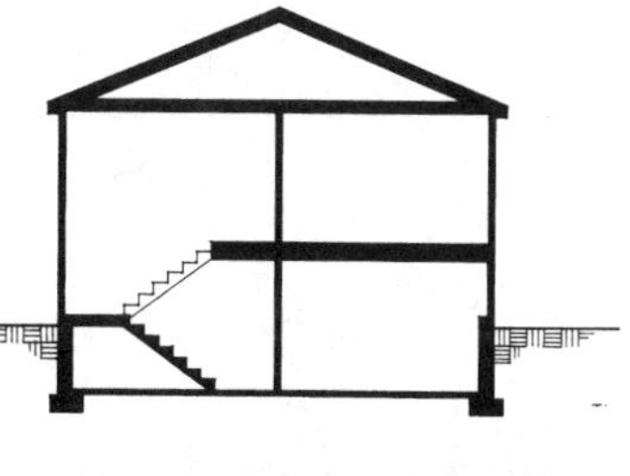

		Labor-Hours	Cost Per Square Foot Of Living Area		
			Mat.	Labor	Total
1 Site Work	Excavation for lower level, 4′ deep. Site preparation for slab.	.029		.61	.61
2 Foundation	Continuous reinforced concrete footing 8″ deep x 18″ wide, dampproofed and insulated reinforced concrete foundation wall, 8″ thick, 4′ deep; 4″ concrete slab on 4″ crushed stone base and polyethylene vapor barrier, trowel finish.	.066	2.33	3.13	5.46
3 Framing	Exterior walls - 2″ x 4″ wood studs, 16″ O.C.; 1/2″ plywood sheathing; 2″ x 6″ rafters 16″ O.C. with 1/2″ plywood sheathing, 4 in 12 pitch; 2″ x 6″ ceiling joists 16″ O.C.; 2″ x 8″ floor joists 16″ O.C. with 5/8″ plywood subfloor; 1/2″ plywood subfloor on 1″ x 2″ wood sleepers 16″ O.C.	.118	5.91	6.53	12.44
4 Exterior Walls	Horizontal beveled wood siding; building paper; 3-1/2″ batt insulation; wood double hung windows; 3 flush solid core wood exterior doors; storms and screens.	.091	6.89	3.57	10.46
5 Roofing	25 year asphalt shingles; #15 felt building paper; aluminum gutters, downspouts, drip edge and flashings.	.024	.48	.94	1.42
6 Interiors	Walls and ceilings, 1/2″ taped and finished gypsum wallboard, primed and painted with 2 coats; painted baseboard and trim, finished hardwood floor 40%, carpet with 1/2″ underlayment 40%, vinyl tile with 1/2″ underlayment 15%, ceramic tile with 1/2″ underlayment 5%; hollow core and louvered doors.	.217	11.21	11.08	22.29
7 Specialties	Average grade kitchen cabinets - 14 L.F. wall and base with solid surface counter top and kitchen sink; 40 gallon electric water heater.	.021	1.48	.68	2.16
8 Mechanical	1 lavatory, white, wall hung; 1 water closet, white; 1 bathtub with shower, enameled steel, white; gas fired warm air heat.	.061	2.27	2.22	4.49
9 Electrical	200 Amp. service; romex wiring; incandescent lighting fixtures, switches, receptacles.	.039	.89	1.29	2.18
10 Overhead	Contractor's overhead and profit and plans.		5.34	5.10	10.44
		Total	36.80	35.15	**71.95**

- **Simple design from standard plans**
- **Single family — 1 full bath, 1 kitchen**
- **No basement**
- **Asphalt shingles on roof**
- **Hot air heat**
- **Gypsum wallboard interior finishes**
- **Materials and workmanship are average**

Note: The illustration shown may contain some optional components (for example: garages and/or fireplaces) whose costs are shown in the modifications, adjustments, & alternatives below or at the end of the square foot section.

Base cost per square foot of living area

Exterior Wall	1200	1500	1800	2100	2400	2700	3000	3400	3800	4200	4600
	Living Area										
Wood Siding - Wood Frame	88.05	81.45	76.45	72.20	69.40	67.85	66.05	64.45	61.65	59.40	58.35
Brick Veneer - Wood Frame	93.25	86.15	80.70	76.05	73.05	71.30	69.30	67.60	64.60	62.20	61.00
Stucco on Wood Frame	87.50	81.00	76.00	71.85	69.05	67.55	65.75	64.15	61.40	59.10	58.10
Solid Masonry	101.30	93.45	87.30	82.05	78.65	76.75	74.45	72.50	69.20	66.45	65.15
Finished Basement, Add*	18.40	18.10	17.25	16.70	16.35	16.10	15.75	15.55	15.25	14.95	14.80
Unfinished Basement, Add*	7.15	6.65	6.10	5.75	5.50	5.40	5.20	5.05	4.85	4.70	4.60

*Basement under middle level only.

Modifications

Add to the total cost

Upgrade Kitchen Cabinets	$ + 3157
Solid Surface Countertops (Included)	
Full Bath - including plumbing, wall and floor finishes	+ 4261
Half Bath - including plumbing, wall and floor finishes	+ 2660
One Car Attached Garage	+ 9490
One Car Detached Garage	+ 12,424
Fireplace & Chimney	+ 4670

Adjustments

For multi family - add to total cost

Additional Kitchen	$ + 5105
Additional Bath	+ 4261
Additional Entry & Exit	+ 1276
Separate Heating	+ 1248
Separate Electric	+ 1575

For Townhouse/Rowhouse - Multiply cost per square foot by

Inner Unit	.90
End Unit	.95

Alternatives

Add to or deduct from the cost per square foot of living area

Cedar Shake Roof	+ 1.6
Clay Tile Roof	+ 2.9
Slate Roof	+ 5.6
Upgrade Walls to Skim Coat Plaster	+ .3
Upgrade Ceilings to Textured Finish	+ .4
Air Conditioning, in Heating Ductwork	+ 1.4
In Separate Ductwork	+ 4.1
Heating Systems, Hot Water	+ 1.3
Heat Pump	+ 2.2
Electric Heat	– .4
Not Heated	– 2.3

Additional upgrades or components

Kitchen Cabinets & Countertops	Page 9
Bathroom Vanities	9
Fireplaces & Chimneys	9
Windows, Skylights & Dormers	9
Appliances	9
Breezeways & Porches	9
Finished Attic	9
Garages	9
Site Improvements	9
Wings & Ells	5

Important: See the Reference Section for Location Factors (to adjust for your city) and Estimating Form

Average Tri-Level

Living Area 2400 S.F.
Perimeter 163 L.F.

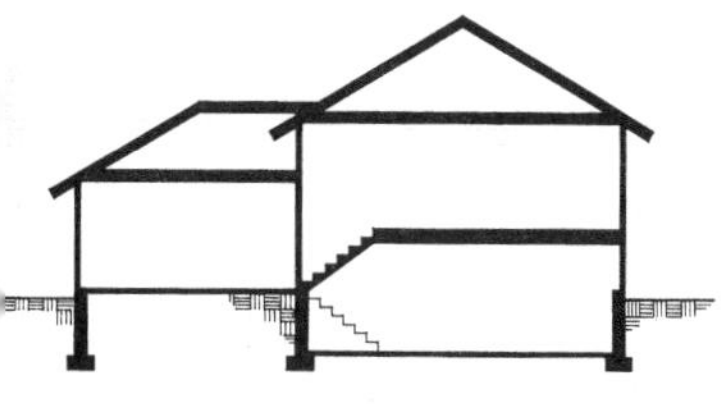

SQUARE FOOT COSTS

		Labor-Hours	Cost Per Square Foot Of Living Area		
			Mat.	Labor	Total
1 Site Work	Site preparation for slab; 4′ deep trench excavation for foundation wall, excavation for lower level, 4′ deep.	.029		.50	.50
2 Foundation	Continuous reinforced concrete footing 8″ deep x 18″ wide; dampproofed and insulated reinforced concrete foundation wall, 8″ thick, 4′ deep; 4″ concrete slab on 4″ crushed stone base and polyethylene vapor barrier, trowel finish.	.080	2.66	3.40	6.06
3 Framing	Exterior walls - 2″ x 4″ wood studs, 16″ O.C.; 1/2″ plywood sheathing; 2″ x 6″ rafters 16″ O.C. with 1/2″ plywood sheathing, 4 in 12 pitch; 2″ x 6″ ceiling joists 16″ O.C.; 2″ x 8″ floor joists 16″ O.C. with 5/8″ plywood subfloor; 1/2″ plywood subfloor on 1″ x 2″ wood sleepers 16″ O.C.	.124	5.58	6.13	11.71
4 Exterior Walls	Horizontal beveled wood siding: building paper; 3-1/2″ batt insulation; wood double hung windows; 3 flush solid core wood exterior doors; storms and screens.	.083	5.95	3.12	9.07
5 Roofing	25 year asphalt shingles; #15 felt building paper; aluminum gutters, downspouts, drip edge and flashings.	.032	.64	1.24	1.88
6 Interiors	Walls and ceilings, 1/2″ taped and finished gypsum wallboard, primed and painted with 2 coats; painted baseboard and trim, finished hardwood floor 40%, carpet with 1/2″ underlayment 40%, vinyl tile with 1/2″ underlayment 15%, ceramic tile with 1/2″ underlayment 5%; hollow core and louvered doors.	.186	11.33	10.81	22.14
7 Specialties	Average grade kitchen cabinets - 14 L.F. wall and base with solid surface counter top and kitchen sink; 40 gallon electric water heater.	.012	1.24	.55	1.79
8 Mechanical	1 lavatory, white, wall hung; 1 water closet, white; 1 bathtub with shower, enameled steel, white; gas fired warm air heat.	.059	2.01	2.11	4.12
9 Electrical	200 Amp. service; romex wiring; incandescent lighting fixtures, switches, receptacles.	.036	.83	1.22	2.05
10 Overhead	Contractor's overhead and profit and plans.		5.16	4.92	10.08
		Total	35.40	34.00	**69.40**

Carlinville Public Library
P.O. Box 17
Carlinville, Illinois 62626

RESIDENTIAL | Solid Wall | 1 Story

- **Post and beam frame**
- **Log exterior walls**
- **Simple design from standard plans**
- **Single family — 1 full bath, 1 kitchen**
- **No basement**
- **Asphalt shingles on roof**
- **Hot air heat**
- **Gypsum wallboard interior finishes**
- **Materials and workmanship are average**

Note: The illustration shown may contain some optional components (for example: garages and/or fireplaces) whose costs are shown in the modifications, adjustments, & alternatives below or at the end of the square foot section.

Base cost per square foot of living area

Exterior Wall	Living Area 600	800	1000	1200	1400	1600	1800	2000	2400	2800	3200
6" Log - Solid Wall	108.90	98.40	90.20	83.80	78.60	74.20	72.00	70.50	65.70	61.90	59.90
8" Log - Solid Wall	108.00	97.60	89.50	83.10	78.00	73.60	71.50	70.00	65.30	61.30	59.40
Finished Basement, Add	25.00	23.50	22.40	21.50	20.70	20.20	19.90	19.50	18.80	18.40	18.00
Unfinished Basement, Add	10.70	9.70	9.10	8.40	7.90	7.60	7.40	7.10	6.80	6.40	6.20

Modifications

Add to the total cost

Upgrade Kitchen Cabinets	$ + 3157
Solid Surface Countertops (Included)	
Full Bath - including plumbing, wall and floor finishes	+ 4261
Half Bath - including plumbing, wall and floor finishes	+ 2660
One Car Attached Garage	+ 9490
One Car Detached Garage	+ 12,424
Fireplace & Chimney	+ 4185

Adjustments

For multi family - add to total cost

Additional Kitchen	$ + 5105
Additional Bath	+ 4261
Additional Entry & Exit	+ 1276
Separate Heating	+ 1248
Separate Electric	+ 1575

For Townhouse/Rowhouse - Multiply cost per square foot by

Inner Unit	.92
End Unit	.96

Alternatives

Add to or deduct from the cost per square foot of living area

Cedar Shake Roof	+ 2.3
Air Conditioning, in Heating Ductwork	+ 2.8
In Separate Ductwork	+ 5.5
Heating Systems, Hot Water	+ 1.5
Heat Pump	+ 1.8
Electric Heat	– .7
Not Heated	– 2.6

Additional upgrades or components

Kitchen Cabinets & Countertops	Page 9
Bathroom Vanities	9
Fireplaces & Chimneys	9
Windows, Skylights & Dormers	9
Appliances	9
Breezeways & Porches	9
Finished Attic	9
Garages	9
Site Improvements	9
Wings & Ells	5

Important: See the Reference Section for Location Factors (to adjust for your city) and Estimating Form

Solid Wall 1 Story

Living Area 1600 S.F.
Perimeter 163 L.F.

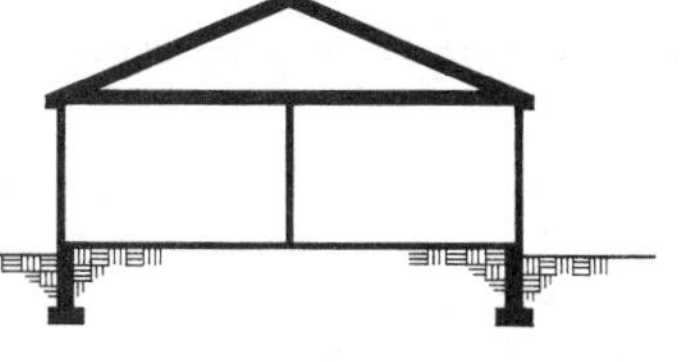

SQUARE FOOT COSTS

		Labor-Hours	Cost Per Square Foot Of Living Area		
			Mat.	Labor	Total
1 Site Work	Site preparation for slab; 4′ deep trench excavation for foundation wall.	.048		.82	.88
2 Foundation	Continuous reinforced concrete footing 8″ deep x 18″ wide; dampproofed and insulated reinforced concrete foundation wall, 8″ thick, 4′ deep; 4″ concrete slab on 4″ crushed stone base and polyethylene vapor barrier, trowel finish.	.113	4.09	5.10	9.19
3 Framing	Exterior walls - Precut traditional log home. Handicrafted white cedar or pine logs. Delivery included.	.201	12.22	8.68	20.90
4 Exterior Walls					
5 Roofing	25 year asphalt shingles; #15 felt building paper; aluminum gutters, downspouts, drip edge and flashings.	.047	1.02	1.78	2.80
6 Interiors	Walls and ceilings, 1/2″ taped and finished gypsum wallboard, primed and painted with 2 coats; painted baseboard and trim, finished hardwood floor 40%, carpet with 1/2″ underlayment 40%, vinyl tile with 1/2″ underlayment 15%, ceramic tile with 1/2″ underlayment 5%; hollow core and louvered doors.	.232	8.89	8.72	17.61
7 Specialties	Average grade kitchen cabinets - 14 L.F. wall and base with solid surface counter top and kitchen sink; 40 gallon electric water heater.	.009	1.85	.62	2.47
8 Mechanical	1 lavatory, white, wall hung; 1 water closet, white; 1 bathtub with shower, enameled steel, white; gas fired warm air heat.	.098	2.91	2.29	5.20
9 Electrical	200 Amp. service; romex wiring; incandescent lighting fixtures, switches, receptacles.	.041	.93	1.37	2.30
10 Overhead	Contractor's overhead and profit and plans.		7.52	5.33	12.85
		Total	39.43	34.77	**74.20**

- **Post and beam frame**
- **Log exterior walls**
- **Simple design from standard plans**
- **Single family — 1 full bath, 1 kitchen**
- **No basement**
- **Asphalt shingles on roof**
- **Hot air heat**
- **Gypsum wallboard interior finishes**
- **Materials and workmanship are average**

Note: The illustration shown may contain some optional components (for example: garages and/or fireplaces) whose costs are shown in the modifications, adjustments, & alternatives below or at the end of the square foot section.

Base cost per square foot of living area

Exterior Wall	Living Area 1000	1200	1400	1600	1800	2000	2200	2600	3000	3400	3800
6" Log-Solid	109.00	89.80	85.70	82.70	79.30	76.20	73.90	69.50	64.50	62.80	61.10
8" Log-Solid	104.70	93.90	89.30	86.40	82.90	79.30	77.20	72.50	67.50	65.70	63.70
Finished Basement, Add	14.50	14.00	13.50	13.20	12.90	12.70	12.50	12.00	11.70	11.60	11.30
Unfinished Basement, Add	5.80	5.50	5.20	4.90	4.80	4.50	4.50	4.20	3.90	3.80	3.70

Modifications

Add to the total cost

Upgrade Kitchen Cabinets	$ + 3157
Solid Surface Countertops (Included)	
Full Bath - including plumbing, wall and floor finishes	+ 4261
Half Bath - including plumbing, wall and floor finishes	+ 2660
One Car Attached Garage	+ 9490
One Car Detached Garage	+ 12,424
Fireplace & Chimney	+ 4670

Adjustments

For multi family - add to total cost

Additional Kitchen	$ + 5105
Additional Bath	+ 4261
Additional Entry & Exit	+ 1276
Separate Heating	+ 1248
Separate Electric	+ 1575

For Townhouse/Rowhouse - Multiply cost per square foot by

Inner Unit	.92
End Unit	.96

Alternatives

Add to or deduct from the cost per square foot of living area

Cedar Shake Roof	+ 1.6
Air Conditioning, in Heating Ductwork	+ 1.7
In Separate Ductwork	+ 4.3
Heating Systems, Hot Water	+ 1.3
Heat Pump	+ 2.1
Electric Heat	– .5
Not Heated	– 2.4

Additional upgrades or components

Kitchen Cabinets & Countertops	Page 9
Bathroom Vanities	9
Fireplaces & Chimneys	9
Windows, Skylights & Dormers	9
Appliances	9
Breezeways & Porches	9
Finished Attic	9
Garages	9
Site Improvements	9
Wings & Ells	5

Important: See the Reference Section for Location Factors (to adjust for your city) and Estimating Form

Solid Wall 2 Story

Living Area 2000 S.F.
Perimeter 135 L.F.

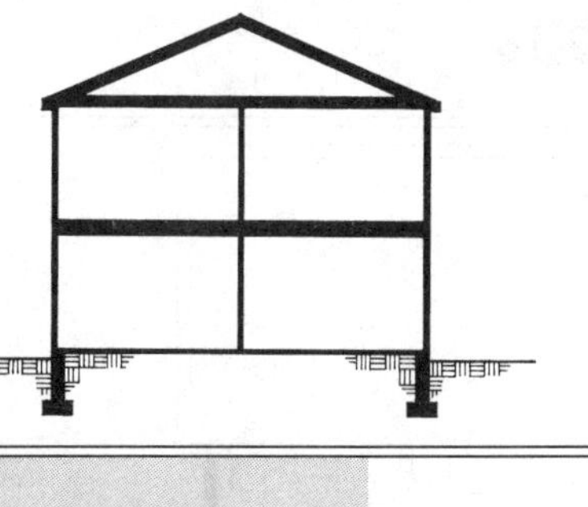

		Labor-Hours	Cost Per Square Foot Of Living Area		
			Mat.	Labor	Total
1 Site Work	Site preparation for slab; 4′ deep trench excavation for foundation wall.	.034		.65	.65
2 Foundation	Continuous reinforced concrete footing 8″ deep x 18″ wide; dampproofed and insulated reinforced concrete foundation wall, 8″ thick, 4′ deep; 4″ concrete slab on 4″ crushed stone base and polyethylene vapor barrier, trowel finish.	.066	2.45	3.10	5.55
3 Framing	Exterior walls - Precut traditional log home. Handicrafted white cedar or pine logs. Delivery included.	.232	14.86	9.87	24.73
4 Exterior Walls					
5 Roofing	25 year asphalt shingles; #15 felt building paper; aluminum gutters, downspouts, drip edge and flashings.	.024	.50	.92	1.42
6 Interiors	Walls and ceilings, 1/2″ taped and finished gypsum wallboard, primed and painted with 2 coats; painted baseboard and trim, finished hardwood floor 40%, carpet with 1/2″ underlayment 40%, vinyl tile with 1/2″ underlayment 15%, ceramic tile with 1/2″ underlayment 5%; hollow core and louvered doors.	.225	10.42	10.23	20.65
7 Specialties	Average grade kitchen cabinets - 14 L.F. wall and base with solid surface counter top and kitchen sink; 40 gallon electric water heater.	.021	1.46	.51	1.97
8 Mechanical	1 lavatory, white, wall hung; 1 water closet, white; 1 bathtub with shower, enameled steel, white; gas fired warm air heat.	.060	2.57	2.10	4.67
9 Electrical	200 Amp. service; romex wiring; incandescent lighting fixtures, switches, receptacles.	.039	.85	1.23	2.08
10 Overhead	Contractor's overhead and profit and plans.		7.36	7.12	14.48
		Total	40.47	35.73	**76.20**

1 Story — Base cost per square foot of living area

Exterior Wall	Living Area 50	100	200	300	400	500	600	700
Wood Siding - Wood Frame	146.90	113.85	99.50	84.80	80.15	77.25	75.35	76.10
Brick Veneer - Wood Frame	155.15	117.10	100.60	82.45	77.05	73.75	71.60	72.20
Stucco on Wood Frame	145.05	112.45	98.30	83.95	79.30	76.50	74.65	75.45
Solid Masonry	191.70	145.90	126.20	102.60	96.10	92.15	90.90	90.90
Finished Basement, Add	44.05	36.60	33.00	26.90	25.65	24.95	24.45	24.15
Unfinished Basement, Add	20.75	15.80	13.75	10.30	9.60	9.15	8.90	8.70

1-1/2 Story — Base cost per square foot of living area

Exterior Wall	Living Area 100	200	300	400	500	600	700	800
Wood Siding - Wood Frame	118.20	95.40	81.85	74.25	70.10	68.15	65.70	64.85
Brick Veneer - Wood Frame	172.15	127.10	105.05	91.95	85.20	81.70	77.90	76.35
Stucco on Wood Frame	154.95	113.35	93.65	83.05	77.00	73.95	70.60	69.05
Solid Masonry	196.50	146.60	121.35	104.65	96.95	92.70	88.35	86.70
Finished Basement, Add	29.80	27.00	24.55	21.90	21.15	20.65	20.20	20.15
Unfinished Basement, Add	13.05	11.00	9.60	8.10	7.70	7.40	7.20	7.10

2 Story — Base cost per square foot of living area

Exterior Wall	Living Area 100	200	400	600	800	1000	1200	1400
Wood Siding - Wood Frame	117.60	88.75	76.05	64.15	60.05	57.55	55.85	56.80
Brick Veneer - Wood Frame	175.15	121.25	96.45	77.85	71.35	67.35	64.80	65.00
Stucco on Wood Frame	155.95	107.50	85.00	70.15	64.50	61.00	58.70	59.10
Solid Masonry	202.40	140.80	112.75	88.65	81.05	76.45	73.45	73.35
Finished Basement, Add	23.55	19.90	18.00	15.05	14.40	14.05	13.80	13.60
Unfinished Basement, Add	10.55	8.05	6.95	5.30	4.90	4.75	4.60	4.50

Base costs do not include bathroom or kitchen facilities. Use Modifications/Adjustments/Alternatives on pages 93-96 where appropriate.

Important: See the Reference Section for Location Factors (to adjust for your city) and Estimating Forms

Custom Class

1 Story

1-1/2 Story

2 Story

2-1/2 Story

Bi-Level

Tri-Level

- **A distinct residence from designer's plans**
- **Single family – 1 full bath, 1 half bath, 1 kitchen**
- **No basement**
- **Asphalt shingles on roof**
- **Forced hot air heat/air conditioning**
- **Gypsum wallboard interior finishes**
- **Materials and workmanship are above average**

Note: The illustration shown may contain some optional components (for example: garages and/or fireplaces) whose costs are shown in the modifications, adjustments, & alternatives below or at the end of the square foot section.

Base cost per square foot of living area

Exterior Wall	Living Area 800	1000	1200	1400	1600	1800	2000	2400	2800	3200	3600
Wood Siding - Wood Frame	134.20	121.95	112.15	104.25	98.90	95.95	92.45	85.90	81.10	77.85	74.20
Brick Veneer - Wood Frame	150.70	137.85	127.45	119.00	113.35	110.20	106.45	99.50	94.45	90.95	87.10
Stone Veneer - Wood Frame	155.20	141.95	131.15	122.30	116.50	113.25	109.25	102.15	96.80	93.15	89.15
Solid Masonry	157.30	143.80	132.85	123.85	117.95	114.70	110.60	103.30	97.90	94.15	90.10
Finished Basement, Add	42.55	42.30	40.45	38.95	37.95	37.45	36.65	35.55	34.70	33.95	33.30
Unfinished Basement, Add	18.00	17.00	16.10	15.40	14.90	14.65	14.25	13.70	13.30	12.95	12.65

Modifications

Add to the total cost

Upgrade Kitchen Cabinets	$ + 910
Solid Surface Countertops (Included)	
Full Bath - including plumbing, wall and floor finishes	+ 5050
Half Bath - including plumbing, wall and floor finishes	+ 3152
Two Car Attached Garage	+ 19,382
Two Car Detached Garage	+ 22,032
Fireplace & Chimney	+ 4355

Adjustments

For multi family - add to total cost

Additional Kitchen	$ + 11,494
Additional Full Bath & Half Bath	+ 8202
Additional Entry & Exit	+ 1276
Separate Heating & Air Conditioning	+ 4978
Separate Electric	+ 1575

For Townhouse/Rowhouse - Multiply cost per square foot by

Inner Unit	.90
End Unit	.95

Alternatives

Add to or deduct from the cost per square foot of living area

Cedar Shake Roof	+ 1.9
Clay Tile Roof	+ 3.7
Slate Roof	+ 7.4
Upgrade Ceilings to Textured Finish	+ .4
Air Conditioning, in Heating Ductwork	Base Syster
Heating Systems, Hot Water	+ 1.5
Heat Pump	+ 1.8
Electric Heat	– 2.0
Not Heated	– 3.3

Additional upgrades or components

Kitchen Cabinets & Countertops	Page 9
Bathroom Vanities	9
Fireplaces & Chimneys	9
Windows, Skylights & Dormers	9
Appliances	9
Breezeways & Porches	9
Finished Attic	9
Garages	9
Site Improvements	9
Wings & Ells	7

Important: See the Reference Section for Location Factors (to adjust for your city) and Estimating Form

Custom 1 Story

Living Area 2400 S.F.
Perimeter 207 L.F.

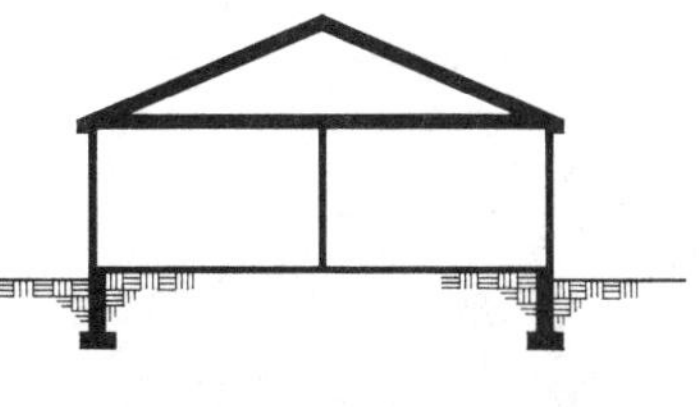

		Labor-Hours	Cost Per Square Foot Of Living Area		
			Mat.	Labor	Total
1 Site Work	Site preparation for slab; 4' deep trench excavation for foundation wall.	.028		.60	.60
2 Foundation	Continuous reinforced concrete footing 8" deep x 18" wide; dampproofed and insulated reinforced concrete foundation wall, 8" thick, 4' deep; 4" concrete slab on 4" crushed stone base and polyethylene vapor barrier, trowel finish.	.113	4.43	5.60	10.03
3 Framing	Exterior walls - 2" x 6" wood studs, 16" O.C.; 1/2" plywood sheathing; 2" x 8" rafters 16" O.C. with 1/2" plywood sheathing, 4 in 12 pitch; 2" x 6" ceiling joists 16" O.C.; 5/8" plywood subfloor on 1" x 3" wood sleepers 16" O.C.	.190	3.87	5.17	9.04
4 Exterior Walls	Horizontal beveled wood siding; building paper; 6" batt insulation; wood double hung windows; 3 solid core wood exterior doors; storms and screens.	.085	7.11	2.60	9.71
5 Roofing	30 year asphalt shingles; #15 felt building paper; aluminum gutters, downspouts and drip edge; copper flashings.	.082	2.82	2.73	5.55
6 Interiors	Walls and ceilings - 5/8" gypsum wallboard, skim coat plaster, painted with primer and 2 coats; hardwood baseboard and trim, sanded and finished; hardwood floor 70%, ceramic tile with underlayment 20%, vinyl tile with underlayment 10%; wood panel interior doors, primed and painted with 2 coats.	.292	12.46	9.84	22.30
7 Specialties	Custom grade kitchen cabinets - 20 L.F. wall and base with solid surface counter top and kitchen sink; 4 L.F. bathroom vanity; 75 gallon electric water heater, medicine cabinet.	.019	3.94	.96	4.90
8 Mechanical	Gas fired warm air heat/air conditioning; one full bath including: bathtub, corner shower, built in lavatory and water closet; one 1/2 bath including: built in lavatory and water closet.	.092	4.69	2.44	7.13
9 Electrical	200 Amp. service; romex wiring; fluorescent and incandescent lighting fixtures, switches, receptacles.	.039	.94	1.38	2.32
10 Overhead	Contractor's overhead and profit and design.		8.04	6.28	14.32
		Total	48.30	37.60	**85.90**

- **A distinct residence from designer's plans**
- **Single family – 1 full bath, 1 half bath, 1 kitchen**
- **No basement**
- **Asphalt shingles on roof**
- **Forced hot air heat/air conditioning**
- **Gypsum wallboard interior finishes**
- **Materials and workmanship are above average**

Note: The illustration shown may contain some optional components (for example: garages and/or fireplaces) whose costs are shown in the modifications, adjustments, & alternatives below or at the end of the square foot section.

Base cost per square foot of living area

	Living Area										
Exterior Wall	1000	1200	1400	1600	1800	2000	2400	2800	3200	3600	4000
Wood Siding - Wood Frame	122.40	114.10	108.15	101.05	97.00	93.00	85.30	**81.80**	78.65	76.10	72.75
Brick Veneer - Wood Frame	129.55	120.80	114.60	106.90	102.50	98.25	89.90	86.20	82.70	80.05	76.35
Stone Veneer - Wood Frame	134.30	125.25	118.80	110.75	106.15	101.65	92.95	89.05	85.40	82.60	78.75
Solid Masonry	136.45	127.30	120.80	112.50	107.85	103.25	94.35	90.40	86.60	83.85	79.90
Finished Basement, Add	28.25	28.45	27.60	26.45	25.80	25.20	24.10	23.55	22.90	22.55	22.10
Unfinished Basement, Add	12.05	11.50	11.20	10.60	10.35	10.00	9.50	9.20	8.90	8.80	8.55

Modifications

Add to the total cost

Upgrade Kitchen Cabinets	$ + 910
Solid Surface Countertops (Included)	
Full Bath - including plumbing, wall and floor finishes	+ 5050
Half Bath - including plumbing, wall and floor finishes	+ 3152
Two Car Attached Garage	+ 19,382
Two Car Detached Garage	+ 22,032
Fireplace & Chimney	+ 4915

Adjustments

For multi family - add to total cost

Additional Kitchen	$ + 11,494
Additional Full Bath & Half Bath	+ 8202
Additional Entry & Exit	+ 1276
Separate Heating & Air Conditioning	+ 4978
Separate Electric	+ 1575

For Townhouse/Rowhouse - Multiply cost per square foot by

Inner Unit	.90
End Unit	.95

Alternatives

Add to or deduct from the cost per square foot of living area

Cedar Shake Roof	+ 1.4
Clay Tile Roof	+ 2.6
Slate Roof	+ 5.3
Upgrade Ceilings to Textured Finish	+ .4
Air Conditioning, in Heating Ductwork	Base System
Heating Systems, Hot Water	+ 1.4
Heat Pump	+ 1.9
Electric Heat	– 1.7
Not Heated	– 3.0

Additional upgrades or components

Kitchen Cabinets & Countertops	Page 9
Bathroom Vanities	9
Fireplaces & Chimneys	9
Windows, Skylights & Dormers	9
Appliances	9
Breezeways & Porches	9
Finished Attic	9
Garages	9
Site Improvements	9
Wings & Ells	7

Important: See the Reference Section for Location Factors (to adjust for your city) and Estimating Form

Custom 1-1/2 Story

Living Area 2800 S.F.
Perimeter 175 L.F.

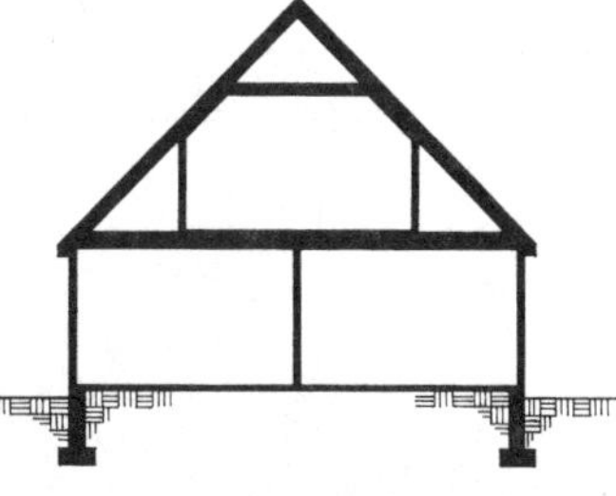

		Labor-Hours	Cost Per Square Foot Of Living Area		
			Mat.	Labor	Total
1 Site Work	Site preparation for slab; 4' deep trench excavation for foundation wall.	.028		.51	.51
2 Foundation	Continuous reinforced concrete footing 8" deep x 18" wide; dampproofed and insulated reinforced concrete foundation wall, 8" thick, 4' deep; 4" concrete slab on 4" crushed stone base and polyethylene vapor barrier, trowel finish.	.065	3.07	4.00	7.07
3 Framing	Exterior walls - 2" x 6" wood studs, 16" O.C.; 1/2" plywood sheathing; 2" x 8" rafters 16" O.C. with 1/2" plywood sheathing, 8 in 12 pitch; 2" x 10" floor joists 16" O.C. with 5/8" plywood subfloor; 5/8" plywood subfloor on 1" x 3" wood sleepers 16" O.C.	.192	4.77	5.44	10.21
4 Exterior Walls	Horizontal beveled wood siding; building paper; 6" batt insulation; wood double hung windows; 3 solid core wood exterior doors; storms and screens.	.064	7.10	2.63	9.73
5 Roofing	30 year asphalt shingles; #15 felt building paper; aluminum gutters, downspouts and drip edge; copper flashings.	.048	1.77	1.70	3.47
6 Interiors	Walls and ceilings - 5/8" gypsum wallboard, skim coat plaster, painted with primer and 2 coats; hardwood baseboard and trim, sanded and finished; hardwood floor 70%, ceramic tile with underlayment 20%, vinyl tile with underlayment 10%; wood panel interior doors, primed and painted with 2 coats.	.259	13.63	10.80	24.43
7 Specialties	Custom grade kitchen cabinets - 20 L.F. wall and base with solid surface counter top and kitchen sink; 4 L.F. bathroom vanity; 75 gallon electric water heater, medicine cabinet.	.030	3.36	.81	4.17
8 Mechanical	Gas fired warm air heat/air conditioning; one full bath including: bathtub, corner shower, built in lavatory and water closet; one 1/2 bath including: built in lavatory and water closet.	.084	4.07	2.28	6.35
9 Electrical	200 Amp. service; romex wiring; fluorescent and incandescent lighting fixtures, switches, receptacles.	.038	.90	1.32	2.22
10 Overhead	Contractor's overhead and profit and design.		7.73	5.91	13.64
		Total	46.40	35.40	**81.80**

RESIDENTIAL | Custom | 2 Story

- **A distinct residence from designer's plans**
- **Single family – 1 full bath, 1 half bath, 1 kitchen**
- **No basement**
- **Asphalt shingles on roof**
- **Forced hot air heat/air conditioning**
- **Gypsum wallboard interior finishes**
- **Materials and workmanship are above average**

Note: The illustration shown may contain some optional components (for example: garages and/or fireplaces) whose costs are shown in the modifications, adjustments, & alternatives below or at the end of the square foot section.

Base cost per square foot of living area

Exterior Wall	Living Area 1200	1400	1600	1800	2000	2400	2800	3200	3600	4000	4400
Wood Siding - Wood Frame	114.90	108.20	103.55	99.35	94.85	88.30	82.75	79.15	77.00	74.75	72.80
Brick Veneer - Wood Frame	122.55	115.40	110.35	105.80	101.05	93.90	87.85	83.90	81.60	79.00	77.00
Stone Veneer - Wood Frame	127.60	120.10	114.85	110.00	105.15	97.55	91.15	87.05	84.65	81.90	79.75
Solid Masonry	129.95	122.30	116.95	112.05	107.05	99.25	92.70	88.50	86.05	83.20	81.00
Finished Basement, Add	22.75	22.90	22.25	21.60	21.20	20.25	19.50	19.00	18.70	18.30	18.10
Unfinished Basement, Add	9.70	9.30	9.00	8.70	8.50	8.10	7.70	7.50	7.35	7.15	7.05

Modifications

Add to the total cost

Upgrade Kitchen Cabinets	$ + 910
Solid Surface Countertops (Included)	
Full Bath - including plumbing, wall and floor finishes	+ 5050
Half Bath - including plumbing, wall and floor finishes	+ 3152
Two Car Attached Garage	+ 19,382
Two Car Detached Garage	+ 22,032
Fireplace & Chimney	+ 4915

Adjustments

For multi family - add to total cost

Additional Kitchen	$ + 11,494
Additional Full Bath & Half Bath	+ 8202
Additional Entry & Exit	+ 1276
Separate Heating & Air Conditioning	+ 4978
Separate Electric	+ 1575

For Townhouse/Rowhouse - Multiply cost per square foot by

Inner Unit	.87
End Unit	.93

Alternatives

Add to or deduct from the cost per square foot of living area

Cedar Shake Roof	+ .98
Clay Tile Roof	+ 1.85
Slate Roof	+ 3.72
Upgrade Ceilings to Textured Finish	+ .40
Air Conditioning, in Heating Ductwork	Base System
Heating Systems, Hot Water	+ 1.42
Heat Pump	+ 2.14
Electric Heat	– 1.76
Not Heated	– 2.88

Additional upgrades or components

Kitchen Cabinets & Countertops	Page 93
Bathroom Vanities	94
Fireplaces & Chimneys	94
Windows, Skylights & Dormers	94
Appliances	95
Breezeways & Porches	95
Finished Attic	95
Garages	96
Site Improvements	96
Wings & Ells	74

Important: See the Reference Section for Location Factors (to adjust for your city) and Estimating Forms

Custom 2 Story

Living Area 2800 S.F.
Perimeter 156 L.F.

		Labor-Hours	Cost Per Square Foot Of Living Area		
			Mat.	Labor	Total
1 Site Work	Site preparation for slab; 4' deep trench excavation for foundation wall.	.024		.51	.51
2 Foundation	Continuous reinforced concrete footing 8" deep x 18" wide; dampproofed and insulated reinforced concrete foundation wall, 8" thick, 4' deep; 4" concrete slab on 4" crushed stone base and polyethylene vapor barrier, trowel finish.	.058	2.59	3.42	6.01
3 Framing	Exterior walls - 2" x 6" wood studs, 16" O.C.; 1/2" plywood sheathing; 2" x 8" rafters 16" O.C. with 1/2" plywood sheathing, 6 in 12 pitch; 2" x 8" ceiling joists 16" O.C.; 2" x 10" floor joists 16" O.C. with 5/8" plywood subfloor; 5/8" plywood subfloor on 1" x 3" wood sleepers 16" O.C.	.159	5.17	5.62	10.79
4 Exterior Walls	Horizontal beveled wood siding; building paper; 6" batt insulation; wood double hung windows; 3 solid core wood exterior doors; storms and screens.	.091	8.06	3.02	11.08
5 Roofing	30 year asphalt shingles; #15 felt building paper; aluminum gutters, downspouts and drip edge; copper flashings.	.042	1.41	1.37	2.78
6 Interiors	Walls and ceilings - 5/8" gypsum wallboard, skim coat plaster, painted with primer and 2 coats; hardwood baseboard and trim, sanded and finished; hardwood floor 70%, ceramic tile with underlayment 20%, vinyl tile with underlayment 10%; wood panel interior doors, primed and painted with 2 coats.	.271	13.80	11.06	24.86
7 Specialties	Custom grade kitchen cabinets - 20 L.F. wall and base with solid surface counter top and kitchen sink; 4 L.F. bathroom vanity; 75 gallon electric water heater, medicine cabinet.	.028	3.36	.81	4.17
8 Mechanical	Gas fired warm air heat/air conditioning; one full bath including: bathtub, corner shower; built in lavatory and water closet; one 1/2 bath including: built in lavatory and water closet.	.078	4.19	2.32	6.51
9 Electrical	200 Amp. service; romex wiring; fluorescent and incandescent lighting fixtures, switches, receptacles.	.038	.90	1.32	2.22
10 Overhead	Contractor's overhead and profit and design.		7.92	5.90	13.82
		Total	47.40	35.35	**82.75**

RESIDENTIAL | Custom | 2-1/2 Story

- **A distinct residence from designer's plans**
- **Single family — 1 full bath, 1 half bath, 1 kitchen**
- **No basement**
- **Asphalt shingles on roof**
- **Forced hot air heat/air conditioning**
- **Gypsum wallboard interior finishes**
- **Materials and workmanship are above average**

Note: The illustration shown may contain some optional components (for example: garages and/or fireplaces) whose costs are shown in the modifications, adjustments, & alternatives below or at the end of the square foot section.

Base cost per square foot of living area

Exterior Wall	Living Area 1500	1800	2100	2400	2800	3200	3600	4000	4500	5000	5500
Wood Siding - Wood Frame	113.80	102.65	96.20	92.25	87.10	82.35	79.65	75.40	73.25	71.20	69.20
Brick Veneer - Wood Frame	121.80	110.00	102.75	98.50	93.10	87.75	84.85	80.15	77.70	75.55	73.30
Stone Veneer - Wood Frame	127.00	114.80	107.10	102.60	97.05	91.30	88.15	83.25	80.65	78.30	75.90
Solid Masonry	129.45	117.10	109.05	104.50	98.80	92.95	89.75	84.75	82.00	79.60	77.20
Finished Basement, Add	18.05	18.00	17.00	16.60	16.15	15.55	15.20	14.75	14.45	14.25	13.95
Unfinished Basement, Add	7.75	7.35	6.90	6.65	6.45	6.15	6.05	5.80	5.70	5.55	5.45

Modifications

Add to the total cost

Upgrade Kitchen Cabinets	$ + 910
Solid Surface Countertops (Included)	
Full Bath - including plumbing, wall and floor finishes	+ 5050
Half Bath - including plumbing, wall and floor finishes	+ 3152
Two Car Attached Garage	+ 19,382
Two Car Detached Garage	+ 22,032
Fireplace & Chimney	+ 5555

Adjustments

For multi family - add to total cost

Additional Kitchen	$ + 11,494
Additional Full Bath & Half Bath	+ 8202
Additional Entry & Exit	+ 1276
Separate Heating & Air Conditioning	+ 4978
Separate Electric	+ 1575

For Townhouse/Rowhouse - Multiply cost per square foot by

Inner Unit	.87
End Unit	.94

Alternatives

Add to or deduct from the cost per square foot of living area

Cedar Shake Roof	+ .85
Clay Tile Roof	+ 1.60
Slate Roof	+ 3.22
Upgrade Ceilings to Textured Finish	+ .40
Air Conditioning, in Heating Ductwork	Base System
Heating Systems, Hot Water	+ 1.28
Heat Pump	+ 2.20
Electric Heat	– 3.14
Not Heated	– 2.88

Additional upgrades or components

Kitchen Cabinets & Countertops	Page 93
Bathroom Vanities	94
Fireplaces & Chimneys	94
Windows, Skylights & Dormers	94
Appliances	95
Breezeways & Porches	95
Finished Attic	95
Garages	96
Site Improvements	96
Wings & Ells	74

Custom 2-1/2 Story

Living Area 3200 S.F.
Perimeter 150 L.F.

SQUARE FOOT COSTS

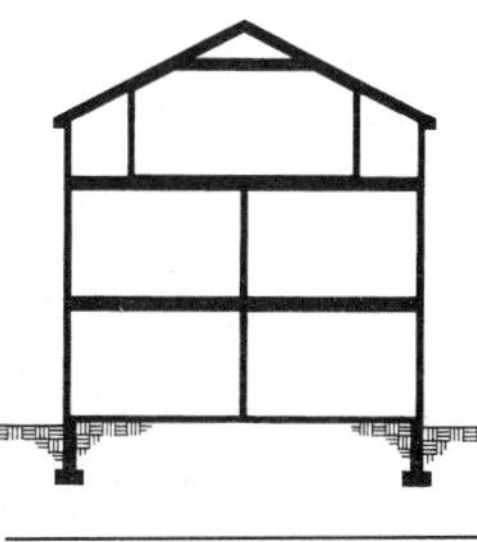

		Labor-Hours	Cost Per Square Foot Of Living Area		
			Mat.	Labor	Total
1 Site Work	Site preparation for slab; 4′ deep trench excavation for foundation wall.	.048		.45	.45
2 Foundation	Continuous reinforced concrete footing 8″ deep x 18″ wide; dampproofed and insulated reinforced concrete foundation wall, 8″ thick, 4′ deep; 4″ concrete slab on 4″ crushed stone base and polyethylene vapor barrier, trowel finish.	.063	2.14	2.85	4.99
3 Framing	Exterior walls - 2″ x 6″ wood studs, 16″ O.C.; 1/2″ plywood sheathing; 2″ x 8″ rafters 16″ O.C. with 1/2″ plywood sheathing, 6 in 12 pitch; 2″ x 8″ ceiling joists 16″ O.C.; 2″ x 10″ floor joists 16″ O.C. with 5/8″ plywood subfloor; 5/8″ plywood subfloor on 1″ x 3″ wood sleepers 16″ O.C.	.177	5.54	5.85	11.39
4 Exterior Walls	Horizontal beveled wood siding; building paper; 6″ batt insulation; wood double hung windows; 3 solid core wood exterior doors; storms and screens.	.134	8.36	3.14	11.50
5 Roofing	30 year asphalt shingles; #15 felt building paper; aluminum gutters, downspouts and drip edge; copper flashings.	.032	1.09	1.05	2.14
6 Interiors	Walls and ceilings - 5/8″ gypsum wallboard, skim coat plaster, painted with primer and 2 coats; hardwood baseboard and trim, sanded and finished; hardwood floor 70%, ceramic tile with underlayment 20%, vinyl tile with underlayment 10%; wood panel interior doors, primed and painted with 2 coats.	.354	14.53	11.77	26.30
7 Specialties	Custom grade kitchen cabinets - 20 L.F. wall and base with solid surface counter top and kitchen sink; 4 L.F. bathroom vanity; 75 gallon electric water heater, medicine cabinet.	.053	2.96	.72	3.68
8 Mechanical	Gas fired warm air heat/air conditioning; one full bath including: bathtub, corner shower; built in lavatory and water closet; one 1/2 bath including: built in lavatory and water closet.	.104	3.81	2.24	6.05
9 Electrical	200 Amp. service; romex wiring; fluorescent and incandescent lighting fixtures, switches, receptacles.	.048	.87	1.28	2.15
10 Overhead	Contractor's overhead and profit and design.		7.85	5.85	13.70
		Total	47.15	35.20	**82.35**

RESIDENTIAL | Custom | 3 Story

- **A distinct residence from designer's plans**
- **Single family – 1 full bath, 1 half bath, 1 kitchen**
- **No basement**
- **Asphalt shingles on roof**
- **Forced hot air heat/air conditioning**
- **Gypsum wallboard interior finishes**
- **Materials and workmanship are above average**

Note: The illustration shown may contain some optional components (for example: garages and/or fireplaces) whose costs are shown in the modifications, adjustments, & alternatives below or at the end of the square foot section.

Base cost per square foot of living area

Exterior Wall	Living Area 1500	1800	2100	2500	3000	3500	4000	4500	5000	5500	6000
Wood Siding - Wood Frame	113.35	102.30	97.05	92.65	85.80	82.45	78.25	73.75	72.20	70.55	68.80
Brick Veneer - Wood Frame	121.65	110.00	104.20	99.45	92.05	88.35	83.55	78.75	76.95	75.15	73.10
Stone Veneer - Wood Frame	127.10	115.05	108.85	103.95	96.10	92.20	87.10	82.00	80.15	78.15	75.90
Solid Masonry	129.65	117.35	111.10	106.05	97.95	93.95	88.70	83.45	81.55	79.55	77.30
Finished Basement, Add	15.90	15.80	15.15	14.75	14.10	13.65	13.20	12.75	12.60	12.45	12.25
Unfinished Basement, Add	6.75	6.40	6.15	5.95	5.65	5.45	5.20	5.05	4.95	4.85	4.75

Modifications

Add to the total cost

Upgrade Kitchen Cabinets	$ + 910
Solid Surface Countertops (Included)	
Full Bath - including plumbing, wall and floor finishes	+ 5050
Half Bath - including plumbing, wall and floor finishes	+ 3152
Two Car Attached Garage	+ 19,382
Two Car Detached Garage	+ 22,032
Fireplace & Chimney	+ 5555

Adjustments

For multi family - add to total cost

Additional Kitchen	$ + 11,494
Additional Full Bath & Half Bath	+ 8202
Additional Entry & Exit	+ 1276
Separate Heating & Air Conditioning	+ 4978
Separate Electric	+ 1575

For Townhouse/Rowhouse - Multiply cost per square foot by

Inner Unit	.85
End Unit	.93

Alternatives

Add to or deduct from the cost per square foot of living area

Cedar Shake Roof	+ .6
Clay Tile Roof	+ 1.2
Slate Roof	+ 2.4
Upgrade Ceilings to Textured Finish	+ .4
Air Conditioning, in Heating Ductwork	Base System
Heating Systems, Hot Water	+ 1.2
Heat Pump	+ 2.2
Electric Heat	– 3.1
Not Heated	– 2.7

Additional upgrades or components

Kitchen Cabinets & Countertops	Page 9
Bathroom Vanities	9
Fireplaces & Chimneys	9
Windows, Skylights & Dormers	9
Appliances	9
Breezeways & Porches	9
Finished Attic	9
Garages	9
Site Improvements	9
Wings & Ells	7

Important: See the Reference Section for Location Factors (to adjust for your city) and Estimating Forms

Custom 3 Story

Living Area 3000 S.F.
Perimeter 135 L.F.

		Labor-Hours	Cost Per Square Foot Of Living Area		
			Mat.	Labor	Total
1 Site Work	Site preparation for slab; 4′ deep trench excavation for foundation wall.	.048		.48	.48
2 Foundation	Continuous reinforced concrete footing 8″ deep x 18″ wide; dampproofed and insulated reinforced concrete foundation wall, 8″ thick, 4′ deep; 4″ concrete slab on 4″ crushed stone base and polyethylene vapor barrier, trowel finish.	.060	1.99	2.69	4.68
3 Framing	Exterior walls - 2″ x 6″ wood studs, 16″ O.C.; 1/2″ plywood sheathing; 2″ x 8″ rafters 16″ O.C. with 1/2″ plywood sheathing, 6 in 12 pitch; 2″ x 8″ ceiling joists 16″ O.C.; 2″ x 10″ floor joists 16″ O.C. with 5/8″ plywood subfloor; 5/8″ plywood subfloor on 1″ x 3″ wood sleepers 16″ O.C.	.191	5.83	6.04	11.87
4 Exterior Walls	Horizontal beveled wood siding; building paper; 6″ batt insulation; wood double hung windows; 3 solid core wood exterior doors; storms and screens.	.150	9.56	3.62	13.18
5 Roofing	30 year asphalt shingles; #15 felt building paper; aluminum gutters, downspouts and drip edge; copper flashings.	.028	.94	.91	1.85
6 Interiors	Walls and ceilings - 5/8″ gypsum wallboard, skim coat plaster, painted with primer and 2 coats; hardwood baseboard and trim, sanded and finished; hardwood floor 70%, ceramic tile with underlayment 20%, vinyl tile with underlayment 10%; wood panel interior doors, primed and painted with 2 coats.	.409	14.94	12.18	27.12
7 Specialties	Custon grade kitchen cabinets - 20 L.F. wall and base with solid surface counter top and kitchen sink; 4 L.F. bathroom vanity; 75 gallon electric water heater, medicine cabinet.	.053	3.13	.77	3.90
8 Mechanical	Gas fired warm air heat/air conditioning; one full bath including: bathtub, corner shower; built in lavatory and water closet; one 1/2 bath including: built in lavatory and water closet.	.105	3.98	2.27	6.25
9 Electrical	200 Amp. service; romex wiring; fluorescent and incandescent lighting fixtures, switches, receptacles.	.048	.88	1.30	2.18
10 Overhead	Contractor's overhead and profit and design.		8.25	6.04	14.29
		Total	49.50	36.30	**85.80**

- **A distinct residence from designer's plans**
- **Single family — 1 full bath, 1 half bath, 1 kitchen**
- **No basement**
- **Asphalt shingles on roof**
- **Forced hot air heat/air conditioning**
- **Gypsum wallboard interior finishes**
- **Materials and workmanship are above average**

Note: The illustration shown may contain some optional components (for example: garages and/or fireplaces) whose costs are shown in the modifications, adjustments, & alternatives below or at the end of the square foot section.

Base cost per square foot of living area

Exterior Wall	Living Area 1200	1400	1600	1800	2000	2400	2800	3200	3600	4000	4400
Wood Siding - Wood Frame	108.80	102.50	98.05	94.20	89.80	83.85	78.65	75.30	73.35	71.25	69.45
Brick Veneer - Wood Frame	114.50	107.85	103.15	99.00	94.45	88.00	82.50	78.90	76.80	74.50	72.60
Stone Veneer - Wood Frame	118.30	111.40	106.55	102.20	97.55	90.80	84.95	81.25	79.05	76.65	74.70
Solid Masonry	120.05	113.00	108.15	103.65	98.95	92.05	86.15	82.30	80.10	77.60	75.60
Finished Basement, Add	22.75	22.90	22.25	21.60	21.20	20.25	19.50	19.00	18.70	18.30	18.10
Unfinished Basement, Add	9.70	9.30	9.00	8.70	8.50	8.10	7.70	7.50	7.35	7.15	7.05

Modifications

Add to the total cost

Upgrade Kitchen Cabinets	$ + 910
Solid Surface Countertops (Included)	
Full Bath - including plumbing, wall and floor finishes	+ 5050
Half Bath - including plumbing, wall and floor finishes	+ 3152
Two Car Attached Garage	+ 19,382
Two Car Detached Garage	+ 22,032
Fireplace & Chimney	+ 4355

Adjustments

For multi family - add to total cost

Additional Kitchen	$ + 11,494
Additional Full Bath & Half Bath	+ 8202
Additional Entry & Exit	+ 1276
Separate Heating & Air Conditioning	+ 4978
Separate Electric	+ 1575

For Townhouse/Rowhouse - Multiply cost per square foot by

Inner Unit	.89
End Unit	.95

Alternatives

Add to or deduct from the cost per square foot of living area

Cedar Shake Roof	+ 1.9
Clay Tile Roof	+ 3.7
Slate Roof	+ 3.7
Upgrade Ceilings to Textured Finish	+ .4
Air Conditioning, in Heating Ductwork	Base Syster
Heating Systems, Hot Water	+ 1.4
Heat Pump	+ 2.1
Electric Heat	– 1.7
Not Heated	– 2.7

Additional upgrades or components

Kitchen Cabinets & Countertops	Page 9
Bathroom Vanities	9
Fireplaces & Chimneys	9
Windows, Skylights & Dormers	9
Appliances	9
Breezeways & Porches	9
Finished Attic	9
Garages	9
Site Improvements	9
Wings & Ells	7

Important: See the Reference Section for Location Factors (to adjust for your city) and Estimating Form

Custom Bi-Level

Living Area 2800 S.F.
Perimeter 156 L.F.

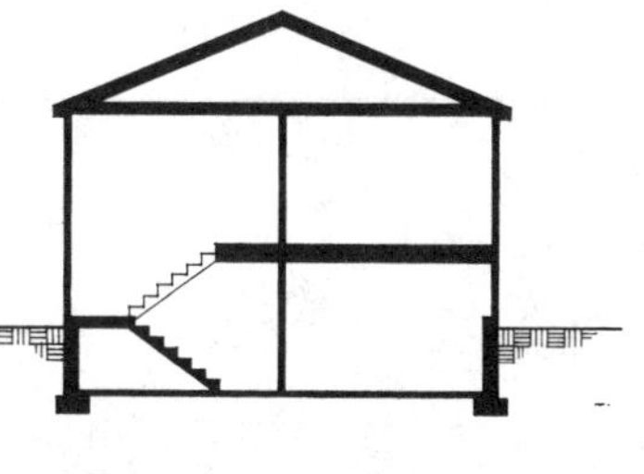

		Labor-Hours	Cost Per Square Foot Of Living Area		
			Mat.	Labor	Total
1 Site Work	Excavation for lower level, 4′ deep. Site preparation for slab.	.024		.51	.51
2 Foundation	Continuous reinforced concrete footing 8″ deep x 18″ wide; dampproofed and insulated reinforced concrete foundation wall, 8″ thick, 4′ deep; 4″ concrete slab on 4″ crushed stone base and polyethylene vapor barrier, trowel finish.	.058	2.59	3.42	6.01
3 Framing	Exterior walls - 2″ x 6″ wood studs, 16″ O.C.; 1/2″ plywood sheathing; 2″ x 8″ rafters 16″ O.C. with 1/2″ plywood sheathing, 6 in 12 pitch; 2″ x 8″ ceiling joists 16″ O.C.; 2″ x 10″ floor joists 16″ O.C. with 5/8″ plywood subfloor; 5/8″ plywood subfloor on 1″ x 3″ wood sleepers 16″ O.C.	.147	4.89	5.34	10.23
4 Exterior Walls	Horizontal beveled wood siding; building paper; 6″ batt insulation; wood double hung windows; 3 solid core wood exterior doors; storms and screens.	.079	6.31	2.35	8.66
5 Roofing	30 year asphalt shingles; #15 felt building paper; aluminum gutters, downspouts and drip edge; copper flashings.	.033	1.41	1.37	2.78
6 Interiors	Walls and ceilings - 5/8″ gypsum wallboard, skim coat plaster, painted with primer and 2 coats; hardwood baseboard and trim, sanded and finished; hardwood floor 70%, ceramic tile with underlayment 20%, vinyl tile with underlayment 10%; wood panel interior doors, primed and painted with 2 coats.	.257	13.67	10.78	24.45
7 Specialties	Custom grade kitchen cabinets - 20 L.F. wall and base with solid surface counter top and kitchen sink; 4 L.F. bathroom vanity; 75 gallon electric water heater, medicine cabinet.	.028	3.36	.81	4.17
8 Mechanical	Gas fired warm air heat/air conditioning; one full bath including: bathtub, corner shower, built in lavatory and water closet; one 1/2 bath including: built in lavatory and water closet.	.078	4.19	2.32	6.51
9 Electrical	200 Amp. service; romex wiring; fluorescent and incandescent lighting fixtures, switches, receptacles.	.038	.90	1.32	2.22
10 Overhead	Contractor's overhead and profit and design.		7.48	5.63	13.11
		Total	44.80	33.85	**78.65**

RESIDENTIAL | Custom | Tri-Level

- **A distinct residence from designer's plans**
- **Single family – 1 full bath, 1 half bath, 1 kitchen**
- **No basement**
- **Asphalt shingles on roof**
- **Forced hot air heat/air conditioning**
- **Gypsum wallboard interior finishes**
- **Materials and workmanship are above average**

Note: The illustration shown may contain some optional components (for example: garages and/or fireplaces) whose costs are shown in the modifications, adjustments, & alternatives below or at the end of the square foot section.

Base cost per square foot of living area

Exterior Wall	1200	1500	1800	2100	2400	2800	3200	3600	4000	4500	5000
	Living Area										
Wood Siding - Wood Frame	111.80	101.90	94.35	88.15	84.00	80.95	77.40	73.60	72.05	68.45	66.45
Brick Veneer - Wood Frame	117.50	107.05	99.00	92.35	88.00	84.75	80.95	76.90	75.25	71.35	69.30
Stone Veneer - Wood Frame	121.25	110.45	102.10	95.15	90.60	87.30	83.30	79.00	77.35	73.30	71.10
Solid Masonry	123.00	112.00	103.55	96.45	91.80	88.50	84.35	80.10	78.30	74.20	71.90
Finished Basement, Add*	28.30	28.20	27.00	25.95	25.30	24.85	24.20	23.65	23.40	22.90	22.55
Unfinished Basement, Add*	12.00	11.35	10.75	10.25	9.95	9.75	9.45	9.15	9.05	8.80	8.60

*Basement under middle level only.

Modifications

Add to the total cost

Upgrade Kitchen Cabinets	$ + 910
Solid Surface Countertops (Included)	
Full Bath - including plumbing, wall and floor finishes	+ 5050
Half Bath - including plumbing, wall and floor finishes	+ 3152
Two Car Attached Garage	+ 19,382
Two Car Detached Garage	+ 22,032
Fireplace & Chimney	+ 4915

Adjustments

For multi family - add to total cost

Additional Kitchen	$ + 11,494
Additional Full Bath & Half Bath	+ 8202
Additional Entry & Exit	+ 1276
Separate Heating & Air Conditioning	+ 4978
Separate Electric	+ 1575

For Townhouse/Rowhouse - Multiply cost per square foot by

Inner Unit	.87
End Unit	.94

Alternatives

Add to or deduct from the cost per square foot of living area

Cedar Shake Roof	+ 1.4
Clay Tile Roof	+ 2.6
Slate Roof	+ 5.3
Upgrade Ceilings to Textured Finish	+ .4
Air Conditioning, in Heating Ductwork	Base Syste
Heating Systems, Hot Water	+ 1.3
Heat Pump	+ 2.2
Electric Heat	– 1.5
Not Heated	– 2.7

Additional upgrades or components

Kitchen Cabinets & Countertops	Page 9
Bathroom Vanities	9
Fireplaces & Chimneys	9
Windows, Skylights & Dormers	9
Appliances	9
Breezeways & Porches	9
Finished Attic	9
Garages	9
Site Improvements	9
Wings & Ells	7

Important: See the Reference Section for Location Factors (to adjust for your city) and Estimating Form

Custom Tri-Level

Living Area 3200 S.F.
Perimeter 198 L.F.

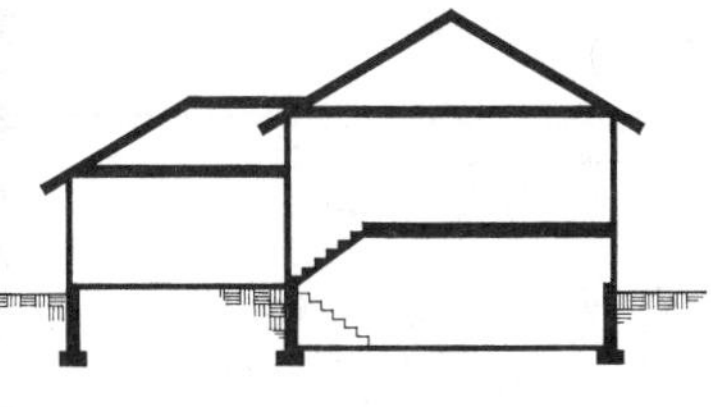

		Labor-Hours	Cost Per Square Foot Of Living Area		
			Mat.	Labor	Total
1 Site Work	Site preparation for slab; 4′ deep trench excavation for foundation wall, excavation for lower level, 4′ deep.	.023		.45	.45
2 Foundation	Continuous reinforced concrete footing 8″ deep x 18″ wide; dampproofed and insulated reinforced concrete foundation wall, 8″ thick, 4′ deep; 4″ concrete slab on 4″ crushed stone base and polyethylene vapor barrier, trowel finish.	.073	3.08	3.94	7.02
3 Framing	Exterior walls - 2″ x 6″ wood studs, 16″ O.C.; 1/2″ plywood sheathing; 2″ x 8″ rafters 16″ O.C. with 1/2″ plywood sheathing, 6 in 12 pitch; 2″ x 8″ ceiling joists 16″ O.C.; 2″ x 10″ floor joists 16″ O.C. with 5/8″ plywood subfloor; 5/8″ plywood subfloor on 1″ x 3″ wood sleepers 16″ O.C.	.162	4.48	5.25	9.73
4 Exterior Walls	Horizontal beveled wood siding; building paper; 6″ batt insulation; wood double hung windows; 3 solid core wood exterior doors; storms and screens.	.076	6.10	2.25	8.35
5 Roofing	30 year asphalt shingles; #15 felt building paper; aluminum gutters, downspouts and drip edge; copper flashings.	.045	1.88	1.82	3.70
6 Interiors	Walls and ceilings - 5/8″ gypsum wallboard, skim coat plaster, painted with primer and 2 coats; hardwood baseboard and trim, sanded and finished; hardwood floor 70%, ceramic tile with underlayment 20%, vinyl tile with underlayment 10%; wood panel interior doors, primed and painted with 2 coats.	.242	13.08	10.28	23.36
7 Specialties	Custom grade kitchen cabinets - 20 L.F. wall and base with solid surface counter top and kitchen sink; 4 L.F. bathroom vanity; 75 gallon electric water heater, medicine cabinet.	.026	2.96	.72	3.68
8 Mechanical	Gas fired warm air heat/air conditioning; one full bath including: bathtub, corner shower, built in lavatory and water closet; one 1/2 bath including: built in lavatory and water closet.	.073	3.81	2.24	6.05
9 Electrical	200 Amp. service; romex wiring; fluorescent and incandescent lighting fixtures, switches, receptacles.	.036	.87	1.28	2.15
10 Overhead	Contractor's overhead and profit and design.		7.24	5.67	12.91
		Total	43.50	33.90	**77.40**

1 Story — Base cost per square foot of living area

Exterior Wall	Living Area 50	100	200	300	400	500	600	700
Wood Siding - Wood Frame	172.05	134.00	117.45	100.10	94.65	91.50	89.30	89.95
Brick Veneer - Wood Frame	191.40	147.80	128.95	107.75	101.60	97.90	95.45	95.90
Stone Veneer - Wood Frame	204.10	156.85	136.50	112.80	106.10	102.10	99.45	99.80
Solid Masonry	210.05	161.15	140.05	115.15	108.20	104.10	101.30	101.55
Finished Basement, Add	66.00	56.60	51.40	42.75	41.00	39.95	39.25	38.75
Unfinished Basement, Add	49.20	33.95	26.90	20.75	19.10	18.10	17.45	17.00

1-1/2 Story — Base cost per square foot of living area

Exterior Wall	Living Area 100	200	300	400	500	600	700	800
Wood Siding - Wood Frame	137.20	112.60	97.60	89.05	84.45	82.35	79.55	78.75
Brick Veneer - Wood Frame	154.50	126.40	109.15	98.00	92.70	90.15	87.00	86.10
Stone Veneer - Wood Frame	165.80	135.45	116.65	103.85	98.15	95.30	91.85	90.90
Solid Masonry	171.10	139.70	120.25	106.65	100.75	97.70	94.15	93.10
Finished Basement, Add	44.30	41.10	37.60	33.85	32.80	32.15	31.45	31.40
Unfinished Basement, Add	29.45	22.40	19.05	16.30	15.30	14.65	14.05	13.90

2 Story — Base cost per square foot of living area

Exterior Wall	Living Area 100	200	400	600	800	1000	1200	1400
Wood Siding - Wood Frame	136.95	104.75	90.55	77.30	72.65	69.90	68.10	69.00
Brick Veneer - Wood Frame	156.25	118.55	102.10	84.95	79.60	76.35	74.20	74.95
Stone Veneer - Wood Frame	168.95	127.60	109.60	89.95	84.10	80.55	78.25	78.85
Solid Masonry	174.90	131.85	113.20	92.35	86.20	82.60	80.10	80.60
Finished Basement, Add	33.05	28.30	25.75	21.40	20.55	20.00	19.70	19.45
Unfinished Basement, Add	24.60	17.05	13.45	10.40	9.60	9.05	8.75	8.50

Base costs do not include bathroom or kitchen facilities. Use Modifications/Adjustments/Alternatives on pages 93-96 where appropriate.

Important: See the Reference Section for Location Factors (to adjust for your city) and Estimating Forms

Luxury Class

1 Story

1-1/2 Story

2 Story

2-1/2 Story

Bi-Level

Tri-Level

RESIDENTIAL | Luxury | 1 Story

- Unique residence built from an architect's plan
- Single family – 1 full bath, 1 half bath, 1 kitchen
- No basement
- Cedar shakes on roof
- Forced hot air heat/air conditioning
- Gypsum wallboard interior finishes
- Many special features
- Extraordinary materials and workmanship

Note: The illustration shown may contain some optional components (for example: garages and/or fireplaces) whose costs are shown in the modifications, adjustments, & alternatives below or at the end of the square foot section.

Base cost per square foot of living area

Exterior Wall	Living Area 1000	1200	1400	1600	1800	2000	2400	2800	3200	3600	4000
Wood Siding - Wood Frame	155.40	143.85	134.40	128.10	124.50	120.30	112.55	107.00	103.15	99.10	95.70
Brick Veneer - Wood Frame	162.55	150.25	140.20	133.55	129.80	125.25	117.10	111.20	107.05	102.70	99.15
Solid Brick	173.50	160.10	149.20	141.95	137.85	132.90	124.10	117.60	113.00	108.20	104.30
Solid Stone	173.15	159.85	148.90	141.70	137.65	132.70	123.80	117.45	112.80	108.05	104.10
Finished Basement, Add	42.30	45.45	43.65	42.50	41.75	40.75	39.40	38.40	37.50	36.70	36.05
Unfinished Basement, Add	18.95	17.85	17.00	16.45	16.10	15.60	15.00	14.50	14.00	13.65	13.30

Modifications

Add to the total cost

Upgrade Kitchen Cabinets	$ + 1200
Solid Surface Countertops (Included)	
Full Bath - including plumbing, wall and floor finishes	+ 5858
Half Bath - including plumbing, wall and floor finishes	+ 3657
Two Car Attached Garage	+ 22,219
Two Car Detached Garage	+ 25,057
Fireplace & Chimney	+ 6105

Adjustments

For multi family - add to total cost

Additional Kitchen	$ + 14,855
Additional Full Bath & Half Bath	+ 9515
Additional Entry & Exit	+ 2114
Separate Heating & Air Conditioning	+ 4978
Separate Electric	+ 1575

For Townhouse/Rowhouse - Multiply cost per square foot by

Inner Unit	.90
End Unit	.95

Alternatives

Add to or deduct from the cost per square foot of living area

Heavyweight Asphalt Shingles	– 1.9
Clay Tile Roof	+ 1.7
Slate Roof	+ 5.6
Upgrade Ceilings to Textured Finish	+ .4
Air Conditioning, in Heating Ductwork	Base Syste
Heating Systems, Hot Water	+ 1.6
Heat Pump	+ 1.9
Electric Heat	– 1.7
Not Heated	– 3.5

Additional upgrades or components

Kitchen Cabinets & Countertops	Page 9
Bathroom Vanities	9
Fireplaces & Chimneys	9
Windows, Skylights & Dormers	9
Appliances	9
Breezeways & Porches	9
Finished Attic	9
Garages	9
Site Improvements	9
Wings & Ells	9

Important: See the Reference Section for Location Factors (to adjust for your city) and Estimating Form

Luxury 1 Story

Living Area 2800 S.F.
Perimeter 219 L.F.

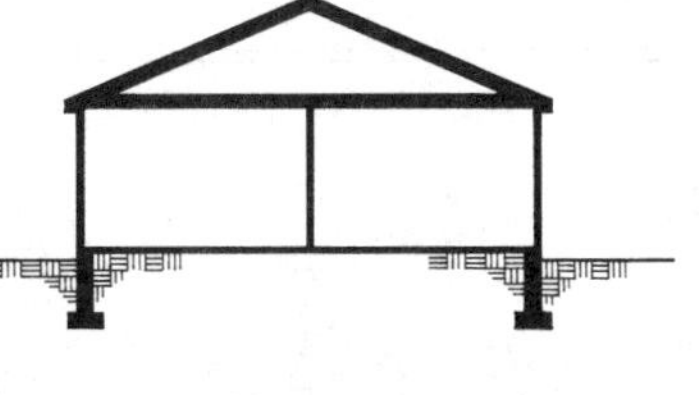

		Labor-Hours	Cost Per Square Foot Of Living Area		
			Mat.	Labor	Total
1 Site Work	Site preparation for slab; 4′ deep trench excavation for foundation wall.	.028		.55	.55
2 Foundation	Continuous reinforced concrete footing 8″ deep x 18″ wide; dampproofed and insulated reinforced concrete foundation wall, 12″ thick, 4′ deep; 4″ concrete slab on 4″ crushed stone base and polyethylene vapor barrier, trowel finish.	.098	5.27	5.81	11.08
3 Framing	Exterior walls - 2″ x 6″ wood studs, 16″ O.C.; 5/8″ plywood sheathing; 2″ x 10″ rafters 16″ O.C. with 5/8″ plywood sheathing, 6 in 12 pitch; 2″ x 8″ ceiling joists 16″ O.C.; 5/8″ plywood subfloor on 1″ x 3″ wood sleepers 16″ O.C.	.260	11.12	10.61	21.73
4 Exterior Walls	Horizontal beveled wood siding; building paper; 6″ batt insulation; wood double hung windows; 3 solid core wood exterior doors; storms and screens.	.204	6.82	2.56	9.38
5 Roofing	Red cedar shingles; #15 felt building paper; aluminum gutters, downspouts and drip edge; copper flashings.	.082	3.30	3.19	6.49
6 Interiors	Walls and ceilings - 5/8″ gypsum wallboard, skim coat plaster, painted with primer and 2 coats; hardwood baseboard and trim, sanded and finished; hardwood floor 70%, ceramic tile with underlayment 20%, vinyl tile with underlayment 10%; wood panel interior doors, primed and painted with 2 coats.	.287	10.82	10.82	21.64
7 Specialties	Luxury grade kitchen cabinets - 25 L.F. wall and base with solid surface counter top and kitchen sink; 6 L.F. bathroom vanity; 75 gallon electric water heater; medicine cabinet.	.052	4.41	1.04	5.45
8 Mechanical	Gas fired warm air heat/air conditioning; one full bath including: bathtub, corner shower; built in lavatory and water closet; one 1/2 bath including: built in lavatory and water closet.	.078	4.79	2.55	7.34
9 Electrical	200 Amp. service; romex wiring; fluorescent and incandescent lighting fixtures; intercom, switches, receptacles.	.044	1.06	1.59	2.65
10 Overhead	Contractor's overhead and profit and architect's fees.		11.41	9.28	20.69
		Total	59.00	48.00	**107.00**

- **Unique residence built from an architect's plan**
- **Single family – 1 full bath, 1 half bath, 1 kitchen**
- **No basement**
- **Cedar shakes on roof**
- **Forced hot air heat/air conditioning**
- **Gypsum wallboard interior finishes**
- **Many special features**
- **Extraordinary materials and workmanship**

Note: The illustration shown may contain some optional components (for example: garages and/or fireplaces) whose costs are shown in the modifications, adjustments, & alternatives below or at the end of the square foot section.

Base cost per square foot of living area

Exterior Wall	Living Area 1000	1200	1400	1600	1800	2000	2400	2800	3200	3600	4000
Wood Siding - Wood Frame	143.80	133.75	126.55	118.15	113.30	108.45	99.45	95.25	91.55	88.65	84.65
Brick Veneer - Wood Frame	152.05	141.55	133.90	124.80	119.60	114.40	104.70	100.30	96.20	93.10	88.80
Solid Brick	164.70	153.35	145.20	135.10	129.35	123.70	112.95	108.10	103.35	100.05	95.25
Solid Stone	164.30	152.95	144.85	134.80	129.05	123.40	112.65	107.80	103.10	99.80	95.05
Finished Basement, Add	29.80	32.30	31.30	29.90	29.15	28.40	26.95	26.30	25.50	25.10	24.55
Unfinished Basement, Add	13.50	12.95	12.50	11.80	11.45	11.10	10.45	10.15	9.75	9.55	9.30

Modifications

Add to the total cost

Upgrade Kitchen Cabinets	$ + 1200
Solid Surface Countertops (Included)	
Full Bath - including plumbing, wall and floor finishes	+ 5858
Half Bath - including plumbing, wall and floor finishes	+ 3657
Two Car Attached Garage	+ 22,219
Two Car Detached Garage	+ 25,057
Fireplace & Chimney	+ 6695

Adjustments

For multi family - add to total cost

Additional Kitchen	$ + 14,855
Additional Full Bath & Half Bath	+ 9515
Additional Entry & Exit	+ 2114
Separate Heating & Air Conditioning	+ 4978
Separate Electric	+ 1575

For Townhouse/Rowhouse - Multiply cost per square foot by

Inner Unit	.90
End Unit	.95

Alternatives

Add to or deduct from the cost per square foot of living area

Heavyweight Asphalt Shingles	– 1.4
Clay Tile Roof	+ 1.2
Slate Roof	+ 4.1
Upgrade Ceilings to Textured Finish	+ .4
Air Conditioning, in Heating Ductwork	Base Syste
Heating Systems, Hot Water	+ 1.5
Heat Pump	+ 2.1
Electric Heat	– 1.7
Not Heated	– 3.2

Additional upgrades or components

Kitchen Cabinets & Countertops	Page 9
Bathroom Vanities	9
Fireplaces & Chimneys	9
Windows, Skylights & Dormers	9
Appliances	9
Breezeways & Porches	9
Finished Attic	9
Garages	9
Site Improvements	9
Wings & Ells	9

Important: See the Reference Section for Location Factors (to adjust for your city) and Estimating Form

Luxury 1-1/2 Story

Living Area 2800 S.F.
Perimeter 175 L.F.

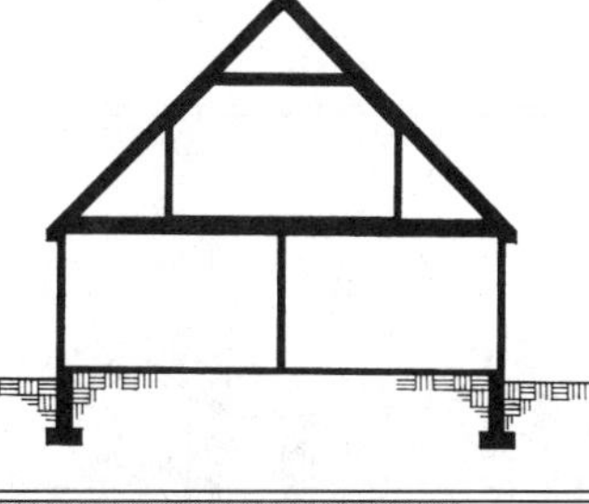

		Labor-Hours	Cost Per Square Foot Of Living Area		
			Mat.	Labor	Total
1 Site Work	Site preparation for slab; 4' deep trench excavation for foundation wall.	.025		.55	.55
2 Foundation	Continuous reinforced concrete footing 8" deep x 18" wide; dampproofed and insulated reinforced concrete foundation wall, 12" thick, 4' deep; 4" concrete slab on 4" crushed stone base and polyethylene vapor barrier, trowel finish.	.066	3.77	4.38	8.15
3 Framing	Exterior walls - 2" x 6" wood studs, 16" O.C.; 5/8" plywood sheathing; 2" x 10" rafters 16" O.C. with 5/8" plywood sheathing, 8 in 12 pitch; 2" x 8" ceiling joists 16" O.C.; 2" x 12" floor joists 16" O.C. with 5/8" plywood subfloor; 5/8" plywood subfloor on 1" x 3" wood sleepers 16" O.C.	.189	6.69	6.88	13.57
4 Exterior Walls	Horizontal beveled wood siding; building paper; 6" batt insulation; wood double hung windows; 3 solid core wood exterior doors; storms and screens.	.174	7.50	2.83	10.33
5 Roofing	Red cedar shingles; #15 felt building paper; aluminum gutters, downspouts and drip edge; copper flashings.	.065	2.06	1.99	4.05
6 Interiors	Walls and ceilings - 5/8" gypsum wallboard, skim coat plaster, painted with primer and 2 coats; hardwood baseboard and trim, sanded and finished; hardwood floor 70%, ceramic tile with underlayment 20%, vinyl tile with underlayment 10%; wood panel interior doors, primed and painted with 2 coats.	.260	12.40	12.34	24.74
7 Specialties	Luxury grade kitchen cabinets - 25 L.F. wall and base with solid surface counter top and kitchen sink; 6 L.F. bathroom vanity; 75 gallon electric water heater; medicine cabinet.	.062	4.41	1.04	5.45
8 Mechanical	Gas fired warm air heat/air conditioning; one full bath including: bathtub, corner shower; built in lavatory and water closet; one 1/2 bath including: built in lavatory and water closet.	.080	4.79	2.55	7.34
9 Electrical	200 Amp. service; romex wiring; fluorescent and incandescent lighting fixtures; intercom, switches, receptacles.	.044	1.06	1.59	2.65
10 Overhead	Contractor's overhead and profit and architect's fees.		10.22	8.20	18.42
		Total	52.90	42.35	**95.25**

- **Unique residence built from an architect's plan**
- **Single family – 1 full bath, 1 half bath, 1 kitchen**
- **No basement**
- **Cedar shakes on roof**
- **Forced hot air heat/air conditioning**
- **Gypsum wallboard interior finishes**
- **Many special features**
- **Extraordinary materials and workmanship**

Note: The illustration shown may contain some optional components (for example: garages and/or fireplaces) whose costs are shown in the modifications, adjustments, & alternatives below or at the end of the square foot section.

Base cost per square foot of living area

Exterior Wall	Living Area 1200	1400	1600	1800	2000	2400	2800	3200	3600	4000	4400
Wood Siding - Wood Frame	133.40	125.40	119.75	114.85	109.40	101.80	95.25	91.05	88.60	85.90	83.55
Brick Veneer - Wood Frame	142.20	133.55	127.60	122.20	116.55	108.20	101.05	96.55	93.90	90.85	88.35
Solid Brick	155.75	146.20	139.65	133.55	127.50	118.05	110.10	105.00	101.95	98.45	95.75
Solid Stone	155.30	145.80	139.25	133.15	127.15	117.80	109.75	104.75	101.75	98.30	95.50
Finished Basement, Add	23.95	25.95	25.20	24.45	23.90	22.80	21.80	21.25	20.85	20.45	20.20
Unfinished Basement, Add	10.90	10.45	10.05	9.70	9.45	8.95	8.50	8.25	8.05	7.85	7.75

Modifications

Add to the total cost

Upgrade Kitchen Cabinets	$ + 1200
Solid Surface Countertops (Included)	
Full Bath - including plumbing, wall and floor finishes	+ 5858
Half Bath - including plumbing, wall and floor finishes	+ 3657
Two Car Attached Garage	+ 22,219
Two Car Detached Garage	+ 25,057
Fireplace & Chimney	+ 6695

Adjustments

For multi family - add to total cost

Additional Kitchen	$ + 14,855
Additional Full Bath & Half Bath	+ 9515
Additional Entry & Exit	+ 2114
Separate Heating & Air Conditioning	+ 4978
Separate Electric	+ 1575

For Townhouse/Rowhouse - Multiply cost per square foot by

Inner Unit	.86
End Unit	.93

Alternatives

Add to or deduct from the cost per square foot of living area

Heavyweight Asphalt Shingles	– .9
Clay Tile Roof	+ .8
Slate Roof	+ 2.8
Upgrade Ceilings to Textured Finish	+ .4
Air Conditioning, in Heating Ductwork	Base Syste
Heating Systems, Hot Water	+ 1.5
Heat Pump	+ 2.2
Electric Heat	– 1.5
Not Heated	– 3.1

Additional upgrades or components

Kitchen Cabinets & Countertops	Page 9
Bathroom Vanities	9
Fireplaces & Chimneys	9
Windows, Skylights & Dormers	9
Appliances	9
Breezeways & Porches	9
Finished Attic	9
Garages	9
Site Improvements	9
Wings & Ells	9

Important: See the Reference Section for Location Factors (to adjust for your city) and Estimating Forn

Luxury 2 Story

Living Area 3200 S.F.
Perimeter 163 L.F.

		Labor-Hours	Cost Per Square Foot Of Living Area		
			Mat.	Labor	Total
1 Site Work	Site preparation for slab; 4′ deep trench excavation for foundation wall.	.024		.49	.49
2 Foundation	Continuous reinforced concrete footing 8″ deep x 18″ wide; dampproofed and insulated reinforced concrete foundation wall, 12″ thick, 4′ deep; 4″ concrete slab on 4″ crushed stone base and polyethylene vapor barrier, trowel finish.	.058	3.04	3.57	6.61
3 Framing	Exterior walls - 2″ x 6″ wood studs, 16″ O.C.; 5/8″ plywood sheathing; 2″ x 10″ rafters 16″ O.C. with 5/8″ plywood sheathing, 6 in 12 pitch; 2″ x 8″ ceiling joists 16″ O.C.; 2″ x 12″ floor joists 16″ O.C. with 5/8″ plywood subfloor; 5/8″ plywood subfloor on 1″ x 3″ wood sleepers 16″ O.C.	.193	6.74	6.82	13.56
4 Exterior Walls	Horizontal beveled wood siding, building paper; 6″ batt insulation; wood double hung windows; 3 solid core wood exterior doors; storms and screens.	.247	7.90	3.03	10.93
5 Roofing	Red cedar shingles; #15 felt building paper; aluminum gutters, downspouts and drip edge; copper flashings.	.049	1.65	1.59	3.24
6 Interiors	Walls and ceilings - 5/8″ gypsum wallboard, skim coat plaster, painted with primer and 2 coats; hardwood baseboard and trim, sanded and finished; hardwood floor 70%, ceramic tile with underlayment 20%, vinyl tile with underlayment 10%; wood panel interior doors, primed and painted with 2 coats.	.252	12.19	12.26	24.45
7 Specialties	Luxury grade kitchen cabinets - 25 L.F. wall and base with solid surface counter top and kitchen sink; 6 L.F. bathroom vanity; 75 gallon electric water heater; medicine cabinet.	.057	3.87	.91	4.78
8 Mechanical	Gas fired warm air heat/air conditioning; one full bath including: bathtub, corner shower; built in lavatory and water closet; one 1/2 bath including: built in lavatory and water closet.	.071	4.36	2.45	6.81
9 Electrical	200 Amp. service; romex wiring; fluorescent and incandescent lighting fixtures; intercom, switches, receptacles.	.042	1.03	1.55	2.58
10 Overhead	Contractor's overhead and profit and architect's fee.		9.77	7.83	17.60
		Total	50.55	40.50	**91.05**

- **Unique residence built from an architect's plan**
- **Single family — 1 full bath, 1 half bath, 1 kitchen**
- **No basement**
- **Cedar shakes on roof**
- **Forced hot air heat/air conditioning**
- **Gypsum wallboard interior finishes**
- **Many special features**
- **Extraordinary materials and workmanship**

Note: The illustration shown may contain some optional components (for example: garages and/or fireplaces) whose costs are shown in the modifications, adjustments, & alternatives below or at the end of the square foot section.

Base cost per square foot of living area

Exterior Wall	Living Area 1500	1800	2100	2500	3000	3500	4000	4500	5000	5500	6000
Wood Siding - Wood Frame	130.60	117.60	109.95	104.40	96.70	91.10	85.70	83.15	80.80	78.50	76.00
Brick Veneer - Wood Frame	139.70	126.05	117.50	111.50	103.15	96.95	91.20	88.30	85.70	83.20	80.45
Solid Brick	153.80	138.95	129.10	122.50	113.05	106.00	99.60	96.20	93.25	90.45	87.35
Solid Stone	153.40	138.65	128.70	122.15	112.80	105.70	99.35	95.95	93.00	90.15	87.15
Finished Basement, Add	19.05	20.50	19.35	18.70	17.85	17.05	16.55	16.15	15.90	15.60	15.30
Unfinished Basement, Add	8.75	8.30	7.75	7.45	7.05	6.70	6.45	6.25	6.05	5.95	5.85

Modifications

Add to the total cost

Upgrade Kitchen Cabinets	$ + 1200
Solid Surface Countertops (Included)	
Full Bath - including plumbing, wall and floor finishes	+ 5858
Half Bath - including plumbing, wall and floor finishes	+ 3657
Two Car Attached Garage	+ 22,219
Two Car Detached Garage	+ 25,057
Fireplace & Chimney	+ 7320

Adjustments

For multi family - add to total cost

Additional Kitchen	$ + 14,855
Additional Full Bath & Half Bath	+ 9515
Additional Entry & Exit	+ 2114
Separate Heating & Air Conditioning	+ 4978
Separate Electric	+ 1575

For Townhouse/Rowhouse - Multiply cost per square foot by

Inner Unit	.86
End Unit	.93

Alternatives

Add to or deduct from the cost per square foot of living area

Heavyweight Asphalt Shingles	– .8
Clay Tile Roof	+ .7
Slate Roof	+ 2.4
Upgrade Ceilings to Textured Finish	+ .4
Air Conditioning, in Heating Ductwork	Base Syste
Heating Systems, Hot Water	+ 1.3
Heat Pump	+ 2.3
Electric Heat	– 3.1
Not Heated	– 3.1

Additional upgrades or components

Kitchen Cabinets & Countertops	Page 9
Bathroom Vanities	9
Fireplaces & Chimneys	9
Windows, Skylights & Dormers	9
Appliances	9
Breezeways & Porches	9
Finished Attic	9
Garages	9
Site Improvements	9
Wings & Ells	9

Luxury 2-1/2 Story

Living Area 3000 S.F.
Perimeter 148 L.F.

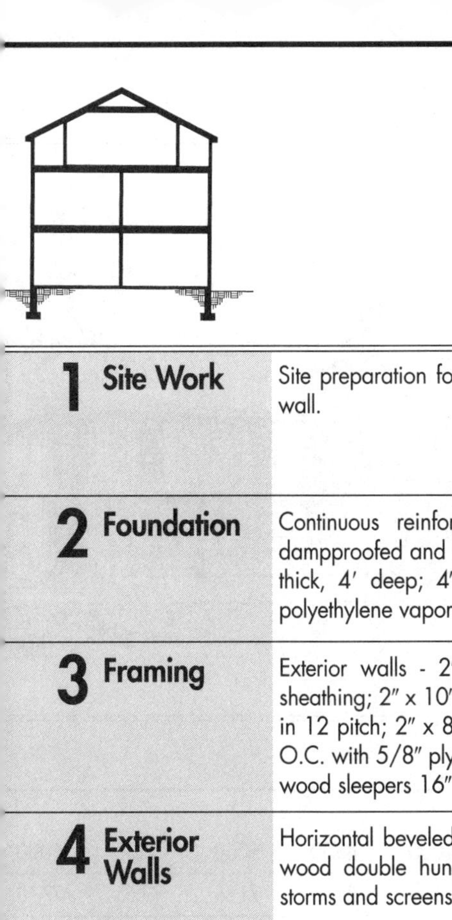

		Labor-Hours	Cost Per Square Foot Of Living Area		
			Mat.	Labor	Total
1 Site Work	Site preparation for slab; 4′ deep trench excavation for foundation wall.	.055		.52	.52
2 Foundation	Continuous reinforced concrete footing 8″ deep x 18″ wide; dampproofed and insulated reinforced concrete foundation wall, 12″ thick, 4′ deep; 4″ concrete slab on 4″ crushed stone base and polyethylene vapor barrier, trowel finish.	.067	2.64	3.24	5.88
3 Framing	Exterior walls - 2″ x 6″ wood studs, 16″ O.C.; 5/8″ plywood sheathing; 2″ x 10″ rafters 16″ O.C. with 5/8″ plywood sheathing, 6 in 12 pitch; 2″ x 8″ ceiling joists 16″ O.C.; 2″ x 12″ floor joists 16″ O.C. with 5/8″ plywood subfloor; 5/8″ plywood subfloor on 1″ x 3″ wood sleepers 16″ O.C.	.209	6.96	7.02	13.98
4 Exterior Walls	Horizontal beveled wood siding; building paper; 6″ batt insulation; wood double hung windows; 3 solid core wood exterior doors; storms and screens.	.405	9.25	3.54	12.79
5 Roofing	Red cedar shingles; #15 felt building paper; aluminum gutters, downspouts and drip edge; copper flashings.	.039	1.27	1.22	2.49
6 Interiors	Walls and ceilings - 5/8″ gypsum wallboard, skim coat plaster, painted with primer and 2 coats; hardwood baseboard and trim, sanded and finished; hardwood floor 70%, ceramic tile with underlayment 20%, vinyl tile with underlayment 10%; wood panel interior doors, primed and painted with 2 coats.	.341	13.81	13.74	27.55
7 Specialties	Luxury grade kitchen cabinets - 25 L.F. wall and base with solid surface counter top and kitchen sink; 6 L.F. bathroom vanity; 75 gallon electric water heater; medicine cabinet.	.119	4.12	.97	5.09
8 Mechanical	Gas fired warm air heat/air conditioning; one full bath including: bathtub, corner shower; built in lavatory and water closet; one 1/2 bath including: built in lavatory and water closet.	.103	4.57	2.50	7.07
9 Electrical	200 Amp. service; romex wiring; fluorescent and incandescent lighting fixtures; intercom, switches, receptacles.	.054	1.04	1.57	2.61
10 Overhead	Contractor's overhead and profit and architect's fee.		10.49	8.23	18.72
		Total	54.15	42.55	**96.70**

- **Unique residence built from an architect's plan**
- **Single family — 1 full bath, 1 half bath, 1 kitchen**
- **No basement**
- **Cedar shakes on roof**
- **Forced hot air heat/air conditioning**
- **Gypsum wallboard interior finishes**
- **Many special features**
- **Extraordinary materials and workmanship**

Note: The illustration shown may contain some optional components (for example: garages and/or fireplaces) whose costs are shown in the modifications, adjustments, & alternatives below or at the end of the square foot section.

Base cost per square foot of living area

Exterior Wall	Living Area 1500	1800	2100	2500	3000	3500	4000	4500	5000	5500	6000
Wood Siding - Wood Frame	129.80	117.00	110.60	105.35	97.45	93.45	88.60	83.55	81.65	79.65	77.60
Brick Veneer - Wood Frame	139.30	125.80	118.80	113.20	104.60	100.15	94.75	89.15	87.10	84.95	82.55
Solid Brick	153.90	139.30	131.40	125.30	115.60	110.50	104.20	97.90	95.50	93.00	90.20
Solid Stone	153.45	138.85	131.00	124.95	115.20	110.20	103.85	97.60	95.25	92.75	90.00
Finished Basement, Add	16.75	18.05	17.30	16.75	15.95	15.45	14.85	14.40	14.15	13.90	13.60
Unfinished Basement, Add	7.65	7.25	6.90	6.65	6.30	6.10	5.80	5.55	5.45	5.35	5.20

Modifications

Add to the total cost

Upgrade Kitchen Cabinets	$ + 1200
Solid Surface Countertops (Included)	
Full Bath - including plumbing, wall and floor finishes	+ 5858
Half Bath - including plumbing, wall and floor finishes	+ 3657
Two Car Attached Garage	+ 22,219
Two Car Detached Garage	+ 25,057
Fireplace & Chimney	+ 7320

Adjustments

For multi family - add to total cost

Additional Kitchen	$ + 14,855
Additional Full Bath & Half Bath	+ 9515
Additional Entry & Exit	+ 2114
Separate Heating & Air Conditioning	+ 4978
Separate Electric	+ 1575

For Townhouse/Rowhouse - Multiply cost per square foot by

Inner Unit	.84
End Unit	.92

Alternatives

Add to or deduct from the cost per square foot of living area

Heavyweight Asphalt Shingles	– .6
Clay Tile Roof	+ .5
Slate Roof	+ 1.8
Upgrade Ceilings to Textured Finish	+ .4
Air Conditioning, in Heating Ductwork	Base Syste
Heating Systems, Hot Water	+ 1.3
Heat Pump	+ 2.3
Electric Heat	– 3.1
Not Heated	– 3.0

Additional upgrades or components

Kitchen Cabinets & Countertops	Page 9
Bathroom Vanities	9
Fireplaces & Chimneys	9
Windows, Skylights & Dormers	9
Appliances	9
Breezeways & Porches	9
Finished Attic	9
Garages	9
Site Improvements	9
Wings & Ells	9

Luxury 3 Story

Living Area 3000 S.F.
Perimeter 135 L.F.

		Labor-Hours	Cost Per Square Foot Of Living Area		
			Mat.	Labor	Total
1 Site Work	Site preparation for slab; 4′ deep trench excavation for foundation wall.	.055		.52	.52
2 Foundation	Continuous reinforced concrete footing 8″ deep x 18″ wide; dampproofed and insulated reinforced concrete foundation wall, 12″ thick, 4′ deep; 4″ concrete slab on 4″ crushed stone base and polyethylene vapor barrier, trowel finish.	.063	2.39	2.94	5.33
3 Framing	Exterior walls - 2″ x 6″ wood studs, 16″ O.C.; 5/8″ plywood sheathing; 2″ x 10″ rafters 16″ O.C. with 5/8″ plywood sheathing, 6 in 12 pitch; 2″ x 8″ ceiling joists 16″ O.C.; 2″ x 12″ floor joists 16″ O.C. with 5/8″ plywood subfloor; 5/8″ plywood subfloor on 1″ x 3″ wood sleepers 16″ O.C.	.225	7.07	7.13	14.20
4 Exterior Walls	Horizontal beveled wood siding; building paper; 6″ batt insulation; wood double hung windows; 3 solid core wood exterior doors; storms and screens.	.454	10.09	3.88	13.97
5 Roofing	Red cedar shingles; #15 felt building paper; aluminum gutters, downspouts and drip edge; copper flashings.	.034	1.10	1.06	2.16
6 Interiors	Walls and ceilings - 5/8″ gypsum wallboard, skim coat plaster, painted with primer and 2 coats; hardwood baseboard and trim, sanded and finished; hardwood floor 70%, ceramic tile with underlayment 20%, vinyl tile with underlayment 10%; wood panel interior doors, primed and painted with 2 coats.	.390	13.83	13.84	27.67
7 Specialties	Luxury grade kitchen cabinets - 25 L.F. wall and base with solid surface counter top and kitchen sink; 6 L.F. bathroom vanity; 75 gallon electric water heater; medicine cabinet.	.119	4.12	.97	5.09
8 Mechanical	Gas fired warm air heat/air conditioning; one full bath including: bathtub, corner shower; built in lavatory and water closet; one 1/2 bath including: built in lavatory and water closet.	.103	4.57	2.50	7.07
9 Electrical	200 Amp. service; romex wiring; fluorescent and incandescent lighting fixtures; intercom, switches, receptacles.	.053	1.04	1.57	2.61
10 Overhead	Contractor's overhead and profit and architect's fees.		10.59	8.24	18.83
		Total	54.80	42.65	**97.45**

- **Unique residence built from an architect's plan**
- **Single family — 1 full bath, 1 half bath, 1 kitchen**
- **No basement**
- **Cedar shakes on roof**
- **Forced hot air heat/air conditioning**
- **Gypsum wallboard interior finishes**
- **Many special features**
- **Extraordinary materials and workmanship**

Note: The illustration shown may contain some optional components (for example: garages and/or fireplaces) whose costs are shown in the modifications, adjustments, & alternatives below or at the end of the square foot section.

Base cost per square foot of living area

	Living Area										
Exterior Wall	**1200**	**1400**	**1600**	**1800**	**2000**	**2400**	**2800**	**3200**	**3600**	**4000**	**4400**
Wood Siding - Wood Frame	126.45	118.95	113.55	108.95	103.75	96.70	90.75	86.70	84.50	81.95	79.80
Brick Veneer - Wood Frame	133.05	125.10	119.40	114.50	109.10	101.50	95.10	90.80	88.40	85.70	83.35
Solid Brick	143.15	134.60	128.45	122.95	117.30	108.95	101.85	97.15	94.45	91.35	88.90
Solid Stone	142.85	134.30	128.15	122.75	117.05	108.65	101.65	96.95	94.30	91.25	88.75
Finished Basement, Add	23.95	25.95	25.20	24.45	23.90	22.80	21.80	21.25	20.85	20.45	20.20
Unfinished Basement, Add	10.90	10.45	10.05	9.70	9.45	8.95	8.50	8.25	8.05	7.85	7.75

Modifications

Add to the total cost

Upgrade Kitchen Cabinets	$ + 1200
Solid Surface Countertops (Included)	
Full Bath - including plumbing, wall and floor finishes	+ 5858
Half Bath - including plumbing, wall and floor finishes	+ 3657
Two Car Attached Garage	+ 22,219
Two Car Detached Garage	+ 25,057
Fireplace & Chimney	+ 6105

Adjustments

For multi family - add to total cost

Additional Kitchen	$ + 14,855
Additional Full Bath & Half Bath	+ 9515
Additional Entry & Exit	+ 2114
Separate Heating & Air Conditioning	+ 4978
Separate Electric	+ 1575

For Townhouse/Rowhouse - Multiply cost per square foot by

Inner Unit	.89
End Unit	.94

Alternatives

Add to or deduct from the cost per square foot of living area

Heavyweight Asphalt Shingles	– 1.9
Clay Tile Roof	+ 1.7
Slate Roof	+ 5.6
Upgrade Ceilings to Textured Finish	+ .4
Air Conditioning, in Heating Ductwork	Base Syste
Heating Systems, Hot Water	+ 1.5
Heat Pump	+ 2.2
Electric Heat	– 1.5
Not Heated	– 3.1

Additional upgrades or components

Kitchen Cabinets & Countertops	Page 9
Bathroom Vanities	9
Fireplaces & Chimneys	9
Windows, Skylights & Dormers	9
Appliances	9
Breezeways & Porches	9
Finished Attic	9
Garages	9
Site Improvements	9
Wings & Ells	9

Important: See the Reference Section for Location Factors (to adjust for your city) and Estimating Form

Luxury Bi-Level

Living Area 3200 S.F.
Perimeter 163 L.F.

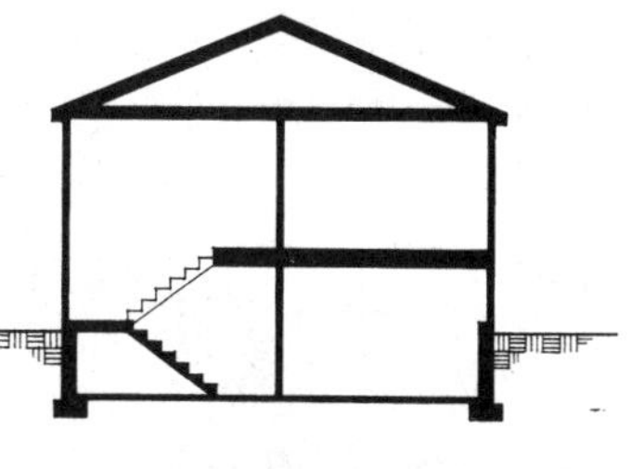

		Labor-Hours	Cost Per Square Foot Of Living Area		
			Mat.	Labor	Total
1 Site Work	Excavation for lower level, 4' deep. Site preparation for slab.	.024		.49	.49
2 Foundation	Continuous reinforced concrete footing 8" deep x 18" wide; dampproofed and insulated reinforced concrete foundation wall, 12" thick, 4' deep; 4" concrete slab on 4" crushed stone base and polyethylene vapor barrier, trowel finish.	.058	3.04	3.57	6.61
3 Framing	Exterior walls - 2" x 6" wood studs, 16" O.C.; 5/8" plywood sheathing; 2" x 10" rafters 16" O.C. with 5/8" plywood sheathing, 6 in 12 pitch; 2" x 8" ceiling joists 16" O.C.; 2" x 12" floor joists 16" O.C. with 5/8" plywood subfloor; 5/8" plywood subfloor on 1" x 3" wood sleepers 16" O.C.	.232	6.42	6.48	12.90
4 Exterior Walls	Horizontal beveled wood siding; building paper; 6" batt insulation; wood double hung windows; 3 solid core wood exterior doors; storms and screens.	.185	6.18	2.34	8.52
5 Roofing	Red cedar shingles: #15 felt building paper; aluminum gutters, downspouts and drip edge; copper flashings.	.042	1.65	1.59	3.24
6 Interiors	Walls and ceilings - 5/8" gypsum wallboard, skim coat plaster, painted with primer and 2 coats; hardwood baseboard and trim, sanded and finished; hardwood floor 70%, ceramic tile with underlayment 20%; vinyl tile with underlayment 10%; wood panel interior doors, primed and painted with 2 coats.	.238	12.04	11.96	24.00
7 Specialties	Luxury grade kitchen cabinets - 25 L.F. wall and base with solid surface counter top and kitchen sink; 6 L.F. bathroom vanity; 75 gallon electric water heater; medicine cabinet.	.056	3.87	.91	4.78
8 Mechanical	Gas fired warm air heat/air conditioning; one full bath including: bathtub, corner shower; built in lavatory and water closet; one 1/2 bath including: built in lavatory and water closet.	.071	4.36	2.45	6.81
9 Electrical	200 Amp. service; romex wiring; fluorescent and incandescent lighting fixtures; intercom, switches, receptacles.	.042	1.03	1.55	2.58
10 Overhead	Contractor's overhead and profit and architect's fees.		9.26	7.51	16.77
		Total	47.85	38.85	**86.70**

- **Unique residence built from an architect's plan**
- **Single family — 1 full bath, 1 half bath, 1 kitchen**
- **No basement**
- **Cedar shakes on roof**
- **Forced hot air heat/air conditioning**
- **Gypsum wallboard interior finishes**
- **Many special features**
- **Extraordinary materials and workmanship**

Note: The illustration shown may contain some optional components (for example: garages and/or fireplaces) whose costs are shown in the modifications, adjustments, & alternatives below or at the end of the square foot section.

Base cost per square foot of living area

Exterior Wall	Living Area 1500	1800	2100	2400	2800	3200	3600	4000	4500	5000	5500
Wood Siding - Wood Frame	119.15	110.20	102.90	98.00	94.45	90.20	85.85	84.00	79.80	77.50	74.95
Brick Veneer - Wood Frame	125.05	115.55	107.75	102.55	98.85	94.30	89.65	87.70	83.15	80.65	78.00
Solid Brick	134.20	123.85	115.20	109.60	105.55	100.55	95.45	93.30	88.40	85.55	82.60
Solid Stone	133.90	123.60	115.00	109.40	105.35	100.35	95.20	93.15	88.20	85.45	82.50
Finished Basement, Add*	28.20	30.40	29.15	28.35	27.80	27.00	26.30	26.00	25.35	24.90	24.45
Unfinished Basement, Add*	12.60	11.85	11.35	10.90	10.70	10.30	9.95	9.80	9.50	9.25	9.10

*Basement under middle level only.

Modifications

Add to the total cost

Upgrade Kitchen Cabinets	$ + 1200
Solid Surface Countertops (Included)	
Full Bath - including plumbing, wall and floor finishes	+ 5858
Half Bath - including plumbing, wall and floor finishes	+ 3657
Two Car Attached Garage	+ 22,219
Two Car Detached Garage	+ 25,057
Fireplace & Chimney	+ 6695

Adjustments

For multi family - add to total cost

Additional Kitchen	$ + 14,855
Additional Full Bath & Half Bath	+ 9515
Additional Entry & Exit	+ 2114
Separate Heating & Air Conditioning	+ 4978
Separate Electric	+ 1575

For Townhouse/Rowhouse - Multiply cost per square foot by

Inner Unit	.86
End Unit	.93

Alternatives

Add to or deduct from the cost per square foot of living area

Heavyweight Asphalt Shingles	- 1.4
Clay Tile Roof	+ 1.2
Slate Roof	+ 4.1
Upgrade Ceilings to Textured Finish	+ .4
Air Conditioning, in Heating Ductwork	Base Syster
Heating Systems, Hot Water	+ 1.4
Heat Pump	+ 2.3
Electric Heat	- 1.4
Not Heated	- 3.0

Additional upgrades or components

Kitchen Cabinets & Countertops	Page 9
Bathroom Vanities	9
Fireplaces & Chimneys	9
Windows, Skylights & Dormers	9
Appliances	9
Breezeways & Porches	9
Finished Attic	9
Garages	9
Site Improvements	9
Wings & Ells	9

Important: See the Reference Section for Location Factors (to adjust for your city) and Estimating Form

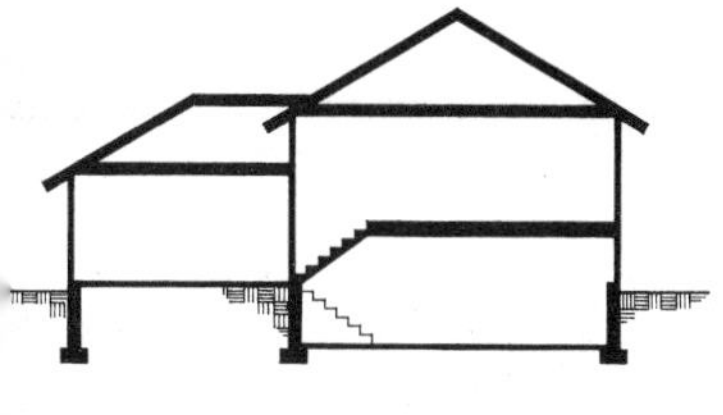

Luxury Tri-Level

Living Area 3600 S.F.
Perimeter 207 L.F.

SQUARE FOOT COSTS

		Labor-Hours	Cost Per Square Foot Of Living Area		
			Mat.	Labor	Total
1 Site Work	Site preparation for slab; 4′ deep trench excavation for foundation wall, excavation for lower level, 4′ deep.	.021		.43	.43
2 Foundation	Continuous reinforced concrete footing 8″ deep x 18″ wide; dampproofed and insulated reinforced concrete foundation wall, 12″ thick, 4′ deep; 4″ concrete slab on 4″ crushed stone base and polyethylene vapor barrier, trowel finish.	.109	3.65	4.12	7.77
3 Framing	Exterior walls - 2″ x 6″ wood studs, 16″ O.C.; 5/8″ plywood sheathing; 2″ x 10″ rafters 16″ O.C. with 5/8″ plywood sheathing, 6 in 12 pitch; 2″ x 8″ ceiling joists 16″ O.C.; 2″ x 12″ floor joists 16″ O.C. with 5/8″ plywood subfloor; 5/8″ plywood subfloor on 1″ x 3″ wood sleepers 16″ O.C.	.204	6.37	6.48	12.85
4 Exterior Walls	Horizontal beveled wood siding; building paper; 6″ batt insulation; wood double hung windows; 3 solid core wood exterior doors; storms and screens.	.181	5.96	2.25	8.21
5 Roofing	Red cedar shingles; #15 felt building paper; aluminum gutters, downspouts and drip edge; copper flashings.	.056	2.19	2.13	4.32
6 Interiors	Walls and ceilings - 5/8″ gypsum wallboard, skim coat plaster, painted with primer and 2 coats; hardwood baseboard and trim, sanded and finished; hardwood floor 70%, ceramic tile with underlayment 20%, vinyl tile with underlayment 10%; wood panel interior doors, primed and painted with 2 coats.	.217	11.33	11.18	22.51
7 Specialties	Luxury grade kitchen cabinets - 25 L.F. wall and base with solid surface counter top and kitchen sink; 6 L.F. bathroom vanity; 75 gallon electric water heater; medicine cabinet.	.048	3.44	.83	4.27
8 Mechanical	Gas fired warm air heat/air conditioning; one full bath including: bathtub, corner shower; built in lavatory and water closet; one 1/2 bath including: built in lavatory and water closet.	.057	4.02	2.37	6.39
9 Electrical	200 Amp. service; romex wiring; fluorescent and incandescent lighting fixtures; intercom, switches, receptacles.	.039	1.00	1.51	2.51
10 Overhead	Contractor's overhead and profit and architect's fees.		9.09	7.50	16.59
		Total	47.05	38.80	**85.85**

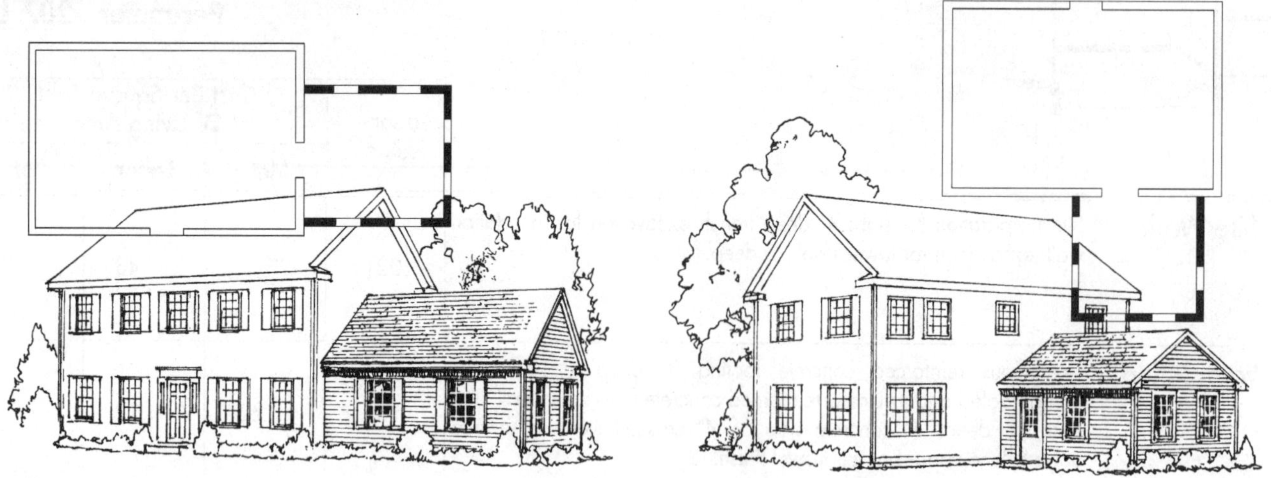

1 Story — Base cost per square foot of living area

Exterior Wall	Living Area 50	100	200	300	400	500	600	700
Wood Siding - Wood Frame	191.90	149.25	130.55	111.00	104.90	101.25	98.85	99.65
Brick Veneer - Wood Frame	214.05	165.05	143.80	119.80	112.80	108.60	105.80	106.40
Solid Brick	248.20	189.40	164.10	133.35	125.00	120.00	116.65	116.90
Solid Stone	247.10	188.60	163.45	132.85	124.60	119.65	116.35	116.45
Finished Basement, Add	72.80	66.40	59.80	48.85	46.65	45.30	44.45	43.80
Unfinished Basement, Add	37.75	29.45	26.00	20.20	19.05	18.35	17.90	17.55

1-1/2 Story — Base cost per square foot of living area

Exterior Wall	Living Area 100	200	300	400	500	600	700	800
Wood Siding - Wood Frame	152.20	124.60	107.75	97.95	92.85	90.50	87.30	86.45
Brick Veneer - Wood Frame	172.00	140.40	120.95	108.20	102.35	99.45	95.85	94.85
Solid Brick	202.45	164.80	141.25	124.05	117.00	113.20	108.90	107.75
Solid Stone	201.55	164.00	140.60	123.55	116.55	112.80	108.50	107.35
Finished Basement, Add	48.20	47.90	43.45	38.60	37.30	36.40	35.55	35.50
Unfinished Basement, Add	24.30	20.75	18.45	15.90	15.25	14.80	14.35	14.30

2 Story — Base cost per square foot of living area

Exterior Wall	Living Area 100	200	400	600	800	1000	1200	1400
Wood Siding - Wood Frame	149.95	113.70	97.80	82.70	77.50	74.45	72.40	73.45
Brick Veneer - Wood Frame	172.05	129.50	110.95	91.50	85.40	81.85	79.40	80.25
Solid Brick	206.15	153.85	131.30	105.00	97.65	93.20	90.20	90.70
Solid Stone	205.15	153.05	130.60	104.55	97.25	92.80	89.85	90.30
Finished Basement, Add	36.40	33.30	29.95	24.40	23.40	22.70	22.25	21.95
Unfinished Basement, Add	18.90	14.75	13.00	10.10	9.55	9.20	9.00	8.80

Base costs do not include bathroom or kitchen facilities. Use Modifications/Adjustments/Alternatives on pages 93-96 where appropriate.

Kitchen cabinets - Base units, hardwood (Cost per Unit)

	Economy	Average	Custom	Luxury
24" deep, 35" high,				
One top drawer,				
One door below				
12" wide	$130	$173	$230	$305
15" wide	170	226	300	395
18" wide	183	244	325	425
21" wide	191	254	340	445
24" wide	216	288	385	505
Four drawers				
12" wide	278	370	490	650
15" wide	220	293	390	515
18" wide	244	325	430	570
24" wide	263	350	465	615
Two top drawers,				
Two doors below				
27" wide	236	315	420	550
30" wide	255	340	450	595
33" wide	263	350	465	615
36" wide	270	360	480	630
42" wide	293	390	520	685
48" wide	311	415	550	725
Range or sink base				
(Cost per unit)				
Two doors below				
30" wide	214	285	380	500
33" wide	229	305	405	535
36" wide	236	315	420	550
42" wide	255	340	450	595
48" wide	266	355	470	620
Corner Base Cabinet				
(Cost per unit)				
36" wide	341	455	605	795
Lazy Susan *(Cost per unit)*				
With revolving door	334	445	590	780

Kitchen cabinets - Wall cabinets, hardwood (Cost per Unit)

	Economy	Average	Custom	Luxury
12" deep, 2 doors				
12" high				
30" wide	$140	$187	$250	$ 325
36" wide	160	213	285	375
15" high				
30" wide	146	195	260	340
33" wide	161	215	285	375
36" wide	164	218	290	380
24" high				
30" wide	177	236	315	415
36" wide	194	259	345	455
42" wide	212	283	375	495
30" high, 1 door				
12" wide	128	170	225	300
15" wide	143	190	255	335
18" wide	154	205	275	360
24" wide	171	228	305	400
30" high, 2 doors				
27" wide	207	276	365	485
30" wide	201	268	355	470
36" wide	229	305	405	535
42" wide	244	325	430	570
48" wide	274	365	485	640
Corner wall, 30" high				
24" wide	147	196	260	345
30" wide	172	229	305	400
36" wide	185	246	325	430
Broom closet				
84" high, 24" deep				
18" wide	371	495	660	865
Oven Cabinet				
84" high, 24" deep				
27" wide	529	705	940	1235

Kitchen countertops (Cost per L.F.)

	Economy	Average	Custom	Luxury
Solid Surface				
24" wide, no backsplash	$ 86	$115	$155	$200
with backsplash	94	125	165	220
Stock plastic laminate, 24" wide				
with backsplash	15	21	25	35
Custom plastic laminate, no splash				
7/8" thick, alum. molding	23	30	40	55
1-1/4" thick, no splash	25	34	45	60
Marble				
1/2" - 3/4" thick w/splash	41	55	70	95
Maple, laminated				
1-1/2" thick w/splash	60	80	105	140
Stainless steel				
(per S.F.)	110	146	195	255
Cutting blocks, recessed				
16" x 20" x 1" (each)	81	108	145	190

Vanity bases *(Cost per Unit)*

	Economy	Average	Custom	Luxury
2 door, 30" high, 21" deep				
24" wide	$170	$227	$300	$395
30" wide	198	264	350	460
36" wide	259	345	460	605
48" wide	308	410	545	720

Solid surface vanity tops *(Cost Each)*

	Economy	Average	Custom	Luxury
Center bowl				
22" x 25"	$285	$308	$332	$359
22" x 31"	325	351	379	409
22" x 37"	375	405	437	472
22" x 49"	465	502	542	586

Fireplaces & Chimneys *(Cost per Unit)*

	1-1/2 Story	2 Story	3 Story
Economy (prefab metal)			
Exterior chimney & 1 fireplace	$4200	$4640	$5085
Interior chimney & 1 fireplace	4020	4475	4685
Average (masonry)			
Exterior chimney & 1 fireplace	4185	4670	5305
Interior chimney & 1 fireplace	3920	4400	4790
For more than 1 flue, add	305	515	855
For more than 1 fireplace, add	2965	2965	2965
Custom (masonry)			
Exterior chimney & 1 fireplace	4355	4915	5555
Interior chimney & 1 fireplace	4085	4620	4980
For more than 1 flue, add	340	590	805
For more than 1 fireplace, add	3130	3130	3130
Luxury (masonry)			
Exterior chimney & 1 fireplace	6105	6695	7320
Interior chimney & 1 fireplace	5825	6370	6740
For more than 1 flue, add	505	840	1175
For more than 1 fireplace, add	4810	4810	4810

Windows and Skylights *(Cost Each)*

	Economy	Average	Custom	Luxury
Fixed Picture Windows				
3'-6" x 4'-0"	$ 454	$ 491	$ 530	$ 572
4'-0" x 6'-0"	652	704	760	821
5'-0" x 6'-0"	922	995	1075	1161
6'-0" x 6'-0"	943	1019	1100	1188
Bay/Bow Windows				
8'-0" x 5'-0"	1200	1296	1400	1512
10'-0" x 5'-0"	1329	1435	1550	1674
10'-0" x 6'-0"	2122	2292	2475	2673
12'-0" x 6'-0"	2658	2870	3100	3348
Palladian Windows				
3'-2" x 6'-4"		1667	1800	1944
4'-0" x 6'-0"		1667	1800	1944
5'-5" x 6'-10"		2060	2225	2403
8'-0" x 6'-0"		2477	2675	2889
Skylights				
46" x 21-1/2"	350	378	479	517
46" x 28"	385	416	471	509
57" x 44"	479	517	637	688

Dormers *(Cost/S.F. of plan area)*

	Economy	Average	Custom	Luxury
Framing and Roofing Only				
Gable dormer, 2" x 6" roof frame	$21	$24	$27	$43
2" x 8" roof frame	23	25	28	45
Shed dormer, 2" x 6" roof frame	14	15	17	28
2" x 8" roof frame	15	16	18	29
2" x 10" roof frame	16	18	20	30

Appliances (Cost per Unit)

	Economy	Average	Custom	Luxury
Range				
30" free standing, 1 oven	$ 310	$1030	$1390	$1750
2 oven	1675	1713	1731	1750
30" built-in, 1 oven	550	1163	1469	1775
2 oven	1200	1575	1763	1950
21" free standing				
1 oven	320	370	395	420
Counter Top Ranges				
4 burner standard	264	477	584	690
As above with griddle	570	698	761	825
Microwave Oven	180	395	503	610
Combination Range, Refrigerator, Sink				
30" wide	1025	1538	1794	2050
60" wide	2400	2760	2940	3120
72" wide	2725	3134	3339	3543
Comb. Range, Refrig., Sink, Microwave Oven & Ice Maker	4168	4793	5106	5418
Compactor				
4 to 1 compaction	520	573	599	625
Deep Freeze				
15 to 23 C.F.	475	568	614	660
30 C.F.	885	993	1046	1100
Dehumidifier, portable, auto.				
15 pint	164	189	201	213
30 pint	183	211	224	238
Washing Machine, automatic	425	950	1213	1475
Water Heater				
Electric, glass lined				
30 gal.	365	450	493	535
80 gal.	750	963	1069	1175
Water Heater, Gas, glass lined				
30 gal.	520	643	704	765
50 gal.	815	1008	1104	1200
Water Softener, automatic				
30 grains/gal.	615	708	754	800
100 grains/gal.	835	961	1023	1086
Dishwasher, built-in				
2 cycles	385	463	501	540
4 or more cycles	400	448	471	495
Dryer, automatic	500	825	988	1150
Garage Door Opener	320	385	418	450
Garbage Disposal	91	148	177	205
Heater, Electric, built-in				
1250 watt ceiling type	169	209	229	249
1250 watt wall type	202	235	252	268
Wall type w/blower				
1500 watt	228	317	361	405
3000 watt	405	446	466	486
Hood For Range, 2 speed				
30" wide	124	460	627	795
42" wide	330	550	660	770
Humidifier, portable				
7 gal. per day	164	189	201	213
15 gal. per day	197	227	241	256
Ice Maker, automatic				
13 lb. per day	445	512	546	579
51 lb. per day	1275	1467	1562	1658
Refrigerator, no frost				
10-12 C.F.	530	680	755	830
14-16 C.F.	540	583	604	625
18-20 C.F.	620	835	943	1050
21-29 C.F.	780	1603	2014	2425
Sump Pump, 1/3 H.P.	214	295	335	375

Breezeway (Cost per S.F.)

Class	Type	Area (S.F.)			
		50	100	150	200
Economy	Open	$ 19.10	$ 16.25	$13.65	$13.40
	Enclosed	91.60	70.80	58.75	51.45
Average	Open	23.65	20.80	18.20	16.55
	Enclosed	100.40	74.80	61.20	53.85
Custom	Open	33.70	29.65	25.80	23.65
	Enclosed	138.45	103.00	84.20	74.00
Luxury	Open	34.80	30.50	27.50	26.65
	Enclosed	140.10	103.90	84.50	75.25

Porches (Cost per S.F.)

Class	Type	Area (S.F.)				
		25	50	100	200	300
Economy	Open	$ 54.75	$ 36.65	$28.60	$24.25	$20.70
	Enclosed	109.40	76.25	57.60	44.95	38.50
Average	Open	66.45	42.30	32.45	27.00	27.00
	Enclosed	131.30	88.95	67.30	52.10	44.15
Custom	Open	86.10	57.10	43.00	37.55	33.65
	Enclosed	171.10	116.90	88.95	69.25	59.65
Luxury	Open	92.15	60.25	44.45	39.95	35.60
	Enclosed	180.30	126.65	94.00	72.95	62.85

Finished attic (Cost per S.F.)

Class	Area (S.F.)				
	400	500	600	800	1000
Economy	$14.65	$14.10	$13.55	$13.30	$12.85
Average	22.50	21.95	21.45	21.15	20.60
Custom	27.45	26.80	26.25	25.80	25.30
Luxury	34.55	33.75	32.95	32.20	31.65

Alarm system (Cost per System)

	Burglar Alarm	Smoke Detector
Economy	$ 370	$ 57
Average	425	70
Custom	720	140
Luxury	1075	172

Sauna, prefabricated
(Cost per unit, including heater and controls—7' high)

Size	Cost
6' x 4'	$ 4275
6' x 5'	4800
6' x 6'	5125
6' x 9'	6350
8' x 10'	8350
8' x 12'	9800
10' x 12'	10400

Garages *

(Costs include exterior wall systems comparable with the quality of the residence. Included in the cost is an allowance for one personnel door, manual overhead door(s) and electrical fixture.)

Class	Type									
	Detached			Attached			Built-in		Basement	
Economy	One Car	Two Car	Three Car	One Car	Two Car	Three Car	One Car	Two Car	One Car	Two Car
Wood	$11,433	$17,426	$23,418	$ 8870	$15,261	$21,254	$1600	$3200	$11,680	$1499
Masonry	15,732	22,806	29,879	11,560	19,032	26,106	2128	4256		
Average										
Wood	12,424	18,665	24,906	9490	16,130	22,371	1722	3443	1336	1835
Masonry	15,886	22,998	30,110	11,656	19,167	26,279	2147	3723		
Custom										
Wood	14,438	22,032	29,626	11,248	19,382	26,976	3322	3592	1976	3115
Masonry	17,690	26,102	34,514	13,284	22,235	30,647	3722	4391		
Luxury										
Wood	16,139	25,057	33,976	12,761	22,219	31,138	3384	3716	2687	4208
Masonry	20,612	30,655	40,699	15,560	26,143	36,187	3934	4814		

*See the Introduction to this section for definitions of garage types.

Swimming pools *(Cost per S.F.)*

Residential	(includes equipment)
In-ground	$ 19.35 - 48.50
Deck equipment	1.30
Paint pool, preparation & 3 coats (epoxy)	3.00
Rubber base paint	2.78
Pool Cover	0.72
Swimming Pool Heaters	(Cost per unit)
(not including wiring, external piping, base or pad)	
Gas	
155 MBH	$ 3050
190 MBH	4075
500 MBH	9950
Electric	
15 KW 7200 gallon pool	2250
24 KW 9600 gallon pool	3000
54 KW 24,000 gallon pool	4475

Wood and coal stoves

Wood Only	
Free Standing (minimum)	$ 1275
Fireplace Insert (minimum)	1285
Coal Only	
Free Standing	$ 1452
Fireplace Insert	1590
Wood and Coal	
Free Standing	$ 2987
Fireplace Insert	3060

Sidewalks *(Cost per S.F.)*

Concrete, 3000 psi with wire mesh	4" thick	$ 2.92
	5" thick	3.56
	6" thick	3.99
Precast concrete patio blocks (natural)	2" thick	8.50
Precast concrete patio blocks (colors)	2" thick	9.00
Flagstone, bluestone	1" thick	12.25
Flagstone, bluestone	1-1/2" thick	16.05
Slate (natural, irregular)	3/4" thick	12.00
Slate (random rectangular)	1/2" thick	17.80
Seeding		
Fine grading & seeding includes lime, fertilizer & seed	per S.Y.	1.93
Lawn Sprinkler System	per S.F.	.76

Fencing *(Cost per L.F.)*

Chain Link, 4' high, galvanized	$ 13.75
Gate, 4' high (each)	144.00
Cedar Picket, 3' high, 2 rail	10.25
Gate (each)	146.00
3 Rail, 4' high	12.30
Gate (each)	155.00
Cedar Stockade, 3 Rail, 6' high	12.50
Gate (each)	155.00
Board & Battens, 2 sides 6' high, pine	17.85
6' high, cedar	25.50
No. 1 Cedar, basketweave, 6' high	14.30
Gate, 6' high (each)	178.00

Carport *(Cost per S.F.)*

Economy	$ 6.89
Average	10.44
Custom	15.57
Luxury	17.68

Assemblies Section

Table of Contents

Table No. — Page

Site Work 101
-04 Footing Excavation 102
-08 Foundation Excavation 104
-12 Utility Trenching 106
-16 Sidewalk 108
-20 Driveway 110
-24 Septic 112
-60 Chain Link Fence 114
-64 Wood Fence 115

2 Foundations 117
2-04 Footing 118
2-08 Block Wall 120
2-12 Concrete Wall 122
2-16 Wood Wall Foundation 124
2-20 Floor Slab 126

3 Framing 129
3-02 Floor (Wood) 130
3-04 Floor (Wood) 132
3-06 Floor (Wood) 134
3-08 Exterior Wall 136
3-12 Gable End Roof 138
3-16 Truss Roof 140
3-20 Hip Roof 142
3-24 Gambrel Roof 144
3-28 Mansard Roof 146
3-32 Shed/Flat Roof 148
3-40 Gable Dormer 150
3-44 Shed Dormer 152
3-48 Partition 154

4 Exterior Walls 157
4-02 Masonry Block Wall 158
4-04 Brick/Stone Veneer 160
4-08 Wood Siding 162
4-12 Shingle Siding 164
4-16 Metal & Plastic Siding 166
4-20 Insulation 168
4-28 Double Hung Window 170
4-32 Casement Window 172
4-36 Awning Window 174
4-40 Sliding Window 176
4-44 Bow/Bay Window 178
4-48 Fixed Window 180
4-52 Entrance Door 182
4-53 Sliding Door 184
4-56 Residential Overhead Door 186
4-58 Aluminum Window 188
4-60 Storm Door & Window 190
4-64 Shutters/Blinds 191

5 Roofing 193
5-04 Gable End Roofing 194
5-08 Hip Roof Roofing 196
5-12 Gambrel Roofing 198
5-16 Mansard Roofing 200
5-20 Shed Roofing 202
5-24 Gable Dormer Roofing 204
5-28 Shed Dormer Roofing 206
5-32 Skylight/Skywindow 208
5-34 Built-up Roofing 210

6 Interiors 213
6-04 Drywall & Thincoat Wall 214
6-08 Drywall & Thincoat Ceiling 216
6-12 Plaster & Stucco Wall 218
6-16 Plaster & Stucco Ceiling 220
6-18 Suspended Ceiling 222
6-20 Interior Door 224
6-24 Closet Door 226
6-60 Carpet 228
6-64 Flooring 229
6-90 Stairways 230

7 Specialties 233
7-08 Kitchen 234
7-12 Appliances 236
7-16 Bath Accessories 237
7-24 Masonry Fireplace 238
7-30 Prefabricated Fireplace 240
7-32 Greenhouse 242
7-36 Swimming Pool 243
7-40 Wood Deck 244

8 Mechanical 247
8-04 Two Fixture Lavatory 248
8-12 Three Fixture Bathroom 250
8-16 Three Fixture Bathroom 252
8-20 Three Fixture Bathroom 254
8-24 Three Fixture Bathroom 256
8-28 Three Fixture Bathroom 258
8-32 Three Fixture Bathroom 260
8-36 Four Fixture Bathroom 262
8-40 Four Fixture Bathroom 264
8-44 Five Fixture Bathroom 266
8-60 Gas Fired Heating/Cooling 268
8-64 Oil Fired Heating/Cooling 270
8-68 Hot Water Heating 272
8-80 Rooftop Heating/Cooling 274

9 Electrical 277
9-10 Electric Service 278
9-20 Electric Heating 279
9-30 Wiring Devices 280
9-40 Light Fixtures 281

How to Use the Assemblies Cost Tables

The following is a detailed explanation of a sample Assemblies Cost Table. Included are an illustration and accompanying system descriptions. Additionally, related systems and price sheets may be included. Next to each bold number below is the item being described with the appropriate component of the sample entry following in parenthesis. General contractors should add an additional markup to the figures shown in the Assemblies section. Note: Throughout this section, the words assembly and system are used interchangeably.

System Identification (3 Framing 12)

Each Assemblies section has been assigned a unique identification number, component category, system numbe and system description.

3 | FRAMING **12 Gable End Roof Framing Systems**

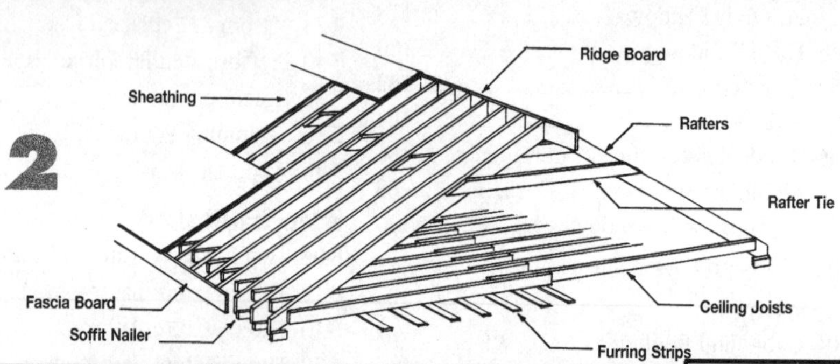

System Description	QUAN.	UNIT	LABOR HOURS	COST PER S.F. MAT.	INST.	TOTAL
2" X 6" RAFTERS, 16" O.C., 4/12 PITCH						
Rafters, 2" x 6", 16" O.C., 4/12 pitch	1.170	L.F.	.019	.69	.76	1.45
Ceiling joists, 2" x 4", 16" O.C.	1.000	L.F.	.013	.37	.52	.89
Ridge board, 2" x 6"	.050	L.F.	.002	.03	.07	.10
Fascia board, 2" x 6"	.100	L.F.	.005	.07	.22	.29
Rafter tie, 1" x 4", 4' O.C.	.060	L.F.	.001	.03	.05	.08
Soffit nailer (outrigger), 2" x 4", 24" O.C.	.170	L.F.	.004	.06	.18	.24
Sheathing, exterior, plywood, CDX, 1/2" thick	1.170	S.F.	.013	.80	.55	1.35
Furring strips, 1" x 3", 16" O.C.	1.000	L.F.	.023	.31	.93	1.24
TOTAL		S.F.	.080	2.36	3.28	5.64
2" X 8" RAFTERS, 16" O.C., 4/12 PITCH						
Rafters, 2" x 8", 16" O.C., 4/12 pitch	1.170	L.F.	.020	1.06	.81	1.87
Ceiling joists, 2" x 6", 16" O.C.	1.000	L.F.	.013	.59	.52	1.11
Ridge board, 2" x 8"	.050	L.F.	.002	.05	.07	.12
Fascia board, 2" x 8"	.100	L.F.	.007	.09	.29	.38
Rafter tie, 1" x 4", 4' O.C.	.060	L.F.	.001	[illegible]	[illegible]	.08
Soffit nailer (outrigger), 2" x 4", 24" O.C.	.170	L.F.	.004	[illegible]	[illegible]	[illegible]
Sheathing, exterior, plywood, CDX, 1/2" thick	1.170	S.F.	.013	.80	.55	[illegible]
Furring strips, 1" x 3", 16" O.C.	1.000	L.F.	.023	.31	.93	[illegible]
TOTAL		S.F.	.083	2.99	3.40	6.39

The cost of this system is based on the square foot of plan area.
All quantities have been adjusted accordingly.

Description	QUAN.	UNIT	LABOR HOURS	COST PER S.F. MAT.	INST.	TOTAL

2 *Illustration*

At the top of most assembly pages is an illustration with individual components labeled. Elements involved in the total system function are shown.

3 *System Description (2" x 6" Rafters, 16" O.C., 4/12 Pitch)*

The components of a typical system are listed separately to show what has been included in the development of the total system price. Each page includes a brief outline of any special conditions to be used when pricing a system. Alternative components can be found on the opposite page. Simply insert any chosen new element into the chart to develop a custom system.

4 *Quantities for Each Component*

Each material in a system is shown with the quantity required for the system unit. For example, there are 1.170 L.F. of rafter per S.F. of plan area.

5 *Unit of Measure for Each Component*

The abbreviated designation indicates the unit of measure, as defined by industry standards, upon which the individual component has been priced. In this example, items are priced by the linear foot (L.F.) or the square foot (S.F.).

6 *Labor Hours*

This is the amount of time it takes to install the quantity of the individual component.

7 *Unit of Measure (Cost per S.F.)*

In the three right-hand columns, each cost figure is adjusted to agree with the unit of measure for the entire system. In this case, cost per S.F. is the common unit of measure.

8 *Labor Hours (.083)*

The labor hours column shows the amount of time necessary to install the system per the unit of measure. For example, it takes .083 labor hours to install one square foot (plan area) of this roof framing system.

9 *Materials (2.99)*

This column contains the material cost of each element. These cost figures include 10% for profit.

10 *Installation (3.40)*

This column contains labor and equipment costs. Labor rates include bare cost and the installing contractor's overhead and profit. On the average, the labor cost will be 69.5% over the bare labor cost. Equipment costs include 10% for profit.

11 *Totals (6.39)*

The figure in this column is the sum of the material and installation costs.

12 *Work Sheet*

Using the selective price sheet on the page opposite each system, it is possible to create estimates with alternative items for any number of systems.

Division 1
Site Work

No part of this publication may be reproduced, stored in a retrieval system, or transmitted in any form or by any means without prior written permission of Reed Construction Data.

04 Footing Excavation Systems

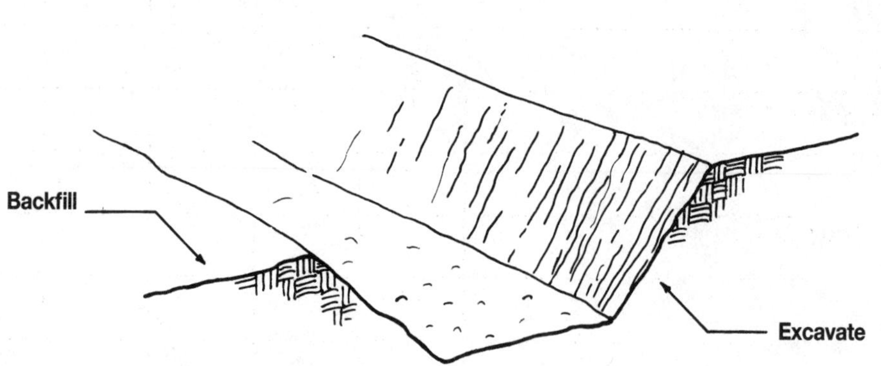

System Description	QUAN.	UNIT	LABOR HOURS	COST EACH		
				MAT.	INST.	TOTAL
BUILDING, 24′ X 38′, 4′ DEEP						
Cut & chip light trees to 6″ diam.	.190	Acre	9.120		503.50	503.50
Excavate, backhoe	174.000	C.Y.	4.641		341.04	341.04
Backfill, dozer, 4″ lifts, no compaction	87.000	C.Y.	.580		95.70	95.70
Rough grade, dozer, 30′ from building	87.000	C.Y.	.580		95.70	95.70
TOTAL		Ea.	14.921		1,035.94	1,035.94
BUILDING, 26′ X 46′, 4′ DEEP						
Cut & chip light trees to 6″ diam.	.210	Acre	10.080		556.50	556.50
Excavate, backhoe	201.000	C.Y.	5.361		393.96	393.96
Backfill, dozer, 4″ lifts, no compaction	100.000	C.Y.	.667		110	110
Rough grade, dozer, 30′ from building	100.000	C.Y.	.667		110	110
TOTAL		Ea.	16.775		1,170.46	1,170.46
BUILDING, 26′ X 60′, 4′ DEEP						
Cut & chip light trees to 6″ diam.	.240	Acre	11.520		636	636
Excavate, backhoe	240.000	C.Y.	6.401		470.40	470.40
Backfill, dozer, 4″ lifts, no compaction	120.000	C.Y.	.800		132	132
Rough grade, dozer, 30′ from building	120.000	C.Y.	.800		132	132
TOTAL		Ea.	19.521		1,370.40	1,370.40
BUILDING, 30′ X 66′, 4′ DEEP						
Cut & chip light trees to 6″ diam.	.260	Acre	12.480		689	689
Excavate, backhoe	268.000	C.Y.	7.148		525.28	525.28
Backfill, dozer, 4″ lifts, no compaction	134.000	C.Y.	.894		147.40	147.40
Rough grade, dozer, 30′ from building	134.000	C.Y.	.894		147.40	147.40
TOTAL		Ea.	21.416		1,509.08	1,509.08

The costs in this system are on a cost each basis.
Quantities are based on 1′-0″ clearance on each side of footing.

Description	QUAN.	UNIT	LABOR HOURS	COST EACH		
				MAT.	INST.	TOTAL

Footing Excavation Price Sheet	QUAN.	UNIT	LABOR HOURS	COST EACH		
				MAT.	INST.	TOTAL
Clear and grub, medium brush, 30′ from building, 24′ x 38′	.190	Acre	9.120		505	505
26′ x 46′	.210	Acre	10.080		555	555
26′ x 60′	.240	Acre	11.520		635	635
30′ x 66′	.260	Acre	12.480		690	690
Light trees, to 6″ dia. cut & chip, 24′ x 38′	.190	Acre	9.120		505	505
26′ x 46′	.210	Acre	10.080		555	555
26′ x 60′	.240	Acre	11.520		635	635
30′ x 66′	.260	Acre	12.480		690	690
Medium trees, to 10″ dia. cut & chip, 24′ x 38′	.190	Acre	13.029		720	720
26′ x 46′	.210	Acre	14.400		790	790
26′ x 60′	.240	Acre	16.457		905	905
30′ x 66′	.260	Acre	17.829		980	980
Excavation, footing, 24′ x 38′, 2′ deep	68.000	C.Y.	.906		134	134
4′ deep	174.000	C.Y.	2.319		340	340
8′ deep	384.000	C.Y.	5.119		750	750
26′ x 46′, 2′ deep	79.000	C.Y.	1.053		155	155
4′ deep	201.000	C.Y.	2.679		395	395
8′ deep	404.000	C.Y.	5.385		790	790
26′ x 60′, 2′ deep	94.000	C.Y.	1.253		185	185
4′ deep	240.000	C.Y.	3.199		470	470
8′ deep	483.000	C.Y.	6.438		950	950
30′ x 66′, 2′ deep	105.000	C.Y.	1.400		206	206
4′ deep	268.000	C.Y.	3.572		525	525
8′ deep	539.000	C.Y.	7.185		1,050	1,050
Backfill, 24′ x 38′, 2″ lifts, no compaction	34.000	C.Y.	.227		37.50	37.50
Compaction, air tamped, add	34.000	C.Y.	2.267		315	315
4″ lifts, no compaction	87.000	C.Y.	.580		95.50	95.50
Compaction, air tamped, add	87.000	C.Y.	5.800		805	805
8″ lifts, no compaction	192.000	C.Y.	1.281		211	211
Compaction, air tamped, add	192.000	C.Y.	12.801		1,750	1,750
26′ x 46′, 2″ lifts, no compaction	40.000	C.Y.	.267		44	44
Compaction, air tamped, add	40.000	C.Y.	2.667		370	370
4″ lifts, no compaction	100.000	C.Y.	.667		110	110
Compaction, air tamped, add	100.000	C.Y.	6.667		925	925
8″ lifts, no compaction	202.000	C.Y.	1.347		223	223
Compaction, air tamped, add	202.000	C.Y.	13.467		1,850	1,850
26′ x 60′, 2″ lifts, no compaction	47.000	C.Y.	.313		51.50	51.50
Compaction, air tamped, add	47.000	C.Y.	3.133		435	435
4″ lifts, no compaction	120.000	C.Y.	.800		132	132
Compaction, air tamped, add	120.000	C.Y.	8.000		1,100	1,100
8″ lifts, no compaction	242.000	C.Y.	1.614		266	266
Compaction, air tamped, add	242.000	C.Y.	16.134		2,225	2,225
30′ x 66′, 2″ lifts, no compaction	53.000	C.Y.	.354		58.50	58.50
Compaction, air tamped, add	53.000	C.Y.	3.534		490	490
4″ lifts, no compaction	134.000	C.Y.	.894		148	148
Compaction, air tamped, add	134.000	C.Y.	8.934		1,250	1,250
8″ lifts, no compaction	269.000	C.Y.	1.794		296	296
Compaction, air tamped, add	269.000	C.Y.	17.934		2,475	2,475
Rough grade, 30′ from building, 24′ x 38′	87.000	C.Y.	.580		95.50	95.50
26′ x 46′	100.000	C.Y.	.667		110	110
26′ x 60′	120.000	C.Y.	.800		132	132
30′ x 66′	134.000	C.Y.	.894		148	148

08 Foundation Excavation Systems

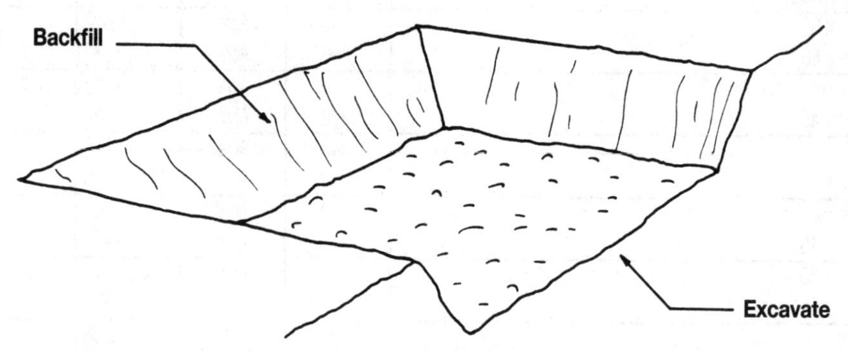

System Description	QUAN.	UNIT	LABOR HOURS	COST EACH MAT.	COST EACH INST.	COST EACH TOTAL
BUILDING, 24′ X 38′, 8′ DEEP						
Clear & grub, dozer, medium brush, 30′ from building	.190	Acre	2.027		197.60	197.60
Excavate, track loader, 1-1/2 C.Y. bucket	550.000	C.Y.	7.860		643.50	643.50
Backfill, dozer, 8″ lifts, no compaction	180.000	C.Y.	1.201		198	198
Rough grade, dozer, 30′ from building	280.000	C.Y.	1.868		308	308
TOTAL		Ea.	12.956		1,347.10	1,347.10
BUILDING, 26′ X 46′, 8′ DEEP						
Clear & grub, dozer, medium brush, 30′ from building	.210	Acre	2.240		218.40	218.40
Excavate, track loader, 1-1/2 C.Y. bucket	672.000	C.Y.	9.603		786.24	786.24
Backfill, dozer, 8″ lifts, no compaction	220.000	C.Y.	1.467		242	242
Rough grade, dozer, 30′ from building	340.000	C.Y.	2.268		374	374
TOTAL		Ea.	15.578		1,620.64	1,620.64
BUILDING, 26′ X 60′, 8′ DEEP						
Clear & grub, dozer, medium brush, 30′ from building	.240	Acre	2.560		249.60	249.60
Excavate, track loader, 1-1/2 C.Y. bucket	829.000	C.Y.	11.846		969.93	969.93
Backfill, dozer, 8″ lifts, no compaction	270.000	C.Y.	1.801		297	297
Rough grade, dozer, 30′ from building	420.000	C.Y.	2.801		462	462
TOTAL		Ea.	19.008		1,978.53	1,978.53
BUILDING, 30′ X 66′, 8′ DEEP						
Clear & grub, dozer, medium brush, 30′ from building	.260	Acre	2.773		270.40	270.40
Excavate, track loader, 1-1/2 C.Y. bucket	990.000	C.Y.	14.147		1,158.30	1,158.30
Backfill dozer, 8″ lifts, no compaction	320.000	C.Y.	2.134		352	352
Rough grade, dozer, 30′ from building	500.000	C.Y.	3.335		550	550
TOTAL		Ea.	22.389		2,330.70	2,330.70

The costs in this system are on a cost each basis.
Quantities are based on 1′-0″ clearance beyond footing projection.

Description	QUAN.	UNIT	LABOR HOURS	COST EACH MAT.	COST EACH INST.	COST EACH TOTAL

Important: See the Reference Section for critical supporting data - Reference Nos., Crews & Location Factors

Foundation Excavation Price Sheet

	QUAN.	UNIT	LABOR HOURS	COST EACH MAT.	INST.	TOTAL
Clear & grub, medium brush, 30′ from building, 24′ x 38′	.190	Acre	2.027		198	198
26′ x 46′	.210	Acre	2.240		219	219
26′ x 60′	.240	Acre	2.560		250	250
30′ x 66′	.260	Acre	2.773		271	271
Light trees, to 6″ dia. cut & chip, 24′ x 38′	.190	Acre	9.120		505	505
26′ x 46′	.210	Acre	10.080		555	555
26′ x 60′	.240	Acre	11.520		635	635
30′ x 66′	.260	Acre	12.480		690	690
Medium trees, to 10″ dia. cut & chip, 24′ x 38′	.190	Acre	13.029		720	720
26′ x 46′	.210	Acre	14.400		790	790
26′ x 60′	.240	Acre	16.457		905	905
30′ x 66′	.260	Acre	17.829		980	980
Excavation, basement, 24′ x 38′, 2′ deep	98.000	C.Y.	1.400		115	115
4′ deep	220.000	C.Y.	3.144		257	257
8′ deep	550.000	C.Y.	7.860		645	645
26′ x 46′, 2′ deep	123.000	C.Y.	1.758		144	144
4′ deep	274.000	C.Y.	3.915		320	320
8′ deep	672.000	C.Y.	9.603		785	785
26′ x 60′, 2′ deep	157.000	C.Y.	2.244		184	184
4′ deep	345.000	C.Y.	4.930		405	405
8′ deep	829.000	C.Y.	11.846		970	970
30′ x 66′, 2′ deep	192.000	C.Y.	2.744		225	225
4′ deep	419.000	C.Y.	5.988		490	490
8′ deep	990.000	C.Y.	14.147		1,150	1,150
Backfill, 24′ x 38′, 2″ lifts, no compaction	32.000	C.Y.	.213		35.50	35.50
Compaction, air tamped, add	32.000	C.Y.	2.133		295	295
4″ lifts, no compaction	72.000	C.Y.	.480		79	79
Compaction, air tamped, add	72.000	C.Y.	4.800		660	660
8″ lifts, no compaction	180.000	C.Y.	1.201		198	198
Compaction, air tamped, add	180.000	C.Y.	12.001		1,650	1,650
26′ x 46′, 2″ lifts, no compaction	40.000	C.Y.	.267		44	44
Compaction, air tamped, add	40.000	C.Y.	2.667		370	370
4″ lifts, no compaction	90.000	C.Y.	.600		99	99
Compaction, air tamped, add	90.000	C.Y.	6.000		835	835
8″ lifts, no compaction	220.000	C.Y.	1.467		242	242
Compacton, air tamped, add	220.000	C.Y.	14.667		2,025	2,025
26′ x 60′, 2″ lifts, no compaction	50.000	C.Y.	.334		55	55
Compaction, air tamped, add	50.000	C.Y.	3.334		460	460
4″ lifts, no compaction	110.000	C.Y.	.734		121	121
Compaction, air tamped, add	110.000	C.Y.	7.334		1,025	1,025
8″ lifts, no compaction	270.000	C.Y.	1.801		297	297
Compaction, air tamped, add	270.000	C.Y.	18.001		2,475	2,475
30′ x 66′, 2″ lifts, no compaction	60.000	C.Y.	.400		66	66
Compaction, air tamped, add	60.000	C.Y.	4.000		550	550
4″ lifts, no compaction	130.000	C.Y.	.867		143	143
Compaction, air tamped, add	130.000	C.Y.	8.667		1,200	1,200
8″ lifts, no compaction	320.000	C.Y.	2.134		350	350
Compaction, air tamped, add	320.000	C.Y.	21.334		2,950	2,950
Rough grade, 30′ from building, 24′ x 38′	280.000	C.Y.	1.868		310	310
26′ x 46′	340.000	C.Y.	2.268		375	375
26′ x 60′	420.000	C.Y.	2.801		465	465
30′ x 66′	500.000	C.Y.	3.335		550	550

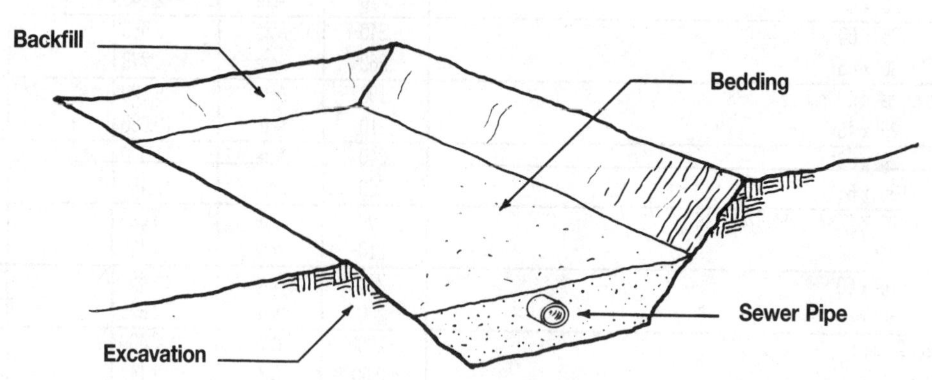

System Description		QUAN.	UNIT	LABOR HOURS	COST PER L.F. MAT.	INST.	TOTAL
2′ DEEP							
Excavation, backhoe		.296	C.Y.	.032		1.56	1.56
Bedding, sand		.111	C.Y.	.044	1.44	1.50	2.94
Utility, sewer, 6″ cast iron		1.000	L.F.	.283	13.03	10.84	23.87
Backfill, incl. compaction		.185	C.Y.	.044		1.29	1.29
	TOTAL		L.F.	.403	14.47	15.19	29.66
4′ DEEP							
Excavation, backhoe		.889	C.Y.	.095		4.66	4.66
Bedding, sand		.111	C.Y.	.044	1.44	1.50	2.94
Utility, sewer, 6″ cast iron		1.000	L.F.	.283	13.03	10.84	23.87
Backfill, incl. compaction		.778	C.Y.	.183		5.41	5.41
	TOTAL		L.F.	.605	14.47	22.41	36.88
6′ DEEP							
Excavation, backhoe		1.770	C.Y.	.189		9.30	9.30
Bedding, sand		.111	C.Y.	.044	1.44	1.50	2.94
Utility, sewer, 6″ cast iron		1.000	L.F.	.283	13.03	10.84	23.87
Backfill, incl. compaction		1.660	C.Y.	.391		11.54	11.54
	TOTAL		L.F.	.907	14.47	33.18	47.65
8′ DEEP							
Excavation, backhoe		2.960	C.Y.	.316		15.54	15.54
Bedding, sand		.111	C.Y.	.044	1.44	1.50	2.94
Utility, sewer, 6″ cast iron		1.000	L.F.	.283	13.03	10.84	23.87
Backfill, incl. compaction		2.850	C.Y.	.671		19.81	19.81
	TOTAL		L.F.	1.314	14.47	47.69	62.16

The costs in this system are based on a cost per linear foot of trench, and based on 2′ wide at bottom of trench up to 6′ deep.

Description	QUAN.	UNIT	LABOR HOURS	COST PER L.F. MAT.	INST.	TOTAL

Important: See the Reference Section for critical supporting data - Reference Nos., Crews & Location Factors

Utility Trenching Price Sheet	QUAN.	UNIT	LABOR HOURS	COST PER C.Y.		
				MAT.	INST.	TOTAL
Excavation, bottom of trench 2′ wide, 2′ deep	.296	C.Y.	.032		1.56	1.56
4′ deep	.889	C.Y.	.095		4.66	4.66
6′ deep	1.770	C.Y.	.142		7.35	7.35
8′ deep	2.960	C.Y.	.105		14.25	14.25
Bedding, sand, bottom of trench 2′ wide, no compaction, pipe, 2″ diameter	.070	C.Y.	.028	.91	.95	1.86
4″ diameter	.084	C.Y.	.034	1.09	1.13	2.22
6″ diameter	.105	C.Y.	.042	1.37	1.42	2.79
8″ diameter	.122	C.Y.	.049	1.59	1.65	3.24
Compacted, pipe, 2″ diameter	.074	C.Y.	.030	.96	1	1.96
4″ diameter	.092	C.Y.	.037	1.20	1.24	2.44
6″ diameter	.111	C.Y.	.044	1.44	1.50	2.94
8″ diameter	.129	C.Y.	.052	1.68	1.75	3.43
3/4″ stone, bottom of trench 2′ wide, pipe, 4″ diameter	.082	C.Y.	.033	1.07	1.11	2.18
6″ diameter	.099	C.Y.	.040	1.29	1.34	2.63
3/8″ stone, bottom of trench 2′ wide, pipe, 4″ diameter	.084	C.Y.	.034	1.09	1.13	2.22
6″ diameter	.102	C.Y.	.041	1.33	1.38	2.71
Utilities, drainage & sewerage, corrugated plastic, 6″ diameter	1.000	L.F.	.069	3.78	2.10	5.88
8″ diameter	1.000	L.F.	.072	6.40	2.19	8.59
Bituminous fiber, 4″ diameter	1.000	L.F.	.064	2.11	1.96	4.07
6″ diameter	1.000	L.F.	.069	3.78	2.10	5.88
8″ diameter	1.000	L.F.	.072	6.40	2.19	8.59
Concrete, non-reinforced, 6″ diameter	1.000	L.F.	.181	4.54	6.60	11.14
8″ diameter	1.000	L.F.	.214	5	7.75	12.75
PVC, SDR 35, 4″ diameter	1.000	L.F.	.064	2.11	1.96	4.07
6″ diameter	1.000	L.F.	.069	3.78	2.10	5.88
8″ diameter	1.000	L.F.	.072	6.40	2.19	8.59
Vitrified clay, 4″ diameter	1.000	L.F.	.091	1.87	2.77	4.64
6″ diameter	1.000	L.F.	.120	3.12	3.67	6.79
8″ diameter	1.000	L.F.	.140	4.43	5.35	9.78
Gas & service, polyethylene, 1-1/4″ diameter	1.000	L.F.	.059	.96	2.09	3.05
Steel sched.40, 1″ diameter	1.000	L.F.	.107	2.90	4.79	7.69
2″ diameter	1.000	L.F.	.114	4.56	5.15	9.71
Sub-drainage, PVC, perforated, 3″ diameter	1.000	L.F.	.064	2.11	1.96	4.07
4″ diameter	1.000	L.F.	.064	2.11	1.96	4.07
5″ diameter	1.000	L.F.	.069	3.78	2.10	5.88
6″ diameter	1.000	L.F.	.069	3.78	2.10	5.88
Porous wall concrete, 4″ diameter	1.000	L.F.	.072	2.15	2.19	4.34
Vitrified clay, perforated, 4″ diameter	1.000	L.F.	.120	2.20	4.35	6.55
6″ diameter	1.000	L.F.	.152	3.64	5.55	9.19
Water service, copper, type K, 3/4″	1.000	L.F.	.083	2.56	3.69	6.25
1″ diameter	1.000	L.F.	.093	3.36	4.12	7.48
PVC, 3/4″	1.000	L.F.	.121	.90	5.35	6.25
1″ diameter	1.000	L.F.	.134	.99	5.95	6.94
Backfill, bottom of trench 2′ wide no compact, 2′ deep, pipe, 2″ diameter	.226	L.F.	.053		1.57	1.57
4″ diameter	.212	L.F.	.050		1.47	1.47
6″ diameter	.185	L.F.	.044		1.29	1.29
4′ deep, pipe, 2″ diameter	.819	C.Y.	.193		5.70	5.70
4″ diameter	.805	C.Y.	.189		5.60	5.60
6″ diameter	.778	C.Y.	.183		5.40	5.40
6′ deep, pipe, 2″ diameter	1.700	C.Y.	.400		11.80	11.80
4″ diameter	1.690	C.Y.	.398		11.75	11.75
6″ diameter	1.660	C.Y.	.391		11.55	11.55
8′ deep, pipe, 2″ diameter	2.890	C.Y.	.680		20	20
4″ diameter	2.870	C.Y.	.675		19.95	19.95
6″ diameter	2.850	C.Y.	.671		19.80	19.80

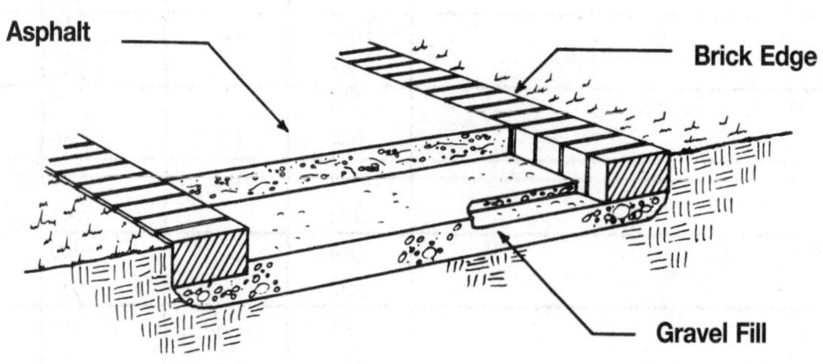

System Description	QUAN.	UNIT	LABOR HOURS	COST PER S.F. MAT.	COST PER S.F. INST.	COST PER S.F. TOTAL
ASPHALT SIDEWALK SYSTEM, 3′ WIDE WALK						
Gravel fill, 4″ deep	1.000	S.F.	.001	.28	.03	.31
Compact fill	.012	C.Y.			.01	.01
Handgrade	1.000	S.F.	.004		.13	.13
Walking surface, bituminous paving, 2″ thick	1.000	S.F.	.007	.45	.25	.70
Edging, brick, laid on edge	.670	L.F.	.079	1.37	2.87	4.24
TOTAL		S.F.	.091	2.10	3.29	5.39
CONCRETE SIDEWALK SYSTEM, 3′ WIDE WALK						
Gravel fill, 4″ deep	1.000	S.F.	.001	.28	.03	.31
Compact fill	.012	C.Y.			.01	.01
Handgrade	1.000	S.F.	.004		.13	.13
Walking surface, concrete, 4″ thick	1.000	S.F.	.040	1.49	1.43	2.92
Edging, brick, laid on edge	.670	L.F.	.079	1.37	2.87	4.24
TOTAL		S.F.	.124	3.14	4.47	7.61
PAVERS, BRICK SIDEWALK SYSTEM, 3′ WIDE WALK						
Sand base fill, 4″ deep	1.000	S.F.	.001	.35	.06	.41
Compact fill	.012	C.Y.			.01	.01
Handgrade	1.000	S.F.	.004		.13	.13
Walking surface, brick pavers	1.000	S.F.	.160	2.74	5.80	8.54
Edging, redwood, untreated, 1″ x 4″	.670	L.F.	.032	1.61	1.32	2.93
TOTAL		S.F.	.197	4.70	7.32	12.02

The costs in this system are based on a cost per square foot of sidewalk area. Concrete used is 3000 p.s.i.

Description	QUAN.	UNIT	LABOR HOURS	COST PER S.F. MAT.	COST PER S.F. INST.	COST PER S.F. TOTAL

 Important: See the Reference Section for critical supporting data - Reference Nos., Crews & Location Factors

Sidewalk Price Sheet	QUAN.	UNIT	LABOR HOURS	COST PER S.F. MAT.	INST.	TOTAL
Base, crushed stone, 3″ deep	1.000	S.F.	.001	.32	.07	.39
6″ deep	1.000	S.F.	.001	.65	.08	.73
9″ deep	1.000	S.F.	.002	.95	.10	1.05
12″ deep	1.000	S.F.	.002	1.53	.12	1.65
Bank run gravel, 6″ deep	1.000	S.F.	.001	.42	.05	.47
9″ deep	1.000	S.F.	.001	.62	.07	.69
12″ deep	1.000	S.F.	.001	.85	.09	.94
Compact base, 3″ deep	.009	C.Y.	.001		.01	.01
6″ deep	.019	C.Y.	.001		.02	.02
9″ deep	.028	C.Y.	.001		.03	.03
Handgrade	1.000	S.F.	.004		.13	.13
Surface, brick, pavers dry joints, laid flat, running bond	1.000	S.F.	.160	2.74	5.80	8.54
Basket weave	1.000	S.F.	.168	3.04	6.10	9.14
Herringbone	1.000	S.F.	.174	3.04	6.30	9.34
Laid on edge, running bond	1.000	S.F.	.229	2.59	8.30	10.89
Mortar jts. laid flat, running bond	1.000	S.F.	.192	3.29	6.95	10.24
Basket weave	1.000	S.F.	.202	3.65	7.30	10.95
Herringbone	1.000	S.F.	.209	3.65	7.55	11.20
Laid on edge, running bond	1.000	S.F.	.274	3.11	9.95	13.06
Bituminous paving, 1-1/2″ thick	1.000	S.F.	.006	.33	.18	.51
2″ thick	1.000	S.F.	.007	.45	.25	.70
2-1/2″ thick	1.000	S.F.	.008	.57	.27	.84
Sand finish, 3/4″ thick	1.000	S.F.	.001	.21	.09	.30
1″ thick	1.000	S.F.	.001	.26	.11	.37
Concrete, reinforced, broom finish, 4″ thick	1.000	S.F.	.040	1.49	1.43	2.92
5″ thick	1.000	S.F.	.044	1.98	1.58	3.56
6″ thick	1.000	S.F.	.047	2.31	1.68	3.99
Crushed stone, white marble, 3″ thick	1.000	S.F.	.009	.21	.28	.49
Bluestone, 3″ thick	1.000	S.F.	.009	.23	.28	.51
Flagging, bluestone, 1″	1.000	S.F.	.198	5.10	7.15	12.25
1-1/2″	1.000	S.F.	.188	9.25	6.80	16.05
Slate, natural cleft, 3/4″	1.000	S.F.	.174	5.70	6.30	12
Random rect., 1/2″	1.000	S.F.	.152	12.30	5.50	17.80
Granite blocks	1.000	S.F.	.174	6.90	6.30	13.20
Edging, corrugated aluminum, 4″, 3′ wide walk	.666	L.F.	.008	.27	.33	.60
4′ wide walk	.500	L.F.	.006	.21	.25	.46
6″, 3′ wide walk	.666	L.F.	.010	.34	.39	.73
4′ wide walk	.500	L.F.	.007	.26	.30	.56
Redwood-cedar-cypress, 1″ x 4″, 3′ wide walk	.666	L.F.	.021	.81	.87	1.68
4′ wide walk	.500	L.F.	.016	.61	.65	1.26
2″ x 4″, 3′ wide walk	.666	L.F.	.032	1.61	1.32	2.93
4′ wide walk	.500	L.F.	.024	1.21	.99	2.20
Brick, dry joints, 3′ wide walk	.666	L.F.	.079	1.37	2.87	4.24
4′ wide walk	.500	L.F.	.059	1.03	2.15	3.18
Mortar joints, 3′ wide walk	.666	L.F.	.095	1.65	3.45	5.10
4′ wide walk	.500	L.F.	.071	1.23	2.57	3.80

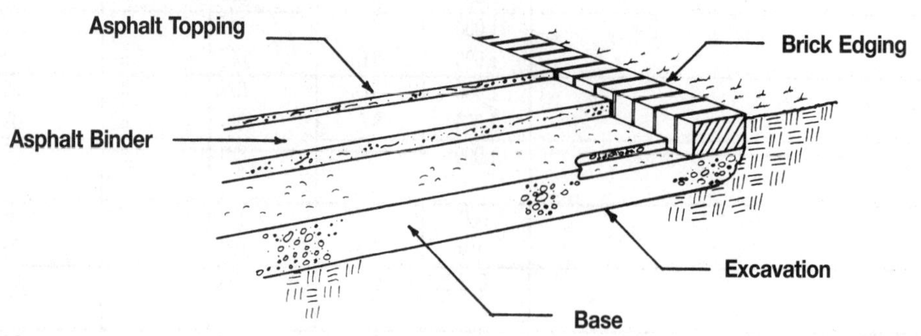

System Description	QUAN.	UNIT	LABOR HOURS	COST PER S.F. MAT.	COST PER S.F. INST.	COST PER S.F. TOTAL
ASPHALT DRIVEWAY TO 10′ WIDE						
Excavation, driveway to 10′ wide, 6″ deep	.019	C.Y.			.03	.03
Base, 6″ crushed stone	1.000	S.F.	.001	.65	.08	.73
Handgrade base	1.000	S.F.	.004		.13	.13
2″ thick base	1.000	S.F.	.002	.45	.14	.59
1″ topping	1.000	S.F.	.001	.26	.11	.37
Edging, brick pavers	.200	L.F.	.024	.41	.86	1.27
TOTAL		S.F.	.032	1.77	1.35	3.12
CONCRETE DRIVEWAY TO 10′ WIDE						
Excavation, driveway to 10′ wide, 6″ deep	.019	C.Y.			.03	.03
Base, 6″ crushed stone	1.000	S.F.	.001	.65	.08	.73
Handgrade base	1.000	S.F.	.004		.13	.13
Surface, concrete, 4″ thick	1.000	S.F.	.040	1.49	1.43	2.92
Edging, brick pavers	.200	L.F.	.024	.41	.86	1.27
TOTAL		S.F.	.069	2.55	2.53	5.08
PAVERS, BRICK DRIVEWAY TO 10′ WIDE						
Excavation, driveway to 10′ wide, 6″ deep	.019	C.Y.			.03	.03
Base, 6″ sand	1.000	S.F.	.001	.55	.10	.65
Handgrade base	1.000	S.F.	.004		.13	.13
Surface, pavers, brick laid flat, running bond	1.000	S.F.	.160	2.74	5.80	8.54
Edging, redwood, untreated, 2″ x 4″	.200	L.F.	.010	.48	.40	.88
TOTAL		S.F.	.175	3.77	6.46	10.23

Description	QUAN.	UNIT	LABOR HOURS	COST PER S.F. MAT.	COST PER S.F. INST.	COST PER S.F. TOTAL

Important: See the Reference Section for critical supporting data - Reference Nos., Crews & Location Factors

Driveway Price Sheet	QUAN.	UNIT	LABOR HOURS	COST PER S.F. MAT.	INST.	TOTAL
Excavation, by machine, 10′ wide, 6″ deep	.019	C.Y.	.001		.03	.03
12″ deep	.037	C.Y.	.001		.05	.05
18″ deep	.055	C.Y.	.001		.08	.08
20′ wide, 6″ deep	.019	C.Y.	.001		.03	.03
12″ deep	.037	C.Y.	.001		.05	.05
18″ deep	.055	C.Y.	.001		.08	.08
Base, crushed stone, 10′ wide, 3″ deep	1.000	S.F.	.001	.33	.04	.37
6″ deep	1.000	S.F.	.001	.65	.08	.73
9″ deep	1.000	S.F.	.002	.95	.10	1.05
20′ wide, 3″ deep	1.000	S.F.	.001	.33	.04	.37
6″ deep	1.000	S.F.	.001	.65	.08	.73
9″ deep	1.000	S.F.	.002	.95	.10	1.05
Bank run gravel, 10′ wide, 3″ deep	1.000	S.F.	.001	.21	.03	.24
6″ deep	1.000	S.F.	.001	.42	.05	.47
9″ deep	1.000	S.F.	.001	.62	.07	.69
20′ wide, 3″ deep	1.000	S.F.	.001	.21	.03	.24
6″ deep	1.000	S.F.	.001	.42	.05	.47
9″ deep	1.000	S.F.	.001	.62	.07	.69
Handgrade, 10′ wide	1.000	S.F.	.004		.13	.13
20′ wide	1.000	S.F.	.004		.13	.13
Surface, asphalt, 10′ wide, 3/4″ topping, 1″ base	1.000	S.F.	.002	.55	.20	.75
2″ base	1.000	S.F.	.003	.66	.23	.89
1″ topping, 1″ base	1.000	S.F.	.002	.60	.22	.82
2″ base	1.000	S.F.	.003	.71	.25	.96
20′ wide, 3/4″ topping, 1″ base	1.000	S.F.	.002	.55	.20	.75
2″ base	1.000	S.F.	.003	.66	.23	.89
1″ topping, 1″ base	1.000	S.F.	.002	.60	.22	.82
2″ base	1.000	S.F.	.003	.71	.25	.96
Concrete, 10′ wide, 4″ thick	1.000	S.F.	.040	1.49	1.43	2.92
6″ thick	1.000	S.F.	.047	2.31	1.68	3.99
20′ wide, 4″ thick	1.000	S.F.	.040	1.49	1.43	2.92
6″ thick	1.000	S.F.	.047	2.31	1.68	3.99
Paver, brick 10′ wide dry joints, running bond, laid flat	1.000	S.F.	.160	2.74	5.80	8.54
Laid on edge	1.000	S.F.	.229	2.59	8.30	10.89
Mortar joints, laid flat	1.000	S.F.	.192	3.29	6.95	10.24
Laid on edge	1.000	S.F.	.274	3.11	9.95	13.06
20′ wide, running bond, dry jts., laid flat	1.000	S.F.	.160	2.74	5.80	8.54
Laid on edge	1.000	S.F.	.229	2.59	8.30	10.89
Mortar joints, laid flat	1.000	S.F.	.192	3.29	6.95	10.24
Laid on edge	1.000	S.F.	.274	3.11	9.95	13.06
Crushed stone, 10′ wide, white marble, 3″	1.000	S.F.	.009	.21	.28	.49
Bluestone, 3″	1.000	S.F.	.009	.23	.28	.51
20′ wide, white marble, 3″	1.000	S.F.	.009	.21	.28	.49
Bluestone, 3″	1.000	S.F.	.009	.23	.28	.51
Soil cement, 10′ wide	1.000	S.F.	.007	.22	.66	.88
20′ wide	1.000	S.F.	.007	.22	.66	.88
Granite blocks, 10′ wide	1.000	S.F.	.174	6.90	6.30	13.20
20′ wide	1.000	S.F.	.174	6.90	6.30	13.20
Asphalt block, solid 1-1/4″ thick	1.000	S.F.	.119	4.64	4.29	8.93
Solid 3″ thick	1.000	S.F.	.123	6.50	4.46	10.96
Edging, brick, 10′ wide	.200	L.F.	.024	.41	.86	1.27
20′ wide	.100	L.F.	.012	.21	.43	.64
Redwood, untreated 2″ x 4″, 10′ wide	.200	L.F.	.010	.48	.40	.88
20′ wide	.100	L.F.	.005	.24	.20	.44
Granite, 4 1/2″ x 12″ straight, 10′ wide	.200	L.F.	.032	1.04	1.63	2.67
20′ wide	.100	L.F.	.016	.52	.82	1.34
Finishes, asphalt sealer, 10′ wide	1.000	S.F.	.023	.51	.70	1.21
20′ wide	1.000	S.F.	.023	.51	.70	1.21
Concrete, exposed aggregate 10′ wide	1.000	S.F.	.013	4.49	.47	4.96
20′ wide	1.000	S.F.	.013	4.49	.47	4.96

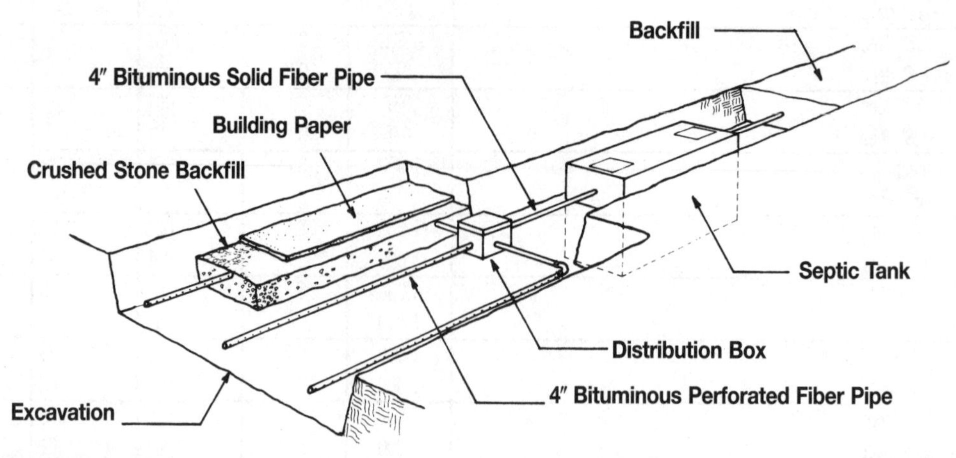

System Description	QUAN.	UNIT	LABOR HOURS	COST EACH		
				MAT.	INST.	TOTAL
SEPTIC SYSTEM WITH 1000 S.F. LEACHING FIELD, 1000 GALLON TANK						
Tank, 1000 gallon, concrete	1.000	Ea.	3.500	615	134	749
Distribution box, concrete	1.000	Ea.	1.000	114	29.50	143.50
4″ PVC pipe	25.000	L.F.	1.600	52.75	49	101.75
Tank and field excavation	119.000	C.Y.	13.130		888.93	888.93
Crushed stone backfill	76.000	C.Y.	12.160	1,976	511.48	2,487.48
Backfill with excavated material	36.000	C.Y.	.240		39.60	39.60
Building paper	125.000	S.Y.	2.430	45	101.25	146.25
4″ PVC perforated pipe	145.000	L.F.	9.280	305.95	284.20	590.15
4″ pipe fittings	2.000	Ea.	1.939	19.90	77	96.90
TOTAL		Ea.	45.279	3,128.60	2,114.96	5,243.56
SEPTIC SYSTEM WITH 2 LEACHING PITS, 1000 GALLON TANK						
Tank, 1000 gallon, concrete	1.000	Ea.	3.500	615	134	749
Distribution box, concrete	1.000	Ea.	1.000	114	29.50	143.50
4″ PVC pipe	75.000	L.F.	4.800	158.25	147	305.25
Excavation for tank only	20.000	C.Y.	2.207		149.40	149.40
Crushed stone backfill	10.000	C.Y.	1.600	260	67.30	327.30
Backfill with excavated material	55.000	C.Y.	.367		60.50	60.50
Pits, 6′ diameter, including excavation and stone backfill	2.000	Ea.		1,420		1,420
TOTAL		Ea.	13.474	2,567.25	587.70	3,154.95

The costs in this system include all necessary piping and excavation.

Description	QUAN.	UNIT	LABOR HOURS	COST EACH		
				MAT.	INST.	TOTAL

Important: See the Reference Section for critical supporting data - Reference Nos., Crews & Location Factors

Septic Systems Price Sheet	QUAN.	UNIT	LABOR HOURS	COST EACH		
				MAT.	INST.	TOTAL
Tank, precast concrete, 1000 gallon	1.000	Ea.	3.500	615	134	749
2000 gallon	1.000	Ea.	5.600	1,225	214	1,439
Distribution box, concrete, 5 outlets	1.000	Ea.	1.000	114	29.50	143.50
12 outlets	1.000	Ea.	2.000	310	59	369
4″ pipe, PVC, solid	25.000	L.F.	1.600	53	49	102
Tank and field excavation, 1000 S.F. field	119.000	C.Y.	6.565		890	890
2000 S.F. field	190.000	C.Y.	10.482		1,425	1,425
Tank excavation only, 1000 gallon tank	20.000	C.Y.	1.103		150	150
2000 gallon tank	32.000	C.Y.	1.765		239	239
Backfill, crushed stone 1000 S.F. field	76.000	C.Y.	12.160	1,975	510	2,485
2000 S.F. field	140.000	C.Y.	22.400	3,650	940	4,590
Backfill with excavated material, 1000 S.F. field	36.000	C.Y.	.240		39.50	39.50
2000 S.F. field	60.000	C.Y.	.400		66	66
6′ diameter pits	55.000	C.Y.	.367		60.50	60.50
3′ diameter pits	42.000	C.Y.	.280		46.50	46.50
Building paper, 1000 S.F. field	125.000	S.Y.	2.376	44	99	143
2000 S.F. field	250.000	S.Y.	4.860	90	203	293
4″ pipe, PVC, perforated, 1000 S.F. field	145.000	L.F.	9.280	305	284	589
2000 S.F. field	265.000	L.F.	16.960	560	520	1,080
Pipe fittings, bituminous fiber, 1000 S.F. field	2.000	Ea.	1.939	19.90	77	96.90
2000 S.F. field	4.000	Ea.	3.879	40	154	194
Leaching pit, including excavation and stone backfill, 3′ diameter	1.000	Ea.		535		535
6′ diameter	1.000	Ea.		710		710

60 Chain Link Fence Price Sheet

System Description	QUAN.	UNIT	LABOR HOURS	COST PER UNIT MAT.	INST.	TOTAL
Chain link fence						
Galv.9ga. wire, 1-5/8"post 10'O.C., 1-3/8"top rail, 2"corner post, 3'hi	1.000	L.F.	.130	6.35	3.97	10.32
4' high	1.000	L.F.	.141	9.45	4.32	13.77
6' high	1.000	L.F.	.209	10.70	6.40	17.10
Add for gate 3' wide 1-3/8" frame 3' high	1.000	Ea.	2.000	57	61	118
4' high	1.000	Ea.	2.400	70.50	73.50	144
6' high	1.000	Ea.	2.400	127	73.50	200.50
Add for gate 4' wide 1-3/8" frame 3' high	1.000	Ea.	2.667	66.50	81.50	148
4' high	1.000	Ea.	2.667	87	81.50	168.50
6' high	1.000	Ea.	3.000	161	91.50	252.50
Alum.9ga. wire, 1-5/8"post, 10'O.C., 1-3/8"top rail, 2"corner post,3'hi	1.000	L.F.	.130	7.60	3.97	11.57
4' high	1.000	L.F.	.141	8.70	4.32	13.02
6' high	1.000	L.F.	.209	11.15	6.40	17.55
Add for gate 3' wide 1-3/8" frame 3' high	1.000	Ea.	2.000	74.50	61	135.50
4' high	1.000	Ea.	2.400	101	73.50	174.50
6' high	1.000	Ea.	2.400	152	73.50	225.50
Add for gate 4' wide 1-3/8" frame 3' high	1.000	Ea.	2.400	101	73.50	174.50
4' high	1.000	Ea.	2.667	135	81.50	216.50
6' high	1.000	Ea.	3.000	211	91.50	302.50
Vinyl 9ga. wire, 1-5/8"post 10'O.C., 1-3/8"top rail, 2"corner post,3'hi	1.000	L.F.	.130	6.75	3.97	10.72
4' high	1.000	L.F.	.141	11.10	4.32	15.42
6' high	1.000	L.F.	.209	12.65	6.40	19.05
Add for gate 3' wide 1-3/8" frame 3' high	1.000	Ea.	2.000	84.50	61	145.50
4' high	1.000	Ea.	2.400	110	73.50	183.50
6' high	1.000	Ea.	2.400	169	73.50	242.50
Add for gate 4' wide 1-3/8" frame 3' high	1.000	Ea.	2.400	115	73.50	188.50
4' high	1.000	Ea.	2.667	152	81.50	233.50
6' high	1.000	Ea.	3.000	220	91.50	311.50
Tennis court, chain link fence, 10' high						
Galv.11ga.wire, 2"post 10'O.C., 1-3/8"top rail, 2-1/2"corner post	1.000	L.F.	.253	16.90	7.75	24.65
Add for gate 3' wide 1-3/8" frame	1.000	Ea.	2.400	211	73.50	284.50
Alum.11ga.wire, 2"post 10'O.C., 1-3/8"top rail, 2-1/2"corner post	1.000	L.F.	.253	23.50	7.75	31.25
Add for gate 3' wide 1-3/8" frame	1.000	Ea.	2.400	270	73.50	343.50
Vinyl 11ga.wire,2"post 10' O.C.,1-3/8"top rail,2-1/2"corner post	1.000	L.F.	.253	20.50	7.75	28.25
Add for gate 3' wide 1-3/8" frame	1.000	Ea.	2.400	305	73.50	378.50
Railings, commercial						
Aluminum balcony rail, 1-1/2" posts with pickets	1.000	L.F.	.164	44	8.70	52.70
With expanded metal panels	1.000	L.F.	.164	56.50	8.70	65.20
With porcelain enamel panel inserts	1.000	L.F.	.164	50.50	8.70	59.20
Mild steel, ornamental rounded top rail	1.000	L.F.	.164	49.50	8.70	58.20
As above, but pitch down stairs	1.000	L.F.	.183	53.50	9.70	63.20
Steel pipe, welded, 1-1/2" round, painted	1.000	L.F.	.160	14.50	8.50	23
Galvanized	1.000	L.F.	.160	20.50	8.50	29
Residential, stock units, mild steel, deluxe	1.000	L.F.	.102	11.30	5.40	16.70
Economy	1.000	L.F.	.102	8.45	5.40	13.85

64 Wood Fence Price Sheet

System Description	QUAN.	UNIT	LABOR HOURS	COST PER UNIT MAT.	INST.	TOTAL
Basketweave, 3/8"x4" boards, 2"x4" stringers on spreaders, 4"x4" posts						
No. 1 cedar, 6' high	1.000	L.F.	.150	8.85	5.45	14.30
Treated pine, 6' high	1.000	L.F.	.160	10.75	5.80	16.55
Board fence, 1"x4" boards, 2"x4" rails, 4"x4" posts						
Preservative treated, 2 rail, 3' high	1.000	L.F.	.166	6.55	6	12.55
4' high	1.000	L.F.	.178	7.20	6.45	13.65
3 rail, 5' high	1.000	L.F.	.185	8.15	6.75	14.90
6' high	1.000	L.F.	.192	9.30	7	16.30
Western cedar, No. 1, 2 rail, 3' high	1.000	L.F.	.166	7.15	6	13.15
3 rail, 4' high	1.000	L.F.	.178	8.50	6.45	14.95
5' high	1.000	L.F.	.185	9.80	6.75	16.55
6' high	1.000	L.F.	.192	10.75	7	17.75
No. 1 cedar, 2 rail, 3' high	1.000	L.F.	.166	10.75	6	16.75
4' high	1.000	L.F.	.178	12.25	6.45	18.70
3 rail, 5' high	1.000	L.F.	.185	14.15	6.75	20.90
6' high	1.000	L.F.	.192	15.75	7	22.75
Shadow box, 1"x6" boards, 2"x4" rails, 4"x4" posts						
Fir, pine or spruce, treated, 3 rail, 6' high	1.000	L.F.	.160	12.05	5.80	17.85
No. 1 cedar, 3 rail, 4' high	1.000	L.F.	.185	14.85	6.75	21.60
6' high	1.000	L.F.	.192	18.30	7	25.30
Open rail, split rails, No. 1 cedar, 2 rail, 3' high	1.000	L.F.	.150	5.95	5.45	11.40
3 rail, 4' high	1.000	L.F.	.160	8	5.80	13.80
No. 2 cedar, 2 rail, 3' high	1.000	L.F.	.150	4.63	5.45	10.08
3 rail, 4' high	1.000	L.F.	.160	5.30	5.80	11.10
Open rail, rustic rails, No. 1 cedar, 2 rail, 3' high	1.000	L.F.	.150	3.71	5.45	9.16
3 rail, 4' high	1.000	L.F.	.160	4.98	5.80	10.78
No. 2 cedar, 2 rail, 3' high	1.000	L.F.	.150	3.55	5.45	9
3 rail, 4' high	1.000	L.F.	.160	3.76	5.80	9.56
Rustic picket, molded pine pickets, 2 rail, 3' high	1.000	L.F.	.171	5.25	6.25	11.50
3 rail, 4' high	1.000	L.F.	.197	6.05	7.20	13.25
No. 1 cedar, 2 rail, 3' high	1.000	L.F.	.171	7.15	6.25	13.40
3 rail, 4' high	1.000	L.F.	.197	8.20	7.20	15.40
Picket fence, fir, pine or spruce, preserved, treated						
2 rail, 3' high	1.000	L.F.	.171	4.58	6.25	10.83
3 rail, 4' high	1.000	L.F.	.185	5.40	6.75	12.15
Western cedar, 2 rail, 3' high	1.000	L.F.	.171	5.70	6.25	11.95
3 rail, 4' high	1.000	L.F.	.185	5.85	6.75	12.60
No. 1 cedar, 2 rail, 3' high	1.000	L.F.	.171	11.45	6.25	17.70
3 rail, 4' high	1.000	L.F.	.185	13.30	6.75	20.05
Stockade, No. 1 cedar, 3-1/4" rails, 6' high	1.000	L.F.	.150	10.80	5.45	16.25
8' high	1.000	L.F.	.155	13.95	5.65	19.60
No. 2 cedar, treated rails, 6' high	1.000	L.F.	.150	10.80	5.45	16.25
Treated pine, treated rails, 6' high	1.000	L.F.	.150	10.55	5.45	16
Gates, No. 2 cedar, picket, 3'-6" wide 4' high	1.000	Ea.	2.667	57	97	154
No. 2 cedar, rustic round, 3' wide, 3' high	1.000	Ea.	2.667	73	97	170
No. 2 cedar, stockade screen, 3'-6" wide, 6' high	1.000	Ea.	3.000	63.50	109	172.50
General, wood, 3'-6" wide, 4' high	1.000	Ea.	2.400	55	87	142
6' high	1.000	Ea.	3.000	69	109	178

Division 2
Foundations

o part of this publication may be reproduced, stored in a retrieval system, or transmitted in any form
r by any means without prior written permission of Reed Construction Data.

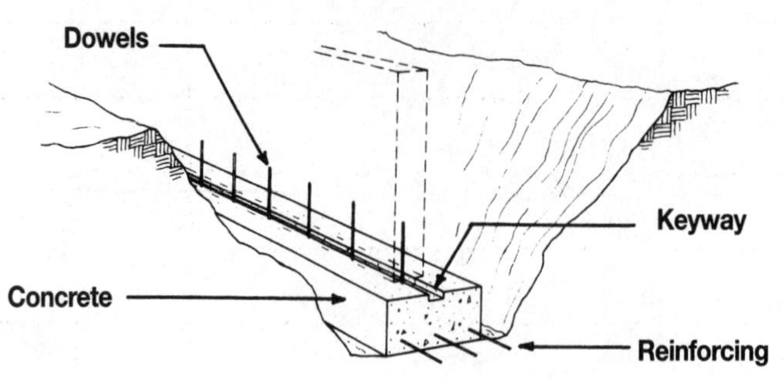

System Description	QUAN.	UNIT	LABOR HOURS	COST PER L.F.		
				MAT.	INST.	TOTAL
8″ THICK BY 18″ WIDE FOOTING						
Concrete, 3000 psi	.040	C.Y.		3.56		3.56
Place concrete, direct chute	.040	C.Y.	.016		.52	.52
Forms, footing, 4 uses	1.330	SFCA	.103	.78	3.62	4.40
Reinforcing, 1/2″ diameter bars, 2 each	1.380	Lb.	.011	.63	.48	1.11
Keyway, 2″ x 4″, beveled, 4 uses	1.000	L.F.	.015	.20	.61	.81
Dowels, 1/2″ diameter bars, 2′ long, 6′ O.C.	.166	Ea.	.006	.11	.25	.36
TOTAL		L.F.	.151	5.28	5.48	10.76
12″ THICK BY 24″ WIDE FOOTING						
Concrete, 3000 psi	.070	C.Y.		6.23		6.23
Place concrete, direct chute	.070	C.Y.	.028		.91	.91
Forms, footing, 4 uses	2.000	SFCA	.155	1.18	5.44	6.62
Reinforcing, 1/2″ diameter bars, 2 each	1.380	Lb.	.011	.63	.48	1.11
Keyway, 2″ x 4″, beveled, 4 uses	1.000	L.F.	.015	.20	.61	.81
Dowels, 1/2″ diameter bars, 2′ long, 6′ O.C.	.166	Ea.	.006	.11	.25	.36
TOTAL		L.F.	.215	8.35	7.69	16.04
12″ THICK BY 36″ WIDE FOOTING						
Concrete, 3000 psi	.110	C.Y.		9.79		9.79
Place concrete, direct chute	.110	C.Y.	.044		1.43	1.43
Forms, footing, 4 uses	2.000	SFCA	.155	1.18	5.44	6.62
Reinforcing, 1/2″ diameter bars, 2 each	1.380	Lb.	.011	.63	.48	1.11
Keyway, 2″ x 4″, beveled, 4 uses	1.000	L.F.	.015	.20	.61	.81
Dowels, 1/2″ diameter bars, 2′ long, 6′ O.C.	.166	Ea.	.006	.11	.25	.36
TOTAL		L.F.	.231	11.91	8.21	20.12

The footing costs in this system are on a cost per linear foot basis

Description	QUAN.	UNIT	LABOR HOURS	COST PER S.F.		
				MAT.	INST.	TOTAL

Footing Price Sheet	QUAN.	UNIT	LABOR HOURS	COST PER L.F. MAT.	INST.	TOTAL
Concrete, 8″ thick by 18″ wide footing						
2000 psi concrete	.040	C.Y.		3.42		3.42
2500 psi concrete	.040	C.Y.		3.50		3.50
3000 psi concrete	.040	C.Y.		3.56		3.56
3500 psi concrete	.040	C.Y.		3.60		3.60
4000 psi concrete	.040	C.Y.		3.70		3.70
12″ thick by 24″ wide footing						
2000 psi concrete	.070	C.Y.		6		6
2500 psi concrete	.070	C.Y.		6.15		6.15
3000 psi concrete	.070	C.Y.		6.25		6.25
3500 psi concrete	.070	C.Y.		6.30		6.30
4000 psi concrete	.070	C.Y.		6.50		6.50
12″ thick by 36″ wide footing						
2000 psi concrete	.110	C.Y.		9.40		9.40
2500 psi concrete	.110	C.Y.		9.65		9.65
3000 psi concrete	.110	C.Y.		9.80		9.80
3500 psi concrete	.110	C.Y.		9.90		9.90
4000 psi concrete	.110	C.Y.		10.20		10.20
Place concrete, 8″ thick by 18″ wide footing, direct chute	.040	C.Y.	.016		.52	.52
Pumped concrete	.040	C.Y.	.017		.77	.77
Crane & bucket	.040	C.Y.	.032		1.51	1.51
12″ thick by 24″ wide footing, direct chute	.070	C.Y.	.028		.91	.91
Pumped concrete	.070	C.Y.	.030		1.35	1.35
Crane & bucket	.070	C.Y.	.056		2.64	2.64
12″ thick by 36″ wide footing, direct chute	.110	C.Y.	.044		1.43	1.43
Pumped concrete	.110	C.Y.	.047		2.12	2.12
Crane & bucket	.110	C.Y.	.088		4.15	4.15
Forms, 8″ thick footing, 1 use	1.330	SFCA	.140	.21	4.93	5.14
4 uses	1.330	SFCA	.103	.78	3.62	4.40
12″ thick footing, 1 use	2.000	SFCA	.211	.32	7.40	7.72
4 uses	2.000	SFCA	.155	1.18	5.45	6.63
Reinforcing, 3/8″ diameter bar, 1 each	.400	Lb.	.003	.18	.14	.32
2 each	.800	Lb.	.006	.37	.28	.65
3 each	1.200	Lb.	.009	.55	.42	.97
1/2″ diameter bar, 1 each	.700	Lb.	.005	.32	.25	.57
2 each	1.380	Lb.	.011	.63	.48	1.11
3 each	2.100	Lb.	.016	.97	.74	1.71
5/8″ diameter bar, 1 each	1.040	Lb.	.008	.48	.36	.84
2 each	2.080	Lb.	.016	.96	.73	1.69
Keyway, beveled, 2″ x 4″, 1 use	1.000	L.F.	.030	.40	1.22	1.62
2 uses	1.000	L.F.	.023	.30	.92	1.22
2″ x 6″, 1 use	1.000	L.F.	.032	.60	1.30	1.90
2 uses	1.000	L.F.	.024	.45	.98	1.43
Dowels, 2 feet long, 6′ O.C., 3/8″ bar	.166	Ea.	.005	.06	.23	.29
1/2″ bar	.166	Ea.	.006	.11	.25	.36
5/8″ bar	.166	Ea.	.006	.17	.28	.45
3/4″ bar	.166	Ea.	.006	.17	.28	.45

08 Block Wall Systems

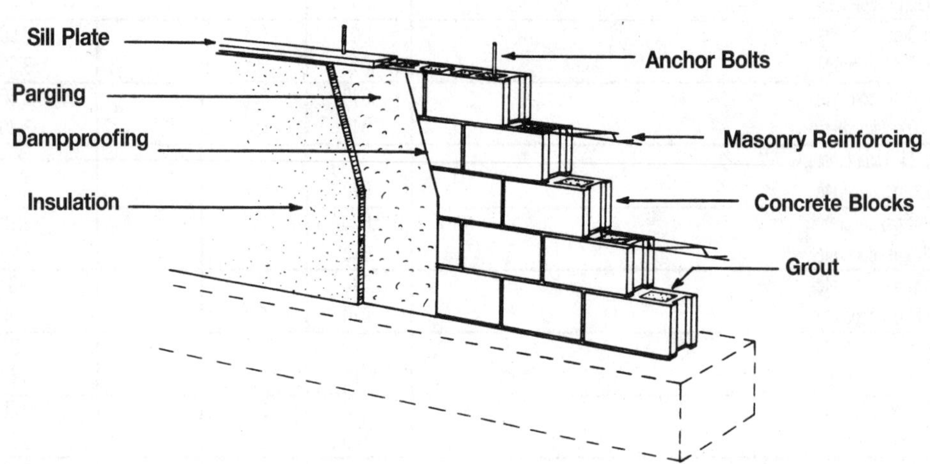

System Description	QUAN.	UNIT	LABOR HOURS	COST PER S.F. MAT.	COST PER S.F. INST.	COST PER S.F. TOTAL
8″ WALL, GROUTED, FULL HEIGHT						
Concrete block, 8″ x 16″ x 8″	1.000	S.F.	.094	1.97	3.50	5.47
Masonry reinforcing, every second course	.750	L.F.	.002	.14	.08	.22
Parging, plastering with portland cement plaster, 1 coat	1.000	S.F.	.014	.23	.54	.77
Dampproofing, bituminous coating, 1 coat	1.000	S.F.	.012	.08	.45	.53
Insulation, 1″ rigid polystyrene	1.000	S.F.	.010	.42	.41	.83
Grout, solid, pumped	1.000	S.F.	.059	.94	2.16	3.10
Anchor bolts, 1/2″ diameter, 8″ long, 4′ O.C.	.060	Ea.	.002	.04	.10	.14
Sill plate, 2″ x 4″, treated	.250	L.F.	.007	.14	.30	.44
TOTAL		S.F.	.200	3.96	7.54	11.50
12″ WALL, GROUTED, FULL HEIGHT						
Concrete block, 8″ x 16″ x 12″	1.000	S.F.	.160	2.76	5.80	8.56
Masonry reinforcing, every second course	.750	L.F.	.003	.17	.12	.29
Parging, plastering with portland cement plaster, 1 coat	1.000	S.F.	.014	.23	.54	.77
Dampproofing, bituminous coating, 1 coat	1.000	S.F.	.012	.08	.45	.53
Insulation, 1″ rigid polystyrene	1.000	S.F.	.010	.42	.41	.83
Grout, solid, pumped	1.000	S.F.	.063	1.53	2.29	3.82
Anchor bolts, 1/2″ diameter, 8″ long, 4′ O.C.	.060	Ea.	.002	.04	.10	.14
Sill plate, 2″ x 4″, treated	.250	L.F.	.007	.14	.30	.44
TOTAL		S.F.	.271	5.37	10.01	15.38

The costs in this system are based on a square foot of wall. Do not subtract for window or door openings.

Description	QUAN.	UNIT	LABOR HOURS	COST PER S.F. MAT.	COST PER S.F. INST.	COST PER S.F. TOTAL

Important: See the Reference Section for critical supporting data - Reference Nos., Crews & Location Factors

Block Wall Systems	QUAN.	UNIT	LABOR HOURS	COST PER S.F.		
				MAT.	INST.	TOTAL
Concrete, block, 8″ x 16″ x, 6″ thick	1.000	S.F.	.089	1.84	3.27	5.11
8″ thick	1.000	S.F.	.093	1.97	3.50	5.47
10″ thick	1.000	S.F.	.095	2.68	4.25	6.93
12″ thick	1.000	S.F.	.122	2.76	5.80	8.56
Solid block, 8″ x 16″ x, 6″ thick	1.000	S.F.	.091	1.99	3.38	5.37
8″ thick	1.000	S.F.	.096	2.84	3.59	6.43
10″ thick	1.000	S.F.	.096	2.84	3.59	6.43
12″ thick	1.000	S.F.	.126	4.14	4.97	9.11
Masonry reinforcing, wire strips, to 8″ wide, every course	1.500	L.F.	.004	.29	.17	.46
Every 2nd course	.750	L.F.	.002	.14	.08	.22
Every 3rd course	.500	L.F.	.001	.10	.06	.16
Every 4th course	.400	L.F.	.001	.08	.04	.12
Wire strips to 12″ wide, every course	1.500	L.F.	.006	.33	.24	.57
Every 2nd course	.750	L.F.	.003	.17	.12	.29
Every 3rd course	.500	L.F.	.002	.11	.08	.19
Every 4th course	.400	L.F.	.002	.09	.06	.15
Parging, plastering with portland cement plaster, 1 coat	1.000	S.F.	.014	.23	.54	.77
2 coats	1.000	S.F.	.022	.36	.83	1.19
Dampproofing, bituminous, brushed on, 1 coat	1.000	S.F.	.012	.08	.45	.53
2 coats	1.000	S.F.	.016	.11	.60	.71
Sprayed on, 1 coat	1.000	S.F.	.010	.08	.36	.44
2 coats	1.000	S.F.	.016	.16	.60	.76
Troweled on, 1/16″ thick	1.000	S.F.	.016	.18	.60	.78
1/8″ thick	1.000	S.F.	.020	.32	.75	1.07
1/2″ thick	1.000	S.F.	.023	1.03	.86	1.89
Insulation, rigid, fiberglass, 1.5#/C.F., unfaced						
1-1/2″ thick R 6.2	1.000	S.F.	.008	.45	.33	.78
2″ thick R 8.5	1.000	S.F.	.008	.52	.33	.85
3″ thick R 13	1.000	S.F.	.010	.63	.41	1.04
Foamglass, 1-1/2″ thick R 2.64	1.000	S.F.	.010	1.33	.41	1.74
2″ thick R 5.26	1.000	S.F.	.011	3.21	.45	3.66
Perlite, 1″ thick R 2.77	1.000	S.F.	.010	.29	.41	.70
2″ thick R 5.55	1.000	S.F.	.011	.55	.45	1
Polystyrene, extruded, 1″ thick R 5.4	1.000	S.F.	.010	.42	.41	.83
2″ thick R 10.8	1.000	S.F.	.011	1.14	.45	1.59
Molded 1″ thick R 3.85	1.000	S.F.	.010	.19	.41	.60
2″ thick R 7.7	1.000	S.F.	.011	.57	.45	1.02
Grout, concrete block cores, 6″ thick	1.000	S.F.	.044	.71	1.63	2.34
8″ thick	1.000	S.F.	.059	.94	2.16	3.10
10″ thick	1.000	S.F.	.061	1.24	2.22	3.46
12″ thick	1.000	S.F.	.063	1.53	2.29	3.82
Anchor bolts, 2′ on center, 1/2″ diameter, 8″ long	.120	Ea.	.005	.07	.20	.27
12″ long	.120	Ea.	.005	.15	.21	.36
3/4″ diameter, 8″ long	.120	Ea.	.006	.18	.25	.43
12″ long	.120	Ea.	.006	.23	.26	.49
4′ on center, 1/2″ diameter, 8″ long	.060	Ea.	.002	.04	.10	.14
12″ long	.060	Ea.	.003	.08	.10	.18
3/4″ diameter, 8″ long	.060	Ea.	.003	.09	.12	.21
12″ long	.060	Ea.	.003	.11	.13	.24
Sill plates, treated, 2″ x 4″	.250	L.F.	.007	.14	.30	.44
4″ x 4″	.250	L.F.	.007	.37	.28	.65

12 Concrete Wall Systems

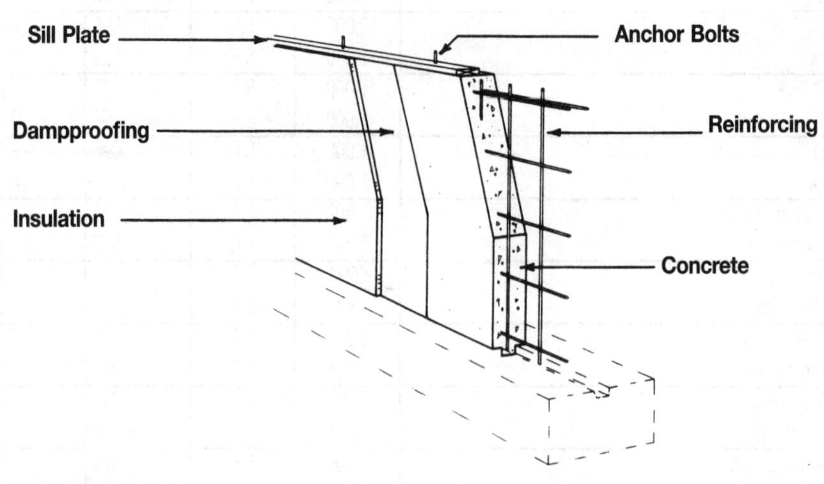

System Description	QUAN.	UNIT	LABOR HOURS	COST PER S.F.		
				MAT.	INST.	TOTAL
8″ THICK, POURED CONCRETE WALL						
Concrete, 8″ thick , 3000 psi	.025	C.Y.		2.23		2.23
Forms, prefabricated plywood, 4 uses per month	2.000	SFCA	.076	1.22	2.72	3.94
Reinforcing, light	.670	Lb.	.004	.31	.16	.47
Placing concrete, direct chute	.025	C.Y.	.013		.43	.43
Dampproofing, brushed on, 2 coats	1.000	S.F.	.016	.11	.60	.71
Rigid insulation, 1″ polystyrene	1.000	S.F.	.010	.42	.41	.83
Anchor bolts, 1/2″ diameter, 12″ long, 4′ O.C.	.060	Ea.	.003	.08	.10	.18
Sill plates, 2″ x 4″, treated	.250	L.F.	.007	.14	.30	.44
TOTAL		S.F.	.129	4.51	4.72	9.23
12″ THICK, POURED CONCRETE WALL						
Concrete, 12″ thick, 3000 psi	.040	C.Y.		3.56		3.56
Forms, prefabricated plywood, 4 uses per month	2.000	SFCA	.076	1.22	2.72	3.94
Reinforcing, light	1.000	Lb.	.005	.46	.24	.70
Placing concrete, direct chute	.040	C.Y.	.019		.62	.62
Dampproofing, brushed on, 2 coats	1.000	S.F.	.016	.11	.60	.71
Rigid insulation, 1″ polystyrene	1.000	S.F.	.010	.42	.41	.83
Anchor bolts, 1/2″ diameter, 12″ long, 4′ O.C.	.060	Ea.	.003	.08	.10	.18
Sill plates, 2″ x 4″ treated	.250	L.F.	.007	.14	.30	.44
TOTAL		S.F.	.136	5.99	4.99	10.98

The costs in this system are based on sq. ft. of wall. Do not subtract for window and door openings. The costs assume a 4′ high wall.

Description	QUAN.	UNIT	LABOR HOURS	COST PER S.F.		
				MAT.	INST.	TOTAL

Important: See the Reference Section for critical supporting data - Reference Nos., Crews & Location Factors

Concrete Wall Price Sheet	QUAN.	UNIT	LABOR HOURS	COST PER S.F.		
				MAT.	INST.	TOTAL
ormwork, prefabricated plywood, 1 use per month	2.000	SFCA	.081	3.70	2.90	6.60
4 uses per month	2.000	SFCA	.076	1.22	2.72	3.94
Job built forms, 1 use per month	2.000	SFCA	.320	5.20	11.40	16.60
4 uses per month	2.000	SFCA	.221	1.94	7.90	9.84
einforcing, 8″ wall, light reinforcing	.670	Lb.	.004	.31	.16	.47
Heavy reinforcing	1.500	Lb.	.008	.69	.36	1.05
10″ wall, light reinforcing	.850	Lb.	.005	.39	.20	.59
Heavy reinforcing	2.000	Lb.	.011	.92	.48	1.40
12″ wall light reinforcing	1.000	Lb.	.005	.46	.24	.70
Heavy reinforcing	2.250	Lb.	.012	1.04	.54	1.58
lacing concrete, 8″ wall, direct chute	.025	C.Y.	.013		.43	.43
Pumped concrete	.025	C.Y.	.016		.72	.72
Crane & bucket	.025	C.Y.	.023		1.06	1.06
10″ wall, direct chute	.030	C.Y.	.016		.52	.52
Pumped concrete	.030	C.Y.	.019		.87	.87
Crane & bucket	.030	C.Y.	.027		1.27	1.27
12″ wall, direct chute	.040	C.Y.	.019		.62	.62
Pumped concrete	.040	C.Y.	.023		1.05	1.05
Crane & bucket	.040	C.Y.	.032		1.51	1.51
ampproofing, bituminous, brushed on, 1 coat	1.000	S.F.	.012	.08	.45	.53
2 coats	1.000	S.F.	.016	.11	.60	.71
Sprayed on, 1 coat	1.000	S.F.	.010	.08	.36	.44
2 coats	1.000	S.F.	.016	.16	.60	.76
Troweled on, 1/16″ thick	1.000	S.F.	.016	.18	.60	.78
1/8″ thick	1.000	S.F.	.020	.32	.75	1.07
1/2″ thick	1.000	S.F.	.023	1.03	.86	1.89
nsulation rigid, fiberglass, 1.5#/C.F., unfaced						
1-1/2″ thick, R 6.2	1.000	S.F.	.008	.45	.33	.78
2″ thick, R 8.3	1.000	S.F.	.008	.52	.33	.85
3″ thick, R 12.4	1.000	S.F.	.010	.63	.41	1.04
Foamglass, 1-1/2″ thick R 2.64	1.000	S.F.	.010	1.33	.41	1.74
2″ thick R 5.26	1.000	S.F.	.011	3.21	.45	3.66
Perlite, 1″ thick R 2.77	1.000	S.F.	.010	.29	.41	.70
2″ thick R 5.55	1.000	S.F.	.011	.55	.45	1
Polystyrene, extruded, 1″ thick R 5.40	1.000	S.F.	.010	.42	.41	.83
2″ thick R 10.8	1.000	S.F.	.011	1.14	.45	1.59
Molded, 1″ thick R 3.85	1.000	S.F.	.010	.19	.41	.60
2″ thick R 7.70	1.000	S.F.	.011	.57	.45	1.02
nchor bolts, 2′ on center, 1/2″ diameter, 8″ long	.120	Ea.	.005	.07	.20	.27
12″ long	.120	Ea.	.005	.15	.21	.36
3/4″ diameter, 8″ long	.120	Ea.	.006	.18	.25	.43
12″ long	.120	Ea.	.006	.23	.26	.49
Sill plates, treated lumber, 2″ x 4″	.250	L.F.	.007	.14	.30	.44
4″ x 4″	.250	L.F.	.007	.37	.28	.65

16 | Wood Wall Foundation Systems

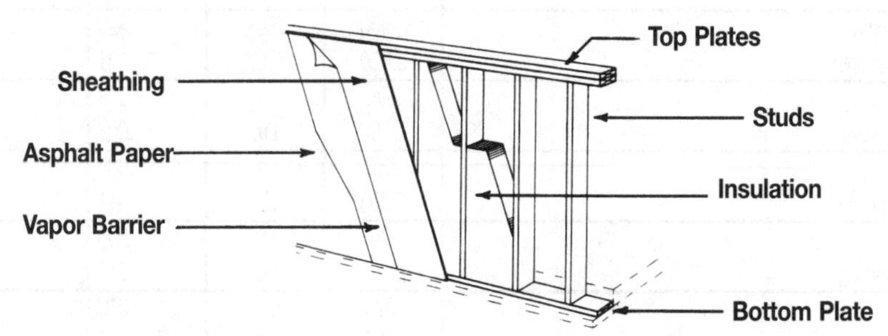

System Description	QUAN.	UNIT	LABOR HOURS	COST PER S.F. MAT.	COST PER S.F. INST.	COST PER S.F. TOTAL
2″ X 4″ STUDS, 16″ O.C., WALL						
Studs, 2″ x 4″, 16″ O.C., treated	1.000	L.F.	.015	.54	.59	1.13
Plates, double top plate, single bottom plate, treated, 2″ x 4″	.750	L.F.	.011	.41	.44	.85
Sheathing, 1/2″, exterior grade, CDX, treated	1.000	S.F.	.014	.88	.58	1.46
Asphalt paper, 15# roll	1.100	S.F.	.002	.04	.10	.14
Vapor barrier, 4 mil polyethylene	1.000	S.F.	.002	.03	.09	.12
Insulation, batts, fiberglass, 3-1/2″ thick, R 11	1.000	S.F.	.005	.30	.20	.50
TOTAL		S.F.	.049	2.20	2	4.20
2″ X 6″ STUDS, 16″ O.C., WALL						
Studs, 2″ x 6″, 16″ O.C., treated	1.000	L.F.	.016	.85	.65	1.50
Plates, double top plate, single bottom plate, treated, 2″ x 6″	.750	L.F.	.012	.64	.49	1.13
Sheathing, 5/8″ exterior grade, CDX, treated	1.000	S.F.	.015	1.30	.62	1.92
Asphalt paper, 15# roll	1.100	S.F.	.002	.04	.10	.14
Vapor barrier, 4 mil polyethylene	1.000	S.F.	.002	.03	.09	.12
Insulation, batts, fiberglass, 6″ thick, R 19	1.000	S.F.	.006	.39	.24	.63
TOTAL		S.F.	.053	3.25	2.19	5.44
2″ X 8″ STUDS, 16″ O.C., WALL						
Studs, 2″ x 8″, 16″ O.C. treated	1.000	L.F.	.018	1.10	.72	1.82
Plates, double top plate, single bottom plate, treated, 2″ x 8″	.750	L.F.	.013	.83	.54	1.37
Sheathing, 3/4″ exterior grade, CDX, treated	1.000	S.F.	.016	1.51	.67	2.18
Asphalt paper, 15# roll	1.100	S.F.	.002	.04	.10	.14
Vapor barrier, 4 mil polyethylene	1.000	S.F.	.002	.03	.09	.12
Insulation, batts, fiberglass, 9″ thick, R 30	1.000	S.F.	.006	.66	.24	.90
TOTAL		S.F.	.057	4.17	2.36	6.53

The costs in this system are based on a sq. ft. of wall area. Do not Subtract for window or door openings. The costs assume a 4′ high wall.

Description	QUAN.	UNIT	LABOR HOURS	COST PER S.F. MAT.	COST PER S.F. INST.	COST PER S.F. TOTAL

Wood Wall Foundation Price Sheet	QUAN.	UNIT	LABOR HOURS	COST PER S.F. MAT.	INST.	TOTAL
Studs, treated, 2" x 4", 12" O.C.	1.250	L.F.	.018	.68	.74	1.42
16" O.C.	1.000	L.F.	.015	.54	.59	1.13
2" x 6", 12" O.C.	1.250	L.F.	.020	1.06	.81	1.87
16" O.C.	1.000	L.F.	.016	.85	.65	1.50
2" x 8", 12" O.C.	1.250	L.F.	.022	1.38	.90	2.28
16" O.C.	1.000	L.F.	.018	1.10	.72	1.82
Plates, treated double top single bottom, 2" x 4"	.750	L.F.	.011	.41	.44	.85
2" x 6"	.750	L.F.	.012	.64	.49	1.13
2" x 8"	.750	L.F.	.013	.83	.54	1.37
Sheathing, treated exterior grade CDX, 1/2" thick	1.000	S.F.	.014	.88	.58	1.46
5/8" thick	1.000	S.F.	.015	1.30	.62	1.92
3/4" thick	1.000	S.F.	.016	1.51	.67	2.18
Asphalt paper, 15# roll	1.100	S.F.	.002	.04	.10	.14
Vapor barrier, polyethylene, 4 mil	1.000	S.F.	.002	.02	.09	.11
10 mil	1.000	S.F.	.002	.06	.09	.15
Insulation, rigid, fiberglass, 1.5#/C.F., unfaced	1.000	S.F.	.008	.34	.33	.67
1-1/2" thick, R 6.2	1.000	S.F.	.008	.45	.33	.78
2" thick, R 8.3	1.000	S.F.	.008	.52	.33	.85
3" thick, R 12.4	1.000	S.F.	.010	.64	.42	1.06
Foamglass 1 1/2" thick, R 2.64	1.000	S.F.	.010	1.33	.41	1.74
2" thick, R 5.26	1.000	S.F.	.011	3.21	.45	3.66
Perlite 1" thick, R 2.77	1.000	S.F.	.010	.29	.41	.70
2" thick, R 5.55	1.000	S.F.	.011	.55	.45	1
Polystyrene, extruded, 1" thick, R 5.40	1.000	S.F.	.010	.42	.41	.83
2" thick, R 10.8	1.000	S.F.	.011	1.14	.45	1.59
Molded 1" thick, R 3.85	1.000	S.F.	.010	.19	.41	.60
2" thick, R 7.7	1.000	S.F.	.011	.57	.45	1.02
Non rigid, batts, fiberglass, paper backed, 3-1/2" thick roll, R 11	1.000	S.F.	.005	.30	.20	.50
6", R 19	1.000	S.F.	.006	.39	.24	.63
9", R 30	1.000	S.F.	.006	.66	.24	.90
12", R 38	1.000	S.F.	.006	.84	.24	1.08
Mineral fiber, paper backed, 3-1/2", R 13	1.000	S.F.	.005	.31	.20	.51
6", R 19	1.000	S.F.	.005	.42	.20	.62
10", R 30	1.000	S.F.	.006	.63	.24	.87

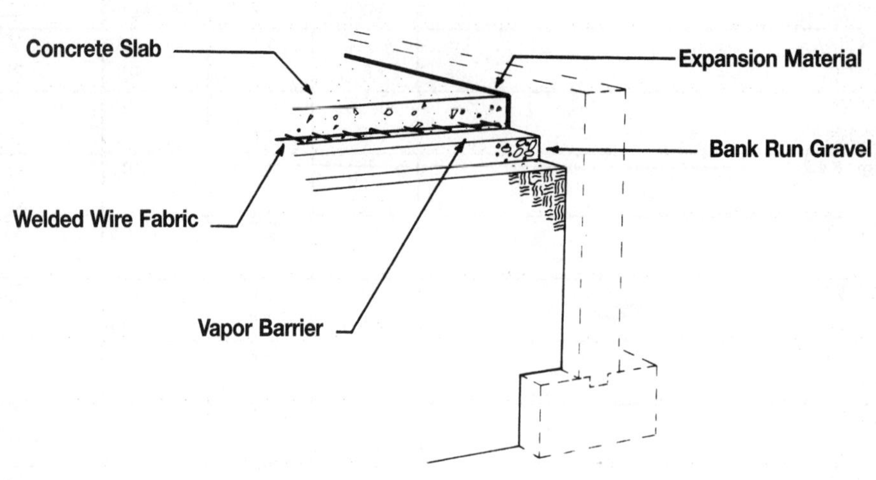

System Description	QUAN.	UNIT	LABOR HOURS	COST PER S.F.		
				MAT.	INST.	TOTAL
4″ THICK SLAB						
Concrete, 4″ thick, 3000 psi concrete	.012	C.Y.		1.07		1.07
Place concrete, direct chute	.012	C.Y.	.005		.17	.17
Bank run gravel, 4″ deep	1.000	S.F.	.001	.32	.04	.36
Polyethylene vapor barrier, .006″ thick	1.000	S.F.	.002	.03	.09	.12
Edge forms, expansion material	.100	L.F.	.005	.03	.19	.22
Welded wire fabric, 6 x 6, 10/10 (W1.4/W1.4)	1.100	S.F.	.005	.23	.23	.46
Steel trowel finish	1.000	S.F.	.015		.54	.54
TOTAL		S.F.	.033	1.68	1.26	2.94
6″ THICK SLAB						
Concrete, 6″ thick, 3000 psi concrete	.019	C.Y.		1.69		1.69
Place concrete, direct chute	.019	C.Y.	.008		.27	.27
Bank run gravel, 4″ deep	1.000	S.F.	.001	.32	.04	.36
Polyethylene vapor barrier, .006″ thick	1.000	S.F.	.002	.03	.09	.12
Edge forms, expansion material	.100	L.F.	.005	.03	.19	.22
Welded wire fabric, 6 x 6, 10/10 (W1.4/W1.4)	1.100	S.F.	.005	.23	.23	.46
Steel trowel finish	1.000	S.F.	.015		.54	.54
TOTAL		S.F.	.036	2.30	1.36	3.66

The slab costs in this section are based on a cost per square foot of floor area.

Description	QUAN.	UNIT	LABOR HOURS	COST PER S.F.		
				MAT.	INST.	TOTAL

Floor Slab Price Sheet	QUAN.	UNIT	LABOR HOURS	COST PER S.F.		
				MAT.	INST.	TOTAL
;oncrete, 4″ thick slab, 2000 psi concrete	.012	C.Y.		1.03		1.03
2500 psi concrete	.012	C.Y.		1.05		1.05
3000 psi concrete	.012	C.Y.		1.07		1.07
3500 psi concrete	.012	C.Y.		1.08		1.08
4000 psi concrete	.012	C.Y.		1.11		1.11
4500 psi concrete	.012	C.Y.		1.13		1.13
5″ thick slab, 2000 psi concrete	.015	C.Y.		1.28		1.28
2500 psi concrete	.015	C.Y.		1.31		1.31
3000 psi concrete	.015	C.Y.		1.34		1.34
3500 psi concrete	.015	C.Y.		1.35		1.35
4000 psi concrete	.015	C.Y.		1.39		1.39
4500 psi concrete	.015	C.Y.		1.42		1.42
6″ thick slab, 2000 psi concrete	.019	C.Y.		1.62		1.62
2500 psi concrete	.019	C.Y.		1.66		1.66
3000 psi concrete	.019	C.Y.		1.69		1.69
3500 psi concrete	.019	C.Y.		1.71		1.71
4000 psi concrete	.019	C.Y.		1.76		1.76
4500 psi concrete	.019	C.Y.		1.80		1.80
'lace concrete, 4″ slab, direct chute	.012	C.Y.	.005		.17	.17
Pumped concrete	.012	C.Y.	.006		.27	.27
Crane & bucket	.012	C.Y.	.008		.37	.37
5″ slab, direct chute	.015	C.Y.	.007		.21	.21
Pumped concrete	.015	C.Y.	.007		.34	.34
Crane & bucket	.015	C.Y.	.010		.46	.46
6″ slab, direct chute	.019	C.Y.	.008		.27	.27
Pumped concrete	.019	C.Y.	.009		.42	.42
Crane & bucket	.019	C.Y.	.012		.58	.58
;ravel, bank run, 4″ deep	1.000	S.F.	.001	.32	.04	.36
6″ deep	1.000	S.F.	.001	.42	.05	.47
9″ deep	1.000	S.F.	.001	.62	.07	.69
12″ deep	1.000	S.F.	.001	.85	.09	.94
3/4″ crushed stone, 3″ deep	1.000	S.F.	.001	.33	.04	.37
6″ deep	1.000	S.F.	.001	.65	.08	.73
9″ deep	1.000	S.F.	.002	.95	.10	1.05
12″ deep	1.000	S.F.	.002	1.53	.12	1.65
/apor barrier polyethylene, .004″ thick	1.000	S.F.	.002	.02	.09	.11
.006″ thick	1.000	S.F.	.002	.03	.09	.12
:dge forms, expansion material, 4″ thick slab	.100	L.F.	.004	.02	.12	.14
6″ thick slab	.100	L.F.	.005	.03	.19	.22
Velded wire fabric 6 x 6, 10/10 (W1.4/W1.4)	1.100	S.F.	.005	.23	.23	.46
6 x 6, 6/6 (W2.9/W2.9)	1.100	S.F.	.006	.40	.28	.68
4 x 4, 10/10 (W1.4/W1.4)	1.100	S.F.	.006	.21	.26	.47
Finish concrete, screed finish	1.000	S.F.	.009		.33	.33
Float finish	1.000	S.F.	.011		.41	.41
Steel trowel, for resilient floor	1.000	S.F.	.013		.49	.49
For finished floor	1.000	S.F.	.015		.54	.54

Division 3 Framing

No part of this publication may be reproduced, stored in a retrieval system, or transmitted in any form or by any means without prior written permission of Reed Construction Data.

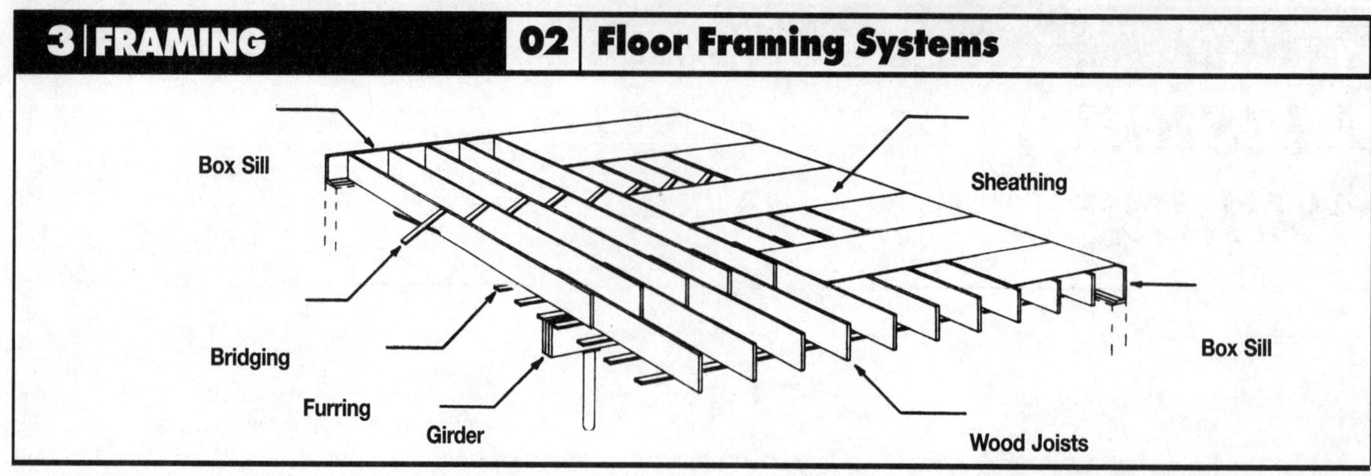

System Description	QUAN.	UNIT	LABOR HOURS	COST PER S.F.		
				MAT.	INST.	TOTAL
2″ X 8″, 16″ O.C.						
Wood joists, 2″ x 8″, 16″ O.C.	1.000	L.F.	.015	.91	.59	1.50
Bridging, 1″ x 3″, 6′ O.C.	.080	Pr.	.005	.04	.20	.24
Box sills, 2″ x 8″	.150	L.F.	.002	.14	.09	.23
Concrete filled steel column, 4″ diameter	.125	L.F.	.002	.11	.10	.21
Girder, built up from three 2″ x 8″	.125	L.F.	.013	.34	.54	.88
Sheathing, plywood, subfloor, 5/8″ CDX	1.000	S.F.	.012	.86	.48	1.34
Furring, 1″ x 3″, 16″ O.C.	1.000	L.F.	.023	.31	.93	1.24
TOTAL		S.F.	.072	2.71	2.93	5.64
2″ X 10″, 16″ O.C.						
Wood joists, 2″ x 10″, 16″ OC	1.000	L.F.	.018	1.29	.72	2.01
Bridging, 1″ x 3″, 6′ OC	.080	Pr.	.005	.04	.20	.24
Box sills, 2″ x 10″	.150	L.F.	.003	.19	.11	.30
Girder, built up from three 2″ x 10″	.125	L.F.	.002	.11	.10	.21
Sheathing, plywood, subfloor, 5/8″ CDX	1.000	S.F.	.012	.86	.48	1.34
Furring, 1″ x 3″,16″ OC	1.000	L.F.	.023	.31	.93	1.24
TOTAL		S.F.	.077	3.29	3.12	6.41
2″ X 12″, 16″ O.C.						
Wood joists, 2″ x 12″, 16″ O.C.	1.000	L.F.	.018	1.76	.75	2.51
Bridging, 1″ x 3″, 6′ O.C.	.080	Pr.	.005	.04	.20	.24
Box sills, 2″ x 12″	.150	L.F.	.003	.26	.11	.37
Concrete filled steel column, 4″ diameter	.125	L.F.	.002	.11	.10	.21
Girder, built up from three 2″ x 12″	.125	L.F.	.015	.66	.62	1.28
Sheathing, plywood, subfloor, 5/8″ CDX	1.000	S.F.	.012	.86	.48	1.34
Furring, 1″ x 3″, 16″ O.C.	1.000	L.F.	.023	.31	.93	1.24
TOTAL		S.F.	.078	4	3.19	7.19

Floor costs on this page are given on a cost per square foot basis.

Description	QUAN.	UNIT	LABOR HOURS	COST PER S.F.		
				MAT.	INST.	TOTAL

Floor Framing Price Sheet (Wood)

	QUAN.	UNIT	LABOR HOURS	COST PER S.F. MAT.	INST.	TOTAL
Joists, #2 or better, pine, 2″ x 4″, 12″ O.C.	1.250	L.F.	.016	.46	.65	1.11
16″ O.C.	1.000	L.F.	.013	.37	.52	.89
2″ x 6″, 12″ O.C.	1.250	L.F.	.016	.74	.65	1.39
16″ O.C.	1.000	L.F.	.013	.59	.52	1.11
2″ x 8″, 12″ O.C.	1.250	L.F.	.018	1.14	.74	1.88
16″ O.C.	1.000	L.F.	.015	.91	.59	1.50
2″ x 10″, 12″ O.C.	1.250	L.F.	.022	1.61	.90	2.51
16″ O.C.	1.000	L.F.	.018	1.29	.72	2.01
2″x 12″, 12″ O.C.	1.250	L.F.	.023	2.20	.94	3.14
16″ O.C.	1.000	L.F.	.018	1.76	.75	2.51
Bridging, wood 1″ x 3″, joists 12″ O.C.	.100	Pr.	.006	.05	.25	.30
16″ O.C.	.080	Pr.	.005	.04	.20	.24
Metal, galvanized, joists 12″ O.C.	.100	Pr.	.006	.11	.25	.36
16″ O.C.	.080	Pr.	.005	.09	.20	.29
Compression type, joists 12″ O.C.	.100	Pr.	.004	.13	.16	.29
16″ O.C.	.080	Pr.	.003	.10	.13	.23
Box sills, #2 or better pine, 2″ x 4″	.150	L.F.	.002	.06	.08	.14
2″ x 6″	.150	L.F.	.002	.09	.08	.17
2″ x 8″	.150	L.F.	.002	.14	.09	.23
2″ x 10″	.150	L.F.	.003	.19	.11	.30
2″ x 12″	.150	L.F.	.003	.26	.11	.37
Girders, including lally columns, 3 pieces spiked together, 2″ x 8″	.125	L.F.	.015	.45	.64	1.09
2″ x 10″	.125	L.F.	.016	.60	.68	1.28
2″ x 12″	.125	L.F.	.017	.77	.72	1.49
Solid girders, 3″ x 8″	.040	L.F.	.004	.23	.17	.40
3″ x 10″	.040	L.F.	.004	.26	.17	.43
3″ x 12″	.040	L.F.	.004	.29	.18	.47
4″ x 8″	.040	L.F.	.004	.23	.18	.41
4″ x 10″	.040	L.F.	.004	.26	.19	.45
4″ x 12″	.040	L.F.	.004	.31	.20	.51
Steel girders, bolted & including fabrication, wide flange shapes						
12″ deep, 14#/l.f.	.040	L.F.	.003	.69	.21	.90
10″ deep, 15#/l.f.	.040	L.F.	.003	.69	.21	.90
8″ deep, 10#/l.f.	.040	L.F.	.003	.46	.21	.67
6″ deep, 9#/l.f.	.040	L.F.	.003	.42	.21	.63
5″ deep, 16#/l.f.	.040	L.F.	.003	.69	.21	.90
Sheathing, plywood exterior grade CDX, 1/2″ thick	1.000	S.F.	.011	.68	.47	1.15
5/8″ thick	1.000	S.F.	.012	.86	.48	1.34
3/4″ thick	1.000	S.F.	.013	1.03	.52	1.55
Boards, 1″ x 8″ laid regular	1.000	S.F.	.016	1.23	.65	1.88
Laid diagonal	1.000	S.F.	.019	1.23	.77	2
1″ x 10″ laid regular	1.000	S.F.	.015	1.50	.59	2.09
Laid diagonal	1.000	S.F.	.018	1.50	.72	2.22
Furring, 1″ x 3″, 12″ O.C.	1.250	L.F.	.029	.39	1.16	1.55
16″ O.C.	1.000	L.F.	.023	.31	.93	1.24
24″ O.C.	.750	L.F.	.017	.23	.70	.93

04 Floor Framing Systems

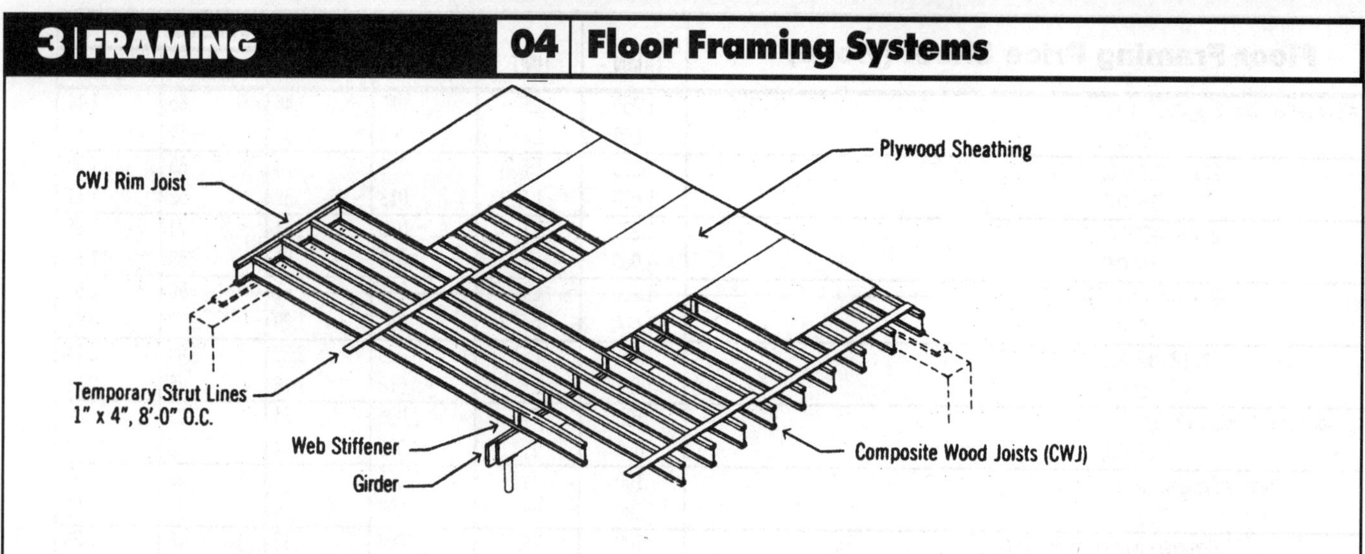

System Description	QUAN.	UNIT	LABOR HOURS	COST PER S.F.		
				MAT.	INST.	TOTAL
9-1/2″ COMPOSITE WOOD JOISTS, 16″ O.C.						
CWJ, 9-1/2″, 16″ O.C., 15′ span	1.000	L.F.	.018	1.63	.73	2.36
Temp. strut line, 1″ x 4″, 8′ O.C.	.160	L.F.	.003	.07	.13	.20
CWJ rim joist, 9-1/2″	.150	L.F.	.003	.24	.11	.35
Concrete filled steel column, 4″ diameter	.125	L.F.	.002	.11	.10	.21
Girder, built up from three 2″ x 8″	.125	L.F.	.013	.34	.54	.88
Sheathing, plywood, subfloor, 5/8″ CDX	1.000	S.F.	.012	.86	.48	1.34
TOTAL		S.F.	.051	3.25	2.09	5.34
11-1/2″ COMPOSITE WOOD JOISTS, 16″ O.C.						
CWJ, 11-1/2″, 16″ O.C., 18′ span	1.000	L.F.	.018	1.73	.74	2.47
Temp. strut line, 1″ x 4″, 8′ O.C.	.160	L.F.	.003	.07	.13	.20
CWJ rim joist, 11-1/2″	.150	L.F.	.003	.26	.11	.37
Concrete filled steel column, 4″ diameter	.125	L.F.	.002	.11	.10	.21
Girder, built up from three 2″ x 10″	.125	L.F.	.014	.49	.58	1.07
Sheathing, plywood, subfloor, 5/8″ CDX	1.000	S.F.	.012	.86	.48	1.34
TOTAL		S.F.	.052	3.52	2.14	5.66
14″ COMPOSITE WOOD JOISTS, 16″ O.C.						
CWJ, 14″, 16″ O.C., 22′ span	1.000	L.F.	.020	1.98	.80	2.78
Temp. strut line, 1″ x 4″, 8′ O.C.	.160	L.F.	.003	.07	.13	.20
CWJ rim joist, 14″	.150	L.F.	.003	.30	.12	.42
Concrete filled steel column, 4″ diameter	.600	L.F.	.002	.11	.10	.21
Girder, built up from three 2″ x 12″	.600	L.F.	.015	.66	.62	1.28
Sheathing, plywood, subfloor, 5/8″ CDX	1.000	S.F.	.012	.86	.48	1.34
TOTAL		S.F.	.055	3.98	2.25	6.23

Floor costs on this page are given on a cost per square foot basis.

Description	QUAN.	UNIT	LABOR HOURS	COST PER S.F.		
				MAT.	INST.	TOTAL

Important: See the Reference Section for critical supporting data - Reference Nos., Crews & Location Factors

Floor Framing Price Sheet (Wood)	QUAN.	UNIT	LABOR HOURS	COST PER S.F.		
				MAT.	INST.	TOTAL
Composite wood joist 9-1/2" deep, 12" O.C.	1.250	L.F.	.022	2.03	.91	2.94
16" O.C.	1.000	L.F.	.018	1.63	.73	2.36
11-1/2" deep, 12" O.C.	1.250	L.F.	.023	2.16	.93	3.09
16" O.C.	1.000	L.F.	.018	1.73	.74	2.47
14" deep, 12" O.C.	1.250	L.F.	.024	2.47	.99	3.46
16" O.C.	1.000	L.F.	.020	1.98	.80	2.78
16 " deep, 12" O.C.	1.250	L.F.	.026	3.50	1.04	4.54
16" O.C.	1.000	L.F.	.021	2.80	.84	3.64
CWJ rim joist, 9-1/2"	.150	L.F.	.003	.24	.11	.35
11-1/2"	.150	L.F.	.003	.26	.11	.37
14"	.150	L.F.	.003	.30	.12	.42
16"	.150	L.F.	.003	.42	.13	.55
Girders, including lally columns, 3 pieces spiked together, 2" x 8"	.125	L.F.	.015	.45	.64	1.09
2" x 10"	.125	L.F.	.016	.60	.68	1.28
2" x 12"	.125	L.F.	.017	.77	.72	1.49
Solid girders, 3" x 8"	.040	L.F.	.004	.23	.17	.40
3" x 10"	.040	L.F.	.004	.26	.17	.43
3" x 12"	.040	L.F.	.004	.29	.18	.47
4" x 8"	.040	L.F.	.004	.23	.18	.41
4" x 10"	.040	L.F.	.004	.26	.19	.45
4" x 12"	.040	L.F.	.004	.31	.20	.51
Steel girders, bolted & including fabrication, wide flange shapes						
12" deep, 14#/l.f.	.040	L.F.	.061	15.90	4.81	20.71
10" deep, 15#/l.f.	.040	L.F.	.067	17.35	5.25	22.60
8" deep, 10#/l.f.	.040	L.F.	.067	11.55	5.25	16.80
6" deep, 9#/l.f.	.040	L.F.	.067	10.40	5.25	15.65
5" deep, 16#/l.f.	.040	L.F.	.064	16.80	5.05	21.85
Sheathing, plywood exterior grade CDX, 1/2" thick	1.000	S.F.	.011	.68	.47	1.15
5/8" thick	1.000	S.F.	.012	.86	.48	1.34
3/4" thick	1.000	S.F.	.013	1.03	.52	1.55
Boards, 1" x 8" laid regular	1.000	S.F.	.016	1.23	.65	1.88
Laid diagonal	1.000	S.F.	.019	1.23	.77	2
1" x 10" laid regular	1.000	S.F.	.015	1.50	.59	2.09
Laid diagonal	1.000	S.F.	.018	1.50	.72	2.22
Furring, 1" x 3", 12" O.C.	1.250	L.F.	.029	.39	1.16	1.55
16" O.C.	1.000	L.F.	.023	.31	.93	1.24
24" O.C.	.750	L.F.	.017	.23	.70	.93

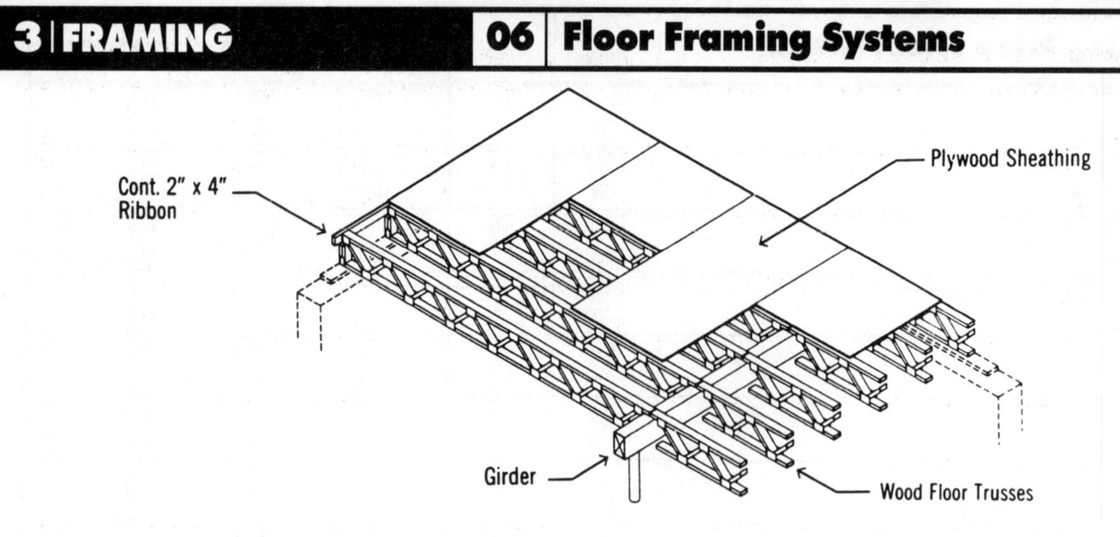

System Description	QUAN.	UNIT	LABOR HOURS	COST PER S.F. MAT.	INST.	TOTAL
12" OPEN WEB JOISTS, 16" O.C.						
OWJ 12", 16" O.C., 21' span	1.000	L.F.	.018	1.85	.74	2.59
Continuous ribbing, 2" x 4"	.150	L.F.	.002	.06	.08	.14
Concrete filled steel column, 4" diameter	.125	L.F.	.002	.11	.10	.21
Girder, built up from three 2" x 8"	.125	L.F.	.013	.34	.54	.88
Sheathing, plywood, subfloor, 5/8" CDX	1.000	S.F.	.012	.86	.48	1.34
Furring, 1" x 3", 16" O.C.	1.000	L.F.	.023	.31	.93	1.24
TOTAL		S.F.	.070	3.53	2.87	6.40
14" OPEN WEB WOOD JOISTS, 16" O.C.						
OWJ 14", 16" O.C., 22' span	1.000	L.F.	.020	2.15	.80	2.95
Continuous ribbing, 2" x 4"	.150	L.F.	.002	.06	.08	.14
Concrete filled steel column, 4" diameter	.125	L.F.	.002	.11	.10	.21
Girder, built up from three 2" x 10"	.125	L.F.	.014	.49	.58	1.07
Sheathing, plywood, subfloor, 5/8" CDX	1.000	S.F.	.012	.86	.48	1.34
Furring, 1" x 3",16" O.C.	1.000	L.F.	.023	.31	.93	1.24
TOTAL		S.F.	.073	3.98	2.97	6.95
16" OPEN WEB WOOD JOISTS, 16" O.C.						
OWJ 16", 16" O.C., 24' span	1.000	L.F.	.021	2.23	.84	3.07
Continuous ribbing, 2" x 4"	.150	L.F.	.002	.06	.08	.14
Concrete filled steel column, 4" diameter	.125	L.F.	.002	.11	.10	.21
Girder, built up from three 2" x 12"	.125	L.F.	.015	.66	.62	1.28
Sheathing, plywood, subfloor, 5/8" CDX	1.000	S.F.	.012	.86	.48	1.34
Furring, 1" x 3", 16" O.C.	1.000	L.F.	.023	.31	.93	1.24
TOTAL		S.F.	.075	4.23	3.05	7.28

Floor costs on this page are given on a cost per square foot basis.

Description	QUAN.	UNIT	LABOR HOURS	COST PER S.F. MAT.	INST.	TOTAL

 Important: See the Reference Section for critical supporting data - Reference Nos., Crews & Location Factors

Floor Framing Price Sheet (Wood)	QUAN.	UNIT	LABOR HOURS	COST PER S.F.		
				MAT.	INST.	TOTAL
Open web joists, 12″ deep, 12″ O.C.	1.250	L.F.	.023	2.31	.93	3.24
16″ O.C.	1.000	L.F.	.018	1.85	.74	2.59
14″ deep, 12″ O.C.	1.250	L.F.	.024	2.69	.99	3.68
16″ O.C.	1.000	L.F.	.020	2.15	.80	2.95
16 ″ deep, 12″ O.C.	1.250	L.F.	.026	2.78	1.04	3.82
16″ O.C.	1.000	L.F.	.021	2.23	.84	3.07
18″ deep, 12″ O.C.	1.250	L.F.	.027	2.81	1.10	3.91
16″ O.C.	1.000	L.F.	.022	2.25	.88	3.13
Continuous ribbing, 2″ x 4″	.150	L.F.	.002	.06	.08	.14
2″ x 6″	.150	L.F.	.002	.09	.08	.17
2″ x 8″	.150	L.F.	.002	.14	.09	.23
2″ x 10″	.150	L.F.	.003	.19	.11	.30
2″ x 12″	.150	L.F.	.003	.26	.11	.37
Girders, including lally columns, 3 pieces spiked together, 2″ x 8″	.125	L.F.	.015	.45	.64	1.09
2″ x 10″	.125	L.F.	.016	.60	.68	1.28
2″ x 12″	.125	L.F.	.017	.77	.72	1.49
Solid girders, 3″ x 8″	.040	L.F.	.004	.23	.17	.40
3″ x 10″	.040	L.F.	.004	.26	.17	.43
3″ x 12″	.040	L.F.	.004	.29	.18	.47
4″ x 8″	.040	L.F.	.004	.23	.18	.41
4″ x 10″	.040	L.F.	.004	.26	.19	.45
4″ x 12″	.040	L.F.	.004	.31	.20	.51
Steel girders, bolted & including fabrication, wide flange shapes						
12″ deep, 14#/l.f.	.040	L.F.	.061	15.90	4.81	20.71
10″ deep, 15#/l.f.	.040	L.F.	.067	17.35	5.25	22.60
8″ deep, 10#/l.f.	.040	L.F.	.067	11.55	5.25	16.80
6″ deep, 9#/l.f.	.040	L.F.	.067	10.40	5.25	15.65
5″ deep, 16#/l.f.	.040	L.F.	.064	16.80	5.05	21.85
Sheathing, plywood exterior grade CDX, 1/2″ thick	1.000	S.F.	.011	.68	.47	1.15
5/8″ thick	1.000	S.F.	.012	.86	.48	1.34
3/4″ thick	1.000	S.F.	.013	1.03	.52	1.55
Boards, 1″ x 8″ laid regular	1.000	S.F.	.016	1.23	.65	1.88
Laid diagonal	1.000	S.F.	.019	1.23	.77	2
1″ x 10″ laid regular	1.000	S.F.	.015	1.50	.59	2.09
Laid diagonal	1.000	S.F.	.018	1.50	.72	2.22
Furring, 1″ x 3″, 12″ O.C.	1.250	L.F.	.029	.39	1.16	1.55
16″ O.C.	1.000	L.F.	.023	.31	.93	1.24
24″ O.C.	.750	L.F.	.017	.23	.70	.93

08 Exterior Wall Framing Systems

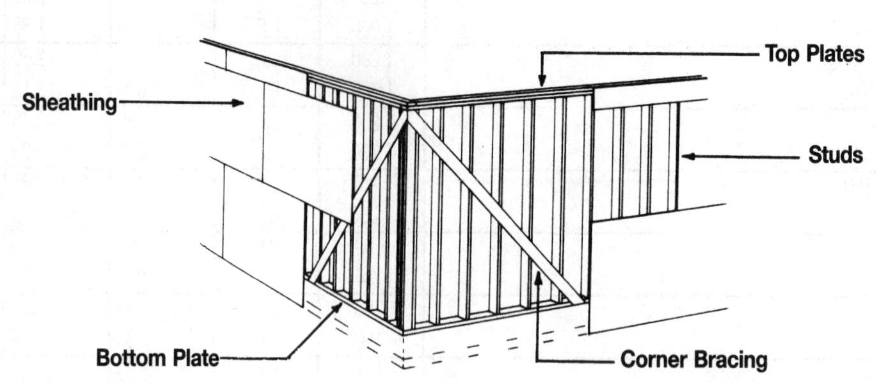

System Description	QUAN.	UNIT	LABOR HOURS	COST PER S.F. MAT.	COST PER S.F. INST.	COST PER S.F. TOTAL
2″ X 4″, 16″ O.C.						
2″ x 4″ studs, 16″ O.C.	1.000	L.F.	.015	.37	.59	.96
Plates, 2″ x 4″, double top, single bottom	.375	L.F.	.005	.14	.22	.36
Corner bracing, let-in, 1″ x 6″	.063	L.F.	.003	.04	.14	.18
Sheathing, 1/2″ plywood, CDX	1.000	S.F.	.011	.68	.47	1.15
TOTAL		S.F.	.034	1.23	1.42	2.65
2″ X 4″, 24″ O.C.						
2″ x 4″ studs, 24″ O.C.	.750	L.F.	.011	.28	.44	.72
Plates, 2″ x 4″, double top, single bottom	.375	L.F.	.005	.14	.22	.36
Corner bracing, let-in, 1″ x 6″	.063	L.F.	.002	.04	.09	.13
Sheathing, 1/2″ plywood, CDX	1.000	S.F.	.011	.68	.47	1.15
TOTAL		S.F.	.029	1.14	1.22	2.36
2″ X 6″, 16″ O.C.						
2″ x 6″ studs, 16″ O.C.	1.000	L.F.	.016	.59	.65	1.24
Plates, 2″ x 6″, double top, single bottom	.375	L.F.	.006	.22	.24	.46
Corner bracing, let-in, 1″ x 6″	.063	L.F.	.003	.04	.14	.18
Sheathing, 1/2″ plywood, CDX	1.000	S.F.	.014	.68	.58	1.26
TOTAL		S.F.	.039	1.53	1.61	3.14
2″ X 6″, 24″ O.C.						
2″ x 6″ studs, 24″ O.C.	.750	L.F.	.012	.44	.49	.93
Plates, 2″ x 6″, double top, single bottom	.375	L.F.	.006	.22	.24	.46
Corner bracing, let-in, 1″ x 6″	.063	L.F.	.002	.04	.09	.13
Sheathing, 1/2″ plywood, CDX	1.000	S.F.	.011	.68	.47	1.15
TOTAL		S.F.	.031	1.38	1.29	2.67

The wall costs on this page are given in cost per square foot of wall.
For window and door openings see below.

Description	QUAN.	UNIT	LABOR HOURS	COST PER S.F. MAT.	COST PER S.F. INST.	COST PER S.F. TOTAL

Exterior Wall Framing Price Sheet	QUAN.	UNIT	LABOR HOURS	COST PER S.F. MAT.	INST.	TOTAL
tuds, #2 or better, 2″ x 4″, 12″ O.C.	1.250	L.F.	.018	.46	.74	1.20
16″ O.C.	1.000	L.F.	.015	.37	.59	.96
24″ O.C.	.750	L.F.	.011	.28	.44	.72
32″ O.C.	.600	L.F.	.009	.22	.35	.57
2″ x 6″, 12″ O.C.	1.250	L.F.	.020	.74	.81	1.55
16″ O.C.	1.000	L.F.	.016	.59	.65	1.24
24″ O.C.	.750	L.F.	.012	.44	.49	.93
32″ O.C.	.600	L.F.	.010	.35	.39	.74
2″ x 8″, 12″ O.C.	1.250	L.F.	.025	1.50	1.03	2.53
16″ O.C.	1.000	L.F.	.020	1.20	.82	2.02
24″ O.C.	.750	L.F.	.015	.90	.62	1.52
32″ O.C.	.600	L.F.	.012	.72	.49	1.21
lates, #2 or better, double top, single bottom, 2″ x 4″	.375	L.F.	.005	.14	.22	.36
2″ x 6″	.375	L.F.	.006	.22	.24	.46
2″ x 8″	.375	L.F.	.008	.45	.31	.76
orner bracing, let-in 1″ x 6″ boards, studs, 12″ O.C.	.070	L.F.	.004	.04	.15	.19
16″ O.C.	.063	L.F.	.003	.04	.14	.18
24″ O.C.	.063	L.F.	.002	.04	.09	.13
32″ O.C.	.057	L.F.	.002	.04	.08	.12
Let-in steel ("T" shape), studs, 12″ O.C.	.070	L.F.	.001	.04	.04	.08
16″ O.C.	.063	L.F.	.001	.03	.04	.07
24″ O.C.	.063	L.F.	.001	.03	.03	.06
32″ O.C.	.057	L.F.	.001	.03	.03	.06
Sheathing, plywood CDX, 3/8″ thick	1.000	S.F.	.010	.72	.43	1.15
1/2″ thick	1.000	S.F.	.011	.68	.47	1.15
5/8″ thick	1.000	S.F.	.012	.86	.50	1.36
3/4″ thick	1.000	S.F.	.013	1.03	.54	1.57
Boards, 1″ x 6″, laid regular	1.000	S.F.	.025	1.38	1	2.38
Laid diagonal	1.000	S.F.	.027	1.38	1.11	2.49
1″ x 8″, laid regular	1.000	S.F.	.021	1.23	.85	2.08
Laid diagonal	1.000	S.F.	.025	1.23	1	2.23
Wood fiber, regular, no vapor barrier, 1/2″ thick	1.000	S.F.	.013	.61	.54	1.15
5/8″ thick	1.000	S.F.	.013	.79	.54	1.33
Asphalt impregnated 25/32″ thick	1.000	S.F.	.013	.31	.54	.85
1/2″ thick	1.000	S.F.	.013	.20	.54	.74
Polystyrene, regular, 3/4″ thick	1.000	S.F.	.010	.42	.41	.83
2″ thick	1.000	S.F.	.011	1.14	.45	1.59
Fiberglass, foil faced, 1″ thick	1.000	S.F.	.008	.88	.33	1.21
2″ thick	1.000	S.F.	.009	1.49	.37	1.86

Window & Door Openings	QUAN.	UNIT	LABOR HOURS	COST EACH MAT.	INST.	TOTAL
The following costs are to be added to the total costs of the wall for each opening. Do not subtract the area of the openings.						
Headers, 2″ x 6″ double, 2′ long	4.000	L.F.	.178	2.36	7.25	9.61
3′ long	6.000	L.F.	.267	3.54	10.85	14.39
4′ long	8.000	L.F.	.356	4.72	14.50	19.22
5′ long	10.000	L.F.	.444	5.90	18.10	24
2″ x 8″ double, 4′ long	8.000	L.F.	.376	7.30	15.35	22.65
5′ long	10.000	L.F.	.471	9.10	19.20	28.30
6′ long	12.000	L.F.	.565	10.90	23	33.90
8′ long	16.000	L.F.	.753	14.55	30.50	45.05
2″ x 10″ double, 4′ long	8.000	L.F.	.400	10.30	16.30	26.60
6′ long	12.000	L.F.	.600	15.50	24.50	40
8′ long	16.000	L.F.	.800	20.50	32.50	53
10′ long	20.000	L.F.	1.000	26	41	67
2″ x 12″ double, 8′ long	16.000	L.F.	.853	28	34.50	62.50
12′ long	24.000	L.F.	1.280	42	52	94

12 Gable End Roof Framing Systems

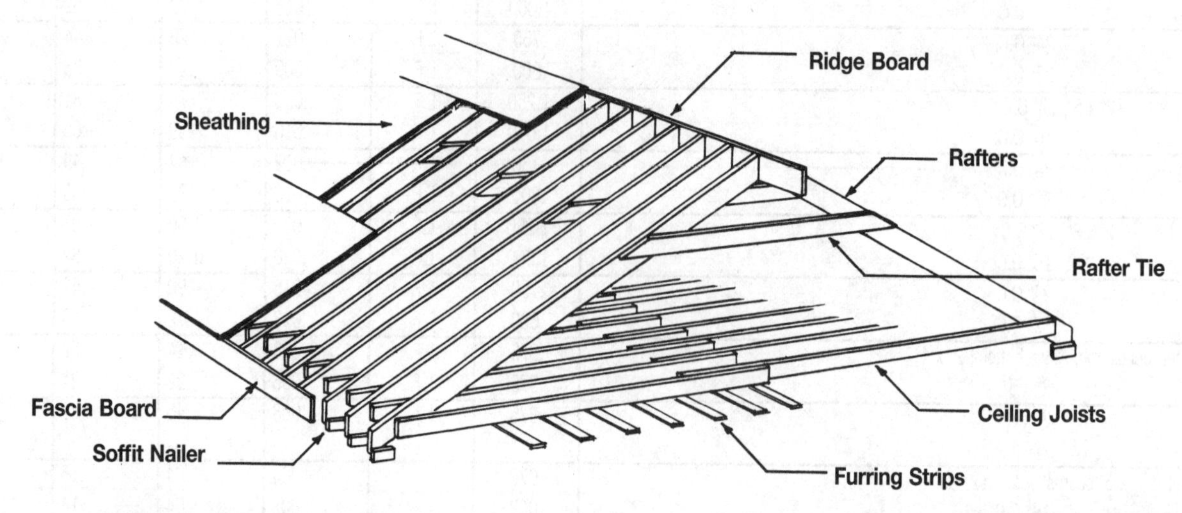

System Description	QUAN.	UNIT	LABOR HOURS	COST PER S.F.		
				MAT.	INST.	TOTAL
2″ X 6″ RAFTERS, 16″ O.C., 4/12 PITCH						
Rafters, 2″ x 6″, 16″ O.C., 4/12 pitch	1.170	L.F.	.019	.69	.76	1.45
Ceiling joists, 2″ x 4″, 16″ O.C.	1.000	L.F.	.013	.37	.52	.89
Ridge board, 2″ x 6″	.050	L.F.	.002	.03	.07	.10
Fascia board, 2″ x 6″	.100	L.F.	.005	.07	.22	.29
Rafter tie, 1″ x 4″, 4′ O.C.	.060	L.F.	.001	.03	.05	.08
Soffit nailer (outrigger), 2″ x 4″, 24″ O.C.	.170	L.F.	.004	.06	.18	.24
Sheathing, exterior, plywood, CDX, 1/2″ thick	1.170	S.F.	.013	.80	.55	1.35
Furring strips, 1″ x 3″, 16″ O.C.	1.000	L.F.	.023	.31	.93	1.24
TOTAL		S.F.	.080	2.36	3.28	5.64
2″ X 8″ RAFTERS, 16″ O.C., 4/12 PITCH						
Rafters, 2″ x 8″, 16″ O.C., 4/12 pitch	1.170	L.F.	.020	1.06	.81	1.87
Ceiling joists, 2″ x 6″, 16″ O.C.	1.000	L.F.	.013	.59	.52	1.11
Ridge board, 2″ x 8″	.050	L.F.	.002	.05	.07	.12
Fascia board, 2″ x 8″	.100	L.F.	.007	.09	.29	.38
Rafter tie, 1″ x 4″, 4′ O.C.	.060	L.F.	.001	.03	.05	.08
Soffit nailer (outrigger), 2″ x 4″, 24″ O.C.	.170	L.F.	.004	.06	.18	.24
Sheathing, exterior, plywood, CDX, 1/2″ thick	1.170	S.F.	.013	.80	.55	1.35
Furring strips, 1″ x 3″, 16″ O.C.	1.000	L.F.	.023	.31	.93	1.24
TOTAL		S.F.	.083	2.99	3.40	6.39

The cost of this system is based on the square foot of plan area.
All quantities have been adjusted accordingly.

Description	QUAN.	UNIT	LABOR HOURS	COST PER S.F.		
				MAT.	INST.	TOTAL

Important: See the Reference Section for critical supporting data - Reference Nos., Crews & Location Factors

Gable End Roof Framing Price Sheet	QUAN.	UNIT	LABOR HOURS	COST PER S.F.		
				MAT.	INST.	TOTAL
Rafters, #2 or better, 16" O.C., 2" x 6", 4/12 pitch	1.170	L.F.	.019	.69	.76	1.45
8/12 pitch	1.330	L.F.	.027	.78	1.09	1.87
2" x 8", 4/12 pitch	1.170	L.F.	.020	1.06	.81	1.87
8/12 pitch	1.330	L.F.	.028	1.21	1.16	2.37
2" x 10", 4/12 pitch	1.170	L.F.	.030	1.51	1.22	2.73
8/12 pitch	1.330	L.F.	.043	1.72	1.76	3.48
24" O.C., 2" x 6", 4/12 pitch	.940	L.F.	.015	.55	.61	1.16
8/12 pitch	1.060	L.F.	.021	.63	.87	1.50
2" x 8", 4/12 pitch	.940	L.F.	.016	.86	.65	1.51
8/12 pitch	1.060	L.F.	.023	.96	.92	1.88
2" x 10", 4/12 pitch	.940	L.F.	.024	1.21	.98	2.19
8/12 pitch	1.060	L.F.	.034	1.37	1.40	2.77
Ceiling joist, #2 or better, 2" x 4", 16" O.C.	1.000	L.F.	.013	.37	.52	.89
24" O.C.	.750	L.F.	.010	.28	.39	.67
2" x 6", 16" O.C.	1.000	L.F.	.013	.59	.52	1.11
24" O.C.	.750	L.F.	.010	.44	.39	.83
2" x 8", 16" O.C.	1.000	L.F.	.015	.91	.59	1.50
24" O.C.	.750	L.F.	.011	.68	.44	1.12
2" x 10", 16" O.C.	1.000	L.F.	.018	1.29	.72	2.01
24" O.C.	.750	L.F.	.013	.97	.54	1.51
Ridge board, #2 or better, 1" x 6"	.050	L.F.	.001	.05	.05	.10
1" x 8"	.050	L.F.	.001	.07	.06	.13
1" x 10"	.050	L.F.	.002	.08	.07	.15
2" x 6"	.050	L.F.	.002	.03	.07	.10
2" x 8"	.050	L.F.	.002	.05	.07	.12
2" x 10"	.050	L.F.	.002	.06	.08	.14
Fascia board, #2 or better, 1" x 6"	.100	L.F.	.004	.05	.16	.21
1" x 8"	.100	L.F.	.005	.06	.19	.25
1" x 10"	.100	L.F.	.005	.07	.21	.28
2" x 6"	.100	L.F.	.006	.07	.23	.30
2" x 8"	.100	L.F.	.007	.09	.29	.38
2" x 10"	.100	L.F.	.004	.26	.14	.40
Rafter tie, #2 or better, 4' O.C., 1" x 4"	.060	L.F.	.001	.03	.05	.08
1" x 6"	.060	L.F.	.001	.03	.05	.08
2" x 4"	.060	L.F.	.002	.04	.07	.11
2" x 6"	.060	L.F.	.002	.05	.08	.13
Soffit nailer (outrigger), 2" x 4", 16" O.C.	.220	L.F.	.006	.08	.23	.31
24" O.C.	.170	L.F.	.004	.06	.18	.24
2" x 6", 16" O.C.	.220	L.F.	.006	.09	.26	.35
24" O.C.	.170	L.F.	.005	.07	.21	.28
Sheathing, plywood CDX, 4/12 pitch, 3/8" thick.	1.170	S.F.	.012	.84	.50	1.34
1/2" thick	1.170	S.F.	.013	.80	.55	1.35
5/8" thick	1.170	S.F.	.014	1.01	.59	1.60
8/12 pitch, 3/8"	1.330	S.F.	.014	.96	.57	1.53
1/2" thick	1.330	S.F.	.015	.90	.63	1.53
5/8" thick	1.330	S.F.	.016	1.14	.67	1.81
Boards, 4/12 pitch roof, 1" x 6"	1.170	S.F.	.026	1.61	1.05	2.66
1" x 8"	1.170	S.F.	.021	1.44	.88	2.32
8/12 pitch roof, 1" x 6"	1.330	S.F.	.029	1.84	1.20	3.04
1" x 8"	1.330	S.F.	.024	1.64	1	2.64
Furring, 1" x 3", 12" O.C.	1.200	L.F.	.027	.37	1.12	1.49
16" O.C.	1.000	L.F.	.023	.31	.93	1.24
24" O.C.	.800	L.F.	.018	.25	.74	.99

3 FRAMING

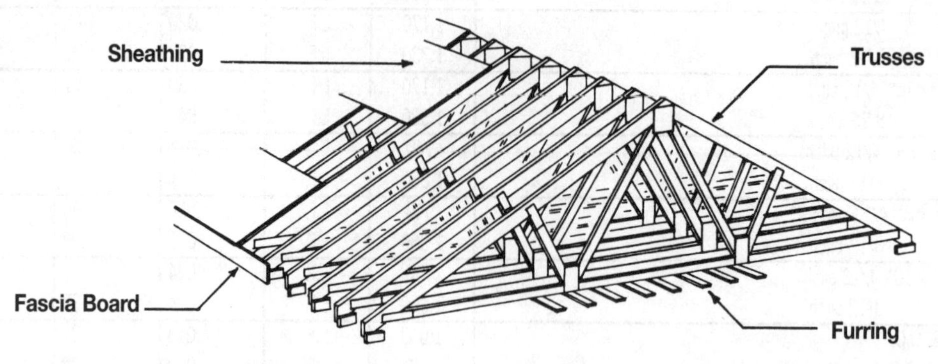

System Description	QUAN.	UNIT	LABOR HOURS	COST PER S.F. MAT.	COST PER S.F. INST.	COST PER S.F. TOTAL
TRUSS, 16" O.C., 4/12 PITCH, 1' OVERHANG, 26' SPAN						
Truss, 40# loading, 16" O.C., 4/12 pitch, 26' span	.030	Ea.	.021	2.34	1.12	3.46
Fascia board, 2" x 6"	.100	L.F.	.005	.07	.22	.29
Sheathing, exterior, plywood, CDX, 1/2" thick	1.170	S.F.	.013	.80	.55	1.35
Furring, 1" x 3", 16" O.C.	1.000	L.F.	.023	.31	.93	1.24
TOTAL		S.F.	.062	3.52	2.82	6.34
TRUSS, 16" O.C., 8/12 PITCH, 1' OVERHANG, 26' SPAN						
Truss, 40# loading, 16" O.C., 8/12 pitch, 26' span	.030	Ea.	.023	2.78	1.22	4
Fascia board, 2" x 6"	.100	L.F.	.005	.07	.22	.29
Sheathing, exterior, plywood, CDX, 1/2" thick	1.330	S.F.	.015	.90	.63	1.53
Furring, 1" x 3", 16" O.C.	1.000	L.F.	.023	.31	.93	1.24
TOTAL		S.F.	.066	4.06	3	7.06
TRUSS, 24" O.C., 4/12 PITCH, 1' OVERHANG, 26' SPAN						
Truss, 40# loading, 24" O.C., 4/12 pitch, 26' span	.020	Ea.	.014	1.56	.74	2.30
Fascia board, 2" x 6"	.100	L.F.	.005	.07	.22	.29
Sheathing, exterior, plywood, CDX, 1/2" thick	1.170	S.F.	.013	.80	.55	1.35
Furring, 1" x 3", 16" O.C.	1.000	L.F.	.023	.31	.93	1.24
TOTAL		S.F.	.055	2.74	2.44	5.18
TRUSS, 24" O.C., 8/12 PITCH, 1' OVERHANG, 26' SPAN						
Truss, 40# loading, 24" O.C., 8/12 pitch, 26' span	.020	Ea.	.015	1.85	.82	2.67
Fascia board, 2" x 6"	.100	L.F.	.005	.07	.22	.29
Sheathing, exterior, plywood, CDX, 1/2" thick	1.330	S.F.	.015	.90	.63	1.53
Furring, 1" x 3", 16" O.C.	1.000	L.F.	.023	.31	.93	1.24
TOTAL		S.F.	.058	3.13	2.60	5.73

The cost of this system is based on the square foot of plan area.
A one foot overhang is included.

Description	QUAN.	UNIT	LABOR HOURS	COST PER S.F. MAT.	COST PER S.F. INST.	COST PER S.F. TOTAL

Important: See the Reference Section for critical supporting data - Reference Nos., Crews & Location Factors

Truss Roof Framing Price Sheet

Truss Roof Framing Price Sheet	QUAN.	UNIT	LABOR HOURS	COST PER S.F.		
				MAT.	INST.	TOTAL
russ, 40# loading, including 1′ overhang, 4/12 pitch, 24′ span, 16″ O.C.	.033	Ea.	.022	1.85	1.15	3
24″ O.C.	.022	Ea.	.015	1.23	.77	2
26′ span, 16″ O.C.	.030	Ea.	.021	2.34	1.12	3.46
24″ O.C.	.020	Ea.	.014	1.56	.74	2.30
28′ span, 16″ O.C.	.027	Ea.	.020	1.84	1.08	2.92
24″ O.C.	.019	Ea.	.014	1.29	.76	2.05
32′ span, 16″ O.C.	.024	Ea.	.019	2.29	1.02	3.31
24″ O.C.	.016	Ea.	.013	1.53	.67	2.20
36′ span, 16″ O.C.	.022	Ea.	.019	2.62	1.01	3.63
24″ O.C.	.015	Ea.	.013	1.79	.69	2.48
8/12 pitch, 24′ span, 16″ O.C.	.033	Ea.	.024	2.82	1.27	4.09
24″ O.C.	.022	Ea.	.016	1.88	.85	2.73
26′ span, 16″ O.C.	.030	Ea.	.023	2.78	1.22	4
24″ O.C.	.020	Ea.	.015	1.85	.82	2.67
28′ span, 16″ O.C.	.027	Ea.	.022	2.69	1.17	3.86
24″ O.C.	.019	Ea.	.016	1.89	.82	2.71
32′ span, 16″ O.C.	.024	Ea.	.021	2.83	1.12	3.95
24″ O.C.	.016	Ea.	.014	1.89	.75	2.64
36′ span, 16″ O.C.	.022	Ea.	.021	3.06	1.14	4.20
24″ O.C.	.015	Ea.	.015	2.09	.77	2.86
ascia board, #2 or better, 1″ x 6″	.100	L.F.	.004	.05	.16	.21
1″ x 8″	.100	L.F.	.005	.06	.19	.25
1″ x 10″	.100	L.F.	.005	.07	.21	.28
2″ x 6″	.100	L.F.	.006	.07	.23	.30
2″ x 8″	.100	L.F.	.007	.09	.29	.38
2″ x 10″	.100	L.F.	.009	.13	.36	.49
Sheathing, plywood CDX, 4/12 pitch, 3/8″ thick	1.170	S.F.	.012	.84	.50	1.34
1/2″ thick	1.170	S.F.	.013	.80	.55	1.35
5/8″ thick	1.170	S.F.	.014	1.01	.59	1.60
8/12 pitch, 3/8″ thick	1.330	S.F.	.014	.96	.57	1.53
1/2″ thick	1.330	S.F.	.015	.90	.63	1.53
5/8″ thick	1.330	S.F.	.016	1.14	.67	1.81
Boards, 4/12 pitch, 1″ x 6″	1.170	S.F.	.026	1.61	1.05	2.66
1″ x 8″	1.170	S.F.	.021	1.44	.88	2.32
8/12 pitch, 1″ x 6″	1.330	S.F.	.029	1.84	1.20	3.04
1″ x 8″	1.330	S.F.	.024	1.64	1	2.64
Furring, 1″ x 3″, 12″ O.C.	1.200	L.F.	.027	.37	1.12	1.49
16″ O.C.	1.000	L.F.	.023	.31	.93	1.24
24″ O.C.	.800	L.F.	.018	.25	.74	.99

20 Hip Roof Framing Systems

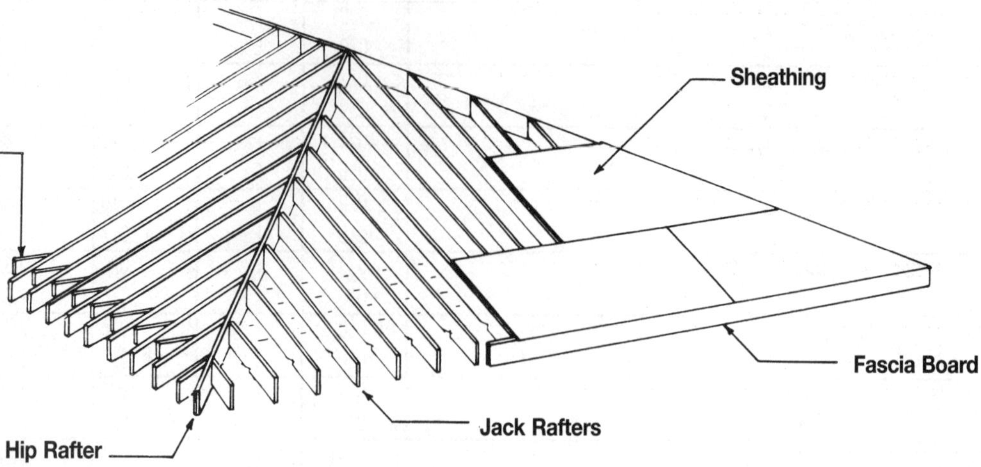

Sheathing

Ceiling Joists

Fascia Board

Jack Rafters

Hip Rafter

System Description	QUAN.	UNIT	LABOR HOURS	COST PER S.F. MAT.	COST PER S.F. INST.	COST PER S.F. TOTAL
2″ X 6″, 16″ O.C., 4/12 PITCH						
Hip rafters, 2″ x 8″, 4/12 pitch	.160	L.F.	.004	.15	.15	.30
Jack rafters, 2″ x 6″, 16″ O.C., 4/12 pitch	1.430	L.F.	.038	.84	1.56	2.40
Ceiling joists, 2″ x 6″, 16″ O.C.	1.000	L.F.	.013	.59	.52	1.11
Fascia board, 2″ x 8″	.220	L.F.	.016	.20	.64	.84
Soffit nailer (outrigger), 2″ x 4″, 24″ O.C.	.220	L.F.	.006	.08	.23	.31
Sheathing, 1/2″ exterior plywood, CDX	1.570	S.F.	.018	1.07	.74	1.81
Furring strips, 1″ x 3″, 16″ O.C.	1.000	L.F.	.023	.31	.93	1.24
TOTAL		S.F.	.118	3.24	4.77	8.01
2″ X 8″, 16″ O.C., 4/12 PITCH						
Hip rafters, 2″ x 10″, 4/12 pitch	.160	L.F.	.004	.21	.18	.39
Jack rafters, 2″ x 8″, 16″ O.C., 4/12 pitch	1.430	L.F.	.047	1.30	1.90	3.20
Ceiling joists, 2″ x 6″, 16″ O.C.	1.000	L.F.	.013	.59	.52	1.11
Fascia board, 2″ x 8″	.220	L.F.	.012	.15	.49	.64
Soffit nailer (outrigger), 2″ x 4″, 24″ O.C.	.220	L.F.	.006	.08	.23	.31
Sheathing, 1/2″ exterior plywood, CDX	1.570	S.F.	.018	1.07	.74	1.81
Furring strips, 1″ x 3″, 16″ O.C.	1.000	L.F.	.023	.31	.93	1.24
TOTAL		S.F.	.123	3.71	4.99	8.70

The cost of this system is based on S.F. of plan area. Measurement is area under the hip roof only. See gable roof system for added costs.

Description	QUAN.	UNIT	LABOR HOURS	COST PER S.F. MAT.	COST PER S.F. INST.	COST PER S.F. TOTAL

Important: See the Reference Section for critical supporting data - Reference Nos., Crews & Location Factors

Hip Roof Framing Price Sheet	QUAN.	UNIT	LABOR HOURS	COST PER S.F.		
				MAT.	INST.	TOTAL
Hip rafters, #2 or better, 2" x 6", 4/12 pitch	.160	L.F.	.003	.09	.14	.23
8/12 pitch	.210	L.F.	.006	.12	.23	.35
2" x 8", 4/12 pitch	.160	L.F.	.004	.15	.15	.30
8/12 pitch	.210	L.F.	.006	.19	.25	.44
2" x 10", 4/12 pitch	.160	L.F.	.004	.21	.18	.39
8/12 pitch roof	.210	L.F.	.008	.27	.31	.58
Jack rafters, #2 or better, 16" O.C., 2" x 6", 4/12 pitch	1.430	L.F.	.038	.84	1.56	2.40
8/12 pitch	1.800	L.F.	.061	1.06	2.47	3.53
2" x 8", 4/12 pitch	1.430	L.F.	.047	1.30	1.90	3.20
8/12 pitch	1.800	L.F.	.075	1.64	3.04	4.68
2" x 10", 4/12 pitch	1.430	L.F.	.051	1.84	2.07	3.91
8/12 pitch	1.800	L.F.	.082	2.32	3.35	5.67
24" O.C., 2" x 6", 4/12 pitch	1.150	L.F.	.031	.68	1.25	1.93
8/12 pitch	1.440	L.F.	.048	.85	1.97	2.82
2" x 8", 4/12 pitch	1.150	L.F.	.038	1.05	1.53	2.58
8/12 pitch	1.440	L.F.	.060	1.31	2.43	3.74
2" x 10", 4/12 pitch	1.150	L.F.	.041	1.48	1.67	3.15
8/12 pitch	1.440	L.F.	.066	1.86	2.68	4.54
Ceiling joists, #2 or better, 2" x 4", 16" O.C.	1.000	L.F.	.013	.37	.52	.89
24" O.C.	.750	L.F.	.010	.28	.39	.67
2" x 6", 16" O.C.	1.000	L.F.	.013	.59	.52	1.11
24" O.C.	.750	L.F.	.010	.44	.39	.83
2" x 8", 16" O.C.	1.000	L.F.	.015	.91	.59	1.50
24" O.C.	.750	L.F.	.011	.68	.44	1.12
2" x 10", 16" O.C.	1.000	L.F.	.018	1.29	.72	2.01
24" O.C.	.750	L.F.	.013	.97	.54	1.51
Fascia board, #2 or better, 1" x 6"	.220	L.F.	.009	.11	.35	.46
1" x 8"	.220	L.F.	.010	.13	.41	.54
1" x 10"	.220	L.F.	.011	.14	.46	.60
2" x 6"	.220	L.F.	.013	.16	.51	.67
2" x 8"	.220	L.F.	.016	.20	.64	.84
2" x 10"	.220	L.F.	.020	.28	.80	1.08
Soffit nailer (outrigger), 2" x 4", 16" O.C.	.280	L.F.	.007	.10	.29	.39
24" O.C.	.220	L.F.	.006	.08	.23	.31
2" x 8", 16" O.C.	.280	L.F.	.007	.19	.27	.46
24" O.C.	.220	L.F.	.005	.15	.22	.37
Sheathing, plywood CDX, 4/12 pitch, 3/8" thick	1.570	S.F.	.016	1.13	.68	1.81
1/2" thick	1.570	S.F.	.018	1.07	.74	1.81
5/8" thick	1.570	S.F.	.019	1.35	.79	2.14
8/12 pitch, 3/8" thick	1.900	S.F.	.020	1.37	.82	2.19
1/2" thick	1.900	S.F.	.022	1.29	.89	2.18
5/8" thick	1.900	S.F.	.023	1.63	.95	2.58
Boards, 4/12 pitch, 1" x 6" boards	1.450	S.F.	.032	2	1.31	3.31
1" x 8" boards	1.450	S.F.	.027	1.78	1.09	2.87
8/12 pitch, 1" x 6" boards	1.750	S.F.	.039	2.42	1.58	4
1" x 8" boards	1.750	S.F.	.032	2.15	1.31	3.46
Furring, 1" x 3", 12" O.C.	1.200	L.F.	.027	.37	1.12	1.49
16" O.C.	1.000	L.F.	.023	.31	.93	1.24
24" O.C.	.800	L.F.	.018	.25	.74	.99

24 Gambrel Roof Framing Systems

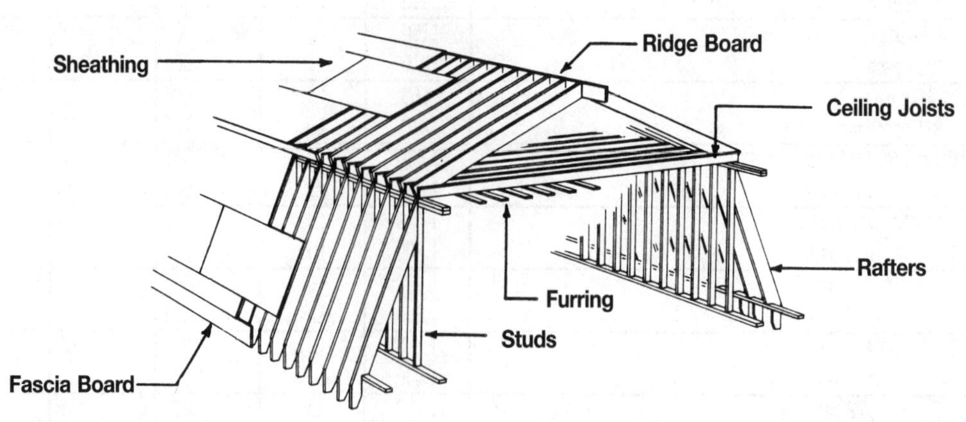

System Description	QUAN.	UNIT	LABOR HOURS	COST PER S.F. MAT.	COST PER S.F. INST.	COST PER S.F. TOTAL
2" X 6" RAFTERS, 16" O.C.						
Roof rafters, 2" x 6", 16" O.C.	1.430	L.F.	.029	.84	1.17	2.01
Ceiling joists, 2" x 6", 16" O.C.	.710	L.F.	.009	.42	.37	.79
Stud wall, 2" x 4", 16" O.C., including plates	.790	L.F.	.012	.29	.51	.80
Furring strips, 1" x 3", 16" O.C.	.710	L.F.	.016	.22	.66	.88
Ridge board, 2" x 8"	.050	L.F.	.002	.05	.07	.12
Fascia board, 2" x 6"	.100	L.F.	.006	.07	.23	.30
Sheathing, exterior grade plywood, 1/2" thick	1.450	S.F.	.017	.99	.68	1.67
TOTAL		S.F.	.091	2.88	3.69	6.57
2" X 8" RAFTERS, 16" O.C.						
Roof rafters, 2" x 8", 16" O.C.	1.430	L.F.	.031	1.30	1.24	2.54
Ceiling joists, 2" x 6", 16" O.C.	.710	L.F.	.009	.42	.37	.79
Stud wall, 2" x 4", 16" O.C., including plates	.790	L.F.	.012	.29	.51	.80
Furring strips, 1" x 3", 16" O.C.	.710	L.F.	.016	.22	.66	.88
Ridge board, 2" x 8"	.050	L.F.	.002	.05	.07	.12
Fascia board, 2" x 8"	.100	L.F.	.007	.09	.29	.38
Sheathing, exterior grade plywood, 1/2" thick	1.450	S.F.	.017	.99	.68	1.67
TOTAL		S.F.	.094	3.36	3.82	7.18

The cost of this system is based on the square foot of plan area on the first floor.

Description	QUAN.	UNIT	LABOR HOURS	COST PER S.F. MAT.	COST PER S.F. INST.	COST PER S.F. TOTAL

Important: See the Reference Section for critical supporting data - Reference Nos., Crews & Location Factors

Gambrel Roof Framing Price Sheet	QUAN.	UNIT	LABOR HOURS	COST PER S.F.		
				MAT.	INST.	TOTAL
Roof rafters, #2 or better, 2" x 6", 16" O.C.	1.430	L.F.	.029	.84	1.17	2.01
24" O.C.	1.140	L.F.	.023	.67	.93	1.60
2" x 8", 16" O.C.	1.430	L.F.	.031	1.30	1.24	2.54
24" O.C.	1.140	L.F.	.024	1.04	.99	2.03
2" x 10", 16" O.C.	1.430	L.F.	.046	1.84	1.89	3.73
24" O.C.	1.140	L.F.	.037	1.47	1.50	2.97
Ceiling joist, #2 or better, 2" x 4", 16" O.C.	.710	L.F.	.009	.26	.37	.63
24" O.C.	.570	L.F.	.007	.21	.30	.51
2" x 6", 16" O.C.	.710	L.F.	.009	.42	.37	.79
24" O.C.	.570	L.F.	.007	.34	.30	.64
2" x 8", 16" O.C.	.710	L.F.	.010	.65	.42	1.07
24" O.C.	.570	L.F.	.008	.52	.34	.86
Stud wall, #2 or better, 2" x 4", 16" O.C.	.790	L.F.	.012	.29	.51	.80
24" O.C.	.630	L.F.	.010	.23	.40	.63
2" x 6", 16" O.C.	.790	L.F.	.014	.47	.58	1.05
24" O.C.	.630	L.F.	.011	.37	.46	.83
Furring, 1" x 3", 16" O.C.	.710	L.F.	.016	.22	.66	.88
24" O.C.	.590	L.F.	.013	.18	.55	.73
Ridge board, #2 or better, 1" x 6"	.050	L.F.	.001	.05	.05	.10
1" x 8"	.050	L.F.	.001	.07	.06	.13
1" x 10"	.050	L.F.	.002	.08	.07	.15
2" x 6"	.050	L.F.	.002	.03	.07	.10
2" x 8"	.050	L.F.	.002	.05	.07	.12
2" x 10"	.050	L.F.	.002	.06	.08	.14
Fascia board, #2 or better, 1" x 6"	.100	L.F.	.004	.05	.16	.21
1" x 8"	.100	L.F.	.005	.06	.19	.25
1" x 10"	.100	L.F.	.005	.07	.21	.28
2" x 6"	.100	L.F.	.006	.07	.23	.30
2" x 8"	.100	L.F.	.007	.09	.29	.38
2" x 10"	.100	L.F.	.009	.13	.36	.49
Sheathing, plywood, exterior grade CDX, 3/8" thick	1.450	S.F.	.015	1.04	.62	1.66
1/2" thick	1.450	S.F.	.017	.99	.68	1.67
5/8" thick	1.450	S.F.	.018	1.25	.73	1.98
3/4" thick	1.450	S.F.	.019	1.49	.78	2.27
Boards, 1" x 6", laid regular	1.450	S.F.	.032	2	1.31	3.31
Laid diagonal	1.450	S.F.	.036	2	1.45	3.45
1" x 8", laid regular	1.450	S.F.	.027	1.78	1.09	2.87
Laid diagonal	1.450	S.F.	.032	1.78	1.31	3.09

Hip Rafters
Sheathing
Ridge Board
Rafters
Top Plates
Furring
Ceiling Joists
Rafters
Bottom Plate

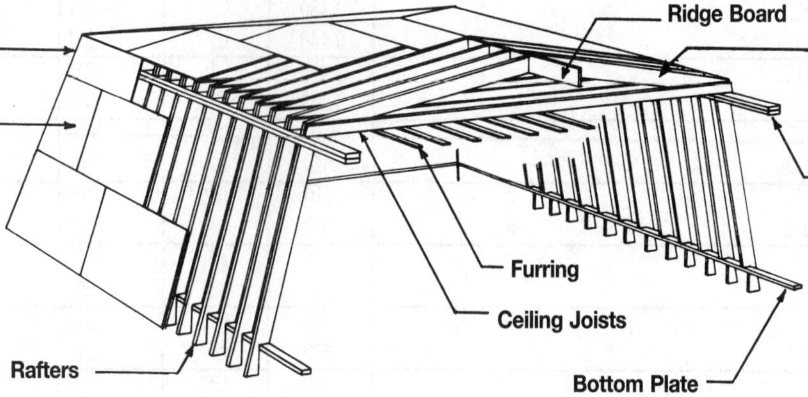

System Description	QUAN.	UNIT	LABOR HOURS	COST PER S.F.		
				MAT.	INST.	TOTAL
2" X 6" RAFTERS, 16" O.C.						
Roof rafters, 2" x 6", 16" O.C.	1.210	L.F.	.033	.71	1.34	2.05
Rafter plates, 2" x 6", double top, single bottom	.364	L.F.	.010	.21	.40	.61
Ceiling joists, 2" x 4", 16" O.C.	.920	L.F.	.012	.34	.48	.82
Hip rafter, 2" x 6"	.070	L.F.	.002	.04	.09	.13
Jack rafter, 2" x 6", 16" O.C.	1.000	L.F.	.039	.59	1.59	2.18
Ridge board, 2" x 6"	.018	L.F.	.001	.01	.02	.03
Sheathing, exterior grade plywood, 1/2" thick	2.210	S.F.	.025	1.50	1.04	2.54
Furring strips, 1" x 3", 16" O.C.	.920	L.F.	.021	.29	.86	1.15
TOTAL		S.F.	.143	3.69	5.82	9.51
2" X 8" RAFTERS, 16" O.C.						
Roof rafters, 2" x 8", 16" O.C.	1.210	L.F.	.036	1.10	1.46	2.56
Rafter plates, 2" x 8", double top, single bottom	.364	L.F.	.011	.33	.44	.77
Ceiling joists, 2" x 6", 16" O.C.	.920	L.F.	.012	.54	.48	1.02
Hip rafter, 2" x 8"	.070	L.F.	.002	.06	.10	.16
Jack rafter, 2" x 8", 16" O.C.	1.000	L.F.	.048	.91	1.95	2.86
Ridge board, 2" x 8"	.018	L.F.	.001	.02	.03	.05
Sheathing, exterior grade plywood, 1/2" thick	2.210	S.F.	.025	1.50	1.04	2.54
Furring strips, 1" x 3", 16" O.C.	.920	L.F.	.021	.29	.86	1.15
TOTAL		S.F.	.156	4.75	6.36	11.11

The cost of this system is based on the square foot of plan area.

Description	QUAN.	UNIT	LABOR HOURS	COST PER S.F.		
				MAT.	INST.	TOTAL

Important: See the Reference Section for critical supporting data - Reference Nos., Crews & Location Factors

Mansard Roof Framing Price Sheet	QUAN.	UNIT	LABOR HOURS	COST PER S.F.		
				MAT.	INST.	TOTAL
oof rafters, #2 or better, 2″ x 6″, 16″ O.C.	1.210	L.F.	.033	.71	1.34	2.05
24″ O.C.	.970	L.F.	.026	.57	1.08	1.65
2″ x 8″, 16″ O.C.	1.210	L.F.	.036	1.10	1.46	2.56
24″ O.C.	.970	L.F.	.029	.88	1.17	2.05
2″ x 10″, 16″ O.C.	1.210	L.F.	.046	1.56	1.85	3.41
24″ O.C.	.970	L.F.	.037	1.25	1.48	2.73
after plates, #2 or better double top single bottom, 2″ x 6″	.364	L.F.	.010	.21	.40	.61
2″ x 8″	.364	L.F.	.011	.33	.44	.77
2″ x 10″	.364	L.F.	.014	.47	.56	1.03
eiling joist, #2 or better, 2″ x 4″, 16″ O.C.	.920	L.F.	.012	.34	.48	.82
24″ O.C.	.740	L.F.	.009	.27	.38	.65
2″ x 6″, 16″ O.C.	.920	L.F.	.012	.54	.48	1.02
24″ O.C.	.740	L.F.	.009	.44	.38	.82
2″ x 8″, 16″ O.C.	.920	L.F.	.013	.84	.54	1.38
24″ O.C.	.740	L.F.	.011	.67	.44	1.11
lip rafter, #2 or better, 2″ x 6″	.070	L.F.	.002	.04	.09	.13
2″ x 8″	.070	L.F.	.002	.06	.10	.16
2″ x 10″	.070	L.F.	.003	.09	.12	.21
ack rafter, #2 or better, 2″ x 6″, 16″ O.C.	1.000	L.F.	.039	.59	1.59	2.18
24″ O.C.	.800	L.F.	.031	.47	1.27	1.74
2″ x 8″, 16″ O.C.	1.000	L.F.	.048	.91	1.95	2.86
24″ O.C.	.800	L.F.	.038	.73	1.56	2.29
idge board, #2 or better, 1″ x 6″	.018	L.F.	.001	.02	.02	.04
1″ x 8″	.018	L.F.	.001	.02	.02	.04
1″ x 10″	.018	L.F.	.001	.03	.02	.05
2″ x 6″	.018	L.F.	.001	.01	.02	.03
2″ x 8″	.018	L.F.	.001	.02	.03	.05
2″ x 10″	.018	L.F.	.001	.02	.03	.05
heathing, plywood exterior grade CDX, 3/8″ thick	2.210	S.F.	.023	1.59	.95	2.54
1/2″ thick	2.210	S.F.	.025	1.50	1.04	2.54
5/8″ thick	2.210	S.F.	.027	1.90	1.11	3.01
3/4″ thick	2.210	S.F.	.029	2.28	1.19	3.47
Boards, 1″ x 6″, laid regular	2.210	S.F.	.049	3.05	1.99	5.04
Laid diagonal	2.210	S.F.	.054	3.05	2.21	5.26
1″ x 8″, laid regular	2.210	S.F.	.040	2.72	1.66	4.38
Laid diagonal	2.210	S.F.	.049	2.72	1.99	4.71
urring, 1″ x 3″, 12″ O.C.	1.150	L.F.	.026	.36	1.07	1.43
24″ O.C.	.740	L.F.	.017	.23	.69	.92

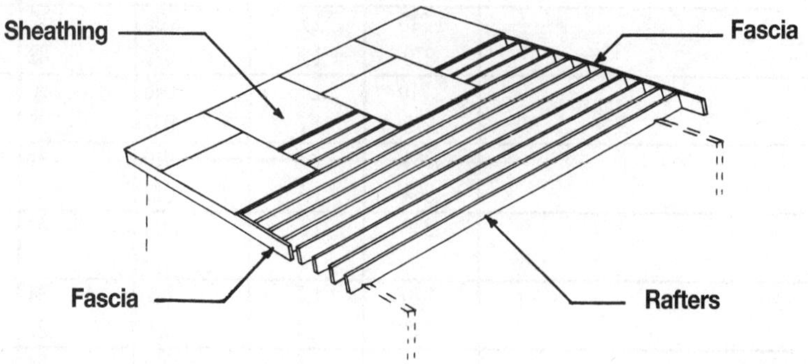

32 Shed/Flat Roof Framing Systems

Sheathing
Fascia
Fascia
Rafters

System Description	QUAN.	UNIT	LABOR HOURS	COST PER S.F.		
				MAT.	INST.	TOTAL
2″ X 6″,16″ O.C., 4/12 PITCH						
Rafters, 2″ x 6″, 16″ O.C., 4/12 pitch	1.170	L.F.	.019	.69	.76	1.45
Fascia, 2″ x 6″	.100	L.F.	.006	.07	.23	.30
Bridging, 1″ x 3″, 6′ O.C.	.080	Pr.	.005	.04	.20	.24
Sheathing, exterior grade plywood, 1/2″ thick	1.230	S.F.	.014	.84	.58	1.42
TOTAL		S.F.	.044	1.64	1.77	3.41
2″ X 6″, 24″ O.C., 4/12 PITCH						
Rafters, 2″ x 6″, 24″ O.C., 4/12 pitch	.940	L.F.	.015	.55	.61	1.16
Fascia, 2″ x 6″	.100	L.F.	.006	.07	.23	.30
Bridging, 1″ x 3″, 6′ O.C.	.060	Pr.	.004	.03	.15	.18
Sheathing, exterior grade plywood, 1/2″ thick	1.230	S.F.	.014	.84	.58	1.42
TOTAL		S.F.	.039	1.49	1.57	3.06
2″ X 8″, 16″ O.C., 4/12 PITCH						
Rafters, 2″ x 8″, 16″ O.C., 4/12 pitch	1.170	L.F.	.020	1.06	.81	1.87
Fascia, 2″ x 8″	.100	L.F.	.007	.09	.29	.38
Bridging, 1″ x 3″, 6′ O.C.	.080	Pr.	.005	.04	.20	.24
Sheathing, exterior grade plywood, 1/2″ thick	1.230	S.F.	.014	.84	.58	1.42
TOTAL		S.F.	.046	2.03	1.88	3.91
2″ X 8″, 24″ O.C., 4/12 PITCH						
Rafters, 2″ x 8″, 24″ O.C., 4/12 pitch	.940	L.F.	.016	.86	.65	1.51
Fascia, 2″ x 8″	.100	L.F.	.007	.09	.29	.38
Bridging, 1″ x 3″, 6′ O.C.	.060	Pr.	.004	.03	.15	.18
Sheathing, exterior grade plywood, 1/2″ thick	1.230	S.F.	.014	.84	.58	1.42
TOTAL		S.F.	.041	1.82	1.67	3.49

The cost of this system is based on the square foot of plan area.
A 1′ overhang is assumed. No ceiling joists or furring are included.

Description	QUAN.	UNIT	LABOR HOURS	COST PER S.F.		
				MAT.	INST.	TOTAL

Important: See the Reference Section for critical supporting data - Reference Nos., Crews & Location Factors

Shed/Flat Roof Framing Price Sheet

	QUAN.	UNIT	LABOR HOURS	COST PER S.F.		
				MAT.	INST.	TOTAL
afters, #2 or better, 16″ O.C., 2″ x 4″, 0 - 4/12 pitch	1.170	L.F.	.014	.51	.57	1.08
5/12 - 8/12 pitch	1.330	L.F.	.020	.59	.82	1.41
2″ x 6″, 0 - 4/12 pitch	1.170	L.F.	.019	.69	.76	1.45
5/12 - 8/12 pitch	1.330	L.F.	.027	.78	1.09	1.87
2″ x 8″, 0 - 4/12 pitch	1.170	L.F.	.020	1.06	.81	1.87
5/12 - 8/12 pitch	1.330	L.F.	.028	1.21	1.16	2.37
2″ x 10″, 0 - 4/12 pitch	1.170	L.F.	.030	1.51	1.22	2.73
5/12 - 8/12 pitch	1.330	L.F.	.043	1.72	1.76	3.48
24″ O.C., 2″ x 4″, 0 - 4/12 pitch	.940	L.F.	.011	.42	.46	.88
5/12 - 8/12 pitch	1.060	L.F.	.021	.63	.87	1.50
2″ x 6″, 0 - 4/12 pitch	.940	L.F.	.015	.55	.61	1.16
5/12 - 8/12 pitch	1.060	L.F.	.021	.63	.87	1.50
2″ x 8″, 0 - 4/12 pitch	.940	L.F.	.016	.86	.65	1.51
5/12 - 8/12 pitch	1.060	L.F.	.023	.96	.92	1.88
2″ x 10″, 0 - 4/12 pitch	.940	L.F.	.024	1.21	.98	2.19
5/12 - 8/12 pitch	1.060	L.F.	.034	1.37	1.40	2.77
ascia, #2 or better,, 1″ x 4″	.100	L.F.	.003	.04	.12	.16
1″ x 6″	.100	L.F.	.004	.05	.16	.21
1″ x 8″	.100	L.F.	.005	.06	.19	.25
1″ x 10″	.100	L.F.	.005	.07	.21	.28
2″ x 4″	.100	L.F.	.005	.06	.20	.26
2″ x 6″	.100	L.F.	.006	.07	.23	.30
2″ x 8″	.100	L.F.	.007	.09	.29	.38
2″ x 10″	.100	L.F.	.009	.13	.36	.49
ridging, wood 6′ O.C., 1″ x 3″, rafters, 16″ O.C.	.080	Pr.	.005	.04	.20	.24
24″ O.C.	.060	Pr.	.004	.03	.15	.18
Metal, galvanized, rafters, 16″ O.C.	.080	Pr.	.005	.09	.20	.29
24″ O.C.	.060	Pr.	.003	.08	.14	.22
Compression type, rafters, 16″ O.C.	.080	Pr.	.003	.10	.13	.23
24″ O.C.	.060	Pr.	.002	.08	.10	.18
heathing, plywood, exterior grade, 3/8″ thick, flat 0 - 4/12 pitch	1.230	S.F.	.013	.89	.53	1.42
5/12 - 8/12 pitch	1.330	S.F.	.014	.96	.57	1.53
1/2″ thick, flat 0 - 4/12 pitch	1.230	S.F.	.014	.84	.58	1.42
5/12 - 8/12 pitch	1.330	S.F.	.015	.90	.63	1.53
5/8″ thick, flat 0 - 4/12 pitch	1.230	S.F.	.015	1.06	.62	1.68
5/12 - 8/12 pitch	1.330	S.F.	.016	1.14	.67	1.81
3/4″ thick, flat 0 - 4/12 pitch	1.230	S.F.	.016	1.27	.66	1.93
5/12 - 8/12 pitch	1.330	S.F.	.018	1.37	.72	2.09
Boards, 1″ x 6″, laid regular, flat 0 - 4/12 pitch	1.230	S.F.	.027	1.70	1.11	2.81
5/12 - 8/12 pitch	1.330	S.F.	.041	1.84	1.66	3.50
Laid diagonal, flat 0 - 4/12 pitch	1.230	S.F.	.030	1.70	1.23	2.93
5/12 - 8/12 pitch	1.330	S.F.	.044	1.84	1.81	3.65
1″ x 8″, laid regular, flat 0 - 4/12 pitch	1.230	S.F.	.022	1.51	.92	2.43
5/12 - 8/12 pitch	1.330	S.F.	.034	1.64	1.37	3.01
Laid diagonal, flat 0 - 4/12 pitch	1.230	S.F.	.027	1.51	1.11	2.62
5/12 - 8/12 pitch	1.330	S.F.	.044	1.84	1.81	3.65

40 Gable Dormer Framing Systems

Valley Rafter
Ridge Board
Sheathing
Rafters
Fascia Board
Studs & Plates
Headers
Trimmer Rafters

System Description	QUAN.	UNIT	LABOR HOURS	COST PER S.F. MAT.	COST PER S.F. INST.	COST PER S.F. TOTAL
2″ X 6″, 16″ O.C.						
Dormer rafter, 2″ x 6″, 16″ O.C.	1.330	L.F.	.036	.78	1.48	2.26
Ridge board, 2″ x 6″	.280	L.F.	.009	.17	.36	.53
Trimmer rafters, 2″ x 6″	.880	L.F.	.014	.52	.57	1.09
Wall studs & plates, 2″ x 4″, 16″ O.C.	3.160	L.F.	.056	1.17	2.28	3.45
Fascia, 2″ x 6″	.220	L.F.	.012	.15	.49	.64
Valley rafter, 2″ x 6″, 16″ O.C.	.280	L.F.	.009	.17	.36	.53
Cripple rafter, 2″ x 6″, 16″ O.C.	.560	L.F.	.022	.33	.89	1.22
Headers, 2″ x 6″, doubled	.670	L.F.	.030	.40	1.21	1.61
Ceiling joist, 2″ x 4″, 16″ O.C.	1.000	L.F.	.013	.37	.52	.89
Sheathing, exterior grade plywood, 1/2″ thick	3.610	S.F.	.041	2.45	1.70	4.15
TOTAL		S.F.	.242	6.51	9.86	16.37
2″ X 8″, 16″ O.C.						
Dormer rafter, 2″ x 8″, 16″ O.C.	1.330	L.F.	.039	1.21	1.61	2.82
Ridge board, 2″ x 8″	.280	L.F.	.010	.25	.41	.66
Trimmer rafter, 2″ x 8″	.880	L.F.	.015	.80	.61	1.41
Wall studs & plates, 2″ x 4″, 16″ O.C.	3.160	L.F.	.056	1.17	2.28	3.45
Fascia, 2″ x 8″	.220	L.F.	.016	.20	.64	.84
Valley rafter, 2″ x 8″, 16″ O.C.	.280	L.F.	.010	.25	.39	.64
Cripple rafter, 2″ x 8″, 16″ O.C.	.560	L.F.	.027	.51	1.09	1.60
Headers, 2″ x 8″, doubled	.670	L.F.	.032	.61	1.29	1.90
Ceiling joist, 2″ x 4″, 16″ O.C.	1.000	L.F.	.013	.37	.52	.89
Sheathing,, exterior grade plywood, 1/2″ thick	3.610	S.F.	.041	2.45	1.70	4.15
TOTAL		S.F.	.259	7.82	10.54	18.36

The cost in this system is based on the square foot of plan area.
The measurement being the plan area of the dormer only.

Description	QUAN.	UNIT	LABOR HOURS	COST PER S.F. MAT.	COST PER S.F. INST.	COST PER S.F. TOTAL

Gable Dormer Framing Price Sheet

Gable Dormer Framing Price Sheet	QUAN.	UNIT	LABOR HOURS	COST PER S.F.		
				MAT.	INST.	TOTAL
ormer rafters, #2 or better, 2″ x 4″, 16″ O.C.	1.330	L.F.	.029	.63	1.18	1.81
24″ O.C.	1.060	L.F.	.023	.50	.94	1.44
2″ x 6″, 16″ O.C.	1.330	L.F.	.036	.78	1.48	2.26
24″ O.C.	1.060	L.F.	.029	.63	1.18	1.81
2″ x 8″, 16″ O.C.	1.330	L.F.	.039	1.21	1.61	2.82
24″ O.C.	1.060	L.F.	.031	.96	1.28	2.24
idge board, #2 or better, 1″ x 4″	.280	L.F.	.006	.22	.24	.46
1″ x 6″	.280	L.F.	.007	.28	.31	.59
1″ x 8″	.280	L.F.	.008	.38	.33	.71
2″ x 4″	.280	L.F.	.007	.13	.29	.42
2″ x 6″	.280	L.F.	.009	.17	.36	.53
2″ x 8″	.280	L.F.	.010	.25	.41	.66
immer rafters, #2 or better, 2″ x 4″	.880	L.F.	.011	.42	.46	.88
2″ x 6″	.880	L.F.	.014	.52	.57	1.09
2″ x 8″	.880	L.F.	.015	.80	.61	1.41
2″ x 10″	.880	L.F.	.022	1.14	.92	2.06
/all studs & plates, #2 or better, 2″ x 4″ studs, 16″ O.C.	3.160	L.F.	.056	1.17	2.28	3.45
24″ O.C.	2.800	L.F.	.050	1.04	2.02	3.06
2″ x 6″ studs, 16″ O.C.	3.160	L.F.	.063	1.86	2.59	4.45
24″ O.C.	2.800	L.F.	.056	1.65	2.30	3.95
ascia, #2 or better, 1″ x 4″	.220	L.F.	.006	.08	.26	.34
1″ x 6″	.220	L.F.	.008	.10	.32	.42
1″ x 8″	.220	L.F.	.009	.12	.37	.49
2″ x 4″	.220	L.F.	.011	.14	.44	.58
2″ x 6″	.220	L.F.	.014	.17	.55	.72
2″ x 8″	.220	L.F.	.016	.20	.64	.84
alley rafter, #2 or better, 2″ x 4″	.280	L.F.	.007	.13	.29	.42
2″ x 6″	.280	L.F.	.009	.17	.36	.53
2″ x 8″	.280	L.F.	.010	.25	.39	.64
2″ x 10″	.280	L.F.	.012	.36	.48	.84
ripple rafter, #2 or better, 2″ x 4″, 16″ O.C.	.560	L.F.	.018	.27	.72	.99
24″ O.C.	.450	L.F.	.014	.21	.57	.78
2″ x 6″, 16″ O.C.	.560	L.F.	.022	.33	.89	1.22
24″ O.C.	.450	L.F.	.018	.27	.72	.99
2″ x 8″, 16″ O.C.	.560	L.F.	.027	.51	1.09	1.60
24″ O.C.	.450	L.F.	.021	.41	.88	1.29
leaders, #2 or better double header, 2″ x 4″	.670	L.F.	.024	.32	.98	1.30
2″ x 6″	.670	L.F.	.030	.40	1.21	1.61
2″ x 8″	.670	L.F.	.032	.61	1.29	1.90
2″ x 10″	.670	L.F.	.034	.86	1.37	2.23
eiling joist, #2 or better, 2″ x 4″, 16″ O.C.	1.000	L.F.	.013	.37	.52	.89
24″ O.C.	.800	L.F.	.010	.30	.42	.72
2″ x 6″, 16″ O.C.	1.000	L.F.	.013	.59	.52	1.11
24″ O.C.	.800	L.F.	.010	.47	.42	.89
heathing, plywood exterior grade, 3/8″ thick	3.610	S.F.	.038	2.60	1.55	4.15
1/2″ thick	3.610	S.F.	.041	2.45	1.70	4.15
5/8″ thick	3.610	S.F.	.044	3.10	1.81	4.91
3/4″ thick	3.610	S.F.	.048	3.72	1.95	5.67
Boards, 1″ x 6″, laid regular	3.610	S.F.	.089	4.98	3.61	8.59
Laid diagonal	3.610	S.F.	.099	4.98	4.01	8.99
1″ x 8″, laid regular	3.610	S.F.	.076	4.44	3.07	7.51
Laid diagonal	3.610	S.F.	.089	4.44	3.61	8.05

3 FRAMING

44 Shed Dormer Framing Systems

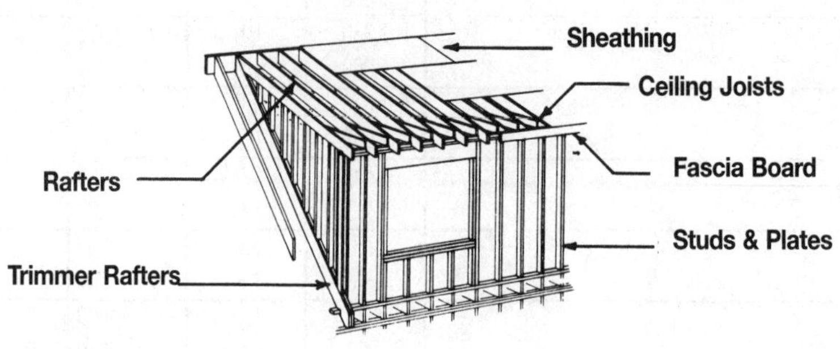

System Description	QUAN.	UNIT	LABOR HOURS	COST PER S.F.		
				MAT.	INST.	TOTAL
2″ X 6″ RAFTERS, 16″ O.C.						
Dormer rafter, 2″ x 6″, 16″ O.C.	1.080	L.F.	.029	.64	1.20	1.84
Trimmer rafter, 2″ x 6″	.400	L.F.	.006	.24	.26	.50
Studs & plates, 2″ x 4″, 16″ O.C.	2.750	L.F.	.049	1.02	1.98	3
Fascia, 2″ x 6″	.250	L.F.	.014	.17	.55	.72
Ceiling joist, 2″ x 4″, 16″ O.C.	1.000	L.F.	.013	.37	.52	.89
Sheathing, exterior grade plywood, CDX, 1/2″ thick	2.940	S.F.	.034	2	1.38	3.38
TOTAL		S.F.	.145	4.44	5.89	10.33
2″ X 8″ RAFTERS, 16″ O.C.						
Dormer rafter, 2″ x 8″, 16″ O.C.	1.080	L.F.	.032	.98	1.31	2.29
Trimmer rafter, 2″ x 8″	.400	L.F.	.007	.36	.28	.64
Studs & plates, 2″ x 4″, 16″ O.C.	2.750	L.F.	.049	1.02	1.98	3
Fascia, 2″ x 8″	.250	L.F.	.018	.23	.73	.96
Ceiling joist, 2″ x 6″, 16″ O.C.	1.000	L.F.	.013	.59	.52	1.11
Sheathing, exterior grade plywood, CDX, 1/2″ thick	2.940	S.F.	.034	2	1.38	3.38
TOTAL		S.F.	.153	5.18	6.20	11.38
2″ X 10″ RAFTERS, 16″ O.C.						
Dormer rafter, 2″ x 10″, 16″ O.C.	1.080	L.F.	.041	1.39	1.65	3.04
Trimmer rafter, 2″ x 10″	.400	L.F.	.010	.52	.42	.94
Studs & plates, 2″ x 4″, 16″ O.C.	2.750	L.F.	.049	1.02	1.98	3
Fascia, 2″ x 10″	.250	L.F.	.022	.32	.91	1.23
Ceiling joist, 2″ x 6″, 16″ O.C.	1.000	L.F.	.013	.59	.52	1.11
Sheathing, exterior grade plywood, CDX, 1/2″ thick	2.940	S.F.	.034	2	1.38	3.38
TOTAL		S.F.	.169	5.84	6.86	12.70

The cost in this system is based on the square foot of plan area.
The measurement is the plan area of the dormer only.

Description	QUAN.	UNIT	LABOR HOURS	COST PER S.F.		
				MAT.	INST.	TOTAL

Shed Dormer Framing Price Sheet	QUAN.	UNIT	LABOR HOURS	COST PER S.F.		
				MAT.	INST.	TOTAL
rmer rafters, #2 or better, 2″ x 4″, 16″ O.C.	1.080	L.F.	.023	.51	.96	1.47
24″ O.C.	.860	L.F.	.019	.41	.76	1.17
2″ x 6″, 16″ O.C.	1.080	L.F.	.029	.64	1.20	1.84
24″ O.C.	.860	L.F.	.023	.51	.95	1.46
2″ x 8″, 16″ O.C.	1.080	L.F.	.032	.98	1.31	2.29
24″ O.C.	.860	L.F.	.025	.78	1.04	1.82
2″ x 10″, 16″ O.C.	1.080	L.F.	.041	1.39	1.65	3.04
24″ O.C.	.860	L.F.	.032	1.11	1.32	2.43
mmer rafter, #2 or better, 2″ x 4″	.400	L.F.	.005	.19	.21	.40
2″ x 6″	.400	L.F.	.006	.24	.26	.50
2″ x 8″	.400	L.F.	.007	.36	.28	.64
2″ x 10″	.400	L.F.	.010	.52	.42	.94
uds & plates, #2 or better, 2″ x 4″, 16″ O.C.	2.750	L.F.	.049	1.02	1.98	3
24″ O.C.	2.200	L.F.	.039	.81	1.58	2.39
2″ x 6″, 16″ O.C.	2.750	L.F.	.055	1.62	2.26	3.88
24″ O.C.	2.200	L.F.	.044	1.30	1.80	3.10
scia, #2 or better, 1″ x 4″	.250	L.F.	.006	.08	.26	.34
1″ x 6″	.250	L.F.	.008	.10	.32	.42
1″ x 8″	.250	L.F.	.009	.12	.37	.49
2″ x 4″	.250	L.F.	.011	.14	.44	.58
2″ x 6″	.250	L.F.	.014	.17	.55	.72
2″ x 8″	.250	L.F.	.018	.23	.73	.96
eiling joist, #2 or better, 2″ x 4″, 16″ O.C.	1.000	L.F.	.013	.37	.52	.89
24″ O.C.	.800	L.F.	.010	.30	.42	.72
2″ x 6″, 16″ O.C.	1.000	L.F.	.013	.59	.52	1.11
24″ O.C.	.800	L.F.	.010	.47	.42	.89
2″ x 8″, 16″ O.C.	1.000	L.F.	.015	.91	.59	1.50
24″ O.C.	.800	L.F.	.012	.73	.47	1.20
heathing, plywood exterior grade, 3/8″ thick	2.940	S.F.	.031	2.12	1.26	3.38
1/2″ thick	2.940	S.F.	.034	2	1.38	3.38
5/8″ thick	2.940	S.F.	.036	2.53	1.47	4
3/4″ thick	2.940	S.F.	.039	3.03	1.59	4.62
Boards, 1″ x 6″, laid regular	2.940	S.F.	.072	4.06	2.94	7
Laid diagonal	2.940	S.F.	.080	4.06	3.26	7.32
1″ x 8″, laid regular	2.940	S.F.	.062	3.62	2.50	6.12
Laid diagonal	2.940	S.F.	.072	3.62	2.94	6.56

Window Openings	QUAN.	UNIT	LABOR HOURS	COST EACH		
				MAT.	INST.	TOTAL
he following are to be added to the total cost of the dormers for window openings. Do not subtract window area from the stud wall quantities.						
eaders, 2″ x 6″ doubled, 2′ long	4.000	L.F.	.178	2.36	7.25	9.61
3′ long	6.000	L.F.	.267	3.54	10.85	14.39
4′ long	8.000	L.F.	.356	4.72	14.50	19.22
5′ long	10.000	L.F.	.444	5.90	18.10	24
2″ x 8″ doubled, 4′ long	8.000	L.F.	.376	7.30	15.35	22.65
5′ long	10.000	L.F.	.471	9.10	19.20	28.30
6′ long	12.000	L.F.	.565	10.90	23	33.90
8′ long	16.000	L.F.	.753	14.55	30.50	45.05
2″ x 10″ doubled, 4′ long	8.000	L.F.	.400	10.30	16.30	26.60
6′ long	12.000	L.F.	.600	15.50	24.50	40
8′ long	16.000	L.F.	.800	20.50	32.50	53
10′ long	20.000	L.F.	1.000	26	41	67

48 Partition Framing Systems

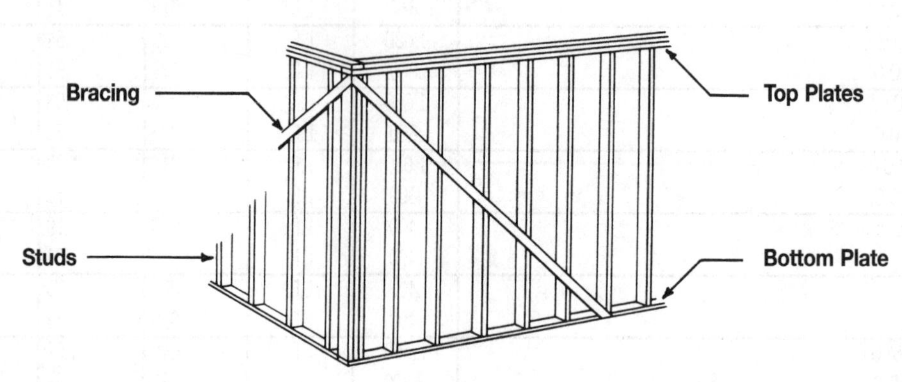

System Description	QUAN.	UNIT	LABOR HOURS	COST PER S.F.		
				MAT.	INST.	TOTAL
2" X 4", 16" O.C.						
2" x 4" studs, #2 or better, 16" O.C.	1.000	L.F.	.015	.37	.59	.96
Plates, double top, single bottom	.375	L.F.	.005	.14	.22	.36
Cross bracing, let-in, 1" x 6"	.080	L.F.	.004	.05	.17	.22
TOTAL		S.F.	.024	.56	.98	1.54
2" X 4", 24" O.C.						
2" x 4" studs, #2 or better, 24" O.C.	.800	L.F.	.012	.30	.47	.77
Plates, double top, single bottom	.375	L.F.	.005	.14	.22	.36
Cross bracing, let-in, 1" x 6"	.080	L.F.	.003	.05	.11	.16
TOTAL		S.F.	.020	.49	.80	1.29
2" X 6", 16" O.C.						
2" x 6" studs, #2 or better, 16" O.C.	1.000	L.F.	.016	.59	.65	1.24
Plates, double top, single bottom	.375	L.F.	.006	.22	.24	.46
Cross bracing, let-in, 1" x 6"	.080	L.F.	.004	.05	.17	.22
TOTAL		S.F.	.026	.86	1.06	1.92
2" X 6", 24" O.C.						
2" x 6" studs, #2 or better, 24" O.C.	.800	L.F.	.013	.47	.52	.99
Plates, double top, single bottom	.375	L.F.	.006	.22	.24	.46
Cross bracing, let-in, 1" x 6"	.080	L.F.	.003	.05	.11	.16
TOTAL		S.F.	.022	.74	.87	1.61

The costs in this system are based on a square foot of wall area. Do not subtract for door or window openings.

Description	QUAN.	UNIT	LABOR HOURS	COST PER S.F.		
				MAT.	INST.	TOTAL

Partition Framing Price Sheet	QUAN.	UNIT	LABOR HOURS	COST PER S.F.		
				MAT.	INST.	TOTAL
Wood studs, #2 or better, 2″ x 4″, 12″ O.C.	1.250	L.F.	.018	.46	.74	1.20
16″ O.C.	1.000	L.F.	.015	.37	.59	.96
24″ O.C.	.800	L.F.	.012	.30	.47	.77
32″ O.C.	.650	L.F.	.009	.24	.38	.62
2″ x 6″, 12″ O.C.	1.250	L.F.	.020	.74	.81	1.55
16″ O.C.	1.000	L.F.	.016	.59	.65	1.24
24″ O.C.	.800	L.F.	.013	.47	.52	.99
32″ O.C.	.650	L.F.	.010	.38	.42	.80
Plates, #2 or better double top single bottom, 2″ x 4″	.375	L.F.	.005	.14	.22	.36
2″ x 6″	.375	L.F.	.006	.22	.24	.46
2″ x 8″	.375	L.F.	.005	.34	.22	.56
Cross bracing, let-in, 1″ x 6″ boards studs, 12″ O.C.	.080	L.F.	.005	.06	.22	.28
16″ O.C.	.080	L.F.	.004	.05	.17	.22
24″ O.C.	.080	L.F.	.003	.05	.11	.16
32″ O.C.	.080	L.F.	.002	.04	.09	.13
Let-in steel (T shaped) studs, 12″ O.C.	.080	L.F.	.001	.05	.06	.11
16″ O.C.	.080	L.F.	.001	.04	.04	.08
24″ O.C.	.080	L.F.	.001	.04	.04	.08
32″ O.C.	.080	L.F.	.001	.03	.03	.06
Steel straps studs, 12″ O.C.	.080	L.F.	.001	.06	.05	.11
16″ O.C.	.080	L.F.	.001	.06	.04	.10
24″ O.C.	.080	L.F.	.001	.06	.04	.10
32″ O.C.	.080	L.F.	.001	.06	.04	.10
Metal studs, load bearing 24″ O.C., 20 ga. galv., 2-1/2″ wide	1.000	S.F.	.015	.73	.61	1.34
3-5/8″ wide	1.000	S.F.	.015	.86	.62	1.48
4″ wide	1.000	S.F.	.016	.90	.64	1.54
6″ wide	1.000	S.F.	.016	1.15	.65	1.80
16 ga., 2-1/2″ wide	1.000	S.F.	.017	.84	.70	1.54
3-5/8″ wide	1.000	S.F.	.017	1.01	.71	1.72
4″ wide	1.000	S.F.	.018	1.06	.73	1.79
6″ wide	1.000	S.F.	.018	1.33	.74	2.07
Non-load bearing 24″ O.C., 25 ga. galv., 1-5/8″ wide	1.000	S.F.	.011	.25	.43	.68
2-1/2″ wide	1.000	S.F.	.011	.28	.43	.71
3-5/8″ wide	1.000	S.F.	.011	.33	.44	.77
4″ wide	1.000	S.F.	.011	.43	.44	.87
6″ wide	1.000	S.F.	.011	.50	.45	.95
20 ga., 2-1/2″ wide	1.000	S.F.	.013	.54	.54	1.08
3-5/8″ wide	1.000	S.F.	.014	.62	.55	1.17
4″ wide	1.000	S.F.	.014	.70	.55	1.25
6″ wide	1.000	S.F.	.014	.88	.56	1.44

Window & Door Openings	QUAN.	UNIT	LABOR HOURS	COST EACH		
				MAT.	INST.	TOTAL
The following costs are to be added to the total costs of the walls. Do not subtract openings from total wall area.						
Headers, 2″ x 6″ double, 2′ long	4.000	L.F.	.178	2.36	7.25	9.61
3′ long	6.000	L.F.	.267	3.54	10.85	14.39
4′ long	8.000	L.F.	.356	4.72	14.50	19.22
5′ long	10.000	L.F.	.444	5.90	18.10	24
2″ x 8″ double, 4′ long	8.000	L.F.	.376	7.30	15.35	22.65
5′ long	10.000	L.F.	.471	9.10	19.20	28.30
6′ long	12.000	L.F.	.565	10.90	23	33.90
8′ long	16.000	L.F.	.753	14.55	30.50	45.05
2″ x 10″ double, 4′ long	8.000	L.F.	.400	10.30	16.30	26.60
6′ long	12.000	L.F.	.600	15.50	24.50	40
8′ long	16.000	L.F.	.800	20.50	32.50	53
10′ long	20.000	L.F.	1.000	26	41	67
2″ x 12″ double, 8′ long	16.000	L.F.	.853	28	34.50	62.50
12′ long	24.000	L.F.	1.280	42	52	94

Division 4
Exterior Walls

No part of this publication may be reproduced, stored in a retrieval system, or transmitted in any form or by any means without prior written permission of Reed Construction Data.

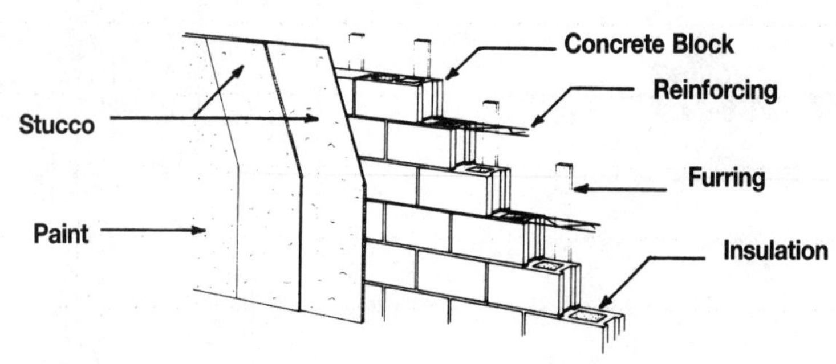

System Description	QUAN.	UNIT	LABOR HOURS	COST PER S.F. MAT.	COST PER S.F. INST.	COST PER S.F. TOTAL
6" THICK CONCRETE BLOCK WALL						
6" thick concrete block, 6" x 8" x 16"	1.000	S.F.	.100	1.54	3.72	5.26
Masonry reinforcing, truss strips every other course	.625	L.F.	.002	.12	.07	.19
Furring, 1" x 3", 16" O.C.	1.000	L.F.	.016	.31	.66	.97
Masonry insulation, poured vermiculite	1.000	S.F.	.013	.61	.54	1.15
Stucco, 2 coats	1.000	S.F.	.069	.20	2.57	2.77
Masonry paint, 2 coats	1.000	S.F.	.016	.19	.57	.76
TOTAL		S.F.	.216	2.97	8.13	11.10
8" THICK CONCRETE BLOCK WALL						
8" thick concrete block, 8" x 8" x 16"	1.000	S.F.	.107	1.67	3.97	5.64
Masonry reinforcing, truss strips every other course	.625	L.F.	.002	.12	.07	.19
Furring, 1" x 3", 16" O.C.	1.000	L.F.	.016	.31	.66	.97
Masonry insulation, poured vermiculite	1.000	S.F.	.018	.80	.72	1.52
Stucco, 2 coats	1.000	S.F.	.069	.20	2.57	2.77
Masonry paint, 2 coats	1.000	S.F.	.016	.19	.57	.76
TOTAL		S.F.	.228	3.29	8.56	11.85
12" THICK CONCRETE BLOCK WALL						
12" thick concrete block, 12" x 8" x 16"	1.000	S.F.	.141	2.44	5.10	7.54
Masonry reinforcing, truss strips every other course	.625	L.F.	.003	.14	.10	.24
Furring, 1" x 3", 16" O.C.	1.000	L.F.	.016	.31	.66	.97
Masonry insulation, poured vermiculite	1.000	S.F.	.026	1.18	1.06	2.24
Stucco, 2 coats	1.000	S.F.	.069	.20	2.57	2.77
Masonry paint, 2 coats	1.000	S.F.	.016	.19	.57	.76
TOTAL		S.F.	.271	4.46	10.06	14.52

Costs for this system are based on a square foot of wall area. Do not subtract for window openings.

Description	QUAN.	UNIT	LABOR HOURS	COST PER S.F. MAT.	COST PER S.F. INST.	COST PER S.F. TOTAL

Important: See the Reference Section for critical supporting data - Reference Nos., Crews & Location Factors

Masonry Block Price Sheet	QUAN.	UNIT	LABOR HOURS	COST PER S.F.		
				MAT.	INST.	TOTAL
Block concrete, 8" x 16" regular, 4" thick	1.000	S.F.	.093	1.04	3.46	4.50
6" thick	1.000	S.F.	.100	1.54	3.72	5.26
8" thick	1.000	S.F.	.107	1.67	3.97	5.64
10" thick	1.000	S.F.	.111	2.37	4.13	6.50
12" thick	1.000	S.F.	.141	2.44	5.10	7.54
Solid block, 4" thick	1.000	S.F.	.096	1.43	3.59	5.02
6" thick	1.000	S.F.	.104	1.69	3.87	5.56
8" thick	1.000	S.F.	.111	2.54	4.13	6.67
10" thick	1.000	S.F.	.133	3.44	4.82	8.26
12" thick	1.000	S.F.	.148	3.82	5.35	9.17
Lightweight, 4" thick	1.000	S.F.	.093	1.04	3.46	4.50
6" thick	1.000	S.F.	.100	1.54	3.72	5.26
8" thick	1.000	S.F.	.107	1.67	3.97	5.64
10" thick	1.000	S.F.	.111	2.37	4.13	6.50
12" thick	1.000	S.F.	.141	2.44	5.10	7.54
Split rib profile, 4" thick	1.000	S.F.	.116	2.35	4.31	6.66
6" thick	1.000	S.F.	.123	2.70	4.58	7.28
8" thick	1.000	S.F.	.131	3.11	4.96	8.07
10" thick	1.000	S.F.	.157	3.31	5.65	8.96
12" thick	1.000	S.F.	.175	3.68	6.30	9.98
Masonry reinforcing, wire truss strips, every course, 8" block	1.375	L.F.	.004	.26	.15	.41
12" block	1.375	L.F.	.006	.30	.22	.52
Every other course, 8" block	.625	L.F.	.002	.12	.07	.19
12" block	.625	L.F.	.003	.14	.10	.24
Furring, wood, 1" x 3", 12" O.C.	1.250	L.F.	.020	.39	.83	1.22
16" O.C.	1.000	L.F.	.016	.31	.66	.97
24" O.C.	.800	L.F.	.013	.25	.53	.78
32" O.C.	.640	L.F.	.010	.20	.42	.62
Steel, 3/4" channels, 12" O.C.	1.250	L.F.	.034	.24	1.22	1.46
16" O.C.	1.000	L.F.	.030	.22	1.08	1.30
24" O.C.	.800	L.F.	.023	.15	.82	.97
32" O.C.	.640	L.F.	.018	.12	.66	.78
Masonry insulation, vermiculite or perlite poured 4" thick	1.000	S.F.	.009	.39	.35	.74
6" thick	1.000	S.F.	.013	.60	.54	1.14
8" thick	1.000	S.F.	.018	.80	.72	1.52
10" thick	1.000	S.F.	.021	.97	.87	1.84
12" thick	1.000	S.F.	.026	1.18	1.06	2.24
Block inserts polystyrene, 6" thick	1.000	S.F.		.94		.94
8" thick	1.000	S.F.		.94		.94
10" thick	1.000	S.F.		1.10		1.10
12" thick	1.000	S.F.		1.16		1.16
Stucco, 1 coat	1.000	S.F.	.057	.16	2.11	2.27
2 coats	1.000	S.F.	.069	.20	2.57	2.77
3 coats	1.000	S.F.	.081	.23	3.02	3.25
Painting, 1 coat	1.000	S.F.	.011	.12	.40	.52
2 coats	1.000	S.F.	.016	.19	.57	.76
Primer & 1 coat	1.000	S.F.	.013	.20	.47	.67
2 coats	1.000	S.F.	.018	.27	.65	.92
Lath, metal lath expanded 2.5 lb/S.Y., painted	1.000	S.F.	.010	.31	.37	.68
Galvanized	1.000	S.F.	.012	.34	.41	.75

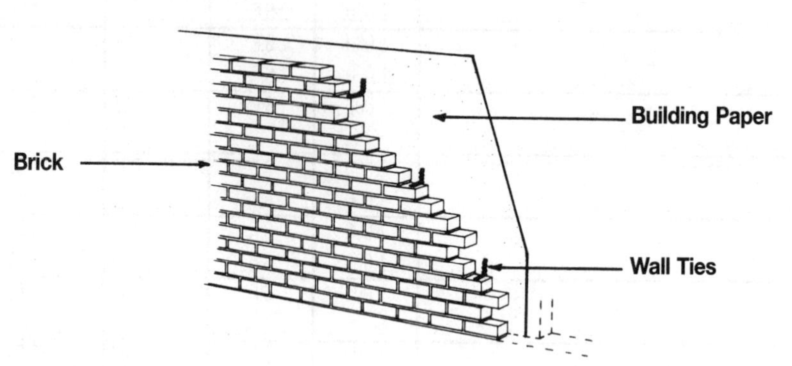

System Description	QUAN.	UNIT	LABOR HOURS	COST PER S.F.		
				MAT.	INST.	TOTAL
SELECT COMMON BRICK						
Brick, select common, running bond	1.000	S.F.	.174	3.17	6.45	9.62
Wall ties, 7/8" x 7", 22 gauge	1.000	Ea.	.008	.06	.32	.38
Building paper, spunbonded polypropylene	1.100	S.F.	.002	.12	.09	.21
Trim, pine, painted	.125	L.F.	.004	.09	.16	.25
TOTAL		S.F.	.188	3.44	7.02	10.46
RED FACED COMMON BRICK						
Brick, common, red faced, running bond	1.000	S.F.	.182	3.17	6.75	9.92
Wall ties, 7/8" x 7", 22 gauge	1.000	Ea.	.008	.06	.32	.38
Building paper, spundbonded polypropylene	1.100	S.F.	.002	.12	.09	.21
Trim, pine, painted	.125	L.F.	.004	.09	.16	.25
TOTAL		S.F.	.196	3.44	7.32	10.76
BUFF OR GREY FACE BRICK						
Brick, buff or grey	1.000	S.F.	.182	3.36	6.75	10.11
Wall ties, 7/8" x 7", 22 gauge	1.000	Ea.	.008	.06	.32	.38
Building paper, spundbonded polypropylene	1.100	S.F.	.002	.12	.09	.21
Trim, pine, painted	.125	L.F.	.004	.09	.16	.25
TOTAL		S.F.	.196	3.63	7.32	10.95
STONE WORK, ROUGH STONE, AVERAGE						
Field stone veneer	1.000	S.F.	.223	5.16	8.31	13.47
Wall ties, 7/8" x 7", 22 gauge	1.000	Ea.	.008	.06	.32	.38
Building paper, spundbonded polypropylene	1.000	S.F.	.002	.12	.09	.21
Trim, pine, painted	.125	L.F.	.004	.09	.16	.25
TOTAL		S.F.	.237	5.43	8.88	14.31

The costs in this system are based on a square foot of wall area. Do not subtract area for window & door openings.

Description	QUAN.	UNIT	LABOR HOURS	COST PER S.F.		
				MAT.	INST.	TOTAL

EXTERIOR WALLS 4

Important: See the Reference Section for critical supporting data - Reference Nos., Crews & Location Factors

Brick/Stone Veneer Price Sheet	QUAN.	UNIT	LABOR HOURS	COST PER S.F. MAT.	INST.	TOTAL
Brick						
Select common, running bond	1.000	S.F.	.174	3.17	6.45	9.62
Red faced, running bond	1.000	S.F.	.182	3.17	6.75	9.92
Buff or grey faced, running bond	1.000	S.F.	.182	3.36	6.75	10.11
Header every 6th course	1.000	S.F.	.216	3.69	8.05	11.74
English bond	1.000	S.F.	.286	4.73	10.65	15.38
Flemish bond	1.000	S.F.	.195	3.34	7.25	10.59
Common bond	1.000	S.F.	.267	4.21	9.90	14.11
Stack bond	1.000	S.F.	.182	3.36	6.75	10.11
Jumbo, running bond	1.000	S.F.	.092	3.98	3.42	7.40
Norman, running bond	1.000	S.F.	.125	4.32	4.65	8.97
Norwegian, running bond	1.000	S.F.	.107	3.25	3.97	7.22
Economy, running bond	1.000	S.F.	.129	3.90	4.80	8.70
Engineer, running bond	1.000	S.F.	.154	3.26	5.70	8.96
Roman, running bond	1.000	S.F.	.160	5.10	5.95	11.05
Utility, running bond	1.000	S.F.	.089	3.70	3.31	7.01
Glazed, running bond	1.000	S.F.	.190	9.15	7.10	16.25
Stone work, rough stone, average	1.000	S.F.	.179	5.15	8.30	13.45
Maximum	1.000	S.F.	.267	7.70	12.40	20.10
Wall ties, galvanized, corrugated 7/8″ x 7″, 22 gauge	1.000	Ea.	.008	.06	.32	.38
16 gauge	1.000	Ea.	.008	.19	.32	.51
Cavity wall, every 3rd course 6″ long Z type, 1/4″ diameter	1.330	L.F.	.010	.41	.41	.82
3/16″ diameter	1.330	L.F.	.010	.21	.41	.62
8″ long, Z type, 1/4″ diameter	1.330	L.F.	.010	.44	.41	.85
3/16″ diameter	1.330	L.F.	.010	.24	.41	.65
Building paper, aluminum and kraft laminated foil, 1 side	1.000	S.F.	.002	.04	.09	.13
2 sides	1.000	S.F.	.002	.07	.09	.16
#15 asphalt paper	1.100	S.F.	.002	.04	.10	.14
Polyethylene, .002″ thick	1.000	S.F.	.002	.01	.09	.10
.004″ thick	1.000	S.F.	.002	.02	.09	.11
.006″ thick	1.000	S.F.	.002	.03	.09	.12
.010″ thick	1.000	S.F.	.002	.06	.09	.15
Trim, 1″ x 4″, cedar	.125	L.F.	.005	.21	.20	.41
Fir	.125	L.F.	.005	.09	.20	.29
Redwood	.125	L.F.	.005	.21	.20	.41
White pine	.125	L.F.	.005	.09	.20	.29

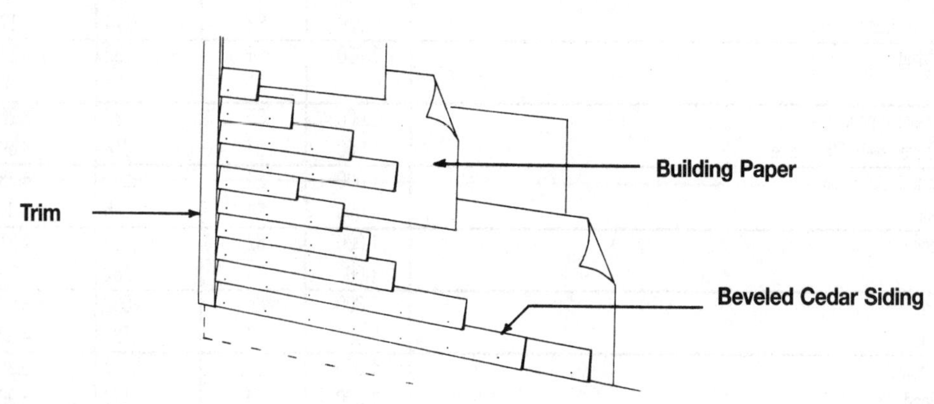

System Description	QUAN.	UNIT	LABOR HOURS	COST PER S.F. MAT.	COST PER S.F. INST.	COST PER S.F. TOTAL
1/2" X 6" BEVELED CEDAR SIDING, "A" GRADE						
1/2" x 6" beveled cedar siding	1.000	S.F.	.032	3.61	1.30	4.91
Building wrap, spundbonded polypropylene	1.100	S.F.	.002	.12	.09	.21
Trim, cedar	.125	L.F.	.005	.21	.20	.41
Paint, primer & 2 coats	1.000	S.F.	.017	.18	.61	.79
TOTAL		S.F.	.056	4.12	2.20	6.32
1/2" X 8" BEVELED CEDAR SIDING, "A" GRADE						
1/2" x 8" beveled cedar siding	1.000	S.F.	.029	3.21	1.19	4.40
Building wrap, spundbonded polypropylene	1.100	S.F.	.002	.12	.09	.21
Trim, cedar	.125	L.F.	.005	.21	.20	.41
Paint, primer & 2 coats	1.000	S.F.	.017	.18	.61	.79
TOTAL		S.F.	.053	3.72	2.09	5.81
1" X 4" TONGUE & GROOVE, REDWOOD, VERTICAL GRAIN						
Redwood, clear, vertical grain, 1" x 10"	1.000	S.F.	.018	3.62	.75	4.37
Building wrap, spunbonded polypropylene	1.100	S.F.	.002	.12	.09	.21
Trim, redwood	.125	L.F.	.005	.21	.20	.41
Sealer, 1 coat, stain, 1 coat	1.000	S.F.	.013	.12	.48	.60
TOTAL		S.F.	.038	4.07	1.52	5.59
1" X 6" TONGUE & GROOVE, REDWOOD, VERTICAL GRAIN						
Redwood, clear, vertical grain, 1" x 10"	1.000	S.F.	.019	3.73	.77	4.50
Building wrap, spunbonded polypropylene	1.100	S.F.	.002	.12	.09	.21
Trim, redwood	.125	L.F.	.005	.21	.20	.41
Sealer, 1 coat, stain, 1 coat	1.000	S.F.	.013	.12	.48	.60
TOTAL		S.F.	.039	4.18	1.54	5.72

The costs in this system are based on a square foot of wall area.
Do not subtract area for door or window openings.

Description	QUAN.	UNIT	LABOR HOURS	COST PER S.F. MAT.	COST PER S.F. INST.	COST PER S.F. TOTAL

Important: See the Reference Section for critical supporting data - Reference Nos., Crews & Location Factors

Wood Siding Price Sheet	QUAN.	UNIT	LABOR HOURS	COST PER S.F.		
				MAT.	INST.	TOTAL
Siding, beveled cedar, "A" grade, 1/2" x 6"	1.000	S.F.	.028	3.61	1.30	4.91
1/2" x 8"	1.000	S.F.	.023	3.21	1.19	4.40
"B" grade, 1/2" x 6"	1.000	S.F.	.032	4.01	1.44	5.45
1/2" x 8"	1.000	S.F.	.029	3.57	1.32	4.89
Clear grade, 1/2" x 6"	1.000	S.F.	.028	4.51	1.63	6.14
1/2" x 8"	1.000	S.F.	.023	4.01	1.49	5.50
Redwood, clear vertical grain, 1/2" x 6"	1.000	S.F.	.036	2.88	1.45	4.33
1/2" x 8"	1.000	S.F.	.032	2.33	1.30	3.63
Clear all heart vertical grain, 1/2" x 6"	1.000	S.F.	.028	3.20	1.61	4.81
1/2" x 8"	1.000	S.F.	.023	2.59	1.44	4.03
Siding board & batten, cedar, "B" grade, 1" x 10"	1.000	S.F.	.031	2.22	1.25	3.47
1" x 12"	1.000	S.F.	.031	2.22	1.25	3.47
Redwood, clear vertical grain, 1" x 6"	1.000	S.F.	.043	2.76	1.98	4.74
1" x 8"	1.000	S.F.	.018	2.52	1.74	4.26
White pine, #2 & better, 1" x 10"	1.000	S.F.	.029	.74	1.19	1.93
1" x 12"	1.000	S.F.	.029	.74	1.19	1.93
Siding vertical, tongue & groove, cedar "B" grade, 1" x 4"	1.000	S.F.	.033	2	.75	2.75
1" x 6"	1.000	S.F.	.024	2.06	.77	2.83
1" x 8"	1.000	S.F.	.024	2.12	.80	2.92
1" x 10"	1.000	S.F.	.021	2.18	.82	3
"A" grade, 1" x 4"	1.000	S.F.	.033	1.83	.69	2.52
1" x 6"	1.000	S.F.	.024	1.88	.71	2.59
1" x 8"	1.000	S.F.	.024	1.93	.73	2.66
1" x 10"	1.000	S.F.	.021	1.98	.75	2.73
Clear vertical grain, 1" x 4"	1.000	S.F.	.033	1.69	.64	2.33
1" x 6"	1.000	S.F.	.024	1.73	.65	2.38
1" x 8"	1.000	S.F.	.024	1.77	.67	2.44
1" x 10"	1.000	S.F.	.021	1.82	.68	2.50
Redwood, clear vertical grain, 1" x 4"	1.000	S.F.	.033	3.62	.75	4.37
1" x 6"	1.000	S.F.	.024	3.73	.77	4.50
1" x 8"	1.000	S.F.	.024	3.83	.80	4.63
1" x 10"	1.000	S.F.	.021	3.95	.82	4.77
Clear all heart vertical grain, 1" x 4"	1.000	S.F.	.033	3.32	.69	4.01
1" x 6"	1.000	S.F.	.024	3.41	.71	4.12
1" x 8"	1.000	S.F.	.024	3.50	.73	4.23
1" x 10"	1.000	S.F.	.021	3.59	.75	4.34
White pine, 1" x 10"	1.000	S.F.	.024	.77	.82	1.59
Siding plywood, texture 1-11 cedar, 3/8" thick	1.000	S.F.	.024	1.19	.97	2.16
5/8" thick	1.000	S.F.	.024	2.62	.97	3.59
Redwood, 3/8" thick	1.000	S.F.	.024	1.19	.97	2.16
5/8" thick	1.000	S.F.	.024	1.98	.97	2.95
Fir, 3/8" thick	1.000	S.F.	.024	.64	.97	1.61
5/8" thick	1.000	S.F.	.024	1.11	.97	2.08
Southern yellow pine, 3/8" thick	1.000	S.F.	.024	.64	.97	1.61
5/8" thick	1.000	S.F.	.024	.89	.97	1.86
Hard board, 7/16" thick primed, plain finish	1.000	S.F.	.025	1.14	1	2.14
Board finish	1.000	S.F.	.023	.80	.93	1.73
Polyvinyl coated, 3/8" thick	1.000	S.F.	.021	.97	.87	1.84
5/8" thick	1.000	S.F.	.024	.89	.97	1.86
Paper, #15 asphalt felt	1.100	S.F.	.002	.04	.10	.14
Trim, cedar	.125	L.F.	.005	.21	.20	.41
Fir	.125	L.F.	.005	.09	.20	.29
Redwood	.125	L.F.	.005	.21	.20	.41
White pine	.125	L.F.	.005	.09	.20	.29
Painting, primer, & 1 coat	1.000	S.F.	.013	.12	.48	.60
2 coats	1.000	S.F.	.017	.18	.61	.79
Stain, sealer, & 1 coat	1.000	S.F.	.017	.09	.62	.71
2 coats	1.000	S.F.	.019	.15	.67	.82

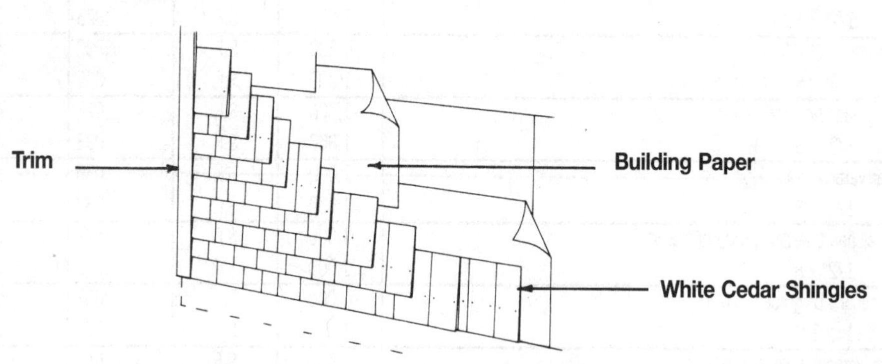

System Description	QUAN.	UNIT	LABOR HOURS	COST PER S.F.		
				MAT.	INST.	TOTAL
WHITE CEDAR SHINGLES, 5″ EXPOSURE						
White cedar shingles, 16″ long, grade "A", 5″ exposure	1.000	S.F.	.033	1.35	1.36	2.71
Building wrap, spunbonded polypropylene	1.100	S.F.	.002	.12	.09	.21
Trim, cedar	.125	S.F.	.005	.21	.20	.41
Paint, primer & 1 coat	1.000	S.F.	.017	.09	.62	.71
TOTAL		S.F.	.057	1.77	2.27	4.04
NO. 1 PERFECTIONS, 5-1/2″ EXPOSURE						
No. 1 perfections, red cedar, 5-1/2″ exposure	1.000	S.F.	.029	1.76	1.19	2.95
Building wrap, spunbonded polypropylene	1.100	S.F.	.002	.12	.09	.21
Trim, cedar	.125	S.F.	.005	.21	.20	.41
Stain, sealer & 1 coat	1.000	S.F.	.017	.09	.62	.71
TOTAL		S.F.	.053	2.18	2.10	4.28
RESQUARED & REBUTTED PERFECTIONS, 5-1/2″ EXPOSURE						
Resquared & rebutted perfections, 5-1/2″ exposure	1.000	S.F.	.027	2.19	1.09	3.28
Building wrap, spunbonded polypropylene	1.100	S.F.	.002	.12	.09	.21
Trim, cedar	.125	S.F.	.005	.21	.20	.41
Stain, sealer & 1 coat	1.000	S.F.	.017	.09	.62	.71
TOTAL		S.F.	.051	2.61	2	4.61
HAND-SPLIT SHAKES, 8-1/2″ EXPOSURE						
Hand-split red cedar shakes, 18″ long, 8-1/2″ exposure	1.000	S.F.	.040	1.07	1.63	2.70
Building wrap, spunbonded polypropylene	1.100	S.F.	.002	.12	.09	.21
Trim, cedar	.125	S.F.	.005	.21	.20	.41
Stain, sealer & 1 coat	1.000	S.F.	.017	.09	.62	.71
TOTAL		S.F.	.064	1.49	2.54	4.03

The costs in this system are based on a square foot of wall area.
Do not subtract area for door or window openings.

Description	QUAN.	UNIT	LABOR HOURS	COST PER S.F.		
				MAT.	INST.	TOTAL

Shingle Siding Price Sheet	QUAN.	UNIT	LABOR HOURS	COST PER S.F. MAT.	INST.	TOTAL
hingles wood, white cedar 16" long, "A" grade, 5" exposure	1.000	S.F.	.033	1.35	1.36	2.71
7" exposure	1.000	S.F.	.030	1.22	1.22	2.44
8-1/2" exposure	1.000	S.F.	.032	.77	1.30	2.07
10" exposure	1.000	S.F.	.028	.67	1.14	1.81
"B" grade, 5" exposure	1.000	S.F.	.040	1.21	1.63	2.84
7" exposure	1.000	S.F.	.028	.85	1.14	1.99
8-1/2" exposure	1.000	S.F.	.024	.73	.98	1.71
10" exposure	1.000	S.F.	.020	.61	.82	1.43
Fire retardant, "A" grade, 5" exposure	1.000	S.F.	.033	1.68	1.36	3.04
7" exposure	1.000	S.F.	.028	1.08	1.14	2.22
8-1/2" exposure	1.000	S.F.	.032	1.16	1.31	2.47
10" exposure	1.000	S.F.	.025	.90	1.02	1.92
Fire retardant, 5" exposure	1.000	S.F.	.029	2.08	1.19	3.27
7" exposure	1.000	S.F.	.036	1.61	1.45	3.06
8-1/2" exposure	1.000	S.F.	.032	1.44	1.31	2.75
10" exposure	1.000	S.F.	.025	1.12	1.02	2.14
Resquared & rebutted, 5-1/2" exposure	1.000	S.F.	.027	2.19	1.09	3.28
7" exposure	1.000	S.F.	.024	1.97	.98	2.95
8-1/2" exposure	1.000	S.F.	.021	1.75	.87	2.62
10" exposure	1.000	S.F.	.019	1.53	.76	2.29
Fire retardant, 5" exposure	1.000	S.F.	.027	2.51	1.09	3.60
7" exposure	1.000	S.F.	.024	2.25	.98	3.23
8-1/2" exposure	1.000	S.F.	.021	2	.87	2.87
10" exposure	1.000	S.F.	.023	1.35	.93	2.28
Hand-split, red cedar, 24" long, 7" exposure	1.000	S.F.	.045	2.13	1.82	3.95
8-1/2" exposure	1.000	S.F.	.038	1.82	1.56	3.38
10" exposure	1.000	S.F.	.032	1.52	1.30	2.82
12" exposure	1.000	S.F.	.026	1.22	1.04	2.26
Fire retardant, 7" exposure	1.000	S.F.	.045	2.59	1.82	4.41
8-1/2" exposure	1.000	S.F.	.038	2.22	1.56	3.78
10" exposure	1.000	S.F.	.032	1.85	1.30	3.15
12" exposure	1.000	S.F.	.026	1.48	1.04	2.52
18" long, 5" exposure	1.000	S.F.	.068	1.82	2.77	4.59
7" exposure	1.000	S.F.	.048	1.28	1.96	3.24
8-1/2" exposure	1.000	S.F.	.040	1.07	1.63	2.70
10" exposure	1.000	S.F.	.036	.96	1.47	2.43
Fire retardant, 5" exposure	1.000	S.F.	.068	2.38	2.77	5.15
7" exposure	1.000	S.F.	.048	1.68	1.96	3.64
8-1/2" exposure	1.000	S.F.	.040	1.40	1.63	3.03
10" exposure	1.000	S.F.	.036	1.26	1.47	2.73
Paper, #15 asphalt felt	1.100	S.F.	.002	.04	.09	.13
Trim, cedar	.125	S.F.	.005	.21	.20	.41
Fir	.125	S.F.	.005	.09	.20	.29
Redwood	.125	S.F.	.005	.21	.20	.41
White pine	.125	S.F.	.005	.09	.20	.29
Painting, primer, & 1 coat	1.000	S.F.	.013	.12	.48	.60
2 coats	1.000	S.F.	.017	.18	.61	.79
Staining, sealer, & 1 coat	1.000	S.F.	.017	.09	.62	.71
2 coats	1.000	S.F.	.019	.15	.67	.82

4 EXTERIOR WALLS

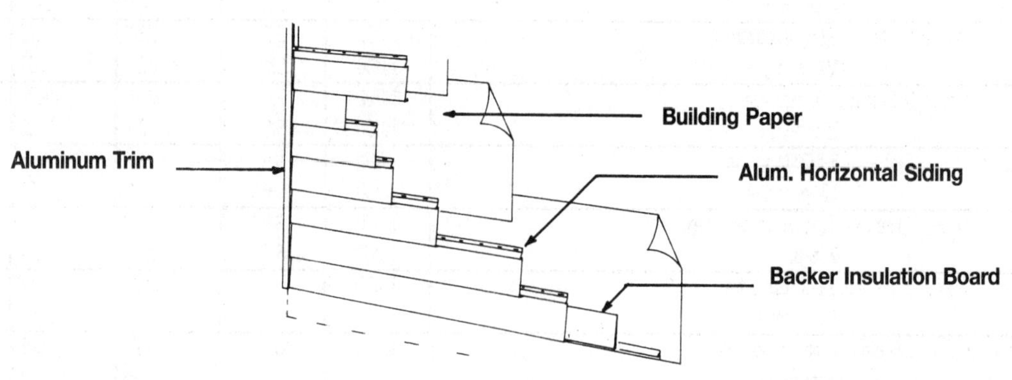

System Description	QUAN.	UNIT	LABOR HOURS	COST PER S.F. MAT.	INST.	TOTAL
ALUMINUM CLAPBOARD SIDING, 8″ WIDE, WHITE						
Aluminum horizontal siding, 8″ clapboard	1.000	S.F.	.031	1.38	1.27	2.65
Backer, insulation board	1.000	S.F.	.008	.45	.33	.78
Trim, aluminum	.600	L.F.	.016	.62	.64	1.26
Building wrap, spunbonded polypropylene	1.100	S.F.	.002	.12	.09	.21
TOTAL		S.F.	.057	2.57	2.33	4.90
ALUMINUM VERTICAL BOARD & BATTEN, WHITE						
Aluminum vertical board & batten	1.000	S.F.	.027	1.54	1.11	2.65
Backer insulation board	1.000	S.F.	.008	.45	.33	.78
Trim, aluminum	.600	L.F.	.016	.62	.64	1.26
Building wrap, spunbonded polypropylene	1.100	S.F.	.002	.12	.09	.21
TOTAL		S.F.	.053	2.73	2.17	4.90
VINYL CLAPBOARD SIDING, 8″ WIDE, WHITE						
PVC vinyl horizontal siding, 8″ clapboard	1.000	S.F.	.032	.68	1.32	2
Backer, insulation board	1.000	S.F.	.008	.45	.33	.78
Trim, vinyl	.600	L.F.	.014	.45	.57	1.02
Building wrap, spunbonded polypropylene	1.100	S.F.	.002	.12	.09	.21
TOTAL		S.F.	.056	1.70	2.31	4.01
VINYL VERTICAL BOARD & BATTEN, WHITE						
PVC vinyl vertical board & batten	1.000	S.F.	.029	1.51	1.19	2.70
Backer, insulation board	1.000	S.F.	.008	.45	.33	.78
Trim, vinyl	.600	L.F.	.014	.45	.57	1.02
Building wrap, spunbonded polypropylene	1.100	S.F.	.002	.12	.09	.21
TOTAL		S.F.	.053	2.53	2.18	4.71

The costs in this system are on a square foot of wall basis.
subtract openings from wall area.

Description	QUAN.	UNIT	LABOR HOURS	COST PER S.F. MAT.	INST.	TOTAL

Important: See the Reference Section for critical supporting data - Reference Nos., Crews & Location Factors

Metal & Plastic Siding Price Sheet	QUAN.	UNIT	LABOR HOURS	COST PER S.F. MAT.	INST.	TOTAL
ding, aluminum, .024″ thick, smooth, 8″ wide, white	1.000	S.F.	.031	1.38	1.27	2.65
Color	1.000	S.F.	.031	1.47	1.27	2.74
Double 4″ pattern, 8″ wide, white	1.000	S.F.	.031	1.31	1.27	2.58
Color	1.000	S.F.	.031	1.40	1.27	2.67
Double 5″ pattern, 10″ wide, white	1.000	S.F.	.029	1.31	1.19	2.50
Color	1.000	S.F.	.029	1.40	1.19	2.59
Embossed, single, 8″ wide, white	1.000	S.F.	.031	1.63	1.27	2.90
Color	1.000	S.F.	.031	1.72	1.27	2.99
Double 4″ pattern, 8″ wide, white	1.000	S.F.	.031	1.49	1.27	2.76
Color	1.000	S.F.	.031	1.58	1.27	2.85
Double 5″ pattern, 10″ wide, white	1.000	S.F.	.029	1.49	1.19	2.68
Color	1.000	S.F.	.029	1.58	1.19	2.77
Alum siding with insulation board, smooth, 8″ wide, white	1.000	S.F.	.031	1.32	1.27	2.59
Color	1.000	S.F.	.031	1.41	1.27	2.68
Double 4″ pattern, 8″ wide, white	1.000	S.F.	.031	1.30	1.27	2.57
Color	1.000	S.F.	.031	1.39	1.27	2.66
Double 5″ pattern, 10″ wide, white	1.000	S.F.	.029	1.30	1.19	2.49
Color	1.000	S.F.	.029	1.39	1.19	2.58
Embossed, single, 8″ wide, white	1.000	S.F.	.031	1.52	1.27	2.79
Color	1.000	S.F.	.031	1.61	1.27	2.88
Double 4″ pattern, 8″ wide, white	1.000	S.F.	.031	1.54	1.27	2.81
Color	1.000	S.F.	.031	1.63	1.27	2.90
Double 5″ pattern, 10″ wide, white	1.000	S.F.	.029	1.54	1.19	2.73
Color	1.000	S.F.	.029	1.63	1.19	2.82
Aluminum, shake finish, 10″ wide, white	1.000	S.F.	.029	1.63	1.19	2.82
Color	1.000	S.F.	.029	1.72	1.19	2.91
Aluminum, vertical, 12″ wide, white	1.000	S.F.	.027	1.54	1.11	2.65
Color	1.000	S.F.	.027	1.63	1.11	2.74
Vinyl siding, 8″ wide, smooth, white	1.000	S.F.	.032	.68	1.32	2
Color	1.000	S.F.	.032	.77	1.32	2.09
10″ wide, Dutch lap, smooth, white	1.000	S.F.	.029	.70	1.19	1.89
Color	1.000	S.F.	.029	.79	1.19	1.98
Double 4″ pattern, 8″ wide, white	1.000	S.F.	.032	.61	1.32	1.93
Color	1.000	S.F.	.032	.70	1.32	2.02
Double 5″ pattern, 10″ wide, white	1.000	S.F.	.029	.62	1.19	1.81
Color	1.000	S.F.	.029	.71	1.19	1.90
Embossed, single, 8″ wide, white	1.000	S.F.	.032	.73	1.32	2.05
Color	1.000	S.F.	.032	.82	1.32	2.14
10″ wide, white	1.000	S.F.	.029	.74	1.19	1.93
Color	1.000	S.F.	.029	.83	1.19	2.02
Double 4″ pattern, 8″ wide, white	1.000	S.F.	.032	.65	1.32	1.97
Color	1.000	S.F.	.032	.74	1.32	2.06
Double 5″ pattern, 10″ wide, white	1.000	S.F.	.029	.67	1.19	1.86
Color	1.000	S.F.	.029	.76	1.19	1.95
Vinyl, shake finish, 10″ wide, white	1.000	S.F.	.029	2.09	1.19	3.28
Color	1.000	S.F.	.029	2.18	1.19	3.37
Vinyl, vertical, double 5″ pattern, 10″ wide, white	1.000	S.F.	.029	1.51	1.19	2.70
Color	1.000	S.F.	.029	1.60	1.19	2.79
Backer board, installed in siding panels 8″ or 10″ wide	1.000	S.F.	.008	.45	.33	.78
4′ x 8′ sheets, polystyrene, 3/4″ thick	1.000	S.F.	.010	.42	.41	.83
4′ x 8′ fiberboard, plain	1.000	S.F.	.008	.45	.33	.78
Trim, aluminum, white	.600	L.F.	.016	.62	.64	1.26
Color	.600	L.F.	.016	.67	.64	1.31
Vinyl, white	.600	L.F.	.014	.45	.57	1.02
Color	.600	L.F.	.014	.44	.58	1.02
Paper, #15 asphalt felt	1.100	S.F.	.002	.04	.10	.14
Kraft paper, plain	1.100	S.F.	.002	.04	.10	.14
Foil backed	1.100	S.F.	.002	.08	.10	.18

EXTERIOR WALLS 4

Description	QUAN.	UNIT	LABOR HOURS	COST PER S.F. MAT.	INST.	TOTAL
Poured insulation, cellulose fiber, R3.8 per inch (1" thick)	1.000	S.F.	.003	.04	.14	.18
Fiberglass , R4.0 per inch (1" thick)	1.000	S.F.	.003	.04	.14	.18
Mineral wool, R3.0 per inch (1" thick)	1.000	S.F.	.003	.03	.14	.17
Polystyrene, R4.0 per inch (1" thick)	1.000	S.F.	.003	.20	.14	.34
Vermiculite, R2.7 per inch (1" thick)	1.000	S.F.	.003	.15	.14	.29
Perlite, R2.7 per inch (1" thick)	1.000	S.F.	.003	.15	.14	.29
Reflective insulation, aluminum foil reinforced with scrim	1.000	S.F.	.004	.15	.17	.32
Reinforced with woven polyolefin	1.000	S.F.	.004	.19	.17	.36
With single bubble air space, R8.8	1.000	S.F.	.005	.30	.22	.52
With double bubble air space, R9.8	1.000	S.F.	.005	.32	.22	.54
Rigid insulation, fiberglass, unfaced,						
1-1/2" thick, R6.2	1.000	S.F.	.008	.45	.33	.78
2" thick, R8.3	1.000	S.F.	.008	.52	.33	.85
2-1/2" thick, R10.3	1.000	S.F.	.010	.63	.41	1.04
3" thick, R12.4	1.000	S.F.	.010	.63	.41	1.04
Foil faced, 1" thick, R4.3	1.000	S.F.	.008	.88	.33	1.21
1-1/2" thick, R6.2	1.000	S.F.	.008	1.19	.33	1.52
2" thick, R8.7	1.000	S.F.	.009	1.49	.37	1.86
2-1/2" thick, R10.9	1.000	S.F.	.010	1.76	.41	2.17
3" thick, R13.0	1.000	S.F.	.010	1.91	.41	2.32
Foam glass, 1-1/2" thick R2.64	1.000	S.F.	.010	1.33	.41	1.74
2" thick R5.26	1.000	S.F.	.011	3.21	.45	3.66
Perlite, 1" thick R2.77	1.000	S.F.	.010	.29	.41	.70
2" thick R5.55	1.000	S.F.	.011	.55	.45	1
Polystyrene, extruded, blue, 2.2#/C.F., 3/4" thick R4	1.000	S.F.	.010	.42	.41	.83
1-1/2" thick R8.1	1.000	S.F.	.011	.83	.45	1.28
2" thick R10.8	1.000	S.F.	.011	1.14	.45	1.59
Molded bead board, white, 1" thick R3.85	1.000	S.F.	.010	.19	.41	.60
1-1/2" thick, R5.6	1.000	S.F.	.011	.46	.45	.91
2" thick, R7.7	1.000	S.F.	.011	.57	.45	1.02
Non-rigid insulation, batts						
Fiberglass, kraft faced, 3-1/2" thick, R11, 11" wide	1.000	S.F.	.005	.30	.20	.50
15" wide	1.000	S.F.	.005	.30	.20	.50
23" wide	1.000	S.F.	.005	.30	.20	.50
6" thick, R19, 11" wide	1.000	S.F.	.006	.39	.24	.63
15" wide	1.000	S.F.	.006	.39	.24	.63
23" wide	1.000	S.F.	.006	.39	.24	.63
9" thick, R30, 15" wide	1.000	S.F.	.006	.66	.24	.90
23" wide	1.000	S.F.	.006	.66	.24	.90
12" thick, R38, 15" wide	1.000	S.F.	.006	.84	.24	1.08
23" wide	1.000	S.F.	.006	.84	.24	1.08
Fiberglass, foil faced, 3-1/2" thick, R11, 15" wide	1.000	S.F.	.005	.44	.20	.64
23" wide	1.000	S.F.	.005	.44	.20	.64
6" thick, R19, 15" thick	1.000	S.F.	.005	.45	.20	.65
23" wide	1.000	S.F.	.005	.45	.20	.65
9" thick, R30, 15" wide	1.000	S.F.	.006	.78	.24	1.02
23" wide	1.000	S.F.	.006	.78	.24	1.02

Important: See the Reference Section for critical supporting data - Reference Nos., Crews & Location Factors

Insulation Systems	QUAN.	UNIT	LABOR HOURS	COST PER S.F.		
				MAT.	INST.	TOTAL
Non-rigid insulation batts						
Fiberglass unfaced, 3-1/2" thick, R11, 15" wide	1.000	S.F.	.005	.23	.20	.43
23" wide	1.000	S.F.	.005	.23	.20	.43
6" thick, R19, 15" wide	1.000	S.F.	.006	.37	.24	.61
23" wide	1.000	S.F.	.006	.37	.24	.61
9" thick, R19, 15" wide	1.000	S.F.	.007	.66	.28	.94
23" wide	1.000	S.F.	.007	.66	.28	.94
12" thick, R38, 15" wide	1.000	S.F.	.007	.84	.28	1.12
23" wide	1.000	S.F.	.007	.84	.28	1.12
Mineral fiber batts, 3" thick, R11	1.000	S.F.	.005	.31	.20	.51
3-1/2" thick, R13	1.000	S.F.	.005	.31	.20	.51
6" thick, R19	1.000	S.F.	.005	.42	.20	.62
6-1/2" thick, R22	1.000	S.F.	.005	.42	.20	.62
10" thick, R30	1.000	S.F.	.006	.63	.24	.87

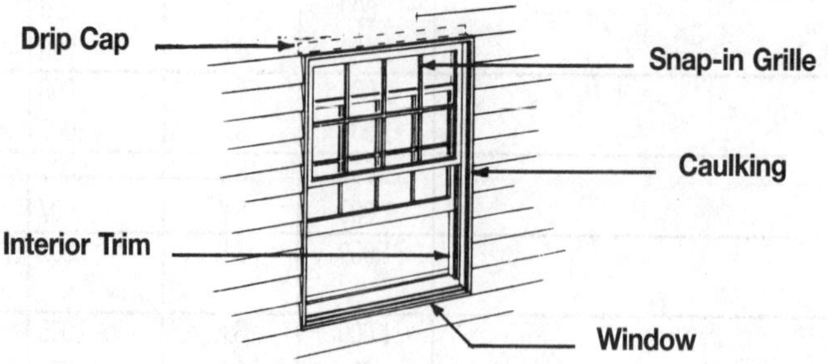

Drip Cap
Snap-in Grille
Caulking
Interior Trim
Window

System Description	QUAN.	UNIT	LABOR HOURS	COST EACH MAT.	COST EACH INST.	COST EACH TOTAL
BUILDER'S QUALITY WOOD WINDOW 2′ X 3′, DOUBLE HUNG						
Window, primed, builder's quality, 2′ x 3′, insulating glass	1.000	Ea.	.800	198	32.50	230.50
Trim, interior casing	11.000	L.F.	.367	10.34	14.96	25.30
Paint, interior & exterior, primer & 2 coats	2.000	Face	1.778	2.10	64	66.10
Caulking	10.000	L.F.	.323	1.70	13.30	15
Snap-in grille	1.000	Set	.333	46	13.60	59.60
Drip cap, metal	2.000	L.F.	.040	.56	1.64	2.20
TOTAL		Ea.	3.641	258.70	140	398.70
PLASTIC CLAD WOOD WINDOW 3′ X 4′, DOUBLE HUNG						
Window, plastic clad, premium, 3′ x 4′, insulating glass	1.000	Ea.	.889	355	36	391
Trim, interior casing	15.000	L.F.	.500	14.10	20.40	34.50
Paint, interior, primer & 2 coats	1.000	Face	.889	1.05	32	33.05
Caulking	14.000	L.F.	.452	2.38	18.62	21
Snap-in grille	1.000	Set	.333	46	13.60	59.60
TOTAL		Ea.	3.063	418.53	120.62	539.15
METAL CLAD WOOD WINDOW, 3′ X 5′, DOUBLE HUNG						
Window, metal clad, deluxe, 3′ x 5′, insulating glass	1.000	Ea.	1.000	297	41	338
Trim, interior casing	17.000	L.F.	.567	15.98	23.12	39.10
Paint, interior, primer & 2 coats	1.000	Face	.889	1.05	32	33.05
Caulking	16.000	L.F.	.516	2.72	21.28	24
Snap-in grille	1.000	Set	.235	128	9.60	137.60
Drip cap, metal	3.000	L.F.	.060	.84	2.46	3.30
TOTAL		Ea.	3.267	445.59	129.46	575.05

The cost of this system is on a cost per each window basis.

Description	QUAN.	UNIT	LABOR HOURS	COST EACH MAT.	COST EACH INST.	COST EACH TOTAL

Double Hung Window Price Sheet	QUAN.	UNIT	LABOR HOURS	COST EACH MAT.	COST EACH INST.	COST EACH TOTAL
Windows, double-hung, builder's quality, 2′ x 3′, single glass	1.000	Ea.	.800	202	32.50	234.50
Insulating glass	1.000	Ea.	.800	198	32.50	230.50
3′ x 4′, single glass	1.000	Ea.	.889	269	36	305
Insulating glass	1.000	Ea.	.889	258	36	294
4′ x 4′-6″, single glass	1.000	Ea.	1.000	305	41	346
Insulating glass	1.000	Ea.	1.000	330	41	371
Plastic clad premium insulating glass, 2′-6″ x 3′	1.000	Ea.	.800	215	32.50	247.50
3′ x 3′-6″	1.000	Ea.	.800	253	32.50	285.50
3′ x 4′	1.000	Ea.	.889	355	36	391
3′ x 4′-6″	1.000	Ea.	.889	292	36	328
3′ x 5′	1.000	Ea.	1.000	305	41	346
3′-6″ x 6′	1.000	Ea.	1.000	355	41	396
Metal clad deluxe insulating glass, 2′-6″ x 3′	1.000	Ea.	.800	204	32.50	236.50
3′ x 3′-6″	1.000	Ea.	.800	242	32.50	274.50
3′ x 4′	1.000	Ea.	.889	257	36	293
3′ x 4′-6″	1.000	Ea.	.889	278	36	314
3′ x 5′	1.000	Ea.	1.000	297	41	338
3′-6″ x 6′	1.000	Ea.	1.000	360	41	401
Trim, interior casing, window 2′ x 3′	11.000	L.F.	.367	10.35	14.95	25.30
2′-6″ x 3′	12.000	L.F.	.400	11.30	16.30	27.60
3′ x 3′-6″	14.000	L.F.	.467	13.15	19.05	32.20
3′ x 4′	15.000	L.F.	.500	14.10	20.50	34.60
3′ x 4′-6″	16.000	L.F.	.533	15.05	22	37.05
3′ x 5′	17.000	L.F.	.567	16	23	39
3′-6″ x 6′	20.000	L.F.	.667	18.80	27	45.80
4′ x 4′-6″	18.000	L.F.	.600	16.90	24.50	41.40
Paint or stain, interior or exterior, 2′ x 3′ window, 1 coat	1.000	Face	.444	.38	15.90	16.28
2 coats	1.000	Face	.727	.77	26	26.77
Primer & 1 coat	1.000	Face	.727	.70	26	26.70
Primer & 2 coats	1.000	Face	.889	1.05	32	33.05
3′ x 4′ window, 1 coat	1.000	Face	.667	.79	24	24.79
2 coats	1.000	Face	.667	.89	24	24.89
Primer & 1 coat	1.000	Face	.727	1.07	26	27.07
Primer & 2 coats	1.000	Face	.889	1.05	32	33.05
4′ x 4′-6″ window, 1 coat	1.000	Face	.667	.79	24	24.79
2 coats	1.000	Face	.667	.89	24	24.89
Primer & 1 coat	1.000	Face	.727	1.07	26	27.07
Primer & 2 coats	1.000	Face	.889	1.05	32	33.05
Caulking, window, 2′ x 3′	10.000	L.F.	.323	1.70	13.30	15
2′-6″ x 3′	11.000	L.F.	.355	1.87	14.65	16.52
3′ x 3′-6″	13.000	L.F.	.419	2.21	17.30	19.51
3′ x 4′	14.000	L.F.	.452	2.38	18.60	20.98
3′ x 4′-6″	15.000	L.F.	.484	2.55	19.95	22.50
3′ x 5′	16.000	L.F.	.516	2.72	21.50	24.22
3′-6″ x 6′	19.000	L.F.	.613	3.23	25.50	28.73
4′ x 4′-6″	17.000	L.F.	.548	2.89	22.50	25.39
Grilles, glass size to, 16″ x 24″ per sash	1.000	Set	.333	46	13.60	59.60
32″ x 32″ per sash	1.000	Set	.235	128	9.60	137.60
Drip cap, aluminum, 2′ long	2.000	L.F.	.040	.56	1.64	2.20
3′ long	3.000	L.F.	.060	.84	2.46	3.30
4′ long	4.000	L.F.	.080	1.12	3.28	4.40
Wood, 2′ long	2.000	L.F.	.067	1.88	2.72	4.60
3′ long	3.000	L.F.	.100	2.82	4.08	6.90
4′ long	4.000	L.F.	.133	3.76	5.45	9.21

32 Casement Window Systems

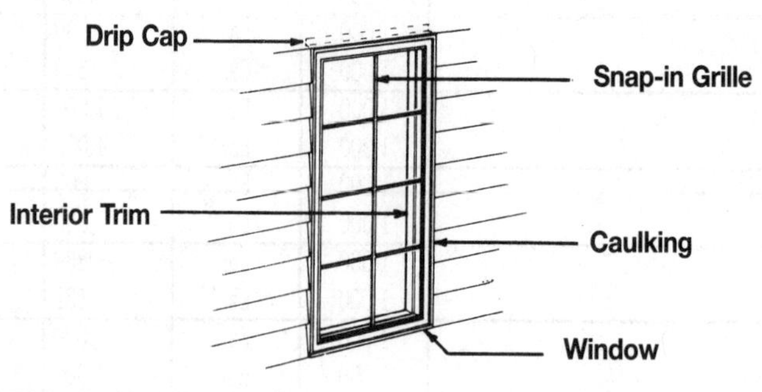

System Description	QUAN.	UNIT	LABOR HOURS	COST EACH		
				MAT.	INST.	TOTAL
BUILDER'S QUALITY WINDOW, WOOD, 2′ BY 3′, CASEMENT						
Window, primed, builder's quality, 2′ x 3′, insulating glass	1.000	Ea.	.800	310	32.50	342.50
Trim, interior casing	11.000	L.F.	.367	10.34	14.96	25.30
Paint, interior & exterior, primer & 2 coats	2.000	Face	1.778	2.10	64	66.10
Caulking	10.000	L.F.	.323	1.70	13.30	15
Snap-in grille	1.000	Ea.	.267	26	10.85	36.85
Drip cap, metal	2.000	L.F.	.040	.56	1.64	2.20
TOTAL		Ea.	3.575	350.70	137.25	487.95
PLASTIC CLAD WOOD WINDOW, 2′ X 4′, CASEMENT						
Window, plastic clad, premium, 2′ x 4′, insulating glass	1.000	Ea.	.889	305	36	341
Trim, interior casing	13.000	L.F.	.433	12.22	17.68	29.90
Paint, interior, primer & 2 coats	1.000	Ea.	.889	1.05	32	33.05
Caulking	12.000	L.F.	.387	2.04	15.96	18
Snap-in grille	1.000	Ea.	.267	26	10.85	36.85
TOTAL		Ea.	2.865	346.31	112.49	458.80
METAL CLAD WOOD WINDOW, 2′ X 5′, CASEMENT						
Window, metal clad, deluxe, 2′ x 5′, insulating glass	1.000	Ea.	1.000	263	41	304
Trim, interior casing	15.000	L.F.	.500	14.10	20.40	34.50
Paint, interior, primer & 2 coats	1.000	Ea.	.889	1.05	32	33.05
Caulking	14.000	L.F.	.452	2.38	18.62	21
Snap-in grille	1.000	Ea.	.250	37	10.20	47.20
Drip cap, metal	12.000	L.F.	.040	.56	1.64	2.20
TOTAL		Ea.	3.131	318.09	123.86	441.95

The cost of this system is on a cost per each window basis.

Description	QUAN.	UNIT	LABOR HOURS	COST EACH		
				MAT.	INST.	TOTAL

Important: See the Reference Section for critical supporting data - Reference Nos., Crews & Location Factors

Casement Window Price Sheet

	QUAN.	UNIT	LABOR HOURS	COST EACH MAT.	INST.	TOTAL
indow, casement, builders quality, 2′ x 3′, single glass	1.000	Ea.	.800	193	32.50	225.50
Insulating glass	1.000	Ea.	.800	310	32.50	342.50
2′ x 4′-6″, single glass	1.000	Ea.	.727	725	29.50	754.50
Insulating glass	1.000	Ea.	.727	670	29.50	699.50
2′ x 6′, single glass	1.000	Ea.	.889	1,050	36	1,086
Insulating glass	1.000	Ea.	.889	930	36	966
Plastic clad premium insulating glass, 2′ x 3′	1.000	Ea.	.800	187	32.50	219.50
2′ x 4′	1.000	Ea.	.889	208	36	244
2′ x 5′	1.000	Ea.	1.000	284	41	325
2′ x 6′	1.000	Ea.	1.000	360	41	401
Metal clad deluxe insulating glass, 2′ x 3′	1.000	Ea.	.800	193	32.50	225.50
2′ x 4′	1.000	Ea.	.889	232	36	268
2′ x 5′	1.000	Ea.	1.000	264	41	305
2′ x 6′	1.000	Ea.	1.000	305	41	346
rim, interior casing, window 2′ x 3′	11.000	L.F.	.367	10.35	14.95	25.30
2′ x 4′	13.000	L.F.	.433	12.20	17.70	29.90
2′ x 4′-6″	14.000	L.F.	.467	13.15	19.05	32.20
2′ x 5′	15.000	L.F.	.500	14.10	20.50	34.60
2′ x 6′	17.000	L.F.	.567	16	23	39
aint or stain, interior or exterior, 2′ x 3′ window, 1 coat	1.000	Face	.444	.38	15.90	16.28
2 coats	1.000	Face	.727	.77	26	26.77
Primer & 1 coat	1.000	Face	.727	.70	26	26.70
Primer & 2 coats	1.000	Face	.889	1.05	32	33.05
2′ x 4′ window, 1 coat	1.000	Face	.444	.38	15.90	16.28
2 coats	1.000	Face	.727	.77	26	26.77
Primer & 1 coat	1.000	Face	.727	.70	26	26.70
Primer & 2 coats	1.000	Face	.889	1.05	32	33.05
2′ x 6′ window, 1 coat	1.000	Face	.667	.79	24	24.79
2 coats	1.000	Face	.667	.89	24	24.89
Primer & 1 coat	1.000	Face	.727	1.07	26	27.07
Primer & 2 coats	1.000	Face	.889	1.05	32	33.05
aulking, window, 2′ x 3′	10.000	L.F.	.323	1.70	13.30	15
2′ x 4′	12.000	L.F.	.387	2.04	15.95	17.99
2′ x 4′-6″	13.000	L.F.	.419	2.21	17.30	19.51
2′ x 5′	14.000	L.F.	.452	2.38	18.60	20.98
2′ x 6′	16.000	L.F.	.516	2.72	21.50	24.22
rilles, glass size, to 20″ x 36″	1.000	Ea.	.267	26	10.85	36.85
To 20″ x 56″	1.000	Ea.	.250	37	10.20	47.20
rip cap, metal, 2′ long	2.000	L.F.	.040	.56	1.64	2.20
Wood, 2′ long	2.000	L.F.	.067	1.88	2.72	4.60

36 Awning Window Systems

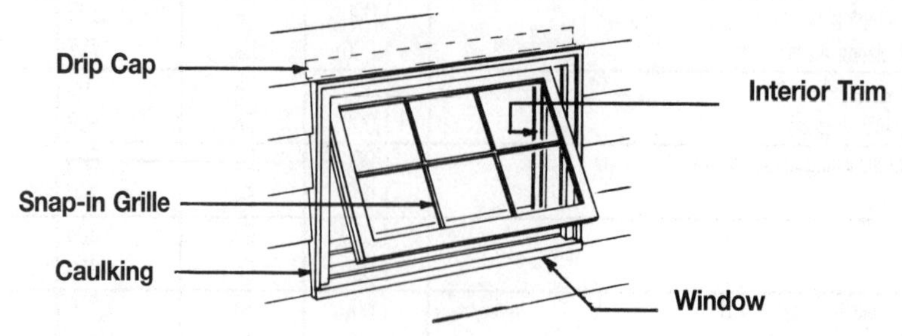

System Description	QUAN.	UNIT	LABOR HOURS	COST EACH MAT.	COST EACH INST.	COST EACH TOTAL
BUILDER'S QUALITY WINDOW, WOOD, 34" X 22", AWNING						
Window, builder quality, 34" x 22", insulating glass	1.000	Ea.	.800	243	32.50	275.50
Trim, interior casing	10.500	L.F.	.350	9.87	14.28	24.15
Paint, interior & exterior, primer & 2 coats	2.000	Face	1.778	2.10	64	66.10
Caulking	9.500	L.F.	.306	1.62	12.64	14.26
Snap-in grille	1.000	Ea.	.267	21	10.85	31.85
Drip cap, metal	3.000	L.F.	.060	.84	2.46	3.30
TOTAL		Ea.	3.561	278.43	136.73	415.16
PLASTIC CLAD WOOD WINDOW, 40" X 28", AWNING						
Window, plastic clad, premium, 40" x 28", insulating glass	1.000	Ea.	.889	315	36	351
Trim interior casing	13.500	L.F.	.450	12.69	18.36	31.05
Paint, interior, primer & 2 coats	1.000	Face	.889	1.05	32	33.05
Caulking	12.500	L.F.	.403	2.13	16.63	18.76
Snap-in grille	1.000	Ea.	.267	21	10.85	31.85
TOTAL		Ea.	2.898	351.87	113.84	465.71
METAL CLAD WOOD WINDOW, 48" X 36", AWNING						
Window, metal clad, deluxe, 48" x 36", insulating glass	1.000	Ea.	1.000	340	41	381
Trim, interior casing	15.000	L.F.	.500	14.10	20.40	34.50
Paint, interior, primer & 2 coats	1.000	Face	.889	1.05	32	33.05
Caulking	14.000	L.F.	.452	2.38	18.62	21
Snap-in grille	1.000	Ea.	.250	30.50	10.20	40.70
Drip cap, metal	4.000	L.F.	.080	1.12	3.28	4.40
TOTAL		Ea.	3.171	389.15	125.50	514.65

The cost of this system is on a cost per each window basis.

Description	QUAN.	UNIT	LABOR HOURS	COST EACH MAT.	COST EACH INST.	COST EACH TOTAL

Awning Window Price Sheet

	QUAN.	UNIT	LABOR HOURS	COST EACH MAT.	COST EACH INST.	COST EACH TOTAL
indows, awning, builder's quality, 34″ x 22″, insulated glass	1.000	Ea.	.800	230	32.50	262.50
Low E glass	1.000	Ea.	.800	243	32.50	275.50
40″ x 28″, insulated glass	1.000	Ea.	.889	290	36	326
Low E glass	1.000	Ea.	.889	310	36	346
48″ x 36″, insulated glass	1.000	Ea.	1.000	425	41	466
Low E glass	1.000	Ea.	1.000	445	41	486
Plastic clad premium insulating glass, 34″ x 22″	1.000	Ea.	.800	245	32.50	277.50
40″ x 22″	1.000	Ea.	.800	267	32.50	299.50
36″ x 28″	1.000	Ea.	.889	285	36	321
36″ x 36″	1.000	Ea.	.889	315	36	351
48″ x 28″	1.000	Ea.	1.000	340	41	381
60″ x 36″	1.000	Ea.	1.000	495	41	536
Metal clad deluxe insulating glass, 34″ x 22″	1.000	Ea.	.800	229	32.50	261.50
40″ x 22″	1.000	Ea.	.800	268	32.50	300.50
36″ x 25″	1.000	Ea.	.889	249	36	285
40″ x 30″	1.000	Ea.	.889	310	36	346
48″ x 28″	1.000	Ea.	1.000	320	41	361
60″ x 36″	1.000	Ea.	1.000	340	41	381
rim, interior casing window, 34″ x 22″	10.500	L.F.	.350	9.85	14.30	24.15
40″ x 22″	11.500	L.F.	.383	10.80	15.65	26.45
36″ x 28″	12.500	L.F.	.417	11.75	17	28.75
40″ x 28″	13.500	L.F.	.450	12.70	18.35	31.05
48″ x 28″	14.500	L.F.	.483	13.65	19.70	33.35
48″ x 36″	15.000	L.F.	.500	14.10	20.50	34.60
aint or stain, interior or exterior, 34″ x 22″, 1 coat	1.000	Face	.444	.38	15.90	16.28
2 coats	1.000	Face	.727	.77	26	26.77
Primer & 1 coat	1.000	Face	.727	.70	26	26.70
Primer & 2 coats	1.000	Face	.889	1.05	32	33.05
36″ x 28″, 1 coat	1.000	Face	.444	.38	15.90	16.28
2 coats	1.000	Face	.727	.77	26	26.77
Primer & 1 coat	1.000	Face	.727	.70	26	26.70
Primer & 2 coats	1.000	Face	.889	1.05	32	33.05
48″ x 36″, 1 coat	1.000	Face	.667	.79	24	24.79
2 coats	1.000	Face	.667	.89	24	24.89
Primer & 1 coat	1.000	Face	.727	1.07	26	27.07
Primer & 2 coats	1.000	Face	.889	1.05	32	33.05
aulking, window, 34″ x 22″	9.500	L.F.	.306	1.62	12.65	14.27
40″ x 22″	10.500	L.F.	.339	1.79	13.95	15.74
36″ x 28″	11.500	L.F.	.371	1.96	15.30	17.26
40″ x 28″	12.500	L.F.	.403	2.13	16.65	18.78
48″ x 28″	13.500	L.F.	.436	2.30	17.95	20.25
48″ x 36″	14.000	L.F.	.452	2.38	18.60	20.98
irilles, glass size, to 28″ by 16″	1.000	Ea.	.267	21	10.85	31.85
To 44″ by 24″	1.000	Ea.	.250	30.50	10.20	40.70
rip cap, aluminum, 3′ long	3.000	L.F.	.060	.84	2.46	3.30
3′-6″ long	3.500	L.F.	.070	.98	2.87	3.85
4′ long	4.000	L.F.	.080	1.12	3.28	4.40
Wood, 3′ long	3.000	L.F.	.100	2.82	4.08	6.90
3′-6″ long	3.500	L.F.	.117	3.29	4.76	8.05
4′ long	4.000	L.F.	.133	3.76	5.45	9.21

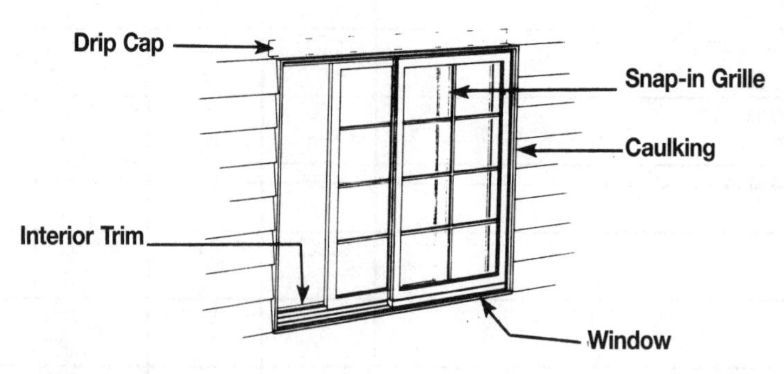

System Description	QUAN.	UNIT	LABOR HOURS	COST EACH		
				MAT.	INST.	TOTAL
BUILDER'S QUALITY WOOD WINDOW, 3′ X 2′, SLIDING						
Window, primed, builder's quality, 3′ x 2′, insul. glass	1.000	Ea.	.800	200	32.50	232.50
Trim, interior casing	11.000	L.F.	.367	10.34	14.96	25.30
Paint, interior & exterior, primer & 2 coats	2.000	Face	1.778	2.10	64	66.10
Caulking	10.000	L.F.	.323	1.70	13.30	15
Snap-in grille	1.000	Set	.333	25.50	13.60	39.10
Drip cap, metal	3.000	L.F.	.060	.84	2.46	3.30
TOTAL		Ea.	3.661	240.48	140.82	381.30
PLASTIC CLAD WOOD WINDOW, 4′ X 3′-6″, SLIDING						
Window, plastic clad, premium, 4′ x 3′-6″, insulating glass	1.000	Ea.	.889	705	36	741
Trim, interior casing	16.000	L.F.	.533	15.04	21.76	36.80
Paint, interior, primer & 2 coats	1.000	Face	.889	1.05	32	33.05
Caulking	17.000	L.F.	.548	2.89	22.61	25.50
Snap-in grille	1.000	Set	.333	25.50	13.60	39.10
TOTAL		Ea.	3.192	749.48	125.97	875.45
METAL CLAD WOOD WINDOW, 6′ X 5′, SLIDING						
Window, metal clad, deluxe, 6′ x 5′, insulating glass	1.000	Ea.	1.000	710	41	751
Trim, interior casing	23.000	L.F.	.767	21.62	31.28	52.90
Paint, interior, primer & 2 coats	1.000	Face	.889	1.05	32	33.05
Caulking	22.000	L.F.	.710	3.74	29.26	33
Snap-in grille	1.000	Set	.364	38	14.80	52.80
Drip cap, metal	6.000	L.F.	.120	1.68	4.92	6.60
TOTAL		Ea.	3.850	776.09	153.26	929.35

The cost of this system is on a cost per each window basis.

Description	QUAN.	UNIT	LABOR HOURS	COST EACH		
				MAT.	INST.	TOTAL

EXTERIOR WALLS 4

Important: See the Reference Section for critical supporting data - Reference Nos., Crews & Location Factors

Sliding Window Price Sheet	QUAN.	UNIT	LABOR HOURS	COST EACH		
				MAT.	INST.	TOTAL
indows, sliding, builder's quality, 3′ x 3′, single glass	1.000	Ea.	.800	158	32.50	190.50
Insulating glass	1.000	Ea.	.800	200	32.50	232.50
4′ x 3′-6″, single glass	1.000	Ea.	.889	188	36	224
Insulating glass	1.000	Ea.	.889	236	36	272
6′ x 5′, single glass	1.000	Ea.	1.000	345	41	386
Insulating glass	1.000	Ea.	1.000	415	41	456
Plastic clad premium insulating glass, 3′ x 3′	1.000	Ea.	.800	565	32.50	597.50
4′ x 3′-6″	1.000	Ea.	.889	705	36	741
5′ x 4′	1.000	Ea.	.889	850	36	886
6′ x 5′	1.000	Ea.	1.000	1,025	41	1,066
Metal clad deluxe insulating glass, 3′ x 3′	1.000	Ea.	.800	320	32.50	352.50
4′ x 3′-6″	1.000	Ea.	.889	390	36	426
5′ x 4′	1.000	Ea.	.889	470	36	506
6′ x 5′	1.000	Ea.	1.000	710	41	751
im, interior casing, window 3′ x 2′	11.000	L.F.	.367	10.35	14.95	25.30
3′ x 3′	13.000	L.F.	.433	12.20	17.70	29.90
4′ x 3′-6″	16.000	L.F.	.533	15.05	22	37.05
5′ x 4′	19.000	L.F.	.633	17.85	26	43.85
6′ x 5′	23.000	L.F.	.767	21.50	31.50	53
aint or stain, interior or exterior, 3′ x 2′ window, 1 coat	1.000	Face	.444	.38	15.90	16.28
2 coats	1.000	Face	.727	.77	26	26.77
Primer & 1 coat	1.000	Face	.727	.70	26	26.70
Primer & 2 coats	1.000	Face	.889	1.05	32	33.05
4′ x 3′-6″ window, 1 coat	1.000	Face	.667	.79	24	24.79
2 coats	1.000	Face	.667	.89	24	24.89
Primer & 1 coat	1.000	Face	.727	1.07	26	27.07
Primer & 2 coats	1.000	Face	.889	1.05	32	33.05
6′ x 5′ window, 1 coat	1.000	Face	.889	2.17	32	34.17
2 coats	1.000	Face	1.333	3.97	47.50	51.47
Primer & 1 coat	1.000	Face	1.333	3.76	47.50	51.26
Primer & 2 coats	1.000	Face	1.600	5.60	57	62.60
aulking, window, 3′ x 2′	10.000	L.F.	.323	1.70	13.30	15
3′ x 3′	12.000	L.F.	.387	2.04	15.95	17.99
4′ x 3′-6″	15.000	L.F.	.484	2.55	19.95	22.50
5′ x 4′	18.000	L.F.	.581	3.06	24	27.06
6′ x 5′	22.000	L.F.	.710	3.74	29.50	33.24
rilles, glass size, to 14″ x 36″	1.000	Set	.333	25.50	13.60	39.10
To 36″ x 36″	1.000	Set	.364	38	14.80	52.80
rip cap, aluminum, 3′ long	3.000	L.F.	.060	.84	2.46	3.30
4′ long	4.000	L.F.	.080	1.12	3.28	4.40
5′ long	5.000	L.F.	.100	1.40	4.10	5.50
6′ long	6.000	L.F.	.120	1.68	4.92	6.60
Wood, 3′ long	3.000	L.F.	.100	2.82	4.08	6.90
4′ long	4.000	L.F.	.133	3.76	5.45	9.21
5′ long	5.000	L.F.	.167	4.70	6.80	11.50
6′ long	6.000	L.F.	.200	5.65	8.15	13.80

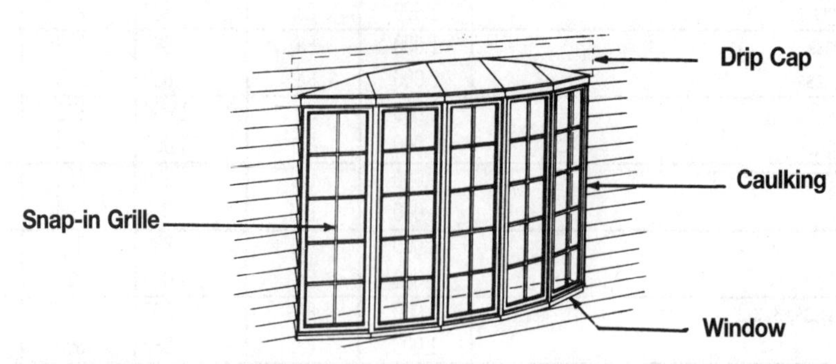

System Description	QUAN.	UNIT	LABOR HOURS	COST EACH MAT.	COST EACH INST.	COST EACH TOTAL
AWNING TYPE BOW WINDOW, BUILDER'S QUALITY, 8′ X 5′						
Window, primed, builder's quality, 8′ x 5′, insulating glass	1.000	Ea.	1.600	1,325	65	1,390
Trim, interior casing	27.000	L.F.	.900	25.38	36.72	62.10
Paint, interior & exterior, primer & 1 coat	2.000	Face	3.200	11.20	114	125.20
Drip cap, vinyl	1.000	Ea.	.533	82.50	21.50	104
Caulking	26.000	L.F.	.839	4.42	34.58	39
Snap-in grilles	1.000	Set	1.067	104	43.40	147.40
TOTAL		Ea.	8.139	1,552.50	315.20	1,867.70
CASEMENT TYPE BOW WINDOW, PLASTIC CLAD, 10′ X 6′						
Window, plastic clad, premium, 10′ x 6′, insulating glass	1.000	Ea.	2.286	1,875	93	1,968
Trim, interior casing	33.000	L.F.	1.100	31.02	44.88	75.90
Paint, interior, primer & 1 coat	1.000	Face	1.778	2.10	64	66.10
Drip cap, vinyl	1.000	Ea.	.615	90	25	115
Caulking	32.000	L.F.	1.032	5.44	42.56	48
Snap-in grilles	1.000	Set	1.333	130	54.25	184.25
TOTAL		Ea.	8.144	2,133.56	323.69	2,457.25
DOUBLE HUNG TYPE, METAL CLAD, 9′ X 5′						
Window, metal clad, deluxe, 9′ x 5′, insulating glass	1.000	Ea.	2.667	1,200	109	1,309
Trim, interior casing	29.000	L.F.	.967	27.26	39.44	66.70
Paint, interior, primer & 1 coat	1.000	Face	1.778	2.10	64	66.10
Drip cap, vinyl	1.000	Set	.615	90	25	115
Caulking	28.000	L.F.	.903	4.76	37.24	42
Snap-in grilles	1.000	Set	1.067	104	43.40	147.40
TOTAL		Ea.	7.997	1,428.12	318.08	1,746.20

The cost of this system is on a cost per each window basis.

Description	QUAN.	UNIT	LABOR HOURS	COST EACH MAT.	COST EACH INST.	COST EACH TOTAL

Bow/Bay Window Price Sheet	QUAN.	UNIT	LABOR HOURS	COST EACH		
				MAT.	INST.	TOTAL
Windows, bow awning type, builder's quality, 8′ x 5′, insulating glass	1.000	Ea.	1.600	1,625	65	1,690
Low E glass	1.000	Ea.	1.600	1,325	65	1,390
12′ x 6′, insulating glass	1.000	Ea.	2.667	1,375	109	1,484
Low E glass	1.000	Ea.	2.667	1,450	109	1,559
Plastic clad premium insulating glass, 6′ x 4′	1.000	Ea.	1.600	1,100	65	1,165
9′ x 4′	1.000	Ea.	2.000	1,475	81.50	1,556.50
10′ x 5′	1.000	Ea.	2.286	2,375	93	2,468
12′ x 6′	1.000	Ea.	2.667	3,000	109	3,109
Metal clad deluxe insulating glass, 6′ x 4′	1.000	Ea.	1.600	940	65	1,005
9′ x 4′	1.000	Ea.	2.000	1,325	81.50	1,406.50
10′ x 5′	1.000	Ea.	2.286	1,825	93	1,918
12′ x 6′	1.000	Ea.	2.667	2,525	109	2,634
Bow casement type, builder's quality, 8′ x 5′, single glass	1.000	Ea.	1.600	1,550	65	1,615
Insulating glass	1.000	Ea.	1.600	1,875	65	1,940
12′ x 6′, single glass	1.000	Ea.	2.667	1,950	109	2,059
Insulating glass	1.000	Ea.	2.667	2,025	109	2,134
Plastic clad premium insulating glass, 8′ x 5′	1.000	Ea.	1.600	1,250	65	1,315
10′ x 5′	1.000	Ea.	2.000	1,775	81.50	1,856.50
10′ x 6′	1.000	Ea.	2.286	1,875	93	1,968
12′ x 6′	1.000	Ea.	2.667	2,225	109	2,334
Metal clad deluxe insulating glass, 8′ x 5′	1.000	Ea.	1.600	1,375	65	1,440
10′ x 5′	1.000	Ea.	2.000	1,475	81.50	1,556.50
10′ x 6′	1.000	Ea.	2.286	1,750	93	1,843
12′ x 6′	1.000	Ea.	2.667	2,425	109	2,534
Bow, double hung type, builder's quality, 8′ x 4′, single glass	1.000	Ea.	1.600	1,100	65	1,165
Insulating glass	1.000	Ea.	1.600	1,175	65	1,240
9′ x 5′, single glass	1.000	Ea.	2.667	1,175	109	1,284
Insulating glass	1.000	Ea.	2.667	1,250	109	1,359
Plastic clad premium insulating glass, 7′ x 4′	1.000	Ea.	1.600	1,125	65	1,190
8′ x 4′	1.000	Ea.	2.000	1,150	81.50	1,231.50
8′ x 5′	1.000	Ea.	2.286	1,200	93	1,293
9′ x 5′	1.000	Ea.	2.667	1,250	109	1,359
Metal clad deluxe insulating glass, 7′ x 4′	1.000	Ea.	1.600	1,050	65	1,115
8′ x 4′	1.000	Ea.	2.000	1,075	81.50	1,156.50
8′ x 5′	1.000	Ea.	2.286	1,125	93	1,218
9′ x 5′	1.000	Ea.	2.667	1,200	109	1,309
Trim, interior casing, window 7′ x 4′	1.000	Ea.	.767	21.50	31.50	53
8′ x 5′	1.000	Ea.	.900	25.50	36.50	62
10′ x 6′	1.000	Ea.	1.100	31	45	76
12′ x 6′	1.000	Ea.	1.233	35	50.50	85.50
Paint or stain, interior, or exterior, 7′ x 4′ window, 1 coat	1.000	Face	.889	2.17	32	34.17
Primer & 1 coat	1.000	Face	1.333	3.76	47.50	51.26
8′ x 5′ window, 1 coat	1.000	Face	.889	2.17	32	34.17
Primer & 1 coat	1.000	Face	1.333	3.76	47.50	51.26
10′ x 6′ window, 1 coat	1.000	Face	1.333	1.58	48	49.58
Primer & 1 coat	1.000	Face	1.778	2.10	64	66.10
12′ x 6′ window, 1 coat	1.000	Face	1.778	4.34	64	68.34
Primer & 1 coat	1.000	Face	2.667	7.50	95	102.50
Drip cap, vinyl moulded window, 7′ long	1.000	Ea.	.533	82.50	21.50	104
8′ long	1.000	Ea.	.533	82.50	21.50	104
10′ long	1.000	Ea.	.615	90	25	115
12′ long	1.000	Ea.	.615	90	25	115
Caulking, window, 7′ x 4′	1.000	Ea.	.710	3.74	29.50	33.24
8′ x 5′	1.000	Ea.	.839	4.42	34.50	38.92
10′ x 6′	1.000	Ea.	1.032	5.45	42.50	47.95
12′ x 6′	1.000	Ea.	1.161	6.10	48	54.10
Grilles, window, 7′ x 4′	1.000	Set	.800	78	32.50	110.50
8′ x 5′	1.000	Set	1.067	104	43.50	147.50
10′ x 6′	1.000	Set	1.333	130	54.50	184.50
12′ x 6′	1.000	Set	1.600	156	65	221

48 Fixed Window Systems

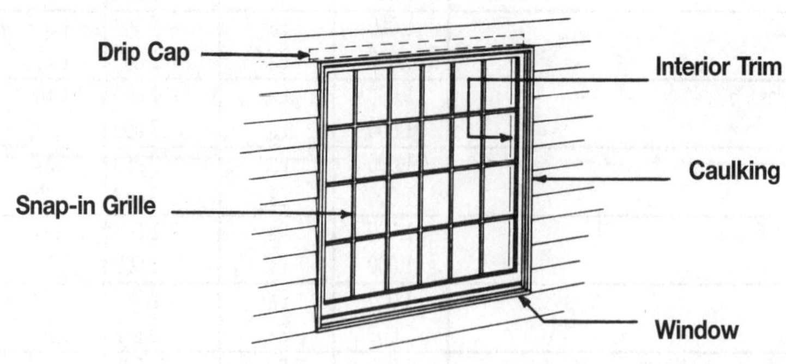

System Description	QUAN.	UNIT	LABOR HOURS	COST EACH		
				MAT.	INST.	TOTAL
BUILDER'S QUALITY PICTURE WINDOW, 4′ X 4′						
Window, primed, builder's quality, 4′ x 4′, insulating glass	1.000	Ea.	1.333	320	54.50	374.50
Trim, interior casing	17.000	L.F.	.567	15.98	23.12	39.10
Paint, interior & exterior, primer & 2 coats	2.000	Face	1.778	2.10	64	66.10
Caulking	16.000	L.F.	.516	2.72	21.28	24
Snap-in grille	1.000	Ea.	.267	146	10.85	156.85
Drip cap, metal	4.000	L.F.	.080	1.12	3.28	4.40
TOTAL		Ea.	4.541	487.92	177.03	664.95
PLASTIC CLAD WOOD WINDOW, 4′-6″ X 6′-6″						
Window, plastic clad, prem., 4′-6″ x 6′-6″, insul. glass	1.000	Ea.	1.455	700	59.50	759.50
Trim, interior casing	23.000	L.F.	.767	21.62	31.28	52.90
Paint, interior, primer & 2 coats	1.000	Face	.889	1.05	32	33.05
Caulking	22.000	L.F.	.710	3.74	29.26	33
Snap-in grille	1.000	Ea.	.267	146	10.85	156.85
TOTAL		Ea.	4.088	872.41	162.89	1,035.30
METAL CLAD WOOD WINDOW, 6′-6″ X 6′-6″						
Window, metal clad, deluxe, 6′-6″ x 6′-6″, insulating glass	1.000	Ea.	1.600	565	65	630
Trim interior casing	27.000	L.F.	.900	25.38	36.72	62.10
Paint, interior, primer & 2 coats	1.000	Face	1.600	5.60	57	62.60
Caulking	26.000	L.F.	.839	4.42	34.58	39
Snap-in grille	1.000	Ea.	.267	146	10.85	156.85
Drip cap, metal	6.500	L.F.	.130	1.82	5.33	7.15
TOTAL		Ea.	5.336	748.22	209.48	957.70

The cost of this system is on a cost per each window basis.

Description	QUAN.	UNIT	LABOR HOURS	COST EACH		
				MAT.	INST.	TOTAL

Fixed Window Price Sheet	QUAN.	UNIT	LABOR HOURS	COST EACH		
				MAT.	INST.	TOTAL
Window-picture, builder's quality, 4′ x 4′, single glass	1.000	Ea.	1.333	295	54.50	349.50
Insulating glass	1.000	Ea.	1.333	320	54.50	374.50
4′ x 4′-6″, single glass	1.000	Ea.	1.455	295	59.50	354.50
Insulating glass	1.000	Ea.	1.455	325	59.50	384.50
5′ x 4′, single glass	1.000	Ea.	1.455	365	59.50	424.50
Insulating glass	1.000	Ea.	1.455	410	59.50	469.50
6′ x 4′-6″, single glass	1.000	Ea.	1.600	465	65	530
Insulating glass	1.000	Ea.	1.600	520	65	585
Plastic clad premium insulating glass, 4′ x 4′	1.000	Ea.	1.333	475	54.50	529.50
4′-6″ x 6′-6″	1.000	Ea.	1.455	700	59.50	759.50
5′-6″ x 6′-6″	1.000	Ea.	1.600	1,000	65	1,065
6′-6″ x 6′-6″	1.000	Ea.	1.600	1,025	65	1,090
Metal clad deluxe insulating glass, 4′ x 4′	1.000	Ea.	1.333	305	54.50	359.50
4′-6″ x 6′-6″	1.000	Ea.	1.455	445	59.50	504.50
5′-6″ x 6′-6″	1.000	Ea.	1.600	490	65	555
6′-6″ x 6′-6″	1.000	Ea.	1.600	565	65	630
Trim, interior casing, window 4′ x 4′	17.000	L.F.	.567	16	23	39
4′-6″ x 4′-6″	19.000	L.F.	.633	17.85	26	43.85
5′-0″ x 4′-0″	19.000	L.F.	.633	17.85	26	43.85
4′-6″ x 6′-6″	23.000	L.F.	.767	21.50	31.50	53
5′-6″ x 6′-6″	25.000	L.F.	.833	23.50	34	57.50
6′-6″ x 6′-6″	27.000	L.F.	.900	25.50	36.50	62
Paint or stain, interior or exterior, 4′ x 4′ window, 1 coat	1.000	Face	.667	.79	24	24.79
2 coats	1.000	Face	.667	.89	24	24.89
Primer & 1 coat	1.000	Face	.727	1.07	26	27.07
Primer & 2 coats	1.000	Face	.889	1.05	32	33.05
4′-6″ x 6′-6″ window, 1 coat	1.000	Face	.667	.79	24	24.79
2 coats	1.000	Face	.667	.89	24	24.89
Primer & 1 coat	1.000	Face	.727	1.07	26	27.07
Primer & 2 coats	1.000	Face	.889	1.05	32	33.05
6′-6″ x 6′-6″ window, 1 coat	1.000	Face	.889	2.17	32	34.17
2 coats	1.000	Face	1.333	3.97	47.50	51.47
Primer & 1 coat	1.000	Face	1.333	3.76	47.50	51.26
Primer & 2 coats	1.000	Face	1.600	5.60	57	62.60
Caulking, window, 4′ x 4′	1.000	Ea.	.516	2.72	21.50	24.22
4′-6″ x 4′-6″	1.000	Ea.	.581	3.06	24	27.06
5′-0″ x 4′-0″	1.000	Ea.	.581	3.06	24	27.06
4′-6″ x 6′-6″	1.000	Ea.	.710	3.74	29.50	33.24
5′-6″ x 6′-6″	1.000	Ea.	.774	4.08	32	36.08
6′-6″ x 6′-6″	1.000	Ea.	.839	4.42	34.50	38.92
Grilles, glass size, to 48″ x 48″	1.000	Ea.	.267	146	10.85	156.85
To 60″ x 68″	1.000	Ea.	.286	107	11.65	118.65
Drip cap, aluminum, 4′ long	4.000	L.F.	.080	1.12	3.28	4.40
4′-6″ long	4.500	L.F.	.090	1.26	3.69	4.95
5′ long	5.000	L.F.	.100	1.40	4.10	5.50
6′ long	6.000	L.F.	.120	1.68	4.92	6.60
Wood, 4′ long	4.000	L.F.	.133	3.76	5.45	9.21
4′-6″ long	4.500	L.F.	.150	4.23	6.10	10.33
5′ long	5.000	L.F.	.167	4.70	6.80	11.50
6′ long	6.000	L.F.	.200	5.65	8.15	13.80

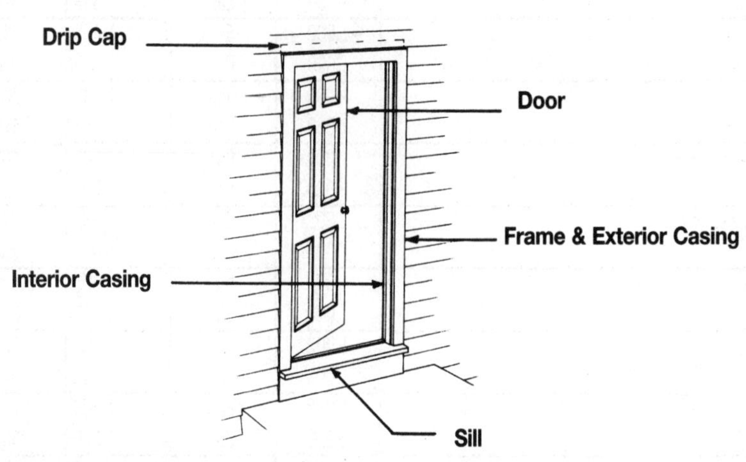

System Description	QUAN.	UNIT	LABOR HOURS	COST EACH		
				MAT.	INST.	TOTAL
COLONIAL, 6 PANEL, 3′ X 6′-8″, WOOD						
Door, 3′ x 6′-8″ x 1-3/4″ thick, pine, 6 panel colonial	1.000	Ea.	1.067	390	43.50	433.50
Frame, 5-13/16″ deep, incl. exterior casing & drip cap	17.000	L.F.	.725	141.95	29.58	171.53
Interior casing, 2-1/2″ wide	18.000	L.F.	.600	16.92	24.48	41.40
Sill, 8/4 x 8″ deep	3.000	L.F.	.480	43.95	19.50	63.45
Butt hinges, brass, 4-1/2″ x 4-1/2″	1.500	Pr.		20.40		20.40
Lockset	1.000	Ea.	.571	34.50	23.50	58
Weatherstripping, metal, spring type, bronze	1.000	Set	1.053	17.70	43	60.70
Paint, interior & exterior, primer & 2 coats	2.000	Face	1.778	11.10	64	75.10
TOTAL		Ea.	6.274	676.52	247.56	924.08
SOLID CORE BIRCH, FLUSH, 3′ X 6′-8″						
Door, 3′ x 6′-8″, 1-3/4″ thick, birch, flush solid core	1.000	Ea.	1.067	104	43.50	147.50
Frame, 5-13/16″ deep, incl. exterior casing & drip cap	17.000	L.F.	.725	141.95	29.58	171.53
Interior casing, 2-1/2″ wide	18.000	L.F.	.600	16.92	24.48	41.40
Sill, 8/4 x 8″ deep	3.000	L.F.	.480	43.95	19.50	63.45
Butt hinges, brass, 4-1/2″ x 4-1/2″	1.500	Pr.		20.40		20.40
Lockset	1.000	Ea.	.571	34.50	23.50	58
Weatherstripping, metal, spring type, bronze	1.000	Set	1.053	17.70	43	60.70
Paint, Interior & exterior, primer & 2 coats	2.000	Face	1.778	10.30	64	74.30
TOTAL		Ea.	6.274	389.72	247.56	637.28

These systems are on a cost per each door basis.

Description	QUAN.	UNIT	LABOR HOURS	COST EACH		
				MAT.	INST.	TOTAL

Important: See the Reference Section for critical supporting data - Reference Nos., Crews & Location Factors

Entrance Door Price Sheet	QUAN.	UNIT	LABOR HOURS	COST EACH		
				MAT.	INST.	TOTAL
Door exterior wood 1-3/4" thick, pine, dutch door, 2'-8" x 6'-8" minimum	1.000	Ea.	1.333	655	54.50	709.50
Maximum	1.000	Ea.	1.600	690	65	755
3'-0" x 6'-8", minimum	1.000	Ea.	1.333	680	54.50	734.50
Maximum	1.000	Ea.	1.600	740	65	805
Colonial, 6 panel, 2'-8" x 6'-8"	1.000	Ea.	1.000	365	41	406
3'-0" x 6'-8"	1.000	Ea.	1.067	390	43.50	433.50
8 panel, 2'-6" x 6'-8"	1.000	Ea.	1.000	645	41	686
3'-0" x 6'-8"	1.000	Ea.	1.067	550	43.50	593.50
Flush, birch, solid core, 2'-8" x 6'-8"	1.000	Ea.	1.000	100	41	141
3'-0" x 6'-8"	1.000	Ea.	1.067	104	43.50	147.50
Porch door, 2'-8" x 6'-8"	1.000	Ea.	1.000	259	41	300
3'-0" x 6'-8"	1.000	Ea.	1.067	260	43.50	303.50
Hand carved mahogany, 2'-8" x 6'-8"	1.000	Ea.	1.067	505	43.50	548.50
3'-0" x 6'-8"	1.000	Ea.	1.067	540	43.50	583.50
Rosewood, 2'-8" x 6'-8"	1.000	Ea.	1.067	785	43.50	828.50
3'-0" x 6-8"	1.000	Ea.	1.067	815	43.50	858.50
Door, metal clad wood 1-3/8" thick raised panel, 2'-8" x 6'-8"	1.000	Ea.	1.067	273	43.50	316.50
3'-0" x 6'-8"	1.000	Ea.	1.067	271	43.50	314.50
Deluxe metal door, 2'-8" x 6'-8"	1.000	Ea.	1.231	420	50	470
3'-0" x 6'-8"	1.000	Ea.	1.231	415	50	465
Frame, pine, including exterior trim & drip cap, 5/4, x 4-9/16" deep	17.000	L.F.	.725	91	29.50	120.50
5-13/16" deep	17.000	L.F.	.725	142	29.50	171.50
6-9/16" deep	17.000	L.F.	.725	140	29.50	169.50
Safety glass lites, add	1.000	Ea.		68		68
Interior casing, 2'-8" x 6'-8" door	18.000	L.F.	.600	16.90	24.50	41.40
3'-0" x 6'-8" door	19.000	L.F.	.633	17.85	26	43.85
Sill, oak, 8/4 x 8" deep	3.000	L.F.	.480	44	19.50	63.50
8/4 x 10" deep	3.000	L.F.	.533	59.50	22	81.50
Butt hinges, steel plated, 4-1/2" x 4-1/2", plain	1.500	Pr.		20.50		20.50
Ball bearing	1.500	Pr.		46		46
Bronze, 4-1/2" x 4-1/2", plain	1.500	Pr.		23.50		23.50
Ball bearing	1.500	Pr.		48		48
Lockset, minimum	1.000	Ea.	.571	34.50	23.50	58
Maximum	1.000	Ea.	1.000	145	41	186
Weatherstripping, metal, interlocking, zinc	1.000	Set	2.667	14.30	109	123.30
Bronze	1.000	Set	2.667	22.50	109	131.50
Spring type, bronze	1.000	Set	1.053	17.70	43	60.70
Rubber, minimum	1.000	Set	1.053	4.87	43	47.87
Maximum	1.000	Set	1.143	5.55	46.50	52.05
Felt minimum	1.000	Set	.571	2.26	23.50	25.76
Maximum	1.000	Set	.615	2.43	25	27.43
Paint or stain, flush door, interior or exterior, 1 coat	2.000	Face	.941	3.76	33.50	37.26
2 coats	2.000	Face	1.455	7.50	52	59.50
Primer & 1 coat	2.000	Face	1.455	6.80	52	58.80
Primer & 2 coats	2.000	Face	1.778	10.30	64	74.30
Paneled door, interior & exterior, 1 coat	2.000	Face	1.143	4.02	41	45.02
2 coats	2.000	Face	2.000	8.05	72	80.05
Primer & 1 coat	2.000	Face	1.455	7.30	52	59.30
Primer & 2 coats	2.000	Face	1.778	11.10	64	75.10

53 Sliding Door Systems

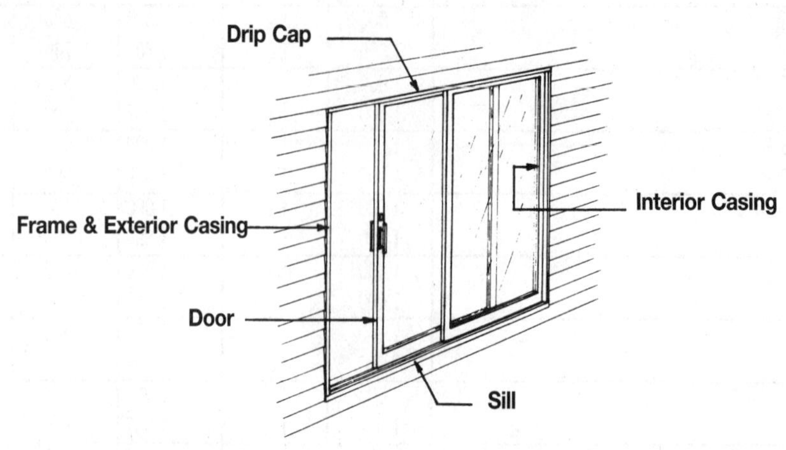

System Description	QUAN.	UNIT	LABOR HOURS	COST EACH MAT.	COST EACH INST.	COST EACH TOTAL
WOOD SLIDING DOOR, 8′ WIDE, PREMIUM						
Wood, 5/8″ thick tempered insul. glass, 8′ wide, premium	1.000	Ea.	5.333	1,250	217	1,467
Interior casing	22.000	L.F.	.733	20.68	29.92	50.60
Exterior casing	22.000	L.F.	.733	20.68	29.92	50.60
Sill, oak, 8/4 x 8″ deep	8.000	L.F.	1.280	117.20	52	169.20
Drip cap	8.000	L.F.	.160	2.24	6.56	8.80
Paint, interior & exterior, primer & 2 coats	2.000	Face	2.816	15.84	100.32	116.16
TOTAL		Ea.	11.055	1,426.64	435.72	1,862.36
ALUMINUM SLIDING DOOR, 8′ WIDE, PREMIUM						
Aluminum, 5/8″ tempered insul. glass, 8′ wide, premium	1.000	Ea.	5.333	1,425	217	1,642
Interior casing	22.000	L.F.	.733	20.68	29.92	50.60
Exterior casing	22.000	L.F.	.733	20.68	29.92	50.60
Sill, oak, 8/4 x 8″ deep	8.000	L.F.	1.280	117.20	52	169.20
Drip cap	8.000	L.F.	.160	2.24	6.56	8.80
Paint, interior & exterior, primer & 2 coats	2.000	Face	2.816	15.84	100.32	116.16
TOTAL		Ea.	11.055	1,601.64	435.72	2,037.36

The cost of this system is on a cost per each door basis.

Description	QUAN.	UNIT	LABOR HOURS	COST EACH MAT.	COST EACH INST.	COST EACH TOTAL

Sliding Door Price Sheet	QUAN.	UNIT	LABOR HOURS	MAT.	INST.	TOTAL
ding door, wood, 5/8" thick, tempered insul. glass, 6' wide, premium	1.000	Ea.	4.000	1,075	163	1,238
Economy	1.000	Ea.	4.000	765	163	928
8'wide, wood premium	1.000	Ea.	5.333	1,250	217	1,467
Economy	1.000	Ea.	5.333	885	217	1,102
12' wide, wood premium	1.000	Ea.	6.400	2,975	261	3,236
Economy	1.000	Ea.	6.400	2,025	261	2,286
Aluminum, 5/8" thick, tempered insul. glass, 6'wide, premium	1.000	Ea.	4.000	1,375	163	1,538
Economy	1.000	Ea.	4.000	720	163	883
8'wide, premium	1.000	Ea.	5.333	1,425	217	1,642
Economy	1.000	Ea.	5.333	1,200	217	1,417
12' wide, premium	1.000	Ea.	6.400	2,375	261	2,636
Economy	1.000	Ea.	6.400	1,500	261	1,761
terior casing, 6' wide door	20.000	L.F.	.667	18.80	27	45.80
8' wide door	22.000	L.F.	.733	20.50	30	50.50
12' wide door	26.000	L.F.	.867	24.50	35.50	60
xterior casing, 6' wide door	20.000	L.F.	.667	18.80	27	45.80
8' wide door	22.000	L.F.	.733	20.50	30	50.50
12' wide door	26.000	L.F.	.867	24.50	35.50	60
ll, oak, 8/4 x 8" deep, 6' wide door	6.000	L.F.	.960	88	39	127
8' wide door	8.000	L.F.	1.280	117	52	169
12' wide door	12.000	L.F.	1.920	176	78	254
8/4 x 10" deep, 6' wide door	6.000	L.F.	1.067	119	43.50	162.50
8' wide door	8.000	L.F.	1.422	158	58	216
12' wide door	12.000	L.F.	2.133	237	87	324
rip cap, 6' wide door	6.000	L.F.	.120	1.68	4.92	6.60
8' wide door	8.000	L.F.	.160	2.24	6.55	8.79
12' wide door	12.000	L.F.	.240	3.36	9.85	13.21
aint or stain, interior & exterior, 6' wide door, 1 coat	2.000	Face	1.600	4.80	57.50	62.30
2 coats	2.000	Face	1.600	4.80	57.50	62.30
Primer & 1 coat	2.000	Face	1.778	9.60	63	72.60
Primer & 2 coats	2.000	Face	2.560	14.40	91	105.40
8' wide door, 1 coat	2.000	Face	1.760	5.30	63.50	68.80
2 coats	2.000	Face	1.760	5.30	63.50	68.80
Primer & 1 coat	2.000	Face	1.955	10.55	69.50	80.05
Primer & 2 coats	2.000	Face	2.816	15.85	100	115.85
12' wide door, 1 coat	2.000	Face	2.080	6.25	75	81.25
2 coats	2.000	Face	2.080	6.25	75	81.25
Primer & 1 coat	2.000	Face	2.311	12.50	82	94.50
Primer & 2 coats	2.000	Face	3.328	18.70	119	137.70
Aluminum door, trim only, interior & exterior, 6' door, 1 coat	2.000	Face	.800	2.40	29	31.40
2 coats	2.000	Face	.800	2.40	29	31.40
Primer & 1 coat	2.000	Face	.889	4.80	31.50	36.30
Primer & 2 coats	2.000	Face	1.280	7.20	45.50	52.70
8' wide door, 1 coat	2.000	Face	.880	2.64	31.50	34.14
2 coats	2.000	Face	.880	2.64	31.50	34.14
Primer & 1 coat	2.000	Face	.978	5.30	35	40.30
Primer & 2 coats	2.000	Face	1.408	7.90	50	57.90
12' wide door, 1 coat	2.000	Face	1.040	3.12	37.50	40.62
2 coats	2.000	Face	1.040	3.12	37.50	40.62
Primer & 1 coat	2.000	Face	1.155	6.25	41	47.25
Primer & 2 coats	2.000	Face	1.664	9.35	59.50	68.85

56 Residential Garage Door Systems

System Description	QUAN.	UNIT	LABOR HOURS	COST EACH		
				MAT.	INST.	TOTAL
OVERHEAD, SECTIONAL GARAGE DOOR, 9′ X 7′						
Wood, overhead sectional door, std., incl. hardware, 9′ x 7′	1.000	Ea.	2.000	485	81.50	566.50
Jamb & header blocking, 2″ x 6″	25.000	L.F.	.901	14.75	36.75	51.50
Exterior trim	25.000	L.F.	.833	23.50	34	57.50
Paint, interior & exterior, primer & 2 coats	2.000	Face	3.556	22.20	128	150.20
Weatherstripping, molding type	1.000	Set	.767	21.62	31.28	52.90
Drip cap	9.000	L.F.	.180	2.52	7.38	9.90
TOTAL		Ea.	8.237	569.59	318.91	888.50
OVERHEAD, SECTIONAL GARAGE DOOR, 16′ X 7′						
Wood, overhead sectional, std., incl. hardware, 16′ x 7′	1.000	Ea.	2.667	980	109	1,089
Jamb & header blocking, 2″ x 6″	30.000	L.F.	1.081	17.70	44.10	61.80
Exterior trim	30.000	L.F.	1.000	28.20	40.80	69
Paint, interior & exterior, primer & 2 coats	2.000	Face	5.333	33.30	192	225.30
Weatherstripping, molding type	1.000	Set	1.000	28.20	40.80	69
Drip cap	16.000	L.F.	.320	4.48	13.12	17.60
TOTAL		Ea.	11.401	1,091.88	439.82	1,531.70
OVERHEAD, SWING-UP TYPE, GARAGE DOOR, 16′ X 7′						
Wood, overhead, swing-up, std., incl. hardware, 16′ x 7′	1.000	Ea.	2.667	670	109	779
Jamb & header blocking, 2″ x 6″	30.000	L.F.	1.081	17.70	44.10	61.80
Exterior trim	30.000	L.F.	1.000	28.20	40.80	69
Paint, interior & exterior, primer & 2 coats	2.000	Face	5.333	33.30	192	225.30
Weatherstripping, molding type	1.000	Set	1.000	28.20	40.80	69
Drip cap	16.000	L.F.	.320	4.48	13.12	17.60
TOTAL		Ea.	11.401	781.88	439.82	1,221.70

This system is on a cost per each door basis.

Description	QUAN.	UNIT	LABOR HOURS	COST EACH		
				MAT.	INST.	TOTAL

Important: See the Reference Section for critical supporting data - Reference Nos., Crews & Location Factors

Resi Garage Door Price Sheet	QUAN.	UNIT	LABOR HOURS	COST EACH		
				MAT.	INST.	TOTAL
verhead, sectional, including hardware, fiberglass, 9′ x 7′, standard	1.000	Ea.	3.030	620	123	743
Deluxe	1.000	Ea.	3.030	790	123	913
16′ x 7′, standard	1.000	Ea.	2.667	1,050	109	1,159
Deluxe	1.000	Ea.	2.667	1,325	109	1,434
Hardboard, 9′ x 7′, standard	1.000	Ea.	2.000	410	81.50	491.50
Deluxe	1.000	Ea.	2.000	525	81.50	606.50
16′ x 7′, standard	1.000	Ea.	2.667	770	109	879
Deluxe	1.000	Ea.	2.667	900	109	1,009
Metal, 9′ x 7′, standard	1.000	Ea.	3.030	495	123	618
Deluxe	1.000	Ea.	2.000	640	81.50	721.50
16′ x 7′, standard	1.000	Ea.	5.333	630	217	847
Deluxe	1.000	Ea.	2.667	940	109	1,049
Wood, 9′ x 7′, standard	1.000	Ea.	2.000	485	81.50	566.50
Deluxe	1.000	Ea.	2.000	1,400	81.50	1,481.50
16′ x 7′, standard	1.000	Ea.	2.667	980	109	1,089
Deluxe	1.000	Ea.	2.667	2,050	109	2,159
Overhead swing-up type including hardware, fiberglass, 9′ x 7′, standard	1.000	Ea.	2.000	670	81.50	751.50
Deluxe	1.000	Ea.	2.000	740	81.50	821.50
16′ x 7′, standard	1.000	Ea.	2.667	845	109	954
Deluxe	1.000	Ea.	2.667	915	109	1,024
Hardboard, 9′ x 7′, standard	1.000	Ea.	2.000	320	81.50	401.50
Deluxe	1.000	Ea.	2.000	425	81.50	506.50
16′ x 7′, standard	1.000	Ea.	2.667	450	109	559
Deluxe	1.000	Ea.	2.667	670	109	779
Metal, 9′ x 7′, standard	1.000	Ea.	2.000	355	81.50	436.50
Deluxe	1.000	Ea.	2.000	630	81.50	711.50
16′ x 7′, standard	1.000	Ea.	2.667	555	109	664
Deluxe	1.000	Ea.	2.667	890	109	999
Wood, 9′ x 7′, standard	1.000	Ea.	2.000	385	81.50	466.50
Deluxe	1.000	Ea.	2.000	685	81.50	766.50
16′ x 7′, standard	1.000	Ea.	2.667	670	109	779
Deluxe	1.000	Ea.	2.667	945	109	1,054
amb & header blocking, 2″ x 6″, 9′ x 7′ door	25.000	L.F.	.901	14.75	37	51.75
16′ x 7′ door	30.000	L.F.	1.081	17.70	44	61.70
2″ x 8″, 9′ x 7′ door	25.000	L.F.	1.000	23	41	64
16′ x 7′ door	30.000	L.F.	1.200	27.50	49	76.50
xterior trim, 9′ x 7′ door	25.000	L.F.	.833	23.50	34	57.50
16′ x 7′ door	30.000	L.F.	1.000	28	41	69
aint or stain, interior & exterior, 9′ x 7′ door, 1 coat	1.000	Face	2.286	8.05	82	90.05
2 coats	1.000	Face	4.000	16.10	144	160.10
Primer & 1 coat	1.000	Face	2.909	14.55	104	118.55
Primer & 2 coats	1.000	Face	3.556	22	128	150
16′ x 7′ door, 1 coat	1.000	Face	3.429	12.05	123	135.05
2 coats	1.000	Face	6.000	24	216	240
Primer & 1 coat	1.000	Face	4.364	22	156	178
Primer & 2 coats	1.000	Face	5.333	33.50	192	225.50
Veatherstripping, molding type, 9′ x 7′ door	1.000	Set	.767	21.50	31.50	53
16′ x 7′ door	1.000	Set	1.000	28	41	69
rip cap, 9′ door	9.000	L.F.	.180	2.52	7.40	9.92
16′ door	16.000	L.F.	.320	4.48	13.10	17.58
arage door opener, economy	1.000	Ea.	1.000	279	41	320
Deluxe, including remote control	1.000	Ea.	1.000	410	41	451

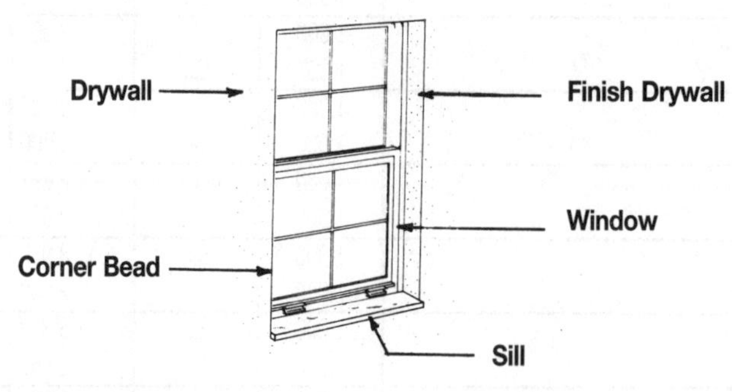

System Description	QUAN.	UNIT	LABOR HOURS	COST EACH		
				MAT.	INST.	TOTAL
SINGLE HUNG, 2′ X 3′ OPENING						
Window, 2′ x 3′ opening, enameled, insulating glass	1.000	Ea.	1.600	191	79	270
Blocking, 1″ x 3″ furring strip nailers	10.000	L.F.	.146	3.10	5.90	9
Drywall, 1/2″ thick, standard	5.000	S.F.	.040	1.20	1.65	2.85
Corner bead, 1″ x 1″, galvanized steel	8.000	L.F.	.160	1.04	6.56	7.60
Finish drywall, tape and finish corners inside and outside	16.000	L.F.	.269	1.12	11.04	12.16
Sill, slate	2.000	L.F.	.400	17.30	14.30	31.60
TOTAL		Ea.	2.615	214.76	118.45	333.21
SLIDING, 3′ X 2′ OPENING						
Window, 3′ x 2′ opening, enameled, insulating glass	1.000	Ea.	1.600	197	79	276
Blocking, 1″ x 3″ furring strip nailers	10.000	L.F.	.146	3.10	5.90	9
Drywall, 1/2″ thick, standard	5.000	S.F.	.040	1.20	1.65	2.85
Corner bead, 1″ x 1″, galvanized steel	7.000	L.F.	.140	.91	5.74	6.65
Finish drywall, tape and finish corners inside and outside	14.000	L.F.	.236	.98	9.66	10.64
Sill, slate	3.000	L.F.	.600	25.95	21.45	47.40
TOTAL		Ea.	2.762	229.14	123.40	352.54
AWNING, 3′-1″ X 3′-2″						
Window, 3′-1″ x 3′-2″ opening, enameled, insul. glass	1.000	Ea.	1.600	212	79	291
Blocking, 1″ x 3″ furring strip, nailers	12.500	L.F.	.182	3.88	7.38	11.26
Drywall, 1/2″ thick, standard	4.500	S.F.	.036	1.08	1.49	2.57
Corner bead, 1″ x 1″, galvanized steel	9.250	L.F.	.185	1.20	7.59	8.79
Finish drywall, tape and finish corners, inside and outside	18.500	L.F.	.312	1.30	12.77	14.07
Sill, slate	3.250	L.F.	.650	28.11	23.24	51.35
TOTAL		Ea.	2.965	247.57	131.47	379.04

Description	QUAN.	UNIT	LABOR HOURS	COST PER S.F.		
				MAT.	INST.	TOTAL

Aluminum Window Price Sheet	QUAN.	UNIT	LABOR HOURS	COST EACH		
				MAT.	INST.	TOTAL
Window, aluminum, awning, 3′-1″ x 3′-2″, standard glass	1.000	Ea.	1.600	234	79	313
Insulating glass	1.000	Ea.	1.600	212	79	291
4′-5″ x 5′-3″, standard glass	1.000	Ea.	2.000	330	98.50	428.50
Insulating glass	1.000	Ea.	2.000	370	98.50	468.50
Casement, 3′-1″ x 3′-2″, standard glass	1.000	Ea.	1.600	330	79	409
Insulating glass	1.000	Ea.	1.600	315	79	394
Single hung, 2′ x 3′, standard glass	1.000	Ea.	1.600	158	79	237
Insulating glass	1.000	Ea.	1.600	191	79	270
2′-8″ x 6′-8″, standard glass	1.000	Ea.	2.000	335	98.50	433.50
Insulating glass	1.000	Ea.	2.000	430	98.50	528.50
3′-4″ x 5′-0″, standard glass	1.000	Ea.	1.778	216	87.50	303.50
Insulating glass	1.000	Ea.	1.778	305	87.50	392.50
Sliding, 3′ x 2′, standard glass	1.000	Ea.	1.600	177	79	256
Insulating glass	1.000	Ea.	1.600	197	79	276
5′ x 3′, standard glass	1.000	Ea.	1.778	225	87.50	312.50
Insulating glass	1.000	Ea.	1.778	315	87.50	402.50
8′ x 4′, standard glass	1.000	Ea.	2.667	325	131	456
Insulating glass	1.000	Ea.	2.667	520	131	651
Blocking, 1″ x 3″ furring, opening 3′ x 2′	10.000	L.F.	.146	3.10	5.90	9
3′ x 3′	12.500	L.F.	.182	3.88	7.40	11.28
3′ x 5′	16.000	L.F.	.233	4.96	9.45	14.41
4′ x 4′	16.000	L.F.	.233	4.96	9.45	14.41
4′ x 5′	18.000	L.F.	.262	5.60	10.60	16.20
4′ x 6′	20.000	L.F.	.291	6.20	11.80	18
4′ x 8′	24.000	L.F.	.349	7.45	14.15	21.60
6′-8″ x 2′-8″	19.000	L.F.	.276	5.90	11.20	17.10
Drywall, 1/2″ thick, standard, opening 3′ x 2′	5.000	S.F.	.040	1.20	1.65	2.85
3′ x 3′	6.000	S.F.	.048	1.44	1.98	3.42
3′ x 5′	8.000	S.F.	.064	1.92	2.64	4.56
4′ x 4′	8.000	S.F.	.064	1.92	2.64	4.56
4′ x 5′	9.000	S.F.	.072	2.16	2.97	5.13
4′ x 6′	10.000	S.F.	.080	2.40	3.30	5.70
4′ x 8′	12.000	S.F.	.096	2.88	3.96	6.84
6′-8″ x 2′	9.500	S.F.	.076	2.28	3.14	5.42
Corner bead, 1″ x 1″, galvanized steel, opening 3′ x 2′	7.000	L.F.	.140	.91	5.75	6.66
3′ x 3′	9.000	L.F.	.180	1.17	7.40	8.57
3′ x 5′	11.000	L.F.	.220	1.43	9	10.43
4′ x 4′	12.000	L.F.	.240	1.56	9.85	11.41
4′ x 5′	13.000	L.F.	.260	1.69	10.65	12.34
4′ x 6′	14.000	L.F.	.280	1.82	11.50	13.32
4′ x 8′	16.000	L.F.	.320	2.08	13.10	15.18
6′-8″ x 2′	15.000	L.F.	.300	1.95	12.30	14.25
Tape and finish corners, inside and outside, opening 3′ x 2′	14.000	L.F.	.204	.98	9.65	10.63
3′ x 3′	18.000	L.F.	.262	1.26	12.40	13.66
3′ x 5′	22.000	L.F.	.320	1.54	15.20	16.74
4′ x 4′	24.000	L.F.	.349	1.68	16.55	18.23
4′ x 5′	26.000	L.F.	.378	1.82	17.95	19.77
4′ x 6′	28.000	L.F.	.407	1.96	19.30	21.26
4′ x 8′	32.000	L.F.	.466	2.24	22	24.24
6′-8″ x 2′	30.000	L.F.	.437	2.10	20.50	22.60
Sill, slate, 2′ long	2.000	L.F.	.400	17.30	14.30	31.60
3′ long	3.000	L.F.	.600	26	21.50	47.50
4′ long	4.000	L.F.	.800	34.50	28.50	63
Wood, 1-5/8″ x 6-1/4″, 2′ long	2.000	L.F.	.128	16.70	5.20	21.90
3′ long	3.000	L.F.	.192	25	7.85	32.85
4′ long	4.000	L.F.	.256	33.50	10.45	43.95

60 Storm Door & Window Systems

System Description	QUAN.	UNIT	LABOR HOURS	COST EACH		
				MAT.	INST.	TOTAL
Storm door, aluminum, combination, storm & screen, anodized, 2'-6" x 6'-8"	1.000	Ea.	1.067	168	43.50	211.50
2'-8" x 6'-8"	1.000	Ea.	1.143	194	46.50	240.50
3'-0" x 6'-8"	1.000	Ea.	1.143	194	46.50	240.50
Mill finish, 2'-6" x 6'-8"	1.000	Ea.	1.067	224	43.50	267.50
2'-8" x 6'-8"	1.000	Ea.	1.143	224	46.50	270.50
3'-0" x 6'-8"	1.000	Ea.	1.143	243	46.50	289.50
Painted, 2'-6" x 6'-8"	1.000	Ea.	1.067	223	43.50	266.50
2'-8" x 6'-8"	1.000	Ea.	1.143	227	46.50	273.50
3'-0" x 6'-8"	1.000	Ea.	1.143	237	46.50	283.50
Wood, combination, storm & screen, crossbuck, 2'-6" x 6'-9"	1.000	Ea.	1.455	320	59.50	379.50
2'-8" x 6'-9"	1.000	Ea.	1.600	281	65	346
3'-0" x 6'-9"	1.000	Ea.	1.778	287	72.50	359.50
Full lite, 2'-6" x 6'-9"	1.000	Ea.	1.455	292	59.50	351.50
2'-8" x 6'-9"	1.000	Ea.	1.600	292	65	357
3'-0" x 6'-9"	1.000	Ea.	1.778	300	72.50	372.50
Windows, aluminum, combination storm & screen, basement, 1'-10" x 1'-0"	1.000	Ea.	.533	30.50	21.50	52
2'-9" x 1'-6"	1.000	Ea.	.533	33.50	21.50	55
3'-4" x 2'-0"	1.000	Ea.	.533	40.50	21.50	62
Double hung, anodized, 2'-0" x 3'-5"	1.000	Ea.	.533	79.50	21.50	101
2'-6" x 5'-0"	1.000	Ea.	.571	106	23.50	129.50
4'-0" x 6'-0"	1.000	Ea.	.640	225	26	251
Painted, 2'-0" x 3'-5"	1.000	Ea.	.533	94.50	21.50	116
2'-6" x 5'-0"	1.000	Ea.	.571	152	23.50	175.50
4'-0" x 6'-0"	1.000	Ea.	.640	272	26	298
Fixed window, anodized, 4'-6" x 4'-6"	1.000	Ea.	.640	121	26	147
5'-8" x 4'-6"	1.000	Ea.	.800	137	32.50	169.50
Painted, 4'-6" x 4'-6"	1.000	Ea.	.640	121	26	147
5'-8" x 4'-6"	1.000	Ea.	.800	137	32.50	169.50

Important: See the Reference Section for critical supporting data - Reference Nos., Crews & Location Factors

64 Shutters/Blinds Systems

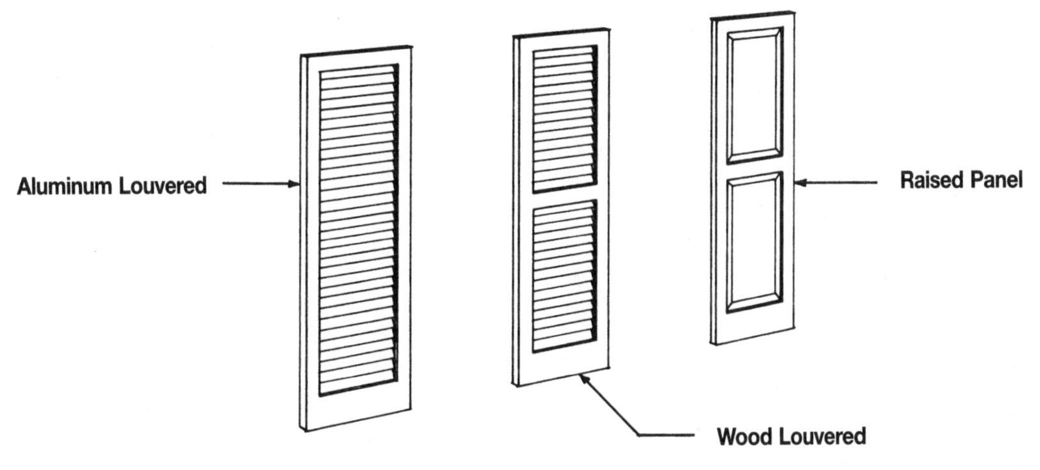

System Description	QUAN.	UNIT	LABOR HOURS	COST PER PAIR MAT.	INST.	TOTAL
Shutters, exterior blinds, aluminum, louvered, 1′-4″ wide, 3″-0″ long	1.000	Set	.800	46	32.50	78.50
4′-0″ long	1.000	Set	.800	55	32.50	87.50
5′-4″ long	1.000	Set	.800	72.50	32.50	105
6′-8″ long	1.000	Set	.889	92.50	36	128.50
Wood, louvered, 1′-2″ wide, 3′-3″ long	1.000	Set	.800	87.50	32.50	120
4′-7″ long	1.000	Set	.800	118	32.50	150.50
5′-3″ long	1.000	Set	.800	134	32.50	166.50
1′-6″ wide, 3′-3″ long	1.000	Set	.800	93	32.50	125.50
4′-7″ long	1.000	Set	.800	130	32.50	162.50
Polystyrene, solid raised panel, 3′-3″ wide, 3′-0″ long	1.000	Set	.800	182	32.50	214.50
3′-11″ long	1.000	Set	.800	198	32.50	230.50
5′-3″ long	1.000	Set	.800	260	32.50	292.50
6′-8″ long	1.000	Set	.889	293	36	329
Polystyrene, louvered, 1′-2″ wide, 3′-3″ long	1.000	Set	.800	44	32.50	76.50
4′-7″ long	1.000	Set	.800	54.50	32.50	87
5′-3″ long	1.000	Set	.800	58	32.50	90.50
6′-8″ long	1.000	Set	.889	95.50	36	131.50
Vinyl, louvered, 1′-2″ wide, 4′-7″ long	1.000	Set	.720	51.50	29.50	81
1′-4″ x 6′-8″ long	1.000	Set	.889	87	36	123

Division 5 Roofing

No part of this publication may be reproduced, stored in a retrieval system, or transmitted in any form or by any means without prior written permission of Reed Construction Data.

04 Gable End Roofing Systems

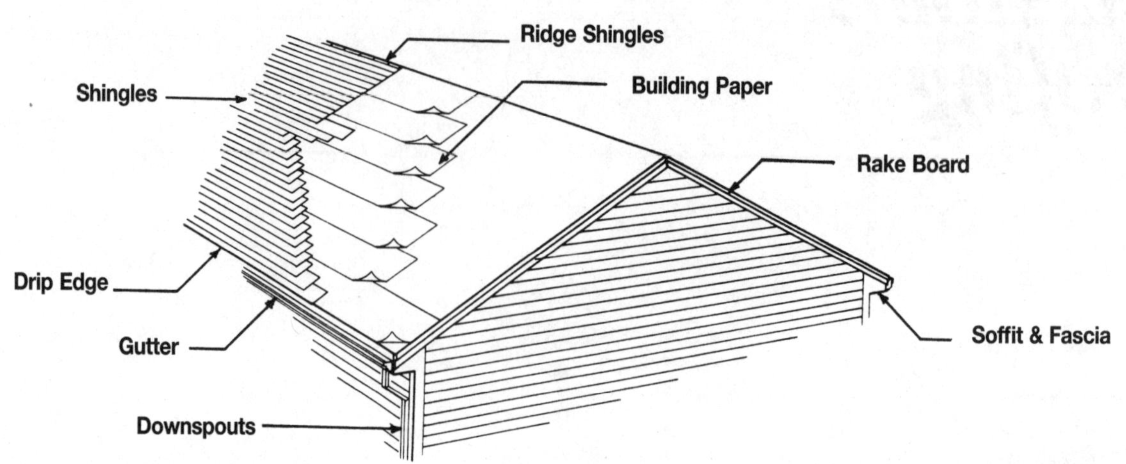

System Description	QUAN.	UNIT	LABOR HOURS	COST PER S.F. MAT.	COST PER S.F. INST.	COST PER S.F. TOTAL
ASPHALT, ROOF SHINGLES, CLASS A						
Shingles, inorganic class A, 210-235 lb./sq., 4/12 pitch	1.160	S.F.	.017	.41	.65	1.06
Drip edge, metal, 5″ wide	.150	L.F.	.003	.04	.12	.16
Building paper, #15 felt	1.300	S.F.	.002	.05	.07	.12
Ridge shingles, asphalt	.042	L.F.	.001	.04	.04	.08
Soffit & fascia, white painted aluminum, 1′ overhang	.083	L.F.	.012	.20	.49	.69
Rake trim, 1″ x 6″	.040	L.F.	.002	.04	.07	.11
Rake trim, prime and paint	.040	L.F.	.002	.01	.06	.07
Gutter, seamless, aluminum painted	.083	L.F.	.006	.10	.25	.35
Downspouts, aluminum painted	.035	L.F.	.002	.04	.07	.11
TOTAL		S.F.	.047	.93	1.82	2.75
WOOD, CEDAR SHINGLES NO. 1 PERFECTIONS, 18″ LONG						
Shingles, wood, cedar, No. 1 perfections, 4/12 pitch	1.160	S.F.	.035	2.11	1.43	3.54
Drip edge, metal, 5″ wide	.150	L.F.	.003	.04	.12	.16
Building paper, #15 felt	1.300	S.F.	.002	.05	.07	.12
Ridge shingles, cedar	.042	L.F.	.001	.11	.05	.16
Soffit & fascia, white painted aluminum, 1′ overhang	.083	L.F.	.012	.20	.49	.69
Rake trim, 1″ x 6″	.040	L.F.	.002	.04	.07	.11
Rake trim, prime and paint	.040	L.F.	.002	.01	.06	.07
Gutter, seamless, aluminum, painted	.083	L.F.	.006	.10	.25	.35
Downspouts, aluminum, painted	.035	L.F.	.002	.04	.07	.11
TOTAL		S.F.	.065	2.70	2.61	5.31

The prices in these systems are based on a square foot of plan area.
All quantities have been adjusted accordingly.

Description	QUAN.	UNIT	LABOR HOURS	COST PER S.F. MAT.	COST PER S.F. INST.	COST PER S.F. TOTAL

Gable End Roofing Price Sheet	QUAN.	UNIT	LABOR HOURS	COST PER S.F.		
				MAT.	INST.	TOTAL
Shingles, asphalt, inorganic, class A, 210-235 lb./sq., 4/12 pitch	1.160	S.F.	.017	.41	.65	1.06
8/12 pitch	1.330	S.F.	.019	.44	.71	1.15
Laminated, multi-layered, 240-260 lb./sq., 4/12 pitch	1.160	S.F.	.021	.57	.80	1.37
8/12 pitch	1.330	S.F.	.023	.62	.86	1.48
Premium laminated, multi-layered, 260-300 lb./sq., 4/12 pitch	1.160	S.F.	.027	.72	1.03	1.75
8/12 pitch	1.330	S.F.	.030	.78	1.11	1.89
Clay tile, Spanish tile, red, 4/12 pitch	1.160	S.F.	.053	3.43	2.02	5.45
8/12 pitch	1.330	S.F.	.058	3.72	2.18	5.90
Mission tile, red, 4/12 pitch	1.160	S.F.	.083	8.35	3.16	11.51
8/12 pitch	1.330	S.F.	.090	9.05	3.42	12.47
French tile, red, 4/12 pitch	1.160	S.F.	.071	7.55	2.69	10.24
8/12 pitch	1.330	S.F.	.077	8.20	2.91	11.11
Slate, Buckingham, Virginia, black, 4/12 pitch	1.160	S.F.	.055	7.25	2.08	9.33
8/12 pitch	1.330	S.F.	.059	7.85	2.25	10.10
Vermont, black or grey, 4/12 pitch	1.160	S.F.	.055	4.50	2.08	6.58
8/12 pitch	1.330	S.F.	.059	4.88	2.25	7.13
Wood, No. 1 red cedar, 5X, 16″ long, 5″ exposure, 4/12 pitch	1.160	S.F.	.038	2.15	1.56	3.71
8/12 pitch	1.330	S.F.	.042	2.33	1.69	4.02
Fire retardant, 4/12 pitch	1.160	S.F.	.038	2.55	1.56	4.11
8/12 pitch	1.330	S.F.	.042	2.76	1.69	4.45
18″ long, No.1 perfections, 5″ exposure, 4/12 pitch	1.160	S.F.	.035	2.11	1.43	3.54
8/12 pitch	1.330	S.F.	.038	2.29	1.55	3.84
Fire retardant, 4/12 pitch	1.160	S.F.	.035	2.49	1.43	3.92
8/12 pitch	1.330	S.F.	.038	2.70	1.55	4.25
Resquared & rebutted, 18″ long, 6″ exposure, 4/12 pitch	1.160	S.F.	.032	2.63	1.31	3.94
8/12 pitch	1.330	S.F.	.035	2.85	1.42	4.27
Fire retardant, 4/12 pitch	1.160	S.F.	.032	3.01	1.31	4.32
8/12 pitch	1.330	S.F.	.035	3.26	1.42	4.68
Wood shakes hand split, 24″ long, 10″ exposure, 4/12 pitch	1.160	S.F.	.038	1.82	1.56	3.38
8/12 pitch	1.330	S.F.	.042	1.98	1.69	3.67
Fire retardant, 4/12 pitch	1.160	S.F.	.038	2.22	1.56	3.78
8/12 pitch	1.330	S.F.	.042	2.41	1.69	4.10
18″ long, 8″ exposure, 4/12 pitch	1.160	S.F.	.048	1.28	1.96	3.24
8/12 pitch	1.330	S.F.	.052	1.39	2.12	3.51
Fire retardant, 4/12 pitch	1.160	S.F.	.048	1.68	1.96	3.64
8/12 pitch	1.330	S.F.	.052	1.82	2.12	3.94
Drip edge, metal, 5″ wide	.150	L.F.	.003	.04	.12	.16
8″ wide	.150	L.F.	.003	.05	.12	.17
Building paper, #15 asphalt felt	1.300	S.F.	.002	.05	.07	.12
Ridge shingles, asphalt	.042	L.F.	.001	.04	.04	.08
Clay	.042	L.F.	.002	.40	.06	.46
Slate	.042	L.F.	.002	.39	.06	.45
Wood, shingles	.042	L.F.	.001	.11	.05	.16
Shakes	.042	L.F.	.001	.11	.05	.16
Soffit & fascia, aluminum, vented, 1′ overhang	.083	L.F.	.012	.20	.49	.69
2′ overhang	.083	L.F.	.013	.29	.54	.83
Vinyl, vented, 1′ overhang	.083	L.F.	.011	.14	.45	.59
2′ overhang	.083	L.F.	.012	.19	.49	.68
Wood, board fascia, plywood soffit, 1′ overhang	.083	L.F.	.004	.02	.13	.15
2′ overhang	.083	L.F.	.006	.03	.20	.23
Rake trim, painted, 1″ x 6″	.040	L.F.	.004	.05	.13	.18
1″ x 8″	.040	L.F.	.004	.08	.14	.22
Gutter, 5″ box, aluminum, seamless, painted	.083	L.F.	.006	.10	.25	.35
Vinyl	.083	L.F.	.006	.09	.25	.34
Downspout, 2″ x 3″, aluminum, one story house	.035	L.F.	.001	.04	.07	.11
Two story house	.060	L.F.	.003	.06	.11	.17
Vinyl, one story house	.035	L.F.	.002	.04	.07	.11
Two story house	.060	L.F.	.003	.06	.11	.17

08 Hip Roof Roofing Systems

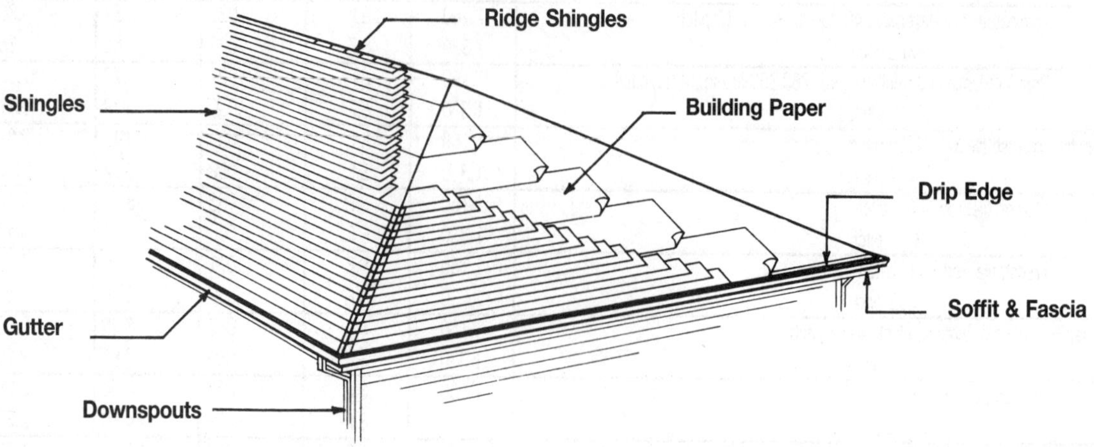

System Description	QUAN.	UNIT	LABOR HOURS	COST PER S.F.		
				MAT.	INST.	TOTAL
ASPHALT, ROOF SHINGLES, CLASS A						
Shingles, inorganic, class A, 210-235 lb./sq. 4/12 pitch	1.570	S.F.	.023	.54	.87	1.41
Drip edge, metal, 5″ wide	.122	L.F.	.002	.03	.10	.13
Building paper, #15 asphalt felt	1.800	S.F.	.002	.07	.09	.16
Ridge shingles, asphalt	.075	L.F.	.002	.07	.07	.14
Soffit & fascia, white painted aluminum, 1′ overhang	.120	L.F.	.017	.29	.71	1
Gutter, seamless, aluminum, painted	.120	L.F.	.008	.14	.36	.50
Downspouts, aluminum, painted	.035	L.F.	.002	.04	.07	.11
TOTAL		S.F.	.056	1.18	2.27	3.45
WOOD, CEDAR SHINGLES, NO. 1 PERFECTIONS, 18″ LONG						
Shingles, red cedar, No. 1 perfections, 5″ exp., 4/12 pitch	1.570	S.F.	.047	2.82	1.90	4.72
Drip edge, metal, 5″ wide	.122	L.F.	.002	.03	.10	.13
Building paper, #15 asphalt felt	1.800	S.F.	.002	.07	.09	.16
Ridge shingles, wood, cedar	.075	L.F.	.002	.20	.09	.29
Soffit & fascia, white painted aluminum, 1′ overhang	.120	L.F.	.017	.29	.71	1
Gutter, seamless, aluminum, painted	.120	L.F.	.008	.14	.36	.50
Downspouts, aluminum, painted	.035	L.F.	.002	.04	.07	.11
TOTAL		S.F.	.080	3.59	3.32	6.91

The prices in these systems are based on a square foot of plan area.
All quantities have been adjusted accordingly.

Description	QUAN.	UNIT	LABOR HOURS	COST PER S.F.		
				MAT.	INST.	TOTAL

Hip Roof - Roofing Price Sheet	QUAN.	UNIT	LABOR HOURS	COST PER S.F.		
				MAT.	INST.	TOTAL
ingles, asphalt, inorganic, class A, 210-235 lb./sq., 4/12 pitch	1.570	S.F.	.023	.54	.87	1.41
8/12 pitch	1.850	S.F.	.028	.65	1.04	1.69
Laminated, multi-layered, 240-260 lb./sq., 4/12 pitch	1.570	S.F.	.028	.76	1.06	1.82
8/12 pitch	1.850	S.F.	.034	.90	1.26	2.16
Prem. laminated, multi-layered, 260-300 lb./sq., 4/12 pitch	1.570	S.F.	.037	.96	1.37	2.33
8/12 pitch	1.850	S.F.	.043	1.14	1.62	2.76
Clay tile, Spanish tile, red, 4/12 pitch	1.570	S.F.	.071	4.58	2.69	7.27
8/12 pitch	1.850	S.F.	.084	5.45	3.19	8.64
Mission tile, red, 4/12 pitch	1.570	S.F.	.111	11.10	4.21	15.31
8/12 pitch	1.850	S.F.	.132	13.20	5	18.20
French tile, red, 4/12 pitch	1.570	S.F.	.095	10.10	3.58	13.68
8/12 pitch	1.850	S.F.	.113	11.95	4.26	16.21
Slate, Buckingham, Virginia, black, 4/12 pitch	1.570	S.F.	.073	9.70	2.77	12.47
8/12 pitch	1.850	S.F.	.087	11.50	3.29	14.79
Vermont, black or grey, 4/12 pitch	1.570	S.F.	.073	6	2.77	8.77
8/12 pitch	1.850	S.F.	.087	7.15	3.29	10.44
Wood, red cedar, No.1 5X, 16″ long, 5″ exposure, 4/12 pitch	1.570	S.F.	.051	2.86	2.08	4.94
8/12 pitch	1.850	S.F.	.061	3.40	2.47	5.87
Fire retardant, 4/12 pitch	1.570	S.F.	.051	3.39	2.08	5.47
8/12 pitch	1.850	S.F.	.061	4.03	2.47	6.50
18″ long, No.1 perfections, 5″ exposure, 4/12 pitch	1.570	S.F.	.047	2.82	1.90	4.72
8/12 pitch	1.850	S.F.	.055	3.34	2.26	5.60
Fire retardant, 4/12 pitch	1.570	S.F.	.047	3.32	1.90	5.22
8/12 pitch	1.850	S.F.	.055	3.94	2.26	6.20
Resquared & rebutted, 18″ long, 6″ exposure, 4/12 pitch	1.570	S.F.	.043	3.50	1.74	5.24
8/12 pitch	1.850	S.F.	.051	4.16	2.07	6.23
Fire retardant, 4/12 pitch	1.570	S.F.	.043	4	1.74	5.74
8/12 pitch	1.850	S.F.	.051	4.76	2.07	6.83
Wood shakes hand split, 24″ long, 10″ exposure, 4/12 pitch	1.570	S.F.	.051	2.43	2.08	4.51
8/12 pitch	1.850	S.F.	.061	2.89	2.47	5.36
Fire retardant, 4/12 pitch	1.570	S.F.	.051	2.96	2.08	5.04
8/12 pitch	1.850	S.F.	.061	3.52	2.47	5.99
18″ long, 8″ exposure, 4/12 pitch	1.570	S.F.	.064	1.71	2.61	4.32
8/12 pitch	1.850	S.F.	.076	2.03	3.10	5.13
Fire retardant, 4/12 pitch	1.570	S.F.	.064	2.24	2.61	4.85
8/12 pitch	1.850	S.F.	.076	2.66	3.10	5.76
rip edge, metal, 5″ wide	.122	L.F.	.002	.03	.10	.13
8″ wide	.122	L.F.	.002	.04	.10	.14
uilding paper, #15 asphalt felt	1.800	S.F.	.002	.07	.09	.16
idge shingles, asphalt	.075	L.F.	.002	.07	.07	.14
Clay	.075	L.F.	.003	.71	.11	.82
Slate	.075	L.F.	.003	.70	.11	.81
Wood, shingles	.075	L.F.	.002	.20	.09	.29
Shakes	.075	L.F.	.002	.20	.09	.29
offit & fascia, aluminum, vented, 1′ overhang	.120	L.F.	.017	.29	.71	1
2′ overhang	.120	L.F.	.019	.42	.78	1.20
Vinyl, vented, 1′ overhang	.120	L.F.	.016	.20	.65	.85
2′ overhang	.120	L.F.	.017	.28	.71	.99
Wood, board fascia, plywood soffit, 1′ overhang	.120	L.F.	.004	.02	.13	.15
2′ overhang	.120	L.F.	.006	.03	.20	.23
utter, 5″ box, aluminum, seamless, painted	.120	L.F.	.008	.14	.36	.50
Vinyl	.120	L.F.	.009	.13	.36	.49
Downspout, 2″ x 3″, aluminum, one story house	.035	L.F.	.002	.04	.07	.11
Two story house	.060	L.F.	.003	.06	.11	.17
Vinyl, one story house	.035	L.F.	.001	.04	.07	.11
Two story house	.060	L.F.	.003	.06	.11	.17

5 ROOFING

12 Gambrel Roofing Systems

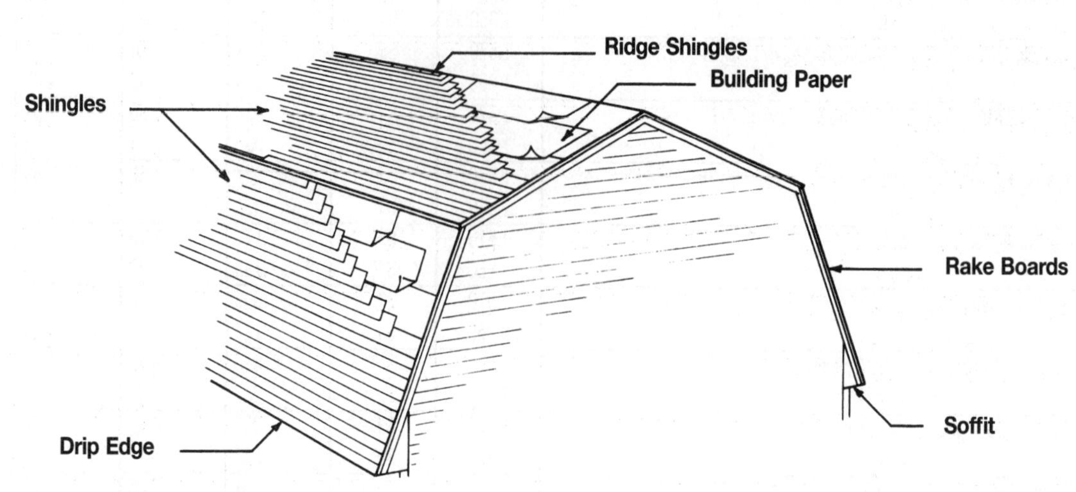

System Description	QUAN.	UNIT	LABOR HOURS	COST PER S.F. MAT.	COST PER S.F. INST.	COST PER S.F. TOTAL
ASPHALT, ROOF SHINGLES, CLASS A						
Shingles, asphalt, inorganic, class A, 210-235 lb./sq.	1.450	S.F.	.022	.51	.82	1.33
Drip edge, metal, 5″ wide	.146	L.F.	.003	.04	.12	.16
Building paper, #15 asphalt felt	1.500	S.F.	.002	.06	.08	.14
Ridge shingles, asphalt	.042	L.F.	.001	.04	.04	.08
Soffit & fascia, painted aluminum, 1′ overhang	.083	L.F.	.012	.20	.49	.69
Rake trim, 1″ x 6″	.063	L.F.	.003	.07	.10	.17
Rake trim, prime and paint	.063	L.F.	.003	.02	.10	.12
Gutter, seamless, alumunum, painted	.083	L.F.	.006	.10	.25	.35
Downspouts, aluminum, painted	.042	L.F.	.002	.05	.08	.13
TOTAL		S.F.	.054	1.09	2.08	3.17
WOOD, CEDAR SHINGLES, NO. 1 PERFECTIONS, 18″ LONG						
Shingles, wood, red cedar, No. 1 perfections, 5″ exposure	1.450	S.F.	.044	2.64	1.79	4.43
Drip edge, metal, 5″ wide	.146	L.F.	.003	.04	.12	.16
Building paper, #15 asphalt felt	1.500	S.F.	.002	.06	.08	.14
Ridge shingles, wood	.042	L.F.	.001	.11	.05	.16
Soffit & fascia, white painted aluminum, 1′ overhang	.083	L.F.	.012	.20	.49	.69
Rake trim, 1″ x 6″	.063	L.F.	.003	.07	.10	.17
Rake trim, prime and paint	.063	L.F.	.001	.01	.05	.06
Gutter, seamless, aluminum, painted	.083	L.F.	.006	.10	.25	.35
Downspouts, aluminum, painted	.042	L.F.	.002	.05	.08	.13
TOTAL		S.F.	.074	3.28	3.01	6.29

The prices in this system are based on a square foot of plan area.
All quantities have been adjusted accordingly.

Description	QUAN.	UNIT	LABOR HOURS	COST PER S.F. MAT.	COST PER S.F. INST.	COST PER S.F. TOTAL

Important: See the Reference Section for critical supporting data - Reference Nos., Crews & Location Factors

Gambrel Roofing Price Sheet	QUAN.	UNIT	LABOR HOURS	COST PER S.F.		
				MAT.	INST.	TOTAL
hingles, asphalt, standard, inorganic, class A, 210-235 lb./sq.	1.450	S.F.	.022	.51	.82	1.33
Laminated, multi-layered, 240-260 lb./sq.	1.450	S.F.	.027	.71	1	1.71
Premium laminated, multi-layered, 260-300 lb./sq.	1.450	S.F.	.034	.90	1.28	2.18
Slate, Buckingham, Virginia, black	1.450	S.F.	.069	9.10	2.60	11.70
Vermont, black or grey	1.450	S.F.	.069	5.65	2.60	8.25
Wood, red cedar, No.1 5X, 16″ long, 5″ exposure, plain	1.450	S.F.	.048	2.69	1.95	4.64
Fire retardant	1.450	S.F.	.048	3.19	1.95	5.14
18″ long, No.1 perfections, 6″ exposure, plain	1.450	S.F.	.044	2.64	1.79	4.43
Fire retardant	1.450	S.F.	.044	3.11	1.79	4.90
Resquared & rebutted, 18″ long, 6″ exposure, plain	1.450	S.F.	.040	3.29	1.64	4.93
Fire retardant	1.450	S.F.	.040	3.75	1.64	5.39
Shakes, hand split, 24″ long, 10″ exposure, plain	1.450	S.F.	.048	2.28	1.95	4.23
Fire retardant	1.450	S.F.	.048	2.78	1.95	4.73
18″ long, 8″ exposure, plain	1.450	S.F.	.060	1.61	2.45	4.06
Fire retardant	1.450	S.F.	.060	2.11	2.45	4.56
rip edge, metal, 5″ wide	.146	L.F.	.003	.04	.12	.16
8″ wide	.146	L.F.	.003	.05	.12	.17
uilding paper, #15 asphalt felt	1.500	S.F.	.002	.06	.08	.14
idge shingles, asphalt	.042	L.F.	.001	.04	.04	.08
Slate	.042	L.F.	.002	.39	.06	.45
Wood, shingles	.042	L.F.	.001	.11	.05	.16
Shakes	.042	L.F.	.001	.11	.05	.16
offit & fascia, aluminum, vented, 1′ overhang	.083	L.F.	.012	.20	.49	.69
2′ overhang	.083	L.F.	.013	.29	.54	.83
Vinyl vented, 1′ overhang	.083	L.F.	.011	.14	.45	.59
2′ overhang	.083	L.F.	.012	.19	.49	.68
Wood board fascia, plywood soffit, 1′ overhang	.083	L.F.	.004	.02	.13	.15
2′ overhang	.083	L.F.	.006	.03	.20	.23
ake trim, painted, 1″ x 6″	.063	L.F.	.006	.09	.20	.29
1″ x 8″	.063	L.F.	.007	.11	.27	.38
utter, 5″ box, aluminum, seamless, painted	.083	L.F.	.006	.10	.25	.35
Vinyl	.083	L.F.	.006	.09	.25	.34
ownspout 2″ x 3″, aluminum, one story house	.042	L.F.	.002	.04	.08	.12
Two story house	.070	L.F.	.003	.07	.13	.20
Vinyl, one story house	.042	L.F.	.002	.04	.08	.12
Two story house	.070	L.F.	.003	.07	.13	.20

16 Mansard Roofing Systems

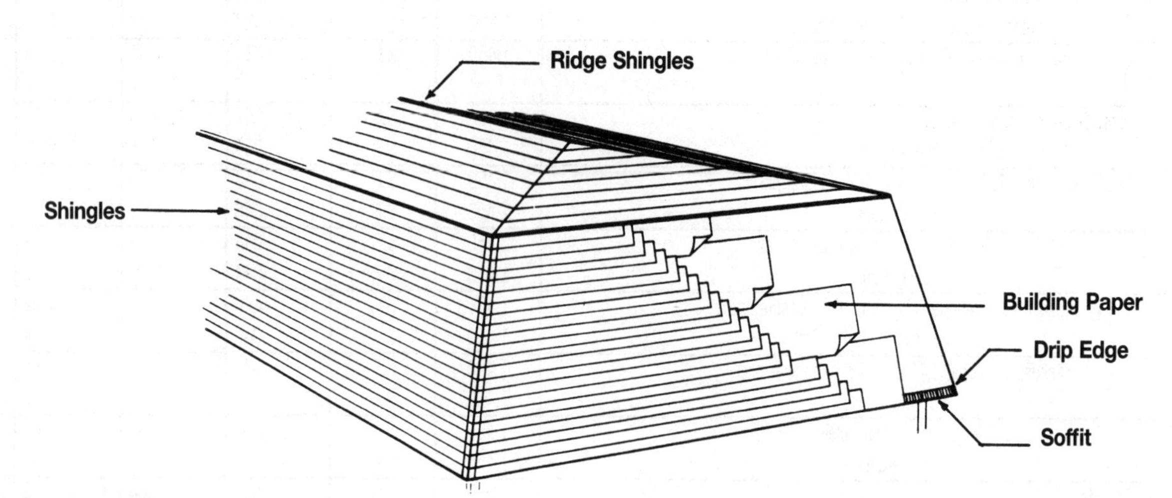

System Description	QUAN.	UNIT	LABOR HOURS	COST PER S.F. MAT.	COST PER S.F. INST.	COST PER S.F. TOTAL
ASPHALT, ROOF SHINGLES, CLASS A						
Shingles, standard inorganic class A 210-235 lb./sq.	2.210	S.F.	.032	.75	1.20	1.95
Drip edge, metal, 5″ wide	.122	L.F.	.002	.03	.10	.13
Building paper, #15 asphalt felt	2.300	S.F.	.003	.09	.12	.21
Ridge shingles, asphalt	.090	L.F.	.002	.08	.08	.16
Soffit & fascia, white painted aluminum, 1′ overhang	.122	L.F.	.018	.29	.73	1.02
Gutter, seamless, aluminum, painted	.122	L.F.	.008	.15	.36	.51
Downspouts, aluminum, painted	.042	L.F.	.002	.05	.08	.13
TOTAL		S.F.	.067	1.44	2.67	4.11
WOOD, CEDAR SHINGLES, NO. 1 PERFECTIONS, 18″ LONG						
Shingles, wood, red cedar, No. 1 perfections, 5″ exposure	2.210	S.F.	.064	3.87	2.62	6.49
Drip edge, metal, 5″ wide	.122	L.F.	.002	.03	.10	.13
Building paper, #15 asphalt felt	2.300	S.F.	.003	.09	.12	.21
Ridge shingles, wood	.090	L.F.	.003	.24	.10	.34
Soffit & fascia, white painted aluminum, 1′ overhang	.122	L.F.	.018	.29	.73	1.02
Gutter, seamless, aluminum, painted	.122	L.F.	.008	.15	.36	.51
Downspouts, aluminum, painted	.042	L.F.	.002	.05	.08	.13
TOTAL		S.F.	.100	4.72	4.11	8.83

The prices in these systems are based on a square foot of plan area.
All quantities have been adjusted accordingly.

Description	QUAN.	UNIT	LABOR HOURS	COST PER S.F. MAT.	COST PER S.F. INST.	COST PER S.F. TOTAL

Important: See the Reference Section for critical supporting data - Reference Nos., Crews & Location Factors

Mansard Roofing Price Sheet	QUAN.	UNIT	LABOR HOURS	COST PER S.F. MAT.	INST.	TOTAL
Shingles, asphalt, standard, inorganic, class A, 210-235 lb./sq.	2.210	S.F.	.032	.75	1.20	1.95
Laminated, multi-layered, 240-260 lb./sq.	2.210	S.F.	.039	1.05	1.46	2.51
Premium laminated, multi-layered, 260-300 lb./sq.	2.210	S.F.	.050	1.32	1.88	3.20
Slate Buckingham, Virginia, black	2.210	S.F.	.101	13.30	3.81	17.11
Vermont, black or grey	2.210	S.F.	.101	8.25	3.81	12.06
Wood, red cedar, No.1 5X, 16″ long, 5″ exposure, plain	2.210	S.F.	.070	3.94	2.86	6.80
Fire retardant	2.210	S.F.	.070	4.67	2.86	7.53
18″ long, No.1 perfections 6″ exposure, plain	2.210	S.F.	.064	3.87	2.62	6.49
Fire retardant	2.210	S.F.	.064	4.56	2.62	7.18
Resquared & rebutted, 18″ long, 6″ exposure, plain	2.210	S.F.	.059	4.82	2.40	7.22
Fire retardant	2.210	S.F.	.059	5.50	2.40	7.90
Shakes, hand split, 24″ long 10″ exposure, plain	2.210	S.F.	.070	3.34	2.86	6.20
Fire retardant	2.210	S.F.	.070	4.07	2.86	6.93
18″ long, 8″ exposure, plain	2.210	S.F.	.088	2.35	3.59	5.94
Fire retardant	2.210	S.F.	.088	3.08	3.59	6.67
Drip edge, metal, 5″ wide	.122	S.F.	.002	.03	.10	.13
8″ wide	.122	S.F.	.002	.04	.10	.14
Building paper, #15 asphalt felt	2.300	S.F.	.003	.09	.12	.21
Ridge shingles, asphalt	.090	L.F.	.002	.08	.08	.16
Slate	.090	L.F.	.004	.84	.14	.98
Wood, shingles	.090	L.F.	.003	.24	.10	.34
Shakes	.090	L.F.	.003	.24	.10	.34
Soffit & fascia, aluminum vented, 1′ overhang	.122	L.F.	.018	.29	.73	1.02
2′ overhang	.122	L.F.	.020	.43	.79	1.22
Vinyl vented, 1′ overhang	.122	L.F.	.016	.20	.66	.86
2′ overhang	.122	L.F.	.018	.29	.73	1.02
Wood board fascia, plywood soffit, 1′ overhang	.122	L.F.	.013	.33	.50	.83
2′ overhang	.122	L.F.	.019	.45	.76	1.21
Gutter, 5″ box, aluminum, seamless, painted	.122	L.F.	.008	.15	.36	.51
Vinyl	.122	L.F.	.009	.13	.36	.49
Downspout 2″ x 3″, aluminum, one story house	.042	L.F.	.002	.04	.08	.12
Two story house	.070	L.F.	.003	.07	.13	.20
Vinyl, one story house	.042	L.F.	.002	.04	.08	.12
Two story house	.070	L.F.	.003	.07	.13	.20

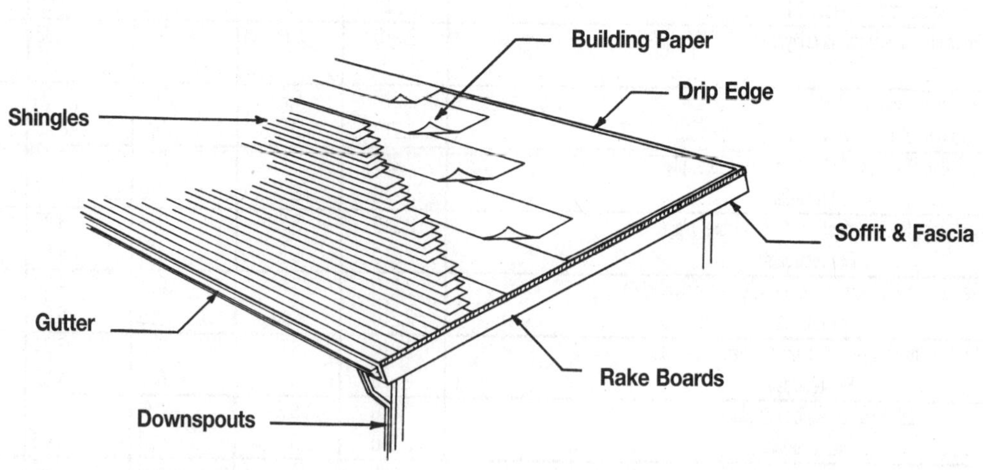

System Description	QUAN.	UNIT	LABOR HOURS	COST PER S.F.		
				MAT.	INST.	TOTAL
ASPHALT, ROOF SHINGLES, CLASS A						
Shingles, inorganic class A 210-235 lb./sq. 4/12 pitch	1.230	S.F.	.019	.44	.71	1.15
Drip edge, metal, 5″ wide	.100	L.F.	.002	.02	.08	.10
Building paper, #15 asphalt felt	1.300	S.F.	.002	.05	.07	.12
Soffit & fascia, white painted aluminum, 1′ overhang	.080	L.F.	.012	.19	.48	.67
Rake trim, 1″ x 6″	.043	L.F.	.002	.05	.07	.12
Rake trim, prime and paint	.043	L.F.	.002	.01	.07	.08
Gutter, seamless, aluminum, painted	.040	L.F.	.003	.05	.12	.17
Downspouts, painted aluminum	.020	L.F.	.001	.02	.04	.06
TOTAL		S.F.	.043	.83	1.64	2.47
WOOD, CEDAR SHINGLES, NO. 1 PERFECTIONS, 18″ LONG						
Shingles, red cedar, No. 1 perfections, 5″ exp., 4/12 pitch	1.230	S.F.	.035	2.11	1.43	3.54
Drip edge, metal, 5″ wide	.100	L.F.	.002	.02	.08	.10
Building paper, #15 asphalt felt	1.300	S.F.	.002	.05	.07	.12
Soffit & fascia, white painted aluminum, 1′ overhang	.080	L.F.	.012	.19	.48	.67
Rake trim, 1″ x 6″	.043	L.F.	.002	.05	.07	.12
Rake trim, prime and paint	.043	L.F.	.001	.01	.03	.04
Gutter, seamless, aluminum, painted	.040	L.F.	.003	.05	.12	.17
Downspouts, painted aluminum	.020	L.F.	.001	.02	.04	.06
TOTAL		S.F.	.058	2.50	2.32	4.82

The prices in these systems are based on a square foot of plan area.
All quantities have been adjusted accordingly.

Description	QUAN.	UNIT	LABOR HOURS	COST PER S.F.		
				MAT.	INST.	TOTAL

Shed Roofing Price Sheet	QUAN.	UNIT	LABOR HOURS	COST PER S.F. MAT.	INST.	TOTAL
Shingles, asphalt, inorganic, class A, 210-235 lb./sq., 4/12 pitch	1.230	S.F.	.017	.41	.65	1.06
8/12 pitch	1.330	S.F.	.019	.44	.71	1.15
Laminated, multi-layered, 240-260 lb./sq. 4/12 pitch	1.230	S.F.	.021	.57	.80	1.37
8/12 pitch	1.330	S.F.	.023	.62	.86	1.48
Premium laminated, multi-layered, 260-300 lb./sq. 4/12 pitch	1.230	S.F.	.027	.72	1.03	1.75
8/12 pitch	1.330	S.F.	.030	.78	1.11	1.89
Clay tile, Spanish tile, red, 4/12 pitch	1.230	S.F.	.053	3.43	2.02	5.45
8/12 pitch	1.330	S.F.	.058	3.72	2.18	5.90
Mission tile, red, 4/12 pitch	1.230	S.F.	.083	8.35	3.16	11.51
8/12 pitch	1.330	S.F.	.090	9.05	3.42	12.47
French tile, red, 4/12 pitch	1.230	S.F.	.071	7.55	2.69	10.24
8/12 pitch	1.330	S.F.	.077	8.20	2.91	11.11
Slate, Buckingham, Virginia, black, 4/12 pitch	1.230	S.F.	.055	7.25	2.08	9.33
8/12 pitch	1.330	S.F.	.059	7.85	2.25	10.10
Vermont, black or grey, 4/12 pitch	1.230	S.F.	.055	4.50	2.08	6.58
8/12 pitch	1.330	S.F.	.059	4.88	2.25	7.13
Wood, red cedar, No.1 5X, 16″ long, 5″ exposure, 4/12 pitch	1.230	S.F.	.038	2.15	1.56	3.71
8/12 pitch	1.330	S.F.	.042	2.33	1.69	4.02
Fire retardant, 4/12 pitch	1.230	S.F.	.038	2.55	1.56	4.11
8/12 pitch	1.330	S.F.	.042	2.76	1.69	4.45
18″ long, 6″ exposure, 4/12 pitch	1.230	S.F.	.035	2.11	1.43	3.54
8/12 pitch	1.330	S.F.	.038	2.29	1.55	3.84
Fire retardant, 4/12 pitch	1.230	S.F.	.035	2.49	1.43	3.92
8/12 pitch	1.330	S.F.	.038	2.70	1.55	4.25
Resquared & rebutted, 18″ long, 6″ exposure, 4/12 pitch	1.230	S.F.	.032	2.63	1.31	3.94
8/12 pitch	1.330	S.F.	.035	2.85	1.42	4.27
Fire retardant, 4/12 pitch	1.230	S.F.	.032	3.01	1.31	4.32
8/12 pitch	1.330	S.F.	.035	3.26	1.42	4.68
Wood shakes, hand split, 24″ long, 10″ exposure, 4/12 pitch	1.230	S.F.	.038	1.82	1.56	3.38
8/12 pitch	1.330	S.F.	.042	1.98	1.69	3.67
Fire retardant, 4/12 pitch	1.230	S.F.	.038	2.22	1.56	3.78
8/12 pitch	1.330	S.F.	.042	2.41	1.69	4.10
18″ long, 8″ exposure, 4/12 pitch	1.230	S.F.	.048	1.28	1.96	3.24
8/12 pitch	1.330	S.F.	.052	1.39	2.12	3.51
Fire retardant, 4/12 pitch	1.230	S.F.	.048	1.68	1.96	3.64
8/12 pitch	1.330	S.F.	.052	1.82	2.12	3.94
Drip edge, metal, 5″ wide	.100	L.F.	.002	.02	.08	.10
8″ wide	.100	L.F.	.002	.04	.08	.12
Building paper, #15 asphalt felt	1.300	S.F.	.002	.05	.07	.12
Soffit & fascia, aluminum vented, 1′ overhang	.080	L.F.	.012	.19	.48	.67
2′ overhang	.080	L.F.	.013	.28	.52	.80
Vinyl vented, 1′ overhang	.080	L.F.	.011	.13	.44	.57
2′ overhang	.080	L.F.	.012	.19	.48	.67
Wood board fascia, plywood soffit, 1′ overhang	.080	L.F.	.010	.23	.37	.60
2′ overhang	.080	L.F.	.014	.31	.56	.87
Rake, trim, painted, 1″ x 6″	.043	L.F.	.004	.06	.14	.20
1″ x 8″	.043	L.F.	.004	.06	.14	.20
Gutter, 5″ box, aluminum, seamless, painted	.040	L.F.	.003	.05	.12	.17
Vinyl	.040	L.F.	.003	.04	.12	.16
Downspout 2″ x 3″, aluminum, one story house	.020	L.F.	.001	.02	.04	.06
Two story house	.020	L.F.	.001	.04	.06	.10
Vinyl, one story house	.020	L.F.	.001	.02	.04	.06
Two story house	.020	L.F.	.001	.04	.06	.10

5 ROOFING

24 Gable Dormer Roofing Systems

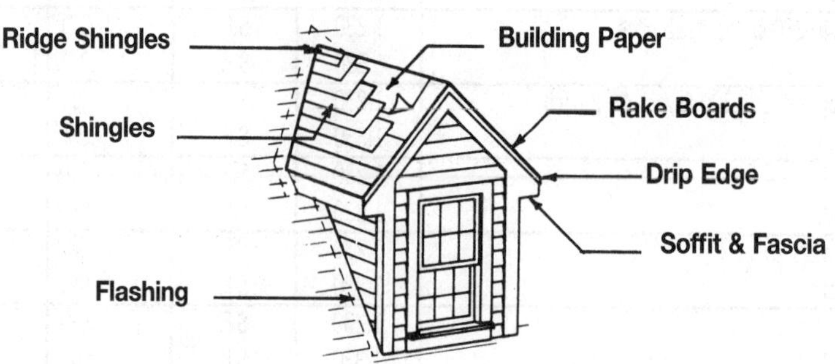

System Description	QUAN.	UNIT	LABOR HOURS	COST PER S.F.		
				MAT.	INST.	TOTAL
ASPHALT, ROOF SHINGLES, CLASS A						
Shingles, standard inorganic class A 210-235 lb./sq	1.400	S.F.	.020	.48	.76	1.24
Drip edge, metal, 5″ wide	.220	L.F.	.004	.05	.18	.23
Building paper, #15 asphalt felt	1.500	S.F.	.002	.06	.08	.14
Ridge shingles, asphalt	.280	L.F.	.007	.24	.25	.49
Soffit & fascia, aluminum, vented	.220	L.F.	.032	.52	1.31	1.83
Flashing, aluminum, mill finish, .013″ thick	1.500	S.F.	.083	.59	3.11	3.70
TOTAL		S.F.	.148	1.94	5.69	7.63
WOOD, CEDAR, NO. 1 PERFECTIONS						
Shingles, red cedar, No.1 perfections, 18″ long, 5″ exp.	1.400	S.F.	.041	2.46	1.67	4.13
Drip edge, metal, 5″ wide	.220	L.F.	.004	.05	.18	.23
Building paper, #15 asphalt felt	1.500	S.F.	.002	.06	.08	.14
Ridge shingles, wood	.280	L.F.	.008	.74	.32	1.06
Soffit & fascia, aluminum, vented	.220	L.F.	.032	.52	1.31	1.83
Flashing, aluminum, mill finish, .013″ thick	1.500	S.F.	.083	.59	3.11	3.70
TOTAL		S.F.	.170	4.42	6.67	11.09
SLATE, BUCKINGHAM, BLACK						
Shingles, Buckingham, Virginia, black	1.400	S.F.	.064	8.47	2.42	10.89
Drip edge, metal, 5″ wide	.220	L.F.	.004	.05	.18	.23
Building paper, #15 asphalt felt	1.500	S.F.	.002	.06	.08	.14
Ridge shingles, slate	.280	L.F.	.011	2.62	.42	3.04
Soffit & fascia, aluminum, vented	.220	L.F.	.032	.52	1.31	1.83
Flashing, copper, 16 oz.	1.500	S.F.	.104	4.32	3.92	8.24
TOTAL		S.F.	.217	16.04	8.33	24.37

The prices in these systems are based on a square foot of plan area under the dormer roof.

Description	QUAN.	UNIT	LABOR HOURS	COST PER S.F.		
				MAT.	INST.	TOTAL

Important: See the Reference Section for critical supporting data - Reference Nos., Crews & Location Factors

Gable Dormer Roofing Price Sheet	QUAN.	UNIT	LABOR HOURS	COST PER S.F.		
				MAT.	INST.	TOTAL
hingles, asphalt, standard, inorganic, class A, 210-235 lb./sq.	1.400	S.F.	.020	.48	.76	1.24
Laminated, multi-layered, 240-260 lb./sq.	1.400	S.F.	.025	.67	.93	1.60
Premium laminated, multi-layered, 260-300 lb./sq.	1.400	S.F.	.032	.84	1.20	2.04
Clay tile, Spanish tile, red	1.400	S.F.	.062	4	2.35	6.35
Mission tile, red	1.400	S.F.	.097	9.75	3.68	13.43
French tile, red	1.400	S.F.	.083	8.80	3.14	11.94
Slate Buckingham, Virginia, black	1.400	S.F.	.064	8.45	2.42	10.87
Vermont, black or grey	1.400	S.F.	.064	5.25	2.42	7.67
Wood, red cedar, No.1 5X, 16" long, 5" exposure	1.400	S.F.	.045	2.51	1.82	4.33
Fire retardant	1.400	S.F.	.045	2.97	1.82	4.79
18" long, No.1 perfections, 5" exposure	1.400	S.F.	.041	2.46	1.67	4.13
Fire retardant	1.400	S.F.	.041	2.90	1.67	4.57
Resquared & rebutted, 18" long, 5" exposure	1.400	S.F.	.037	3.07	1.53	4.60
Fire retardant	1.400	S.F.	.037	3.51	1.53	5.04
Shakes hand split, 24" long, 10" exposure	1.400	S.F.	.045	2.13	1.82	3.95
Fire retardant	1.400	S.F.	.045	2.59	1.82	4.41
18" long, 8" exposure	1.400	S.F.	.056	1.50	2.28	3.78
Fire retardant	1.400	S.F.	.056	1.96	2.28	4.24
rip edge, metal, 5" wide	.220	L.F.	.004	.05	.18	.23
8" wide	.220	L.F.	.004	.08	.18	.26
uilding paper, #15 asphalt felt	1.500	S.F.	.002	.06	.08	.14
idge shingles, asphalt	.280	L.F.	.007	.24	.25	.49
Clay	.280	L.F.	.011	2.66	.42	3.08
Slate	.280	L.F.	.011	2.62	.42	3.04
Wood	.280	L.F.	.008	.74	.32	1.06
offit & fascia, aluminum, vented	.220	L.F.	.032	.52	1.31	1.83
Vinyl, vented	.220	L.F.	.029	.36	1.20	1.56
Wood, board fascia, plywood soffit	.220	L.F.	.026	.62	1.01	1.63
lashing, aluminum, .013" thick	1.500	S.F.	.083	.59	3.11	3.70
.032" thick	1.500	S.F.	.083	1.76	3.11	4.87
.040" thick	1.500	S.F.	.083	2.40	3.11	5.51
.050" thick	1.500	S.F.	.083	3.03	3.11	6.14
Copper, 16 oz.	1.500	S.F.	.104	4.32	3.92	8.24
20 oz.	1.500	S.F.	.109	6.45	4.08	10.53
24 oz.	1.500	S.F.	.114	7.75	4.28	12.03
32 oz.	1.500	S.F.	.120	10.30	4.50	14.80

28 Shed Dormer Roofing Systems

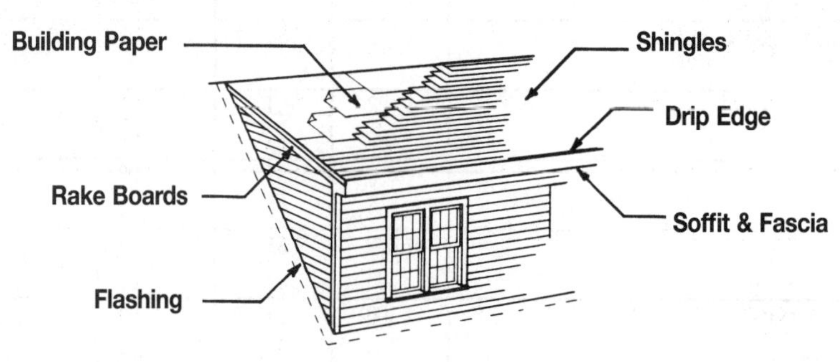

System Description	QUAN.	UNIT	LABOR HOURS	COST PER S.F. MAT.	COST PER S.F. INST.	COST PER S.F. TOTAL
ASPHALT, ROOF SHINGLES, CLASS A						
Shingles, standard inorganic class A 210-235 lb./sq.	1.100	S.F.	.016	.37	.60	.97
Drip edge, aluminum, 5″ wide	.250	L.F.	.005	.07	.21	.28
Building paper, #15 asphalt felt	1.200	S.F.	.002	.05	.06	.11
Soffit & fascia, aluminum, vented, 1′ overhang	.250	L.F.	.036	.60	1.49	2.09
Flashing, aluminum, mill finish, 0.013″ thick	.800	L.F.	.044	.31	1.66	1.97
TOTAL		S.F.	.103	1.40	4.02	5.42
WOOD, CEDAR, NO. 1 PERFECTIONS, 18″ LONG						
Shingles, wood, red cedar, #1 perfections, 5″ exposure	1.100	S.F.	.032	1.94	1.31	3.25
Drip edge, aluminum, 5″ wide	.250	L.F.	.005	.07	.21	.28
Building paper, #15 asphalt felt	1.200	S.F.	.002	.05	.06	.11
Soffit & fascia, aluminum, vented, 1′ overhang	.250	L.F.	.036	.60	1.49	2.09
Flashing, aluminum, mill finish, 0.013″ thick	.800	L.F.	.044	.31	1.66	1.97
TOTAL		S.F.	.119	2.97	4.73	7.70
SLATE, BUCKINGHAM, BLACK						
Shingles, slate, Buckingham, black	1.100	S.F.	.050	6.66	1.90	8.56
Drip edge, aluminum, 5″ wide	.250	L.F.	.005	.07	.21	.28
Building paper, #15 asphalt felt	1.200	S.F.	.002	.05	.06	.11
Soffit & fascia, aluminum, vented, 1′ overhang	.250	L.F.	.036	.60	1.49	2.09
Flashing, copper, 16 oz.	.800	L.F.	.056	2.30	2.09	4.39
TOTAL		S.F.	.149	9.68	5.75	15.43

The prices in this system are based on a square foot of plan area under the dormer roof.

Description	QUAN.	UNIT	LABOR HOURS	COST PER S.F. MAT.	COST PER S.F. INST.	COST PER S.F. TOTAL

Important: See the Reference Section for critical supporting data - Reference Nos., Crews & Location Factors

Shed Dormer Roofing Price Sheet	QUAN.	UNIT	LABOR HOURS	COST PER S.F.		
				MAT.	INST.	TOTAL
hingles, asphalt, standard, inorganic, class A, 210-235 lb./sq.	1.100	S.F.	.016	.37	.60	.97
Laminated, multi-layered, 240-260 lb./sq.	1.100	S.F.	.020	.52	.73	1.25
Premium laminated, multi-layered, 260-300 lb./sq.	1.100	S.F.	.025	.66	.94	1.60
Clay tile, Spanish tile, red	1.100	S.F.	.049	3.15	1.85	5
Mission tile, red	1.100	S.F.	.077	7.65	2.89	10.54
French tile, red	1.100	S.F.	.065	6.95	2.46	9.41
Slate Buckingham, Virginia, black	1.100	S.F.	.050	6.65	1.90	8.55
Vermont, black or grey	1.100	S.F.	.050	4.13	1.90	6.03
Wood, red cedar, No. 1 5X, 16″ long, 5″ exposure	1.100	S.F.	.035	1.97	1.43	3.40
Fire retardant	1.100	S.F.	.035	2.33	1.43	3.76
18″ long, No.1 perfections, 5″ exposure	1.100	S.F.	.032	1.94	1.31	3.25
Fire retardant	1.100	S.F.	.032	2.29	1.31	3.60
Resquared & rebutted, 18″ long, 5″ exposure	1.100	S.F.	.029	2.41	1.20	3.61
Fire retardant	1.100	S.F.	.029	2.76	1.20	3.96
Shakes hand split, 24″ long, 10″ exposure	1.100	S.F.	.035	1.67	1.43	3.10
Fire retardant	1.100	S.F.	.035	2.03	1.43	3.46
18″ long, 8″ exposure	1.100	S.F.	.044	1.18	1.79	2.97
Fire retardant	1.100	S.F.	.044	1.54	1.79	3.33
rip edge, metal, 5″ wide	.250	L.F.	.005	.07	.21	.28
8″ wide	.250	L.F.	.005	.09	.21	.30
uilding paper, #15 asphalt felt	1.200	S.F.	.002	.05	.06	.11
offit & fascia, aluminum, vented	.250	L.F.	.036	.60	1.49	2.09
Vinyl, vented	.250	L.F.	.033	.41	1.36	1.77
Wood, board fascia, plywood soffit	.250	L.F.	.030	.71	1.15	1.86
lashing, aluminum, .013″ thick	.800	L.F.	.044	.31	1.66	1.97
.032″ thick	.800	L.F.	.044	.94	1.66	2.60
.040″ thick	.800	L.F.	.044	1.28	1.66	2.94
.050″ thick	.800	L.F.	.044	1.62	1.66	3.28
Copper, 16 oz.	.800	L.F.	.056	2.30	2.09	4.39
20 oz.	.800	L.F.	.058	3.43	2.18	5.61
24 oz.	.800	L.F.	.061	4.12	2.28	6.40
32 oz.	.800	L.F.	.064	5.50	2.40	7.90

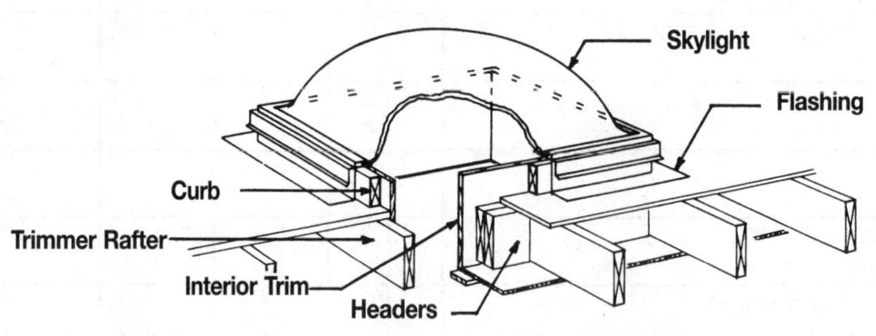

System Description	QUAN.	UNIT	LABOR HOURS	COST EACH MAT.	COST EACH INST.	COST EACH TOTAL
SKYLIGHT, FIXED, 32″ X 32″						
Skylight, fixed bubble, insulating, 32″ x 32″	1.000	Ea.	1.422	107.73	52.62	160.35
Trimmer rafters, 2″ x 6″	28.000	L.F.	.448	16.52	18.20	34.72
Headers, 2″ x 6″	6.000	L.F.	.267	3.54	10.86	14.40
Curb, 2″ x 4″	12.000	L.F.	.154	4.44	6.24	10.68
Flashing, aluminum, .013″ thick	13.500	S.F.	.745	5.27	27.95	33.22
Trim, stock pine, 11/16″ x 2-1/2″	12.000	L.F.	.400	11.28	16.32	27.60
Trim primer coat, oil base, brushwork	12.000	L.F.	.148	.36	5.28	5.64
Trim paint, 1 coat, brushwork	12.000	L.F.	.148	.36	5.28	5.64
TOTAL		Ea.	3.732	149.50	142.75	292.25
SKYLIGHT, FIXED, 48″ X 48″						
Skylight, fixed bubble, insulating, 48″ x 48″	1.000	Ea.	1.296	148	48	196
Trimmer rafters, 2″ x 6″	28.000	L.F.	.448	16.52	18.20	34.72
Headers, 2″ x 6″	8.000	L.F.	.356	4.72	14.48	19.20
Curb, 2″ x 4″	16.000	L.F.	.205	5.92	8.32	14.24
Flashing, aluminum, .013″ thick	16.000	S.F.	.883	6.24	33.12	39.36
Trim, stock pine, 11/16″ x 2-1/2″	16.000	L.F.	.533	15.04	21.76	36.80
Trim primer coat, oil base, brushwork	16.000	L.F.	.197	.48	7.04	7.52
Trim paint, 1 coat, brushwork	16.000	L.F.	.197	.48	7.04	7.52
TOTAL		Ea.	4.115	197.40	157.96	355.36
SKYWINDOW, OPERATING, 24″ X 48″						
Skywindow, operating, thermopane glass, 24″ x 48″	1.000	Ea.	3.200	600	118	718
Trimmer rafters, 2″ x 6″	28.000	L.F.	.448	16.52	18.20	34.72
Headers, 2″ x 6″	8.000	L.F.	.267	3.54	10.86	14.40
Curb, 2″ x 4″	14.000	L.F.	.179	5.18	7.28	12.46
Flashing, aluminum, .013″ thick	14.000	S.F.	.772	5.46	28.98	34.44
Trim, stock pine, 11/16″ x 2-1/2″	14.000	L.F.	.467	13.16	19.04	32.20
Trim primer coat, oil base, brushwork	14.000	L.F.	.172	.42	6.16	6.58
Trim paint, 1 coat, brushwork	14.000	L.F.	.172	.42	6.16	6.58
TOTAL		Ea.	5.677	644.70	214.68	859.38

The prices in these systems are on a cost each basis.

Description	QUAN.	UNIT	LABOR HOURS	COST EACH MAT.	COST EACH INST.	COST EACH TOTAL

Skylight/Skywindow Price Sheet	QUAN.	UNIT	LABOR HOURS	COST EACH MAT.	COST EACH INST.	COST EACH TOTAL
kylight, fixed bubble insulating, 24″ x 24″	1.000	Ea.	.800	60.50	29.50	90
32″ x 32″	1.000	Ea.	1.422	108	52.50	160.50
32″ x 48″	1.000	Ea.	.864	98.50	32	130.50
48″ x 48″	1.000	Ea.	1.296	148	48	196
Ventilating bubble insulating, 36″ x 36″	1.000	Ea.	2.667	420	98.50	518.50
52″ x 52″	1.000	Ea.	2.667	625	98.50	723.50
28″ x 52″	1.000	Ea.	3.200	490	118	608
36″ x 52″	1.000	Ea.	3.200	530	118	648
Skywindow, operating, thermopane glass, 24″ x 48″	1.000	Ea.	3.200	600	118	718
32″ x 48″	1.000	Ea.	3.556	630	132	762
rimmer rafters, 2″ x 6″	28.000	L.F.	.448	16.50	18.20	34.70
2″ x 8″	28.000	L.F.	.472	25.50	19.30	44.80
2″ x 10″	28.000	L.F.	.711	36	29	65
eaders, 24″ window, 2″ x 6″	4.000	L.F.	.178	2.36	7.25	9.61
2″ x 8″	4.000	L.F.	.188	3.64	7.70	11.34
2″ x 10″	4.000	L.F.	.200	5.15	8.15	13.30
32″ window, 2″ x 6″	6.000	L.F.	.267	3.54	10.85	14.39
2″ x 8″	6.000	L.F.	.282	5.45	11.50	16.95
2″ x 10″	6.000	L.F.	.300	7.75	12.25	20
48″ window, 2″ x 6″	8.000	L.F.	.356	4.72	14.50	19.22
2″ x 8″	8.000	L.F.	.376	7.30	15.35	22.65
2″ x 10″	8.000	L.F.	.400	10.30	16.30	26.60
urb, 2″ x 4″, skylight, 24″ x 24″	8.000	L.F.	.102	2.96	4.16	7.12
32″ x 32″	12.000	L.F.	.154	4.44	6.25	10.69
32″ x 48″	14.000	L.F.	.179	5.20	7.30	12.50
48″ x 48″	16.000	L.F.	.205	5.90	8.30	14.20
lashing, aluminum .013″ thick, skylight, 24″ x 24″	9.000	S.F.	.497	3.51	18.65	22.16
32″ x 32″	13.500	S.F.	.745	5.25	28	33.25
32″ x 48″	14.000	S.F.	.772	5.45	29	34.45
48″ x 48″	16.000	S.F.	.883	6.25	33	39.25
Copper 16 oz., skylight, 24″ x 24″	9.000	S.F.	.626	26	23.50	49.50
32″ x 32″	13.500	S.F.	.939	39	35	74
32″ x 48″	14.000	S.F.	.974	40.50	36.50	77
48″ x 48″	16.000	S.F.	1.113	46	42	88
rim, interior casing painted, 24″ x 24″	8.000	L.F.	.347	8.30	13.75	22.05
32″ x 32″	12.000	L.F.	.520	12.50	20.50	33
32″ x 48″	14.000	L.F.	.607	14.55	24	38.55
48″ x 48″	16.000	L.F.	.693	16.65	27.50	44.15

34 Built-up Roofing Systems

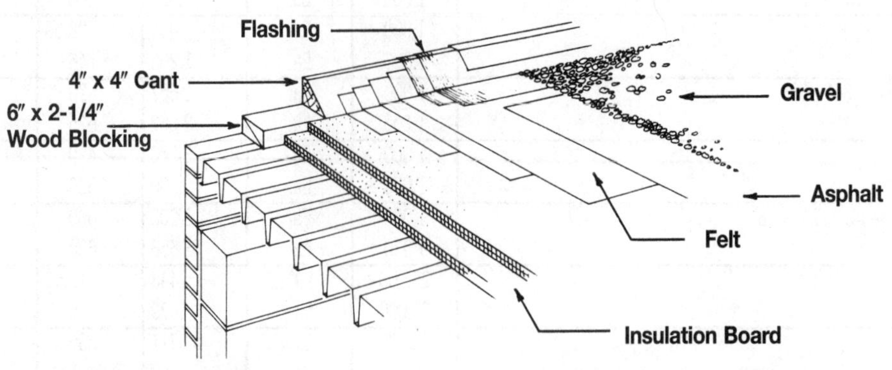

System Description	QUAN.	UNIT	LABOR HOURS	COST PER S.F. MAT.	COST PER S.F. INST.	COST PER S.F. TOTAL
ASPHALT, ORGANIC, 4-PLY, INSULATED DECK						
Membrane, asphalt, 4-plies #15 felt, gravel surfacing	1.000	S.F.	.025	.71	1.05	1.76
Insulation board, 2-layers of 1-1/16" glass fiber	2.000	S.F.	.016	1.68	.60	2.28
Wood blocking, 2" x 6"	.040	L.F.	.004	.07	.18	.25
Treated 4" x 4" cant strip	.040	L.F.	.001	.06	.04	.10
Flashing, aluminum, 0.040" thick	.050	S.F.	.003	.08	.10	.18
TOTAL		S.F.	.049	2.60	1.97	4.57
ASPHALT, INORGANIC, 3-PLY, INSULATED DECK						
Membrane, asphalt, 3-plies type IV glass felt, gravel surfacing	1.000	S.F.	.028	.66	1.15	1.81
Insulation board, 2-layers of 1-1/16" glass fiber	2.000	S.F.	.016	1.68	.60	2.28
Wood blocking, 2" x 6"	.040	L.F.	.004	.07	.18	.25
Treated 4" x 4" cant strip	.040	L.F.	.001	.06	.04	.10
Flashing, aluminum, 0.040" thick	.050	S.F.	.003	.08	.10	.18
TOTAL		S.F.	.052	2.55	2.07	4.62
COAL TAR, ORGANIC, 4-PLY, INSULATED DECK						
Membrane, coal tar, 4-plies #15 felt, gravel surfacing	1.000	S.F.	.027	1.17	1.10	2.27
Insulation board, 2-layers of 1-1/16" glass fiber	2.000	S.F.	.016	1.68	.60	2.28
Wood blocking, 2" x 6"	.040	L.F.	.004	.07	.18	.25
Treated 4" x 4" cant strip	.040	L.F.	.001	.06	.04	.10
Flashing, aluminum, 0.040" thick	.050	S.F.	.003	.08	.10	.18
TOTAL		S.F.	.051	3.06	2.02	5.08
COAL TAR, INORGANIC, 3-PLY, INSULATED DECK						
Membrane, coal tar, 3-plies type IV glass felt, gravel surfacing	1.000	S.F.	.029	.96	1.21	2.17
Insulation board, 2-layers of 1-1/16" glass fiber	2.000	S.F.	.016	1.68	.60	2.28
Wood blocking, 2" x 6"	.040	L.F.	.004	.07	.18	.25
Treated 4" x 4" cant strip	.040	L.F.	.001	.06	.04	.10
Flashing, aluminum, 0.040" thick	.050	S.F.	.003	.08	.10	.18
TOTAL		S.F.	.053	2.85	2.13	4.98

Built-Up Roofing Price Sheet	QUAN.	UNIT	LABOR HOURS	COST PER S.F.		
				MAT.	INST.	TOTAL
embrane, asphalt, 4-plies #15 organic felt, gravel surfacing	1.000	S.F.	.025	.71	1.05	1.76
Asphalt base sheet & 3-plies #15 asphalt felt	1.000	S.F.	.025	.54	1.05	1.59
3-plies type IV glass fiber felt	1.000	S.F.	.028	.66	1.15	1.81
4-plies type IV glass fiber felt	1.000	S.F.	.028	.80	1.15	1.95
Coal tar, 4-plies #15 organic felt, gravel surfacing	1.000	S.F.	.027			
4-plies tarred felt	1.000	S.F.	.027	1.17	1.10	2.27
3-plies type IV glass fiber felt	1.000	S.F.	.029	.96	1.21	2.17
4-plies type IV glass fiber felt	1.000	S.F.	.027	1.33	1.10	2.43
Roll, asphalt, 1-ply #15 organic felt, 2-plies mineral surfaced	1.000	S.F.	.021	.43	.85	1.28
3-plies type IV glass fiber, 1-ply mineral surfaced	1.000	S.F.	.022	.67	.92	1.59
sulation boards, glass fiber, 1-1/16″ thick	1.000	S.F.	.008	.84	.30	1.14
2-1/16″ thick	1.000	S.F.	.010	1.23	.37	1.60
2-7/16″ thick	1.000	S.F.	.010	1.41	.37	1.78
Expanded perlite, 1″ thick	1.000	S.F.	.010	.34	.37	.71
1-1/2″ thick	1.000	S.F.	.010	.43	.37	.80
2″ thick	1.000	S.F.	.011	.68	.43	1.11
Fiberboard, 1″ thick	1.000	S.F.	.010	.37	.37	.74
1-1/2″ thick	1.000	S.F.	.010	.55	.37	.92
2″ thick	1.000	S.F.	.010	.75	.37	1.12
Extruded polystyrene, 15 PSI compressive strength, 2″ thick R10	1.000	S.F.	.006	.39	.24	.63
3″ thick R15	1.000	S.F.	.008	.75	.30	1.05
4″ thick R20	1.000	S.F.	.008	1.18	.30	1.48
Tapered for drainage	1.000	S.F.	.005	.39	.20	.59
40 PSI compressive strength, 1″ thick R5	1.000	S.F.	.005	.39	.20	.59
2″ thick R10	1.000	S.F.	.006	.76	.24	1
3″ thick R15	1.000	S.F.	.008	1.11	.30	1.41
4″ thick R20	1.000	S.F.	.008	1.49	.30	1.79
Fiberboard high density, 1/2″ thick R1.3	1.000	S.F.	.008	.22	.30	.52
1″ thick R2.5	1.000	S.F.	.010	.40	.37	.77
1 1/2″ thick R3.8	1.000	S.F.	.010	.65	.37	1.02
Polyisocyanurate, 1 1/2″ thick R10.87	1.000	S.F.	.006	.39	.24	.63
2″ thick R14.29	1.000	S.F.	.007	.50	.27	.77
3 1/2″ thick R25	1.000	S.F.	.008	.81	.30	1.11
Tapered for drainage	1.000	S.F.	.006	.42	.21	.63
Expanded polystyrene, 1″ thick	1.000	S.F.	.005	.21	.20	.41
2″ thick R10	1.000	S.F.	.006	.45	.24	.69
3″ thick	1.000	S.F.	.006	.77	.24	1.01
ood blocking, treated, 6″ x 2″ & 4″ x 4″ cant	.040	L.F.	.002	.09	.10	.19
6″ x 4-1/2″ & 4″ x 4″ cant	.040	L.F.	.005	.15	.22	.37
6″ x 5″ & 4″ x 4″ cant	.040	L.F.	.007	.18	.28	.46
ashing, aluminum, 0.019″ thick	.050	S.F.	.003	.04	.10	.14
0.032″ thick	.050	S.F.	.003	.06	.10	.16
0.040″ thick	.050	S.F.	.003	.08	.10	.18
Copper sheets, 16 oz., under 500 lbs.	.050	S.F.	.003	.14	.13	.27
Over 500 lbs.	.050	S.F.	.003	.16	.10	.26
20 oz., under 500 lbs.	.050	S.F.	.004	.21	.14	.35
Over 500 lbs.	.050	S.F.	.003	.20	.10	.30
Stainless steel, 32 gauge	.050	S.F.	.003	.12	.10	.22
28 gauge	.050	S.F.	.003	.15	.10	.25
26 gauge	.050	S.F.	.003	.18	.10	.28
24 gauge	.050	S.F.	.003	.23	.10	.33

Division 6
Interiors

No part of this publication may be reproduced, stored in a retrieval system, or transmitted in any form or by any means without prior written permission of Reed Construction Data.

04 Drywall & Thincoat Wall Systems

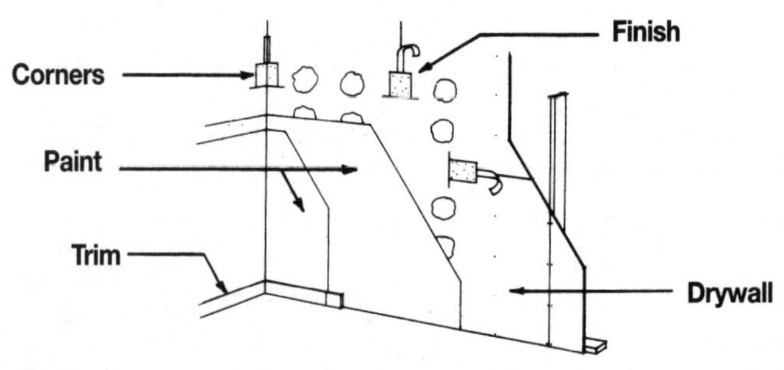

System Description	QUAN.	UNIT	LABOR HOURS	COST PER S.F.		
				MAT.	INST.	TOTAL
1/2" DRYWALL, TAPED & FINISHED						
Gypsum wallboard, 1/2" thick, standard	1.000	S.F.	.008	.24	.33	.57
Finish, taped & finished joints	1.000	S.F.	.008	.03	.33	.36
Corners, taped & finished, 32 L.F. per 12' x 12' room	.083	L.F.	.002	.01	.06	.07
Painting, primer & 2 coats	1.000	S.F.	.011	.15	.38	.53
Paint trim, to 6" wide, primer + 1 coat enamel	.125	L.F.	.001	.01	.05	.06
Trim, baseboard	.125	L.F.	.005	.22	.20	.42
TOTAL		S.F.	.035	.66	1.35	2.01
THINCOAT, SKIM-COAT, ON 1/2" BACKER DRYWALL						
Gypsum wallboard, 1/2" thick, thincoat backer	1.000	S.F.	.008	.24	.33	.57
Thincoat plaster	1.000	S.F.	.011	.08	.41	.49
Corners, taped & finished, 32 L.F. per 12' x 12' room	.083	L.F.	.002	.01	.06	.07
Painting, primer & 2 coats	1.000	S.F.	.011	.15	.38	.53
Paint trim, to 6" wide, primer + 1 coat enamel	.125	L.F.	.001	.01	.05	.06
Trim, baseboard	.125	L.F.	.005	.22	.20	.42
TOTAL		S.F.	.038	.71	1.43	2.14
5/8" DRYWALL, TAPED & FINISHED						
Gypsum wallboard, 5/8" thick, standard	1.000	S.F.	.008	.26	.33	.59
Finish, taped & finished joints	1.000	S.F.	.008	.03	.33	.36
Corners, taped & finished, 32 L.F. per 12' x 12' room	.083	L.F.	.002	.01	.06	.07
Painting, primer & 2 coats	1.000	S.F.	.011	.15	.38	.53
Trim, baseboard	.125	L.F.	.005	.22	.20	.42
Paint trim, to 6" wide, primer + 1 coat enamel	.125	L.F.	.001	.01	.05	.06
TOTAL		S.F.	.035	.68	1.35	2.03

The costs in this system are based on a square foot of wall.
Do not deduct for openings.

Description	QUAN.	UNIT	LABOR HOURS	COST PER S.F.		
				MAT.	INST.	TOTAL

Important: See the Reference Section for critical supporting data - Reference Nos., Crews & Location Factors

Drywall & Thincoat Wall Price Sheet	QUAN.	UNIT	LABOR HOURS	COST PER S.F.		
				MAT.	INST.	TOTAL
ypsum wallboard, 1/2" thick, standard	1.000	S.F.	.008	.24	.33	.57
Fire resistant	1.000	S.F.	.008	.24	.33	.57
Water resistant	1.000	S.F.	.008	.24	.33	.57
5/8" thick, standard	1.000	S.F.	.008	.26	.33	.59
Fire resistant	1.000	S.F.	.008	.28	.33	.61
Water resistant	1.000	S.F.	.008	.29	.33	.62
Gypsum wallboard backer for thincoat system, 1/2" thick	1.000	S.F.	.008	.24	.33	.57
5/8" thick	1.000	S.F.	.008	.26	.33	.59
ypsum wallboard, taped & finished	1.000	S.F.	.008	.03	.33	.36
Texture spray	1.000	S.F.	.010	.04	.36	.40
Thincoat plaster, including tape	1.000	S.F.	.011	.08	.41	.49
ypsum wallboard corners, taped & finished, 32 L.F. per 4′ x 4′ room	.250	L.F.	.004	.02	.17	.19
6′ x 6′ room	.110	L.F.	.002	.01	.08	.09
10′ x 10′ room	.100	L.F.	.001	.01	.07	.08
12′ x 12′ room	.083	L.F.	.001	.01	.06	.07
16′ x 16′ room	.063	L.F.	.001		.04	.04
Thincoat system, 32 L.F. per 4′ x 4′ room	.250	L.F.	.003	.02	.11	.13
6′ x 6′ room	.110	L.F.	.001	.01	.04	.05
10′ x 10′ room	.100	L.F.	.001	.01	.04	.05
12′ x 12′ room	.083	L.F.	.001	.01	.03	.04
16′ x 16′ room	.063	L.F.	.001	.01	.02	.03
ainting, primer, & 1 coat	1.000	S.F.	.008	.10	.29	.39
& 2 coats	1.000	S.F.	.011	.15	.38	.53
Wallpaper, $7/double roll	1.000	S.F.	.013	.32	.44	.76
$17/double roll	1.000	S.F.	.015	.72	.53	1.25
$40/double roll	1.000	S.F.	.018	1.68	.65	2.33
Tile, ceramic adhesive thin set, 4 1/4" x 4 1/4" tiles	1.000	S.F.	.084	2.30	2.76	5.06
6" x 6" tiles	1.000	S.F.	.080	2.94	2.62	5.56
Pregrouted sheets	1.000	S.F.	.067	4.65	2.18	6.83
rim, painted or stained, baseboard	.125	L.F.	.006	.23	.25	.48
Base shoe	.125	L.F.	.005	.15	.22	.37
Chair rail	.125	L.F.	.005	.16	.20	.36
Cornice molding	.125	L.F.	.004	.10	.17	.27
Cove base, vinyl	.125	L.F.	.003	.06	.12	.18
aneling, not including furring or trim						
Plywood, prefinished, 1/4" thick, 4′ x 8′ sheets, vert. grooves						
Birch faced, minimum	1.000	S.F.	.032	.87	1.30	2.17
Average	1.000	S.F.	.038	1.33	1.55	2.88
Maximum	1.000	S.F.	.046	1.94	1.86	3.80
Mahogany, African	1.000	S.F.	.040	2.48	1.63	4.11
Philippine (lauan)	1.000	S.F.	.032	1.07	1.30	2.37
Oak or cherry, minimum	1.000	S.F.	.032	2.08	1.30	3.38
Maximum	1.000	S.F.	.040	3.19	1.63	4.82
Rosewood	1.000	S.F.	.050	4.52	2.04	6.56
Teak	1.000	S.F.	.040	3.19	1.63	4.82
Chestnut	1.000	S.F.	.043	4.71	1.74	6.45
Pecan	1.000	S.F.	.040	2.04	1.63	3.67
Walnut, minimum	1.000	S.F.	.032	2.72	1.30	4.02
Maximum	1.000	S.F.	.040	5.15	1.63	6.78

08 Drywall & Thincoat Ceiling Systems

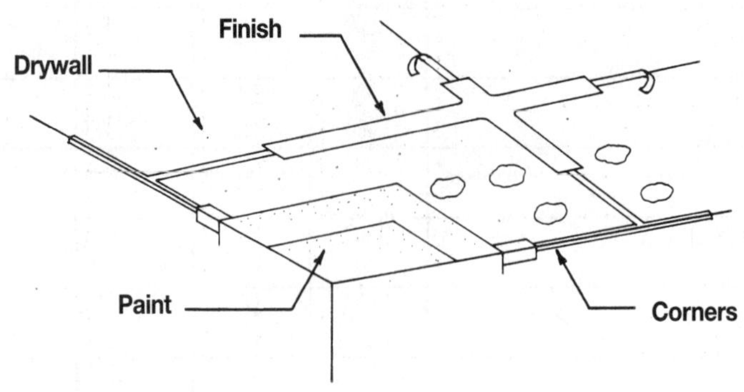

System Description	QUAN.	UNIT	LABOR HOURS	COST PER S.F. MAT.	INST.	TOTAL
1/2" SHEETROCK, TAPED & FINISHED						
Gypsum wallboard, 1/2" thick, standard	1.000	S.F.	.008	.24	.33	.57
Finish, taped & finished	1.000	S.F.	.008	.03	.33	.36
Corners, taped & finished, 12′ x 12′ room	.333	L.F.	.006	.02	.23	.25
Paint, primer & 2 coats	1.000	S.F.	.011	.15	.38	.53
TOTAL		S.F.	.033	.44	1.27	1.71
THINCOAT, SKIM COAT ON 1/2" GYPSUM WALLBOARD						
Gypsum wallboard, 1/2" thick, thincoat backer	1.000	S.F.	.008	.24	.33	.57
Thincoat plaster	1.000	S.F.	.011	.08	.41	.49
Corners, taped & finished, 12′ x 12′ room	.333	L.F.	.006	.02	.23	.25
Paint, primer & 2 coats	1.000	S.F.	.011	.15	.38	.53
TOTAL		S.F.	.036	.49	1.35	1.84
WATER-RESISTANT GYPSUM WALLBOARD, 1/2" THICK, TAPED & FINISHED						
Gypsum wallboard, 1/2" thick, water-resistant	1.000	S.F.	.008	.24	.33	.57
Finish, taped & finished	1.000	S.F.	.008	.03	.33	.36
Corners, taped & finished, 12′ x 12′ room	.333	L.F.	.006	.02	.23	.25
Paint, primer & 2 coats	1.000	S.F.	.011	.15	.38	.53
TOTAL		S.F.	.033	.44	1.27	1.71
5/8" GYPSUM WALLBOARD, TAPED & FINISHED						
Gypsum wallboard, 5/8" thick, standard	1.000	S.F.	.008	.26	.33	.59
Finish, taped & finished	1.000	S.F.	.008	.03	.33	.36
Corners, taped & finished, 12′ x 12′ room	.333	L.F.	.006	.02	.23	.25
Paint, primer & 2 coats	1.000	S.F.	.011	.15	.38	.53
TOTAL		S.F.	.033	.46	1.27	1.73

The costs in this system are based on a square foot of ceiling.

Description	QUAN.	UNIT	LABOR HOURS	COST PER S.F. MAT.	INST.	TOTAL

Drywall & Thincoat Ceilings

	QUAN.	UNIT	LABOR HOURS	COST PER S.F. MAT.	INST.	TOTAL
psum wallboard ceilings, 1/2" thick, standard	1.000	S.F.	.008	.24	.33	.57
Fire resistant	1.000	S.F.	.008	.24	.33	.57
Water resistant	1.000	S.F.	.008	.24	.33	.57
5/8" thick, standard	1.000	S.F.	.008	.26	.33	.59
Fire resistant	1.000	S.F.	.008	.28	.33	.61
Water resistant	1.000	S.F.	.008	.29	.33	.62
Gypsum wallboard backer for thincoat ceiling system, 1/2" thick	1.000	S.F.	.016	.48	.66	1.14
5/8" thick	1.000	S.F.	.016	.50	.66	1.16
psum wallboard ceilings, taped & finished	1.000	S.F.	.008	.03	.33	.36
Texture spray	1.000	S.F.	.010	.04	.36	.40
Thincoat plaster	1.000	S.F.	.011	.08	.41	.49
rners taped & finished, 4' x 4' room	1.000	L.F.	.015	.07	.69	.76
6' x 6' room	.667	L.F.	.010	.05	.46	.51
10' x 10' room	.400	L.F.	.006	.03	.28	.31
12' x 12' room	.333	L.F.	.005	.02	.23	.25
16' x 16' room	.250	L.F.	.003	.01	.13	.14
Thincoat system, 4' x 4' room	1.000	L.F.	.011	.08	.41	.49
6' x 6' room	.667	L.F.	.007	.05	.27	.32
10' x 10' room	.400	L.F.	.004	.03	.16	.19
12' x 12' room	.333	L.F.	.004	.03	.14	.17
16' x 16' room	.250	L.F.	.002	.01	.08	.09
inting, primer & 1 coat	1.000	S.F.	.008	.10	.29	.39
& 2 coats	1.000	S.F.	.011	.15	.38	.53
Wallpaper, double roll, solid pattern, avg. workmanship	1.000	S.F.	.013	.32	.44	.76
Basic pattern, avg. workmanship	1.000	S.F.	.015	.72	.53	1.25
Basic pattern, quality workmanship	1.000	S.F.	.018	1.68	.65	2.33
Tile, ceramic adhesive thin set, 4 1/4" x 4 1/4" tiles	1.000	S.F.	.084	2.30	2.76	5.06
6" x 6" tiles	1.000	S.F.	.080	2.94	2.62	5.56
Pregrouted sheets	1.000	S.F.	.067	4.65	2.18	6.83

Plaster & Stucco Wall Systems

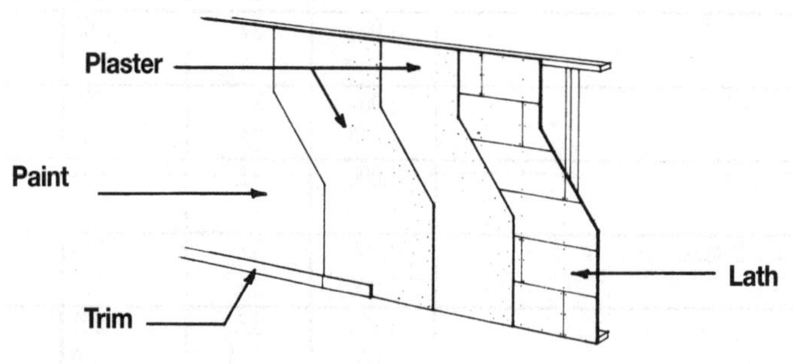

System Description	QUAN.	UNIT	LABOR HOURS	COST PER S.F. MAT.	INST.	TOTAL
PLASTER ON GYPSUM LATH						
Plaster, gypsum or perlite, 2 coats	1.000	S.F.	.053	.38	1.99	2.37
Lath, 3/8″ gypsum	1.000	S.F.	.010	.50	.37	.87
Corners, expanded metal, 32 L.F. per 12′ x 12′ room	.083	L.F.	.002	.01	.07	.08
Painting, primer & 2 coats	1.000	S.F.	.011	.15	.38	.53
Paint trim, to 6″ wide, primer + 1 coat enamel	.125	L.F.	.001	.01	.05	.06
Trim, baseboard	.125	L.F.	.005	.22	.20	.42
TOTAL		S.F.	.082	1.27	3.06	4.33
PLASTER ON METAL LATH						
Plaster, gypsum or perlite, 2 coats	1.000	S.F.	.053	.38	1.99	2.37
Lath, 2.5 Lb. diamond, metal	1.000	S.F.	.010	.31	.37	.68
Corners, expanded metal, 32 L.F. per 12′ x 12′ room	.083	L.F.	.002	.01	.07	.08
Painting, primer & 2 coats	1.000	S.F.	.011	.15	.38	.53
Paint trim, to 6″ wide, primer + 1 coat enamel	.125	L.F.	.001	.01	.05	.06
Trim, baseboard	.125	L.F.	.005	.22	.20	.42
TOTAL		S.F.	.082	1.08	3.06	4.14
STUCCO ON METAL LATH						
Stucco, 2 coats	1.000	S.F.	.041	.24	1.53	1.77
Lath, 2.5 Lb. diamond, metal	1.000	S.F.	.010	.31	.37	.68
Corners, expanded metal, 32 L.F. per 12′ x 12′ room	.083	L.F.	.002	.01	.07	.08
Painting, primer & 2 coats	1.000	S.F.	.011	.15	.38	.53
Paint trim, to 6″ wide, primer + 1 coat enamel	.125	L.F.	.001	.01	.05	.06
Trim, baseboard	.125	L.F.	.005	.22	.20	.42
TOTAL		S.F.	.070	.94	2.60	3.54

The costs in these systems are based on a per square foot of wall area.
Do not deduct for openings.

Description	QUAN.	UNIT	LABOR HOURS	COST PER S.F. MAT.	INST.	TOTAL

Important: See the Reference Section for critical supporting data - Reference Nos., Crews & Location Factors

Plaster & Stucco Wall Price Sheet	QUAN.	UNIT	LABOR HOURS	COST PER S.F. MAT.	INST.	TOTAL
aster, gypsum or perlite, 2 coats	1.000	S.F.	.053	.38	1.99	2.37
3 coats	1.000	S.F.	.065	.55	2.41	2.96
th, gypsum, standard, 3/8″ thick	1.000	S.F.	.010	.50	.37	.87
1/2″ thick	1.000	S.F.	.013	.50	.45	.95
Fire resistant, 3/8″ thick	1.000	S.F.	.013	.43	.45	.88
1/2″ thick	1.000	S.F.	.014	.55	.49	1.04
Metal, diamond, 2.5 Lb.	1.000	S.F.	.010	.31	.37	.68
3.4 Lb.	1.000	S.F.	.012	.40	.42	.82
Rib, 2.75 Lb.	1.000	S.F.	.012	.33	.42	.75
3.4 Lb.	1.000	S.F.	.013	.48	.45	.93
rners, expanded metal, 32 L.F. per 4′ x 4′ room	.250	L.F.	.005	.03	.21	.24
6′ x 6′ room	.110	L.F.	.002	.01	.09	.10
10′ x 10′ room	.100	L.F.	.002	.01	.08	.09
12′ x 12′ room	.083	L.F.	.002	.01	.07	.08
16′ x 16′ room	.063	L.F.	.001	.01	.05	.06
inting, primer & 1 coats	1.000	S.F.	.008	.10	.29	.39
Primer & 2 coats	1.000	S.F.	.011	.15	.38	.53
Wallpaper, low price double roll	1.000	S.F.	.013	.32	.44	.76
Medium price double roll	1.000	S.F.	.015	.72	.53	1.25
High price double roll	1.000	S.F.	.018	1.68	.65	2.33
Tile, ceramic thin set, 4-1/4″ x 4-1/4″ tiles	1.000	S.F.	.084	2.30	2.76	5.06
6″ x 6″ tiles	1.000	S.F.	.080	2.94	2.62	5.56
Pregrouted sheets	1.000	S.F.	.067	4.65	2.18	6.83
im, painted or stained, baseboard	.125	L.F.	.006	.23	.25	.48
Base shoe	.125	L.F.	.005	.15	.22	.37
Chair rail	.125	L.F.	.005	.16	.20	.36
Cornice molding	.125	L.F.	.004	.10	.17	.27
Cove base, vinyl	.125	L.F.	.003	.06	.12	.18
aneling not including furring or trim						
Plywood, prefinished, 1/4″ thick, 4′ x 8′ sheets, vert. grooves						
Birch faced, minimum	1.000	S.F.	.032	.87	1.30	2.17
Average	1.000	S.F.	.038	1.33	1.55	2.88
Maximum	1.000	S.F.	.046	1.94	1.86	3.80
Mahogany, African	1.000	S.F.	.040	2.48	1.63	4.11
Philippine (lauan)	1.000	S.F.	.032	1.07	1.30	2.37
Oak or cherry, minimum	1.000	S.F.	.032	2.08	1.30	3.38
Maximum	1.000	S.F.	.040	3.19	1.63	4.82
Rosewood	1.000	S.F.	.050	4.52	2.04	6.56
Teak	1.000	S.F.	.040	3.19	1.63	4.82
Chestnut	1.000	S.F.	.043	4.71	1.74	6.45
Pecan	1.000	S.F.	.040	2.04	1.63	3.67
Walnut, minimum	1.000	S.F.	.032	2.72	1.30	4.02
Maximum	1.000	S.F.	.040	5.15	1.63	6.78

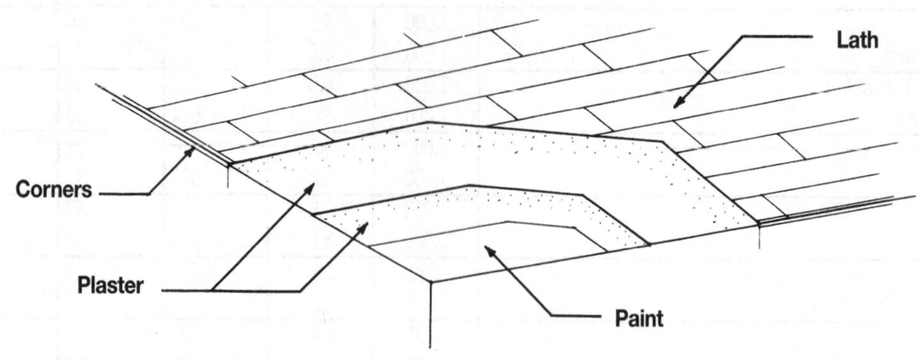

System Description	QUAN.	UNIT	LABOR HOURS	COST PER S.F.		
				MAT.	INST.	TOTAL
PLASTER ON GYPSUM LATH						
Plaster, gypsum or perlite, 2 coats	1.000	S.F.	.061	.38	2.26	2.64
Gypsum lath, plain or perforated, nailed, 3/8" thick	1.000	S.F.	.010	.50	.37	.87
Gypsum lath, ceiling installation adder	1.000	S.F.	.004		.15	.15
Corners, expanded metal, 12′ x 12′ room	.330	L.F.	.007	.04	.27	.31
Painting, primer & 2 coats	1.000	S.F.	.011	.15	.38	.53
TOTAL		S.F.	.093	1.07	3.43	4.50
PLASTER ON METAL LATH						
Plaster, gypsum or perlite, 2 coats	1.000	S.F.	.061	.38	2.26	2.64
Lath, 2.5 Lb. diamond, metal	1.000	S.F.	.012	.31	.42	.73
Corners, expanded metal, 12′ x 12′ room	.330	L.F.	.007	.04	.27	.31
Painting, primer & 2 coats	1.000	S.F.	.011	.15	.38	.53
TOTAL		S.F.	.091	.88	3.33	4.21
STUCCO ON GYPSUM LATH						
Stucco, 2 coats	1.000	S.F.	.041	.24	1.53	1.77
Gypsum lath, plain or perforated, nailed, 3/8" thick	1.000	S.F.	.010	.50	.37	.87
Gypsum lath, ceiling installation adder	1.000	S.F.	.004		.15	.15
Corners, expanded metal, 12′ x 12′ room	.330	L.F.	.007	.04	.27	.31
Painting, primer & 2 coats	1.000	S.F.	.011	.15	.38	.53
TOTAL		S.F.	.073	.93	2.70	3.63
STUCCO ON METAL LATH						
Stucco, 2 coats	1.000	S.F.	.041	.24	1.53	1.77
Lath, 2.5 Lb. diamond, metal	1.000	S.F.	.012	.31	.42	.73
Corners, expanded metal, 12′ x 12′ room	.330	L.F.	.007	.04	.27	.31
Painting, primer & 2 coats	1.000	S.F.	.011	.15	.38	.53
TOTAL		S.F.	.071	.74	2.60	3.34

The costs in these systems are based on a square foot of ceiling area.

Description	QUAN.	UNIT	LABOR HOURS	COST PER S.F.		
				MAT.	INST.	TOTAL

 Important: See the Reference Section for critical supporting data - Reference Nos., Crews & Location Factors

Plaster & Stucco Ceiling Price Sheet	QUAN.	UNIT	LABOR HOURS	COST PER S.F. MAT.	INST.	TOTAL
aster, gypsum or perlite, 2 coats	1.000	S.F.	.061	.38	2.26	2.64
3 coats	1.000	S.F.	.065	.55	2.41	2.96
ath, gypsum, standard, 3/8" thick	1.000	S.F.	.014	.50	.52	1.02
1/2" thick	1.000	S.F.	.015	.50	.55	1.05
Fire resistant, 3/8" thick	1.000	S.F.	.017	.43	.60	1.03
1/2" thick	1.000	S.F.	.018	.55	.64	1.19
Metal, diamond, 2.5 Lb.	1.000	S.F.	.012	.31	.42	.73
3.4 Lb.	1.000	S.F.	.015	.40	.53	.93
Rib, 2.75 Lb.	1.000	S.F.	.012	.33	.42	.75
3.4 Lb.	1.000	S.F.	.013	.48	.45	.93
orners expanded metal, 4' x 4' room	1.000	L.F.	.020	.13	.82	.95
6' x 6' room	.667	L.F.	.013	.09	.55	.64
10' x 10' room	.400	L.F.	.008	.05	.33	.38
12' x 12' room	.333	L.F.	.007	.04	.27	.31
16' x 16' room	.250	L.F.	.004	.02	.15	.17
ainting, primer & 1 coat	1.000	S.F.	.008	.10	.29	.39
Primer & 2 coats	1.000	S.F.	.011	.15	.38	.53

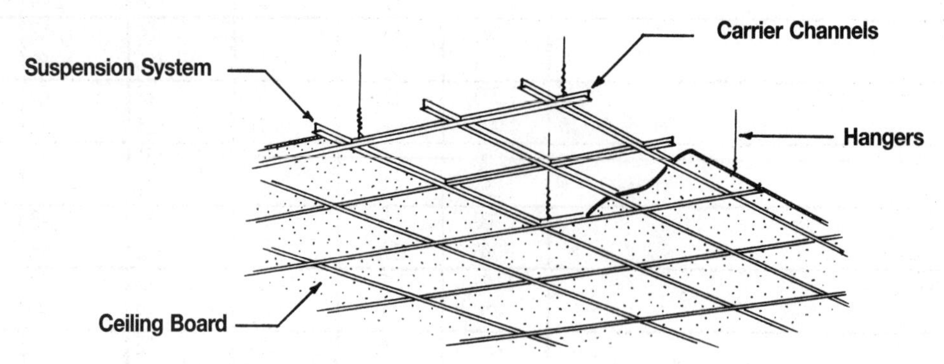

System Description	QUAN.	UNIT	LABOR HOURS	COST PER S.F. MAT.	COST PER S.F. INST.	COST PER S.F. TOTAL
2′ X 2′ GRID, FILM FACED FIBERGLASS, 5/8″ THICK						
Suspension system, 2′ x 2′ grid, T bar	1.000	S.F.	.012	.67	.50	1.17
Ceiling board, film faced fiberglass, 5/8″ thick	1.000	S.F.	.013	.56	.52	1.08
Carrier channels, 1-1/2″ x 3/4″	1.000	S.F.	.017	.14	.69	.83
Hangers, #12 wire	1.000	S.F.	.002	.06	.07	.13
TOTAL		S.F.	.044	1.43	1.78	3.21
2′ X 4′ GRID, FILM FACED FIBERGLASS, 5/8″ THICK						
Suspension system, 2′ x 4′ grid, T bar	1.000	S.F.	.010	.54	.41	.95
Ceiling board, film faced fiberglass, 5/8″ thick	1.000	S.F.	.013	.56	.52	1.08
Carrier channels, 1-1/2″ x 3/4″	1.000	S.F.	.017	.14	.69	.83
Hangers, #12 wire	1.000	S.F.	.002	.06	.07	.13
TOTAL		S.F.	.042	1.30	1.69	2.99
2′ X 2′ GRID, MINERAL FIBER, REVEAL EDGE, 1″ THICK						
Suspension system, 2′ x 2′ grid, T bar	1.000	S.F.	.012	.67	.50	1.17
Ceiling board, mineral fiber, reveal edge, 1″ thick	1.000	S.F.	.013	1.32	.54	1.86
Carrier channels, 1-1/2″ x 3/4″	1.000	S.F.	.017	.14	.69	.83
Hangers, #12 wire	1.000	S.F.	.002	.06	.07	.13
TOTAL		S.F.	.044	2.19	1.80	3.99
2′ X 4′ GRID, MINERAL FIBER, REVEAL EDGE, 1″ THICK						
Suspension system, 2′ x 4′ grid, T bar	1.000	S.F.	.010	.54	.41	.95
Ceiling board, mineral fiber, reveal edge, 1″ thick	1.000	S.F.	.013	1.32	.54	1.86
Carrier channels, 1-1/2″ x 3/4″	1.000	S.F.	.017	.14	.69	.83
Hangers, #12 wire	1.000	S.F.	.002	.06	.07	.13
TOTAL		S.F.	.042	2.06	1.71	3.77

Description	QUAN.	UNIT	LABOR HOURS	COST PER S.F. MAT.	COST PER S.F. INST.	COST PER S.F. TOTAL

Suspended Ceiling Price Sheet	QUAN.	UNIT	LABOR HOURS	COST PER S.F.		
				MAT.	INST.	TOTAL
uspension systems, T bar, 2′ x 2′ grid	1.000	S.F.	.012	.67	.50	1.17
2′ x 4′ grid	1.000	S.F.	.010	.54	.41	.95
Concealed Z bar, 12″ module	1.000	S.F.	.015	.48	.63	1.11
eiling boards, fiberglass, film faced, 2′ x 2′ or 2′ x 4′, 5/8″ thick	1.000	S.F.	.013	.56	.52	1.08
3/4″ thick	1.000	S.F.	.013	1.25	.54	1.79
3″ thick thermal R11	1.000	S.F.	.018	1.39	.72	2.11
Glass cloth faced, 3/4″ thick	1.000	S.F.	.016	1.78	.65	2.43
1″ thick	1.000	S.F.	.016	1.97	.67	2.64
1-1/2″ thick, nubby face	1.000	S.F.	.017	2.45	.69	3.14
Mineral fiber boards, 5/8″ thick, aluminum face 2′ x 2′	1.000	S.F.	.013	1.62	.54	2.16
2′ x 4′	1.000	S.F.	.012	1.09	.50	1.59
Standard faced, 2′ x 2′ or 2′ x 4′	1.000	S.F.	.012	.68	.48	1.16
Plastic coated face, 2′ x 2′ or 2′ x 4′	1.000	S.F.	.020	1.06	.82	1.88
Fire rated, 2 hour rating, 5/8″ thick	1.000	S.F.	.012	.92	.48	1.40
Tegular edge, 2′ x 2′ or 2′ x 4′, 5/8″ thick, fine textured	1.000	S.F.	.013	1.14	.69	1.83
Rough textured	1.000	S.F.	.015	1.49	.69	2.18
3/4″ thick, fine textured	1.000	S.F.	.016	1.62	.72	2.34
Rough textured	1.000	S.F.	.018	1.83	.72	2.55
Luminous panels, prismatic, acrylic	1.000	S.F.	.020	1.90	.82	2.72
Polystyrene	1.000	S.F.	.020	.97	.82	1.79
Flat or ribbed, acrylic	1.000	S.F.	.020	3.31	.82	4.13
Polystyrene	1.000	S.F.	.020	2.27	.82	3.09
Drop pan, white, acrylic	1.000	S.F.	.020	4.85	.82	5.67
Polystyrene	1.000	S.F.	.020	4.06	.82	4.88
Carrier channels, 4′-0″ on center, 3/4″ x 1-1/2″	1.000	S.F.	.017	.14	.69	.83
1-1/2″ x 3-1/2″	1.000	S.F.	.017	.35	.69	1.04
Hangers, #12 wire	1.000	S.F.	.002	.06	.07	.13

6 INTERIORS

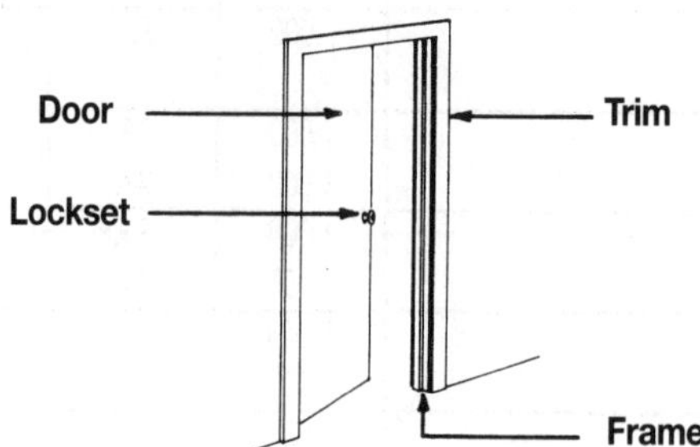

System Description	QUAN.	UNIT	LABOR HOURS	COST EACH		
				MAT.	INST.	TOTAL
LAUAN, FLUSH DOOR, HOLLOW CORE						
Door, flush, lauan, hollow core, 2′-8″ wide x 6′-8″ high	1.000	Ea.	.889	33	36	69
Frame, pine, 4-5/8″ jamb	17.000	L.F.	.725	102.85	29.58	132.43
Trim, stock pine, 11/16″ x 2-1/2″	34.000	L.F.	1.133	31.96	46.24	78.20
Paint trim, to 6″ wide, primer + 1 coat enamel	34.000	L.F.	.340	3.40	12.24	15.64
Butt hinges, chrome, 3-1/2″ x 3-1/2″	1.500	Pr.		36.75		36.75
Lockset, passage	1.000	Ea.	.500	15.45	20.50	35.95
Prime door & frame, oil, brushwork	2.000	Face	1.600	4.66	57	61.66
Paint door and frame, oil, 2 coats	2.000	Face	2.667	7.56	95	102.56
TOTAL		Ea.	7.854	235.63	296.56	532.19
BIRCH, FLUSH DOOR, HOLLOW CORE						
Door, flush, birch, hollow core, 2′-8″ wide x 6′-8″ high	1.000	Ea.	.889	47.50	36	83.50
Frame, pine, 4-5/8″ jamb	17.000	L.F.	.725	102.85	29.58	132.43
Trim, stock pine, 11/16″ x 2-1/2″	34.000	L.F.	1.133	31.96	46.24	78.20
Butt hinges, chrome, 3-1/2″ x 3-1/2″	1.500	Pr.		36.75		36.75
Lockset, passage	1.000	Ea.	.500	15.45	20.50	35.95
Prime door & frame, oil, brushwork	2.000	Face	1.600	4.66	57	61.66
Paint door and frame, oil, 2 coats	2.000	Face	2.667	7.56	95	102.56
TOTAL		Ea.	7.514	246.73	284.32	531.05
RAISED PANEL, SOLID, PINE DOOR						
Door, pine, raised panel, 2′-8″ wide x 6′-8″ high	1.000	Ea.	.889	156	36	192
Frame, pine, 4-5/8″ jamb	17.000	L.F.	.725	102.85	29.58	132.43
Trim, stock pine, 11/16″ x 2-1/2″	34.000	L.F.	1.133	31.96	46.24	78.20
Butt hinges, bronze, 3-1/2″ x 3-1/2″	1.500	Pr.		42		42
Lockset, passage	1.000	Ea.	.500	15.45	20.50	35.95
Prime door & frame, oil, brushwork	2.000	Face	1.600	4.66	57	61.66
Paint door and frame, oil, 2 coats	2.000	Face	2.667	7.56	95	102.56
TOTAL		Ea.	7.514	360.48	284.32	644.80

The costs in these systems are based on a cost per each door.

Description	QUAN.	UNIT	LABOR HOURS	COST EACH		
				MAT.	INST.	TOTAL

Interior Door Price Sheet	QUAN.	UNIT	LABOR HOURS	COST EACH		
				MAT.	INST.	TOTAL
oor, hollow core, lauan 1-3/8" thick, 6'-8" high x 1'-6" wide	1.000	Ea.	.889	29.50	36	65.50
2'-0" wide	1.000	Ea.	.889	28	36	64
2'-6" wide	1.000	Ea.	.889	31.50	36	67.50
2'-8" wide	1.000	Ea.	.889	33	36	69
3'-0" wide	1.000	Ea.	.941	35	38.50	73.50
Birch 1-3/8" thick, 6'-8" high x 1'-6" wide	1.000	Ea.	.889	37	36	73
2'-0" wide	1.000	Ea.	.889	41	36	77
2'-6" wide	1.000	Ea.	.889	46	36	82
2'-8" wide	1.000	Ea.	.889	47.50	36	83.50
3'-0" wide	1.000	Ea.	.941	52	38.50	90.50
Louvered pine 1-3/8" thick, 6'-8" high x 1'-6" wide	1.000	Ea.	.842	98.50	34.50	133
2'-0" wide	1.000	Ea.	.889	124	36	160
2'-6" wide	1.000	Ea.	.889	135	36	171
2'-8" wide	1.000	Ea.	.889	143	36	179
3'-0" wide	1.000	Ea.	.941	153	38.50	191.50
Paneled pine 1-3/8" thick, 6'-8" high x 1'-6" wide	1.000	Ea.	.842	111	34.50	145.50
2'-0" wide	1.000	Ea.	.889	128	36	164
2'-6" wide	1.000	Ea.	.889	144	36	180
2'-8" wide	1.000	Ea.	.889	156	36	192
3'-0" wide	1.000	Ea.	.941	163	38.50	201.50
Frame, pine, 1'-6" thru 2'-0" wide door, 3-5/8" deep	16.000	L.F.	.683	70.50	28	98.50
4-5/8" deep	16.000	L.F.	.683	97	28	125
5-5/8" deep	16.000	L.F.	.683	62	28	90
2'-6" thru 3'0" wide door, 3-5/8" deep	17.000	L.F.	.725	75	29.50	104.50
4-5/8" deep	17.000	L.F.	.725	103	29.50	132.50
5-5/8" deep	17.000	L.F.	.725	66	29.50	95.50
Trim, casing, painted, both sides, 1'-6" thru 2'-6" wide door	32.000	L.F.	1.855	32	71.50	103.50
2'-6" thru 3'-0" wide door	34.000	L.F.	1.971	34	76	110
Butt hinges 3-1/2" x 3-1/2", steel plated, chrome	1.500	Pr.		37		37
Bronze	1.500	Pr.		42		42
Locksets, passage, minimum	1.000	Ea.	.500	15.45	20.50	35.95
Maximum	1.000	Ea.	.575	17.75	23.50	41.25
Privacy, miniumum	1.000	Ea.	.625	19.30	25.50	44.80
Maximum	1.000	Ea.	.675	21	27.50	48.50
Paint 2 sides, primer & 2 cts., flush door, 1'-6" to 2'-0" wide	2.000	Face	5.547	14.15	198	212.15
2'-6" thru 3'-0" wide	2.000	Face	6.933	17.65	248	265.65
Louvered door, 1'-6" thru 2'-0" wide	2.000	Face	6.400	13.60	229	242.60
2'-6" thru 3'-0" wide	2.000	Face	8.000	17	286	303
Paneled door, 1'-6" thru 2'-0" wide	2.000	Face	6.400	13.60	229	242.60
2'-6" thru 3'-0" wide	2.000	Face	8.000	17	286	303

Closet Door Systems

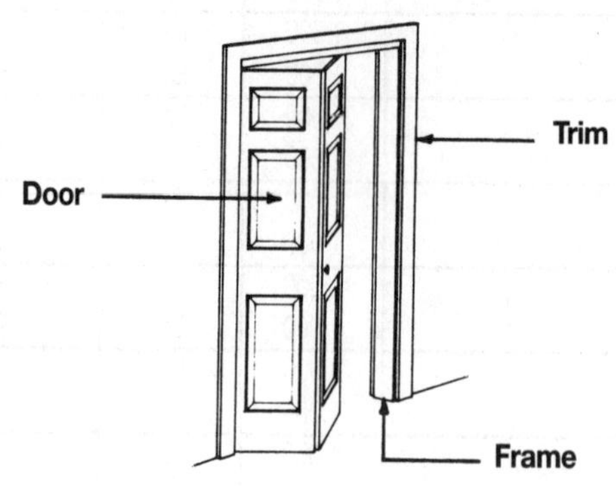

System Description		QUAN.	UNIT	LABOR HOURS	COST EACH		
					MAT.	INST.	TOTAL
BI-PASSING, FLUSH, LAUAN, HOLLOW CORE, 4′-0″ X 6′-8″							
Door, flush, lauan, hollow core, 4′-0″ x 6′-8″ opening		1.000	Ea.	1.333	172	54.50	226.50
Frame, pine, 4-5/8″ jamb		18.000	L.F.	.768	108.90	31.32	140.22
Trim, stock pine, 11/16″ x 2-1/2″		36.000	L.F.	1.200	33.84	48.96	82.80
Prime door & frame, oil, brushwork		2.000	Face	1.600	4.66	57	61.66
Paint door and frame, oil, 2 coats		2.000	Face	2.667	7.56	95	102.56
	TOTAL		Ea.	7.568	326.96	286.78	613.74
BI-PASSING, FLUSH, BIRCH, HOLLOW CORE, 6′-0″ X 6′-8″							
Door, flush, birch, hollow core, 6′-0″ x 6′-8″ opening		1.000	Ea.	1.600	253	65	318
Frame, pine, 4-5/8″ jamb		19.000	L.F.	.811	114.95	33.06	148.01
Trim, stock pine, 11/16″ x 2-1/2″		38.000	L.F.	1.267	35.72	51.68	87.40
Prime door & frame, oil, brushwork		2.000	Face	2.000	5.83	71.25	77.08
Paint door and frame, oil, 2 coats		2.000	Face	3.333	9.45	118.75	128.20
	TOTAL		Ea.	9.011	418.95	339.74	758.69
BI-FOLD, PINE, PANELED, 3′-0″ X 6′-8″							
Door, pine, paneled, 3′-0″ x 6′-8″ opening		1.000	Ea.	1.231	162	50	212
Frame, pine, 4-5/8″ jamb		17.000	L.F.	.725	102.85	29.58	132.43
Trim, stock pine, 11/16″ x 2-1/2″		34.000	L.F.	1.133	31.96	46.24	78.20
Prime door & frame, oil, brushwork		2.000	Face	1.600	4.66	57	61.66
Paint door and frame, oil, 2 coats		2.000	Face	2.667	7.56	95	102.56
	TOTAL		Ea.	7.356	309.03	277.82	586.85
BI-FOLD, PINE, LOUVERED, 6′-0″ X 6′-8″							
Door, pine, louvered, 6′-0″ x 6′-8″ opening		1.000	Ea.	1.600	225	65	290
Frame, pine, 4-5/8″ jamb		19.000	L.F.	.811	114.95	33.06	148.01
Trim, stock pine, 11/16″ x 2-1/2″		38.000	L.F.	1.267	35.72	51.68	87.40
Prime door & frame, oil, brushwork		2.500	Face	2.000	5.83	71.25	77.08
Paint door and frame, oil, 2 coats		2.500	Face	3.333	9.45	118.75	128.20
	TOTAL		Ea.	9.011	390.95	339.74	730.69

The costs in this system are based on a cost per each door.

Description	QUAN.	UNIT	LABOR HOURS	COST EACH		
				MAT.	INST.	TOTAL

Important: See the Reference Section for critical supporting data - Reference Nos., Crews & Location Factors

Closet Door Price Sheet

Closet Door Price Sheet	QUAN.	UNIT	LABOR HOURS	COST EACH MAT.	COST EACH INST.	COST EACH TOTAL
oors, bi-passing, pine, louvered, 4′-0″ x 6′-8″ opening	1.000	Ea.	1.333	415	54.50	469.50
6′-0″ x 6′-8″ opening	1.000	Ea.	1.600	510	65	575
Paneled, 4′-0″ x 6′-8″ opening	1.000	Ea.	1.333	400	54.50	454.50
6′-0″ x 6′-8″ opening	1.000	Ea.	1.600	490	65	555
Flush, birch, hollow core, 4′-0″ x 6′-8″ opening	1.000	Ea.	1.333	211	54.50	265.50
6′-0″ x 6′-8″ opening	1.000	Ea.	1.600	253	65	318
Flush, lauan, hollow core, 4′-0″ x 6′-8″ opening	1.000	Ea.	1.333	172	54.50	226.50
6′-0″ x 6′-8″ opening	1.000	Ea.	1.600	201	65	266
Bi-fold, pine, louvered, 3′-0″ x 6′-8″ opening	1.000	Ea.	1.231	162	50	212
6′-0″ x 6′-8″ opening	1.000	Ea.	1.600	225	65	290
Paneled, 3′-0″ x 6′-8″ opening	1.000	Ea.	1.231	162	50	212
6′-0″ x 6′-8″ opening	1.000	Ea.	1.600	225	65	290
Flush, birch, hollow core, 3′-0″ x 6′-8″ opening	1.000	Ea.	1.231	54.50	50	104.50
6′-0″ x 6′-8″ opening	1.000	Ea.	1.600	108	65	173
Flush, lauan, hollow core, 3′-0″ x 6′8″ opening	1.000	Ea.	1.231	188	50	238
6′-0″ x 6′-8″ opening	1.000	Ea.	1.600	310	65	375
rame pine, 3′-0″ door, 3-5/8″ deep	17.000	L.F.	.725	75	29.50	104.50
4-5/8″ deep	17.000	L.F.	.725	103	29.50	132.50
5-5/8″ deep	17.000	L.F.	.725	66	29.50	95.50
4′-0″ door, 3-5/8″ deep	18.000	L.F.	.768	79.50	31.50	111
4-5/8″ deep	18.000	L.F.	.768	109	31.50	140.50
5-5/8″ deep	18.000	L.F.	.768	70	31.50	101.50
6′-0″ door, 3-5/8″ deep	19.000	L.F.	.811	84	33	117
4-5/8″ deep	19.000	L.F.	.811	115	33	148
5-5/8″ deep	19.000	L.F.	.811	73.50	33	106.50
rim both sides, painted 3′-0″ x 6′-8″ door	34.000	L.F.	1.971	34	76	110
4′-0″ x 6′-8″ door	36.000	L.F.	2.086	36	80.50	116.50
6′-0″ x 6′-8″ door	38.000	L.F.	2.203	38	85	123
aint 2 sides, primer & 2 cts., flush door & frame, 3′ x 6′-8″ opng	2.000	Face	2.914	9.15	114	123.15
4′-0″ x 6′-8″ opening	2.000	Face	3.886	12.20	152	164.20
6′-0″ x 6′-8″ opening	2.000	Face	4.857	15.30	190	205.30
Paneled door & frame, 3′-0″ x 6′-8″ opening	2.000	Face	6.000	12.75	215	227.75
4′-0″ x 6′-8″ opening	2.000	Face	8.000	17	286	303
6′-0″ x 6′-8″ opening	2.000	Face	10.000	21.50	360	381.50
Louvered door & frame, 3′-0″ x 6′-8″ opening	2.000	Face	6.000	12.75	215	227.75
4′-0″ x 6′-8″ opening	2.000	Face	8.000	17	286	303
6′-0″ x 6′-8″ opening	2.000	Face	10.000	21.50	360	381.50

System Description	QUAN.	UNIT	LABOR HOURS	COST PER S.F. MAT.	INST.	TOTAL
Carpet, direct glue-down, nylon, level loop, 26 oz.	1.000	S.F.	.018	2.01	.66	2.67
32 oz.	1.000	S.F.	.018	2.84	.66	3.50
40 oz.	1.000	S.F.	.018	4.23	.66	4.89
Nylon, plush, 20 oz.	1.000	S.F.	.018	1.33	.66	1.99
24 oz.	1.000	S.F.	.018	1.41	.66	2.07
30 oz.	1.000	S.F.	.018	2.10	.66	2.76
42 oz.	1.000	S.F.	.022	2.60	.80	3.40
48 oz.	1.000	S.F.	.022	3.65	.80	4.45
54 oz.	1.000	S.F.	.022	4.12	.80	4.92
Olefin, 15 oz.	1.000	S.F.	.018	.71	.66	1.37
22 oz.	1.000	S.F.	.018	.85	.66	1.51
Tile, foam backed, needle punch	1.000	S.F.	.014	3.12	.52	3.64
Tufted loop or shag	1.000	S.F.	.014	1.32	.52	1.84
Wool, 36 oz., level loop	1.000	S.F.	.018	9.35	.66	10.01
32 oz., patterned	1.000	S.F.	.020	9.20	.74	9.94
48 oz., patterned	1.000	S.F.	.020	9.40	.74	10.14
Padding, sponge rubber cushion, minimum	1.000	S.F.	.006	.39	.22	.61
Maximum	1.000	S.F.	.006	1.01	.22	1.23
Felt, 32 oz. to 56 oz., minimum	1.000	S.F.	.006	.44	.22	.66
Maximum	1.000	S.F.	.006	.82	.22	1.04
Bonded urethane, 3/8" thick, minimum	1.000	S.F.	.006	.48	.22	.70
Maximum	1.000	S.F.	.006	.83	.22	1.05
Prime urethane, 1/4" thick, minimum	1.000	S.F.	.006	.28	.22	.50
Maximum	1.000	S.F.	.006	.51	.22	.73
Stairs, for stairs, add to above carpet prices	1.000	Riser	.267		9.85	9.85
Underlayment plywood, 3/8" thick	1.000	S.F.	.011	.96	.43	1.39
1/2" thick	1.000	S.F.	.011	1.17	.45	1.62
5/8" thick	1.000	S.F.	.011	1.34	.47	1.81
3/4" thick	1.000	S.F.	.012	1.54	.50	2.04
Particle board, 3/8" thick	1.000	S.F.	.011	.41	.43	.84
1/2" thick	1.000	S.F.	.011	.43	.45	.88
5/8" thick	1.000	S.F.	.011	.54	.47	1.01
3/4" thick	1.000	S.F.	.012	.61	.50	1.11
Hardboard, 4' x 4', 0.215" thick	1.000	S.F.	.011	.43	.43	.86

System Description	QUAN.	UNIT	LABOR HOURS	COST PER S.F. MAT.	COST PER S.F. INST.	COST PER S.F. TOTAL
esilient flooring, asphalt tile on concrete, 1/8″ thick						
Color group B	1.000	S.F.	.020	1.13	.74	1.87
Color group C & D	1.000	S.F.	.020	1.24	.74	1.98
Asphalt tile on wood subfloor, 1/8″ thick						
Color group B	1.000	S.F.	.020	1.34	.74	2.08
Color group C & D	1.000	S.F.	.020	1.45	.74	2.19
Vinyl composition tile, 12″ x 12″, 1/16″ thick	1.000	S.F.	.016	.87	.59	1.46
Embossed	1.000	S.F.	.016	1.10	.59	1.69
Marbleized	1.000	S.F.	.016	1.10	.59	1.69
Plain	1.000	S.F.	.016	1.22	.59	1.81
.080″ thick, embossed	1.000	S.F.	.016	1.11	.59	1.70
Marbleized	1.000	S.F.	.016	1.23	.59	1.82
Plain	1.000	S.F.	.016	1.80	.59	2.39
1/8″ thick, marbleized	1.000	S.F.	.016	1.18	.59	1.77
Plain	1.000	S.F.	.016	2.21	.59	2.80
Vinyl tile, 12″ x 12″, .050″ thick, minimum	1.000	S.F.	.016	1.94	.59	2.53
Maximum	1.000	S.F.	.016	3.80	.59	4.39
1/8″ thick, minimum	1.000	S.F.	.016	2.45	.59	3.04
Maximum	1.000	S.F.	.016	5.30	.59	5.89
1/8″ thick, solid colors	1.000	S.F.	.016	3.92	.59	4.51
Florentine pattern	1.000	S.F.	.016	4.51	.59	5.10
Marbleized or travertine pattern	1.000	S.F.	.016	9.25	.59	9.84
Vinyl sheet goods, backed, .070″ thick, minimum	1.000	S.F.	.032	2.28	1.18	3.46
Maximum	1.000	S.F.	.040	2.90	1.47	4.37
.093″ thick, minimum	1.000	S.F.	.035	2.45	1.28	3.73
Maximum	1.000	S.F.	.040	3.50	1.47	4.97
.125″ thick, minimum	1.000	S.F.	.035	2.76	1.28	4.04
Maximum	1.000	S.F.	.040	4.38	1.47	5.85
Wood, oak, finished in place, 25/32″ x 2-1/2″ clear	1.000	S.F.	.074	4.06	2.72	6.78
Select	1.000	S.F.	.074	3.58	2.72	6.30
No. 1 common	1.000	S.F.	.074	4.85	2.72	7.57
Prefinished, oak, 2-1/2″ wide	1.000	S.F.	.047	6.80	1.92	8.72
3-1/4″ wide	1.000	S.F.	.043	8.75	1.76	10.51
Ranch plank, oak, random width	1.000	S.F.	.055	8.45	2.25	10.70
Parquet, 5/16″ thick, finished in place, oak, minimum	1.000	S.F.	.077	4.29	2.84	7.13
Maximum	1.000	S.F.	.107	6.60	4.06	10.66
Teak, minimum	1.000	S.F.	.077	5.60	2.84	8.44
Maximum	1.000	S.F.	.107	9.20	4.06	13.26
Sleepers, treated, 16″ O.C., 1″ x 2″	1.000	S.F.	.007	.11	.28	.39
1″ x 3″	1.000	S.F.	.008	.21	.33	.54
2″ x 4″	1.000	S.F.	.011	.54	.43	.97
2″ x 6″	1.000	S.F.	.012	.85	.50	1.35
Subfloor, plywood, 1/2″ thick	1.000	S.F.	.011	.68	.43	1.11
5/8″ thick	1.000	S.F.	.012	.86	.48	1.34
3/4″ thick	1.000	S.F.	.013	1.03	.52	1.55
Ceramic tile, color group 2, 1″ x 1″	1.000	S.F.	.087	4.70	2.86	7.56
2″ x 2″ or 2″ x 1″	1.000	S.F.	.084	4.48	2.76	7.24
Color group 1, 8″ x 8″		S.F.	.064	3.45	2.09	5.54
12″ x 12″		S.F.	.049	4.33	1.61	5.94
16″ x 16″		S.F.	.029	5.45	.95	6.40

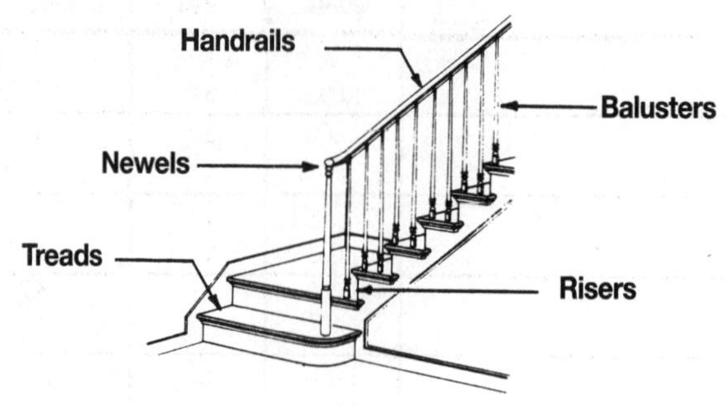

System Description	QUAN.	UNIT	LABOR HOURS	COST EACH		
				MAT.	INST.	TOTAL
7 RISERS, OAK TREADS, BOX STAIRS						
Treads, oak, 1-1/4" x 10" wide, 3' long	6.000	Ea.	2.667	534	108.60	642.60
Risers, 3/4" thick, beech	7.000	Ea.	2.625	133.35	107.10	240.45
Balusters, birch, 30" high	12.000	Ea.	3.429	83.40	139.80	223.20
Newels, 3-1/4" wide	2.000	Ea.	2.286	84	93	177
Handrails, oak laminated	7.000	L.F.	.933	231	38.15	269.15
Stringers, 2" x 10", 3 each	21.000	L.F.	.306	7.77	12.39	20.16
TOTAL		Ea.	12.246	1,073.52	499.04	1,572.56
14 RISERS, OAK TREADS, BOX STAIRS						
Treads, oak, 1-1/4" x 10" wide, 3' long	13.000	Ea.	5.778	1,157	235.30	1,392.30
Risers, 3/4" thick, beech	14.000	Ea.	5.250	266.70	214.20	480.90
Balusters, birch, 30" high	26.000	Ea.	7.428	180.70	302.90	483.60
Newels, 3-1/4" wide	2.000	Ea.	2.286	84	93	177
Handrails, oak, laminated	14.000	L.F.	1.867	462	76.30	538.30
Stringers, 2" x 10", 3 each	42.000	L.F.	5.169	54.18	210	264.18
TOTAL		Ea.	27.778	2,204.58	1,131.70	3,336.28
14 RISERS, PINE TREADS, BOX STAIRS						
Treads, pine, 9-1/2" x 3/4" thick	13.000	Ea.	5.778	273	235.30	508.30
Risers, 3/4" thick, pine	14.000	Ea.	5.091	141.12	207.48	348.60
Balusters, pine, 30" high	26.000	Ea.	7.428	105.82	302.90	408.72
Newels, 3-1/4" wide	2.000	Ea.	2.286	84	93	177
Handrails, oak, laminated	14.000	L.F.	1.867	462	76.30	538.30
Stringers, 2" x 10", 3 each	42.000	L.F.	5.169	54.18	210	264.18
TOTAL		Ea.	27.619	1,120.12	1,124.98	2,245.10

Description	QUAN.	UNIT	LABOR HOURS	COST EACH		
				MAT.	INST.	TOTAL

Important: See the Reference Section for critical supporting data - Reference Nos., Crews & Location Factors

Stairway Price Sheet	QUAN.	UNIT	LABOR HOURS	COST EACH MAT.	INST.	TOTAL
eads, oak, 1-1/16" x 9-1/2", 3' long, 7 riser stair	6.000	Ea.	2.667	535	109	644
14 riser stair	13.000	Ea.	5.778	1,150	235	1,385
1-1/16" x 11-1/2", 3' long, 7 riser stair	6.000	Ea.	2.667	192	109	301
14 riser stair	13.000	Ea.	5.778	415	235	650
Pine, 3/4" x 9-1/2", 3' long, 7 riser stair	6.000	Ea.	2.667	126	109	235
14 riser stair	13.000	Ea.	5.778	273	235	508
3/4" x 11-1/4", 3' long, 7 riser stair	6.000	Ea.	2.667	153	109	262
14 riser stair	13.000	Ea.	5.778	330	235	565
sers, oak, 3/4" x 7-1/2" high, 7 riser stair	7.000	Ea.	2.625	141	107	248
14 riser stair	14.000	Ea.	5.250	281	214	495
Beech, 3/4" x 7-1/2" high, 7 riser stair	7.000	Ea.	2.625	133	107	240
14 riser stair	14.000	Ea.	5.250	267	214	481
aluster, turned, 30" high, pine, 7 riser stair	12.000	Ea.	3.429	49	140	189
14 riser stair	26.000	Ea.	7.428	106	305	411
30" birch, 7 riser stair	12.000	Ea.	3.429	83.50	140	223.50
14 riser stair	26.000	Ea.	7.428	181	305	486
42" pine, 7 riser stair	12.000	Ea.	3.556	65	145	210
14 riser stair	26.000	Ea.	7.704	140	315	455
42" birch, 7 riser stair	12.000	Ea.	3.556	133	145	278
14 riser stair	26.000	Ea.	7.704	289	315	604
lewels, 3-1/4" wide, starting, 7 riser stair	2.000	Ea.	2.286	84	93	177
14 riser stair	2.000	Ea.	2.286	84	93	177
Landing, 7 riser stair	2.000	Ea.	3.200	216	130	346
14 riser stair	2.000	Ea.	3.200	216	130	346
landrails, oak, laminated, 7 riser stair	7.000	L.F.	.933	231	38	269
14 riser stair	14.000	L.F.	1.867	460	76.50	536.50
Stringers, fir, 2" x 10" 7 riser stair	21.000	L.F.	2.585	27	105	132
14 riser stair	42.000	L.F.	5.169	54	210	264
2" x 12", 7 riser stair	21.000	L.F.	2.585	37	105	142
14 riser stair	42.000	L.F.	5.169	74	210	284

Special Stairways	QUAN.	UNIT	LABOR HOURS	COST EACH MAT.	INST.	TOTAL
Basement stairs, open risers	1.000	Flight	4.000	655	163	818
Spiral stairs, oak, 4'-6" diameter, prefabricated, 9' high	1.000	Flight	10.667	4,850	435	5,285
Aluminum, 5'-0" diameter stock unit	1.000	Flight	9.956	3,575	525	4,100
Custom unit	1.000	Flight	9.956	6,800	525	7,325
Cast iron, 4'-0" diameter, minimum	1.000	Flight	9.956	3,200	525	3,725
Maximum	1.000	Flight	17.920	4,350	955	5,305
Steel, industrial, pre-erected, 3'-6" wide, bar rail	1.000	Flight	7.724	3,250	615	3,865
Picket rail	1.000	Flight	7.724	3,650	615	4,265

Division 7
Specialties

No part of this publication may be reproduced, stored in a retrieval system, or transmitted in any form or by any means without prior written permission of Reed Construction Data.

System Description	QUAN.	UNIT	LABOR HOURS	COST PER L.F.		
				MAT.	INST.	TOTAL
KITCHEN, ECONOMY GRADE						
Top cabinets, economy grade	1.000	L.F.	.171	32.32	6.88	39.20
Bottom cabinets, economy grade	1.000	L.F.	.256	48.48	10.32	58.80
Square edge, plastic face countertop	1.000		.267	24	10.85	34.85
Blocking, wood, 2″ x 4″	1.000	L.F.	.032	.37	1.30	1.67
Soffit, framing, wood, 2″ x 4″	4.000	L.F.	.071	1.48	2.88	4.36
Soffit drywall	2.000	S.F.	.047	.60	1.94	2.54
Drywall painting	2.000	S.F.	.013	.10	.54	.64
TOTAL		L.F.	.857	107.35	34.71	142.06
AVERAGE GRADE						
Top cabinets, average grade	1.000	L.F.	.213	40.40	8.60	49
Bottom cabinets, average grade	1.000	L.F.	.320	60.60	12.90	73.50
Solid surface countertop, solid color	1.000	L.F.	.800	65	32.50	97.50
Blocking, wood, 2″ x 4″	1.000	L.F.	.032	.37	1.30	1.67
Soffit framing, wood, 2″ x 4″	4.000	L.F.	.071	1.48	2.88	4.36
Soffit drywall	2.000	S.F.	.047	.60	1.94	2.54
Drywall painting	2.000	S.F.	.013	.10	.54	.64
TOTAL		L.F.	1.496	168.55	60.66	229.21
CUSTOM GRADE						
Top cabinets, custom grade	1.000	L.F.	.256	105.60	10.40	116
Bottom cabinets, custom grade	1.000	L.F.	.384	158.40	15.60	174
Solid surface countertop, premium patterned color	1.000	L.F.	1.067	112	43.50	155.50
Blocking, wood, 2″ x 4″	1.000	L.F.	.032	.37	1.30	1.67
Soffit framing, wood, 2″ x 4″	4.000	L.F.	.071	1.48	2.88	4.36
Soffit drywall	2.000	S.F.	.047	.60	1.94	2.54
Drywall painting	2.000	S.F.	.013	.10	.54	.64
TOTAL		L.F.	1.870	378.55	76.16	454.71

Description	QUAN.	UNIT	LABOR HOURS	COST PER L.F.		
				MAT.	INST.	TOTAL

Kitchen Price Sheet	QUAN.	UNIT	LABOR HOURS	COST PER L.F.		
				MAT.	INST.	TOTAL
op cabinets, economy grade	1.000	L.F.	.171	32.50	6.90	39.40
Average grade	1.000	L.F.	.213	40.50	8.60	49.10
Custom grade	1.000	L.F.	.256	106	10.40	116.40
ottom cabinets, economy grade	1.000	L.F.	.256	48.50	10.30	58.80
Average grade	1.000	L.F.	.320	60.50	12.90	73.40
Custom grade	1.000	L.F.	.384	158	15.60	173.60
ounter top, laminated plastic, 7/8″ thick, no splash	1.000	L.F.	.267	19.25	10.85	30.10
With backsplash	1.000	L.F.	.267	25.50	10.85	36.35
1-1/4″ thick, no splash	1.000	L.F.	.286	22	11.65	33.65
With backsplash	1.000	L.F.	.286	27.50	11.65	39.15
Post formed, laminated plastic	1.000	L.F.	.267	9.90	10.85	20.75
Marble, with backsplash, minimum	1.000	L.F.	.471	35	19.35	54.35
Maximum	1.000	L.F.	.615	88	25.50	113.50
Maple, solid laminated, no backsplash	1.000	L.F.	.286	57.50	11.65	69.15
With backsplash	1.000	L.F.	.286	68.50	11.65	80.15
locking, wood, 2″ x 4″	1.000	L.F.	.032	.37	1.30	1.67
2″ x 6″	1.000	L.F.	.036	.59	1.47	2.06
2″ x 8″	1.000	L.F.	.040	.91	1.63	2.54
offit framing, wood, 2″ x 3″	4.000	L.F.	.064	1.16	2.60	3.76
2″ x 4″	4.000	L.F.	.071	1.48	2.88	4.36
offit, drywall, painted	2.000	S.F.	.060	.70	2.48	3.18
Paneling, standard	2.000	S.F.	.064	1.74	2.60	4.34
Deluxe	2.000	S.F.	.091	3.88	3.72	7.60
Sinks, porcelain on cast iron, single bowl, 21″ x 24″	1.000	Ea.	10.334	340	410	750
21″ x 30″	1.000	Ea.	10.334	390	410	800
Double bowl, 20″ x 32″	1.000	Ea.	10.810	430	430	860
Stainless steel, single bowl, 16″ x 20″	1.000	Ea.	10.334	480	410	890
22″ x 25″	1.000	Ea.	10.334	520	410	930
Double bowl, 20″ x 32″	1.000	Ea.	10.810	276	430	706

Kitchen Price Sheet	QUAN.	UNIT	LABOR HOURS	COST PER L.F. MAT.	INST.	TOTAL
Range, free standing, minimum	1.000	Ea.	3.600	340	136	476
Maximum	1.000	Ea.	6.000	1,700	207	1,907
Built-in, minimum	1.000	Ea.	3.333	565	149	714
Maximum	1.000	Ea.	10.000	1,525	415	1,940
Counter top range, 4-burner, minimum	1.000	Ea.	3.333	279	149	428
Maximum	1.000	Ea.	4.667	645	208	853
Compactor, built-in, minimum	1.000	Ea.	2.215	470	92.50	562.50
Maximum	1.000	Ea.	3.282	530	137	667
Dishwasher, built-in, minimum	1.000	Ea.	6.735	360	299	659
Maximum	1.000	Ea.	9.235	410	410	820
Garbage disposer, minimum	1.000	Ea.	2.810	82.50	125	207.50
Maximum	1.000	Ea.	2.810	196	125	321
Microwave oven, minimum	1.000	Ea.	2.615	108	117	225
Maximum	1.000	Ea.	4.615	445	206	651
Range hood, ducted, minimum	1.000	Ea.	4.658	81	196	277
Maximum	1.000	Ea.	5.991	695	251	946
Ductless, minimum	1.000	Ea.	2.615	74.50	111	185.50
Maximum	1.000	Ea.	3.948	690	166	856
Refrigerator, 16 cu.ft., minimum	1.000	Ea.	2.000	605	59	664
Maximum	1.000	Ea.	3.200	970	94	1,064
16 cu.ft. with icemaker, minimum	1.000	Ea.	4.210	765	151	916
Maximum	1.000	Ea.	5.410	1,125	186	1,311
19 cu.ft., minimum	1.000	Ea.	2.667	750	78.50	828.50
Maximum	1.000	Ea.	4.667	1,325	137	1,462
19 cu.ft. with icemaker, minimum	1.000	Ea.	5.143	975	178	1,153
Maximum	1.000	Ea.	7.143	1,525	237	1,762
Sinks, porcelain on cast iron single bowl, 21″ x 24″	1.000	Ea.	10.334	340	410	750
21″ x 30″	1.000	Ea.	10.334	390	410	800
Double bowl, 20″ x 32″	1.000	Ea.	10.810	430	430	860
Stainless steel, single bowl 16″ x 20″	1.000	Ea.	10.334	480	410	890
22″ x 25″	1.000	Ea.	10.334	520	410	930
Double bowl, 20″ x 32″	1.000	Ea.	10.810	276	430	706
Water heater, electric, 30 gallon	1.000	Ea.	3.636	310	161	471
40 gallon	1.000	Ea.	4.000	330	177	507
Gas, 30 gallon	1.000	Ea.	4.000	425	177	602
75 gallon	1.000	Ea.	5.333	820	236	1,056
Wall, packaged terminal heater/air conditioner cabinet, wall sleeve, louver, electric heat, thermostat, manual changeover, 208V		Ea.				
6000 BTUH cooling, 8800 BTU heating	1.000	Ea.	2.667	1,050	107	1,157
9000 BTUH cooling, 13,900 BTU heating	1.000	Ea.	3.200	1,100	128	1,228
12,000 BTUH cooling, 13,900 BTU heating	1.000	Ea.	4.000	1,225	160	1,385
15,000 BTUH cooling, 13,900 BTU heating	1.000	Ea.	5.333	1,450	213	1,663

Important: See the Reference Section for critical supporting data - Reference Nos., Crews & Location Factors

System Description	QUAN.	UNIT	LABOR HOURS	COST EACH MAT.	COST EACH INST.	COST EACH TOTAL
Curtain rods, stainless, 1″ diameter, 3′ long	1.000	Ea.	.615	33.50	25	58.50
5′ long	1.000	Ea.	.615	33.50	25	58.50
Grab bar, 1″ diameter, 12″ long	1.000	Ea.	.283	17.45	11.55	29
36″ long	1.000	Ea.	.340	18.30	13.85	32.15
1-1/4″ diameter, 12″ long	1.000	Ea.	.333	20.50	13.60	34.10
36″ long	1.000	Ea.	.400	21.50	16.30	37.80
1-1/2″ diameter, 12″ long	1.000	Ea.	.383	23.50	15.65	39.15
36″ long	1.000	Ea.	.460	24.50	18.75	43.25
Mirror, 18″ x 24″	1.000	Ea.	.400	72	16.30	88.30
72″ x 24″	1.000	Ea.	1.333	216	54.50	270.50
Medicine chest with mirror, 18″ x 24″	1.000	Ea.	.400	125	16.30	141.30
36″ x 24″	1.000	Ea.	.600	188	24.50	212.50
Toilet tissue dispenser, surface mounted, minimum	1.000	Ea.	.267	11.90	10.85	22.75
Maximum	1.000	Ea.	.400	17.85	16.30	34.15
Flush mounted, minimum	1.000	Ea.	.293	13.10	11.95	25.05
Maximum	1.000	Ea.	.427	19.05	17.35	36.40
Towel bar, 18″ long, minimum	1.000	Ea.	.278	31	11.30	42.30
Maximum	1.000	Ea.	.348	38.50	14.15	52.65
24″ long, minimum	1.000	Ea.	.313	34.50	12.75	47.25
Maximum	1.000	Ea.	.383	42.50	15.55	58.05
36″ long, minimum	1.000	Ea.	.381	62.50	15.50	78
Maximum	1.000	Ea.	.419	69	17.05	86.05

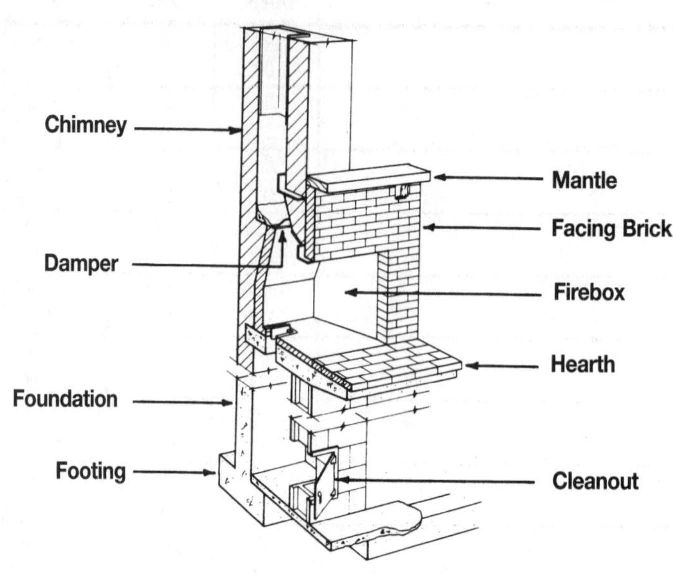

System Description	QUAN.	UNIT	LABOR HOURS	COST EACH MAT.	COST EACH INST.	COST EACH TOTAL
MASONRY FIREPLACE						
Footing, 8″ thick, concrete, 4′ x 7′	.700	C.Y.	2.110	89.60	74.21	163.81
Foundation, concrete block, 32″ x 60″ x 4′ deep	1.000	Ea.	5.275	110.40	196.20	306.60
Fireplace, brick firebox, 30″ x 29″ opening	1.000	Ea.	40.000	420	1,450	1,870
Damper, cast iron, 30″ opening	1.000	Ea.	1.333	78	55	133
Facing brick, standard size brick, 6′ x 5′	30.000	S.F.	5.217	95.10	193.50	288.60
Hearth, standard size brick, 3′ x 6′	1.000	Ea.	8.000	154	290	444
Chimney, standard size brick, 8″ x 12″ flue, one story house	12.000	V.L.F.	12.000	330	432	762
Mantle, 4″ x 8″, wood	6.000	L.F.	1.333	33.90	54.30	88.20
Cleanout, cast iron, 8″ x 8″	1.000	Ea.	.667	32	27.50	59.50
TOTAL		Ea.	75.935	1,343	2,772.71	4,115.71

The costs in this system are on a cost each basis.

Description	QUAN.	UNIT	LABOR HOURS	COST EACH MAT.	COST EACH INST.	COST EACH TOTAL

Masonry Fireplace Price Sheet	QUAN.	UNIT	LABOR HOURS	COST EACH		
				MAT.	INST.	TOTAL
oting 8″ thick, 3′ x 6′	.440	C.Y.	1.326	56.50	46.50	103
4′ x 7′	.700	C.Y.	2.110	89.50	74	163.50
5′ x 8′	1.000	C.Y.	3.014	128	106	234
1′ thick, 3′ x 6′	.670	C.Y.	2.020	86	71	157
4′ x 7′	1.030	C.Y.	3.105	132	109	241
5′ x 8′	1.480	C.Y.	4.461	189	156	345
oundation-concrete block, 24″ x 48″, 4′ deep	1.000	Ea.	4.267	88.50	157	245.50
8′ deep	1.000	Ea.	8.533	177	315	492
24″ x 60″, 4′ deep	1.000	Ea.	4.978	103	183	286
8′ deep	1.000	Ea.	9.956	206	365	571
32″ x 48″, 4′ deep	1.000	Ea.	4.711	97.50	173	270.50
8′ deep	1.000	Ea.	9.422	195	345	540
32″ x 60″, 4′ deep	1.000	Ea.	5.333	110	196	306
8′ deep	1.000	Ea.	10.845	224	400	624
32″ x 72″, 4′ deep	1.000	Ea.	6.133	127	226	353
8′ deep	1.000	Ea.	12.267	254	450	704
ireplace, brick firebox 30″ x 29″ opening	1.000	Ea.	40.000	420	1,450	1,870
48″ x 30″ opening	1.000	Ea.	60.000	630	2,175	2,805
Steel fire box with registers, 25″ opening	1.000	Ea.	26.667	830	980	1,810
48″ opening	1.000	Ea.	44.000	1,250	1,625	2,875
amper, cast iron, 30″ opening	1.000	Ea.	1.333	78	55	133
36″ opening	1.000	Ea.	1.556	91	64	155
Steel, 30″ opening	1.000	Ea.	1.333	70	55	125
36″ opening	1.000	Ea.	1.556	81.50	64	145.50
acing for fireplace, standard size brick, 6′ x 5′	30.000	S.F.	5.217	95	194	289
7′ x 5′	35.000	S.F.	6.087	111	226	337
8′ x 6′	48.000	S.F.	8.348	152	310	462
Fieldstone, 6′ x 5′	30.000	S.F.	5.217	425	194	619
7′ x 5′	35.000	S.F.	6.087	495	226	721
8′ x 6′	48.000	S.F.	8.348	680	310	990
Sheetrock on metal, studs, 6′ x 5′	30.000	S.F.	.980	21.50	40	61.50
7′ x 5′	35.000	S.F.	1.143	25	47	72
8′ x 6′	48.000	S.F.	1.568	34.50	64.50	99
earth, standard size brick, 3′ x 6′	1.000	Ea.	8.000	154	290	444
3′ x 7′	1.000	Ea.	9.280	179	335	514
3′ x 8′	1.000	Ea.	10.640	205	385	590
Stone, 3′ x 6′	1.000	Ea.	8.000	166	290	456
3′ x 7′	1.000	Ea.	9.280	193	335	528
3′ x 8′	1.000	Ea.	10.640	221	385	606
:himney, standard size brick , 8″ x 12″ flue, one story house	12.000	V.L.F.	12.000	330	430	760
Two story house	20.000	V.L.F.	20.000	550	720	1,270
Aantle wood, beams, 4″ x 8″	6.000	L.F.	1.333	34	54.50	88.50
4″ x 10″	6.000	L.F.	1.371	41	56	97
Ornate, prefabricated, 6′ x 3′-6″ opening, minimum	1.000	Ea.	1.600	153	65	218
Maximum	1.000	Ea.	1.600	190	65	255
:leanout, door and frame, cast iron, 8″ x 8″	1.000	Ea.	.667	32	27.50	59.50
12″ x 12″	1.000	Ea.	.800	36.50	33	69.50

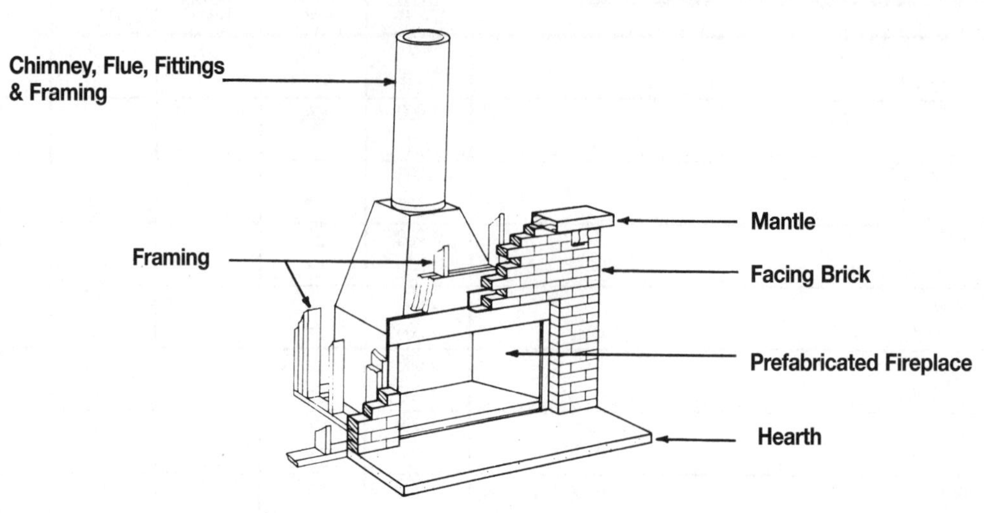

System Description	QUAN.	UNIT	LABOR HOURS	COST EACH		
				MAT.	INST.	TOTAL
PREFABRICATED FIREPLACE						
Prefabricated fireplace, metal, minimum	1.000	Ea.	6.154	1,250	251	1,501
Framing, 2″ x 4″ studs, 6′ x 5′	35.000	L.F.	.509	12.95	20.65	33.60
Fire resistant gypsum drywall, unfinished	40.000	S.F.	.320	9.60	13.20	22.80
Drywall finishing adder	40.000	S.F.	.320	1.20	13.20	14.40
Facing, brick, standard size brick, 6′ x 5′	30.000	S.F.	5.217	95.10	193.50	288.60
Hearth, standard size brick, 3′ x 6′	1.000	Ea.	8.000	154	290	444
Chimney, one story house, framing, 2″ x 4″ studs	80.000	L.F.	1.164	29.60	47.20	76.80
Sheathing, plywood, 5/8″ thick	32.000	S.F.	.758	83.84	31.04	114.88
Flue, 10″ metal, insulated pipe	12.000	V.L.F.	4.000	294	160.20	454.20
Fittings, ceiling support	1.000	Ea.	.667	129	26.50	155.50
Fittings, joist shield	1.000	Ea.	.667	73.50	26.50	100
Fittings, roof flashing	1.000	Ea.	.667	149	26.50	175.50
Mantle beam, wood, 4″ x 8″	6.000	L.F.	1.333	33.90	54.30	88.20
TOTAL		Ea.	29.776	2,315.69	1,153.79	3,469.48

The costs in this system are on a cost each basis.

Description	QUAN.	UNIT	LABOR HOURS	COST EACH		
				MAT.	INST.	TOTAL

Prefabricated Fireplace Price Sheet	QUAN.	UNIT	LABOR HOURS	COST EACH MAT.	INST.	TOTAL
efabricated fireplace, minimum	1.000	Ea.	6.154	1,250	251	1,501
Average	1.000	Ea.	8.000	1,500	325	1,825
Maximum	1.000	Ea.	8.889	3,700	360	4,060
aming, 2″ x 4″ studs, fireplace, 6′ x 5′	35.000	L.F.	.509	12.95	20.50	33.45
7′ x 5′	40.000	L.F.	.582	14.80	23.50	38.30
8′ x 6′	45.000	L.F.	.655	16.65	26.50	43.15
ıeetrock, 1/2″ thick, fireplace, 6′ x 5′	40.000	S.F.	.640	10.80	26.50	37.30
7′ x 5′	45.000	S.F.	.720	12.15	29.50	41.65
8′ x 6′	50.000	S.F.	.800	13.50	33	46.50
acing for fireplace, brick, 6′ x 5′	30.000	S.F.	5.217	95	194	289
7′ x 5′	35.000	S.F.	6.087	111	226	337
8′ x 6′	48.000	S.F.	8.348	152	310	462
Fieldstone, 6′ x 5′	30.000	S.F.	5.217	460	194	654
7′ x 5′	35.000	S.F.	6.087	535	226	761
8′ x 6′	48.000	S.F.	8.348	735	310	1,045
earth, standard size brick, 3′ x 6′	1.000	Ea.	8.000	154	290	444
3′ x 7′	1.000	Ea.	9.280	179	335	514
3′ x 8′	1.000	Ea.	10.640	205	385	590
Stone, 3′ x 6′	1.000	Ea.	8.000	166	290	456
3′ x 7′	1.000	Ea.	9.280	193	335	528
3′ x 8′	1.000	Ea.	10.640	221	385	606
himney, framing, 2″ x 4″, one story house	80.000	L.F.	1.164	29.50	47	76.50
Two story house	120.000	L.F.	1.746	44.50	71	115.50
heathing, plywood, 5/8″ thick	32.000	S.F.	.758	84	31	115
Stucco on plywood	32.000	S.F.	1.125	36	44.50	80.50
ue, 10″ metal pipe, insulated, one story house	12.000	V.L.F.	4.000	294	160	454
Two story house	20.000	V.L.F.	6.667	490	267	757
ittings, ceiling support	1.000	Ea.	.667	129	26.50	155.50
ittings joist sheild, one story house	1.000	Ea.	.667	73.50	26.50	100
Two story house	2.000	Ea.	1.333	147	53	200
ittings roof flashing	1.000	Ea.	.667	149	26.50	175.50
lantle, wood beam, 4″ x 8″	6.000	L.F.	1.333	34	54.50	88.50
4″ x 10″	6.000	L.F.	1.371	41	56	97
Ornate prefabricated, 6′ x 3′-6″ opening, minimum	1.000	Ea.	1.600	153	65	218
Maximum	1.000	Ea.	1.600	190	65	255

System Description	QUAN.	UNIT	LABOR HOURS	COST EACH		
				MAT.	INST.	TOTAL
Economy, lean to, shell only, not including 2′ stub wall, fndtn, flrs, heat						
4′ x 16′	1.000	Ea.	26.212	2,150	1,075	3,225
4′ x 24′	1.000	Ea.	30.259	2,475	1,225	3,700
6′ x 10′	1.000	Ea.	16.552	1,800	675	2,475
6′ x 16′	1.000	Ea.	23.034	2,500	940	3,440
6′ x 24′	1.000	Ea.	29.793	3,250	1,225	4,475
8′ x 10′	1.000	Ea.	22.069	2,400	900	3,300
8′ x 16′	1.000	Ea.	38.400	4,175	1,575	5,750
8′ x 24′	1.000	Ea.	49.655	5,400	2,025	7,425
Free standing, 8′ x 8′	1.000	Ea.	17.356	2,775	705	3,480
8′ x 16′	1.000	Ea.	30.211	4,850	1,225	6,075
8′ x 24′	1.000	Ea.	39.051	6,275	1,600	7,875
10′ x 10′	1.000	Ea.	18.824	3,350	765	4,115
10′ x 16′	1.000	Ea.	24.095	4,300	980	5,280
10′ x 24′	1.000	Ea.	31.624	5,625	1,275	6,900
14′ x 10′	1.000	Ea.	20.741	4,200	845	5,045
14′ x 16′	1.000	Ea.	24.889	5,050	1,025	6,075
14′ x 24′	1.000	Ea.	33.349	6,750	1,350	8,100
Standard,lean to,shell only,not incl.2′stub wall, fndtn,flrs, heat 4′x10′	1.000	Ea.	28.235	2,300	1,150	3,450
4′ x 16′	1.000	Ea.	39.341	3,225	1,600	4,825
4′ x 24′	1.000	Ea.	45.412	3,725	1,850	5,575
6′ x 10′	1.000	Ea.	24.827	2,700	1,025	3,725
6′ x 16′	1.000	Ea.	34.538	3,750	1,400	5,150
6′ x 24′	1.000	Ea.	44.689	4,850	1,825	6,675
8′ x 10′	1.000	Ea.	33.103	3,600	1,350	4,950
8′ x 16′	1.000	Ea.	57.600	6,275	2,350	8,625
8′ x 24′	1.000	Ea.	74.482	8,100	3,050	11,150
Free standing, 8′ x 8′	1.000	Ea.	26.034	4,175	1,050	5,225
8′ x 16′	1.000	Ea.	45.316	7,275	1,850	9,125
8′ x 24′	1.000	Ea.	58.577	9,400	2,375	11,775
10′ x 10′	1.000	Ea.	28.236	5,025	1,150	6,175
10′ x 16′	1.000	Ea.	36.142	6,425	1,475	7,900
10′ x 24′	1.000	Ea.	47.436	8,450	1,925	10,375
14′ x 10′	1.000	Ea.	31.112	6,300	1,275	7,575
14′ x 16′	1.000	Ea.	37.334	7,550	1,525	9,075
14′ x 24′	1.000	Ea.	50.030	10,100	2,050	12,150
Deluxe,lean to,shell only,not incl.2′stub wall, fndtn, flrs or heat, 4′x10′	1.000	Ea.	20.645	4,000	840	4,840
4′ x 16′	1.000	Ea.	33.032	6,400	1,350	7,750
4′ x 24′	1.000	Ea.	49.548	9,600	2,025	11,625
6′ x 10′	1.000	Ea.	30.968	6,000	1,250	7,250
6′ x 16′	1.000	Ea.	49.548	9,600	2,025	11,625
6′ x 24′	1.000	Ea.	74.323	14,400	3,025	17,425
8′ x 10′	1.000	Ea.	41.290	8,000	1,675	9,675
8′ x 16′	1.000	Ea.	66.065	12,800	2,700	15,500
8′ x 24′	1.000	Ea.	99.097	19,200	4,025	23,225
Freestanding, 8′ x 8′	1.000	Ea.	18.618	5,500	760	6,260
8′ x 16′	1.000	Ea.	37.236	11,000	1,525	12,525
8′ x 24′	1.000	Ea.	55.855	16,500	2,275	18,775
10′ x 10′	1.000	Ea.	29.091	8,600	1,175	9,775
10′ x 16′	1.000	Ea.	46.546	13,800	1,900	15,700
10′ x 24′	1.000	Ea.	69.818	20,600	2,850	23,450
14′ x 10′	1.000	Ea.	40.727	12,000	1,650	13,650
14′ x 16′	1.000	Ea.	65.164	19,300	2,650	21,950
14′ x 24′	1.000	Ea.	97.746	28,900	3,975	32,875

System Description	QUAN.	UNIT	LABOR HOURS	COST EACH MAT.	COST EACH INST.	COST EACH TOTAL
vimming pools, vinyl lined, metal sides, sand bottom, 12′ x 28′	1.000	Ea.	50.177	3,125	2,050	5,175
12′ x 32′	1.000	Ea.	55.366	3,450	2,275	5,725
12′ x 36′	1.000	Ea.	60.061	3,750	2,475	6,225
16′ x 32′	1.000	Ea.	66.798	4,175	2,750	6,925
16′ x 36′	1.000	Ea.	71.190	4,450	2,925	7,375
16′ x 40′	1.000	Ea.	74.703	4,650	3,075	7,725
20′ x 36′	1.000	Ea.	77.860	4,850	3,200	8,050
20′ x 40′	1.000	Ea.	82.135	5,125	3,375	8,500
20′ x 44′	1.000	Ea.	90.348	5,650	3,725	9,375
24′ x 40′	1.000	Ea.	98.562	6,150	4,050	10,200
24′ x 44′	1.000	Ea.	108.418	6,775	4,450	11,225
24′ x 48′	1.000	Ea.	118.274	7,375	4,850	12,225
Vinyl lined, concrete sides, 12′ x 28′	1.000	Ea.	79.447	4,950	3,275	8,225
12′ x 32′	1.000	Ea.	88.818	5,550	3,650	9,200
12′ x 36′	1.000	Ea.	97.656	6,100	4,025	10,125
16′ x 32′	1.000	Ea.	111.393	6,950	4,575	11,525
16′ x 36′	1.000	Ea.	121.354	7,575	5,000	12,575
16′ x 40′	1.000	Ea.	130.445	8,150	5,375	13,525
28′ x 36′	1.000	Ea.	140.585	8,775	5,775	14,550
20′ x 40′	1.000	Ea.	149.336	9,325	6,150	15,475
20′ x 44′	1.000	Ea.	164.270	10,300	6,775	17,075
24′ x 40′	1.000	Ea.	179.203	11,200	7,375	18,575
24′ x 44′	1.000	Ea.	197.124	12,300	8,125	20,425
24′ x 48′	1.000	Ea.	215.044	13,400	8,850	22,250
Gunite, bottom and sides, 12′ x 28′	1.000	Ea.	129.767	6,500	5,325	11,825
12′ x 32′	1.000	Ea.	142.164	7,125	5,850	12,975
12′ x 36′	1.000	Ea.	153.028	7,675	6,275	13,950
16′ x 32′	1.000	Ea.	167.743	8,400	6,900	15,300
16′ x 36′	1.000	Ea.	176.421	8,850	7,275	16,125
16′ x 40′	1.000	Ea.	182.368	9,125	7,500	16,625
20′ x 36′	1.000	Ea.	187.949	9,425	7,725	17,150
20′ x 40′	1.000	Ea.	179.200	12,400	7,375	19,775
20′ x 44′	1.000	Ea.	197.120	13,700	8,125	21,825
24′ x 40′	1.000	Ea.	215.040	14,900	8,850	23,750
24′ x 44′	1.000	Ea.	273.244	13,700	11,300	25,000
24′ x 48′	1.000	Ea.	298.077	14,900	12,300	27,200

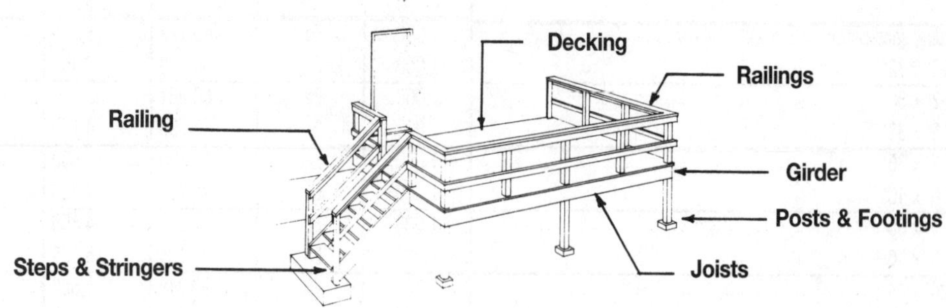

System Description	QUAN.	UNIT	LABOR HOURS	COST PER S.F. MAT.	COST PER S.F. INST.	COST PER S.F. TOTAL
8′ X 12′ DECK, PRESSURE TREATED LUMBER, JOISTS 16″ O.C.						
Decking, 2″ x 6″ lumber	2.080	L.F.	.027	1.23	1.08	2.31
Lumber preservative	2.080	L.F.		.27		.27
Joists, 2″ x 8″, 16″ O.C.	1.000	L.F.	.015	.91	.59	1.50
Lumber preservative	1.000	L.F.		.17		.17
Girder, 2″ x 10″	.125	L.F.	.002	.16	.09	.25
Lumber preservative	.125	L.F.		.03		.03
Hand excavation for footings	.250	L.F.	.006		.18	.18
Concrete footings	.250	L.F.	.006	.26	.21	.47
4″ x 4″ Posts	.250	L.F.	.010	.35	.42	.77
Lumber preservative	.250	L.F.		.04		.04
Framing, pressure treated wood stairs, 3′ wide, 8 closed risers	1.000	Set	.080	1.65	3.25	4.90
Railings, 2″ x 4″	1.000	L.F.	.026	.37	1.05	1.42
Lumber preservative	1.000	L.F.		.09		.09
TOTAL		S.F.	.172	5.53	6.87	12.40
12′ X 16′ DECK, PRESSURE TREATED LUMBER, JOISTS 24″ O.C.						
Decking, 2″ x 6″	2.080	L.F.	.027	1.23	1.08	2.31
Lumber preservative	2.080	L.F.		.27		.27
Joists, 2″ x 10″, 24″ O.C.	.800	L.F.	.014	1.03	.58	1.61
Lumber preservative	.800	L.F.		.17		.17
Girder, 2″ x 10″	.083	L.F.	.001	.11	.06	.17
Lumber preservative	.083	L.F.		.02		.02
Hand excavation for footings	.122	L.F.	.006		.18	.18
Concrete footings	.122	L.F.	.006	.26	.21	.47
4″ x 4″ Posts	.122	L.F.	.005	.17	.20	.37
Lumber preservative	.122	L.F.		.02		.02
Framing, pressure treated wood stairs, 3′ wide, 8 closed risers	1.000	Set	.040	.83	1.63	2.46
Railings, 2″ x 4″	.670	L.F.	.017	.25	.70	.95
Lumber preservative	.670	L.F.		.06		.06
TOTAL		S.F.	.116	4.42	4.64	9.06
12′ X 24′ DECK, REDWOOD OR CEDAR, JOISTS 16″ O.C.						
Decking, 2″ x 6″ redwood	2.080	L.F.	.027	7.55	1.08	8.63
Joists, 2″ x 10″, 16″ O.C.	1.000	L.F.	.018	6.95	.72	7.67
Girder, 2″ x 10″	.083	L.F.	.001	.58	.06	.64
Hand excavation for footings	.111	L.F.	.006		.18	.18
Concrete footings	.111	L.F.	.006	.26	.21	.47
Lumber preservative	.111	L.F.		.02		.02
Post, 4″ x 4″, including concrete footing	.111	L.F.	.009	.73	.36	1.09
Framing, redwood or cedar stairs, 3′ wide, 8 closed risers	1.000	Set	.028	2	1.14	3.14
Railings, 2″ x 4″	.540	L.F.	.005	1.31	.19	1.50
TOTAL		S.F.	.100	19.40	3.94	23.34

The costs in this system are on a square foot basis.

Important: See the Reference Section for critical supporting data - Reference Nos., Crews & Location Factors

Wood Deck Price Sheet	QUAN.	UNIT	LABOR HOURS	COST PER S.F.		
				MAT.	INST.	TOTAL
ecking, treated lumber, 1″ x 4″	3.430	L.F.	.031	2.52	1.28	3.80
1″ x 6″	2.180	L.F.	.033	2.65	1.34	3.99
2″ x 4″	3.200	L.F.	.041	1.47	1.66	3.13
2″ x 6″	2.080	L.F.	.027	1.50	1.08	2.58
Redwood or cedar,, 1″ x 4″	3.430	L.F.	.035	2.95	1.41	4.36
1″ x 6″	2.180	L.F.	.036	3.08	1.47	4.55
2″ x 4″	3.200	L.F.	.028	8	1.14	9.14
2″ x 6″	2.080	L.F.	.027	7.55	1.08	8.63
oists for deck, treated lumber, 2″ x 8″, 16″ O.C.	1.000	L.F.	.015	1.08	.59	1.67
24″ O.C.	.800	L.F.	.012	.87	.47	1.34
2″ x 10″, 16″ O.C.	1.000	L.F.	.018	1.51	.72	2.23
24″ O.C.	.800	L.F.	.014	1.20	.58	1.78
Redwood or cedar, 2″ x 8″, 16″ O.C.	1.000	L.F.	.015	4.83	.59	5.42
24″ O.C.	.800	L.F.	.012	3.86	.47	4.33
2″ x 10″, 16″ O.C.	1.000	L.F.	.018	6.95	.72	7.67
24″ O.C.	.800	L.F.	.014	5.55	.58	6.13
irder for joists, treated lumber, 2″ x 10″, 8′ x 12′ deck	.125	L.F.	.002	.19	.09	.28
12′ x 16′ deck	.083	L.F.	.001	.13	.06	.19
12′ x 24′ deck	.083	L.F.	.001	.13	.06	.19
Redwood or cedar, 2″ x 10″, 8′ x 12′ deck	.125	L.F.	.002	.87	.09	.96
12′ x 16′ deck	.083	L.F.	.001	.58	.06	.64
12′ x 24′ deck	.083	L.F.	.001	.58	.06	.64
osts, 4″ x 4″, including concrete footing, 8′ x 12′ deck	.250	S.F.	.022	.65	.81	1.46
12′ x 16′ deck	.122	L.F.	.017	.45	.59	1.04
12′ x 24′ deck	.111	L.F.	.017	.43	.58	1.01
tairs 2″ x 10″ stringers, treated lumber, 8′ x 12′ deck	1.000	Set	.020	1.65	3.25	4.90
12′ x 16′ deck	1.000	Set	.012	.83	1.63	2.46
12′ x 24′ deck	1.000	Set	.008	.58	1.14	1.72
Redwood or cedar, 8′ x 12′ deck	1.000	Set	.040	5.70	3.25	8.95
12′ x 16′ deck	1.000	Set	.020	2.85	1.63	4.48
12′ x 24′ deck	1.000	Set	.012	2	1.14	3.14
ailings 2″ x 4″, treated lumber, 8′ x 12′ deck	1.000	L.F.	.026	.46	1.05	1.51
12′ x 16′ deck	.670	L.F.	.017	.31	.70	1.01
12′ x 24′ deck	.540	L.F.	.014	.25	.57	.82
Redwood or cedar, 8′ x 12′ deck	1.000	L.F.	.009	2.43	.35	2.78
12′ x 16′ deck	.670	L.F.	.006	1.60	.23	1.83
12′ x 24′ deck	.540	L.F.	.005	1.31	.19	1.50

Division 8
Mechanical

No part of this publication may be reproduced, stored in a retrieval system, or transmitted in any form or by any means without prior written permission of Reed Construction Data.

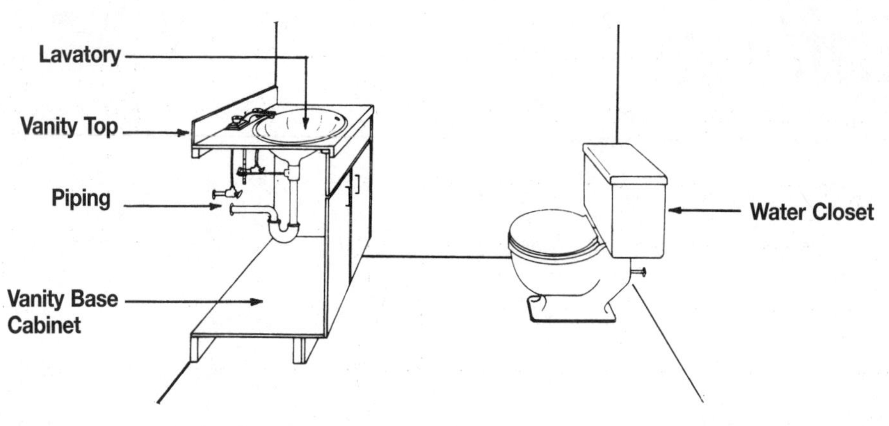

System Description	QUAN.	UNIT	LABOR HOURS	COST EACH		
				MAT.	INST.	TOTAL
LAVATORY INSTALLED WITH VANITY, PLUMBING IN 2 WALLS						
Water closet, floor mounted, 2 piece, close coupled, white	1.000	Ea.	3.019	179	120	299
Rough-in, vent, 2" diameter DWV piping	1.000	Ea.	.955	24	38	62
Waste, 4" diameter DWV piping	1.000	Ea.	.828	30	33	63
Supply, 1/2" diameter type "L" copper supply piping	1.000	Ea.	.593	9.90	26.28	36.18
Lavatory, 20" x 18", P.E. cast iron white	1.000	Ea.	2.500	227	99.50	326.50
Rough-in, vent, 1-1/2" diameter DWV piping	1.000	Ea.	.901	23.40	36	59.40
Waste, 2" diameter DWV piping	1.000	Ea.	.955	24	38	62
Supply, 1/2" diameter type "L" copper supply piping	1.000	Ea.	.988	16.50	43.80	60.30
Piping, supply, 1/2" diameter type "L" copper supply piping	10.000	L.F.	.988	16.50	43.80	60.30
Waste, 4" diameter DWV piping	7.000	L.F.	1.931	70	77	147
Vent, 2" diameter DWV piping	12.000	L.F.	2.866	72	114	186
Vanity base cabinet, 2 door, 30" wide	1.000	Ea.	1.000	223	41	264
Vanity top, plastic & laminated, square edge	2.670	L.F.	.712	82.77	28.97	111.74
TOTAL		Ea.	18.236	998.07	739.35	1,737.42
LAVATORY WITH WALL-HUNG LAVATORY, PLUMBING IN 2 WALLS						
Water closet, floor mounted, 2 piece close coupled, white	1.000	Ea.	3.019	179	120	299
Rough-in, vent, 2" diameter DWV piping	1.000	Ea.	.955	24	38	62
Waste, 4" diameter DWV piping	1.000	Ea.	.828	30	33	63
Supply, 1/2" diameter type "L" copper supply piping	1.000	Ea.	.593	9.90	26.28	36.18
Lavatory, 20" x 18", P.E. cast iron, wall hung, white	1.000	Ea.	2.000	261	79.50	340.50
Rough-in, vent, 1-1/2" diameter DWV piping	1.000	Ea.	.901	23.40	36	59.40
Waste, 2" diameter DWV piping	1.000	Ea.	.955	24	38	62
Supply, 1/2" diameter type "L" copper supply piping	1.000	Ea.	.988	16.50	43.80	60.30
Piping, supply, 1/2" diameter type "L" copper supply piping	10.000	L.F.	.988	16.50	43.80	60.30
Waste, 4" diameter DWV piping	7.000	L.F.	1.931	70	77	147
Vent, 2" diameter DWV piping	12.000	L.F.	2.866	72	114	186
Carrier, steel for studs, no arms	1.000	Ea.	1.143	38.50	50.50	89
TOTAL		Ea.	17.167	764.80	699.88	1,464.68

Description	QUAN.	UNIT	LABOR HOURS	COST EACH		
				MAT.	INST.	TOTAL

 Important: See the Reference Section for critical supporting data - Reference Nos., Crews & Location Factors

Two Fixture Lavatory Price Sheet	QUAN.	UNIT	LABOR HOURS	COST EACH		
				MAT.	INST.	TOTAL
ater closet, close coupled standard 2 piece, white	1.000	Ea.	3.019	179	120	299
Color	1.000	Ea.	3.019	215	120	335
One piece elongated bowl, white	1.000	Ea.	3.019	545	120	665
Color	1.000	Ea.	3.019	680	120	800
Low profile, one piece elongated bowl, white	1.000	Ea.	3.019	745	120	865
Color	1.000	Ea.	3.019	970	120	1,090
ough-in for water closet						
1/2" copper supply, 4" cast iron waste, 2" cast iron vent	1.000	Ea.	2.376	64	97.50	161.50
4" PVC waste, 2" PVC vent	1.000	Ea.	2.678	26	109	135
4" copper waste , 2" copper vent	1.000	Ea.	2.520	99.50	106	205.50
3" cast iron waste, 1-1/2" cast iron vent	1.000	Ea.	2.244	57	92	149
3" PVC waste, 1-1/2" PVC vent	1.000	Ea.	2.388	23	102	125
3" copper waste, 1-1/2" copper vent	1.000	Ea.	2.524	94	103	197
1/2" PVC supply, 4" PVC waste, 2" PVC vent	1.000	Ea.	2.974	31.50	122	153.50
3" PVC waste, 1-1/2" PVC vent	1.000	Ea.	2.684	28.50	115	143.50
1/2" steel supply, 4" cast iron waste, 2" cast iron vent	1.000	Ea.	2.545	64	105	169
4" cast iron waste, 2" steel vent	1.000	Ea.	2.590	61.50	107	168.50
4" PVC waste, 2" PVC vent	1.000	Ea.	2.847	26.50	117	143.50
avatory, vanity top mounted, P.E. on cast iron 20" x 18" white	1.000	Ea.	2.500	227	99.50	326.50
Color	1.000	Ea.	2.500	256	99.50	355.50
Steel, enameled 10" x 17" white	1.000	Ea.	2.759	142	110	252
Color	1.000	Ea.	2.500	148	99.50	247.50
Vitreous china 20" x 16", white	1.000	Ea.	2.963	244	118	362
Color	1.000	Ea.	2.963	244	118	362
Wall hung, P.E. on cast iron, 20" x 18", white	1.000	Ea.	2.000	261	79.50	340.50
Color	1.000	Ea.	2.000	297	79.50	376.50
Vitreous china 19" x 17", white	1.000	Ea.	2.286	204	91	295
Color	1.000	Ea.	2.286	204	91	295
ough-in supply waste and vent for lavatory						
1/2" copper supply, 2" cast iron waste, 1-1/2" cast iron vent	1.000	Ea.	2.844	64	118	182
2" PVC waste, 1-1/2" PVC vent	1.000	Ea.	2.962	27.50	126	153.50
2" copper waste, 1-1/2" copper vent	1.000	Ea.	2.308	63.50	102	165.50
1-1/2" PVC waste, 1-1/4" PVC vent	1.000	Ea.	2.639	27	117	144
1-1/2" copper waste, 1-1/4" copper vent	1.000	Ea.	2.114	52.50	93.50	146
1/2" PVC supply, 2" PVC waste, 1-1/2" PVC vent	1.000	Ea.	3.456	36.50	148	184.50
1-1/2" PVC waste, 1-1/4" PVC vent	1.000	Ea.	3.133	36	139	175
1/2" steel supply, 2" cast iron waste, 1-1/2" cast iron vent	1.000	Ea.	3.126	64	131	195
2" cast iron waste, 2" steel vent	1.000	Ea.	3.225	62	134	196
2" PVC waste, 1-1/2" PVC vent	1.000	Ea.	3.244	28	139	167
1-1/2" PVC waste, 1-1/4" PVC vent	1.000	Ea.	2.921	27	130	157
iping, supply, 1/2" copper, type "L"	10.000	L.F.	.988	16.50	44	60.50
1/2" steel	10.000	L.F.	1.270	16.70	56.50	73.20
1/2" PVC	10.000	L.F.	1.482	25.50	65.50	91
Waste, 4" cast iron	7.000	L.F.	1.931	70	77	147
4" copper	7.000	L.F.	2.800	147	112	259
4" PVC	7.000	L.F.	2.333	24	93	117
Vent, 2" cast iron	12.000	L.F.	2.866	72	114	186
2" copper	12.000	L.F.	2.182	80.50	96.50	177
2" PVC	12.000	L.F.	3.254	17.90	130	147.90
2" steel	12.000	Ea.	3.000	64	119	183
anity base cabinet, 2 door, 24" x 30"	1.000	Ea.	1.000	223	41	264
24" x 36"	1.000	Ea.	1.200	298	49	347
anity top, laminated plastic, square edge 25" x 32"	2.670	L.F.	.712	83	29	112
25" x 38"	3.170	L.F.	.845	98.50	34.50	133
Post formed, laminated plastic, 25" x 32"	2.670	L.F.	.712	26.50	29	55.50
25" x 38"	3.170	L.F.	.845	31.50	34.50	66
Cultured marble, 25" x 32" with bowl	1.000	Ea.	2.500	169	99.50	268.50
25" x 38" with bowl	1.000	Ea.	2.500	196	99.50	295.50
Carrier for lavatory, steel for studs	1.000	Ea.	1.143	38.50	50.50	89
Wood 2" x 8" blocking	1.330	L.F.	.053	1.21	2.17	3.38

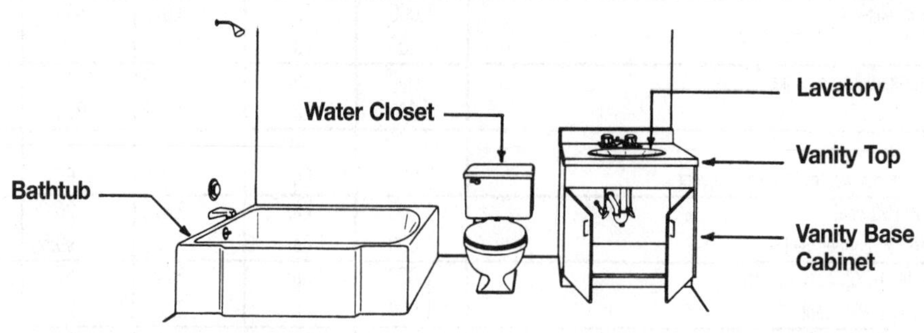

System Description	QUAN.	UNIT	LABOR HOURS	COST EACH		
				MAT.	INST.	TOTAL
BATHROOM INSTALLED WITH VANITY						
Water closet, floor mounted, 2 piece, close coupled, white	1.000	Ea.	3.019	179	120	299
Rough-in, waste, 4" diameter DWV piping	1.000	Ea.	.828	30	33	63
Vent, 2" diameter DWV piping	1.000	Ea.	.955	24	38	62
Supply, 1/2" diameter type "L" copper supply piping	1.000	Ea.	.593	9.90	26.28	36.18
Lavatory, 20" x 18", P.E. cast iron with accessories, white	1.000	Ea.	2.500	227	99.50	326.50
Rough-in, supply, 1/2" diameter type "L" copper supply piping	1.000	Ea.	.988	16.50	43.80	60.30
Waste, 1-1/2" diameter DWV piping	1.000	Ea.	1.803	46.80	72	118.80
Bathtub, P.E. cast iron, 5' long with accessories, white	1.000	Ea.	3.636	460	145	605
Rough-in, waste, 4" diameter DWV piping	1.000	Ea.	.828	30	33	63
Vent, 1-1/2" diameter DWV piping	1.000	Ea.	.593	20	26.20	46.20
Supply, 1/2" diameter type "L" copper supply piping	1.000	Ea.	.988	16.50	43.80	60.30
Piping, supply, 1/2" diameter type "L" copper supply piping	20.000	L.F.	1.975	33	87.60	120.60
Waste, 4" diameter DWV piping	9.000	L.F.	2.483	90	99	189
Vent, 2" diameter DWV piping	6.000	L.F.	1.500	32.10	59.70	91.80
Vanity base cabinet, 2 door, 30" wide	1.000	Ea.	1.000	223	41	264
Vanity top, plastic laminated square edge	2.670	L.F.	.712	64.08	28.97	93.05
TOTAL		Ea.	24.401	1,501.88	996.85	2,498.73
BATHROOM WITH WALL HUNG LAVATORY						
Water closet, floor mounted, 2 piece, close coupled, white	1.000	Ea.	3.019	179	120	299
Rough-in, vent, 2" diameter DWV piping	1.000	Ea.	.955	24	38	62
Waste, 4" diameter DWV piping	1.000	Ea.	.828	30	33	63
Supply, 1/2" diameter type "L" copper supply piping	1.000	Ea.	.593	9.90	26.28	36.18
Lavatory, 20" x 18" P.E. cast iron, wall hung, white	1.000	Ea.	2.000	261	79.50	340.50
Rough-in, waste, 1-1/2" diameter DWV piping	1.000	Ea.	1.803	46.80	72	118.80
Supply, 1/2" diameter type "L" copper supply piping	1.000	Ea.	.988	16.50	43.80	60.30
Bathtub, P.E. cast iron, 5' long with accessories, white	1.000	Ea.	3.636	460	145	605
Rough-in, waste, 4" diameter DWV piping	1.000	Ea.	.828	30	33	63
Supply, 1/2" diameter type "L" copper supply piping	1.000	Ea.	.988	16.50	43.80	60.30
Vent, 1-1/2" diameter DWV piping	1.000	Ea.	1.482	50	65.50	115.50
Piping, supply, 1/2" diameter type "L" copper supply piping	20.000	L.F.	1.975	33	87.60	120.60
Waste, 4" diameter DWV piping	9.000	L.F.	2.483	90	99	189
Vent, 2" diameter DWV piping	6.000	L.F.	1.500	32.10	59.70	91.80
Carrier, steel, for studs, no arms	1.000	Ea.	1.143	38.50	50.50	89
TOTAL		Ea.	24.221	1,317.30	996.68	2,313.98

The costs in this system are a cost each basis, all necessary piping is included.

Important: See the Reference Section for critical supporting data - Reference Nos., Crews & Location Factors

Three Fixture Bathroom Price Sheet	QUAN.	UNIT	LABOR HOURS	COST EACH		
				MAT.	INST.	TOTAL
ter closet, close coupled standard 2 piece, white	1.000	Ea.	3.019	179	120	299
Color	1.000	Ea.	3.019	215	120	335
One piece, elongated bowl, white	1.000	Ea.	3.019	545	120	665
Color	1.000	Ea.	3.019	680	120	800
Low profile, one piece elongated bowl, white	1.000	Ea.	3.019	745	120	865
Color	1.000	Ea.	3.019	970	120	1,090
ugh-in, for water closet						
1/2″ copper supply, 4″ cast iron waste, 2″ cast iron vent	1.000	Ea.	2.376	64	97.50	161.50
4″ PVC/DWV waste, 2″ PVC vent	1.000	Ea.	2.678	26	109	135
4″ copper waste, 2″ copper vent	1.000	Ea.	2.520	99.50	106	205.50
3″ cast iron waste, 1-1/2″ cast iron vent	1.000	Ea.	2.244	57	92	149
3″ PVC waste, 1-1/2″ PVC vent	1.000	Ea.	2.388	23	102	125
3″ copper waste, 1-1/2″ copper vent	1.000	Ea.	2.014	66	85.50	151.50
1/2″ PVC supply, 4″ PVC waste, 2″ PVC vent	1.000	Ea.	2.974	31.50	122	153.50
3″ PVC waste, 1-1/2″ PVC supply	1.000	Ea.	2.684	28.50	115	143.50
1/2″ steel supply, 4″ cast iron waste, 2″ cast iron vent	1.000	Ea.	2.545	64	105	169
4″ cast iron waste, 2″ steel vent	1.000	Ea.	2.590	61.50	107	168.50
4″ PVC waste, 2″ PVC vent	1.000	Ea.	2.847	26.50	117	143.50
vatory, wall hung, P.E. cast iron 20″ x 18″, white	1.000	Ea.	2.000	261	79.50	340.50
Color	1.000	Ea.	2.000	297	79.50	376.50
Vitreous china 19″ x 17″, white	1.000	Ea.	2.286	204	91	295
Color	1.000	Ea.	2.286	204	91	295
Lavatory, for vanity top, P.E. cast iron 20″ x 18″″, white	1.000	Ea.	2.500	227	99.50	326.50
Color	1.000	Ea.	2.500	256	99.50	355.50
Steel, enameled 20″ x 17″, white	1.000	Ea.	2.759	142	110	252
Color	1.000	Ea.	2.500	148	99.50	247.50
Vitreous china 20″ x 16″, white	1.000	Ea.	2.963	244	118	362
Color	1.000	Ea.	2.963	244	118	362
ough-in, for lavatory						
1/2″ copper supply, 1-1/2″ C.I. waste, 1-1/2″ C.I. vent	1.000	Ea.	2.791	63.50	116	179.50
1-1/2″ PVC waste, 1-1/4″ PVC vent	1.000	Ea.	2.639	27	117	144
1/2″ steel supply, 1-1/4″ cast iron waste, 1-1/4″ steel vent	1.000	Ea.	2.890	54	121	175
1-1/4″ PVC@ waste, 1-1/4″ PVC vent	1.000	Ea.	2.794	27	124	151
1/2″ PVC supply, 1-1/2″ PVC waste, 1-1/2″ PVC vent	1.000	Ea.	3.260	36	144	180
athtub, P.E. cast iron, 5′ long corner with fittings, white	1.000	Ea.	3.636	460	145	605
Color	1.000	Ea.	3.636	500	145	645
ough-in, for bathtub						
1/2″ copper supply, 4″ cast iron waste, 1-1/2″ copper vent	1.000	Ea.	2.409	66.50	103	169.50
4″ PVC waste, 1-1/2″ PVC vent	1.000	Ea.	2.877	32	123	155
1/2″ steel supply, 4″ cast iron waste, 1-1/2″ steel vent	1.000	Ea.	2.898	63	121	184
4″ PVC waste, 1-1/2″ PVC vent	1.000	Ea.	3.159	32.50	136	168.50
1/2″ PVC supply, 4″ PVC waste, 1-1/2″ PVC vent	1.000	Ea.	3.371	41	145	186
ping, supply 1/2″ copper	20.000	L.F.	1.975	33	87.50	120.50
1/2″ steel	20.000	L.F.	2.540	33.50	113	146.50
1/2″ PVC	20.000	L.F.	2.963	51	131	182
iping, waste, 4″ cast iron no hub	9.000	L.F.	2.483	90	99	189
4″ PVC/DWV	9.000	L.F.	3.000	31	120	151
4″ copper/DWV	9.000	L.F.	3.600	189	144	333
iping, vent 2″ cast iron no hub	6.000	L.F.	1.433	36	57	93
2″ copper/DWV	6.000	L.F.	1.091	40	48.50	88.50
2″ PVC/DWV	6.000	L.F.	1.627	8.95	65	73.95
2″ steel, galvanized	6.000	L.F.	1.500	32	59.50	91.50
anity base cabinet, 2 door, 24″ x 30″	1.000	Ea.	1.000	223	41	264
24″ x 36″	1.000	Ea.	1.200	298	49	347
anity top, laminated plastic square edge 25″ x 32″	2.670	L.F.	.712	64	29	93
25″ x 38″	3.160	L.F.	.843	76	34.50	110.50
Cultured marble, 25″ x 32″, with bowl	1.000	Ea.	2.500	169	99.50	268.50
25″ x 38″, with bowl	1.000	Ea.	2.500	196	99.50	295.50
arrier, for lavatory, steel for studs, no arms	1.000	Ea.	1.143	38.50	50.50	89
Wood, 2″ x 8″ blocking	1.300	L.F.	.052	1.18	2.12	3.30

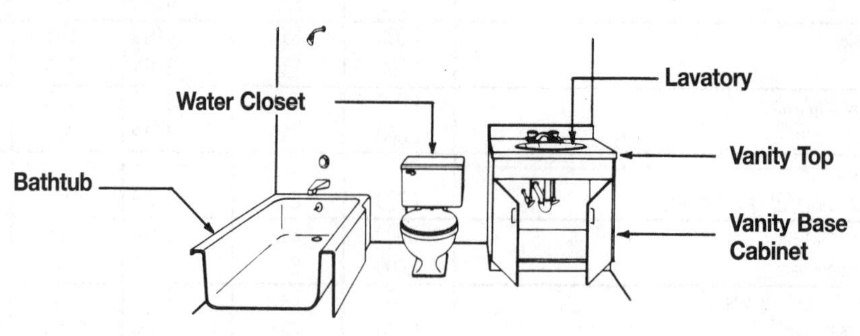

System Description	QUAN.	UNIT	LABOR HOURS	COST EACH		
				MAT.	INST.	TOTAL
BATHROOM WITH LAVATORY INSTALLED IN VANITY						
Water closet, floor mounted, 2 piece, close coupled, white	1.000	Ea.	3.019	179	120	299
Rough-in, waste, 4" diameter DWV piping	1.000	Ea.	.828	30	33	63
Vent, 2" diameter DWV piping	1.000	Ea.	.955	24	38	62
Supply, 1/2" diameter type "L" copper supply piping	1.000	Ea.	.593	9.90	26.28	36.18
Lavatory, 20" x 18", P.E. cast iron with accessories, white	1.000	Ea.	2.500	227	99.50	326.50
Rough-in, waste, 1-1/2" diameter DWV piping	1.000	Ea.	1.803	46.80	72	118.80
Supply, 1/2" diameter type "L" copper supply piping	1.000	Ea.	.988	16.50	43.80	60.30
Bathtub, P.E. cast iron 5' long with accessories, white	1.000	Ea.	3.636	460	145	605
Rough-in, waste, 4" diameter DWV piping	1.000	Ea.	.828	30	33	63
Vent, 1-1/2" diameter DWV piping	1.000	Ea.	.593	20	26.20	46.20
Supply, 1/2" diameter type "L" copper supply piping	1.000	Ea.	.988	16.50	43.80	60.30
Piping, supply, 1/2" diameter type "L" copper supply piping	10.000	L.F.	.988	16.50	43.80	60.30
Waste, 4" diameter DWV piping	6.000	L.F.	1.655	60	66	126
Vent, 2" diameter DWV piping	6.000	L.F.	1.500	32.10	59.70	91.80
Vanity base cabinet, 2 door, 30" wide	1.000	Ea.	1.000	223	41	264
Vanity top, plastic laminated square edge	2.670	L.F.	.712	64.08	28.97	93.05
TOTAL		Ea.	22.586	1,455.38	920.05	2,375.43
BATHROOM WITH WALL HUNG LAVATORY						
Water closet, floor mounted, 2 piece, close coupled, white	1.000	Ea.	3.019	179	120	299
Rough-in, vent, 2" diameter DWV piping	1.000	Ea.	.955	24	38	62
Waste, 4" diameter DWV piping	1.000	Ea.	.828	30	33	63
Supply, 1/2" diameter type "L" copper supply piping	1.000	Ea.	.593	9.90	26.28	36.18
Lavatory, 20" x 18" P.E. cast iron, wall hung, white	1.000	Ea.	2.000	261	79.50	340.50
Rough-in, waste, 1-1/2" diameter DWV piping	1.000	Ea.	1.803	46.80	72	118.80
Supply, 1/2" diameter type "L" copper supply piping	1.000	Ea.	.988	16.50	43.80	60.30
Bathtub, P.E. cast iron, 5' long with accessories, white	1.000	Ea.	3.636	460	145	605
Rough-in, waste, 4" diameter DWV piping	1.000	Ea.	.828	30	33	63
Supply, 1/2" diameter type "L" copper supply piping	1.000	Ea.	.988	16.50	43.80	60.30
Vent, 1-1/2" diameter DWV piping	1.000	Ea.	.593	20	26.20	46.20
Piping, supply, 1/2" diameter type "L" copper supply piping	10.000	L.F.	.988	16.50	43.80	60.30
Waste, 4" diameter DWV piping	6.000	L.F.	1.655	60	66	126
Vent, 2" diameter DWV piping	6.000	L.F.	1.500	32.10	59.70	91.80
Carrier, steel, for studs, no arms	1.000	Ea.	1.143	38.50	50.50	89
TOTAL		Ea.	21.517	1,240.80	880.58	2,121.38

The costs in this system are on a cost each basis. All necessary piping is included.

Three Fixture Bathroom Price Sheet	QUAN.	UNIT	LABOR HOURS	COST EACH		
				MAT.	INST.	TOTAL
ater closet, close coupled standard 2 piece, white	1.000	Ea.	3.019	179	120	299
Color	1.000	Ea.	3.019	215	120	335
One piece elongated bowl, white	1.000	Ea.	3.019	545	120	665
Color	1.000	Ea.	3.019	680	120	800
Low profile, one piece elongated bowl, white	1.000	Ea.	3.019	745	120	865
Color	1.000	Ea.	3.019	970	120	1,090
ugh-in for water closet						
1/2″ copper supply, 4″ cast iron waste, 2″ cast iron vent	1.000	Ea.	2.376	64	97.50	161.50
4″ PVC/DWV waste, 2″ PVC vent	1.000	Ea.	2.678	26	109	135
4″ carrier waste, 2″ copper vent	1.000	Ea.	2.520	99.50	106	205.50
3″ cast iron waste, 1-1/2″ cast iron vent	1.000	Ea.	2.244	57	92	149
3″ PVC waste, 1-1/2″ PVC vent	1.000	Ea.	2.388	23	102	125
3″ copper waste, 1-1/2″ copper vent	1.000	Ea.	2.014	66	85.50	151.50
1/2″ PVC supply, 4″ PVC waste, 2″ PVC vent	1.000	Ea.	2.974	31.50	122	153.50
3″ PVC waste, 1-1/2″ PVC supply	1.000	Ea.	2.684	28.50	115	143.50
1/2″ steel supply, 4″ cast iron waste, 2″ cast iron vent	1.000	Ea.	2.545	64	105	169
4″ cast iron waste, 2″ steel vent	1.000	Ea.	2.590	61.50	107	168.50
4″ PVC waste, 2″ PVC vent	1.000	Ea.	2.847	26.50	117	143.50
avatory, wall hung, PE cast iron 20″ x 18″, white	1.000	Ea.	2.000	261	79.50	340.50
Color	1.000	Ea.	2.000	297	79.50	376.50
Vitreous china 19″ x 17″, white	1.000	Ea.	2.286	204	91	295
Color	1.000	Ea.	2.286	204	91	295
Lavatory, for vanity top, PE cast iron 20″ x 18″, white	1.000	Ea.	2.500	227	99.50	326.50
Color	1.000	Ea.	2.500	256	99.50	355.50
Steel enameled 20″ x 17″, white	1.000	Ea.	2.759	142	110	252
Color	1.000	Ea.	2.500	148	99.50	247.50
Vitreous china 20″ x 16″, white	1.000	Ea.	2.963	244	118	362
Color	1.000	Ea.	2.963	244	118	362
ough-in for lavatory						
1/2″ copper supply, 1-1/2″ cast iron waste, 1-1/2″ cast iron vent	1.000	Ea.	2.791	63.50	116	179.50
1-1/2″ PVC waste, 1-1/4″ PVC vent	1.000	Ea.	2.639	27	117	144
1/2″ steel supply, 1-1/4″ cast iron waste, 1-1/4″ steel vent	1.000	Ea.	2.890	54	121	175
1-1/4″ PVC waste, 1-1/4″ PVC vent	1.000	Ea.	2.794	27	124	151
1/2″ PVC supply, 1-1/2″ PVC waste, 1-1/2″ PVC vent	1.000	Ea.	3.260	36	144	180
athtub, PE cast iron, 5′ long corner with fittings, white	1.000	Ea.	3.636	460	145	605
Color	1.000	Ea.	3.636	500	145	645
ough-in for bathtub						
1/2″ copper supply, 4″ cast iron waste, 1-1/2″ copper vent	1.000	Ea.	2.409	66.50	103	169.50
4″ PVC waste, 1/2″ PVC vent	1.000	Ea.	2.877	32	123	155
1/2″ steel supply, 4″ cast iron waste, 1-1/2″ steel vent	1.000	Ea.	2.898	63	121	184
4″ PVC waste, 1-1/2″ PVC vent	1.000	Ea.	3.159	32.50	136	168.50
1/2″ PVC supply, 4″ PVC waste, 1-1/2″ PVC vent	1.000	Ea.	3.371	41	145	186
iping supply, 1/2″ copper	10.000	L.F.	.988	16.50	44	60.50
1/2″ steel	10.000	L.F.	1.270	16.70	56.50	73.20
1/2″ PVC	10.000	L.F.	1.482	25.50	65.50	91
iping waste, 4″ cast iron no hub	6.000	L.F.	1.655	60	66	126
4″ PVC/DWV	6.000	L.F.	2.000	20.50	80	100.50
4″ copper/DWV	6.000	L.F.	2.400	126	95.50	221.50
iping vent 2″ cast iron no hub	6.000	L.F.	1.433	36	57	93
2″ copper/DWV	6.000	L.F.	1.091	40	48.50	88.50
2″ PVC/DWV	6.000	L.F.	1.627	8.95	65	73.95
2″ steel, galvanized	6.000	L.F.	1.500	32	59.50	91.50
anity base cabinet, 2 door, 24″ x 30″	1.000	Ea.	1.000	223	41	264
24″ x 36″	1.000	Ea.	1.200	298	49	347
anity top, laminated plastic square edge 25″ x 32″	2.670	L.F.	.712	64	29	93
25″ x 38″	3.160	L.F.	.843	76	34.50	110.50
Cultured marble, 25″ x 32″, with bowl	1.000	Ea.	2.500	169	99.50	268.50
25″ x 38″, with bowl	1.000	Ea.	2.500	196	99.50	295.50
Carrier, for lavatory, steel for studs, no arms	1.000	Ea.	1.143	38.50	50.50	89
Wood, 2″ x 8″ blocking	1.300	L.F.	.052	1.18	2.12	3.30

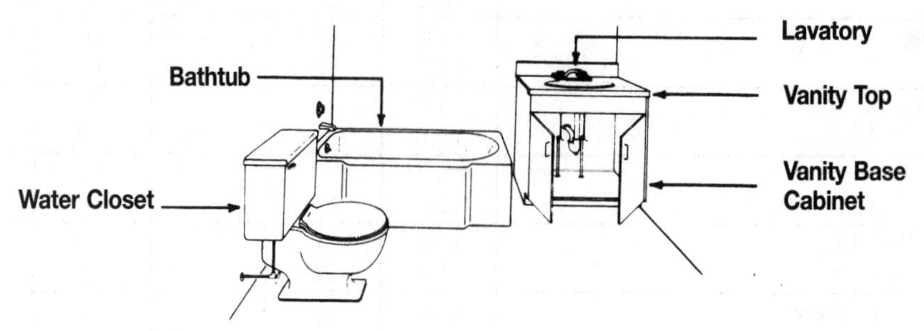

System Description	QUAN.	UNIT	LABOR HOURS	COST EACH		
				MAT.	INST.	TOTAL
BATHROOM WITH LAVATORY INSTALLED IN VANITY						
Water closet, floor mounted, 2 piece, close coupled, white	1.000	Ea.	3.019	179	120	299
Rough-in, vent, 2" diameter DWV piping	1.000	Ea.	.955	24	38	62
Waste, 4" diameter DWV piping	1.000	Ea.	.828	30	33	63
Supply, 1/2" diameter type "L" copper supply piping	1.000	Ea.	.593	9.90	26.28	36.18
Lavatory, 20" x 18", PE cast iron with accessories, white	1.000	Ea.	2.500	227	99.50	326.50
Rough-in, vent, 1-1/2" diameter DWV piping	1.000	Ea.	1.803	46.80	72	118.80
Supply, 1/2" diameter type "L" copper supply piping	1.000	Ea.	.988	16.50	43.80	60.30
Bathtub, P.E. cast iron, 5' long with accessories, white	1.000	Ea.	3.636	460	145	605
Rough-in, waste, 4" diameter DWV piping	1.000	Ea.	.828	30	33	63
Supply, 1/2" diameter type "L" copper supply piping	1.000	Ea.	.988	16.50	43.80	60.30
Vent, 1-1/2" diameter DWV piping	1.000	Ea.	.593	20	26.20	46.20
Piping, supply, 1/2" diameter type "L" copper supply piping	32.000	L.F.	3.161	52.80	140.16	192.96
Waste, 4" diameter DWV piping	12.000	L.F.	3.310	120	132	252
Vent, 2" diameter DWV piping	6.000	L.F.	1.500	32.10	59.70	91.80
Vanity base cabinet, 2 door, 30" wide	1.000	Ea.	1.000	223	41	264
Vanity top, plastic laminated square edge	2.670	L.F.	.712	64.08	28.97	93.05
TOTAL		Ea.	26.414	1,551.68	1,082.41	2,634.09
BATHROOM WITH WALL HUNG LAVATORY						
Water closet, floor mounted, 2 piece, close coupled, white	1.000	Ea.	3.019	179	120	299
Rough-in, vent, 2" diameter DWV piping	1.000	Ea.	.955	24	38	62
Waste, 4" diameter DWV piping	1.000	Ea.	.828	30	33	63
Supply, 1/2" diameter type "L" copper supply piping	1.000	Ea.	.593	9.90	26.28	36.18
Lavatory, 20" x 18" P.E. cast iron, wall hung, white	1.000	Ea.	2.000	261	79.50	340.50
Rough-in, waste, 1-1/2" diameter DWV piping	1.000	Ea.	1.803	46.80	72	118.80
Supply, 1/2" diameter type "L" copper supply piping	1.000	Ea.	.988	16.50	43.80	60.30
Bathtub, P.E. cast iron, 5' long with accessories, white	1.000	Ea.	3.636	460	145	605
Rough-in, waste, 4" diameter DWV piping	1.000	Ea.	.828	30	33	63
Supply, 1/2" diameter type "L" copper supply piping	1.000	Ea.	.988	16.50	43.80	60.30
Vent, 1-1/2" diameter DWV piping	1.000	Ea.	.593	20	26.20	46.20
Piping, supply, 1/2" diameter type "L" copper supply piping	32.000	L.F.	3.161	52.80	140.16	192.96
Waste, 4" diameter DWV piping	12.000	L.F.	3.310	120	132	252
Vent, 2" diameter DWV piping	6.000	L.F.	1.500	32.10	59.70	91.80
Carrier steel, for studs, no arms	1.000	Ea.	1.143	38.50	50.50	89
TOTAL		Ea.	25.345	1,337.10	1,042.94	2,380.04

The costs in this system are on a cost each basis. All necessary piping is included.

Three Fixture Bathroom Price Sheet	QUAN.	UNIT	LABOR HOURS	COST EACH		
				MAT.	INST.	TOTAL
ater closet, close coupled, standard 2 piece, white	1.000	Ea.	3.019	179	120	299
Color	1.000	Ea.	3.019	215	120	335
One piece, elongated bowl, white	1.000	Ea.	3.019	545	120	665
Color	1.000	Ea.	3.019	680	120	800
Low profile, one piece, elongated bowl, white	1.000	Ea.	3.019	745	120	865
Color	1.000	Ea.	3.019	970	120	1,090
ugh-in, for water closet						
1/2″ copper supply, 4″ cast iron waste, 2″ cast iron vent	1.000	Ea.	2.376	64	97.50	161.50
4″ PVC/DWV waste, 2″ PVC vent	1.000	Ea.	2.678	26	109	135
4″ copper waste, 2″ copper vent	1.000	Ea.	2.520	99.50	106	205.50
3″ cast iron waste, 1-1/2″ cast iron vent	1.000	Ea.	2.244	57	92	149
3″ PVC waste, 1-1/2″ PVC vent	1.000	Ea.	2.388	23	102	125
3″ copper waste, 1-1/2″ copper vent	1.000	Ea.	2.014	66	85.50	151.50
1/2″ PVC supply, 4″ PVC waste, 2″ PVC vent	1.000	Ea.	2.974	31.50	122	153.50
3″ PVC waste, 1-1/2″ PVC supply	1.000	Ea.	2.684	28.50	115	143.50
1/2″ steel supply, 4″ cast iron waste, 2″ cast iron vent	1.000	Ea.	2.545	64	105	169
4″ cast iron waste, 2″ steel vent	1.000	Ea.	2.590	61.50	107	168.50
4″ PVC waste, 2″ PVC vent	1.000	Ea.	2.847	26.50	117	143.50
avatory wall hung, P.E. cast iron, 20″ x 18″, white	1.000	Ea.	2.000	261	79.50	340.50
Color	1.000	Ea.	2.000	297	79.50	376.50
Vitreous china, 19″ x 17″, white	1.000	Ea.	2.286	204	91	295
Color	1.000	Ea.	2.286	204	91	295
Lavatory, for vanity top, P.E., cast iron, 20″ x 18″, white	1.000	Ea.	2.500	227	99.50	326.50
Color	1.000	Ea.	2.500	256	99.50	355.50
Steel, enameled, 20″ x 17″, white	1.000	Ea.	2.759	142	110	252
Color	1.000	Ea.	2.500	148	99.50	247.50
Vitreous china, 20″ x 16″, white	1.000	Ea.	2.963	244	118	362
Color	1.000	Ea.	2.963	244	118	362
ough-in, for lavatory						
1/2″ copper supply, 1-1/2″ C.I. waste, 1-1/2″ C.I. vent	1.000	Ea.	2.791	63.50	116	179.50
1-1/2″ PVC waste, 1-1/4″ PVC vent	1.000	Ea.	2.639	27	117	144
1/2″ steel supply, 1-1/4″ cast iron waste, 1-1/4″ steel vent	1.000	Ea.	2.890	54	121	175
1-1/4″ PVC waste, 1-1/4″ PVC vent	1.000	Ea.	2.794	27	124	151
1/2″ PVC supply, 1-1/2″ PVC waste, 1-1/2″ PVC vent	1.000	Ea.	3.260	36	144	180
athtub, P.E. cast iron, 5′ long corner with fittings, white	1.000	Ea.	3.636	460	145	605
Color	1.000	Ea.	3.636	500	145	645
ough-in, for bathtub						
1/2″ copper supply, 4″ cast iron waste, 1-1/2″ copper vent	1.000	Ea.	2.409	66.50	103	169.50
4″ PVC waste, 1/2″ PVC vent	1.000	Ea.	2.877	32	123	155
1/2″ steel supply, 4″ cast iron waste, 1-1/2″ steel vent	1.000	Ea.	2.898	63	121	184
4″ PVC waste, 1-1/2″ PVC vent	1.000	Ea.	3.159	32.50	136	168.50
1/2″ PVC supply, 4″ PVC waste, 1-1/2″ PVC vent	1.000	Ea.	3.371	41	145	186
iping, supply, 1/2″ copper	32.000	L.F.	3.161	53	140	193
1/2″ steel	32.000	L.F.	4.063	53.50	181	234.50
1/2″ PVC	32.000	L.F.	4.741	81.50	210	291.50
iping, waste, 4″ cast iron no hub	12.000	L.F.	3.310	120	132	252
4″ PVC/DWV	12.000	L.F.	4.000	41.50	160	201.50
4″ copper/DWV	12.000	L.F.	4.800	252	191	443
Piping, vent, 2″ cast iron no hub	6.000	L.F.	1.433	36	57	93
2″ copper/DWV	6.000	L.F.	1.091	40	48.50	88.50
2″ PVC/DWV	6.000	L.F.	1.627	8.95	65	73.95
2″ steel, galvanized	6.000	L.F.	1.500	32	59.50	91.50
Vanity base cabinet, 2 door, 24″ x 30″	1.000	Ea.	1.000	223	41	264
24″ x 36″	1.000	Ea.	1.200	298	49	347
Vanity top, laminated plastic square edge, 25″ x 32″	2.670	L.F.	.712	64	29	93
25″ x 38″	3.160	L.F.	.843	76	34.50	110.50
Cultured marble, 25″ x 32″, with bowl	1.000	Ea.	2.500	169	99.50	268.50
25″ x 38″, with bowl	1.000	Ea.	2.500	196	99.50	295.50
Carrier, for lavatory, steel for studs, no arms	1.000	Ea.	1.143	38.50	50.50	89
Wood, 2″ x 8″ blocking	1.300	L.F.	.052	1.18	2.12	3.30

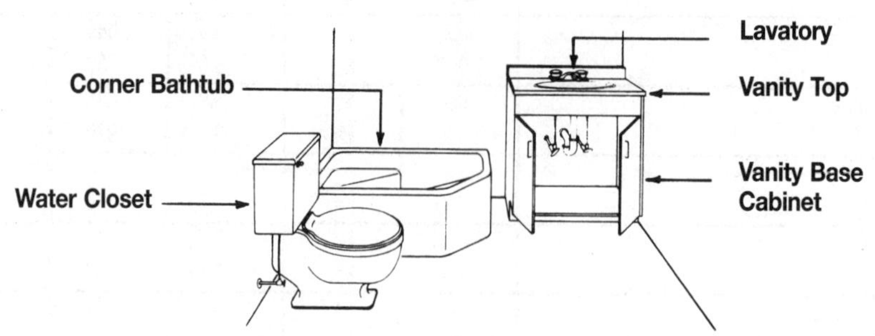

System Description	QUAN.	UNIT	LABOR HOURS	COST EACH MAT.	COST EACH INST.	COST EACH TOTAL
BATHROOM WITH LAVATORY INSTALLED IN VANITY						
Water closet, floor mounted, 2 piece, close coupled, white	1.000	Ea.	3.019	179	120	299
Rough-in, vent, 2" diameter DWV piping	1.000	Ea.	.955	24	38	62
Waste, 4" diameter DWV piping	1.000	Ea.	.828	30	33	63
Supply, 1/2" diameter type "L" copper supply piping	1.000	Ea.	.593	9.90	26.28	36.18
Lavatory, 20" x 18", P.E. cast iron with fittings, white	1.000	Ea.	2.500	227	99.50	326.50
Rough-in, waste, 1-1/2" diameter DWV piping	1.000	Ea.	1.803	46.80	72	118.80
Supply, 1/2" diameter type "L" copper supply piping	1.000	Ea.	.988	16.50	43.80	60.30
Bathtub, P.E. cast iron, corner with fittings, white	1.000	Ea.	3.636	1,650	145	1,795
Rough-in, waste, 4" diameter DWV piping	1.000	Ea.	.828	30	33	63
Supply, 1/2" diameter type "L" copper supply piping	1.000	Ea.	.988	16.50	43.80	60.30
Vent, 1-1/2" diameter DWV piping	1.000	Ea.	.593	20	26.20	46.20
Piping, supply, 1/2" diameter type "L" copper supply piping	32.000	L.F.	3.161	52.80	140.16	192.96
Waste, 4" diameter DWV piping	12.000	L.F.	3.310	120	132	252
Vent, 2" diameter DWV piping	6.000	L.F.	1.500	32.10	59.70	91.80
Vanity base cabinet, 2 door, 30" wide	1.000	Ea.	1.000	223	41	264
Vanity top, plastic laminated, square edge	2.670	L.F.	.712	82.77	28.97	111.74
TOTAL		Ea.	26.414	2,760.37	1,082.41	3,842.78
BATHROOM WITH WALL HUNG LAVATORY						
Water closet, floor mounted, 2 piece, close coupled, white	1.000	Ea.	3.019	179	120	299
Rough-in, vent, 2" diameter DWV piping	1.000	Ea.	.955	24	38	62
Waste, 4" diameter DWV piping	1.000	Ea.	.828	30	33	63
Supply, 1/2" diameter type "L" copper supply piping	1.000	Ea.	.593	9.90	26.28	36.18
Lavatory, 20" x 18", P.E. cast iron, with fittings, white	1.000	Ea.	2.000	261	79.50	340.50
Rough-in, waste, 1-1/2" diameter DWV piping	1.000	Ea.	1.803	46.80	72	118.80
Supply, 1/2" diameter type "L" copper supply piping	1.000	Ea.	.988	16.50	43.80	60.30
Bathtub, P.E. cast iron, corner, with fittings, white	1.000	Ea.	3.636	1,650	145	1,795
Rough-in, waste, 4" diameter DWV piping	1.000	Ea.	.828	30	33	63
Supply, 1/2" diameter type "L" copper supply piping	1.000	Ea.	.988	16.50	43.80	60.30
Vent, 1-1/2" diameter DWV piping	1.000	Ea.	.593	20	26.20	46.20
Piping, supply, 1/2" diameter type "L" copper supply piping	32.000	L.F.	3.161	52.80	140.16	192.96
Waste, 4" diameter DWV piping	12.000	L.F.	3.310	120	132	252
Vent, 2" diameter DWV piping	6.000	L.F.	1.500	32.10	59.70	91.80
Carrier, steel, for studs, no arms	1.000	Ea.	1.143	38.50	50.50	89
TOTAL		Ea.	25.345	2,527.10	1,042.94	3,570.04

The costs in this system are on a cost each basis. All necessary piping is included.

Three Fixture Bathroom Price Sheet	QUAN.	UNIT	LABOR HOURS	COST EACH		
				MAT.	INST.	TOTAL
ater closet, close coupled, standard 2 piece, white	1.000	Ea.	3.019	179	120	299
Color	1.000	Ea.	3.019	215	120	335
One piece elongated bowl, white	1.000	Ea.	3.019	545	120	665
Color	1.000	Ea.	3.019	680	120	800
Low profile, one piece elongated bowl, white	1.000	Ea.	3.019	745	120	865
Color	1.000	Ea.	3.019	970	120	1,090
ough-in, for water closet						
1/2" copper supply, 4" cast iron waste, 2" cast iron vent	1.000	Ea.	2.376	64	97.50	161.50
4" PVC/DWV waste, 2" PVC vent	1.000	Ea.	2.678	26	109	135
4" copper waste, 2" copper vent	1.000	Ea.	2.520	99.50	106	205.50
3" cast iron waste, 1-1/2" cast iron vent	1.000	Ea.	2.244	57	92	149
3" PVC waste, 1-1/2" PVC vent	1.000	Ea.	2.388	23	102	125
3" copper waste, 1-1/2" copper vent	1.000	Ea.	2.014	66	85.50	151.50
1/2" PVC supply, 4" PVC waste, 2" PVC vent	1.000	Ea.	2.974	31.50	122	153.50
3" PVC waste, 1-1/2" PVC supply	1.000	Ea.	2.684	28.50	115	143.50
1/2" steel supply, 4" cast iron waste, 2" cast iron vent	1.000	Ea.	2.545	64	105	169
4" cast iron waste, 2" steel vent	1.000	Ea.	2.590	61.50	107	168.50
4" PVC waste, 2" PVC vent	1.000	Ea.	2.847	26.50	117	143.50
avatory, wall hung P.E. cast iron 20" x 18", white	1.000	Ea.	2.000	261	79.50	340.50
Color	1.000	Ea.	2.000	297	79.50	376.50
Vitreous china 19" x 17", white	1.000	Ea.	2.286	204	91	295
Color	1.000	Ea.	2.286	204	91	295
Lavatory, for vanity top, P.E., cast iron, 20" x 18", white	1.000	Ea.	2.500	227	99.50	326.50
Color	1.000	Ea.	2.500	256	99.50	355.50
Steel enameled 20" x 17", white	1.000	Ea.	2.759	142	110	252
Color	1.000	Ea.	2.500	148	99.50	247.50
Vitreous china 20" x 16", white	1.000	Ea.	2.963	244	118	362
Color	1.000	Ea.	2.963	244	118	362
ough-in, for lavatory						
1/2" copper supply, 1-1/2" cast iron waste, 1-1/2" cast iron vent	1.000	Ea.	2.791	63.50	116	179.50
1-1/2" PVC waste, 1-1/4" PVC vent	1.000	Ea.	2.639	27	117	144
1/2" steel supply, 1-1/4" cast iron waste, 1-1/4" steel vent	1.000	Ea.	2.890	54	121	175
1-1/4" PVC waste, 1-1/4" PVC vent	1.000	Ea.	2.794	27	124	151
1/2" PVC supply, 1-1/2" PVC waste, 1-1/2" PVC vent	1.000	Ea.	3.260	36	144	180
Bathtub, P.E. cast iron, corner with fittings, white	1.000	Ea.	3.636	1,650	145	1,795
Color	1.000	Ea.	4.000	1,850	159	2,009
Rough-in, for bathtub						
1/2" copper supply, 4" cast iron waste, 1-1/2" copper vent	1.000	Ea.	2.409	66.50	103	169.50
4" PVC waste, 1-1/2" PVC vent	1.000	Ea.	2.877	32	123	155
1/2" steel supply, 4" cast iron waste, 1-1/2" steel vent	1.000	Ea.	2.898	63	121	184
4" PVC waste, 1-1/2" PVC vent	1.000	Ea.	3.159	32.50	136	168.50
1/2" PVC supply, 4" PVC waste, 1-1/2" PVC vent	1.000	Ea.	3.371	41	145	186
Piping, supply, 1/2" copper	32.000	L.F.	3.161	53	140	193
1/2" steel	32.000	L.F.	4.063	53.50	181	234.50
1/2" PVC	32.000	L.F.	4.741	81.50	210	291.50
Piping, waste, 4" cast iron, no hub	12.000	L.F.	3.310	120	132	252
4" PVC/DWV	12.000	L.F.	4.000	41.50	160	201.50
4" copper/DWV	12.000	L.F.	4.800	252	191	443
Piping, vent 2" cast iron, no hub	6.000	L.F.	1.433	36	57	93
2" copper/DWV	6.000	L.F.	1.091	40	48.50	88.50
2" PVC/DWV	6.000	L.F.	1.627	8.95	65	73.95
2" steel, galvanized	6.000	L.F.	1.500	32	59.50	91.50
Vanity base cabinet, 2 door, 24" x 30"	1.000	Ea.	1.000	223	41	264
24" x 36"	1.000	Ea.	1.200	298	49	347
Vanity top, laminated plastic square edge 25" x 32"	2.670	L.F.	.712	83	29	112
25" x 38"	3.160	L.F.	.843	98	34.50	132.50
Cultured marble, 25" x 32", with bowl	1.000	Ea.	2.500	169	99.50	268.50
25" x 38", with bowl	1.000	Ea.	2.500	196	99.50	295.50
Carrier, for lavatory, steel for studs, no arms	1.000	Ea.	1.143	38.50	50.50	89
Wood, 2" x 8" blocking	1.300	L.F.	.053	1.21	2.17	3.38

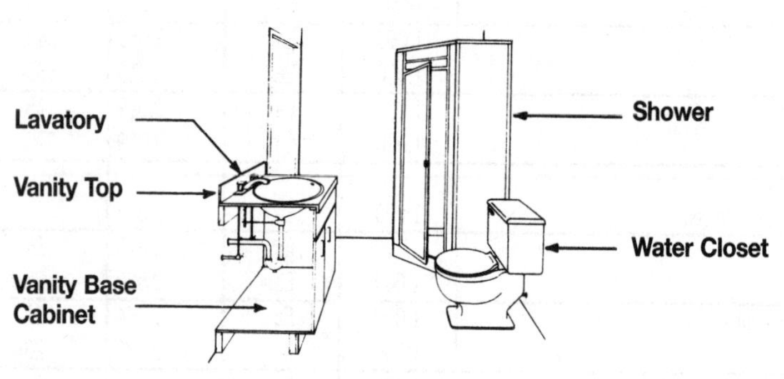

System Description	QUAN.	UNIT	LABOR HOURS	COST EACH		
				MAT.	INST.	TOTAL
BATHROOM WITH SHOWER, LAVATORY INSTALLED IN VANITY						
Water closet, floor mounted, 2 piece, close coupled, white	1.000	Ea.	3.019	179	120	299
Rough-in, vent, 2" diameter DWV piping	1.000	Ea.	.955	24	38	62
Waste, 4" diameter DWV piping	1.000	Ea.	.828	30	33	63
Supply, 1/2" diameter type "L" copper supply piping	1.000	Ea.	.593	9.90	26.28	36.18
Lavatory, 20" x 18" P.E. cast iron with fittings, white	1.000	Ea.	2.500	227	99.50	326.50
Rough-in, waste, 1-1/2" diameter DWV piping	1.000	Ea.	1.803	46.80	72	118.80
Supply, 1/2" diameter type "L" copper supply piping	1.000	Ea.	.988	16.50	43.80	60.30
Shower, steel enameled, stone base, corner, white	1.000	Ea.	8.000	360	320	680
Rough-in, vent, 1-1/2" diameter DWV piping	1.000	Ea.	.225	5.85	9	14.85
Waste, 2" diameter DWV piping	1.000	Ea.	1.433	36	57	93
Supply, 1/2" diameter type "L" copper supply piping	1.000	Ea.	1.580	26.40	70.08	96.48
Piping, supply, 1/2" diameter type "L" copper supply piping	36.000	L.F.	4.148	69.30	183.96	253.26
Waste, 4" diameter DWV piping	7.000	L.F.	2.759	100	110	210
Vent, 2" diameter DWV piping	6.000	L.F.	2.250	48.15	89.55	137.70
Vanity base 2 door, 30" wide	1.000	Ea.	1.000	223	41	264
Vanity top, plastic laminated, square edge	2.170	L.F.	.712	68.09	28.97	97.06
TOTAL		Ea.	32.793	1,469.99	1,342.14	2,812.13
BATHROOM WITH SHOWER, WALL HUNG LAVATORY						
Water closet, floor mounted, close coupled	1.000	Ea.	3.019	179	120	299
Rough-in, vent, 2" diameter DWV piping	1.000	Ea.	.955	24	38	62
Waste, 4" diameter DWV piping	1.000	Ea.	.828	30	33	63
Supply, 1/2" diameter type "L" copper supply piping	1.000	Ea.	.593	9.90	26.28	36.18
Lavatory, 20" x 18" P.E. cast iron with fittings, white	1.000	Ea.	2.000	261	79.50	340.50
Rough-in, waste, 1-1/2" diameter DWV piping	1.000	Ea.	1.803	46.80	72	118.80
Supply, 1/2" diameter type "L" copper supply piping	1.000	Ea.	.988	16.50	43.80	60.30
Shower, steel enameled, stone base, white	1.000	Ea.	8.000	360	320	680
Rough-in, vent, 1-1/2" diameter DWV piping	1.000	Ea.	.225	5.85	9	14.85
Waste, 2" diameter DWV piping	1.000	Ea.	1.433	36	57	93
Supply, 1/2" diameter type "L" copper supply piping	1.000	Ea.	1.580	26.40	70.08	96.48
Piping, supply, 1/2" diameter type "L" copper supply piping	36.000	L.F.	4.148	69.30	183.96	253.26
Waste, 4" diameter DWV piping	7.000	L.F.	2.759	100	110	210
Vent, 2" diameter DWV piping	6.000	L.F.	2.250	48.15	89.55	137.70
Carrier, steel, for studs, no arms	1.000	Ea.	1.143	38.50	50.50	89
TOTAL		Ea.	31.724	1,251.40	1,302.67	2,554.07

The costs in this system are on a cost each basis. All necessary piping is included.

Important: See the Reference Section for critical supporting data - Reference Nos., Crews & Location Factors

Three Fixture Bathroom Price Sheet	QUAN.	UNIT	LABOR HOURS	COST EACH		
				MAT.	INST.	TOTAL
ater closet, close coupled, standard 2 piece, white	1.000	Ea.	3.019	179	120	299
Color	1.000	Ea.	3.019	215	120	335
One piece elongated bowl, white	1.000	Ea.	3.019	545	120	665
Color	1.000	Ea.	3.019	680	120	800
Low profile, one piece elongated bowl, white	1.000	Ea.	3.019	745	120	865
Color	1.000	Ea.	3.019	970	120	1,090
ough-in, for water closet						
1/2" copper supply, 4" cast iron waste, 2" cast iron vent	1.000	Ea.	2.376	64	97.50	161.50
4" PVC/DWV waste, 2" PVC vent	1.000	Ea.	2.678	26	109	135
4" copper waste, 2" copper vent	1.000	Ea.	2.520	99.50	106	205.50
3" cast iron waste, 1-1/2" cast iron vent	1.000	Ea.	2.244	57	92	149
3" PVC waste, 1-1/2" PVC vent	1.000	Ea.	2.388	23	102	125
3" copper waste, 1-1/2" copper vent	1.000	Ea.	2.014	66	85.50	151.50
1/2" PVC supply, 4" PVC waste, 2" PVC vent	1.000	Ea.	2.974	31.50	122	153.50
3" PVC waste, 1-1/2" PVC supply	1.000	Ea.	2.684	28.50	115	143.50
1/2" steel supply, 4" cast iron waste, 2" cast iron vent	1.000	Ea.	2.545	64	105	169
4" cast iron waste, 2" steel vent	1.000	Ea.	2.590	61.50	107	168.50
4" PVC waste, 2" PVC vent	1.000	Ea.	2.847	26.50	117	143.50
avatory, wall hung, P.E. cast iron 20" x 18", white	1.000	Ea.	2.000	261	79.50	340.50
Color	1.000	Ea.	2.000	297	79.50	376.50
Vitreous china 19" x 17", white	1.000	Ea.	2.286	204	91	295
Color	1.000	Ea.	2.286	204	91	295
Lavatory, for vanity top, P.E. cast iron 20" x 18", white	1.000	Ea.	2.500	227	99.50	326.50
Color	1.000	Ea.	2.500	256	99.50	355.50
Steel enameled 20" x 17", white	1.000	Ea.	2.759	142	110	252
Color	1.000	Ea.	2.500	148	99.50	247.50
Vitreous china 20" x 16", white	1.000	Ea.	2.963	244	118	362
Color	1.000	Ea.	2.963	244	118	362
ough-in, for lavatory						
1/2" copper supply, 1-1/2" cast iron waste, 1-1/2" cast iron vent	1.000	Ea.	2.791	63.50	116	179.50
1-1/2" PVC waste, 1-1/2" PVC vent	1.000	Ea.	2.639	27	117	144
1/2" steel supply, 1-1/4" cast iron waste, 1-1/4" steel vent	1.000	Ea.	2.890	54	121	175
1-1/4" PVC waste, 1-1/4" PVC vent	1.000	Ea.	2.921	27	130	157
1/2" PVC supply, 1-1/2" PVC waste, 1-1/2" PVC vent	1.000	Ea.	3.260	36	144	180
Shower, steel enameled stone base, 32" x 32", white	1.000	Ea.	8.000	360	320	680
Color	1.000	Ea.	7.822	845	310	1,155
36" x 36" white	1.000	Ea.	8.889	960	355	1,315
Color	1.000	Ea.	8.889	960	355	1,315
ough-in, for shower						
1/2" copper supply, 4" cast iron waste, 1-1/2" copper vent	1.000	Ea.	3.238	68.50	136	204.50
4" PVC waste, 1-1/2" PVC vent	1.000	Ea.	3.429	36.50	145	181.50
1/2" steel supply, 4" cast iron waste, 1-1/2" steel vent	1.000	Ea.	3.665	67	155	222
4" PVC waste, 1-1/2" PVC vent	1.000	Ea.	3.881	37	165	202
1/2" PVC supply, 4" PVC waste, 1-1/2" PVC vent	1.000	Ea.	4.219	51	179	230
Piping, supply, 1/2" copper	36.000	L.F.	4.148	69.50	184	253.50
1/2" steel	36.000	L.F.	5.333	70	237	307
1/2" PVC	36.000	L.F.	6.222	107	275	382
Piping, waste, 4" cast iron no hub	7.000	L.F.	2.759	100	110	210
4" PVC/DWV	7.000	L.F.	3.333	34.50	133	167.50
4" copper/DWV	7.000	L.F.	4.000	210	160	370
Piping, vent, 2" cast iron no hub	6.000	L.F.	2.149	54	85.50	139.50
2" copper/DWV	6.000	L.F.	1.636	60.50	72.50	133
2" PVC/DWV	6.000	L.F.	2.441	13.40	97	110.40
2" steel, galvanized	6.000	L.F.	2.250	48	89.50	137.50
Vanity base cabinet, 2 door, 24" x 30"	1.000	Ea.	1.000	223	41	264
24" x 36"	1.000	Ea.	1.200	298	49	347
Vanity top, laminated plastic square edge, 25" x 32"	2.170	L.F.	.712	68	29	97
25" x 38"	2.670	L.F.	.845	81	34.50	115.50
Carrier, for lavatory, steel for studs, no arms	1.000	Ea.	1.143	38.50	50.50	89
Wood, 2" x 8" blocking	1.300	L.F.	.052	1.18	2.12	3.30

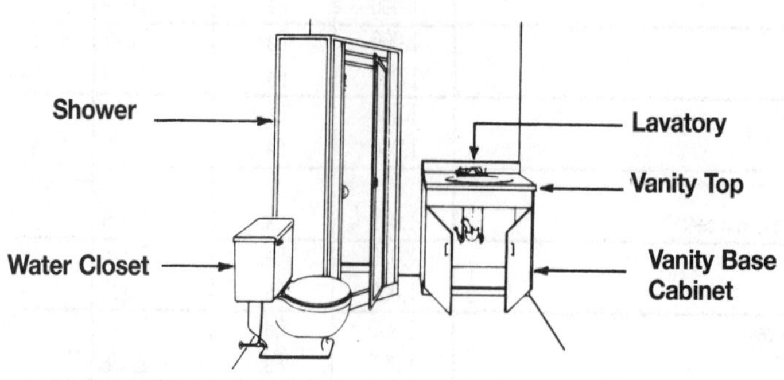

System Description	QUAN.	UNIT	LABOR HOURS	COST EACH		
				MAT.	INST.	TOTAL
BATHROOM WITH LAVATORY INSTALLED IN VANITY						
Water closet, floor mounted, 2 piece, close coupled, white	1.000	Ea.	3.019	179	120	299
Rough-in, vent, 2" diameter DWV piping	1.000	Ea.	.955	24	38	62
Waste, 4" diameter DWV piping	1.000	Ea.	.828	30	33	63
Supply, 1/2" diameter type "L" copper supply piping	1.000	Ea.	.593	9.90	26.28	36.18
Lavatory, 20" x 18", P.E. cast iron with fittings, white	1.000	Ea.	2.500	227	99.50	326.50
Rough-in, waste, 1-1/2" diameter DWV piping	1.000	Ea.	1.803	46.80	72	118.80
Supply, 1/2" diameter type "L" copper supply piping	1.000	Ea.	.988	16.50	43.80	60.30
Shower, steel enameled, stone base, corner, white	1.000	Ea.	8.000	360	320	680
Rough-in, vent, 1-1/2" diameter DWV piping	1.000	Ea.	.225	5.85	9	14.85
Waste, 2" diameter DWV piping	1.000	Ea.	1.433	36	57	93
Supply, 1/2" diameter type "L" copper supply piping	1.000	Ea.	1.580	26.40	70.08	96.48
Piping, supply, 1/2" diameter type "L" copper supply piping	36.000	L.F.	3.556	59.40	157.68	217.08
Waste, 4" diameter DWV piping	7.000	L.F.	1.931	70	77	147
Vent, 2" diameter DWV piping	6.000	L.F.	1.500	32.10	59.70	91.80
Vanity base, 2 door, 30" wide	1.000	Ea.	1.000	223	41	264
Vanity top, plastic laminated, square edge	2.670	L.F.	.712	64.08	28.97	93.05
TOTAL		Ea.	30.623	1,410.03	1,253.01	2,663.04
BATHROOM, WITH WALL HUNG LAVATORY						
Water closet, floor mounted, 2 piece, close coupled, white	1.000	Ea.	3.019	179	120	299
Rough-in, vent, 2" diameter DWV piping	1.000	Ea.	.955	24	38	62
Waste, 4" diameter DWV piping	1.000	Ea.	.828	30	33	63
Supply, 1/2" diameter type "L" copper supply piping	1.000	Ea.	.593	9.90	26.28	36.18
Lavatory, wall hung, 20" x 18" P.E. cast iron with fittings, white	1.000	Ea.	2.000	261	79.50	340.50
Rough-in, waste, 1-1/2" diameter DWV piping	1.000	Ea.	1.803	46.80	72	118.80
Supply, 1/2" diameter type "L" copper supply piping	1.000	Ea.	.988	16.50	43.80	60.30
Shower, steel enameled, stone base, corner, white	1.000	Ea.	8.000	360	320	680
Rough-in, waste, 1-1/2" diameter DWV piping	1.000	Ea.	.225	5.85	9	14.85
Waste, 2" diameter DWV piping	1.000	Ea.	1.433	36	57	93
Supply, 1/2" diameter type "L" copper supply piping	1.000	Ea.	1.580	26.40	70.08	96.48
Piping, supply, 1/2" diameter type "L" copper supply piping	36.000	L.F.	3.556	59.40	157.68	217.08
Waste, 4" diameter DWV piping	7.000	L.F.	1.931	70	77	147
Vent, 2" diameter DWV piping	6.000	L.F.	1.500	32.10	59.70	91.80
Carrier, steel, for studs, no arms	1.000	Ea.	1.143	38.50	50.50	89
TOTAL		Ea.	29.554	1,195.45	1,213.54	2,408.99

The costs in this system are on a cost each basis. All necessary piping is included.

Important: See the Reference Section for critical supporting data - Reference Nos., Crews & Location Factors

Three Fixture Bathroom Price Sheet	QUAN.	UNIT	LABOR HOURS	COST EACH		
				MAT.	INST.	TOTAL
ater closet, close coupled, standard 2 piece, white	1.000	Ea.	3.019	179	120	299
Color	1.000	Ea.	3.019	215	120	335
One piece elongated bowl, white	1.000	Ea.	3.019	545	120	665
Color	1.000	Ea.	3.019	680	120	800
Low profile one piece elongated bowl, white	1.000	Ea.	3.019	745	120	865
Color	1.000	Ea.	3.623	970	120	1,090
ugh-in, for water closet						
1/2" copper supply, 4" cast iron waste, 2" cast iron vent	1.000	Ea.	2.376	64	97.50	161.50
4" P.V.C./DWV waste, 2" PVC vent	1.000	Ea.	2.678	26	109	135
4" copper waste, 2" copper vent	1.000	Ea.	2.520	99.50	106	205.50
3" cast iron waste, 1-1/2" cast iron vent	1.000	Ea.	2.244	57	92	149
3" PVC waste, 1-1/2" PVC vent	1.000	Ea.	2.388	23	102	125
3" copper waste, 1-1/2" copper vent	1.000	Ea.	2.014	66	85.50	151.50
1/2" P.V.C. supply, 4" P.V.C. waste, 2" P.V.C. vent	1.000	Ea.	2.974	31.50	122	153.50
3" P.V.C. waste, 1-1/2" P.V.C. vent	1.000	Ea.	2.684	28.50	115	143.50
1/2" steel supply, 4" cast iron waste, 2" cast iron vent	1.000	Ea.	2.545	64	105	169
4" cast iron waste, 2" steel vent	1.000	Ea.	2.590	61.50	107	168.50
4" P.V.C. waste, 2" P.V.C. vent	1.000	Ea.	2.847	26.50	117	143.50
avatory, wall hung P.E. cast iron 20" x 18", white	1.000	Ea.	2.000	261	79.50	340.50
Color	1.000	Ea.	2.000	297	79.50	376.50
Vitreous china 19" x 17", white	1.000	Ea.	2.286	204	91	295
Color	1.000	Ea.	2.286	204	91	295
Lavatory, for vanity top P.E. cast iron 20" x 18", white	1.000	Ea.	2.500	227	99.50	326.50
Color	1.000	Ea.	2.500	256	99.50	355.50
Steel enameled 20" x 17", white	1.000	Ea.	2.759	142	110	252
Color	1.000	Ea.	2.500	148	99.50	247.50
Vitreous china 20" x 16", white	1.000	Ea.	2.963	244	118	362
Color	1.000	Ea.	2.963	244	118	362
ough-in, for lavatory						
1/2" copper supply, 1-1/2" cast iron waste, 1-1/2" cast iron vent	1.000	Ea.	2.791	63.50	116	179.50
1-1/2" P.V.C. waste, 1-1/2" P.V.C. vent	1.000	Ea.	2.639	27	117	144
1/2" steel supply, 1-1/2" cast iron waste, 1-1/4" steel vent	1.000	Ea.	2.890	54	121	175
1-1/2" P.V.C. waste, 1-1/4" P.V.C. vent	1.000	Ea.	2.921	27	130	157
1/2" P.V.C. supply, 1-1/2" P.V.C. waste, 1-1/2" P.V.C. vent	1.000	Ea.	3.260	36	144	180
hower, steel enameled stone base, 32" x 32", white	1.000	Ea.	8.000	360	320	680
Color	1.000	Ea.	7.822	845	310	1,155
36" x 36", white	1.000	Ea.	8.889	960	355	1,315
Color	1.000	Ea.	8.889	960	355	1,315
lough-in, for shower						
1/2" copper supply, 2" cast iron waste, 1-1/2" copper vent	1.000	Ea.	3.161	67.50	134	201.50
2" P.V.C. waste, 1-1/2" P.V.C. vent	1.000	Ea.	3.429	36.50	145	181.50
1/2" steel supply, 2" cast iron waste, 1-1/2" steel vent	1.000	Ea.	3.887	91	164	255
2" P.V.C. waste, 1-1/2" P.V.C. vent	1.000	Ea.	3.881	37	165	202
1/2" P.V.C. supply, 2" P.V.C. waste, 1-1/2" P.V.C. vent	1.000	Ea.	4.219	51	179	230
iping, supply, 1/2" copper	36.000	L.F.	3.556	59.50	158	217.50
1/2" steel	36.000	L.F.	4.571	60	203	263
1/2" P.V.C.	36.000	L.F.	5.333	91.50	236	327.50
Waste, 4" cast iron, no hub	7.000	L.F.	1.931	70	77	147
4" P.V.C./DWV	7.000	L.F.	2.333	24	93	117
4" copper/DWV	7.000	L.F.	2.800	147	112	259
Vent, 2" cast iron, no hub	6.000	L.F.	1.091	40	48.50	88.50
2" copper/DWV	6.000	L.F.	1.091	40	48.50	88.50
2" P.V.C./DWV	6.000	L.F.	1.627	8.95	65	73.95
2" steel, galvanized	6.000	L.F.	1.500	32	59.50	91.50
Vanity base cabinet, 2 door, 24" x 30"	1.000	Ea.	1.000	223	41	264
24" x 36"	1.000	Ea.	1.200	298	49	347
Vanity top, laminated plastic square edge, 25" x 32"	2.670	L.F.	.712	64	29	93
25" x 38"	3.170	L.F.	.845	76	34.50	110.50
Carrier , for lavatory, steel, for studs, no arms	1.000	Ea.	1.143	38.50	50.50	89
Wood, 2" x 8" blocking	1.300	L.F.	.052	1.18	2.12	3.30

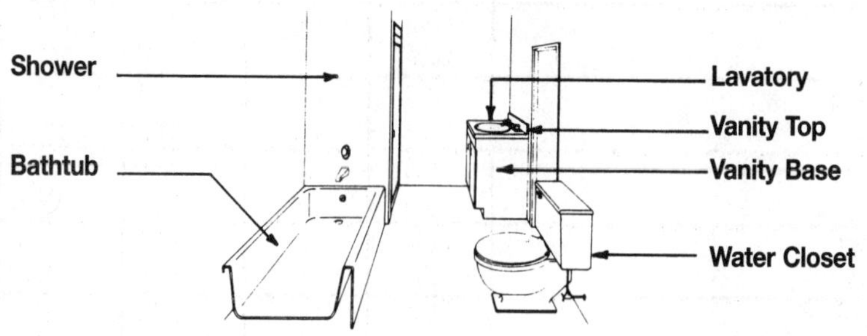

System Description	QUAN.	UNIT	LABOR HOURS	COST EACH MAT.	INST.	TOTAL
BATHROOM WITH LAVATORY INSTALLED IN VANITY						
Water closet, floor mounted, 2 piece, close coupled, white	1.000	Ea.	3.019	179	120	299
Rough-in, vent, 2″ diameter DWV piping	1.000	Ea.	.955	24	38	62
Waste, 4″ diameter DWV piping	1.000	Ea.	.828	30	33	63
Supply, 1/2″ diameter type "L" copper supply piping	1.000	Ea.	.593	9.90	26.28	36.18
Lavatory, 20″ x 18″ P.E. cast iron with fittings, white	1.000	Ea.	2.500	227	99.50	326.50
Shower, steel, enameled, stone base, corner, white	1.000	Ea.	8.889	880	355	1,235
Rough-in, waste, 1-1/2″ diameter DWV piping	2.000	Ea.	4.507	117	180	297
Supply, 1/2″ diameter type "L" copper supply piping	2.000	Ea.	3.161	52.80	140.16	192.96
Bathtub, P.E. cast iron, 5′ long with fittings, white	1.000	Ea.	3.636	460	145	605
Rough-in, waste, 4″ diameter DWV piping	1.000	Ea.	.828	30	33	63
Supply, 1/2″ diameter type "L" copper supply piping	1.000	Ea.	.988	16.50	43.80	60.30
Vent, 1-1/2″ diameter DWV piping	1.000	Ea.	.593	20	26.20	46.20
Piping, supply, 1/2″ diameter type "L" copper supply piping	42.000	L.F.	4.148	69.30	183.96	253.26
Waste, 4″ diameter DWV piping	10.000	L.F.	2.759	100	110	210
Vent, 2″ diameter DWV piping	13.000	L.F.	3.250	69.55	129.35	198.90
Vanity base, 2 doors, 30″ wide	1.000	Ea.	1.000	223	41	264
Vanity top, plastic laminated, square edge	2.670	L.F.	.712	64.08	28.97	93.05
TOTAL		Ea.	42.366	2,572.13	1,733.22	4,305.35
BATHROOM WITH WALL HUNG LAVATORY						
Water closet, floor mounted, 2 piece, close coupled, white	1.000	Ea.	3.019	179	120	299
Rough-in, vent, 2″ diameter DWV piping	1.000	Ea.	.955	24	38	62
Waste, 4″ diameter DWV piping	1.000	Ea.	.828	30	33	63
Supply, 1/2″ diameter type "L" copper supply piping	1.000	Ea.	.593	9.90	26.28	36.18
Lavatory, 20″ x 18″ P.E. cast iron with fittings, white	1.000	Ea.	2.000	261	79.50	340.50
Shower, steel enameled, stone base, corner , white	1.000	Ea.	8.889	880	355	1,235
Rough-in, waste, 1-1/2″ diameter DWV piping	2.000	Ea.	4.507	117	180	297
Supply, 1/2″ diameter type "L" copper supply piping	2.000	Ea.	3.161	52.80	140.16	192.96
Bathtub, P.E. cast iron, 5′ long with fittings, white	1.000	Ea.	3.636	460	145	605
Rough-in, waste, 4″ diameter DWV piping	1.000	Ea.	.828	30	33	63
Supply, 1/2″ diameter type "L" copper supply piping	1.000	Ea.	.988	16.50	43.80	60.30
Vent, 1-1/2″ diameter copper DWV piping	1.000	Ea.	.593	20	26.20	46.20
Piping, supply, 1/2″ diameter type "L" copper supply piping	42.000	L.F.	4.148	69.30	183.96	253.26
Waste, 4″ diameter DWV piping	10.000	L.F.	2.759	100	110	210
Vent, 2″ diameter DWV piping	13.000	L.F.	3.250	69.55	129.35	198.90
Carrier, steel, for studs, no arms	1.000	Ea.	1.143	38.50	50.50	89
TOTAL		Ea.	41.297	2,357.55	1,693.75	4,051.30

The costs in this system are on a cost each basis. All necessary piping is included.

Important: See the Reference Section for critical supporting data - Reference Nos., Crews & Location Factors

Four Fixture Bathroom Price Sheet	QUAN.	UNIT	LABOR HOURS	COST EACH		
				MAT.	INST.	TOTAL
ater closet, close coupled, standard 2 piece, white	1.000	Ea.	3.019	179	120	299
Color	1.000	Ea.	3.019	215	120	335
One piece elongated bowl, white	1.000	Ea.	3.019	545	120	665
Color	1.000	Ea.	3.019	680	120	800
Low profile, one piece elongated bowl, white	1.000	Ea.	3.019	745	120	865
Color	1.000	Ea.	3.019	970	120	1,090
1/2″ copper supply, 4″ cast iron waste, 2″ cast iron vent	1.000	Ea.	2.376	64	97.50	161.50
4″ PVC/DWV waste, 2″ PVC vent	1.000	Ea.	2.678	26	109	135
4″ copper waste, 2″ copper vent	1.000	Ea.	2.520	99.50	106	205.50
3″ cast iron waste, 1-1/2″ cast iron vent	1.000	Ea.	2.244	57	92	149
3″ P.V.C. waste, 1-1/2″ P.V.C. vent	1.000	Ea.	2.388	23	102	125
3″ copper waste, 1-1/2″ copper vent	1.000	Ea.	2.014	66	85.50	151.50
1/2″ P.V.C. supply, 4″ P.V.C. waste, 2″ P.V.C. vent	1.000	Ea.	2.974	31.50	122	153.50
3″ P.V.C. waste, 1-1/2″ P.V.C. vent	1.000	Ea.	2.684	28.50	115	143.50
1/2″ steel supply, 4″ cast iron waste, 2″ cast iron vent	1.000	Ea.	2.545	64	105	169
4″ cast iron waste, 2″ steel vent	1.000	Ea.	2.590	61.50	107	168.50
4″ P.V.C. waste, 2″ P.V.C. vent	1.000	Ea.	2.847	26.50	117	143.50
avatory, wall hung P.E. cast iron 20″ x 18″, white	1.000	Ea.	2.000	261	79.50	340.50
Color	1.000	Ea.	2.000	297	79.50	376.50
Vitreous china 19″ x 17″, white	1.000	Ea.	2.286	204	91	295
Color	1.000	Ea.	2.286	204	91	295
Lavatory for vanity top, P.E. cast iron 20″ x 18″, white	1.000	Ea.	2.500	227	99.50	326.50
Color	1.000	Ea.	2.500	256	99.50	355.50
Steel enameled, 20″ x 17″, white	1.000	Ea.	2.759	142	110	252
Color	1.000	Ea.	2.500	148	99.50	247.50
Vitreous china 20″ x 16″, white	1.000	Ea.	2.963	244	118	362
Color	1.000	Ea.	2.963	244	118	362
hower, steel enameled stone base, 36″ square, white	1.000	Ea.	8.889	880	355	1,235
Color	1.000	Ea.	8.889	895	355	1,250
ough-in, for lavatory or shower						
1/2″ copper supply, 1-1/2″ cast iron waste, 1-1/2″ cast iron vent	1.000	Ea.	3.834	85	160	245
1-1/2″ P.V.C. waste, 1-1/4″ P.V.C. vent	1.000	Ea.	3.675	39.50	163	202.50
1/2″ steel supply, 1-1/4″ cast iron waste, 1-1/4″ steel vent	1.000	Ea.	4.103	76	173	249
1-1/4″ P.V.C. waste, 1-1/4″ P.V.C. vent	1.000	Ea.	3.937	40	175	215
1/2″ P.V.C. supply, 1-1/2″ P.V.C. waste, 1-1/2″ P.V.C. vent	1.000	Ea.	4.592	54	203	257
athtub, P.E. cast iron, 5′ long with fittings, white	1.000	Ea.	3.636	460	145	605
Color	1.000	Ea.	3.636	500	145	645
Steel, enameled 5′ long with fittings, white	1.000	Ea.	2.909	330	116	446
Color	1.000	Ea.	2.909	330	116	446
Rough-in, for bathtub						
1/2″ copper supply, 4″ cast iron waste, 1-1/2″ copper vent	1.000	Ea.	2.409	66.50	103	169.50
4″ P.V.C. waste, 1-1/2″ P.V.C. vent	1.000	Ea.	2.877	32	123	155
1/2″ steel supply, 4″ cast iron waste, 1-1/2″ steel vent	1.000	Ea.	2.898	63	121	184
4″ P.V.C. waste, 1-1/2″ P.V.C. vent	1.000	Ea.	3.159	32.50	136	168.50
1/2″ P.V.C. supply, 4″ P.V.C. waste, 1-1/2″ P.V.C. vent	1.000	Ea.	3.371	41	145	186
Piping, supply, 1/2″ copper	42.000	L.F.	4.148	69.50	184	253.50
1/2″ steel	42.000	L.F.	5.333	70	237	307
1/2″ P.V.C.	42.000	L.F.	6.222	107	275	382
Waste, 4″ cast iron, no hub	10.000	L.F.	2.759	100	110	210
4″ P.V.C./DWV	10.000	L.F.	3.333	34.50	133	167.50
4″ copper/DWV	10.000	Ea.	4.000	210	160	370
Vent 2″ cast iron, no hub	13.000	L.F.	3.105	78	124	202
2″ copper/DWV	13.000	L.F.	2.364	87	105	192
2″ P.V.C./DWV	13.000	L.F.	3.525	19.35	140	159.35
2″ steel, galvanized	13.000	L.F.	3.250	69.50	129	198.50
Vanity base cabinet, 2 doors, 30″ wide	1.000	Ea.	1.000	223	41	264
Vanity top, plastic laminated, square edge	2.670	L.F.	.712	64	29	93
Carrier, steel for studs, no arms	1.000	Ea.	1.143	38.50	50.50	89
Wood, 2″ x 8″ blocking	1.300	L.F.	.052	1.18	2.12	3.30

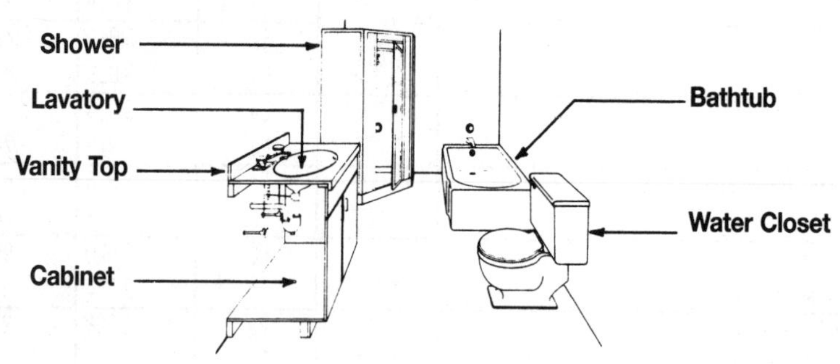

System Description	QUAN.	UNIT	LABOR HOURS	COST EACH MAT.	INST.	TOTAL
BATHROOM WITH LAVATORY INSTALLED IN VANITY						
Water closet, floor mounted, 2 piece, close coupled, white	1.000	Ea.	3.019	179	120	299
Rough-in, vent, 2″ diameter DWV piping	1.000	Ea.	.955	24	38	62
Waste, 4″ diameter DWV piping	1.000	Ea.	.828	30	33	63
Supply, 1/2″ diameter type "L" copper supply piping	1.000	Ea.	.593	9.90	26.28	36.18
Lavatory, 20″ x 18″ P.E. cast iron with fittings, white	1.000	Ea.	2.500	227	99.50	326.50
Shower, steel, enameled, stone base, corner, white	1.000	Ea.	8.889	880	355	1,235
Rough-in, waste, 1-1/2″ diameter DWV piping	2.000	Ea.	4.507	117	180	297
Supply, 1/2″ diameter type "L" copper supply piping	2.000	Ea.	3.161	52.80	140.16	192.96
Bathtub, P.E. cast iron, 5′ long with fittings, white	1.000	Ea.	3.636	460	145	605
Rough-in, waste, 4″ diameter DWV piping	1.000	Ea.	.828	30	33	63
Supply, 1/2″ diameter type "L" copper supply piping	1.000	Ea.	.988	16.50	43.80	60.30
Vent, 1-1/2″ diameter DWV piping	1.000	Ea.	.593	20	26.20	46.20
Piping, supply, 1/2″ diameter type "L" copper supply piping	42.000	L.F.	4.939	82.50	219	301.50
Waste, 4″ diameter DWV piping	10.000	L.F.	4.138	150	165	315
Vent, 2″ diameter DWV piping	13.000	L.F.	4.500	96.30	179.10	275.40
Vanity base, 2 doors, 30″ wide	1.000	Ea.	1.000	223	41	264
Vanity top, plastic laminated, square edge	2.670	L.F.	.712	68.09	28.97	97.06
TOTAL		Ea.	45.786	2,666.09	1,873.01	4,539.10
BATHROOM WITH WALL HUNG LAVATORY						
Water closet, floor mounted, 2 piece, close coupled, white	1.000	Ea.	3.019	179	120	299
Rough-in, vent, 2″ diameter DWV piping	1.000	Ea.	.955	24	38	62
Waste, 4″ diameter DWV piping	1.000	Ea.	.828	30	33	63
Supply, 1/2″ diameter type "L" copper supply piping	1.000	Ea.	.593	9.90	26.28	36.18
Lavatory, 20″ x 18″ P.E. cast iron with fittings, white	1.000	Ea.	2.000	261	79.50	340.50
Shower, steel enameled, stone base, corner, white	1.000	Ea.	8.889	880	355	1,235
Rough-in, waste, 1-1/2″ diameter DWV piping	2.000	Ea.	4.507	117	180	297
Supply, 1/2″ diameter type "L" copper supply piping	2.000	Ea.	3.161	52.80	140.16	192.96
Bathtub, P.E. cast iron, 5″ long with fittings, white	1.000	Ea.	3.636	460	145	605
Rough-in, waste, 4″ diameter DWV piping	1.000	Ea.	.828	30	33	63
Supply, 1/2″ diameter type "L" copper supply piping	1.000	Ea.	.988	16.50	43.80	60.30
Vent, 1-1/2″ diameter DWV piping	1.000	Ea.	.593	20	26.20	46.20
Piping, supply, 1/2″ diameter type "L" copper supply piping	42.000	L.F.	4.939	82.50	219	301.50
Waste, 4″ diameter DWV piping	10.000	L.F.	4.138	150	165	315
Vent, 2″ diameter DWV piping	13.000	L.F.	4.500	96.30	179.10	275.40
Carrier, steel for studs, no arms	1.000	Ea.	1.143	38.50	50.50	89
TOTAL		Ea.	44.717	2,447.50	1,833.54	4,281.04

The costs in this system are on a cost each basis. All necessary piping is included

Important: See the Reference Section for critical supporting data - Reference Nos., Crews & Location Factors

Four Fixture Bathroom Price Sheet	QUAN.	UNIT	LABOR HOURS	COST EACH		
				MAT.	INST.	TOTAL
Water closet, close coupled, standard 2 piece, white	1.000	Ea.	3.019	179	120	299
Color	1.000	Ea.	3.019	215	120	335
One piece, elongated bowl, white	1.000	Ea.	3.019	545	120	665
Color	1.000	Ea.	3.019	680	120	800
Low profile, one piece elongated bowl, white	1.000	Ea.	3.019	745	120	865
Color	1.000	Ea.	3.019	970	120	1,090
Rough-in, for water closet						
1/2″ copper supply, 4″ cast iron waste, 2″ cast iron vent	1.000	Ea.	2.376	64	97.50	161.50
4″ PVC/DWV waste, 2″ PVC vent	1.000	Ea.	2.678	26	109	135
4″ copper waste, 2″ copper vent	1.000	Ea.	2.520	99.50	106	205.50
3″ cast iron waste, 1-1/2″ cast iron vent	1.000	Ea.	2.244	57	92	149
3″ PVC waste, 1-1/2″ PVC vent	1.000	Ea.	2.388	23	102	125
3″ PVC waste, 1-1/2″ PVC vent	1.000	Ea.	2.014	66	85.50	151.50
1/2″ PVC supply, 4″ PVC waste, 2″ PVC vent	1.000	Ea.	2.974	31.50	122	153.50
3″ PVC waste, 1-1/2″ PVC vent	1.000	Ea.	2.684	28.50	115	143.50
1/2″ steel supply, 4″ cast iron waste, 2″ cast iron vent	1.000	Ea.	2.545	64	105	169
4″ cast iron waste, 2″ steel vent	1.000	Ea.	2.590	61.50	107	168.50
4″ PVC waste, 2″ PVC vent	1.000	Ea.	2.847	26.50	117	143.50
Lavatory wall hung, P.E. cast iron 20″ x 18″, white	1.000	Ea.	2.000	261	79.50	340.50
Color	1.000	Ea.	2.000	297	79.50	376.50
Vitreous china 19″ x 17″, white	1.000	Ea.	2.286	204	91	295
Color	1.000	Ea.	2.286	204	91	295
Lavatory for vanity top, P.E. cast iron, 20″ x 18″, white	1.000	Ea.	2.500	227	99.50	326.50
Color	1.000	Ea.	2.500	256	99.50	355.50
Steel, enameled 20″ x 17″, white	1.000	Ea.	2.759	142	110	252
Color	1.000	Ea.	2.500	148	99.50	247.50
Vitreous china 20″ x 16″, white	1.000	Ea.	2.963	244	118	362
Color	1.000	Ea.	2.963	244	118	362
Shower, steel enameled, stone base 36″ square, white	1.000	Ea.	8.889	880	355	1,235
Color	1.000	Ea.	8.889	895	355	1,250
Rough-in, for lavatory and shower						
1/2″ copper supply, 1-1/2″ cast iron waste, 1-1/2″ cast iron vent	1.000	Ea.	7.668	170	320	490
1-1/2″ PVC waste, 1-1/4″ PVC vent	1.000	Ea.	7.352	79	325	404
1/2″ steel supply, 1-1/4″ cast iron waste, 1-1/4″ steel vent	1.000	Ea.	8.205	152	345	497
1-1/4″ PVC waste, 1-1/4″ PVC vent	1.000	Ea.	7.873	79.50	350	429.50
1/2″ PVC supply, 1-1/2″ PVC waste, 1-1/2″ PVC vent	1.000	Ea.	9.185	108	405	513
Bathtub, P.E. cast iron, 5′ long with fittings, white	1.000	Ea.	3.636	460	145	605
Color	1.000	Ea.	3.636	500	145	645
Steel enameled, 5′ long with fittings, white	1.000	Ea.	2.909	330	116	446
Color	1.000	Ea.	2.909	330	116	446
Rough-in, for bathtub						
1/2″ copper supply, 4″ cast iron waste, 1-1/2″ copper vent	1.000	Ea.	2.409	66.50	103	169.50
4″ PVC waste, 1-1/2″ PVC vent	1.000	Ea.	2.877	32	123	155
1/2″ steel supply, 4″ cast iron waste, 1-1/2″ steel vent	1.000	Ea.	2.898	63	121	184
4″ PVC waste, 1-1/2″ PVC vent	1.000	Ea.	3.159	32.50	136	168.50
1/2″ PVC supply, 4″ PVC waste, 1-1/2″ PVC vent	1.000	Ea.	3.371	41	145	186
Piping supply, 1/2″ copper	42.000	L.F.	4.148	69.50	184	253.50
1/2″ steel	42.000	L.F.	5.333	70	237	307
1/2″ PVC	42.000	L.F.	6.222	107	275	382
Piping, waste, 4″ cast iron, no hub	10.000	L.F.	3.586	130	143	273
4″ PVC/DWV	10.000	L.F.	4.333	44.50	173	217.50
4″ copper/DWV	10.000	L.F.	5.200	273	207	480
Piping, vent, 2″ cast iron, no hub	13.000	L.F.	3.105	78	124	202
2″ copper/DWV	13.000	L.F.	2.364	87	105	192
2″ PVC/DWV	13.000	L.F.	3.525	19.35	140	159.35
2″ steel, galvanized	13.000	L.F.	3.250	69.50	129	198.50
Vanity base cabinet, 2 doors, 30″ wide	1.000	Ea.	1.000	223	41	264
Vanity top, plastic laminated, square edge	3.160	L.F.	.843	76	34.50	110.50
Carrier, steel, for studs, no arms	1.000	Ea.	1.143	38.50	50.50	89
Wood, 2″ x 8″ blocking	1.300	L.F.	.052	1.18	2.12	3.30

44 Five Fixture Bathroom Systems

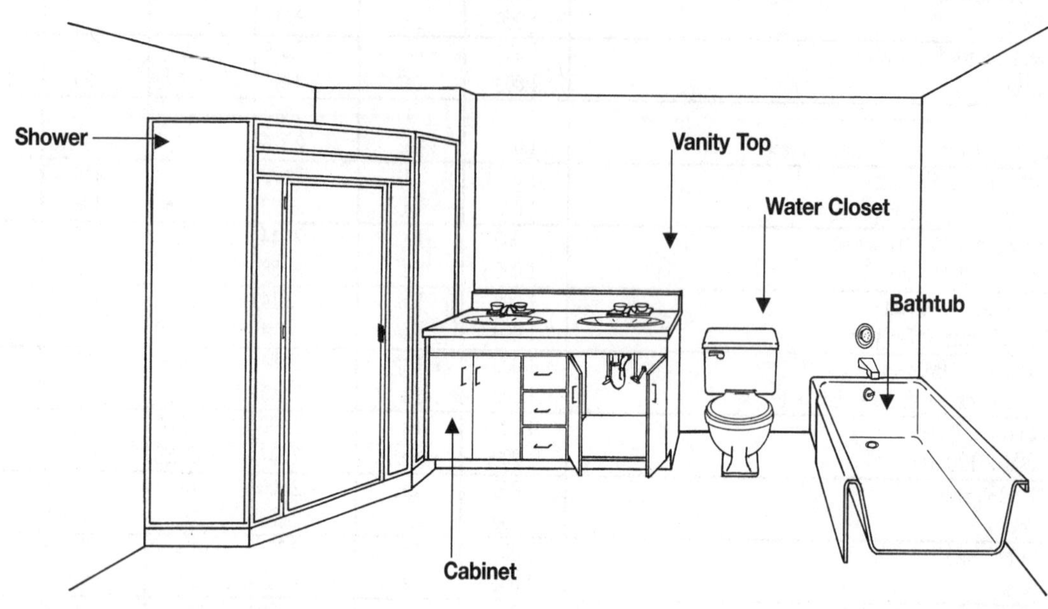

System Description	QUAN.	UNIT	LABOR HOURS	COST EACH		
				MAT.	INST.	TOTAL
BATHROOM WITH SHOWER, BATHTUB, LAVATORIES IN VANITY						
Water closet, floor mounted, 1 piece combination, white	1.000	Ea.	3.019	745	120	865
Rough-in, vent, 2" diameter DWV piping	1.000	Ea.	.955	24	38	62
Waste, 4" diameter DWV piping	1.000	Ea.	.828	30	33	63
Supply, 1/2" diameter type "L" copper supply piping	1.000	Ea.	.593	9.90	26.28	36.18
Lavatory, 20" x 16", vitreous china oval, with fittings, white	2.000	Ea.	5.926	488	236	724
Shower, steel enameled, stone base, corner, white	1.000	Ea.	8.889	880	355	1,235
Rough-in, waste, 1-1/2" diameter DWV piping	3.000	Ea.	5.408	140.40	216	356.40
Supply, 1/2" diameter type "L" copper supply piping	3.000	Ea.	2.963	49.50	131.40	180.90
Bathtub, P.E. cast iron, 5' long with fittings, white	1.000	Ea.	3.636	460	145	605
Rough-in, waste, 4" diameter DWV piping	1.000	Ea.	1.103	40	44	84
Supply, 1/2" diameter type "L" copper supply piping	1.000	Ea.	.988	16.50	43.80	60.30
Vent, 1-1/2" diameter copper DWV piping	1.000	Ea.	.593	20	26.20	46.20
Piping, supply, 1/2" diameter type "L" copper supply piping	42.000	L.F.	4.148	69.30	183.96	253.26
Waste, 4" diameter DWV piping	10.000	L.F.	2.759	100	110	210
Vent, 2" diameter DWV piping	13.000	L.F.	3.250	69.55	129.35	198.90
Vanity base, 2 door, 24" x 48"	1.000	Ea.	1.400	355	57	412
Vanity top, plastic laminated, square edge	4.170	L.F.	1.112	100.08	45.24	145.32
TOTAL		Ea.	47.570	3,597.23	1,940.23	5,537.46

The costs in this system are on a cost each basis. All necessary piping is included

Description	QUAN.	UNIT	LABOR HOURS	COST EACH		
				MAT.	INST.	TOTAL

Important: See the Reference Section for critical supporting data - Reference Nos., Crews & Location Factors

Five Fixture Bathroom Price Sheet	QUAN.	UNIT	LABOR HOURS	COST EACH MAT.	INST.	TOTAL
'ater closet, close coupled, standard 2 piece, white	1.000	Ea.	3.019	179	120	299
Color	1.000	Ea.	3.019	215	120	335
One piece elongated bowl, white	1.000	Ea.	3.019	545	120	665
Color	1.000	Ea.	3.019	680	120	800
Low profile, one piece elongated bowl, white	1.000	Ea.	3.019	745	120	865
Color	1.000	Ea.	3.019	970	120	1,090
ough-in, supply, waste and vent for water closet						
1/2" copper supply, 4" cast iron waste, 2" cast iron vent	1.000	Ea.	2.376	64	97.50	161.50
4" P.V.C./DWV waste, 2" P.V.C. vent	1.000	Ea.	2.678	26	109	135
4" copper waste, 2" copper vent	1.000	Ea.	2.520	99.50	106	205.50
3" cast iron waste, 1-1/2" cast iron vent	1.000	Ea.	2.244	57	92	149
3" P.V.C. waste, 1-1/2" P.V.C. vent	1.000	Ea.	2.388	23	102	125
3" copper waste, 1-1/2" copper vent	1.000	Ea.	2.014	66	85.50	151.50
1/2" P.V.C. supply, 4" P.V.C. waste, 2" P.V.C. vent	1.000	Ea.	2.974	31.50	122	153.50
3" P.V.C. waste, 1-1/2" P.V.C. supply	1.000	Ea.	2.684	28.50	115	143.50
1/2" steel supply, 4" cast iron waste, 2" cast iron vent	1.000	Ea.	2.545	64	105	169
4" cast iron waste, 2" steel vent	1.000	Ea.	2.590	61.50	107	168.50
4" P.V.C. waste, 2" P.V.C. vent	1.000	Ea.	2.847	26.50	117	143.50
avatory, wall hung, P.E. cast iron 20" x 18", white	2.000	Ea.	4.000	520	159	679
Color	2.000	Ea.	4.000	595	159	754
Vitreous china, 19" x 17", white	2.000	Ea.	4.571	410	182	592
Color	2.000	Ea.	4.571	410	182	592
Lavatory, for vanity top, P.E. cast iron, 20" x 18", white	2.000	Ea.	5.000	455	199	654
Color	2.000	Ea.	5.000	510	199	709
Steel enameled 20" x 17", white	2.000	Ea.	5.517	284	220	504
Color	2.000	Ea.	5.000	296	199	495
Vitreous china 20" x 16", white	2.000	Ea.	5.926	490	236	726
Color	2.000	Ea.	5.926	490	236	726
Shower, steel enameled, stone base 36" square, white	1.000	Ea.	8.889	880	355	1,235
Color	1.000	Ea.	8.889	895	355	1,250
Rough-in, for lavatory or shower						
1/2" copper supply, 1-1/2" cast iron waste, 1-1/2" cast iron vent	3.000	Ea.	8.371	190	345	535
1-1/2" P.V.C. waste, 1-1/4" P.V.C. vent	3.000	Ea.	7.916	81	350	431
1/2" steel supply, 1-1/4" cast iron waste, 1-1/4" steel vent	3.000	Ea.	8.670	163	365	528
1-1/4" P.V.C. waste, 1-1/4" P.V.C. vent	3.000	Ea.	8.381	81.50	370	451.50
1/2" P.V.C. supply, 1-1/2" P.V.C. waste, 1-1/2" P.V.C. vent	3.000	Ea.	9.778	108	435	543
Bathtub, P.E. cast iron 5' long with fittings, white	1.000	Ea.	3.636	460	145	605
Color	1.000	Ea.	3.636	500	145	645
Steel, enameled 5' long with fittings, white	1.000	Ea.	2.909	330	116	446
Color	1.000	Ea.	2.909	330	116	446
Rough-in, for bathtub						
1/2" copper supply, 4" cast iron waste, 1-1/2" copper vent	1.000	Ea.	2.684	76.50	114	190.50
4" P.V.C. waste, 1-1/2" P.V.C. vent	1.000	Ea.	3.210	35.50	136	171.50
1/2" steel supply, 4" cast iron waste, 1-1/2" steel vent	1.000	Ea.	3.173	73	132	205
4" P.V.C. waste, 1-1/2" P.V.C. vent	1.000	Ea.	3.492	35.50	149	184.50
1/2" P.V.C. supply, 4" P.V.C. waste, 1-1/2" P.V.C. vent	1.000	Ea.	3.704	44.50	158	202.50
Piping, supply, 1/2" copper	42.000	L.F.	4.148	69.50	184	253.50
1/2" steel	42.000	L.F.	5.333	70	237	307
1/2" P.V.C.	42.000	L.F.	6.222	107	275	382
Piping, waste, 4" cast iron, no hub	10.000	L.F.	2.759	100	110	210
4" P.V.C./DWV	10.000	L.F.	3.333	34.50	133	167.50
4" copper/DWV	10.000	L.F.	4.000	210	160	370
Piping, vent, 2" cast iron, no hub	13.000	L.F.	3.105	78	124	202
2" copper/DWV	13.000	L.F.	2.364	87	105	192
2" P.V.C./DWV	13.000	L.F.	3.525	19.35	140	159.35
2" steel, galvanized	13.000	L.F.	3.250	69.50	129	198.50
Vanity base cabinet, 2 doors, 24" x 48"	1.000	Ea.	1.400	355	57	412
Vanity top, plastic laminated, square edge	4.170	L.F.	1.112	100	45	145
Carrier, steel, for studs, no arms	1.000	Ea.	1.143	38.50	50.50	89
Wood, 2" x 8" blocking	1.300	L.F.	.052	1.18	2.12	3.30

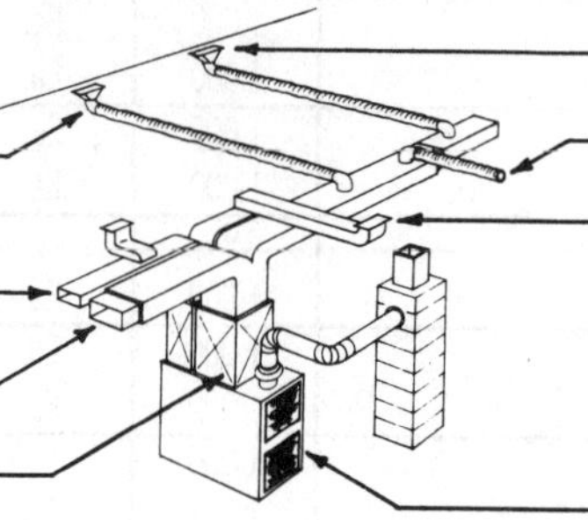

Gas Heating/Cooling Systems

Floor Registers
Register Elbows
Lateral Ducts
Return Air Grille
Return Air Duct
Supply Duct
Plenum
Furnace

System Description	QUAN.	UNIT	LABOR HOURS	COST PER SYSTEM MAT.	INST.	TOTAL
HEATING ONLY, GAS FIRED HOT AIR, ONE ZONE, 1200 S.F. BUILDING						
Furnace, gas, up flow	1.000	Ea.	5.000	720	201	921
Intermittent pilot	1.000	Ea.		145		145
Supply duct, rigid fiberglass	176.000	S.F.	12.068	128.48	501.60	630.08
Return duct, sheet metal, galvanized	158.000	Lb.	16.137	180.12	671.50	851.62
Lateral ducts, 6″ flexible fiberglass	144.000	L.F.	8.862	247.68	355.68	603.36
Register, elbows	12.000	Ea.	3.200	318	128.40	446.40
Floor registers, enameled steel	12.000	Ea.	3.000	234.60	133.80	368.40
Floor grille, return air	2.000	Ea.	.727	51	32.40	83.40
Thermostat	1.000	Ea.	1.000	30	44.50	74.50
Plenum	1.000	Ea.	1.000	67.50	40	107.50
TOTAL		System	50.994	2,122.38	2,108.88	4,231.26
HEATING/COOLING, GAS FIRED FORCED AIR, ONE ZONE, 1200 S.F. BUILDING						
Furnace, including plenum, compressor, coil	1.000	Ea.	14.720	3,956	588.80	4,544.80
Intermittent pilot	1.000	Ea.		145		145
Supply duct, rigid fiberglass	176.000	S.F.	12.068	128.48	501.60	630.08
Return duct, sheet metal, galvanized	158.000	Lb.	16.137	180.12	671.50	851.62
Lateral duct, 6″ flexible fiberglass	144.000	L.F.	8.862	247.68	355.68	603.36
Register elbows	12.000	Ea.	3.200	318	128.40	446.40
Floor registers, enameled steel	12.000	Ea.	3.000	234.60	133.80	368.40
Floor grille return air	2.000	Ea.	.727	51	32.40	83.40
Thermostat	1.000	Ea.	1.000	30	44.50	74.50
Refrigeration piping, 25 ft. (pre-charged)	1.000	Ea.		203		203
TOTAL		System	59.714	5,493.88	2,456.68	7,950.56

The costs in these systems are based on complete system basis. For larger buildings use the price sheet on the opposite page.

Description	QUAN.	UNIT	LABOR HOURS	COST PER SYSTEM MAT.	INST.	TOTAL

Gas Heating/Cooling Price Sheet	QUAN.	UNIT	LABOR HOURS	COST EACH MAT.	INST.	TOTAL
rnace, heating only, 100 MBH, area to 1200 S.F.	1.000	Ea.	5.000	720	201	921
120 MBH, area to 1500 S.F.	1.000	Ea.	5.000	720	201	921
160 MBH, area to 2000 S.F.	1.000	Ea.	5.714	965	229	1,194
200 MBH, area to 2400 S.F.	1.000	Ea.	6.154	2,200	247	2,447
Heating/cooling, 100 MBH heat, 36 MBH cool, to 1200 S.F.	1.000	Ea.	16.000	4,300	640	4,940
120 MBH heat, 42 MBH cool, to 1500 S.F.	1.000	Ea.	18.462	4,575	770	5,345
144 MBH heat, 47 MBH cool, to 2000 S.F.	1.000	Ea.	20.000	5,300	830	6,130
200 MBH heat, 60 MBH cool, to 2400 S.F.	1.000	Ea.	34.286	5,550	1,425	6,975
termittent pilot, 100 MBH furnace	1.000	Ea.		145		145
200 MBH furnace	1.000	Ea.		145		145
upply duct, rectangular, area to 1200 S.F., rigid fiberglass	176.000	S.F.	12.068	128	500	628
Sheet metal insulated	228.000	Lb.	31.331	355	1,275	1,630
Area to 1500 S.F., rigid fiberglass	176.000	S.F.	12.068	128	500	628
Sheet metal insulated	228.000	Lb.	31.331	355	1,275	1,630
Area to 2400 S.F., rigid fiberglass	205.000	S.F.	14.057	150	585	735
Sheet metal insulated	271.000	Lb.	37.048	420	1,500	1,920
Round flexible, insulated 6″ diameter, to 1200 S.F.	156.000	L.F.	9.600	268	385	653
To 1500 S.F.	184.000	L.F.	11.323	315	455	770
8″ diameter, to 2000 S.F.	269.000	L.F.	23.911	580	960	1,540
To 2400 S.F.	248.000	L.F.	22.045	535	885	1,420
eturn duct, sheet metal galvanized, to 1500 S.F.	158.000	Lb.	16.137	180	670	850
To 2400 S.F.	191.000	Lb.	19.507	218	810	1,028
ateral ducts, flexible round 6″ insulated, to 1200 S.F.	144.000	L.F.	8.862	248	355	603
To 1500 S.F.	172.000	L.F.	10.585	296	425	721
To 2000 S.F.	261.000	L.F.	16.062	450	645	1,095
To 2400 S.F.	300.000	L.F.	18.462	515	740	1,255
Spiral steel insulated, to 1200 S.F.	144.000	L.F.	20.067	335	780	1,115
To 1500 S.F.	172.000	L.F.	23.952	400	930	1,330
To 2000 S.F.	261.000	L.F.	36.352	610	1,400	2,010
To 2400 S.F.	300.000	L.F.	41.825	700	1,625	2,325
Rectangular sheet metal galvanized insulated, to 1200 S.F.	228.000	Lb.	39.056	450	1,575	2,025
To 1500 S.F.	344.000	Lb.	53.966	620	2,175	2,795
To 2000 S.F.	522.000	Lb.	81.926	940	3,300	4,240
To 2400 S.F.	600.000	Lb.	94.189	1,075	3,800	4,875
egister elbows, to 1500 S.F.	12.000	Ea.	3.200	320	128	448
To 2400 S.F.	14.000	Ea.	3.733	370	150	520
loor registers, enameled steel w/damper, to 1500 S.F.	12.000	Ea.	3.000	235	134	369
To 2400 S.F.	14.000	Ea.	4.308	320	192	512
Return air grille, area to 1500 S.F. 12″ x 12″	2.000	Ea.	.727	51	32.50	83.50
Area to 2400 S.F. 8″ x 16″	2.000	Ea.	.444	46	19.80	65.80
Area to 2400 S.F. 8″ x 16″	2.000	Ea.	.727	51	32.50	83.50
16″ x 16″	1.000	Ea.	.364	37	16.20	53.20
Thermostat, manual, 1 set back	1.000	Ea.	1.000	30	44.50	74.50
Electric, timed, 1 set back	1.000	Ea.	1.000	89	44.50	133.50
2 set back	1.000	Ea.	1.000	196	44.50	240.50
Plenum, heating only, 100 M.B.H.	1.000	Ea.	1.000	67.50	40	107.50
120 MBH	1.000	Ea.	1.000	67.50	40	107.50
160 MBH	1.000	Ea.	1.000	67.50	40	107.50
200 MBH	1.000	Ea.	1.000	67.50	40	107.50
Refrigeration piping, 3/8″	25.000	L.F.		20		20
3/4″	25.000	L.F.		40.50		40.50
7/8″	25.000	L.F.		47		47
Refrigerant piping, 25 ft. (precharged)	1.000	Ea.		203		203
Diffusers, ceiling, 6″ diameter, to 1500 S.F.	10.000	Ea.	4.444	186	198	384
To 2400 S.F.	12.000	Ea.	6.000	240	270	510
Floor, aluminum, adjustable, 2-1/4″ x 12″ to 1500 S.F.	12.000	Ea.	3.000	170	134	304
To 2400 S.F.	14.000	Ea.	3.500	199	156	355
Side wall, aluminum, adjustable, 8″ x 4″, to 1500 S.F.	12.000	Ea.	3.000	380	134	514
5″ x 10″ to 2400 S.F.	12.000	Ea.	3.692	500	164	664

64 Oil Fired Heating/Cooling Systems

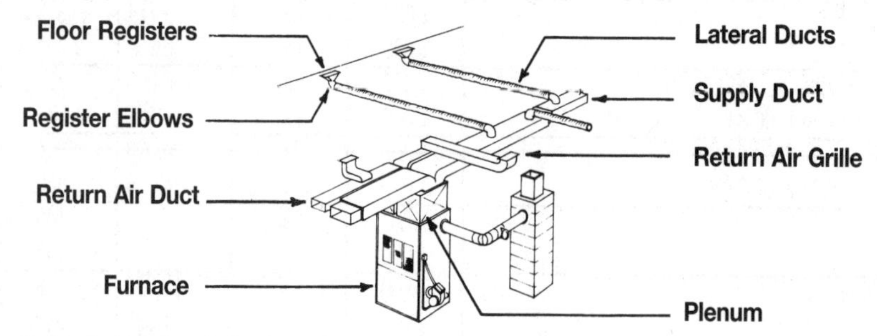

System Description	QUAN.	UNIT	LABOR HOURS	COST PER SYSTEM		
				MAT.	INST.	TOTAL
HEATING ONLY, OIL FIRED HOT AIR, ONE ZONE, 1200 S.F. BUILDING						
Furnace, oil fired, atomizing gun type burner	1.000	Ea.	4.571	850	183	1,033
3/8" diameter copper supply pipe	1.000	Ea.	2.759	34.50	122.10	156.60
Shut off valve	1.000	Ea.	.333	7.70	14.75	22.45
Oil tank, 275 gallon, on legs	1.000	Ea.	3.200	340	128	468
Supply duct, rigid fiberglass	176.000	S.F.	12.068	128.48	501.60	630.08
Return duct, sheet metal, galvanized	158.000	Lb.	16.137	180.12	671.50	851.62
Lateral ducts, 6" flexible fiberglass	144.000	L.F.	8.862	247.68	355.68	603.36
Register elbows	12.000	Ea.	3.200	318	128.40	446.40
Floor register, enameled steel	12.000	Ea.	3.000	234.60	133.80	368.40
Floor grille, return air	2.000	Ea.	.727	51	32.40	83.40
Thermostat	1.000	Ea.	1.000	30	44.50	74.50
TOTAL		System	55.857	2,422.08	2,315.73	4,737.81
HEATING/COOLING, OIL FIRED, FORCED AIR, ONE ZONE, 1200 S.F. BUILDING						
Furnace, including plenum, compressor, coil	1.000	Ea.	16.000	4,575	640	5,215
3/8" diameter copper supply pipe	1.000	Ea.	2.759	34.50	122.10	156.60
Shut off valve	1.000	Ea.	.333	7.70	14.75	22.45
Oil tank, 275 gallon on legs	1.000	Ea.	3.200	340	128	468
Supply duct, rigid fiberglass	176.000	S.F.	12.068	128.48	501.60	630.08
Return duct, sheet metal, galvanized	158.000	Lb.	16.137	180.12	671.50	851.62
Lateral ducts, 6" flexible fiberglass	144.000	L.F.	8.862	247.68	355.68	603.36
Register elbows	12.000	Ea.	3.200	318	128.40	446.40
Floor registers, enameled steel	12.000	Ea.	3.000	234.60	133.80	368.40
Floor grille, return air	2.000	Ea.	.727	51	32.40	83.40
Refrigeration piping (precharged)	25.000	L.F.		203		203
TOTAL		System	66.286	6,320.08	2,728.23	9,048.31

Description	QUAN.	UNIT	LABOR HOURS	COST EACH		
				MAT.	INST.	TOTAL

Important: See the Reference Section for critical supporting data - Reference Nos., Crews & Location Factors

Oil Fired Heating/Cooling	QUAN.	UNIT	LABOR HOURS	COST EACH		
				MAT.	INST.	TOTAL
rnace, heating, 95.2 MBH,area to 1200 S.F.	1.000	Ea.	4.706	870	189	1,059
123.2 MBH, area to 1500 S.F.	1.000	Ea.	5.000	1,200	201	1,401
151.2 MBH, area to 2000 S.F.	1.000	Ea.	5.333	1,325	214	1,539
200 MBH, area to 2400 S.F.	1.000	Ea.	6.154	2,325	247	2,572
Heating/cooling, 95.2 MBH heat, 36 MBH cool, to 1200 S.F.	1.000	Ea.	16.000	4,575	640	5,215
112 MBH heat, 42 MBH cool, to 1500 S.F.	1.000	Ea.	24.000	6,875	960	7,835
151 MBH heat, 47 MBH cool, to 2000 S.F.	1.000	Ea.	20.800	5,950	830	6,780
184.8 MBH heat, 60 MBH cool, to 2400 S.F.	1.000	Ea.	24.000	6,300	1,000	7,300
il piping to furnace, 3/8″ dia., copper	1.000	Ea.	3.412	134	150	284
il tank, on legs above ground, 275 gallons	1.000	Ea.	3.200	340	128	468
550 gallons	1.000	Ea.	5.926	1,525	237	1,762
Below ground, 275 gallons	1.000	Ea.	3.200	340	128	468
550 gallons	1.000	Ea.	5.926	1,525	237	1,762
1000 gallons	1.000	Ea.	6.400	2,425	256	2,681
upply duct, rectangular, area to 1200 S.F., rigid fiberglass	176.000	S.F.	12.068	128	500	628
Sheet metal, insulated	228.000	Lb.	31.331	355	1,275	1,630
Area to 1500 S.F., rigid fiberglass	176.000	S.F.	12.068	128	500	628
Sheet metal, insulated	228.000	Lb.	31.331	355	1,275	1,630
Area to 2400 S.F., rigid fiberglass	205.000	S.F.	14.057	150	585	735
Sheet metal, insulated	271.000	Lb.	37.048	420	1,500	1,920
Round flexible, insulated, 6″ diameter to 1200 S.F.	156.000	L.F.	9.600	268	385	653
To 1500 S.F.	184.000	L.F.	11.323	315	455	770
8″ diameter to 2000 S.F.	269.000	L.F.	23.911	580	960	1,540
To 2400 S.F.	269.000	L.F.	22.045	535	885	1,420
eturn duct, sheet metal galvanized, to 1500 S.F.	158.000	Lb.	16.137	180	670	850
To 2400 S.F.	191.000	Lb.	19.507	218	810	1,028
ateral ducts, flexible round, 6″, insulated to 1200 S.F.	144.000	L.F.	8.862	248	355	603
To 1500 S.F.	172.000	L.F.	10.585	296	425	721
To 2000 S.F.	261.000	L.F.	16.062	450	645	1,095
To 2400 S.F.	300.000	L.F.	18.462	515	740	1,255
Spiral steel, insulated to 1200 S.F.	144.000	L.F.	20.067	335	780	1,115
To 1500 S.F.	172.000	L.F.	23.952	400	930	1,330
To 2000 S.F.	261.000	L.F.	36.352	610	1,400	2,010
To 2400 S.F.	300.000	L.F.	41.825	700	1,625	2,325
Rectangular sheet metal galvanized insulated, to 1200 S.F.	288.000	Lb.	45.183	520	1,825	2,345
To 1500 S.F.	344.000	Lb.	53.966	620	2,175	2,795
To 2000 S.F.	522.000	Lb.	81.926	940	3,300	4,240
To 2400 S.F.	600.000	Lb.	94.189	1,075	3,800	4,875
Register elbows, to 1500 S.F.	12.000	Ea.	3.200	320	128	448
To 2400 S.F.	14.000	Ea.	3.733	370	150	520
Floor registers, enameled steel w/damper, to 1500 S.F.	12.000	Ea.	3.000	235	134	369
To 2400 S.F.	14.000	Ea.	4.308	320	192	512
Return air grille, area to 1500 S.F., 12″ x 12″	2.000	Ea.	.727	51	32.50	83.50
12″ x 24″	1.000	Ea.	.444	46	19.80	65.80
Area to 2400 S.F., 8″ x 16″	2.000	Ea.	.727	51	32.50	83.50
16″ x 16″	1.000	Ea.	.364	37	16.20	53.20
Thermostat, manual, 1 set back	1.000	Ea.	1.000	30	44.50	74.50
Electric, timed, 1 set back	1.000	Ea.	1.000	89	44.50	133.50
2 set back	1.000	Ea.	1.000	196	44.50	240.50
Refrigeration piping, 3/8″	25.000	L.F.		20		20
3/4″	25.000	L.F.		40.50		40.50
Diffusers, ceiling, 6″ diameter, to 1500 S.F.	10.000	Ea.	4.444	186	198	384
To 2400 S.F.	12.000	Ea.	6.000	240	270	510
Floor, aluminum, adjustable, 2-1/4″ x 12″ to 1500 S.F.	12.000	Ea.	3.000	170	134	304
To 2400 S.F.	14.000	Ea.	3.500	199	156	355
Side wall, aluminum, adjustable, 8″ x 4″, to 1500 S.F.	12.000	Ea.	3.000	380	134	514
5″ x 10″ to 2400 S.F.	12.000	Ea.	3.692	500	164	664

68 Hot Water Heating Systems

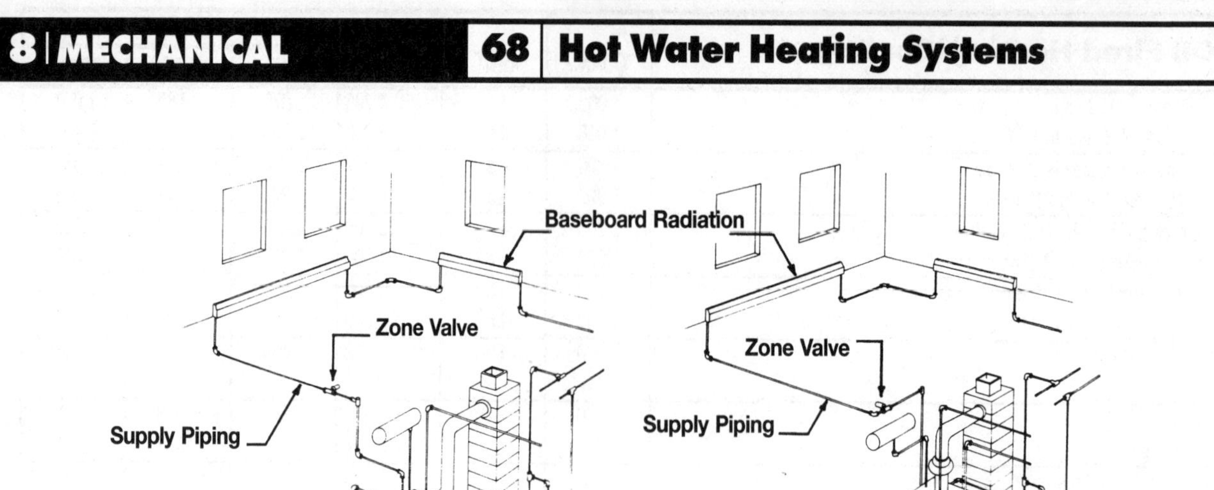

System Description	QUAN.	UNIT	LABOR HOURS	COST EACH		
				MAT.	INST.	TOTAL
OIL FIRED HOT WATER HEATING SYSTEM, AREA TO 1200 S.F.						
Boiler package, oil fired, 97 MBH, area to 1200 S.F. building	1.000	Ea.	15.000	1,400	580	1,980
3/8" diameter copper supply pipe	1.000	Ea.	2.759	34.50	122.10	156.60
Shut off valve	1.000	Ea.	.333	7.70	14.75	22.45
Oil tank, 275 gallon, with black iron filler pipe	1.000	Ea.	3.200	340	128	468
Supply piping, 3/4" copper tubing	176.000	L.F.	18.526	425.92	820.16	1,246.08
Supply fittings, copper 3/4"	36.000	Ea.	15.158	42.12	671.40	713.52
Supply valves, 3/4"	2.000	Ea.	.800	117	35.40	152.40
Baseboard radiation, 3/4"	106.000	L.F.	35.333	368.88	1,415.10	1,783.98
Zone valve	1.000	Ea.	.400	74.50	17.80	92.30
TOTAL		Ea.	91.509	2,810.62	3,804.71	6,615.33
OIL FIRED HOT WATER HEATING SYSTEM, AREA TO 2400 S.F.						
Boiler package, oil fired, 225 MBH, area to 2400 S.F. building	1.000	Ea.	19.704	3,525	760	4,285
3/8" diameter copper supply pipe	1.000	Ea.	2.759	34.50	122.10	156.60
Shut off valve	1.000	Ea.	.333	7.70	14.75	22.45
Oil tank, 550 gallon, with black iron pipe filler pipe	1.000	Ea.	5.926	1,525	237	1,762
Supply piping, 3/4" copper tubing	228.000	L.F.	23.999	551.76	1,062.48	1,614.24
Supply fittings, copper	46.000	Ea.	19.368	53.82	857.90	911.72
Supply valves	2.000	Ea.	.800	117	35.40	152.40
Baseboard radiation	212.000	L.F.	70.666	737.76	2,830.20	3,567.96
Zone valve	1.000	Ea.	.400	74.50	17.80	92.30
TOTAL		Ea.	143.955	6,627.04	5,937.63	12,564.67

The costs in this system are on a cost each basis. the costs represent total cost for the system based on a gross square foot of plan area.

Description	QUAN.	UNIT	LABOR HOURS	COST EACH		
				MAT.	INST.	TOTAL

Important: See the Reference Section for critical supporting data - Reference Nos., Crews & Location Factors

Hot Water Heating Price Sheet

	QUAN.	UNIT	LABOR HOURS	COST EACH		
				MAT.	INST.	TOTAL
iler, oil fired, 97 MBH, area to 1200 S.F.	1.000	Ea.	15.000	1,400	580	1,980
118 MBH, area to 1500 S.F.	1.000	Ea.	16.506	2,725	635	3,360
161 MBH, area to 2000 S.F.	1.000	Ea.	18.405	3,425	710	4,135
215 MBH, area to 2400 S.F.	1.000	Ea.	19.704	3,525	760	4,285
piping, (valve & filter), 3/8″ copper	1.000	Ea.	3.289	77.50	144	221.50
1/4″ copper	1.000	Ea.	3.242	56	143	199
l tank, filler pipe and cap on legs, 275 gallon	1.000	Ea.	3.200	340	128	468
550 gallon	1.000	Ea.	5.926	1,525	237	1,762
Buried underground, 275 gallon	1.000	Ea.	3.200	340	128	468
550 gallon	1.000	Ea.	5.926	1,525	237	1,762
1000 gallon	1.000	Ea.	6.400	2,425	256	2,681
ipply piping copper, area to 1200 S.F., 1/2″ tubing	176.000	L.F.	17.384	290	770	1,060
3/4″ tubing	176.000	L.F.	18.526	425	820	1,245
Area to 1500 S.F., 1/2″ tubing	186.000	L.F.	18.371	305	815	1,120
3/4″ tubing	186.000	L.F.	19.578	450	865	1,315
Area to 2000 S.F., 1/2″ tubing	204.000	L.F.	20.149	335	895	1,230
3/4″ tubing	204.000	L.F.	21.473	495	950	1,445
Area to 2400 S.F., 1/2″ tubing	228.000	L.F.	22.520	375	1,000	1,375
3/4″ tubing	228.000	L.F.	23.999	550	1,050	1,600
ipply pipe fittings copper, area to 1200 S.F., 1/2″	36.000	Ea.	14.400	19.10	635	654.10
3/4″	36.000	Ea.	15.158	42	670	712
Area to 1500 S.F., 1/2″	40.000	Ea.	16.000	21	710	731
3/4″	40.000	Ea.	16.842	47	745	792
Area to 2000 S.F., 1/2″	44.000	Ea.	17.600	23.50	780	803.50
3/4″	44.000	Ea.	18.526	51.50	820	871.50
Area to 2400, S.F., 1/2″	46.000	Ea.	18.400	24.50	815	839.50
3/4″	46.000	Ea.	19.368	54	860	914
upply valves, 1/2″ pipe size	2.000	Ea.	.667	86	29.50	115.50
3/4″	2.000	Ea.	.800	117	35.50	152.50
aseboard radiation, area to 1200 S.F., 1/2″ tubing	106.000	L.F.	28.267	625	1,125	1,750
3/4″ tubing	106.000	L.F.	35.333	370	1,425	1,795
Area to 1500 S.F., 1/2″ tubing	134.000	L.F.	35.734	790	1,425	2,215
3/4″ tubing	134.000	L.F.	44.666	465	1,800	2,265
Area to 2000 S.F., 1/2″ tubing	178.000	L.F.	47.467	1,050	1,900	2,950
3/4″ tubing	178.000	L.F.	59.333	620	2,375	2,995
Area to 2400 S.F., 1/2″ tubing	212.000	L.F.	56.534	1,250	2,250	3,500
3/4″ tubing	212.000	L.F.	70.666	740	2,825	3,565
one valves, 1/2″ tubing	1.000	Ea.	.400	74.50	17.80	92.30
3/4″ tubing	1.000	Ea.	.400	79.50	17.80	97.30

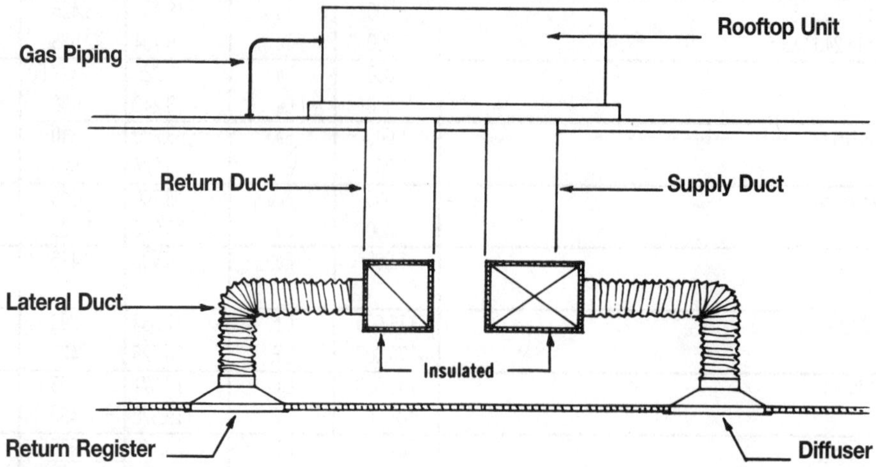

System Description	QUAN.	UNIT	LABOR HOURS	COST EACH		
				MAT.	INST.	TOTAL
ROOFTOP HEATING/COOLING UNIT, AREA TO 2000 S.F.						
Rooftop unit, single zone, electric cool, gas heat, to 2000 s.f.	1.000	Ea.	28.521	4,450	1,150	5,600
Gas piping	34.500	L.F.	5.207	82.11	231.15	313.26
Duct, supply and return, galvanized steel	38.000	Lb.	3.881	43.32	161.50	204.82
Insulation, ductwork	33.000	S.F.	1.508	18.15	57.09	75.24
Lateral duct, flexible duct 12" diameter, insulated	72.000	L.F.	11.520	234	460.80	694.80
Diffusers	4.000	Ea.	4.571	1,240	204	1,444
Return registers	1.000	Ea.	.727	123	32.50	155.50
TOTAL		Ea.	55.935	6,190.58	2,297.04	8,487.62
ROOFTOP HEATING/COOLING UNIT, AREA TO 5000 S.F.						
Rooftop unit, single zone, electric cool, gas heat, to 5000 s.f.	1.000	Ea.	42.032	13,100	1,625	14,725
Gas piping	86.250	L.F.	13.019	205.28	577.88	783.16
Duct supply and return, galvanized steel	95.000	Lb.	9.702	108.30	403.75	512.05
Insulation, ductwork	82.000	S.F.	3.748	45.10	141.86	186.96
Lateral duct, flexible duct, 12" diameter, insulated	180.000	L.F.	28.800	585	1,152	1,737
Diffusers	10.000	Ea.	11.429	3,100	510	3,610
Return registers	3.000	Ea.	2.182	369	97.50	466.50
TOTAL		Ea.	110.912	17,512.68	4,507.99	22,020.67

Description	QUAN.	UNIT	LABOR HOURS	COST EACH		
				MAT.	INST.	TOTAL

Important: See the Reference Section for critical supporting data - Reference Nos., Crews & Location Factors

Rooftop Price Sheet	QUAN.	UNIT	LABOR HOURS	COST EACH		
				MAT.	INST.	TOTAL
oftop unit, single zone, electric cool, gas heat to 2000 S.F.	1.000	Ea.	28.521	4,450	1,150	5,600
Area to 3000 S.F.	1.000	Ea.	35.982	8,350	1,375	9,725
Area to 5000 S.F.	1.000	Ea.	42.032	13,100	1,625	14,725
Area to 10000 S.F.	1.000	Ea.	68.376	27,000	2,725	29,725
as piping, area 2000 through 4000 S.F.	34.500	L.F.	5.207	82	231	313
Area 5000 to 10000 S.F.	86.250	L.F.	13.019	205	580	785
uct, supply and return, galvanized steel, to 2000 S.F.	38.000	Lb.	3.881	43.50	162	205.50
Area to 3000 S.F.	57.000	Lb.	5.821	65	242	307
Area to 5000 S.F.	95.000	Lb.	9.702	108	405	513
Area to 10000 S.F.	190.000	Lb.	19.405	217	810	1,027
Rigid fiberglass, area to 2000 S.F.	33.000	S.F.	2.263	24	94	118
Area to 3000 S.F.	49.000	S.F.	3.360	36	140	176
Area to 5000 S.F.	82.000	S.F.	5.623	60	234	294
Area to 10000 S.F.	164.000	S.F.	11.245	120	465	585
sulation, supply and return, blanket type, area to 2000 S.F.	33.000	S.F.	1.508	18.15	57	75.15
Area to 3000 S.F.	49.000	S.F.	2.240	27	85	112
Area to 5000 S.F.	82.000	S.F.	3.748	45	142	187
Area to 10000 S.F.	164.000	S.F.	7.496	90	284	374
ateral ducts, flexible round, 12″ insulated, to 2000 S.F.	72.000	L.F.	11.520	234	460	694
Area to 3000 S.F.	108.000	L.F.	17.280	350	690	1,040
Area to 5000 S.F.	180.000	L.F.	28.800	585	1,150	1,735
Area to 10000 S.F.	360.000	L.F.	57.600	1,175	2,300	3,475
Rectangular, galvanized steel, to 2000 S.F.	239.000	Lb.	24.409	272	1,025	1,297
Area to 3000 S.F.	360.000	Lb.	36.767	410	1,525	1,935
Area to 5000 S.F.	599.000	Lb.	61.176	685	2,550	3,235
Area to 10000 S.F.	998.000	Lb.	101.926	1,150	4,250	5,400
iffusers, ceiling, 1 to 4 way blow, 24″ x 24″, to 2000 S.F.	4.000	Ea.	4.571	1,250	204	1,454
Area to 3000 S.F.	6.000	Ea.	6.857	1,850	305	2,155
Area to 5000 S.F.	10.000	Ea.	11.429	3,100	510	3,610
Area to 10000 S.F.	20.000	Ea.	22.857	6,200	1,025	7,225
eturn grilles, 24″ x 24″, to 2000 S.F.	1.000	Ea.	.727	123	32.50	155.50
Area to 3000 S.F.	2.000	Ea.	1.455	246	65	311
Area to 5000 S.F.	3.000	Ea.	2.182	370	97.50	467.50
Area to 10000 S.F.	5.000	Ea.	3.636	615	163	778

Division 9
Electrical

o part of this publication may be reproduced, stored in a retrieval system, or transmitted in any form
r by any means without prior written permission of Reed Construction Data.

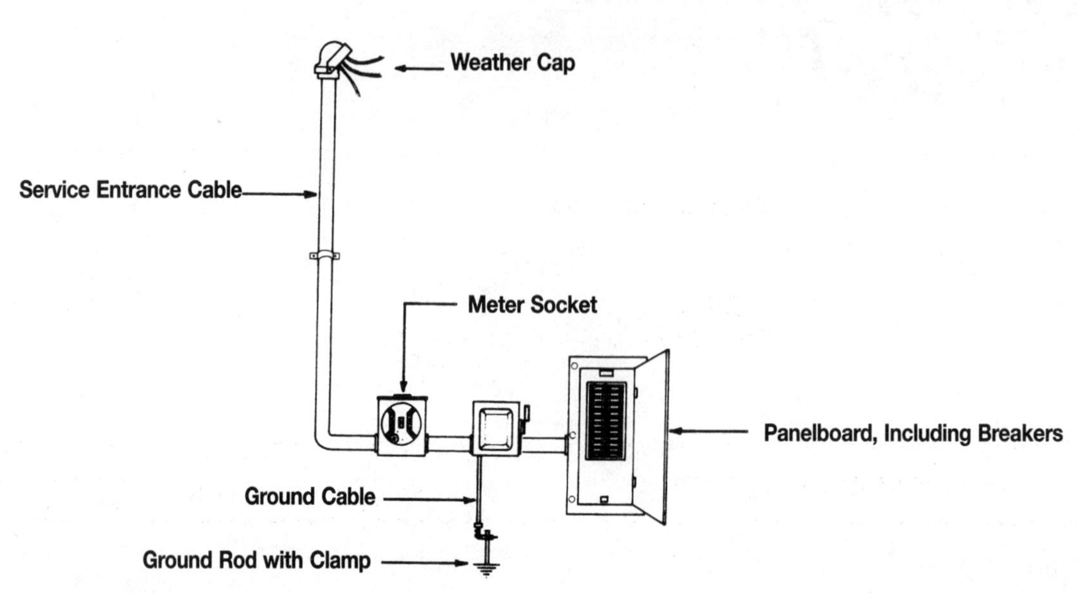

System Description	QUAN.	UNIT	LABOR HOURS	COST EACH MAT.	COST EACH INST.	COST EACH TOTAL
100 AMP SERVICE						
Weather cap	1.000	Ea.	.667	9.50	29.50	39
Service entrance cable	10.000	L.F.	.762	30.90	33.90	64.80
Meter socket	1.000	Ea.	2.500	37.50	111	148.50
Ground rod with clamp	1.000	Ea.	1.455	14.85	64.50	79.35
Ground cable	5.000	L.F.	.250	8.20	11.10	19.30
Panel board, 12 circuit	1.000	Ea.	6.667	195	296	491
TOTAL		Ea.	12.301	295.95	546	841.95
200 AMP SERVICE						
Weather cap	1.000	Ea.	1.000	29	44.50	73.50
Service entrance cable	10.000	L.F.	1.143	58.50	51	109.50
Meter socket	1.000	Ea.	4.211	56.50	187	243.50
Ground rod with clamp	1.000	Ea.	1.818	34.50	81	115.50
Ground cable	10.000	L.F.	.500	16.40	22.20	38.60
3/4" EMT	5.000	L.F.	.308	5.80	13.70	19.50
Panel board, 24 circuit	1.000	Ea.	12.308	520	455	975
TOTAL		Ea.	21.288	720.70	854.40	1,575.10
400 AMP SERVICE						
Weather cap	1.000	Ea.	2.963	380	132	512
Service entrance cable	180.000	L.F.	5.760	289.80	255.60	545.40
Meter socket	1.000	Ea.	4.211	56.50	187	243.50
Ground rod with clamp	1.000	Ea.	2.000	90.50	89	179.50
Ground cable	20.000	L.F.	.485	20.60	21.60	42.20
3/4" greenfield	20.000	L.F.	1.000	11.20	44.40	55.60
Current transformer cabinet	1.000	Ea.	6.154	152	274	426
Panel board, 42 circuit	1.000	Ea.	33.333	2,650	1,475	4,125
TOTAL		Ea.	55.906	3,650.60	2,478.60	6,129.20

Important: See the Reference Section for critical supporting data - Reference Nos., Crews & Location Factors

20 Electric Perimeter Heating Systems

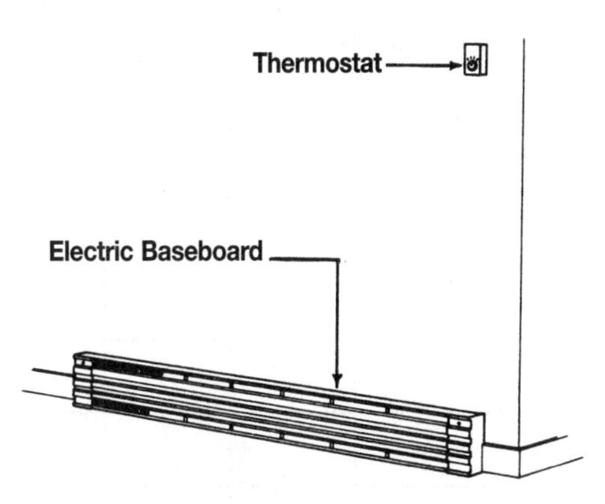

System Description	QUAN.	UNIT	LABOR HOURS	COST EACH MAT.	COST EACH INST.	COST EACH TOTAL
′ BASEBOARD HEATER						
Electric baseboard heater, 4′ long	1.000	Ea.	1.194	45	53	98
Thermostat, integral	1.000	Ea.	.500	19.80	22	41.80
Romex, 12-3 with ground	40.000	L.F.	1.600	18.40	71.20	89.60
Panel board breaker, 20 Amp	1.000	Ea.	.300	7.95	13.35	21.30
TOTAL		Ea.	3.594	91.15	159.55	250.70
′ BASEBOARD HEATER						
Electric baseboard heater, 6′ long	1.000	Ea.	1.600	59.50	71	130.50
Thermostat, integral	1.000	Ea.	.500	19.80	22	41.80
Romex, 12-3 with ground	40.000	L.F.	1.600	18.40	71.20	89.60
Panel board breaker, 20 Amp	1.000	Ea.	.400	10.60	17.80	28.40
TOTAL		Ea.	4.100	108.30	182	290.30
′ BASEBOARD HEATER						
Electric baseboard heater, 8′ long	1.000	Ea.	2.000	75	89	164
Thermostat, integral	1.000	Ea.	.500	19.80	22	41.80
Romex, 12-3 with ground	40.000	L.F.	1.600	18.40	71.20	89.60
Panel board breaker, 20 Amp	1.000	Ea.	.500	13.25	22.25	35.50
TOTAL		Ea.	4.600	126.45	204.45	330.90
.0′ BASEBOARD HEATER						
Electric baseboard heater, 10′ long	1.000	Ea.	2.424	92.50	108	200.50
Thermostat, integral	1.000	Ea.	.500	19.80	22	41.80
Romex, 12-3 with ground	40.000	L.F.	1.600	18.40	71.20	89.60
Panel board breaker, 20 Amp	1.000	Ea.	.750	19.88	33.38	53.26
TOTAL		Ea.	5.274	150.58	234.58	385.16

The costs in this system are on a cost each basis and include all necessary conduit fittings.

Description	QUAN.	UNIT	LABOR HOURS	COST EACH MAT.	COST EACH INST.	COST EACH TOTAL

System Description	QUAN.	UNIT	LABOR HOURS	COST EACH		
				MAT.	INST.	TOTAL
Air conditioning receptacles						
Using non-metallic sheathed cable	1.000	Ea.	.800	17	35.50	52.50
Using BX cable	1.000	Ea.	.964	29.50	43	72.50
Using EMT conduit	1.000	Ea.	1.194	40	53	93
Disposal wiring						
Using non-metallic sheathed cable	1.000	Ea.	.889	12.85	39.50	52.35
Using BX cable	1.000	Ea.	1.067	24.50	47.50	72
Using EMT conduit	1.000	Ea.	1.333	36.50	59.50	96
Dryer circuit						
Using non-metallic sheathed cable	1.000	Ea.	1.455	38.50	64.50	103
Using BX cable	1.000	Ea.	1.739	50.50	77.50	128
Using EMT conduit	1.000	Ea.	2.162	55	96	151
Duplex receptacles						
Using non-metallic sheathed cable	1.000	Ea.	.615	17	27.50	44.50
Using BX cable	1.000	Ea.	.741	29.50	33	62.50
Using EMT conduit	1.000	Ea.	.920	40	41	81
Exhaust fan wiring						
Using non-metallic sheathed cable	1.000	Ea.	.800	15.45	35.50	50.95
Using BX cable	1.000	Ea.	.964	27.50	43	70.50
Using EMT conduit	1.000	Ea.	1.194	38.50	53	91.50
Furnace circuit & switch						
Using non-metallic sheathed cable	1.000	Ea.	1.333	21	59.50	80.50
Using BX cable	1.000	Ea.	1.600	35	71	106
Using EMT conduit	1.000	Ea.	2.000	44.50	89	133.50
Ground fault						
Using non-metallic sheathed cable	1.000	Ea.	1.000	46	44.50	90.50
Using BX cable	1.000	Ea.	1.212	58.50	54	112.50
Using EMT conduit	1.000	Ea.	1.481	78.50	66	144.50
Heater circuits						
Using non-metallic sheathed cable	1.000	Ea.	1.000	16.80	44.50	61.30
Using BX cable	1.000	Ea.	1.212	25.50	54	79.50
Using EMT conduit	1.000	Ea.	1.481	36	66	102
Lighting wiring						
Using non-metallic sheathed cable	1.000	Ea.	.500	16.90	22	38.90
Using BX cable	1.000	Ea.	.602	25.50	26.50	52
Using EMT conduit	1.000	Ea.	.748	34.50	33	67.50
Range circuits						
Using non-metallic sheathed cable	1.000	Ea.	2.000	75	89	164
Using BX cable	1.000	Ea.	2.424	112	108	220
Using EMT conduit	1.000	Ea.	2.963	79.50	132	211.50
Switches, single pole						
Using non-metallic sheathed cable	1.000	Ea.	.500	15.45	22	37.45
Using BX cable	1.000	Ea.	.602	27.50	26.50	54
Using EMT conduit	1.000	Ea.	.748	38.50	33	71.50
Switches, 3-way						
Using non-metallic sheathed cable	1.000	Ea.	.667	20	29.50	49.50
Using BX cable	1.000	Ea.	.800	31.50	35.50	67
Using EMT conduit	1.000	Ea.	1.333	50.50	59.50	110
Water heater						
Using non-metallic sheathed cable	1.000	Ea.	1.600	24	71	95
Using BX cable	1.000	Ea.	1.905	41.50	84.50	126
Using EMT conduit	1.000	Ea.	2.353	41.50	105	146.50
Weatherproof receptacle						
Using non-metallic sheathed cable	1.000	Ea.	1.333	110	59.50	169.50
Using BX cable	1.000	Ea.	1.600	119	71	190
Using EMT conduit	1.000	Ea.	2.000	130	89	219

DESCRIPTION	QUAN.	UNIT	LABOR HOURS	COST EACH MAT.	COST EACH INST.	COST EACH TOTAL
orescent strip, 4′ long, 1 light, average	1.000	Ea.	.941	31	42	73
Deluxe	1.000	Ea.	1.129	37	50.50	87.50
2 lights, average	1.000	Ea.	1.000	33.50	44.50	78
Deluxe	1.000	Ea.	1.200	40	53.50	93.50
8′ long, 1 light, average	1.000	Ea.	1.194	46.50	53	99.50
Deluxe	1.000	Ea.	1.433	56	63.50	119.50
2 lights, average	1.000	Ea.	1.290	56	57.50	113.50
Deluxe	1.000	Ea.	1.548	67	69	136
Surface mounted, 4′ x 1′, economy	1.000	Ea.	.914	65	41	106
Average	1.000	Ea.	1.143	81	51	132
Deluxe	1.000	Ea.	1.371	97	61	158
4′ x 2′, economy	1.000	Ea.	1.208	82.50	53.50	136
Average	1.000	Ea.	1.509	103	67	170
Deluxe	1.000	Ea.	1.811	124	80.50	204.50
Recessed, 4′x 1′, 2 lamps, economy	1.000	Ea.	1.123	42.50	50	92.50
Average	1.000	Ea.	1.404	53	62.50	115.50
Deluxe	1.000	Ea.	1.684	63.50	75	138.50
4′ x 2′, 4′ lamps, economy	1.000	Ea.	1.362	51	60.50	111.50
Average	1.000	Ea.	1.702	64	75.50	139.50
Deluxe	1.000	Ea.	2.043	77	90.50	167.50
candescent, exterior, 150W, single spot	1.000	Ea.	.500	18.60	22	40.60
Double spot	1.000	Ea.	1.167	75.50	51.50	127
Recessed, 100W, economy	1.000	Ea.	.800	51	35.50	86.50
Average	1.000	Ea.	1.000	64	44.50	108.50
Deluxe	1.000	Ea.	1.200	77	53.50	130.50
150W, economy	1.000	Ea.	.800	75	35.50	110.50
Average	1.000	Ea.	1.000	93.50	44.50	138
Deluxe	1.000	Ea.	1.200	112	53.50	165.50
Surface mounted, 60W, economy	1.000	Ea.	.800	39.50	35.50	75
Average	1.000	Ea.	1.000	44.50	44.50	89
Deluxe	1.000	Ea.	1.194	60.50	53	113.50
etal halide, recessed 2′ x 2′ 250W	1.000	Ea.	2.500	320	111	431
2′ x 2′, 400W	1.000	Ea.	2.759	360	123	483
Surface mounted, 2′ x 2′, 250W	1.000	Ea.	2.963	310	132	442
2′ x 2′, 400W	1.000	Ea.	3.333	365	148	513
High bay, single, unit, 400W	1.000	Ea.	3.478	370	155	525
Twin unit, 400W	1.000	Ea.	5.000	740	222	962
Low bay, 250W	1.000	Ea.	2.500	370	111	481

Unit Price Section

able of Contents

General Requirements 287

100 Summary 288
200 Price & Payment Procedures 288
300 Administrative Requirements 288
500 Temporary Facilities & Controls 289
590 Equipment Rental 292
700 Execution Requirements 302

Site Construction 303

050 Basic Site Materials & Methods 304
200 Site Preparation 304
300 Earthwork 308
400 Tunneling, Boring & Jacking 310
500 Utility Services 311
600 Drainage & Containment 312
700 Bases, Ballasts, Pavements, & Appurtenances 313
800 Site Improvements & Amenities 316
900 Planting 320

Concrete 325

050 Basic Conc. Mat. & Methods 326
100 Concrete Forms & Accessories 326
200 Concrete Reinforcement 328
300 Cast-In-Place Concrete 329
400 Precast Concrete 331
900 Concrete Restor. & Cleaning 332

Masonry 333

050 Basic Masonry Mat. & Methods 334
200 Masonry Units 336
500 Refractories 336
800 Masonry Assemblies 336
900 Masonry Restoration & Cleaning 341

Metals 343

050 Basic Materials & Methods 347
100 Structural Metal Framing 347
300 Metal Deck 348
400 Cold Formed Metal Framing 349
500 Metal Fabrications 358

Wood & Plastics 361

050 Basic Wd. & Plastic Mat. & Meth. 362
100 Rough Carpentry 370
200 Finish Carpentry 384
400 Architectural Woodwork 388
600 Plastic Fabrications 396

Thermal & Moisture Protection 397

07050 Basic Materials & Methods 398
07100 Dampproofing & Waterproofing 398
07200 Thermal Protection 399
07300 Shingles, Roof Tiles & Roof Cov. 403
07400 Roofing & Siding Panels 402
07500 Membrane Roofing 408
07600 Flashing & Sheet Metal 409
07700 Roof Specialties & Accessories 411
07900 Joint Sealers 414

Doors & Windows 415

08050 Basic Materials & Methods 416
08100 Metal Doors & Frames 417
08200 Wood & Plastic Doors 419
08300 Specialty Doors 425
08500 Windows 426
08600 Skylights 436
08700 Hardware 437
08800 Glazing 440

Finishes 441

09050 Basic Materials & Methods 442
09100 Metal Support Assemblies 442
09200 Plaster & Gypsum Board 444
09300 Tile 448
09400 Terrazzo 450
09500 Ceilings 451
09600 Flooring 452
09700 Wall Finishes 455
09900 Paints & Coatings 456

Specialties 469

10150 Compartments & Cubicles 470
10200 Louvers & Vents 470
10300 Fireplaces & Stoves 471
10340 Manufactured Exterior Specialties 472
10350 Flagpoles 473
10520 Fire Protection Specialties 473
10530 Protective Covers 473
10550 Postal Specialties 474
10670 Storage Shelving 474
10800 Toilet, Bath, & Laundry Access. 475

Equipment 477

11010 Maintenance Equipment 478
11400 Food Service Equipment 478
11450 Residential Equipment 478

Furnishings 481

12300 Manufactured Casework 482
12400 Furnishings & Access. 482

Special Construction 485

13030 Special Purpose Rooms 486
13100 Lightning Protection 486
13120 Pre-Engineered Structures 486
13150 Swimming Pools 487
13200 Storage Tanks 488
13280 Hazardous Mat. Remediation 489
13800 Building Automation & Control 490

Conveying Systems 493

14200 Elevators 494

Mechanical 495

15050 Basic Materials & Methods 496
15100 Building Services Piping 497
15400 Plumbing Fixtures & Equipment 506
15500 Heat Generation Equipment 509
15700 Heating, Ventilation, & A/C Equip. 512
15800 Air Distribution 514

Electrical 517

16050 Basic Elect. Mat. & Methods 518
16100 Wiring Methods 519
16200 Electrical Power 527
16400 Low-Voltage Distribution 528
16500 Lighting 529
16800 Sound & Video 531

How to Use the Unit Price Pages

The following is a detailed explanation of a sample entry in the Unit Price Section. Next to each bold number below is the item being described with appropriate component of the sample entry following in parenthesis. Some prices are listed as bare costs, others as costs that include overhead and profit of the installing contractor. In most cases, if the work is to be subcontracted, the general contractor will need to add an additional markup (RSMeans suggests using 10%) to the figures in the column "Total Incl. O&P."

1 *Division Number/Title (03300/Cast-In-Place Concrete)*

Use the Unit Price Section Table of Contents to locate specific items. The sections are classified according to the CSI MasterFormat (1995 Edition).

2 *Line Numbers (03310 240 3900)*

Each unit price line item has been assigned a unique 12-digit code based on the CSI MasterFormat classification.

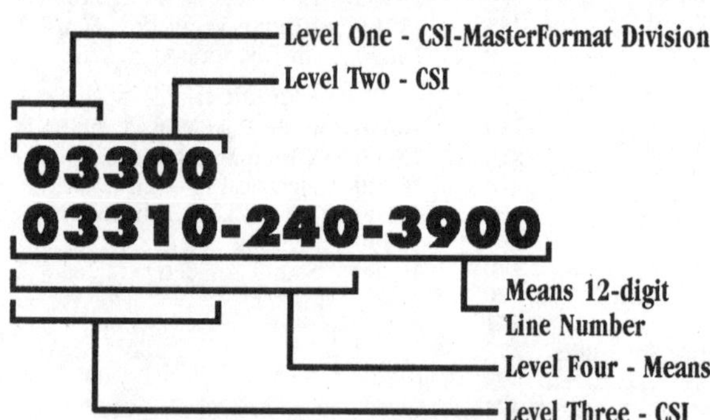

3 *Description (Concrete-In-Place, etc.)*

Each line item is described in detail. Sub-items and additional sizes are indented beneath the appropriate line items. The first line or two after the main item (in boldfac may contain descriptive information that pertains to all lir items beneath this boldface listing.

4 *Reference Number Information*

R03310 -010 You'll see reference numbers shown in bold rectangles at the beginning of some sections. These refer to related items in the Reference Section, visually identified by a vertical gray ba on the edge of pages.

The relation may be: (1) an estimating procedure that should be read before estimating, (2) an alternate pricing method, or (3) technical information.

The "R" designates the Reference Section. The numbers refer to the MasterFormat classification system.

It is strongly recommended that you review all reference numbers that appear within the section in which you are working.

Note: Not all reference numbers appear in all Mean publications.

03300 | Cast-In-Place Concrete

		03310 | Structural Concrete	CREW	DAILY OUTPUT	LABOR-HOURS	UNIT	2005 BARE COSTS MAT.	LABOR	EQUIP.	TOTAL	TOTAL INCL O&P	
240	0010	**CONCRETE IN PLACE** Including forms (4 uses), reinforcing										240
	0050	steel and finishing unless otherwise indicated										
	0500	Chimney foundations, industrial, minimum	C-14C	32.22	3.476	C.Y.	112	78	.73	190.73	257	
	0510	Maximum		23.71	4.724		119	106	.99	225.99	310	
	3800	Footings, spread under 1 C.Y.		38.07	2.942		162	66	.62	228.62	291	
	3850	Over 5 C.Y.		81.04	1.382		226	31	.29	257.29	300	
	3900	Footings, strip, 18" x 9", unreinforced	C-14C	40	2.800	C.Y.	101	62.50	.59	164.09	219	
	3920	18" x 9", reinforced		35	3.200		120	71.50	.67	192.17	255	
	3925	20" x 10", unreinforced		45	2.489		98	55.50	.52	154.02	204	
	3930	20" x 10", reinforced		40	2.800		114	62.50	.59	177.09	233	
	3935	24" x 12", unreinforced		55	2.036		97	45.50	.43	142.93	185	
	3940	24" x 12", reinforced		48	2.333		113	52	.49	165.49	214	

5 *Crew (C-14C)*

The "Crew" column designates the typical trade or crew used to install the item. If an installation can be accomplished by one trade and requires no power equipment, that trade and the number of workers are listed (for example, "2 Carpenters"). If an installation requires a composite crew, a crew code designation is listed (for example, "C-14C"). You'll find full details on all composite crews in the Crew Listings.

- For a complete list of all trades utilized in this book and their abbreviations, see the inside back cover.

Crews

Crew No.	Bare Costs		Incl. Subs O & P		Cost Per Labor-Hour	
Crew C-14C	Hr.	Daily	Hr.	Daily	Bare Costs	Incl. O&P
1 Carpenter Foreman (out)	$26.00	$208.00	$44.10	$352.80	$22.38	$38.19
6 Carpenters	24.00	1152.00	40.75	1956.00		
2 Rodmen (reinf.)	25.45	407.20	45.55	728.80		
4 Laborers	17.35	555.20	29.45	942.40		
1 Cement Finisher	23.00	184.00	37.10	296.80		
1 Gas Engine Vibrator		24.00		26.40	.21	.24
112 L.H., Daily Totals		$2530.40		$4303.20	$22.59	$38.43

6 *Productivity: Daily Output (40)/Labor-Hours (2.800)*

The "Daily Output" represents the typical number of units the designated crew will install in a normal 8-hour day. To find out the number of days the given crew would require to complete the installation, divide your quantity by the daily output. For example:

Quantity	÷	Daily Output	=	Duration
100 C.Y.	÷	40/ Crew Day	=	2.5 Crew Days

The "Labor-Hours" figure represents the number of labor-hours required to install one unit of work. To find out the number of labor-hours required for your particular task, multiply the quantity of the item times the number of labor-hours shown. For example:

Quantity	x	Productivity Rate	=	Duration
100 C.Y.	x	2.800 Labor-Hours/ C.Y.	=	280 Labor-Hours

7 *Unit (C.Y.)*

The abbreviated designation indicates the unit of measure upon which the price, production, and crew are based (C.Y. = Cubic Yard). For a complete listing of abbreviations refer to the Abbreviations Listing in the Reference Section of this book.

8 *Bare Costs:*

Mat. (Bare Material Cost) (101)

The unit material cost is the "bare" material cost with no overhead and profit included. *Costs shown reflect national average material prices for January of the current year and include delivery to the job site. No sales taxes are included.*

Labor (62.50)

The unit labor cost is derived by multiplying bare labor-hour costs for Crew C-14C by labor-hour units. The bare labor-hour cost is found in the Crew Section under C-14C. (If a trade is listed, the hourly labor cost—the wage rate—is found on the inside back cover.)

Labor-Hour Cost Crew C-14C	x	Labor-Hour Units	=	Labor
$22.38	x	2.800	=	$62.50

Equip. (Equipment) (.59)

Equipment costs for each crew are listed in the description of each crew. Tools or equipment whose value justifies purchase or ownership by a contractor are considered overhead as shown on the inside back cover. The unit equipment cost is derived by multiplying the bare equipment hourly cost by the labor-hour units.

Equipment Cost Crew C-14C	x	Labor-Hour Units	=	Equip.
.21	x	2.800	=	.59

Total (164.09)

The total of the bare costs is the arithmetic total of the three previous columns: mat., labor, and equip.

Material	+	Labor	+	Equip.	=	Total
$101	+	$62.50	+	$.59	=	$164.09

9 *Total Costs Including O&P*

This figure is the sum of the bare material cost plus 10% for profit; the bare labor cost plus total overhead and profit (per the inside back cover or, if a crew is listed, from the crew listings); and the bare equipment cost plus 10% for profit.

Material is Bare Material Cost + 10% = 101 + 10.10	=	$111.10
Labor for Crew C-14C = Labor-Hour Cost (38.19) x Labor-Hour Units (2.800)	=	$106.93
Equip. is Bare Equip. Cost + 10% = .59 + .06	=	$ .65
Total (Rounded)	=	$219

Division 1 General Requirements

'stimating Tips

ıe General Requirements of any contract are ·ry important to both the bidder and the vner. These lay the ground rules under hich the contract will be executed and have significant influence on the cost of perations. Therefore, it is extremely important thoroughly read and understand the General equirements both before preparing an timate and when the estimate is complete, to certain that nothing in the contract is verlooked. Caution should be exercised when pplying items listed in Division 1 to an timate. Many of the items are included in the nit prices listed in the other divisions such as ark-ups on labor and company overhead.

1200 Price & Payment Procedures

When estimating historic preservation projects (depending on the condition of the existing structure and the owner's requirements), a 15-20% contingency or allowance is recommended, regardless of the stage of the drawings.

1300 Administrative Requirements

Before determining a final cost estimate, it is a good practice to review all the items listed in subdivision 01300 to make final adjustments for items that may need customizing to specific job conditions.

Historic preservation projects may require specialty labor and methods, as well as extra time to protect existing materials that must be preserved and/or restored. Some additional expenses may be incurred in architectural fees for facility surveys and other special inspections and analyses.

Requirements for initial and periodic submittals can represent a significant cost to the General Requirements of a job. Thoroughly check the submittal specifications when estimating a project to determine any costs that should be included.

01400 Quality Requirements

- All projects will require some degree of Quality Control. This cost is not included in the unit cost of construction listed in each division. Depending upon the terms of the contract, the various costs of inspection and testing can be the responsibility of either the owner or the contractor. Be sure to include the required costs in your estimate.

01500 Temporary Facilities & Controls

- Barricades, access roads, safety nets, scaffolding, security and many more requirements for the execution of a safe project are elements of direct cost. These costs can easily be overlooked when preparing an estimate. When looking through the major classifications of this subdivision, determine which items apply to each division in your estimate.

01590 Equipment Rental

- This subdivision contains transportation, handling, storage, protection and product options and substitutions. Listed in this cost manual are average equipment rental rates for all types of equipment. This is useful information when estimating the time and materials requirement of any particular operation in order to establish a unit or total cost.
- A good rule of thumb is that weekly rental is 3 times daily rental and that monthly rental is 3 times weekly rental.
- The figures in the column for Crew Equipment Cost represent the rental rate used in determining the daily cost of equipment in a crew. It is calculated by dividing the weekly rate by 5 days and adding the hourly operating cost times 8 hours.

01740 Execution Requirements

- When preparing an estimate, read the specifications to determine the requirements for Contract Closeout thoroughly. Final cleaning, record documentation, operation and maintenance data, warranties and bonds, and spare parts and maintenance materials can all be elements of cost for the completion of a contract. Do not overlook these in your estimate.

01830 Operations & Maintenance

- If maintenance and repair are included in your contract, they require special attention. To estimate the cost to remove and replace any unit usually requires a site visit to determine the accessibility and the specific difficulty at that location. Obstructions, dust control, safety, and often overtime hours must be considered when preparing your estimate.

Reference Numbers

Reference numbers are shown in bold squares at the beginning of some major classifications. These numbers refer to related items in the Reference Section. The reference information may be an estimating procedure, an alternate pricing method or technical information.

Note: Not all subdivisions listed here necessarily appear in this publication.

Io part of this publication may be reproduced, stored in a retrieval system, or transmitted in any form r by any means without prior written permission of Reed Construction Data.

01100 | Summary

01103 | Models & Renderings

			CREW	DAILY OUTPUT	LABOR-HOURS	UNIT	2005 BARE COSTS MAT.	LABOR	EQUIP.	TOTAL	TOTAL INCL O&P
500	0010	**RENDERINGS** Color, matted, 20" x 30", eye level,									
	0050	Average				Ea.	2,575			2,575	2,825

01107 | Professional Consultant

			CREW	DAILY OUTPUT	LABOR-HOURS	UNIT	2005 BARE COSTS MAT.	LABOR	EQUIP.	TOTAL	TOTAL INCL O&P
100	0011	**ARCHITECTURAL FEES** R01107-010									
	0020	For new construction									
	0060	Minimum				Project					4.90%
	0090	Maximum									16%
	0100	For alteration work, to $500,000, add to fee									50%
	0150	Over $500,000, add to fee									25%
200	0011	**CONSTRUCTION MANAGEMENT FEES**									
	0060	For work to $10,000				Project					10%
	0070	To $25,000									9%
	0090	To $100,000									6%
700	0010	**SURVEYING** Conventional, topographical, minimum	A-7	3.30	7.273	Acre	16.30	183	18.40	217.70	350
	0100	Maximum	A-8	.60	53.333		49	1,325	101	1,475	2,400
	0300	Lot location and lines, minimum, for large quantities	A-7	2	12		25.50	305	30.50	361	570
	0320	Average	"	1.25	19.200		46	485	48.50	579.50	920
	0400	Maximum, for small quantities	A-8	1	32		73.50	795	61	929.50	1,475
	0600	Monuments, 3' long	A-7	10	2.400	Ea.	20	60.50	6.05	86.55	131
	0800	Property lines, perimeter, cleared land	"	1,000	.024	L.F.	.03	.61	.06	.70	1.12
	0900	Wooded land	A-8	875	.037	"	.05	.91	.07	1.03	1.66
	1100	Crew for layout of building, trenching or pipe laying, 2 person crew	A-6	1	16	Day		380	60.50	440.50	700
	1200	3 person crew	A-7	1	24	"		605	60.50	665.50	1,100

01200 | Price & Payment Procedures

01250 | Contract Modification Procedures

			CREW	DAILY OUTPUT	LABOR-HOURS	UNIT	2005 BARE COSTS MAT.	LABOR	EQUIP.	TOTAL	TOTAL INCL O&P
200	0010	**CONTINGENCIES** for estimate at conceptual stage				Project					20%
	0150	Final working drawing stage				"					3%

01290 | Payment Procedures

			CREW	DAILY OUTPUT	LABOR-HOURS	UNIT	2005 BARE COSTS MAT.	LABOR	EQUIP.	TOTAL	TOTAL INCL O&P
800	0010	**TAXES** Sales tax, State, average R01100-090				%	4.65%				
	0050	Maximum					7%				
	0200	Social Security, on first $87,900 of wages R01100-100						7.65%			
	0300	Unemployment, MA, combined Federal and State, minimum						2.10%			
	0350	Average						6.20%			
	0400	Maximum						8%			

01300 | Administrative Requirements

01310 | Project Management/Coordination

			CREW	DAILY OUTPUT	LABOR-HOURS	UNIT	2005 BARE COSTS MAT.	LABOR	EQUIP.	TOTAL	TOTAL INCL O&P
150	0010	**PERMITS** Rule of thumb, most cities, minimum				Job					.50%
	0100	Maximum				"					2%

01300 | Administrative Requirements

01310 | Project Management/Coordination

			CREW	DAILY OUTPUT	LABOR-HOURS	UNIT	2005 BARE COSTS MAT.	LABOR	EQUIP.	TOTAL	TOTAL INCL O&P	
350	0010	**INSURANCE** Builders risk, standard, minimum R01100-040				Job					.22%	350
	0050	Maximum									.59%	
	0200	All-risk type, minimum R01100-060									.25%	
	0250	Maximum				↓					.62%	
	0400	Contractor's equipment floater, minimum				Value					.50%	
	0450	Maximum				"					1.50%	
	0600	Public liability, average				Job					1.55%	
	0800	Workers' compensation & employer's liability, average										
	0850	by trade, carpentry, general				Payroll		18.51%				
	0900	Clerical						.60%				
	0950	Concrete						15.79%				
	1000	Electrical						6.40%				
	1050	Excavation						10.34%				
	1100	Glazing						13.82%				
	1150	Insulation						15.24%				
	1200	Lathing						10.71%				
	1250	Masonry						14.97%				
	1300	Painting & decorating						12.89%				
	1350	Pile driving						22.94%				
	1400	Plastering						14.62%				
	1450	Plumbing						7.78%				
	1500	Roofing						31.75%				
	1550	Sheet metal work (HVAC)						11.09%				
	1600	Steel erection, structural						38.86%				
	1650	Tile work, interior ceramic						9.63%				
	1700	Waterproofing, brush or hand caulking						7.27%				
	1800	Wrecking						40.51%				
	2000	Range of 35 trades in 50 states, excl. wrecking, min.						2.50%				
	2100	Average						16.20%				
	2200	Maximum				↓		110.10%				

01500 | Temporary Facilities & Controls

01540 | Construction Aids

			CREW	DAILY OUTPUT	LABOR-HOURS	UNIT	2005 BARE COSTS MAT.	LABOR	EQUIP.	TOTAL	TOTAL INCL O&P	
550	0010	**PUMP STAGING**, Aluminum										550
	1300	System in place, 50' working height, per use based on 50 uses	2 Carp	84.80	.189	C.S.F.	5.30	4.53		9.83	13.50	
	1400	100 uses R01540-200		84.80	.189		2.65	4.53		7.18	10.60	
	1500	150 uses	↓	84.80	.189	↓	1.77	4.53		6.30	9.65	
750	0010	**SCAFFOLDING** R01540-100										750
	0015	Steel tubular, reg, rent/mo, no plank, incl erect or dismantle										
	0090	Building exterior, wall face, 1 to 5 stories, 6'-4" x 5' frames	3 Carp	24	1	C.S.F.	24.50	24		48.50	67.50	
	0200	6 to 12 stories	4 Carp	21.20	1.509		24.50	36		60.50	88	
	0310	13 to 20 stories	5 Carp	20	2		24.50	48		72.50	108	
	0460	Building interior, wall face area, up to 16' high	3 Carp	25	.960		24.50	23		47.50	65.50	
	0560	16' to 40' high		23	1.043	↓	24.50	25		49.50	69	
	0800	Building interior floor area, up to 30' high	↓	312	.077	C.C.F.	2.57	1.85		4.42	5.95	
	0900	Over 30' high	4 Carp	275	.116	"	2.57	2.79		5.36	7.55	
	0910	Steel tubular, heavy duty shoring, buy										
	0920	Frames 5' high 2' wide				Ea.	82.50			82.50	91	
	0925	5' high 4' wide				↓	93.50			93.50	103	

01540 | Construction Aids			CREW	DAILY OUTPUT	LABOR-HOURS	UNIT	2005 BARE COSTS MAT.	LABOR	EQUIP.	TOTAL	TOTAL INCL O&P
750	0930	6′ high 2′ wide R01540-100				Ea.	94.50			94.50	104
	0935	6′ high 4′ wide				↓	111			111	122
	0940	Accessories									
	0945	Cross braces				Ea.	17.60			17.60	19.35
	0950	U-head, 8″ x 8″					19.25			19.25	21
	0955	J-head, 4″ x 8″					14.10			14.10	15.50
	0960	Base plate, 8″ x 8″					15.60			15.60	17.20
	0965	Leveling jack				↓	33.50			33.50	37
	1000	Steel tubular, regular, buy									
	1100	Frames 3′ high 5′ wide				Ea.	64			64	70
	1150	5′ high 5′ wide					73.50			73.50	81
	1200	6′-4″ high 5′ wide					92.50			92.50	102
	1350	7′-6″ high 6′ wide					160			160	175
	1500	Accessories cross braces					16.50			16.50	18.15
	1550	Guardrail post					16.50			16.50	18.15
	1600	Guardrail 7′ section					8			8	8.80
	1650	Screw jacks & plates					26.50			26.50	29
	1700	Sidearm brackets					31			31	34
	1750	8″ casters					36.50			36.50	40
	1800	Plank 2″ x 10″ x 16′-0″					47			47	51.50
	1900	Stairway section					270			270	296
	1910	Stairway starter bar					32			32	35
	1920	Stairway inside handrail					58.50			58.50	64
	1930	Stairway outside handrail					80.50			80.50	88.50
	1940	Walk-thru frame guardrail				↓	40.50			40.50	44.50
	2000	Steel tubular, regular, rent/mo.									
	2100	Frames 3′ high 5′ wide				Ea.	3.75			3.75	4.13
	2150	5′ high 5′ wide					3.75			3.75	4.13
	2200	6′-4″ high 5′ wide					3.75			3.75	4.13
	2250	7′-6″ high 6′ wide					7			7	7.70
	2500	Accessories, cross braces					.60			.60	.66
	2550	Guardrail post					1			1	1.10
	2600	Guardrail 7′ section					.75			.75	.83
	2650	Screw jacks & plates					1.50			1.50	1.65
	2700	Sidearm brackets					1.50			1.50	1.65
	2750	8″ casters					6			6	6.60
	2800	Outrigger for rolling tower					3			3	3.30
	2850	Plank 2″ x 10″ x 16′-0″					5			5	5.50
	2900	Stairway section					10			10	11
	2910	Stairway starter bar					.10			.10	.11
	2920	Stairway inside handrail					5			5	5.50
	2930	Stairway outside handrail					5			5	5.50
	2940	Walk-thru frame guardrail				↓	2			2	2.20
	3000	Steel tubular, heavy duty shoring, rent/mo.									
	3250	5′ high 2′ & 4′ wide				Ea.	5			5	5.50
	3300	6′ high 2′ & 4′ wide					5			5	5.50
	3500	Accessories, cross braces					1			1	1.10
	3600	U - head, 8″ x 8″					1			1	1.10
	3650	J - head, 4″ x 8″					1			1	1.10
	3700	Base plate, 8″ x 8″					1			1	1.10
	3750	Leveling jack					2			2	2.20
	5700	Planks, 2x10x16′-0″, labor only, erect or remove to 50′ H	3 Carp	144	.167			4		4	6.80
	5800	Over 50′ high	4 Carp	160	.200			4.80		4.80	8.15
	6820	Erect or dismantle frames, 1st tier	4 Clab	45	.711			12.35		12.35	21
	6830	2nd tier		93	.344			5.95		5.95	10.15
	6840	3rd tier	↓	87	.368	↓		6.40		6.40	10.85

01540	Construction Aids	Crew	Daily Output	Labor-Hours	Unit	2005 Bare Costs Mat.	Labor	Equip.	Total	Total Incl O&P	
6850	4th tier R01540-100	4 Clab	75	.427	Ea.		7.40		7.40	12.55	750
0010	**STAGING AIDS** and fall protection equipment										760
0100	Sidewall staging bracket, tubular, buy				Ea.	30.50			30.50	33.50	
0110	Cost each per day, based on 250 days use				Day	.12			.12	.13	
0200	Guard post, buy				Ea.	15			15	16.50	
0210	Cost each per day, based on 250 days use				Day	.06			.06	.07	
0300	End guard chains, buy per pair				Pair	25			25	27.50	
0310	Cost per set per day, based on 250 days use				Day	.12			.12	.13	
1010	Cost each per day, based on 250 days use				″	.03			.03	.03	
1100	Wood bracket, buy				Ea.	13.50			13.50	14.85	
1110	Cost each per day, based on 250 days use				Day	.05			.05	.06	
2010	Cost per pair per day, based on 250 days use				″	.34			.34	.37	
2100	Steel siderail jack, buy per pair				Pair	63			63	69.50	
2110	Cost per pair per day, based on 250 days use				Day	.25			.25	.28	
3010	Cost each per day, based on 250 days use				″	.17			.17	.19	
3100	Aluminum scaffolding plank, 20″ wide x 24′ long, buy				Ea.	690			690	760	
3110	Cost each per day, based on 250 days use				Day	2.76			2.76	3.04	
4010	Cost each per day, based on 250 days use				″	.83			.83	.91	
4100	Rope for safety line, 5/8″ x 100′ nylon, buy				Ea.	41			41	45	
4110	Cost each per day, based on 250 days use				Day	.16			.16	.18	
4200	Permanent U-Bolt roof anchor, buy				Ea.	31			31	34	
4300	Temporary (one use) roof ridge anchor, buy				″	24			24	26.50	
5000	Installation (setup and removal) of staging aids										
5010	Sidewall staging bracket	2 Carp	64	.250	Ea.		6		6	10.20	
5020	Guard post with 2 wood rails	″	64	.250			6		6	10.20	
5030	End guard chains, set	1 Carp	64	.125			3		3	5.10	
5100	Roof shingling bracket		96	.083			2		2	3.40	
5200	Ladder jack	↓	64	.125			3		3	5.10	
5300	Wood plank, 2x10x16′	2 Carp	80	.200			4.80		4.80	8.15	
5310	Aluminum scaffold plank, 20″ x 24′	″	40	.400			9.60		9.60	16.30	
5410	Safety rope	1 Carp	40	.200			4.80		4.80	8.15	
5420	Permanent U-Bolt roof anchor (install only)	2 Carp	40	.400			9.60		9.60	16.30	
5430	Temporary roof ridge anchor (install only)	1 Carp	64	.125	↓		3		3	5.10	
0010	**TARPAULINS** Cotton duck, 10 oz. to 13.13 oz. per S.Y., minimum				S.F.	.49			.49	.54	800
0050	Maximum					.58			.58	.64	
0200	Reinforced polyethylene 3 mils thick, white					.11			.11	.12	
0300	4 mils thick, white, clear or black					.14			.14	.15	
0730	Polyester reinforced w/ integral fastening system 11 mils thick				↓	1.07			1.07	1.18	
0010	**SMALL TOOLS** As % of contractor's work, minimum				Total					.50%	820
0100	Maximum	″			″					2%	

01590 | Equipment Rental

		01590 Equipment Rental	UNIT	HOURLY OPER. COST	RENT PER DAY	RENT PER WEEK	RENT PER MONTH	CREW EQUIPMENT COST/DAY
100	0010	**CONCRETE EQUIPMENT RENTAL**						
	0100	without operators						
	0150	For batch plant, see div. 01590 500						
	0200	Bucket, concrete lightweight, 1/2 C.Y.	Ea.	.55	15.35	46	138	13.60
	0300	1 C.Y.		.60	18.65	56	168	16
	0400	1 1/2 C.Y.		.75	25.50	76	228	21.20
	0500	2 C.Y.		.80	30.50	91	273	24.60
	0580	8 C.Y.		4.40	200	600	1,800	155.20
	0600	Cart, concrete, self propelled, operator walking, 10 C.F.		2.10	55	165	495	49.80
	0700	Operator riding, 18 C.F.		3.35	83.50	250	750	76.80
	0800	Conveyer for concrete, portable, gas, 16" wide, 26' long		7.35	117	350	1,050	128.80
	0900	46' long		7.70	142	425	1,275	146.60
	1000	56' long		7.85	150	450	1,350	152.80
	1100	Core drill, electric, 2 1/2 H.P., 1" to 8" bit diameter		1.61	62.50	187	560	50.30
	1150	11 HP, 8" to 18" cores		6.61	83.50	250.80	750	103.05
	1200	Finisher, concrete floor, gas, riding trowel, 48" diameter		4.95	86.50	260	780	91.60
	1300	Gas, manual, 3 blade, 36" trowel		1.05	16.65	50	150	18.40
	1400	4 blade, 48" trowel		1.50	21	63	189	24.60
	1500	Float, hand operated (Bull float) 48" wide		.08	13.35	40	120	8.65
	1570	Curb builder, 14 H.P., gas, single screw		9.50	195	585	1,750	193
	1590	Double screw		9.90	228	685	2,050	216.20
	1600	Grinder, concrete and terrazzo, electric, floor		2.10	88.50	266	800	70
	1700	Wall grinder		1.05	44.50	133	400	35
	1800	Mixer, powered, mortar and concrete, gas, 6 C.F., 18 H.P.		5.25	102	305	915	103
	1900	10 C.F., 25 H.P.		6.40	120	360	1,075	123.20
	2000	16 C.F.		6.70	143	430	1,300	139.60
	2100	Concrete, stationary, tilt drum, 2 C.Y.		5.25	200	600	1,800	162
	2120	Pump, concrete, truck mounted 4" line 80' boom		21.65	885	2,655	7,975	704.20
	2140	5" line, 110' boom		28.05	1,175	3,510	10,500	926.40
	2160	Mud jack, 50 C.F. per hr.		4.82	115	346	1,050	107.75
	2180	225 C.F. per hr.		7.56	154	461.60	1,375	152.80
	2190	Shotcrete pump rig, 12 CY/hr		11.10	225	675	2,025	223.80
	2600	Saw, concrete, manual, gas, 18 H.P.		3.60	35	105	315	49.80
	2650	Self propelled, gas, 30 H.P.		7.05	96.50	290	870	114.40
	2700	Vibrators, concrete, electric, 60 cycle, 2 H.P.		.37	10.65	32	96	9.35
	2800	3 H.P.		.56	15.65	47	141	13.90
	2900	Gas engine, 5 H.P.		.95	18.35	55	165	18.60
	3000	8 H.P.		1.35	22	66	198	24
	3050	Vibrating screed, gas engine, 8 H.P.		1.68	49.50	148	445	43.05
	3100	Concrete transit mixer, hydraulic drive						
	3120	6 x 4, 250 H.P., 8 C.Y., rear discharge		34.95	550	1,650	4,950	609.60
	3200	Front discharge		40.65	670	2,015	6,050	728.20
	3300	6 x 6, 285 H.P., 12 C.Y., rear discharge		40.60	670	2,010	6,025	726.80
	3400	Front discharge	↓	41.55	680	2,040	6,125	740.40
200	0010	**EARTHWORK EQUIPMENT RENTAL** Without operators						
	0040	Aggregate spreader, push type 8' to 12' wide	Ea.	1.80	34.50	103	310	35
	0045	Tailgate type, 8' wide	"	1.75	31.50	95	285	33
	0050	Augers for vertical drilling						
	0055	Earth auger, truck mounted, for fence & sign posts	Ea.	8.30	485	1,455	4,375	357.40
	0060	For borings and monitoring wells		29.20	620	1,855	5,575	604.60
	0070	Earth auger, portable, trailer mounted		1.55	22.50	67	201	25.80
	0075	Earth auger, truck mounted, for caissons, water wells, utility poles		153.10	3,250	9,770	29,300	3,179
	0080	Auger, horizontal boring machine, 12" to 36" diameter, 45 H.P.		15.85	182	545	1,625	235.80
	0090	12" to 48" diameter, 65 H.P.		22.30	325	970	2,900	372.40
	0095	Auger, for fence posts, gas engine, hand held		.30	4.33	13	39	5
	0100	Excavator, diesel hydraulic, crawler mounted, 1/2 C.Y. cap.		15.80	345	1,030	3,100	332.40
	0120	5/8 C.Y. capacity		19.30	465	1,390	4,175	432.40
	0140	3/4 C.Y. capacity	↓	22.50	490	1,470	4,400	474

01590 | Equipment Rental 200

	01590 Equipment Rental	UNIT	HOURLY OPER. COST	RENT PER DAY	RENT PER WEEK	RENT PER MONTH	CREW EQUIPMENT COST/DAY
0150	1 C.Y. capacity	Ea.	27.10	570	1,705	5,125	557.80
0200	1 1/2 C.Y. capacity		33.55	755	2,260	6,775	720.40
0300	2 C.Y. capacity		42.35	950	2,855	8,575	909.80
0320	2 1/2 C.Y. capacity		55.75	1,275	3,845	11,500	1,215
0340	3 1/2 C.Y. capacity		94.45	2,100	6,320	19,000	2,020
0341	Attachments						
0342	Bucket thumbs		2.50	208	625	1,875	145
0345	Grapples		1	200	600.40	1,800	128.10
0350	Gradall type, truck mounted, 3 ton @ 15' radius, 5/8 C.Y.		38.35	885	2,655	7,975	837.80
0370	1 C.Y. capacity		44.25	1,025	3,085	9,250	971
0400	Backhoe loader, 40 to 45 H.P., 5/8 C.Y. capacity		8.40	177	530	1,600	173.20
0450	45 H.P. to 60 H.P., 3/4 C.Y. capacity		10.60	218	655	1,975	215.80
0460	80 H.P., 1 1/4 C.Y. capacity		13.30	250	750	2,250	256.40
0470	112 H.P., 1 1/2 C.Y. capacity		18.10	390	1,165	3,500	377.80
0480	Attachments						
0482	Compactor, 20,000 lb		4.25	117	350	1,050	104
0485	Hydraulic hammer, 750 ft lbs		1.95	68.50	205	615	56.60
0486	Hydraulic hammer, 1200 ft lbs		4	133	400	1,200	112
0500	Brush chipper, gas engine, 6" cutter head, 35 H.P.		5.90	93.50	280	840	103.20
0550	12" cutter head, 130 H.P.		9.35	150	450	1,350	164.80
0600	15" cutter head, 165 H.P.		13.40	158	475	1,425	202.20
0750	Bucket, clamshell, general purpose, 3/8 C.Y.		1	35	105	315	29
0800	1/2 C.Y.		1.10	41.50	125	375	33.80
0850	3/4 C.Y.		1.25	51.50	155	465	41
0900	1 C.Y.		1.30	55	165	495	43.40
0950	1 1/2 C.Y.		2.05	75	225	675	61.40
1000	2 C.Y.		2.15	85	255	765	68.20
1010	Bucket, dragline, medium duty, 1/2 C.Y.		.60	22.50	67	201	18.20
1020	3/4 C.Y.		.60	23.50	70	210	18.80
1030	1 C.Y.		.65	25.50	76	228	20.40
1040	1 1/2 C.Y.		.95	38.50	115	345	30.60
1050	2 C.Y.		1.05	43.50	130	390	34.40
1070	3 C.Y.		1.55	58.50	175	525	47.40
1200	Compactor, manually guided 2 drum vibratory smooth roller, 7.5 H.P.		4.60	143	430	1,300	122.80
1250	Rammer compactor, gas, 1000 lb. blow		1.60	36.50	110	330	34.80
1300	Vibratory plate, gas, 18" plate, 3000 lb. blow		1.50	22	66	198	25.20
1350	21" plate, 5000 lb. blow		1.80	37.50	112	335	36.80
1370	Curb builder/extruder, 14 H.P., gas, single screw		9.50	197	590	1,775	194
1390	Double screw		9.90	228	685	2,050	216.20
1500	Disc harrow attachment, for tractor	↓	.36	60.50	181	545	39.10
1750	Extractor, piling, see lines 2500 to 2750						
1810	Feller buncher, shearing & accumulating trees, 100 H.P.	Ea.	19.80	455	1,370	4,100	432.40
1860	Grader, self propelled, 25,000 lb.		18.20	400	1,205	3,625	386.60
1910	30,000 lb.		20.70	485	1,455	4,375	456.60
1920	40,000 lb.		30.50	730	2,190	6,575	682
1930	55,000 lb.		40.25	1,025	3,080	9,250	938
1950	Hammer, pavement demo., hyd., gas, self prop., 1000 to 1250 lb.		20.15	390	1,170	3,500	395.20
2000	Diesel 1300 to 1500 lb.		27.80	590	1,765	5,300	575.40
2050	Pile driving hammer, steam or air, 4150 ft. lb. @ 225 BPM		6.35	275	825	2,475	215.80
2100	8750 ft. lb. @ 145 BPM		8.25	450	1,350	4,050	336
2150	15,000 ft. lb. @ 60 BPM		8.60	485	1,455	4,375	359.80
2200	24,450 ft. lb. @ 111 BPM	↓	11.35	535	1,605	4,825	411.80
2250	Leads, 15,000 ft. lb. hammers	L.F.	.03	1.69	5.06	15.20	1.25
2300	24,450 ft. lb. hammers and heavier	"	.05	2.67	8	24	2
2350	Diesel type hammer, 22,400 ft. lb.	Ea.	24	620	1,860	5,575	564
2400	41,300 ft. lb.		32.20	675	2,025	6,075	662.60
2450	141,000 ft. lb.		60.90	1,475	4,430	13,300	1,373
2500	Vib. elec. hammer/extractor, 200 KW diesel generator, 34 H.P.	↓	26.25	655	1,965	5,900	603

01590 | Equipment Rental

		Description	UNIT	HOURLY OPER. COST	RENT PER DAY	RENT PER WEEK	RENT PER MONTH	CREW EQUIPMENT COST/DAY
200	2550	80 H.P.	Ea.	44.80	960	2,885	8,650	935.40
	2600	150 H.P.		64	1,475	4,455	13,400	1,403
	2700	Extractor, steam or air, 700 ft. lb.		14.50	410	1,235	3,700	363
	2750	1000 ft. lb.		16.60	515	1,545	4,625	441.80
	2800	Log chipper, up to 22" diam, 600 H.P.		38.53	1,250	3,770	11,300	1,062
	2850	Logger, for skidding & stacking logs, 150 H.P.		33.80	790	2,375	7,125	745.40
	2900	Rake, spring tooth, with tractor		7.99	213	638	1,925	191.50
	3000	Roller, vibratory, tandem, smooth drum, 20 H.P.		5.05	112	335	1,000	107.40
	3050	35 H.P.		7.15	203	610	1,825	179.20
	3100	Towed type vibratory compactor, smooth drum, 50 H.P.		31.95	560	1,680	5,050	591.60
	3150	Sheepsfoot, 50 H.P.		32.65	580	1,745	5,225	610.20
	3170	Landfill compactor, 220 HP		46.75	1,150	3,470	10,400	1,068
	3200	Pneumatic tire roller, 80 H.P.		8.35	292	875	2,625	241.80
	3250	120 H.P.		13.60	505	1,515	4,550	411.80
	3300	Sheepsfoot vibratory roller, 200 H.P.		35.35	855	2,560	7,675	794.80
	3320	340 H.P.		48.90	1,225	3,670	11,000	1,125
	3350	Smooth drum vibratory roller, 75 H.P.		13.60	395	1,180	3,550	344.80
	3400	125 H.P.		17.15	495	1,480	4,450	433.20
	3410	Rotary mower, brush, 60", with tractor		11.35	230	690	2,075	228.80
	3450	Scrapers, towed type, 9 to 12 C.Y. capacity		4.12	196	589	1,775	150.75
	3500	12 to 17 C.Y. capacity		1.51	262	785.60	2,350	169.20
	3550	Scrapers, self propelled, 4 x 4 drive, 2 engine, 14 C.Y. capacity		82.40	1,500	4,465	13,400	1,552
	3600	2 engine, 24 C.Y. capacity		119.45	2,275	6,805	20,400	2,317
	3640	32 44 C.Y. capacity		142.65	2,650	7,935	23,800	2,728
	3650	Self loading, 11 C.Y. capacity		40.55	825	2,475	7,425	819.40
	3700	22 C.Y. capacity		76.40	1,725	5,185	15,600	1,648
	3710	Screening plant 110 H.P. w/ 5' x 10' screen		21.75	370	1,105	3,325	395
	3720	5' x 16' screen		23.80	465	1,390	4,175	468.40
	3850	Shovels, see Cranes division 01590 600						
	3860	Shovel/backhoe bucket, 1/2 C.Y.	Ea.	1.80	55	165	495	47.40
	3870	3/4 C.Y.		1.85	61.50	185	555	51.80
	3880	1 C.Y.		1.95	71.50	215	645	58.60
	3890	1 1/2 C.Y.		2.10	85	255	765	67.80
	3910	3 C.Y.		2.40	118	355	1,075	90.20
	3950	Stump chipper, 18" deep, 30 H.P.		4.56	57	171	515	70.70
	4110	Tractor, crawler, with bulldozer, torque converter, diesel 80 H.P.		16.10	310	930	2,800	314.80
	4150	105 H.P.		22.10	460	1,385	4,150	453.80
	4200	140 H.P.		27.25	590	1,770	5,300	572
	4260	200 H.P.		40.95	985	2,960	8,875	919.60
	4310	300 H.P.		53.10	1,275	3,850	11,600	1,195
	4360	410 H.P.		71.95	1,575	4,740	14,200	1,524
	4370	500 H.P.		95.25	2,125	6,365	19,100	2,035
	4380	700 H.P.		142.20	3,325	10,005	30,000	3,139
	4400	Loader, crawler, torque conv., diesel, 1 1/2 C.Y., 80 H.P.		14.95	320	965	2,900	312.60
	4450	1 1/2 to 1 3/4 C.Y., 95 H.P.		17.35	385	1,150	3,450	368.80
	4510	1 3/4 to 2 1/4 C.Y., 130 H.P.		23.60	605	1,815	5,450	551.80
	4530	2 1/2 to 3 1/4 C.Y., 190 H.P.		35.50	840	2,520	7,550	788
	4560	3 1/2 to 5 C.Y., 275 H.P.		46.95	1,200	3,590	10,800	1,094
	4610	Tractor loader, wheel, torque conv., 4 x 4, 1 to 1 1/4 C.Y., 65 H.P.		9.85	192	575	1,725	193.80
	4620	1 1/2 to 1 3/4 C.Y., 80 H.P.		12.50	235	705	2,125	241
	4650	1 3/4 to 2 C.Y., 100 H.P.		14.20	278	835	2,500	280.60
	4710	2 1/2 to 3 1/2 C.Y., 130 H.P.		15.25	305	915	2,750	305
	4730	3 to 4 1/2 C.Y., 170 H.P.		20.95	465	1,390	4,175	445.60
	4760	5 1/4 to 5 3/4 C.Y., 270 H.P.		34.55	705	2,120	6,350	700.40
	4810	7 to 8 C.Y., 375 H.P.		58.95	1,250	3,755	11,300	1,223
	4870	12 1/2 C.Y., 690 H.P.		82.30	1,950	5,860	17,600	1,830
	4880	Wheeled, skid steer, 10 C.F., 30 H.P. gas		8.80	138	415	1,250	153.40
	4890	1 C.Y., 78 H.P., diesel		10.35	192	575	1,725	197.80

	01590 | Equipment Rental	UNIT	HOURLY OPER. COST	RENT PER DAY	RENT PER WEEK	RENT PER MONTH	CREW EQUIPMENT COST/DAY	
4891	Attachments for all skid steer loaders							200
4892	Auger	Ea.	.40	66.50	200	600	43.20	
4893	Backhoe		.65	109	326	980	70.40	
4894	Broom		.63	105	314	940	67.85	
4895	Forks		.22	37.50	112	335	24.15	
4896	Grapple		.50	83	249	745	53.80	
4897	Concrete hammer		.98	163	489	1,475	105.65	
4898	Tree spade		1	167	500	1,500	108	
4899	Trencher		.68	114	342	1,025	73.85	
4900	Trencher, chain, boom type, gas, operator walking, 12 H.P.		2.75	43.50	130	390	48	
4910	Operator riding, 40 H.P.		8.55	242	725	2,175	213.40	
5000	Wheel type, diesel, 4′ deep, 12″ wide		46.90	735	2,205	6,625	816.20	
5100	Diesel, 6′ deep, 20″ wide		63.75	1,575	4,750	14,300	1,460	
5150	Ladder type, diesel, 5′ deep, 8″ wide		23.10	705	2,115	6,350	607.80	
5200	Diesel, 8′ deep, 16″ wide		56.80	1,600	4,830	14,500	1,420	
5210	Tree spade, self propelled		9.60	267	800	2,400	236.80	
5250	Truck, dump, tandem, 12 ton payload		20.45	272	815	2,450	326.60	
5300	Three axle dump, 16 ton payload		27.95	420	1,265	3,800	476.60	
5350	Dump trailer only, rear dump, 16 1/2 C.Y.		4.25	115	345	1,025	103	
5400	20 C.Y.		4.60	132	395	1,175	115.80	
5450	Flatbed, single axle, 1 1/2 ton rating		11.85	56.50	170	510	128.80	
5500	3 ton rating		15.15	91.50	275	825	176.20	
5550	Off highway rear dump, 25 ton capacity		40.40	980	2,945	8,825	912.20	
5600	35 ton capacity		41.30	1,000	3,015	9,050	933.40	
5610	50 ton capacity		53.15	1,250	3,780	11,300	1,181	
5620	65 ton capacity		56.85	1,375	4,105	12,300	1,276	
5630	100 ton capacity		73	1,775	5,325	16,000	1,649	
6000	Vibratory plow, 25 H.P., walking	↓	3.90	58.50	175	525	66.20	
0010	**GENERAL EQUIPMENT RENTAL** Without operators							400
0150	Aerial lift, scissor type, to 15′ high, 1000 lb. cap., electric	Ea.	2.30	41.50	125	375	43.40	
0160	To 25′ high, 2000 lb. capacity		2.70	60	180	540	57.60	
0170	Telescoping boom to 40′ high, 500 lb. capacity, gas		11.45	268	805	2,425	252.60	
0180	To 45′ high, 500 lb. capacity		12.30	310	925	2,775	283.40	
0190	To 60′ high, 600 lb. capacity		14.25	410	1,230	3,700	360	
0195	Air compressor, portable, 6.5 CFM, electric		.42	14.65	44	132	12.15	
0196	Gasoline		.51	22	66	198	17.30	
0200	Air compressor, portable, gas engine, 60 C.F.M.		4.50	32.50	97	291	55.40	
0300	160 C.F.M.		7.80	41.50	125	375	87.40	
0400	Diesel engine, rotary screw, 250 C.F.M.		7.80	83.50	250	750	112.40	
0500	365 C.F.M.		10.35	107	320	960	146.80	
0550	450 C.F.M.		12.55	127	380	1,150	176.40	
0600	600 C.F.M.		21.35	212	635	1,900	297.80	
0700	750 C.F.M.		23.05	222	665	2,000	317.40	
0800	For silenced models, small sizes, add		3%	5%	5%	5%		
0900	Large sizes, add	↓	5%	7%	7%	7%		
0920	Air tools and accessories							
0930	Breaker, pavement, 60 lb.	Ea.	.40	11	33	99	9.80	
0940	80 lb.		.40	12.65	38	114	10.80	
0950	Drills, hand (jackhammer) 65 lb.		.45	15	45	135	12.60	
0960	Track or wagon, swing boom, 4″ drifter		35.05	640	1,920	5,750	664.40	
0970	5″ drifter		47.70	735	2,205	6,625	822.60	
0975	Track mounted quarry drill, 6″ diameter drill		50.45	795	2,380	7,150	879.60	
0980	Dust control per drill		.78	12	36	108	13.45	
0990	Hammer, chipping, 12 lb.		.40	21.50	64	192	16	
1000	Hose, air with couplings, 50′ long, 3/4″ diameter		.03	5.65	17	51	3.65	
1100	1″ diameter		.03	5.65	17	51	3.65	
1200	1 1/2″ diameter		.04	7.35	22	66	4.70	
1300	2″ diameter	↓	.10	17.35	52	156	11.20	

01590 | Equipment Rental

			UNIT	HOURLY OPER. COST	RENT PER DAY	RENT PER WEEK	RENT PER MONTH	CREW EQUIPMENT COST/DAY
400	1400	2 1/2" diameter	Ea.	.12	19.65	59	177	12.75
	1410	3" diameter		.16	27.50	82	246	17.70
	1450	Drill, steel, 7/8" x 2'		.05	8	24	72	5.20
	1460	7/8" x 6'		.06	9.35	28	84	6.10
	1520	Moil points		.03	4.33	13	39	2.85
	1525	Pneumatic nailer w/accessories		.41	27	81	243	19.50
	1530	Sheeting driver for 60 lb. breaker		.10	6.95	20.80	62.50	4.95
	1540	For 90 lb. breaker		.15	10	30	90	7.20
	1550	Spade, 25 lb.		.35	6.35	19	57	6.60
	1560	Tamper, single, 35 lb.		.58	38.50	115	345	27.65
	1570	Triple, 140 lb.		.87	57.50	173	520	41.55
	1580	Wrenches, impact, air powered, up to 3/4" bolt		.25	7.35	22	66	6.40
	1590	Up to 1 1/4" bolt		.35	15.35	46	138	12
	1600	Barricades, barrels, reflectorized, 1 to 50 barrels		.02	2.60	7.80	23.50	1.70
	1610	100 to 200 barrels		.01	1.93	5.80	17.40	1.25
	1620	Barrels with flashers, 1 to 50 barrels		.02	3.27	9.80	29.50	2.10
	1630	100 to 200 barrels		.02	2.60	7.80	23.50	1.70
	1640	Barrels with steady burn type C lights		.03	4.33	13	39	2.85
	1650	Illuminated board, trailer mounted, with generator		.65	117	350	1,050	75.20
	1670	Portable barricade, stock, with flashers, 1 to 6 units		.02	3.27	9.80	29.50	2.10
	1680	25 to 50 units		.02	3.03	9.10	27.50	2
	1690	Butt fusion machine, electric		22.35	425	1,270	3,800	432.80
	1695	Electro fusion machine		8.85	168	505	1,525	171.80
	1700	Carts, brick, hand powered, 1000 lb. capacity		.30	49.50	148	445	32
	1800	Gas engine, 1500 lb., 7 1/2' lift		2.77	94	282	845	78.55
	1822	Dehumidifier, medium, 6 lb/hr, 150 CFM		.68	41.50	124	370	30.25
	1824	Large, 18 lb/hr, 600 CFM		1.36	82.50	248	745	60.50
	1830	Distributor, asphalt, trailer mtd, 2000 gal., 38 H.P. diesel		7.35	243	730	2,200	204.80
	1840	3000 gal., 38 H.P. diesel		8.35	278	835	2,500	233.80
	1850	Drill, rotary hammer, electric, 1 1/2" diameter		.40	24.50	73	219	17.80
	1860	Carbide bit for above		.03	5.35	16	48	3.45
	1865	Rotary, crawler, 250 H.P.		87.35	1,600	4,835	14,500	1,666
	1870	Emulsion sprayer, 65 gal., 5 H.P. gas engine		1.80	72	216	650	57.60
	1880	200 gal., 5 H.P. engine		4.85	120	360	1,075	110.80
	1920	Floodlight, mercury vapor, or quartz, on tripod						
	1930	1000 watt	Ea.	.30	12	36	108	9.60
	1940	2000 watt		.52	22	66	198	17.35
	1950	Floodlights, trailer mounted with generator, 1 300 watt light		2.45	65	195	585	58.60
	1960	2 1000 watt lights		3.30	108	325	975	91.40
	2000	4 300 watt lights		2.75	76.50	230	690	68
	2020	Forklift, wheeled, for brick, 18', 3000 lb., 2 wheel drive, gas		13.70	177	530	1,600	215.60
	2040	28', 4000 lb., 4 wheel drive, diesel		11.75	232	695	2,075	233
	2050	For rough terrain, 8000 lb., 16' lift, 68 H.P.		15.70	360	1,075	3,225	340.60
	2060	For plant, 4 T. capacity, 80 H.P., 2 wheel drive, gas		7.80	83.50	250	750	112.40
	2080	10 T. capacity, 120 H.P., 2 wheel drive, diesel		11.95	157	470	1,400	189.60
	2100	Generator, electric, gas engine, 1.5 KW to 3 KW		1.75	15	45	135	23
	2200	5 KW		2.40	21.50	65	195	32.20
	2300	10 KW		3.95	46.50	140	420	59.60
	2400	25 KW		7.65	71.50	215	645	104.20
	2500	Diesel engine, 20 KW		5.95	61.50	185	555	84.60
	2600	50 KW		11.05	68.50	205	615	129.40
	2700	100 KW		16.20	78.50	235	705	176.60
	2800	250 KW		45.65	142	425	1,275	450.20
	2850	Hammer, hydraulic, for mounting on boom, to 500 ft. lb.		1.75	63.50	190	570	52
	2860	1000 ft. lb.		3.10	102	305	915	85.80
	2900	Heaters, space, oil or electric, 50 MBH		.92	9.65	29	87	13.15
	3000	100 MBH		1.66	13.65	41	123	21.50
	3100	300 MBH		5.33	35	105	315	63.65

01590 | Equipment Rental

400

	01590 Equipment Rental	UNIT	HOURLY OPER. COST	RENT PER DAY	RENT PER WEEK	RENT PER MONTH	CREW EQUIPMENT COST/DAY
3150	500 MBH	Ea.	10.70	50	150	450	115.60
3200	Hose, water, suction with coupling, 20' long, 2" diameter		.02	4	12	36	2.55
3210	3" diameter		.03	6.65	20	60	4.25
3220	4" diameter		.03	8	24	72	5.05
3230	6" diameter		.10	20	60	180	12.80
3240	8" diameter		.25	41.50	125	375	27
3250	Discharge hose with coupling, 50' long, 2" diameter		.01	3	9	27	1.90
3260	3" diameter		.01	4	12	36	2.50
3270	4" diameter		.02	5.65	17	51	3.55
3280	6" diameter		.06	14	42	126	8.90
3290	8" diameter		.31	52	156	470	33.70
3295	Insulation blower		.11	7	21	63	5.10
3300	Ladders, extension type, 16' to 36' long		.17	28.50	86	258	18.55
3400	40' to 60' long		.26	36.50	110	330	24.10
3405	Lance for cutting concrete		2.78	105	315	945	85.25
3407	Lawn mower, rotary, 22", 5HP		.97	27	81	243	23.95
3408	48" self propelled		3.06	100	300	900	84.50
3410	Level, laser type, for pipe and sewer leveling		1.27	84.50	253	760	60.75
3430	Electronic		.76	50.50	151	455	36.30
3440	Laser type, rotating beam for grade control		1.04	69	207	620	49.70
3460	Builders level with tripod and rod		.07	12.35	37	111	7.95
3500	Light towers, towable, with diesel generator, 2000 watt		2.75	76.50	230	690	68
3600	4000 watt		3.30	108	325	975	91.40
3700	Mixer, powered, plaster and mortar, 6 C.F., 7 H.P.		1.30	26.50	80	240	26.40
3800	10 C.F., 9 H.P.		1.55	41.50	125	375	37.40
3850	Nailer, pneumatic		.41	27	81	243	19.50
3900	Paint sprayers complete, 8 CFM		.69	45.50	137	410	32.90
4000	17 CFM		1.10	73.50	220	660	52.80
4020	Pavers, bituminous, rubber tires, 8' wide, 50 H.P., diesel		26.80	775	2,320	6,950	678.40
4030	10' wide, 150 H.P.		59.45	1,300	3,895	11,700	1,255
4050	Crawler, 8' wide, 100 H.P., diesel		55.60	1,450	4,335	13,000	1,312
4060	10' wide, 150 H.P.		66.95	1,875	5,590	16,800	1,654
4070	Concrete paver, 12' to 24' wide, 250 H.P.		57.90	1,325	3,995	12,000	1,262
4080	Placer spreader trimmer, 24' wide, 300 H.P.		87.60	2,025	6,045	18,100	1,910
4100	Pump, centrifugal gas pump, 1 1/2", 4 MGPH		2.60	36.50	110	330	42.80
4200	2", 8 MGPH		3.25	41.50	125	375	51
4300	3", 15 MGPH		3.45	43.50	130	390	53.60
4400	6", 90 MGPH		15.75	152	455	1,375	217
4500	Submersible electric pump, 1 1/4", 55 GPM		.35	17	51	153	13
4600	1 1/2", 83 GPM		.41	19.65	59	177	15.10
4700	2", 120 GPM		.43	24.50	73	219	18.05
4800	3", 300 GPM		.90	33.50	100	300	27.20
4900	4", 560 GPM		6.16	135	405	1,225	130.30
5000	6", 1590 GPM		9.27	200	600	1,800	194.15
5100	Diaphragm pump, gas, single, 1 1/2" diameter		.78	38.50	116	350	29.45
5200	2" diameter		2.70	48.50	145	435	50.60
5300	3" diameter		2.75	48.50	145	435	51
5400	Double, 4" diameter		3.55	76.50	230	690	74.40
5500	Trash pump, self priming, gas, 2" diameter		2.55	26.50	80	240	36.40
5600	Diesel, 4" diameter		6.05	53.50	160	480	80.40
5650	Diesel, 6" diameter		18.35	112	335	1,000	213.80
5655	Grout Pump		8.30	58.50	175	525	101.40
5660	Rollers, see division 01590 200						
5700	Salamanders, L.P. gas fired, 100,000 BTU	Ea.	1.66	10.65	32	96	19.70
5705	50,000 BTU		1.25	7.65	23	69	14.60
5720	Sandblaster, portable, open top, 3 C.F. capacity		.40	20	60	180	15.20
5730	6 C.F. capacity		.65	29	87	261	22.60
5740	Accessories for above		.11	18.35	55	165	11.90

01590 | Equipment Rental

			UNIT	HOURLY OPER. COST	RENT PER DAY	RENT PER WEEK	RENT PER MONTH	CREW EQUIPMENT COST/DAY
400	5750	Sander, floor	Ea.	.72	17	51	153	15.95
	5760	Edger		.71	25	75	225	20.70
	5800	Saw, chain, gas engine, 18" long		1.20	16	48	144	19.20
	5900	36" long		.55	48.50	145	435	33.40
	5950	60" long		.55	50	150	450	34.40
	6000	Masonry, table mounted, 14" diameter, 5 H.P.		1.26	54.50	163	490	42.70
	6050	Portable cut off, 8 H.P.		1.30	28.50	85	255	27.40
	6100	Circular, hand held, electric, 7 1/4" diameter		.20	7.35	22	66	6
	6200	12" diameter		.27	11	33	99	8.75
	6250	Wall saw, w/hydraulic power, 10 H.P		2.04	95	284.80	855	73.30
	6275	Shot blaster, walk behind, 20" wide		1.42	545	1,635	4,900	338.35
	6300	Steam cleaner, 100 gallons per hour		2.25	61.50	185	555	55
	6310	200 gallons per hour		3	75	225	675	69
	6340	Tar Kettle/Pot, 400 gallon		2.69	51.50	155	465	52.50
	6350	Torch, cutting, acetylene oxygen, 150' hose		1.50	10	30	90	18
	6360	Hourly operating cost includes tips and gas		8.10				64.80
	6410	Toilet, portable chemical		.11	17.65	53	159	11.50
	6420	Recycle flush type		.13	21.50	65	195	14.05
	6430	Toilet, fresh water flush, garden hose,		.15	24.50	73	219	15.80
	6440	Hoisted, non flush, for high rise		.13	21.50	64	192	13.85
	6450	Toilet, trailers, minimum		.22	36.50	110	330	23.75
	6460	Maximum		.66	110	330	990	71.30
	6465	Tractor, farm with attachment	↓	10.25	222	665	2,000	215
	6470	Trailer, office, see division 01520 500						
	6500	Trailers, platform, flush deck, 2 axle, 25 ton capacity	Ea.	4.25	90	270	810	88
	6600	40 ton capacity		5.55	125	375	1,125	119.40
	6700	3 axle, 50 ton capacity		6	138	415	1,250	131
	6800	75 ton capacity		7.45	182	545	1,625	168.60
	6810	Trailer mounted cable reel for H.V. line work		4.47	213	639	1,925	163.55
	6820	Trailer mounted cable tensioning rig		8.75	415	1,250	3,750	320
	6830	Cable pulling rig	↓	55.44	2,375	7,120	21,400	1,868
	6850	Trailer, storage, see division 01520 500						
	6900	Water tank, engine driven discharge, 5000 gallons	Ea.	5.60	120	360	1,075	116.80
	6925	10,000 gallons		7.70	168	505	1,525	162.60
	6950	Water truck, off highway, 6000 gallons		49.25	700	2,095	6,275	813
	7010	Tram car for H.V. line work, powered, 2 conductor		5.73	116	347	1,050	115.25
	7020	Transit (builder's level) with tripod		.07	12.35	37	111	7.95
	7030	Trench box, 3000 lbs. 6'x8'		.67	90.50	272	815	59.75
	7040	7200 lbs. 6'x20'		.82	136	409	1,225	88.35
	7050	8000 lbs., 8' x 16'		.88	147	442	1,325	95.45
	7060	9500 lbs., 8'x20'		1.19	199	597	1,800	128.90
	7065	11,000 lbs., 8'x24'		1.34	224	671	2,025	144.90
	7070	12,000 lbs., 10' x 20'		1.77	294	883	2,650	190.75
	7100	Truck, pickup, 3/4 ton, 2 wheel drive		5.75	53.50	160	480	78
	7200	4 wheel drive		5.90	63.50	190	570	85.20
	7250	Crew carrier, 9 passenger		5.44	92	276.40	830	98.80
	7290	Tool van, 24,000 G.V.W.		8.40	91.50	274.80	825	122.15
	7300	Tractor, 4 x 2, 30 ton capacity, 195 H.P.		13.65	165	495	1,475	208.20
	7410	250 H.P.		18.20	243	730	2,200	291.60
	7500	6 x 2, 40 ton capacity, 240 H.P.		17.45	270	810	2,425	301.60
	7600	6 x 4, 45 ton capacity, 240 H.P.		22.20	293	880	2,650	353.60
	7620	Vacuum truck, hazardous material, 2500 gallon		6.51	300	902	2,700	232.50
	7625	5,000 gallon		11.37	400	1,202.80	3,600	331.50
	7640	Tractor, with A frame, boom and winch, 225 H.P.		15.75	223	670	2,000	260
	7650	Vacuum, H.E.P.A., 16 gal., wet/dry		.27	24	72	216	16.55
	7655	55 gal, wet/dry		.60	36	108	325	26.40
	7660	Water tank, portable		1	9.35	28	84	13.60
	7690	Large production vacuum loader, 3150 CFM	↓	15	600	1,800	5,400	480

01590 | Equipment Rental

			UNIT	HOURLY OPER. COST	RENT PER DAY	RENT PER WEEK	RENT PER MONTH	CREW EQUIPMENT COST/DAY
400	7700	Welder, electric, 200 amp	Ea.	3.79	60.50	181	545	66.50
	7800	300 amp		5.28	64	192	575	80.65
	7900	Gas engine, 200 amp		5.70	28	84	252	62.40
	8000	300 amp		7.70	32.50	98	294	81.20
	8100	Wheelbarrow, any size		.06	10.65	32	96	6.90
	8200	Wrecking ball, 4000 lb.		1.85	68.50	205	615	55.80
500	0010	**HIGHWAY EQUIPMENT RENTAL**						
	0050	Asphalt batch plant, portable drum mixer, 100 ton/hr.	Ea.	55.75	1,275	3,850	11,600	1,216
	0060	200 ton/hr.		61.85	1,350	4,045	12,100	1,304
	0070	300 ton/hr.		72.05	1,600	4,785	14,400	1,533
	0100	Backhoe attachment, long stick, up to 185 HP, 10.5' long		.30	20	60	180	14.40
	0140	Up to 250 HP, 12' long		.33	21.50	65	195	15.65
	0180	Over 250 HP, 15' long		.43	28.50	85	255	20.45
	0200	Special dipper arm, up to 100 HP, 32' long		.88	58.50	175	525	42.05
	0240	Over 100 HP, 33' long		1.10	73.50	220	660	52.80
	0300	Concrete batch plant, portable, electric, 200 CY/Hr		10.40	570	1,715	5,150	426.20
	0500	Grader attachment, ripper/scarifier, rear mounted						
	0520	Up to 135 HP	Ea.	2.75	58.50	175	525	57
	0540	Up to 180 HP		3.30	75	225	675	71.40
	0580	Up to 250 HP		3.65	85	255	765	80.20
	0700	Pvmt. removal bucket, for hyd. excavator, up to 90 HP		1.40	43.50	130	390	37.20
	0740	Up to 200 HP		1.60	65	195	585	51.80
	0780	Over 200 HP		1.70	76.50	230	690	59.60
	0900	Aggregate spreader, self propelled, 187 HP		37.15	790	2,370	7,100	771.20
	1000	Chemical spreader, 3 C.Y.		2.20	61.50	185	555	54.60
	1900	Hammermill, traveling, 250 HP		39.98	1,700	5,120	15,400	1,344
	2000	Horizontal borer, 3" diam, 13 HP gas driven		3.95	51.50	155	465	62.60
	2200	Hydromulchers, gas power, 3000 gal., for truck mounting		10.50	187	560	1,675	196
	2400	Joint & crack cleaner, walk behind, 25 HP		2.05	46.50	140	420	44.40
	2500	Filler, trailer mounted, 400 gal., 20 HP		6	182	545	1,625	157
	3000	Paint striper, self propelled, double line, 30 HP		5.15	155	465	1,400	134.20
	3200	Post drivers, 6" I Beam frame, for truck mounting		8.15	415	1,250	3,750	315.20
	3400	Road sweeper, self propelled, 8' wide, 90 HP		23.50	430	1,285	3,850	445
	4000	Road mixer, self propelled, 130 HP		29.05	570	1,715	5,150	575.40
	4100	310 HP		54	1,850	5,570	16,700	1,546
	4200	Cold mix paver, incl pug mill and bitumen tank,						
	4220	165 HP	Ea.	66.95	1,875	5,600	16,800	1,656
	4250	Paver, asphalt, wheel or crawler, 130 H.P., diesel		65.65	1,775	5,325	16,000	1,590
	4300	Paver, road widener, gas 1' to 6', 67 HP		29.65	630	1,890	5,675	615.20
	4400	Diesel, 2' to 14', 88 HP		39.95	955	2,870	8,600	893.60
	4600	Slipform pavers, curb and gutter, 2 track, 75 HP		24.85	640	1,920	5,750	582.80
	4700	4 track, 165 HP		34.75	745	2,235	6,700	725
	4800	Median barrier, 215 HP		35.35	775	2,325	6,975	747.80
	4901	Trailer, low bed, 75 ton capacity		7.90	178	535	1,600	170.20
	5000	Road planer, walk behind, 10" cutting width, 10 HP		2	26	78	234	31.60
	5100	Self propelled, 12" cutting width, 64 HP		5.30	195	585	1,750	159.40
	5200	Pavement profiler, 4' to 6' wide, 450 HP		146.05	2,675	8,040	24,100	2,776
	5300	8' to 10' wide, 750 HP		233.70	4,000	11,980	35,900	4,266
	5400	Roadway plate, steel, 1"x8'x20'		.06	10	30	90	6.50
	5600	Stabilizer,self propelled, 150 HP		27.40	540	1,625	4,875	544.20
	5700	310 HP		44.55	1,150	3,445	10,300	1,045
	5800	Striper, thermal, truck mounted 120 gal. paint, 150 H.P.		32.40	485	1,450	4,350	549.20
	6000	Tar kettle, 330 gal., trailer mounted		2.37	36.50	110	330	40.95
	7000	Tunnel locomotive, diesel, 8 to 12 ton		20.95	540	1,615	4,850	490.60
	7005	Electric, 10 ton		20.65	610	1,830	5,500	531.20
	7010	Muck cars, 1/2 C.Y. capacity		1.50	20	60	180	24
	7020	1 C.Y. capacity		1.70	28	84	252	30.40
	7030	2 C.Y. capacity		1.85	32.50	98	294	34.40

01590 | Equipment Rental

			UNIT	HOURLY OPER. COST	RENT PER DAY	RENT PER WEEK	RENT PER MONTH	CREW EQUIPMENT COST/DAY
500	7040	Side dump, 2 C.Y. capacity	Ea.	2	40	120	360	40
	7050	3 C.Y. capacity		2.70	45	135	405	48.60
	7060	5 C.Y. capacity		3.85	58.50	175	525	65.80
	7100	Ventilating blower for tunnel, 7 1/2 H.P.		1.22	44	132	395	36.15
	7110	10 H.P.		1.41	45.50	137	410	38.70
	7120	20 H.P.		2.30	56.50	170	510	52.40
	7140	40 H.P.		4.02	83.50	250	750	82.15
	7160	60 H.P.		6.09	127	380	1,150	124.70
	7175	75 H.P.		7.79	168	505	1,525	163.30
	7180	200 H.P.		17.45	245	735	2,200	286.60
	7800	Windrow loader, elevating	↓	33.40	875	2,620	7,850	791.20
600	0010	**LIFTING AND HOISTING EQUIPMENT RENTAL**						
	0100	without operators						
	0120	Aerial lift truck, 2 person, to 80'	Ea.	18.60	580	1,740	5,225	496.80
	0140	Boom work platform, 40' snorkel		9.80	203	610	1,825	200.40
	0150	Crane, flatbed mntd, 3 ton cap.		12.45	183	550	1,650	209.60
	0200	Crane, climbing, 106' jib, 6000 lb. capacity, 410 FPM		43.35	1,350	4,070	12,200	1,161
	0300	101' jib, 10,250 lb. capacity, 270 FPM	↓	48.75	1,725	5,150	15,500	1,420
	0400	Tower, static, 130' high, 106' jib,						
	0500	6200 lb. capacity at 400 FPM	Ea.	46.50	1,575	4,700	14,100	1,312
	0600	Crawler mounted, lattice boom, 1/2 C.Y., 15 tons at 12' radius		20.76	570	1,710	5,125	508.10
	0700	3/4 C.Y., 20 tons at 12' radius		27.68	680	2,045	6,125	630.45
	0800	1 C.Y., 25 tons at 12' radius		36.90	895	2,690	8,075	833.20
	0900	1 1/2 C.Y., 40 tons at 12' radius		41.15	930	2,795	8,375	888.20
	1000	2 C.Y., 50 tons at 12' radius		51.35	1,325	3,975	11,900	1,206
	1100	3 C.Y., 75 tons at 12' radius		49.10	1,300	3,885	11,700	1,170
	1200	100 ton capacity, 60' boom		62.80	1,700	5,085	15,300	1,519
	1300	165 ton capacity, 60' boom		92.20	2,175	6,560	19,700	2,050
	1400	200 ton capacity, 70' boom		98.10	2,350	7,015	21,000	2,188
	1500	350 ton capacity, 80' boom		141.90	3,325	9,945	29,800	3,124
	1600	Truck mounted, lattice boom, 6 x 4, 20 tons at 10' radius		27.76	960	2,875	8,625	797.10
	1700	25 tons at 10' radius		29.66	1,025	3,070	9,200	851.30
	1800	8 x 4, 30 tons at 10' radius		36.66	1,075	3,260	9,775	945.30
	1900	40 tons at 12' radius		32.74	1,150	3,450	10,400	951.90
	2000	8 x 4, 60 tons at 15' radius		39.81	1,250	3,770	11,300	1,072
	2050	82 tons at 15' radius		46.50	1,500	4,470	13,400	1,266
	2100	90 tons at 15' radius		44.08	1,625	4,860	14,600	1,325
	2200	115 tons at 15' radius		50.55	1,800	5,400	16,200	1,484
	2300	150 tons at 18' radius		43.10	1,825	5,500	16,500	1,445
	2350	165 tons at 18' radius		77.25	2,150	6,470	19,400	1,912
	2400	Truck mounted, hydraulic, 12 ton capacity		32.45	570	1,715	5,150	602.60
	2500	25 ton capacity		32.60	595	1,780	5,350	616.80
	2550	33 ton capacity		33.45	625	1,870	5,600	641.60
	2560	40 ton capacity		31.85	595	1,790	5,375	612.80
	2600	55 ton capacity		47.05	865	2,595	7,775	895.40
	2700	80 ton capacity		56	910	2,735	8,200	995
	2720	100 ton capacity		80.75	2,325	6,945	20,800	2,035
	2740	120 ton capacity		85.55	2,525	7,550	22,700	2,194
	2760	150 ton capacity		104.65	3,175	9,535	28,600	2,744
	2800	Self propelled, 4 x 4, with telescoping boom, 5 ton		15.50	320	965	2,900	317
	2900	12 1/2 ton capacity		25.10	510	1,525	4,575	505.80
	3000	15 ton capacity		25.70	585	1,755	5,275	556.60
	3050	20 ton capacity		26.80	610	1,835	5,500	581.40
	3100	25 ton capacity		27.25	595	1,785	5,350	575
	3150	40 ton capacity		45.20	845	2,540	7,625	869.60
	3200	Derricks, guy, 20 ton capacity, 60' boom, 75' mast		12.72	330	996	3,000	300.95
	3300	100' boom, 115' mast	↓	20.60	570	1,710	5,125	506.80

01590 | Equipment Rental

		UNIT	HOURLY OPER. COST	RENT PER DAY	RENT PER WEEK	RENT PER MONTH	CREW EQUIPMENT COST/DAY	
3400	Stiffleg, 20 ton capacity, 70' boom, 37' mast	Ea.	14.71	425	1,280	3,850	373.70	600
3500	100' boom, 47' mast		23.19	695	2,080	6,250	601.50	
3550	Helicopter, small, lift to 1250 lbs. maximum, w/pilot		67.85	2,675	8,050	24,200	2,153	
3600	Hoists, chain type, overhead, manual, 3/4 ton		.10	2	6	18	2	
3900	10 ton		.60	13	39	117	12.60	
4000	Hoist and tower, 5000 lb. cap., portable electric, 40' high		4.19	191	574	1,725	148.30	
4100	For each added 10' section, add		.09	15	45	135	9.70	
4200	Hoist and single tubular tower, 5000 lb. electric, 100' high		5.66	267	801	2,400	205.50	
4300	For each added 6' 6" section, add		.15	25	75	225	16.20	
4400	Hoist and double tubular tower, 5000 lb., 100' high		6.06	294	882	2,650	224.90	
4500	For each added 6' 6" section, add		.17	28.50	85	255	18.35	
4550	Hoist and tower, mast type, 6000 lb., 100' high		6.56	305	915	2,750	235.50	
4570	For each added 10' section, add		.11	18.35	55	165	11.90	
4600	Hoist and tower, personnel, electric, 2000 lb., 100' @ 125 FPM		13.52	815	2,440	7,325	596.15	
4700	3000 lb., 100' @ 200 FPM		15.40	915	2,750	8,250	673.20	
4800	3000 lb., 150' @ 300 FPM		17.10	1,025	3,090	9,275	754.80	
4900	4000 lb., 100' @ 300 FPM		17.73	1,050	3,150	9,450	771.85	
5000	6000 lb., 100' @ 275 FPM		19.14	1,100	3,300	9,900	813.10	
5100	For added heights up to 500', add	L.F.	.01	1.67	5	15	1.10	
5200	Jacks, hydraulic, 20 ton	Ea.	.05	6.35	19	57	4.20	
5500	100 ton	"	.30	18.65	56	168	13.60	
6000	Jacks, hydraulic, climbing with 50' jackrods							
6010	and control consoles, minimum 3 mo. rental							
6100	30 ton capacity	Ea.	1.65	110	329	985	79	
6150	For each added 10' jackrod section, add		.05	3.33	10	30	2.40	
6300	50 ton capacity		2.65	177	530	1,600	127.20	
6350	For each added 10' jackrod section, add		.06	4	12	36	2.90	
6500	125 ton capacity		6.95	465	1,390	4,175	333.60	
6550	For each added 10' jackrod section, add		.48	31.50	95	285	22.85	
6600	Cable jack, 10 ton capacity with 200' cable		1.38	91.50	275	825	66.05	
6650	For each added 50' of cable, add		.15	10	30	90	7.20	
0010	**WELLPOINT EQUIPMENT RENTAL** See also division 02240							700
0020	Based on 2 months rental							
0100	Combination jetting & wellpoint pump, 60 H.P. diesel	Ea.	9.17	272	817	2,450	236.75	
0200	High pressure gas jet pump, 200 H.P., 300 psi	"	16.39	233	698	2,100	270.70	
0300	Discharge pipe, 8" diameter	L.F.	.01	.44	1.32	3.96	.35	
0350	12" diameter		.01	.65	1.96	5.90	.45	
0400	Header pipe, flows up to 150 G.P.M., 4" diameter		.01	.40	1.20	3.60	.30	
0500	400 G.P.M., 6" diameter		.01	.47	1.42	4.26	.35	
0600	800 G.P.M., 8" diameter		.01	.65	1.96	5.90	.45	
0700	1500 G.P.M., 10" diameter		.01	.69	2.06	6.20	.50	
0800	2500 G.P.M., 12" diameter		.02	1.30	3.89	11.65	.95	
0900	4500 G.P.M., 16" diameter		.02	1.66	4.98	14.95	1.15	
0950	For quick coupling aluminum and plastic pipe, add		.03	1.72	5.15	15.45	1.25	
1100	Wellpoint, 25' long, with fittings & riser pipe, 1 1/2" or 2" diameter	Ea.	.05	3.43	10.28	31	2.45	
1200	Wellpoint pump, diesel powered, 4" diameter, 20 H.P.		4.45	157	471	1,425	129.80	
1300	6" diameter, 30 H.P.		5.81	195	584	1,750	163.30	
1400	8" suction, 40 H.P.		7.91	267	801	2,400	223.50	
1500	10" suction, 75 H.P.		10.86	310	936	2,800	274.10	
1600	12" suction, 100 H.P.		16.25	500	1,500	4,500	430	
1700	12" suction, 175 H.P.		21.61	550	1,650	4,950	502.90	
0010	**MARINE EQUIPMENT RENTAL**							800
0200	Barge, 400 Ton, 30' wide x 90' long	Ea.	16.15	240	720	2,150	273.20	
0240	800 Ton, 45' wide x 90' long		26.65	345	1,030	3,100	419.20	
2000	Tugboat, diesel, 100 HP		16.95	168	505	1,525	236.60	
2040	250 HP		31.45	315	945	2,825	440.60	
2080	380 HP		73.10	925	2,780	8,350	1,141	

		01740 | Cleaning	CREW	DAILY OUTPUT	LABOR-HOURS	UNIT	2005 BARE COSTS MAT.	LABOR	EQUIP.	TOTAL	TOTAL INCL O&P
500	0010	**CLEANING UP** After job completion, allow, minimum				Job					.30%
	0040	Maximum				"					1%

For information about Means Estimating Seminars, see yellow pages 12 and 13 in back of book

Division 2 Site Construction

stimating Tips

2200 Site Preparation

If possible visit the site and take an inventory of the type, quantity and size of the trees. Certain trees may have a landscape resale value or firewood value. Stump disposal can be very expensive, particularly if they cannot be buried at the site. Consider using a bulldozer in lieu of hand cutting trees.

Estimators should visit the site to determine the need for haul road, access, storage of materials, and security considerations. When estimating for access roads on unstable soil, consider using a geotextile stabilization fabric. It can greatly reduce the quantity of crushed stone or gravel. Sites of limited size and access can cause cost overruns due to lost productivity. Theft and damage is another consideration if the location is isolated. A temporary fence or security guards may be required. Investigate the site thoroughly.

2210 Subsurface Investigation

n preparing estimates on structures involving arthwork or foundations, all information oncerning soil characteristics should be btained. Look particularly for hazardous vaste, evidence of prior dumping of debris, nd previous stream beds.

2220 Selective Demolition

he costs shown for selective demolition do ot include rubbish handling or disposal. These tems should be estimated separately using leans data or other sources.

- Historic preservation often requires that the contractor remove materials from the existing structure, rehab them and replace them. The estimator must be aware of any related measures and precautions that must be taken when doing selective demolition, and cutting and patching. Requirements may include special handling and storage, as well as security.
- In addition to Section 02220, you can find selective demolition items in each division. Example: Roofing demolition is in division 7.

02300 Earthwork

- Estimating the actual cost of performing earthwork requires careful consideration of the variables involved. This includes items such as type of soil, whether or not water will be encountered, dewatering, whether or not banks need bracing, disposal of excavated earth, length of haul to fill or spoil sites, etc. If the project has large quantities of cut or fill, consider raising or lowering the site to reduce costs while paying close attention to the effect on site drainage and utilities if doing this.
- If the project has large quantities of fill, creating a borrow pit on the site can significantly lower the costs.
- It is very important to consider what time of year the project is scheduled for completion. Bad weather can create large cost overruns from dewatering, site repair and lost productivity from cold weather.

02500 Utility Services
02600 Drainage & Containment

- Never assume that the water, sewer and drainage lines will go in at the early stages of the project. Consider the site access needs before dividing the site in half with open trenches, loose pipe, and machinery obstructions. Always inspect the site to establish that the site drawings are complete. Check off all existing utilities on your drawing as you locate them. If you find any discrepancies, mark up the site plan for further research. Differing site conditions can be very costly if discovered later in the project.
- See also Section 02955 for restoration of pipe where removal/replacement may be undesirable. Use of new types of piping materials can reduce the overall project cost. Owners/design engineers should consider the installing construction as a valuable source of current information on piping products that could lead to significant utility cost savings.

02700 Bases, Ballasts, Pavements/Appurtenances

- When estimating paving, keep in mind the project schedule. If an asphaltic paving project is in a colder climate and runs through to the spring, consider placing the base course in the autumn, then topping it in the spring just prior to completion. This could save considerable costs in spring repair. Keep in mind that prices for asphalt and concrete are generally higher in the cold seasons.
- See also Sections 02960/02965.

02900 Planting

- The timing of planting and guarantee specifications often dictate the costs for establishing tree and shrub growth and a stand of grass or ground cover. Establish the work performance schedule to coincide with the local planting season. Maintenance and growth guarantees can add from 20% to 100% to the total landscaping cost. The cost to replace trees and shrubs can be as high as 5% of the total cost depending on the planting zone, soil conditions and time of year.

02960 & 02965 Flexible Pavement Surfacing Recovery

- Recycling of asphalt pavement is becoming very popular and is an alternative to removal and replacement of asphalt pavement. It can be a good value engineering proposal if removed pavement can be recycled either at the site or another site that is reasonably close to the project site.

Reference Numbers

Reference numbers are shown in bold squares at the beginning of some major classifications. These numbers refer to related items in the Reference Section. The reference information may be an estimating procedure, an alternate pricing method or technical information.

Note: Not all subdivisions listed here necessarily appear in this publication.

o part of this publication may be reproduced, stored in a retrieval system, or transmitted in any form r by any means without prior written permission of Reed Construction Data.

02050 | Basic Site Materials & Methods

	02055 Soils		CREW	DAILY OUTPUT	LABOR-HOURS	UNIT	2005 BARE COSTS MAT.	LABOR	EQUIP.	TOTAL	TOTAL INCL O&P
150	0010	**BORROW**									
	0020	Spread, 200 H.P. dozer, no compaction, 2 mi. RT haul									
	0200	Common borrow	B-15	600	.047	C.Y.	6	.96	3.12	10.08	11.60

	02060 Aggregate		CREW	DAILY OUTPUT	LABOR-HOURS	UNIT	MAT.	LABOR	EQUIP.	TOTAL	TOTAL INCL O&P
150	0010	**BORROW**									
	0020	Spread, with 200 H.P. dozer, no compaction, 2 mi RT haul									
	0100	Bank run gravel	B-15	600	.047	C.Y.	17.80	.96	3.12	21.88	24.50
	0300	Crushed stone, (1.40 tons per CY), 1-1/2"		600	.047		27.50	.96	3.12	31.58	35
	0320	3/4"		600	.047		27.50	.96	3.12	31.58	35
	0340	1/2"		600	.047		26.50	.96	3.12	30.58	34.50
	0360	3/8"		600	.047		26.50	.96	3.12	30.58	34
	0400	Sand, washed, concrete		600	.047		26.50	.96	3.12	30.58	34
	0500	Dead or bank sand		600	.047		4.13	.96	3.12	8.21	9.55

	02080 Utility Materials		CREW	DAILY OUTPUT	LABOR-HOURS	UNIT	MAT.	LABOR	EQUIP.	TOTAL	TOTAL INCL O&P
400	0010	**UTILITY BOXES** Precast concrete, 6" thick									
	0050	5' x 10' x 6' high, I.D.	B-13	2	24	Ea.	1,525	455	310	2,290	2,775
	0350	Hand hole, precast concrete, 1-1/2" thick									
	0400	1'-0" x 2'-0" x 1'-9", I.D., light duty	B-1	4	6	Ea.	260	108		368	470
	0450	4'-6" x 3'-2" x 2'-0", O.D., heavy duty	B-6	3	8	"	800	153	72	1,025	1,225

02200 | Site Preparation

	02210 Subsurface Investigation		CREW	DAILY OUTPUT	LABOR-HOURS	UNIT	2005 BARE COSTS MAT.	LABOR	EQUIP.	TOTAL	TOTAL INCL O&P
120	0010	**BORING AND EXPLORATORY DRILLING**									
	0020	Borings, initial field stake out & determination of elevations	A-6	1	16	Day		380	60.50	440.50	700
	0100	Drawings showing boring details				Total		185		185	270
	0200	Report and recommendations from P.E.						415		415	595
	0300	Mobilization and demobilization, minimum	B-55	4	4			73	195	268	335
	0350	For over 100 miles, per added mile		450	.036	Mile		.65	1.74	2.39	3
	0600	Auger holes in earth, no samples, 2-1/2" diameter		78.60	.204	L.F.		3.70	9.95	13.65	17.15
	0800	Cased borings in earth, with samples, 2-1/2" diameter		55.50	.288	"	14.20	5.25	14.05	33.50	40
	1400	Drill rig and crew with truck mounted auger		1	16	Day		291	780	1,071	1,350
	1500	For inner city borings add, minimum									10%
	1510	Maximum									20%

	02220 Site Demolition		CREW	DAILY OUTPUT	LABOR-HOURS	UNIT	MAT.	LABOR	EQUIP.	TOTAL	TOTAL INCL O&P
110	0010	**BUILDING DEMOLITION** Large urban projects, incl. 20 mi. haul (R02220-510)									
	0500	Small bldgs, or single bldgs, no salvage included, steel	B-3	14,800	.003	C.F.		.06	.12	.18	.24
	0600	Concrete	"	11,300	.004	"		.08	.15	.23	.31
	0605	Concrete, plain	B-5	33	1.212	C.Y.		23	28	51	70
	0610	Reinforced		25	1.600			30.50	37	67.50	92.50
	0615	Concrete walls		34	1.176			22.50	27.50	50	68
	0620	Elevated slabs		26	1.538			29.50	36	65.50	89
	0650	Masonry	B-3	14,800	.003	C.F.		.06	.12	.18	.24

02220 | Site Demolition

02220	Line	Description	Crew	Daily Output	Labor-Hours	Unit	2005 Bare Costs Mat.	2005 Bare Costs Labor	2005 Bare Costs Equip.	2005 Bare Costs Total	Total Incl O&P
110	0700	Wood (R02220-510)	B-3	14,800	.003	C.F.		.06	.12	.18	.24
	1000	Single family, one story house, wood, minimum				Ea.				2,525	2,975
	1020	Maximum								4,400	5,275
	1200	Two family, two story house, wood, minimum								3,300	3,950
	1220	Maximum								6,375	7,700
	1300	Three family, three story house, wood, minimum								4,400	5,275
	1320	Maximum								7,700	9,250
	1400	Gutting building, see division 02225-400									
130	0010	**BLDG. FOOTINGS AND FOUNDATIONS DEMOLITION** (R02220-510)									
	0200	Floors, concrete slab on grade,									
	0240	4" thick, plain concrete	B-9C	500	.080	S.F.		1.42	.28	1.70	2.72
	0280	Reinforced, wire mesh		470	.085			1.51	.30	1.81	2.89
	0300	Rods		400	.100			1.78	.35	2.13	3.40
	0400	6" thick, plain concrete		375	.107			1.89	.38	2.27	3.62
	0420	Reinforced, wire mesh		340	.118			2.09	.42	2.51	4
	0440	Rods		300	.133			2.37	.47	2.84	4.54
	1000	Footings, concrete, 1' thick, 2' wide	B-5	300	.133	L.F.		2.54	3.10	5.64	7.70
	1080	1'-6" thick, 2' wide		250	.160			3.05	3.72	6.77	9.25
	1120	3' wide		200	.200			3.81	4.65	8.46	11.50
	1200	Average reinforcing, add								10%	10%
	2000	Walls, block, 4" thick	1 Clab	180	.044	S.F.		.77		.77	1.31
	2040	6" thick		170	.047			.82		.82	1.39
	2080	8" thick		150	.053			.93		.93	1.57
	2100	12" thick		150	.053			.93		.93	1.57
	2400	Concrete, plain concrete, 6" thick	B-9	160	.250			4.44	.89	5.33	8.50
	2420	8" thick		140	.286			5.05	1.01	6.06	9.70
	2440	10" thick		120	.333			5.90	1.18	7.08	11.35
	2500	12" thick		100	.400			7.10	1.42	8.52	13.60
	2600	For average reinforcing, add								10%	10%
	4000	For congested sites or small quantities, add up to								200%	200%
	4200	Add for disposal, on site	B-11A	232	.069	C.Y.		1.42	3.96	5.38	6.75
	4250	To five miles	B-30	220	.109	"		2.29	7.60	9.89	12.15
220	0010	**FENCING DEMOLITION** (R02220-510)									
	1600	Fencing, barbed wire, 3 strand	2 Clab	430	.037	L.F.		.65		.65	1.10
	1650	5 strand	"	280	.057			.99		.99	1.68
	1700	Chain link, posts & fabric, remove only, 8' to 10' high	B-6	445	.054			1.03	.48	1.51	2.26
	1750	Remove and reset	"	70	.343			6.55	3.08	9.63	14.40
240	0010	**MINOR SITE DEMOLITION** (R02220-510)									
	0015	No hauling, abandon catch basin or manhole	B-6	7	3.429	Ea.		65.50	31	96.50	144
	0020	Remove existing catch basin or manhole, masonry		4	6			115	54	169	253
	0030	Catch basin or manhole frames and covers, stored		13	1.846			35.50	16.60	52.10	78
	0040	Remove and reset		7	3.429			65.50	31	96.50	144
	1000	Masonry walls, block or tile, solid, remove	B-5	1,800	.022	C.F.		.42	.52	.94	1.28
	1100	Cavity wall		2,200	.018			.35	.42	.77	1.04
	1200	Brick, solid		900	.044			.85	1.03	1.88	2.57
	1300	With block back-up		1,130	.035			.67	.82	1.49	2.04
	1400	Stone, with mortar		900	.044			.85	1.03	1.88	2.57
	1500	Dry set		1,500	.027			.51	.62	1.13	1.54
	2900	Pipe removal, sewer/water, no excavation, 12" diameter	B-6	175	.137	L.F.		2.63	1.23	3.86	5.75
	2960	21"-24" diameter		120	.200	"		3.83	1.80	5.63	8.45
	4000	Sidewalk removal, bituminous, 2-1/2" thick		325	.074	S.Y.		1.42	.66	2.08	3.10
	4050	Brick, set in mortar		185	.130			2.49	1.17	3.66	5.45
	4100	Concrete, plain, 4"		160	.150			2.88	1.35	4.23	6.30
	4200	Mesh reinforced		150	.160			3.07	1.44	4.51	6.75
	5000	Slab on grade removal, plain	B-5	45	.889	C.Y.		16.95	20.50	37.45	51

02200 | Site Preparation

02220 | Site Demolition

				CREW	DAILY OUTPUT	LABOR-HOURS	UNIT	2005 BARE COSTS MAT.	LABOR	EQUIP.	TOTAL	TOTAL INCL O&P
240	5100	Mesh reinforced	R02220-510	B-5	33	1.212	C.Y.		23	28	51	70
	5200	Rod reinforced		↓	25	1.600			30.50	37	67.50	92.50
	5500	For congested sites or small quantities, add up to									200%	200%
	5550	For disposal on site, add		B-11A	232	.069			1.42	3.96	5.38	6.75
	5600	To 5 miles, add		B-34D	76	.105	↓		2.06	5.50	7.56	9.50
250	0010	**DEMOLISH, REMOVE PAVEMENT AND CURB**	R02220-510									
	5010	Pavement removal, bituminous roads, 3″ thick		B-38	690	.035	S.Y.		.67	.48	1.15	1.64
	5050	4″ to 6″ thick			420	.057			1.10	.78	1.88	2.70
	5100	Bituminous driveways			640	.037			.72	.51	1.23	1.77
	5200	Concrete to 6″ thick, hydraulic hammer, mesh reinforced			255	.094			1.80	1.29	3.09	4.44
	5300	Rod reinforced		↓	200	.120	↓		2.30	1.64	3.94	5.65
	5600	With hand held air equipment, bituminous, to 6″ thick		B-39	1,900	.025	S.F.		.45	.07	.52	.84
	5700	Concrete to 6″ thick, no reinforcing			1,600	.030			.53	.09	.62	1
	5800	Mesh reinforced			1,400	.034			.61	.10	.71	1.14
	5900	Rod reinforced		↓	765	.063	↓		1.11	.19	1.30	2.08
	6000	Curbs, concrete, plain		B-6	360	.067	L.F.		1.28	.60	1.88	2.80
	6100	Reinforced			275	.087			1.67	.78	2.45	3.67
	6200	Granite			360	.067			1.28	.60	1.88	2.80
	6300	Bituminous		↓	528	.045	↓		.87	.41	1.28	1.91
310	0010	**SELECTIVE DEMOLITION, CUTOUT**	R02220-510									
	0020	Concrete, elev. slab, light reinforcement, under 6 CF		B-9C	65	.615	C.F.		10.90	2.18	13.08	21
	0050	Light reinforcing, over 6 C.F.		″	75	.533	″		9.45	1.89	11.34	18.10
	0200	Slab on grade to 6″ thick, not reinforced, under 8 S.F.		B-9	85	.471	S.F.		8.35	1.67	10.02	16.05
	0250	8 - 16 S.F.		″	175	.229	″		4.06	.81	4.87	7.80
	0255	For over 16 SF see 02220-130-0400										
	0600	Walls, not reinforced, under 6 C.F.		B-9	60	.667	C.F.		11.85	2.36	14.21	22.50
	0650	6 - 12 C.F.		″	80	.500	″		8.90	1.77	10.67	17
	0655	For over 12 CF see 02220-130-2500										
	1000	Concrete, elevated slab, bar reinforced, under 6 C.F.		B-9C	45	.889	C.F.		15.80	3.15	18.95	30.50
	1050	Bar reinforced, over 6 C.F.		″	50	.800	″		14.20	2.83	17.03	27
	1200	Slab on grade to 6″ thick, bar reinforced, under 8 S.F.		B-9	75	.533	S.F.		9.45	1.89	11.34	18.10
	1250	8 - 16 S.F.		″	150	.267	″		4.73	.94	5.67	9.10
	1255	For over 16 SF see 02220-130-0440										
	1400	Walls, bar reinforced, under 6 C.F.		B-9C	50	.800	C.F.		14.20	2.83	17.03	27
	1450	6 - 12 CF		″	70	.571	″		10.15	2.02	12.17	19.40
	1455	For over 12 CF see 02220-130-2500 & 2600										
	2000	Brick, to 4 S.F. opening, not including toothing										
	2040	4″ thick		B-9C	30	1.333	Ea.		23.50	4.72	28.22	45
	2060	8″ thick			18	2.222			39.50	7.85	47.35	75.50
	2080	12″ thick			10	4			71	14.15	85.15	137
	2400	Concrete block, to 4 S.F. opening, 2″ thick			35	1.143			20.50	4.05	24.55	39
	2420	4″ thick			30	1.333			23.50	4.72	28.22	45
	2440	8″ thick			27	1.481			26.50	5.25	31.75	50.50
	2460	12″ thick		↓	24	1.667			29.50	5.90	35.40	56.50
	2600	Gypsum block, to 4 S.F. opening, 2″ thick		B-9	80	.500			8.90	1.77	10.67	17
	2620	4″ thick			70	.571			10.15	2.02	12.17	19.40
	2640	8″ thick			55	.727			12.90	2.57	15.47	25
	2800	Terra cotta, to 4 S.F. opening, 4″ thick			70	.571			10.15	2.02	12.17	19.40
	2840	8″ thick			65	.615			10.90	2.18	13.08	21
	2880	12″ thick		↓	50	.800	↓		14.20	2.83	17.03	27
	3000	Toothing masonry cutouts, brick, soft old mortar		1 Brhe	40	.200	V.L.F.		3.77		3.77	6.25
	3100	Hard mortar			30	.267			5.05		5.05	8.35
	3200	Block, soft old mortar			70	.114			2.15		2.15	3.58
	3400	Hard mortar		↓	50	.160	↓		3.02		3.02	5
	6000	Walls, interior, not including re-framing,										

2 SITE CONSTRUCTION

02220 | Site Demolition

		02220 Site Demolition	Crew	Daily Output	Labor-Hours	Unit	2005 Bare Costs Mat.	Labor	Equip.	Total	Total Incl O&P
310	6010	openings to 5 S.F. (R02220-510)									
	6100	Drywall to 5/8″ thick	1 Clab	24	.333	Ea.		5.80		5.80	9.80
	6200	Paneling to 3/4″ thick		20	.400			6.95		6.95	11.80
	6300	Plaster, on gypsum lath		20	.400			6.95		6.95	11.80
	6340	On wire lath		14	.571			9.90		9.90	16.85
	7000	Wood frame, not including re-framing, openings to 5 S.F.									
	7200	Floors, sheathing and flooring to 2″ thick	1 Clab	5	1.600	Ea.		28		28	47
	7310	Roofs, sheathing to 1″ thick, not including roofing		6	1.333			23		23	39.50
	7410	Walls, sheathing to 1″ thick, not including siding		7	1.143			19.85		19.85	33.50
330	0010	**SELECTIVE DEMOLITION, DUMP CHARGES** (R02220-510)									
	0020	Dump charges, typical urban city, tipping fees only									
	0100	Building construction materials				Ton					70
	0200	Trees, brush, lumber									50
	0300	Rubbish only									60
	0500	Reclamation station, usual charge									85
340	0010	**SELECTIVE DEMOLITION, GUTTING** (R02220-510)									
	0020	Building interior, including disposal, dumpter fees not included									
	0500	Residential building									
	0560	Minimum	B-16	400	.080	SF Flr.		1.47	1.19	2.66	3.80
	0580	Maximum	″	360	.089	″		1.64	1.32	2.96	4.22
	0900	Commercial building									
	1000	Minimum	B-16	350	.091	SF Flr.		1.68	1.36	3.04	4.34
	1020	Maximum	″	250	.128	″		2.36	1.91	4.27	6.10
350	0010	**SELECTIVE DEMOLITION, RUBBISH HANDLING** (R02220-510)									
	0020	The following are to be added to the demolition prices									
	0400	Chute, circular, prefabricated steel, 18″ diameter	B-1	40	.600	L.F.	28	10.80		38.80	49.50
	0440	30″ diameter	″	30	.800	″	37.50	14.40		51.90	66
	0700	10 C.Y. capacity (4 Tons)				Week					375
	0725	Dumpster, weekly rental, 1 dump/week, 20 C.Y. capacity (8 Tons)									440
	0800	30 C.Y. capacity (10 Tons)									665
	0840	40 C.Y. capacity (13 Tons)									805
	1000	Dust partition, 6 mil polyethylene, 1″ x 3″ frame	2 Carp	2,000	.008	S.F.	.17	.19		.36	.52
	1080	2″ x 4″ frame	″	2,000	.008	″	.28	.19		.47	.64
	2000	Load, haul, and dump, 50′ haul	2 Clab	24	.667	C.Y.		11.55		11.55	19.65
	2040	100′ haul		16.50	.970			16.80		16.80	28.50
	2080	Over 100′ haul, add per 100 L.F.		35.50	.451			7.80		7.80	13.25
	2120	In elevators, per 10 floors, add		140	.114			1.98		1.98	3.37
	3000	Loading & trucking, including 2 mile haul, chute loaded	B-16	45	.711			13.10	10.60	23.70	33.50
	3040	Hand loading truck, 50′ haul	″	48	.667			12.25	9.95	22.20	31.50
	3080	Machine loading truck	B-17	120	.267			5.15	4.52	9.67	13.55
	5000	Haul, per mile, up to 8 C.Y. truck	B-34B	1,165	.007			.13	.41	.54	.67
	5100	Over 8 C.Y. truck	″	1,550	.005			.10	.31	.41	.51

02230 | Site Clearing

		02230 Site Clearing	Crew	Daily Output	Labor-Hours	Unit	2005 Bare Costs Mat.	Labor	Equip.	Total	Total Incl O&P
100	0010	**CLEAR AND GRUB**									
	0020	Cut & chip light trees to 6″ diam.	B-7	1	48	Acre		900	1,025	1,925	2,650
	0150	Grub stumps and remove	B-30	2	12			252	835	1,087	1,350
	0200	Cut & chip medium, trees to 12″ diam.	B-7	.70	68.571			1,300	1,450	2,750	3,775
	0250	Grub stumps and remove	B-30	1	24			505	1,675	2,180	2,675
	0300	Cut & chip heavy, trees to 24″ diam.	B-7	.30	160			3,000	3,400	6,400	8,825
	0350	Grub stumps and remove	B-30	.50	48			1,000	3,350	4,350	5,350
	0400	If burning is allowed, reduce cut & chip									40%
200	0010	**SELECTIVE CLEARING**									
	0020	Clearing brush with brush saw	A-1C	.25	32	Acre		555	77	632	1,025

02200 | Site Preparation

02230 | Site Clearing

		02230 Site Clearing	CREW	DAILY OUTPUT	LABOR-HOURS	UNIT	MAT.	LABOR	EQUIP.	TOTAL	TOTAL INCL O&P
							2005 BARE COSTS				
200	0100	By hand	1 Clab	.12	66.667	Acre		1,150		1,150	1,975
	0300	With dozer, ball and chain, light clearing	B-11A	2	8			165	460	625	780
	0400	Medium clearing	"	1.50	10.667			220	615	835	1,050
500	0010	**STRIPPING & STOCKPILING OF SOIL**									
	1400	Loam or topsoil, remove and stockpile on site									
	1420	6″ deep, 200′ haul	B-10B	865	.009	C.Y.		.22	1.06	1.28	1.53
	1430	300′ haul		520	.015			.37	1.77	2.14	2.55
	1440	500′ haul		225	.036			.85	4.09	4.94	5.90
	1450	Alternate method: 6″ deep, 200′ haul		5,090	.002	S.Y.		.04	.18	.22	.26
	1460	500′ haul		1,325	.006	"		.14	.69	.83	1

02300 | Earthwork

02305 | Equipment

		02305 Equipment	CREW	DAILY OUTPUT	LABOR-HOURS	UNIT	MAT.	LABOR	EQUIP.	TOTAL	TOTAL INCL O&P
							2005 BARE COSTS				
250	0010	**MOBILIZATION OR DEMOB.** (One or the other, unless noted)									
	0015	Up to 25 mi haul dist (50 mi round trip for mob/demob crew)									
	0020	Dozer, loader, backhoe, excav., grader, paver, roller, 70 to 150 H.P.	B-34N	4	2	Ea.		39	112	151	188
	0900	Shovel or dragline, 3/4 C.Y.	B-34K	3.60	2.222			43.50	146	189.50	233
	1100	Small equipment, placed in rear of, or towed by pickup truck	A-3A	8	1			19.05	10.65	29.70	43
	1150	Equip up to 70 HP, on flatbed trailer behind pickup truck	A-3D	4	2			38	43.50	81.50	111
	2000	Mob & demob truck-mounted crane up to 75 ton, driver only	1 EQHV	3.60	2.222			55		55	90.50
	2200	Crawler-mounted, up to 75 ton	A-3F	2	8			177	278	455	600
	2500	For each additional 5 miles haul distance, add						10%	10%		
	3000	For large pieces of equipment, allow for assembly/knockdown									
	3100	For mob/demob of micro-tunneling equip, see section 02441-400									

02310 | Grading

		02310 Grading	CREW	DAILY OUTPUT	LABOR-HOURS	UNIT	MAT.	LABOR	EQUIP.	TOTAL	TOTAL INCL O&P
100	0010	**FINISH GRADING**									
	0012	Finish grading area to be paved with grader, small area	B-11L	400	.040	S.Y.		.83	1.14	1.97	2.64
	0100	Large area		2,000	.008			.17	.23	.40	.53
	0200	Grade subgrade for base course, roadways		3,500	.005			.09	.13	.22	.30
	1020	For large parking lots	B-32C	5,000	.010			.20	.31	.51	.68
	1050	For small irregular areas	"	2,000	.024			.50	.78	1.28	1.70
	1100	Fine grade for slab on grade, machine	B-11L	1,040	.015			.32	.44	.76	1.01
	1150	Hand grading	B-18	700	.034			.62	.05	.67	1.11
	1200	Fine grade granular base for sidewalks and bikeways	B-62	1,200	.020			.38	.13	.51	.78
	2550	Hand grade select gravel	2 Clab	60	.267	C.S.F.		4.63		4.63	7.85
	3000	Hand grade select gravel, including compaction, 4″ deep	B-18	555	.043	S.Y.		.78	.07	.85	1.39
	3100	6″ deep		400	.060			1.08	.09	1.17	1.93
	3120	8″ deep		300	.080			1.44	.12	1.56	2.59
	3300	Finishing grading slopes, gentle	B-11L	8,900	.002			.04	.05	.09	.12
	3310	Steep slopes	"	7,100	.002			.05	.06	.11	.15

02315 | Excavation and Fill

		02315 Excavation and Fill	CREW	DAILY OUTPUT	LABOR-HOURS	UNIT	MAT.	LABOR	EQUIP.	TOTAL	TOTAL INCL O&P
110	0010	**BACKFILL, GENERAL**									
	0015	By hand, no compaction, light soil	1 Clab	14	.571	C.Y.		9.90		9.90	16.85

	02315 | Excavation and Fill		CREW	DAILY OUTPUT	LABOR-HOURS	UNIT	2005 BARE COSTS MAT.	LABOR	EQUIP.	TOTAL	TOTAL INCL O&P	
110	0100	Heavy soil	1 Clab	11	.727	L.C.Y.		12.60		12.60	21.50	110
	0300	Compaction in 6″ layers, hand tamp, add to above	↓	20.60	.388	E.C.Y.		6.75		6.75	11.45	
	0500	Air tamp, add	B-9D	190	.211			3.74	.93	4.67	7.40	
	0600	Vibrating plate, add	A-1D	60	.133			2.31	.42	2.73	4.39	
	0800	Compaction in 12″ layers, hand tamp, add to above	1 Clab	34	.235	↓		4.08		4.08	6.95	
	1300	Dozer backfilling, bulk, up to 300′ haul, no compaction	B-10B	1,200	.007	L.C.Y.		.16	.77	.93	1.10	
	1400	Air tamped, add	B-11B	80	.200	E.C.Y.		4.02	2.30	6.32	9.25	
320	0010	**COMPACTION, STRUCTURAL**										320
	0020	Steel wheel tandem roller, 5 tons	B-10E	8	1	Hr.		24	13.45	37.45	54.50	
	0050	Air tamp, 6″ to 8″ lifts, common fill	B-9	250	.160	E.C.Y.		2.84	.57	3.41	5.45	
	0060	Select fill	″	300	.133			2.37	.47	2.84	4.54	
	0600	Vibratory plate, 8″ lifts, common fill	A-1D	200	.040			.69	.13	.82	1.32	
	0700	Select fill	″	216	.037	↓		.64	.12	.76	1.22	
424	0010	**EXCAVATING, BULK BANK MEASURE** Common earth piled										424
	0020	For loading onto trucks, add								15%	15%	
	0200	Backhoe, hydraulic, crawler mtd., 1 C.Y. cap. = 75 C.Y./hr.	B-12A	600	.027	B.C.Y.		.56	.93	1.49	1.96	
	0310	Wheel mounted, 1/2 C.Y. cap. = 30 C.Y./hr.	B-12E	240	.067			1.40	1.39	2.79	3.86	
	1200	Front end loader, track mtd., 1-1/2 C.Y. cap. = 70 C.Y./hr.	B-10N	560	.014			.34	.56	.90	1.17	
	1500	Wheel mounted, 3/4 C.Y. cap. = 45 C.Y./hr.	B-10R	360	.022	↓		.53	.54	1.07	1.46	
	8000	For hauling excavated material, see div. 02315-490										
462	0010	**EXCAVATION, STRUCTURAL**										462
	0015	Hand, pits to 6′ deep, sandy soil	1 Clab	8	1	C.Y.		17.35		17.35	29.50	
	0100	Heavy soil or clay		4	2	B.C.Y.		34.50		34.50	59	
	1100	Hand loading trucks from stock pile, sandy soil		12	.667			11.55		11.55	19.65	
	1300	Heavy soil or clay	↓	8	1	↓		17.35		17.35	29.50	
	1500	For wet or muck hand excavation, add to above				%				50%	50%	
490	0010	**HAULING**, excavated or borrow, loose cubic yards										490
	0012	no loading included, highway haulers										
	0020	6 C.Y. dump truck, 1/4 mile round trip, 5.0 loads/hr.	B-34A	195	.041	L.C.Y.		.80	1.68	2.48	3.18	
	0200	4 mile round trip, 1.8 loads/hr.	″	70	.114			2.23	4.67	6.90	8.85	
	0310	12 C.Y. dump truck, 1/4 mile round trip 3.7 loads/hr.	B-34B	288	.028			.54	1.66	2.20	2.72	
	0500	4 mile round trip, 1.6 loads/hr.	″	125	.064	↓		1.25	3.81	5.06	6.25	
520	0010	**FILL**, spread dumped material, no compaction										520
	0020	By dozer, no compaction	B-10B	1,000	.008	C.Y.		.19	.92	1.11	1.32	
	0100	By hand	1 Clab	12	.667	L.C.Y.		11.55		11.55	19.65	
	0500	Gravel fill, compacted, under floor slabs, 4″ deep	B-37	10,000	.005	S.F.	.16	.09	.01	.26	.33	
	0600	6″ deep		8,600	.006		.24	.10	.01	.35	.45	
	0700	9″ deep		7,200	.007		.39	.12	.01	.52	.66	
	0800	12″ deep		6,000	.008	↓	.55	.15	.02	.72	.88	
	1000	Alternate pricing method, 4″ deep		120	.400	E.C.Y.	11.80	7.45	.90	20.15	26.50	
	1100	6″ deep		160	.300		11.80	5.60	.67	18.07	23	
	1200	9″ deep		200	.240		11.80	4.46	.54	16.80	21	
	1300	12″ deep	↓	220	.218	↓	11.80	4.06	.49	16.35	20.50	
610	0010	**EXCAVATING, TRENCH** or continuous footing, common earth										610
	0050	1′ to 4′ deep, 3/8 C.Y. tractor loader/backhoe	B-11C	150	.107	B.C.Y.		2.20	1.44	3.64	5.25	
	0060	1/2 C.Y. tractor loader/backhoe	B-11M	200	.080			1.65	1.28	2.93	4.16	
	0090	4′ to 6′ deep, 1/2 C.Y. tractor loader/backhoe	″	200	.080			1.65	1.28	2.93	4.16	
	0100	5/8 C.Y. hydraulic backhoe	B-12Q	250	.064			1.35	1.73	3.08	4.15	
	0300	1/2 C.Y. hydraulic excavator, truck mounted	B-12J	200	.080			1.68	4.19	5.87	7.40	
	1400	By hand with pick and shovel 2′ to 6′ deep, light soil	1 Clab	8	1			17.35		17.35	29.50	
	1500	Heavy soil	″	4	2	↓		34.50		34.50	59	
620	0010	**EXCAVATING, UTILITY TRENCH** Common earth										620
	0050	Trenching with chain trencher, 12 H.P., operator walking										

SITE CONSTRUCTION 2

02300 | Earthwork

		02315 | Excavation and Fill	CREW	DAILY OUTPUT	LABOR-HOURS	UNIT	2005 BARE COSTS MAT.	LABOR	EQUIP.	TOTAL	TOTAL INCL O&P	
620	0100	4" wide trench, 12" deep	B-53	800	.010	L.F.		.17	.06	.23	.36	620
	1000	Backfill by hand including compaction, add										
	1050	4" wide trench, 12" deep	A-1G	800	.010	L.F.		.17	.04	.21	.34	
640	0010	**UTILITY BEDDING** For pipe and conduit, not incl. compaction										640
	0050	Crushed or screened bank run gravel	B-6	150	.160	L.C.Y.	20	3.07	1.44	24.51	28.50	
	0100	Crushed stone 3/4" to 1/2"	↓	150	.160	↓	27.50	3.07	1.44	32.01	36.50	
	0200	Sand, dead or bank	↓	150	.160	↓	4.13	3.07	1.44	8.64	11.25	
	0500	Compacting bedding in trench	A-1D	90	.089	E.C.Y.		1.54	.28	1.82	2.93	
	0600	If material source exceeds 2 miles, add for extra mileage.				Ton						
	0610	See 02315-490-0010 for hauling mileage add.				"						

		02360 | Soil Treatment	CREW	DAILY OUTPUT	LABOR-HOURS	UNIT	MAT.	LABOR	EQUIP.	TOTAL	TOTAL INCL O&P	
200	0010	**TERMITE PRETREATMENT**										200
	0020	Slab and walls, residential	1 Skwk	1,200	.007	SF Flr.	.28	.16		.44	.58	
	0400	Insecticides for termite control, minimum	↓	14.20	.563	Gal.	11.60	13.45		25.05	36	
	0500	Maximum	↓	11	.727	"	19.85	17.40		37.25	51.50	

		02370 | Erosion & Sedimentation Control	CREW	DAILY OUTPUT	LABOR-HOURS	UNIT	MAT.	LABOR	EQUIP.	TOTAL	TOTAL INCL O&P	
700	0010	**SYNTHETIC EROSION CONTROL**										700
	0020	Jute mesh, 100 SY per roll, 4' wide, stapled	B-80A	2,400	.010	S.Y.	.69	.17	.07	.93	1.13	
	0100	Plastic netting, stapled, 2" x 1" mesh, 20 mil	B-1	2,500	.010	↓	.63	.17		.80	.98	
	0200	Polypropylene mesh, stapled, 6.5 oz./S.Y.	↓	2,500	.010	↓	1.30	.17		1.47	1.72	
	0300	Tobacco netting, or jute mesh #2, stapled	↓	2,500	.010	↓	.07	.17		.24	.37	
	1000	Silt fence, polypropylene, 3' high, ideal conditions	2 Clab	1,600	.010	L.F.	.32	.17		.49	.64	
	1100	Adverse conditions	"	950	.017	"	.32	.29		.61	.85	

02400 | Tunneling, Boring & Jacking

		02441 | Microtunneling	CREW	DAILY OUTPUT	LABOR-HOURS	UNIT	2005 BARE COSTS MAT.	LABOR	EQUIP.	TOTAL	TOTAL INCL O&P	
400	0010	**MICROTUNNELING** Not including excavation, backfill, shoring,										400
	0020	or dewatering, average 50'/day, slurry method										
	0100	24" to 48" outside diameter, minimum				L.F.					640	
	0110	Adverse conditions, add				%					50%	
	1000	Rent microtunneling machine, average monthly lease				Month					85,500	
	1010	Operating technician				Day					640	
	1100	Mobilization and demobilization, minimum				Job					42,800	
	1110	Maximum				"					430,000	

02500 | Utility Services

02510 | Water Distribution		CREW	DAILY OUTPUT	LABOR-HOURS	UNIT	2005 BARE COSTS MAT.	LABOR	EQUIP.	TOTAL	TOTAL INCL O&P	
0010	**TAPPING, CROSSES AND SLEEVES**										710
4000	Drill and tap pressurized main (labor only)										
4100	6" main, 1" to 2" service	Q-1	3	5.333	Ea.		129		129	213	
4150	8" main, 1" to 2" service	"	2.75	5.818	"		141		141	232	
4500	Tap and insert gate valve										
4600	8" main, 4" branch	B-21	3.20	8.750	Ea.		166	49.50	215.50	335	
4650	6" branch		2.70	10.370			197	58.50	255.50	395	
4651	Piping, drill, tap & insert gate valve, 8" main, 6" branch		2.70	10.370			197	58.50	255.50	395	
4700	10" Main, 4" branch		2.70	10.370			197	58.50	255.50	395	
4750	6" branch		2.35	11.915			226	67.50	293.50	455	
4800	12" main, 6" branch		2.35	11.915			226	67.50	293.50	455	
0010	**WATER SUPPLY, DUCTILE IRON PIPE** cement lined										730
0020	Not including excavation or backfill										
2000	Pipe, class 50 water piping, 18' lengths										
2020	Mechanical joint, 4" diameter	B-21A	200	.200	L.F.	12.50	4.40	2.53	19.43	24	
2040	6" diameter		160	.250		14.50	5.50	3.16	23.16	28.50	
3000	Tyton, Push-on joint, 4" diameter		400	.100		7.35	2.20	1.27	10.82	13.15	
3020	6" diameter		333.33	.120		8.40	2.64	1.52	12.56	15.25	
8000	Fittings, mechanical joint										
8006	90° bend, 4" diameter	B-20A	16	2	Ea.	126	42.50		168.50	209	
8020	6" diameter		12.80	2.500		169	53.50		222.50	275	
8200	Wye or tee, 4" diameter		10.67	2.999		153	64		217	275	
8220	6" diameter		8.53	3.751		192	80		272	345	
8398	45° bends, 4" diameter		16	2		133	42.50		175.50	217	
8400	6" diameter		12.80	2.500		146	53.50		199.50	250	
8450	Decreaser, 6" x 4" diameter		14.22	2.250		104	48		152	195	
8460	8" x 6" diameter		11.64	2.749		213	58.50		271.50	335	
8550	Butterfly valves with boxes, cast iron										
8560	4" diameter	B-20	6	4	Ea.	760	72		832	960	
9600	Steel sleeve and tap, 4" diameter		3	8		430	144		574	715	
9620	6" diameter		2	12		505	216		721	920	
0010	**WATER SUPPLY, POLYVINYL CHLORIDE PIPE**										750
2100	AWWA Class 160, S.D.R. 26, 1-1/2" diameter	B-20	750	.032	L.F.	.45	.58		1.03	1.48	
2120	2" diameter		686	.035		1.09	.63		1.72	2.27	
2140	2-1/2" diameter		500	.048		1.62	.87		2.49	3.25	
2160	3" diameter		430	.056		2.32	1.01		3.33	4.26	
2180	4" diameter		375	.064		3.78	1.15		4.93	6.10	
02520 | Wells											
0010	**WELLS & ACCESSORIES**, domestic										510
0100	Drilled, 4" to 6" diameter	B-23	120	.333	L.F.		5.90	28	33.90	41	
1500	Pumps, installed in wells to 100' deep, 4" submersible										
1520	3/4 H.P.	Q-1	2.66	6.015	Ea.	390	146		536	670	
1600	1 H.P.	"	2.29	6.987	"	415	169		584	735	
02530 | Sanitary Sewerage											
0010	**SEWAGE COLLECTION, CONCRETE PIPE**										730
0020	See 02630-530 for sewage/drainage collection, concrete pipe										
0010	**SEWAGE COLLECTION, POLYVINYL CHLORIDE PIPE**										780
0020	Not including excavation or backfill										
2000	10' lengths, S.D.R. 35, B&S, 4" diameter	B-20	375	.064	L.F.	1.92	1.15		3.07	4.07	
2040	6" diameter		350	.069		3.44	1.24		4.68	5.90	
2080	8" diameter		335	.072		5.80	1.29		7.09	8.60	
2120	10" diameter	B-21	330	.085		8.75	1.61	.48	10.84	12.90	

02500 | Utility Services

	02530 | Sanitary Sewerage		CREW	DAILY OUTPUT	LABOR-HOURS	UNIT	2005 BARE COSTS MAT.	LABOR	EQUIP.	TOTAL	TOTAL INCL O&P
780	4000	Piping, DWV PVC,no exc/bkfill, 10' L, Sch 40, 4" dia	B-20	375	.064	L.F.	1.88	1.15		3.03	4.03
	4010	6" dia		350	.069		4.06	1.24		5.30	6.55
	4020	8" dia		335	.072		12.50	1.29		13.79	15.95
	02540 | Septic Tank Systems										
400	0010	**SEPTIC TANKS**									
	0015	Septic tanks, not incl exc or piping, precast, 1,000 gal	B-21	8	3.500	Ea.	560	66.50	19.80	646.30	750
	0100	2,000 gallon		5	5.600		1,100	106	31.50	1,237.50	1,450
	0600	High density polyethylene, 1,000 gallon		6	4.667		865	88.50	26.50	980	1,125
	0700	1,500 gallon		4	7		1,125	133	39.50	1,297.50	1,500
	1000	Distribution boxes, concrete, 7 outlets	2 Clab	16	1		104	17.35		121.35	144
	1100	9 outlets	"	8	2		281	34.50		315.50	370
	1150	Leaching field chambers, 13' x 3'-7" x 1'-4", standard	B-13	16	3		565	57	38.50	660.50	760
	1420	Leaching pit, 6', dia, 3' deep complete					645			645	710
	2200	Excavation for septic tank, 3/4 C.Y. backhoe	B-12F	145	.110	C.Y.		2.32	3.27	5.59	7.45
	2400	4' trench for disposal field, 3/4 C.Y. backhoe	"	335	.048	L.F.		1.01	1.42	2.43	3.24
	2600	Gravel fill, run of bank	B-6	150	.160	C.Y.	16.15	3.07	1.44	20.66	24.50
	2800	Crushed stone, 3/4"	"	150	.160	"	24	3.07	1.44	28.51	32.50
	02550 | Piped Energy Distribution										
464	0010	**PIPING, GAS SERVICE & DISTRIBUTION, POLYETHYLENE**									
	0020	not including excavation or backfill									
	1000	60 psi coils, comp cplg @ 100', 1/2" diameter, SDR 9.3	B-20A	608	.053	L.F.	.47	1.12		1.59	2.39
	1040	1-1/4" diameter, SDR 11		544	.059		.87	1.25		2.12	3.05
	1100	2" diameter, SDR 11		488	.066		1.08	1.40		2.48	3.52
	1160	3" diameter, SDR 11		408	.078		2.25	1.67		3.92	5.25
	1500	60 PSI 40' joints with coupling, 3" diameter, SDR 11	B-21A	408	.098		2.25	2.16	1.24	5.65	7.40
	1540	4" diameter, SDR 11		352	.114		5.15	2.50	1.44	9.09	11.45
	1600	6" diameter, SDR 11		328	.122		16.10	2.68	1.54	20.32	24
	1640	8" diameter, SDR 11		272	.147		22	3.23	1.86	27.09	31.50
466	0010	**PIPING, GAS SERVICE & DISTRIBUTION, STEEL**									
	0020	not including excavation or backfill, tar coated and wrapped									
	4000	Schedule 40, plain end									
	4040	1" diameter	Q-4	300	.107	L.F.	2.64	2.73	.27	5.64	7.70
	4080	2" diameter	"	280	.114	"	4.14	2.93	.29	7.36	9.70

02600 | Drainage & Containment

	02620 | Subdrainage		CREW	DAILY OUTPUT	LABOR-HOURS	UNIT	2005 BARE COSTS MAT.	LABOR	EQUIP.	TOTAL	TOTAL INCL O&P
610	0010	**PIPING, SUBDRAINAGE, CONCRETE**									
	0021	Not including excavation and backfill									
	3000	Porous wall concrete underdrain, std. strength, 4" diameter	B-20	335	.072	L.F.	1.95	1.29		3.24	4.34
	3020	6" diameter	"	315	.076		2.54	1.37		3.91	5.10
	3040	8" diameter	B-21	310	.090		3.13	1.71	.51	5.35	6.90
620	0010	**PIPING, SUBDRAINAGE, CORRUGATED METAL**									
	0021	Not including excavation and backfill									

MEANS CostWorks® 2005

Maximize your estimating & budgeting efforts quickly & efficiently . . .

RSMeans

Introducing Means CostWorks Online – your preeminent Cost Data resource is now available in an easy-to-use online format as well as on CD.

Means CostWorks 2005 in CD or Online formats – Join the thousands of industry professionals who rely on RSMeans Cost Data each year!

Means CostWorks 2005 CD offers 18 detailed cost titles that cover every job costing requirement you can imagine . . . it turns your desktop or laptop PC into a comprehensive cost library. Each title addresses a specific area of the commercial or residential building and maintenance industry.

Plus you get the productivity-enhancing tools to really put this valuable information to work.

Save hundreds on specially priced CD packages! Choose the one that's right for you . . .

Builder's Package

Includes:

- Residential Cost Data
- Light Commercial Cost Data
- Site Work & Landscape Cost Data
- Concrete & Masonry Cost Data
- Open Shop Building Construction Cost Data

Just $380.95

Save over $300 off the price of buying each book separately!

Building Professional's Package

Includes:

- Building Construction Cost Data
- Mechanical Cost Data
- Plumbing Cost Data
- Electrical Cost Data
- Square Foot Costs

Just $445.95

Save over $250 off the price of buying each book separately!

Facility Manager's Package

Includes:

- Facilities Construction Cost Data
- Facilities Maintenance & Repair Cost Data
- Repair & Remodeling Cost Data
- Building Construction Cost Data

Just $572.95

Save over $300 off the price of buying each book separately!

Design Professional's Package

Includes:

- Building Construction Cost Data
- Square Foot Costs
- Interior Cost Data
- Assemblies Cost Data

Just $445.95

Save over $200 off the price of buying each book separately!

Go to www.rsmeans.com/costworks/web/index.asp to learn about other packages *available exclusively online* with a subscription to *Means CostWorks Online 2005*!

02600 | Drainage & Containment

02620 | Subdrainage		CREW	DAILY OUTPUT	LABOR-HOURS	UNIT	2005 BARE COSTS MAT.	LABOR	EQUIP.	TOTAL	TOTAL INCL O&P	
2010	Aluminum, perforated										620
2020	6″ diameter, 18 ga.	B-14	380	.126	L.F.	2.76	2.35	.57	5.68	7.65	
2200	8″ diameter, 16 ga.		370	.130		3.94	2.41	.58	6.93	9.05	
2220	10″ diameter, 16 ga.	↓	360	.133	↓	4.93	2.48	.60	8.01	10.25	
3000	Uncoated galvanized, perforated										
3020	6″ diameter, 18 ga.	B-20	380	.063	L.F.	4.37	1.14		5.51	6.75	
3200	8″ diameter, 16 ga.	″	370	.065		6	1.17		7.17	8.60	
3220	10″ diameter, 16 ga.	B-21	360	.078		9	1.48	.44	10.92	12.85	
3240	12″ diameter, 16 ga.	″	285	.098	↓	9.45	1.86	.56	11.87	14.15	
4000	Steel, perforated, asphalt coated										
4020	6″ diameter 18 ga.	B-20	380	.063	L.F.	3.49	1.14		4.63	5.75	
4030	8″ diameter 18 ga	″	370	.065		5.45	1.17		6.62	8	
4040	10″ diameter 16 ga	B-21	360	.078		6.30	1.48	.44	8.22	9.85	
4050	12″ diameter 16 ga		285	.098		7.20	1.86	.56	9.62	11.70	
4060	18″ diameter 16 ga	↓	205	.137	↓	9.85	2.59	.77	13.21	16.05	

02630 | Storm Drainage		CREW	DAILY OUTPUT	LABOR-HOURS	UNIT	MAT.	LABOR	EQUIP.	TOTAL	TOTAL INCL O&P	
0010	**CATCH BASIN GRATES AND FRAMES** not including footing, excavation										110
1600	Frames & covers, C.I., 24″ square, 500 lb.	B-6	7.80	3.077	Ea.	239	59	27.50	325.50	390	
0010	**STORM DRAINAGE MANHOLES, FRAMES & COVERS** not including										400
0020	footing, excavation, backfill (See line items for frame & cover)										
0050	Brick, 4′ inside diameter, 4′ deep	D-1	1	16	Ea.	320	350		670	930	
1110	Precast, 4′ I.D., 4′ deep	B-22	4.10	7.317	″	675	142	58	875	1,050	
0010	**SEWAGE/DRAINAGE COLLECTION, CONCRETE PIPE**										530
0020	Not including excavation or backfill										
1020	8″ diameter	B-14	224	.214	L.F.	4.54	3.98	.96	9.48	12.75	
1030	10″ diameter	″	216	.222	″	5.05	4.13	1	10.18	13.60	
3780	Concrete slotted pipe, class 4 mortar joint										
3800	12″ diameter	B-21	168	.167	L.F.	13.15	3.16	.94	17.25	21	
3840	18″ diameter	″	152	.184	″	20.50	3.50	1.04	25.04	29.50	
3900	Class 4 O-ring										
3940	12″ diameter	B-21	168	.167	L.F.	13.75	3.16	.94	17.85	21.50	
3960	18″ diameter	″	152	.184	″	18.45	3.50	1.04	22.99	27.50	

02700 | Bases, Ballasts, Pavements & Appurtenances

02710 | Bound Base Courses		CREW	DAILY OUTPUT	LABOR-HOURS	UNIT	2005 BARE COSTS MAT.	LABOR	EQUIP.	TOTAL	TOTAL INCL O&P	
0010	**ASPHALT-TREATED PERMEABLE BASE COURSE** for roadways										200
0020	and large paved areas										
0700	Liquid application to gravel base, asphalt emulsion	B-45	6,000	.003	Gal.	2.92	.05	.04	3.01	3.33	
0800	Prime and seal, cut back asphalt		6,000	.003	″	3.45	.05	.04	3.54	3.91	
1000	Macadam penetration crushed stone, 2 gal. per S.Y., 4″ thick		6,000	.003	S.Y.	5.85	.05	.04	5.94	6.50	
1100	6″ thick, 3 gal. per S.Y.		4,000	.004		8.75	.07	.06	8.88	9.85	
1200	8″ thick, 4 gal. per S.Y.	↓	3,000	.005	↓	11.70	.10	.08	11.88	13.10	
8900	For small and irregular areas, add						50%	50%			

02720 | Unbound Base Courses & Ballasts		CREW	DAILY OUTPUT	LABOR-HOURS	UNIT	MAT.	LABOR	EQUIP.	TOTAL	TOTAL INCL O&P	
0010	**AGGREGRATE BASE COURSE** For roadways and large areas										200
0051	3/4″ stone compacted to 3″ deep	B-36	36,000	.001	S.F.	.29	.02	.03	.34	.39	

	02720 | Unbound Base Courses & Ballasts		CREW	DAILY OUTPUT	LABOR-HOURS	UNIT	2005 BARE COSTS MAT.	LABOR	EQUIP.	TOTAL	TOTAL INCL O&P
200	0101	6″ deep	B-36	35,100	.001	S.F.	.59	.02	.03	.64	.73
	0201	9″ deep		25,875	.002		.86	.03	.04	.93	1.05
	0305	12″ deep		21,150	.002		1.39	.04	.05	1.48	1.65
	0306	Crushed 1-1/2″ stone base, compacted to 4″ deep		47,000	.001		.05	.02	.02	.09	.12
	0307	6″ deep		35,100	.001		.70	.02	.03	.75	.84
	0308	8″ deep		27,000	.001		.93	.03	.04	1	1.12
	0309	12″ deep		16,200	.002		1.39	.05	.07	1.51	1.69
	0350	Bank run gravel, spread and compacted									
	0371	6″ deep	B-32	54,000	.001	S.F.	.39	.01	.03	.43	.47
	0391	9″ deep		39,600	.001		.56	.02	.04	.62	.69
	0401	12″ deep		32,400	.001		.77	.02	.05	.84	.94
	6900	For small and irregular areas, add						50%	50%		
	02740 | Flexible Pavement										
315	0010	**PAVING** Asphaltic concrete R02065-300									
	0020	6″ stone base, 2″ binder course, 1″ topping	B-25C	9,000	.005	S.F.	1.38	.11	.20	1.69	1.92
	0300	Binder course, 1-1/2″ thick		35,000	.001		.31	.03	.05	.39	.45
	0400	2″ thick		25,000	.002		.41	.04	.07	.52	.59
	0500	3″ thick		15,000	.003		.63	.06	.12	.81	.93
	0600	4″ thick		10,800	.004		.82	.09	.16	1.07	1.23
	0800	Sand finish course, 3/4″ thick		41,000	.001		.19	.02	.04	.25	.30
	0900	1″ thick		34,000	.001		.23	.03	.05	.31	.37
	1000	Fill pot holes, hot mix, 2″ thick	B-16	4,200	.008		.42	.14	.11	.67	.82
	1100	4″ thick		3,500	.009		.61	.17	.14	.92	1.10
	1120	6″ thick		3,100	.010		.82	.19	.15	1.16	1.39
	1140	Cold patch, 2″ thick	B-51	3,000	.016		.50	.29	.04	.83	1.09
	1160	4″ thick		2,700	.018		.95	.32	.05	1.32	1.64
	1180	6″ thick		1,900	.025		1.48	.45	.07	2	2.47
	02750 | Rigid Pavement										
300	0010	**PLAIN CEMENT CONCRETE PAVEMENT**									
	0015	Including joints, finishing and curing									
	0021	Fixed form, 12′ pass, unreinforced, 6″ thick	B-26	18,000	.005	S.F.	2.49	.10	.13	2.72	3.04
	0101	8″ thick	″	13,500	.007		3.35	.13	.18	3.66	4.09
	0701	Finishing, broom finish small areas	2 Cefi	1,215	.013			.30		.30	.49
	02770 | Curbs and Gutters										
100	0010	**BITUMINOUS CONCRETE CURBS**									
	0012	Curbs, asphaltic, machine formed, 8″ wide, 6″ high, 40 L.F./ton	B-27	1,000	.032	L.F.	.62	.57	.22	1.41	1.89
	0100	8″ wide, 8″ high, 30 L.F. per ton		900	.036		.71	.63	.24	1.58	2.12
	0150	Asphaltic berm, 12″ W, 3″-6″ H, 35 L.F./ton, before pavement		700	.046		.95	.82	.31	2.08	2.78
	0200	12″ W, 1-1/2″ to 4″ H, 60 L.F. per ton, laid with pavement	B-2	1,050	.038		.58	.68		1.26	1.79
300	0010	**CEMENT CONCRETE CURBS**									
	0300	Concrete, wood forms, 6″ x 18″, straight	C-2A	500	.096	L.F.	2.66	2.21		4.87	6.65
	0400	6″ x 18″, radius	″	200	.240		2.77	5.55		8.32	12.35
	0550	Precast, 6″ x 18″, straight	B-29	700	.069		7.35	1.30	1.20	9.85	11.60
	0600	6″ x 18″, radius	″	325	.148		8.40	2.79	2.58	13.77	16.80
500	0010	**STONE CURBS**									
	1000	Granite, split face, straight, 5″ x 16″	D-13	500	.096	L.F.	9.45	2.18	1.01	12.64	15.15
	1100	6″ x 18″	″	450	.107		12.40	2.43	1.12	15.95	18.90
	1300	Radius curbing, 6″ x 18″, over 10′ radius	B-29	260	.185		15.20	3.49	3.22	21.91	26
	1400	Corners, 2′ radius		80	.600	Ea.	51	11.35	10.45	72.80	86.50
	1600	Edging, 4-1/2″ x 12″, straight		300	.160	L.F.	4.72	3.03	2.79	10.54	13.35

02770 | Curbs and Gutters		CREW	DAILY OUTPUT	LABOR-HOURS	UNIT	2005 BARE COSTS MAT.	LABOR	EQUIP.	TOTAL	TOTAL INCL O&P	
1800	Curb inlets, (guttermouth) straight	B-29	41	1.171	Ea.	113	22	20.50	155.50	185	500
2000	Indian granite (belgian block)										
2100	Jumbo, 10-1/2″ x 7-1/2″ x 4″, grey	D-1	150	.107	L.F.	1.72	2.32		4.04	5.75	
2150	Pink		150	.107		2.24	2.32		4.56	6.30	
2200	Regular, 9″ x 4-1/2″ x 4-1/2″, grey		160	.100		1.56	2.18		3.74	5.35	
2250	Pink		160	.100		2.14	2.18		4.32	6	
2300	Cubes, 4″ x 4″ x 4″, grey		175	.091		1.49	1.99		3.48	4.94	
2350	Pink		175	.091		1.56	1.99		3.55	5	
2400	6″ x 6″ x 6″, pink		155	.103		3.86	2.25		6.11	8	
2500	Alternate pricing method for indian granite										
2550	Jumbo, 10-1/2″ x 7-1/2″ x 4″ (30lb), grey				Ton	98			98	108	
2600	Pink					130			130	143	
2650	Regular, 9″ x 4-1/2″ x 4-1/2″ (20lb), grey					110			110	121	
2700	Pink					150			150	165	
2750	Cubes, 4″ x 4″ x 4″ (5lb), grey					180			180	198	
2800	Pink					200			200	220	
2850	6″ x 6″ x 6″ (25lb), pink					150			150	165	
2900	For pallets, add					16.50			16.50	18.15	
02775 | Sidewalks											
0010	**SIDEWALKS, DRIVEWAYS, & PATIOS** No base										275
0021	Asphaltic concrete, 2″ thick	B-37	6,480	.007	S.F.	.41	.14	.02	.57	.70	
0101	2-1/2″ thick	″	5,950	.008	″	.51	.15	.02	.68	.84	
0300	Concrete, 3000 psi, CIP, 6 x 6 - W1.4 x W1.4 mesh,										
0310	broomed finish, no base, 4″ thick	B-24	600	.040	S.F.	1.35	.86		2.21	2.92	
0350	5″ thick		545	.044		1.80	.94		2.74	3.56	
0400	6″ thick		510	.047		2.10	1.01		3.11	3.99	
0450	For bank run gravel base, 4″ thick, add	B-18	2,500	.010		.41	.17	.01	.59	.77	
0520	8″ thick, add	″	1,600	.015		.84	.27	.02	1.13	1.41	
1000	Crushed stone, 1″ thick, white marble	2 Clab	1,700	.009		.19	.16		.35	.49	
1050	Bluestone	″	1,700	.009		.21	.16		.37	.51	
1700	Redwood, prefabricated, 4′ x 4′ sections	2 Carp	316	.051		8.10	1.22		9.32	10.95	
1750	Redwood planks, 1″ thick, on sleepers	″	240	.067		5.65	1.60		7.25	8.95	
2250	Stone dust, 4″ thick	B-62	900	.027	S.Y.	2.88	.51	.17	3.56	4.22	
02780 | Unit Pavers											
0010	**ASPHALT BLOCKS**										100
0020	Rectangular, 6″ x 12″ x 1-1/4″, w/bed & neopr. adhesive	D-1	135	.119	S.F.	4.22	2.58		6.80	8.95	
0100	3″ thick		130	.123		5.90	2.68		8.58	10.95	
0300	Hexagonal tile, 8″ wide, 1-1/4″ thick		135	.119		4.22	2.58		6.80	8.95	
0400	2″ thick		130	.123		5.90	2.68		8.58	10.95	
0500	Square, 8″ x 8″, 1-1/4″ thick		135	.119		4.22	2.58		6.80	8.95	
0600	2″ thick		130	.123		5.90	2.68		8.58	10.95	
0010	**BRICK PAVING** 4″ x 8″ x 1-1/2″, without joints (4.5 brick/S.F.)	D-1	110	.145	S.F.	2.43	3.17		5.60	7.95	200
0100	Grouted, 3/8″ joint (3.9 brick/S.F.)		90	.178		2.78	3.87		6.65	9.50	
0200	4″ x 8″ x 2-1/4″, without joints (4.5 bricks/S.F.)		110	.145		3.08	3.17		6.25	8.65	
0300	Grouted, 3/8″ joint (3.9 brick/S.F.)		90	.178		2.84	3.87		6.71	9.55	
0500	Bedding, asphalt, 3/4″ thick	B-25	5,130	.017		.34	.33	.39	1.06	1.37	
0540	Course washed sand bed, 1″ thick	B-18	5,000	.005		.18	.09	.01	.28	.36	
0580	Mortar, 1″ thick	D-1	300	.053		.29	1.16		1.45	2.25	
0620	2″ thick		200	.080		.58	1.74		2.32	3.54	
1500	Brick on 1″ thick sand bed laid flat, 4.5 per S.F.		100	.160		2.49	3.48		5.97	8.55	
2000	Brick pavers, laid on edge, 7.2 per S.F.		70	.229		2.36	4.98		7.34	10.90	
0010	**STONE PAVERS**										600
1100	Flagging, bluestone, irregular, 1″ thick,	D-1	81	.198	S.F.	4.63	4.30		8.93	12.25	

02700 | Bases, Ballasts, Pavements & Appurtenances

	02780 | Unit Pavers		CREW	DAILY OUTPUT	LABOR-HOURS	UNIT	2005 BARE COSTS MAT.	LABOR	EQUIP.	TOTAL	TOTAL INCL O&P
600	1150	Snapped random rectangular, 1″ thick	D-1	92	.174	S.F.	7	3.79		10.79	14
	1200	1-1/2″ thick		85	.188		8.45	4.10		12.55	16.05
	1250	2″ thick		83	.193		9.80	4.20		14	17.80
	1300	Slate, natural cleft, irregular, 3/4″ thick		92	.174		5.15	3.79		8.94	12
	1310	1″ thick		85	.188		6	4.10		10.10	13.40
	1351	Random rectangular, gauged, 1/2″ thick		105	.152		11.15	3.32		14.47	17.80
	1400	Random rectangular, butt joint, gauged, 1/4″ thick		150	.107		12	2.32		14.32	17.10
	1450	For sand rubbed finish, add					5.60			5.60	6.15
	1500	For interior setting, add								25%	25%
	1550	Granite blocks, 3-1/2″ x 3-1/2″ x 3-1/2″	D-1	92	.174	S.F.	6.25	3.79		10.04	13.20
700	0010	**STEPS** Incl. excav., borrow & concrete base, where applicable									
	0100	Brick steps	B-24	35	.686	LF Riser	9.10	14.70		23.80	34.50
	0200	Railroad ties	2 Clab	25	.640		2.89	11.10		13.99	22
	0300	Bluestone treads, 12″ x 2″ or 12″ x 1-1/2″	B-24	30	.800		22	17.15		39.15	52.50
	0600	Precast concrete, see division 03480-800									
	4025	Steel edge strips, incl. stakes, 1/4″ x 5″	B-1	390	.062	L.F.	3.35	1.11		4.46	5.55
	4050	Edging, landscape timber or railroad ties, 6″ x 8″	2 Carp	170	.094	″	2.56	2.26		4.82	6.65

	02785 | Flexible Pavement Coating		CREW	DAILY OUTPUT	LABOR-HOURS	UNIT	MAT.	LABOR	EQUIP.	TOTAL	TOTAL INCL O&P
250	0010	**FOG SEAL**									
	0012	Sealcoating, 2 coat coal tar pitch emulsion over 10,000 SY	B-45	5,000	.003	S.Y.	.46	.06	.05	.57	.66
	0030	1000 to 10,000 S.Y.	″	3,000	.005		.46	.10	.08	.64	.77
	0100	Under 1000 S.Y.	B-1	1,050	.023		.46	.41		.87	1.21
	0300	Petroleum resistant, over 10,000 S.Y.	B-45	5,000	.003		.57	.06	.05	.68	.78
	0320	1000 to 10,000 S.Y.	″	3,000	.005		.57	.10	.08	.75	.89
	0400	Under 1000 S.Y.	B-1	1,050	.023		.57	.41		.98	1.33

02800 | Site Improvements and Amenities

	02810 | Irrigation System		CREW	DAILY OUTPUT	LABOR-HOURS	UNIT	2005 BARE COSTS MAT.	LABOR	EQUIP.	TOTAL	TOTAL INCL O&P
300	0010	**SPRINKLER IRRIGATION SYSTEM** For lawns									
	0800	Residential system, custom, 1″ supply	B-20	2,000	.012	S.F.	.27	.22		.49	.67
	0900	1-1/2″ supply	″	1,800	.013	″	.32	.24		.56	.76
	02815 | Outdoor Fountains										
100	0010	**YARD FOUNTAINS**									
	0100	Outdoor fountain, 48″ high with bowl and figures	2 Clab	2	8	Ea.	159	139		298	410
	02820 | Fences & Gates										
120	0010	**CHAIN LINK FENCE** 11 ga. wire									
	0020	1-5/8″ post 10′O.C.,1-3/8″ top rail,2″ corner post galv. stl., 3′ high	B-1	185	.130	L.F.	5.75	2.34		8.09	10.30
	0050	4′ high		170	.141		8.60	2.54		11.14	13.75
	0100	6′ high		115	.209		9.70	3.76		13.46	17.10
	0150	Add for gate 3′ wide, 1-3/8″ frame 3′ high		12	2	Ea.	51.50	36		87.50	118
	0170	4′ high		10	2.400		64	43.50		107.50	144
	0190	6′ high		10	2.400		115	43.50		158.50	201
	0200	Add for gate 4′ wide, 1-3/8″ frame 3′ high		9	2.667		60.50	48		108.50	148

02820 Fences & Gates		Crew	Daily Output	Labor-Hours	Unit	2005 Bare Costs Mat.	Labor	Equip.	Total	Total Incl O&P	
0220	4′ high	B-1	9	2.667	Ea.	79	48		127	169	120
0240	6′ high		8	3		146	54		200	253	
0350	Aluminized steel, 9 ga. wire, 3′ high		185	.130	L.F.	6.95	2.34		9.29	11.55	
0380	4′ high		170	.141		7.90	2.54		10.44	13	
0400	6′ high		115	.209		10.15	3.76		13.91	17.55	
0450	Add for gate 3′ wide, 1-3/8″ frame 3′ high		12	2	Ea.	67.50	36		103.50	136	
0470	4′ high		10	2.400		92	43.50		135.50	175	
0490	6′ high		10	2.400		138	43.50		181.50	226	
0500	Add for gate 4′ wide, 1-3/8″ frame 3′ high		10	2.400		92	43.50		135.50	175	
0520	4′ high		9	2.667		123	48		171	217	
0540	6′ high		8	3		192	54		246	305	
0620	Vinyl covered 9 ga. wire, 3′ high		185	.130	L.F.	6.15	2.34		8.49	10.70	
0640	4′ high		170	.141		10.10	2.54		12.64	15.40	
0660	6′ high		115	.209		11.50	3.76		15.26	19.05	
0720	Add for gate 3′ wide, 1-3/8″ frame 3′ high		12	2	Ea.	77	36		113	146	
0740	4′ high		10	2.400		100	43.50		143.50	184	
0760	6′ high		10	2.400		154	43.50		197.50	243	
0780	Add for gate 4′ wide, 1-3/8″ frame 3′ high		10	2.400		104	43.50		147.50	189	
0800	4′ high		9	2.667		138	48		186	234	
0820	6′ high		8	3		200	54		254	310	
0860	Tennis courts, 11 ga. wire, 2 1/2″ post 10′ O.C., 1-5/8″ top rail										
0900	2-1/2″ corner post, 10′ high	B-1	95	.253	L.F.	15.35	4.55		19.90	24.50	
0920	12′ high		80	.300	″	18.40	5.40		23.80	29.50	
1000	Add for gate 3′ wide, 1-5/8″ frame 10′ high		10	2.400	Ea.	192	43.50		235.50	285	
1040	Aluminized, 11 ga. wire 10′ high		95	.253	L.F.	21.50	4.55		26.05	31.50	
1100	12′ high		80	.300	″	22.50	5.40		27.90	33.50	
1140	Add for gate 3′ wide, 1-5/8″ frame, 10′ high		10	2.400	Ea.	246	43.50		289.50	345	
1250	Vinyl covered 11 ga. wire, 10′ high		95	.253	L.F.	18.40	4.55		22.95	28.50	
1300	12′ high		80	.300	″	21.50	5.40		26.90	32.50	
1400	Add for gate 3′ wide, 1-3/8″ frame, 10′ high		10	2.400	Ea.	277	43.50		320.50	380	
0010	**FENCE, VINYL**, white, steel reinforced, stainless steel fasteners										300
0020	Picket, 4″ x 4″ posts @ 6′ - 0″ OC, 3′ high	B-1	140	.171	L.F.	15.90	3.09		18.99	23	
0030	4′ high		130	.185		18.30	3.33		21.63	25.50	
0040	5′ high		120	.200		21	3.60		24.60	29	
0100	Board (semi-privacy), 5″ x 5″ posts @ 7′ - 6″ OC, 5′ high		130	.185		22	3.33		25.33	29.50	
0120	6′ high		125	.192		25	3.46		28.46	33.50	
0200	Basketweave, 5″ x 5″ posts @ 7′ - 6″ OC, 5′ high		160	.150		19.30	2.70		22	25.50	
0220	6′ high		150	.160		23	2.88		25.88	30	
0300	Privacy, 5″ x 5″ posts @ 7′ - 6″ OC, 5′ high		130	.185		23.50	3.33		26.83	31.50	
0320	6′ high		150	.160		27	2.88		29.88	34.50	
0350	Gate, 5′ high		9	2.667	Ea.	330	48		378	445	
0360	6′ high		9	2.667		340	48		388	455	
0400	For posts set in concrete, add		25	.960		6.30	17.30		23.60	36.50	
0010	**FENCES, MISC. METAL**										410
0012	Chicken wire, posts @ 4′, 1″ mesh, 4′ high	B-80C	410	.059	L.F.	1.63	1.05	.33	3.01	3.92	
0100	2″ mesh, 6′ high		350	.069		1.47	1.23	.38	3.08	4.11	
0200	Galv. steel, 12 ga., 2″ x 4″ mesh, posts 5′ O.C., 3′ high		300	.080		2.21	1.43	.45	4.09	5.35	
0300	5′ high		300	.080		2.95	1.43	.45	4.83	6.15	
0400	14 ga., 1″ x 2″ mesh, 3′ high		300	.080		2.35	1.43	.45	4.23	5.50	
0500	5′ high		300	.080		3.25	1.43	.45	5.13	6.50	
1000	Kennel fencing, 1-1/2″ mesh, 6′ long, 3′-6″ wide, 6′-2″ high	2 Clab	4	4	Ea.	370	69.50		439.50	525	
1050	12′ long		4	4		445	69.50		514.50	605	
1200	Top covers, 1-1/2″ mesh, 6′ long		15	1.067		75	18.50		93.50	114	
1250	12′ long		12	1.333		120	23		143	172	

SITE CONSTRUCTION 2

	02820 | Fences & Gates		CREW	DAILY OUTPUT	LABOR-HOURS	UNIT	2005 BARE COSTS MAT.	LABOR	EQUIP.	TOTAL	TOTAL INCL O&P
510	0010	**FENCE, WOOD** Basket weave, 3/8" x 4" boards, 2" x 4"									
	0020	stringers on spreaders, 4" x 4" posts									
	0050	No. 1 cedar, 6' high	B-80C	160	.150	L.F.	8.05	2.69	.84	11.58	14.30
	0070	Treated pine, 6' high		150	.160		9.75	2.87	.89	13.51	16.55
	0090	Vertical weave 6' high		145	.166		11.75	2.97	.92	15.64	18.95
	0200	Board fence, 1" x 4" boards, 2" x 4" rails, 4" x 4" post									
	0220	Preservative treated, 2 rail, 3' high	B-80C	145	.166	L.F.	5.95	2.97	.92	9.84	12.55
	0240	4' high		135	.178		6.55	3.19	.99	10.73	13.65
	0260	3 rail, 5' high		130	.185		7.40	3.31	1.03	11.74	14.90
	0300	6' high		125	.192		8.45	3.44	1.07	12.96	16.30
	0320	No. 2 grade western cedar, 2 rail, 3' high		145	.166		6.50	2.97	.92	10.39	13.15
	0340	4' high		135	.178		7.70	3.19	.99	11.88	14.95
	0360	3 rail, 5' high		130	.185		8.90	3.31	1.03	13.24	16.55
	0400	6' high		125	.192		9.75	3.44	1.07	14.26	17.75
	0420	No. 1 grade cedar, 2 rail, 3' high		145	.166		9.80	2.97	.92	13.69	16.75
	0440	4' high		135	.178		11.10	3.19	.99	15.28	18.70
	0460	3 rail, 5' high		130	.185		12.85	3.31	1.03	17.19	21
	0500	6' high		125	.192		14.35	3.44	1.07	18.86	22.50
	0540	Shadow box, 1" x 6" board, 2" x 4" rail, 4" x 4"post									
	0560	Pine, pressure treated, 3 rail, 6' high	B-80C	150	.160	L.F.	10.95	2.87	.89	14.71	17.85
	0600	Gate, 3'-6" wide		8	3	Ea.	63	54	16.75	133.75	178
	0620	No. 1 cedar, 3 rail, 4' high		130	.185	L.F.	13.50	3.31	1.03	17.84	21.50
	0640	6' high		125	.192		16.65	3.44	1.07	21.16	25.50
	0860	Open rail fence, split rails, 2 rail 3' high, no. 1 cedar		160	.150		5.40	2.69	.84	8.93	11.40
	0870	No. 2 cedar		160	.150		4.21	2.69	.84	7.74	10.10
	0880	3 rail, 4' high, no. 1 cedar		150	.160		7.30	2.87	.89	11.06	13.80
	0890	No. 2 cedar		150	.160		4.80	2.87	.89	8.56	11.10
	0920	Rustic rails, 2 rail 3' high, no. 1 cedar		160	.150		3.37	2.69	.84	6.90	9.15
	0930	No. 2 cedar		160	.150		3.23	2.69	.84	6.76	9
	0940	3 rail, 4' high		150	.160		4.53	2.87	.89	8.29	10.80
	0950	No. 2 cedar		150	.160		3.42	2.87	.89	7.18	9.55
	0960	Picket fence, gothic, pressure treated pine									
	1000	2 rail, 3' high	B-80C	140	.171	L.F.	4.16	3.07	.96	8.19	10.85
	1020	3 rail, 4' high		130	.185	"	4.90	3.31	1.03	9.24	12.15
	1040	Gate, 3'-6" wide		9	2.667	Ea.	44	48	14.90	106.90	145
	1060	No. 2 cedar, 2 rail, 3' high		140	.171	L.F.	5.20	3.07	.96	9.23	11.95
	1100	3 rail, 4' high		130	.185	"	5.30	3.31	1.03	9.64	12.60
	1120	Gate, 3'-6" wide		9	2.667	Ea.	52	48	14.90	114.90	154
	1140	No. 1 cedar, 2 rail 3' high		140	.171	L.F.	10.40	3.07	.96	14.43	17.70
	1160	3 rail, 4' high		130	.185		12.10	3.31	1.03	16.44	20
	1200	Rustic picket, molded pine, 2 rail, 3' high		140	.171		4.77	3.07	.96	8.80	11.50
	1220	No. 1 cedar, 2 rail, 3' high		140	.171		6.50	3.07	.96	10.53	13.40
	1240	Stockade fence, no. 1 cedar, 3-1/4" rails, 6' high		160	.150		9.80	2.69	.84	13.33	16.25
	1260	8' high		155	.155		12.70	2.77	.86	16.33	19.60
	1300	No. 2 cedar, treated wood rails, 6' high		160	.150		9.80	2.69	.84	13.33	16.25
	1320	Gate, 3'-6" wide		8	3	Ea.	57.50	54	16.75	128.25	172
	1360	Treated pine, treated rails, 6' high		160	.150	L.F.	9.60	2.69	.84	13.13	16
	1400	8' high		150	.160	"	14.40	2.87	.89	18.16	21.50
520	0010	**FENCE, WOOD RAIL** Picket, No. 2 cedar, Gothic, 2 rail, 3' high	B-1	160	.150	L.F.	5.15	2.70		7.85	10.25
	0050	Gate, 3'-6" wide	B-80C	9	2.667	Ea.	44.50	48	14.90	107.40	146
	0400	3 rail, 4' high		150	.160	L.F.	5.90	2.87	.89	9.66	12.30
	0500	Gate, 3'-6" wide		9	2.667	Ea.	53.50	48	14.90	116.40	155
	1200	Stockade, No. 2 cedar, treated wood rails, 6' high		160	.150	L.F.	6.45	2.69	.84	9.98	12.50
	1250	Gate, 3' wide		9	2.667	Ea.	52.50	48	14.90	115.40	155
	1300	No. 1 cedar, 3-1/4" cedar rails, 6' high		160	.150	L.F.	16.15	2.69	.84	19.68	23
	1500	Gate, 3' wide		9	2.667	Ea.	134	48	14.90	196.90	245

02820 | Fences & Gates

	Description	Crew	Daily Output	Labor-Hours	Unit	2005 Bare Costs Mat.	Labor	Equip.	Total	Total Incl O&P	
2700	Prefabricated redwood or cedar, 4′ high	B-80C	160	.150	L.F.	12.20	2.69	.84	15.73	18.90	520
2800	6′ high		150	.160		16.20	2.87	.89	19.96	23.50	
3300	Board, shadow box, 1″ x 6″, treated pine, 6′ high		160	.150		9.50	2.69	.84	13.03	15.90	
3400	No. 1 cedar, 6′ high		150	.160		18.75	2.87	.89	22.51	26.50	
3900	Basket weave, No. 1 cedar, 6′ high		160	.150		18.55	2.69	.84	22.08	26	
4200	Gate, 3′-6″ wide	B-1	9	2.667	Ea.	59	48		107	147	
5000	Fence rail, redwood, 2″ x 4″, merch grade 8′	″	2,400	.010	L.F.	1.01	.18		1.19	1.42	

02830 | Retaining Walls

	Description	Crew	Daily Output	Labor-Hours	Unit	2005 Bare Costs Mat.	Labor	Equip.	Total	Total Incl O&P	
0010	**CAST IN PLACE RETAINING WALLS**										100
1800	Concrete gravity wall with vertical face including excavation & backfill										
1850	No reinforcing										
1900	6′ high, level embankment	C-17C	36	2.306	L.F.	59.50	56	10.40	125.90	172	
2000	33° slope embankment	″	32	2.594	″	54	63	11.65	128.65	179	
2800	Reinforced concrete cantilever, incl. excavation, backfill & reinf.										
2900	6′ high, 33° slope embankment	C-17C	35	2.371	L.F.	54	57.50	10.65	122.15	169	
0010	**INTERLOCKING SEGMENTAL RETAINING WALLS**										200
7100	Segmental Retaining Wall system, incl pins, and void fill										
7120	base not included										
7140	Large unit, 8″ high x 18″ wide x 20″ deep, 3 plane split	B-62	300	.080	S.F.	8.55	1.53	.51	10.59	12.60	
7150	straight split		300	.080		8.55	1.53	.51	10.59	12.60	
7160	Medium, ltwt, 8″ high x 18″ wide x 12″ deep, 3 plane split		400	.060		7.55	1.15	.38	9.08	10.65	
7170	straight split		400	.060		7.55	1.15	.38	9.08	10.65	
7180	Small unit, 4″ x 18″ x 10″ deep, 3 plane split		400	.060		9.45	1.15	.38	10.98	12.75	
7190	straight split		400	.060		9.45	1.15	.38	10.98	12.75	
7200	Cap unit, 3 plane split		300	.080		10.60	1.53	.51	12.64	14.85	
7210	Cap unit, st split		300	.080		10.60	1.53	.51	12.64	14.85	
7260	For reinforcing, add								2.20	2.75	
8000	For higher walls, add components as necessary										
0010	**METAL BIN RETAINING WALLS**, Aluminized steel bin, excavation										600
0020	and backfill not included, 10′ wide										
0100	4′ high, 5.5′ deep	B-13	650	.074	S.F.	16.30	1.40	.95	18.65	21.50	
0200	8′ high, 5.5′ deep		615	.078		18.75	1.48	1	21.23	24	
0300	10′ high, 7.7′ deep		580	.083		19.75	1.57	1.06	22.38	25.50	
0400	12′ high, 7.7′ deep		530	.091		21.50	1.71	1.16	24.37	27.50	
0500	16′ high, 7.7′ deep		515	.093		22.50	1.76	1.20	25.46	29.50	
0010	**STONE RETAINING WALLS**										800
0015	Including excavation, concrete footing and										
0020	stone 3′ below grade. Price is exposed face area.										
0200	Decorative random stone, to 6′ high, 1′-6″ thick, dry set	D-1	35	.457	S.F.	35	9.95		44.95	55	
0300	Mortar set		40	.400		35	8.70		43.70	53	
0500	Cut stone, to 6′ high, 1′-6″ thick, dry set		35	.457		35	9.95		44.95	55	
0600	Mortar set		40	.400		35	8.70		43.70	53	
0800	Random stone, 6′ to 10′ high, 2′ thick, dry set		45	.356		35	7.75		42.75	51.50	
0900	Mortar set		50	.320		35	6.95		41.95	50	
1100	Cut stone, 6′ to 10′ high, 2′ thick, dry set		45	.356		35	7.75		42.75	51.50	
1200	Mortar set		50	.320		35	6.95		41.95	50	

02870 | Site Furnishings

	Description	Crew	Daily Output	Labor-Hours	Unit	2005 Bare Costs Mat.	Labor	Equip.	Total	Total Incl O&P	
0010	**BENCHES**										310
0012	Seating, benches, park, precast conc, w/backs, wood rails, 4′ long	2 Clab	5	3.200	Ea.	320	55.50		375.50	445	

02800 | Site Improvements and Amenities

	02870 | Site Furnishings		CREW	DAILY OUTPUT	LABOR-HOURS	UNIT	2005 BARE COSTS MAT.	LABOR	EQUIP.	TOTAL	TOTAL INCL O&P
310	0100	8′ long	2 Clab	4	4	Ea.	670	69.50		739.50	855
	0500	Steel barstock pedestals w/backs, 2″ x 3″ wood rails, 4′ long		10	1.600		770	28		798	895
	0510	8′ long		7	2.286		910	39.50		949.50	1,075
	0800	Cast iron pedestals, back & arms, wood slats, 4′ long		8	2		290	34.50		324.50	380
	0820	8′ long		5	3.200		845	55.50		900.50	1,025
	1700	Steel frame, fir seat, 10′ long	↓	10	1.600	↓	176	28		204	241

02900 | Planting

	02905 | Plants, Planting, Transplanting		CREW	DAILY OUTPUT	LABOR-HOURS	UNIT	2005 BARE COSTS MAT.	LABOR	EQUIP.	TOTAL	TOTAL INCL O&P
725	0010	PLANTING									
	0012	Moving shrubs on site, 12″ ball	B-62	28	.857	Ea.		16.45	5.50	21.95	33.50
	0100	24″ ball	″	22	1.091			21	6.95	27.95	42.50
	0300	Moving trees on site, 36″ ball	B-6	3.75	6.400			123	57.50	180.50	270
	0400	60″ ball	″	1	24	↓		460	216	676	1,000

	02910 | Plant Preparation		CREW	DAILY OUTPUT	LABOR-HOURS	UNIT	2005 BARE COSTS MAT.	LABOR	EQUIP.	TOTAL	TOTAL INCL O&P
500	0010	MULCHING									
	0100	Aged barks, 3″ deep, hand spread	1 Clab	100	.080	S.Y.	2.03	1.39		3.42	4.59
	0150	Skid steer loader	B-63	13.50	2.963	M.S.F.	226	51.50	11.40	288.90	350
	0200	Hay, 1″ deep, hand spread	1 Clab	475	.017	S.Y.	.56	.29		.85	1.12
	0250	Power mulcher, small	B-64	180	.089	M.S.F.	62	1.62	1.29	64.91	72.50
	0350	Large	B-65	530	.030	″	62	.55	.59	63.14	70
	0400	Humus peat, 1″ deep, hand spread	1 Clab	700	.011	S.Y.	2.13	.20		2.33	2.68
	0450	Push spreader	″	2,500	.003	″	2.13	.06		2.19	2.43
	0550	Tractor spreader	B-66	700	.011	M.S.F.	237	.26	.25	237.51	261
	0600	Oat straw, 1″ deep, hand spread	1 Clab	475	.017	S.Y.	.32	.29		.61	.85
	0650	Power mulcher, small	B-64	180	.089	M.S.F.	35.50	1.62	1.29	38.41	43
	0700	Large	B-65	530	.030	″	35.50	.55	.59	36.64	40.50
	0750	Add for asphaltic emulsion	B-45	1,770	.009	Gal.	1.74	.17	.13	2.04	2.34
	0800	Peat moss, 1″ deep, hand spread	1 Clab	900	.009	S.Y.	1.69	.15		1.84	2.12
	0850	Push spreader	″	2,500	.003	″	1.69	.06		1.75	1.95
	0950	Tractor spreader	B-66	700	.011	M.S.F.	188	.26	.25	188.51	208
	1000	Polyethylene film, 6 mil.	2 Clab	2,000	.008	S.Y.	.16	.14		.30	.42
	1100	Redwood nuggets, 3″ deep, hand spread	1 Clab	150	.053	″	4.17	.93		5.10	6.15
	1150	Skid steer loader	B-63	13.50	2.963	M.S.F.	465	51.50	11.40	527.90	610
	1200	Stone mulch, hand spread, ceramic chips, economy	1 Clab	125	.064	S.Y.	6.10	1.11		7.21	8.65
	1250	Deluxe	″	95	.084	″	9.40	1.46		10.86	12.85
	1300	Granite chips	B-1	10	2.400	C.Y.	30.50	43.50		74	107
	1400	Marble chips		10	2.400		115	43.50		158.50	200
	1600	Pea gravel		28	.857		56	15.45		71.45	87.50
	1700	Quartz	↓	10	2.400	↓	148	43.50		191.50	236
	1800	Tar paper, 15 Lb. felt	1 Clab	800	.010	S.Y.	.31	.17		.48	.63
	1900	Wood chips, 2″ deep, hand spread	″	220	.036	″	1.74	.63		2.37	2.98
	1950	Skid steer loader	B-63	20.30	1.970	M.S.F.	193	34	7.55	234.55	279
720	0010	PLANT BED PREPARATION, SHRUB & TREE									
	0100	Backfill planting pit, by hand, on site topsoil	2 Clab	18	.889	C.Y.		15.40		15.40	26
	0200	Prepared planting mix	″	24	.667			11.55		11.55	19.65
	0300	Skid steer loader, on site topsoil	B-62	340	.071	↓		1.35	.45	1.80	2.77

02900 | Planting

02910 | Plant Preparation		Crew	Daily Output	Labor-Hours	Unit	2005 Bare Costs Mat.	Labor	Equip.	Total	Total Incl O&P	
0400	Prepared planting mix	B-62	410	.059	C.Y.		1.12	.37	1.49	2.29	720
1000	Excavate planting pit, by hand, sandy soil	2 Clab	16	1			17.35		17.35	29.50	
1100	Heavy soil or clay	"	8	2			34.50		34.50	59	
1200	1/2 C.Y. backhoe, sandy soil	B-11C	150	.107			2.20	1.44	3.64	5.25	
1300	Heavy soil or clay	"	115	.139			2.87	1.88	4.75	6.85	
2000	Mix planting soil, incl. loam, manure, peat, by hand	2 Clab	60	.267		37	4.63		41.63	49	
2100	Skid steer loader	B-62	150	.160		37	3.07	1.02	41.09	47.50	
3000	Pile sod, skid steer loader	"	2,800	.009	S.Y.		.16	.05	.21	.34	
3100	By hand	2 Clab	400	.040			.69		.69	1.18	
4000	Remove sod, F.E. loader	B-10S	2,000	.004			.10	.12	.22	.29	
4100	Sod cutter	B-12K	3,200	.005			.11	.30	.41	.51	
4200	By hand	2 Clab	240	.067			1.16		1.16	1.96	
0010	**LOAM & TOPSOIL**										810
0300	Fine grade, base course for paving, see div. 02720-200										
0701	Furnish and place, truck dumped, unscreened, 4″ deep	B-10S	12,000	.001	S.F.	.35	.02	.02	.39	.44	
0801	6″ deep	"	7,400	.001	"	.38	.03	.03	.44	.49	
0900	Fine grading and seeding, incl. lime, fertilizer & seed,										
1001	With equipment	B-14	9,000	.005	S.F.	.06	.10	.02	.18	.26	
02915 | Shrub and Tree Transplanting											
0010	**GROUND COVER** Plants, pachysandra, in prepared beds	B-1	15	1.600	C	25	29		54	76.50	200
0200	Vinca minor, 1 yr, bare root		12	2	"	26	36		62	89.50	
0600	Stone chips, in 50 lb. bags, Georgia marble		520	.046	Bag	2.35	.83		3.18	4	
0700	Onyx gemstone		260	.092		17	1.66		18.66	21.50	
0800	Quartz		260	.092		6.35	1.66		8.01	9.80	
0900	Pea gravel, truckload lots		28	.857	Ton	24.50	15.45		39.95	53	
02920 | Lawns & Grasses											
0010	**SEEDING, GENERAL** R02920-500										310
0020	Mechanical seeding, 215 lb./acre	B-66	1.50	5.333	Acre	510	122	115	747	885	
0101	$2.00/lb., 44 lb./M.S.Y.	1 Clab	13,950	.001	S.F.	.02	.01		.03	.04	
0300	Fine grading and seeding incl. lime, fertilizer & seed,										
0310	with equipment	B-14	1,000	.048	S.Y.	.16	.89	.22	1.27	1.93	
0600	Limestone hand push spreader, 50 lbs. per M.S.F.	1 Clab	180	.044	M.S.F.	3.38	.77		4.15	5.05	
0800	Grass seed hand push spreader, 4.5 lbs. per M.S.F.	"	180	.044	"	16.65	.77		17.42	19.65	
0010	**SODDING**										400
0020	Sodding, 1″ deep, bluegrass sod, on level ground, over 8 MSF	B-63	22	1.818	M.S.F.	217	31.50	7	255.50	300	
0200	4 M.S.F.		17	2.353		243	41	9.05	293.05	345	
0300	1000 S.F.		13.50	2.963		265	51.50	11.40	327.90	390	
0500	Sloped ground, over 8 M.S.F.		6	6.667		217	116	25.50	358.50	465	
0600	4 M.S.F.		5	8		243	139	30.50	412.50	535	
0700	1000 S.F.		4	10		265	174	38.50	477.50	630	
1000	Bent grass sod, on level ground, over 6 M.S.F.		20	2		485	34.50	7.70	527.20	600	
1100	3 M.S.F.		18	2.222		540	38.50	8.55	587.05	670	
1200	Sodding 1000 S.F. or less		14	2.857		615	49.50	10.95	675.45	775	
1500	Sloped ground, over 6 M.S.F.		15	2.667		485	46.50	10.25	541.75	625	
1600	3 M.S.F.		13.50	2.963		540	51.50	11.40	602.90	695	
1700	1000 S.F.		12	3.333		615	58	12.80	685.80	790	
02930 | Exterior Plants											
0010	**SHRUBS AND TREES** Evergreen, in prepared beds, B & B										310
0100	Arborvitae pyramidal, 4′-5′	B-17	30	1.067	Ea.	42.50	20.50	18.10	81.10	101	
0150	Globe, 12″-15″	B-1	96	.250		10.85	4.51		15.36	19.55	
0300	Cedar, blue, 8′-10′	B-17	18	1.778		179	34	30	243	288	

	02930	Exterior Plants	CREW	DAILY OUTPUT	LABOR-HOURS	UNIT	2005 BARE COSTS MAT.	LABOR	EQUIP.	TOTAL	TOTAL INCL O&P
310	0500	Hemlock, canadian, 2-1/2′-3′	B-1	36	.667	Ea.	22.50	12		34.50	45.50
	0550	Holly, Savannah, 8′ - 10′ H		9.68	2.479		515	44.50		559.50	645
	0600	Juniper, andorra, 18″-24″		80	.300		15.60	5.40		21	26.50
	0620	Wiltoni, 15″-18″		80	.300		14.80	5.40		20.20	25.50
	0640	Skyrocket, 4-1/2′-5′	B-17	55	.582		47	11.20	9.85	68.05	81
	0660	Blue pfitzer, 2′-2-1/2′	B-1	44	.545		24.50	9.85		34.35	43.50
	0680	Ketleerie, 2-1/2′-3′		50	.480		31	8.65		39.65	49
	0700	Pine, black, 2-1/2′-3′		50	.480		38.50	8.65		47.15	56.50
	0720	Mugo, 18″-24″		60	.400		33.50	7.20		40.70	49.50
	0740	White, 4′-5′	B-17	75	.427		49.50	8.20	7.25	64.95	76
	0800	Spruce, blue, 18″-24″	B-1	60	.400		37.50	7.20		44.70	53.50
	0840	Norway, 4′-5′	B-17	75	.427		84	8.20	7.25	99.45	114
	0900	Yew, denisforma, 12″-15″	B-1	60	.400		22.50	7.20		29.70	37
	1000	Capitata, 18″-24″		30	.800		19.25	14.40		33.65	45.50
	1100	Hicksi, 2′-2-1/2′		30	.800		28	14.40		42.40	55.50
320	0010	**SHRUBS** Broadleaf evergreen, planted in prepared beds									
	0100	Andromeda, 15″-18″, container	B-1	96	.250	Ea.	22	4.51		26.51	31.50
	0200	Azalea, 15″ - 18″, container		96	.250		24.50	4.51		29.01	34
	0300	Barberry, 9″-12″, container		130	.185		9.75	3.33		13.08	16.40
	0400	Boxwood, 15″-18″, B & B		96	.250		26	4.51		30.51	36
	0500	Euonymus, emerald gaiety, 12″ to 15″, container		115	.209		15.75	3.76		19.51	24
	0600	Holly, 15″-18″, B & B		96	.250		15.35	4.51		19.86	24.50
	0900	Mount laurel, 18″ - 24″, B & B		80	.300		48.50	5.40		53.90	62.50
	1000	Paxistema, 9 - 12″ high		130	.185		15.75	3.33		19.08	23
	1100	Rhododendron, 18″-24″, container		48	.500		27	9		36	45
	1200	Rosemary, 1 gal container		600	.040		57.50	.72		58.22	64.50
	2000	Deciduous, amelanchier, 2′-3′, B & B		57	.421		78	7.60		85.60	99
	2100	Azalea, 15″-18″, B & B		96	.250		21	4.51		25.51	30.50
	2300	Bayberry, 2′-3′, B & B		57	.421		24	7.60		31.60	39.50
	2600	Cotoneaster, 15″-18″, B & B		80	.300		14.25	5.40		19.65	25
	2800	Dogwood, 3′-4′, B & B	B-17	40	.800		22.50	15.40	13.55	51.45	66
	2900	Euonymus, alatus compacta, 15″ to 18″, container	B-1	80	.300		18.75	5.40		24.15	29.50
	3200	Forsythia, 2′-3′, container	″	60	.400		16.50	7.20		23.70	30.50
	3300	Hibiscus, 3′-4′, B & B	B-17	75	.427		12.90	8.20	7.25	28.35	36
	3400	Honeysuckle, 3′-4′, B & B	B-1	60	.400		18.40	7.20		25.60	32.50
	3500	Hydrangea, 2′-3′, B & B	″	57	.421		21.50	7.60		29.10	36.50
	3600	Lilac, 3′-4′, B & B	B-17	40	.800		22	15.40	13.55	50.95	65.50
	3900	Privet, bare root, 18″-24″	B-1	80	.300		11.40	5.40		16.80	21.50
	4100	Quince, 2′-3′, B & B	″	57	.421		18.95	7.60		26.55	34
	4200	Russian olive, 3′-4′, B & B	B-17	75	.427		20.50	8.20	7.25	35.95	44
	4400	Spirea, 3′-4′, B & B	B-1	70	.343		23.50	6.20		29.70	36.50
	4500	Viburnum, 3′-4′, B & B	B-17	40	.800		24	15.40	13.55	52.95	67.50
410	0010	**TREES** Deciduous, in prep. beds, balled & burlapped (B&B)									
	0100	Ash, 2″ caliper	B-17	8	4	Ea.	110	77	68	255	325
	0200	Beech, 5′-6′		50	.640		215	12.35	10.85	238.20	269
	0300	Birch, 6′-8′, 3 stems		20	1.600		119	31	27	177	212
	0500	Crabapple, 6′-8′		20	1.600		155	31	27	213	252
	0600	Dogwood, 4′-5′		40	.800		66	15.40	13.55	94.95	113
	0700	Eastern redbud 4′-5′		40	.800		129	15.40	13.55	157.95	183
	0800	Elm, 8′-10′		20	1.600		109	31	27	167	202
	0900	Ginkgo, 6′-7′		24	1.333		159	25.50	22.50	207	243
	1000	Hawthorn, 8′-10′, 1″ caliper		20	1.600		121	31	27	179	216
	1100	Honeylocust, 10′-12′, 1-1/2″ caliper		10	3.200		145	61.50	54	260.50	325
	1300	Larch, 8′		32	1		93.50	19.25	16.95	129.70	154

	02930 | Exterior Plants		CREW	DAILY OUTPUT	LABOR-HOURS	UNIT	2005 BARE COSTS MAT.	LABOR	EQUIP.	TOTAL	TOTAL INCL O&P	
	1400	Linden, 8′-10′, 1″ caliper	B-17	20	1.600	Ea.	101	31	27	159	193	410
	1500	Magnolia, 4′-5′		20	1.600		69.50	31	27	127.50	158	
	1600	Maple, red, 8′-10′, 1-1/2″ caliper		10	3.200		154	61.50	54	269.50	335	
	1700	Mountain ash, 8′-10′, 1″ caliper		16	2		159	38.50	34	231.50	277	
	1800	Oak, 2-1/2″-3″ caliper		6	5.333		240	103	90.50	433.50	535	
	2100	Planetree, 9′-11′, 1-1/4″ caliper		10	3.200		99.50	61.50	54	215	272	
	2200	Plum, 6′-8′, 1″ caliper		20	1.600		88	31	27	146	179	
	2300	Poplar, 9′-11′, 1-1/4″ caliper		10	3.200		46	61.50	54	161.50	213	
	2500	Sumac, 2′-3′		75	.427		22	8.20	7.25	37.45	45.50	
	2700	Tulip, 5′-6′		40	.800		48	15.40	13.55	76.95	93.50	
	2800	Willow, 6′-8′, 1″ caliper		20	1.600		59	31	27	117	147	

	02945 | Planting Accessories		CREW	DAILY OUTPUT	LABOR-HOURS	UNIT	MAT.	LABOR	EQUIP.	TOTAL	TOTAL INCL O&P	
120	0010	**EDGING**										120
	0050	Aluminum alloy, including stakes, 1/8″ x 4″, mill finish	B-1	390	.062	L.F.	2.21	1.11		3.32	4.31	
	0051	Black paint		390	.062		2.56	1.11		3.67	4.70	
	0052	Black anodized		390	.062		2.96	1.11		4.07	5.15	
	0100	Brick, set horizontally, 1-1/2 bricks per L.F.	D-1	370	.043		.93	.94		1.87	2.59	
	0150	Set vertically, 3 bricks per L.F.	″	135	.119		2.70	2.58		5.28	7.25	
	0200	Corrugated aluminum, roll, 4″ wide	1 Carp	650	.012		.37	.30		.67	.91	
	0250	6″ wide	″	550	.015		.46	.35		.81	1.10	
	0600	Railroad ties, 6″ x 8″	2 Carp	170	.094		2.56	2.26		4.82	6.65	
	0650	7″ x 9″	″	136	.118		2.84	2.82		5.66	7.90	
	0700	Redwood										
	0750	2″ x 4″	2 Carp	330	.048	L.F.	2.20	1.16		3.36	4.40	
	0800	Steel edge strips, incl. stakes, 1/4″ x 5″	B-1	390	.062		3.35	1.11		4.46	5.55	
	0850	3/16″ x 4″	″	390	.062		2.65	1.11		3.76	4.79	
300	0010	**PLANTERS** Concrete, sandblasted, precast, 48″ diameter, 24″ high	2 Clab	15	1.067	Ea.	555	18.50		573.50	640	300
	0300	Fiberglass, circular, 36″ diameter, 24″ high		15	1.067		385	18.50		403.50	455	
	1200	Wood, square, 48″ side, 24″ high		15	1.067		860	18.50		878.50	980	
	1300	Circular, 48″ diameter, 30″ high		10	1.600		710	28		738	825	
	1600	Planter/bench, 72″		5	3.200		2,775	55.50		2,830.50	3,150	
510	0010	**TREE GUYING**										510
	0015	Tree guying Including stakes, guy wire and wrap										
	0100	Less than 3″ caliper, 2 stakes	2 Clab	35	.457	Ea.	15.10	7.95		23.05	30	
	0200	3″ to 4″ caliper, 3 stakes	″	21	.762	″	17.85	13.20		31.05	42	
	1000	Including arrowhead anchor, cable, turnbuckles and wrap										
	1100	Less than 3″ caliper, 3″ anchors	2 Clab	20	.800	Ea.	47.50	13.90		61.40	75.50	
	1200	3″ to 6″ caliper, 4″ anchors		15	1.067		68	18.50		86.50	107	
	1300	6″ caliper, 6″ anchors		12	1.333		84	23		107	132	
	1400	8″ caliper, 8″ anchors		9	1.778		96.50	31		127.50	159	

For information about Means Estimating Seminars, see yellow pages 12 and 13 in back of book

Division Notes

		CREW	DAILY OUTPUT	LABOR-HOURS	UNIT	2005 BARE COSTS MAT.	LABOR	EQUIP.	TOTAL	TOTAL INCL O&P

Division 3 Concrete

stimating Tips

eneral

Carefully check all the plans and specifications. Concrete often appears on drawings other than structural drawings, including mechanical and electrical drawings for equipment pads. The cost of cutting and patching is often difficult to estimate. See Subdivisions 02220 and 03055 for demolition costs.

Always obtain concrete prices from suppliers near the job site. A volume discount can often be negotiated depending upon competition in the area. Remember to add for waste, particularly for slabs and footings on grade.

3100 Concrete Forms & Accessories

A primary cost for concrete construction is forming. Most jobs today are constructed with prefabricated forms. The selection of the forms best suited for the job and the total square feet of forms required for efficient concrete forming and placing are key elements in estimating concrete construction. Enough forms must be available for erection to make efficient use of the concrete placing equipment and crew.

- Concrete accessories for forming and placing depend upon the systems used. Study the plans and specifications to assure that all special accessory requirements have been included in the cost estimate such as anchor bolts, inserts and hangers.

03200 Concrete Reinforcement

- Ascertain that the reinforcing steel supplier has included all accessories, cutting, bending and an allowance for lapping, splicing and waste. A good rule of thumb is 10% for lapping, splicing and waste. Also, 10% waste should be allowed for welded wire fabric.

03300 Cast-in-Place Concrete

- When estimating structural concrete, pay particular attention to requirements for concrete additives, curing methods and surface treatments. Special consideration for climate, hot or cold, must be included in your estimate. Be sure to include requirements for concrete placing equipment and concrete finishing.

03400 Precast Concrete
03500 Cementitious Decks & Toppings

- The cost of hauling precast concrete structural members is often an important factor. For this reason, it is important to get a quote from the nearest supplier. It may become economically feasible to set up precasting beds on the site if the hauling costs are prohibitive.

Reference Numbers

Reference numbers are shown in bold squares at the beginning of some major classifications. These numbers refer to related items in the Reference Section. The reference information may be an estimating procedure, an alternate pricing method or technical information.

Note: Not all subdivisions listed here necessarily appear in this publication.

o part of this publication may be reproduced, stored in a retrieval system, or transmitted in any form r by any means without prior written permission of Reed Construction Data.

03050 | Basic Concrete Materials & Methods

	03055 | Selective Demolition		CREW	DAILY OUTPUT	LABOR-HOURS	UNIT	2005 BARE COSTS MAT.	LABOR	EQUIP.	TOTAL	TOTAL INCL O&P
870	0010	**WINTER PROTECTION** For heated ready mix, add, minimum				C.Y.	4.50			4.50	4.95
	0050	Maximum				″	5.65			5.65	6.20
	0100	Temporary heat to protect concrete, 24 hours, minimum	2 Clab	50	.320	M.S.F.	94	5.55		99.55	112
	0200	Temporary shelter for slab on grade, wood frame/polyethylene sheeting									
	0201	Build or remove, minimum	2 Carp	10	1.600	M.S.F.	257	38.50		295.50	350
	0210	Maximum	″	3	5.333	″	310	128		438	555
	0300	See also Division 03390-200									

03100 | Concrete Forms & Accessories

	03110 | Structural C.I.P. Forms		CREW	DAILY OUTPUT	LABOR-HOURS	UNIT	2005 BARE COSTS MAT.	LABOR	EQUIP.	TOTAL	TOTAL INCL O&P
410	0010	**FORMS IN PLACE, COLUMNS**									
	1500	Round fiber tube, 1 use, 8″ diameter	C-1	155	.206	L.F.	1.51	4.28		5.79	8.90
	1550	10″ diameter		155	.206		1.88	4.28		6.16	9.30
	1600	12″ diameter		150	.213		2.17	4.42		6.59	9.90
	1700	16″ diameter		140	.229		3.58	4.74		8.32	12
	5000	Job-built plywood, 8″ x 8″ columns, 1 use		165	.194	SFCA	2.11	4.02		6.13	9.10
	5500	12″ x 12″ columns, 1 use		180	.178		2.12	3.69		5.81	8.60
	7500	Steel framed plywood, 4 use per mo., rent, 8″ x 8″		340	.094		3.01	1.95		4.96	6.60
	7550	10″ x 10″		350	.091		2.40	1.90		4.30	5.85
	7600	12″ x 12″		370	.086		2.81	1.79		4.60	6.15
430	0010	**FORMS IN PLACE, FOOTINGS** Continuous wall, plywood, 1 use	C-1	375	.085	SFCA	2.31	1.77		4.08	5.55
	0150	4 use	″	485	.066	″	.75	1.37		2.12	3.15
	1500	Keyway, 4 use, tapered wood, 2″ x 4″	1 Carp	530	.015	L.F.	.18	.36		.54	.81
	1550	2″ x 6″	″	500	.016	″	.27	.38		.65	.95
	5000	Spread footings, job-built lumber, 1 use	C-1	305	.105	SFCA	1.66	2.18		3.84	5.50
	5150	4 use	″	414	.077	″	.54	1.60		2.14	3.31
435	0010	**FORMS IN PLACE, GRADE BEAM** Job-built plywood, 1 use	C-2	530	.091	SFCA	1.56	1.91		3.47	4.96
	0150	4 use	″	605	.079	″	.51	1.67		2.18	3.40
445	0010	**FORMS IN PLACE, SLAB ON GRADE**									
	1000	Bulkhead forms w/keyway, wood, 6″ high, 1 use	C-1	510	.063	L.F.	.80	1.30		2.10	3.09
	1400	Bulkhead form for slab, 4-1/2″ high, exp metal, incl keyway & stakes		1,200	.027		.70	.55		1.25	1.71
	1410	5-1/2″ high		1,100	.029		.78	.60		1.38	1.88
	1420	7-1/2″ high		960	.033		.94	.69		1.63	2.20
	2000	Curb forms, wood, 6″ to 12″ high, on grade, 1 use		215	.149	SFCA	2.19	3.09		5.28	7.65
	2150	4 use		275	.116	″	.71	2.41		3.12	4.87
	3000	Edge forms, wood, 4 use, on grade, to 6″ high		600	.053	L.F.	.27	1.11		1.38	2.17
	3050	7″ to 12″ high		435	.074	SFCA	.74	1.52		2.26	3.40
	4000	For slab blockouts, to 12″ high, 1 use		200	.160	L.F.	.73	3.32		4.05	6.45
	4100	Plastic (extruded), to 6″ high, multiple use, on grade		800	.040	″	.37	.83		1.20	1.82
	8760	Void form, corrugated fiberboard, 6″ x 12″, 10′ long		240	.133	S.F.	9.20	2.76		11.96	14.80
450	0010	**FORMS IN PLACE, STAIRS** (Slant length x width), 1 use	C-2	165	.291	S.F.	3.66	6.15		9.81	14.45
	0150	4 use		190	.253		1.19	5.35		6.54	10.35
	2000	Stairs, cast on sloping ground (length x width), 1 use		220	.218		1.36	4.60		5.96	9.30
	2100	4 use		240	.200		.44	4.22		4.66	7.65
455	0010	**FORMS IN PLACE, WALLS**									
	0100	Box out for wall openings, to 16″ thick, to 10 S.F.	C-2	24	2	Ea.	24.50	42		66.50	98.50
	0150	Over 10 S.F. (use perimeter)	″	280	.171	L.F.	2.07	3.61		5.68	8.45
	0250	Brick shelf, 4″ w, add to wall forms, use wall area abv shelf									

03100 | Concrete Forms & Accessories

		03110 | Structural C.I.P. Forms	CREW	DAILY OUTPUT	LABOR-HOURS	UNIT	2005 BARE COSTS MAT.	LABOR	EQUIP.	TOTAL	TOTAL INCL O&P	
5	0260	1 use	C-2	240	.200	SFCA	2.18	4.22		6.40	9.55	455
	0350	4 use		300	.160	″	.87	3.37		4.24	6.65	
	0500	Bulkhead, with keyway, 1 use, 2 piece		265	.181	L.F.	2.76	3.82		6.58	9.55	
	0550	3 piece		175	.274		3.48	5.80		9.28	13.65	
	0600	Bulkhead forms with keyway, 1 piece expanded metal, 8″ wall	C-1	1,000	.032		.94	.66		1.60	2.16	
	0610	10″ wall	″	800	.040		1.62	.83		2.45	3.19	
	2000	Wall, below grade, job-built plywood, to 8′ high, 1 use	C-2	300	.160	SFCA	2.37	3.37		5.74	8.30	
	2150	4 use		435	.110		.88	2.33		3.21	4.92	
	2400	Over 8′ to 16′ high, 1 use		280	.171		4.51	3.61		8.12	11.10	
	2420	2 use		345	.139		1.89	2.93		4.82	7.05	
	2430	3 use		375	.128		1.57	2.70		4.27	6.30	
	2440	4 use		395	.122		1.40	2.56		3.96	5.90	
	2445	Exterior wall, 8′ to 16′ high, 1 use		280	.171		2.18	3.61		5.79	8.55	
	2550	4 use		395	.122		.70	2.56		3.26	5.10	
	3000	For architectural finish, add		1,820	.026		5.40	.56		5.96	6.85	
	7800	Modular prefabricated plywood, to 8′ high, 1 use		1,180	.041		1.68	.86		2.54	3.30	
	7860	4 use		1,260	.038		.55	.80		1.35	1.97	
	8000	To 16′ high, 1 use		715	.067		2.19	1.42		3.61	4.81	
	8060	4 use		790	.061		.72	1.28		2	2.96	
	8100	Over 16′ high, 1 use		715	.067		2.63	1.42		4.05	5.30	
	8160	4 use		790	.061		.88	1.28		2.16	3.13	
00	0010	**SCAFFOLDING** See division 01540-750										800
		03150 | Concrete Accessories										
80	0010	**ACCESSORIES, ANCHOR BOLTS** J-type, incl. nut and washer										080
	0020	1/2″ diameter, 6″ long	1 Carp	90	.089	Ea.	.94	2.13		3.07	4.65	
	0050	10″ long		85	.094		1.06	2.26		3.32	5	
	0100	12″ long		85	.094		1.17	2.26		3.43	5.15	
	0200	5/8″ diameter, 12″ long		80	.100		1.18	2.40		3.58	5.40	
	0250	18″ long		70	.114		1.39	2.74		4.13	6.20	
	0300	24″ long		60	.133		1.60	3.20		4.80	7.20	
	0350	3/4″ diameter, 8″ long		80	.100		1.39	2.40		3.79	5.60	
	0400	12″ long		70	.114		1.74	2.74		4.48	6.55	
	0450	18″ long		60	.133		2.26	3.20		5.46	7.95	
	0500	24″ long		50	.160		2.96	3.84		6.80	9.75	
085	0012	**ANCHOR BOLTS** See division 04080-070										085
160	0010	**ACCESSORIES, CHAMFER STRIPS**										160
	5000	Wood, 1/2″ wide	1 Carp	535	.015	L.F.	.09	.36		.45	.71	
	5200	3/4″ wide		525	.015		.10	.37		.47	.73	
	5400	1″ wide		515	.016		.14	.37		.51	.78	
170	0010	**ACCESSORIES, COLUMN FORM**										170
	1000	Column clamps, adjustable to 24″ x 24″, buy				Set	83.50			83.50	91.50	
	1400	Rent per month				″	8.55			8.55	9.45	
250	0010	**EXPANSION JOINT** Keyed, cold, 24 ga, incl. stakes, 3-1/2″ high	1 Carp	200	.040	L.F.	.58	.96		1.54	2.27	250
	0050	4-1/2″ high		200	.040		.70	.96		1.66	2.40	
	0100	5-1/2″ high		195	.041		.78	.98		1.76	2.53	
	2000	Premolded, bituminous fiber, 1/2″ x 6″		375	.021		.39	.51		.90	1.30	
	2050	1″ x 12″		300	.027		1.37	.64		2.01	2.60	
	2500	Neoprene sponge, closed cell, 1/2″ x 6″		375	.021		1.32	.51		1.83	2.32	
	2550	1″ x 12″		300	.027		6.05	.64		6.69	7.80	
	5000	For installation in walls, add						75%				

03100 | Concrete Forms & Accessories

	03150	Concrete Accessories	CREW	DAILY OUTPUT	LABOR-HOURS	UNIT	2005 BARE COSTS MAT.	LABOR	EQUIP.	TOTAL	TOTAL INCL O&P
250	5250	For installation in boxouts, add						25%			
400	0010	**ACCESSORIES, INSERTS**									
	1000	All size nut insert, 5/8" & 3/4", incl. nut	1 Carp	84	.095	Ea.	3.39	2.29		5.68	7.60
	2000	Continuous slotted, 1-5/8" x 1-3/8"									
	2100	12 ga., 3" long	1 Carp	65	.123	Ea.	3.08	2.95		6.03	8.40
	2150	6" long		65	.123		3.98	2.95		6.93	9.40
	2200	8 ga., 12" long		65	.123		9.55	2.95		12.50	15.50
	2300	36" long		60	.133		22	3.20		25.20	29.50
620	0010	**ACCESSORIES, SLEEVES AND CHASES**									
	0100	Plastic, 1 use, 9" long, 2" diameter	1 Carp	100	.080	Ea.	.53	1.92		2.45	3.84
	0150	4" diameter		90	.089		1.56	2.13		3.69	5.35
	0200	6" diameter		75	.107		2.75	2.56		5.31	7.40
640	0010	**ACCESSORIES, SNAP TIES, FLAT WASHER**, 4-3/4" L&W									
	0100	3000 lb., to 8"				C	96			96	106
	0250	16"					122			122	135
	0300	18"					121			121	133
	0500	With plastic cone, to 8"					89			89	97.50
	0600	11" & 12"					105			105	116
	0650	16"					111			111	122
	0700	18"					116			116	128
850	0010	**ACCESSORIES, WALL AND FOUNDATION**									
	0020	Coil tie system									
	0700	1-1/4", 36,000 lb., to 8"				C	830			830	915
	1200	1-1/4" diameter x 3" long				"	1,625			1,625	1,775
	4200	30" long				Ea.	4.20			4.20	4.62
	4250	36" long				"	4.78			4.78	5.25
860	0010	**WATERSTOP** PVC, ribbed 3/16" thick, 4" wide	1 Carp	155	.052	L.F.	.78	1.24		2.02	2.96
	0050	6" wide		145	.055		1.34	1.32		2.66	3.72
	0500	Ribbed, PVC, with center bulb, 9" wide, 3/16" thick		135	.059		2	1.42		3.42	4.61
	0550	3/8" thick		130	.062		3.03	1.48		4.51	5.85

03200 | Concrete Reinforcement

	03210	Reinforcing Steel	CREW	DAILY OUTPUT	LABOR-HOURS	UNIT	2005 BARE COSTS MAT.	LABOR	EQUIP.	TOTAL	TOTAL INCL O&P
600	0010	**REINFORCING IN PLACE** A615 Grade 60, incl. access. labor									
	0502	Footings, #4 to #7	4 Rodm	4,200	.008	Lb.	.42	.19		.61	.81
	0550	#8 to #18		3.60	8.889	Ton	720	226		946	1,200
	0702	Walls, #3 to #7		6,000	.005	Lb.	.42	.14		.56	.70
	0750	#8 to #18		4	8	Ton	760	204		964	1,200
	2400	Dowels, 2 feet long, deformed, #3	2 Rodm	520	.031	Ea.	.33	.78		1.11	1.76
	2410	#4		480	.033		.59	.85		1.44	2.17
	2420	#5		435	.037		.92	.94		1.86	2.69
	2430	#6		360	.044		1.32	1.13		2.45	3.47
	2600	Dowel sleeves for CIP concrete, 2-part system									
	2610	Sleeve base, plastic, for #5 bar, fasten to edge form	1 Rodm	200	.040	Ea.	.33	1.02		1.35	2.18
	2615	Sleeve, plastic, for #5 bar x 9" long, snap onto base		400	.020		.96	.51		1.47	1.97
	2620	Sleeve base, for #6 bar		175	.046		.33	1.16		1.49	2.44
	2625	Sleeve, for #6 bar		350	.023		1.02	.58		1.60	2.16

03200 | Concrete Reinforcement

	03210 Reinforcing Steel		CREW	DAILY OUTPUT	LABOR-HOURS	UNIT	2005 BARE COSTS MAT.	LABOR	EQUIP.	TOTAL	TOTAL INCL O&P	
0	2630	Sleeve base, for #8 bar	1 Rodm	150	.053	Ea.	.40	1.36		1.76	2.87	600
	2635	Sleeve, for #8 bar		300	.027		1.11	.68		1.79	2.43	
	2700	Dowel caps, visual warning only, plastic, #3 to #8	2 Rodm	800	.020		.24	.51		.75	1.17	
	2720	#7 to #14		750	.021		.56	.54		1.10	1.59	
	2750	Impalement protective, plastic, #3 to #7		800	.020		1.74	.51		2.25	2.82	
	2760	#7 to #11		775	.021		2.12	.53		2.65	3.27	
	2770	#11 to #16		750	.021		2.35	.54		2.89	3.56	
	3000	For epoxy dowel anchoring, see Div 05090-300										

	03220 Welded Wire Fabric		CREW	DAILY OUTPUT	LABOR-HOURS	UNIT	MAT.	LABOR	EQUIP.	TOTAL	TOTAL INCL O&P	
00	0011	**WELDED WIRE FABRIC** Sheets, 6 x 6 - W1.4 x W1.4 (10 x 10)	2 Rodm	3,500	.005	S.F.	.19	.12		.31	.42	200
	0301	6 x 6 - W2.9 x W2.9 (6 x 6) 42 lb. per C.S.F.		2,900	.006		.33	.14		.47	.61	
	0501	4 x 4 - W1.4 x W1.4 (10 x 10) 31 lb. per C.S.F.		3,100	.005		.18	.13		.31	.43	
	0750	Rolls										
	0901	2 x 2 - #12 galv. for gunite reinforcing	2 Rodm	650	.025	S.F.	.32	.63		.95	1.47	

	03240 Fibrous Reinforcing		CREW	DAILY OUTPUT	LABOR-HOURS	UNIT	MAT.	LABOR	EQUIP.	TOTAL	TOTAL INCL O&P	
00	0010	**FIBROUS REINFORCING**										300
	0100	Synthetic fibers, add to concrete				Lb.	3.87			3.87	4.26	
	0110	1-1/2 lb. per C.Y.				C.Y.	6			6	6.60	
	0150	Steel fibers, add to concrete				Lb.	.44			.44	.48	
	0155	25 lb. per C.Y.				C.Y.	11			11	12.10	
	0160	50 lb. per C.Y.					22			22	24	
	0170	75 lb. per C.Y.					34			34	37.50	
	0180	100 lb. per C.Y.					44			44	48.50	

03300 | Cast-In-Place Concrete

	03310 Structural Concrete		CREW	DAILY OUTPUT	LABOR-HOURS	UNIT	2005 BARE COSTS MAT.	LABOR	EQUIP.	TOTAL	TOTAL INCL O&P	
220	0010	**CONCRETE, READY MIX** Normal weight										220
	0020	2000 psi				C.Y.	77.50			77.50	85.50	
	0100	2500 psi					79.50			79.50	87.50	
	0150	3000 psi					81			81	89	
	0200	3500 psi					82			82	90	
	0300	4000 psi					84			84	92.50	
	0350	4500 psi					86			86	94.50	
	0400	5000 psi					90			90	99	
	0411	6000 psi					103			103	113	
	0412	8000 psi					167			167	184	
	0413	10,000 psi					238			238	261	
	0414	12,000 psi					287			287	315	
	1000	For high early strength cement, add					10%					
	2000	For all lightweight aggregate, add					45%					
240	0010	**CONCRETE IN PLACE** Including forms (4 uses), reinforcing										240
	0050	steel and finishing unless otherwise indicated										
	0500	Chimney foundations, industrial, minimum	C-14C	32.22	3.476	C.Y.	112	78	.73	190.73	257	
	0510	Maximum		23.71	4.724		119	106	.99	225.99	310	
	3800	Footings, spread under 1 C.Y.		38.07	2.942		162	66	.62	228.62	291	
	3850	Over 5 C.Y.		81.04	1.382		226	31	.29	257.29	300	

03300 | Cast-In-Place Concrete

03310 | Structural Concrete

		03310 Structural Concrete	CREW	DAILY OUTPUT	LABOR-HOURS	UNIT	2005 BARE COSTS MAT.	LABOR	EQUIP.	TOTAL	TOTAL INCL O&P
240	3900	Footings, strip, 18″ x 9″, unreinforced	C-14C	40	2.800	C.Y.	101	62.50	.59	164.09	219
	3920	18″ x 9″, reinforced		35	3.200		120	71.50	.67	192.17	255
	3925	20″ x 10″, unreinforced		45	2.489		98	55.50	.52	154.02	204
	3930	20″ x 10″, reinforced		40	2.800		114	62.50	.59	177.09	233
	3935	24″ x 12″, unreinforced		55	2.036		97	45.50	.43	142.93	185
	3940	24″ x 12″, reinforced		48	2.333		113	52	.49	165.49	214
	3945	36″ x 12″, unreinforced		70	1.600		93	36	.34	129.34	163
	3950	36″ x 12″, reinforced		60	1.867		108	42	.39	150.39	190
	4000	Foundation mat, under 10 C.Y.		38.67	2.896		164	65	.61	229.61	293
	4050	Over 20 C.Y.		56.40	1.986		144	44.50	.42	188.92	234
	4520	Handicap access ramp, railing both sides, 3′ wide	C-14H	14.58	3.292	L.F.	167	76.50	1.65	245.15	315
	4525	5′ wide		12.22	3.928		175	91.50	1.96	268.46	350
	4530	With 6″ curb and rails both sides, 3′ wide		8.55	5.614		174	131	2.81	307.81	415
	4535	5′ wide		7.31	6.566		178	153	3.28	334.28	460
	4650	Slab on grade, not including finish, 4″ thick	C-14E	60.75	1.449	C.Y.	104	33	.39	137.39	172
	4700	6″ thick	″	92	.957	″	98	22	.26	120.26	146
	4751	Slab on grade, incl. troweled finish, not incl. forms									
	4760	or reinforcing, over 10,000 S.F., 4″ thick	C-14F	3,425	.021	S.F.	1.07	.45	.01	1.53	1.91
	4820	6″ thick	″	3,350	.021	″	1.56	.46	.01	2.03	2.47
	5000	Slab on grade, incl. textured finish, not incl. forms									
	5001	or reinforcing, 4″ thick	C-14G	2,873	.019	S.F.	1.05	.41	.01	1.47	1.84
	5010	6″ thick		2,590	.022		1.65	.45	.01	2.11	2.56
	5020	8″ thick		2,320	.024		2.15	.50	.01	2.66	3.20
	6203	Retaining walls, gravity, 4′ high				C.Y.	103			103	113
	6800	Stairs, not including safety treads, free standing, 3′-6″ wide	C-14H	83	.578	LF Nose	14.80	13.45	.29	28.54	39.50
	6850	Cast on ground		125	.384	″	8.70	8.95	.19	17.84	25
	7000	Stair landings, free standing		200	.240	S.F.	3.13	5.60	.12	8.85	13.05
	7050	Cast on ground		475	.101	″	1.84	2.35	.05	4.24	6.10
700	0010	**PLACING CONCRETE** and vibrating, including labor & equipment									
	1900	Footings, continuous, shallow, direct chute	C-6	120	.400	C.Y.		7.45	.40	7.85	12.95
	1950	Pumped	C-20	150	.427			8.15	5	13.15	19.20
	2000	With crane and bucket	C-7	90	.800			15.45	10.65	26.10	38
	2400	Footings, spread, under 1 C.Y., direct chute	C-6	55	.873			16.25	.87	17.12	28.50
	2600	Over 5 C.Y., direct chute		120	.400			7.45	.40	7.85	12.95
	2900	Foundation mats, over 20 C.Y., direct chute		350	.137			2.55	.14	2.69	4.44
	4300	Slab on grade, 4″ thick, direct chute		110	.436			8.15	.44	8.59	14.15
	4350	Pumped	C-20	130	.492			9.40	5.80	15.20	22
	4400	With crane and bucket	C-7	110	.655			12.65	8.75	21.40	30.50
	4900	Walls, 8″ thick, direct chute	C-6	90	.533			9.95	.53	10.48	17.30
	4950	Pumped	C-20	100	.640			12.25	7.50	19.75	29
	5000	With crane and bucket	C-7	80	.900			17.40	12	29.40	42
	5050	12″ thick, direct chute	C-6	100	.480			8.95	.48	9.43	15.55
	5100	Pumped	C-20	110	.582			11.15	6.85	18	26
	5200	With crane and bucket	C-7	90	.800			15.45	10.65	26.10	38
	5600	Wheeled concrete dumping, add to placing costs above									
	5610	Walking cart, 50′ haul, add	C-18	32	.281	C.Y.		4.94	1.56	6.50	10.10
	5620	150′ haul, add		24	.375			6.60	2.07	8.67	13.50
	5700	250′ haul, add		18	.500			8.80	2.77	11.57	17.95
	5800	Riding cart, 50′ haul, add	C-19	80	.112			1.98	.96	2.94	4.42
	5810	150′ haul, add		60	.150			2.64	1.28	3.92	5.90
	5900	250′ haul, add		45	.200			3.51	1.71	5.22	7.85
		03350 Concrete Finishing									
300	0010	**FINISHING FLOORS** Monolithic, screed finish	1 Cefi	900	.009	S.F.		.20		.20	.33
	0100	Screed and bull float (darby) finish		725	.011			.25		.25	.41

03300 | Cast-In-Place Concrete

03350 | Concrete Finishing

		03350 Concrete Finishing	CREW	DAILY OUTPUT	LABOR-HOURS	UNIT	2005 BARE COSTS MAT.	LABOR	EQUIP.	TOTAL	TOTAL INCL O&P	
00	0150	Screed, float, and broom finish	1 Cefi	630	.013	S.F.		.29		.29	.47	300
	0200	Screed, float, and hand trowel		600	.013			.31		.31	.49	
	0250	Machine trowel		550	.015			.33		.33	.54	
	1600	Exposed local aggregate finish, minimum		625	.013		4.08	.29		4.37	4.96	
	1650	Maximum		465	.017		12.90	.40		13.30	14.85	
25	0010	**CONTROL JOINT**, concrete floor slab										325
	0100	Sawcut in green concrete										
	0120	1″ depth	C-27	2,000	.008	L.F.		.18	.06	.24	.36	
	0140	1-1/2″ depth		1,800	.009			.20	.06	.26	.40	
	0160	2″ depth		1,600	.010			.23	.07	.30	.45	
	0200	Clean out control joint of debris	C-28	6,000	.001			.03		.03	.05	
	0300	Joint sealant										
	0320	Backer rod, polyethylene, 1/4″ diameter	1 Cefi	460	.017	L.F.	.02	.40		.42	.67	
	0340	Sealant, polyurethane										
	0360	1/4″ x 1/4″ (308 LF/Gal)	1 Cefi	270	.030	L.F.	.15	.68		.83	1.27	
	0380	1/4″ x 1/2″ (154 LF/Gal)	″	255	.031	″	.31	.72		1.03	1.50	
50	0010	**FINISHING WALLS** Break ties and patch voids	1 Cefi	540	.015	S.F.	.03	.34		.37	.58	350
	0050	Burlap rub with grout	″	450	.018		.03	.41		.44	.69	
	0300	Bush hammer, green concrete	B-39	1,000	.048		.03	.85	.14	1.02	1.63	
	0350	Cured concrete	″	650	.074		.03	1.31	.22	1.56	2.49	
00	0010	**SLAB TEXTURE STAMPING**, buy										600
	0020	Approx. 3 S.F.- 5 S.F. each, minimum				Ea.	45			45	49.50	
	0030	Average				″	49.50			49.50	54.50	
	0120	Per S.F. of tool, average				S.F.	54			54	59.50	
	0200	Commonly used chemicals for texture systems										
	0210	Hardeners w/colors average				S.F.	.45			.45	.50	
	0220	Release agents w/colors, average					.18			.18	.20	
	0225	Clear, average					.14			.14	.15	
	0230	Sealers, clear, average					.12			.12	.13	
	0240	Colors, average					.15			.15	.17	

03390 | Concrete Curing

		03390 Concrete Curing	CREW	DAILY OUTPUT	LABOR-HOURS	UNIT	MAT.	LABOR	EQUIP.	TOTAL	TOTAL INCL O&P	
200	0011	**CURING** With burlap, 4 uses assumed, 7.5 oz.	2 Clab	5,500	.003	S.F.	.07	.05		.12	.17	200
	0101	10 oz.		5,500	.003		.11	.05		.16	.21	
	0201	With waterproof curing paper, 2 ply, reinforced		7,000	.002		.06	.04		.10	.14	
	0301	With sprayed membrane curing compound		9,500	.002		.05	.03		.08	.11	
	0710	Electrically, heated pads, 15 watts/S.F., 20 uses, minimum					.17			.17	.18	
	0800	Maximum					.28			.28	.31	

03400 | Precast Concrete

03450 | Plant-Precast Architectural Concrete

		03450 Plant-Precast Architectural Concrete	CREW	DAILY OUTPUT	LABOR-HOURS	UNIT	2005 BARE COSTS MAT.	LABOR	EQUIP.	TOTAL	TOTAL INCL O&P	
855	0010	**PRECAST WINDOW SILLS**										855
	0600	Precast concrete, 4″ tapers to 3″, 9″ wide	D-1	70	.229	L.F.	9.45	4.98		14.43	18.65	
	0650	11″ wide		60	.267		12.50	5.80		18.30	23.50	
	0700	13″ wide, 3 1/2″ tapers to 2 1/2″, 12″ wall		50	.320		13.50	6.95		20.45	26.50	

CONCRETE 3

03400 | Precast Concrete

		03480 | Precast Concrete Specialties	CREW	DAILY OUTPUT	LABOR-HOURS	UNIT	2005 BARE COSTS MAT.	LABOR	EQUIP.	TOTAL	TOTAL INCL O&P
400	0010	**LINTELS**									
	0800	Precast concrete, 4" wide, 8" high, to 5' long	D-10	28	1.429	Ea.	24	32	18.05	74.05	100
	0850	5'-12' long		24	1.667		68	37.50	21	126.50	160
	1000	6" wide, 8" high, to 5' long		26	1.538		29.50	34.50	19.45	83.45	112
	1050	5'-12' long	↓	22	1.818	↓	84	41	23	148	186
800	0010	**STAIRS**, Precast concrete treads on steel stringers, 3' wide	C-12	75	.640	Riser	64.50	14.95	8.05	87.50	105
	0300	Front entrance, 5' wide with 48" platform, 2 risers		16	3	Flight	340	70	37.50	447.50	535
	0350	5 risers		12	4		530	93.50	50	673.50	800
	0500	6' wide, 2 risers	↓	15	3.200		360	74.50	40	474.50	565
	1200	Basement entrance stairs, steel bulkhead doors, minimum	B-51	22	2.182		700	39	5.85	744.85	845
	1250	Maximum	"	11	4.364	↓	1,000	78.50	11.70	1,090.20	1,250

03900 | Concrete Restoration & Cleaning

		03930 | Concrete Rehabilitation	CREW	DAILY OUTPUT	LABOR-HOURS	UNIT	2005 BARE COSTS MAT.	LABOR	EQUIP.	TOTAL	TOTAL INCL O&P
400	0012	**FLOOR PATCHING** 1/4" thick, small areas, regular	1 Cefi	170	.047	S.F.	2.96	1.08		4.04	5
	0100	Epoxy	"	100	.080	"	4.11	1.84		5.95	7.50

For information about Means Estimating Seminars, see yellow pages 12 and 13 in back of book

Division 4 Masonry

Estimating Tips

04050 Basic Masonry Materials Methods

The terms *mortar* and *grout* are often used interchangeably, and incorrectly. Mortar is used to bed masonry units, seal the entry of air and moisture, provide architectural appearance, and allow for size variations in the units. Grout is used primarily in reinforced masonry construction and is used to bond the masonry to the reinforcing steel. Common mortar types are M(2500 psi), S(1800 psi), N(750 psi), and O(350 psi), and conform to ASTM C270. Grout is either fine or coarse, conforms to ASTM C476, and in-place strengths generally exceed 2500 psi. Mortar and grout are different components of masonry construction and are placed by entirely different methods. An estimator should be aware of their unique uses and costs.

Waste, specifically the loss/droppings of mortar and the breakage of brick and block, is included in all masonry assemblies in this division. A factor of 25% is added for mortar and 3% for brick and concrete masonry units.

Scaffolding or staging is not included in any of the Division 4 costs. Refer to section 01540 for scaffolding and staging costs.

04800 Masonry Assemblies

- The most common types of unit masonry are brick and concrete masonry. The major classifications of brick are building brick (ASTM C62), facing brick (ASTM C216) and glazed brick, fire brick and pavers. Many varieties of texture and appearance can exist within these classifications, and the estimator would be wise to check local custom and availability within the project area. On repair and remodeling jobs, matching the existing brick may be the most important criteria.
- Brick and concrete block are priced by the piece and then converted into a price per square foot of wall. Openings less than two square feet are generally ignored by the estimator because any savings in units used is offset by the cutting and trimming required.
- It is often difficult and expensive to find and purchase small lots of historic brick. Costs can vary widely. Many design issues affect costs, selection of mortar mix, and repairs or replacement of masonry materials. Cleaning techniques must be reflected in the estimate.
- All masonry walls, whether interior or exterior, require bracing. The cost of bracing walls during construction should be included by the estimator and this bracing must remain in place until permanent bracing is complete. Permanent bracing of masonry walls is accomplished by masonry itself, in the form of pilasters or abutting wall corners, or by anchoring the walls to the structural frame. Accessories in the form of anchors, anchor slots and ties are used, but their supply and installation can be by different trades. For instance, anchor slots on spandrel beams and columns are supplied and welded in place by the steel fabricator, but the ties from the slots into the masonry are installed by the bricklayer. Regardless of the installation method the estimator must be certain that these accessories are accounted for in pricing.

Reference Numbers

Reference numbers are shown in bold squares at the beginning of some major classifications. These numbers refer to related items in the Reference Section. The reference information may be an estimating procedure, an alternate pricing method or technical information.

Note: Not all subdivisions listed here necessarily appear in this publication.

No part of this publication may be reproduced, stored in a retrieval system, or transmitted in any form or by any means without prior written permission of Reed Construction Data.

04055 | Selective Demolition		CREW	DAILY OUTPUT	LABOR-HOURS	UNIT	2005 BARE COSTS MAT.	LABOR	EQUIP.	TOTAL	TOTAL INCL O&P
110	0010 SELECTIVE DEMOLITION, MASONRY R02220-510									
	0300 Concrete block walls, unreinforced, 2″ thick	2 Clab	1,200	.013	S.F.		.23		.23	.39
	0310 4″ thick		1,150	.014			.24		.24	.41
	0320 6″ thick		1,100	.015			.25		.25	.43
	0330 8″ thick		1,050	.015			.26		.26	.45
	0340 10″ thick		1,000	.016			.28		.28	.47
	0360 12″ thick		950	.017			.29		.29	.50
	0380 Reinforced alternate courses, 2″ thick		1,130	.014			.25		.25	.42
	0390 4″ thick		1,080	.015			.26		.26	.44
	0400 6″ thick		1,035	.015			.27		.27	.46
	0410 8″ thick		990	.016			.28		.28	.48
	0420 10″ thick		940	.017			.30		.30	.50
	0430 12″ thick		890	.018			.31		.31	.53
	0440 Reinforced alternate courses & vertically 48″ OC, 4″ thick		900	.018			.31		.31	.52
	0450 6″ thick		850	.019			.33		.33	.55
	0460 8″ thick		800	.020			.35		.35	.59
	0480 10″ thick		750	.021			.37		.37	.63
	0490 12″ thick		700	.023			.40		.40	.67
	1000 Chimney, 16″ x 16″, soft old mortar	1 Clab	55	.145	C.F.		2.52		2.52	4.28
	1020 Hard mortar		40	.200			3.47		3.47	5.90
	1030 16″ x 20″, soft old mortar		55	.145			2.52		2.52	4.28
	1040 Hard mortar		40	.200			3.47		3.47	5.90
	1050 16″ x 24″, soft old mortar		55	.145			2.52		2.52	4.28
	1060 Hard mortar		40	.200			3.47		3.47	5.90
	1080 20″ x 20″, soft old mortar		55	.145			2.52		2.52	4.28
	1100 Hard mortar		40	.200			3.47		3.47	5.90
	1110 20″ x 24″, soft old mortar		55	.145			2.52		2.52	4.28
	1120 Hard mortar		40	.200			3.47		3.47	5.90
	1140 20″ x 32″, soft old mortar		55	.145			2.52		2.52	4.28
	1160 Hard mortar		40	.200			3.47		3.47	5.90
	1200 48″ x 48″, soft old mortar		55	.145			2.52		2.52	4.28
	1220 Hard mortar		40	.200			3.47		3.47	5.90
	1250 Metal, high temp steel jacket, 24″ diameter	E-2	130	.369	V.L.F.		9.50	10.20	19.70	29
	1260 60″ diameter	″	60	.800			20.50	22	42.50	63.50
	2000 Columns, 8″ x 8″, soft old mortar	1 Clab	48	.167			2.89		2.89	4.91
	2020 Hard mortar		40	.200			3.47		3.47	5.90
	2060 16″ x 16″, soft old mortar		16	.500			8.70		8.70	14.75
	2100 Hard mortar		14	.571			9.90		9.90	16.85
	2140 24″ x 24″, soft old mortar		8	1			17.35		17.35	29.50
	2160 Hard mortar		6	1.333			23		23	39.50
	2200 36″ x 36″, soft old mortar		4	2			34.50		34.50	59
	2220 Hard mortar		3	2.667			46.50		46.50	78.50
	2230 Alternate pricing method, soft old mortar		30	.267	C.F.		4.63		4.63	7.85
	2240 Hard mortar		23	.348	″		6.05		6.05	10.25
	3000 Copings, precast or masonry, to 8″ wide									
	3020 Soft old mortar	1 Clab	180	.044	L.F.		.77		.77	1.31
	3040 Hard mortar	″	160	.050	″		.87		.87	1.47
	3100 To 12″ wide									
	3120 Soft old mortar	1 Clab	160	.050	L.F.		.87		.87	1.47
	3140 Hard mortar	″	140	.057	″		.99		.99	1.68
	4000 Fireplace, brick, 30″ x 24″ opening									
	4020 Soft old mortar	1 Clab	2	4	Ea.		69.50		69.50	118
	4040 Hard mortar		1.25	6.400			111		111	188
	4100 Stone, soft old mortar		1.50	5.333			92.50		92.50	157
	4120 Hard mortar		1	8			139		139	236
	5000 Veneers, brick, soft old mortar		140	.057	S.F.		.99		.99	1.68

04055 | Selective Demolition

			CREW	DAILY OUTPUT	LABOR-HOURS	UNIT	2005 BARE COSTS MAT.	LABOR	EQUIP.	TOTAL	TOTAL INCL O&P	
110	5020	Hard mortar (R02220-510)	1 Clab	125	.064	S.F.		1.11		1.11	1.88	110
	5100	Granite and marble, 2" thick		180	.044			.77		.77	1.31	
	5120	4" thick		170	.047			.82		.82	1.39	
	5140	Stone, 4" thick		180	.044			.77		.77	1.31	
	5160	8" thick		175	.046			.79		.79	1.35	
	5400	Alternate pricing method, stone, 4" thick		60	.133	C.F.		2.31		2.31	3.93	
	5420	8" thick		85	.094	"		1.63		1.63	2.77	

04060 | Masonry Mortar

			CREW	DAILY OUTPUT	LABOR-HOURS	UNIT	MAT.	LABOR	EQUIP.	TOTAL	TOTAL INCL O&P	
200	0010	**CEMENT** Gypsum 80 lb. bag, T.L. lots (R04060-100)				Bag	23			23	25.50	200
	0050	L.T.L. lots					23			23	25.50	
	0100	Masonry, 70 lb. bag, T.L. lots					6.10			6.10	6.70	
	0150	L.T.L. lots					6.45			6.45	7.10	
	0200	White, 70 lb. bag, T.L. lots					16.35			16.35	18	
	0250	L.T.L. lots					17.45			17.45	19.20	

04070 | Masonry Grout

			CREW	DAILY OUTPUT	LABOR-HOURS	UNIT	MAT.	LABOR	EQUIP.	TOTAL	TOTAL INCL O&P	
420	0010	**GROUTING** Bond bms. & lintels, 8" dp., pumped, not incl. block										420
	0200	Concrete block cores, solid, 4" thk., by hand, 0.067 C.F./S.F. of wall	D-8	1,100	.036	S.F.	.22	.81		1.03	1.59	
	0210	6" thick, pumped, 0.175 C.F. per S.F.	D-4	720	.056		.58	1.10	.19	1.87	2.68	
	0250	8" thick, pumped, 0.258 C.F. per S.F.		680	.059		.85	1.16	.20	2.21	3.10	
	0300	10" thick, pumped, 0.340 C.F. per S.F.		660	.061		1.12	1.20	.20	2.52	3.46	
	0350	12" thick, pumped, 0.422 C.F. per S.F.		640	.063		1.39	1.23	.21	2.83	3.82	

04080 | Anchorage & Reinforcement

			CREW	DAILY OUTPUT	LABOR-HOURS	UNIT	MAT.	LABOR	EQUIP.	TOTAL	TOTAL INCL O&P	
070	0010	**ANCHOR BOLTS** Hooked type with nut and washer, 1/2" diam., 8" long	1 Bric	200	.040	Ea.	.55	.99		1.54	2.25	070
	0030	12" long		190	.042		1.17	1.04		2.21	3.02	
	0060	3/4" diameter, 8" long		160	.050		1.39	1.24		2.63	3.59	
	0070	12" long		150	.053		1.74	1.32		3.06	4.10	
200	0010	**REINFORCING** Steel bars A615, placed horiz., #3 & #4 bars (R04080-500)	1 Bric	450	.018	Lb.	.40	.44		.84	1.17	200
	0050	Placed vertical, #3 & #4 bars		350	.023		.40	.56		.96	1.38	
	0060	#5 & #6 bars		650	.012		.40	.30		.70	.95	
	0200	Joint reinforcing, regular truss, to 6" wide, mill std galvanized		30	.267	C.L.F.	11.90	6.60		18.50	24	
	0250	12" wide		20	.400		14.40	9.90		24.30	32.50	
	0400	Cavity truss with drip section, to 6" wide		30	.267		13.60	6.60		20.20	26	
	0450	12" wide		20	.400		15.45	9.90		25.35	33.50	
650	0010	**WALL TIES** To brick veneer, galv., corrugated, 7/8" x 7", 22 Ga.	1 Bric	10.50	.762	C	5.75	18.80		24.55	38	650
	0100	24 Ga.		10.50	.762		5.25	18.80		24.05	37.50	
	0150	16 Ga.		10.50	.762		17.65	18.80		36.45	51	
	0200	Buck anchors, galv., corrugated, 16 gauge, 2" bend, 8" x 2"		10.50	.762		115	18.80		133.80	159	
	0250	8" x 3"		10.50	.762		107	18.80		125.80	150	
	0680	Cavity wall, Z type, galvanized, 6" long, 1/4" diameter		10.50	.762		28.50	18.80		47.30	63	
	0850	3/16" diameter		10.50	.762		16.80	18.80		35.60	50	
	0855	8" long, 1/4" diameter		10.50	.762		31	18.80		49.80	65.50	
	1000	Rectangular type, galvanized, 1/4" diameter, 2" x 6"		10.50	.762		30.50	18.80		49.30	65.50	
	1050	4" x 6"		10.50	.762		34.50	18.80		53.30	69.50	
	1100	3/16" diameter, 2" x 6"		10.50	.762		21.50	18.80		40.30	55	
	1150	4" x 6"		10.50	.762		23.50	18.80		42.30	57.50	
	1500	Rigid partition anchors, plain, 8" long, 1" x 1/8"		10.50	.762		67	18.80		85.80	106	
	1550	1" x 1/4"		10.50	.762		132	18.80		150.80	177	
	1580	1-1/2" x 1/8"		10.50	.762		93	18.80		111.80	134	
	1600	1-1/2" x 1/4"		10.50	.762		221	18.80		239.80	275	
	1650	2" x 1/8"		10.50	.762		116	18.80		134.80	160	

04050 | Basic Masonry Materials and Methods

04080 | Anchorage & Reinforcement

		04080 Anchorage & Reinforcement	CREW	DAILY OUTPUT	LABOR-HOURS	UNIT	2005 BARE COSTS MAT.	LABOR	EQUIP.	TOTAL	TOTAL INCL O&P
650	1700	2" x 1/4"	1 Bric	10.50	.762	C	315	18.80		333.80	375

04200 | Masonry Units

04210 | Clay Masonry Units

		04210 Clay Masonry Units	CREW	DAILY OUTPUT	LABOR-HOURS	UNIT	2005 BARE COSTS MAT.	LABOR	EQUIP.	TOTAL	TOTAL INCL O&P
100	0010	**COMMON BUILDING BRICK** C62, TL lots, material only R04210-120									
	0020	Standard, minimum				M	298			298	330
	0050	Average (select)				"	355			355	390
300	0010	**FACE BRICK** C216, TL lots, material only R04210-120									
	0300	Standard modular, 4" x 2-2/3" x 8", minimum				M	375			375	410
	0350	Maximum					465			465	510
	2170	For less than truck load lots, add					10			10	11
	2180	For buff or gray brick, add					15			15	16.50

04500 | Refractories

04550 | Flue Liners

		04550 Flue Liners	CREW	DAILY OUTPUT	LABOR-HOURS	UNIT	2005 BARE COSTS MAT.	LABOR	EQUIP.	TOTAL	TOTAL INCL O&P
250	0010	**FLUE LINING** Including mortar joints, 8" x 8"	D-1	125	.128	V.L.F.	3.87	2.79		6.66	8.90
	0100	8" x 12"		103	.155		5.45	3.38		8.83	11.65
	0200	12" x 12"		93	.172		6.60	3.75		10.35	13.50
	0300	12" x 18"		84	.190		11.45	4.15		15.60	19.45
	0400	18" x 18"		75	.213		16.15	4.65		20.80	25.50
	0500	20" x 20"		66	.242		34	5.30		39.30	46.50
	0600	24" x 24"		56	.286		43.50	6.20		49.70	58
	1000	Round, 18" diameter		66	.242		25	5.30		30.30	36.50
	1100	24" diameter		47	.340		39	7.40		46.40	55.50

04800 | Masonry Assemblies

04810 | Unit Masonry Assemblies

		04810 Unit Masonry Assemblies	CREW	DAILY OUTPUT	LABOR-HOURS	UNIT	2005 BARE COSTS MAT.	LABOR	EQUIP.	TOTAL	TOTAL INCL O&P
040	0010	**ADOBE BRICK** Semi-stabilized, with cement mortar									
	0060	Brick, 10" x 4" x 14" , 2.6/S.F.	D-8	560	.071	S.F.	2.13	1.60		3.73	5
	0080	12" x 4" x 16", 2.3/S.F.		580	.069		3.49	1.54		5.03	6.40
	0100	10" x 4" x 16", 2.3/S.F.		590	.068		3.17	1.52		4.69	6

04810 | Unit Masonry Assemblies

	Line	04810 Unit Masonry Assemblies	CREW	DAILY OUTPUT	LABOR-HOURS	UNIT	2005 BARE COSTS MAT.	LABOR	EQUIP.	TOTAL	TOTAL INCL O&P	
040	0120	8" x 4" x 16", 2.3/S.F.	D-8	560	.071	S.F.	2.43	1.60		4.03	5.35	040
	0140	4" x 4" x 16", 2.3/S.F.		540	.074		2.55	1.66		4.21	5.55	
	0160	6" x 4" x 16", 2.3/S.F.		540	.074		1.94	1.66		3.60	4.89	
	0180	4" x 4" x 12", 3.0/S.F.		520	.077		2.42	1.72		4.14	5.50	
	0200	8" x 4" x 12", 3.0/S.F.	↓	520	.077	↓	2.63	1.72		4.35	5.75	
050	0010	**AUTOCLAVED AERATED CONCRETE BLOCK** Scaffolding not incl										050
	0050	Solid, 4" x 12" x 24", incl mortar	D-8	600	.067	S.F.	1.84	1.49		3.33	4.50	
	0060	6" x 12" x 24"		600	.067		2.47	1.49		3.96	5.20	
	0070	8" x 8" x 24"		575	.070		3.10	1.56		4.66	6	
	0080	10" x 12" x 24"		575	.070		4.04	1.56		5.60	7.05	
	0090	12" x 12" x 24"	↓	550	.073	↓	4.71	1.63		6.34	7.90	
100	0010	**BRICK VENEER** Scaffolding not included, truck load lots										100
	0015	Material costs incl. 3% brick and 25% mortar waste										
	2000	Standard, sel. common, 4" x 2-2/3" x 8", (6.75/S.F.) R04210-100	D-8	230	.174	S.F.	2.88	3.89		6.77	9.60	
	2020	Standard, red, 4" x 2-2/3" x 8", running bond (6.75/SF)		220	.182		2.88	4.07		6.95	9.90	
	2050	Full header every 6th course (7.88/S.F.) R04210-120		185	.216		3.35	4.83		8.18	11.75	
	2100	English, full header every 2nd course (10.13/S.F.)		140	.286		4.30	6.40		10.70	15.40	
	2150	Flemish, alternate header every course (9.00/S.F.) R04210-180		150	.267		3.82	5.95		9.77	14.10	
	2200	Flemish, alt. header every 6th course (7.13/S.F.)		205	.195		3.04	4.36		7.40	10.60	
	2250	Full headers throughout (13.50/S.F.) R04210-500		105	.381		5.70	8.50		14.20	20.50	
	2300	Rowlock course (13.50/S.F.)		100	.400		5.70	8.95		14.65	21	
	2350	Rowlock stretcher (4.50/S.F.)		310	.129		1.93	2.89		4.82	6.95	
	2400	Soldier course (6.75/S.F.)		200	.200		2.88	4.47		7.35	10.60	
	2450	Sailor course (4.50/S.F.)		290	.138		1.93	3.08		5.01	7.30	
	2600	Buff or gray face, running bond, (6.75/S.F.)		220	.182		3.06	4.07		7.13	10.10	
	2700	Glazed face brick, running bond		210	.190		8.30	4.26		12.56	16.25	
	2750	Full header every 6th course (7.88/S.F.)		170	.235		9.70	5.25		14.95	19.40	
	3000	Jumbo, 6" x 4" x 12" running bond (3.00/S.F.)		435	.092		3.62	2.06		5.68	7.40	
	3050	Norman, 4" x 2-2/3" x 12" running bond, (4.5/S.F.)		320	.125		3.93	2.80		6.73	8.95	
	3100	Norwegian, 4" x 3-1/5" x 12" (3.75/S.F.)		375	.107		2.95	2.39		5.34	7.20	
	3150	Economy, 4" x 4" x 8" (4.50/S.F.)		310	.129		3.54	2.89		6.43	8.70	
	3200	Engineer, 4" x 3-1/5" x 8" (5.63/S.F.)		260	.154		2.97	3.44		6.41	8.95	
	3250	Roman, 4" x 2" x 12" (6.00/S.F.)		250	.160		4.63	3.58		8.21	11.05	
	3300	SCR, 6" x 2-2/3" x 12" (4.50/S.F.)		310	.129		4.24	2.89		7.13	9.45	
	3350	Utility, 4" x 4" x 12" (3.00/S.F.)	↓	450	.089	↓	3.37	1.99		5.36	7	
	3400	For cavity wall construction, add						15%				
	3450	For stacked bond, add						10%				
	3500	For interior veneer construction, add						15%				
	3550	For curved walls, add						30%				
110	0010	**OVERSIZED BRICK**, scaffolding not included										110
	0100	Oversized brick, 4" x 4" x 12"	D-8	118	.339	S.F.	.29	7.60		7.89	12.90	
	0110	4" x 4" x 16"		155	.258		.29	5.75		6.04	9.90	
	0120	4" x 8" x 16"	↓	155	.258	↓	.29	5.75		6.04	9.90	
160	0010	**CHIMNEY** See Div. 03310-240 for foundation, add to prices below										160
	0100	Brick, 16" x 16", 8" flue, scaff. not incl.	D-1	18.20	.879	V.L.F.	15.90	19.15		35.05	49.50	
	0150	16" x 20" with one 8" x 12" flue		16	1		25	22		47	63.50	
	0200	16" x 24" with two 8" x 8" flues		14	1.143		36.50	25		61.50	82	
	0250	20" x 20" with one 12" x 12" flue		13.70	1.168		28	25.50		53.50	73	
	0300	20" x 24" with two 8" x 12" flues		12	1.333		41	29		70	93.50	
	0350	20" x 32" with two 12" x 12" flues	↓	10	1.600	↓	48	35		83	111	
170	0010	**COLUMNS** Face brick, includes mortar, scaffolding not included										170
	0050	8" x 8", 9 brick per course	D-1	56	.286	V.L.F.	3.74	6.20		9.94	14.45	
	0100	12" x 8", 13.5 brick		37	.432		5.60	9.40		15	22	
	0200	12" x 12", 20 brick	↓	25	.640	↓	8.30	13.95		22.25	32	

04810 | Unit Masonry Assemblies

		04810 Unit Masonry Assemblies	CREW	DAILY OUTPUT	LABOR-HOURS	UNIT	2005 BARE COSTS MAT.	LABOR	EQUIP.	TOTAL	TOTAL INCL O&P
170	0300	16″ x 12″, 27 brick	D-1	19	.842	V.L.F.	11.25	18.35		29.60	43
	0400	16″ x 16″, 36 brick		14	1.143		14.95	25		39.95	58
	0500	20″ x 16″, 45 brick		11	1.455		18.70	31.50		50.20	73
	0600	20″ x 20″, 56 brick		9	1.778		23.50	38.50		62	90
172	0010	**CONCRETE BLOCK, BACK-UP,** C90, 2000 psi R04220-200									
	0020	Normal weight, 8″ x 16″ units, tooled joint 1 side									
	0050	Not-reinforced, 2000 psi, 2″ thick	D-8	475	.084	S.F.	.83	1.88		2.71	4.04
	0200	4″ thick		460	.087		.98	1.94		2.92	4.31
	0300	6″ thick		440	.091		1.44	2.03		3.47	4.97
	0350	8″ thick		400	.100		1.56	2.24		3.80	5.45
	0400	10″ thick		330	.121		2.19	2.71		4.90	6.90
	0450	12″ thick	D-9	310	.155		2.26	3.37		5.63	8.10
	1000	Reinforced, alternate courses, 4″ thick	D-8	450	.089		1.08	1.99		3.07	4.50
	1100	6″ thick		430	.093		1.55	2.08		3.63	5.15
	1150	8″ thick		395	.101		1.67	2.26		3.93	5.60
	1200	10″ thick		320	.125		2.30	2.80		5.10	7.20
	1250	12″ thick	D-9	300	.160		2.38	3.48		5.86	8.40
175	0010	**CONCRETE BLOCK BOND BEAM** C90, 2000 psi									
	0020	Not including grout or reinforcing									
	0130	8″ high, 8″ thick	D-8	565	.071	L.F.	1.64	1.58		3.22	4.44
	0150	12″ thick	D-9	510	.094	″	2.28	2.05		4.33	5.90
182	0010	**CONCRETE BLOCK, DECORATIVE** C90, 2000 psi									
	5000	Split rib profile units, 1″ deep ribs, 8 ribs									
	5100	8″ x 16″ x 4″ thick	D-8	345	.116	S.F.	2.14	2.59		4.73	6.65
	5150	6″ thick		325	.123		2.45	2.75		5.20	7.30
	5200	8″ thick		300	.133		2.83	2.98		5.81	8.05
	5250	12″ thick	D-9	275	.175		3.34	3.80		7.14	10
	5400	For special deeper colors, 4″ thick, add					.80			.80	.88
	5450	12″ thick, add					.69			.69	.76
	5600	For white, 4″ thick, add					.80			.80	.88
	5650	6″ thick, add					.80			.80	.88
	5700	8″ thick, add					.75			.75	.82
	5750	12″ thick, add					.69			.69	.76
184	0010	**CONCRETE BLOCK, EXTERIOR** C90, 2000 psi									
	0020	Reinforced alt courses, tooled joints 2 sides									
	0100	Normal weight, 8″ x 16″ x 6″ thick	D-8	395	.101	S.F.	1.67	2.26		3.93	5.60
	0200	8″ thick		360	.111		2.48	2.48		4.96	6.85
	0250	10″ thick		290	.138		3	3.08		6.08	8.45
	0300	12″ thick	D-9	250	.192		3.01	4.18		7.19	10.25
186	0010	**CONCRETE BLOCK FOUNDATION WALL** C90/C145									
	0050	Normal-weight, cut joints, horiz joint reinf, no vert reinf									
	0200	Hollow, 8″ x 16″ x 6″ thick	D-8	455	.088	S.F.	1.67	1.97		3.64	5.10
	0250	8″ thick		425	.094		1.80	2.10		3.90	5.45
	0300	10″ thick		350	.114		2.43	2.56		4.99	6.95
	0350	12″ thick	D-9	300	.160		2.51	3.48		5.99	8.55
	0500	Solid, 8″ x 16″ block, 6″ thick	D-8	440	.091		1.81	2.03		3.84	5.35
	0550	8″ thick	″	415	.096		2.59	2.16		4.75	6.45
	0600	12″ thick	D-9	350	.137		3.76	2.99		6.75	9.10
188	0010	**CONCRETE BLOCK INSULATION INSERTS**									
	0100	Inserts, styrofoam, plant installed, add to block prices									
	0200	8″ x 16″ units, 6″ thick				S.F.	.85			.85	.94
	0250	8″ thick					.85			.85	.94
	0300	10″ thick					1			1	1.10
	0350	12″ thick					1.05			1.05	1.16

04810 | Unit Masonry Assemblies

Section	Line	Description	Crew	Daily Output	Labor-Hours	Unit	2005 Bare Costs Mat.	Labor	Equip.	Total	Total Incl O&P
190	0010	**CONCRETE BLOCK, LINTELS** C90, normal weight									
	0100	Including grout and horizontal reinforcing									
	0200	8" x 8" x 8", 1 #4 bar	D-4	300	.133	L.F.	3.63	2.63	.45	6.71	8.85
	0250	2 #4 bars		295	.136		3.81	2.67	.46	6.94	9.15
	1000	12" x 8" x 8", 1 #4 bar		275	.145		5.10	2.87	.49	8.46	10.95
	1150	2 #5 bars		270	.148		5.45	2.92	.50	8.87	11.45
210	0010	**CONCRETE BLOCK, PARTITIONS**, scaffolding not included									
	1000	Lightweight block, tooled joints, 2 sides, hollow									
	1100	Not reinforced, 8" x 16" x 4" thick	D-8	440	.091	S.F.	1.12	2.03		3.15	4.61
	1150	6" thick		410	.098		1.54	2.18		3.72	5.30
	1200	8" thick		385	.104		1.89	2.32		4.21	5.95
	1250	10" thick		370	.108		2.49	2.42		4.91	6.75
	1300	12" thick	D-9	350	.137		2.92	2.99		5.91	8.20
	4000	Regular block, tooled joints, 2 sides, hollow									
	4100	Not reinforced, 8" x 16" x 4" thick	D-8	430	.093	S.F.	.94	2.08		3.02	4.50
	4150	6" thick		400	.100		1.40	2.24		3.64	5.25
	4200	8" thick		375	.107		1.52	2.39		3.91	5.65
	4250	10" thick		360	.111		2.15	2.48		4.63	6.50
	4300	12" thick	D-9	340	.141		2.22	3.07		5.29	7.55
250	0010	**COPING** Stock units									
	0050	Precast concrete, 10" wide, 4" tapers to 3-1/2", 8" wall	D-1	75	.213	L.F.	15	4.65		19.65	24.50
	0100	12" wide, 3-1/2" tapers to 3", 10" wall		70	.229		10.50	4.98		15.48	19.85
	0150	16" wide, 4" tapers to 3-1/2", 14" wall		60	.267		14.50	5.80		20.30	25.50
	0300	Limestone for 12" wall, 4" thick		90	.178		13.50	3.87		17.37	21.50
	0350	6" thick		80	.200		15.75	4.36		20.11	24.50
	0500	Marble, to 4" thick, no wash, 9" wide		90	.178		19.25	3.87		23.12	27.50
	0550	12" wide		80	.200		29	4.36		33.36	39.50
	0700	Terra cotta, 9" wide		90	.178		4.75	3.87		8.62	11.70
	0800	Aluminum, for 12" wall		80	.200		11.25	4.36		15.61	19.65
260	0010	**CORNICES** Brick, on existing building									
	0110	Face bricks, 12 brick/S.F., minimum	D-1	30	.533	SF Face	5	11.60		16.60	25
	0150	15 brick/S.F., maximum	"	23	.696	"	6	15.15		21.15	31.50
325	0010	**GLASS BLOCK** R04060-200									
	0150	8" x 8"	D-8	160	.250	S.F.	10.15	5.60		15.75	20.50
	0160	end block		160	.250		27.50	5.60		33.10	39.50
	0170	90 deg corner		160	.250		28	5.60		33.60	40
	0180	45 deg corner		160	.250		13.85	5.60		19.45	24.50
	0200	12" x 12"		175	.229		11.90	5.10		17	21.50
	0210	4" x 8"		160	.250		7.75	5.60		13.35	17.80
	0220	6" x 8"		160	.250		9.40	5.60		15	19.65
	0700	For solar reflective blocks, add					100%				
	1000	Thinline, plain, 3-1/8" thick, under 1,000 S.F., 6" x 6"	D-8	115	.348	S.F.	13.05	7.80		20.85	27.50
	1050	8" x 8"		160	.250		7.70	5.60		13.30	17.75
	1400	For cleaning block after installation (both sides), add		1,000	.040		.10	.89		.99	1.60
650	0016	**WALLS** R04210-500									
	0800	4" wall, face, 4" x 2-2/3" x 8"	D-8	215	.186	S.F.	2.83	4.16		6.99	10
	0850	4" thick, as back up, 6.75 bricks per S.F.		240	.167		2.31	3.73		6.04	8.75
	0900	8" thick wall, 13.50 brick per S.F.		135	.296		4.73	6.65		11.38	16.20
	1000	12" thick wall, 20.25 bricks per S.F.		95	.421		7.10	9.40		16.50	23.50
	1050	16" thick wall, 27.00 bricks per S.F.		75	.533		9.60	11.95		21.55	30.50
	1200	Reinforced, 4" x 2-2/3" x 8" , 4" wall		205	.195		2.31	4.36		6.67	9.80
	1250	8" thick wall, 13.50 brick per S.F.		130	.308		4.73	6.90		11.63	16.65
	1300	12" thick wall, 20.25 bricks per S.F.		90	.444		7.10	9.95		17.05	24.50
	1350	16" thick wall, 27.00 bricks per S.F.		70	.571		9.60	12.80		22.40	32

	04850 | Stone Assemblies		CREW	DAILY OUTPUT	LABOR-HOURS	UNIT	2005 BARE COSTS MAT.	LABOR	EQUIP.	TOTAL	TOTAL INCL O&P
300	0010	**GRANITE** Cut to size									
	2450	For radius under 5′, add				L.F.	100%				
	2500	Steps, copings, etc., finished on more than one surface									
	2550	Minimum	D-10	50	.800	C.F.	76.50	18	10.10	104.60	125
	2600	Maximum	″	50	.800	″	122	18	10.10	150.10	176
	2800	Pavers, 4″ x 4″ x 4″ blocks, split face and joints									
	2850	Minimum	D-11	80	.300	S.F.	11.20	6.70		17.90	23.50
	2900	Maximum	″	80	.300	″	22.50	6.70		29.20	35.50
	3500	Curbing, city street type, See Division 02770-300									
400	0010	**LIMESTONE,** Cut to size									
	0020	Veneer facing panels									
	0500	Texture finish, light stick, 4-1/2″ thick, 5′x 12′	D-4	300	.133	S.F.	31	2.63	.45	34.08	39
	0750	5″ thick, 5′ x 14′ panels	D-10	275	.145		33.50	3.27	1.84	38.61	44.50
	1000	Sugarcube finish, 2″ Thick, 3′ x 5′ panels		275	.145		18.35	3.27	1.84	23.46	27.50
	1050	3″ Thick, 4′ x 9′ panels		275	.145		22.50	3.27	1.84	27.61	32.50
	1200	4″ Thick, 5′ x 11′ panels		275	.145		26.50	3.27	1.84	31.61	36.50
	1400	Sugarcube, textured finish, 4-1/2″ thick, 5′ x 12′		275	.145		29	3.27	1.84	34.11	39.50
	1450	5″ thick, 5′ x 14′ panels		275	.145	↓	34	3.27	1.84	39.11	45
	2000	Coping, sugarcube finish, top & 2 sides		30	1.333	C.F.	37	30	16.85	83.85	109
	2100	Sills, lintels, jambs, trim, stops, sugarcube finish, average		20	2		55	45	25.50	125.50	163
	2150	Detailed		20	2	↓	55	45	25.50	125.50	163
	2300	Steps, extra hard, 14″ wide, 6″ rise	↓	50	.800	L.F.	19.50	18	10.10	47.60	62.50
	3000	Quoins, plain finish, 6″x12″x12″	D-12	25	1.280	Ea.	50	27.50		77.50	101
	3050	6″x16″x24″	″	25	1.280	″	66.50	27.50		94	119
500	0011	**MARBLE**, ashlar, split face, 4″ + or - thick, random									
	0040	lengths 1′ to 4′ & heights 2″ to 7-1/2″, average	D-8	175	.229	S.F.	14.55	5.10		19.65	24.50
	0100	Base, polished, 3/4″ or 7/8″ thick, polished, 6″ high	D-10	65	.615	L.F.	13.25	13.85	7.80	34.90	46
	1000	Facing, polished finish, cut to size, 3/4″ to 7/8″ thick									
	1050	Average	D-10	130	.308	S.F.	19.45	6.90	3.89	30.24	37.50
	1100	Maximum	″	130	.308	″	45	6.90	3.89	55.79	65.50
	2200	Window sills, 6″ x 3/4″ thick	D-1	85	.188	L.F.	7.30	4.10		11.40	14.85
	2500	Flooring, polished tiles, 12″ x 12″ x 3/8″ thick									
	2510	Thin set, average	D-11	90	.267	S.F.	9	5.95		14.95	19.80
	2600	Maximum		90	.267		32.50	5.95		38.45	46
	2700	Mortar bed, average		65	.369		9.15	8.25		17.40	24
	2740	Maximum	↓	65	.369		30	8.25		38.25	46.50
	2780	Travertine, 3/8″ thick, average	D-10	130	.308		12.40	6.90	3.89	23.19	29.50
	2790	Maximum	″	130	.308	↓	30.50	6.90	3.89	41.29	49.50
	3500	Thresholds, 3′ long, 7/8″ thick, 4″ to 5″ wide, plain	D-12	24	1.333	Ea.	14.05	28.50		42.55	63
	3550	Beveled		24	1.333	″	16.30	28.50		44.80	65.50
	3700	Window stools, polished, 7/8″ thick, 5″ wide	↓	85	.376	L.F.	13.10	8.05		21.15	28
600	0011	**ROUGH STONE WALL**, Dry									
	0100	Random fieldstone, under 18″ thick	D-12	60	.533	C.F.	8.05	11.45		19.50	28
	0150	Over 18″ thick	″	63	.508	″	9.70	10.90		20.60	29
700	0011	**SANDSTONE OR BROWNSTONE**									
	0100	Sawed face veneer, 2-1/2″ thick, to 2′ x 4′ panels	D-10	130	.308	S.F.	17.75	6.90	3.89	28.54	35.50
	0150	4′ thick, to 3′-6″ x 8′panels		100	.400		17.75	9	5.05	31.80	40
	0300	Split face, random sizes	↓	100	.400	↓	10.65	9	5.05	24.70	32
	0350	Cut stone trim (limestone)									
	0360	Ribbon stone, 4″ thick, 5′ pieces	D-8	120	.333	Ea.	131	7.45		138.45	156
	0370	Cove stone, 4″ thick, 5′ pieces		105	.381		131	8.50		139.50	158
	0380	Cornice stone, 10″ to 12″ wide	↓	90	.444	↓	162	9.95		171.95	195

04800 | Masonry Assemblies

04850 | Stone Assemblies

		CREW	DAILY OUTPUT	LABOR-HOURS	UNIT	2005 BARE COSTS MAT.	LABOR	EQUIP.	TOTAL	TOTAL INCL O&P	
0390	Band stone, 4″ thick, 5′ pieces	D-8	145	.276	Ea.	83.50	6.15		89.65	102	700
0410	Window and door trim, 3″ to 4″ wide		160	.250		71	5.60		76.60	87.50	
0420	Key stone, 18″ long		60	.667		74.50	14.90		89.40	107	
0010	**SLATE** Pennsylvania, blue gray to gray black; Vermont,										800
3500	Stair treads, sand finish, 1″ thick x 12″ wide										
3600	3 L.F. to 6 L.F.	D-10	120	.333	L.F.	16.30	7.50	4.22	28.02	35	
3700	Ribbon, sand finish, 1″ thick x 12″ wide										
3750	To 6 L.F.	D-10	120	.333	L.F.	10.75	7.50	4.22	22.47	29	
0010	**WINDOW SILL** Bluestone, thermal top, 10″ wide, 1-1/2″ thick	D-1	85	.188	S.F.	13.50	4.10		17.60	21.50	900
0050	2″ thick		75	.213	″	15.75	4.65		20.40	25	
0100	Cut stone, 5″ x 8″ plain		48	.333	L.F.	10.20	7.25		17.45	23.50	
0200	Face brick on edge, brick, 8″ wide		80	.200		2.15	4.36		6.51	9.60	
0400	Marble, 9″ wide, 1″ thick		85	.188		7.50	4.10		11.60	15.05	
0900	Slate, colored, unfading, honed, 12″ wide, 1″ thick		85	.188		15.25	4.10		19.35	23.50	
0950	2″ thick		70	.229		21.50	4.98		26.48	32	

04880 | Masonry Fireplaces

		CREW	DAILY OUTPUT	LABOR-HOURS	UNIT	MAT.	LABOR	EQUIP.	TOTAL	TOTAL INCL O&P	
0010	**FIREPLACE** For prefabricated fireplace, see div. 10305-100										600
0100	Brick fireplace, not incl. foundations or chimneys										
0110	30″ x 29″ opening, incl. chamber, plain brickwork	D-1	.40	40	Ea.	380	870		1,250	1,875	
0200	Fireplace box only (110 brick)	″	2	8	″	125	174		299	430	
0300	For elaborate brickwork and details, add					35%	35%				
0400	For hearth, brick & stone, add	D-1	2	8	Ea.	140	174		314	445	
0410	For steel angle, damper, cleanouts, add		4	4		98	87		185	253	
0600	Plain brickwork, incl. metal circulator		.50	32		725	695		1,420	1,950	
0800	Face brick only, standard size, 8″ x 2-2/3″ x 4″		.30	53.333	M	380	1,150		1,530	2,350	
0900	Stone fireplace, fieldstone, add				SF Face	10			10	11	
1000	Cut stone, add				″	11			11	12.10	

04900 | Masonry Restoration and Cleaning

04910 | Unit Masonry Restoration

		CREW	DAILY OUTPUT	LABOR-HOURS	UNIT	2005 BARE COSTS MAT.	LABOR	EQUIP.	TOTAL	TOTAL INCL O&P	
0010	**POINTING MASONRY**										720
0300	Cut and repoint brick, hard mortar, running bond	1 Bric	80	.100	S.F.	.27	2.47		2.74	4.40	
0320	Common bond		77	.104		.27	2.57		2.84	4.56	
0360	Flemish bond		70	.114		.28	2.82		3.10	5	
0400	English bond		65	.123		.28	3.04		3.32	5.35	
0600	Soft old mortar, running bond		100	.080		.27	1.98		2.25	3.59	
0620	Common bond		96	.083		.27	2.06		2.33	3.71	
0640	Flemish bond		90	.089		.28	2.20		2.48	3.96	
0680	English bond		82	.098		.28	2.41		2.69	4.32	
0700	Stonework, hard mortar		140	.057	L.F.	.35	1.41		1.76	2.74	
0720	Soft old mortar		160	.050	″	.35	1.24		1.59	2.45	
1000	Repoint, mask and grout method, running bond		95	.084	S.F.	.35	2.08		2.43	3.85	
1020	Common bond		90	.089		.35	2.20		2.55	4.04	
1040	Flemish bond		86	.093		.35	2.30		2.65	4.21	

04900 | Masonry Restoration and Cleaning

	04910 Unit Masonry Restoration		CREW	DAILY OUTPUT	LABOR-HOURS	UNIT	2005 BARE COSTS MAT.	LABOR	EQUIP.	TOTAL	TOTAL INCL O&P
720	1060	English bond	1 Bric	77	.104	S.F.	.35	2.57		2.92	4.66
	2000	Scrub coat, sand grout on walls, minimum		120	.067		2.82	1.65		4.47	5.85
	2020	Maximum	↓	98	.082	↓	2.03	2.02		4.05	5.60

	04930 Unit Masonry Cleaning		CREW	DAILY OUTPUT	LABOR-HOURS	UNIT	MAT.	LABOR	EQUIP.	TOTAL	TOTAL INCL O&P
900	0010	**BRICK WASHING** Acid, smooth brick R04930 -100	1 Bric	560	.014	S.F.	.02	.35		.37	.62
	0050	Rough brick		400	.020		.03	.49		.52	.86
	0060	Stone, acid wash	↓	600	.013	↓	.04	.33		.37	.59
	1000	Muriatic acid, price per gallon in 5 gallon lots				Gal.	4.90			4.90	5.40

For information about Means Estimating Seminars, see yellow pages 12 and 13 in back of book

Division 5
Metals

stimating Tips

050 Basic Metal aterials & Methods

Nuts, bolts, washers, connection angles and plates can add a significant amount to both the tonnage of a structural steel job as well as the estimated cost. As a rule of thumb add 10% to the total weight to account for these accessories.

Type 2 steel construction, commonly referred to as "simple construction," consists generally of field bolted connections with lateral bracing supplied by other elements of the building, such as masonry walls or x-bracing. The estimator should be aware, however, that shop connections may be accomplished by welding or bolting. The method may be particular to the fabrication shop and may have an impact on the estimated cost.

200 Metal Joists

In any given project the total weight of open web steel joists is determined by the loads to be supported and the design. However, economies can be realized in minimizing the amount of labor used to place the joists. This is done by maximizing the joist spacing and therefore minimizing the number of joists required to be installed on the job. Certain spacings and locations may be required by the design, but in other cases maximizing the spacing and keeping it as uniform as possible will keep the costs down.

05300 Metal Deck

- The takeoff and estimating of metal deck involves more than simply the area of the floor or roof and the type of deck specified or shown on the drawings. Many different sizes and types of openings may exist. Small openings for individual pipes or conduits may be drilled after the floor/roof is installed, but larger openings may require special deck lengths as well as reinforcing or structural support. The estimator should determine who will be supplying this reinforcing. Additionally, some deck terminations are part of the deck package, such as screed angles and pour stops, and others will be part of the steel contract, such as angles attached to structural members and cast-in-place angles and plates. The estimator must ensure that all pieces are accounted for in the complete estimate.

05500 Metal Fabrications

- The most economical steel stairs are those that use common materials, standard details and most importantly, a uniform and relatively simple method of field assembly. Commonly available A36 channels and plates are very good choices for the main stringers of the stairs, as are angles and tees for the carrier members. Risers and treads are usually made by specialty shops, and it is most economical to use a typical detail in as many places as possible. The stairs should be pre-assembled and shipped directly to the site. The field connections should be simple and straightforward to be accomplished efficiently and with a minimum of equipment and labor.

Reference Numbers

Reference numbers are shown in bold squares at the beginning of some major classifications. These numbers refer to related items in the Reference Section. The reference information may be an estimating procedure, an alternate pricing method or technical information.

Note: Not all subdivisions listed here necessarily appear in this publication.

part of this publication may be reproduced, stored in a retrieval system, or transmitted in any form by any means without prior written permission of Reed Construction Data.

5 METALS

05090 | Metal Fastenings

			Crew	Daily Output	Labor-Hours	Unit	2005 Bare Costs Mat.	Labor	Equip.	Total	Total Incl O&P
150	0010	**BOLTS & HEX NUTS** Steel, A307									
	0100	1/4" diameter, 1/2" long	1 Sswk	140	.057	Ea.	.06	1.46		1.52	2.88
	0200	1" long		140	.057		.07	1.46		1.53	2.89
	0300	2" long		130	.062		.09	1.57		1.66	3.13
	0400	3" long		130	.062		.13	1.57		1.70	3.18
	0500	4" long		120	.067		.15	1.70		1.85	3.45
	0600	3/8" diameter, 1" long		130	.062		.11	1.57		1.68	3.15
	0700	2" long		130	.062		.14	1.57		1.71	3.18
	0800	3" long		120	.067		.18	1.70		1.88	3.49
	0900	4" long		120	.067		.23	1.70		1.93	3.54
	1000	5" long		115	.070		.28	1.78		2.06	3.74
	1100	1/2" diameter, 1-1/2" long		120	.067		.21	1.70		1.91	3.52
	1200	2" long		120	.067		.24	1.70		1.94	3.55
	1300	4" long		115	.070		.36	1.78		2.14	3.83
	1400	6" long		110	.073		.49	1.86		2.35	4.13
	1500	8" long		105	.076		.64	1.95		2.59	4.46
	1600	5/8" diameter, 1-1/2" long		120	.067		.40	1.70		2.10	3.73
	1700	2" long		120	.067		.44	1.70		2.14	3.77
	1800	4" long		115	.070		.61	1.78		2.39	4.10
	1900	6" long		110	.073		.77	1.86		2.63	4.44
	2000	8" long		105	.076		1.12	1.95		3.07	4.99
	2100	10" long		100	.080		1.39	2.04		3.43	5.45
	2200	3/4" diameter, 2" long		120	.067		.63	1.70		2.33	3.98
	2300	4" long		110	.073		.88	1.86		2.74	4.56
	2400	6" long		105	.076		1.12	1.95		3.07	4.99
	2500	8" long		95	.084		1.66	2.15		3.81	6
	2600	10" long		85	.094		2.16	2.40		4.56	7
	2700	12" long		80	.100		2.52	2.56		5.08	7.70
	2800	1" diameter, 3" long		105	.076		1.65	1.95		3.60	5.55
	2900	6" long		90	.089		2.54	2.27		4.81	7.15
	3000	12" long		75	.107		4.78	2.73		7.51	10.50
	3100	For galvanized, add					75%				
	3200	For stainless, add					350%				
300	0010	**CHEMICAL ANCHORS**, Includes layout & drilling									
	1430	Chemical anchor, w/rod & epoxy cartridge, 3/4" diam. x 9-1/2" long	B-89A	27	.593	Ea.	11.65	12.25	3.82	27.72	37.50
	1435	1" diameter x 11-3/4" long		24	.667		22.50	13.75	4.29	40.54	53
	1440	1-1/4" diameter x 14" long		21	.762		43	15.70	4.91	63.61	79
	1445	1-3/4" diameter x 15" long		20	.800		81	16.50	5.15	102.65	123
	1450	18" long		17	.941		97.50	19.40	6.05	122.95	147
	1455	2" diameter x 18" long		16	1		124	20.50	6.45	150.95	179
	1460	24" long		15	1.067		162	22	6.85	190.85	224
	1500	Chemical anchoring, epoxy cartridge, excludes layout, drilling, fastener									
	1530	For fastener 3/4" dia x 6" embedment	B-89A	27	.593	Ea.	4.58	12.25	3.82	20.65	30
	1535	1" dia x 8" embedment		24	.667		6.85	13.75	4.29	24.89	36
	1540	1-1/4" dia x 10" embedment		21	.762		13.75	15.70	4.91	34.36	47
	1545	1-3/4" dia x 12" embedment		20	.800		23	16.50	5.15	44.65	58.50
	1550	14" embedment		17	.941		27.50	19.40	6.05	52.95	69.50
	1555	2" dia x 12" embedment		16	1		36.50	20.50	6.45	63.45	82.50
	1560	18" embedment		15	1.067		46	22	6.85	74.85	95.50
340	0010	**DRILLING** For anchors, up to 4" deep, incl. bit and layout									
	0050	in concrete or brick walls and floors, no anchor									
	0100	Holes, 1/4" diameter	1 Carp	75	.107	Ea.	.08	2.56		2.64	4.44
	0150	For each additional inch of depth, add		430	.019		.02	.45		.47	.78
	0200	3/8" diameter		63	.127		.08	3.05		3.13	5.25
	0250	For each additional inch of depth, add		340	.024		.02	.56		.58	.98

05090	Metal Fastenings	CREW	DAILY OUTPUT	LABOR-HOURS	UNIT	2005 BARE COSTS MAT.	LABOR	EQUIP.	TOTAL	TOTAL INCL O&P	
0300	1/2" diameter	1 Carp	50	.160	Ea.	.08	3.84		3.92	6.60	340
0350	For each additional inch of depth, add		250	.032		.02	.77		.79	1.32	
0400	5/8" diameter		48	.167		.15	4		4.15	6.95	
0450	For each additional inch of depth, add		240	.033		.04	.80		.84	1.40	
0500	3/4" diameter		45	.178		.18	4.27		4.45	7.45	
0550	For each additional inch of depth, add		220	.036		.04	.87		.91	1.53	
0600	7/8" diameter		43	.186		.21	4.47		4.68	7.85	
0650	For each additional inch of depth, add		210	.038		.05	.91		.96	1.61	
0700	1" diameter		40	.200		.24	4.80		5.04	8.40	
0750	For each additional inch of depth, add		190	.042		.06	1.01		1.07	1.79	
0800	1-1/4" diameter		38	.211		.35	5.05		5.40	9	
0850	For each additional inch of depth, add		180	.044		.09	1.07		1.16	1.91	
0900	1-1/2" diameter		35	.229		.53	5.50		6.03	9.90	
0950	For each additional inch of depth, add		165	.048		.13	1.16		1.29	2.12	
1000	For ceiling installations, add						40%				
1100	Drilling & layout for drywall/plaster walls, up to 1" deep, no anchor										
1200	Holes, 1/4" diameter	1 Carp	150	.053	Ea.	.01	1.28		1.29	2.18	
1300	3/8" diameter		140	.057		.01	1.37		1.38	2.34	
1400	1/2" diameter		130	.062		.01	1.48		1.49	2.52	
1500	3/4" diameter		120	.067		.02	1.60		1.62	2.74	
1600	1" diameter		110	.073		.03	1.75		1.78	2.99	
1700	1-1/4" diameter		100	.080		.04	1.92		1.96	3.31	
1800	1-1/2" diameter		90	.089		.07	2.13		2.20	3.69	
1900	For ceiling installations, add						40%				
1910	Drilling & layout for steel, up to 1/4" deep, no anchor										
1920	Holes, 1/4" diameter	1 Sswk	112	.071	Ea.	.11	1.83		1.94	3.64	
1925	For each additional 1/4" depth, add		336	.024		.11	.61		.72	1.29	
1930	3/8" diameter		104	.077		.13	1.97		2.10	3.93	
1935	For each additional 1/4" depth, add		312	.026		.13	.66		.79	1.40	
1940	1/2" diameter		96	.083		.14	2.13		2.27	4.27	
1945	For each additional 1/4" depth, add		288	.028		.14	.71		.85	1.53	
1950	5/8" diameter		88	.091		.24	2.32		2.56	4.74	
1955	For each additional 1/4" depth, add		264	.030		.24	.77		1.01	1.75	
1960	3/4" diameter		80	.100		.26	2.56		2.82	5.20	
1965	For each additional 1/4" depth, add		240	.033		.26	.85		1.11	1.93	
1970	7/8" diameter		72	.111		.31	2.84		3.15	5.85	
1975	For each additional 1/4" depth, add		216	.037		.31	.95		1.26	2.17	
1980	1" diameter		64	.125		.35	3.19		3.54	6.55	
1985	For each additional 1/4" depth, add		192	.042		.35	1.06		1.41	2.44	
1990	For drilling up, add						40%				
0010	**EXPANSION ANCHORS** & shields										380
0100	Bolt anchors for concrete, brick or stone, no layout and drilling										
0200	Expansion shields, zinc, 1/4" diameter, 1-5/16" long, single	1 Carp	90	.089	Ea.	1.01	2.13		3.14	4.73	
0300	1-3/8" long, double		85	.094		1.11	2.26		3.37	5.05	
0500	2" long, double		80	.100		2.05	2.40		4.45	6.35	
0700	2-1/2" long, double		75	.107		2.65	2.56		5.21	7.25	
0900	2-3/4" long, double		70	.114		3.93	2.74		6.67	9	
1100	3-15/16" long, double		65	.123		7.80	2.95		10.75	13.55	
2100	Hollow wall anchors for gypsum wall board, plaster or tile										
2500	3/16" diameter, short	1 Carp	150	.053	Ea.	.55	1.28		1.83	2.78	
3000	Toggle bolts, bright steel, 1/8" diameter, 2" long		85	.094		.25	2.26		2.51	4.12	
3100	4" long		80	.100		.37	2.40		2.77	4.49	
3200	3/16" diameter, 3" long		80	.100		.42	2.40		2.82	4.54	
3300	6" long		75	.107		.58	2.56		3.14	4.99	
3400	1/4" diameter, 3" long		75	.107		.47	2.56		3.03	4.87	
3500	6" long		70	.114		.66	2.74		3.40	5.40	

	05090	Metal Fastenings	CREW	DAILY OUTPUT	LABOR-HOURS	UNIT	2005 BARE COSTS MAT.	LABOR	EQUIP.	TOTAL	TOTAL INCL O&P
380	3600	3/8" diameter, 3" long	1 Carp	70	.114	Ea.	.90	2.74		3.64	5.65
	3700	6" long		60	.133		1.58	3.20		4.78	7.20
	3800	1/2" diameter, 4" long		60	.133		2.33	3.20		5.53	8
	3900	6" long		50	.160		3.85	3.84		7.69	10.75
	4000	Nailing anchors									
	4100	Nylon nailing anchor, 1/4" diameter, 1" long	1 Carp	3.20	2.500	C	18	60		78	122
	4200	1-1/2" long		2.80	2.857		23	68.50		91.50	142
	4300	2" long		2.40	3.333		38.50	80		118.50	179
	4400	Metal nailing anchor, 1/4" diameter, 1" long		3.20	2.500		27.50	60		87.50	133
	4500	1-1/2" long		2.80	2.857		37.50	68.50		106	158
	4600	2" long		2.40	3.333		48	80		128	189
	5000	Screw anchors for concrete, masonry,									
	5100	stone & tile, no layout or drilling included									
	5200	Jute fiber, #6, #8, & #10, 1" long	1 Carp	240	.033	Ea.	.24	.80		1.04	1.62
	5300	#12, 1-1/2" long		200	.040		.35	.96		1.31	2.02
	5400	#14, 2" long		160	.050		.55	1.20		1.75	2.65
	5500	#16, 2" long		150	.053		.57	1.28		1.85	2.80
	5600	#20, 2" long		140	.057		.91	1.37		2.28	3.33
	5700	Lag screw shields, 1/4" diameter, short		90	.089		.44	2.13		2.57	4.10
	5800	Long		85	.094		.51	2.26		2.77	4.40
	5900	3/8" diameter, short		85	.094		.80	2.26		3.06	4.72
	6000	Long		80	.100		.94	2.40		3.34	5.10
	6100	1/2" diameter, short		80	.100		1.11	2.40		3.51	5.30
	6200	Long		75	.107		1.39	2.56		3.95	5.90
	6300	3/4" diameter, short		70	.114		3.12	2.74		5.86	8.10
	6400	Long		65	.123		3.78	2.95		6.73	9.15
	6600	Lead, #6 & #8, 3/4" long		260	.031		.16	.74		.90	1.43
	6700	#10 - #14, 1-1/2" long		200	.040		.23	.96		1.19	1.88
	6800	#16 & #18, 1-1/2" long		160	.050		.31	1.20		1.51	2.38
	6900	Plastic, #6 & #8, 3/4" long		260	.031		.10	.74		.84	1.36
	7000	#8 & #10, 7/8" long		240	.033		.04	.80		.84	1.40
	7100	#10 & #12, 1" long		220	.036		.13	.87		1	1.62
	7200	#14 & #16, 1-1/2" long		160	.050		.07	1.20		1.27	2.12
	8950	Self-drilling concrete screw, hex washer head, 3/16" dia x 1-3/4" long		300	.027		.32	.64		.96	1.44
	8960	2-1/4" long		250	.032		.49	.77		1.26	1.84
	8970	Phillips flat head, 3/16" dia x 1-3/4" long		300	.027		.33	.64		.97	1.45
	8980	2-1/4" long		250	.032		.48	.77		1.25	1.83
460	0010	**LAG SCREWS**									
	0020	Steel, 1/4" diameter, 2" long	1 Carp	200	.040	Ea.	.08	.96		1.04	1.72
	0100	3/8" diameter, 3" long		150	.053		.23	1.28		1.51	2.42
	0200	1/2" diameter, 3" long		130	.062		.38	1.48		1.86	2.93
	0300	5/8" diameter, 3" long		120	.067		.75	1.60		2.35	3.55
580	0010	**POWDER ACTUATED** Tools & fasteners									
	0020	Stud driver, .22 caliber, buy, minimum				Ea.	310			310	340
	0100	Maximum				"	495			495	545
	0300	Powder charges for above, low velocity				C	16.60			16.60	18.25
	0400	Standard velocity					23.50			23.50	26
	0600	Drive pins & studs, 1/4" & 3/8" diam., to 3" long, minimum					11.85			11.85	13.05
	0700	Maximum					46.50			46.50	51
600	0010	**RIVETS**									
	0100	Aluminum rivet & mandrel, 1/2" grip length x 1/8" diameter				C	5			5	5.50
	0200	3/16" diameter					7.70			7.70	8.45
	0300	Aluminum rivet, steel mandrel, 1/8" diameter					7.30			7.30	8.05

05050 | Basic Metal Materials & Methods

05090 | Metal Fastenings

	05090 Metal Fastenings	Crew	Daily Output	Labor-Hours	Unit	2005 Bare Costs Mat.	2005 Bare Costs Labor	2005 Bare Costs Equip.	2005 Bare Costs Total	Total Incl O&P	
0400	3/16" diameter				C	6.70			6.70	7.35	600
0500	Copper rivet, steel mandrel, 1/8" diameter					6.65			6.65	7.30	
0600	Monel rivet, steel mandrel, 1/8" diameter					23.50			23.50	26	
0700	3/16" diameter					68			68	75	
0800	Stainless rivet & mandrel, 1/8" diameter					11.90			11.90	13.05	
0900	3/16" diameter					22			22	24.50	
1000	Stainless rivet, steel mandrel, 1/8" diameter					9.25			9.25	10.20	
1100	3/16" diameter					16.75			16.75	18.45	
1200	Steel rivet and mandrel, 1/8" diameter					5.75			5.75	6.30	
1300	3/16" diameter					8.60			8.60	9.50	
1400	Hand riveting tool, minimum				Ea.	111			111	122	
1500	Maximum					212			212	233	
1600	Power riveting tool, minimum					800			800	880	
1700	Maximum					2,025			2,025	2,225	

05100 | Structural Metal Framing

05120 | Structural Steel

	05120 Structural Steel	Crew	Daily Output	Labor-Hours	Unit	2005 Bare Costs Mat.	2005 Bare Costs Labor	2005 Bare Costs Equip.	2005 Bare Costs Total	Total Incl O&P	
0010	**CEILING SUPPORTS**										220
1000	Entrance door/folding partition supports	E-4	60	.533	L.F.	17.50	13.90	1.35	32.75	47.50	
1100	Linear accelerator door supports		14	2.286		79.50	59.50	5.80	144.80	209	
1200	Lintels or shelf angles, hung, exterior hot dipped galv.		267	.120		11.95	3.12	.30	15.37	19.50	
1250	Two coats primer paint instead of galv.		267	.120		10.35	3.12	.30	13.77	17.75	
1400	Monitor support, ceiling hung, expansion bolted		4	8	Ea.	277	208	20.50	505.50	730	
1450	Hung from pre-set inserts		6	5.333		298	139	13.55	450.55	615	
1600	Motor supports for overhead doors		4	8		141	208	20.50	369.50	580	
1700	Partition support for heavy folding partitions, without pocket		24	1.333	L.F.	40	34.50	3.39	77.89	115	
1750	Supports at pocket only		12	2.667		79.50	69.50	6.75	155.75	229	
2000	Rolling grilles & fire door supports		34	.941		34	24.50	2.39	60.89	87.50	
2100	Spider-leg light supports, expansion bolted to ceiling slab		8	4	Ea.	114	104	10.15	228.15	335	
2150	Hung from pre-set inserts		12	2.667	"	123	69.50	6.75	199.25	276	
2400	Toilet partition support		36	.889	L.F.	40	23	2.26	65.26	91	
2500	X-ray travel gantry support		12	2.667	"	137	69.50	6.75	213.25	291	
0010	**COLUMNS, LIGHTWEIGHT**										250
1000	Lightweight units (lally), 3-1/2" diameter	E-2	780	.062	L.F.	2.72	1.58	1.70	6	7.85	
1050	4" diameter	"	900	.053	"	3.99	1.37	1.47	6.83	8.60	
8000	Lally columns, to 8', 3-1/2" diameter	2 Carp	24	.667	Ea.	22	16		38	51	
8080	4" diameter	"	20	.800	"	32	19.20		51.20	67.50	
0010	**COLUMNS, STRUCTURAL**										260
0020	Shop fab'd for 100-ton, 1-2 story project, bolted conn's.										
0800	Steel, concrete filled, extra strong pipe, 3-1/2" diameter	E-2	660	.073	L.F.	29	1.87	2.01	32.88	37.50	
0830	4" diameter		780	.062		32	1.58	1.70	35.28	40.50	
0890	5" diameter		1,020	.047		38.50	1.21	1.30	41.01	45.50	
0930	6" diameter		1,200	.040		51	1.03	1.10	53.13	59	
1100	For galvanizing, add				Lb.	.21			.21	.24	
1300	For web ties, angles, etc., add per added lb.	1 Sswk	945	.008		.88	.22		1.10	1.38	
1500	Steel pipe, extra strong, no concrete, 3" to 5" diameter	E-2	16,000	.003		.88	.08	.08	1.04	1.20	
1600	6" to 12" diameter		14,000	.003		.88	.09	.09	1.06	1.23	

METALS 5

05100 | Structural Metal Framing

05120		Structural Steel	CREW	DAILY OUTPUT	LABOR-HOURS	UNIT	2005 BARE COSTS MAT.	LABOR	EQUIP.	TOTAL	TOTAL INCL O&P
260	2400	Structural tubing, rect, 5" to 6" wide, light section	E-2	11,200	.004	Lb.	.88	.11	.12	1.11	1.30
	2700	12" x 8" x 1/2" thk wall		24,000	.002		.88	.05	.06	.99	1.12
	2800	Heavy section	↓	32,000	.002	↓	.88	.04	.04	.96	1.08
	8090	For projects 75 to 99 tons, add				All	10%				
	8092	50 to 74 tons, add					20%				
	8094	25 to 49 tons, add					30%	10%			
	8096	10 to 24 tons, add					50%	25%			
	8098	2 to 9 tons, add					75%	50%			
	8099	Less than 2 tons, add				↓	100%	100%			
480	0010	**LINTELS**									
	0020	Plain steel angles, under 500 lb.	1 Bric	550	.015	Lb.	.67	.36		1.03	1.34
	0100	500 to 1000 lb.		640	.013	"	.66	.31		.97	1.23
	2000	Steel angles, 3-1/2" x 3", 1/4" thick, 2'-6" long		47	.170	Ea.	9.45	4.20		13.65	17.40
	2100	4'-6" long		26	.308		17	7.60		24.60	31.50
	2600	4" x 3-1/2", 1/4" thick, 5'-0" long		21	.381		21.50	9.40		30.90	39.50
	2700	9'-0" long	↓	12	.667	↓	39	16.45		55.45	70.50
	3500	For precast concrete lintels, see div. 03480-400									
720	0010	**STRUCTURAL STEEL** Bolted, incl. fabrication									
	0050	Beams, W 6 x 9	E-2	720	.067	L.F.	9.45	1.72	1.84	13.01	15.65
	0100	W 8 x 10		720	.067		10.50	1.72	1.84	14.06	16.80
	0200	Columns, W 6 x 15		540	.089		17.05	2.29	2.45	21.79	26
	0250	W 8 x 31	↓	540	.089	↓	35.50	2.29	2.45	40.24	46
	7990	For projects 75 to 99 tons, add				All	10%				
	7992	50 to 75 tons, add					20%				
	7994	25 to 49 tons, add					30%	10%			
	7996	10 to 24 tons, add					50%	25%			
	7998	2 to 9 tons, add					75%	50%			
	7999	Less than 2 tons, add				↓	100%	100%			

05300 | Metal Deck

05310		Steel Deck	CREW	DAILY OUTPUT	LABOR-HOURS	UNIT	2005 BARE COSTS MAT.	LABOR	EQUIP.	TOTAL	TOTAL INCL O&P
300	0010	**METAL DECKING** Steel decking R05310-100									
	1900	For multi-story or congested site, add						50%			
	2100	Open type, galv., 1-1/2" deep wide rib, 22 gauge, under 50 squares	E-4	4,500	.007	S.F.	1.47	.19	.02	1.68	1.99
	2600	20 gauge, under 50 squares		3,865	.008		1.73	.22	.02	1.97	2.34
	2900	18 gauge, under 50 squares		3,800	.008		2.24	.22	.02	2.48	2.90
	3050	16 gauge, under 50 squares		3,700	.009		3.01	.23	.02	3.26	3.76
	3700	4-1/2" deep, long span roof, over 50 squares, 20 gauge		2,700	.012		3.78	.31	.03	4.12	4.79
	6100	Slab form, steel, 28 gauge, 9/16" deep, uncoated		4,000	.008		.98	.21	.02	1.21	1.50
	6200	Galvanized		4,000	.008		.87	.21	.02	1.10	1.38
	6220	24 gauge, 1" deep, uncoated		3,900	.008		1.07	.21	.02	1.30	1.61
	6240	Galvanized		3,900	.008		1.26	.21	.02	1.49	1.82
	6300	24 gauge, 1-5/16" deep, uncoated		3,800	.008		1.14	.22	.02	1.38	1.69
	6400	Galvanized		3,800	.008		1.34	.22	.02	1.58	1.91
	6500	22 gauge, 1-5/16" deep, uncoated		3,700	.009		1.43	.23	.02	1.68	2.02
	6600	Galvanized		3,700	.009		1.46	.23	.02	1.71	2.06
	6700	22 gauge, 2" deep uncoated	↓	3,600	.009	↓	1.89	.23	.02	2.14	2.55

05300 | Metal Deck

	05310 Steel Deck		CREW	DAILY OUTPUT	LABOR-HOURS	UNIT	2005 BARE COSTS MAT.	LABOR	EQUIP.	TOTAL	TOTAL INCL O&P	
300	6800	Galvanized	E-4	3,600	.009	S.F.	1.85	.23	.02	2.10	2.51	300

R05310-100

05400 | Cold-Formed Metal Framing

	05410 Load-Bearing Metal Studs		CREW	DAILY OUTPUT	LABOR-HOURS	UNIT	2005 BARE COSTS MAT.	LABOR	EQUIP.	TOTAL	TOTAL INCL O&P	
100	0010	**BRACING**, shear wall X-bracing, per 10′ x 10′ bay, one face										100
	0120	Metal strap, 20 ga x 4″ wide	2 Carp	18	.889	Ea.	22.50	21.50		44	61	
	0130	6″ wide		18	.889		35	21.50		56.50	74.50	
	0160	18 ga x 4″ wide		16	1		32.50	24		56.50	76.50	
	0170	6″ wide	↓	16	1	↓	47.50	24		71.50	93.50	
	0410	Continuous strap bracing, per horizontal row on both faces										
	0420	Metal strap, 20 ga x 2″ wide, studs 12″ O.C.	1 Carp	7	1.143	C.L.F.	57	27.50		84.50	110	
	0430	16″ O.C.		8	1		57	24		81	104	
	0440	24″ O.C.		10	.800		57	19.20		76.20	95.50	
	0450	18 ga x 2″ wide, studs 12″ O.C.		6	1.333		79	32		111	142	
	0460	16″ O.C.		7	1.143		79	27.50		106.50	134	
	0470	24″ O.C.	↓	8	1	↓	79	24		103	128	
120	0010	**BRIDGING**, solid between studs w/ 1-1/4″ leg track, per stud bay										120
	0200	Studs 12″ O.C., 18 ga x 2-1/2″ wide	1 Carp	125	.064	Ea.	.97	1.54		2.51	3.67	
	0210	3-5/8″ wide		120	.067		1.17	1.60		2.77	4.01	
	0220	4″ wide		120	.067		1.24	1.60		2.84	4.08	
	0230	6″ wide		115	.070		1.61	1.67		3.28	4.61	
	0240	8″ wide		110	.073		2.05	1.75		3.80	5.20	
	0300	16 ga x 2-1/2″ wide		115	.070		1.21	1.67		2.88	4.17	
	0310	3-5/8″ wide		110	.073		1.47	1.75		3.22	4.57	
	0320	4″ wide		110	.073		1.57	1.75		3.32	4.69	
	0330	6″ wide		105	.076		2.02	1.83		3.85	5.30	
	0340	8″ wide		100	.080		2.57	1.92		4.49	6.10	
	1200	Studs 16″ O.C., 18 ga x 2-1/2″ wide		125	.064		1.24	1.54		2.78	3.97	
	1210	3-5/8″ wide		120	.067		1.50	1.60		3.10	4.37	
	1220	4″ wide		120	.067		1.58	1.60		3.18	4.46	
	1230	6″ wide		115	.070		2.06	1.67		3.73	5.10	
	1240	8″ wide		110	.073		2.62	1.75		4.37	5.85	
	1300	16 ga x 2-1/2″ wide		115	.070		1.55	1.67		3.22	4.55	
	1310	3-5/8″ wide		110	.073		1.88	1.75		3.63	5.05	
	1320	4″ wide		110	.073		2.01	1.75		3.76	5.15	
	1330	6″ wide		105	.076		2.59	1.83		4.42	5.95	
	1340	8″ wide		100	.080		3.30	1.92		5.22	6.90	
	2200	Studs 24″ O.C., 18 ga x 2-1/2″ wide		125	.064		1.79	1.54		3.33	4.58	
	2210	3-5/8″ wide		120	.067		2.17	1.60		3.77	5.10	
	2220	4″ wide		120	.067		2.29	1.60		3.89	5.25	
	2230	6″ wide		115	.070		2.98	1.67		4.65	6.10	
	2240	8″ wide		110	.073		3.80	1.75		5.55	7.15	
	2300	16 ga x 2-1/2″ wide		115	.070		2.24	1.67		3.91	5.30	
	2310	3-5/8″ wide		110	.073		2.72	1.75		4.47	5.95	
	2320	4″ wide		110	.073		2.91	1.75		4.66	6.15	
	2330	6″ wide		105	.076		3.75	1.83		5.58	7.20	
	2340	8″ wide	↓	100	.080	↓	4.77	1.92		6.69	8.50	
	3000	Continuous bridging, per row										

	05410	Load-Bearing Metal Studs	CREW	DAILY OUTPUT	LABOR-HOURS	UNIT	2005 BARE COSTS MAT.	LABOR	EQUIP.	TOTAL	TOTAL INCL O&P
120	3100	16 ga x 1-1/2" channel thru studs 12" O.C.	1 Carp	6	1.333	C.L.F.	54	32		86	114
	3110	16" O.C.		7	1.143		54	27.50		81.50	106
	3120	24" O.C.		8.80	.909		54	22		76	96.50
	4100	2" x 2" angle x 18 ga, studs 12" O.C.		7	1.143		79	27.50		106.50	134
	4110	16" O.C.		9	.889		79	21.50		100.50	123
	4120	24" O.C.		12	.667		79	16		95	114
	4200	16 ga, studs 12" O.C.		5	1.600		100	38.50		138.50	175
	4210	16" O.C.		7	1.143		100	27.50		127.50	157
	4220	24" O.C.		10	.800		100	19.20		119.20	143
300	0010	**FRAMING,** boxed headers/beams									
	0200	Double, 18 ga x 6" deep	2 Carp	220	.073	L.F.	5.60	1.75		7.35	9.10
	0210	8" deep		210	.076		6.20	1.83		8.03	9.95
	0220	10" deep		200	.080		7.55	1.92		9.47	11.55
	0230	12 " deep		190	.084		8.30	2.02		10.32	12.55
	0300	16 ga x 8" deep		180	.089		7.15	2.13		9.28	11.45
	0310	10" deep		170	.094		8.60	2.26		10.86	13.30
	0320	12 " deep		160	.100		9.35	2.40		11.75	14.40
	0400	14 ga x 10" deep		140	.114		10	2.74		12.74	15.65
	0410	12 " deep		130	.123		10.95	2.95		13.90	17.05
	1210	Triple, 18 ga x 8" deep		170	.094		9	2.26		11.26	13.75
	1220	10" deep		165	.097		10.85	2.33		13.18	15.85
	1230	12 " deep		160	.100		11.95	2.40		14.35	17.25
	1300	16 ga x 8" deep		145	.110		10.40	2.65		13.05	15.90
	1310	10" deep		140	.114		12.40	2.74		15.14	18.30
	1320	12 " deep		135	.119		13.55	2.84		16.39	19.75
	1400	14 ga x 10" deep		115	.139		13.65	3.34		16.99	20.50
	1410	12 " deep		110	.145		15.10	3.49		18.59	22.50
400	0010	**FRAMING, STUD WALLS** w/ top & bottom track, no openings,									
	0020	headers, beams, bridging or bracing									
	4100	8' high walls, 18 ga x 2-1/2" wide, studs 12" O.C.	2 Carp	54	.296	L.F.	9.45	7.10		16.55	22.50
	4110	16" O.C.		77	.208		7.55	4.99		12.54	16.75
	4120	24" O.C.		107	.150		5.65	3.59		9.24	12.30
	4130	3-5/8" wide, studs 12" O.C.		53	.302		11.20	7.25		18.45	24.50
	4140	16" O.C.		76	.211		8.95	5.05		14	18.45
	4150	24" O.C.		105	.152		6.70	3.66		10.36	13.60
	4160	4" wide, studs 12" O.C.		52	.308		11.70	7.40		19.10	25.50
	4170	16" O.C.		74	.216		9.35	5.20		14.55	19.10
	4180	24" O.C.		103	.155		7.05	3.73		10.78	14.10
	4190	6" wide, studs 12" O.C.		51	.314		14.90	7.55		22.45	29
	4200	16" O.C.		73	.219		11.95	5.25		17.20	22
	4210	24" O.C.		101	.158		9	3.80		12.80	16.35
	4220	8" wide, studs 12" O.C.		50	.320		18.20	7.70		25.90	33
	4230	16" O.C.		72	.222		14.65	5.35		20	25
	4240	24" O.C.		100	.160		11.05	3.84		14.89	18.65
	4300	16 ga x 2-1/2" wide, studs 12" O.C.		47	.340		11.05	8.15		19.20	26
	4310	16" O.C.		68	.235		8.75	5.65		14.40	19.20
	4320	24" O.C.		94	.170		6.45	4.09		10.54	14.05
	4330	3-5/8" wide, studs 12" O.C.		46	.348		13.25	8.35		21.60	29
	4340	16" O.C.		66	.242		10.50	5.80		16.30	21.50
	4350	24" O.C.		92	.174		7.75	4.17		11.92	15.60
	4360	4" wide, studs 12" O.C.		45	.356		13.95	8.55		22.50	30
	4370	16" O.C.		65	.246		11.05	5.90		16.95	22
	4380	24" O.C.		90	.178		8.15	4.27		12.42	16.20
	4390	6" wide, studs 12" O.C.		44	.364		17.55	8.75		26.30	34
	4400	16" O.C.		64	.250		13.90	6		19.90	25.50

05410 | Load-Bearing Metal Studs

			CREW	DAILY OUTPUT	LABOR-HOURS	UNIT	2005 BARE COSTS MAT.	LABOR	EQUIP.	TOTAL	TOTAL INCL O&P	
400	4410	24" O.C.	2 Carp	88	.182	L.F.	10.30	4.36		14.66	18.75	400
	4420	8" wide, studs 12" O.C.		43	.372		21.50	8.95		30.45	38.50	
	4430	16" O.C.		63	.254		17.15	6.10		23.25	29	
	4440	24" O.C.		86	.186		12.75	4.47		17.22	21.50	
	5100	10' high walls, 18 ga x 2-1/2" wide, studs 12" O.C.		54	.296		11.35	7.10		18.45	24.50	
	5110	16" O.C.		77	.208		8.95	4.99		13.94	18.30	
	5120	24" O.C.		107	.150		6.60	3.59		10.19	13.35	
	5130	3-5/8" wide, studs 12" O.C.		53	.302		13.45	7.25		20.70	27	
	5140	16" O.C.		76	.211		10.65	5.05		15.70	20.50	
	5150	24" O.C.		105	.152		7.85	3.66		11.51	14.80	
	5160	4" wide, studs 12" O.C.		52	.308		14.05	7.40		21.45	28	
	5170	16" O.C.		74	.216		11.15	5.20		16.35	21	
	5180	24" O.C.		103	.155		8.20	3.73		11.93	15.35	
	5190	6" wide, studs 12" O.C.		51	.314		17.85	7.55		25.40	32.50	
	5200	16" O.C.		73	.219		14.15	5.25		19.40	24.50	
	5210	24" O.C.		101	.158		10.45	3.80		14.25	17.95	
	5220	8" wide, studs 12" O.C.		50	.320		22	7.70		29.70	37	
	5230	16" O.C.		72	.222		17.30	5.35		22.65	28	
	5240	24" O.C.		100	.160		12.85	3.84		16.69	20.50	
	5300	16 ga x 2-1/2" wide, studs 12" O.C.		47	.340		13.35	8.15		21.50	28.50	
	5310	16" O.C.		68	.235		10.45	5.65		16.10	21	
	5320	24" O.C.		94	.170		7.60	4.09		11.69	15.30	
	5330	3-5/8" wide, studs 12" O.C.		46	.348		16.05	8.35		24.40	32	
	5340	16" O.C.		66	.242		12.60	5.80		18.40	24	
	5350	24" O.C.		92	.174		9.15	4.17		13.32	17.15	
	5360	4" wide, studs 12" O.C.		45	.356		16.85	8.55		25.40	33	
	5370	16" O.C.		65	.246		13.25	5.90		19.15	24.50	
	5380	24" O.C.		90	.178		9.60	4.27		13.87	17.80	
	5390	6" wide, studs 12" O.C.		44	.364		21	8.75		29.75	38.50	
	5400	16" O.C.		64	.250		16.65	6		22.65	28.50	
	5410	24" O.C.		88	.182		12.10	4.36		16.46	20.50	
	5420	8" wide, studs 12" O.C.		43	.372		26	8.95		34.95	43.50	
	5430	16" O.C.		63	.254		20.50	6.10		26.60	33	
	5440	24" O.C.		86	.186		14.95	4.47		19.42	24	
	6190	12' high walls, 18 ga x 6" wide, studs 12" O.C.		41	.390		21	9.35		30.35	39	
	6200	16" O.C.		58	.276		16.40	6.60		23	29.50	
	6210	24" O.C.		81	.198		11.95	4.74		16.69	21	
	6220	8" wide, studs 12" O.C.		40	.400		25.50	9.60		35.10	44.50	
	6230	16" O.C.		57	.281		20	6.75		26.75	33.50	
	6240	24" O.C.		80	.200		14.65	4.80		19.45	24.50	
	6390	16 ga x 6" wide, studs 12" O.C.		35	.457		25	10.95		35.95	46	
	6400	16" O.C.		51	.314		19.35	7.55		26.90	34.50	
	6410	24" O.C.		70	.229		13.90	5.50		19.40	24.50	
	6420	8" wide, studs 12" O.C.		34	.471		30.50	11.30		41.80	52.50	
	6430	16" O.C.		50	.320		24	7.70		31.70	39	
	6440	24" O.C.		69	.232		17.15	5.55		22.70	28.50	
	6530	14 ga x 3-5/8" wide, studs 12" O.C.		34	.471		23.50	11.30		34.80	45	
	6540	16" O.C.		48	.333		18.35	8		26.35	33.50	
	6550	24" O.C.		65	.246		13.15	5.90		19.05	24.50	
	6560	4" wide, studs 12" O.C.		33	.485		25	11.65		36.65	47.50	
	6570	16" O.C.		47	.340		19.35	8.15		27.50	35.50	
	6580	24" O.C.		64	.250		13.90	6		19.90	25.50	
	6730	12 ga x 3-5/8" wide, studs 12" O.C.		31	.516		32.50	12.40		44.90	57	
	6740	16" O.C.		43	.372		25	8.95		33.95	42.50	
	6750	24" O.C.		59	.271		17.65	6.50		24.15	30.50	
	6760	4" wide, studs 12" O.C.	↓	30	.533	↓	35	12.80		47.80	60	

	05410 | Load-Bearing Metal Studs		CREW	DAILY OUTPUT	LABOR-HOURS	UNIT	2005 BARE COSTS MAT.	LABOR	EQUIP.	TOTAL	TOTAL INCL O&P
400	6770	16" O.C.	2 Carp	42	.381	L.F.	27	9.15		36.15	45
	6780	24" O.C.		58	.276		18.95	6.60		25.55	32.50
	7390	16' high walls, 16 ga x 6" wide, studs 12" O.C.		33	.485		32	11.65		43.65	55
	7400	16" O.C.		48	.333		25	8		33	41
	7410	24" O.C.		67	.239		17.55	5.75		23.30	29
	7420	8" wide, studs 12" O.C.		32	.500		39.50	12		51.50	63.50
	7430	16" O.C.		47	.340		30.50	8.15		38.65	47.50
	7440	24" O.C.		66	.242		21.50	5.80		27.30	33.50
	7560	14 ga x 4" wide, studs 12" O.C.		31	.516		32	12.40		44.40	56.50
	7570	16" O.C.		45	.356		25	8.55		33.55	42
	7580	24" O.C.		61	.262		17.55	6.30		23.85	30
	7590	6" wide, studs 12" O.C.		30	.533		40.50	12.80		53.30	66
	7600	16" O.C.		44	.364		31.50	8.75		40.25	49.50
	7610	24" O.C.		60	.267		22	6.40		28.40	35.50
	7760	12 ga x 4" wide, studs 12" O.C.		29	.552		45.50	13.25		58.75	72.50
	7770	16" O.C.		40	.400		35	9.60		44.60	55
	7780	24" O.C.		55	.291		24.50	7		31.50	38.50
	7790	6" wide, studs 12" O.C.		28	.571		57.50	13.70		71.20	86.50
	7800	16" O.C.		39	.410		44	9.85		53.85	65
	7810	24" O.C.		54	.296		30.50	7.10		37.60	45.50
	8590	20' high walls, 14 ga x 6" wide, studs 12" O.C.		29	.552		49.50	13.25		62.75	77
	8600	16" O.C.		42	.381		38	9.15		47.15	57.50
	8610	24" O.C.		57	.281		27	6.75		33.75	41
	8620	8" wide, studs 12" O.C.		28	.571		60.50	13.70		74.20	90.50
	8630	16" O.C.		41	.390		47	9.35		56.35	67.50
	8640	24" O.C.		56	.286		33	6.85		39.85	47.50
	8790	12 ga x 6" wide, studs 12" O.C.		27	.593		71	14.20		85.20	102
	8800	16" O.C.		37	.432		54	10.40		64.40	77
	8810	24" O.C.		51	.314		37.50	7.55		45.05	54
	8820	8" wide, studs 12" O.C.		26	.615		86.50	14.75		101.25	120
	8830	16" O.C.		36	.444		66	10.65		76.65	90.50
	8840	24" O.C.	↓	50	.320	↓	45.50	7.70		53.20	63

	05420 | Cold-Formed Metal Joists		CREW	DAILY OUTPUT	LABOR-HOURS	UNIT	MAT.	LABOR	EQUIP.	TOTAL	TOTAL INCL O&P
100	0010	**BRACING**, continuous, per row, top & bottom									
	0120	Flat strap, 20 ga x 2" wide, joists at 12" O.C.	1 Carp	4.67	1.713	C.L.F.	60	41		101	136
	0130	16" O.C.		5.33	1.501		57.50	36		93.50	125
	0140	24" O.C.		6.66	1.201		55.50	29		84.50	110
	0150	18 ga x 2" wide, joists at 12" O.C.		4	2		78.50	48		126.50	168
	0160	16" O.C.		4.67	1.713		77	41		118	155
	0170	24" O.C.	↓	5.33	1.501	↓	75.50	36		111.50	144
120	0010	**BRIDGING**, solid between joists w/ 1-1/4" leg track, per joist bay									
	0230	Joists 12" O.C., 18 ga track x 6" wide	1 Carp	80	.100	Ea.	1.61	2.40		4.01	5.85
	0240	8" wide		75	.107		2.05	2.56		4.61	6.60
	0250	10" wide		70	.114		2.52	2.74		5.26	7.45
	0260	12" wide		65	.123		2.92	2.95		5.87	8.20
	0330	16 ga track x 6" wide		70	.114		2.02	2.74		4.76	6.90
	0340	8" wide		65	.123		2.57	2.95		5.52	7.85
	0350	10" wide		60	.133		3.17	3.20		6.37	8.95
	0360	12" wide		55	.145		3.64	3.49		7.13	9.95
	0440	14 ga track x 8" wide		60	.133		3.24	3.20		6.44	9
	0450	10" wide		55	.145		3.98	3.49		7.47	10.30
	0460	12" wide		50	.160		4.59	3.84		8.43	11.55
	0550	12 ga track x 10" wide		45	.178		5.85	4.27		10.12	13.70
	0560	12" wide	↓	40	.200	↓	6.60	4.80		11.40	15.45

05420 | Cold-Formed Metal Joists

				DAILY	LABOR-		2005 BARE COSTS				TOTAL	
		05420 Cold-Formed Metal Joists	CREW	OUTPUT	HOURS	UNIT	MAT.	LABOR	EQUIP.	TOTAL	INCL O&P	
120	1230	16" O.C., 18 ga track x 6" wide	1 Carp	80	.100	Ea.	2.06	2.40		4.46	6.35	120
	1240	8" wide		75	.107		2.62	2.56		5.18	7.25	
	1250	10" wide		70	.114		3.23	2.74		5.97	8.20	
	1260	12" wide		65	.123		3.75	2.95		6.70	9.10	
	1330	16 ga track x 6" wide		70	.114		2.59	2.74		5.33	7.50	
	1340	8" wide		65	.123		3.30	2.95		6.25	8.65	
	1350	10" wide		60	.133		4.06	3.20		7.26	9.90	
	1360	12" wide		55	.145		4.67	3.49		8.16	11.10	
	1440	14 ga track x 8" wide		60	.133		4.16	3.20		7.36	10	
	1450	10" wide		55	.145		5.10	3.49		8.59	11.55	
	1460	12" wide		50	.160		5.90	3.84		9.74	13	
	1550	12 ga track x 10" wide		45	.178		7.50	4.27		11.77	15.50	
	1560	12" wide		40	.200		8.50	4.80		13.30	17.50	
	2230	24" O.C., 18 ga track x 6" wide		80	.100		2.98	2.40		5.38	7.35	
	2240	8" wide		75	.107		3.80	2.56		6.36	8.50	
	2250	10" wide		70	.114		4.68	2.74		7.42	9.80	
	2260	12" wide		65	.123		5.40	2.95		8.35	10.95	
	2330	16 ga track x 6" wide		70	.114		3.75	2.74		6.49	8.80	
	2340	8" wide		65	.123		4.77	2.95		7.72	10.25	
	2350	10" wide		60	.133		5.85	3.20		9.05	11.90	
	2360	12" wide		55	.145		6.75	3.49		10.24	13.40	
	2440	14 ga track x 8" wide		60	.133		6	3.20		9.20	12.05	
	2450	10" wide		55	.145		7.40	3.49		10.89	14.05	
	2460	12" wide		50	.160		8.50	3.84		12.34	15.85	
	2550	12 ga track x 10" wide		45	.178		10.85	4.27		15.12	19.15	
	2560	12" wide	↓	40	.200	↓	12.25	4.80		17.05	21.50	
200	0010	**FRAMING, BAND JOIST** (track) fastened to bearing wall										200
	0220	18 ga track x 6" deep	2 Carp	1,000	.016	L.F.	1.31	.38		1.69	2.09	
	0230	8" deep		920	.017		1.67	.42		2.09	2.55	
	0240	10" deep		860	.019		2.06	.45		2.51	3.02	
	0320	16 ga track x 6" deep		900	.018		1.65	.43		2.08	2.53	
	0330	8" deep		840	.019		2.10	.46		2.56	3.09	
	0340	10" deep		780	.021		2.58	.49		3.07	3.68	
	0350	12" deep		740	.022		2.97	.52		3.49	4.15	
	0430	14 ga track x 8" deep		750	.021		2.65	.51		3.16	3.78	
	0440	10" deep		720	.022		3.24	.53		3.77	4.48	
	0450	12" deep		700	.023		3.75	.55		4.30	5.05	
	0540	12 ga track x 10" deep		670	.024		4.77	.57		5.34	6.20	
	0550	12" deep	↓	650	.025	↓	5.40	.59		5.99	6.95	
300	0010	**FRAMING, BOXED HEADERS/BEAMS**										300
	0200	Double, 18 ga x 6" deep	2 Carp	220	.073	L.F.	5.60	1.75		7.35	9.10	
	0210	8" deep		210	.076		6.20	1.83		8.03	9.95	
	0220	10" deep		200	.080		7.55	1.92		9.47	11.55	
	0230	12" deep		190	.084		8.30	2.02		10.32	12.55	
	0300	16 ga x 8" deep		180	.089		7.15	2.13		9.28	11.45	
	0310	10" deep		170	.094		8.60	2.26		10.86	13.30	
	0320	12" deep		160	.100		9.35	2.40		11.75	14.40	
	0400	14 ga x 10" deep		140	.114		10	2.74		12.74	15.65	
	0410	12" deep		130	.123		10.95	2.95		13.90	17.05	
	0500	12 ga x 10" deep		110	.145		13.20	3.49		16.69	20.50	
	0510	12" deep		100	.160		14.65	3.84		18.49	22.50	
	1210	Triple, 18 ga x 8" deep		170	.094		9	2.26		11.26	13.75	
	1220	10" deep		165	.097		10.85	2.33		13.18	15.85	
	1230	12" deep		160	.100		11.95	2.40		14.35	17.25	
	1300	16 ga x 8" deep	↓	145	.110	↓	10.40	2.65		13.05	15.90	

	05420	Cold-Formed Metal Joists	CREW	DAILY OUTPUT	LABOR-HOURS	UNIT	2005 BARE COSTS MAT.	LABOR	EQUIP.	TOTAL	TOTAL INCL O&P
300	1310	10" deep	2 Carp	140	.114	L.F.	12.40	2.74		15.14	18.30
	1320	12" deep		135	.119		13.55	2.84		16.39	19.75
	1400	14 ga x 10" deep		115	.139		14.50	3.34		17.84	21.50
	1410	12" deep		110	.145		15.95	3.49		19.44	23.50
	1500	12 ga x 10" deep		90	.178		19.35	4.27		23.62	29
	1510	12" deep		85	.188		21.50	4.52		26.02	31
410	0010	**FRAMING, JOISTS,** no band joists (track), web stiffeners, headers,									
	0020	beams, bridging or bracing									
	0030	Joists (2" flange) and fasteners, materials only									
	0220	18 ga x 6" deep				L.F.	1.73			1.73	1.91
	0230	8" deep					2.06			2.06	2.26
	0240	10" deep					2.42			2.42	2.66
	0320	16 ga x 6" deep					2.12			2.12	2.33
	0330	8" deep					2.54			2.54	2.80
	0340	10" deep					2.97			2.97	3.27
	0350	12" deep					3.37			3.37	3.71
	0430	14 ga x 8" deep					3.21			3.21	3.53
	0440	10" deep					3.71			3.71	4.08
	0450	12" deep					4.21			4.21	4.63
	0540	12 ga x 10" deep					5.40			5.40	5.95
	0550	12" deep					6.15			6.15	6.75
	1010	Installation of joists to band joists, beams & headers, labor only									
	1220	18 ga x 6" deep	2 Carp	110	.145	Ea.		3.49		3.49	5.95
	1230	8" deep		90	.178			4.27		4.27	7.25
	1240	10" deep		80	.200			4.80		4.80	8.15
	1320	16 ga x 6" deep		95	.168			4.04		4.04	6.85
	1330	8" deep		70	.229			5.50		5.50	9.30
	1340	10" deep		60	.267			6.40		6.40	10.85
	1350	12" deep		55	.291			7		7	11.85
	1430	14 ga x 8" deep		65	.246			5.90		5.90	10.05
	1440	10" deep		45	.356			8.55		8.55	14.50
	1450	12" deep		35	.457			10.95		10.95	18.65
	1540	12 ga x 10" deep		40	.400			9.60		9.60	16.30
	1550	12" deep		30	.533			12.80		12.80	21.50
500	0010	**FRAMING, WEB STIFFENERS** at joist bearing, fabricated from									
	0020	stud piece (1-5/8" flange) to stiffen joist (2" flange)									
	2120	For 6" deep joist, with 18 ga x 2-1/2" stud	1 Carp	120	.067	Ea.	2.09	1.60		3.69	5
	2130	3-5/8" stud		110	.073		2.30	1.75		4.05	5.50
	2140	4" stud		105	.076		2.22	1.83		4.05	5.55
	2150	6" stud		100	.080		2.44	1.92		4.36	5.95
	2160	8" stud		95	.084		2.51	2.02		4.53	6.20
	2220	8" deep joist, with 2-1/2" stud		120	.067		2.29	1.60		3.89	5.25
	2230	3-5/8" stud		110	.073		2.48	1.75		4.23	5.70
	2240	4" stud		105	.076		2.43	1.83		4.26	5.75
	2250	6" stud		100	.080		2.68	1.92		4.60	6.20
	2260	8" stud		95	.084		2.88	2.02		4.90	6.60
	2320	10" deep joist, with 2-1/2" stud		110	.073		3.23	1.75		4.98	6.50
	2330	3-5/8" stud		100	.080		3.53	1.92		5.45	7.15
	2340	4" stud		95	.084		3.50	2.02		5.52	7.30
	2350	6" stud		90	.089		3.81	2.13		5.94	7.80
	2360	8" stud		85	.094		3.86	2.26		6.12	8.10
	2420	12" deep joist, with 2-1/2" stud		110	.073		3.42	1.75		5.17	6.70
	2430	3-5/8" stud		100	.080		3.70	1.92		5.62	7.35
	2440	4" stud		95	.084		3.63	2.02		5.65	7.40
	2450	6" stud		90	.089		4	2.13		6.13	8
	2460	8" stud		85	.094		4.30	2.26		6.56	8.55

05420 | Cold-Formed Metal Joists

	05420 Cold-Formed Metal Joists	CREW	DAILY OUTPUT	LABOR-HOURS	UNIT	2005 BARE COSTS MAT.	LABOR	EQUIP.	TOTAL	TOTAL INCL O&P	
3130	For 6″ deep joist, with 16 ga x 3-5/8″ stud	1 Carp	100	.080	Ea.	2.42	1.92		4.34	5.90	500
3140	4″ stud		95	.084		2.39	2.02		4.41	6.05	
3150	6″ stud		90	.089		2.62	2.13		4.75	6.50	
3160	8″ stud		85	.094		2.76	2.26		5.02	6.90	
3230	8″ deep joist, with 3-5/8″ stud		100	.080		2.68	1.92		4.60	6.20	
3240	4″ stud		95	.084		2.62	2.02		4.64	6.30	
3250	6″ stud		90	.089		2.91	2.13		5.04	6.80	
3260	8″ stud		85	.094		3.11	2.26		5.37	7.25	
3330	10″ deep joist, with 3-5/8″ stud		85	.094		3.67	2.26		5.93	7.85	
3340	4″ stud		80	.100		3.73	2.40		6.13	8.20	
3350	6″ stud		75	.107		4.06	2.56		6.62	8.80	
3360	8″ stud		70	.114		4.22	2.74		6.96	9.30	
3430	12″ deep joist, with 3-5/8″ stud		85	.094		4	2.26		6.26	8.25	
3440	4″ stud		80	.100		3.92	2.40		6.32	8.40	
3450	6″ stud		75	.107		4.34	2.56		6.90	9.15	
3460	8″ stud		70	.114		4.64	2.74		7.38	9.75	
4230	For 8″ deep joist, with 14 ga x 3-5/8″ stud		90	.089		3.48	2.13		5.61	7.45	
4240	4″ stud		85	.094		3.54	2.26		5.80	7.75	
4250	6″ stud		80	.100		3.84	2.40		6.24	8.30	
4260	8″ stud		75	.107		4.11	2.56		6.67	8.85	
4330	10″ deep joist, with 3-5/8″ stud		75	.107		4.88	2.56		7.44	9.70	
4340	4″ stud		70	.114		4.83	2.74		7.57	9.95	
4350	6″ stud		65	.123		5.30	2.95		8.25	10.85	
4360	8″ stud		60	.133		5.55	3.20		8.75	11.55	
4430	12″ deep joist, with 3-5/8″ stud		75	.107		5.20	2.56		7.76	10.05	
4440	4″ stud		70	.114		5.30	2.74		8.04	10.45	
4450	6″ stud		65	.123		5.75	2.95		8.70	11.30	
4460	8″ stud		60	.133		6.15	3.20		9.35	12.20	
5330	For 10″ deep joist, with 12 ga x 3-5/8″ stud		65	.123		5.15	2.95		8.10	10.65	
5340	4″ stud		60	.133		5.30	3.20		8.50	11.30	
5350	6″ stud		55	.145		5.85	3.49		9.34	12.35	
5360	8″ stud		50	.160		6.40	3.84		10.24	13.55	
5430	12″ deep joist, with 3-5/8″ stud		65	.123		5.70	2.95		8.65	11.25	
5440	4″ stud		60	.133		5.60	3.20		8.80	11.60	
5450	6″ stud		55	.145		6.35	3.49		9.84	12.95	
5460	8″ stud	↓	50	.160	↓	7.35	3.84		11.19	14.55	

05425 | Cold-Formed Roof Framing

	05425 Cold-Formed Roof Framing	CREW	DAILY OUTPUT	LABOR-HOURS	UNIT	MAT.	LABOR	EQUIP.	TOTAL	TOTAL INCL O&P	
0010	**FRAMING, BRACING**										100
0020	Continuous bracing, per row										
0100	16 ga x 1-1/2″ channel thru rafters/trusses @ 16″ O.C.	1 Carp	4.50	1.778	C.L.F.	54	42.50		96.50	132	
0120	24″ O.C.		6	1.333		54	32		86	114	
0300	2″ x 2″ angle x 18 ga, rafters/trusses @ 16″ O.C.		6	1.333		79	32		111	142	
0320	24″ O.C.		8	1		79	24		103	128	
0400	16 ga, rafters/trusses @ 16″ O.C.		4.50	1.778		100	42.50		142.50	183	
0420	24″ O.C.	↓	6.50	1.231	↓	100	29.50		129.50	160	
0010	**FRAMING, BRIDGING**										200
0020	Solid, between rafters w/ 1-1/4″ leg track, per rafter bay										
1200	Rafters 16″ O.C., 18 ga x 4″ deep	1 Carp	60	.133	Ea.	1.58	3.20		4.78	7.20	
1210	6″ deep		57	.140		2.06	3.37		5.43	7.95	
1220	8″ deep		55	.145		2.62	3.49		6.11	8.85	
1230	10″ deep		52	.154		3.23	3.69		6.92	9.80	
1240	12″ deep		50	.160		3.75	3.84		7.59	10.60	
2200	24″ O.C., 18 ga x 4″ deep		60	.133		2.29	3.20		5.49	7.95	
2210	6″ deep		57	.140		2.98	3.37		6.35	9	
2220	8″ deep	↓	55	.145	↓	3.80	3.49		7.29	10.10	

	05425 | Cold-Formed Roof Framing		CREW	DAILY OUTPUT	LABOR-HOURS	UNIT	2005 BARE COSTS MAT.	LABOR	EQUIP.	TOTAL	TOTAL INCL O&P
200	2230	10″ deep	1 Carp	52	.154	Ea.	4.68	3.69		8.37	11.40
	2240	12″ deep		50	.160		5.40	3.84		9.24	12.45
500	0010	**FRAMING, PARAPETS**									
	0100	3′ high installed on 1st story, 18 ga x 4″ wide studs, 12″ O.C.	2 Carp	100	.160	L.F.	5.85	3.84		9.69	12.95
	0110	16″ O.C.		150	.107		4.98	2.56		7.54	9.85
	0120	24″ O.C.		200	.080		4.11	1.92		6.03	7.80
	0200	6″ wide studs, 12″ O.C.		100	.160		7.50	3.84		11.34	14.75
	0210	16″ O.C.		150	.107		6.40	2.56		8.96	11.40
	0220	24″ O.C.		200	.080		5.30	1.92		7.22	9.05
	1100	Installed on 2nd story, 18 ga x 4″ wide studs, 12″ O.C.		95	.168		5.85	4.04		9.89	13.30
	1110	16″ O.C.		145	.110		4.98	2.65		7.63	10
	1120	24″ O.C.		190	.084		4.11	2.02		6.13	7.95
	1200	6″ wide studs, 12″ O.C.		95	.168		7.50	4.04		11.54	15.10
	1210	16″ O.C.		145	.110		6.40	2.65		9.05	11.55
	1220	24″ O.C.		190	.084		5.30	2.02		7.32	9.25
	2100	Installed on gable, 18 ga x 4″ wide studs, 12″ O.C.		85	.188		5.85	4.52		10.37	14.10
	2110	16″ O.C.		130	.123		4.98	2.95		7.93	10.50
	2120	24″ O.C.		170	.094		4.11	2.26		6.37	8.35
	2200	6″ wide studs, 12″ O.C.		85	.188		7.50	4.52		12.02	15.90
	2210	16″ O.C.		130	.123		6.40	2.95		9.35	12.05
	2220	24″ O.C.		170	.094		5.30	2.26		7.56	9.65
550	0010	**FRAMING, ROOF RAFTERS**									
	0100	Boxed ridge beam, double, 18 ga x 6″ deep	2 Carp	160	.100	L.F.	5.60	2.40		8	10.25
	0110	8″ deep		150	.107		6.20	2.56		8.76	11.20
	0120	10″ deep		140	.114		7.55	2.74		10.29	12.95
	0130	12″ deep		130	.123		8.30	2.95		11.25	14.10
	0200	16 ga x 6″ deep		150	.107		6.35	2.56		8.91	11.35
	0210	8″ deep		140	.114		7.15	2.74		9.89	12.50
	0220	10″ deep		130	.123		8.60	2.95		11.55	14.45
	0230	12″ deep		120	.133		9.35	3.20		12.55	15.75
	1100	Rafters, 2″ flange, material only, 18 ga x 6″ deep					1.73			1.73	1.91
	1110	8″ deep					2.06			2.06	2.26
	1120	10″ deep					2.42			2.42	2.66
	1130	12″ deep					2.81			2.81	3.10
	1200	16 ga x 6″ deep					2.12			2.12	2.33
	1210	8″ deep					2.54			2.54	2.80
	1220	10″ deep					2.97			2.97	3.27
	1230	12″ deep					3.37			3.37	3.71
	2100	Installation only, ordinary rafter to 4:12 pitch, 18 ga x 6″ deep	2 Carp	35	.457	Ea.		10.95		10.95	18.65
	2110	8″ deep		30	.533			12.80		12.80	21.50
	2120	10″ deep		25	.640			15.35		15.35	26
	2130	12″ deep		20	.800			19.20		19.20	32.50
	2200	16 ga x 6″ deep		30	.533			12.80		12.80	21.50
	2210	8″ deep		25	.640			15.35		15.35	26
	2220	10″ deep		20	.800			19.20		19.20	32.50
	2230	12″ deep		15	1.067			25.50		25.50	43.50
	8100	Add to labor, ordinary rafters on steep roofs						25%			
	8110	Dormers & complex roofs						50%			
	8200	Hip & valley rafters to 4:12 pitch						25%			
	8210	Steep roofs						50%			
	8220	Dormers & complex roofs						75%			
	8300	Hip & valley jack rafters to 4:12 pitch						50%			
	8310	Steep roofs						75%			
	8320	Dormers & complex roofs						100%			

05425 | Cold-Formed Roof Framing

	Line	Description	Crew	Daily Output	Labor-Hours	Unit	2005 Bare Costs Mat.	Labor	Equip.	Total	Total Incl O&P
600	0010	**FRAMING, ROOF TRUSSES**									
	0020	Fabrication of trusses on ground, Fink (W) or King Post, to 4:12 pitch									
	0120	18 ga x 4" chords, 16' span	2 Carp	12	1.333	Ea.	65.50	32		97.50	127
	0130	20' span		11	1.455		82	35		117	150
	0140	24' span		11	1.455		98.50	35		133.50	168
	0150	28' span		10	1.600		115	38.50		153.50	191
	0160	32' span		10	1.600		131	38.50		169.50	209
	0250	6" chords, 28' span		9	1.778		145	42.50		187.50	233
	0260	32' span		9	1.778		166	42.50		208.50	255
	0270	36' span		8	2		186	48		234	287
	0280	40' span		8	2		207	48		255	310
	1120	5:12 to 8:12 pitch, 18 ga x 4" chords, 16' span		10	1.600		75	38.50		113.50	148
	1130	20' span		9	1.778		93.50	42.50		136	176
	1140	24' span		9	1.778		112	42.50		154.50	197
	1150	28' span		8	2		131	48		179	226
	1160	32' span		8	2		150	48		198	247
	1250	6" chords, 28' span		7	2.286		166	55		221	275
	1260	32' span		7	2.286		189	55		244	300
	1270	36' span		6	2.667		213	64		277	345
	1280	40' span		6	2.667		237	64		301	370
	2120	9:12 to 12:12 pitch, 18 ga x 4" chords, 16' span		8	2		93.50	48		141.50	185
	2130	20' span		7	2.286		117	55		172	222
	2140	24' span		7	2.286		140	55		195	247
	2150	28' span		6	2.667		164	64		228	289
	2160	32' span		6	2.667		187	64		251	315
	2250	6" chords, 28' span		5	3.200		207	77		284	360
	2260	32' span		5	3.200		237	77		314	390
	2270	36' span		4	4		266	96		362	455
	2280	40' span		4	4		296	96		392	490
	5120	Erection only of roof trusses, to 4:12 pitch, 16' span	F-6	48	.833			17.90	12.55	30.45	44
	5130	20' span		46	.870			18.70	13.10	31.80	46
	5140	24' span		44	.909			19.55	13.70	33.25	48
	5150	28' span		42	.952			20.50	14.35	34.85	50.50
	5160	32' span		40	1			21.50	15.05	36.55	52.50
	5170	36' span		38	1.053			22.50	15.85	38.35	55.50
	5180	40' span		36	1.111			24	16.75	40.75	59
	5220	5:12 to 8:12 pitch, 16' span		42	.952			20.50	14.35	34.85	50.50
	5230	20' span		40	1			21.50	15.05	36.55	52.50
	5240	24' span		38	1.053			22.50	15.85	38.35	55.50
	5250	28' span		36	1.111			24	16.75	40.75	59
	5260	32' span		34	1.176			25.50	17.75	43.25	62
	5270	36' span		32	1.250			27	18.85	45.85	66
	5280	40' span		30	1.333			28.50	20	48.50	70.50
	5320	9:12 to 12:12 pitch, 16' span		36	1.111			24	16.75	40.75	59
	5330	20' span		34	1.176			25.50	17.75	43.25	62
	5340	24' span		32	1.250			27	18.85	45.85	66
	5350	28' span		30	1.333			28.50	20	48.50	70.50
	5360	32' span		28	1.429			30.50	21.50	52	75.50
	5370	36' span		26	1.538			33	23	56	81
	5380	40' span		24	1.667			36	25	61	88
650	0010	**FRAMING, SOFFITS & CANOPIES**									
	0130	Continuous ledger track @ wall, studs @ 16" O.C., 18 ga x 4" wide	2 Carp	535	.030	L.F.	1.06	.72		1.78	2.38
	0140	6" wide		500	.032		1.38	.77		2.15	2.81
	0150	8" wide		465	.034		1.75	.83		2.58	3.32
	0160	10" wide		430	.037		2.16	.89		3.05	3.89
	0230	Studs @ 24" O.C., 18 ga x 4" wide		800	.020		1.01	.48		1.49	1.93

05400 | Cold-Formed Metal Framing

	05425	Cold-Formed Roof Framing	CREW	DAILY OUTPUT	LABOR-HOURS	UNIT	2005 BARE COSTS MAT.	LABOR	EQUIP.	TOTAL	TOTAL INCL O&P
650	0240	6″ wide	2 Carp	750	.021	L.F.	1.31	.51		1.82	2.31
	0250	8″ wide		700	.023		1.67	.55		2.22	2.77
	0260	10″ wide		650	.025		2.06	.59		2.65	3.26
	1000	Horizontal soffit and canopy members, material only									
	1030	1-5/8″ flange studs, 18 ga x 4″ deep				L.F.	1.40			1.40	1.54
	1040	6″ deep					1.78			1.78	1.95
	1050	8″ deep					2.15			2.15	2.36
	1140	2″ flange joists, 18 ga x 6″ deep					1.98			1.98	2.18
	1150	8″ deep					2.35			2.35	2.59
	1160	10″ deep					2.76			2.76	3.04
	4030	Installation only, 18 ga, 1-5/8″ flange x 4″ deep	2 Carp	130	.123	Ea.		2.95		2.95	5
	4040	6″ deep		110	.145			3.49		3.49	5.95
	4050	8″ deep		90	.178			4.27		4.27	7.25
	4140	2″ flange, 18 ga x 6″ deep		110	.145			3.49		3.49	5.95
	4150	8″ deep		90	.178			4.27		4.27	7.25
	4160	10″ deep		80	.200			4.80		4.80	8.15
	6010	Clips to attach facia to rafter tails, 2″ x 2″ x 18 ga angle	1 Carp	120	.067		.94	1.60		2.54	3.75
	6020	16 ga angle	″	100	.080		1.18	1.92		3.10	4.56

05500 | Metal Fabrications

	05517	Metal Stairs	CREW	DAILY OUTPUT	LABOR-HOURS	UNIT	2005 BARE COSTS MAT.	LABOR	EQUIP.	TOTAL	TOTAL INCL O&P
350	0010	**FIRE ESCAPE STAIRS**									
	0020	One story, disappearing, stainless steel	2 Sswk	20	.800	V.L.F.	175	20.50		195.50	232
	0100	Portable ladder				Ea.	53			53	58.50
700	0010	**STAIR**, shop fabricated, steel stringers, safety nosing on treads									
	1700	Pre-erected, steel pan tread, 3′-6″ wide, 2 line pipe rail	E-2	87	.552	Riser	211	14.20	15.25	240.45	277
	1800	With flat bar picket rail	″	87	.552		237	14.20	15.25	266.45	305
	1810	Spiral aluminum, 5′-0″ diameter, stock units	E-4	45	.711		232	18.50	1.81	252.31	293
	1820	Custom units		45	.711		440	18.50	1.81	460.31	520
	1900	Spiral, cast iron, 4′-0″ diameter, ornamental, minimum		45	.711		207	18.50	1.81	227.31	265
	1920	Maximum		25	1.280		283	33.50	3.25	319.75	380

	05520	Handrails & Railings	CREW	DAILY OUTPUT	LABOR-HOURS	UNIT	MAT.	LABOR	EQUIP.	TOTAL	TOTAL INCL O&P
700	0010	**RAILING, PIPE**, shop fabricated									
	0020	Aluminum, 2 rail, satin finish, 1-1/4″ diameter	E-4	160	.200	L.F.	14.60	5.20	.51	20.31	26.50
	0030	Clear anodized		160	.200		18.10	5.20	.51	23.81	30.50
	0040	Dark anodized		160	.200		20.50	5.20	.51	26.21	33
	0080	1-1/2″ diameter, satin finish		160	.200		17.50	5.20	.51	23.21	30
	0090	Clear anodized		160	.200		19.50	5.20	.51	25.21	32
	0100	Dark anodized		160	.200		21.50	5.20	.51	27.21	34.50
	0140	Aluminum, 3 rail, 1-1/4″ diam., satin finish		137	.234		22.50	6.10	.59	29.19	37
	0150	Clear anodized		137	.234		28	6.10	.59	34.69	43.50
	0160	Dark anodized		137	.234		31	6.10	.59	37.69	46.50
	0200	1-1/2″ diameter, satin finish		137	.234		27	6.10	.59	33.69	42
	0210	Clear anodized		137	.234		30.50	6.10	.59	37.19	46
	0220	Dark anodized		137	.234		33.50	6.10	.59	40.19	49.50
	0500	Steel, 2 rail, on stairs, primed, 1-1/4″ diameter		160	.200		12	5.20	.51	17.71	24

05520 Handrails & Railings		Crew	Daily Output	Labor-Hours	Unit	2005 Bare Costs Mat.	Labor	Equip.	Total	Total Incl O&P	
0520	1-1/2" diameter	E-4	160	.200	L.F.	13.20	5.20	.51	18.91	25	700
0540	Galvanized, 1-1/4" diameter		160	.200		16.65	5.20	.51	22.36	29	
0560	1-1/2" diameter		160	.200		18.65	5.20	.51	24.36	31	
0580	Steel, 3 rail, primed, 1-1/4" diameter		137	.234		17.90	6.10	.59	24.59	32	
0600	1-1/2" diameter		137	.234		19	6.10	.59	25.69	33.50	
0620	Galvanized, 1-1/4" diameter		137	.234		25	6.10	.59	31.69	40	
0640	1-1/2" diameter		137	.234		29.50	6.10	.59	36.19	44.50	
0700	Stainless steel, 2 rail, 1-1/4" diam. #4 finish		137	.234		43.50	6.10	.59	50.19	60	
0720	High polish		137	.234		69.50	6.10	.59	76.19	89	
0740	Mirror polish		137	.234		87.50	6.10	.59	94.19	108	
0760	Stainless steel, 3 rail, 1-1/2" diam., #4 finish		120	.267		65.50	6.95	.68	73.13	86	
0770	High polish		120	.267		108	6.95	.68	115.63	133	
0780	Mirror finish		120	.267		131	6.95	.68	138.63	158	
0900	Wall rail, alum. pipe, 1-1/4" diam., satin finish		213	.150		8.35	3.91	.38	12.64	17.15	
0905	Clear anodized		213	.150		10.20	3.91	.38	14.49	19.15	
0910	Dark anodized		213	.150		12.35	3.91	.38	16.64	21.50	
0915	1-1/2" diameter, satin finish		213	.150		9.25	3.91	.38	13.54	18.15	
0920	Clear anodized		213	.150		11.65	3.91	.38	15.94	21	
0925	Dark anodized		213	.150		14.35	3.91	.38	18.64	24	
0930	Steel pipe, 1-1/4" diameter, primed		213	.150		7.30	3.91	.38	11.59	15.95	
0935	Galvanized		213	.150		10.55	3.91	.38	14.84	19.55	
0940	1-1/2" diameter		176	.182		7.50	4.74	.46	12.70	17.90	
0945	Galvanized		213	.150		10.60	3.91	.38	14.89	19.60	
0955	Stainless steel pipe, 1-1/2" diam., #4 finish		107	.299		34.50	7.80	.76	43.06	54	
0960	High polish		107	.299		70.50	7.80	.76	79.06	93.50	
0965	Mirror polish	↓	107	.299	↓	83	7.80	.76	91.56	107	
05580 Formed Metal Fabrications											
0010	LAMP POSTS										600
0020	Aluminum, 7' high, stock units, post only	1 Carp	16	.500	Ea.	33.50	12		45.50	57.50	
0100	Mild steel, plain	"	16	.500	"	32	12		44	56	

or information about Means Estimating Seminars, see yellow pages 12 and 13 in back of book

Division Notes

		CREW	DAILY OUTPUT	LABOR-HOURS	UNIT	2005 BARE COSTS MAT.	LABOR	EQUIP.	TOTAL	TOTAL INCL O&P

Division 6 Wood & Plastics

Estimating Tips

06050 Basic Wood & Plastic Materials & Methods

- Common to any wood framed structure are the accessory connector items such as screws, nails, adhesives, hangers, connector plates, straps, angles and holdowns. For typical wood framed buildings, such as residential projects, the aggregate total for these items can be significant, especially in areas where seismic loading is a concern. For floor and wall framing, the material cost is based on 10 to 25 lbs. per MBF. Holdowns, hangers and other connectors should be taken off by the piece.

06100 Rough Carpentry

- Lumber is a traded commodity and therefore sensitive to supply and demand in the marketplace. Even in "budgetary" estimating of wood framed projects, it is advisable to call local suppliers for the latest market pricing.
- Common quantity units for wood framed projects are "thousand board feet" (MBF). A board foot is a volume of wood, 1″ x 1′ x 1′, or 144 cubic inches. Board foot quantities are generally calculated using nominal material dimensions—dressed sizes are ignored. Board foot per lineal foot of any stick of lumber can be calculated by dividing the nominal cross sectional area by 12. As an example, 2,000 lineal feet of 2 x 12 equates to 4 MBF by dividing the nominal area, 2 x 12, by 12, which equals 2, and multiplying by 2,000 to give 4,000 board feet. This simple rule applies to all nominal dimensioned lumber.
- Waste is an issue of concern at the quantity takeoff for any area of construction. Framing lumber is sold in even foot lengths, i.e., 10′, 12′, 14′, 16′, and depending on spans, wall heights and the grade of lumber, waste is inevitable. A rule of thumb for lumber waste is 5% to 10% depending on material quality and the complexity of the framing.
- Wood in various forms and shapes is used in many projects, even where the main structural framing is steel, concrete or masonry. Plywood as a back-up partition material and 2x boards used as blocking and cant strips around roof edges are two common examples. The estimator should ensure that the costs of all wood materials are included in the final estimate.

06200 Finish Carpentry

- It is necessary to consider the grade of workmanship when estimating labor costs for erecting millwork and interior finish. In practice, there are three grades: premium, custom and economy. The Means daily output for base and case moldings is in the range of 200 to 250 L.F. per carpenter per day. This is appropriate for most average custom grade projects. For premium projects an adjustment to productivity of 25% to 50% should be made depending on the complexity of the job.

Reference Numbers

Reference numbers are shown in bold squares at the beginning of some major classifications. These numbers refer to related items in the Reference Section. The reference information may be an estimating procedure, an alternate pricing method or technical information.

Note: Not all subdivisions listed here necessarily appear in this publication.

No part of this publication may be reproduced, stored in a retrieval system, or transmitted in any form or by any means without prior written permission of Reed Construction Data.

06052 | Selective Demolition											
			DAILY	LABOR-		2005 BARE COSTS				TOTAL	
		CREW	OUTPUT	HOURS	UNIT	MAT.	LABOR	EQUIP.	TOTAL	INCL O&P	
110 0010	**SELECTIVE DEMOLITION, WOOD FRAMING** R02220 -510										110
0100	Timber connector, nailed, small	1 Clab	96	.083	Ea.		1.45		1.45	2.45	
0110	Medium		60	.133			2.31		2.31	3.93	
0120	Large		48	.167			2.89		2.89	4.91	
0130	Bolted, small		48	.167			2.89		2.89	4.91	
0140	Medium		32	.250			4.34		4.34	7.35	
0150	Large		24	.333			5.80		5.80	9.80	
2958	Beams, 2″ x 6″	2 Clab	1,100	.015	L.F.		.25		.25	.43	
2960	2″ x 8″		825	.019			.34		.34	.57	
2965	2″ x 10″		665	.024			.42		.42	.71	
2970	2″ x 12″		550	.029			.50		.50	.86	
2972	2″ x 14″		470	.034			.59		.59	1	
2975	4″ x 8″	B-1	413	.058			1.05		1.05	1.78	
2980	4″ x 10″		330	.073			1.31		1.31	2.22	
2985	4″ x 12″		275	.087			1.57		1.57	2.67	
3000	6″ x 8″		275	.087			1.57		1.57	2.67	
3040	6″ x 10″		220	.109			1.97		1.97	3.34	
3080	6″ x 12″		185	.130			2.34		2.34	3.97	
3120	8″ x 12″		140	.171			3.09		3.09	5.25	
3160	10″ x 12″		110	.218			3.93		3.93	6.65	
3162	Alternate pricing method		1.10	21.818	M.B.F.		395		395	665	
3170	Blocking, in 16″ OC wall framing, 2″ x 4″	1 Clab	600	.013	L.F.		.23		.23	.39	
3172	2″ x 6″		400	.020			.35		.35	.59	
3174	In 24″ OC wall framing, 2″ x 4″		600	.013			.23		.23	.39	
3176	2″ x 6″		400	.020			.35		.35	.59	
3178	Alt method, wood blocking removal from wood framimg		.40	20	M.B.F.		345		345	590	
3179	Wood blocking removal from steel framimg		.36	22.222	″		385		385	655	
3180	Bracing, let in, 1″ x 3″, studs 16″ OC		1,050	.008	L.F.		.13		.13	.22	
3181	Studs 24″ OC		1,080	.007			.13		.13	.22	
3182	1″ x 4″, studs 16″ OC		1,050	.008			.13		.13	.22	
3183	Studs 24″ OC		1,080	.007			.13		.13	.22	
3184	1″ x 6″, studs 16″ OC		1,050	.008			.13		.13	.22	
3185	Studs 24″ OC		1,080	.007			.13		.13	.22	
3186	2″ x 3″, studs 16″ OC		800	.010			.17		.17	.29	
3187	Studs 24″ OC		830	.010			.17		.17	.28	
3188	2″ x 4″, studs 16″ OC		800	.010			.17		.17	.29	
3189	Studs 24″ OC		830	.010			.17		.17	.28	
3190	2″ x 6″, studs 16″ OC		800	.010			.17		.17	.29	
3191	Studs 24″ OC		830	.010			.17		.17	.28	
3192	2″ x 8″, studs 16″ OC		800	.010			.17		.17	.29	
3193	Studs 24″ OC		830	.010			.17		.17	.28	
3194	″T″ shaped metal bracing, studs at 16″ OC		1,060	.008			.13		.13	.22	
3195	Studs at 24″ OC		1,200	.007			.12		.12	.20	
3196	Metal straps, studs at 16″ OC		1,200	.007			.12		.12	.20	
3197	Studs at 24″ OC		1,240	.006			.11		.11	.19	
3200	Columns, round, 8′ to 14′ tall		40	.200	Ea.		3.47		3.47	5.90	
3202	Dimensional lumber sizes	2 Clab	1.10	14.545	M.B.F.		252		252	430	
3250	Blocking, between joists	1 Clab	320	.025	Ea.		.43		.43	.74	
3252	Bridging, metal strap, between joists		320	.025	Pr.		.43		.43	.74	
3254	Wood, between joists		320	.025	″		.43		.43	.74	
3260	Door buck, studs, header & access, 8′ high 2″ x 4″ wall, 3′ wide		32	.250	Ea.		4.34		4.34	7.35	
3261	4′ wide		32	.250			4.34		4.34	7.35	
3262	5′ wide		32	.250			4.34		4.34	7.35	
3263	6′ wide		32	.250			4.34		4.34	7.35	
3264	8′ wide		30	.267			4.63		4.63	7.85	
3265	10′ wide		30	.267			4.63		4.63	7.85	

06052 | Selective Demolition

R02220-510

110

	06052 Selective Demolition	CREW	DAILY OUTPUT	LABOR-HOURS	UNIT	MAT.	LABOR	EQUIP.	TOTAL	TOTAL INCL O&P
3266	12′ wide	1 Clab	30	.267	Ea.		4.63		4.63	7.85
3267	2″ x 6″ wall, 3′ wide		32	.250			4.34		4.34	7.35
3268	4′ wide		32	.250			4.34		4.34	7.35
3269	5′ wide		32	.250			4.34		4.34	7.35
3270	6′ wide		32	.250			4.34		4.34	7.35
3271	8′ wide		30	.267			4.63		4.63	7.85
3272	10′ wide		30	.267			4.63		4.63	7.85
3273	12′ wide		30	.267			4.63		4.63	7.85
3274	Window buck, studs, header & access, 8′ high 2″ x 4″ wall, 2′ wide		24	.333			5.80		5.80	9.80
3275	3′ wide		24	.333			5.80		5.80	9.80
3276	4′ wide		24	.333			5.80		5.80	9.80
3277	5′ wide		24	.333			5.80		5.80	9.80
3278	6′ wide		24	.333			5.80		5.80	9.80
3279	7′ wide		24	.333			5.80		5.80	9.80
3280	8′ wide		22	.364			6.30		6.30	10.70
3281	10′ wide		22	.364			6.30		6.30	10.70
3282	12′ wide		22	.364			6.30		6.30	10.70
3283	2″ x 6″ wall, 2′ wide		24	.333			5.80		5.80	9.80
3284	3′ wide		24	.333			5.80		5.80	9.80
3285	4′ wide		24	.333			5.80		5.80	9.80
3286	5′ wide		24	.333			5.80		5.80	9.80
3287	6′ wide		24	.333			5.80		5.80	9.80
3288	7′ wide		24	.333			5.80		5.80	9.80
3289	8′ wide		22	.364			6.30		6.30	10.70
3290	10′ wide		22	.364			6.30		6.30	10.70
3291	12′ wide		22	.364			6.30		6.30	10.70
3400	Fascia boards, 1″ x 6″		500	.016	L.F.		.28		.28	.47
3440	1″ x 8″		450	.018			.31		.31	.52
3480	1″ x 10″		400	.020			.35		.35	.59
3490	2″ x 6″		450	.018			.31		.31	.52
3500	2″ x 8″		400	.020			.35		.35	.59
3510	2″ x 10″		350	.023			.40		.40	.67
3610	Furring, on wood walls or ceiling		4,000	.002	S.F.		.03		.03	.06
3620	On masonry or concrete walls or ceiling		1,200	.007	"		.12		.12	.20
3800	Headers over openings, 2 @ 2″ x 6″		110	.073	L.F.		1.26		1.26	2.14
3840	2 @ 2″ x 8″		100	.080			1.39		1.39	2.36
3880	2 @ 2″ x 10″		90	.089			1.54		1.54	2.62
3885	Alternate pricing method		.26	30.651	M.B.F.		530		530	905
3920	Joists, 1″ x 4″		1,250	.006	L.F.		.11		.11	.19
3930	1″ x 6″		1,135	.007			.12		.12	.21
3950	1″ x 10″		895	.009			.16		.16	.26
3960	1″ x 12″		765	.010			.18		.18	.31
4200	2″ x 4″	2 Clab	1,000	.016			.28		.28	.47
4230	2″ x 6″		970	.016			.29		.29	.49
4240	2″ x 8″		940	.017			.30		.30	.50
4250	2″ x 10″		910	.018			.31		.31	.52
4280	2″ x 12″		880	.018			.32		.32	.54
4281	2″ x 14″		850	.019			.33		.33	.55
4282	Composite joists, 9-1/2″		960	.017			.29		.29	.49
4283	11-7/8″		930	.017			.30		.30	.51
4284	14″		897	.018			.31		.31	.53
4285	16″		865	.018			.32		.32	.54
4290	Wood joists, alternate pricing method		1.50	10.667	M.B.F.		185		185	315
4500	Open web joist, 12″ deep		500	.032	L.F.		.56		.56	.94
4505	14″ deep		475	.034			.58		.58	.99
4510	16″ deep		450	.036			.62		.62	1.05

06052 | Selective Demolition

		Description	CREW	DAILY OUTPUT	LABOR-HOURS	UNIT	MAT.	LABOR	EQUIP.	TOTAL	TOTAL INCL O&P
							2005 BARE COSTS				
110	4520	18" deep (R02220-510)	2 Clab	425	.038	L.F.		.65		.65	1.11
	4530	24" deep		400	.040			.69		.69	1.18
	4550	Ledger strips, 1" x 2"	1 Clab	1,200	.007			.12		.12	.20
	4560	1" x 3"		1,200	.007			.12		.12	.20
	4570	1" x 4"		1,200	.007			.12		.12	.20
	4580	2" x 2"		1,100	.007			.13		.13	.21
	4590	2" x 4"		1,000	.008			.14		.14	.24
	4600	2" x 6"		1,000	.008			.14		.14	.24
	4601	2" x 8 or 2" x 10"		800	.010			.17		.17	.29
	4602	4" x 6"		600	.013			.23		.23	.39
	4604	4" x 8"		450	.018			.31		.31	.52
	5400	Posts, 4" x 4"	2 Clab	800	.020			.35		.35	.59
	5405	4" x 6"		550	.029			.50		.50	.86
	5410	4" x 8"		440	.036			.63		.63	1.07
	5420	4" x 6"	1 Clab	270	.030			.51		.51	.87
	5425	4" x 10"	2 Clab	390	.041			.71		.71	1.21
	5430	4" x 12"		350	.046			.79		.79	1.35
	5440	6" x 6"		400	.040			.69		.69	1.18
	5445	6" x 8"		350	.046			.79		.79	1.35
	5450	6" x 10"		320	.050			.87		.87	1.47
	5455	6" x 12"		290	.055			.96		.96	1.62
	5480	8" x 8"		300	.053			.93		.93	1.57
	5500	10" x 10"		240	.067			1.16		1.16	1.96
	5660	Tongue and groove floor planks		2	8	M.B.F.		139		139	236
	5750	Rafters, ordinary, 16" OC, 2" x 4"		880	.018	S.F.		.32		.32	.54
	5755	2" x 6"		840	.019			.33		.33	.56
	5760	2" x 8"		820	.020			.34		.34	.57
	5770	2" x 10"		820	.020			.34		.34	.57
	5780	2" x 12"		810	.020			.34		.34	.58
	5785	24" OC, 2" x 4"		1,170	.014			.24		.24	.40
	5786	2" x 6"		1,117	.014			.25		.25	.42
	5787	2" x 8"		1,091	.015			.25		.25	.43
	5788	2" x 10"		1,091	.015			.25		.25	.43
	5789	2" x 12"		1,077	.015			.26		.26	.44
	5795	Rafters, ordinary, 2" x 4" (alternate method)		862	.019	L.F.		.32		.32	.55
	5800	2" x 6" (alternate method)		850	.019			.33		.33	.55
	5840	2" x 8" (alternate method)		837	.019			.33		.33	.56
	5855	2" x 10" (alternate method)		825	.019			.34		.34	.57
	5865	2" x 12" (alternate method)		812	.020			.34		.34	.58
	5870	Sill plate, 2" x 4"	1 Clab	1,170	.007			.12		.12	.20
	5871	2" x 6"		780	.010			.18		.18	.30
	5872	2" x 8"		586	.014			.24		.24	.40
	5873	Alternate pricing method		.78	10.256	M.B.F.		178		178	300
	5885	Ridge board, 1" x 4"	2 Clab	900	.018	L.F.		.31		.31	.52
	5886	1" x 6"		875	.018			.32		.32	.54
	5887	1" x 8"		850	.019			.33		.33	.55
	5888	1" x 10"		825	.019			.34		.34	.57
	5889	1" x 12"		800	.020			.35		.35	.59
	5890	2" x 4"		900	.018			.31		.31	.52
	5892	2" x 6"		875	.018			.32		.32	.54
	5894	2" x 8"		850	.019			.33		.33	.55
	5896	2" x 10"		825	.019			.34		.34	.57
	5898	2" x 12"		800	.020			.35		.35	.59
	6050	Rafter tie, 1" x 4"		1,250	.013			.22		.22	.38
	6052	1" x 6"		1,135	.014			.24		.24	.42
	6054	2" x 4"		1,000	.016			.28		.28	.47

06052 | Selective Demolition		CREW	DAILY OUTPUT	LABOR-HOURS	UNIT	2005 BARE COSTS MAT.	LABOR	EQUIP.	TOTAL	TOTAL INCL O&P
6056	2″ x 6″ R02220 -510	2 Clab	970	.016	L.F.		.29		.29	.49
6070	Sleepers, on concrete, 1″ x 2″	1 Clab	4,700	.002			.03		.03	.05
6075	1″ x 3″		4,000	.002			.03		.03	.06
6080	2″ x 4″		3,000	.003			.05		.05	.08
6085	2″ x 6″	↓	2,600	.003	↓		.05		.05	.09
6086	Sheathing from roof, 5/16″	2 Clab	1,600	.010	S.F.		.17		.17	.29
6088	3/8″		1,525	.010			.18		.18	.31
6090	1/2″		1,400	.011			.20		.20	.34
6092	5/8″		1,300	.012			.21		.21	.36
6094	3/4″		1,200	.013			.23		.23	.39
6096	Board sheathing from roof		1,400	.011			.20		.20	.34
6100	Sheathing, from walls, 1/4″		1,200	.013			.23		.23	.39
6110	5/16″		1,175	.014			.24		.24	.40
6120	3/8″		1,150	.014			.24		.24	.41
6130	1/2″		1,125	.014			.25		.25	.42
6140	5/8″		1,100	.015			.25		.25	.43
6150	3/4″		1,075	.015			.26		.26	.44
6152	Board sheathing from walls		1,500	.011			.19		.19	.31
6158	Subfloor, with boards		1,050	.015			.26		.26	.45
6160	Plywood, 1/2″ thick		768	.021			.36		.36	.61
6162	5/8″ thick		760	.021			.37		.37	.62
6164	3/4″thick		750	.021			.37		.37	.63
6165	1-1/8″ thick	↓	720	.022			.39		.39	.65
6166	Underlayment, particle board, 3/8″ thick	1 Clab	780	.010			.18		.18	.30
6168	1/2″ thick		768	.010			.18		.18	.31
6170	5/8″ thick		760	.011			.18		.18	.31
6172	3/4″ thick	↓	750	.011	↓		.19		.19	.31
6200	Stairs and stringers, minimum	2 Clab	40	.400	Riser		6.95		6.95	11.80
6240	Maximum	″	26	.615	″		10.70		10.70	18.10
6300	Components, tread	1 Clab	110	.073	Ea.		1.26		1.26	2.14
6320	Riser		80	.100	″		1.74		1.74	2.95
6390	Stringer, 2″ x 10″		260	.031	L.F.		.53		.53	.91
6400	2″ x 12″		260	.031			.53		.53	.91
6410	3″ x 10″		250	.032			.56		.56	.94
6420	3″ x 12″	↓	250	.032			.56		.56	.94
6590	Wood studs, 2″ x 3″	2 Clab	3,076	.005			.09		.09	.15
6600	2″ x 4″		2,000	.008			.14		.14	.24
6640	2″ x 6″	↓	1,600	.010	↓		.17		.17	.29
6720	Wall framing, including studs plates and blocking, 2″ x 4″	1 Clab	600	.013	S.F.		.23		.23	.39
6740	2″ x 6″		480	.017	″		.29		.29	.49
6750	Headers, 2″ x 4″		1,125	.007	L.F.		.12		.12	.21
6755	2″ x 6″		1,125	.007			.12		.12	.21
6760	2″ x 8″		1,050	.008			.13		.13	.22
6765	2″ x 10″		1,050	.008			.13		.13	.22
6770	2″ x 12″		1,000	.008			.14		.14	.24
6780	4″ x 10″		525	.015			.26		.26	.45
6785	4″ x 12″		500	.016			.28		.28	.47
6790	6″ x 8″		560	.014			.25		.25	.42
6795	6″ x 10″		525	.015			.26		.26	.45
6797	6″ x 12″	↓	500	.016	↓		.28		.28	.47
7000	Trusses									
7050	12′ span	2 Clab	74	.216	Ea.		3.75		3.75	6.35
7150	24′ span	F-3	66	.606			13.05	9.15	22.20	32
7200	26′ span		64	.625			13.50	9.40	22.90	33
7250	28′ span		62	.645			13.90	9.70	23.60	34
7300	30′ span	↓	58	.690	↓		14.90	10.40	25.30	36.50

110

		06052 | Selective Demolition	CREW	DAILY OUTPUT	LABOR-HOURS	UNIT	2005 BARE COSTS MAT.	LABOR	EQUIP.	TOTAL	TOTAL INCL O&P
110	7350	32′ span (R02220-510)	F-3	56	.714	Ea.		15.40	10.75	26.15	38
	7400	34′ span		54	.741			16	11.15	27.15	39.50
	7450	36′ span		52	.769			16.60	11.60	28.20	41
	8000	Soffit, T & G wood	1 Clab	520	.015	S.F.		.27		.27	.45
	8010	Hardboard, vinyl or aluminum	"	640	.013			.22		.22	.37
	8030	Plywood	2 Carp	315	.051			1.22		1.22	2.07
	9500	See Div. 02220-350 for rubbish handling									
120	0010	**SELECTIVE DEMOLITION, MILLWORK AND TRIM** (R02220-510)									
	1000	Cabinets, wood, base cabinets, per L.F.	2 Clab	80	.200	L.F.		3.47		3.47	5.90
	1020	Wall cabinets, per L.F.		80	.200			3.47		3.47	5.90
	1100	Steel, painted, base cabinets		60	.267			4.63		4.63	7.85
	1500	Counter top, minimum		200	.080			1.39		1.39	2.36
	1510	Maximum		120	.133			2.31		2.31	3.93
	2000	Paneling, 4′ x 8′ sheets		2,000	.008	S.F.		.14		.14	.24
	2100	Boards, 1″ x 4″		700	.023			.40		.40	.67
	2120	1″ x 6″		750	.021			.37		.37	.63
	2140	1″ x 8″		800	.020			.35		.35	.59
	3000	Trim, baseboard, to 6″ wide		1,200	.013	L.F.		.23		.23	.39
	3040	Greater than 6″ and up to 12″ wide		1,000	.016			.28		.28	.47
	3100	Ceiling trim		1,000	.016			.28		.28	.47
	3120	Chair rail		1,200	.013			.23		.23	.39
	3140	Railings with balusters		240	.067			1.16		1.16	1.96
	3160	Wainscoting		700	.023	S.F.		.40		.40	.67
	4000	Curtain rod	1 Clab	80	.100	L.F.		1.74		1.74	2.95
	9000	Minimum labor/equipment charge	"	4	2	Job		34.50		34.50	59
		06070 | Lumber Treatment									
400	0011	**LUMBER TREATMENT**									
	0400	Fire retardant, wet				M.B.F.	291			291	320
	0500	KDAT					277			277	305
	0700	Salt treated, water borne, .40 lb. retention					133			133	146
	0800	Oil borne, 8 lb. retention					156			156	171
	1000	Kiln dried lumber, 1″ & 2″ thick, softwoods					88.50			88.50	97.50
	1100	Hardwoods					94.50			94.50	104
	1500	For small size 1″ stock, add					11.95			11.95	13.15
	1700	For full size rough lumber, add					20%				
600	0010	**PLYWOOD TREATMENT** Fire retardant, 1/4″ thick				M.S.F.	222			222	244
	0030	3/8″ thick					244			244	268
	0050	1/2″ thick					261			261	287
	0070	5/8″ thick					277			277	305
	0100	3/4″ thick					305			305	335
	0200	For KDAT, add					66.50			66.50	73
	0500	Salt treated water borne, .25 lb., wet, 1/4″ thick					122			122	134
	0530	3/8″ thick					127			127	140
	0550	1/2″ thick					133			133	146
	0570	5/8″ thick					145			145	159
	0600	3/4″ thick					150			150	165
	0800	For KDAT add					66.50			66.50	73
	0900	For .40 lb., per C.F. retention, add					55.50			55.50	61
	1000	For certification stamp, add					32.50			32.50	36
		06090 | Wood & Plastic Fastenings									
600	0010	**NAILS** Prices of material only, based on 50# box purchase, copper, plain				Lb.	4.99			4.99	5.50
	0400	Stainless steel, plain					5.75			5.75	6.30

06090 | Wood & Plastic Fastenings		CREW	DAILY OUTPUT	LABOR-HOURS	UNIT	2005 BARE COSTS MAT.	LABOR	EQUIP.	TOTAL	TOTAL INCL O&P	
0500	Box, 3d to 20d, bright				Lb.	1.32			1.32	1.45	600
0520	Galvanized					1.44			1.44	1.58	
0600	Common, 3d to 60d, plain					1.02			1.02	1.12	
0700	Galvanized					1.39			1.39	1.53	
0800	Aluminum					3.38			3.38	3.72	
1000	Annular or spiral thread, 4d to 60d, plain					1.60			1.60	1.76	
1200	Galvanized					1.80			1.80	1.98	
1400	Drywall nails, plain					.79			.79	.87	
1600	Galvanized					1.54			1.54	1.69	
1800	Finish nails, 4d to 10d, plain					.94			.94	1.03	
2000	Galvanized					1.32			1.32	1.45	
2100	Aluminum					4.10			4.10	4.51	
2300	Flooring nails, hardened steel, 2d to 10d, plain					1.40			1.40	1.54	
2400	Galvanized					2.25			2.25	2.48	
2500	Gypsum lath nails, 1-1/8", 13 ga. flathead, blued					1.54			1.54	1.69	
2600	Masonry nails, hardened steel, 3/4" to 3" long, plain					1.48			1.48	1.63	
2700	Galvanized					1.87			1.87	2.06	
2900	Roofing nails, threaded, galvanized					1.35			1.35	1.49	
3100	Aluminum					4.80			4.80	5.30	
3300	Compressed lead head, threaded, galvanized					1.50			1.50	1.65	
3600	Siding nails, plain shank, galvanized					1.45			1.45	1.60	
3800	Aluminum					4.11			4.11	4.52	
5000	Add to prices above for cement coating					.10			.10	.11	
5200	Zinc or tin plating					.13			.13	.14	
5500	Vinyl coated sinkers, 8d to 16d				↓	.57			.57	.63	
0010	**NAILS** mat. only, for pneumatic tools, framing, per carton of 5000, 2"				Ea.	37			37	41	650
0100	2-3/8"					42.50			42.50	46.50	
0200	Per carton of 4000, 3"					37.50			37.50	41.50	
0300	3-1/4"					40			40	44	
0400	Per carton of 5000, 2-3/8", galv.					57.50			57.50	63.50	
0500	Per carton of 4000, 3", galv.					65			65	71.50	
0600	3-1/4", galv.					80.50			80.50	88.50	
0700	Roofing, per carton of 7200, 1"					35			35	38.50	
0800	1-1/4"					32.50			32.50	36	
0900	1-1/2"					37.50			37.50	41.50	
1000	1-3/4"				↓	45.50			45.50	50	
0010	**SHEET METAL SCREWS** Steel, standard, #8 x 3/4", plain				C	2.81			2.81	3.09	700
0100	Galvanized					3.44			3.44	3.78	
0300	#10 x 1", plain					3.76			3.76	4.14	
0400	Galvanized					4.35			4.35	4.79	
1500	Self-drilling, with washers, (pinch point) #8 x 3/4", plain					6.10			6.10	6.70	
1600	Galvanized					6.10			6.10	6.70	
1800	#10 x 3/4", plain					6.10			6.10	6.70	
1900	Galvanized					6.10			6.10	6.70	
3000	Stainless steel w/aluminum or neoprene washers, #14 x 1", plain					18.35			18.35	20	
3100	#14 x 2", plain				↓	25			25	27.50	
0010	**WOOD SCREWS** #8, 1" long, steel				C	3.36			3.36	3.70	750
0100	Brass					11.20			11.20	12.35	
0200	#8, 2" long, steel					3.76			3.76	4.14	
0300	Brass					11.70			11.70	12.90	
0400	#10, 1" long, steel					4.30			4.30	4.73	
0500	Brass					23			23	25	
0600	#10, 2" long, steel					7.65			7.65	8.45	
0700	Brass				↓	40.50			40.50	44.50	

	06090 | Wood & Plastic Fastenings		CREW	DAILY OUTPUT	LABOR-HOURS	UNIT	2005 BARE COSTS MAT.	LABOR	EQUIP.	TOTAL	TOTAL INCL O&P
750	0800	#10, 3" long, steel				C	11.95			11.95	13.15
	1000	#12, 2" long, steel					4.89			4.89	5.40
	1100	Brass					16.30			16.30	17.95
	1500	#12, 3" long, steel					16.20			16.20	17.85
	2000	#12, 4" long, steel					29			29	32
800	0010	**TIMBER CONNECTORS** Add up cost of each part for total									
	0020	cost of connection									
	0100	Connector plates, steel, with bolts, straight	2 Carp	75	.213	Ea.	18.75	5.10		23.85	29
	0110	Tee		50	.320		27.50	7.70		35.20	43.50
	0120	T- Strap, 14 gauge12" x 8" x 2"		50	.320		27.50	7.70		35.20	43.50
	0150	Anchor plate, 7 gauge, 9" x 7"		75	.213		18.75	5.10		23.85	29
	0200	Bolts, machine, sq. hd. with nut & washer, 1/2" diameter, 4" long	1 Carp	140	.057		.36	1.37		1.73	2.73
	0300	7-1/2" long		130	.062		.63	1.48		2.11	3.20
	0500	3/4" diameter, 7-1/2" long		130	.062		1.66	1.48		3.14	4.34
	0610	Machine bolts, w/ nut, washer, 3/4" dia, 15" L, HD's & beam hangers		95	.084		3.05	2.02		5.07	6.80
	0720	Machine bolts, sq. hd. w/nut & wash		150	.053	Lb.	2.14	1.28		3.42	4.52
	0800	Drilling bolt holes in timber, 1/2" diameter		450	.018	Inch		.43		.43	.72
	0900	1" diameter		350	.023	"		.55		.55	.93
	1100	Framing anchors, 2 or 3 dimensional, 10 gauge, no nails incl.		175	.046	Ea.	.46	1.10		1.56	2.37
	1150	Framing anchors, 18 gauge, 4 1/2" x 2 3/4"		175	.046		.46	1.10		1.56	2.37
	1160	Framing anchors, 18 gauge, 4 1/2" x 3"		175	.046		.46	1.10		1.56	2.37
	1170	Clip anchors plates, 18 gauge, 12" x 1 1/8"		175	.046		.46	1.10		1.56	2.37
	1250	Holdowns, 3 gauge base, 10 gauge body		8	1		15.60	24		39.60	58
	1260	Holdowns, 7 gauge 11 1/16" x 3 1/4"		8	1		15.60	24		39.60	58
	1270	Holdowns, 7 gauge 14 3/8" x 3 1/8"		8	1		15.60	24		39.60	58
	1275	Holdowns, 12 gauge 8" x 2 1/2"		8	1		15.60	24		39.60	58
	1300	Joist and beam hangers, 18 ga. galv., for 2" x 4" joist		175	.046		.58	1.10		1.68	2.50
	1400	2" x 6" to 2" x 10" joist		165	.048		.67	1.16		1.83	2.72
	1600	16 ga. galv., 3" x 6" to 3" x 10" joist		160	.050		2.64	1.20		3.84	4.94
	1700	3" x 10" to 3" x 14" joist		160	.050		3.06	1.20		4.26	5.40
	1800	4" x 6" to 4" x 10" joist		155	.052		2.32	1.24		3.56	4.65
	1900	4" x 10" to 4" x 14" joist		155	.052		3.14	1.24		4.38	5.55
	2000	Two-2" x 6" to two-2" x 10" joists		150	.053		2.53	1.28		3.81	4.95
	2100	Two-2" x 10" to two-2" x 14" joists		150	.053		2.53	1.28		3.81	4.95
	2300	3/16" thick, 6" x 8" joist		145	.055		5.50	1.32		6.82	8.30
	2400	6" x 10" joist		140	.057		6.50	1.37		7.87	9.50
	2500	6" x 12" joist		135	.059		7.80	1.42		9.22	11
	2700	1/4" thick, 6" x 14" joist		130	.062		9.70	1.48		11.18	13.20
	2900	Plywood clips, extruded aluminum H clip, for 3/4" panels					.15			.15	.17
	3000	Galvanized 18 ga. back-up clip					.14			.14	.15
	3200	Post framing, 16 ga. galv. for 4" x 4" base, 2 piece	1 Carp	130	.062		5.45	1.48		6.93	8.50
	3300	Cap		130	.062		2.67	1.48		4.15	5.45
	3500	Rafter anchors, 18 ga. galv., 1-1/2" wide, 5-1/4" long		145	.055		.46	1.32		1.78	2.76
	3600	10-3/4" long		145	.055		.91	1.32		2.23	3.25
	3800	Shear plates, 2-5/8" diameter		120	.067		1.63	1.60		3.23	4.51
	3900	4" diameter		115	.070		3.75	1.67		5.42	6.95
	4000	Sill anchors, embedded in concrete or block, 18-5/8" long		115	.070		1.10	1.67		2.77	4.05
	4100	Spike grids, 4" x 4", flat or curved		120	.067		.41	1.60		2.01	3.17
	4400	Split rings, 2-1/2" diameter		120	.067		1.32	1.60		2.92	4.17
	4500	4" diameter		110	.073		2.04	1.75		3.79	5.20
	4550	Tie plate, 20 gauge, 7" x 3 1/8"		110	.073		2.04	1.75		3.79	5.20
	4560	Tie plate, 20 gauge, 5" x 4 1/8"		110	.073		2.04	1.75		3.79	5.20
	4575	Twist straps, 18 gauge, 12" x 1 1/4"		110	.073		2.04	1.75		3.79	5.20
	4580	Twist straps, 18 gauge, 16" x 1 1/4"		110	.073		2.04	1.75		3.79	5.20
	4600	Strap ties, 20 ga., 2 -1/16" wide, 12 13/16" long		180	.044		1.19	1.07		2.26	3.12

06090 | Wood & Plastic Fastenings		CREW	DAILY OUTPUT	LABOR-HOURS	UNIT	2005 BARE COSTS MAT.	LABOR	EQUIP.	TOTAL	TOTAL INCL O&P
4700	Strap ties, 16 ga., 1-3/8" wide, 12" long	1 Carp	180	.044	Ea.	1.19	1.07		2.26	3.12
4800	24" long		160	.050		1.64	1.20		2.84	3.84
5000	Toothed rings, 2-5/8" or 4" diameter		90	.089		1.13	2.13		3.26	4.86
5200	Truss plates, nailed, 20 gauge, up to 32' span		17	.471	Truss	8.15	11.30		19.45	28
5400	Washers, 2" x 2" x 1/8"				Ea.	.26			.26	.29
5500	3" x 3" x 3/16"				"	.67			.67	.74
6101	Beam hangers, polymer painted									
6102	Bolted, 3 ga., (W x H x L)									
6104	3-1/4" x 9" x 12" top flange	1 Carp	1	8	C	6,375	192		6,567	7,350
6106	5-1/4" x 9" x 12" top flange		1	8		6,650	192		6,842	7,625
6108	5-1/4" x 11" x 11-3/4" top flange		1	8		14,900	192		15,092	16,700
6110	6-7/8" x 9" x 12" top flange		1	8		6,850	192		7,042	7,875
6112	6-7/8" x 11" x 13-1/2" top flange		1	8		15,700	192		15,892	17,600
6114	8-7/8" x 11" x 15-1/2" top flange		1	8		16,800	192		16,992	18,800
6116	Nailed, 3 ga., (W x H x L)									
6118	3-1/4" x 10-1/2" x 10" top flange	1 Carp	1.80	4.444	C	4,400	107		4,507	5,025
6120	3-1/4" x 10-1/2" x 12" top flange		1.80	4.444		5,100	107		5,207	5,800
6122	5-1/4" x 9-1/2" x 10" top flange		1.80	4.444		4,750	107		4,857	5,400
6124	5-1/4" x 9-1/2" x 12" top flange		1.80	4.444		5,400	107		5,507	6,100
6126	5-1/2" x 9-1/2" x 12" top flange		1.80	4.444		4,750	107		4,857	5,400
6128	6-7/8" x 8-1/2" x 12" top flange		1.80	4.444		4,925	107		5,032	5,575
6130	7-1/2" x 8-1/2" x 12" top flange		1.80	4.444		5,075	107		5,182	5,750
6132	8-7/8" x 7-1/2" x 14" top flange		1.80	4.444		5,625	107		5,732	6,350
6201	Beam and purlin hangers, galvanized, 12 ga.									
6202	Purlin or joist size, 3" x 8"	1 Carp	1.70	4.706	C	805	113		918	1,075
6204	3" x 10"		1.70	4.706		905	113		1,018	1,175
6206	3" x 12"		1.65	4.848		1,075	116		1,191	1,375
6208	3" x 14"		1.65	4.848		1,250	116		1,366	1,575
6210	3" x 16"		1.65	4.848		1,425	116		1,541	1,775
6212	4" x 8"		1.65	4.848		805	116		921	1,075
6214	4" x 10"		1.65	4.848		940	116		1,056	1,225
6216	4" x 12"		1.60	5		1,000	120		1,120	1,300
6218	4" x 14"		1.60	5		1,100	120		1,220	1,400
6220	4" x 16"		1.60	5		1,250	120		1,370	1,575
6224	6" x 10"		1.55	5.161		1,375	124		1,499	1,700
6226	6" x 12"		1.55	5.161		1,475	124		1,599	1,825
6228	6" x 14"		1.50	5.333		1,600	128		1,728	1,975
6230	6" x 16"		1.50	5.333		1,825	128		1,953	2,225
6300	Column bases									
6302	4 x 4, 16 ga.	1 Carp	1.80	4.444	C	1,125	107		1,232	1,400
6306	7 ga.		1.80	4.444		2,075	107		2,182	2,475
6314	6 x 6, 16 ga.		1.75	4.571		1,425	110		1,535	1,750
6318	7 ga.		1.75	4.571		2,925	110		3,035	3,400
6326	8 x 8, 7 ga.		1.65	4.848		4,875	116		4,991	5,575
6330	8 x 10, 7 ga.		1.65	4.848		5,200	116		5,316	5,925
6590	Joist hangers, heavy duty 12 ga., galvanized									
6592	2" x 4"	1 Carp	1.75	4.571	C	915	110		1,025	1,175
6594	2" x 6"		1.65	4.848		980	116		1,096	1,275
6595	2" x 6", 16 gauge		1.65	4.848		980	116		1,096	1,275
6596	2" x 8"		1.65	4.848		1,050	116		1,166	1,350
6597	2" x 8", 16 gauge		1.65	4.848		1,050	116		1,166	1,350
6598	2" x 10"		1.65	4.848		1,125	116		1,241	1,425
6600	2" x 12"		1.65	4.848		1,250	116		1,366	1,575
6622	(2) 2" x 6"		1.60	5		1,225	120		1,345	1,550
6624	(2) 2" x 8"		1.60	5		1,325	120		1,445	1,650
6626	(2) 2" x 10"		1.55	5.161		1,475	124		1,599	1,825

800

06050 | Basic Wood / Plastic Materials / Methods

	06090 | Wood & Plastic Fastenings		CREW	DAILY OUTPUT	LABOR-HOURS	UNIT	2005 BARE COSTS MAT.	LABOR	EQUIP.	TOTAL	TOTAL INCL O&P
800	6628	(2) 2" x 12"	1 Carp	1.55	5.161	C	1,800	124		1,924	2,175
	6890	Purlin hangers, painted									
	6892	12 ga., 2" x 6"	1 Carp	1.80	4.444	C	1,125	107		1,232	1,400
	6894	2" x 8"		1.80	4.444		1,175	107		1,282	1,475
	6896	2" x 10"		1.80	4.444		1,200	107		1,307	1,500
	6898	2" x 12"		1.75	4.571		1,300	110		1,410	1,625
	6934	(2) 2" x 6"		1.70	4.706		1,125	113		1,238	1,450
	6936	(2) 2" x 8"		1.70	4.706		1,250	113		1,363	1,575
	6938	(2) 2" x 10"		1.70	4.706		1,375	113		1,488	1,700
	6940	(2) 2" x 12"	↓	1.65	4.848	↓	1,475	116		1,591	1,825
825	0010	**ROUGH HARDWARE** Average % of carpentry material, minimum					.50%				
	0200	Maximum					1.50%				
	0210	In seismic or hurricane areas, up to					10%				
850	0010	**BRACING**									
	0302	Let-in, "T" shaped, 22 ga. galv. steel, studs at 16" O.C.	1 Carp	580	.014	L.F.	.45	.33		.78	1.06
	0402	Studs at 24" O.C.		600	.013		.45	.32		.77	1.04
	0502	16 ga. galv. steel straps, studs at 16" O.C.		600	.013		.67	.32		.99	1.28
	0602	Studs at 24" O.C.	↓	620	.013	↓	.67	.31		.98	1.27

06100 | Rough Carpentry

	06110 | Wood Framing		CREW	DAILY OUTPUT	LABOR-HOURS	UNIT	2005 BARE COSTS MAT.	LABOR	EQUIP.	TOTAL	TOTAL INCL O&P
100	0010	**BLOCKING** R06100-010									
	1950	Miscellaneous, to wood construction									
	2000	2" x 4"	1 Carp	250	.032	L.F.	.34	.77		1.11	1.67
	2005	Pneumatic nailed		305	.026		.34	.63		.97	1.44
	2050	2" x 6"		222	.036		.54	.87		1.41	2.06
	2055	Pneumatic nailed		271	.030		.54	.71		1.25	1.79
	2100	2" x 8"		200	.040		.83	.96		1.79	2.54
	2105	Pneumatic nailed		244	.033		.83	.79		1.62	2.25
	2150	2" x 10"		178	.045		1.18	1.08		2.26	3.12
	2155	Pneumatic nailed		217	.037		1.18	.88		2.06	2.79
	2200	2" x 12"		151	.053		1.60	1.27		2.87	3.92
	2205	Pneumatic nailed	↓	185	.043	↓	1.60	1.04		2.64	3.52
	2300	To steel construction									
	2320	2" x 4"	1 Carp	208	.038	L.F.	.34	.92		1.26	1.94
	2340	2" x 6"		180	.044		.54	1.07		1.61	2.40
	2360	2" x 8"		158	.051		.83	1.22		2.05	2.97
	2380	2" x 10"		136	.059		1.18	1.41		2.59	3.69
	2400	2" x 12"	↓	109	.073	↓	1.60	1.76		3.36	4.75
150	0012	**BRACING** Let-in, with 1" x 6" boards, studs @ 16" O.C.	1 Carp	150	.053	L.F.	.58	1.28		1.86	2.80
	0202	Studs @ 24" O.C.	"	230	.035	"	.58	.83		1.41	2.05
200	0012	**BRIDGING** Wood, for joists 16" O.C., 1" x 3"	1 Carp	130	.062	Pr.	.42	1.48		1.90	2.98
	0017	Pneumatic nailed		170	.047		.42	1.13		1.55	2.39
	0102	2" x 3" bridging		130	.062		.40	1.48		1.88	2.95
	0107	Pneumatic nailed		170	.047		.40	1.13		1.53	2.36
	0302	Steel, galvanized, 18 ga., for 2" x 10" joists at 12" O.C.		130	.062		.98	1.48		2.46	3.59
	0402	24" O.C.	↓	140	.057	↓	1.18	1.37		2.55	3.63

		06110 | Wood Framing	CREW	DAILY OUTPUT	LABOR-HOURS	UNIT	2005 BARE COSTS MAT.	LABOR	EQUIP.	TOTAL	TOTAL INCL O&P	
0	0602	For 2" x 14" joists at 16" O.C.	1 Carp	130	.062	Pr.	1.16	1.48		2.64	3.78	200
	0902	Compression type, 16" O.C., 2" x 8" joists		200	.040		1.15	.96		2.11	2.90	
	1002	2" x 12" joists		200	.040		1.15	.96		2.11	2.90	
5	0010	**FRAMING, BEAMS & GIRDERS** R06100-010 R06110-030										505
	1002	Single, 2" x 6"	2 Carp	700	.023	L.F.	.54	.55		1.09	1.52	
	1007	Pneumatic nailed		812	.020		.54	.47		1.01	1.39	
	1022	2" x 8"		650	.025		.83	.59		1.42	1.91	
	1027	Pneumatic nailed		754	.021		.83	.51		1.34	1.77	
	1042	2" x 10"		600	.027		1.18	.64		1.82	2.38	
	1047	Pneumatic nailed		696	.023		1.18	.55		1.73	2.23	
	1062	2" x 12"		550	.029		1.60	.70		2.30	2.95	
	1067	Pneumatic nailed		638	.025		1.60	.60		2.20	2.78	
	1082	2" x 14"		500	.032		2.16	.77		2.93	3.67	
	1087	Pneumatic nailed		580	.028		2.16	.66		2.82	3.49	
	1102	3" x 8"		550	.029		2.59	.70		3.29	4.04	
	1122	3" x 10"		500	.032		3.26	.77		4.03	4.89	
	1142	3" x 12"		450	.036		3.92	.85		4.77	5.75	
	1162	3" x 14"		400	.040		4.73	.96		5.69	6.85	
	1170	4" x 6"	F-3	1,100	.036		2	.78	.55	3.33	4.12	
	1182	4" x 8"		1,000	.040		2.69	.86	.60	4.15	5.05	
	1202	4" x 10"		950	.042		3.37	.91	.63	4.91	5.95	
	1222	4" x 12"		900	.044		4.40	.96	.67	6.03	7.20	
	1242	4" x 14"		850	.047		4.95	1.02	.71	6.68	7.95	
	2002	Double, 2" x 6"	2 Carp	625	.026		1.07	.61		1.68	2.22	
	2007	Pneumatic nailed		725	.022		1.07	.53		1.60	2.08	
	2022	2" x 8"		575	.028		1.65	.67		2.32	2.95	
	2027	Pneumatic nailed		667	.024		1.65	.58		2.23	2.80	
	2042	2" x 10"		550	.029		2.35	.70		3.05	3.78	
	2047	Pneumatic nailed		638	.025		2.35	.60		2.95	3.61	
	2062	2" x 12"		525	.030		3.20	.73		3.93	4.76	
	2067	Pneumatic nailed		610	.026		3.20	.63		3.83	4.59	
	2082	2" x 14"		475	.034		4.32	.81		5.13	6.10	
	2087	Pneumatic nailed		551	.029		4.32	.70		5.02	5.95	
	3002	Triple, 2" x 6"		550	.029		1.61	.70		2.31	2.96	
	3007	Pneumatic nailed		638	.025		1.61	.60		2.21	2.79	
	3022	2" x 8"		525	.030		2.48	.73		3.21	3.97	
	3027	Pneumatic nailed		609	.026		2.48	.63		3.11	3.80	
	3042	2" x 10"		500	.032		3.53	.77		4.30	5.20	
	3047	Pneumatic nailed		580	.028		3.53	.66		4.19	5	
	3062	2" x 12"		475	.034		4.80	.81		5.61	6.65	
	3067	Pneumatic nailed		551	.029		4.80	.70		5.50	6.50	
	3082	2" x 14"		450	.036		6.50	.85		7.35	8.55	
	3087	Pneumatic nailed		522	.031		6.50	.74		7.24	8.35	
10	0010	**FRAMING, CEILINGS**										510
	6002	Suspended, 2" x 3"	2 Carp	1,000	.016	L.F.	.26	.38		.64	.94	
	6052	2" x 4"		900	.018		.34	.43		.77	1.09	
	6102	2" x 6"		800	.020		.54	.48		1.02	1.41	
	6152	2" x 8"		650	.025		.83	.59		1.42	1.91	
15	0010	**FRAMING, COLUMNS**										515
	0101	4" x 4"	2 Carp	390	.041	L.F.	1.25	.98		2.23	3.05	
	0151	4" x 6"		275	.058		2	1.40		3.40	4.57	
	0201	4" x 8"		220	.073		2.69	1.75		4.44	5.90	
	0251	6" x 6"		215	.074		4.45	1.79		6.24	7.90	
	0301	6" x 8"		175	.091		5.85	2.19		8.04	10.20	

06110 | Wood Framing

			CREW	DAILY OUTPUT	LABOR-HOURS	UNIT	2005 BARE COSTS MAT.	LABOR	EQUIP.	TOTAL	TOTAL INCL O&P
515	0351	6″ x 10″	2 Carp	150	.107	L.F.	9.50	2.56		12.06	14.75
520	0010	**FRAMING, HEAVY** Mill timber, beams, single 6″ x 10″	2 Carp	1.10	14.545	M.B.F.	2,125	350		2,475	2,950
	0100	Single 8″ x 16″		1.20	13.333	″	2,675	320		2,995	3,500
	0202	Built from 2″ lumber, multiple 2″ x 14″		900	.018	B.F.	.93	.43		1.36	1.74
	0212	Built from 3″ lumber, multiple 3″ x 6″		700	.023		1.28	.55		1.83	2.34
	0222	Multiple 3″ x 8″		800	.020		1.30	.48		1.78	2.24
	0232	Multiple 3″ x 10″		900	.018		1.30	.43		1.73	2.15
	0242	Multiple 3″ x 12″		1,000	.016		1.31	.38		1.69	2.09
	0252	Built from 4″ lumber, multiple 4″ x 6″		800	.020		1	.48		1.48	1.92
	0262	Multiple 4″ x 8″		900	.018		1.01	.43		1.44	1.83
	0272	Multiple 4″ x 10″		1,000	.016		1.01	.38		1.39	1.76
	0282	Multiple 4″ x 12″		1,100	.015		1.10	.35		1.45	1.80
	0292	Columns, structural grade, 1500f, 4″ x 4″		450	.036	L.F.	2.31	.85		3.16	3.99
	0302	6″ x 6″		225	.071		6.35	1.71		8.06	9.90
	0402	8″ x 8″		240	.067		12.35	1.60		13.95	16.25
	0502	10″ x 10″		90	.178		19.65	4.27		23.92	29
	0602	12″ x 12″		70	.229		28.50	5.50		34	40.50
	0802	Floor planks, 2″ thick, T & G, 2″ x 6″		1,050	.015	B.F.	1.86	.37		2.23	2.67
	0902	2″ x 10″		1,100	.015		1.86	.35		2.21	2.64
	1102	3″ thick, 3″ x 6″		1,050	.015		1.34	.37		1.71	2.09
	1202	3″ x 10″		1,100	.015		1.34	.35		1.69	2.06
	1402	Girders, structural grade, 12″ x 12″		800	.020		1.92	.48		2.40	2.93
	1502	10″ x 16″		1,000	.016		1.87	.38		2.25	2.71
	2050	Roof planks, see division 06150-600									
	2302	Roof purlins, 4″ thick, structural grade	2 Carp	1,050	.015	B.F.	1.31	.37		1.68	2.06
	2502	Roof trusses, add timber connectors, division 06090-800	″	450	.036	″	1.28	.85		2.13	2.86
530	0010	**FRAMING, JOISTS**									
	2002	Joists, 2″ x 4″	2 Carp	1,250	.013	L.F.	.34	.31		.65	.89
	2007	Pneumatic nailed		1,438	.011		.34	.27		.61	.82
	2100	2″ x 6″		1,250	.013		.54	.31		.85	1.11
	2105	Pneumatic nailed		1,438	.011		.54	.27		.81	1.04
	2152	2″ x 8″		1,100	.015		.83	.35		1.18	1.50
	2157	Pneumatic nailed		1,265	.013		.83	.30		1.13	1.43
	2202	2″ x 10″		900	.018		1.18	.43		1.61	2.01
	2207	Pneumatic nailed		1,035	.015		1.18	.37		1.55	1.92
	2252	2″ x 12″		875	.018		1.60	.44		2.04	2.51
	2257	Pneumatic nailed		1,006	.016		1.60	.38		1.98	2.41
	2302	2″ x 14″		770	.021		2.16	.50		2.66	3.22
	2307	Pneumatic nailed		886	.018		2.16	.43		2.59	3.11
	2352	3″ x 6″		925	.017		1.92	.42		2.34	2.82
	2402	3″ x 10″		780	.021		3.26	.49		3.75	4.43
	2452	3″ x 12″		600	.027		3.92	.64		4.56	5.40
	2502	4″ x 6″		800	.020		2	.48		2.48	3.02
	2552	4″ x 10″		600	.027		3.37	.64		4.01	4.80
	2602	4″ x 12″		450	.036		4.40	.85		5.25	6.30
	2607	Sister joist, 2″ x 6″		800	.020		.54	.48		1.02	1.41
	2608	Pneumatic nailed		960	.017		.54	.40		.94	1.27
	3000	Composite wood joist 9-1/2″ deep		.90	17.778	M.L.F.	1,475	425		1,900	2,350
	3010	11-1/2″ deep		.88	18.182		1,550	435		1,985	2,475
	3020	14″ deep		.82	19.512		1,775	470		2,245	2,775
	3030	16″ deep		.78	20.513		2,550	490		3,040	3,625
	4000	Open web joist 12″ deep		.88	18.182		1,675	435		2,110	2,600
	4010	14″ deep		.82	19.512		1,950	470		2,420	2,950
	4020	16″ deep		.78	20.513		2,025	490		2,515	3,050

06110 | Wood Framing

	Description	Crew	Daily Output	Labor-Hours	Unit	2005 Bare Costs Mat.	Labor	Equip.	Total	Total Incl O&P	
4030	18" deep	2 Carp	.74	21.622	M.L.F.	2,050	520		2,570	3,125	530
6000	Composite rim joist, 1-1/4" x 9-1/2"		90	.178		1,825	4.27		1,829.27	2,000	
6010	1-1/4" x 11-1/2"		.88	18.182		2,100	435		2,535	3,075	
6020	1-1/4" x 14-1/2"		.82	19.512		2,575	470		3,045	3,625	
6030	1-1/4" x 16-1/2"		.78	20.513		2,925	490		3,415	4,050	
0010	**FRAMING, MISCELLANEOUS**										545
2002	Firestops, 2" x 4"	2 Carp	780	.021	L.F.	.34	.49		.83	1.21	
2007	Pneumatic nailed		952	.017		.34	.40		.74	1.06	
2102	2" x 6"		600	.027		.54	.64		1.18	1.68	
2107	Pneumatic nailed		732	.022		.54	.52		1.06	1.48	
5002	Nailers, treated, wood construction, 2" x 4"		800	.020		.49	.48		.97	1.36	
5007	Pneumatic nailed		960	.017		.49	.40		.89	1.22	
5102	2" x 6"		750	.021		.77	.51		1.28	1.72	
5107	Pneumatic nailed		900	.018		.77	.43		1.20	1.57	
5122	2" x 8"		700	.023		1	.55		1.55	2.03	
5127	Pneumatic nailed		840	.019		1	.46		1.46	1.88	
5202	Steel construction, 2" x 4"		750	.021		.49	.51		1	1.41	
5222	2" x 6"		700	.023		.77	.55		1.32	1.78	
5242	2" x 8"		650	.025		1	.59		1.59	2.10	
7002	Rough bucks, treated, for doors or windows, 2" x 6"		400	.040		.77	.96		1.73	2.48	
7007	Pneumatic nailed		480	.033		.77	.80		1.57	2.21	
7102	2" x 8"		380	.042		1	1.01		2.01	2.82	
7107	Pneumatic nailed		456	.035		1	.84		1.84	2.53	
8001	Stair stringers, 2" x 10"		130	.123		1.18	2.95		4.13	6.30	
8101	2" x 12"		130	.123		1.60	2.95		4.55	6.75	
8151	3" x 10"		125	.128		3.26	3.07		6.33	8.80	
8201	3" x 12"		125	.128		3.92	3.07		6.99	9.50	
8870	Composite LSL, 1-1/4" x 11-1/2"		130	.123		2.11	2.95		5.06	7.30	
8880	1-1/4" x 14-1/2"		130	.123		2.57	2.95		5.52	7.85	
0010	**PARTITIONS** Wood stud with single bottom plate and										550
0020	double top plate, no waste, std. & better lumber										
0182	2" x 4" studs, 8' high, studs 12" O.C.	2 Carp	80	.200	L.F.	4.10	4.80		8.90	12.65	
0187	12" O.C., pneumatic nailed		96	.167		4.10	4		8.10	11.30	
0202	16" O.C.		100	.160		3.35	3.84		7.19	10.20	
0207	16" O.C., pneumatic nailed		120	.133		3.35	3.20		6.55	9.15	
0302	24" O.C.		125	.128		2.61	3.07		5.68	8.05	
0307	24" O.C., pneumatic nailed		150	.107		2.61	2.56		5.17	7.20	
0382	10' high, studs 12" O.C.		80	.200		4.84	4.80		9.64	13.50	
0387	12" O.C., pneumatic nailed		96	.167		4.84	4		8.84	12.15	
0402	16" O.C.		100	.160		3.91	3.84		7.75	10.80	
0407	16" O.C., pneumatic nailed		120	.133		3.91	3.20		7.11	9.75	
0502	24" O.C.		125	.128		2.98	3.07		6.05	8.50	
0507	24" O.C., pneumatic nailed		150	.107		2.98	2.56		5.54	7.65	
0582	12' high, studs 12" O.C.		65	.246		5.60	5.90		11.50	16.20	
0587	12" O.C., pneumatic nailed		78	.205		5.60	4.92		10.52	14.50	
0602	16" O.C.		80	.200		4.47	4.80		9.27	13.05	
0607	16" O.C., pneumatic nailed		96	.167		4.47	4		8.47	11.70	
0700	24" O.C.		100	.160		3.05	3.84		6.89	9.85	
0705	24" O.C., pneumatic nailed		120	.133		3.05	3.20		6.25	8.80	
0782	2" x 6" studs, 8' high, studs 12" O.C.		70	.229		6.50	5.50		12	16.45	
0787	12" O.C., pneumatic nailed		84	.190		6.50	4.57		11.07	14.90	
0802	16" O.C.		90	.178		5.30	4.27		9.57	13.10	
0807	16" O.C., pneumatic nailed		108	.148		5.30	3.56		8.86	11.90	
0902	24" O.C.		115	.139		4.13	3.34		7.47	10.20	
0907	24" O.C., pneumatic nailed		138	.116		4.13	2.78		6.91	9.25	

	06110 | Wood Framing		CREW	DAILY OUTPUT	LABOR-HOURS	UNIT	2005 BARE COSTS MAT.	LABOR	EQUIP.	TOTAL	TOTAL INCL O&P
550	0982	10′ high, studs 12″ O.C.	2 Carp	70	.229	L.F.	7.70	5.50		13.20	17.75
	0987	12″ O.C., pneumatic nailed		84	.190		7.70	4.57		12.27	16.20
	1002	16″ O.C.		90	.178		6.20	4.27		10.47	14.05
	1007	16″ O.C., pneumatic nailed		108	.148		6.20	3.56		9.76	12.85
	1102	24″ O.C.		115	.139		4.73	3.34		8.07	10.85
	1107	24″ O.C., pneumatic nailed		138	.116		4.73	2.78		7.51	9.90
	1182	12′ high, studs 12″ O.C.		55	.291		8.85	7		15.85	21.50
	1187	12″ O.C., pneumatic nailed		66	.242		8.85	5.80		14.65	19.65
	1202	16″ O.C.		70	.229		7.10	5.50		12.60	17.10
	1207	16″ O.C., pneumatic nailed		84	.190		7.10	4.57		11.67	15.55
	1302	24″ O.C.		90	.178		5.30	4.27		9.57	13.10
	1307	24″ O.C., pneumatic nailed		108	.148		5.30	3.56		8.86	11.90
	1402	For horizontal blocking, 2″ x 4″, add		600	.027		.37	.64		1.01	1.50
	1502	2″ x 6″, add		600	.027		.59	.64		1.23	1.74
	1600	For openings, add	↓	250	.064	↓		1.54		1.54	2.61
	1702	Headers for above openings, material only, add				B.F.	.68			.68	.75
552	0010	**FRAMING, PORCH OR DECK**									
	0100	Treated lumber, posts or columns, 4″ x 4″	2 Carp	390	.041	L.F.	1.38	.98		2.36	3.19
	0110	4″ x 6″		275	.058		3.03	1.40		4.43	5.70
	0120	4″ x 8″		220	.073		6	1.75		7.75	9.55
	0130	Girder, single, 4″ x 4″		675	.024		1.38	.57		1.95	2.49
	0140	4″ x 6″		600	.027		3.03	.64		3.67	4.42
	0150	4″ x 8″		525	.030		6	.73		6.73	7.85
	0160	Double, 2″ x 4″		625	.026		1	.61		1.61	2.14
	0170	2″ x 6″		600	.027		1.59	.64		2.23	2.83
	0180	2″ x 8″		575	.028		2.06	.67		2.73	3.40
	0190	2″ x 10″		550	.029		3.04	.70		3.74	4.54
	0200	2″ x 12″		525	.030		3.90	.73		4.63	5.55
	0210	Triple, 2″ x 4″		575	.028		1.50	.67		2.17	2.79
	0220	2″ x 6″		550	.029		2.38	.70		3.08	3.81
	0230	2″ x 8″		525	.030		3.09	.73		3.82	4.64
	0240	2″ x 10″		500	.032		4.56	.77		5.33	6.30
	0250	2″ x 12″		475	.034		5.85	.81		6.66	7.80
	0260	Ledger, bolted 4′ O.C., 2″ x 4″		400	.040		.58	.96		1.54	2.27
	0270	2″ x 6″		395	.041		.87	.97		1.84	2.60
	0280	2″ x 8″		390	.041		1.10	.98		2.08	2.88
	0300	2″ x 12″		380	.042		2.01	1.01		3.02	3.93
	0310	Joists, 2″ x 4″		1,250	.013		.50	.31		.81	1.07
	0320	2″ x 6″		1,250	.013		.79	.31		1.10	1.39
	0330	2″ x 8″		1,100	.015		1.03	.35		1.38	1.72
	0340	2″ x 10″		900	.018		1.52	.43		1.95	2.40
	0350	2″ x 12″	↓	875	.018		1.83	.44		2.27	2.76
	0360	Railings and trim , 1″ x 4″	1 Carp	300	.027		.60	.64		1.24	1.75
	0370	2″ x 2″		300	.027		.30	.64		.94	1.42
	0380	2″ x 4″		300	.027		.49	.64		1.13	1.63
	0390	2″ x 6″		300	.027	↓	.77	.64		1.41	1.94
	0400	Decking, 1″ x 4″		275	.029	S.F.	1.89	.70		2.59	3.27
	0410	2″ x 4″		300	.027		1.67	.64		2.31	2.93
	0420	2″ x 6″		320	.025		1.68	.60		2.28	2.87
	0430	5/4″ x 6″	↓	320	.025	↓	2.71	.60		3.31	4
	0440	Balusters, square, 2″ x 2″	2 Carp	220	.073	L.F.	.31	1.75		2.06	3.30
	0450	Turned, 2″ x 2″		140	.114		.41	2.74		3.15	5.10
	0460	Stair stringer, 2″ x 10″		130	.123		1.52	2.95		4.47	6.70
	0470	2″ x 12″		130	.123		1.83	2.95		4.78	7
	0480	Stair treads, 1″ x 4″		140	.114		.64	2.74		3.38	5.35
	0490	2″ x 4″	↓	140	.114	↓	.50	2.74		3.24	5.20

06110 | Wood Framing

	06110 | Wood Framing		Crew	Daily Output	Labor-Hours	Unit	2005 Bare Costs Mat.	Labor	Equip.	Total	Total Incl O&P	
552	0500	2" x 6"	2 Carp	160	.100	L.F.	.75	2.40		3.15	4.91	552
	0510	5/4" x 6"		160	.100		1.25	2.40		3.65	5.45	
	0520	Turned handrail post, 4" x 4"		64	.250	Ea.	13.05	6		19.05	24.50	
	0530	Lattice panel, 4' x 8'		1,600	.010	S.F.	.66	.24		.90	1.14	
	0540	Cedar, posts or columns, 4" x 4"		390	.041	L.F.	1.34	.98		2.32	3.15	
	0550	4" x 6"		275	.058		2.44	1.40		3.84	5.05	
	0560	4" x 8"		220	.073		5.85	1.75		7.60	9.35	
	0800	Decking, 1" x 4"		550	.029		1.13	.70		1.83	2.44	
	0810	2" x 4"		600	.027		2.33	.64		2.97	3.66	
	0820	2" x 6"		640	.025		4.25	.60		4.85	5.70	
	0830	5/4" x 6"		640	.025		3.16	.60		3.76	4.49	
	0840	Railings and trim, 1" x 4"		600	.027		1.13	.64		1.77	2.34	
	0860	2" x 4"		600	.027		2.33	.64		2.97	3.66	
	0870	2" x 6"		600	.027		4.25	.64		4.89	5.75	
	0920	Stair treads, 1" x 4"		140	.114		1.13	2.74		3.87	5.90	
	0930	2" x 4"		140	.114		2.33	2.74		5.07	7.25	
	0940	2" x 6"		160	.100		4.25	2.40		6.65	8.75	
	0950	5/4" x 6"		160	.100		3.16	2.40		5.56	7.55	
	0980	Redwood, posts or columns, 4" x 4"		390	.041		3.73	.98		4.71	5.80	
	0990	4" x 6"		275	.058		6.75	1.40		8.15	9.75	
	1000	4" x 8"		220	.073		10.05	1.75		11.80	14.05	
	1240	Redwood decking, 1" x 4"	1 Carp	275	.029	S.F.	5.05	.70		5.75	6.75	
	1260	2" x 6"		340	.024		12.55	.56		13.11	14.75	
	1270	5/4" x 6"		320	.025		8.10	.60		8.70	9.90	
	1280	Railings and trim, 1" x 4"	2 Carp	600	.027	L.F.	1.48	.64		2.12	2.72	
	1310	2" x 6"		600	.027		5.80	.64		6.44	7.45	
	1420	Alternative decking, wood / plastic composite, 5/4" x 6"		640	.025		2.67	.60		3.27	3.96	
	1430	Vinyl, 1-1/2" x 5-1/2"		640	.025		3.16	.60		3.76	4.50	
	1440	1" x 4" square edge fir		550	.029		1.10	.70		1.80	2.40	
	1450	1" x 4" tongue and groove fir		450	.036		1.10	.85		1.95	2.66	
	1460	1" x 4" mahogany		550	.029		1.40	.70		2.10	2.73	
	1470	Accessories, joist hangers, 2" x 4"	1 Carp	160	.050	Ea.	.58	1.20		1.78	2.68	
	1480	2" x 6" through 2" x 12"	"	150	.053		.67	1.28		1.95	2.91	
	1530	Post footing, incl excav, backfill, tube form & concrete, 4' deep, 8" dia	F-7	12	2.667		10.10	55		65.10	105	
	1540	10" diameter		11	2.909		14	60		74	117	
	1550	12" diameter		10	3.200		18.40	66		84.40	132	
555	0010	**FRAMING, ROOFS**										555
	2001	Fascia boards, 2" x 8"	2 Carp	225	.071	L.F.	.83	1.71		2.54	3.81	
	2101	2" x 10"		180	.089		1.18	2.13		3.31	4.91	
	5002	Rafters, to 4 in 12 pitch, 2" x 6", ordinary		1,000	.016		.54	.38		.92	1.24	
	5021	On steep roofs		800	.020		.54	.48		1.02	1.41	
	5041	On dormers or complex roofs		590	.027		.54	.65		1.19	1.70	
	5062	2" x 8", ordinary		950	.017		.83	.40		1.23	1.60	
	5081	On steep roofs		750	.021		.83	.51		1.34	1.78	
	5101	On dormers or complex roofs		540	.030		.83	.71		1.54	2.12	
	5122	2" x 10", ordinary		630	.025		1.18	.61		1.79	2.33	
	5141	On steep roofs		495	.032		1.18	.78		1.96	2.61	
	5161	On dormers or complex roofs		425	.038		1.18	.90		2.08	2.82	
	5182	2" x 12", ordinary		575	.028		1.60	.67		2.27	2.89	
	5201	On steep roofs		455	.035		1.60	.84		2.44	3.19	
	5221	On dormers or complex roofs		395	.041		1.60	.97		2.57	3.41	
	5250	Composite rafter, 9-1/2" deep		575	.028		1.47	.67		2.14	2.74	
	5260	11-1/2" deep		575	.028		1.56	.67		2.23	2.84	
	5301	Hip and valley rafters, 2" x 6", ordinary		760	.021		.54	.51		1.05	1.45	
	5321	On steep roofs		585	.027		.54	.66		1.20	1.70	
	5341	On dormers or complex roofs		510	.031		.54	.75		1.29	1.87	

	06110 | Wood Framing		CREW	DAILY OUTPUT	LABOR-HOURS	UNIT	2005 BARE COSTS MAT.	LABOR	EQUIP.	TOTAL	TOTAL INCL O&P
555	5361	2" x 8", ordinary	2 Carp	720	.022	L.F.	.83	.53		1.36	1.82
	5381	On steep roofs		545	.029		.83	.70		1.53	2.11
	5401	On dormers or complex roofs		470	.034		.83	.82		1.65	2.30
	5421	2" x 10", ordinary		570	.028		1.18	.67		1.85	2.43
	5441	On steep roofs		440	.036		1.18	.87		2.05	2.77
	5461	On dormers or complex roofs		380	.042		1.18	1.01		2.19	3.01
	5470										
	5481	Hip and valley rafters, 2" x 12", ordinary	2 Carp	525	.030	L.F.	1.60	.73		2.33	3
	5501	On steep roofs		410	.039		1.60	.94		2.54	3.35
	5521	On dormers or complex roofs		355	.045		1.60	1.08		2.68	3.60
	5541	Hip and valley jacks, 2" x 6", ordinary		600	.027		.54	.64		1.18	1.68
	5561	On steep roofs		475	.034		.54	.81		1.35	1.96
	5581	On dormers or complex roofs		410	.039		.54	.94		1.48	2.18
	5601	2" x 8", ordinary		490	.033		.83	.78		1.61	2.24
	5621	On steep roofs		385	.042		.83	1		1.83	2.60
	5641	On dormers or complex roofs		335	.048		.83	1.15		1.98	2.86
	5661	2" x 10", ordinary		450	.036		1.18	.85		2.03	2.74
	5681	On steep roofs		350	.046		1.18	1.10		2.28	3.15
	5701	On dormers or complex roofs		305	.052		1.18	1.26		2.44	3.43
	5721	2" x 12", ordinary		375	.043		1.60	1.02		2.62	3.50
	5741	On steep roofs		295	.054		1.60	1.30		2.90	3.97
	5762	On dormers or complex roofs		255	.063		1.60	1.51		3.11	4.32
	5781	Rafter tie, 1" x 4", #3		800	.020		.41	.48		.89	1.27
	5791	2" x 4", #3		800	.020		.34	.48		.82	1.19
	5801	Ridge board, #2 or better, 1" x 6"		600	.027		.91	.64		1.55	2.09
	5821	1" x 8"		550	.029		1.22	.70		1.92	2.53
	5841	1" x 10"		500	.032		1.52	.77		2.29	2.97
	5861	2" x 6"		500	.032		.54	.77		1.31	1.89
	5881	2" x 8"		450	.036		.83	.85		1.68	2.36
	5901	2" x 10"		400	.040		1.18	.96		2.14	2.92
	5921	Roof cants, split, 4" x 4"		650	.025		1.25	.59		1.84	2.38
	5941	6" x 6"		600	.027		4.45	.64		5.09	6
	5961	Roof curbs, untreated, 2" x 6"		520	.031		.54	.74		1.28	1.84
	5981	2" x 12"		400	.040		1.60	.96		2.56	3.39
	6001	Sister rafters, 2" x 6"		800	.020		.54	.48		1.02	1.41
	6021	2" x 8"		640	.025		.83	.60		1.43	1.93
	6041	2" x 10"		535	.030		1.18	.72		1.90	2.51
	6061	2" x 12"		455	.035		1.60	.84		2.44	3.19
560	0010	**FRAMING, SILLS**									
	2002	Ledgers, nailed, 2" x 4"	2 Carp	755	.021	L.F.	.34	.51		.85	1.23
	2052	2" x 6"		600	.027		.54	.64		1.18	1.68
	2102	Bolted, not including bolts, 3" x 6"		325	.049		1.92	1.18		3.10	4.12
	2152	3" x 12"		233	.069		3.92	1.65		5.57	7.10
	2602	Mud sills, redwood, construction grade, 2" x 4"		895	.018		2.20	.43		2.63	3.15
	2622	2" x 6"		780	.021		3.30	.49		3.79	4.47
	4002	Sills, 2" x 4"		600	.027		.34	.64		.98	1.46
	4052	2" x 6"		550	.029		.54	.70		1.24	1.78
	4082	2" x 8"		500	.032		.83	.77		1.60	2.21
	4101	2" x 10"		450	.036		1.18	.85		2.03	2.74
	4121	2" x 12"		400	.040		1.60	.96		2.56	3.39
	4202	Treated, 2" x 4"		550	.029		.49	.70		1.19	1.73
	4222	2" x 6"		500	.032		.77	.77		1.54	2.15
	4242	2" x 8"		450	.036		1	.85		1.85	2.55
	4261	2" x 10"		400	.040		1.49	.96		2.45	3.27
	4281	2" x 12"		350	.046		1.91	1.10		3.01	3.96
	4402	4" x 4"		450	.036		1.35	.85		2.20	2.94

06110 | Wood Framing		Crew	Daily Output	Labor-Hours	Unit	2005 Bare Costs Mat.	2005 Bare Costs Labor	2005 Bare Costs Equip.	2005 Bare Costs Total	Total Incl O&P	
4422	4" x 6"	2 Carp	350	.046	L.F.	2.99	1.10		4.09	5.15	560
4462	4" x 8"		300	.053		5.95	1.28		7.23	8.70	
4480	4" x 10"		260	.062		7.40	1.48		8.88	10.65	
0010	**FRAMING, SLEEPERS**										565
0100	On concrete, treated, 1" x 2"	2 Carp	2,350	.007	L.F.	.10	.16		.26	.39	
0150	1" x 3"		2,000	.008		.19	.19		.38	.54	
0200	2" x 4"		1,500	.011		.49	.26		.75	.97	
0250	2" x 6"		1,300	.012		.77	.30		1.07	1.35	
0010	**FRAMING, SOFFITS & CANOPIES**										570
1002	Canopy or soffit framing , 1" x 4"	2 Carp	900	.018	L.F.	.60	.43		1.03	1.38	
1021	1" x 6"		850	.019		.91	.45		1.36	1.77	
1042	1" x 8"		750	.021		1.22	.51		1.73	2.21	
1102	2" x 4"		620	.026		.34	.62		.96	1.42	
1121	2" x 6"		560	.029		.54	.69		1.23	1.75	
1142	2" x 8"		500	.032		.83	.77		1.60	2.21	
1202	3" x 4"		500	.032		1.09	.77		1.86	2.50	
1221	3" x 6"		400	.040		1.92	.96		2.88	3.74	
1242	3" x 10"		300	.053		3.26	1.28		4.54	5.75	
1250											
0010	**FRAMING, TREATED LUMBER**										575
0100	2" x 4"				M.B.F.	730			730	805	
0110	2" x 6"					770			770	850	
0120	2" x 8"					750			750	825	
0130	2" x 10"					890			890	980	
0140	2" x 12"					955			955	1,050	
0200	4" x 4"					1,025			1,025	1,125	
0210	4" x 6"					1,500			1,500	1,650	
0220	4" x 8"					2,225			2,225	2,450	
0010	**FRAMING, WALLS** R06100-010										590
0100	Door buck, studs, header, access, 8' H, 2"x4" wall, 3' W	1 Carp	32	.250	Ea.	14.35	6		20.35	26	
0110	4' wide R06110-030		32	.250		15.40	6		21.40	27	
0120	5' wide		32	.250		19.50	6		25.50	31.50	
0130	6' wide		32	.250		21	6		27	33.50	
0140	8' wide		30	.267		30	6.40		36.40	44.50	
0150	10' wide		30	.267		43.50	6.40		49.90	59	
0160	12' wide		30	.267		63.50	6.40		69.90	81	
0170	2" x 6" wall, 3' wide		32	.250		20.50	6		26.50	32.50	
0180	4' wide		32	.250		22	6		28	34	
0190	5' wide		32	.250		26	6		32	38.50	
0200	6' wide		32	.250		27.50	6		33.50	40.50	
0210	8' wide		30	.267		36.50	6.40		42.90	51	
0220	10' wide		30	.267		50	6.40		56.40	66	
0230	12' wide		30	.267		70	6.40		76.40	88	
0240	Window buck, studs, header & access, 8' high 2" x 4" wall, 2' wide		24	.333		15.05	8		23.05	30	
0250	3' wide		24	.333		17.70	8		25.70	33	
0260	4' wide		24	.333		19.45	8		27.45	35	
0270	5' wide		24	.333		23.50	8		31.50	39.50	
0280	6' wide		24	.333		26.50	8		34.50	42.50	
0300	8' wide		22	.364		37.50	8.75		46.25	56	
0310	10' wide		22	.364		52	8.75		60.75	72	
0320	12' wide		22	.364		74	8.75		82.75	96.50	
0330	2" x 6" wall, 2' wide		24	.333		23	8		31	39	

06110 | Wood Framing			CREW	DAILY OUTPUT	LABOR-HOURS	UNIT	2005 BARE COSTS MAT.	LABOR	EQUIP.	TOTAL	TOTAL INCL O&P
590 0340	3′ wide	R06100-010	1 Carp	24	.333	Ea.	26	8		34	42
0350	4′ wide			24	.333		28	8		36	44
0360	5′ wide	R06110-030		24	.333		32.50	8		40.50	49
0370	6′ wide			24	.333		35.50	8		43.50	52.50
0380	7′ wide			24	.333		44	8		52	62
0390	8′ wide			22	.364		48	8.75		56.75	67.50
0400	10′ wide			22	.364		63	8.75		71.75	84
0410	12′ wide		↓	22	.364	↓	86	8.75		94.75	110
2002	Headers over openings, 2″ x 6″		2 Carp	360	.044	L.F.	.54	1.07		1.61	2.40
2007	2″ x 6″, pneumatic nailed			432	.037		.54	.89		1.43	2.10
2052	2″ x 8″			340	.047		.83	1.13		1.96	2.83
2057	2″ x 8″, pneumatic nailed			408	.039		.83	.94		1.77	2.51
2101	2″ x 10″			320	.050		1.18	1.20		2.38	3.33
2106	2″ x 10″, pneumatic nailed			384	.042		1.18	1		2.18	2.99
2152	2″ x 12″			300	.053		1.60	1.28		2.88	3.93
2157	2″ x 12″, pneumatic nailed			360	.044		1.60	1.07		2.67	3.57
2191	4″ x 10″			240	.067		3.37	1.60		4.97	6.45
2196	4″ x 10″, pneumatic nailed			288	.056		3.37	1.33		4.70	5.95
2202	4″ x 12″			190	.084		4.40	2.02		6.42	8.25
2207	4″ x 12″, pneumatic nailed			228	.070		4.40	1.68		6.08	7.70
2241	6″ x 10″			165	.097		9.50	2.33		11.83	14.35
2246	6″ x 10″, pneumatic nailed			198	.081		9.50	1.94		11.44	13.70
2251	6″ x 12″			140	.114		9.10	2.74		11.84	14.65
2256	6″ x 12″, pneumatic nailed			168	.095		9.10	2.29		11.39	13.90
3000	Radius, 2″ x 6″			270	.059		.84	1.42		2.26	3.33
3010	2″ x 6″, pneumatic nailed			324	.049		.84	1.19		2.03	2.93
3020	2″ x 8″			255	.063		1.29	1.51		2.80	3.98
3030	2″ x 8″, pneumatic nailed			296	.054		1.29	1.30		2.59	3.62
3040	2″ x 10″			240	.067		1.84	1.60		3.44	4.74
3050	2″ x 10″, pneumatic nailed			285	.056		1.84	1.35		3.19	4.31
3060	2″ x 12″			225	.071		2.50	1.71		4.21	5.65
3070	2″ x 12″, pneumatic nailed			270	.059		2.50	1.42		3.92	5.15
5002	Plates, untreated, 2″ x 3″			850	.019		.26	.45		.71	1.06
5007	2″ x 3″, pneumatic nailed			1,020	.016		.26	.38		.64	.93
5022	2″ x 4″			800	.020		.34	.48		.82	1.19
5027	2″ x 4″, pneumatic nailed			960	.017		.34	.40		.74	1.05
5040	2″ x 6″			750	.021		.54	.51		1.05	1.46
5045	2″ x 6″, pneumatic nailed			900	.018		.54	.43		.97	1.31
5061	Treated, 2″ x 3″			850	.019		.49	.45		.94	1.31
5066	2″ x 3″, treated, pneumatic nailed			1,020	.016		.49	.38		.87	1.18
5081	2″ x 4″			800	.020		.49	.48		.97	1.36
5086	2″ x 4″, treated, pneumatic nailed			960	.017		.49	.40		.89	1.22
5101	2″ x 6″			750	.021		.77	.51		1.28	1.72
5106	2″ x 6″, treated, pneumatic nailed			900	.018		.77	.43		1.20	1.57
5122	Studs, 8′ high wall, 2″ x 3″			1,200	.013		.26	.32		.58	.83
5127	2″ x 3″, pneumatic nailed			1,440	.011		.26	.27		.53	.74
5142	2″ x 4″			1,100	.015		.34	.35		.69	.96
5147	2″ x 4″, pneumatic nailed			1,320	.012		.34	.29		.63	.86
5162	2″ x 6″			1,000	.016		.54	.38		.92	1.24
5167	2″ x 6″, pneumatic nailed			1,200	.013		.54	.32		.86	1.13
5182	3″ x 4″			800	.020		1.09	.48		1.57	2.02
5187	3″ x 4″, pneumatic nailed			960	.017		1.09	.40		1.49	1.88
5201	Installed on second story, 2″ x 3″			1,170	.014		.26	.33		.59	.85
5206	2″ x 3″, pneumatic nailed			1,200	.013		.26	.32		.58	.83
5221	2″ x 4″			1,015	.016		.34	.38		.72	1.01
5226	2″ x 4″, pneumatic nailed		↓	1,080	.015	↓	.34	.36		.70	.97

06110 | Wood Framing			CREW	DAILY OUTPUT	LABOR-HOURS	UNIT	2005 BARE COSTS MAT.	LABOR	EQUIP.	TOTAL	TOTAL INCL O&P	
5241	2″ x 6″	R06100-010	2 Carp	890	.018	L.F.	.54	.43		.97	1.32	590
5246	2″ x 6″, pneumatic nailed			1,020	.016		.54	.38		.92	1.23	
5261	3″ x 4″	R06110-030		800	.020		1.09	.48		1.57	2.02	
5266	3″ x 4″, pneumatic nailed			960	.017		1.09	.40		1.49	1.88	
5281	Installed on dormer or gable, 2″ x 3″			1,045	.015		.26	.37		.63	.91	
5286	2″ x 3″, pneumatic nailed			1,254	.013		.26	.31		.57	.81	
5301	2″ x 4″			905	.018		.34	.42		.76	1.09	
5306	2″ x 4″, pneumatic nailed			1,086	.015		.34	.35		.69	.97	
5321	2″ x 6″			800	.020		.54	.48		1.02	1.41	
5326	2″ x 6″, pneumatic nailed			960	.017		.54	.40		.94	1.27	
5341	3″ x 4″			700	.023		1.09	.55		1.64	2.13	
5346	3″ x 4″, pneumatic nailed			840	.019		1.09	.46		1.55	1.98	
5361	6′ high wall, 2″ x 3″			970	.016		.26	.40		.66	.96	
5366	2″ x 3″, pneumatic nailed			1,164	.014		.26	.33		.59	.85	
5381	2″ x 4″			850	.019		.34	.45		.79	1.14	
5386	2″ x 4″, pneumatic nailed			1,020	.016		.34	.38		.72	1.01	
5401	2″ x 6″			740	.022		.54	.52		1.06	1.47	
5406	2″ x 6″, pneumatic nailed			888	.018		.54	.43		.97	1.32	
5421	3″ x 4″			600	.027		1.09	.64		1.73	2.29	
5426	3″ x 4″, pneumatic nailed			720	.022		1.09	.53		1.62	2.11	
5441	Installed on second story, 2″ x 3″			950	.017		.26	.40		.66	.98	
5446	2″ x 3″, pneumatic nailed			1,140	.014		.26	.34		.60	.86	
5461	2″ x 4″			810	.020		.34	.47		.81	1.17	
5466	2″ x 4″, pneumatic nailed			972	.016		.34	.40		.74	1.04	
5481	2″ x 6″			700	.023		.54	.55		1.09	1.52	
5486	2″ x 6″, pneumatic nailed			840	.019		.54	.46		1	1.37	
5501	3″ x 4″			550	.029		1.09	.70		1.79	2.39	
5506	3″ x 4″, pneumatic nailed			660	.024		1.09	.58		1.67	2.19	
5521	Installed on dormer or gable, 2″ x 3″			850	.019		.26	.45		.71	1.06	
5526	2″ x 3″, pneumatic nailed			1,020	.016		.26	.38		.64	.93	
5541	2″ x 4″			720	.022		.34	.53		.87	1.28	
5546	2″ x 4″, pneumatic nailed			864	.019		.34	.44		.78	1.12	
5561	2″ x 6″			620	.026		.54	.62		1.16	1.64	
5566	2″ x 6″, pneumatic nailed			744	.022		.54	.52		1.06	1.47	
5581	3″ x 4″			480	.033		1.09	.80		1.89	2.56	
5586	3″ x 4″, pneumatic nailed			576	.028		1.09	.67		1.76	2.33	
5601	3′ high wall, 2″ x 3″			740	.022		.26	.52		.78	1.17	
5606	2″ x 3″, pneumatic nailed			888	.018		.26	.43		.69	1.02	
5621	2″ x 4″			640	.025		.34	.60		.94	1.39	
5626	2″ x 4″, pneumatic nailed			768	.021		.34	.50		.84	1.22	
5641	2″ x 6″			550	.029		.54	.70		1.24	1.78	
5646	2″ x 6″, pneumatic nailed			660	.024		.54	.58		1.12	1.58	
5661	3″ x 4″			440	.036		1.09	.87		1.96	2.68	
5666	3″ x 4″, pneumatic nailed			528	.030		1.09	.73		1.82	2.43	
5681	Installed on second story, 2″ x 3″			700	.023		.26	.55		.81	1.22	
5686	2″ x 3″, pneumatic nailed			840	.019		.26	.46		.72	1.07	
5701	2″ x 4″			610	.026		.34	.63		.97	1.44	
5706	2″ x 4″, pneumatic nailed			732	.022		.34	.52		.86	1.26	
5721	2″ x 6″			520	.031		.54	.74		1.28	1.84	
5726	2″ x 6″, pneumatic nailed			624	.026		.54	.62		1.16	1.63	
5741	3″ x 4″			430	.037		1.09	.89		1.98	2.72	
5746	3″ x 4″, pneumatic nailed			516	.031		1.09	.74		1.83	2.46	
5761	Installed on dormer or gable, 2″ x 3″			625	.026		.26	.61		.87	1.33	
5766	2″ x 3″, pneumatic nailed			750	.021		.26	.51		.77	1.16	
5781	2″ x 4″			545	.029		.34	.70		1.04	1.57	
5786	2″ x 4″, pneumatic nailed			654	.024		.34	.59		.93	1.37	

	06110 | Wood Framing		CREW	DAILY OUTPUT	LABOR-HOURS	UNIT	2005 BARE COSTS MAT.	LABOR	EQUIP.	TOTAL	TOTAL INCL O&P
590	5801	2″ x 6″ R06100 -010	2 Carp	465	.034	L.F.	.54	.83		1.37	1.99
	5806	2″ x 6″, pneumatic nailed		558	.029		.54	.69		1.23	1.76
	5821	3″ x 4″ R06110 -030		380	.042		1.09	1.01		2.10	2.92
	5826	3″ x 4″, pneumatic nailed		456	.035		1.09	.84		1.93	2.63
	8250	For second story & above, add						5%			
	8300	For dormer & gable, add						15%			
600	0010	**FURRING** Wood strips, 1″ x 2″, on walls, on wood	1 Carp	550	.015	L.F.	.19	.35		.54	.80
	0017	Pneumatic nailed		710	.011		.19	.27		.46	.67
	0302	On masonry		495	.016		.19	.39		.58	.87
	0402	On concrete		260	.031		.19	.74		.93	1.46
	0602	1″ x 3″, on walls, on wood		550	.015		.28	.35		.63	.90
	0607	Pneumatic nailed		710	.011		.28	.27		.55	.77
	0702	On masonry		495	.016		.28	.39		.67	.97
	0802	On concrete		260	.031		.28	.74		1.02	1.56
	0852	On ceilings, on wood		350	.023		.28	.55		.83	1.24
	0857	Pneumatic nailed		450	.018		.28	.43		.71	1.03
	0902	On masonry		320	.025		.28	.60		.88	1.33
	0952	On concrete		210	.038		.28	.91		1.19	1.86
700	0010	**GROUNDS** For casework, 1″ x 2″ wood strips, on wood	1 Carp	330	.024	L.F.	.19	.58		.77	1.20
	0102	On masonry		285	.028		.19	.67		.86	1.35
	0202	On concrete		250	.032		.19	.77		.96	1.51
	0402	For plaster, 3/4″ deep, on wood		450	.018		.19	.43		.62	.93
	0502	On masonry		225	.036		.19	.85		1.04	1.66
	0602	On concrete		175	.046		.19	1.10		1.29	2.07
	0702	On metal lath		200	.040		.19	.96		1.15	1.84
750	0010	**LOG STRUCTURES**									
	0020	Exterior walls, pine, D logs, with double T & G, 6″ x 6″	2 Carp	500	.032	L.F.	3.83	.77		4.60	5.50
	0030	6″ x 8″		375	.043		4.63	1.02		5.65	6.85
	0040	8″ x 6″		375	.043		3.83	1.02		4.85	5.95
	0050	8″ x 7″		322	.050		3.83	1.19		5.02	6.25
	0060	Square/rectangular logs, with double T & G, 6″ x 6″		500	.032		3.83	.77		4.60	5.50
	0160	Upper floor framing, pine, posts/columns, 4″ x 6″		750	.021		2	.51		2.51	3.07
	0180	4″ x 8″		562	.028		2.69	.68		3.37	4.11
	0190	6″ x 6″		500	.032		4.45	.77		5.22	6.20
	0200	6″ x 8″		375	.043		5.85	1.02		6.87	8.20
	0210	8″ x 8″		281	.057		8.05	1.37		9.42	11.15
	0220	8″ x 10″		225	.071		10.10	1.71		11.81	14
	0230	Beams, 4″ x 8″		562	.028		2.69	.68		3.37	4.11
	0240	4″ x 10″		449	.036		3.37	.86		4.23	5.15
	0250	4″ x 12″		375	.043		4.40	1.02		5.42	6.60
	0260	6″ x 8″		375	.043		5.85	1.02		6.87	8.20
	0270	6″ x 10″		300	.053		9.50	1.28		10.78	12.55
	0280	6″ x 12″		250	.064		9.10	1.54		10.64	12.60
	0290	8″ x 10″		225	.071		10.10	1.71		11.81	14
	0300	8″ x 12″		188	.085		9.85	2.04		11.89	14.25
	0310	Joists, 4″ x 8″		562	.028		2.69	.68		3.37	4.11
	0320	4″ x 10″		449	.036		3.37	.86		4.23	5.15
	0330	4″ x 12″		375	.043		4.40	1.02		5.42	6.60
	0340	6″ x 8″		375	.043		5.85	1.02		6.87	8.20
	0350	6″ x 10″		300	.053		9.50	1.28		10.78	12.55
	0390	Decking, 1″ x 6″ T & G		964	.017	S.F.	1.20	.40		1.60	2
	0400	1″ x 8″ T & G		700	.023	″	1.13	.55		1.68	2.17
	0420	Gable end roof framing, rafters, 4″ x 8″		562	.028	L.F.	2.69	.68		3.37	4.11

06150 | Wood Decking		CREW	DAILY OUTPUT	LABOR-HOURS	UNIT	2005 BARE COSTS MAT.	LABOR	EQUIP.	TOTAL	TOTAL INCL O&P	
0010	ROOF DECKS										600
0400	Cedar planks, 3" thick	2 Carp	320	.050	S.F.	6.40	1.20		7.60	9.10	
0500	4" thick		250	.064		8.65	1.54		10.19	12.10	
0702	Douglas fir, 3" thick		320	.050		2.40	1.20		3.60	4.68	
0802	4" thick		250	.064		3.21	1.54		4.75	6.15	
1002	Hemlock, 3" thick		320	.050		2.40	1.20		3.60	4.68	
1102	4" thick		250	.064		3.20	1.54		4.74	6.15	
1302	Western white spruce, 3" thick		320	.050		2.31	1.20		3.51	4.58	
1402	4" thick		250	.064		3.08	1.54		4.62	6	

06160 | Sheathing		CREW	DAILY OUTPUT	LABOR-HOURS	UNIT	MAT.	LABOR	EQUIP.	TOTAL	TOTAL INCL O&P	
0010	SHEATHING Plywood on roof, CDX (R06160-020)										800
0032	5/16" thick	2 Carp	1,600	.010	S.F.	.63	.24		.87	1.10	
0037	Pneumatic nailed		1,952	.008		.63	.20		.83	1.02	
0052	3/8" thick		1,525	.010		.65	.25		.90	1.15	
0057	Pneumatic nailed		1,860	.009		.65	.21		.86	1.07	
0102	1/2" thick		1,400	.011		.62	.27		.89	1.15	
0103	Pneumatic nailed		1,708	.009		.62	.22		.84	1.06	
0202	5/8" thick		1,300	.012		.78	.30		1.08	1.36	
0207	Pneumatic nailed		1,586	.010		.78	.24		1.02	1.27	
0302	3/4" thick		1,200	.013		.94	.32		1.26	1.57	
0307	Pneumatic nailed		1,464	.011		.94	.26		1.20	1.48	
0502	Plywood on walls with exterior CDX, 3/8" thick		1,200	.013		.65	.32		.97	1.26	
0507	Pneumatic nailed		1,488	.011		.65	.26		.91	1.16	
0603	1/2" thick		1,125	.014		.62	.34		.96	1.26	
0608	Pneumatic nailed		1,395	.011		.62	.28		.90	1.15	
0702	5/8" thick		1,050	.015		.78	.37		1.15	1.48	
0707	Pneumatic nailed		1,302	.012		.78	.30		1.08	1.36	
0803	3/4" thick		975	.016		.94	.39		1.33	1.70	
0808	Pneumatic nailed		1,209	.013		.94	.32		1.26	1.57	
1000	For shear wall construction, add						20%				
1200	For structural 1 exterior plywood, add				S.F.	10%					
1402	With boards, on roof 1" x 6" boards, laid horizontal	2 Carp	725	.022		1.25	.53		1.78	2.28	
1502	Laid diagonal		650	.025		1.25	.59		1.84	2.38	
1702	1" x 8" boards, laid horizontal		875	.018		1.11	.44		1.55	1.98	
1802	Laid diagonal		725	.022		1.11	.53		1.64	2.13	
2000	For steep roofs, add						40%				
2200	For dormers, hips and valleys, add					5%	50%				
2402	Boards on walls, 1" x 6" boards, laid regular	2 Carp	650	.025		1.25	.59		1.84	2.38	
2502	Laid diagonal		585	.027		1.25	.66		1.91	2.49	
2702	1" x 8" boards, laid regular		765	.021		1.11	.50		1.61	2.08	
2802	Laid diagonal		650	.025		1.11	.59		1.70	2.23	
2852	Gypsum, weatherproof, 1/2" thick		1,050	.015		.29	.37		.66	.94	
2902	Sealed, 4/10" thick		1,100	.015		.45	.35		.80	1.09	
3000	Wood fiber, regular, no vapor barrier, 1/2" thick		1,200	.013		.55	.32		.87	1.15	
3100	5/8" thick		1,200	.013		.72	.32		1.04	1.33	
3300	No vapor barrier, in colors, 1/2" thick		1,200	.013		.78	.32		1.10	1.40	
3400	5/8" thick		1,200	.013		.96	.32		1.28	1.60	
3600	With vapor barrier one side, white, 1/2" thick		1,200	.013		.55	.32		.87	1.15	
3700	Vapor barrier 2 sides, 1/2" thick		1,200	.013		.77	.32		1.09	1.39	
3800	Asphalt impregnated, 25/32" thick		1,200	.013		.28	.32		.60	.85	
3850	Intermediate, 1/2" thick		1,200	.013		.18	.32		.50	.74	
4000	Wafer board on roof, 1/2" thick		1,455	.011		.62	.26		.88	1.13	
4100	5/8" thick		1,330	.012		.78	.29		1.07	1.35	

	06160 | Sheathing		CREW	DAILY OUTPUT	LABOR-HOURS	UNIT	2005 BARE COSTS MAT.	LABOR	EQUIP.	TOTAL	TOTAL INCL O&P
850	0010	**SUBFLOOR** Plywood, CDX, 1/2″ thick	2 Carp	1,500	.011	SF Flr.	.62	.26		.88	1.11
	0017	Pneumatic nailed		1,860	.009		.62	.21		.83	1.03
	0102	5/8″ thick		1,350	.012		.78	.28		1.06	1.34
	0107	Pneumatic nailed		1,674	.010		.78	.23		1.01	1.25
	0202	3/4″ thick		1,250	.013		.94	.31		1.25	1.55
	0207	Pneumatic nailed		1,550	.010		.94	.25		1.19	1.45
	0302	1-1/8″ thick, 2-4-1 including underlayment		1,050	.015		1.19	.37		1.56	1.93
	0452	1″ x 8″ S4S, laid regular		1,000	.016		1.11	.38		1.49	1.88
	0462	Laid diagonal		850	.019		1.11	.45		1.56	2
	0502	1″ x 10″ S4S, laid regular		1,100	.015		1.36	.35		1.71	2.09
	0602	Laid diagonal		900	.018		1.36	.43		1.79	2.22
	1500	Wafer board, 5/8″ thick		1,330	.012	S.F.	.49	.29		.78	1.03
	1600	3/4″ thick		1,230	.013	″	.55	.31		.86	1.14
900	0011	**UNDERLAYMENT** Plywood, underlayment grade, 3/8″ thick	2 Carp	1,500	.011	SF Flr.	.87	.26		1.13	1.39
	0017	Pneumatic nailed		1,860	.009		.87	.21		1.08	1.31
	0102	1/2″ thick		1,450	.011		1.06	.26		1.32	1.62
	0107	Pneumatic nailed		1,798	.009		1.06	.21		1.27	1.53
	0202	5/8″ thick		1,400	.011		1.22	.27		1.49	1.81
	0207	Pneumatic nailed		1,736	.009		1.22	.22		1.44	1.72
	0302	3/4″ thick		1,300	.012		1.40	.30		1.70	2.04
	0306	Pneumatic nailed		1,612	.010		1.40	.24		1.64	1.94
	0502	Particle board, 3/8″ thick		1,500	.011		.37	.26		.63	.84
	0507	Pneumatic nailed		1,860	.009		.37	.21		.58	.76
	0602	1/2″ thick		1,450	.011		.39	.26		.65	.88
	0607	Pneumatic nailed		1,798	.009		.39	.21		.60	.79
	0802	5/8″ thick		1,400	.011		.49	.27		.76	1.01
	0807	Pneumatic nailed		1,736	.009		.49	.22		.71	.92
	0902	3/4″ thick		1,300	.012		.55	.30		.85	1.11
	0907	Pneumatic nailed		1,612	.010		.55	.24		.79	1.01
	1102	Hardboard, underlayment grade, 4′ x 4′, .215″ thick		1,500	.011		.39	.26		.65	.86

	06170 | Prefabricated Structural Wood		CREW	DAILY OUTPUT	LABOR-HOURS	UNIT	MAT.	LABOR	EQUIP.	TOTAL	TOTAL INCL O&P
200	0010	**LAMINATED BEAMS** Fb 2800									
	0050	3-1/2″ x 18″	F-3	480	.083	L.F.	28	1.80	1.26	31.06	35.50
	0100	5-1/4″ x 11-7/8″		450	.089		27.50	1.92	1.34	30.76	34.50
	0150	5-1/4″ x 16″		360	.111		37	2.40	1.67	41.07	46.50
	0200	5-1/4″ x 18″		290	.138		41.50	2.98	2.08	46.56	53
	0250	5-1/4″ x 24″		220	.182		47	3.92	2.74	53.66	61.50
	0300	7″ x 11-7/8″		320	.125		33.50	2.70	1.88	38.08	43
	0350	7″ x 16″		260	.154		45	3.32	2.32	50.64	57.50
	0400	7″ x 18″		210	.190		50.50	4.11	2.87	57.48	66
	0500	For premium appearance, add to S.F. prices					5%				
	0550	For industrial type, deduct					15%				
	0600	For stain and varnish, add					5%				
	0650	For 3/4″ laminations, add					25%				
550	0010	**LAMINATED ROOF DECK** Pine or hemlock, 3″ thick	2 Carp	425	.038	S.F.	3.02	.90		3.92	4.85
	0100	4″ thick		325	.049		4.02	1.18		5.20	6.45
	0300	Cedar, 3″ thick		425	.038		3.69	.90		4.59	5.60
	0400	4″ thick		325	.049		4.70	1.18		5.88	7.15
	0600	Fir, 3″ thick		425	.038		3.08	.90		3.98	4.92
	0700	4″ thick		325	.049		3.85	1.18		5.03	6.25
600	0010	**STRUCTURAL JOISTS** Fabricated "I" joists with wood flanges,									
	0100	Plywood webs, incl. bridging & blocking, panels 24″ O.C.									

06170 | Prefabricated Structural Wood

	Line	Description	Crew	Daily Output	Labor-Hours	Unit	2005 Bare Costs Mat.	Labor	Equip.	Total	Total Incl O&P	
600	1200	15′ to 24′ span, 50 psf live load	F-5	2,400	.013	SF Flr.	1.67	.28		1.95	2.31	600
	1300	55 psf live load		2,250	.014		1.78	.30		2.08	2.45	
	1400	24′ to 30′ span, 45 psf live load		2,600	.012		2.03	.26		2.29	2.67	
	1500	55 psf live load		2,400	.013		2.90	.28		3.18	3.66	
	1600	Tubular steel open webs, 45 psf, 24″ O.C., 40′ span	F-3	6,250	.006		1.80	.14	.10	2.04	2.32	
	1700	55′ span		7,750	.005		1.75	.11	.08	1.94	2.21	
	1800	70′ span		9,250	.004		2.27	.09	.07	2.43	2.73	
	1900	85 psf live load, 26′ span		2,300	.017		2.11	.38	.26	2.75	3.24	
980	0010	**ROOF TRUSSES**										980
	0020	For timber connectors, see div. 06090-800										
	5000	Common wood, 2″ x 4″ metal plate connected, 24″ O.C., 4/12 slope										
	5010	1′ overhang, 12′ span	F-5	55	.582	Ea.	26.50	12.10		38.60	49.50	
	5050	20′ span (R06170 -100)	F-6	62	.645		51	13.85	9.70	74.55	90	
	5100	24′ span		60	.667		51	14.35	10.05	75.40	91	
	5150	26′ span		57	.702		71	15.10	10.60	96.70	115	
	5200	28′ span		53	.755		61.50	16.20	11.35	89.05	108	
	5240	30′ span		51	.784		67	16.85	11.80	95.65	115	
	5250	32′ span		50	.800		87	17.20	12.05	116.25	138	
	5280	34′ span		48	.833		106	17.90	12.55	136.45	161	
	5350	8/12 pitch, 1′ overhang, 20′ span		57	.702		65.50	15.10	10.60	91.20	109	
	5400	24′ span		55	.727		77.50	15.65	10.95	104.10	124	
	5450	26′ span		52	.769		84	16.55	11.60	112.15	133	
	5500	28′ span		49	.816		90.50	17.55	12.30	120.35	143	
	5550	32′ span		45	.889		107	19.10	13.40	139.50	165	
	5600	36′ span		41	.976		127	21	14.70	162.70	191	
	5650	38′ span		40	1		138	21.50	15.05	174.55	204	
	5700	40′ span		40	1		157	21.50	15.05	193.55	226	

06180 | Glued-Laminated Construction

	Line	Description	Crew	Daily Output	Labor-Hours	Unit	Mat.	Labor	Equip.	Total	Total Incl O&P	
400	0010	**LAMINATED FRAMING** Not including decking										400
	0020	30 lb., short term live load, 15 lb. dead load										
	0200	Straight roof beams, 20′ clear span, beams 8′ O.C.	F-3	2,560	.016	SF Flr.	1.63	.34	.24	2.21	2.62	
	0300	Beams 16′ O.C.		3,200	.013		1.17	.27	.19	1.63	1.95	
	0500	40′ clear span, beams 8′ O.C.		3,200	.013		3.12	.27	.19	3.58	4.09	
	0600	Beams 16′ O.C.		3,840	.010		2.54	.22	.16	2.92	3.34	
	0800	60′ clear span, beams 8′ O.C.	F-4	2,880	.014		5.35	.30	.31	5.96	6.75	
	0900	Beams 16′ O.C.	″	3,840	.010		3.99	.22	.23	4.44	5.05	
	1100	Tudor arches, 30′ to 40′ clear span, frames 8′ O.C.	F-3	1,680	.024		7	.51	.36	7.87	8.95	
	1200	Frames 16′ O.C.	″	2,240	.018		5.45	.39	.27	6.11	6.95	
	1400	50′ to 60′ clear span, frames 8′ O.C.	F-4	2,200	.018		7.55	.39	.41	8.35	9.40	
	1500	Frames 16′ O.C.		2,640	.015		6.40	.33	.34	7.07	7.95	
	1700	Radial arches, 60′ clear span, frames 8′ O.C.		1,920	.021		7.05	.45	.47	7.97	9	
	1800	Frames 16′ O.C.		2,880	.014		5.40	.30	.31	6.01	6.80	
	2000	100′ clear span, frames 8′ O.C.		1,600	.025		7.30	.54	.56	8.40	9.55	
	2100	Frames 16′ O.C.		2,400	.017		6.40	.36	.37	7.13	8.05	
	2300	120′ clear span, frames 8′ O.C.		1,440	.028		9.70	.60	.62	10.92	12.35	
	2400	Frames 16′ O.C.		1,920	.021		8.85	.45	.47	9.77	11	
	2600	Bowstring trusses, 20′ O.C., 40′ clear span	F-3	2,400	.017		4.37	.36	.25	4.98	5.70	
	2700	60′ clear span	F-4	3,600	.011		3.92	.24	.25	4.41	4.98	
	2800	100′ clear span		4,000	.010		5.55	.22	.22	5.99	6.70	
	2900	120′ clear span		3,600	.011		5.95	.24	.25	6.44	7.20	
	3100	For premium appearance, add to S.F. prices					5%					
	3300	For industrial type, deduct					15%					
	3500	For stain and varnish, add					5%					
	3900	For 3/4″ laminations, add to straight					25%					

06100 | Rough Carpentry

06180 | Glued-Laminated Construction

		06180 Glued-Laminated Construction	CREW	DAILY OUTPUT	LABOR-HOURS	UNIT	2005 BARE COSTS MAT.	LABOR	EQUIP.	TOTAL	TOTAL INCL O&P
400	4100	Add to curved				SF Flr.	15%				
	4300	Alternate pricing method: (use nominal footage of									
	4310	components). Straight beams, camber less than 6"	F-3	3.50	11.429	M.B.F.	2,425	247	172	2,844	3,250
	4400	Columns, including hardware		2	20		2,600	430	300	3,330	3,900
	4600	Curved members, radius over 32'		2.50	16		2,650	345	241	3,236	3,775
	4700	Radius 10' to 32'		3	13.333		2,625	288	201	3,114	3,600
	4900	For complicated shapes, add maximum					100%				
	5100	For pressure treating, add to straight					35%				
	5200	Add to curved					45%				
	6000	Laminated veneer members, southern pine or western species									
	6050	1-3/4" wide x 5-1/2" deep	2 Carp	480	.033	L.F.	2.77	.80		3.57	4.41
	6100	9-1/2" deep		480	.033		3.45	.80		4.25	5.15
	6150	14" deep		450	.036		5.15	.85		6	7.10
	6200	18" deep		450	.036		6.95	.85		7.80	9.10
	6300	Parallel strand members, southern pine or western species									
	6350	1-3/4" wide x 9-1/4" deep	2 Carp	480	.033	L.F.	3.27	.80		4.07	4.96
	6400	11-1/4" deep		450	.036		4.02	.85		4.87	5.85
	6450	14" deep		400	.040		4.79	.96		5.75	6.90
	6500	3-1/2" wide x 9-1/4" deep		480	.033		7.95	.80		8.75	10.10
	6550	11-1/4" deep		450	.036		9.85	.85		10.70	12.25
	6600	14" deep		400	.040		11.65	.96		12.61	14.50
	6650	7" wide x 9-1/4" deep		450	.036		16.55	.85		17.40	19.70
	6700	11-1/4" deep		420	.038		20.50	.91		21.41	24.50
	6750	14" deep		400	.040		25	.96		25.96	29

06200 | Finish Carpentry

06220 | Millwork

		06220 Millwork	CREW	DAILY OUTPUT	LABOR-HOURS	UNIT	2005 BARE COSTS MAT.	LABOR	EQUIP.	TOTAL	TOTAL INCL O&P
100	0010	**MILLWORK** Rule of thumb: Milled material cost									
	0020	equals three times cost of lumber									
	1000	Typical finish hardwood milled material									
	1020	1" x 12", custom birch				L.F.	5.50			5.50	6.05
	1040	Cedar					2.73			2.73	3
	1060	Oak					6.20			6.20	6.80
	1080	Redwood					6			6	6.60
	1100	Southern yellow pine					2.10			2.10	2.31
	1120	Sugar pine					2.51			2.51	2.76
	1140	Teak					16.75			16.75	18.40
	1160	Walnut					7.20			7.20	7.95
	1180	White pine					3.33			3.33	3.66
200	0010	**MOLDINGS, BASE**									
	0500	Base, stock pine, 9/16" x 3-1/2"	1 Carp	240	.033	L.F.	1.53	.80		2.33	3.04
	0501	Oak or birch, 9/16" x 3-1/2"		240	.033		3.10	.80		3.90	4.77
	0550	9/16" x 4-1/2"		200	.040		1.63	.96		2.59	3.42
	0561	Base shoe, oak, 3/4" x 1"		240	.033		1.04	.80		1.84	2.50
	0570	Base, prefinished, 2-1/2" x 9/16"		242	.033		1.05	.79		1.84	2.51
	0580	Shoe, prefinished, 3/8" x 5/8"		266	.030		.45	.72		1.17	1.73
	0585	Flooring cant strip, 3/4" x 1/2"		500	.016		.46	.38		.84	1.16
400	0010	**MOLDINGS, CASINGS**									
	0090	Apron, stock pine, 5/8" x 2"	1 Carp	250	.032	L.F.	.99	.77		1.76	2.39

	06220 | Millwork		CREW	DAILY OUTPUT	LABOR-HOURS	UNIT	2005 BARE COSTS MAT.	LABOR	EQUIP.	TOTAL	TOTAL INCL O&P	
400	0110	5/8" x 3-1/2"	1 Carp	220	.036	L.F.	1.19	.87		2.06	2.79	400
	0300	Band, stock pine, 11/16" x 1-1/8"		270	.030		.54	.71		1.25	1.80	
	0350	11/16" x 1-3/4"		250	.032		.86	.77		1.63	2.25	
	0700	Casing, stock pine, 11/16" x 2-1/2"		240	.033		.85	.80		1.65	2.30	
	0701	Oak or birch		240	.033		2.38	.80		3.18	3.98	
	0750	11/16" x 3-1/2"		215	.037		1.21	.89		2.10	2.85	
	0760	Door & window casing, exterior, 1-1/4" x 2"		200	.040		1.64	.96		2.60	3.43	
	0770	Finger jointed, 1-1/4" x 2"		200	.040		1.01	.96		1.97	2.74	
	4600	Mullion casing, stock pine, 5/16" x 2"		200	.040		.72	.96		1.68	2.42	
	4601	Oak or birch, 9/16" x 2-1/2"		200	.040		2.55	.96		3.51	4.44	
	4700	Teak, custom, nominal 1" x 1"		215	.037		1.16	.89		2.05	2.80	
	4800	Nominal 1" x 3"		200	.040		3.50	.96		4.46	5.50	
450	0010	**MOLDINGS, CEILINGS**										450
	0600	Bed, stock pine, 9/16" x 1-3/4"	1 Carp	270	.030	L.F.	.65	.71		1.36	1.93	
	0650	9/16" x 2"		240	.033		.92	.80		1.72	2.37	
	1200	Cornice molding, stock pine, 9/16" x 1-3/4"		330	.024		.66	.58		1.24	1.72	
	1300	9/16" x 2-1/4"		300	.027		.83	.64		1.47	2	
	2400	Cove scotia, stock pine, 9/16" x 1-3/4"		270	.030		.60	.71		1.31	1.87	
	2401	Oak or birch, 9/16" x 1-3/4"		270	.030		1.71	.71		2.42	3.09	
	2500	11/16" x 2-3/4"		255	.031		1.25	.75		2	2.66	
	2600	Crown, stock pine, 9/16" x 3-5/8"		250	.032		1.67	.77		2.44	3.14	
	2700	11/16" x 4-5/8"		220	.036		2.58	.87		3.45	4.32	
500	0010	**MOLDINGS, EXTERIOR**										500
	1500	Cornice, boards, pine, 1" x 2"	1 Carp	330	.024	L.F.	.33	.58		.91	1.35	
	1600	1" x 4"		250	.032		.68	.77		1.45	2.05	
	1700	1" x 6"		250	.032		1.19	.77		1.96	2.61	
	1800	1" x 8"		200	.040		1.46	.96		2.42	3.24	
	1900	1" x 10"		180	.044		1.62	1.07		2.69	3.59	
	2000	1" x 12"		180	.044		1.77	1.07		2.84	3.76	
	2200	Three piece, built-up, pine, minimum		80	.100		2.20	2.40		4.60	6.50	
	2300	Maximum		65	.123		4.85	2.95		7.80	10.35	
	3000	Corner board, sterling pine, 1" x 4"		200	.040		.68	.96		1.64	2.38	
	3100	1" x 6"		200	.040		1	.96		1.96	2.73	
	3200	2" x 6"		165	.048		2.12	1.16		3.28	4.31	
	3300	2" x 8"		165	.048		2.87	1.16		4.03	5.15	
	3350	Fascia, sterling pine, 1" x 6"		250	.032		1	.77		1.77	2.40	
	3370	1" x 8"		225	.036		1.56	.85		2.41	3.17	
	3372	2" x 6"		225	.036		2.12	.85		2.97	3.78	
	3374	2" x 8"		200	.040		2.87	.96		3.83	4.79	
	3376	2" x 10"		180	.044		3.49	1.07		4.56	5.65	
	3395	Grounds, 1" x 1" redwood		300	.027		.19	.64		.83	1.30	
	3400	Trim, exterior, sterling pine, back band		250	.032		.68	.77		1.45	2.05	
	3500	Casing		250	.032		1.80	.77		2.57	3.28	
	3600	Crown		250	.032		1.73	.77		2.50	3.20	
	3700	Porch rail with balusters		22	.364		14.15	8.75		22.90	30.50	
	3800	Screen		395	.020		1.09	.49		1.58	2.03	
	4100	Verge board, sterling pine, 1" x 4"		200	.040		.66	.96		1.62	2.36	
	4200	1" x 6"		200	.040		.99	.96		1.95	2.72	
	4300	2" x 6"		165	.048		1.61	1.16		2.77	3.75	
	4400	2" x 8"		165	.048		2.13	1.16		3.29	4.32	
	4700	For redwood trim, add					200%					
	5000	Casing/fascia, rough-sawn cedar										
	5100	1" x 2"	1 Carp	275	.029	L.F.	.27	.70		.97	1.49	
	5200	1" x 6"		250	.032		.80	.77		1.57	2.18	
	5300	1" x 8"		230	.035		1.03	.83		1.86	2.55	
	5400	2" x 4"		220	.036		1.03	.87		1.90	2.61	

	06220 | Millwork		CREW	DAILY OUTPUT	LABOR-HOURS	UNIT	2005 BARE COSTS MAT.	LABOR	EQUIP.	TOTAL	TOTAL INCL O&P
500	5500	2″ x 6″	1 Carp	220	.036	L.F.	1.57	.87		2.44	3.21
	5600	2″ x 8″		200	.040		2.08	.96		3.04	3.92
	5700	2″ x 10″		180	.044		2.61	1.07		3.68	4.68
	5800	2″ x 12″		170	.047		3.11	1.13		4.24	5.35
700	0010	**MOLDINGS, TRIM**									
	0200	Astragal, stock pine, 11/16″ x 1-3/4″	1 Carp	255	.031	L.F.	.96	.75		1.71	2.34
	0250	1-5/16″ x 2-3/16″		240	.033		2.57	.80		3.37	4.19
	0800	Chair rail, stock pine, 5/8″ x 2-1/2″		270	.030		1.12	.71		1.83	2.44
	0900	5/8″ x 3-1/2″		240	.033		1.62	.80		2.42	3.14
	1000	Closet pole, stock pine, 1-1/8″ diameter		200	.040		.79	.96		1.75	2.50
	1100	Fir, 1-5/8″ diameter		200	.040		1.27	.96		2.23	3.03
	1150	Corner, inside, 5/16″ x 1″		225	.036		.32	.85		1.17	1.80
	1160	Outside, 1-1/16″ x 1-1/16″		240	.033		.88	.80		1.68	2.33
	1161	1-5/16″ x 1-5/16″		240	.033		1.03	.80		1.83	2.49
	3300	Half round, stock pine, 1/4″ x 1/2″		270	.030		.16	.71		.87	1.39
	3350	1/2″ x 1″		255	.031		.39	.75		1.14	1.71
	3400	Handrail, fir, single piece, stock, hardware not included									
	3450	1-1/2″ x 1-3/4″	1 Carp	80	.100	L.F.	1.22	2.40		3.62	5.40
	3470	Pine, 1-1/2″ x 1-3/4″		80	.100		1.22	2.40		3.62	5.40
	3500	1-1/2″ x 2-1/2″		76	.105		1.27	2.53		3.80	5.70
	3600	Lattice, stock pine, 1/4″ x 1-1/8″		270	.030		.23	.71		.94	1.46
	3700	1/4″ x 1-3/4″		250	.032		.35	.77		1.12	1.69
	3800	Miscellaneous, custom, pine, 1″ x 1″		270	.030		.32	.71		1.03	1.56
	3900	1″ x 3″		240	.033		.65	.80		1.45	2.08
	4100	Birch or oak, nominal 1″ x 1″		240	.033		.50	.80		1.30	1.91
	4200	Nominal 1″ x 3″		215	.037		1.71	.89		2.60	3.40
	4400	Walnut, nominal 1″ x 1″		215	.037		.82	.89		1.71	2.42
	4500	Nominal 1″ x 3″		200	.040		2.46	.96		3.42	4.34
	4700	Teak, nominal 1″ x 1″		215	.037		1.16	.89		2.05	2.80
	4800	Nominal 1″ x 3″		200	.040		3.32	.96		4.28	5.30
	4900	Quarter round, stock pine, 1/4″ x 1/4″		275	.029		.14	.70		.84	1.34
	4950	3/4″ x 3/4″		255	.031		.36	.75		1.11	1.68
	5600	Wainscot moldings, 1-1/8″ x 9/16″, 2′ high, minimum		76	.105	S.F.	8.95	2.53		11.48	14.15
	5700	Maximum		65	.123	″	18.30	2.95		21.25	25
800	0010	**MOLDINGS, WINDOW AND DOOR**									
	2800	Door moldings, stock, decorative, 1-1/8″ wide, plain	1 Carp	17	.471	Set	29	11.30		40.30	51
	2900	Detailed		17	.471	″	81	11.30		92.30	108
	2960	Clear pine door jamb, no stops, 11/16″ x 4-9/16″		240	.033	L.F.	2.30	.80		3.10	3.89
	3150	Door trim set, 1 head and 2 sides, pine, 2-1/2 wide		5.90	1.356	Opng.	14.45	32.50		46.95	71.50
	3170	3-1/2″ wide		5.30	1.509	″	20.50	36		56.50	84
	3250	Glass beads, stock pine, 3/8″ x 1/2″		275	.029	L.F.	.36	.70		1.06	1.59
	3270	3/8″ x 7/8″		270	.030		.40	.71		1.11	1.65
	4850	Parting bead, stock pine, 3/8″ x 3/4″		275	.029		.26	.70		.96	1.48
	4870	1/2″ x 3/4″		255	.031		.31	.75		1.06	1.62
	5000	Stool caps, stock pine, 11/16″ x 3-1/2″		200	.040		1.53	.96		2.49	3.31
	5100	1-1/16″ x 3-1/4″		150	.053		2.67	1.28		3.95	5.10
	5300	Threshold, oak, 3′ long, inside, 5/8″ x 3-5/8″		32	.250	Ea.	7.75	6		13.75	18.75
	5400	Outside, 1-1/2″ x 7-5/8″		16	.500	″	35	12		47	59
	5900	Window trim sets, including casings, header, stops,									
	5910	stool and apron, 2-1/2″ wide, minimum	1 Carp	13	.615	Opng.	19.90	14.75		34.65	47
	5950	Average		10	.800		23.50	19.20		42.70	58
	6000	Maximum		6	1.333		58	32		90	118
900	0010	**SOFFITS**									
	0200	Soffits, pine, 1″ x 4″	2 Carp	420	.038	L.F.	.41	.91		1.32	2
	0210	1″ x 6″		420	.038		.58	.91		1.49	2.18
	0220	1″ x 8″		420	.038		.70	.91		1.61	2.32

	06220 | Millwork		CREW	DAILY OUTPUT	LABOR-HOURS	UNIT	2005 BARE COSTS MAT.	LABOR	EQUIP.	TOTAL	TOTAL INCL O&P
900	0230	1″ x 10″	2 Carp	400	.040	L.F.	1.08	.96		2.04	2.82
	0240	1″ x 12″		400	.040		1.56	.96		2.52	3.35
	0250	STK cedar, 1″ x 4″		420	.038		.52	.91		1.43	2.12
	0260	1″ x 6″		420	.038		.83	.91		1.74	2.46
	0270	1″ x 8″		420	.038		1.09	.91		2	2.75
	0280	1″ x 10″		400	.040		1.31	.96		2.27	3.07
	0290	1″ x 12″		400	.040		1.61	.96		2.57	3.40
	1000	Exterior AC plywood, 1/4″ thick		420	.038	S.F.	.78	.91		1.69	2.41
	1050	3/8″ thick		420	.038		.87	.91		1.78	2.51
	1100	1/2″ thick		420	.038		1.06	.91		1.97	2.72
	1150	Polyvinyl chloride, white, solid	1 Carp	230	.035		.66	.83		1.49	2.15
	1160	Perforated	″	230	.035		.66	.83		1.49	2.15
	1170	Accessories, ″J″ channel 5/8″	2 Carp	700	.023	L.F.	.26	.55		.81	1.22

	06250 | Prefinished Paneling		CREW	DAILY OUTPUT	LABOR-HOURS	UNIT	MAT.	LABOR	EQUIP.	TOTAL	TOTAL INCL O&P
200	0010	**PANELING, HARDBOARD**									
	0050	Not incl. furring or trim, hardboard, tempered, 1/8″ thick	2 Carp	500	.032	S.F.	.30	.77		1.07	1.63
	0100	1/4″ thick		500	.032		.38	.77		1.15	1.72
	0300	Tempered pegboard, 1/8″ thick		500	.032		.37	.77		1.14	1.71
	0400	1/4″ thick		500	.032		.42	.77		1.19	1.76
	0600	Untempered hardboard, natural finish, 1/8″ thick		500	.032		.33	.77		1.10	1.66
	0700	1/4″ thick		500	.032		.32	.77		1.09	1.65
	0900	Untempered pegboard, 1/8″ thick		500	.032		.33	.77		1.10	1.66
	1000	1/4″ thick		500	.032		.37	.77		1.14	1.71
	1200	Plastic faced hardboard, 1/8″ thick		500	.032		.55	.77		1.32	1.91
	1300	1/4″ thick		500	.032		.73	.77		1.50	2.10
	1500	Plastic faced pegboard, 1/8″ thick		500	.032		.52	.77		1.29	1.87
	1600	1/4″ thick		500	.032		.64	.77		1.41	2
	1800	Wood grained, plain or grooved, 1/4″ thick, minimum		500	.032		.48	.77		1.25	1.83
	1900	Maximum		425	.038		1.02	.90		1.92	2.65
	2100	Moldings for hardboard, wood or aluminum, minimum		500	.032	L.F.	.33	.77		1.10	1.66
	2200	Maximum		425	.038	″	.92	.90		1.82	2.54
500	0010	**PANELING, PLYWOOD**									
	2400	Plywood, prefinished, 1/4″ thick, 4′ x 8′ sheets									
	2410	with vertical grooves. Birch faced, minimum	2 Carp	500	.032	S.F.	.79	.77		1.56	2.17
	2420	Average		420	.038		1.21	.91		2.12	2.88
	2430	Maximum		350	.046		1.76	1.10		2.86	3.80
	2600	Mahogany, African		400	.040		2.25	.96		3.21	4.11
	2700	Philippine (Lauan)		500	.032		.97	.77		1.74	2.37
	2900	Oak or Cherry, minimum		500	.032		1.89	.77		2.66	3.38
	3000	Maximum		400	.040		2.90	.96		3.86	4.82
	3200	Rosewood		320	.050		4.11	1.20		5.31	6.55
	3400	Teak		400	.040		2.90	.96		3.86	4.82
	3600	Chestnut		375	.043		4.28	1.02		5.30	6.45
	3800	Pecan		400	.040		1.85	.96		2.81	3.67
	3900	Walnut, minimum		500	.032		2.47	.77		3.24	4.02
	3950	Maximum		400	.040		4.68	.96		5.64	6.80
	4000	Plywood, prefinished, 3/4″ thick, stock grades, minimum		320	.050		1.12	1.20		2.32	3.27
	4100	Maximum		224	.071		4.83	1.71		6.54	8.20
	4300	Architectural grade, minimum		224	.071		3.57	1.71		5.28	6.85
	4400	Maximum		160	.100		5.45	2.40		7.85	10.10
	4600	Plywood, ″A″ face, birch, V.C., 1/2″ thick, natural		450	.036		1.69	.85		2.54	3.31
	4700	Select		450	.036		1.85	.85		2.70	3.49
	4900	Veneer core, 3/4″ thick, natural		320	.050		1.78	1.20		2.98	4

6 WOOD & PLASTICS

06200 | Finish Carpentry

		06250 Prefinished Paneling	Crew	Daily Output	Labor-Hours	Unit	2005 Bare Costs Mat.	Labor	Equip.	Total	Total Incl O&P
500	5000	Select	2 Carp	320	.050	S.F.	2.01	1.20		3.21	4.25
	5200	Lumber core, 3/4″ thick, natural		320	.050		2.68	1.20		3.88	4.99
	5500	Plywood, knotty pine, 1/4″ thick, A2 grade		450	.036		1.46	.85		2.31	3.06
	5600	A3 grade		450	.036		1.85	.85		2.70	3.49
	5800	3/4″ thick, veneer core, A2 grade		320	.050		1.90	1.20		3.10	4.13
	5900	A3 grade		320	.050		2.13	1.20		3.33	4.38
	6100	Aromatic cedar, 1/4″ thick, plywood		400	.040		1.87	.96		2.83	3.69
	6200	1/4″ thick, particle board		400	.040		.91	.96		1.87	2.63

		06260 Board Paneling	Crew	Daily Output	Labor-Hours	Unit	Mat.	Labor	Equip.	Total	Total Incl O&P
400	0010	**PANELING, BOARDS**									
	6400	Wood board paneling, 3/4″ thick, knotty pine	2 Carp	300	.053	S.F.	1.36	1.28		2.64	3.67
	6500	Rough sawn cedar		300	.053		1.74	1.28		3.02	4.08
	6700	Redwood, clear, 1″ x 4″ boards		300	.053		4.06	1.28		5.34	6.65
	6900	Aromatic cedar, closet lining, boards		275	.058		3.13	1.40		4.53	5.80

		06270 Closet/Utility Wood Shelving	Crew	Daily Output	Labor-Hours	Unit	Mat.	Labor	Equip.	Total	Total Incl O&P
200	0010	**SHELVING** Pine, clear grade, no edge band, 1″ x 8″	1 Carp	115	.070	L.F.	1.76	1.67		3.43	4.78
	0100	1″ x 10″		110	.073		2.24	1.75		3.99	5.40
	0200	1″ x 12″		105	.076		3.01	1.83		4.84	6.40
	0450	1″ x 18″		95	.084		4.52	2.02		6.54	8.40
	0460	1″ x 24″		85	.094		6	2.26		8.26	10.45
	0600	Plywood, 3/4″ thick with lumber edge, 12″ wide		75	.107		.32	2.56		2.88	4.70
	0700	24″ wide		70	.114		3.12	2.74		5.86	8.10
	0900	Bookcase, clear grade pine, shelves 12″ O.C., 8″ deep, /SF shelf		70	.114	S.F.	5.70	2.74		8.44	10.95
	1000	12″ deep shelves		65	.123	″	9.80	2.95		12.75	15.75
	1200	Adjustable closet rod and shelf, 12″ wide, 3′ long		20	.400	Ea.	8.60	9.60		18.20	26
	1300	8′ long		15	.533	″	21	12.80		33.80	44.50
	1500	Prefinished shelves with supports, stock, 8″ wide		75	.107	L.F.	3.72	2.56		6.28	8.45
	1600	10″ wide		70	.114	″	4.13	2.74		6.87	9.20

06400 | Architectural Woodwork

		06410 Custom Cabinets	Crew	Daily Output	Labor-Hours	Unit	2005 Bare Costs Mat.	Labor	Equip.	Total	Total Incl O&P
100	0010	**CABINETS** Corner china cabinets, stock pine,									
	0020	80″ high, unfinished, minimum	2 Carp	6.60	2.424	Ea.	455	58		513	600
	0100	Maximum	″	4.40	3.636	″	1,000	87.50		1,087.50	1,250
	0700	Kitchen base cabinets, hardwood, not incl. counter tops,									
	0710	24″ deep, 35″ high, prefinished									
	0800	One top drawer, one door below, 12″ wide	2 Carp	24.80	.645	Ea.	132	15.50		147.50	173
	0820	15″ wide		24	.667		181	16		197	226
	0840	18″ wide		23.30	.687		197	16.50		213.50	244
	0860	21″ wide		22.70	.705		205	16.90		221.90	254
	0880	24″ wide		22.30	.717		236	17.20		253.20	288
	1000	Four drawers, 12″ wide		24.80	.645		315	15.50		330.50	370
	1020	15″ wide		24	.667		242	16		258	293
	1040	18″ wide		23.30	.687		269	16.50		285.50	325
	1060	24″ wide		22.30	.717		292	17.20		309.20	350

	06410 | Custom Cabinets		CREW	DAILY OUTPUT	LABOR-HOURS	UNIT	2005 BARE COSTS MAT.	LABOR	EQUIP.	TOTAL	TOTAL INCL O&P	
100	1200	Two top drawers, two doors below, 27″ wide	2 Carp	22	.727	Ea.	261	17.45		278.45	315	100
	1220	30″ wide		21.40	.748		280	17.95		297.95	340	
	1240	33″ wide		20.90	.766		289	18.35		307.35	350	
	1260	36″ wide		20.30	.788		300	18.90		318.90	360	
	1280	42″ wide		19.80	.808		320	19.40		339.40	390	
	1300	48″ wide		18.90	.847		345	20.50		365.50	415	
	1500	Range or sink base, two doors below, 30″ wide		21.40	.748		231	17.95		248.95	285	
	1520	33″ wide		20.90	.766		248	18.35		266.35	305	
	1540	36″ wide		20.30	.788		259	18.90		277.90	315	
	1560	42″ wide		19.80	.808		277	19.40		296.40	340	
	1580	48″ wide	↓	18.90	.847		291	20.50		311.50	355	
	1800	For sink front units, deduct					50			50	55	
	2000	Corner base cabinets, 36″ wide, standard	2 Carp	18	.889		380	21.50		401.50	455	
	2100	Lazy Susan with revolving door	″	16.50	.970	↓	370	23.50		393.50	445	
	4000	Kitchen wall cabinets, hardwood, 12″ deep with two doors										
	4050	12″ high, 30″ wide	2 Carp	24.80	.645	Ea.	145	15.50		160.50	187	
	4100	36″ wide		24	.667		169	16		185	213	
	4400	15″ high, 30″ wide		24	.667		152	16		168	195	
	4420	33″ wide		23.30	.687		170	16.50		186.50	215	
	4440	36″ wide		22.70	.705		172	16.90		188.90	218	
	4450	42″ wide		22.70	.705		200	16.90		216.90	249	
	4700	24″ high, 30″ wide		23.30	.687		189	16.50		205.50	236	
	4720	36″ wide		22.70	.705		209	16.90		225.90	259	
	4740	42″ wide		22.30	.717		231	17.20		248.20	283	
	5000	30″ high, one door, 12″ wide		22	.727		128	17.45		145.45	170	
	5020	15″ wide		21.40	.748		145	17.95		162.95	190	
	5040	18″ wide		20.90	.766		158	18.35		176.35	205	
	5060	24″ wide		20.30	.788		178	18.90		196.90	228	
	5300	Two doors, 27″ wide		19.80	.808		221	19.40		240.40	276	
	5320	30″ wide		19.30	.829		213	19.90		232.90	268	
	5340	36″ wide		18.80	.851		244	20.50		264.50	305	
	5360	42″ wide		18.50	.865		265	21		286	325	
	5380	48″ wide		18.40	.870		299	21		320	365	
	6000	Corner wall, 30″ high, 24″ wide		18	.889		146	21.50		167.50	196	
	6050	30″ wide		17.20	.930		173	22.50		195.50	229	
	6100	36″ wide		16.50	.970		188	23.50		211.50	246	
	6500	Revolving Lazy Susan		15.20	1.053		286	25.50		311.50	360	
	7000	Broom cabinet, 84″ high, 24″ deep, 18″ wide		10	1.600		390	38.50		428.50	495	
	7500	Oven cabinets, 84″ high, 24″ deep, 27″ wide		8	2	↓	565	48		613	705	
	7750	Valance board trim	↓	396	.040	L.F.	8.10	.97		9.07	10.55	
	9000	For deluxe models of all cabinets, add					40%					
	9500	For custom built in place, add					25%	10%				
	9550	Rule of thumb, kitchen cabinets not including										
	9560	appliances & counter top, minimum	2 Carp	30	.533	L.F.	92	12.80		104.80	123	
	9600	Maximum	″	25	.640	″	240	15.35		255.35	290	
210	0010	**CASEWORK, FRAMES**										210
	0050	Base cabinets, counter storage, 36″ high, one bay										
	0100	18″ wide	1 Carp	2.70	2.963	Ea.	103	71		174	234	
	0400	Two bay, 36″ wide		2.20	3.636		157	87.50		244.50	320	
	1100	Three bay, 54″ wide		1.50	5.333		187	128		315	420	
	2800	Book cases, one bay, 7′ high, 18″ wide		2.40	3.333		121	80		201	269	
	3500	Two bay, 36″ wide		1.60	5		176	120		296	395	
	4100	Three bay, 54″ wide		1.20	6.667		291	160		451	590	
	6100	Wall mounted cabinet, one bay, 24″ high, 18″ wide		3.60	2.222		66.50	53.50		120	164	
	6800	Two bay, 36″ wide	↓	2.20	3.636	↓	97	87.50		184.50	255	

	06410	Custom Cabinets	CREW	DAILY OUTPUT	LABOR-HOURS	UNIT	2005 BARE COSTS MAT.	LABOR	EQUIP.	TOTAL	TOTAL INCL O&P
210	7400	Three bay, 54" wide	1 Carp	1.70	4.706	Ea.	121	113		234	325
	8400	30" high, one bay, 18" wide		3.60	2.222		72.50	53.50		126	170
	9000	Two bay, 36" wide		2.15	3.721		96	89.50		185.50	258
	9400	Three bay, 54" wide		1.60	5		120	120		240	335
	9800	Wardrobe, 7' high, single, 24" wide		2.70	2.963		133	71		204	268
	9950	Partition, adjustable shelves & drawers, 48" wide		1.40	5.714		254	137		391	515
220	0010	CABINET DOORS									
	2000	Glass panel, hardwood frame									
	2200	12" wide, 18" high	1 Carp	34	.235	Ea.	22	5.65		27.65	34
	2400	24" high		33	.242		24	5.80		29.80	36
	2600	30" high		32	.250		25.50	6		31.50	38
	2800	36" high		30	.267		29.50	6.40		35.90	43.50
	3000	48" high		23	.348		37.50	8.35		45.85	55.50
	3200	60" high		17	.471		46	11.30		57.30	69.50
	3400	72" high		15	.533		55	12.80		67.80	82
	3600	15" wide x 18" high		33	.242		23	5.80		28.80	35
	3800	24" high		32	.250		24.50	6		30.50	37
	4000	30" high		30	.267		26	6.40		32.40	40
	4250	36" high		28	.286		30.50	6.85		37.35	45
	4300	48" high		22	.364		38.50	8.75		47.25	57.50
	4350	60" high		16	.500		54.50	12		66.50	80.50
	4400	72" high		14	.571		63.50	13.70		77.20	93.50
	4450	18" wide, 18" high		32	.250		24	6		30	36
	4500	24" high		30	.267		24.50	6.40		30.90	38
	4550	30" high		29	.276		26.50	6.60		33.10	40.50
	4600	36" high		27	.296		31.50	7.10		38.60	46.50
	4650	48" high		21	.381		39.50	9.15		48.65	58.50
	4700	60" high		15	.533		55.50	12.80		68.30	82.50
	4750	72" high		13	.615		64.50	14.75		79.25	96
	5000	Hardwood, raised panel									
	5100	12" wide, 18" high	1 Carp	16	.500	Ea.	33.50	12		45.50	57.50
	5150	24" high		15.50	.516		36.50	12.40		48.90	61
	5200	30" high		15	.533		40	12.80		52.80	65.50
	5250	36" high		14	.571		43	13.70		56.70	71
	5300	48" high		11	.727		52.50	17.45		69.95	87.50
	5320	60" high		8	1		70.50	24		94.50	119
	5340	72" high		7	1.143		81	27.50		108.50	136
	5360	15" wide x 18" high		15.50	.516		35	12.40		47.40	59
	5380	24" high		15	.533		38.50	12.80		51.30	64
	5400	30" high		14.50	.552		42.50	13.25		55.75	69
	5420	36" high		13.50	.593		48.50	14.20		62.70	77.50
	5440	48" high		10.50	.762		59	18.30		77.30	95.50
	5460	60" high		7.50	1.067		86	25.50		111.50	139
	5480	72" high		6.50	1.231		97	29.50		126.50	157
	5500	18" wide, 18" high		15	.533		37	12.80		49.80	62
	5550	24" high		14.50	.552		41	13.25		54.25	67.50
	5600	30" high		14	.571		45	13.70		58.70	73
	5650	36" high		13	.615		51.50	14.75		66.25	81.50
	5700	48" high		10	.800		62	19.20		81.20	101
	5750	60" high		7	1.143		89.50	27.50		117	145
	5800	72" high		6	1.333		101	32		133	166
	6000	Plastic laminate on particle board									
	6100	12" wide, 18" high	1 Carp	25	.320	Ea.	15.20	7.70		22.90	30
	6120	24" high		24	.333		19.90	8		27.90	35.50
	6140	30" high		23	.348		24.50	8.35		32.85	41
	6160	36" high		21	.381		30.50	9.15		39.65	49

	06410 | Custom Cabinets		CREW	DAILY OUTPUT	LABOR-HOURS	UNIT	2005 BARE COSTS MAT.	LABOR	EQUIP.	TOTAL	TOTAL INCL O&P	
220	6200	48" high	1 Carp	16	.500	Ea.	40	12		52	64	220
	6250	60" high		13	.615		50.50	14.75		65.25	80.50	
	6300	72" high		12	.667		61	16		77	94	
	6320	15" wide x 18" high		24.50	.327		18.70	7.85		26.55	34	
	6340	24" high		23.50	.340		25.50	8.15		33.65	42.50	
	6360	30" high		22.50	.356		31.50	8.55		40.05	49	
	6380	36" high		20.50	.390		37.50	9.35		46.85	57	
	6400	48" high		15.50	.516		50.50	12.40		62.90	76.50	
	6450	60" high		12.50	.640		63	15.35		78.35	95.50	
	6480	72" high		11.50	.696		76	16.70		92.70	112	
	6500	18" wide, 18" high		24	.333		22	8		30	38	
	6550	24" high		23	.348		31.50	8.35		39.85	48.50	
	6600	30" high		22	.364		37.50	8.75		46.25	56	
	6650	36" high		20	.400		45.50	9.60		55.10	66.50	
	6700	48" high		15	.533		61	12.80		73.80	88.50	
	6750	60" high		12	.667		76	16		92	111	
	6800	72" high	↓	11	.727	↓	90	17.45		107.45	129	
	7000	Plywood, with edge band										
	7010	12" wide, 18" high	1 Carp	27	.296	Ea.	20.50	7.10		27.60	34.50	
	7100	24" high		26	.308		27.50	7.40		34.90	43	
	7120	30" high		25	.320		35	7.70		42.70	51.50	
	7140	36" high		23	.348		42	8.35		50.35	60	
	7180	48" high		18	.444		55	10.65		65.65	78.50	
	7200	60" high		15	.533		69.50	12.80		82.30	98	
	7250	72" high		14	.571		82	13.70		95.70	114	
	7300	15" wide x 18" high		26.50	.302		26.50	7.25		33.75	41.50	
	7350	24" high		25.50	.314		34	7.55		41.55	50.50	
	7400	30" high		24.50	.327		42	7.85		49.85	60	
	7450	36" high		22.50	.356		51	8.55		59.55	70.50	
	7500	48" high		17.50	.457		68	10.95		78.95	93	
	7550	60" high		14.50	.552		85.50	13.25		98.75	117	
	7600	72" high		13.50	.593		102	14.20		116.20	136	
	7650	18" wide, 18" high		26	.308		30	7.40		37.40	45.50	
	7700	24" high		25	.320		40.50	7.70		48.20	57.50	
	7750	30" high	↓	24	.333	↓	51.50	8		59.50	70	
230	0010	**CABINET HARDWARE**										230
	1000	Catches, minimum	1 Carp	235	.034	Ea.	.82	.82		1.64	2.29	
	1020	Average		119.40	.067		2.67	1.61		4.28	5.65	
	1040	Maximum	↓	80	.100	↓	5.05	2.40		7.45	9.70	
	2000	Door/drawer pulls, handles										
	2200	Handles and pulls, projecting, metal, minimum	1 Carp	160	.050	Ea.	3.79	1.20		4.99	6.20	
	2220	Average		95.24	.084		5.05	2.02		7.07	8.95	
	2240	Maximum		68	.118		6.95	2.82		9.77	12.45	
	2300	Wood, minimum		160	.050		3.79	1.20		4.99	6.20	
	2320	Average		95.24	.084		5.05	2.02		7.07	8.95	
	2340	Maximum		68	.118		6.95	2.82		9.77	12.45	
	2600	Flush, metal, minimum		160	.050		3.79	1.20		4.99	6.20	
	2620	Average		95.24	.084		5.05	2.02		7.07	8.95	
	2640	Maximum		68	.118	↓	6.95	2.82		9.77	12.45	
	3000	Drawer tracks/glides, minimum		48	.167	Pr.	6.45	4		10.45	13.85	
	3020	Average		32	.250		10.90	6		16.90	22	
	3040	Maximum		24	.333		18.70	8		26.70	34	
	4000	Cabinet hinges, minimum		160	.050		2.18	1.20		3.38	4.44	
	4020	Average		95.24	.084		3.16	2.02		5.18	6.90	
	4040	Maximum	↓	68	.118	↓	7.65	2.82		10.47	13.20	

6 WOOD & PLASTICS

6 WOOD & PLASTICS

06410 | Custom Cabinets

		06410 Custom Cabinets	CREW	DAILY OUTPUT	LABOR-HOURS	UNIT	2005 BARE COSTS MAT.	LABOR	EQUIP.	TOTAL	TOTAL INCL O&P
240	0010	**DRAWERS**									
	0100	Solid hardwood front									
	1000	4″ high, 12″ wide	1 Carp	17	.471	Ea.	18.10	11.30		29.40	39
	1200	18″ wide		16	.500		25	12		37	48
	1400	24″ wide		15	.533		31.50	12.80		44.30	56.50
	1600	6″ high, 12″ wide		16	.500		20.50	12		32.50	43
	1800	18″ wide		15	.533		27	12.80		39.80	51.50
	2000	24″ wide		14	.571		34	13.70		47.70	61
	2200	9″ high, 12″ wide		15	.533		24	12.80		36.80	47.50
	2400	18″ wide		14	.571		30.50	13.70		44.20	57
	2600	24″ wide		13	.615		37.50	14.75		52.25	66
	2800	Plastic laminate on particle board front									
	3000	4″ high, 12″ wide	1 Carp	17	.471	Ea.	23.50	11.30		34.80	45
	3200	18″ wide		16	.500		27.50	12		39.50	50.50
	3600	24″ wide		15	.533		32.50	12.80		45.30	57.50
	3800	6″ high, 12″ wide		16	.500		27.50	12		39.50	50.50
	4000	18″ wide		15	.533		33.50	12.80		46.30	58.50
	4500	24″ wide		14	.571		41	13.70		54.70	69
	4800	9″ high, 12″ wide		15	.533		32	12.80		44.80	56.50
	5000	18″ wide		14	.571		32	13.70		45.70	59
	5200	24″ wide		13	.615		51.50	14.75		66.25	81.50
	5400	Plywood, flush panel front									
	6000	4″ high, 12″ wide	1 Carp	17	.471	Ea.	24.50	11.30		35.80	46
	6200	18″ wide	″	16	.500	″	29.50	12		41.50	53
400	0010	**VANITIES**									
	8000	Vanity bases, 2 doors, 30″ high, 21″ deep, 24″ wide	2 Carp	20	.800	Ea.	177	19.20		196.20	227
	8050	30″ wide		16	1		203	24		227	264
	8100	36″ wide		13.33	1.200		271	29		300	345
	8150	48″ wide		11.43	1.400		325	33.50		358.50	410
	9000	For deluxe models of all vanities, add to above					40%				
	9500	For custom built in place, add to above					25%	10%			
500	0010	**CABINETS, OUTDOOR**									
	0020	Cabinet, base, sink/range, 36″	2 Carp	20.30	.788	Ea.	1,250	18.90		1,268.90	1,400
	0100	Cabinet, base, 36″		20.30	.788		1,800	18.90		1,818.90	2,025
	0200	Cabinet, filler strip, 1″ x 30″		158	.101		25	2.43		27.43	31.50

06415 | Countertops

		06415 Countertops	CREW	DAILY OUTPUT	LABOR-HOURS	UNIT	2005 BARE COSTS MAT.	LABOR	EQUIP.	TOTAL	TOTAL INCL O&P
100	0010	**COUNTER TOP** Stock, plastic lam., 24″ wide w/backsplash, min.	1 Carp	30	.267	L.F.	8.75	6.40		15.15	20.50
	0100	Maximum		25	.320		15.85	7.70		23.55	30.50
	0300	Custom plastic, 7/8″ thick, aluminum molding, no splash		30	.267		17.50	6.40		23.90	30
	0400	Cove splash		30	.267		23	6.40		29.40	36.50
	0600	1-1/4″ thick, no splash		28	.286		20	6.85		26.85	33.50
	0700	Square splash		28	.286		25	6.85		31.85	39
	0900	Square edge, plastic face, 7/8″ thick, no splash		30	.267		22	6.40		28.40	35
	1000	With splash		30	.267		28	6.40		34.40	42
	1200	For stainless channel edge, 7/8″ thick, add					2.30			2.30	2.53
	1300	1-1/4″ thick, add					2.70			2.70	2.97
	1500	For solid color suede finish, add					2.25			2.25	2.48
	1700	For end splash, add				Ea.	15			15	16.50
	1901	For cut outs, standard, add, minimum	1 Carp	32	.250			6		6	10.20
	2000	Maximum		8	1		5	24		29	46.50
	2010	Cut out in blacksplash for elec. wall outlet		38	.211			5.05		5.05	8.60
	2020	Cut out for sink		20	.400			9.60		9.60	16.30
	2030	Cut out for stove top		18	.444			10.65		10.65	18.10
	2100	Postformed, including backsplash and front edge		30	.267	L.F.	9	6.40		15.40	21

06400 | Architectural Woodwork

		06415 | Countertops	CREW	DAILY OUTPUT	LABOR-HOURS	UNIT	2005 BARE COSTS MAT.	LABOR	EQUIP.	TOTAL	TOTAL INCL O&P	
100	2110	Mitred, add	1 Carp	12	.667	Ea.		16		16	27	100
	2200	Built-in place, 25″ wide, plastic laminate		25	.320	L.F.	12	7.70		19.70	26.50	
	2300	Ceramic tile mosaic		25	.320		26	7.70		33.70	41.50	
	2500	Marble, stock, with splash, 1/2″ thick, minimum	1 Bric	17	.471		32	11.60		43.60	54.50	
	2700	3/4″ thick, maximum	″	13	.615		80	15.20		95.20	114	
	2900	Maple, solid, laminated, 1-1/2″ thick, no splash	1 Carp	28	.286		52.50	6.85		59.35	69	
	3000	With square splash		28	.286		62	6.85		68.85	80	
	3200	Stainless steel		24	.333	S.F.	120	8		128	146	
	3400	Recessed cutting block with trim, 16″ x 20″ x 1″		8	1	Ea.	61	24		85	108	
	3411	Replace cutting block only		16	.500	″	30	12		42	53.50	
		06430 | Stairs & Railings										
500	0010	**RAILING** Custom design, architectural grade, hardwood, minimum	1 Carp	38	.211	L.F.	5.60	5.05		10.65	14.75	500
	0100	Maximum		30	.267		46.50	6.40		52.90	62.50	
	0300	Stock interior railing with spindles 6″ O.C., 4′ long		40	.200		42	4.80		46.80	54	
	0400	8′ long		48	.167		21	4		25	30	
620	0010	**STAIRS, PREFABRICATED**										620
	0100	Box stairs, prefabricated, 3′-0″ wide										
	0110	Oak treads, up to 14 risers	2 Carp	39	.410	Riser	71.50	9.85		81.35	95	
	0600	With pine treads for carpet, up to 14 risers	″	39	.410	″	46	9.85		55.85	67	
	1100	For 4′ wide stairs, add				Flight	25%					
	1550	Stairs, prefabricated stair handrail with balusters	1 Carp	30	.267	L.F.	60.50	6.40		66.90	77.50	
	1700	Basement stairs, prefabricated, pine treads										
	1710	Pine risers, 3′ wide, up to 14 risers	2 Carp	52	.308	Riser	46	7.40		53.40	63	
	4000	Residential, wood, oak treads, prefabricated		1.50	10.667	Flight	930	256		1,186	1,450	
	4200	Built in place		.44	36.364	″	1,800	875		2,675	3,475	
	4400	Spiral, oak, 4′-6″ diameter, unfinished, prefabricated,										
	4500	incl. railing, 9′ high	2 Carp	1.50	10.667	Flight	4,400	256		4,656	5,275	
630	0010	**STAIR PARTS** Balusters, turned, 30″ high, pine, minimum	1 Carp	28	.286	Ea.	3.70	6.85		10.55	15.70	630
	0100	Maximum		26	.308		19	7.40		26.40	33.50	
	0300	30″ high birch balusters, minimum		28	.286		6.30	6.85		13.15	18.60	
	0400	Maximum		26	.308		25.50	7.40		32.90	40.50	
	0600	42″ high, pine balusters, minimum		27	.296		4.90	7.10		12	17.45	
	0700	Maximum		25	.320		27.50	7.70		35.20	43	
	0900	42″ high birch balusters, minimum		27	.296		10.10	7.10		17.20	23	
	1000	Maximum		25	.320		36.50	7.70		44.20	53	
	1050	Baluster, stock pine, 1-1/4″ x 1-1/4″		240	.033	L.F.	4.65	.80		5.45	6.45	
	1100	1-3/4″ x 1-3/4″		220	.036	″	8.25	.87		9.12	10.60	
	1200	Newels, 3-1/4″ wide, starting, minimum		7	1.143	Ea.	38.50	27.50		66	88.50	
	1300	Maximum		6	1.333		300	32		332	385	
	1500	Landing, minimum		5	1.600		98	38.50		136.50	173	
	1600	Maximum		4	2		330	48		378	445	
	1800	Railings, oak, built-up, minimum		60	.133	L.F.	30	3.20		33.20	38.50	
	1900	Maximum		55	.145		43.50	3.49		46.99	54	
	2100	Add for sub rail		110	.073		5	1.75		6.75	8.45	
	2300	Risers, beech, 3/4″ x 7-1/2″ high		64	.125		5.80	3		8.80	11.45	
	2400	Fir, 3/4″ x 7-1/2″ high		64	.125		1.60	3		4.60	6.85	
	2600	Oak, 3/4″ x 7-1/2″ high		64	.125		6.10	3		9.10	11.80	
	2800	Pine, 3/4″ x 7-1/2″ high		66	.121		3.05	2.91		5.96	8.30	
	2850	Skirt board, pine, 1″ x 10″		55	.145		2.90	3.49		6.39	9.15	
	2900	1″ x 12″		52	.154		3.45	3.69		7.14	10.05	
	3000	Treads, oak, 1-1/4″ x 10″ wide, 3′ long		18	.444	Ea.	80.50	10.65		91.15	107	
	3100	4′ long, oak		17	.471		108	11.30		119.30	137	
	3300	1-1/4″ x 11-1/2″ wide, 3′ long, oak		18	.444		29.50	10.65		40.15	50	
	3400	6′ long, oak		14	.571		176	13.70		189.70	217	
	3600	Beech treads, add					40%					

WOOD & PLASTICS 6

		06430 | Stairs & Railings	CREW	DAILY OUTPUT	LABOR-HOURS	UNIT	2005 BARE COSTS MAT.	LABOR	EQUIP.	TOTAL	TOTAL INCL O&P	
630	3800	For mitered return nosings, add				L.F.	3.90			3.90	4.29	63
		06440 | Wood Ornaments										
150	0010	**BEAMS, DECORATIVE** Rough sawn cedar, non-load bearing, 4″ x 4″	2 Carp	180	.089	L.F.	1.35	2.13		3.48	5.10	15
	0100	4″ x 6″		170	.094		2.60	2.26		4.86	6.70	
	0200	4″ x 8″		160	.100		3.34	2.40		5.74	7.75	
	0300	4″ x 10″		150	.107		4.64	2.56		7.20	9.45	
	0400	4″ x 12″		140	.114		5.60	2.74		8.34	10.85	
	0500	8″ x 8″		130	.123		7.85	2.95		10.80	13.65	
	0600	Plastic beam, "hewn finish", 6″ x 2″		240	.067		2.89	1.60		4.49	5.90	
	0601	6″ x 4″		220	.073		3.37	1.75		5.12	6.65	
	1100	Beam connector plates see div. 06090-800										
350	0010	**GRILLES** and panels, hardwood, sanded										35
	0020	2′ x 4′ to 4′ x 8′, custom designs, unfinished, minimum	1 Carp	38	.211	S.F.	12.10	5.05		17.15	22	
	0050	Average		30	.267		26.50	6.40		32.90	40	
	0100	Maximum		19	.421		40.50	10.10		50.60	61.50	
	0300	As above, but prefinished, minimum		38	.211		12.10	5.05		17.15	22	
	0400	Maximum		19	.421		45.50	10.10		55.60	67	
400	0010	**LOUVERS** Redwood, 2′-0″ diameter, full circle	1 Carp	16	.500	Ea.	136	12		148	171	40
	0100	Half circle		16	.500		130	12		142	164	
	0200	Octagonal		16	.500		104	12		116	135	
	0300	Triangular, 5/12 pitch, 5′-0″ at base		16	.500		220	12		232	263	
500	0010	**FIREPLACE MANTELS** 6″ molding, 6′ x 3′-6″ opening, minimum	1 Carp	5	1.600	Opng.	139	38.50		177.50	218	50
	0100	Maximum		5	1.600		173	38.50		211.50	255	
	0300	Prefabricated pine, colonial type, stock, deluxe		2	4		2,175	96		2,271	2,575	
	0400	Economy		3	2.667		375	64		439	520	
550	0010	**FIREPLACE MANTEL BEAMS** Rough texture wood, 4″ x 8″	1 Carp	36	.222	L.F.	5.15	5.35		10.50	14.70	550
	0100	4″ x 10″		35	.229	″	6.20	5.50		11.70	16.10	
	0300	Laminated hardwood, 2-1/4″ x 10-1/2″ wide, 6′ long		5	1.600	Ea.	98.50	38.50		137	173	
	0400	8′ long		5	1.600	″	137	38.50		175.50	215	
	0600	Brackets for above, rough sawn		12	.667	Pr.	9.05	16		25.05	37	
	0700	Laminated		12	.667	″	13.70	16		29.70	42	
700	0010	**COLUMNS**										700
	0050	Aluminum, round colonial, 6″ diameter	2 Carp	80	.200	V.L.F.	18.10	4.80		22.90	28	
	0100	8″ diameter		62.25	.257		25	6.15		31.15	38	
	0200	10″ diameter		55	.291		30	7		37	45	
	0250	Fir, stock units, hollow round, 6″ diameter		80	.200		19.65	4.80		24.45	29.50	
	0300	8″ diameter		80	.200		20	4.80		24.80	30	
	0350	10″ diameter		70	.229		25	5.50		30.50	37	
	0400	Solid turned, to 8′ high, 3-1/2″ diameter		80	.200		7.45	4.80		12.25	16.30	
	0500	4-1/2″ diameter		75	.213		10.60	5.10		15.70	20.50	
	0600	5-1/2″ diameter		70	.229		14.85	5.50		20.35	25.50	
	0800	Square columns, built-up, 5″ x 5″		65	.246		13.80	5.90		19.70	25	
	0900	Solid, 3-1/2″ x 3-1/2″		130	.123		6.35	2.95		9.30	12	
	1600	Hemlock, tapered, T & G, 12″ diam, 10′ high		100	.160		32	3.84		35.84	41.50	
	1700	16′ high		65	.246		56	5.90		61.90	72	
	1900	10′ high, 14″ diameter		100	.160		81.50	3.84		85.34	96	
	2000	18′ high		65	.246		77.50	5.90		83.40	95	
	2200	18″ diameter, 12′ high		65	.246		109	5.90		114.90	130	
	2300	20′ high		50	.320		105	7.70		112.70	129	
	2500	20″ diameter, 14′ high		40	.400		129	9.60		138.60	158	
	2600	20′ high		35	.457		134	10.95		144.95	166	
	2800	For flat pilasters, deduct					33%					
	3000	For splitting into halves, add				Ea.	95			95	105	

6 WOOD & PLASTICS

06400 | Architectural Woodwork

	06440 | Wood Ornaments	CREW	DAILY OUTPUT	LABOR-HOURS	UNIT	2005 BARE COSTS MAT.	LABOR	EQUIP.	TOTAL	TOTAL INCL O&P	
4000	Rough sawn cedar posts, 4″ x 4″	2 Carp	250	.064	V.L.F.	2.58	1.54		4.12	5.45	700
4100	4″ x 6″		235	.068		3.84	1.63		5.47	7	
4200	6″ x 6″		220	.073		5.80	1.75		7.55	9.30	
4300	8″ x 8″		200	.080		5.95	1.92		7.87	9.80	
	06445 | Simulated Wood Ornaments										
0010	**MILLWORK, HIGH DENSITY POLYMER**										100
0100	Base, 9/16″ x 3-3/16″	1 Carp	230	.035	L.F.	1.34	.83		2.17	2.89	
0200	Casing, fluted, 5/8″ x 3-1/4″		215	.037		3.73	.89		4.62	5.60	
0300	Chair rail, 9/16″ x 2-1/4″		260	.031		1.94	.74		2.68	3.38	
0600	Cove, 13/16″ x 3-3/4″		260	.031		3.90	.74		4.64	5.55	
0700	Crown, 3/4″ x 3-13/16″		260	.031		5.50	.74		6.24	7.30	
0800	Half round, 15/16″ x 2″		240	.033		17.40	.80		18.20	20.50	
	06470 | Screen, Blinds & Shutters										
0010	**SHUTTERS, EXTERIOR** Aluminum, louvered, 1′-4″ wide, 3′-0″ long	1 Carp	10	.800	Pr.	42	19.20		61.20	78.50	100
0200	4′-0″ long		10	.800		50	19.20		69.20	87.50	
0300	5′-4″ long		10	.800		66	19.20		85.20	105	
0400	6′-8″ long		9	.889		84	21.50		105.50	129	
1000	Pine, louvered, primed, each 1′-2″ wide, 3′-3″ long		10	.800		79.50	19.20		98.70	120	
1100	4′-7″ long		10	.800		107	19.20		126.20	151	
1250	Each 1′-4″ wide, 3′-0″ long		10	.800		81	19.20		100.20	122	
1350	5′-3″ long		10	.800		122	19.20		141.20	167	
1500	Each 1′-6″ wide, 3′-3″ long		10	.800		84.50	19.20		103.70	126	
1600	4′-7″ long		10	.800		119	19.20		138.20	163	
1620	Hemlock, louvered, 1′-2″ wide, 5′-7″ long		10	.800		130	19.20		149.20	176	
1630	Each 1′-4″ wide, 2′-2″ long		10	.800		81	19.20		100.20	122	
1640	3′-0″ long		10	.800		81	19.20		100.20	122	
1650	3′-3″ long		10	.800		87.50	19.20		106.70	129	
1660	3′-11″ long		10	.800		99.50	19.20		118.70	142	
1670	4′-3″ long		10	.800		97	19.20		116.20	140	
1680	5′-3″ long		10	.800		121	19.20		140.20	166	
1690	5′-11″ long		10	.800		137	19.20		156.20	184	
1700	Door blinds, 6′-9″ long, each 1′-3″ wide		9	.889		138	21.50		159.50	187	
1710	1′-6″ wide		9	.889		148	21.50		169.50	199	
1720	Hemlock, solid raised panel, each 1′-4″ wide, 3′-3″ long		10	.800		131	19.20		150.20	177	
1730	3′-11″ long		10	.800		154	19.20		173.20	202	
1740	4′-3″ long		10	.800		166	19.20		185.20	216	
1750	4′-7″ long		10	.800		179	19.20		198.20	230	
1760	4′-11″ long		10	.800		195	19.20		214.20	247	
1770	5′-11″ long		10	.800		222	19.20		241.20	277	
1800	Door blinds, 6′-9″ long, each 1′-3″ wide		9	.889		250	21.50		271.50	310	
1900	1′-6″ wide		9	.889		272	21.50		293.50	335	
2500	Polystyrene, solid raised panel, each 1′-4″ wide, 3′-3″ long		10	.800		55	19.20		74.20	93	
2600	3′-11″ long		10	.800		63	19.20		82.20	102	
2700	4′-7″ long		10	.800		66.50	19.20		85.70	106	
2800	5′-3″ long		10	.800		74.50	19.20		93.70	115	
2900	6′-8″ long		9	.889		103	21.50		124.50	150	
3500	Polystyrene, solid raised panel, each 3′-3″ wide, 3′-0″ long		10	.800		166	19.20		185.20	215	
3600	3′-11″ long		10	.800		180	19.20		199.20	231	
3700	4′-7″ long		10	.800		213	19.20		232.20	267	
3800	5′-3″ long		10	.800		237	19.20		256.20	293	
3900	6′-8″ long		9	.889		266	21.50		287.50	330	
4500	Polystyrene, louvered, each 1′-2″ wide, 3′-3″ long		10	.800		40	19.20		59.20	76.50	
4600	4′-7″ long		10	.800		49.50	19.20		68.70	87	

06400 | Architectural Woodwork

	06470 | Screen, Blinds & Shutters		CREW	DAILY OUTPUT	LABOR-HOURS	UNIT	2005 BARE COSTS MAT.	LABOR	EQUIP.	TOTAL	TOTAL INCL O&P
100	4750	5′-3″ long	1 Carp	10	.800	Pr.	53	19.20		72.20	90.50
	4850	6′-8″ long		9	.889		87	21.50		108.50	132
	6000	Vinyl, louvered, each 1′-2″ x 4′-7″ long		10	.800		52	19.20		71.20	89.50
	6200	Each 1′-4″ x 6′-8″ long		9	.889		79	21.50		100.50	123
	8000	PVC exterior rolling shutters									
	8100	including crank control	1 Carp	8	1	Ea.	400	24		424	480
	8500	Insulative - 6′ x 6′8″ stock unit	"	8	1	"	570	24		594	670
200	0010	**SHUTTERS, INTERIOR** Wood, louvered,									
	0200	Two panel, 27″ wide, 36″ high	1 Carp	5	1.600	Set	105	38.50		143.50	181
	0300	33″ wide, 36″ high		5	1.600		125	38.50		163.50	203
	0500	47″ wide, 36″ high		5	1.600		182	38.50		220.50	265
	1000	Four panel, 27″ wide, 36″ high		5	1.600		241	38.50		279.50	330
	1100	33″ wide, 36″ high		5	1.600		241	38.50		279.50	330
	1300	47″ wide, 36″ high		5	1.600		320	38.50		358.50	420

06600 | Plastic Fabrications

	06620 | Non-Structural Plastics		CREW	DAILY OUTPUT	LABOR-HOURS	UNIT	2005 BARE COSTS MAT.	LABOR	EQUIP.	TOTAL	TOTAL INCL O&P
810	0010	**SOLID SURFACE COUNTERTOPS**, Acrylic polymer									
	2000	Pricing for order of 1 - 50 L.F.									
	2100	25″ wide, solid colors	2 Carp	20	.800	L.F.	59	19.20		78.20	97.50
	2200	Patterned colors		20	.800		75	19.20		94.20	115
	2300	Premium patterned colors		20	.800		93.50	19.20		112.70	136
	2400	With silicone attached 4″ backsplash, solid colors		19	.842		65	20		85	106
	2500	Patterned colors		19	.842		82	20		102	125
	2600	Premium patterned colors		19	.842		102	20		122	147
	2700	With hard seam attached 4″ backsplash, solid colors		4	4		65	96		161	235
	2800	Patterned colors		15	1.067		82	25.50		107.50	134
	2900	Premium patterned colors		15	1.067		102	25.50		127.50	156
	3800	Sinks, pricing for order of 1 - 50 units									
	3900	Single bowl, hard seamed, solid colors, 13″ x 17″	1 Carp	2	4	Ea.	400	96		496	605
	4000	10″ x 15″		4.55	1.758		184	42		226	275
	4100	Cutouts for sinks		5.25	1.524			36.50		36.50	62
850	0010	**VANITY TOPS**									
	0015	Solid surface, center bowl, 17″ x 19″	1 Carp	12	.667	Ea.	168	16		184	212
	0020	19″ x 25″		12	.667		204	16		220	251
	0030	19″ x 31″		12	.667		247	16		263	299
	0040	19″ x 37″		12	.667		288	16		304	340
	0050	22″ x 25″		10	.800		229	19.20		248.20	285
	0060	22″ x 31″		10	.800		267	19.20		286.20	325
	0070	22″ x 37″		10	.800		310	19.20		329.20	375
	0080	22″ x 43″		10	.800		355	19.20		374.20	425
	0090	22″ x 49″		10	.800		395	19.20		414.20	465
	0110	22″ x 55″		8	1		445	24		469	530
	0120	22″ x 61″		8	1		510	24		534	600
	0220	Double bowl, 22″ x 61″		8	1		575	24		599	670
	0230	Double bowl, 22″ x 73″		8	1		625	24		649	730
	0240	For aggregate colors, add					35%				
	0250	For faucets and fittings see 15410-300									

Division 7
Thermal & Moisture Protection

Estimating Tips

07100 Dampproofing & Waterproofing

- Be sure of the job specifications before pricing this subdivision. The difference in cost between waterproofing and dampproofing can be great. Waterproofing will hold back standing water. Dampproofing prevents the transmission of water vapor. Also included in this section are vapor retarding membranes.

07200 Thermal Protection

- Insulation and fireproofing products are measured by area, thickness, volume or R value. Specifications may only give what the specific R value should be in a certain situation. The estimator may need to choose the type of insulation to meet that R value.

07300 Shingles, Roof Tiles & Roof Coverings
07400 Roofing & Siding Panels

- Many roofing and siding products are bought and sold by the square. One square is equal to an area that measures 100 square feet.
  This simple change in unit of measure could create a large error if the estimator is not observant. Accessories necessary for a complete installation must be figured into any calculations for both material and labor.

07500 Membrane Roofing
07600 Flashing & Sheet Metal
07700 Roof Specialties & Accessories

- The items in these subdivisions compose a roofing system. No one component completes the installation and all must be estimated. Built-up or single ply membrane roofing systems are made up of many products and installation trades. Wood blocking at roof perimeters or penetrations, parapet coverings, reglets, roof drains, gutters, downspouts, sheet metal flashing, skylights, smoke vents or roof hatches all need to be considered along with the roofing material. Several different installation trades will need to work together on the roofing system. Inherent difficulties in the scheduling and coordination of various trades must be accounted for when estimating labor costs.

07900 Joint Sealers

- To complete the weather-tight shell the sealants and caulkings must be estimated. Where different materials meet—at expansion joints, at flashing penetrations, and at hundreds of other locations throughout a construction project—they provide another line of defense against water penetration. Often, an entire system is based on the proper location and placement of caulking or sealants. The detail drawings that are included as part of a set of architectural plans, show typical locations for these materials. When caulking or sealants are shown at typical locations, this means the estimator must include them for all the locations where this detail is applicable. Be careful to keep different types of sealants separate, and remember to consider backer rods and primers if necessary.

Reference Numbers

Reference numbers are shown in bold squares at the beginning of some major classifications. These numbers refer to related items in the Reference Section. The reference information may be an estimating procedure, an alternate pricing method or technical information.

Note: Not all subdivisions listed here necessarily appear in this publication.

No part of this publication may be reproduced, stored in a retrieval system, or transmitted in any form or by any means without prior written permission of Reed Construction Data.

	07060 Selective Demolition		Crew	Daily Output	Labor-Hours	Unit	2005 Bare Costs Mat.	Labor	Equip.	Total	Total Incl O&P
110	0010	**SELECTIVE DEMOLITION, ROOFING AND SIDING** R02220-510									
	1200	Wood, boards, tongue and groove, 2″ x 6″	2 Clab	960	.017	S.F.		.29		.29	.49
	1220	2″ x 10″		1,040	.015			.27		.27	.45
	1280	Standard planks, 1″ x 6″		1,080	.015			.26		.26	.44
	1320	1″ x 8″		1,160	.014			.24		.24	.41
	1340	1″ x 12″		1,200	.013			.23		.23	.39
	1350	Plywood, to 1″ thick		2,000	.008			.14		.14	.24
	1360	Flashing, aluminum	1 Clab	290	.028			.48		.48	.81
	2000	Gutters, aluminum or wood, edge hung	″	240	.033	L.F.		.58		.58	.98
	2010	Remove and reset, aluminum	1 Shee	125	.064			1.70		1.70	2.85
	2020	Remove and reset, vinyl	1 Carp	125	.064			1.54		1.54	2.61
	2100	Built-in	1 Clab	100	.080			1.39		1.39	2.36
	2500	Roof accessories, plumbing vent flashing		14	.571	Ea.		9.90		9.90	16.85
	2600	Adjustable metal chimney flashing		9	.889	″		15.40		15.40	26
	2650	Coping, sheet metal, up to 12″ wide		240	.033	L.F.		.58		.58	.98
	2660	Concrete, up to 12″ wide	2 Clab	160	.100	″		1.74		1.74	2.95
	3000	Roofing, built-up, 5 ply roof, no gravel	B-2	1,600	.025	S.F.		.44		.44	.75
	3100	Gravel removal, minimum		5,000	.008			.14		.14	.24
	3120	Maximum		2,000	.020			.36		.36	.60
	3400	Roof insulation board, up to 2″ thick		3,900	.010			.18		.18	.31
	3450	Roll roofing, cold adhesive	1 Clab	12	.667	Sq.		11.55		11.55	19.65
	4000	Shingles, asphalt strip, 1 layer	B-2	3,500	.011	S.F.		.20		.20	.34
	4100	Slate		2,500	.016			.28		.28	.48
	4300	Wood		2,200	.018			.32		.32	.55
	4500	Skylight to 10 S.F.	1 Clab	8	1	Ea.		17.35		17.35	29.50
	5000	Siding, metal, horizontal		444	.018	S.F.		.31		.31	.53
	5020	Vertical		400	.020			.35		.35	.59
	5200	Wood, boards, vertical		400	.020			.35		.35	.59
	5220	Clapboards, horizontal		380	.021			.37		.37	.62
	5240	Shingles		350	.023			.40		.40	.67
	5260	Textured plywood		725	.011			.19		.19	.32

07100 | Dampproofing and Waterproofing

	07110 Dampproofing		Crew	Daily Output	Labor-Hours	Unit	2005 Bare Costs Mat.	Labor	Equip.	Total	Total Incl O&P
100	0010	**BITUMINOUS ASPHALT COATING** For foundation									
	0030	Brushed on, below grade, 1 coat	1 Rofc	665	.012	S.F.	.07	.24		.31	.53
	0100	2 coat		500	.016		.10	.32		.42	.71
	0300	Sprayed on, below grade, 1 coat, 25.6 S.F./gal.		830	.010		.07	.20		.27	.44
	0400	2 coat, 20.5 S.F./gal.		500	.016		.14	.32		.46	.76
	0600	Troweled on, asphalt with fibers, 1/16″ thick		500	.016		.16	.32		.48	.78
	0700	1/8″ thick		400	.020		.29	.41		.70	1.07
	1000	1/2″ thick		350	.023		.94	.46		1.40	1.89
200	0010	**CEMENT PARGING** 2 coats, 1/2″ thick, regular P.C. R07110-010	D-1	250	.064	S.F.	.17	1.39		1.56	2.51
	0100	Waterproofed Portland cement	″	250	.064	″	.18	1.39		1.57	2.52

	07190 Water Repellents		Crew	Daily Output	Labor-Hours	Unit	2005 Bare Costs Mat.	Labor	Equip.	Total	Total Incl O&P
700	0010	**RUBBER COATING** Water base liquid, roller applied	2 Rofc	7,000	.002	S.F.	.55	.05		.60	.70
	0200	Silicone or stearate, sprayed on CMU, 1 coat	1 Rofc	4,000	.002		.31	.04		.35	.41

07100 | Dampproofing and Waterproofing

07190 | Water Repellents		CREW	DAILY OUTPUT	LABOR-HOURS	UNIT	2005 BARE COSTS MAT.	LABOR	EQUIP.	TOTAL	TOTAL INCL O&P	
0300	2 coats	1 Rofc	3,000	.003	S.F.	.63	.05		.68	.79	700

07200 | Thermal Protection

07210 | Building Insulation		CREW	DAILY OUTPUT	LABOR-HOURS	UNIT	2005 BARE COSTS MAT.	LABOR	EQUIP.	TOTAL	TOTAL INCL O&P	
0010	**BLOWN-IN INSULATION** Ceilings, with open access										150
0020	Cellulose, 3-1/2″ thick, R13	G-4	5,000	.005	S.F.	.13	.09	.04	.26	.34	
0030	5-3/16″ thick, R19		3,800	.006		.21	.11	.06	.38	.48	
0050	6-1/2″ thick, R22		3,000	.008		.26	.14	.07	.47	.61	
1000	Fiberglass, 5″ thick, R11		3,800	.006		.16	.11	.06	.33	.43	
1050	6″ thick, R13		3,000	.008		.22	.14	.07	.43	.56	
1100	8-1/2″ thick, R19		2,200	.011		.31	.20	.10	.61	.78	
1300	12″ thick, R26		1,500	.016		.43	.29	.14	.86	1.12	
2000	Mineral wool, 4″ thick, R12		3,500	.007		.18	.12	.06	.36	.48	
2050	6″ thick, R17		2,500	.010		.20	.17	.09	.46	.61	
2100	9″ thick, R23		1,750	.014		.29	.25	.12	.66	.88	
2500	Wall installation, incl. drilling & patching from outside, two 1″										
2510	diam. holes @ 16″ O.C., top & mid-point of wall, add to above										
2700	For masonry	G-4	415	.058	S.F.	.06	1.04	.52	1.62	2.41	
2800	For wood siding		840	.029		.06	.51	.26	.83	1.22	
2900	For stucco/plaster		665	.036		.06	.65	.33	1.04	1.53	
0010	**FLOOR INSULATION, NONRIGID** Including										350
0020	spring type wire fasteners										
2000	Fiberglass, blankets or batts, paper or foil backing										
2100	1 side, 3-1/2″ thick, R11	1 Carp	700	.011	S.F.	.32	.27		.59	.82	
2150	6″ thick, R19		600	.013		.40	.32		.72	.98	
2200	8-1/2″ thick, R30		550	.015		.65	.35		1	1.31	
0010	**POURED INSULATION** Cellulose fiber, R3.8 per inch	1 Carp	200	.040	C.F.	.48	.96		1.44	2.16	500
0080	Fiberglass wool, R4 per inch		200	.040		.39	.96		1.35	2.06	
0100	Mineral wool, R3 per inch		200	.040		.33	.96		1.29	1.99	
0300	Polystyrene, R4 per inch		200	.040		2.23	.96		3.19	4.08	
0400	Vermiculite or perlite, R2.7 per inch		200	.040		1.65	.96		2.61	3.45	
0010	**MASONRY INSULATION** Vermiculite or perlite, poured										550
0100	In cores of concrete block, 4″ thick wall, .115 CF/SF	D-1	4,800	.003	S.F.	.19	.07		.26	.33	
0700	Foamed in place, urethane in 2-5/8″ cavity	G-2	1,035	.023		.40	.45	.10	.95	1.30	
0800	For each 1″ added thickness, add	″	2,372	.010		.12	.20	.05	.37	.51	
0010	**PERIMETER INSULATION**										600
0600	Polystyrene, expanded, 1″ thick, R4	1 Carp	680	.012	S.F.	.19	.28		.47	.69	
0700	2″ thick, R8	″	675	.012	″	.41	.28		.69	.93	
0011	**REFLECTIVE INSULATION,** aluminum foil on reinforced scrim	1 Carp	1,900	.004	S.F.	.14	.10		.24	.32	700
0101	Reinforced with woven polyolefin		1,900	.004		.17	.10		.27	.36	
0501	With single bubble air space, R8.8		1,500	.005		.27	.13		.40	.52	
0601	With double bubble air space, R9.8		1,500	.005		.29	.13		.42	.54	
0010	**WALL INSULATION, RIGID**										900
0040	Fiberglass, 1.5#/CF, unfaced, 1″ thick, R4.1	1 Carp	1,000	.008	S.F.	.31	.19		.50	.67	
0060	1-1/2″ thick, R6.2		1,000	.008		.41	.19		.60	.78	
0080	2″ thick, R8.3		1,000	.008		.47	.19		.66	.85	

07210 | Building Insulation

		07210 Building Insulation	CREW	DAILY OUTPUT	LABOR-HOURS	UNIT	2005 BARE COSTS MAT.	LABOR	EQUIP.	TOTAL	TOTAL INCL O&P
900	0120	3" thick, R12.4	1 Carp	800	.010	S.F.	.57	.24		.81	1.04
	0370	3#/CF, unfaced, 1" thick, R4.3		1,000	.008		.36	.19		.55	.73
	0390	1-1/2" thick, R6.5		1,000	.008		.69	.19		.88	1.09
	0400	2" thick, R8.7		890	.009		.83	.22		1.05	1.28
	0420	2-1/2" thick, R10.9		800	.010		1.02	.24		1.26	1.53
	0440	3" thick, R13		800	.010		1.21	.24		1.45	1.74
	0520	Foil faced, 1" thick, R4.3		1,000	.008		.80	.19		.99	1.21
	0540	1-1/2" thick, R6.5		1,000	.008		1.08	.19		1.27	1.52
	0560	2" thick, R8.7		890	.009		1.35	.22		1.57	1.86
	0580	2-1/2" thick, R10.9		800	.010		1.60	.24		1.84	2.17
	0600	3" thick, R13		800	.010		1.74	.24		1.98	2.32
	0670	6#/CF, unfaced, 1" thick, R4.3		1,000	.008		.77	.19		.96	1.18
	0690	1-1/2" thick, R6.5		890	.009		1.19	.22		1.41	1.68
	0700	2" thick, R8.7		800	.010		1.68	.24		1.92	2.26
	0721	2-1/2" thick, R10.9		800	.010		1.84	.24		2.08	2.43
	0741	3" thick, R13		730	.011		2.20	.26		2.46	2.87
	0821	Foil faced, 1" thick, R4.3		1,000	.008		1.09	.19		1.28	1.53
	0840	1-1/2" thick, R6.5		890	.009		1.57	.22		1.79	2.10
	0850	2" thick, R8.7		800	.010		2.05	.24		2.29	2.67
	0880	2-1/2" thick, R10.9		800	.010		2.46	.24		2.70	3.12
	0900	3" thick, R13		730	.011		2.94	.26		3.20	3.68
	1500	Foamglass, 1-1/2" thick, R4.5		800	.010		1.21	.24		1.45	1.74
	1550	3" thick, R9		730	.011		2.92	.26		3.18	3.66
	1600	Isocyanurate, 4' x 8' sheet, foil faced, both sides									
	1610	1/2" thick, R3.9	1 Carp	800	.010	S.F.	.29	.24		.53	.73
	1620	5/8" thick, R4.5		800	.010		.30	.24		.54	.74
	1630	3/4" thick, R5.4		800	.010		.24	.24		.48	.67
	1640	1" thick, R7.2		800	.010		.33	.24		.57	.77
	1650	1-1/2" thick, R10.8		730	.011		.35	.26		.61	.84
	1660	2" thick, R14.4		730	.011		.46	.26		.72	.96
	1670	3" thick, R21.6		730	.011		1.11	.26		1.37	1.67
	1680	4" thick, R28.8		730	.011		1.37	.26		1.63	1.96
	1700	Perlite, 1" thick, R2.77		800	.010		.26	.24		.50	.70
	1750	2" thick, R5.55		730	.011		.50	.26		.76	1
	1900	Extruded polystyrene, 25 PSI compressive strength, 1" thick, R5		800	.010		.38	.24		.62	.83
	1940	2" thick R10		730	.011		.75	.26		1.01	1.28
	1960	3" thick, R15		730	.011		1.04	.26		1.30	1.59
	2100	Expanded polystyrene, 1" thick, R3.85		800	.010		.17	.24		.41	.60
	2120	2" thick, R7.69		730	.011		.42	.26		.68	.91
	2140	3" thick, R11.49		730	.011		.52	.26		.78	1.02
950	0010	**WALL OR CEILING INSUL., NON-RIGID**									
	0040	Fiberglass, kraft faced, batts or blankets									
	0061	3-1/2" thick, R11, 11" wide	1 Carp	1,600	.005	S.F.	.27	.12		.39	.50
	0080	15" wide		1,600	.005		.27	.12		.39	.50
	0141	6" thick, R19, 11" wide		1,350	.006		.35	.14		.49	.63
	0201	9" thick, R30, 15" wide		1,350	.006		.60	.14		.74	.90
	0241	12" thick, R38, 15" wide		1,350	.006		.76	.14		.90	1.08
	0400	Fiberglass, foil faced, batts or blankets									
	0420	3-1/2" thick, R11, 15" wide	1 Carp	1,600	.005	S.F.	.40	.12		.52	.64
	0461	6" thick, R19, 15" wide		1,600	.005		.41	.12		.53	.65
	0501	9" thick, R30, 15" wide		1,350	.006		.71	.14		.85	1.02
	0800	Fiberglass, unfaced, batts or blankets									
	0821	3-1/2" thick, R11, 15" wide	1 Carp	1,600	.005	S.F.	.21	.12		.33	.43
	0861	6" thick, R19, 15" wide		1,350	.006		.34	.14		.48	.61
	0901	9" thick, R30, 15" wide		1,150	.007		.60	.17		.77	.94
	0941	12" thick, R38, 15" wide		1,150	.007		.76	.17		.93	1.12

Important: See the Reference Section for critical supporting data - Reference Nos., Crews, & Location Factors

07210	Building Insulation	CREW	DAILY OUTPUT	LABOR-HOURS	UNIT	2005 BARE COSTS MAT.	LABOR	EQUIP.	TOTAL	TOTAL INCL O&P	
1300	Mineral fiber batts, kraft faced										950
1320	3-1/2" thick, R12	1 Carp	1,600	.005	S.F.	.28	.12		.40	.51	
1340	6" thick, R19		1,600	.005		.38	.12		.50	.62	
1380	10" thick, R30		1,350	.006		.57	.14		.71	.87	
1850	Friction fit wire insulation supports, 16" O.C.		960	.008	Ea.	.05	.20		.25	.40	
1900	For foil backing, add				S.F.	.04			.04	.04	
07220	**Roof and Deck Insulation**										
0010	ROOF DECK INSULATION										700
0020	Fiberboard low density, 1/2" thick R1.39	1 Rofc	1,000	.008	S.F.	.19	.16		.35	.51	
0030	1" thick R2.78		800	.010		.34	.20		.54	.74	
0080	1 1/2" thick R4.17		800	.010		.50	.20		.70	.92	
0100	2" thick R5.56		800	.010		.68	.20		.88	1.12	
0110	Fiberboard high density, 1/2" thick R1.3		1,000	.008		.20	.16		.36	.52	
0120	1" thick R2.5		800	.010		.36	.20		.56	.77	
0130	1-1/2" thick R3.8		800	.010		.59	.20		.79	1.02	
0200	Fiberglass, 3/4" thick R2.78		1,000	.008		.46	.16		.62	.81	
0400	15/16" thick R3.70		1,000	.008		.61	.16		.77	.97	
0460	1-1/16" thick R4.17		1,000	.008		.76	.16		.92	1.14	
0600	1-5/16" thick R5.26		1,000	.008		1.05	.16		1.21	1.46	
0650	2-1/16" thick R8.33		800	.010		1.12	.20		1.32	1.60	
0700	2-7/16" thick R10		800	.010		1.28	.20		1.48	1.78	
1650	Perlite, 1/2" thick R1.32		1,050	.008		.27	.15		.42	.59	
1655	3/4" thick R2.08		800	.010		.29	.20		.49	.69	
1660	1" thick R2.78		800	.010		.31	.20		.51	.71	
1670	1-1/2" thick R4.17		800	.010		.39	.20		.59	.80	
1680	2" thick R5.56		700	.011		.62	.23		.85	1.11	
1685	2-1/2" thick R6.67		700	.011		.77	.23		1	1.28	
1700	Polyisocyanurate, 2#/CF density, 3/4" thick, R5.1		1,500	.005		.30	.11		.41	.53	
1705	1" thick R7.14		1,400	.006		.33	.12		.45	.57	
1715	1-1/2" thick R10.87		1,250	.006		.35	.13		.48	.63	
1725	2" thick R14.29		1,100	.007		.45	.15		.60	.77	
1735	2-1/2" thick R16.67		1,050	.008		.50	.15		.65	.84	
1745	3" thick R21.74		1,000	.008		.73	.16		.89	1.10	
1755	3-1/2" thick R25		1,000	.008		.74	.16		.90	1.11	
1765	Tapered for drainage		1,400	.006	B.F.	.38	.12		.50	.63	
1900	Extruded Polystyrene										
1910	15 PSI compressive strength, 1" thick, R5	1 Rofc	1,500	.005	S.F.	.23	.11		.34	.45	
1920	2" thick, R10		1,250	.006		.35	.13		.48	.63	
1930	3" thick R15		1,000	.008		.68	.16		.84	1.05	
1932	4" thick R20		1,000	.008		1.07	.16		1.23	1.48	
1934	Tapered for drainage		1,500	.005	B.F.	.35	.11		.46	.59	
1940	25 PSI compressive strength, 1" thick R5		1,500	.005	S.F.	.50	.11		.61	.75	
1942	2" thick R10		1,250	.006		.97	.13		1.10	1.31	
1944	3" thick R15		1,000	.008		1.47	.16		1.63	1.92	
1946	4" thick R20		1,000	.008		1.96	.16		2.12	2.46	
1948	Tapered for drainage		1,500	.005	B.F.	.40	.11		.51	.64	
1950	40 psi compressive strength, 1" thick R5		1,500	.005	S.F.	.35	.11		.46	.59	
1952	2" thick R10		1,250	.006		.69	.13		.82	1	
1954	3" thick R15		1,000	.008		1.01	.16		1.17	1.41	
1956	4" thick R20		1,000	.008		1.35	.16		1.51	1.79	
1958	Tapered for drainage		1,400	.006	B.F.	.50	.12		.62	.76	
1960	60 PSI compressive strength, 1" thick R5		1,450	.006	S.F.	.42	.11		.53	.67	
1962	2" thick R10		1,200	.007		.75	.14		.89	1.08	
1964	3" thick R15		975	.008		1.12	.17		1.29	1.54	
1966	4" thick R20		950	.008		1.56	.17		1.73	2.04	

07220 | Roof and Deck Insulation

			CREW	DAILY OUTPUT	LABOR-HOURS	UNIT	2005 BARE COSTS MAT.	LABOR	EQUIP.	TOTAL	TOTAL INCL O&P
700	1968	Tapered for drainage	1 Rofc	1,400	.006	B.F.	.61	.12		.73	.88
	2010	Expanded polystyrene, 1#/CF density, 3/4" thick R2.89		1,500	.005	S.F.	.19	.11		.30	.41
	2020	1" thick R3.85		1,500	.005		.19	.11		.30	.41
	2100	2" thick R7.69		1,250	.006		.41	.13		.54	.69
	2110	3" thick R11.49		1,250	.006		.70	.13		.83	1.01
	2120	4" thick R15.38		1,200	.007		.57	.14		.71	.88
	2130	5" thick R19.23		1,150	.007		.71	.14		.85	1.04
	2140	6" thick R23.26		1,150	.007		.83	.14		.97	1.17
	2150	Tapered for drainage		1,500	.005	B.F.	.34	.11		.45	.57
	2400	Composites with 2" EPS									
	2410	1" fiberboard	1 Rofc	950	.008	S.F.	.78	.17		.95	1.18
	2420	7/16" oriented strand board		800	.010		.93	.20		1.13	1.39
	2430	1/2" plywood		800	.010		1	.20		1.20	1.47
	2440	1" perlite		800	.010		.82	.20		1.02	1.27
	2450	Composites with 1 1/2" polyisocyanurate									
	2460	1" fiberboard	1 Rofc	800	.010	S.F.	.84	.20		1.04	1.29
	2470	1" perlite		850	.009		.88	.19		1.07	1.32
	2480	7/16" oriented strand board		800	.010		1.01	.20		1.21	1.48

07240 | Ext. Insulation Finish Systems (EIFS)

			CREW	DAILY OUTPUT	LABOR-HOURS	UNIT	MAT.	LABOR	EQUIP.	TOTAL	TOTAL INCL O&P
100	0010	**EXTERIOR INSULATION FINISH SYSTEM**									
	0095	Field applied, 1" EPS insulation	J-1	295	.136	S.F.	1.97	2.81	.35	5.13	7.20
	0100	With 1/2" cement board sheathing		220	.182		2.72	3.77	.47	6.96	9.75
	0105	2" EPS insulation		295	.136		2.22	2.81	.35	5.38	7.50
	0110	With 1/2" cement board sheathing		220	.182		2.97	3.77	.47	7.21	10.05
	0115	3" EPS insulation		295	.136		2.32	2.81	.35	5.48	7.60
	0120	With 1/2" cement board sheathing		220	.182		3.07	3.77	.47	7.31	10.15
	0125	4" EPS insulation		295	.136		2.64	2.81	.35	5.80	7.95
	0130	With 1/2" cement board sheathing		220	.182		4.15	3.77	.47	8.39	11.30
	0140	Premium finish add		1,265	.032		.28	.66	.08	1.02	1.49
	0150	Heavy duty reinforcement add		914	.044		1.68	.91	.11	2.70	3.48
	0160	2.5#/S.Y. metal lath substrate add	1 Lath	75	.107	S.Y.	2.16	2.35		4.51	6.20
	0170	3.4#/S.Y. metal lath substrate add	"	75	.107	"	2.25	2.35		4.60	6.30
	0180	Color or texture change,	J-1	1,265	.032	S.F.	.73	.66	.08	1.47	1.98
	0190	With substrate leveling base coat	1 Plas	530	.015		.73	.33		1.06	1.35
	0210	With substrate sealing base coat	1 Pord	1,224	.007		.07	.14		.21	.31
	0370	V groove shape in panel face				L.F.	.52			.52	.57
	0380	U groove shape in panel face				"	.69			.69	.76

07260 | Vapor Retarders

			CREW	DAILY OUTPUT	LABOR-HOURS	UNIT	MAT.	LABOR	EQUIP.	TOTAL	TOTAL INCL O&P
100	0011	**BUILDING PAPER** Aluminum and kraft laminated, foil 1 side	1 Carp	3,700	.002	S.F.	.04	.05		.09	.13
	0101	Foil 2 sides		3,700	.002		.06	.05		.11	.16
	0301	Asphalt, two ply, 30#, for subfloors		1,900	.004		.14	.10		.24	.32
	0401	Asphalt felt sheathing paper, 15#		3,700	.002		.03	.05		.08	.13
	0450	Housewrap, exterior, spun bonded polypropylene									
	0470	Small roll	1 Carp	3,800	.002	S.F.	.16	.05		.21	.27
	0480	Large roll	"	4,000	.002	"	.10	.05		.15	.19
	0500	Material only, 3' x 111.1' roll				Ea.	55			55	60.50
	0520	9' x 111.1' roll				"	100			100	110
	0601	Polyethylene vapor barrier, standard, .002" thick	1 Carp	3,700	.002	S.F.	.01	.05		.06	.10
	0701	.004" thick		3,700	.002		.02	.05		.07	.11
	0901	.006" thick		3,700	.002		.03	.05		.08	.12
	1201	.010" thick		3,700	.002		.05	.05		.10	.15
	1501	Red rosin paper, 5 sq rolls, 4 lb per square		3,700	.002		.02	.05		.07	.11
	1601	5 lbs. per square		3,700	.002		.02	.05		.07	.11
	1801	Reinf. waterproof, .002" polyethylene backing, 1 side		3,700	.002		.05	.05		.10	.15

Important: See the Reference Section for critical supporting data - Reference Nos., Crews, & Location Factors

07200 | Thermal Protection

07260 Vapor Retarders		CREW	DAILY OUTPUT	LABOR-HOURS	UNIT	2005 BARE COSTS MAT.	LABOR	EQUIP.	TOTAL	TOTAL INCL O&P	
1901	2 sides	1 Carp	3,700	.002	S.F.	.07	.05		.12	.16	100
3000	Building wrap, spunbonded polyethylene	2 Carp	8,000	.002		.11	.05		.16	.20	

07300 | Shingles, Roof Tiles and Roof Coverings

07310 Shingles		CREW	DAILY OUTPUT	LABOR-HOURS	UNIT	2005 BARE COSTS MAT.	LABOR	EQUIP.	TOTAL	TOTAL INCL O&P	
0010	**ASPHALT SHINGLES**										100
0100	Standard strip shingles										
0150	Inorganic, class A, 210-235 lb/sq	1 Rofc	5.50	1.455	Sq.	31	29.50		60.50	88.50	
0155	Pneumatic nailed		7	1.143		31	23		54	77	
0200	Organic, class C, 235-240 lb/sq		5	1.600		40.50	32.50		73	105	
0205	Pneumatic nailed		6.25	1.280		40.50	26		66.50	92.50	
0250	Standard, laminated multi-layered shingles										
0300	Class A, 240-260 lb/sq	1 Rofc	4.50	1.778	Sq.	43	36		79	114	
0305	Pneumatic nailed		5.63	1.422		43	29		72	101	
0350	Class C, 260-300 lb/square, 4 bundles/square		4	2		49	40.50		89.50	129	
0355	Pneumatic nailed		5	1.600		49	32.50		81.50	114	
0400	Premium, laminated multi-layered shingles										
0450	Class A, 260-300 lb, 4 bundles/sq	1 Rofc	3.50	2.286	Sq.	54.50	46.50		101	146	
0455	Pneumatic nailed		4.37	1.831		54.50	37		91.50	129	
0500	Class C, 300-385 lb/square, 5 bundles/square		3	2.667		65.50	54		119.50	172	
0505	Pneumatic nailed		3.75	2.133		65.50	43.50		109	152	
0800	#15 felt underlayment		64	.125		3.41	2.54		5.95	8.45	
0825	#30 felt underlayment		58	.138		6.95	2.80		9.75	12.80	
0850	Self adhering polyethylene and rubberized asphalt underlayment		22	.364		42	7.40		49.40	59.50	
0900	Ridge shingles		330	.024	L.F.	.79	.49		1.28	1.78	
0905	Pneumatic nailed		412.50	.019	"	.79	.39		1.18	1.60	
1000	For steep roofs (7 to 12 pitch or greater), add						50%				
0010	**FIBER CEMENT** shingles, 16" x 9.35", 500 lb per square	1 Rofc	2.20	3.636	Sq.	244	74		318	405	500
0200	Shakes, 16" x 9.35", 550 lb per square		2.20	3.636	"	221	74		295	380	
0301	Hip & ridge, 4.75 x 14"		100	.080	L.F.	6	1.62		7.62	9.60	
0400	Hexagonal, 16" x 16"		3	2.667	Sq.	165	54		219	282	
0500	Square, 16" x 16"		3	2.667		148	54		202	263	
2000	For steep roofs (7/12 pitch or greater), add						50%				
0010	**SLATE**, Buckingham, Virginia, black R07310 -020										800
0100	3/16" - 1/4" thick	1 Rots	1.75	4.571	Sq.	550	93.50		643.50	780	
0200	1/4" thick		1.75	4.571		580	93.50		673.50	815	
0900	Pennsylvania black, Bangor, #1 clear		1.75	4.571		450	93.50		543.50	670	
1200	Vermont, unfading, green, mottled green		1.75	4.571		345	93.50		438.50	555	
1300	Semi-weathering green & gray		1.75	4.571		330	93.50		423.50	540	
1400	Purple		1.75	4.571		370	93.50		463.50	580	
1500	Black or gray		1.75	4.571		340	93.50		433.50	550	
2700	Ridge shingles, slate		200	.040	L.F.	8.50	.82		9.32	10.85	
0010	**WOOD** 16" No. 1 red cedar shingles, 5" exposure, on roof	1 Carp	2.50	3.200	Sq.	163	77		240	310	980
0015	Pneumatic nailed		3.25	2.462	"	163	59		222	279	
0200	7-1/2" exposure, on walls		2.05	3.902	Sq.	108	93.50		201.50	278	
0205	Pneumatic nailed		2.67	2.996		108	72		180	241	
0300	18" No. 1 red cedar perfections, 5-1/2" exposure, on roof		2.75	2.909		160	70		230	295	
0305	Pneumatic nailed		3.57	2.241		160	54		214	268	

	07310 | Shingles		CREW	DAILY OUTPUT	LABOR-HOURS	UNIT	2005 BARE COSTS MAT.	LABOR	EQUIP.	TOTAL	TOTAL INCL O&P
980	0500	7-1/2" exposure, on walls	1 Carp	2.25	3.556	Sq.	118	85.50		203.50	274
	0505	Pneumatic nailed		2.92	2.740		118	66		184	241
	0600	Resquared, and rebutted, 5-1/2" exposure, on roof		3	2.667		199	64		263	330
	0605	Pneumatic nailed		3.90	2.051		199	49		248	305
	0900	7-1/2" exposure, on walls		2.45	3.265		146	78.50		224.50	294
	0905	Pneumatic nailed	↓	3.18	2.516		146	60.50		206.50	264
	1000	Add to above for fire retardant shingles, 16" long					30			30	33
	1050	18" long				↓	28.50			28.50	31.50
	1060	Preformed ridge shingles	1 Carp	400	.020	L.F.	1.65	.48		2.13	2.64
	1100	Hand-split red cedar shakes, 1/2" thick x 24" long, 10" exp. on roof		2.50	3.200	Sq.	138	77		215	282
	1105	Pneumatic nailed		3.25	2.462		138	59		197	252
	1110	3/4" thick x 24" long, 10" exp. on roof		2.25	3.556		138	85.50		223.50	297
	1115	Pneumatic nailed		2.92	2.740		138	66		204	264
	1200	1/2" thick, 18" long, 8-1/2" exp. on roof		2	4		97.50	96		193.50	270
	1205	Pneumatic nailed		2.60	3.077		97.50	74		171.50	232
	1210	3/4" thick x 18" long, 8 1/2" exp. on roof		1.80	4.444		97.50	107		204.50	288
	1215	Pneumatic nailed		2.34	3.419		97.50	82		179.50	246
	1255	10" exp. on walls		2	4		110	96		206	284
	1260	10" exposure on walls, pneumatic nailed	↓	2.60	3.077		110	74		184	246
	1700	Add to above for fire retardant shakes, 24" long					30			30	33
	1800	18" long				↓	30			30	33
	1810	Ridge shakes	1 Carp	350	.023	L.F.	2.35	.55		2.90	3.52
	2000	White cedar shingles, 16" long, extras, 5" exposure, on roof		2.40	3.333	Sq.	122	80		202	271
	2005	Pneumatic nailed		3.12	2.564		122	61.50		183.50	239
	2050	5" exposure on walls		2	4		122	96		218	298
	2055	Pneumatic nailed		2.60	3.077		122	74		196	260
	2100	7-1/2" exposure, on walls		2	4		87.50	96		183.50	259
	2105	Pneumatic nailed		2.60	3.077		87.50	74		161.50	221
	2150	"B" grade, 5" exposure on walls		2	4		110	96		206	284
	2155	Pneumatic nailed		2.60	3.077		110	74		184	246
	2300	For 15# organic felt underlayment on roof, 1 layer, add		64	.125		3.41	3		6.41	8.85
	2400	2 layers, add	↓	32	.250		6.80	6		12.80	17.70
	2600	For steep roofs (7/12 pitch or greater), add to above				↓		50%			
	3000	Ridge shakes or shingle wood	1 Carp	280	.029	L.F.	2.40	.69		3.09	3.80
	07320 | Roof Tiles										
200	0010	**CLAY TILE** ASTM C1167, GR 1, severe weathering, acces. incl.									
	0200	Lanai tile or Classic tile, 158 pc per sq	1 Rots	1.65	4.848	Sq.	395	99.50		494.50	620
	0300	Americana, 158 pc per sq, most colors		1.65	4.848		510	99.50		609.50	745
	0350	Green, gray or brown		1.65	4.848		490	99.50		589.50	725
	0400	Blue		1.65	4.848		490	99.50		589.50	725
	0600	Spanish tile, 171 pc per sq, red		1.80	4.444		260	91		351	455
	0800	Blend		1.80	4.444		420	91		511	630
	0900	Glazed white		1.80	4.444		500	91		591	720
	1100	Mission tile, 192 pc per sq, machine scored finish, red		1.15	6.957		635	143		778	960
	1700	French tile, 133 pc per sq, smooth finish, red		1.35	5.926		575	121		696	855
	1750	Blue or green		1.35	5.926		685	121		806	980
	1800	Norman black 317 pc per sq		1	8		805	164		969	1,175
	2200	Williamsburg tile, 158 pc per sq, aged cedar		1.35	5.926		490	121		611	765
	2250	Gray or green		1.35	5.926	↓	490	121		611	765
	2350	Ridge shingles, clay tile		200	.040	L.F.	8.65	.82		9.47	11
	2510	One piece mission tile, natural red, 75 pc per square		1.65	4.848	Sq.	133	99.50		232.50	330
	2530	Mission Tile, 134 pc per square	↓	1.15	6.957		178	143		321	460
	3000	For steep roofs (7/12 pitch or greater), add to above				↓		50%			
300	0010	**CONCRETE TILE** Including installation of accessories									
	0020	Corrugated, 13" x 16-1/2", 90 per sq, 950 lb per sq									

07300 | Shingles, Roof Tiles and Roof Coverings

07320 | Roof Tiles

	Description	Crew	Daily Output	Labor-Hours	Unit	2005 Bare Costs Mat.	Labor	Equip.	Total	Total Incl O&P	
0050	Earthtone colors, nailed to wood deck	1 Rots	1.35	5.926	Sq.	91	121		212	325	300
0150	Blues		1.35	5.926		104	121		225	340	
0200	Greens		1.35	5.926		104	121		225	340	
0250	Premium colors		1.35	5.926		153	121		274	390	
0500	Shakes, 13" x 16-1/2", 90 per sq, 950 lb per sq										
0600	All colors, nailed to wood deck	1 Rots	1.50	5.333	Sq.	185	109		294	405	
1500	Accessory pieces, ridge & hip, 10" x 16-1/2", 8 lbs. each				Ea.	2.25			2.25	2.48	
1700	Rake, 6-1/2" x 16-3/4", 9 lbs. each					2.25			2.25	2.48	
1800	Mansard hip, 10" x 16-1/2", 9.2 lbs. each					2.25			2.25	2.48	
1900	Hip starter, 10" x 16-1/2", 10.5 lbs. each					9.50			9.50	10.45	
2000	3 or 4 way apex, 10" each side, 11.5 lbs. each					10.25			10.25	11.30	

07400 | Roofing and Siding Panels

07410 | Metal Roof and Wall Panels

	Description	Crew	Daily Output	Labor-Hours	Unit	2005 Bare Costs Mat.	Labor	Equip.	Total	Total Incl O&P	
0010	**ALUMINUM ROOFING** Corrugated or ribbed, .0155" thick, natural	G-3	1,200	.027	S.F.	.62	.59		1.21	1.67	100
0300	Painted	"	1,200	.027	"	.89	.59		1.48	1.97	

07420 | Plastic Roof and Wall Panels

	Description	Crew	Daily Output	Labor-Hours	Unit	2005 Bare Costs Mat.	Labor	Equip.	Total	Total Incl O&P	
0010	**FIBERGLASS** Corrugated panels, roofing, 8 oz per SF	G-3	1,000	.032	S.F.	2.24	.70		2.94	3.64	770
0100	12 oz per SF		1,000	.032		3.23	.70		3.93	4.73	
0300	Corrugated siding, 6 oz per SF		880	.036		1.94	.80		2.74	3.48	
0400	8 oz per SF		880	.036		2.24	.80		3.04	3.81	
0600	12 oz. siding, textured		880	.036		3.14	.80		3.94	4.80	
0900	Flat panels, 6 oz per SF, clear or colors		880	.036		1.73	.80		2.53	3.25	
1300	8 oz per SF, clear or colors		880	.036		2.24	.80		3.04	3.81	

07460 | Siding

	Description	Crew	Daily Output	Labor-Hours	Unit	2005 Bare Costs Mat.	Labor	Equip.	Total	Total Incl O&P	
0011	**ALUMINUM SIDING**										100
6040	.024 thick smooth white single 8" wide	2 Carp	515	.031	S.F.	1.25	.75		2	2.65	
6060	Double 4" pattern		515	.031		1.19	.75		1.94	2.58	
6080	Double 5" pattern		550	.029		1.19	.70		1.89	2.50	
6120	Embossed white, 8" wide		515	.031		1.48	.75		2.23	2.90	
6140	Double 4" pattern		515	.031		1.35	.75		2.10	2.76	
6160	Double 5" pattern		550	.029		1.35	.70		2.05	2.68	
6170	Vertical, embossed white, 12" wide		590	.027		1.35	.65		2	2.60	
6320	.019 thick, insulated, smooth white, 8" wide		515	.031		1.20	.75		1.95	2.59	
6340	Double 4" pattern		515	.031		1.18	.75		1.93	2.57	
6360	Double 5" pattern		550	.029		1.18	.70		1.88	2.49	
6400	Embossed white, 8" wide		515	.031		1.38	.75		2.13	2.79	
6420	Double 4" pattern		515	.031		1.40	.75		2.15	2.81	
6440	Double 5" pattern		550	.029		1.40	.70		2.10	2.73	
6500	Shake finish 10" wide white		550	.029		1.48	.70		2.18	2.82	
6600	Vertical pattern, 12" wide, white		590	.027		1.40	.65		2.05	2.65	
6640	For colors add					.08			.08	.09	
6700	Accessories, white										
6720	Starter strip 2-1/8"	2 Carp	610	.026	L.F.	.19	.63		.82	1.28	
6740	Sill trim		450	.036		.31	.85		1.16	1.79	

7 THERMAL & MOISTURE PROTECTION

	07460 | Siding		CREW	DAILY OUTPUT	LABOR-HOURS	UNIT	2005 BARE COSTS MAT.	LABOR	EQUIP.	TOTAL	TOTAL INCL O&P
100	6760	Inside corner	2 Carp	610	.026	L.F.	.94	.63		1.57	2.10
	6780	Outside corner post		610	.026		1.60	.63		2.23	2.83
	6800	Door & window trim		440	.036		.30	.87		1.17	1.81
	6820	For colors add					.08			.08	.09
	6900	Soffit & fascia 1′ overhang solid	2 Carp	110	.145		2.16	3.49		5.65	8.35
	6920	Vented		110	.145		2.16	3.49		5.65	8.35
	6940	2′ overhang solid		100	.160		3.17	3.84		7.01	10
	6960	Vented		100	.160		3.17	3.84		7.01	10
300	0010	**FASCIA** Aluminum, reverse board and batten,									
	0100	.032″ thick, colored, no furring included	1 Shee	145	.055	S.F.	2.14	1.46		3.60	4.81
	0200	Residential type, aluminum	1 Carp	200	.040	L.F.	1.15	.96		2.11	2.90
	0300	Steel, galv and enameled, stock, no furring, long panels	1 Shee	145	.055	S.F.	2.23	1.46		3.69	4.91
	0600	Short panels	″	115	.070	″	3.38	1.85		5.23	6.80
500	0010	**FIBER CEMENT SIDING**									
	0020	Lap siding, 5/16″ thick, 6″ wide, smooth texture	2 Carp	415	.039	S.F.	1.01	.93		1.94	2.68
	0025	Woodgrain texture		415	.039		1.01	.93		1.94	2.68
	0030	7-1/2″ wide, smooth texture		425	.038		.88	.90		1.78	2.50
	0035	Woodgrain texture		425	.038		.88	.90		1.78	2.50
	0040	8″ wide, smooth texture		425	.038		.87	.90		1.77	2.49
	0045	Roughsawn texture		425	.038		.87	.90		1.77	2.49
	0050	9-1/2″ wide, smooth texture		440	.036		.84	.87		1.71	2.41
	0055	Woodgrain texture		440	.036		.84	.87		1.71	2.41
	0060	12″ wide, smooth texture		455	.035		.81	.84		1.65	2.33
	0065	Woodgrain texture		455	.035		.81	.84		1.65	2.33
	0070	Panel siding, 5/16″ thick, smooth texture		750	.021		.73	.51		1.24	1.67
	0075	Stucco texture		750	.021		.73	.51		1.24	1.67
	0080	Grooved woodgrain texture		750	.021		.73	.51		1.24	1.67
	0085	V - grooved woodgrain texture		750	.021		.73	.51		1.24	1.67
	0090	Wood starter strip		400	.040	L.F.	.19	.96		1.15	1.84
600	0010	**VINYL SIDING** Solid PVC panels, 8″ to 10″ wide, plain	1 Carp	255	.031	S.F.	.69	.75		1.44	2.04
	2000	Smooth, white, single, 8″ wide	2 Carp	495	.032		.62	.78		1.40	2
	2020	Dutch lap, 10″ wide		550	.029		.64	.70		1.34	1.89
	2100	Double 4″ pattern, 8″ wide		495	.032		.55	.78		1.33	1.93
	2120	Double 5″ pattern, 10″ wide		550	.029		.56	.70		1.26	1.81
	2200	Embossed, white, single, 8″ wide		495	.032		.66	.78		1.44	2.05
	2220	10 ″ wide		550	.029		.67	.70		1.37	1.93
	2300	Double 4″ pattern, 8″ wide		495	.032		.59	.78		1.37	1.97
	2320	5″ pattern, 10″ wide		550	.029		.61	.70		1.31	1.86
	2400	Shake finish, 10″ wide, white		550	.029		1.90	.70		2.60	3.28
	2600	Vertical pattern, double 5″, 10″ wide, white		550	.029		1.37	.70		2.07	2.70
	2620										
	2700	For colors, add				S.F.	.08			.08	.09
	2720	1/4″ extruded polystyrene fan folded insulation	2 Carp	2,000	.008	″	.14	.19		.33	.48
	3000	Accessories, starter strip		700	.023	L.F.	.24	.55		.79	1.19
	3100	″J″ channel, 1/2″		700	.023		.26	.55		.81	1.22
	3120	5/8″		700	.023		.26	.55		.81	1.22
	3140	3/4″		695	.023		.27	.55		.82	1.24
	3160	1″		690	.023		.32	.56		.88	1.30
	3180	1-1/8″		685	.023		.33	.56		.89	1.31
	3190	1-1/4″		680	.024		.35	.56		.91	1.35
	3200	Under sill trim		500	.032		.52	.77		1.29	1.87
	3300	Outside corner post, 3″ face, pocket 5/8″		700	.023		1.14	.55		1.69	2.18
	3320	7/8″		690	.023		1.13	.56		1.69	2.19
	3340	1-1/4″		680	.024		1.19	.56		1.75	2.27
	3400	Inside corner post, pocket 5/8″		700	.023		.61	.55		1.16	1.60

07460 | Siding		CREW	DAILY OUTPUT	LABOR-HOURS	UNIT	2005 BARE COSTS MAT.	LABOR	EQUIP.	TOTAL	TOTAL INCL O&P	
3420	7/8″	2 Carp	690	.023	L.F.	.68	.56		1.24	1.70	600
3440	1-1/4″		680	.024		.67	.56		1.23	1.70	
3500	Door & window trim, 2-1/2″ face, pocket 5/8″		510	.031		.58	.75		1.33	1.92	
3520	7/8″		500	.032		.62	.77		1.39	1.98	
3540	1-1/4″		490	.033		.62	.78		1.40	2.01	
3600	Soffit & fascia, 1′ overhang, solid		120	.133		1.44	3.20		4.64	7.05	
3620	Vented		120	.133		1.48	3.20		4.68	7.10	
3700	2′ overhang, solid		110	.145		2.13	3.49		5.62	8.30	
3720	Vented		110	.145		2.13	3.49		5.62	8.30	
0010	**SOFFIT** Aluminum, residential, stock units, .020″ thick	1 Carp	210	.038	S.F.	1.01	.91		1.92	2.66	750
0100	Baked enamel on steel, 16 or 18 gauge		105	.076		3.74	1.83		5.57	7.20	
0300	Polyvinyl chloride, white, solid		230	.035		.66	.83		1.49	2.15	
0400	Perforated		230	.035		.66	.83		1.49	2.15	
0500	For colors, add					.07			.07	.08	
0010	**STEEL SIDING**, Beveled, vinyl coated, 8″ wide, including fasteners	1 Carp	265	.030	S.F.	1.15	.72		1.87	2.50	800
0050	10″ wide	″	275	.029		1.23	.70		1.93	2.54	
0081	Galv., corrugated or ribbed, on steel frame, 30 gauge	G-3	775	.041		.77	.91		1.68	2.38	
0101	28 gauge		775	.041		.81	.91		1.72	2.42	
0301	26 gauge		775	.041		1.13	.91		2.04	2.77	
0401	24 gauge		775	.041		1.14	.91		2.05	2.78	
0601	22 gauge		775	.041		1.31	.91		2.22	2.97	
0701	Colored, corrugated/ribbed, on steel frame, 10 yr fnsh, 28 ga.		775	.041		1.20	.91		2.11	2.85	
0901	26 gauge		775	.041		1.25	.91		2.16	2.91	
1001	24 gauge		775	.041		1.46	.91		2.37	3.14	
0010	**WOOD SIDING, BOARDS**										900
2000	Board & batten, cedar, "B" grade, 1″ x 10″	1 Carp	400	.020	S.F.	1.98	.48		2.46	3	
2200	Redwood, clear, vertical grain, 1″ x 10″		400	.020		3.59	.48		4.07	4.77	
2400	White pine, #2 & better, 1″ x 10″		400	.020		.70	.48		1.18	1.59	
2410	Board & batten siding, white pine #2, 1″ x 12″		450	.018		.80	.43		1.23	1.60	
3200	Wood, cedar bevel, A grade, 1/2″ x 6″		250	.032		3.28	.77		4.05	4.91	
3300	1/2″ x 8″		275	.029		2.92	.70		3.62	4.40	
3500	3/4″ x 10″, clear grade		300	.027		3.54	.64		4.18	4.98	
3600	"B" grade		300	.027		2.80	.64		3.44	4.17	
3800	Cedar, rough sawn, 1″ x 4″, A grade, natural		240	.033		2.67	.80		3.47	4.30	
3900	Stained		240	.033		3.01	.80		3.81	4.67	
4100	1″ x 12″, board & batten, #3 & Btr., natural		260	.031		2.02	.74		2.76	3.47	
4200	Stained		260	.031		2.35	.74		3.09	3.84	
4400	1″ x 8″ channel siding, #3 & Btr., natural		250	.032		1.96	.77		2.73	3.46	
4500	Stained		250	.032		2.23	.77		3	3.75	
4700	Redwood, clear, beveled, vertical grain, 1/2″ x 4″		200	.040		3.12	.96		4.08	5.05	
4750	1/2″ x 6″		225	.036		2.62	.85		3.47	4.33	
4800	1/2″ x 8″		250	.032		2.12	.77		2.89	3.63	
5000	3/4″ x 10″		300	.027		3.46	.64		4.10	4.90	
5200	Channel siding, 1″ x 10″, B grade		285	.028		2.23	.67		2.90	3.59	
5250	Redwood, T&G boards, B grade, 1″ x 4″	2 Carp	300	.053		2.66	1.28		3.94	5.10	
5270	1″ x 8″	″	375	.043		2.29	1.02		3.31	4.26	
5400	White pine, rough sawn, 1″ x 8″, natural	1 Carp	275	.029		.67	.70		1.37	1.93	
5500	Stained	″	275	.029		.99	.70		1.69	2.28	
5600	Tongue and groove, 1″ x 8″, horizontal	2 Carp	375	.043		.61	1.02		1.63	2.41	
0010	**WOOD PRODUCT SIDING**										950
0030	Lap siding, hardboard, 7/16″ x 8″, primed										
0050	Wood grain texture finish	2 Carp	650	.025	S.F.	1.04	.59		1.63	2.14	
0100	Panels, 7/16″ thick, smooth, textured or grooved, primed		700	.023		.73	.55		1.28	1.73	

07400 | Roofing and Siding Panels

07460 | Siding

		07460 Siding	CREW	DAILY OUTPUT	LABOR-HOURS	UNIT	2005 BARE COSTS MAT.	LABOR	EQUIP.	TOTAL	TOTAL INCL O&P
950	0200	Stained	2 Carp	700	.023	S.F.	.88	.55		1.43	1.90
	0700	Particle board, overlaid, 3/8" thick		750	.021		.63	.51		1.14	1.56
	0900	Plywood, medium density overlaid, 3/8" thick		750	.021		.84	.51		1.35	1.79
	1000	1/2" thick		700	.023		.95	.55		1.50	1.98
	1100	3/4" thick		650	.025		1.28	.59		1.87	2.41
	1600	Texture 1-11, cedar, 5/8" thick, natural		675	.024		2.38	.57		2.95	3.59
	1700	Factory stained		675	.024		2.21	.57		2.78	3.40
	1900	Texture 1-11, fir, 5/8" thick, natural		675	.024		1.01	.57		1.58	2.08
	2000	Factory stained		675	.024		1.08	.57		1.65	2.16
	2050	Texture 1-11, S.Y.P., 5/8" thick, natural		675	.024		.81	.57		1.38	1.86
	2100	Factory stained		675	.024		.88	.57		1.45	1.94
	2200	Rough sawn cedar, 3/8" thick, natural		675	.024		1.08	.57		1.65	2.16
	2300	Factory stained		675	.024		1.20	.57		1.77	2.29
	2500	Rough sawn fir, 3/8" thick, natural		675	.024		.58	.57		1.15	1.61
	2600	Factory stained		675	.024		.65	.57		1.22	1.69
	2800	Redwood, textured siding, 5/8" thick		675	.024		1.80	.57		2.37	2.95

07500 | Membrane Roofing

07510 | Built-Up Bituminous Roofing

		07510 Built-Up Bituminous Roofing	CREW	DAILY OUTPUT	LABOR-HOURS	UNIT	2005 BARE COSTS MAT.	LABOR	EQUIP.	TOTAL	TOTAL INCL O&P
050	0010	**ASPHALT** Coated felt, #30, 2 sq per roll, not mopped	1 Rofc	58	.138	Sq.	6.95	2.80		9.75	12.80
	0200	#15, 4 sq per roll, plain or perforated, not mopped		58	.138		3.41	2.80		6.21	8.90
	0250	Perforated		58	.138		3.41	2.80		6.21	8.90
	0300	Roll roofing, smooth, #65		15	.533		6.35	10.85		17.20	27
	0500	#90		15	.533		17.70	10.85		28.55	39.50
	0520	Mineralized		15	.533		15.35	10.85		26.20	37
	0540	D.C. (Double coverage), 19" selvage edge		10	.800		29.50	16.25		45.75	62.50
	0580	Adhesive (lap cement)				Gal.	3.68			3.68	4.05
300	0010	**BUILT-UP ROOFING**									
	0120	Asphalt flood coat with gravel/slag surfacing, not including									
	0140	Insulation, flashing or wood nailers									
	0200	Asphalt base sheet, 3 plies #15 asphalt felt, mopped	G-1	22	2.545	Sq.	48.50	48.50	13.55	110.55	158
	0350	On nailable decks		21	2.667		53.50	51	14.20	118.70	168
	0500	4 plies #15 asphalt felt, mopped		20	2.800		69	53.50	14.90	137.40	191
	0550	On nailable decks		19	2.947		62.50	56	15.70	134.20	190
	2000	Asphalt flood coat, smooth surface									
	2200	Asphalt base sheet & 3 plies #15 asphalt felt, mopped	G-1	24	2.333	Sq.	52.50	44.50	12.40	109.40	153
	2400	On nailable decks		23	2.435		49	46.50	12.95	108.45	154
	2600	4 plies #15 asphalt felt, mopped		24	2.333		61.50	44.50	12.40	118.40	163
	2700	On nailable decks		23	2.435		58	46.50	12.95	117.45	164
	4500	Coal tar pitch with gravel/slag surfacing									
	4600	4 plies #15 tarred felt, mopped	G-1	21	2.667	Sq.	107	51	14.20	172.20	226
	4800	3 plies glass fiber felt (type IV), mopped	"	19	2.947	"	87.50	56	15.70	159.20	217
400	0010	**CANTS** 4" x 4", treated timber, cut diagonally	1 Rofc	325	.025	L.F.	1.35	.50		1.85	2.41
	0100	Foamglass		325	.025		2.15	.50		2.65	3.29
	0300	Mineral or fiber, trapezoidal, 1"x 4" x 48"		325	.025		.17	.50		.67	1.11
	0400	1-1/2" x 5-5/8" x 48"		325	.025		.29	.50		.79	1.24

07500 | Membrane Roofing

07550 | Modified Bit. Membrane Roofing

		CREW	DAILY OUTPUT	LABOR-HOURS	UNIT	2005 BARE COSTS MAT.	LABOR	EQUIP.	TOTAL	TOTAL INCL O&P	
0010	**MODIFIED BITUMEN ROOFING** R07550-030										500
0020	Base sheet, #15 glass fiber felt, nailed to deck	1 Rofc	58	.138	Sq.	4.85	2.80		7.65	10.50	
0030	Spot mopped to deck	G-1	295	.190		6.50	3.61	1.01	11.12	14.90	
0040	Fully mopped to deck	"	192	.292		8.90	5.55	1.55	16	22	
0050	#15 organic felt, nailed to deck	1 Rofc	58	.138		4.17	2.80		6.97	9.75	
0060	Spot mopped to deck	G-1	295	.190		5.80	3.61	1.01	10.42	14.15	
0070	Fully mopped to deck	"	192	.292		8.20	5.55	1.55	15.30	21	
0080	SBS modified, granule surf cap sheet, polyester rein., mopped										
1500	Glass fiber reinforced, mopped, 160 mils	G-1	2,000	.028	S.F.	.42	.53	.15	1.10	1.60	
1600	Smooth surface cap sheet, mopped, 145 mils		2,100	.027		.42	.51	.14	1.07	1.56	
1700	Smooth surface flashing, 145 mils		1,260	.044		.42	.85	.24	1.51	2.28	
1800	150 mils		1,260	.044		.41	.85	.24	1.50	2.27	
1900	Granular surface flashing, 150 mils		1,260	.044		.45	.85	.24	1.54	2.32	
2000	160 mils		1,260	.044		.66	.85	.24	1.75	2.55	
2100	APP mod., smooth surf. cap sheet, poly. reinf., torched, 160 mils	G-5	2,100	.019		.41	.35	.07	.83	1.18	
2150	170 mils		2,100	.019		.46	.35	.07	.88	1.24	
2200	Granule surface cap sheet, poly. reinf., torched, 180 mils		2,000	.020		.50	.37	.07	.94	1.31	
2250	Smooth surface flashing, torched, 160 mils		1,260	.032		.41	.59	.12	1.12	1.67	
2300	170 mils		1,260	.032		.46	.59	.12	1.17	1.73	
2350	Granule surface flashing, torched, 180 mils		1,260	.032		.50	.59	.12	1.21	1.77	
2400	Fibrated aluminum coating	1 Rofc	3,800	.002		.10	.04		.14	.19	

07580 | Roll Roofing

		CREW	DAILY OUTPUT	LABOR-HOURS	UNIT	MAT.	LABOR	EQUIP.	TOTAL	TOTAL INCL O&P	
0010	**ROLL ROOFING**										200
0100	Asphalt, mineral surface										
0200	1 ply #15 organic felt, 1 ply mineral surfaced										
0300	Selvage roofing, lap 19", nailed & mopped	G-1	27	2.074	Sq.	39	39.50	11.05	89.55	128	
0400	3 plies glass fiber felt (type IV), 1 ply mineral surfaced										
0500	Selvage roofing, lapped 19", mopped	G-1	25	2.240	Sq.	61	42.50	11.90	115.40	159	
0600	Coated glass fiber base sheet, 2 plies of glass fiber										
0700	Felt (type IV), 1 ply mineral surfaced selvage										
0800	Roofing, lapped 19", mopped	G-1	25	2.240	Sq.	66	42.50	11.90	120.40	164	
0900	On nailable decks	"	24	2.333	"	61	44.50	12.40	117.90	163	
1000	3 plies glass fiber felt (type III), 1 ply mineral surfaced										
1100	Selvage roofing, lapped 19", mopped	G-1	25	2.240	Sq.	61	42.50	11.90	115.40	159	

07590 | Roof Maintenance and Repairs

		CREW	DAILY OUTPUT	LABOR-HOURS	UNIT	MAT.	LABOR	EQUIP.	TOTAL	TOTAL INCL O&P	
0010	**ROOF COATINGS** Asphalt				Gal.	2.95			2.95	3.25	300
0800	Glass fibered roof & patching cement, 5 gallon					4.43			4.43	4.87	
1100	Roof patch & flashing cement, 5 gallon					18.05			18.05	19.85	

07600 | Flashing and Sheet Metal

07610 | Sheet Metal Roofing

		CREW	DAILY OUTPUT	LABOR-HOURS	UNIT	2005 BARE COSTS MAT.	LABOR	EQUIP.	TOTAL	TOTAL INCL O&P	
0010	**COPPER ROOFING** Batten seam, over 10 sq, 16 oz, 130 lb/sq	1 Shee	1.10	7.273	Sq.	405	193		598	770	300
0200	18 oz, 145 lb per sq		1	8		450	212		662	850	
0400	Standing seam, over 10 squares, 16 oz, 125 lb per sq		1.30	6.154		390	163		553	700	
0600	18 oz, 140 lb per sq		1.20	6.667		435	177		612	775	

07600 | Flashing and Sheet Metal

		07610 | Sheet Metal Roofing	CREW	DAILY OUTPUT	LABOR-HOURS	UNIT	2005 BARE COSTS MAT.	LABOR	EQUIP.	TOTAL	TOTAL INCL O&P
300	0900	Flat seam, over 10 squares, 16 oz, 115 lb per sq	1 Shee	1.20	6.667	Sq.	355	177		532	690
	1200	For abnormal conditions or small areas, add					25%	100%			
	1300	For lead-coated copper, add					25%				
900	0010	**ZINC** Copper alloy roofing, batten seam, .020" thick	1 Shee	1.20	6.667	Sq.	520	177		697	865
	0100	.027" thick		1.15	6.957		630	185		815	1,000
	0300	.032" thick		1.10	7.273		710	193		903	1,100
	0400	.040" thick		1.05	7.619		835	202		1,037	1,250
	0600	For standing seam construction, deduct					2%				
	0700	For flat seam construction, deduct					3%				
		07620 | Sheet Metal Flashing and Trim									
100	0010	**SHEET METAL CLADDING**									
	0100	Aluminum, up to 6 bends, .032" thick, window casing	1 Carp	180	.044	S.F.	.58	1.07		1.65	2.45
	0200	Window sill		72	.111	L.F.	.58	2.67		3.25	5.15
	0300	Door casing		180	.044	S.F.	.58	1.07		1.65	2.45
	0400	Fascia		250	.032		.58	.77		1.35	1.94
	0500	Rake trim		225	.036		.58	.85		1.43	2.09
	0700	.024" thick, window casing		180	.044		.90	1.07		1.97	2.80
	0800	Window sill		72	.111	L.F.	.90	2.67		3.57	5.50
	0900	Door casing		180	.044	S.F.	.90	1.07		1.97	2.80
	1000	Fascia		250	.032		.90	.77		1.67	2.29
	1100	Rake trim		225	.036		.90	.85		1.75	2.44
	1200	Vinyl coated aluminum, up to 6 bends, window casing		180	.044		.61	1.07		1.68	2.48
	1300	Window sill		72	.111	L.F.	.61	2.67		3.28	5.20
	1400	Door casing		180	.044	S.F.	.61	1.07		1.68	2.48
	1500	Fascia		250	.032		.61	.77		1.38	1.97
	1600	Rake trim		225	.036		.61	.85		1.46	2.12
		07650 | Flexible Flashing									
600	0010	**FLASHING** Aluminum, mill finish, .013" thick	1 Rofc	145	.055	S.F.	.35	1.12		1.47	2.46
	0030	.016" thick		145	.055		.51	1.12		1.63	2.63
	0060	.019" thick		145	.055		.65	1.12		1.77	2.79
	0100	.032" thick		145	.055		1.06	1.12		2.18	3.24
	0200	.040" thick		145	.055		1.45	1.12		2.57	3.67
	0300	.050" thick		145	.055		1.84	1.12		2.96	4.09
	0325	Mill finish 5" x 7" step flashing, .016" thick		1,920	.004	Ea.	.10	.08		.18	.27
	0350	Mill finish 12" x 12" step flashing, .016" thick		1,600	.005	"	.40	.10		.50	.63
	0400	Painted finish, add				S.F.	.24			.24	.26
	1600	Copper, 16 oz, sheets, under 1000 lbs.	1 Rofc	115	.070		2.62	1.41		4.03	5.50
	1900	20 oz sheets, under 1000 lbs.		110	.073		3.90	1.48		5.38	7
	2200	24 oz sheets, under 1000 lbs.		105	.076		4.69	1.55		6.24	8
	2500	32 oz sheets, under 1000 lbs.		100	.080		6.20	1.62		7.82	9.85
	2700	W shape for valleys, 16 oz, 24" wide		100	.080	L.F.	5.90	1.62		7.52	9.50
	2800	Copper, paperbacked 1 side, 2 oz		330	.024	S.F.	.86	.49		1.35	1.86
	2900	3 oz		330	.024		1.12	.49		1.61	2.14
	3100	Paperbacked 2 sides, 2 oz		330	.024		.86	.49		1.35	1.86
	3150	3 oz		330	.024		1.11	.49		1.60	2.13
	3200	5 oz		330	.024		1.66	.49		2.15	2.74
	5800	Lead, 2.5 lb. per SF, up to 12" wide		135	.059		2.94	1.20		4.14	5.45
	5900	Over 12" wide		135	.059		3.25	1.20		4.45	5.80
	6100	Lead-coated copper, fabric-backed, 2 oz		330	.024		1.41	.49		1.90	2.46
	6200	5 oz		330	.024		1.62	.49		2.11	2.69
	6400	Mastic-backed 2 sides, 2 oz		330	.024		1.10	.49		1.59	2.12
	6500	5 oz		330	.024		1.37	.49		1.86	2.42
	6700	Paperbacked 1 side, 2 oz		330	.024		.95	.49		1.44	1.96

Important: See the Reference Section for critical supporting data - Reference Nos., Crews, & Location Factor

07600 | Flashing and Sheet Metal

07650 | Flexible Flashing		CREW	DAILY OUTPUT	LABOR-HOURS	UNIT	2005 BARE COSTS MAT.	LABOR	EQUIP.	TOTAL	TOTAL INCL O&P	
6800	3 oz	1 Rofc	330	.024	S.F.	1.12	.49		1.61	2.14	600
7000	Paperbacked 2 sides, 2 oz		330	.024		.98	.49		1.47	1.99	
7100	5 oz		330	.024		1.60	.49		2.09	2.67	
8500	Shower pan, bituminous membrane, 7 oz		155	.052		1.08	1.05		2.13	3.12	
8550	3 ply copper and fabric, 3 oz		155	.052		1.63	1.05		2.68	3.72	
8600	7 oz		155	.052		3.37	1.05		4.42	5.65	
8650	Copper, 16 oz		100	.080		3.11	1.62		4.73	6.40	
8700	Lead on copper and fabric, 5 oz		155	.052		1.62	1.05		2.67	3.71	
8800	7 oz		155	.052		2.93	1.05		3.98	5.15	
8900	Stainless steel sheets, 32 ga, .010" thick		155	.052		2.16	1.05		3.21	4.31	
9000	28 ga, .015" thick		155	.052		2.68	1.05		3.73	4.88	
9100	26 ga, .018" thick		155	.052		3.25	1.05		4.30	5.50	
9200	24 ga, .025" thick	↓	155	.052	↓	4.22	1.05		5.27	6.55	
9290	For mechanically keyed flashing, add					40%					
9300	Stainless steel, paperbacked 2 sides, .005" thick	1 Rofc	330	.024	S.F.	1.90	.49		2.39	3	
9320	Steel sheets, galvanized, 20 gauge		130	.062		.73	1.25		1.98	3.10	
9340	30 gauge		160	.050		.31	1.02		1.33	2.21	
9400	Terne coated stainless steel, .015" thick, 28 ga		155	.052		4.05	1.05		5.10	6.40	
9500	.018" thick, 26 ga		155	.052		4.57	1.05		5.62	7	
9600	Zinc and copper alloy (brass), .020" thick		155	.052		3.26	1.05		4.31	5.50	
9700	.027" thick		155	.052		4.37	1.05		5.42	6.75	
9800	.032" thick		155	.052		5.10	1.05		6.15	7.55	
9900	.040" thick	↓	155	.052	↓	6.20	1.05		7.25	8.80	

07700 | Roof Specialties and Accessories

07710 | Manufactured Roof Specialties		CREW	DAILY OUTPUT	LABOR-HOURS	UNIT	2005 BARE COSTS MAT.	LABOR	EQUIP.	TOTAL	TOTAL INCL O&P	
0010	**DOWNSPOUTS** Aluminum 2" x 3", .020" thick, embossed	1 Shee	190	.042	L.F.	.68	1.12		1.80	2.63	400
0100	Enameled		190	.042		.97	1.12		2.09	2.95	
0300	Enameled, .024" thick, 2" x 3"		180	.044		1.12	1.18		2.30	3.21	
0400	3" x 4"		140	.057		1.73	1.52		3.25	4.45	
0600	Round, corrugated aluminum, 3" diameter, .020" thick		190	.042		1.04	1.12		2.16	3.02	
0700	4" diameter, .025" thick		140	.057	↓	1.49	1.52		3.01	4.19	
0900	Wire strainer, round, 2" diameter		155	.052	Ea.	1.75	1.37		3.12	4.23	
1000	4" diameter		155	.052		1.82	1.37		3.19	4.30	
1200	Rectangular, perforated, 2" x 3"		145	.055		2.15	1.46		3.61	4.83	
1300	3" x 4"		145	.055	↓	3.10	1.46		4.56	5.85	
1500	Copper, round, 16 oz., stock, 2" diameter		190	.042	L.F.	4.55	1.12		5.67	6.90	
1600	3" diameter		190	.042		4.50	1.12		5.62	6.85	
1800	4" diameter		145	.055		4.91	1.46		6.37	7.85	
1900	5" diameter		130	.062		6.75	1.63		8.38	10.15	
2100	Rectangular, corrugated copper, stock, 2" x 3"		190	.042		3.91	1.12		5.03	6.20	
2200	3" x 4"		145	.055		4.96	1.46		6.42	7.90	
2400	Rectangular, plain copper, stock, 2" x 3"		190	.042		4.78	1.12		5.90	7.15	
2500	3" x 4"		145	.055	↓	6.30	1.46		7.76	9.35	
2700	Wire strainers, rectangular, 2" x 3"		145	.055	Ea.	2.80	1.46		4.26	5.55	
2800	3" x 4"		145	.055		4.44	1.46		5.90	7.35	
3000	Round, 2" diameter		145	.055		2.63	1.46		4.09	5.35	
3100	3" diameter	↓	145	.055	↓	3.67	1.46		5.13	6.50	

	07710 | Manufactured Roof Specialties		DAILY	LABOR-		2005 BARE COSTS				TOTAL
		CREW	OUTPUT	HOURS	UNIT	MAT.	LABOR	EQUIP.	TOTAL	INCL O&P
400	3300 4" diameter	1 Shee	145	.055	Ea.	5.70	1.46		7.16	8.70
	3400 5" diameter		115	.070		8.20	1.85		10.05	12.15
	3600 Lead-coated copper, round, stock, 2" diameter		190	.042	L.F.	5.50	1.12		6.62	7.95
	3700 3" diameter		190	.042		6.35	1.12		7.47	8.85
	3900 4" diameter		145	.055		8.55	1.46		10.01	11.85
	4300 Rectangular, corrugated, stock, 2" x 3"		190	.042		7.35	1.12		8.47	9.95
	4500 Plain, stock, 2" x 3"		190	.042		7.35	1.12		8.47	10
	4600 3" x 4"		145	.055		7.95	1.46		9.41	11.20
	4800 Steel, galvanized, round, corrugated, 2" or 3" diam, 28 ga		190	.042		.75	1.12		1.87	2.71
	4900 4" diameter, 28 gauge		145	.055		1.02	1.46		2.48	3.58
	5700 Rectangular, corrugated, 28 gauge, 2" x 3"		190	.042		.54	1.12		1.66	2.47
	5800 3" x 4"		145	.055		1.50	1.46		2.96	4.11
	6000 Rectangular, plain, 28 gauge, galvanized, 2" x 3"		190	.042		.81	1.12		1.93	2.77
	6100 3" x 4"		145	.055		1.23	1.46		2.69	3.81
	6300 Epoxy painted, 24 gauge, corrugated, 2" x 3"		190	.042		1.05	1.12		2.17	3.04
	6400 3" x 4"		145	.055		1.91	1.46		3.37	4.56
	6600 Wire strainers, rectangular, 2" x 3"		145	.055	Ea.	1.62	1.46		3.08	4.24
	6700 3" x 4"		145	.055		2.61	1.46		4.07	5.35
	6900 Round strainers, 2" or 3" diameter		145	.055		1.20	1.46		2.66	3.78
	7000 4" diameter		145	.055		1.42	1.46		2.88	4.02
	7200 5" diameter		145	.055		2.20	1.46		3.66	4.88
	7300 6" diameter		115	.070		2.63	1.85		4.48	6
	8200 Vinyl, rectangular, 2" x 3"		210	.038	L.F.	.72	1.01		1.73	2.49
	8300 Round, 2-1/2"		220	.036	"	.72	.97		1.69	2.41
450	0010 **DRIP EDGE**, aluminum, .016" thick, 5" wide, mill finish	1 Carp	400	.020	L.F.	.25	.48		.73	1.10
	0100 White finish		400	.020		.22	.48		.70	1.06
	0200 8" wide, mill finish		400	.020		.30	.48		.78	1.15
	0300 Ice belt, 28" wide, mill finish		100	.080		3.42	1.92		5.34	7
	0310 Vented, mill finish		400	.020		1.41	.48		1.89	2.37
	0320 Painted finish		400	.020		1.54	.48		2.02	2.51
	0400 Galvanized, 5" wide		400	.020		.22	.48		.70	1.06
	0500 8" wide, mill finish		400	.020		.33	.48		.81	1.18
	0510 Rake edge, aluminum, 1-1/2" x 1-1/2"		400	.020		.13	.48		.61	.96
	0520 3-1/2" x 1-1/2"		400	.020		.19	.48		.67	1.03
500	0010 **ELBOWS** Aluminum, 2" x 3", embossed	1 Shee	100	.080	Ea.	.91	2.12		3.03	4.56
	0100 Enameled		100	.080		1.76	2.12		3.88	5.50
	0200 3" x 4", .025" thick, embossed		100	.080		3.20	2.12		5.32	7.10
	0300 Enameled		100	.080		3.20	2.12		5.32	7.10
	0400 Round corrugated, 3", embossed, .020" thick		100	.080		1.98	2.12		4.10	5.75
	0500 4", .025" thick		100	.080		2.99	2.12		5.11	6.85
	0600 Copper, 16 oz. round, 2" diameter		100	.080		11.15	2.12		13.27	15.85
	0700 3" diameter		100	.080		5.95	2.12		8.07	10.05
	0800 4" diameter		100	.080		9.05	2.12		11.17	13.50
	1000 2" x 3" corrugated		100	.080		5.15	2.12		7.27	9.25
	1100 3" x 4" corrugated		100	.080		7.65	2.12		9.77	11.95
	1300 Vinyl, 2-1/2" diameter, 45° or 75°		100	.080		2	2.12		4.12	5.75
	1400 Tee Y junction		75	.107		8.50	2.83		11.33	14.10
550	0010 **GRAVEL STOP** Aluminum, .050" thick, 4" face height, mill finish	1 Shee	145	.055	L.F.	3.82	1.46		5.28	6.65
	0080 Duranodic finish		145	.055		3.69	1.46		5.15	6.50
	0100 Painted		145	.055		4.26	1.46		5.72	7.15
	1350 Galv steel, 24 ga., 4" leg, plain, with continuous cleat, 4" face		145	.055		1.50	1.46		2.96	4.11
	1500 Polyvinyl chloride, 6" face height		135	.059		3.28	1.57		4.85	6.25
	1800 Stainless steel, 24 ga., 6" face height		135	.059		7.15	1.57		8.72	10.50
650	0010 **GUTTERS** Aluminum, stock units, 5" box, .027" thick, plain	1 Shee	120	.067	L.F.	1.24	1.77		3.01	4.33
	0020 Inside corner		25	.320	Ea.	5.25	8.50		13.75	20

07700 | Roof Specialties and Accessories

07710 | Manufactured Roof Specialties

	Line	Description	Crew	Daily Output	Labor-Hours	Unit	2005 Bare Costs Mat.	Labor	Equip.	Total	Total Incl O&P
650	0030	Outside corner	1 Shee	25	.320	Ea.	5.25	8.50		13.75	20
	0100	Enameled		120	.067	L.F.	1.09	1.77		2.86	4.17
	0110	Inside corner		25	.320	Ea.	5.30	8.50		13.80	20
	0120	Outside corner		25	.320	″	5.30	8.50		13.80	20
	0300	5″ box type, .032″ thick, plain		120	.067	L.F.	1.22	1.77		2.99	4.31
	0310	Inside corner		25	.320	Ea.	5.20	8.50		13.70	19.95
	0320	Outside corner		25	.320	″	5.45	8.50		13.95	20.50
	0400	Enameled		120	.067	L.F.	1.21	1.77		2.98	4.30
	0410	Inside corner		25	.320	Ea.	5.60	8.50		14.10	20.50
	0420	Outside corner		25	.320	″	5.60	8.50		14.10	20.50
	0600	5″ x 6″ combination fascia & gutter, .032″ thick, enameled		60	.133	L.F.	3.52	3.54		7.06	9.80
	0700	Copper, half round, 16 oz, stock units, 4″ wide		120	.067		4.10	1.77		5.87	7.50
	0900	5″ wide		120	.067		5.35	1.77		7.12	8.85
	1000	6″ wide		115	.070		6.25	1.85		8.10	10
	1200	K type, 16 oz, stock, 4″ wide		120	.067		5.05	1.77		6.82	8.50
	1300	5″ wide		120	.067		5.50	1.77		7.27	9
	1500	Lead coated copper, half round, stock, 4″ wide		120	.067		7.50	1.77		9.27	11.20
	1600	6″ wide		115	.070		10.40	1.85		12.25	14.55
	1800	K type, stock, 4″ wide		120	.067		8.10	1.77		9.87	11.85
	1900	5″ wide		120	.067		8.25	1.77		10.02	12.05
	2100	Stainless steel, half round or box, stock, 4″ wide		120	.067		4.65	1.77		6.42	8.05
	2200	5″ wide		120	.067		5	1.77		6.77	8.45
	2400	Steel, galv, half round or box, 28 ga, 5″ wide, plain		120	.067		1	1.77		2.77	4.07
	2500	Enameled		120	.067		1.12	1.77		2.89	4.20
	2700	26 ga, stock, 5″ wide		120	.067		.95	1.77		2.72	4.02
	2800	6″ wide		120	.067		1.60	1.77		3.37	4.73
	3000	Vinyl, O.G., 4″ wide	1 Carp	110	.073		.85	1.75		2.60	3.90
	3100	5″ wide		110	.073		1	1.75		2.75	4.06
	3200	4″ half round, stock units		110	.073		.68	1.75		2.43	3.71
	3250	Joint connectors				Ea.	1.36			1.36	1.50
	3300	Wood, clear treated cedar, fir or hemlock, 3″ x 4″	1 Carp	100	.080	L.F.	6.30	1.92		8.22	10.15
	3400	4″ x 5″	″	100	.080	″	7.30	1.92		9.22	11.25
700	0010	**GUTTER GUARD** 6″ wide strip, aluminum mesh	1 Carp	500	.016	L.F.	.37	.38		.75	1.06
	0100	Vinyl mesh	″	500	.016	″	.40	.38		.78	1.09

07720 | Roof Accessories

	Line	Description	Crew	Daily Output	Labor-Hours	Unit	Mat.	Labor	Equip.	Total	Total Incl O&P
550	0010	**RIDGE VENT**									
	0100	Aluminum strips, mill finish	1 Rofc	160	.050	L.F.	1.20	1.02		2.22	3.19
	0150	Painted finish		160	.050	″	2.12	1.02		3.14	4.21
	0200	Connectors		48	.167	Ea.	1.91	3.38		5.29	8.35
	0300	End caps		48	.167	″	.93	3.38		4.31	7.25
	0400	Galvanized strips		160	.050	L.F.	2.07	1.02		3.09	4.15
	0430	Molded polyethylene, shingles not included		160	.050	″	2.55	1.02		3.57	4.68
	0440	End plugs		48	.167	Ea.	.93	3.38		4.31	7.25
	0450	Flexible roll, shingles not included		160	.050	L.F.	1.99	1.02		3.01	4.06
560	0010	**SNOW GUARDS**									
	0100	Slate & asphalt shingle roofs	1 Rofc	160	.050	Ea.	8	1.02		9.02	10.65
	0200	Standing seam metal roofs		48	.167		12.25	3.38		15.63	19.75
	0300	Surface mount for metal roofs		48	.167		6.75	3.38		10.13	13.70
865	0010	**VENTS**									
	0100	Soffit or eave, aluminum, mill finish, strips, 2-1/2″ wide	1 Carp	200	.040	L.F.	.30	.96		1.26	1.96
	0200	3″ wide		200	.040		.33	.96		1.29	1.99
	0300	Enamel finish, 3″ wide		200	.040		.45	.96		1.41	2.13
	0400	Mill finish, rectangular, 4″ x 16″		72	.111	Ea.	1.38	2.67		4.05	6.05
	0500	8″ x 16″		72	.111	″	1.65	2.67		4.32	6.35

	07920 | Joint Sealants	CREW	DAILY OUTPUT	LABOR-HOURS	UNIT	2005 BARE COSTS MAT.	LABOR	EQUIP.	TOTAL	TOTAL INCL O&P
800	0010 CAULKING AND SEALANTS									
	0020 Acoustical sealant, elastomeric, cartridges				Ea.	2.21			2.21	2.43
	0100 Acrylic latex caulk, white									
	0200 11 fl. oz cartridge				Ea.	1.88			1.88	2.07
	0500 1/4" x 1/2"	1 Bric	248	.032	L.F.	.15	.80		.95	1.50
	0600 1/2" x 1/2"		250	.032		.31	.79		1.10	1.66
	0800 3/4" x 3/4"		230	.035		.69	.86		1.55	2.19
	0900 3/4" x 1"		200	.040		.92	.99		1.91	2.65
	1000 1" x 1"	↓	180	.044	↓	1.15	1.10		2.25	3.10
	1400 Butyl based, bulk				Gal.	22			22	24.50
	1500 Cartridges				"	27			27	29.50
	1700 Bulk, in place 1/4" x 1/2", 154 L.F./gal.	1 Bric	230	.035	L.F.	.14	.86		1	1.59
	1800 1/2" x 1/2", 77 L.F./gal.	"	180	.044	"	.29	1.10		1.39	2.15
	2000 Latex acrylic based, bulk				Gal.	23			23	25.50
	2100 Cartridges				"	26			26	28.50
	2200 Bulk in place, 1/4" x 1/2", 154 L.F./gal.	1 Bric	230	.035	L.F.	.15	.86		1.01	1.60
	2300 Polysulfide compounds, 1 component, bulk				Gal.	44			44	48
	2400 Cartridges				"	46.50			46.50	51.50
	2600 1 or 2 component, in place, 1/4" x 1/4", 308 L.F./gal.	1 Bric	145	.055	L.F.	.14	1.36		1.50	2.43
	2700 1/2" x 1/4", 154 L.F./gal.		135	.059		.28	1.46		1.74	2.75
	2900 3/4" x 3/8", 68 L.F./gal.		130	.062		.64	1.52		2.16	3.24
	3000 1" x 1/2", 38 L.F./gal.	↓	130	.062	↓	1.15	1.52		2.67	3.80
	3200 Polyurethane, 1 or 2 component				Gal.	47.50			47.50	52.50
	3300 Cartridges				"	46.50			46.50	51.50
	3500 Bulk, in place, 1/4" x 1/4"	1 Bric	150	.053	L.F.	.15	1.32		1.47	2.36
	3600 1/2" x 1/4"		145	.055		.31	1.36		1.67	2.61
	3800 3/4" x 3/8", 68 L.F./gal.		130	.062		.70	1.52		2.22	3.30
	3900 1" x 1/2"	↓	110	.073	↓	1.24	1.80		3.04	4.35
	4100 Silicone rubber, bulk				Gal.	34.50			34.50	38
	4200 Cartridges				"	40.50			40.50	44.50

For information about Means Estimating Seminars, see yellow pages 12 and 13 in back of book

Division 8
Doors & Windows

Estimating Tips

08100 Metal Doors & Frames

Most metal doors and frames look alike, but there may be significant differences among them. When estimating these items be sure to choose the line item that most closely compares to the specification or door schedule requirements regarding:

- type of metal
- metal gauge
- door core material
- fire rating
- finish

08200 Wood & Plastic Doors

Wood and plastic doors vary considerably in price. The primary determinant is the veneer material. Lauan, birch and oak are the most common veneers. Other variables include the following:

- hollow or solid core
- fire rating
- flush or raised panel
- finish

If the specifications require compliance with AWI (Architectural Woodwork Institute) standards or acoustical standards, the cost of the door may increase substantially. All wood doors are priced pre-mortised for hinges and predrilled for cylindrical locksets.

- Frequently doors, frames, and windows are unique in old buildings. Specified replacement units could be stock, custom (similar to the original) or exact reproduction. The estimator should work closely with a window consultant to determine any extra costs that may be associated with the unusual installation requirements.

08300 Specialty Doors

- There are many varieties of special doors, and they are usually priced per each. Add frames, hardware or operators required for a complete installation.

08510 Steel Windows

- Most metal windows are delivered preglazed. However, some metal windows are priced without glass. Refer to 08800 Glazing for glass pricing. The grade C indicates commercial grade windows, usually ASTM C-35.

08550 Wood Windows

- All wood windows are priced preglazed. The two glazing options priced are single pane float glass and insulating glass 1/2″ thick. Add the cost of screens and grills if required.

08700 Hardware

- Hardware costs add considerably to the cost of a door. The most efficient method to determine the hardware requirements for a project is to review the door schedule. This schedule, in conjunction with the specifications, is all you should need to take off the door hardware.
- Door hinges are priced by the pair, with most doors requiring 1-1/2 pairs per door. The hinge prices do not include installation labor because it is included in door installation. Hinges are classified according to the frequency of use.

08800 Glazing

- Different openings require different types of glass. The three most common types are:
  - float
  - tempered
  - insulating
- Most exterior windows are glazed with insulating glass. Entrance doors and window walls, where the glass is less than 18″ from the floor, are generally glazed with tempered glass. Interior windows and some residential windows are glazed with float glass.
- Energy efficient coatings are also available

08900 Glazed Curtain Wall

- Glazed curtain walls consist of the metal tube framing and the glazing material. The cost data in this subdivision is presented for the metal tube framing alone or the composite wall. If your estimate requires a detailed takeoff of the framing, be sure to add the glazing cost.

Reference Numbers

Reference numbers are shown in bold squares at the beginning of some major classifications. These numbers refer to related items in the Reference Section. The reference information may be an estimating procedure, an alternate pricing method or technical information.

Note: Not all subdivisions listed here necessarily appear in this publication.

No part of this publication may be reproduced, stored in a retrieval system, or transmitted in any form or by any means without prior written permission of Reed Construction Data.

08060 | Selective Demolition

				DAILY	LABOR-		2005 BARE COSTS				TOTAL
		08060 Selective Demolition	CREW	OUTPUT	HOURS	UNIT	MAT.	LABOR	EQUIP.	TOTAL	INCL O&P
110	0010	**SELECTIVE DEMOLITION, DOORS** R02220-510									
	0200	Doors, exterior, 1-3/4" thick, single, 3' x 7' high	1 Clab	16	.500	Ea.		8.70		8.70	14.75
	0220	Double, 6' x 7' high		12	.667			11.55		11.55	19.65
	0500	Interior, 1-3/8" thick, single, 3' x 7' high		20	.400			6.95		6.95	11.80
	0520	Double, 6' x 7' high		16	.500			8.70		8.70	14.75
	0700	Bi-folding, 3' x 6'-8" high		20	.400			6.95		6.95	11.80
	0720	6' x 6'-8" high		18	.444			7.70		7.70	13.10
	0900	Bi-passing, 3' x 6'-8" high		16	.500			8.70		8.70	14.75
	0940	6' x 6'-8" high		14	.571			9.90		9.90	16.85
	1500	Remove and reset, minimum	1 Carp	8	1			24		24	41
	1520	Maximum		6	1.333			32		32	54.50
	2000	Frames, including trim, metal		8	1			24		24	41
	2200	Wood	2 Carp	32	.500			12		12	20.50
	2201	Alternate pricing method	1 Carp	200	.040	L.F.		.96		.96	1.63
	3000	Special doors, counter doors	2 Carp	6	2.667	Ea.		64		64	109
	3300	Glass, sliding, including frames		12	1.333			32		32	54.50
	3400	Overhead, commercial, 12' x 12' high		4	4			96		96	163
	3500	Residential, 9' x 7' high		8	2			48		48	81.50
	3540	16' x 7' high		7	2.286			55		55	93
	3600	Remove and reset, minimum		4	4			96		96	163
	3620	Maximum		2.50	6.400			154		154	261
	3660	Remove and reset elec. garage door opener	1 Carp	8	1			24		24	41
	4000	Residential lockset, exterior		30	.267			6.40		6.40	10.85
	4200	Deadbolt lock		32	.250			6		6	10.20
	5020	Remove bay/bow window	2 Carp	6	2.667			64		64	109
	5032	Remove skylight, plstc domes,flush/curb mtd	G-3	395	.081	S.F.		1.78		1.78	3
	9000	Minimum labor/equipment charge	1 Carp	4	2	Job		48		48	81.50
120	0010	**SELECTIVE DEMOLITION, WINDOWS** R02220-510									
	0200	Aluminum, including trim, to 12 S.F.	1 Clab	16	.500	Ea.		8.70		8.70	14.75
	0240	To 25 S.F.		11	.727			12.60		12.60	21.50
	0280	To 50 S.F.		5	1.600			28		28	47
	0320	Storm windows, to 12 S.F.		27	.296			5.15		5.15	8.75
	0360	To 25 S.F.		21	.381			6.60		6.60	11.20
	0400	To 50 S.F.		16	.500			8.70		8.70	14.75
	0500	Screens, incl. aluminum frame, small		20	.400			6.95		6.95	11.80
	0510	Large		16	.500			8.70		8.70	14.75
	0600	Glass, minimum		200	.040	S.F.		.69		.69	1.18
	0620	Maximum		150	.053	"		.93		.93	1.57
	2000	Wood, including trim, to 12 S.F.		22	.364	Ea.		6.30		6.30	10.70
	2020	To 25 S.F.		18	.444			7.70		7.70	13.10
	2060	To 50 S.F.		13	.615			10.70		10.70	18.10
	2065	To 180 S.F.		8	1			17.35		17.35	29.50
	5020	Remove and reset window, minimum	1 Carp	6	1.333			32		32	54.50
	5040	Average		4	2			48		48	81.50
	5080	Maximum		2	4			96		96	163
	9100	Window awning, residential	1 Clab	80	.100	L.F.		1.74		1.74	2.95

08110 Steel Doors and Frames		Crew	Daily Output	Labor-Hours	Unit	2005 Bare Costs Mat.	Labor	Equip.	Total	Total Incl O&P	
0010	**FIRE DOOR** R08110-020										300
0015	Steel, flush, "B" label, 90 minute										
0020	Full panel, 20 ga., 2'-0" x 6'-8"	2 Carp	20	.800	Ea.	203	19.20		222.20	256	
0040	2'-8" x 6'-8"		18	.889		210	21.50		231.50	267	
0060	3'-0" x 6'-8"		17	.941		210	22.50		232.50	270	
0080	3'-0" x 7'-0"		17	.941		218	22.50		240.50	279	
0140	18 ga., 3'-0" x 6'-8"		16	1		232	24		256	296	
0160	2'-8" x 7'-0"		17	.941		245	22.50		267.50	310	
0180	3'-0" x 7'-0"		16	1		239	24		263	305	
0200	4'-0" x 7'-0"		15	1.067		310	25.50		335.50	385	
0210											
0220	For "A" label, 3 hour, 18 ga., use same price as "B" label										
0240	For vision lite, add				Ea.	49.50			49.50	54.50	
0520	Flush, "B" label 90 min., composite, 20 ga., 2'-0" x 6'-8"	2 Carp	18	.889		267	21.50		288.50	330	
0540	2'-8" x 6'-8"		17	.941		272	22.50		294.50	340	
0560	3'-0" x 6'-8"		16	1		275	24		299	340	
0580	3'-0" x 7'-0"		16	1		283	24		307	350	
0640	Flush, "A" label 3 hour, composite, 18 ga., 3'-0" x 6'-8"		15	1.067		232	25.50		257.50	299	
0660	2'-8" x 7'-0"		16	1		243	24		267	310	
0680	3'-0" x 7'-0"		15	1.067		239	25.50		264.50	305	
0700	4'-0" x 7'-0"		14	1.143		310	27.50		337.50	385	
0010	**RESIDENTIAL STEEL DOOR**										600
0020	Prehung, insulated, exterior										
0030	Embossed, full panel, 2'-8" x 6'-8"	2 Carp	17	.941	Ea.	193	22.50		215.50	252	
0040	3'-0" x 6'-8"		15	1.067		189	25.50		214.50	252	
0060	3'-0" x 7'-0"		15	1.067		249	25.50		274.50	320	
0070	5'-4" x 6'-8", double		8	2		375	48		423	495	
0220	Half glass, 2'-8" x 6'-8"		17	.941		230	22.50		252.50	292	
0240	3'-0" x 6'-8"		16	1		234	24		258	298	
0260	3'-0" x 7'-0"		16	1		286	24		310	355	
0270	5'-4" x 6'-8", double		8	2		485	48		533	615	
0720	Raised plastic face, full panel, 2'-8" x 6'-8"		16	1		249	24		273	315	
0740	3'-0" x 6'-8"		15	1.067		251	25.50		276.50	320	
0760	3'-0" x 7'-0"		15	1.067		253	25.50		278.50	320	
0780	5'-4" x 6'-8", double		8	2		470	48		518	595	
0820	Half glass, 2'-8" x 6'-8"		17	.941		278	22.50		300.50	345	
0840	3'-0" x 6'-8"		16	1		282	24		306	350	
0860	3'-0" x 7'-0"		16	1		310	24		334	380	
0880	5'-4" x 6'-8", double		8	2		595	48		643	735	
1320	Flush face, full panel, 2'-6" x 6'-8"		16	1		192	24		216	252	
1340	3'-0" x 6'-8"		15	1.067		195	25.50		220.50	259	
1360	3'-0" x 7'-0"		15	1.067		247	25.50		272.50	315	
1380	5'-4" x 6'-8", double		8	2		405	48		453	525	
1420	Half glass, 2'-8" x 6'-8"		17	.941		241	22.50		263.50	305	
1440	3'-0" x 6'-8"		16	1		245	24		269	310	
1460	3'-0" x 7'-0"		16	1		284	24		308	350	
1480	5'-4" x 6'-8", double		8	2		475	48		523	600	
1500	Sidelight, full lite, 1'-0" x 6'-8" w/grill					199			199	219	
1510	1'-0" x 6'-8" Low E					240			240	264	
1520	1'-0" x 6'-8" half lite					184			184	202	
1530	1'-0" x 6'-8" half lite low E					223			223	246	
2300	Interior, residential, closet, bi-fold, 6'-8" x 2'-0" wide	2 Carp	16	1		136	24		160	190	
2330	3'-0" wide		16	1		152	24		176	209	
2360	4'-0" wide		15	1.067		231	25.50		256.50	299	
2400	5'-0" wide		14	1.143		268	27.50		295.50	340	

8 DOORS & WINDOWS

	08110 | Steel Doors and Frames		CREW	DAILY OUTPUT	LABOR-HOURS	UNIT	2005 BARE COSTS MAT.	LABOR	EQUIP.	TOTAL	TOTAL INCL O&P
600	2420	6′-0″ wide	2 Carp	13	1.231	Ea.	300	29.50		329.50	380
	2510	Bi-passing closet, incl. hardware, no frame or trim incl.									
	2511	Mirrored, metal frame, 6′-8″ x 4′-0″ wide	2 Carp	10	1.600	Opng.	174	38.50		212.50	257
	2512	5′-0″ wide		10	1.600		203	38.50		241.50	289
	2513	6′-0″ wide		10	1.600		231	38.50		269.50	320
	2514	7′-0″ wide		9	1.778		243	42.50		285.50	340
	2515	8′-0″ wide		9	1.778		480	42.50		522.50	605
	2611	Mirrored, metal, 8′-0″ x 4′-0″ wide		10	1.600		273	38.50		311.50	365
	2612	5′-0″ wide		10	1.600		294	38.50		332.50	390
	2613	6′-0″ wide		10	1.600		325	38.50		363.50	420
	2614	7′-0″ wide		9	1.778		350	42.50		392.50	460
	2615	8′-0″ wide		9	1.778		385	42.50		427.50	500
820	0010	**STEEL FRAMES, KNOCK DOWN**									
	0020	16 ga., up to 5-3/4″ jamb depth									
	0025	6′-8″ high, 3′-0″ wide, single	2 Carp	16	1	Ea.	81.50	24		105.50	131
	0028	6′ high, 3′-6″ wide, single		16	1		72.50	24		96.50	121
	0030	6′ high, 4′-0″ wide, single		16	1		72.50	24		96.50	121
	0040	6′-0″ wide, double		14	1.143		95.50	27.50		123	152
	0045	8′-0″ wide, double		14	1.143		99.50	27.50		127	156
	0100	7′-0″ high, 3′-0″ wide, single		16	1		83.50	24		107.50	133
	0110	7′ high, 3′-6″ wide, single		16	1		74.50	24		98.50	123
	0112	7′ high, 4′-0″ wide, single		16	1		127	24		151	180
	0140	6′-0″ wide, double		14	1.143		100	27.50		127.50	157
	0145	8′-0″ wide, double		14	1.143		93.50	27.50		121	150
	1000	16 ga., up to 4-7/8″ deep, 7′-0″ H, 3′-0″ W, single		16	1		81.50	24		105.50	131
	1140	6′-0″ wide, double		14	1.143		100	27.50		127.50	157
	2800	14 ga., up to 3-7/8″ deep, 7′-0″ high, 3′-0″ wide, single		16	1		83.50	24		107.50	133
	2840	6′-0″ wide, double		14	1.143		103	27.50		130.50	160
	3000	14 ga., up to 5-3/4″ deep, 6′-8″ high, 3′-0″ wide, single		16	1		91.50	24		115.50	142
	3002	6′ high, 3′-6″ wide, single		16	1		88	24		112	138
	3005	6′ high, 4′-0″ wide, single		16	1		88	24		112	138
	3600	up to 5-3/4″ jamb depth, 7′-0″ high, 4′-0″ wide, single		15	1.067		69	25.50		94.50	120
	3620	6′-0″ wide, double		12	1.333		108	32		140	174
	3640	8′-0″ wide, double		12	1.333		91	32		123	155
	3700	8′-0″ high, 4′-0″ wide, single		15	1.067		91	25.50		116.50	144
	3740	8′-0″ wide, double		12	1.333		113	32		145	179
	4000	6-3/4″ deep, 7′-0″ high, 4′-0″ wide, single		15	1.067		97	25.50		122.50	151
	4020	6′-0″ wide, double		12	1.333		111	32		143	178
	4040	8′-0″, wide double		12	1.333		129	32		161	197
	4100	8′-0″ high, 4′-0″ wide, single		15	1.067		108	25.50		133.50	163
	4140	8′-0″ wide, double		12	1.333		126	32		158	194
	4400	8-3/4″ deep, 7′-0″ high, 4′-0″ wide, single		15	1.067		101	25.50		126.50	155
	4440	8′-0″ wide, double		12	1.333		134	32		166	202
	4500	8′-0″ high, 4′-0″ wide, single		15	1.067		119	25.50		144.50	174
	4540	8′-0″ wide, double		12	1.333		139	32		171	208
	4900	For welded frames, add					29.50			29.50	32.50
	5400	14 ga., "B" label, up to 5-3/4″ deep, 7′-0″ high, 4′-0″ wide, single	2 Carp	15	1.067		108	25.50		133.50	163
	5440	8′-0″ wide, double		12	1.333		126	32		158	194
	5800	6-3/4″ deep, 7′-0″ high, 4′-0″ wide, single		15	1.067		93.50	25.50		119	147
	5840	8′-0″ wide, double		12	1.333		140	32		172	208
	6200	8-3/4″ deep, 7′-0″ high, 4′-0″ wide, single		15	1.067		116	25.50		141.50	172
	6240	8′-0″ wide, double		12	1.333		149	32		181	219
	6300	For "A" label use same price as "B" label									
	6400	For baked enamel finish, add					30%	15%			
	6500	For galvanizing, add					15%				
	6600	For hospital stop, add				Ea.	176			176	193

08100 | Metal Doors and Frames

08110	Steel Doors and Frames	CREW	DAILY OUTPUT	LABOR-HOURS	UNIT	2005 BARE COSTS MAT.	LABOR	EQUIP.	TOTAL	TOTAL INCL O&P	
7900	Transom lite frames, fixed, add	2 Carp	155	.103	S.F.	28.50	2.48		30.98	35	820
8000	Movable, add	"	130	.123	"	34.50	2.95		37.45	42.50	

08180	Metal Screen & Storm Doors	CREW	DAILY OUTPUT	LABOR-HOURS	UNIT	MAT.	LABOR	EQUIP.	TOTAL	TOTAL INCL O&P	
0010	**STORM DOORS & FRAMES** Aluminum, residential,										100
0020	combination storm and screen										
0400	Clear anodic coating, 6'-8" x 2'-6" wide	2 Carp	15	1.067	Ea.	153	25.50		178.50	212	
0420	2'-8" wide		14	1.143		176	27.50		203.50	241	
0440	3'-0" wide		14	1.143		176	27.50		203.50	241	
0500	For 7' door height, add					5%					
1000	Mill finish, 6'-8" x 2'-6" wide	2 Carp	15	1.067	Ea.	203	25.50		228.50	268	
1020	2'-8" wide		14	1.143		203	27.50		230.50	271	
1040	3'-0" wide		14	1.143		221	27.50		248.50	290	
1100	For 7'-0" door, add					5%					
1500	White painted, 6'-8" x 2'-6" wide	2 Carp	15	1.067		202	25.50		227.50	267	
1520	2'-8" wide		14	1.143		206	27.50		233.50	274	
1540	3'-0" wide		14	1.143		215	27.50		242.50	284	
1541	Storm door, painted, alum., insul., 6'-8" x 2'-6" wide		14	1.143		233	27.50		260.50	305	
1545	2'-8" wide		14	1.143		233	27.50		260.50	305	
1600	For 7'-0" door, add					5%					
1800	Aluminum screen door, minimum, 6'-8" x 2'-8" wide	2 Carp	14	1.143		93.50	27.50		121	150	
1810	3'-0" wide		14	1.143		215	27.50		242.50	284	
1820	Average, 6'-8" x 2'-8" wide		14	1.143		156	27.50		183.50	219	
1830	3'-0" wide		14	1.143		215	27.50		242.50	284	
1840	Maximum, 6'-8" x 2'-8" wide		14	1.143		310	27.50		337.50	390	
1850	3'-0" wide		14	1.143		233	27.50		260.50	305	
2000	Wood door & screen, see division 08210-930										

08200 | Wood and Plastic Doors

08210	Wood Doors	CREW	DAILY OUTPUT	LABOR-HOURS	UNIT	2005 BARE COSTS MAT.	LABOR	EQUIP.	TOTAL	TOTAL INCL O&P	
0010	**PRE-HUNG DOORS**										720
0300	Exterior, wood, comb. storm & screen, 6'-9" x 2'-6" wide	2 Carp	15	1.067	Ea.	266	25.50		291.50	335	
0320	2'-8" wide		15	1.067		266	25.50		291.50	335	
0340	3'-0" wide		15	1.067		273	25.50		298.50	345	
0360	For 7'-0" high door, add					23			23	25.50	
0370	For aluminum storm doors, see division 08180-100										
1600	Entrance door, flush, birch, solid core										
1620	4-5/8" solid jamb, 1-3/4" x 6'-8" x 2'-8" wide	2 Carp	16	1	Ea.	272	24		296	340	
1640	3'-0" wide	"	16	1		279	24		303	345	
1680	For 7'-0" high door, add					16.50			16.50	18.15	
2000	Entrance door, colonial, 6 panel pine										
2020	4-5/8" solid jamb, 1-3/4" x 6'-8" x 2'-8" wide	2 Carp	16	1	Ea.	445	24		469	530	
2040	3'-0" wide	"	16	1		470	24		494	560	
2060	For 7'-0" high door, add					35			35	38	
2200	For 5-5/8" solid jamb, add					22.50			22.50	24.50	
2250	French door, 6'-8" x 4'-0" wide, 1/2" insul. glass and grille	2 Carp	7	2.286	Pr.	1,000	55		1,055	1,200	
2260	5'-0" wide	"	7	2.286	"	1,100	55		1,155	1,325	
2500	Exterior, metal face, insulated, incl. jamb, brickmold and										
2520	threshold, flush, 2'-8" x 6'-8"	2 Carp	16	1	Ea.	178	24		202	236	
2550	3'-0" x 6'-8"	"	16	1	"	181	24		205	240	

	08210 | Wood Doors		Crew	Daily Output	Labor-Hours	Unit	2005 Bare Costs Mat.	Labor	Equip.	Total	Total Incl O&P
720	2990										
	3500	Embossed, 6 panel, 2'-8" x 6'-8"	2 Carp	16	1	Ea.	163	24		187	221
	3550	3'-0" x 6'-8"		16	1		189	24		213	249
	3600	2 narrow lites, 2'-8" x 6'-8"		16	1		228	24		252	292
	3650	3'-0" x 6'-8"		16	1		232	24		256	296
	3700	Half glass, 2'-8" x 6'-8"		16	1		213	24		237	276
	3750	3'-0" x 6'-8"		16	1		215	24		239	278
	3800	2 top lites, 2'-8" x 6'-8"		16	1		207	24		231	269
	3850	3'-0" x 6'-8"		16	1		209	24		233	271
	4000	Interior, passage door, 4-5/8" solid jamb									
	4400	Lauan, flush, solid core, 1-3/8" x 6'-8" x 2'-6" wide	2 Carp	20	.800	Ea.	167	19.20		186.20	217
	4420	2'-8" wide		20	.800		147	19.20		166.20	194
	4440	3'-0" wide		19	.842		151	20		171	202
	4600	Hollow core, 1-3/8" x 6'-8" x 2'-6" wide		20	.800		109	19.20		128.20	153
	4620	2'-8" wide		20	.800		110	19.20		129.20	155
	4640	3'-0" wide		19	.842		112	20		132	158
	4700	For 7'-0" high door, add					21.50			21.50	24
	5000	Birch, flush, solid core, 1-3/8" x 6'-8" x 2'-6" wide	2 Carp	20	.800		156	19.20		175.20	205
	5020	2'-8" wide		20	.800		176	19.20		195.20	227
	5040	3'-0" wide		19	.842		188	20		208	242
	5200	Hollow core, 1-3/8" x 6'-8" x 2'-6" wide		20	.800		129	19.20		148.20	175
	5220	2'-8" wide		20	.800		135	19.20		154.20	182
	5240	3'-0" wide		19	.842		135	20		155	184
	5280	For 7'-0" high door, add					15.45			15.45	17
	5500	Hardboard paneled, 1-3/8" x 6'-8" x 2'-6" wide	2 Carp	20	.800		130	19.20		149.20	176
	5520	2'-8" wide		20	.800		136	19.20		155.20	183
	5540	3'-0" wide		19	.842		135	20		155	183
	6000	Pine paneled, 1-3/8" x 6'-8" x 2'-6" wide		20	.800		223	19.20		242.20	279
	6020	2'-8" wide		20	.800		241	19.20		260.20	298
	6040	3'-0" wide		19	.842		250	20		270	310
	6500	For 5-5/8" solid jamb, add					10.90			10.90	12
	6520	For split jamb, deduct					13.30			13.30	14.65
910	0010	**WOOD DOORS, DECORATOR**									
	3000	Solid wood, 1-3/4" thick stile and rail									
	3020	Mahogany, 3'-0" x 7'-0", minimum	2 Carp	14	1.143	Ea.	925	27.50		952.50	1,075
	3030	Maximum		10	1.600		1,275	38.50		1,313.50	1,475
	3040	3'-6" x 8'-0", minimum		10	1.600		850	38.50		888.50	1,000
	3050	Maximum		8	2		1,550	48		1,598	1,800
	3100	Pine, 3'-0" x 7'-0", minimum		14	1.143		405	27.50		432.50	490
	3110	Maximum		10	1.600		665	38.50		703.50	795
	3120	3'-6" x 8'-0", minimum		10	1.600		625	38.50		663.50	755
	3130	Maximum		8	2		1,050	48		1,098	1,225
	3200	Red oak, 3'-0" x 7'-0", minimum		14	1.143		1,375	27.50		1,402.50	1,575
	3210	Maximum		10	1.600		1,650	38.50		1,688.50	1,900
	3220	3'-6" x 8'-0", minimum		10	1.600		1,525	38.50		1,563.50	1,750
	3230	Maximum		8	2		2,725	48		2,773	3,075
	4000	Hand carved door, mahogany									
	4020	3'-0" x 7'-0", minimum	2 Carp	14	1.143	Ea.	1,375	27.50		1,402.50	1,550
	4030	Maximum		11	1.455		2,700	35		2,735	3,025
	4040	3'-6" x 8'-0", minimum		10	1.600		1,725	38.50		1,763.50	1,975
	4050	Maximum		8	2		2,650	48		2,698	3,000
	4200	Red oak, 3'-0" x 7'-0", minimum		14	1.143		4,200	27.50		4,227.50	4,675
	4210	Maximum		11	1.455		11,500	35		11,535	12,800
	4220	3'-6" x 8'-0", minimum		10	1.600		4,725	38.50		4,763.50	5,275
	4280	For 6'-8" high door, deduct from 7'-0" door					30			30	33
	4400	For custom finish, add					320			320	350

Important: See the Reference Section for critical supporting data - Reference Nos., Crews, & Location Factors

	08210 Wood Doors		Crew	Daily Output	Labor-Hours	Unit	2005 Bare Costs Mat.	Labor	Equip.	Total	Total Incl O&P
910	4600	Side light, mahogany, 7'-0" x 1'-6" wide, minimum	2 Carp	18	.889	Ea.	770	21.50		791.50	885
	4610	Maximum		14	1.143		2,275	27.50		2,302.50	2,550
	4620	8'-0" x 1'-6" wide, minimum		14	1.143		1,450	27.50		1,477.50	1,650
	4630	Maximum		10	1.600		1,650	38.50		1,688.50	1,875
	4640	Side light, oak, 7'-0" x 1'-6" wide, minimum		18	.889		890	21.50		911.50	1,025
	4650	Maximum		14	1.143		1,575	27.50		1,602.50	1,800
	4660	8'-0" x 1-6" wide, minimum		14	1.143		840	27.50		867.50	970
	4670	Maximum		10	1.600		1,575	38.50		1,613.50	1,825
	6520	Interior cafe doors, 2'-6" opening, stock, panel pine		16	1		168	24		192	226
	6540	3'-0" opening		16	1		175	24		199	234
	6550	Louvered pine									
	6560	2'-6" opening	2 Carp	16	1	Ea.	147	24		171	202
	8000	3'-0" opening		16	1		157	24		181	214
	8010	2'-6" opening, hardwood		16	1		260	24		284	325
	8020	3'-0" opening		16	1		286	24		310	355
	8800	Pre-hung doors, see division 08210-720									
920	0010	**WOOD DOORS, PANELED**									
	0020	Interior, six panel, hollow core, 1-3/8" thick									
	0040	Molded hardboard, 2'-0" x 6'-8"	2 Carp	17	.941	Ea.	45.50	22.50		68	88.50
	0060	2'-6" x 6'-8"		17	.941		49	22.50		71.50	92.50
	0070	2'-8" x 6'-8"		17	.941		51.50	22.50		74	95
	0080	3'-0" x 6'-8"		17	.941		54.50	22.50		77	98
	0140	Embossed print, molded hardboard, 2'-0" x 6'-8"		17	.941		49	22.50		71.50	92.50
	0160	2'-6" x 6'-8"		17	.941		49	22.50		71.50	92.50
	0180	3'-0" x 6'-8"		17	.941		54.50	22.50		77	98
	0540	Six panel, solid, 1-3/8" thick, pine, 2'-0" x 6'-8"		15	1.067		119	25.50		144.50	175
	0560	2'-6" x 6'-8"		14	1.143		134	27.50		161.50	194
	0580	3'-0" x 6'-8"		13	1.231		154	29.50		183.50	219
	1020	Two panel, bored rail, solid, 1-3/8" thick, pine, 1'-6" x 6'-8"		16	1		217	24		241	280
	1040	2'-0" x 6'-8"		15	1.067		286	25.50		311.50	360
	1060	2'-6" x 6'-8"		14	1.143		325	27.50		352.50	405
	1340	Two panel, solid, 1-3/8" thick, fir, 2'-0" x 6'-8"		15	1.067		119	25.50		144.50	175
	1360	2'-6" x 6'-8"		14	1.143		134	27.50		161.50	194
	1380	3'-0" x 6'-8"		13	1.231		325	29.50		354.50	410
	1740	Five panel, solid, 1-3/8" thick, fir, 2'-0" x 6'-8"		15	1.067		213	25.50		238.50	279
	1760	2'-6" x 6'-8"		14	1.143		345	27.50		372.50	420
	1780	3'-0" x 6'-8"		13	1.231		345	29.50		374.50	425
930	0010	**WOOD DOORS, RESIDENTIAL**									
	0200	Exterior, combination storm & screen, pine									
	0260	2'-8" wide	2 Carp	10	1.600	Ea.	256	38.50		294.50	345
	0280	3'-0" wide		9	1.778		261	42.50		303.50	360
	0300	7'-1" x 3'-0" wide		9	1.778		284	42.50		326.50	385
	0400	Full lite, 6'-9" x 2'-6" wide		11	1.455		266	35		301	350
	0420	2'-8" wide		10	1.600		266	38.50		304.50	355
	0440	3'-0" wide		9	1.778		273	42.50		315.50	375
	0500	7'-1" x 3'-0" wide		9	1.778		295	42.50		337.50	400
	0700	Dutch door, pine, 1-3/4" x 6'-8" x 2'-8" wide, minimum		12	1.333		595	32		627	710
	0720	Maximum		10	1.600		630	38.50		668.50	755
	0800	3'-0" wide, minimum		12	1.333		620	32		652	735
	0820	Maximum		10	1.600		670	38.50		708.50	805
	1000	Entrance door, colonial, 1-3/4" x 6'-8" x 2'-8" wide		16	1		330	24		354	405
	1020	6 panel pine, 3'-0" wide		15	1.067		355	25.50		380.50	435
	1100	8 panel pine, 2'-8" wide		16	1		585	24		609	685
	1120	3'-0" wide		15	1.067		500	25.50		525.50	595
	1200	For tempered safety glass lites, (min of 2)add					62			62	68

	08210 | Wood Doors	CREW	DAILY OUTPUT	LABOR-HOURS	UNIT	2005 BARE COSTS MAT.	LABOR	EQUIP.	TOTAL	TOTAL INCL O&P
930	1300 Flush, birch, solid core, 1-3/4" x 6'-8" x 2'-8" wide	2 Carp	16	1	Ea.	91	24		115	141
	1320 3'-0" wide		15	1.067		94	25.50		119.50	148
	1350 7'-0" x 2'-8" wide		16	1		98	24		122	149
	1360 3'-0" wide		15	1.067		106	25.50		131.50	161
	1740 Mahogany, 2'-8" x 6'-8"		15	1.067		715	25.50		740.50	830
	1760 3'-0"x 6'-8"		15	1.067		745	25.50		770.50	860
	2700 Interior, closet, bi-fold, w/hardware, no frame or trim incl.									
	2720 Flush, birch, 6'-6" or 6'-8" x 2'-6" wide	2 Carp	13	1.231	Ea.	45.50	29.50		75	100
	2740 3'-0" wide		13	1.231		49.50	29.50		79	105
	2760 4'-0" wide		12	1.333		90	32		122	154
	2780 5'-0" wide		11	1.455		91	35		126	160
	2800 6'-0" wide		10	1.600		98	38.50		136.50	173
	3000 Raised panel pine, 6'-6" or 6'-8" x 2'-6" wide		13	1.231		149	29.50		178.50	214
	3020 3'-0" wide		13	1.231		171	29.50		200.50	238
	3040 4'-0" wide		12	1.333		218	32		250	295
	3060 5'-0" wide		11	1.455		252	35		287	335
	3080 6'-0" wide		10	1.600		283	38.50		321.50	375
	3200 Louvered, pine 6'-6" or 6'-8" x 2'-6" wide		13	1.231		93.50	29.50		123	153
	3220 3'-0" wide		13	1.231		147	29.50		176.50	212
	3240 4'-0" wide		12	1.333		164	32		196	235
	3260 5'-0" wide		11	1.455		185	35		220	263
	3280 6'-0" wide		10	1.600		204	38.50		242.50	290
	4400 Bi-passing closet, incl. hardware and frame, no trim incl.									
	4420 Flush, lauan, 6'-8" x 4'-0" wide	2 Carp	12	1.333	Opng.	156	32		188	227
	4440 5'-0" wide		11	1.455		171	35		206	248
	4460 6'-0" wide		10	1.600		183	38.50		221.50	266
	4600 Flush, birch, 6'-8" x 4'-0" wide		12	1.333		192	32		224	266
	4620 5'-0" wide		11	1.455		193	35		228	272
	4640 6'-0" wide		10	1.600		230	38.50		268.50	320
	4800 Louvered, pine, 6'-8" x 4'-0" wide		12	1.333		380	32		412	470
	4820 5'-0" wide		11	1.455		360	35		395	455
	4840 6'-0" wide		10	1.600		465	38.50		503.50	575
	4900 Mirrored, 6'-8" x 4'-0" wide		12	1.333	Ea.	232	32		264	310
	5000 Paneled, pine, 6'-8" x 4'-0" wide		12	1.333	Opng.	360	32		392	455
	5020 5'-0" wide		11	1.455		375	35		410	470
	5040 6'-0" wide		10	1.600		445	38.50		483.50	555
	5061 Hardboard, 6'-8" x 4'-0" wide		10	1.600		175	38.50		213.50	257
	5062 5'-0" wide		10	1.600		179	38.50		217.50	262
	5063 6'-0" wide		10	1.600		202	38.50		240.50	288
	6100 Folding accordion, closet, including track and frame									
	6121 Vinyl, 2 layer, stock	2 Carp	400	.040	S.F.	2.93	.96		3.89	4.85
	6140 Woven mahogany and vinyl, stock		400	.040		1.87	.96		2.83	3.69
	6160 Wood slats with vinyl overlay, stock		400	.040		9.45	.96		10.41	12.05
	6180 Economy vinyl, stock		400	.040		1.65	.96		2.61	3.45
	6200 Rigid PVC		400	.040		4.40	.96		5.36	6.45
	6220 For custom partition, add					25%	10%			
	7310 Passage doors, flush, no frame included									
	7320 Hardboard, hollow core, 1-3/8" x 6'-8" x 1'-6" wide	2 Carp	18	.889	Ea.	38	21.50		59.50	78
	7330 2'-0" wide		18	.889		38	21.50		59.50	77.50
	7340 2'-6" wide		18	.889		42	21.50		63.50	82
	7350 2'-8" wide		18	.889		44	21.50		65.50	84.50
	7360 3'-0" wide		17	.941		46.50	22.50		69	89.50
	7420 Lauan, hollow core, 1-3/8" x 6'-8" x 1'-6" wide		18	.889		26.50	21.50		48	65.50
	7440 2'-0" wide		18	.889		25.50	21.50		47	64
	7450 2'-4" wide		18	.889		28.50	21.50		50	67.50
	7460 2'-6" wide		18	.889		28.50	21.50		50	67.50

	08210 | Wood Doors	CREW	DAILY OUTPUT	LABOR-HOURS	UNIT	2005 BARE COSTS MAT.	LABOR	EQUIP.	TOTAL	TOTAL INCL O&P	
7480	2′-8″ wide	2 Carp	18	.889	Ea.	30	21.50		51.50	69	930
7500	3′-0″ wide		17	.941		31.50	22.50		54	73.50	
7700	Birch, hollow core, 1-3/8″ x 6′-8″ x 1′-6″ wide		18	.889		34	21.50		55.50	73	
7720	2′-0″ wide		18	.889		37.50	21.50		59	77	
7740	2′-6″ wide		18	.889		41.50	21.50		63	82	
7760	2′-8″ wide		18	.889		43.50	21.50		65	83.50	
7780	3′-0″ wide		17	.941		47	22.50		69.50	90.50	
8000	Pine louvered, 1-3/8″ x 6′-8″ x 1′-6″ wide		19	.842		89.50	20		109.50	133	
8020	2′-0″ wide		18	.889		113	21.50		134.50	160	
8040	2′-6″ wide		18	.889		123	21.50		144.50	171	
8060	2′-8″ wide		18	.889		130	21.50		151.50	179	
8080	3′-0″ wide		17	.941		139	22.50		161.50	192	
8300	Pine paneled, 1-3/8″ x 6′-8″ x 1′-6″ wide		19	.842		101	20		121	146	
8320	2′-0″ wide		18	.889		117	21.50		138.50	164	
8330	2′-4″ wide		18	.889		128	21.50		149.50	177	
8340	2′-6″ wide		18	.889		131	21.50		152.50	180	
8360	2′-8″ wide		18	.889		142	21.50		163.50	192	
8380	3′-0″ wide		17	.941		148	22.50		170.50	202	
8450	French door, pine, 15 lites, 1-3/8″x6′-8″x2′-6″ wide		18	.889		165	21.50		186.50	217	
8470	2′-8″ wide		18	.889		212	21.50		233.50	269	
8490	3′-0″ wide		17	.941		226	22.50		248.50	288	
8550	For over 20 doors, deduct					15%					
0010	**WOOD FRAMES**										960
0400	Exterior frame, incl. ext. trim, pine, 5/4 x 4-9/16″ deep	2 Carp	375	.043	L.F.	4.85	1.02		5.87	7.10	
0420	5-3/16″ deep		375	.043		7.60	1.02		8.62	10.10	
0440	6-9/16″ deep		375	.043		7.50	1.02		8.52	10	
0600	Oak, 5/4 x 4-9/16″ deep		350	.046		8.45	1.10		9.55	11.10	
0620	5-3/16″ deep		350	.046		9.50	1.10		10.60	12.30	
0640	6-9/16″ deep		350	.046		10.55	1.10		11.65	13.45	
0800	Walnut, 5/4 x 4-9/16″ deep		350	.046		9.90	1.10		11	12.75	
0820	5-3/16″ deep		350	.046		14.40	1.10		15.50	17.65	
0840	6-9/16″ deep		350	.046		16.95	1.10		18.05	20.50	
1000	Sills, 8/4 x 8″ deep, oak, no horns		100	.160		11.95	3.84		15.79	19.65	
1020	2″ horns		100	.160		13.30	3.84		17.14	21	
1040	3″ horns		100	.160		15.35	3.84		19.19	23.50	
1100	8/4 x 10″ deep, oak, no horns		90	.178		16.10	4.27		20.37	25	
1120	2″ horns		90	.178		17.95	4.27		22.22	27	
1140	3″ horns		90	.178		19.55	4.27		23.82	29	
2000	Exterior, colonial, frame & trim, 3′ opng., in-swing, minimum		22	.727	Ea.	279	17.45		296.45	335	
2010	Average		21	.762		415	18.30		433.30	485	
2020	Maximum		20	.800		940	19.20		959.20	1,050	
2100	5′-4″ opening, in-swing, minimum		17	.941		315	22.50		337.50	385	
2120	Maximum		15	1.067		940	25.50		965.50	1,075	
2140	Out-swing, minimum		17	.941		325	22.50		347.50	395	
2160	Maximum		15	1.067		975	25.50		1,000.50	1,125	
2400	6′-0″ opening, in-swing, minimum		16	1		300	24		324	375	
2420	Maximum		10	1.600		975	38.50		1,013.50	1,150	
2460	Out-swing, minimum		16	1		325	24		349	395	
2480	Maximum		10	1.600		1,200	38.50		1,238.50	1,400	
2600	For two sidelights, add, minimum		30	.533	Opng.	310	12.80		322.80	365	
2620	Maximum		20	.800	″	995	19.20		1,014.20	1,125	
2700	Custom birch frame, 3′-0″ opening		16	1	Ea.	179	24		203	238	
2750	6′-0″ opening		16	1		270	24		294	340	
2900	Exterior, modern, plain trim, 3′ opng., in-swing, minimum		26	.615		30	14.75		44.75	58.50	
2920	Average		24	.667		36	16		52	66.50	
2940	Maximum		22	.727		44	17.45		61.45	77.50	

	08210 Wood Doors	CREW	DAILY OUTPUT	LABOR-HOURS	UNIT	2005 BARE COSTS MAT.	LABOR	EQUIP.	TOTAL	TOTAL INCL O&P
960	3000 Interior frame, pine, 11/16" x 3-5/8" deep	2 Carp	375	.043	L.F.	4.02	1.02		5.04	6.15
	3020 4-9/16" deep		375	.043		5.50	1.02		6.52	7.80
	3040 5-3/16" deep		375	.043		3.53	1.02		4.55	5.60
	3200 Oak, 11/16" x 3-5/8" deep		350	.046		3.59	1.10		4.69	5.80
	3220 4-9/16" deep		350	.046		3.87	1.10		4.97	6.10
	3240 5-3/16" deep		350	.046		3.98	1.10		5.08	6.25
	3400 Walnut, 11/16" x 3-5/8" deep		350	.046		6.05	1.10		7.15	8.50
	3420 4-9/16" deep		350	.046		6.40	1.10		7.50	8.85
	3440 5-3/16" deep		350	.046		6.60	1.10		7.70	9.15
	3800 Threshold, oak, 5/8" x 3-5/8" deep		200	.080		2.33	1.92		4.25	5.80
	3820 4-5/8" deep		190	.084		2.96	2.02		4.98	6.70
	3840 5-5/8" deep	↓	180	.089	↓	5	2.13		7.13	9.10
	4000 For casing see division 06220-400 & 06220-800									
	08220 Plastic Doors									
100	0010 **FIBERGLASS DOORS**									
	0020 Exterior, fiberglass, door, 2'-8"wide x 6'-8" high	2 Carp	15	1.067	Ea.	289	25.50		314.50	365
	0040 3'-0"wide x 6'-8" high		15	1.067		289	25.50		314.50	365
	0060 3'-0"wide x 7'-0" high		15	1.067		355	25.50		380.50	435
	0080 3'-0"wide x 6'-8" high, with two lites		15	1.067		350	25.50		375.50	430
	0100 3'-0"wide x 7'-0" high, with two lites		15	1.067		415	25.50		440.50	500
	0110 Half glass, 3'-0"wide x 6'-8" high		15	1.067		390	25.50		415.50	475
	0120 3'-0"wide x 6'-8" high, low e		15	1.067		415	25.50		440.50	500
	0130 3'-0"wide x 7'-0" high		15	1.067		455	25.50		480.50	545
	0140 3'-0"wide x 7'-0" high, low e	↓	15	1.067		480	25.50		505.50	575
	0150 Side Lights, 1'-0"wide x 6'-8" high,					265			265	291
	0160 1'-0"wide x 6'-8" high, low e					277			277	305
	0180 1'-0"wide x 6'-8" high, full glass					305			305	335
	0190 1'-0"wide x 6'-8" high, low e				↓	325			325	360
	08260 Sliding Wood and Plastic Doors									
700	0010 **GLASS, SLIDING, VINYL**									
	0012 Vinyl clad, 1" insul. glass, 6'-0" x 6'-10" high	2 Carp	4	4	Opng.	1,150	96		1,246	1,450
	0030 6'-0" x 8'-0" high		4	4	Ea.	1,750	96		1,846	2,100
	0100 8'-0" x 6'-10" high		4	4	Opng.	1,800	96		1,896	2,150
	0500 3 leaf, 9'-0" x 6'-10" high		3	5.333		1,625	128		1,753	2,025
	0600 12'-0" x 6'-10" high	↓	3	5.333	↓	2,025	128		2,153	2,450
900	0010 **GLASS, SLIDING, WOOD**									
	0020 Wood, 5/8" tempered insul. glass, 6' wide, premium	2 Carp	4	4	Ea.	985	96		1,081	1,250
	0100 Economy		4	4		695	96		791	930
	0150 8' wide, wood, premium		3	5.333		1,150	128		1,278	1,475
	0200 Economy		3	5.333		805	128		933	1,100
	0250 12' wide, wood, premium		2.50	6.400		2,700	154		2,854	3,225
	0300 Economy	↓	2.50	6.400	↓	1,825	154		1,979	2,275
	0350 Aluminum, 5/8" tempered insulated glass, 6' wide									
	0400 Premium	2 Carp	4	4	Ea.	1,250	96		1,346	1,550
	0450 Economy		4	4		655	96		751	885
	0500 8' wide, premium		3	5.333		1,300	128		1,428	1,650
	0550 Economy		3	5.333		1,100	128		1,228	1,425
	0600 12' wide, premium		2.50	6.400		2,150	154		2,304	2,625
	0650 Economy	↓	2.50	6.400	↓	1,375	154		1,529	1,750
	1000 Replacement doors, wood									
	1050 6' wide, premium	2 Carp	4	4	Ea.	655	96		751	885

08310 | Access Doors and Panels

	Description	Crew	Daily Output	Labor-Hours	Unit	2005 Bare Costs Mat.	2005 Bare Costs Labor	2005 Bare Costs Equip.	2005 Bare Costs Total	Total Incl O&P	
0010	**BULKHEAD CELLAR DOORS**										150
0020	Steel, not incl. sides, 44″ x 62″	1 Carp	5.50	1.455	Ea.	204	35		239	285	
0100	52″ x 73″		5.10	1.569		227	37.50		264.50	315	
0500	With sides and foundation plates, 57″ x 45″ x 24″		4.70	1.702		266	41		307	365	
0600	42″ x 49″ x 51″		4.30	1.860		320	44.50		364.50	430	

08360 | Overhead Doors

	Description	Crew	Daily Output	Labor-Hours	Unit	2005 Bare Costs Mat.	2005 Bare Costs Labor	2005 Bare Costs Equip.	2005 Bare Costs Total	Total Incl O&P	
0010	**RESIDENTIAL GARAGE DOORS** Including hardware, no frame										600
0050	Hinged, wood, custom, double door, 9′ x 7′	2 Carp	4	4	Ea.	365	96		461	570	
0070	16′ x 7′		3	5.333		625	128		753	900	
0200	Overhead, sectional, incl. hardware, fiberglass, 9′ x 7′, standard		5.28	3.030		565	72.50		637.50	745	
0220	Deluxe		5.28	3.030		720	72.50		792.50	915	
0300	16′ x 7′, standard		6	2.667		965	64		1,029	1,150	
0320	Deluxe		6	2.667		1,200	64		1,264	1,425	
0500	Hardboard, 9′ x 7′, standard		8	2		370	48		418	490	
0520	Deluxe		8	2		480	48		528	605	
0600	16′ x 7′, standard		6	2.667		700	64		764	880	
0620	Deluxe		6	2.667		815	64		879	1,000	
0700	Metal, 9′ x 7′, standard		5.28	3.030		450	72.50		522.50	620	
0720	Deluxe		8	2		585	48		633	720	
0800	16′ x 7′, standard		3	5.333		570	128		698	845	
0820	Deluxe		6	2.667		855	64		919	1,050	
0900	Wood, 9′ x 7′, standard		8	2		440	48		488	565	
0920	Deluxe		8	2		1,275	48		1,323	1,475	
1000	16′ x 7′, standard		6	2.667		890	64		954	1,100	
1020	Deluxe		6	2.667		1,850	64		1,914	2,150	
1800	Door hardware, sectional	1 Carp	4	2		195	48		243	296	
1810	Door tracks only		4	2		95	48		143	186	
1820	One side only		7	1.143		63	27.50		90.50	116	
3000	Swing-up, including hardware, fiberglass, 9′ x 7′, standard	2 Carp	8	2		610	48		658	750	
3020	Deluxe		8	2		670	48		718	820	
3100	16′ x 7′, standard		6	2.667		770	64		834	955	
3120	Deluxe		6	2.667		830	64		894	1,025	
3200	Hardboard, 9′ x 7′, standard		8	2		293	48		341	400	
3220	Deluxe		8	2		390	48		438	505	
3300	16′ x 7′, standard		6	2.667		410	64		474	560	
3320	Deluxe		6	2.667		610	64		674	780	
3400	Metal, 9′ x 7′, standard		8	2		320	48		368	435	
3420	Deluxe		8	2		570	48		618	710	
3500	16′ x 7′, standard		6	2.667		505	64		569	665	
3520	Deluxe		6	2.667		810	64		874	1,000	
3600	Wood, 9′ x 7′, standard		8	2		350	48		398	465	
3620	Deluxe		8	2		620	48		668	765	
3700	16′ x 7′, standard		6	2.667		610	64		674	780	
3720	Deluxe		6	2.667		860	64		924	1,050	
3900	Door hardware only, swing up	1 Carp	4	2		102	48		150	194	
3920	One side only		7	1.143		57	27.50		84.50	110	
4000	For electric operator, economy, add		8	1		254	24		278	320	
4100	Deluxe, including remote control		8	1		370	24		394	450	
4500	For transmitter/receiver control , add to operator				Total	80.50			80.50	88.50	
4600	Transmitters, additional				″	28			28	31	
6000	Replace section, on sectional door, fiberglass, 9′ x 7′	1 Carp	4	2	Ea.	165	48		213	263	
6020	16′ x 7′		3.50	2.286		250	55		305	370	
6200	Hardboard, 9′ x 7′		4	2		85.50	48		133.50	176	
6220	16′ x 7′		3.50	2.286		164	55		219	273	

08300 | Specialty Doors

08360 | Overhead Doors

		08360 Overhead Doors	CREW	DAILY OUTPUT	LABOR-HOURS	UNIT	2005 BARE COSTS MAT.	LABOR	EQUIP.	TOTAL	TOTAL INCL O&P
600	6300	Metal, 9′ x 7′	1 Carp	4	2	Ea.	139	48		187	235
	6320	16′ x 7′	↓	3.50	2.286	↓	228	55		283	345
	6500	Wood, 9′ x 7′		4	2		84	48		132	174
	6520	16′ x 7′	↓	3.50	2.286	↓	163	55		218	273

08500 | Windows

08510 | Steel Windows

		08510 Steel Windows	CREW	DAILY OUTPUT	LABOR-HOURS	UNIT	2005 BARE COSTS MAT.	LABOR	EQUIP.	TOTAL	TOTAL INCL O&P
700	0010	**SCREENS**									
	0020	For metal sash, aluminum or bronze mesh, flat screen	2 Sswk	1,200	.013	S.F.	3.47	.34		3.81	4.48
	0500	Wicket screen, inside window	"	1,000	.016	"	5.30	.41		5.71	6.65
	0600	Residential, aluminum mesh and frame, 2′ x 3′	2 Carp	32	.500	Ea.	11.50	12		23.50	33
	0610	Rescreen		50	.320		8.85	7.70		16.55	23
	0620	3′ x 5′		32	.500		27	12		39	50
	0630	Rescreen		45	.356		23	8.55		31.55	39.50
	0640	4′ x 8′		25	.640		51	15.35		66.35	82
	0650	Rescreen		40	.400		37	9.60		46.60	57
	0660	Patio door		25	.640	↓	135	15.35		150.35	174
	0680	Rescreening	↓	1,600	.010	S.F.	1.13	.24		1.37	1.65
	1000	For solar louvers, add	2 Sswk	160	.100	"	19.35	2.56		21.91	26.50

08520 | Aluminum Windows

		08520 Aluminum Windows	CREW	DAILY OUTPUT	LABOR-HOURS	UNIT	2005 BARE COSTS MAT.	LABOR	EQUIP.	TOTAL	TOTAL INCL O&P
120	0010	**ALUMINUM WINDOWS** Incl. frame and glazing, Commercial grade									
	1000	Stock units, casement, 3′-1″ x 3′-2″ opening	2 Sswk	10	1.600	Ea.	299	41		340	410
	1040	Insulating glass	"	10	1.600		287	41		328	395
	1050	Add for storms					60			60	66
	1600	Projected, with screen, 3′-1″ x 3′-2″ opening	2 Sswk	10	1.600		213	41		254	315
	1650	Insulating glass	"	10	1.600		193	41		234	291
	1700	Add for storms					56.50			56.50	62
	2000	4′-5″ x 5′-3″ opening	2 Sswk	8	2		300	51		351	430
	2050	Insulating glass	"	8	2		335	51		386	470
	2100	Add for storms					78.50			78.50	86.50
	2500	Enamel finish windows, 3′-1″ x 3′-2″	2 Sswk	10	1.600		191	41		232	290
	2550	Insulating glass		10	1.600		201	41		242	300
	2600	4′-5″ x 5′-3″		8	2		287	51		338	415
	2700	Insulating glass		8	2		360	51		411	495
	3000	Single hung, 2′ x 3′ opening, enameled, standard glazed		10	1.600		143	41		184	237
	3100	Insulating glass		10	1.600		174	41		215	270
	3300	2′-8″ x 6′-8″ opening, standard glazed		8	2		305	51		356	435
	3400	Insulating glass		8	2		390	51		441	530
	3700	3′-4″ x 5′-0″ opening, standard glazed		9	1.778		196	45.50		241.50	305
	3800	Insulating glass		9	1.778		276	45.50		321.50	395
	4000	Sliding aluminum, 3′ x 2′ opening, standard glazed		10	1.600		161	41		202	256
	4100	Insulating glass		10	1.600		179	41		220	276
	4300	5′ x 3′ opening, standard glazed		9	1.778		205	45.50		250.50	315
	4400	Insulating glass		9	1.778		286	45.50		331.50	405
	4600	8′ x 4′ opening, standard glazed		6	2.667		295	68		363	455
	4700	Insulating glass		6	2.667		475	68		543	650
	5000	9′ x 5′ opening, standard glazed		4	4		445	102		547	685
	5100	Insulating glass	↓	4	4	↓	715	102		817	980

08500 | Windows

	08520 | Aluminum Windows		CREW	DAILY OUTPUT	LABOR-HOURS	UNIT	2005 BARE COSTS MAT.	LABOR	EQUIP.	TOTAL	TOTAL INCL O&P	
0	5500	Sliding, with thermal barrier and screen, 6′ x 4′, 2 track	2 Sswk	8	2	Ea.	605	51		656	770	120
	5700	4 track		8	2		740	51		791	915	
	6000	For above units with bronze finish, add					12%					
	6200	For installation in concrete openings, add					5%					
0	0010	**JALOUSIES**										500
	0020	Aluminum incl. glazing & screens, stock, 1′-7″ x 3′-2″	2 Sswk	10	1.600	Ea.	131	41		172	223	
	0100	2′-3″ x 4′-0″		10	1.600		187	41		228	285	
	0200	3′-1″ x 2′-0″		10	1.600		147	41		188	241	
	0300	3′-1″ x 5′-3″		10	1.600		267	41		308	370	
	1000	Mullions for above, 2′-0″ long		80	.200		10.05	5.10		15.15	21	
	1100	5′-3″ long		80	.200		17.20	5.10		22.30	29	
	08550 | Wood Windows											
90	0010	**AWNING WINDOW** Including frame, screens and grills										100
	0100	Average quality, builders model, 34″ x 22″, double insulated glass	1 Carp	10	.800	Ea.	209	19.20		228.20	263	
	0200	Low E glass		10	.800		221	19.20		240.20	276	
	0300	40″ x 28″, double insulated glass		9	.889		264	21.50		285.50	325	
	0400	Low E Glass		9	.889		280	21.50		301.50	345	
	0500	48″ x 36″, double insulated glass		8	1		385	24		409	465	
	0600	Low E glass		8	1		405	24		429	485	
	0800	Vinyl clad, premium, double insulated glass, 24″ x 17″		12	.667		178	16		194	223	
	0840	24″ x 28″		11	.727		213	17.45		230.45	264	
	0860	36″ x 17″		11	.727		215	17.45		232.45	267	
	0900	36″ x 40″		10	.800		330	19.20		349.20	400	
	1100	40″ x 22″		10	.800		243	19.20		262.20	300	
	1200	36″ x 28″		9	.889		259	21.50		280.50	320	
	1300	36″ x 36″		9	.889		288	21.50		309.50	350	
	1400	48″ x 28″		8	1		310	24		334	380	
	1500	60″ x 36″		8	1		450	24		474	535	
	2000	Metal clad, deluxe, double insulated glass, 34″ x 22″		10	.800		208	19.20		227.20	262	
	2100	40″ x 22″		10	.800		244	19.20		263.20	300	
	2200	36″ x 25″		9	.889		226	21.50		247.50	285	
	2300	40″ x 30″		9	.889		283	21.50		304.50	345	
	2400	48″ x 28″		8	1		289	24		313	360	
	2500	60″ x 36″		8	1		310	24		334	380	
50	0010	**BOW-BAY WINDOW** Including frame, screens and grills,										150
	0020	end panels operable										
	1000	Bow type, casement, wood, bldrs mdl, 8′ x 5′ dbl insltd glass, 4 panel	2 Carp	10	1.600	Ea.	1,475	38.50		1,513.50	1,700	
	1050	Low E glass		10	1.600		1,200	38.50		1,238.50	1,400	
	1100	10′-0″ x 5′-0″ , double insulated glass, 6 panels		6	2.667		1,250	64		1,314	1,475	
	1200	Low E glass, 6 panels		6	2.667		1,325	64		1,389	1,550	
	1300	Vinyl clad, bldrs model, double insulated glass, 6′-0″ x 4′-0″, 3 panel		10	1.600		1,000	38.50		1,038.50	1,175	
	1340	9′-0″ x 4′-0″, 4 panel		8	2		1,350	48		1,398	1,550	
	1380	10′-0″ x 6′-0″, 5 panels		7	2.286		2,150	55		2,205	2,475	
	1420	12′-0″ x 6′-0″, 6 panels		6	2.667		2,725	64		2,789	3,100	
	1600	Metal clad, casement, bldrs mdl, 6′-0″ x 4′-0″, dbl insltd gls, 3 panels		10	1.600		855	38.50		893.50	1,000	
	1640	9′-0″ x 4′-0″ , 4 panels		8	2		1,200	48		1,248	1,400	
	1680	10′-0″ x 5′-0″, 5 panels		7	2.286		1,650	55		1,705	1,925	
	1720	12′-0″ x 6′-0″, 6 panels		6	2.667		2,300	64		2,364	2,625	
	2000	Bay window, casement, builders model, 8′ x 5′ dbl insul glass, 4 panels		10	1.600		1,425	38.50		1,463.50	1,625	
	2050	Low E glass,		10	1.600		1,725	38.50		1,763.50	1,950	
	2100	12′-0″ x 6′-0″ , double insulated glass, 6 panels		6	2.667		1,775	64		1,839	2,050	
	2200	Low E glass		6	2.667		1,825	64		1,889	2,125	
	2280	6′-0″ x 4′-0″		11	1.455		975	35		1,010	1,125	
	2300	Vinyl clad, premium, double insulated glass, 8′-0″ x 5′-0″		10	1.600		1,150	38.50		1,188.50	1,325	

08550 | Wood Windows

		08550 Wood Windows	CREW	DAILY OUTPUT	LABOR-HOURS	UNIT	2005 BARE COSTS MAT.	LABOR	EQUIP.	TOTAL	TOTAL INCL O&P
150	2340	10'-0" x 5'-0"	2 Carp	8	2	Ea.	1,625	48		1,673	1,850
	2380	10'-0" x 6'-0"		7	2.286		1,700	55		1,755	1,975
	2420	12'-0" x 6'-0"		6	2.667		2,025	64		2,089	2,325
	2430	14'-0" x 3'-0"		7	2.286		1,400	55		1,455	1,625
	2440	14'-0" x 6'-0"		5	3.200		2,200	77		2,277	2,550
	2600	Metal clad, deluxe, dbl insul. glass, 8'-0" x 5'-0" high, 4 panels		10	1.600		1,250	38.50		1,288.50	1,450
	2640	10'-0" x 5'-0" high, 5 panels		8	2		1,350	48		1,398	1,550
	2680	10'-0" x 6'-0" high, 5 panels		7	2.286		1,600	55		1,655	1,850
	2720	12'-0" x 6'-0" high, 6 panels		6	2.667		2,200	64		2,264	2,525
	3000	Double hung, bldrs. model, bay, 8' x 4' high,dbl insulated glass		10	1.600		995	38.50		1,033.50	1,175
	3050	Low E glass		10	1.600		1,075	38.50		1,113.50	1,250
	3100	9'-0" x 5'-0" high, doublel insulated glass		6	2.667		1,075	64		1,139	1,275
	3200	Low E glass		6	2.667		1,125	64		1,189	1,350
	3300	Vinyl clad, premium, double insulated glass, 7'-0" x 4'-6"		10	1.600		1,025	38.50		1,063.50	1,200
	3340	8'-0" x 4'-6"		8	2		1,050	48		1,098	1,225
	3380	8'-0" x 5'-0"		7	2.286		1,100	55		1,155	1,300
	3420	9'-0" x 5'-0"		6	2.667		1,125	64		1,189	1,350
	3600	Metal clad, deluxe, dbl insul. glass, 7'-0" x 4'-0" high		10	1.600		955	38.50		993.50	1,125
	3640	8'-0" x 4'-0" high		8	2		990	48		1,038	1,150
	3680	8'-0" x 5'-0" high		7	2.286		1,025	55		1,080	1,225
	3720	9'-0" x 5'-0" high	↓	6	2.667	↓	1,075	64		1,139	1,300
200	0010	**CASEMENT WINDOW** Including frame, screen, and grills R08550-010									
	0100	Avg. quality, bldrs. model, 2'-0" x 3'-0" H, dbl. insulated glass	1 Carp	10	.800	Ea.	176	19.20		195.20	226
	0150	Low E glass		10	.800		282	19.20		301.20	345
	0200	2'-0" x 4'-6" high, double insulated glass		9	.889		228	21.50		249.50	287
	0250	Low E glass		9	.889		330	21.50		351.50	395
	0300	2'-3" x 6'-0" high, double insulated glass		8	1		395	24		419	475
	0350	Low E glass		8	1		325	24		349	400
	0522	Vinyl clad, premium, double insulated glass, 2'-0" x 3'-0"		10	.800		237	19.20		256.20	294
	0524	2'-0" x 4'-0"		9	.889		277	21.50		298.50	340
	0525	2'-0" x 5'-0"		8	1		315	24		339	390
	0528	2'-0" x 6'-0"	↓	8	1	↓	360	24		384	435
	3020	Vinyl clad, premium, double insulated glass, multiple leaf units									
	3080	Single unit, 1'-6" x 5'-0"	2 Carp	20	.800	Ea.	272	19.20		291.20	330
	3100	2'-0" x 2'-0"		20	.800		182	19.20		201.20	233
	3140	2'-0" x 2'-6"		20	.800		237	19.20		256.20	294
	3220	2'-0" x 3'-6"		20	.800		235	19.20		254.20	291
	3260	2'-0" x 4'-0"		19	.842		277	20		297	340
	3300	2'-0" x 4'-6"		19	.842		272	20		292	335
	3340	2'-0" x 5'-0"		18	.889		315	21.50		336.50	385
	3460	2'-4" x 3'-0"		20	.800		237	19.20		256.20	294
	3500	2'-4" x 4'-0"		19	.842		294	20		314	360
	3540	2'-4" x 5'-0"		18	.889		350	21.50		371.50	420
	3700	Double unit, 2'-8" x 5'-0"		18	.889		490	21.50		511.50	575
	3740	2'-8" x 6'-0"		17	.941		570	22.50		592.50	670
	3840	3'-0" x 4'-6"		18	.889		445	21.50		466.50	525
	3860	3'-0" x 5'-0"		17	.941		580	22.50		602.50	680
	3880	3'-0" x 6'-0"		17	.941		620	22.50		642.50	720
	3980	3'-4" x 2'-6"		19	.842		375	20		395	450
	4000	3'-4" x 3'-0"		12	1.333		375	32		407	470
	4030	3'-4" x 4'-0"		18	.889		460	21.50		481.50	540
	4050	3'-4" x 5'-0"		12	1.333		580	32		612	695
	4100	3'-4" x 6'-0"		11	1.455		620	35		655	740
	4200	3'-6" x 3'-0"		18	.889		375	21.50		396.50	450
	4340	4'-0" x 3'-0"	↓	18	.889	↓	425	21.50		446.50	500

08550 | Wood Windows

			CREW	DAILY OUTPUT	LABOR-HOURS	UNIT	2005 BARE COSTS MAT.	LABOR	EQUIP.	TOTAL	TOTAL INCL O&P	
200	4380	4'-0" x 3'-6" R08550-010	2 Carp	17	.941	Ea.	455	22.50		477.50	540	200
	4420	4'-0" x 4'-0"		16	1		545	24		569	640	
	4460	4'-0" x 4'-4"		16	1		535	24		559	630	
	4540	4'-0" x 5'-0"		16	1		585	24		609	680	
	4580	4'-0" x 6'-0"		15	1.067		665	25.50		690.50	775	
	4740	4'-8" x 3'-0"		18	.889		475	21.50		496.50	560	
	4780	4'-8" x 3'-6"		17	.941		510	22.50		532.50	600	
	4820	4'-8" x 4'-0"		16	1		605	24		629	705	
	4860	4'-8" x 5'-0"		15	1.067		690	25.50		715.50	805	
	4900	4'-8" x 6'-0"		15	1.067		775	25.50		800.50	900	
	5060	Triple unit, 5'-0" x 5'-0"		15	1.067		930	25.50		955.50	1,075	
	5100	5'-6" x 3'-0"		17	.941		640	22.50		662.50	740	
	5140	5'-6" x 3'-6"		16	1		675	24		699	780	
	5180	5'-6" x 4'-6"		15	1.067		760	25.50		785.50	880	
	5220	5'-6" x 5'-6"		15	1.067		1,025	25.50		1,050.50	1,175	
	5300	6'-0" x 4'-6"		15	1.067		760	25.50		785.50	880	
	5850	5'-0" x 3'-0"		12	1.333		675	32		707	795	
	5900	5'-0" x 4'-0"		11	1.455		795	35		830	935	
	6000	5'-0" x 5'-0"		10	1.600		915	38.50		953.50	1,075	
	6100	5'-0" x 5'-6"		10	1.600		980	38.50		1,018.50	1,150	
	6150	5'-0" x 6'-0"		10	1.600		1,025	38.50		1,063.50	1,225	
	6200	6'-0" x 3'-0"		12	1.333		1,050	32		1,082	1,200	
	6250	6'-0" x 3'-4"		12	1.333		675	32		707	795	
	6300	6'-0" x 4'-0"		11	1.455		725	35		760	860	
	6350	6'-0" x 5'-0"		10	1.600		810	38.50		848.50	960	
	6400	6'-0" x 6'-0"		10	1.600		990	38.50		1,028.50	1,150	
	6500	Quadruple unit, 7'-0" x 4'-0"		9	1.778		955	42.50		997.50	1,125	
	6700	8'-0" x 4'-6"		9	1.778		1,250	42.50		1,292.50	1,450	
	6950	6'-8" x 4'-0"		10	1.600		915	38.50		953.50	1,075	
	7000	6'-8" x 6'-0"		10	1.600		1,225	38.50		1,263.50	1,425	
	8100	Metal clad, deluxe, dbl. insul. glass, 2'-0" x 3'-0" high	1 Carp	10	.800		175	19.20		194.20	226	
	8120	2'-0" x 4'-0" high		9	.889		211	21.50		232.50	268	
	8140	2'-0" x 5'-0" high		8	1		240	24		264	305	
	8160	2'-0" x 6'-0" high		8	1		275	24		299	345	
	8200	For multiple leaf units, deduct for stationary sash										
	8220	2' high				Ea.	18.20			18.20	20	
	8240	4'-6" high					21			21	23	
	8260	6' high					28			28	31	
	8300	For installation, add per leaf						15%				
250	0010	**DOUBLE HUNG** Including frame, screens, and grills										250
	0100	Avg. quality, bldrs. model, 2'-0" x 3'-0" high, dbl insul. glass	1 Carp	10	.800	Ea.	183	19.20		202.20	235	
	0150	Low E glass		10	.800		180	19.20		199.20	231	
	0200	3'-0" x 4'-0" high, double insulated glass		9	.889		244	21.50		265.50	305	
	0250	Low E glass		9	.889		234	21.50		255.50	294	
	0300	4'-0" x 4'-6" high, double insulated glass		8	1		277	24		301	345	
	0350	Low E glass		8	1		300	24		324	370	
	1000	Vinyl clad, premium, double insulated glass, 2'-6" x 3'-0"		10	.800		195	19.20		214.20	248	
	1100	3'-0" x 3'-6"		10	.800		230	19.20		249.20	286	
	1200	3'-0" x 4'-0"		9	.889		325	21.50		346.50	390	
	1300	3'-0" x 4'-6"		9	.889		266	21.50		287.50	330	
	1400	3'-0" x 5'-0"		8	1		278	24		302	345	
	1500	3'-6" x 6'-0"		8	1		325	24		349	395	
	2000	Metal clad, deluxe, dbl. insul. glass, 2'-6" x 3'-0" high		10	.800		186	19.20		205.20	237	
	2100	3'-0" x 3'-6" high		10	.800		220	19.20		239.20	275	
	2200	3'-0" x 4'-0" high		9	.889		234	21.50		255.50	293	

	08550 | Wood Windows		CREW	DAILY OUTPUT	LABOR-HOURS	UNIT	2005 BARE COSTS MAT.	LABOR	EQUIP.	TOTAL	TOTAL INCL O&P
250	2300	3′-0″ x 4′-6″ high	1 Carp	9	.889	Ea.	253	21.50		274.50	315
	2400	3′-0″ x 5′-0″ high		8	1		270	24		294	340
	2500	3′-6″ x 6′-0″ high		8	1		330	24		354	400
260	0010	**HALF ROUND WINDOW**, Vinyl clad, double insulated glass, including grill									
	0800	14″ height x 24″ base	2 Carp	9	1.778	Ea.	325	42.50		367.50	435
	1040	15″ height x 25″ base		8	2		330	48		378	440
	1060	16″ height x 28″ base		7	2.286		360	55		415	490
	1080	17″ height x 29″ base		7	2.286		375	55		430	505
	2000	19″ height x 33″ base	1 Carp	6	1.333		400	32		432	495
	2100	20″ height x 35″ base		6	1.333		445	32		477	545
	2200	21″ height x 37″ base		6	1.333		425	32		457	525
	2250	23″ height x 41″ base	2 Carp	6	2.667		465	64		529	620
	2300	26″ height x 48″ base		6	2.667		480	64		544	640
	2350	30″ height x 56″ base		6	2.667		565	64		629	730
	3000	36″ height x 67″base	1 Carp	4	2		975	48		1,023	1,150
	3040	38″ height x 71″ base	2 Carp	5	3.200		900	77		977	1,125
	3050	40″ height x 75″ base	"	5	3.200		1,175	77		1,252	1,425
	5000	Elliptical, 71″ x 16″	1 Carp	11	.727		740	17.45		757.45	845
	5100	95″ x 21″	"	10	.800		1,050	19.20		1,069.20	1,175
650	0010	**PALLADIAN WINDOWS**									
	0020	Vinyl clad, double insulated glass, including frame and grills									
	0040	3′-2″ x 2′-0″ high	2 Carp	11	1.455	Ea.	1,175	35		1,210	1,325
	0060	3′-2″ x 4′-10″		11	1.455		1,325	35		1,360	1,500
	0080	3′-2″ x 6′-4″		10	1.600		1,575	38.50		1,613.50	1,800
	0100	4′-0″ x 4′-0″		10	1.600		1,275	38.50		1,313.50	1,500
	0120	4′-0″ x 5′-4″	3 Carp	10	2.400		1,500	57.50		1,557.50	1,750
	0140	4′-0″ x 6′-0″		9	2.667		1,550	64		1,614	1,800
	0160	4′-0″ x 7′-4″		9	2.667		1,700	64		1,764	1,975
	0180	5′-5″ x 4′-10″		9	2.667		1,700	64		1,764	1,975
	0200	5′-5″ x 6′-10″		9	2.667		1,925	64		1,989	2,225
	0220	5′-5″ x 7′-9″		9	2.667		2,025	64		2,089	2,325
	0240	6′-0″ x 7′-11″		8	3		2,800	72		2,872	3,200
	0260	8′-0″ x 6′-0″		8	3		2,325	72		2,397	2,675
670	0010	**PICTURE WINDOW** Including frame and grills									
	0100	Average quality, bldrs. model, 3′-6″ x 4′-0″ high, dbl insulated glass	2 Carp	12	1.333	Ea.	268	32		300	350
	0150	Low E glass		12	1.333		292	32		324	375
	0200	4′-0″ x 4′-6″ high, double insulated glass		11	1.455		268	35		303	355
	0250	Low E glass		11	1.455		294	35		329	385
	0300	5′-0″ x 4′-0″ high, double insulated glass		11	1.455		330	35		365	425
	0350	Low E glass		11	1.455		370	35		405	470
	0400	6′-0″ x 4′-6″ high, double insulated glass		10	1.600		420	38.50		458.50	530
	0450	Low E glass		10	1.600		475	38.50		513.50	585
	1000	Vinyl clad, premium, dbl. insul. glass, 4′-0″ x 4′-0″		12	1.333		430	32		462	530
	1100	4′-0″ x 6′-0″		11	1.455		635	35		670	760
	1200	5′-0″ x 6′-0″		10	1.600		915	38.50		953.50	1,075
	1300	6′-0″ x 6′-0″		10	1.600		930	38.50		968.50	1,100
	2000	Metal clad, deluxe, dbl. insul. glass, 4′-0″ x 4′-0″ high		12	1.333		275	32		307	360
	2100	4′-0″ x 6′-0″ high		11	1.455		405	35		440	505
	2200	5′-0″ x 6′-0″ high		10	1.600		445	38.50		483.50	555
	2300	6′-0″ x 6′-0″ high		10	1.600		515	38.50		553.50	630
750	0010	**SLIDING WINDOW** Including frame, screen, and grills									
	0100	Average quality, bldrs. model, 3′-0″ x 3′-0″ high, double insulated	1 Carp	10	.800	Ea.	144	19.20		163.20	191
	0120	Low E glass		10	.800		182	19.20		201.20	233
	0200	4′-0″ x 3′-6″ high, double insulated		9	.889		171	21.50		192.50	224

	08550 | Wood Windows		Crew	Daily Output	Labor-Hours	Unit	2005 Bare Costs Mat.	Labor	Equip.	Total	Total Incl O&P
750	0220	Low E glass	1 Carp	9	.889	Ea.	214	21.50		235.50	272
	0300	6′-0″ x 5′-0″ high, double insulated		8	1		315	24		339	385
	0320	Low E glass		8	1		375	24		399	455
	1000	Vinyl clad, premium, dbl. insulated glass, 3′-0″ x 3′-0″		10	.800		515	19.20		534.20	600
	1020	4′-0″ x 1′-11″		11	.727		505	17.45		522.45	585
	1040	4′-0″ x 3′-0″		10	.800		600	19.20		619.20	695
	1050	4′-0″ x 3′-6″		9	.889		640	21.50		661.50	740
	1090	4′-0″ x 5′-0″		9	.889		770	21.50		791.50	885
	1100	5′-0″ x 4′-0″		9	.889		770	21.50		791.50	885
	1120	5′-0″ x 5′-0″		8	1		855	24		879	985
	1140	6′-0″ x 4′-0″		8	1		855	24		879	985
	1150	6′-0″ x 5′-0″		8	1		945	24		969	1,075
	2000	Metal clad, deluxe, double insulated glass, 3′-0″ x 3′-0″ high		10	.800		290	19.20		309.20	355
	2050	4′-0″ x 3′-6″ high		9	.889		355	21.50		376.50	425
	2100	5′-0″ x 4′-0″ high		9	.889		430	21.50		451.50	505
	2150	6′-0″ x 5′-0″ high		8	1		650	24		674	750
760	0010	**TRANSOM WINDOWS**									
	0050	Vinyl clad, premium, double insulated glass, 32″ x 8″	1 Carp	16	.500	Ea.	144	12		156	179
	0100	36″ x 8″		16	.500		154	12		166	190
	0110	36″ x 12″		16	.500		173	12		185	211
	0150	36″ x 48″		12	.667		203	16		219	251
770	0010	**WEATHERSTRIPPING** See division 08720									
780	0010	**TRAPEZOID WINDOWS**									
	0100	Vinyl clad, including frame and exterior trim									
	0900	20″ base x 44″ leg x 53″ leg	2 Carp	13	1.231	Ea.	350	29.50		379.50	430
	1000	24″ base x 90″ leg x 102″ leg		8	2		580	48		628	715
	3000	36″ base x 0″ leg x 22″ leg		12	1.333		370	32		402	460
	3010	36″ base x 4″ leg x 25″ leg		13	1.231		390	29.50		419.50	480
	3050	36″ base x 26″ leg x 48″ leg		9	1.778		405	42.50		447.50	520
	3100	36″ base x 42″ legs, 50″ peak		9	1.778		470	42.50		512.50	595
	3200	36″ base x 60″ leg x 81″ leg		11	1.455		615	35		650	740
	4320	44″ base x 23″ leg x 56″ leg		11	1.455		485	35		520	595
	4350	44″ base x 59″ leg x 92″ leg		10	1.600		730	38.50		768.50	865
	4500	46″ base x 15″ leg x 46″ leg		8	2		365	48		413	485
	4550	46″ base x 16″ leg x 48″ leg		8	2		390	48		438	510
	4600	46″ base x 50″ leg x 80″ leg		7	2.286		585	55		640	740
	6600	66″ base x 12″ leg x 42″ leg		8	2		495	48		543	620
	6650	66″ base x 12″ legs, 28″ peak		9	1.778		400	42.50		442.50	515
	6700	68″ base x 3″ legs, 31″ peak		8	2		500	48		548	630
800	0010	**WINDOW GRILLE OR MUNTIN** Snap-in type									
	0020	Standard pattern interior grills									
	2000	Wood, awning window, glass size 28″ x 16″ high	1 Carp	30	.267	Ea.	19.10	6.40		25.50	32
	2060	44″ x 24″ high		32	.250		28	6		34	40.50
	2100	Casement, glass size, 20″ x 36″ high		30	.267		23.50	6.40		29.90	37
	2180	20″ x 56″ high		32	.250		34	6		40	47
	2200	Double hung, glass size, 16″ x 24″ high		24	.333	Set	41.50	8		49.50	59.50
	2280	32″ x 32″ high		34	.235	″	116	5.65		121.65	138
	2500	Picture, glass size, 48″ x 48″ high		30	.267	Ea.	133	6.40		139.40	157
	2580	60″ x 68″ high		28	.286	″	97.50	6.85		104.35	119
	2600	Sliding, glass size, 14″ x 36″ high		24	.333	Set	23	8		31	39
	2680	36″ x 36″ high		22	.364	″	35	8.75		43.75	53
820	0010	**WOOD SASH** Including glazing but not including trim									
	0050	Custom, 5′-0″ x 4′-0″, 1″ dbl. glazed, 3/16″ thick lites	2 Carp	3.20	5	Ea.	147	120		267	365

08550 | Wood Windows

			CREW	DAILY OUTPUT	LABOR-HOURS	UNIT	2005 BARE COSTS MAT.	LABOR	EQUIP.	TOTAL	TOTAL INCL O&P
820	0100	1/4" thick lites	2 Carp	5	3.200	Ea.	151	77		228	296
	0200	1" thick, triple glazed		5	3.200		345	77		422	510
	0300	7'-0" x 4'-6" high, 1" double glazed, 3/16" thick lites		4.30	3.721		350	89.50		439.50	535
	0400	1/4" thick lites		4.30	3.721		395	89.50		484.50	585
	0500	1" thick, triple glazed		4.30	3.721		455	89.50		544.50	650
	0600	8'-6" x 5'-0" high, 1" double glazed, 3/16" thick lites		3.50	4.571		475	110		585	710
	0700	1/4" thick lites		3.50	4.571		520	110		630	755
	0800	1" thick, triple glazed		3.50	4.571		525	110		635	760
	0900	Window frames only, based on perimeter length				L.F.	2.90			2.90	3.19
	3000	Replacement sash, double hung, double glazing, to 12 S.F.	1 Carp	64	.125	S.F.	16.05	3		19.05	23
	3100	12 S.F. to 20 S.F.		94	.085		16	2.04		18.04	21
	3200	20 S.F. and over		106	.075		13.95	1.81		15.76	18.45
	3800	Triple glazing for above, add					2.13			2.13	2.34
	7000	Sash, single lite, 2'-0" x 2'-0" high	1 Carp	20	.400	Ea.	41	9.60		50.60	61.50
	7050	2'-6" x 2'-0" high		19	.421		44	10.10		54.10	65.50
	7100	2'-6" x 2'-6" high		18	.444		46.50	10.65		57.15	69.50
	7150	3'-0" x 2'-0" high		17	.471		58.50	11.30		69.80	83.50
840	0010	**WOOD SCREENS**									
	0020	Over 3 S.F., 3/4" frames	2 Carp	375	.043	S.F.	3.34	1.02		4.36	5.40
	0100	1-1/8" frames	"	375	.043	"	5.95	1.02		6.97	8.30

08560 | Plastic Windows

			CREW	DAILY OUTPUT	LABOR-HOURS	UNIT	MAT.	LABOR	EQUIP.	TOTAL	TOTAL INCL O&P
100	0010	**VINYL SINGLE HUNG WINDOWS**									
	0100	Grids, low E, J fin, ext. jambs, 21" x 53"	2 Carp	18	.889	Ea.	130	21.50		151.50	179
	0110	21" x 57"		17	.941		133	22.50		155.50	186
	0120	21" x 65"		16	1		139	24		163	194
	0130	25" x 41"		20	.800		123	19.20		142.20	168
	0140	25" x 49"		18	.889		136	21.50		157.50	185
	0150	25" x 57"		17	.941		139	22.50		161.50	192
	0160	25" x 65"		16	1		145	24		169	200
	0170	29" x 41"		18	.889		131	21.50		152.50	180
	0180	29" x 53"		18	.889		140	21.50		161.50	190
	0190	29" x 57"		17	.941		143	22.50		165.50	197
	0200	29" x 65"		16	1		149	24		173	205
	0210	33" x 41"		20	.800		135	19.20		154.20	182
	0220	33" x 53"		18	.889		146	21.50		167.50	196
	0230	33" x 57"		17	.941		149	22.50		171.50	203
	0240	33" x 65"		16	1		155	24		179	211
	0250	37" x 41"		20	.800		143	19.20		162.20	190
	0260	37" x 53"		18	.889		153	21.50		174.50	204
	0270	37" x 57"		17	.941		156	22.50		178.50	211
	0280	37" x 65"		16	1		163	24		187	220
200	0010	**VINYL DOUBLE HUNG WINDOWS**									
	0100	Grids, low E, J fin, ext. jambs, 21" x 53"	2 Carp	18	.889	Ea.	149	21.50		170.50	200
	0102	21" x 37"		18	.889		133	21.50		154.50	182
	0104	21" x 41"		18	.889		136	21.50		157.50	186
	0106	21" x 49"		18	.889		143	21.50		164.50	193
	0110	21" x 57"		17	.941		152	22.50		174.50	206
	0120	21" x 65"		16	1		158	24		182	215
	0128	25" x 37"		20	.800		140	19.20		159.20	187
	0130	25" x 41"		20	.800		143	19.20		162.20	191
	0140	25" x 49"		18	.889		148	21.50		169.50	198
	0145	25" x 53"		18	.889		154	21.50		175.50	205
	0150	25" x 57"		17	.941		154	22.50		176.50	209

08560 | Plastic Windows

			CREW	DAILY OUTPUT	LABOR-HOURS	UNIT	2005 BARE COSTS MAT.	LABOR	EQUIP.	TOTAL	TOTAL INCL O&P
200	0160	25″ x 65″	2 Carp	16	1	Ea.	164	24		188	221
	0162	25″ x 69″		16	1		170	24		194	228
	0164	25″ x 77″		16	1		180	24		204	239
	0168	29″ x 37″		18	.889		145	21.50		166.50	195
	0170	29″ x 41″		18	.889		148	21.50		169.50	198
	0172	29″ x 49″		18	.889		155	21.50		176.50	206
	0180	29″ x 53″		18	.889		158	21.50		179.50	210
	0190	29″ x 57″		17	.941		162	22.50		184.50	217
	0200	29″ x 65″		16	1		168	24		192	226
	0202	29″ x 69″		16	1		174	24		198	233
	0205	29″ x 77″		16	1		184	24		208	244
	0208	33″ x 37″		20	.800		149	19.20		168.20	196
	0210	33″ x 41″		20	.800		152	19.20		171.20	200
	0215	33″ x 49″		20	.800		160	19.20		179.20	209
	0220	33″ x 53″		18	.889		163	21.50		184.50	216
	0230	33″ x 57″		17	.941		167	22.50		189.50	223
	0240	33″ x 65″		16	1		171	24		195	229
	0242	33″ x 69″		16	1		181	24		205	241
	0246	33″ x 77″		16	1		191	24		215	251
	0250	37″ x 41″		20	.800		156	19.20		175.20	204
	0255	37″ x 49″		20	.800		163	19.20		182.20	213
	0260	37″ x 53″		18	.889		170	21.50		191.50	223
	0270	37″ x 57″		17	.941		174	22.50		196.50	230
	0280	37″ x 65″		16	1		178	24		202	237
	0282	37″ x 69″		16	1		233	24		257	298
	0286	37″ x 77″		16	1		244	24		268	310
	0300	Solid vinyl, average quality, double insulated glass, 2′-0″ x 3′-0″	1 Carp	10	.800		127	19.20		146.20	172
	0310	3′-0″ x 4′-0″		9	.889		158	21.50		179.50	210
	0320	4′-0″ x 4′-6″		8	1		190	24		214	250
	0330	Premium, double insulated glass, 2′-6″ x 3′-0″		10	.800		153	19.20		172.20	201
	0340	3′-0″ x 3′-6″		9	.889		177	21.50		198.50	231
	0350	3′-0″ x 4″-0″		9	.889		188	21.50		209.50	243
	0360	3′-0″ x 4′-6″		9	.889		193	21.50		214.50	248
	0370	3′-0″ x 5′-0″		8	1		198	24		222	259
	0380	3′-6″ x 6″-0″		8	1		218	24		242	280
300	0010	**VINYL CASEMENT WINDOWS**									
	0100	Grids, low E, J fin, ext. jambs, 1 lt, 21″ x 41″	2 Carp	20	.800	Ea.	193	19.20		212.20	245
	0110	21″ x 47″		20	.800		210	19.20		229.20	264
	0120	21″ x 53″		20	.800		227	19.20		246.20	283
	0128	24″ x 35″		19	.842		185	20		205	239
	0130	24″ x 41″		19	.842		201	20		221	256
	0140	24″ x 47″		19	.842		218	20		238	275
	0150	24″ x 53″		19	.842		235	20		255	293
	0158	28″ x 35″		19	.842		197	20		217	252
	0160	28″ x 41″		19	.842		213	20		233	269
	0170	28″ x 47″		19	.842		230	20		250	288
	0180	28″ x 53″		19	.842		254	20		274	315
	0184	28″ x 59″		19	.842		258	20		278	320
	0188	Two lites, 33″ x 35″		18	.889		315	21.50		336.50	385
	0190	33″ x 41″		18	.889		340	21.50		361.50	410
	0200	33″ x 47″		18	.889		365	21.50		386.50	435
	0210	33″ x 53″		18	.889		390	21.50		411.50	465
	0212	33″ x 59″		18	.889		415	21.50		436.50	490
	0215	33″ x 72″		18	.889		430	21.50		451.50	505
	0220	41″ x 41″		18	.889		370	21.50		391.50	445

	08560 | Plastic Windows		CREW	DAILY OUTPUT	LABOR-HOURS	UNIT	2005 BARE COSTS MAT.	LABOR	EQUIP.	TOTAL	TOTAL INCL O&P
300	0230	41" x 47"	2 Carp	18	.889	Ea.	395	21.50		416.50	470
	0240	41" x 53"		17	.941		420	22.50		442.50	505
	0242	41" x 59"		17	.941		445	22.50		467.50	530
	0246	41" x 72"		17	.941		465	22.50		487.50	550
	0250	47" x 41"		17	.941		375	22.50		397.50	450
	0260	47" x 47"		17	.941		400	22.50		422.50	480
	0270	47" x 53"		17	.941		425	22.50		447.50	505
	0272	47" x 59"		17	.941		465	22.50		487.50	550
	0280	56" x 41"		15	1.067		400	25.50		425.50	485
	0290	56" x 47"		15	1.067		425	25.50		450.50	510
	0300	56" x 53"		15	1.067		460	25.50		485.50	555
	0302	56" x 59"		15	1.067		480	25.50		505.50	575
	0310	56" x 72"	↓	15	1.067		525	25.50		550.50	620
	0340	Solid vinyl, premium, double insulated glass, 2'-0" x 3'-0" high	1 Carp	10	.800		170	19.20		189.20	220
	0360	2'-0" x 4'-0" high		9	.889		209	21.50		230.50	266
	0380	2'-0" x 5'-0" high	↓	8	1	↓	248	24		272	315
400	0010	**VINYL PICTURE WINDOWS**									
	0100	Grids, low E, J fin, ext. jambs, 33" x 47"	2 Carp	12	1.333	Ea.	194	32		226	268
	0110	35" x 71"		12	1.333		205	32		237	280
	0120	41" x 47"		12	1.333		225	32		257	300
	0130	41" x 71"		12	1.333		244	32		276	325
	0140	47" x 47"		12	1.333		254	32		286	335
	0150	47" x 71"		11	1.455		267	35		302	355
	0160	53" x 47"		11	1.455		250	35		285	335
	0170	53" x 71"		11	1.455		261	35		296	345
	0180	59" x 47"		11	1.455		283	35		318	370
	0190	59" x 71"		11	1.455		305	35		340	395
	0200	71" x 47"		10	1.600		315	38.50		353.50	410
	0210	71" x 71"	↓	10	1.600	↓	330	38.50		368.50	430
500	0010	**VINYL HALF ROUND WINDOW,** including grill, j fin, low E, ext. jambs									
	0100	10" height x 20" base	2 Carp	8	2	Ea.	237	48		285	345
	0110	15" height x 30" base		8	2		335	48		383	450
	0120	17" height x 34" base		7	2.286		355	55		410	485
	0130	19" height x 38" base		7	2.286		405	55		460	540
	0140	19" height x 33" base	↓	7	2.286		345	55		400	470
	0150	24" height x 48" base	1 Carp	6	1.333		415	32		447	510
	0160	25" height x 50" base	"	6	1.333		450	32		482	550
	0170	30" height x 60" base	2 Carp	6	2.667	↓	555	64		619	720

	08580 | Special Function Windows		CREW	DAILY OUTPUT	LABOR-HOURS	UNIT	MAT.	LABOR	EQUIP.	TOTAL	TOTAL INCL O&P
900	0010	**STORM WINDOWS** Aluminum, residential									
	0300	Basement, mill finish, incl. fiberglass screen									
	0320	1'-10" x 1'-0" high	2 Carp	30	.533	Ea.	28	12.80		40.80	52
	0340	2'-9" x 1'-6" high		30	.533		30.50	12.80		43.30	55
	0360	3'-4" x 2'-0" high	↓	30	.533	↓	36.50	12.80		49.30	62
	1600	Double-hung, combination, storm & screen									
	1700	Custom, clear anodic coating, 2'-0" x 3'-5" high	2 Carp	30	.533	Ea.	72	12.80		84.80	101
	1720	2'-6" x 5'-0" high		28	.571		96.50	13.70		110.20	130
	1740	4'-0" x 6'-0" high		25	.640		204	15.35		219.35	251
	1800	White painted, 2'-0" x 3'-5" high		30	.533		86	12.80		98.80	116
	1820	2'-6" x 5'-0" high		28	.571		138	13.70		151.70	176
	1840	4'-0" x 6'-0" high		25	.640		247	15.35		262.35	298
	2000	Average quality, clear anodic coating, 2'-0" x 3'-5" high		30	.533		73.50	12.80		86.30	102
	2020	2'-6" x 5'-0" high	↓	28	.571	↓	89	13.70		102.70	122

08580 | Special Function Windows

		CREW	DAILY OUTPUT	LABOR-HOURS	UNIT	2005 BARE COSTS MAT.	LABOR	EQUIP.	TOTAL	TOTAL INCL O&P	
2040	4′-0″ x 6′-0″ high	2 Carp	25	.640	Ea.	109	15.35		124.35	146	900
2400	White painted, 2′-0″ x 3′-5″ high		30	.533		72	12.80		84.80	101	
2420	2′-6″ x 5′-0″ high		28	.571		79.50	13.70		93.20	111	
2440	4′-0″ x 6′-0″ high		25	.640		87	15.35		102.35	122	
2600	Mill finish, 2′-0″ x 3′-5″ high		30	.533		65.50	12.80		78.30	94	
2620	2′-6″ x 5′-0″ high		28	.571		73.50	13.70		87.20	104	
2640	4′-0″ x 6-8″ high		25	.640		82	15.35		97.35	117	
4000	Picture window, storm, 1 lite, white or bronze finish										
4020	4′-6″ x 4′-6″ high	2 Carp	25	.640	Ea.	110	15.35		125.35	147	
4040	5′-8″ x 4′-6″ high		20	.800		125	19.20		144.20	170	
4400	Mill finish, 4′-6″ x 4′-6″ high		25	.640		110	15.35		125.35	147	
4420	5′-8″ x 4′-6″ high		20	.800		125	19.20		144.20	170	
4600	3 lite, white or bronze finish										
4620	4′-6″ x 4′-6″ high	2 Carp	25	.640	Ea.	134	15.35		149.35	173	
4640	5′-8″ x 4′-6″ high		20	.800		149	19.20		168.20	197	
4800	Mill finish, 4′-6″ x 4′-6″ high		25	.640		118	15.35		133.35	156	
4820	5′-8″ x 4′-6″ high		20	.800		125	19.20		144.20	170	
5000	Sliding glass door, storm 6′ x 6′-8″, standard	1 Glaz	2	4		675	94		769	900	
5100	Economy	″	2	4		310	94		404	495	
6000	Sliding window, storm, 2 lite, white or bronze finish										
6020	3′-4″ x 2′-7″ high	2 Carp	28	.571	Ea.	96	13.70		109.70	130	
6040	4′-4″ x 3′-3″ high		25	.640		131	15.35		146.35	170	
6060	5′-4″ x 6′-0″ high		20	.800		210	19.20		229.20	264	
6400	3 lite, white or bronze finish										
6420	4′-4″ x 3′-3″ high	2 Carp	25	.640	Ea.	152	15.35		167.35	193	
6440	5′-4″ x 6′-0″ high		20	.800		274	19.20		293.20	335	
6460	6′-0″ x 6′-0″ high		18	.889		275	21.50		296.50	340	
6800	Mill finish, 4′-4″ x 3′-3″ high		25	.640		131	15.35		146.35	170	
6820	5′-4″ x 6′-0″ high		20	.800		275	19.20		294.20	340	
6840	6′-0″ x 6-0″ high		18	.889		284	21.50		305.50	350	
9000	Magnetic interior storm window										
9100	3/16″ plate glass	1 Glaz	107	.075	S.F.	3.98	1.75		5.73	7.30	

08590 | Window Restoration & Replace

		CREW	DAILY OUTPUT	LABOR-HOURS	UNIT	MAT.	LABOR	EQUIP.	TOTAL	TOTAL INCL O&P	
0010	**SOLID VINYL REPLACEMENT WINDOWS** R08550-200										600
0020	Double hung, insulated glass, up to 83 united inches	2 Carp	8	2	Ea.	206	48		254	310	
0040	84 to 93		8	2		229	48		277	335	
0060	94 to 101		6	2.667		262	64		326	395	
0080	102 to 111		6	2.667		290	64		354	430	
0100	112 to 120		6	2.667		325	64		389	470	
0120	For each united inch over 120 , add		800	.020	Inch	3.44	.48		3.92	4.60	
0140	Casement windows, one operating sash , 42 to 60 united inches		8	2	Ea.	170	48		218	269	
0160	61 to 70		8	2		193	48		241	294	
0180	71 to 80		8	2		210	48		258	315	
0200	81 to 96		8	2		223	48		271	325	
0220	Two operating sash, 58 to 78 united inches		8	2		340	48		388	455	
0240	79 to 88		8	2		365	48		413	480	
0260	89 to 98		8	2		395	48		443	515	
0280	99 to 108		6	2.667		415	64		479	565	
0300	109 to 121		6	2.667		445	64		509	600	
0320	Three operating sash, 73 to 108 united inches		8	2		535	48		583	670	
0340	109 to 118		8	2		565	48		613	700	
0360	119 to 128		6	2.667		580	64		644	750	
0380	129 to 138		6	2.667		620	64		684	790	
0400	139 to 156		6	2.667		655	64		719	830	
0420	Four operating sash, 98 to 118 united inches		8	2		770	48		818	925	

08500 | Windows

	08590 | Window Restoration & Replace		Crew	Daily Output	Labor-Hours	Unit	2005 Bare Costs Mat.	Labor	Equip.	Total	Total Incl O&P
600	0440	119 to 128 R08550-200	2 Carp	8	2	Ea.	825	48		873	990
	0460	129 to 138		6	2.667		875	64		939	1,075
	0480	139 to 148		6	2.667		920	64		984	1,100
	0500	149 to 168		6	2.667		980	64		1,044	1,175
	0520	169 to 178		6	2.667		1,050	64		1,114	1,275
	0540	For venting unit to fixed unit, deduct					15.50			15.50	17.05
	0560	Fixed picture window, up to 63 united inches	2 Carp	8	2		123	48		171	217
	0580	64 to 83		8	2		146	48		194	242
	0600	84 to 101		8	2		187	48		235	288
	0620	For each united inch over 101, add		900	.018	Inch	2.25	.43		2.68	3.20
	0640	Picture window opt., low E glazing, up to 101 united inches				Ea.	17.50			17.50	19.25
	0660	102 to 124					22.50			22.50	25
	0680	124 and over					34			34	37.50
	0700	Options, low E glazing, up to 101 united inches					8.75			8.75	9.65
	0720	102 to 124					11.25			11.25	12.40
	0740	124 and over					17			17	18.70
	0760	Muntins, between glazing, square, per lite					1.65			1.65	1.82
	0780	Diamond shape, per full or partial diamond					2.60			2.60	2.86
	0800	Celluose fiber insulation, poured into sash balance cavity	1 Carp	36	.222	C.F.	.48	5.35		5.83	9.60
	0820	Silicone caulking at perimeter	"	800	.010	L.F.	.13	.24		.37	.55

08600 | Skylights

	08610 | Roof Windows		Crew	Daily Output	Labor-Hours	Unit	2005 Bare Costs Mat.	Labor	Equip.	Total	Total Incl O&P
600	0010	**METAL ROOF WINDOW** Fixed, high perf tmpd glazing, 46" x 21-1/2"	1 Carp	8	1	Ea.	216	24		240	279
	0100	46" x 28"		8	1		250	24		274	315
	0125	57" x 44"		6	1.333		320	32		352	410
	0130	72" x 28"		7	1.143		320	27.50		347.50	400
	0150	Venting, high performance tempered glazing, 46" x 21-1/2"		8	1		310	24		334	380
	0175	46" x 28"		8	1		298	24		322	370
	0200	57" x 44"		6	1.333		430	32		462	530
	0500	Flashing set for shingled roof, 46" x 21-1/2"		5	1.600		31	38.50		69.50	99
	0525	46" x 28"		5	1.600		32.50	38.50		71	101
	0550	57" x 44"		5	1.600		38	38.50		76.50	107
	0560	72" x 28"		6	1.333		38	32		70	96
	0575	Flashing set for low pitched roof, 46" x 21-1/2"		5	1.600		136	38.50		174.50	215
	0600	46" x 28"		5	1.600		139	38.50		177.50	218
	0625	57" x 44"		5	1.600		158	38.50		196.50	239
	0650	Flashing set for tile roof 46" x 21-1/2"		5	1.600		80.50	38.50		119	154
	0675	46" x 28"		5	1.600		81	38.50		119.50	154
	0700	57" x 44"		5	1.600		92.50	38.50		131	167

	08620 | Unit Skylights		Crew	Daily Output	Labor-Hours	Unit	Mat.	Labor	Equip.	Total	Total Incl O&P
800	0010	**SKYLIGHT** Plastic domes, flush or curb mounted, ten or									
	0100	more units, curb not included									
	0300	Nominal size under 10 S.F., double	G-3	130	.246	S.F.	19.05	5.40		24.45	30
	0400	Single		160	.200		13.80	4.39		18.19	22.50
	0600	10 S.F. to 20 S.F., double		315	.102		16.65	2.23		18.88	22
	0700	Single		395	.081		8.40	1.78		10.18	12.25

08600 | Skylights

	08620 | Unit Skylights		CREW	DAILY OUTPUT	LABOR-HOURS	UNIT	2005 BARE COSTS MAT.	LABOR	EQUIP.	TOTAL	TOTAL INCL O&P	
	0900	20 S.F. to 30 S.F., double	G-3	395	.081	S.F.	15.10	1.78		16.88	19.60	800
	1000	Single		465	.069		10.60	1.51		12.11	14.20	
	1200	30 S.F. to 65 S.F., double		465	.069		11.35	1.51		12.86	15	
	1300	Single		610	.052		14.80	1.15		15.95	18.20	
	1500	For insulated 4" curbs, double, add					25%					
	1600	Single, add					30%					
	1800	For integral insulated 9" curbs, double, add					30%					
	1900	Single, add					40%					
	2120	Ventilating insulated plexiglass dome with										
	2130	curb mounting, 36" x 36"	G-3	12	2.667	Ea.	380	58.50		438.50	520	
	2150	52" x 52"		12	2.667		570	58.50		628.50	725	
	2160	28" x 52"		10	3.200		445	70		515	610	
	2170	36" x 52"		10	3.200		480	70		550	650	
	2180	For electric opening system, add					285			285	315	
	2210	Operating skylight, with thermopane glass, 24" x 48"	G-3	10	3.200		545	70		615	720	
	2220	32" x 48"	"	9	3.556		570	78		648	760	
	2310	Non venting insulated plexiglass dome skylight with										
	2320	Flush mount 22" x 46"	G-3	15.23	2.101	Ea.	310	46		356	420	
	2330	30" x 30"		16	2		285	44		329	390	
	2340	46" x 46"		13.91	2.300		525	50.50		575.50	665	
	2350	Curb mount 22" x 46"		15.23	2.101		272	46		318	380	
	2360	30" x 30"		16	2		260	44		304	360	
	2370	46" x 46"		13.91	2.300		490	50.50		540.50	620	
	2381	Non-insulated flush mount 22" x 46"		15.23	2.101		209	46		255	310	
	2382	30" x 30"		16	2		190	44		234	283	
	2383	46" x 46"		13.91	2.300		355	50.50		405.50	475	
	2384	Curb mount 22" x 46"		15.23	2.101		177	46		223	273	
	2385	30" x 30"		16	2		171	44		215	262	

08700 | Hardware

	08710 | Door Hardware		CREW	DAILY OUTPUT	LABOR-HOURS	UNIT	2005 BARE COSTS MAT.	LABOR	EQUIP.	TOTAL	TOTAL INCL O&P	
	0010	**AVERAGE** Percentage for hardware, total job cost, minimum									.75%	150
	0050	Maximum									3.50%	
	0500	Total hardware for building, average distribution					85%	15%				
	1000	Door hardware, apartment, interior				Door	120			120	132	
	2100	Pocket door				Ea.	120			120	132	
	4000	Door knocker, bright brass	1 Carp	32	.250		40.50	6		46.50	54.50	
	4100	Mail slot, bright brass, 2" x 11"	"	25	.320		53.50	7.70		61.20	72	
	4200	Peep hole, add to price of door					14.80			14.80	16.30	
	0010	**DOORSTOPS** Holder and bumper, floor or wall	1 Carp	32	.250	Ea.	29.50	6		35.50	42.50	340
	1300	Wall bumper, 4" diameter, with rubber pad, aluminum		32	.250		8.90	6		14.90	19.95	
	1600	Door bumper, floor type, aluminum		32	.250		4.58	6		10.58	15.25	
	1900	Plunger type, door mounted		32	.250		24.50	6		30.50	37	
	0010	**ENTRANCE LOCKS** Cylinder, grip handle, deadlocking latch	1 Carp	9	.889	Ea.	113	21.50		134.50	161	400
	0020	Deadbolt		8	1		138	24		162	193	
	0100	Push and pull plate, dead bolt		8	1		131	24		155	185	
	0900	For handicapped lever, add					143			143	158	
	0010	**HINGES** Full mortise, avg. freq., steel base, 4-1/2" x 4-1/2", USP R08700-100				Pr.	19.80			19.80	22	520
	0100	5" x 5", USP					32.50			32.50	36	

DOORS & WINDOWS 8

08710 | Door Hardware

		08710 Door Hardware	CREW	DAILY OUTPUT	LABOR-HOURS	UNIT	2005 BARE COSTS MAT.	LABOR	EQUIP.	TOTAL	TOTAL INCL O&P
520	0200	6" x 6", USP (R08700-100)				Pr.	69.50			69.50	76.50
	0400	Brass base, 4-1/2" x 4-1/2", US10					40.50			40.50	45
	0500	5" x 5", US10					58.50			58.50	64.50
	0600	6" x 6", US10					99.50			99.50	110
	0800	Stainless steel base, 4-1/2" x 4-1/2", US32					62.50			62.50	69
	0900	For non removable pin, add				Ea.	2.27			2.27	2.50
	0910	For floating pin, driven tips, add					2.60			2.60	2.86
	0930	For hospital type tip on pin, add					11.25			11.25	12.35
	0940	For steeple type tip on pin, add					9.85			9.85	10.80
	0950	Full mortise, high frequency, steel base, 3-1/2" x 3-1/2", US26D				Pr.	18.45			18.45	20.50
	1000	4-1/2" x 4-1/2", USP					47.50			47.50	52
	1100	5" x 5", USP					45.50			45.50	50
	1200	6" x 6", USP					109			109	120
	1400	Brass base, 3-1/2" x 3-1/2", US4					37.50			37.50	41.50
	1430	4-1/2" x 4-1/2", US10					65			65	71.50
	1500	5" x 5", US10					96			96	106
	1600	6" x 6", US10					139			139	153
	1800	Stainless steel base, 4-1/2" x 4-1/2", US32					104			104	114
	1810	Stainless steel base, 5" x 4-1/2", US32					147			147	162
	1930	For hospital type tip on pin, add				Ea.	6.70			6.70	7.35
	1950	Full mortise, low frequency, steel base, 3-1/2" x 3-1/2", US26D				Pr.	9.65			9.65	10.65
	2000	4-1/2" x 4-1/2", USP					9			9	9.90
	2100	5" x 5", USP					24			24	26.50
	2200	6" x 6", USP					48.50			48.50	53.50
	2300	4-1/2" x 4-1/2", US3					14.25			14.25	15.70
	2310	5" x 5", US3					35			35	38.50
	2400	Brass bass, 4-1/2" x 4-1/2", US10					34			34	37.50
	2500	5" x 5", US10					52			52	57
	2800	Stainless steel base, 4-1/2" x 4-1/2", US32					58.50			58.50	64.50
550	0010	**KICK PLATE** 6" high, for 3' door, stainless steel	1 Carp	15	.533	Ea.	26	12.80		38.80	50.50
	0500	Bronze	"	15	.533	"	33	12.80		45.80	57.50
650	0010	**LOCKSET** Standard duty, cylindrical, with sectional trim									
	0020	Non-keyed, passage	1 Carp	12	.667	Ea.	40.50	16		56.50	72
	0100	Privacy		12	.667		50.50	16		66.50	82.50
	0400	Keyed, single cylinder function		10	.800		70.50	19.20		89.70	110
	0500	Lever handled, keyed, single cylinder function		10	.800		124	19.20		143.20	170
	1700	Residential, interior door, minimum		16	.500		14.05	12		26.05	36
	1720	Maximum		8	1		37.50	24		61.50	82
	1800	Exterior, minimum		14	.571		31.50	13.70		45.20	58
	1810	Average		8	1		65	24		89	113
	1820	Maximum		8	1		132	24		156	186

08720 | Weatherstripping & Seals

		08720 Weatherstripping & Seals	CREW	DAILY OUTPUT	LABOR-HOURS	UNIT	MAT.	LABOR	EQUIP.	TOTAL	TOTAL INCL O&P
300	0010	**WEATHERSTRIPPING** Window, double hung, 3' x 5', zinc	1 Carp	7.20	1.111	Opng.	11.35	26.50		37.85	58
	0100	Bronze		7.20	1.111		22.50	26.50		49	70
	0200	Vinyl V strip		7	1.143		3.76	27.50		31.26	50.50
	0500	As above but heavy duty, zinc		4.60	1.739		14.50	41.50		56	87
	0600	Bronze		4.60	1.739		25	41.50		66.50	99
	1000	Doors, wood frame, interlocking, for 3' x 7' door, zinc		3	2.667		13	64		77	123
	1100	Bronze		3	2.667		20.50	64		84.50	132
	1300	6' x 7' opening, zinc		2	4		14.20	96		110.20	179
	1400	Bronze		2	4		27	96		123	193
	1500	Vinyl V strip		6.40	1.250	Ea.	7.15	30		37.15	59
	1700	Wood frame, spring type, bronze									
	1800	3' x 7' door	1 Carp	7.60	1.053	Opng.	16.10	25.50		41.60	60.50

08700 | Hardware

Line	08720 Weatherstripping & Seals	Crew	Daily Output	Labor-Hours	Unit	2005 Bare Costs Mat.	Labor	Equip.	Total	Total Incl O&P	
1900	6′ x 7′ door	1 Carp	7	1.143	Opng.	17.60	27.50		45.10	66	300
1920	Felt, 3′ x 7′ door		14	.571		2.05	13.70		15.75	26	
1930	6′ x 7′ door		13	.615		2.21	14.75		16.96	27.50	
1950	Rubber, 3′ x 7′ door		7.60	1.053		4.43	25.50		29.93	48	
1960	6′ x 7′ door		7	1.143		5.05	27.50		32.55	52	
2200	Metal frame, spring type, bronze										
2300	3′ x 7′ door	1 Carp	3	2.667	Opng.	27.50	64		91.50	139	
2400	6′ x 7′ door	″	2.50	3.200	″	38.50	77		115.50	173	
2500	For stainless steel, spring type, add					133%					
2700	Metal frame, extruded sections, 3′ x 7′ door, aluminum	1 Carp	2	4	Opng.	37.50	96		133.50	205	
2800	Bronze		2	4		95.50	96		191.50	268	
3100	6′ x 7′ door, aluminum		1.20	6.667		48	160		208	325	
3200	Bronze		1.20	6.667		112	160		272	395	
3500	Threshold weatherstripping										
3650	Door sweep, flush mounted, aluminum	1 Carp	25	.320	Ea.	11.15	7.70		18.85	25.50	
3700	Vinyl		25	.320	″	13.20	7.70		20.90	27.50	
4000	Astragal for double doors, aluminum		4	2	Opng.	18.20	48		66.20	102	
4100	Bronze		4	2	″	29.50	48		77.50	114	
5000	Garage door bottom weatherstrip, 12′ aluminum, clear		14	.571	Ea.	17.85	13.70		31.55	43	
5010	Bronze		14	.571		68	13.70		81.70	98.50	
5050	Bottom protection, 12′ aluminum, clear		14	.571		21	13.70		34.70	46.50	
5100	Bronze		14	.571		84.50	13.70		98.20	117	
0010	**THRESHOLD** 3′ long door saddles, aluminum	1 Carp	48	.167	L.F.	3.43	4		7.43	10.55	800
0100	Aluminum, 8″ wide, 1/2″ thick		12	.667	Ea.	29	16		45	59	
0500	Bronze		60	.133	L.F.	28.50	3.20		31.70	37	
0600	Bronze, panic threshold, 5″ wide, 1/2″ thick		12	.667	Ea.	57	16		73	89.50	
0700	Rubber, 1/2″ thick, 5-1/2″ wide		20	.400		31.50	9.60		41.10	51	
0800	2-3/4″ wide		20	.400		14.05	9.60		23.65	32	
	08750 Window Hardware										
0010	**WINDOW HARDWARE**										400
1000	Handles, surface mounted, aluminum	1 Carp	24	.333	Ea.	1.90	8		9.90	15.70	
1020	Brass		24	.333		2.20	8		10.20	16	
1040	Chrome		24	.333		2.02	8		10.02	15.80	
1500	Recessed, aluminum		12	.667		1.10	16		17.10	28	
1520	Brass		12	.667		1.22	16		17.22	28.50	
1540	Chrome		12	.667		1.15	16		17.15	28.50	
2000	Latches, aluminum		20	.400		1.58	9.60		11.18	18.05	
2020	Brass		20	.400		1.90	9.60		11.50	18.40	
2040	Chrome		20	.400		1.78	9.60		11.38	18.25	
	08770 Door/Window Accessories										
0010	**DETECTION SYSTEMS** See division 13851										550
0010	**DOOR ACCESSORIES**										560
1000	Knockers, brass, standard	1 Carp	16	.500	Ea.	37	12		49	61	
1100	Deluxe		10	.800		115	19.20		134.20	159	
4000	Security chain, standard		18	.444		6.35	10.65		17	25	
4100	Deluxe		18	.444		38	10.65		48.65	60	

	08810 | Glass		CREW	DAILY OUTPUT	LABOR-HOURS	UNIT	2005 BARE COSTS MAT.	LABOR	EQUIP.	TOTAL	TOTAL INCL O&P
260	0010	**FLOAT GLASS** 3/16" thick, clear, plain	2 Glaz	130	.123	S.F.	4.19	2.89		7.08	9.40
	0200	Tempered, clear		130	.123		4.96	2.89		7.85	10.20
	0300	Tinted		130	.123		6.25	2.89		9.14	11.65
	0600	1/4" thick, clear, plain		120	.133		4.98	3.13		8.11	10.65
	0700	Tinted		120	.133		4.95	3.13		8.08	10.60
	0800	Tempered, clear		120	.133		6.15	3.13		9.28	11.95
	0900	Tinted		120	.133		8.40	3.13		11.53	14.40
	1600	3/8" thick, clear, plain		75	.213		7.95	5		12.95	16.95
	1700	Tinted		75	.213		9.90	5		14.90	19.15
	1800	Tempered, clear		75	.213		12.40	5		17.40	22
	1900	Tinted		75	.213		15.15	5		20.15	25
	2200	1/2" thick, clear, plain		55	.291		15.85	6.80		22.65	28.50
	2300	Tinted		55	.291		17.35	6.80		24.15	30.50
	2400	Tempered, clear		55	.291		18.65	6.80		25.45	32
	2500	Tinted		55	.291		23	6.80		29.80	36.50
	2800	5/8" thick, clear, plain		45	.356		17.35	8.35		25.70	33
	2900	Tempered, clear	↓	45	.356		19.80	8.35		28.15	36
	8900	For low emissivity coating for 3/16" and 1/4" only, add to above				↓	15%				
300	0010	**GLAZING VARIABLES**									
	0600	For glass replacement, add				S.F.		100%			
	0700	For gasket settings, add				L.F.	3.71			3.71	4.08
	0900	For sloped glazing, add				S.F.		25%			
	2000	Fabrication, polished edges, 1/4" thick				Inch	.32			.32	.35
	2100	1/2" thick					.81			.81	.89
	2500	Mitered edges, 1/4" thick					.81			.81	.89
	2600	1/2" thick				↓	1.31			1.31	1.44
460	0010	**INSULATING GLASS** 2 lites 1/8" float, 1/2" thk, under 15 S.F.									
	0100	Tinted	2 Glaz	95	.168	S.F.	10.50	3.95		14.45	18.10
	0280	Double glazed, 5/8" thk unit, 3/16" float, 15-30 S.F., clear		90	.178		8.65	4.17		12.82	16.45
	0400	1" thk, dbl. glazed, 1/4" float, 30-70 S.F., clear		75	.213		12.05	5		17.05	21.50
	0500	Tinted		75	.213		14.85	5		19.85	24.50
	2000	Both lites, light & heat reflective		85	.188		19.80	4.41		24.21	29.50
	2500	Heat reflective, film inside, 1" thick unit, clear		85	.188		17.35	4.41		21.76	26.50
	2600	Tinted		85	.188		18.70	4.41		23.11	28
	3000	Film on weatherside, clear, 1/2" thick unit		95	.168		12.40	3.95		16.35	20
	3100	5/8" thick unit		90	.178		15.10	4.17		19.27	23.50
	3200	1" thick unit	↓	85	.188	↓	17.05	4.41		21.46	26
850	0010	**WINDOW GLASS** Clear float, stops, putty bed, 1/8" thick	2 Glaz	480	.033	S.F.	3.44	.78		4.22	5.05
	0500	3/16" thick, clear		480	.033		4.23	.78		5.01	5.95
	0600	Tinted		480	.033		4.77	.78		5.55	6.55
	0700	Tempered	↓	480	.033	↓	5.75	.78		6.53	7.65

	08830 | Mirrors		CREW	DAILY OUTPUT	LABOR-HOURS	UNIT	MAT.	LABOR	EQUIP.	TOTAL	TOTAL INCL O&P
100	0010	**MIRRORS** No frames, wall type, 1/4" plate glass, polished edge									
	0100	Up to 5 S.F.	2 Glaz	125	.128	S.F.	6.70	3		9.70	12.30
	0200	Over 5 S.F.		160	.100		6.45	2.35		8.80	11
	0500	Door type, 1/4" plate glass, up to 12 S.F.		160	.100		6.95	2.35		9.30	11.55
	1000	Float glass, up to 10 S.F., 1/8" thick		160	.100		3.91	2.35		6.26	8.20
	1100	3/16" thick		150	.107		4.54	2.50		7.04	9.10
	1500	12" x 12" wall tiles, square edge, clear		195	.082		1.59	1.92		3.51	4.93
	1600	Veined		195	.082		4.26	1.92		6.18	7.85
	2010	Bathroom, unframed, laminated	↓	160	.100	↓	11.10	2.35		13.45	16.15

For information about Means Estimating Seminars, see yellow pages 12 and 13 in back of book

Division 9 Finishes

stimating Tips

eneral

Room Finish Schedule: A complete set of plans should contain a room finish schedule. If one is not available, it would be well worth the time and effort to put one together. A room finish schedule should contain the room number, room name (for clarity), floor materials, base materials, wainscot materials, wainscot height, wall materials (for each wall), ceiling materials, ceiling height and special instructions.

Surplus Finishes: Review the specifications to determine if there is any requirement to provide certain amounts of extra materials for the owner's maintenance department. In some cases the owner may require a substantial amount of materials, especially when it is a special order item or long lead time item.

9200 Plaster & Gypsum Board

Lath is estimated by the square yard for both gypsum and metal lath, plus usually 5% allowance for waste. Furring, channels and accessories are measured by the linear foot. An extra foot should be allowed for each accessory miter or stop.

Plaster is also estimated by the square yard. Deductions for openings vary by preference, from zero deduction to 50% of all openings over 2 feet in width. Some estimators deduct a percentage of the total yardage for openings. The estimator should allow one extra square foot for each linear foot of horizontal interior or exterior angle located below the ceiling level. Also, double the areas of small radius work.

Each room should be measured, perimeter times maximum wall height. Floors and ceiling areas are equal to length times width.

Drywall accessories, studs, track, and acoustical caulking are all measured by the linear foot. Drywall taping is figured by the square foot. Gypsum wallboard is estimated by the square foot. No material deductions should be made for door or window openings under 32 S.F. Coreboard can be obtained in a 1″ thickness for solid wall and shaft work. Additions should be made to price out the inside or outside corners.

- Different types of partition construction should be listed separately on the quantity sheets. There may be walls with studs of various widths, double studded, and similar or dissimilar surface materials. Shaft work is usually different construction from surrounding partitions requiring separate quantities and pricing of the work.

09300 Tile
09400 Terrazzo

- Tile and terrazzo areas are taken off on a square foot basis. Trim and base materials are measured by the linear foot. Accent tiles are listed per each. Two basic methods of installation are used. Mud set is approximately 30% more expensive than the thin set. In terrazzo work, be sure to include the linear footage of embedded decorative strips, grounds, machine rubbing and power cleanup.

09600 Flooring

- Wood flooring is available in strip, parquet, or block configuration. The latter two types are set in adhesives with quantities estimated by the square foot. The laying pattern will influence labor costs and material waste. In addition to the material and labor for laying wood floors, the estimator must make allowances for sanding and finishing these areas unless the flooring is prefinished.
- Most of the various types of flooring are all measured on a square foot basis. Base is measured by the linear foot. If adhesive materials are to be quantified, they are estimated at a specified coverage rate by the gallon depending upon the specified type and the manufacturer's recommendations.
- Sheet flooring is measured by the square yard. Roll widths vary, so consideration should be given to use the most economical width, as waste must be figured into the total quantity. Consider also the installation methods available, direct glue down or stretched.

09700 Wall Finishes

- Wall coverings are estimated by the square foot. The area to be covered is measured, length by height of wall above baseboards, to calculate the square footage of each wall. This figure is divided by the number of square feet in the single roll which is being used. Deduct, in full, the areas of openings such as doors and windows. Where a pattern match is required allow 25%-30% waste. One gallon of paste should be sufficient to hang 12 single rolls of light to medium weight paper.

09800 Acoustical Treatment

- Acoustical systems fall into several categories. The takeoff of these materials should be by the square foot of area with a 5% allowance for waste. Do not forget about scaffolding, if applicable, when estimating these systems.

09900 Paints & Coatings

- A major portion of the work in painting involves surface preparation. Be sure to include cleaning, sanding, filling and masking costs in the estimate.
- Painting is one area where bids vary to a greater extent than almost any other section of a project. This arises from the many methods of measuring surfaces to be painted. The estimator should check the plans and specifications carefully to be sure of the required number of coats.
- Protection of adjacent surfaces is not included in painting costs. When considering the method of paint application, an important factor is the amount of protection and masking required. These must be estimated separately and may be the determining factor in choosing the method of application.

Reference Numbers

Reference numbers are shown in bold squares at the beginning of some major classifications. These numbers refer to related items in the Reference Section. The reference information may be an estimating procedure, an alternate pricing method or technical information.

Note: Not all subdivisions listed here necessarily appear in this publication.

o part of this publication may be reproduced, stored in a retrieval system, or transmitted in any form r by any means without prior written permission of Reed Construction Data.

09050 | Basic Material Finishes and Methods

9 FINISHES

09060 | Selective Demolition

		09060 Selective Demolition	CREW	DAILY OUTPUT	LABOR-HOURS	UNIT	2005 BARE COSTS MAT.	LABOR	EQUIP.	TOTAL	TOTAL INCL O&P
110	0010	SELECTIVE DEMOLITION, CEILINGS R02220-510									
	0200	Ceiling, drywall, furred and nailed or screwed	2 Clab	800	.020	S.F.		.35		.35	.59
	1000	Plaster, lime and horse hair, on wood lath, incl. lath		700	.023			.40		.40	.67
	1200	Suspended ceiling, mineral fiber, 2'x2' or 2'x4'		1,500	.011			.19		.19	.31
	1250	On suspension system, incl. system		1,200	.013			.23		.23	.39
	1500	Tile, wood fiber, 12" x 12", glued		900	.018			.31		.31	.52
	1540	Stapled		1,500	.011			.19		.19	.31
	2000	Wood, tongue and groove, 1" x 4"		1,000	.016			.28		.28	.47
	2040	1" x 8"		1,100	.015			.25		.25	.43
	2400	Plywood or wood fiberboard, 4' x 8' sheets		1,200	.013			.23		.23	.39
120	0010	SELECTIVE DEMOLITION, FLOORING R02220-510									
	0200	Brick with mortar	2 Clab	475	.034	S.F.		.58		.58	.99
	0400	Carpet, bonded, including surface scraping		2,000	.008			.14		.14	.24
	0480	Tackless		9,000	.002			.03		.03	.05
	0800	Resilient, sheet goods		1,400	.011			.20		.20	.34
	0900	Vinyl composition tile, 12" x 12"		1,000	.016			.28		.28	.47
	2000	Tile, ceramic, thin set		675	.024			.41		.41	.70
	2020	Mud set		625	.026			.44		.44	.75
	3000	Wood, block, on end	1 Carp	400	.020			.48		.48	.82
	3200	Parquet		450	.018			.43		.43	.72
	3400	Strip flooring, interior, 2-1/4" x 25/32" thick		325	.025			.59		.59	1
	3500	Exterior, porch flooring, 1" x 4"		220	.036			.87		.87	1.48
	3800	Subfloor, tongue and groove, 1" x 6"		325	.025			.59		.59	1
	3820	1" x 8"		430	.019			.45		.45	.76
	3840	1" x 10"		520	.015			.37		.37	.63
	4000	Plywood, nailed		600	.013			.32		.32	.54
	4100	Glued and nailed		400	.020			.48		.48	.82
130	0010	SELECTIVE DEMOLITION, WALLS AND PARTITIONS R02220-510									
	1000	Drywall, nailed, or screwed	1 Clab	1,000	.008	S.F.		.14		.14	.24
	1500	Fiberboard, nailed		900	.009			.15		.15	.26
	1568	Plenum barrier, sheet lead		300	.027			.46		.46	.79
	2200	Metal or wood studs, finish 2 sides, fiberboard	B-1	520	.046			.83		.83	1.41
	2250	Lath and plaster		260	.092			1.66		1.66	2.82
	2300	Plasterboard (drywall)		520	.046			.83		.83	1.41
	2350	Plywood		450	.053			.96		.96	1.63
	3000	Plaster, lime and horsehair, on wood lath	1 Clab	400	.020			.35		.35	.59
	3020	On metal lath	"	335	.024			.41		.41	.70

09100 | Metal Support Assemblies

09110 | Non-Load Bearing Wall Framing

		09110 Non-Load Bearing Wall Framing	CREW	DAILY OUTPUT	LABOR-HOURS	UNIT	2005 BARE COSTS MAT.	LABOR	EQUIP.	TOTAL	TOTAL INCL O&P
100	0010	METAL STUDS AND TRACK									
	1600	Non-load bearing, galv, 8' high, 25 ga. 1-5/8" wide, 16" O.C.	1 Carp	619	.013	S.F.	.33	.31		.64	.89
	1610	24" O.C.		950	.008		.25	.20		.45	.61
	1620	2-1/2" wide, 16" O.C.		613	.013		.37	.31		.68	.93
	1630	24" O.C.		938	.009		.28	.20		.48	.65
	1640	3-5/8" wide, 16" O.C.		600	.013		.43	.32		.75	1.01
	1650	24" O.C.		925	.009		.32	.21		.53	.71
	1660	4" wide, 16" O.C.		594	.013		.55	.32		.87	1.16

09110 | Non-Load Bearing Wall Framing		CREW	DAILY OUTPUT	LABOR-HOURS	UNIT	2005 BARE COSTS MAT.	LABOR	EQUIP.	TOTAL	TOTAL INCL O&P	
1670	24" O.C.	1 Carp	925	.009	S.F.	.42	.21		.63	.81	100
1680	6" wide, 16" O.C.		588	.014		.65	.33		.98	1.27	
1690	24" O.C.		906	.009		.49	.21		.70	.90	
1700	20 ga. studs, 1-5/8" wide, 16" O.C.		494	.016		.62	.39		1.01	1.34	
1710	24" O.C.		763	.010		.46	.25		.71	.94	
1720	2-1/2" wide, 16" O.C.		488	.016		.70	.39		1.09	1.44	
1730	24" O.C.		750	.011		.53	.26		.79	1.01	
1740	3-5/8" wide, 16" O.C.		481	.017		.81	.40		1.21	1.57	
1750	24" O.C.		738	.011		.61	.26		.87	1.11	
1760	4" wide, 16" O.C.		475	.017		.91	.40		1.31	1.70	
1770	24" O.C.		738	.011		.69	.26		.95	1.20	
1780	6" wide, 16" O.C.		469	.017		1.15	.41		1.56	1.96	
1790	24" O.C.		725	.011		.86	.26		1.12	1.40	
2000	Non-load bearing, galv, 10' high, 25 ga. 1-5/8" wide, 16" O.C.		495	.016		.31	.39		.70	1	
2100	24" O.C.		760	.011		.23	.25		.48	.68	
2200	2-1/2" wide, 16" O.C.		490	.016		.35	.39		.74	1.05	
2250	24" O.C.		750	.011		.26	.26		.52	.71	
2300	3-5/8" wide, 16" O.C.		480	.017		.41	.40		.81	1.13	
2350	24" O.C.		740	.011		.30	.26		.56	.77	
2400	4" wide, 16" O.C.		475	.017		.52	.40		.92	1.27	
2450	24" O.C.		740	.011		.39	.26		.65	.87	
2500	6" wide, 16" O.C.		470	.017		.62	.41		1.03	1.37	
2550	24" O.C.		725	.011		.46	.26		.72	.95	
2600	20 ga. studs, 1-5/8" wide, 16" O.C.		395	.020		.58	.49		1.07	1.47	
2650	24" O.C.		610	.013		.43	.31		.74	1	
2700	2-1/2" wide, 16" O.C.		390	.021		.66	.49		1.15	1.57	
2750	24" O.C.		600	.013		.49	.32		.81	1.08	
2800	3-5/8" wide, 16" O.C.		385	.021		.77	.50		1.27	1.70	
2850	24" O.C.		590	.014		.57	.33		.90	1.17	
2900	4" wide, 16" O.C.		380	.021		.87	.51		1.38	1.81	
2950	24" O.C.		590	.014		.64	.33		.97	1.25	
3000	6" wide, 16" O.C.		375	.021		1.09	.51		1.60	2.07	
3050	24" O.C.		580	.014		.80	.33		1.13	1.44	
3060	Non-load bearing, galv, 12' high, 25 ga. 1-5/8" wide, 16" O.C.		413	.019		.29	.46		.75	1.11	
3070	24" O.C.		633	.013		.22	.30		.52	.76	
3080	2-1/2" wide, 16" O.C.		408	.020		.33	.47		.80	1.17	
3090	24" O.C.		625	.013		.24	.31		.55	.79	
3100	3-5/8" wide, 16" O.C.		400	.020		.39	.48		.87	1.25	
3110	24" O.C.		617	.013		.28	.31		.59	.84	
3120	4" wide, 16" O.C.		396	.020		.50	.48		.98	1.37	
3130	24" O.C.		617	.013		.37	.31		.68	.93	
3140	6" wide, 16" O.C.		392	.020		.59	.49		1.08	1.48	
3150	24" O.C.		604	.013		.43	.32		.75	1.02	
3160	20 ga. studs, 1-5/8" wide, 16" O.C.		329	.024		.56	.58		1.14	1.60	
3170	24" O.C.		508	.016		.41	.38		.79	1.09	
3180	2-1/2" wide, 16" O.C.		325	.025		.63	.59		1.22	1.70	
3190	24" O.C.		500	.016		.46	.38		.84	1.16	
3200	3-5/8" wide, 16" O.C.		321	.025		.74	.60		1.34	1.83	
3210	24" O.C.		492	.016		.54	.39		.93	1.25	
3220	4" wide, 16" O.C.		317	.025		.83	.61		1.44	1.94	
3230	24" O.C.		492	.016		.60	.39		.99	1.32	
3240	6" wide, 16" O.C.		313	.026		1.04	.61		1.65	2.18	
3250	24" O.C.	↓	483	.017	↓	.76	.40		1.16	1.50	
5000	Load bearing studs, see division 05410-400										

09100 | Metal Support Assemblies

	09120 | Ceiling Suspension		Crew	Daily Output	Labor-Hours	Unit	2005 Bare Costs Mat.	Labor	Equip.	Total	Total Incl O&P
100	0010	**CEILING SUSPENSION SYSTEMS** For gypsum board or plaster									
	8000	Suspended ceilings, including carriers									
	8200	1-1/2″ carriers, 24″ O.C. with:									
	8300	7/8″ channels, 16″ O.C.	1 Lath	165	.048	S.F.	.43	1.07		1.50	2.20
	8320	24″ O.C.		200	.040		.36	.88		1.24	1.83
	8400	1-5/8″ channels, 16″ O.C.		155	.052		.52	1.14		1.66	2.41
	8420	24″ O.C.		190	.042		.42	.93		1.35	1.96
	8600	2″ carriers, 24″ O.C. with:									
	8700	7/8″ channels, 16″ O.C.	1 Lath	155	.052	S.F.	.41	1.14		1.55	2.29
	8720	24″ O.C.		190	.042		.34	.93		1.27	1.88
	8800	1-5/8″ channels, 16″ O.C.		145	.055		.50	1.21		1.71	2.52
	8820	24″ O.C.		180	.044		.40	.98		1.38	2.03

	09130 | Acoustical Suspension		Crew	Daily Output	Labor-Hours	Unit	Mat.	Labor	Equip.	Total	Total Incl O&P
100	0010	**CEILING SUSPENSION SYSTEMS** For boards and tile									
	0050	Class A suspension system, 15/16″ T bar, 2′ x 4′ grid	1 Carp	800	.010	S.F.	.49	.24		.73	.95
	0300	2′ x 2′ grid	″	650	.012		.61	.30		.91	1.17
	0350	For 9/16″ grid, add					.14			.14	.15
	0360	For fire rated grid, add					.08			.08	.09
	0370	For colored grid, add					.17			.17	.19
	0400	Concealed Z bar suspension system, 12″ module	1 Carp	520	.015		.44	.37		.81	1.11
	0600	1-1/2″ carrier channels, 4′ O.C., add		470	.017		.13	.41		.54	.83
	0650	1-1/2″ x 3-1/2″ channels		470	.017		.32	.41		.73	1.04
	0700	Carrier channels for ceilings with									
	0900	recessed lighting fixtures, add	1 Carp	460	.017	S.F.	.23	.42		.65	.96
	5000	Wire hangers, #12 wire	″	300	.027	Ea.	.85	.64		1.49	2.03

9 FINISHES

09200 | Plaster & Gypsum Board

	09205 | Furring & Lathing		Crew	Daily Output	Labor-Hours	Unit	2005 Bare Costs Mat.	Labor	Equip.	Total	Total Incl O&P
530	0010	**FURRING** Beams & columns, 7/8″ galvanized channels,									
	0030	12″ O.C.	1 Lath	155	.052	S.F.	.24	1.14		1.38	2.11
	0050	16″ O.C.		170	.047		.20	1.04		1.24	1.90
	0070	24″ O.C.		185	.043		.13	.95		1.08	1.69
	0100	Ceilings, on steel, 7/8″ channels, galvanized, 12″ O.C.		210	.038		.22	.84		1.06	1.60
	0300	16″ O.C.		290	.028		.20	.61		.81	1.21
	0400	24″ O.C.		420	.019		.13	.42		.55	.83
	0600	1-5/8″ channels, galvanized, 12″ O.C.		190	.042		.32	.93		1.25	1.85
	0700	16″ O.C.		260	.031		.29	.68		.97	1.42
	0900	24″ O.C.		390	.021		.19	.45		.64	.94
	1000	Walls, 7/8″ channels, galvanized, 12″ O.C.		235	.034		.22	.75		.97	1.46
	1200	16″ O.C.		265	.030		.20	.66		.86	1.30
	1300	24″ O.C.		350	.023		.13	.50		.63	.97
	1500	1-5/8″ channels, galvanized, 12″ O.C.		210	.038		.32	.84		1.16	1.71
	1600	16″ O.C.		240	.033		.29	.73		1.02	1.51
	1800	24″ O.C.		305	.026		.19	.58		.77	1.15
540	0011	**GYPSUM LATH** Plain or perforated, nailed, 3/8″ thick	1 Lath	765	.010	S.F.	.45	.23		.68	.87
	0101	1/2″ thick, nailed		720	.011		.45	.24		.69	.90
	0301	Clipped to steel studs, 3/8″ thick		675	.012		.45	.26		.71	.92
	0401	1/2″ thick		630	.013		.45	.28		.73	.95

09200 | Plaster & Gypsum Board

09205 | Furring & Lathing

	Line	Description	Crew	Daily Output	Labor-Hours	Unit	2005 Bare Costs Mat.	Labor	Equip.	Total	Total Incl O&P	
	0601	Firestop gypsum base, to steel studs, 3/8″ thick	1 Lath	630	.013	S.F.	.39	.28		.67	.88	540
	0701	1/2″ thick		585	.014		.50	.30		.80	1.04	
	0901	Foil back, to steel studs, 3/8″ thick		675	.012		.44	.26		.70	.90	
	1001	1/2″ thick		630	.013		.46	.28		.74	.96	
	1501	For ceiling installations, add		1,950	.004			.09		.09	.15	
	1601	For columns and beams, add		1,550	.005			.11		.11	.18	
	0011	**METAL LATH**										560
	3601	2.5 lb. diamond painted, on wood framing, on walls	1 Lath	765	.010	S.F.	.29	.23		.52	.68	
	3701	On ceilings		675	.012		.29	.26		.55	.73	
	4201	3.4 lb. diamond painted, wired to steel framing, on walls		675	.012		.36	.26		.62	.82	
	4301	On ceilings		540	.015		.36	.33		.69	.93	
	5101	Rib lath, painted, wired to steel, on walls, 2.75 lb.		675	.012		.30	.26		.56	.75	
	5201	3.4 lb.		630	.013		.44	.28		.72	.93	
	5701	Suspended ceiling system, incl. 3.4 lb. diamond lath, painted		135	.059		1.32	1.30		2.62	3.57	
	5801	Galvanized		135	.059		1.36	1.30		2.66	3.62	
	0010	**ACCESSORIES, PLASTER** Casing bead, expanded flange, galvanized	1 Lath	2.70	2.963	C.L.F.	38	65		103	148	700
	0900	Channels, cold rolled, 16 ga., 3/4″ deep, galvanized					22			22	24.50	
	1620	Corner bead, expanded bullnose, 3/4″ radius, #10, galvanized	1 Lath	2.60	3.077		28	67.50		95.50	141	
	1650	#1, galvanized		2.55	3.137		47.50	69		116.50	164	
	1670	Expanded wing, 2-3/4″ wide, galv. #1		2.65	3.019		30	66.50		96.50	141	
	1700	Inside corner, (corner rite) 3″ x 3″, painted		2.60	3.077		25.50	67.50		93	138	
	1750	Strip-ex, 4″ wide, painted		2.55	3.137		22	69		91	137	
	1800	Expansion joint, 3/4″ grounds, limited expansion, galv., 1 piece		2.70	2.963		92.50	65		157.50	207	
	2100	Extreme expansion, galvanized, 2 piece		2.60	3.077		162	67.50		229.50	288	

09210 | Gypsum Plaster

	Line	Description	Crew	Daily Output	Labor-Hours	Unit	Mat.	Labor	Equip.	Total	Total Incl O&P	
	0010	**GYPSUM PLASTER** 80# bag, less than 1 ton				Bag	14.05			14.05	15.45	100
	0302	2 coats, no lath included, on walls	J-1	750	.053	S.F.	.35	1.11	.14	1.60	2.37	
	0402	On ceilings		660	.061		.35	1.26	.16	1.77	2.64	
	0903	3 coats, no lath included, on walls		620	.065		.50	1.34	.17	2.01	2.96	
	1002	On ceilings		560	.071		.81	1.48	.18	2.47	3.55	
	1600	For irregular or curved surfaces, add						30%				
	1800	For columns & beams, add						50%				
	0010	**PERLITE OR VERMICULITE PLASTER** 100 lb. bags Under 200 bags				Bag	13.75			13.75	15.10	500
	0301	2 coats, no lath included, on walls	J-1	830	.048	S.F.	.36	1	.12	1.48	2.20	
	0401	On ceilings		710	.056		.36	1.17	.15	1.68	2.50	
	0901	3 coats, no lath included, on walls		665	.060		.63	1.25	.16	2.04	2.94	
	1001	On ceilings		565	.071		.63	1.47	.18	2.28	3.33	
	1700	For irregular or curved surfaces, add to above				S.Y.		30%				
	1800	For columns and beams, add to above						50%				
	1900	For soffits, add to ceiling prices						40%				
	0010	**THIN COAT** Plaster, 1 coat veneer, not incl. lath	J-1	3,600	.011	S.F.	.07	.23	.03	.33	.49	900
	1000	In 50 lb. bags				Bag	9.65			9.65	10.65	

09220 | Portland Cement Plaster

	Line	Description	Crew	Daily Output	Labor-Hours	Unit	Mat.	Labor	Equip.	Total	Total Incl O&P	
	0011	**STUCCO** 3 coats 1″ thick, float finish, with mesh, on wood frame	J-2	470	.102	S.F.	.54	2.14	.22	2.90	4.37	200
	0101	On masonry construction	J-1	495	.081		.21	1.68	.21	2.10	3.25	
	0151	2 coats, 3/4″ thick, float finish, no lath incl.	″	980	.041		.22	.85	.11	1.18	1.77	
	0301	For trowel finish, add	1 Plas	1,530	.005			.11		.11	.19	
	0600	For coloring and special finish, add, minimum	J-1	685	.058	S.Y.	.36	1.21	.15	1.72	2.58	
	0700	Maximum		200	.200	″	1.26	4.15	.52	5.93	8.85	
	1001	Exterior stucco, with bonding agent, 3 coats, on walls		1,800	.022	S.F.	.33	.46	.06	.85	1.19	
	1201	Ceilings		1,620	.025		.33	.51	.06	.90	1.28	

	09220 | Portland Cement Plaster		CREW	DAILY OUTPUT	LABOR-HOURS	UNIT	2005 BARE COSTS MAT.	LABOR	EQUIP.	TOTAL	TOTAL INCL O&P
200	1301	Beams	J-1	720	.056	S.F.	.33	1.15	.14	1.62	2.44
	1501	Columns		900	.044		.33	.92	.11	1.36	2.02
	1601	Mesh, painted, nailed to wood, 1.8 lb.	1 Lath	540	.015		.46	.33		.79	1.04
	1801	3.6 lb.		495	.016		.37	.36		.73	.98
	1901	Wired to steel, painted, 1.8 lb.		477	.017		.46	.37		.83	1.11
	2101	3.6 lb.		450	.018		.37	.39		.76	1.03

	09250 | Gypsum Board		CREW	DAILY OUTPUT	LABOR-HOURS	UNIT	MAT.	LABOR	EQUIP.	TOTAL	TOTAL INCL O&P
200	0010	**CEMENTITIOUS BACKERBOARD**									
	0070	Cementitious backerboard, on floor, 3′ x 4′x 1/2″ sheets	2 Carp	525	.030	S.F.	1.04	.73		1.77	2.39
	0080	3′ x 5′ x 1/2″ sheets		525	.030		1.04	.73		1.77	2.39
	0090	3′ x 6′ x 1/2″ sheets		525	.030		.76	.73		1.49	2.07
	0100	3′ x 4′x 5/8″ sheets		525	.030		1.08	.73		1.81	2.43
	0110	3′ x 5′ x 5/8″ sheets		525	.030		1.06	.73		1.79	2.41
	0120	3′ x 6′ x 5/8″ sheets		525	.030		1.07	.73		1.80	2.41
	0150	On wall, 3′ x 4′x 1/2″ sheets		350	.046		1.04	1.10		2.14	3.01
	0160	3′ x 5′ x 1/2″ sheets		350	.046		1.04	1.10		2.14	3.01
	0170	3′ x 6′ x 1/2″ sheets		350	.046		.76	1.10		1.86	2.69
	0180	3′ x 4′x 5/8″ sheets		350	.046		1.08	1.10		2.18	3.05
	0190	3′ x 5′ x 5/8″ sheets		350	.046		1.06	1.10		2.16	3.03
	0200	3′ x 6′ x 5/8″ sheets		350	.046		1.07	1.10		2.17	3.03
	0250	On counter, 3′ x 4′x 1/2″ sheets		180	.089		1.04	2.13		3.17	4.77
	0260	3′ x 5′ x 1/2″ sheets		180	.089		1.04	2.13		3.17	4.77
	0270	3′ x 6′ x 1/2″ sheets		180	.089		.76	2.13		2.89	4.45
	0300	3′ x 4′x 5/8″ sheets		180	.089		1.08	2.13		3.21	4.81
	0310	3′ x 5′ x 5/8″ sheets		180	.089		1.06	2.13		3.19	4.79
	0320	3′ x 6′ x 5/8″ sheets		180	.089		1.07	2.13		3.20	4.79
300	0010	**BLUEBOARD** For use with thin coat									
	0100	plaster application (see division 09210-900)									
	1000	3/8″ thick, on walls or ceilings, standard, no finish included	2 Carp	1,900	.008	S.F.	.18	.20		.38	.54
	1100	With thin coat plaster finish		875	.018		.25	.44		.69	1.03
	1400	On beams, columns, or soffits, standard, no finish included		675	.024		.21	.57		.78	1.20
	1450	With thin coat plaster finish		475	.034		.28	.81		1.09	1.68
	3000	1/2″ thick, on walls or ceilings, standard, no finish included		1,900	.008		.21	.20		.41	.57
	3100	With thin coat plaster finish		875	.018		.28	.44		.72	1.06
	3300	Fire resistant, no finish included		1,900	.008		.21	.20		.41	.57
	3400	With thin coat plaster finish		875	.018		.28	.44		.72	1.06
	3450	On beams, columns, or soffits, standard, no finish included		675	.024		.24	.57		.81	1.24
	3500	With thin coat plaster finish		475	.034		.31	.81		1.12	1.71
	3700	Fire resistant, no finish included		675	.024		.24	.57		.81	1.24
	3800	With thin coat plaster finish		475	.034		.31	.81		1.12	1.71
	5000	5/8″ thick, on walls or ceilings, fire resistant, no finish included		1,900	.008		.24	.20		.44	.60
	5100	With thin coat plaster finish		875	.018		.31	.44		.75	1.09
	5500	On beams, columns, or soffits, no finish included		675	.024		.28	.57		.85	1.27
	5600	With thin coat plaster finish		475	.034		.35	.81		1.16	1.75
	6000	For high ceilings, over 8′ high, add		3,060	.005			.13		.13	.21
	6500	For over 3 stories high, add per story		6,100	.003			.06		.06	.11
700	0010	**DRYWALL** Gypsum plasterboard, nailed or screwed (R09250-100)									
	0100	to studs unless otherwise noted									
	0150	3/8″ thick, on walls, standard, no finish included	2 Carp	2,000	.008	S.F.	.20	.19		.39	.55
	0200	On ceilings, standard, no finish included		1,800	.009		.20	.21		.41	.58
	0250	On beams, columns, or soffits, no finish included		675	.024		.20	.57		.77	1.19
	0300	1/2″ thick, on walls, standard, no finish included		2,000	.008		.22	.19		.41	.57
	0350	Taped and finished (level 4 finish)		965	.017		.25	.40		.65	.96
	0390	With compound skim coat (level 5 finish)		775	.021		.28	.50		.78	1.15

09250 | Gypsum Board

	09250 Gypsum Board	CREW	DAILY OUTPUT	LABOR-HOURS	UNIT	2005 BARE COSTS MAT.	LABOR	EQUIP.	TOTAL	TOTAL INCL O&P	
0400	Fire resistant, no finish included R09250-100	2 Carp	2,000	.008	S.F.	.22	.19		.41	.57	700
0450	Taped and finished (level 4 finish)		965	.017		.25	.40		.65	.96	
0490	With compound skim coat (level 5 finish)		775	.021		.28	.50		.78	1.15	
0500	Water resistant, no finish included		2,000	.008		.22	.19		.41	.57	
0550	Taped and finished (level 4 finish)		965	.017		.25	.40		.65	.96	
0590	With compound skim coat (level 5 finish)		775	.021		.28	.50		.78	1.15	
0600	Prefinished, vinyl, clipped to studs		900	.018		.56	.43		.99	1.34	
1000	On ceilings, standard, no finish included		1,800	.009		.22	.21		.43	.60	
1050	Taped and finished (level 4 finish)		765	.021		.25	.50		.75	1.13	
1090	With compound skim coat (level 5 finish)		610	.026		.28	.63		.91	1.38	
1100	Fire resistant, no finish included		1,800	.009		.22	.21		.43	.60	
1150	Taped and finished (level 4 finish)		765	.021		.25	.50		.75	1.13	
1195	With compound skim coat (level 5 finish)		610	.026		.28	.63		.91	1.38	
1200	Water resistant, no finish included		1,800	.009		.22	.21		.43	.60	
1250	Taped and finished (level 4 finish)		765	.021		.25	.50		.75	1.13	
1290	With compound skim coat (level 5 finish)		610	.026		.28	.63		.91	1.38	
1500	On beams, columns, or soffits, standard, no finish included		675	.024		.25	.57		.82	1.25	
1550	Taped and finished (level 4 finish)		475	.034		.25	.81		1.06	1.65	
1590	With compound skim coat (level 5 finish)		540	.030		.28	.71		.99	1.52	
1600	Fire resistant, no finish included		675	.024		.25	.57		.82	1.25	
1650	Taped and finished (level 4 finish)		475	.034		.25	.81		1.06	1.65	
1690	With compound skim coat (level 5 finish)		540	.030		.28	.71		.99	1.52	
1700	Water resistant, no finish included		675	.024		.25	.57		.82	1.25	
1750	Taped and finished (level 4 finish)		475	.034		.25	.81		1.06	1.65	
1790	With compound skim coat (level 5 finish)		540	.030		.28	.71		.99	1.52	
2000	5/8" thick, on walls, standard, no finish included		2,000	.008		.24	.19		.43	.59	
2050	Taped and finished (level 4 finish)		965	.017		.27	.40		.67	.98	
2090	With compound skim coat (level 5 finish)		775	.021		.30	.50		.80	1.17	
2100	Fire resistant, no finish included		2,000	.008		.25	.19		.44	.61	
2150	Taped and finished (level 4 finish)		965	.017		.28	.40		.68	.99	
2195	With compound skim coat (level 5 finish)		775	.021		.31	.50		.81	1.19	
2200	Water resistant, no finish included		2,000	.008		.26	.19		.45	.62	
2250	Taped and finished (level 4 finish)		965	.017		.29	.40		.69	1	
2290	With compound skim coat (level 5 finish)		775	.021		.32	.50		.82	1.20	
2300	Prefinished, vinyl, clipped to studs		900	.018		.65	.43		1.08	1.44	
3000	On ceilings, standard, no finish included		1,800	.009		.24	.21		.45	.62	
3050	Taped and finished (level 4 finish)		765	.021		.27	.50		.77	1.15	
3090	With compound skim coat (level 5 finish)		615	.026		.30	.62		.92	1.39	
3100	Fire resistant, no finish included		1,800	.009		.25	.21		.46	.64	
3150	Taped and finished (level 4 finish)		765	.021		.28	.50		.78	1.16	
3190	With compound skim coat (level 5 finish)		615	.026		.31	.62		.93	1.41	
3200	Water resistant, no finish included		1,800	.009		.26	.21		.47	.65	
3250	Taped and finished (level 4 finish)		765	.021		.29	.50		.79	1.17	
3290	With compound skim coat (level 5 finish)		615	.026		.32	.62		.94	1.42	
3500	On beams, columns, or soffits, no finish included		675	.024		.28	.57		.85	1.27	
3550	Taped and finished (level 4 finish)		475	.034		.31	.81		1.12	1.71	
3590	With compound skim coat (level 5 finish)		380	.042		.35	1.01		1.36	2.11	
3600	Fire resistant, no finish included		675	.024		.29	.57		.86	1.29	
3650	Taped and finished (level 4 finish)		475	.034		.32	.81		1.13	1.72	
3690	With compound skim coat (level 5 finish)		380	.042		.31	1.01		1.32	2.07	
3700	Water resistant, no finish included		675	.024		.30	.57		.87	1.30	
3750	Taped and finished (level 4 finish)		475	.034		.33	.81		1.14	1.74	
3790	With compound skim coat (level 5 finish)		380	.042		.32	1.01		1.33	2.08	
4000	Fireproofing, beams or columns, 2 layers, 1/2" thick, incl finish		330	.048		.47	1.16		1.63	2.50	
4050	5/8" thick		300	.053		.56	1.28		1.84	2.79	
4100	3 layers, 1/2" thick		225	.071		.69	1.71		2.40	3.66	

09200 | Plaster & Gypsum Board

	09250 | Gypsum Board		CREW	DAILY OUTPUT	LABOR-HOURS	UNIT	2005 BARE COSTS MAT.	LABOR	EQUIP.	TOTAL	TOTAL INCL O&P
700	4150	5/8″ thick R09250-100	2 Carp	210	.076	S.F.	.84	1.83		2.67	4.03
	5200	For work over 8′ high, add		3,060	.005			.13		.13	.21
	5270	For textured spray, add	2 Lath	1,600	.010		.04	.22		.26	.40
	5350	For finishing inner corners, add	2 Carp	950	.017	L.F.	.07	.40		.47	.76
	5355	For finishing outer corners, add	″	1,250	.013		.17	.31		.48	.70
	5500	For acoustical sealant, add per bead	1 Carp	500	.016		.03	.38		.41	.68
	5550	Sealant, 1 quart tube				Ea.	4.89			4.89	5.40
	5600	Sound deadening board, 1/4″ gypsum	2 Carp	1,800	.009	S.F.	.22	.21		.43	.60
	5650	1/2″ wood fiber	″	1,800	.009	″	.38	.21		.59	.78

	09270 | Drywall Accessories		CREW	DAILY OUTPUT	LABOR-HOURS	UNIT	MAT.	LABOR	EQUIP.	TOTAL	TOTAL INCL O&P
100	0011	**ACCESSORIES, DRYWALL** Casing bead, galvanized steel	1 Carp	290	.028	L.F.	.17	.66		.83	1.31
	0101	Vinyl		290	.028		.17	.66		.83	1.31
	0401	Corner bead, galvanized steel, 1-1/4″ x 1-1/4″		350	.023		.20	.55		.75	1.15
	0411	1-1/4″ x 1-1/4″, 10′ long		35	.229	Ea.	2	5.50		7.50	11.50
	0601	Vinyl corner bead		400	.020	L.F.	.21	.48		.69	1.05
	0901	Furring channel, galv. steel, 7/8″ deep, standard		260	.031		.23	.74		.97	1.51
	1001	Resilient		260	.031		.23	.74		.97	1.50
	1101	J trim, galvanized steel, 1/2″ wide		300	.027			.64		.64	1.09
	1121	5/8″ wide		300	.027		.16	.64		.80	1.27
	1160	Screws #6 x 1″ A				M	7.15			7.15	7.85
	1170	#6 x 1-5/8″ A				″	9.35			9.35	10.30
	1501	Z stud, galvanized steel, 1-1/2″ wide	1 Carp	260	.031	L.F.	.35	.74		1.09	1.63

	09280 | Gypsum Wallboard Repairs		CREW	DAILY OUTPUT	LABOR-HOURS	UNIT	MAT.	LABOR	EQUIP.	TOTAL	TOTAL INCL O&P
100	0010	**GYPSUM WALLBOARD REPAIRS**									
	0100	Fill and sand, pin / nail holes	1 Carp	960	.008	Ea.		.20		.20	.34
	0110	Screw head pops		480	.017			.40		.40	.68
	0120	Dents, up to 2″ square		48	.167		.01	4		4.01	6.80
	0130	2″ to 4″ square		24	.333		.02	8		8.02	13.60
	0140	Cut square, patch, sand and finish, holes, up to 2″ square		12	.667		.02	16		16.02	27
	0150	2″ to 4″ square		11	.727		.06	17.45		17.51	29.50
	0160	4″ to 8″ square		10	.800		.16	19.20		19.36	32.50
	0170	8″ to 12″ square		8	1		.31	24		24.31	41.50

09300 | Tile

	09310 | Ceramic Tile		CREW	DAILY OUTPUT	LABOR-HOURS	UNIT	2005 BARE COSTS MAT.	LABOR	EQUIP.	TOTAL	TOTAL INCL O&P
100	0010	**CERAMIC TILE**									
	0050	Base, using 1′ x 4″ high pc. with 1″ x 1″ tiles, mud set	D-7	82	.195	L.F.	4.07	3.97		8.04	10.90
	0100	Thin set	″	128	.125		3.87	2.54		6.41	8.35
	0300	For 6″ high base, 1″ x 1″ tile face, add					.64			.64	.70
	0400	For 2″ x 2″ tile face, add to above					.34			.34	.37
	0600	Cove base, 4-1/4″ x 4-1/4″ high, mud set	D-7	91	.176		3.12	3.57		6.69	9.20
	0700	Thin set		128	.125		3.14	2.54		5.68	7.55
	0900	6″ x 4-1/4″ high, mud set		100	.160		2.85	3.25		6.10	8.40
	1000	Thin set		137	.117		2.98	2.37		5.35	7.10
	1200	Sanitary cove base, 6″ x 4-1/4″ high, mud set		93	.172		3.60	3.50		7.10	9.60

09310 | Ceramic Tile			CREW	DAILY OUTPUT	LABOR-HOURS	UNIT	2005 BARE COSTS MAT.	LABOR	EQUIP.	TOTAL	TOTAL INCL O&P	
100	1300	Thin set	D-7	124	.129	L.F.	3.67	2.62		6.29	8.25	100
	1500	6" x 6" high, mud set		84	.190		3.38	3.87		7.25	9.95	
	1600	Thin set		117	.137		3.42	2.78		6.20	8.25	
	1800	Bathroom accessories, average		82	.195	Ea.	9.70	3.97		13.67	17.05	
	1900	Bathtub, 5', rec. 4-1/4" x 4-1/4" tile wainscot, adhesive set 6' high		2.90	5.517		140	112		252	335	
	2100	7' high wainscot		2.50	6.400		160	130		290	385	
	2200	8' high wainscot		2.20	7.273		170	148		318	425	
	2400	Bullnose trim, 4-1/4" x 4-1/4", mud set		82	.195	L.F.	2.69	3.97		6.66	9.35	
	2500	Thin set		128	.125		2.63	2.54		5.17	7	
	2700	6" x 4-1/4" bullnose trim, mud set		84	.190		2.05	3.87		5.92	8.50	
	2800	Thin set		124	.129		2.05	2.62		4.67	6.50	
	3000	Floors, natural clay, random or uniform, thin set, color group 1		183	.087	S.F.	3.69	1.78		5.47	6.90	
	3100	Color group 2		183	.087		3.98	1.78		5.76	7.25	
	3255	Glazed, thin set, 6" x 6", color group 1		200	.080		3.14	1.63		4.77	6.05	
	3260	8" x 8" tile		250	.064		3.14	1.30		4.44	5.55	
	3270	12" x 12" tile		325	.049		3.94	1		4.94	5.95	
	3280	16" x 16" tile		550	.029		4.95	.59		5.54	6.40	
	3285	Border, 6" x 12" tile		275	.058		10	1.18		11.18	12.90	
	3290	3" x 12" tile		200	.080		28.50	1.63		30.13	33.50	
	3300	Porcelain type, 1 color, color group 2, 1" x 1"		183	.087		4.27	1.78		6.05	7.55	
	3310	2" x 2" or 2" x 1", thin set		190	.084		4.07	1.71		5.78	7.25	
	3350	For random blend, 2 colors, add					.79			.79	.87	
	3360	4 colors, add					1.12			1.12	1.23	
	4300	Specialty tile, 4-1/4" x 4-1/4" x 1/2", decorator finish	D-7	183	.087		8.95	1.78		10.73	12.70	
	4500	Add for epoxy grout, 1/16" joint, 1" x 1" tile		800	.020		.55	.41		.96	1.26	
	4600	2" x 2" tile		820	.020		.50	.40		.90	1.19	
	4800	Pregrouted sheets, walls, 4-1/4" x 4-1/4", 6" x 4-1/4"										
	4810	and 8-1/2" x 4-1/4", 4 S.F. sheets, silicone grout	D-7	240	.067	S.F.	4.23	1.36		5.59	6.85	
	5100	Floors, unglazed, 2 S.F. sheets,										
	5110	urethane adhesive	D-7	180	.089	S.F.	4.21	1.81		6.02	7.55	
	5400	Walls, interior, thin set, 4-1/4" x 4-1/4" tile		190	.084		2.09	1.71		3.80	5.05	
	5500	6" x 4-1/4" tile		190	.084		2.35	1.71		4.06	5.35	
	5700	8-1/2" x 4-1/4" tile		190	.084		3.32	1.71		5.03	6.40	
	5800	6" x 6" tile		200	.080		2.67	1.63		4.30	5.55	
	5810	8" x 8" tile		225	.071		3.19	1.45		4.64	5.85	
	5820	12" x 12" tile		300	.053		3.04	1.08		4.12	5.10	
	5830	16" x 16" tile		500	.032		3.29	.65		3.94	4.67	
	6000	Decorated wall tile, 4-1/4" x 4-1/4", minimum		270	.059		3.50	1.20		4.70	5.80	
	6100	Maximum		180	.089		39	1.81		40.81	45.50	
	6600	Crystalline glazed, 4-1/4" x 4-1/4", mud set, plain		100	.160		3.28	3.25		6.53	8.85	
	6700	4-1/4" x 4-1/4", scored tile		100	.160		4.13	3.25		7.38	9.80	
	6900	6"x 6" plain		93	.172		3.31	3.50		6.81	9.30	
	7000	For epoxy grout, 1/16" joints, 4-1/4" tile, add		800	.020		.33	.41		.74	1.01	
	7200	For tile set in dry mortar, add		1,735	.009			.19		.19	.30	
	7300	For tile set in portland cement mortar, add		290	.055			1.12		1.12	1.81	

09330 | Quarry Tile			CREW	DAILY OUTPUT	LABOR-HOURS	UNIT	MAT.	LABOR	EQUIP.	TOTAL	TOTAL INCL O&P	
100	0010	**QUARRY TILE** Base, cove or sanitary, 2" or 5" high, mud set										100
	0100	1/2" thick	D-7	110	.145	L.F.	3.88	2.96		6.84	9.05	
	0300	Bullnose trim, red, mud set, 6" x 6" x 1/2" thick		120	.133		3.81	2.71		6.52	8.55	
	0400	4" x 4" x 1/2" thick		110	.145		4.30	2.96		7.26	9.50	
	0600	4" x 8" x 1/2" thick, using 8" as edge		130	.123		3.78	2.50		6.28	8.20	
	0700	Floors, mud set, 1,000 S.F. lots, red, 4" x 4" x 1/2" thick		120	.133	S.F.	3.89	2.71		6.60	8.65	
	0900	6" x 6" x 1/2" thick		140	.114		2.99	2.32		5.31	7.05	
	1000	4" x 8" x 1/2" thick		130	.123		3.89	2.50		6.39	8.30	

09300 | Tile

		09330 | Quarry Tile	CREW	DAILY OUTPUT	LABOR-HOURS	UNIT	2005 BARE COSTS MAT.	LABOR	EQUIP.	TOTAL	TOTAL INCL O&P
100	1300	For waxed coating, add				S.F.	.62			.62	.68
	1500	For colors other than green, add					.37			.37	.41
	1600	For abrasive surface, add					.44			.44	.48
	1800	Brown tile, imported, 6" x 6" x 3/4"	D-7	120	.133		4.62	2.71		7.33	9.45
	1900	8" x 8" x 1"		110	.145		5.25	2.96		8.21	10.50
	2100	For thin set mortar application, deduct		700	.023			.46		.46	.75
	2700	Stair tread, 6" x 6" x 3/4", plain		50	.320		4.24	6.50		10.74	15.10
	2800	Abrasive		47	.340		4.77	6.90		11.67	16.40
	3000	Wainscot, 6" x 6" x 1/2", thin set, red		105	.152		3.56	3.10		6.66	8.90
	3100	Colors other than green		105	.152	↓	3.97	3.10		7.07	9.35
	3300	Window sill, 6" wide, 3/4" thick		90	.178	L.F.	4.47	3.61		8.08	10.70
	3400	Corners	↓	80	.200	Ea.	5	4.07		9.07	12.05
		09370 | Metal Tile									
100	0010	**METAL TILE** 4' x 4' sheet, 24 ga., tile pattern, nailed									
	0200	Stainless steel	2 Carp	512	.031	S.F.	22	.75		22.75	26
	0400	Aluminized steel	"	512	.031	"	11.90	.75		12.65	14.35

09400 | Terrazzo

		09410 | Portland Cement Terrazzo	CREW	DAILY OUTPUT	LABOR-HOURS	UNIT	2005 BARE COSTS MAT.	LABOR	EQUIP.	TOTAL	TOTAL INCL O&P
100	0010	**PORTLAND CEMENT TERRAZZO**, cast-in-place									
	4300	Stone chips, onyx gemstone, per 50 lb. bag				Bag	12			12	13.20
		09420 | Precast Terrazzo									
900	0010	**TERRAZZO, PRECAST** Base, 6" high, straight	1 Mstz	35	.229	L.F.	9.25	5.20		14.45	18.60
	0100	Cove		30	.267		10.60	6.10		16.70	21.50
	0300	8" high base, straight		30	.267		9.45	6.10		15.55	20
	0400	Cove	↓	25	.320		13.90	7.30		21.20	27
	0600	For white cement, add					.38			.38	.42
	0700	For 16 ga. zinc toe strip, add					1.39			1.39	1.53
	0900	Curbs, 4" x 4" high	1 Mstz	19	.421		26.50	9.60		36.10	45
	1000	8" x 8" high	"	15	.533	↓	31	12.20		43.20	53.50
	1200	Floor tiles, non-slip, 1" thick, 12" x 12"	D-1	29	.552	S.F.	15.85	12		27.85	37.50
	1300	1-1/4" thick, 12" x 12"		29	.552		17.60	12		29.60	39.50
	1500	16" x 16"		23	.696		19.10	15.15		34.25	46
	1600	1-1/2" thick, 16" x 16"	↓	21	.762		17.50	16.60		34.10	47
	4800	Wainscot, 12" x 12" x 1" tiles	1 Mstz	12	.667		5.65	15.25		20.90	31
	4900	16" x 16" x 1-1/2" tiles	"	8	1	↓	12.30	23		35.30	50.50

	09510 | Acoustical Ceilings		CREW	DAILY OUTPUT	LABOR-HOURS	UNIT	2005 BARE COSTS MAT.	LABOR	EQUIP.	TOTAL	TOTAL INCL O&P
700	0010	**SUSPENDED ACOUSTIC CEILING TILES,** Not including									
	0100	suspension system									
	0300	Fiberglass boards, film faced, 2′ x 2′ or 2′ x 4′, 5/8″ thick	1 Carp	625	.013	S.F.	.51	.31		.82	1.08
	0400	3/4″ thick		600	.013		1.14	.32		1.46	1.79
	0500	3″ thick, thermal, R11		450	.018		1.26	.43		1.69	2.11
	0600	Glass cloth faced fiberglass, 3/4″ thick		500	.016		1.62	.38		2	2.43
	0700	1″ thick		485	.016		1.79	.40		2.19	2.64
	0820	1-1/2″ thick, nubby face		475	.017		2.23	.40		2.63	3.14
	1110	Mineral fiber tile, lay-in, 2′ x 2′ or 2′ x 4′, 5/8″ thick, fine texture		625	.013		.43	.31		.74	.99
	1115	Rough textured		625	.013		1.06	.31		1.37	1.69
	1125	3/4″ thick, fine textured		600	.013		1.17	.32		1.49	1.83
	1130	Rough textured		600	.013		1.47	.32		1.79	2.16
	1135	Fissured		600	.013		1.74	.32		2.06	2.45
	1150	Tegular, 5/8″ thick, fine textured		470	.017		1.04	.41		1.45	1.83
	1155	Rough textured		470	.017		1.35	.41		1.76	2.18
	1165	3/4″ thick, fine textured		450	.018		1.47	.43		1.90	2.34
	1170	Rough textured		450	.018		1.66	.43		2.09	2.55
	1175	Fissured	↓	450	.018		2.59	.43		3.02	3.57
	1180	For aluminum face, add					4.72			4.72	5.20
	1185	For plastic film face, add					.75			.75	.83
	1190	For fire rating, add					.35			.35	.39
	1300	Mirror faced panels, 15/16″ thick, 2′ x 2′	1 Carp	500	.016		9.80	.38		10.18	11.45
	1900	Eggcrate, acrylic, 1/2″ x 1/2″ x 1/2″ cubes		500	.016		1.42	.38		1.80	2.21
	2100	Polystyrene eggcrate, 3/8″ x 3/8″ x 1/2″ cubes		510	.016		1.19	.38		1.57	1.95
	2200	1/2″ x 1/2″ x 1/2″ cubes		500	.016		1.60	.38		1.98	2.41
	2400	Luminous panels, prismatic, acrylic		400	.020		1.73	.48		2.21	2.72
	2500	Polystyrene		400	.020		.88	.48		1.36	1.79
	2700	Flat white acrylic		400	.020		3.01	.48		3.49	4.13
	2800	Polystyrene		400	.020		2.06	.48		2.54	3.09
	3000	Drop pan, white, acrylic		400	.020		4.41	.48		4.89	5.65
	3100	Polystyrene		400	.020		3.69	.48		4.17	4.88
	3600	Perforated aluminum sheets, .024″ thick, corrugated, painted		490	.016		1.76	.39		2.15	2.61
	3700	Plain		500	.016		3.07	.38		3.45	4.03
	3750	Wood fiber in cementitious binder, 2′ x 2′ or 4′, painted, 1″ thick		600	.013		1.20	.32		1.52	1.86
	3760	2″ thick		550	.015		1.98	.35		2.33	2.77
	3770	2-1/2″ thick		500	.016		2.71	.38		3.09	3.63
	3780	3″ thick	↓	450	.018	↓	3.04	.43		3.47	4.06
760	0010	**SUSPENDED CEILINGS, COMPLETE** Including standard									
	0100	suspension system but not incl. 1-1/2″ carrier channels									
	0600	Fiberglass ceiling board, 2′ x 4′ x 5/8″, plain faced,	1 Carp	500	.016	S.F.	1	.38		1.38	1.75
	0700	Offices, 2′ x 4′ x 3/4″		380	.021		1.63	.51		2.14	2.65
	1800	Tile, Z bar suspension, 5/8″ mineral fiber tile		150	.053		1.55	1.28		2.83	3.88
	1900	3/4″ mineral fiber tile	↓	150	.053	↓	1.65	1.28		2.93	3.99
900	0010	**CEILING TILE,** Stapled or cemented									
	0100	12″ x 12″ or 12″ x 24″, not including furring									
	0600	Mineral fiber, vinyl coated, 5/8″ thick	1 Carp	1,000	.008	S.F.	1.23	.19		1.42	1.68
	0700	3/4″ thick		1,000	.008		1.35	.19		1.54	1.82
	0900	Fire rated, 3/4″ thick, plain faced		1,000	.008		1.17	.19		1.36	1.62
	1000	Plastic coated face		1,000	.008		1.28	.19		1.47	1.74
	1200	Aluminum faced, 5/8″ thick, plain	↓	1,000	.008		1.12	.19		1.31	1.56
	3300	For flameproofing, add					.09			.09	.10
	3400	For sculptured 3 dimensional, add					.25			.25	.28
	3900	For ceiling primer, add					.12			.12	.13
	4000	For ceiling cement, add				↓	.33			.33	.36

FINISHES 9

09620 | Specialty Flooring

			CREW	DAILY OUTPUT	LABOR-HOURS	UNIT	2005 BARE COSTS MAT.	LABOR	EQUIP.	TOTAL	TOTAL INCL O&P
100	0010	**ATHLETIC FLOORING**									
	3700	Polyethylene, in rolls, no base incl., landscape surfaces	1 Tilf	275	.029	S.F.	2.57	.67		3.24	3.90
	3800	Nylon action surface, 1/8" thick		275	.029		2.76	.67		3.43	4.11
	3900	1/4" thick		275	.029		3.98	.67		4.65	5.45
	4000	3/8" thick		275	.029		5	.67		5.67	6.55

09631 | Brick Flooring

			CREW	DAILY OUTPUT	LABOR-HOURS	UNIT	MAT.	LABOR	EQUIP.	TOTAL	TOTAL INCL O&P
100	0010	**BRICK FLOORING**									
	0020	Acid proof shales, red, 8" x 3-3/4" x 1-1/4" thick	D-7	.43	37.209	M	730	755		1,485	2,025
	0050	2-1/4" thick	D-1	.40	40		795	870		1,665	2,325
	0200	Acid proof clay brick, 8" x 3-3/4" x 2-1/4" thick	"	.40	40		755	870		1,625	2,275
	0260	Cast ceramic, pressed, 4" x 8" x 1/2", unglazed	D-7	100	.160	S.F.	4.94	3.25		8.19	10.70
	0270	Glazed		100	.160		6.60	3.25		9.85	12.50
	0280	Hand molded flooring, 4" x 8" x 3/4", unglazed		95	.168		6.55	3.42		9.97	12.70
	0290	Glazed		95	.168		8.20	3.42		11.62	14.50
	0300	8" hexagonal, 3/4" thick, unglazed		85	.188		7.15	3.83		10.98	14.05
	0310	Glazed		85	.188		12.95	3.83		16.78	20.50
	0450	Acid proof joints, 1/4" wide	D-1	65	.246		1.13	5.35		6.48	10.15
	0500	Pavers, 8" x 4", 1" to 1-1/4" thick, red	D-7	95	.168		2.88	3.42		6.30	8.65
	0510	Ironspot	"	95	.168		4.07	3.42		7.49	10
	0540	1-3/8" to 1-3/4" thick, red	D-1	95	.168		2.78	3.67		6.45	9.15
	0560	Ironspot		95	.168		4.02	3.67		7.69	10.50
	0580	2-1/4" thick, red		90	.178		2.83	3.87		6.70	9.55
	0590	Ironspot		90	.178		4.38	3.87		8.25	11.25
	0800	For sidewalks and patios with pavers, see division 02780-200									
	0870	For epoxy joints, add	D-1	600	.027	S.F.	2.15	.58		2.73	3.34
	0880	For Furan underlayment, add	"	600	.027		1.78	.58		2.36	2.93
	0890	For waxed surface, steam cleaned, add	D-5	1,000	.008		.15	.15		.30	.42

09635 | Marble Flooring

			CREW	DAILY OUTPUT	LABOR-HOURS	UNIT	MAT.	LABOR	EQUIP.	TOTAL	TOTAL INCL O&P
100	0010	**MARBLE** Thin gauge tile, 12" x 6", 3/8", White Carara	D-7	60	.267	S.F.	8.75	5.40		14.15	18.40
	0100	Travertine		60	.267		9.65	5.40		15.05	19.35
	0200	12" x 12" x 3/8", thin set, floors		60	.267		6.70	5.40		12.10	16.15
	0300	On walls		52	.308		8.90	6.25		15.15	19.85

09637 | Stone Flooring

			CREW	DAILY OUTPUT	LABOR-HOURS	UNIT	MAT.	LABOR	EQUIP.	TOTAL	TOTAL INCL O&P
100	0010	**SLATE TILE** Vermont, 6" x 6" x 1/4" thick, thin set	D-7	180	.089	S.F.	4.14	1.81		5.95	7.45
200	0010	**SLATE & STONE FLOORS** See division 02780-600									

09647 | Wood Parquet Flooring

			CREW	DAILY OUTPUT	LABOR-HOURS	UNIT	MAT.	LABOR	EQUIP.	TOTAL	TOTAL INCL O&P
100	0010	**WOOD PARQUET** flooring									
	5200	Parquetry, standard, 5/16" thick, not incl. finish, oak, minimum	1 Carp	160	.050	S.F.	3.21	1.20		4.41	5.55
	5300	Maximum		100	.080		5.30	1.92		7.22	9.10
	5500	Teak, minimum		160	.050		4.40	1.20		5.60	6.90
	5600	Maximum		100	.080		7.70	1.92		9.62	11.70
	5650	13/16" thick, select grade oak, minimum		160	.050		8.50	1.20		9.70	11.40
	5700	Maximum		100	.080		12.90	1.92		14.82	17.40
	5800	Custom parquetry, including finish, minimum		100	.080		14.20	1.92		16.12	18.85
	5900	Maximum		50	.160		18.80	3.84		22.64	27
	6700	Parquetry, prefinished white oak, 5/16" thick, minimum		160	.050		3.50	1.20		4.70	5.90
	6800	Maximum		100	.080		7.15	1.92		9.07	11.15
	7000	Walnut or teak, parquetry, minimum		160	.050		4.77	1.20		5.97	7.30

	09647 | Wood Parquet Flooring		CREW	DAILY OUTPUT	LABOR-HOURS	UNIT	2005 BARE COSTS MAT.	LABOR	EQUIP.	TOTAL	TOTAL INCL O&P	
0	7100	Maximum	1 Carp	100	.080	S.F.	8.30	1.92		10.22	12.40	100
	7200	Acrylic wood parquet blocks, 12″ x 12″ x 5/16″,										
	7210	irradiated, set in epoxy	1 Carp	160	.050	S.F.	6.95	1.20		8.15	9.70	
	09648 | Wood Strip Flooring											
00	0010	**WOOD** Fir, vertical grain, 1″ x 4″, not incl. finish, B & better	1 Carp	255	.031	S.F.	2.44	.75		3.19	3.96	100
	0100	C grade & better		255	.031		2.29	.75		3.04	3.80	
	0300	Flat grain, 1″ x 4″, not incl. finish, B & better		255	.031		2.79	.75		3.54	4.35	
	0400	C & better		255	.031		2.68	.75		3.43	4.23	
	4000	Maple, strip, 25/32″ x 2-1/4″, not incl. finish, select		170	.047		4.65	1.13		5.78	7	
	4100	#2 & better		170	.047		2.95	1.13		4.08	5.15	
	4300	33/32″ x 3-1/4″, not incl. finish, #1 grade		170	.047		3.62	1.13		4.75	5.90	
	4400	#2 & better		170	.047		3.22	1.13		4.35	5.45	
	4600	Oak, white or red, 25/32″ x 2-1/4″, not incl. finish										
	4700	#1 common	1 Carp	170	.047	S.F.	3	1.13		4.13	5.20	
	4900	Select quartered, 2-1/4″ wide		170	.047		2.56	1.13		3.69	4.74	
	5000	Clear		170	.047		3.72	1.13		4.85	6	
	6100	Prefinished, white oak, prime grade, 2-1/4″ wide		170	.047		6.20	1.13		7.33	8.70	
	6200	3-1/4″ wide		185	.043		7.95	1.04		8.99	10.50	
	6400	Ranch plank		145	.055		7.70	1.32		9.02	10.70	
	6500	Hardwood blocks, 9″ x 9″, 25/32″ thick		160	.050		5.15	1.20		6.35	7.70	
	7400	Yellow pine, 3/4″ x 3-1/8″, T & G, C & better, not incl. finish		200	.040		2.28	.96		3.24	4.14	
	7500	Refinish wood floor, sand, 2 cts poly, wax, soft wood, min.	1 Clab	400	.020		.69	.35		1.04	1.35	
	7600	Hard wood, max		130	.062		1.04	1.07		2.11	2.95	
	7800	Sanding and finishing, 2 coats polyurethane		295	.027		.69	.47		1.16	1.56	
	7900	Subfloor and underlayment, see division 06160										
	8015	Transition molding, 2 1/4″ wide, 5′ long	1 Carp	19.20	.417	Ea.	12.90	10		22.90	31	
	8300	Floating floor, wood composition strip, complete.	1 Clab	133	.060	S.F.	3.38	1.04		4.42	5.50	
	8310	Floating floor components, T & G wood composite strips					3.05			3.05	3.35	
	8320	Film					.14			.14	.15	
	8330	Foam					.17			.17	.18	
	8340	Adhesive					.07			.07	.08	
	8350	Installation kit					.16			.16	.18	
	8360	Trim, 2″ wide x 3′ long				L.F.	2.05			2.05	2.26	
	8370	Reducer moulding				″	4.15			4.15	4.57	
200	0010	**RESILIENT BASE**										200
	0800	Base, cove, rubber or vinyl, .080″ thick										
	1100	Standard colors, 2-1/2″ high	1 Tilf	315	.025	L.F.	.46	.58		1.04	1.45	
	1150	4″ high		315	.025		.51	.58		1.09	1.50	
	1200	6″ high		315	.025		.84	.58		1.42	1.86	
	1450	1/8″ thick, standard colors, 2-1/2″ high		315	.025		.54	.58		1.12	1.53	
	1500	4″ high		315	.025		.70	.58		1.28	1.71	
	1550	6″ high		315	.025		.92	.58		1.50	1.95	
	1600	Corners, 2-1/2″ high		315	.025	Ea.	1.17	.58		1.75	2.23	
	1630	4″ high		315	.025		1.22	.58		1.80	2.28	
	1660	6″ high		315	.025		1.59	.58		2.17	2.69	
	09653 | Resilient Sheet Flooring											
100	0010	**RESILIENT SHEET FLOORING**										100
	5900	Rubber, sheet goods, 36″ wide, 1/8″ thick	1 Tilf	120	.067	S.F.	3.54	1.53		5.07	6.35	
	5950	3/16″ thick		100	.080		5.05	1.83		6.88	8.50	
	6000	1/4″ thick		90	.089		5.80	2.04		7.84	9.70	
	8000	Vinyl sheet goods, backed, .065″ thick, minimum		250	.032		2.07	.73		2.80	3.46	
	8050	Maximum		200	.040		2.64	.92		3.56	4.37	

09653 | Resilient Sheet Flooring

		09653 Resilient Sheet Flooring	CREW	DAILY OUTPUT	LABOR-HOURS	UNIT	2005 BARE COSTS MAT.	LABOR	EQUIP.	TOTAL	TOTAL INCL O&P
100	8100	.080" thick, minimum	1 Tilf	230	.035	S.F.	2.23	.80		3.03	3.73
	8150	Maximum		200	.040		3.18	.92		4.10	4.97
	8200	.125" thick, minimum		230	.035		2.51	.80		3.31	4.04
	8250	Maximum		200	.040		3.98	.92		4.90	5.85
	8700	Adhesive cement, 1 gallon does 200 to 300 S.F.				Gal.	16.35			16.35	17.95
	8800	Asphalt primer, 1 gallon per 300 S.F.					9.65			9.65	10.60
	8900	Emulsion, 1 gallon per 140 S.F.					12.25			12.25	13.50
	8950	Latex underlayment, liquid, fortified					33			33	36.50

09658 | Resilient Tile Flooring

		09658 Resilient Tile Flooring	CREW	DAILY OUTPUT	LABOR-HOURS	UNIT	MAT.	LABOR	EQUIP.	TOTAL	TOTAL INCL O&P
100	0010	**RESILIENT TILE FLOORING**									
	2200	Cork tile, standard finish, 1/8" thick	1 Tilf	315	.025	S.F.	3.75	.58		4.33	5.05
	2250	3/16" thick		315	.025		3.74	.58		4.32	5.05
	2300	5/16" thick		315	.025		5	.58		5.58	6.45
	2350	1/2" thick		315	.025		5.75	.58		6.33	7.30
	2500	Urethane finish, 1/8" thick		315	.025		4.72	.58		5.30	6.15
	2550	3/16" thick		315	.025		5.10	.58		5.68	6.60
	2600	5/16" thick		315	.025		6.35	.58		6.93	7.95
	2650	1/2" thick		315	.025		8.95	.58		9.53	10.80
	6050	Tile, marbleized colors, 12" x 12", 1/8" thick		400	.020		4.05	.46		4.51	5.20
	6100	3/16" thick		400	.020		5.45	.46		5.91	6.75
	6300	Special tile, plain colors, 1/8" thick		400	.020		4.39	.46		4.85	5.55
	6350	3/16" thick		400	.020		5.90	.46		6.36	7.25
	7000	Vinyl composition tile, 12" x 12", 1/16" thick		500	.016		.79	.37		1.16	1.46
	7050	Embossed		500	.016		1	.37		1.37	1.69
	7100	Marbleized		500	.016		1	.37		1.37	1.69
	7150	Solid		500	.016		1.11	.37		1.48	1.81
	7200	3/32" thick, embossed		500	.016		1.01	.37		1.38	1.70
	7250	Marbleized		500	.016		1.12	.37		1.49	1.82
	7300	Solid		500	.016		1.64	.37		2.01	2.39
	7350	1/8" thick, marbleized		500	.016		1.07	.37		1.44	1.77
	7400	Solid		500	.016		2.01	.37		2.38	2.80
	7450	Conductive		500	.016		3.76	.37		4.13	4.73
	7500	Vinyl tile, 12" x 12", .050" thick, minimum		500	.016		1.76	.37		2.13	2.53
	7550	Maximum		500	.016		3.45	.37		3.82	4.39
	7600	1/8" thick, minimum		500	.016		2.23	.37		2.60	3.04
	7650	Solid colors		500	.016		4.80	.37		5.17	5.90
	7700	Marbleized or Travertine pattern		500	.016		3.56	.37		3.93	4.51
	7750	Florentine pattern		500	.016		4.10	.37		4.47	5.10
	7800	Maximum		500	.016		8.40	.37		8.77	9.85

09662 | Static Control Flooring

		09662 Static Control Flooring	CREW	DAILY OUTPUT	LABOR-HOURS	UNIT	MAT.	LABOR	EQUIP.	TOTAL	TOTAL INCL O&P
100	0010	**CONDUCTIVE RESILIENT FLOORING**									
	1700	Conductive flooring, rubber tile, 1/8" thick	1 Tilf	315	.025	S.F.	2.88	.58		3.46	4.11
	1800	Homogeneous vinyl tile, 1/8" thick	"	315	.025	"	3.92	.58		4.50	5.25

09680 | Carpet

		09680 Carpet	CREW	DAILY OUTPUT	LABOR-HOURS	UNIT	MAT.	LABOR	EQUIP.	TOTAL	TOTAL INCL O&P
600	0010	**CARPET PAD**, commercial grade									
	9001	Sponge rubber pad, minimum	1 Tilf	1,350	.006	S.F.	.35	.14		.49	.61
	9101	Maximum		1,350	.006		.92	.14		1.06	1.23
	9201	Felt pad, minimum		1,350	.006		.40	.14		.54	.66
	9301	Maximum		1,350	.006		.75	.14		.89	1.04
	9401	Bonded urethane pad, minimum		1,350	.006		.44	.14		.58	.70
	9501	Maximum		1,350	.006		.75	.14		.89	1.05
	9601	Prime urethane pad, minimum		1,350	.006		.25	.14		.39	.50

09600 | Flooring

09680 | Carpet

	Line	Description	CREW	DAILY OUTPUT	LABOR-HOURS	UNIT	2005 BARE COSTS MAT.	LABOR	EQUIP.	TOTAL	TOTAL INCL O&P
600	9701	Maximum	1 Tilf	1,350	.006	S.F.	.47	.14		.61	.73
800	0010	**CARPET** Commercial grades, direct cement									
	0701	Nylon, level loop, 26 oz., light to medium traffic	1 Tilf	445	.018	S.F.	1.83	.41		2.24	2.67
	0901	32 oz., medium traffic		445	.018		2.59	.41		3	3.50
	1101	40 oz., medium to heavy traffic		445	.018		3.84	.41		4.25	4.89
	2101	Nylon, plush, 20 oz., light traffic		445	.018		1.21	.41		1.62	1.99
	2801	24 oz., light to medium traffic		445	.018		1.28	.41		1.69	2.07
	2901	30 oz., medium traffic		445	.018		1.91	.41		2.32	2.76
	3001	36 oz., medium traffic		445	.018		2.41	.41		2.82	3.31
	3101	42 oz., medium to heavy traffic		370	.022		2.37	.50		2.87	3.40
	3201	46 oz., medium to heavy traffic		370	.022		3.32	.50		3.82	4.45
	3301	54 oz., heavy traffic		370	.022		3.75	.50		4.25	4.92
	3501	Olefin, 15 oz., light traffic		445	.018		.65	.41		1.06	1.37
	3651	22 oz., light traffic		445	.018		.77	.41		1.18	1.51
	4501	50 oz., medium to heavy traffic, level loop		445	.018		8.50	.41		8.91	10
	4701	32 oz., medium to heavy traffic, patterned		400	.020		8.35	.46		8.81	9.95
	4901	48 oz., heavy traffic, patterned		400	.020		8.50	.46		8.96	10.15
	5000	For less than full roll, add					25%				
	5100	For small rooms, less than 12′ wide, add						25%			
	5200	For large open areas (no cuts), deduct						25%			
	5600	For bound carpet baseboard, add	1 Tilf	300	.027	L.F.	1.20	.61		1.81	2.30
	5610	For stairs, not incl. price of carpet, add	"	30	.267	Riser		6.10		6.10	9.85
	8950	For tackless, stretched installation, add padding to above									
	9850	For "branded" fiber, add				S.Y.	25%				
900	0010	**CARPET TILE**									
	0100	Tufted nylon, 18″ x 18″, hard back, 20 oz.	1 Tilf	150	.053	S.Y.	20.50	1.22		21.72	24.50
	0110	26 oz.		150	.053		35.50	1.22		36.72	41
	0200	Cushion back, 20 oz.		150	.053		26	1.22		27.22	30.50
	0210	26 oz.		150	.053		40.50	1.22		41.72	46.50

FINISHES 9

09700 | Wall Finishes

09720 | Wall Coverings

	Line	Description	CREW	DAILY OUTPUT	LABOR-HOURS	UNIT	2005 BARE COSTS MAT.	LABOR	EQUIP.	TOTAL	TOTAL INCL O&P
100	0010	**WALL COVERING** Including sizing, add 10%-30% waste at takeoff R09700-700									
	0050	Aluminum foil	1 Pape	275	.029	S.F.	.87	.63		1.50	1.99
	0100	Copper sheets, .025″ thick, vinyl backing		240	.033		4.66	.72		5.38	6.35
	0300	Phenolic backing		240	.033		6.05	.72		6.77	7.85
	0600	Cork tiles, light or dark, 12″ x 12″ x 3/16″		240	.033		2.55	.72		3.27	3.99
	0700	5/16″ thick		235	.034		3.10	.73		3.83	4.61
	0900	1/4″ basketweave		240	.033		4.76	.72		5.48	6.45
	1000	1/2″ natural, non-directional pattern		240	.033		5.35	.72		6.07	7.10
	1100	3/4″ natural, non-directional pattern		240	.033		9	.72		9.72	11.10
	1200	Granular surface, 12″ x 36″, 1/2″ thick		385	.021		1.03	.45		1.48	1.86
	1300	1″ thick		370	.022		1.33	.46		1.79	2.22
	1500	Polyurethane coated, 12″ x 12″ x 3/16″ thick		240	.033		3.21	.72		3.93	4.71
	1600	5/16″ thick		235	.034		4.57	.73		5.30	6.25
	1800	Cork wallpaper, paperbacked, natural		480	.017		1.83	.36		2.19	2.60

09700 | Wall Finishes

		09720 | Wall Coverings	CREW	DAILY OUTPUT	LABOR-HOURS	UNIT	2005 BARE COSTS MAT.	LABOR	EQUIP.	TOTAL	TOTAL INCL O&P
100	1900	Colors R09700-700	1 Pape	480	.017	S.F.	2.26	.36		2.62	3.08
	2100	Flexible wood veneer, 1/32" thick, plain woods		100	.080		1.92	1.72		3.64	4.94
	2200	Exotic woods		95	.084		2.92	1.81		4.73	6.20
	2400	Gypsum-based, fabric-backed, fire									
	2500	resistant for masonry walls, minimum, 21 oz./S.Y.	1 Pape	800	.010	S.F.	.68	.22		.90	1.10
	2600	Average		720	.011		1	.24		1.24	1.49
	2700	Maximum, (small quantities)		640	.013		1.11	.27		1.38	1.66
	2750	Acrylic, modified, semi-rigid PVC, .028" thick	2 Carp	330	.048		.92	1.16		2.08	2.99
	2800	.040" thick	"	320	.050		1.21	1.20		2.41	3.37
	3000	Vinyl wall covering, fabric-backed, lightweight, (12-15 oz./S.Y.)	1 Pape	640	.013		.58	.27		.85	1.08
	3300	Medium weight, type 2, (20-24 oz./S.Y.)		480	.017		.72	.36		1.08	1.38
	3400	Heavy weight, type 3, (28 oz./S.Y.)		435	.018		1.16	.40		1.56	1.93
	3600	Adhesive, 5 gal. lots, (18SY/Gal.)				Gal.	8.95			8.95	9.80
	3700	Wallpaper, average workmanship, solid pattern, low cost paper	1 Pape	640	.013	S.F.	.29	.27		.56	.76
	3900	basic patterns (matching required), avg. cost paper		535	.015		.65	.32		.97	1.25
	4000	Paper at $85per double roll, quality workmanship		435	.018		1.53	.40		1.93	2.33
	4100	Linen wall covering, paper backed									
	4150	Flame treatment, minimum				S.F.	.69			.69	.76
	4180	Maximum					1.26			1.26	1.39
	4200	Grass cloths with lining paper, minimum	1 Pape	400	.020		.63	.43		1.06	1.40
	4300	Maximum	"	350	.023		2.03	.49		2.52	3.04

		09770 | Special Wall Surfaces	CREW	DAILY OUTPUT	LABOR-HOURS	UNIT	MAT.	LABOR	EQUIP.	TOTAL	TOTAL INCL O&P
700	0010	**PANEL SYSTEM**									
	0100	Raised panel, eng. wood core w/ wood veneer, std., paint grade	2 Carp	300	.053	S.F.	9.10	1.28		10.38	12.15
	0110	Oak veneer		300	.053		15.50	1.28		16.78	19.20
	0120	Maple veneer		300	.053		19.80	1.28		21.08	24
	0130	Cherry veneer		300	.053		25	1.28		26.28	29.50
	0300	Class I fire rated, paint grade		300	.053		14.55	1.28		15.83	18.15
	0310	Oak veneer		300	.053		25	1.28		26.28	29.50
	0320	Maple veneer		300	.053		31.50	1.28		32.78	37
	0330	Cherry veneer		300	.053		45	1.28		46.28	51.50
	0510	Beadboard, 5/8" MDF, standard, primed		300	.053		7.15	1.28		8.43	10.05
	0520	Oak veneer, unfinished		300	.053		12.90	1.28		14.18	16.30
	0530	Maple veneer, unfinished		300	.053		15.25	1.28		16.53	18.90
	0610	Rustic paneling, 5/8" MDF, standard, maple veneer, unfinished		300	.053		21.50	1.28		22.78	25.50
	5000	For prefinished paneling, see division 06250-500 & 06250-200									

9 FINISHES

09900 | Paints & Coatings

		09910 | Paints	CREW	DAILY OUTPUT	LABOR-HOURS	UNIT	2005 BARE COSTS MAT.	LABOR	EQUIP.	TOTAL	TOTAL INCL O&P
100	0010	**CABINETS AND CASEWORK**									
	1000	Primer coat, oil base, brushwork	1 Pord	650	.012	S.F.	.05	.27		.32	.49
	2000	Paint, oil base, brushwork, 1 coat		650	.012		.06	.27		.33	.51
	2500	2 coats		400	.020		.12	.44		.56	.85
	3000	Stain, brushwork, wipe off		650	.012		.05	.27		.32	.49
	4000	Shellac, 1 coat, brushwork		650	.012		.06	.27		.33	.50
	4500	Varnish, 3 coats, brushwork, sand after 1st coat		325	.025		.17	.54		.71	1.07
	5000	For latex paint, deduct					10%				

09900 | Paints & Coatings

09910 | Paints

				DAILY	LABOR-		2005 BARE COSTS				TOTAL
			CREW	OUTPUT	HOURS	UNIT	MAT.	LABOR	EQUIP.	TOTAL	INCL O&P
300	0010	**DOORS AND WINDOWS, EXTERIOR** R09910-220									
	0100	Door frames & trim, only									
	0110	Brushwork, primer	1 Pord	512	.016	L.F.	.05	.34		.39	.62
	0120	Finish coat, exterior latex		512	.016		.06	.34		.40	.62
	0130	Primer & 1 coat, exterior latex		300	.027		.11	.58		.69	1.07
	0140	Primer & 2 coats, exterior latex		265	.030		.17	.66		.83	1.26
	0150	Doors, flush, both sides, incl. frame & trim									
	0160	Roll & brush, primer	1 Pord	10	.800	Ea.	3.89	17.40		21.29	33
	0170	Finish coat, exterior latex		10	.800		4.42	17.40		21.82	33.50
	0180	Primer & 1 coat, exterior latex		7	1.143		8.30	25		33.30	50
	0190	Primer & 2 coats, exterior latex		5	1.600		12.75	35		47.75	71
	0200	Brushwork, stain, sealer & 2 coats polyurethane		4	2		15.80	43.50		59.30	89
	0210	Doors, French, both sides, 10-15 lite, incl. frame & trim									
	0220	Brushwork, primer	1 Pord	6	1.333	Ea.	1.95	29		30.95	49.50
	0230	Finish coat, exterior latex		6	1.333		2.21	29		31.21	50
	0240	Primer & 1 coat, exterior latex		3	2.667		4.15	58		62.15	100
	0250	Primer & 2 coats, exterior latex		2	4		6.25	87		93.25	150
	0260	Brushwork, stain, sealer & 2 coats polyurethane		2.50	3.200		5.65	69.50		75.15	120
	0270	Doors, louvered, both sides, incl. frame & trim									
	0280	Brushwork, primer	1 Pord	7	1.143	Ea.	3.89	25		28.89	45.50
	0290	Finish coat, exterior latex		7	1.143		4.42	25		29.42	46
	0300	Primer & 1 coat, exterior latex		4	2		8.30	43.50		51.80	80.50
	0310	Primer & 2 coats, exterior latex		3	2.667		12.50	58		70.50	109
	0320	Brushwork, stain, sealer & 2 coats polyurethane		4.50	1.778		15.80	38.50		54.30	81
	0330	Doors, panel, both sides, incl. frame & trim									
	0340	Roll & brush, primer	1 Pord	6	1.333	Ea.	3.89	29		32.89	52
	0350	Finish coat, exterior latex		6	1.333		4.42	29		33.42	52.50
	0360	Primer & 1 coat, exterior latex		3	2.667		8.30	58		66.30	105
	0370	Primer & 2 coats, exterior latex		2.50	3.200		12.50	69.50		82	128
	0380	Brushwork, stain, sealer & 2 coats polyurethane		3	2.667		15.80	58		73.80	113
	0400	Windows, per ext. side, based on 15 SF									
	0410	1 to 6 lite									
	0420	Brushwork, primer	1 Pord	13	.615	Ea.	.77	13.40		14.17	23
	0430	Finish coat, exterior latex		13	.615		.87	13.40		14.27	23
	0440	Primer & 1 coat, exterior latex		8	1		1.64	22		23.64	38
	0450	Primer & 2 coats, exterior latex		6	1.333		2.46	29		31.46	50
	0460	Stain, sealer & 1 coat varnish		7	1.143		2.24	25		27.24	43.50
	0470	7 to 10 lite									
	0480	Brushwork, primer	1 Pord	11	.727	Ea.	.77	15.80		16.57	27
	0490	Finish coat, exterior latex		11	.727		.87	15.80		16.67	27
	0500	Primer & 1 coat, exterior latex		7	1.143		1.64	25		26.64	43
	0510	Primer & 2 coats, exterior latex		5	1.600		2.46	35		37.46	59.50
	0520	Stain, sealer & 1 coat varnish		6	1.333		2.24	29		31.24	50
	0530	12 lite									
	0540	Brushwork, primer	1 Pord	10	.800	Ea.	.77	17.40		18.17	29.50
	0550	Finish coat, exterior latex		10	.800		.87	17.40		18.27	29.50
	0560	Primer & 1 coat, exterior latex		6	1.333		1.64	29		30.64	49.50
	0570	Primer & 2 coats, exterior latex		5	1.600		2.46	35		37.46	59.50
	0580	Stain, sealer & 1 coat varnish		6	1.333		2.23	29		31.23	50
	0590	For oil base paint, add					10%				
310	0010	**DOORS & WINDOWS, INTERIOR LATEX** R09910-220									
	0100	Doors flush, both sides, incl. frame & trim									
	0110	Roll & brush, primer	1 Pord	10	.800	Ea.	3.38	17.40		20.78	32
	0120	Finish coat, latex		10	.800		3.65	17.40		21.05	32.50
	0130	Primer & 1 coat latex		7	1.143		7.05	25		32.05	49
	0140	Primer & 2 coats latex		5	1.600		10.50	35		45.50	68.50

FINISHES 9

	09910 Paints		CREW	DAILY OUTPUT	LABOR-HOURS	UNIT	2005 BARE COSTS MAT.	LABOR	EQUIP.	TOTAL	TOTAL INCL O&P
310	0160	Spray, both sides, primer (R09910-220)	1 Pord	20	.400	Ea.	3.56	8.70		12.26	18.20
	0170	Finish coat, latex		20	.400		3.83	8.70		12.53	18.50
	0180	Primer & 1 coat latex		11	.727		7.45	15.80		23.25	34
	0190	Primer & 2 coats latex		8	1		11.10	22		33.10	48
	0200	Doors, French, both sides, 10-15 lite, incl. frame & trim									
	0210	Roll & brush, primer	1 Pord	6	1.333	Ea.	1.69	29		30.69	49.50
	0220	Finish coat, latex		6	1.333		1.83	29		30.83	49.50
	0230	Primer & 1 coat latex		3	2.667		3.52	58		61.52	99.50
	0240	Primer & 2 coats latex		2	4		5.25	87		92.25	149
	0260	Doors, louvered, both sides, incl. frame & trim									
	0270	Roll & brush, primer	1 Pord	7	1.143	Ea.	3.38	25		28.38	44.50
	0280	Finish coat, latex		7	1.143		3.65	25		28.65	45
	0290	Primer & 1 coat, latex		4	2		6.85	43.50		50.35	79
	0300	Primer & 2 coats, latex		3	2.667		10.70	58		68.70	107
	0320	Spray, both sides, primer		20	.400		3.56	8.70		12.26	18.20
	0330	Finish coat, latex		20	.400		3.83	8.70		12.53	18.50
	0340	Primer & 1 coat, latex		11	.727		7.45	15.80		23.25	34
	0350	Primer & 2 coats, latex		8	1		11.30	22		33.30	48.50
	0360	Doors, panel, both sides, incl. frame & trim									
	0370	Roll & brush, primer	1 Pord	6	1.333	Ea.	3.56	29		32.56	51.50
	0380	Finish coat, latex		6	1.333		3.65	29		32.65	51.50
	0390	Primer & 1 coat, latex		3	2.667		7.05	58		65.05	103
	0400	Primer & 2 coats, latex		2.50	3.200		10.70	69.50		80.20	126
	0420	Spray, both sides, primer		10	.800		3.56	17.40		20.96	32.50
	0430	Finish coat, latex		10	.800		3.83	17.40		21.23	32.50
	0440	Primer & 1 coat, latex		5	1.600		7.45	35		42.45	65
	0450	Primer & 2 coats, latex		4	2		11.30	43.50		54.80	84
	0460	Windows, per interior side, based on 15 SF									
	0470	1 to 6 lite									
	0480	Brushwork, primer	1 Pord	13	.615	Ea.	.67	13.40		14.07	22.50
	0490	Finish coat, enamel		13	.615		.72	13.40		14.12	23
	0500	Primer & 1 coat enamel		8	1		1.39	22		23.39	37.50
	0510	Primer & 2 coats enamel		6	1.333		2.11	29		31.11	50
	0530	7 to 10 lite									
	0540	Brushwork, primer	1 Pord	11	.727	Ea.	.67	15.80		16.47	26.50
	0550	Finish coat, enamel		11	.727		.72	15.80		16.52	27
	0560	Primer & 1 coat enamel		7	1.143		1.39	25		26.39	42.50
	0570	Primer & 2 coats enamel		5	1.600		2.11	35		37.11	59.50
	0590	12 lite									
	0600	Brushwork, primer	1 Pord	10	.800	Ea.	.67	17.40		18.07	29
	0610	Finish coat, enamel		10	.800		.72	17.40		18.12	29.50
	0620	Primer & 1 coat enamel		6	1.333		1.39	29		30.39	49
	0630	Primer & 2 coats enamel		5	1.600		2.11	35		37.11	59.50
	0650	For oil base paint, add					10%				
320	0010	**DOORS AND WINDOWS, INTERIOR ALKYD (OIL BASE)**									
	0500	Flush door & frame, 3' x 7', oil, primer, brushwork	1 Pord	10	.800	Ea.	2.12	17.40		19.52	31
	1000	Paint, 1 coat		10	.800		2.04	17.40		19.44	30.50
	1200	2 coats		6	1.333		3.43	29		32.43	51.50
	1400	Stain, brushwork, wipe off		18	.444		1.02	9.65		10.67	17
	1600	Shellac, 1 coat, brushwork		25	.320		1.18	6.95		8.13	12.75
	1800	Varnish, 3 coats, brushwork, sand after 1st coat		9	.889		3.61	19.35		22.96	36
	2000	Panel door & frame, 3' x 7', oil, primer, brushwork		6	1.333		1.82	29		30.82	49.50
	2200	Paint, 1 coat		6	1.333		2.04	29		31.04	49.50
	2400	2 coats		3	2.667		5.90	58		63.90	102
	2600	Stain, brushwork, panel door, 3' x 7', not incl. frame		16	.500		1.02	10.90		11.92	19
	2800	Shellac, 1 coat, brushwork		22	.364		1.18	7.90		9.08	14.30

09910 | Paints			CREW	DAILY OUTPUT	LABOR-HOURS	UNIT	2005 BARE COSTS MAT.	LABOR	EQUIP.	TOTAL	TOTAL INCL O&P	
320	3000	Varnish, 3 coats, brushwork, sand after 1st coat	1 Pord	7.50	1.067	Ea.	3.61	23		26.61	42	320
	3020	French door, incl. 3′ x 7′, 6 lites, frame & trim										
	3022	Paint, 1 coat, over existing paint	1 Pord	5	1.600	Ea.	4.08	35		39.08	61.50	
	3024	2 coats, over existing paint		5	1.600		7.90	35		42.90	65.50	
	3026	Primer & 1 coat		3.50	2.286		7.75	49.50		57.25	90	
	3028	Primer & 2 coats		3	2.667		11.80	58		69.80	109	
	3032	Varnish or polyurethane, 1 coat		5	1.600		4.46	35		39.46	62	
	3034	2 coats, sanding between		3	2.667		8.90	58		66.90	105	
	4400	Windows, including frame and trim, per side										
	4600	Colonial type, 6/6 lites, 2′ x 3′, oil, primer, brushwork	1 Pord	14	.571	Ea.	.29	12.45		12.74	21	
	5800	Paint, 1 coat		14	.571		.32	12.45		12.77	21	
	6000	2 coats		9	.889		.63	19.35		19.98	32.50	
	6200	3′ x 5′ opening, 6/6 lites, primer coat, brushwork		12	.667		.72	14.50		15.22	25	
	6400	Paint, 1 coat		12	.667		.80	14.50		15.30	25	
	6600	2 coats		7	1.143		1.56	25		26.56	42.50	
	6800	4′ x 8′ opening, 6/6 lites, primer coat, brushwork		8	1		1.54	22		23.54	37.50	
	7000	Paint, 1 coat		8	1		1.72	22		23.72	38	
	7200	2 coats		5	1.600		3.34	35		38.34	60.50	
	8000	Single lite type, 2′ x 3′, oil base, primer coat, brushwork		33	.242		.29	5.25		5.54	8.95	
	8200	Paint, 1 coat		33	.242		.32	5.25		5.57	9	
	8400	2 coats		20	.400		.63	8.70		9.33	15	
	8600	3′ x 5′ opening, primer coat, brushwork		20	.400		.72	8.70		9.42	15.10	
	8800	Paint, 1 coat		20	.400		.80	8.70		9.50	15.20	
	9000	2 coats		13	.615		1.56	13.40		14.96	23.50	
	9200	4′ x 8′ opening, primer coat, brushwork		14	.571		1.54	12.45		13.99	22	
	9400	Paint, 1 coat		14	.571		1.72	12.45		14.17	22.50	
	9600	2 coats		8	1		3.34	22		25.34	39.50	
400	0010	**FENCES**										400
	0100	Chain link or wire metal, one side, water base										
	0110	Roll & brush, first coat (R09910 -220)	1 Pord	960	.008	S.F.	.06	.18		.24	.36	
	0120	Second coat		1,280	.006		.05	.14		.19	.28	
	0130	Spray, first coat		2,275	.004		.06	.08		.14	.19	
	0140	Second coat		2,600	.003		.06	.07		.13	.17	
	0150	Picket, water base										
	0160	Roll & brush, first coat	1 Pord	865	.009	S.F.	.06	.20		.26	.40	
	0170	Second coat		1,050	.008		.06	.17		.23	.34	
	0180	Spray, first coat		2,275	.004		.06	.08		.14	.20	
	0190	Second coat		2,600	.003		.06	.07		.13	.18	
	0200	Stockade, water base										
	0210	Roll & brush, first coat	1 Pord	1,040	.008	S.F.	.06	.17		.23	.34	
	0220	Second coat		1,200	.007		.06	.15		.21	.31	
	0230	Spray, first coat		2,275	.004		.06	.08		.14	.20	
	0240	Second coat		2,600	.003		.06	.07		.13	.18	
500	0010	**FLOORS, INTERIOR**										500
	0100	Concrete										
	0120	1st coat	1 Pord	975	.008	S.F.	.11	.18		.29	.41	
	0130	2nd coat		1,150	.007		.07	.15		.22	.33	
	0140	3rd coat		1,300	.006		.06	.13		.19	.29	
	0150	Roll, latex, block filler										
	0160	1st coat	1 Pord	2,600	.003	S.F.	.15	.07		.22	.27	
	0170	2nd coat		3,250	.002		.09	.05		.14	.19	
	0180	3rd coat		3,900	.002		.07	.04		.11	.14	
	0190	Spray, latex, block filler										
	0200	1st coat	1 Pord	2,600	.003	S.F.	.13	.07		.20	.25	
	0210	2nd coat		3,250	.002		.07	.05		.12	.17	

	09910	Paints	CREW	DAILY OUTPUT	LABOR-HOURS	UNIT	2005 BARE COSTS MAT.	LABOR	EQUIP.	TOTAL	TOTAL INCL O&P
500	0220	3rd coat	1 Pord	3,900	.002	S.F.	.06	.04		.10	.13
620	0010	**MISCELLANEOUS, EXTERIOR** R09910-220									
	0100	Railing, ext., decorative wood, incl. cap & baluster									
	0110	newels & spindles @ 12″ O.C.									
	0120	Brushwork, stain, sand, seal & varnish									
	0130	First coat	1 Pord	90	.089	L.F.	.45	1.93		2.38	3.67
	0140	Second coat	″	120	.067	″	.45	1.45		1.90	2.87
	0150	Rough sawn wood, 42″ high, 2″x2″ verticals, 6″ O.C.									
	0160	Brushwork, stain, each coat	1 Pord	90	.089	L.F.	.15	1.93		2.08	3.34
	0170	Wrought iron, 1″ rail, 1/2″ sq. verticals									
	0180	Brushwork, zinc chromate, 60″ high, bars 6″ O.C.									
	0190	Primer	1 Pord	130	.062	L.F.	.51	1.34		1.85	2.76
	0200	Finish coat		130	.062		.16	1.34		1.50	2.37
	0210	Additional coat		190	.042		.19	.92		1.11	1.71
	0220	Shutters or blinds, single panel, 2′x4′, paint all sides									
	0230	Brushwork, primer	1 Pord	20	.400	Ea.	.56	8.70		9.26	14.90
	0240	Finish coat, exterior latex		20	.400		.47	8.70		9.17	14.80
	0250	Primer & 1 coat, exterior latex		13	.615		.90	13.40		14.30	23
	0260	Spray, primer		35	.229		.82	4.97		5.79	9.05
	0270	Finish coat, exterior latex		35	.229		.99	4.97		5.96	9.25
	0280	Primer & 1 coat, exterior latex		20	.400		.88	8.70		9.58	15.25
	0290	For louvered shutters, add				S.F.	10%				
	0300	Stair stringers, exterior, metal									
	0310	Roll & brush, zinc chromate, to 14″, each coat	1 Pord	320	.025	L.F.	.05	.54		.59	.95
	0320	Rough sawn wood, 4″ x 12″									
	0330	Roll & brush, exterior latex, each coat	1 Pord	215	.037	L.F.	.07	.81		.88	1.41
	0340	Trellis/lattice, 2″x2″ @ 3″ O.C. with 2″x8″ supports									
	0350	Spray, latex, per side, each coat	1 Pord	475	.017	S.F.	.07	.37		.44	.68
	0450	Decking, Ext., sealer, alkyd, brushwork, sealer coat		1,140	.007		.05	.15		.20	.31
	0460	1st coat		1,140	.007		.06	.15		.21	.31
	0470	2nd coat		1,300	.006		.04	.13		.17	.26
	0500	Paint, alkyd, brushwork, primer coat		1,140	.007		.07	.15		.22	.33
	0510	1st coat		1,140	.007		.07	.15		.22	.33
	0520	2nd coat		1,300	.006		.05	.13		.18	.28
	0600	Sand paint, alkyd, brushwork, 1 coat		150	.053		.09	1.16		1.25	2.01
630	0010	**MISCELLANEOUS, INTERIOR**									
	2400	Floors, conc./wood, oil base, primer/sealer coat, brushwork	2 Pord	1,950	.008	S.F.	.06	.18		.24	.35
	2450	Roller		5,200	.003		.06	.07		.13	.18
	2600	Spray		6,000	.003		.06	.06		.12	.17
	2650	Paint 1 coat, brushwork		1,950	.008		.05	.18		.23	.35
	2800	Roller		5,200	.003		.06	.07		.13	.17
	2850	Spray		6,000	.003		.06	.06		.12	.16
	3000	Stain, wood floor, brushwork, 1 coat		4,550	.004		.05	.08		.13	.18
	3200	Roller		5,200	.003		.05	.07		.12	.17
	3250	Spray		6,000	.003		.05	.06		.11	.16
	3400	Varnish, wood floor, brushwork		4,550	.004		.06	.08		.14	.19
	3450	Roller		5,200	.003		.06	.07		.13	.18
	3600	Spray		6,000	.003		.06	.06		.12	.17
	3800	Grilles, per side, oil base, primer coat, brushwork	1 Pord	520	.015		.10	.33		.43	.66
	3850	Spray		1,140	.007		.10	.15		.25	.36
	3880	Paint 1 coat, brushwork		520	.015		.11	.33		.44	.67
	3900	Spray		1,140	.007		.12	.15		.27	.38
	3920	Paint 2 coats, brushwork		325	.025		.21	.54		.75	1.11
	3940	Spray		650	.012		.24	.27		.51	.70
	4250	Paint 1 coat, brushwork	2 Pord	1,300	.012	L.F.	.11	.27		.38	.56

Important: See the Reference Section for critical supporting data - Reference Nos., Crews, & Location Factors

09910 | Paints

	09910	Paints	CREW	DAILY OUTPUT	LABOR-HOURS	UNIT	2005 BARE COSTS MAT.	LABOR	EQUIP.	TOTAL	TOTAL INCL O&P	
630	4500	Louvers, one side, primer, brushwork	1 Pord	524	.015	S.F.	.06	.33		.39	.61	630
	4520	Paint one coat, brushwork		520	.015		.06	.33		.39	.61	
	4530	Spray		1,140	.007		.06	.15		.21	.32	
	4540	Paint two coats, brushwork		325	.025		.11	.54		.65	1	
	4550	Spray		650	.012		.12	.27		.39	.57	
	4560	Paint three coats, brushwork		270	.030		.16	.64		.80	1.24	
	4570	Spray		500	.016		.18	.35		.53	.77	
	5000	Pipe, to 4″ diameter, primer or sealer coat, oil base, brushwork	2 Pord	1,250	.013	L.F.	.06	.28		.34	.53	
	5100	Spray		2,165	.007		.06	.16		.22	.32	
	5200	Paint 1 coat, brushwork		1,250	.013		.06	.28		.34	.53	
	5300	Spray		2,165	.007		.06	.16		.22	.32	
	5350	Paint 2 coats, brushwork		775	.021		.11	.45		.56	.86	
	5400	Spray		1,240	.013		.13	.28		.41	.60	
	5450	To 8″ diameter, primer or sealer coat, brushwork		620	.026		.12	.56		.68	1.05	
	5500	Spray		1,085	.015		.20	.32		.52	.75	
	5550	Paint 1 coat, brushwork		620	.026		.17	.56		.73	1.11	
	5600	Spray		1,085	.015		.19	.32		.51	.74	
	5650	Paint 2 coats, brushwork		385	.042		.23	.90		1.13	1.74	
	5700	Spray		620	.026		.25	.56		.81	1.20	
	6600	Radiators, per side, primer, brushwork	1 Pord	520	.015	S.F.	.06	.33		.39	.61	
	6620	Paint one coat, brushwork		520	.015		.05	.33		.38	.61	
	6640	Paint two coats, brushwork		340	.024		.11	.51		.62	.96	
	6660	Paint three coats, brushwork		283	.028		.16	.61		.77	1.19	
	7000	Trim, wood, incl. puttying, under 6″ wide										
	7200	Primer coat, oil base, brushwork	1 Pord	650	.012	L.F.	.02	.27		.29	.47	
	7250	Paint, 1 coat, brushwork		650	.012		.03	.27		.30	.47	
	7400	2 coats		400	.020		.05	.44		.49	.78	
	7450	3 coats		325	.025		.08	.54		.62	.97	
	7500	Over 6″ wide, primer coat, brushwork		650	.012		.05	.27		.32	.49	
	7550	Paint, 1 coat, brushwork		650	.012		.05	.27		.32	.50	
	7600	2 coats		400	.020		.10	.44		.54	.83	
	7650	3 coats		325	.025		.15	.54		.69	1.05	
	8000	Cornice, simple design, primer coat, oil base, brushwork		650	.012	S.F.	.05	.27		.32	.49	
	8250	Paint, 1 coat		650	.012		.05	.27		.32	.50	
	8300	2 coats		400	.020		.10	.44		.54	.83	
	8350	Ornate design, primer coat		350	.023		.05	.50		.55	.87	
	8400	Paint, 1 coat		350	.023		.05	.50		.55	.88	
	8450	2 coats		400	.020		.10	.44		.54	.83	
	8600	Balustrades, primer coat, oil base, brushwork		520	.015		.05	.33		.38	.60	
	8650	Paint, 1 coat		520	.015		.05	.33		.38	.61	
	8700	2 coats		325	.025		.10	.54		.64	.99	
	8900	Trusses and wood frames, primer coat, oil base, brushwork		800	.010		.05	.22		.27	.41	
	8950	Spray		1,200	.007		.05	.15		.20	.30	
	9000	Paint 1 coat, brushwork		750	.011		.05	.23		.28	.44	
	9200	Spray		1,200	.007		.06	.15		.21	.31	
	9220	Paint 2 coats, brushwork		500	.016		.10	.35		.45	.68	
	9240	Spray		600	.013		.12	.29		.41	.61	
	9260	Stain, brushwork, wipe off		600	.013		.05	.29		.34	.53	
	9280	Varnish, 3 coats, brushwork		275	.029		.17	.63		.80	1.23	
	9350	For latex paint, deduct					10%					
700	0010	**SIDING EXTERIOR**, Alkyd (oil base)										700
	0450	Steel siding, oil base, paint 1 coat, brushwork	2 Pord	2,015	.008	S.F.	.06	.17		.23	.34	
	0500	Spray		4,550	.004		.08	.08		.16	.22	
	0800	Paint 2 coats, brushwork		1,300	.012		.11	.27		.38	.56	
	1000	Spray		4,550	.004		.14	.08		.22	.28	
	1200	Stucco, rough, oil base, paint 2 coats, brushwork		1,300	.012		.11	.27		.38	.56	

9 FINISHES

	09910 | Paints		CREW	DAILY OUTPUT	LABOR-HOURS	UNIT	2005 BARE COSTS MAT.	LABOR	EQUIP.	TOTAL	TOTAL INCL O&P
700	1400	Roller	2 Pord	1,625	.010	S.F.	.12	.21		.33	.48
	1600	Spray		2,925	.005		.12	.12		.24	.34
	1800	Texture 1-11 or clapboard, oil base, primer coat, brushwork		1,300	.012		.09	.27		.36	.54
	2000	Spray		4,550	.004		.09	.08		.17	.23
	2100	Paint 1 coat, brushwork		1,300	.012		.08	.27		.35	.53
	2200	Spray		4,550	.004		.08	.08		.16	.22
	2400	Paint 2 coats, brushwork		810	.020		.16	.43		.59	.89
	2600	Spray		2,600	.006		.18	.13		.31	.42
	3000	Stain 1 coat, brushwork		1,520	.011		.05	.23		.28	.43
	3200	Spray		5,320	.003		.05	.07		.12	.17
	3400	Stain 2 coats, brushwork		950	.017		.10	.37		.47	.71
	4000	Spray		3,050	.005		.11	.11		.22	.31
	4200	Wood shingles, oil base primer coat, brushwork		1,300	.012		.09	.27		.36	.53
	4400	Spray		3,900	.004		.08	.09		.17	.24
	4600	Paint 1 coat, brushwork		1,300	.012		.07	.27		.34	.51
	4800	Spray		3,900	.004		.08	.09		.17	.24
	5000	Paint 2 coats, brushwork		810	.020		.13	.43		.56	.86
	5200	Spray		2,275	.007		.13	.15		.28	.39
	5800	Stain 1 coat, brushwork		1,500	.011		.05	.23		.28	.43
	6000	Spray		3,900	.004		.05	.09		.14	.20
	6500	Stain 2 coats, brushwork		950	.017		.10	.37		.47	.71
	7000	Spray		2,660	.006		.13	.13		.26	.37
	8000	For latex paint, deduct					10%				
	8100	For work over 12′ H, from pipe scaffolding, add						15%			
	8200	For work over 12′ H, from extension ladder, add						25%			
	8300	For work over 12′ H, from swing staging, add						35%			
710	0010	**SIDING, MISC.** R09910-220									
	0100	Aluminum siding									
	0110	Brushwork, primer	2 Pord	2,275	.007	S.F.	.05	.15		.20	.31
	0120	Finish coat, exterior latex		2,275	.007		.04	.15		.19	.29
	0130	Primer & 1 coat exterior latex		1,300	.012		.10	.27		.37	.55
	0140	Primer & 2 coats exterior latex		975	.016		.14	.36		.50	.74
	0150	Mineral Fiber shingles									
	0160	Brushwork, primer	2 Pord	1,495	.011	S.F.	.09	.23		.32	.48
	0170	Finish coat, industrial enamel		1,495	.011		.10	.23		.33	.49
	0180	Primer & 1 coat enamel		810	.020		.19	.43		.62	.92
	0190	Primer & 2 coats enamel		540	.030		.30	.64		.94	1.38
	0200	Roll, primer		1,625	.010		.10	.21		.31	.46
	0210	Finish coat, industrial enamel		1,625	.010		.11	.21		.32	.47
	0220	Primer & 1 coat enamel		975	.016		.21	.36		.57	.82
	0230	Primer & 2 coats enamel		650	.025		.32	.54		.86	1.23
	0240	Spray, primer		3,900	.004		.08	.09		.17	.24
	0250	Finish coat, industrial enamel		3,900	.004		.09	.09		.18	.25
	0260	Primer & 1 coat enamel		2,275	.007		.17	.15		.32	.44
	0270	Primer & 2 coats enamel		1,625	.010		.26	.21		.47	.64
	0280	Waterproof sealer, first coat		4,485	.004		.07	.08		.15	.20
	0290	Second coat		5,235	.003		.06	.07		.13	.18
	0300	Rough wood incl. shingles, shakes or rough sawn siding									
	0310	Brushwork, primer	2 Pord	1,280	.013	S.F.	.11	.27		.38	.58
	0320	Finish coat, exterior latex		1,280	.013		.07	.27		.34	.53
	0330	Primer & 1 coat exterior latex		960	.017		.18	.36		.54	.80
	0340	Primer & 2 coats exterior latex		700	.023		.25	.50		.75	1.10
	0350	Roll, primer		2,925	.005		.15	.12		.27	.37
	0360	Finish coat, exterior latex		2,925	.005		.08	.12		.20	.29
	0370	Primer & 1 coat exterior latex		1,790	.009		.24	.19		.43	.58
	0380	Primer & 2 coats exterior latex		1,300	.012		.32	.27		.59	.79

09910 | Paints			CREW	DAILY OUTPUT	LABOR-HOURS	UNIT	2005 BARE COSTS MAT.	LABOR	EQUIP.	TOTAL	TOTAL INCL O&P	
710	0390	Spray, primer (R09910-220)	2 Pord	3,900	.004	S.F.	.13	.09		.22	.29	710
	0400	Finish coat, exterior latex		3,900	.004		.06	.09		.15	.22	
	0410	Primer & 1 coat exterior latex		2,600	.006		.19	.13		.32	.43	
	0420	Primer & 2 coats exterior latex		2,080	.008		.26	.17		.43	.55	
	0430	Waterproof sealer, first coat		4,485	.004		.12	.08		.20	.26	
	0440	Second coat	↓	4,485	.004	↓	.07	.08		.15	.20	
	0450	Smooth wood incl. butt, T&G, beveled, drop or B&B siding										
	0460	Brushwork, primer	2 Pord	2,325	.007	S.F.	.08	.15		.23	.34	
	0470	Finish coat, exterior latex		1,280	.013		.07	.27		.34	.53	
	0480	Primer & 1 coat exterior latex		800	.020		.15	.44		.59	.89	
	0490	Primer & 2 coats exterior latex		630	.025		.22	.55		.77	1.15	
	0500	Roll, primer		2,275	.007		.09	.15		.24	.35	
	0510	Finish coat, exterior latex		2,275	.007		.08	.15		.23	.33	
	0520	Primer & 1 coat exterior latex		1,300	.012		.17	.27		.44	.62	
	0530	Primer & 2 coats exterior latex		975	.016		.24	.36		.60	.86	
	0540	Spray, primer		4,550	.004		.07	.08		.15	.21	
	0550	Finish coat, exterior latex		4,550	.004		.06	.08		.14	.20	
	0560	Primer & 1 coat exterior latex		2,600	.006		.14	.13		.27	.37	
	0570	Primer & 2 coats exterior latex		1,950	.008		.20	.18		.38	.51	
	0580	Waterproof sealer, first coat		5,230	.003		.07	.07		.14	.18	
	0590	Second coat	↓	5,980	.003		.07	.06		.13	.17	
	0600	For oil base paint, add				↓	10%					
800	0010	**TRIM, EXTERIOR** (R09910-220)										800
	0100	Door frames & trim (see Doors, interior or exterior)										
	0110	Fascia, latex paint, one coat coverage										
	0120	1″ x 4″, brushwork	1 Pord	640	.013	L.F.	.02	.27		.29	.47	
	0130	Roll		1,280	.006		.02	.14		.16	.24	
	0140	Spray		2,080	.004		.01	.08		.09	.16	
	0150	1″ x 6″ to 1″ x 10″, brushwork		640	.013		.06	.27		.33	.52	
	0160	Roll		1,230	.007		.07	.14		.21	.30	
	0170	Spray		2,100	.004		.05	.08		.13	.19	
	0180	1″ x 12″, brushwork		640	.013		.06	.27		.33	.52	
	0190	Roll		1,050	.008		.07	.17		.24	.34	
	0200	Spray	↓	2,200	.004	↓	.05	.08		.13	.18	
	0210	Gutters & downspouts, metal, zinc chromate paint										
	0220	Brushwork, gutters, 5″, first coat	1 Pord	640	.013	L.F.	.06	.27		.33	.51	
	0230	Second coat		960	.008		.05	.18		.23	.36	
	0240	Third coat		1,280	.006		.04	.14		.18	.27	
	0250	Downspouts, 4″, first coat		640	.013		.06	.27		.33	.51	
	0260	Second coat		960	.008		.05	.18		.23	.36	
	0270	Third coat	↓	1,280	.006	↓	.04	.14		.18	.27	
	0280	Gutters & downspouts, wood										
	0290	Brushwork, gutters, 5″, primer	1 Pord	640	.013	L.F.	.05	.27		.32	.51	
	0300	Finish coat, exterior latex		640	.013		.05	.27		.32	.51	
	0310	Primer & 1 coat exterior latex		400	.020		.11	.44		.55	.84	
	0320	Primer & 2 coats exterior latex		325	.025		.17	.54		.71	1.06	
	0330	Downspouts, 4″, primer		640	.013		.05	.27		.32	.51	
	0340	Finish coat, exterior latex		640	.013		.05	.27		.32	.51	
	0350	Primer & 1 coat exterior latex		400	.020		.11	.44		.55	.84	
	0360	Primer & 2 coats exterior latex	↓	325	.025	↓	.08	.54		.62	.97	
	0370	Molding, exterior, up to 14″ wide										
	0380	Brushwork, primer	1 Pord	640	.013	L.F.	.06	.27		.33	.52	
	0390	Finish coat, exterior latex		640	.013		.06	.27		.33	.52	
	0400	Primer & 1 coat exterior latex		400	.020		.13	.44		.57	.86	
	0410	Primer & 2 coats exterior latex		315	.025		.13	.55		.68	1.05	
	0420	Stain & fill	↓	1,050	.008	↓	.06	.17		.23	.33	

	09910 | Paints		CREW	DAILY OUTPUT	LABOR-HOURS	UNIT	2005 BARE COSTS MAT.	LABOR	EQUIP.	TOTAL	TOTAL INCL O&P
800	0430	Shellac (R09910-220)	1 Pord	1,850	.004	L.F.	.07	.09		.16	.22
	0440	Varnish		1,275	.006		.07	.14		.21	.30
910	0350	**WALLS, MASONRY (CMU), EXTERIOR**									
	0360	Concrete masonry units (CMU), smooth surface									
	0370	Brushwork, latex, first coat	1 Pord	640	.013	S.F.	.04	.27		.31	.49
	0380	Second coat		960	.008		.03	.18		.21	.33
	0390	Waterproof sealer, first coat		736	.011		.07	.24		.31	.47
	0400	Second coat		1,104	.007		.05	.16		.21	.31
	0410	Roll, latex, paint, first coat		1,465	.005		.04	.12		.16	.25
	0420	Second coat		1,790	.004		.03	.10		.13	.19
	0430	Waterproof sealer, first coat		1,680	.005		.07	.10		.17	.25
	0440	Second coat		2,060	.004		.05	.08		.13	.19
	0450	Spray, latex, paint, first coat		1,950	.004		.03	.09		.12	.19
	0460	Second coat		2,600	.003		.03	.07		.10	.14
	0470	Waterproof sealer, first coat		2,245	.004		.09	.08		.17	.23
	0480	Second coat		2,990	.003		.03	.06		.09	.14
	0490	Concrete masonry unit (CMU), porous									
	0500	Brushwork, latex, first coat	1 Pord	640	.013	S.F.	.07	.27		.34	.53
	0510	Second coat		960	.008		.04	.18		.22	.34
	0520	Waterproof sealer, first coat		736	.011		.09	.24		.33	.49
	0530	Second coat		1,104	.007		.04	.16		.20	.31
	0540	Roll latex, first coat		1,465	.005		.05	.12		.17	.26
	0550	Second coat		1,790	.004		.03	.10		.13	.20
	0560	Waterproof sealer, first coat		1,680	.005		.09	.10		.19	.27
	0570	Second coat		2,060	.004		.05	.08		.13	.20
	0580	Spray latex, first coat		1,950	.004		.04	.09		.13	.19
	0590	Second coat		2,600	.003		.03	.07		.10	.14
	0600	Waterproof sealer, first coat		2,245	.004		.07	.08		.15	.21
	0610	Second coat		2,990	.003		.04	.06		.10	.14
920	0010	**WALLS AND CEILINGS, INTERIOR**									
	0100	Concrete, dry wall or plaster, oil base, primer or sealer coat									
	0200	Smooth finish, brushwork	1 Pord	1,150	.007	S.F.	.05	.15		.20	.30
	0240	Roller		2,040	.004		.05	.09		.14	.19
	0300	Sand finish, brushwork		975	.008		.05	.18		.23	.34
	0340	Roller		1,150	.007		.05	.15		.20	.31
	0380	Spray		2,275	.004		.04	.08		.12	.17
	0400	Paint 1 coat, smooth finish, brushwork		1,200	.007		.05	.15		.20	.29
	0440	Roller		1,300	.006		.05	.13		.18	.27
	0480	Spray		2,275	.004		.04	.08		.12	.18
	0500	Sand finish, brushwork		1,050	.008		.05	.17		.22	.32
	0540	Roller		1,600	.005		.05	.11		.16	.23
	0580	Spray		2,100	.004		.04	.08		.12	.19
	0800	Paint 2 coats, smooth finish, brushwork		680	.012		.09	.26		.35	.52
	0840	Roller		800	.010		.10	.22		.32	.47
	0880	Spray		1,625	.005		.08	.11		.19	.27
	0900	Sand finish, brushwork		605	.013		.09	.29		.38	.57
	0940	Roller		1,020	.008		.10	.17		.27	.39
	0980	Spray		1,700	.005		.08	.10		.18	.26
	1200	Paint 3 coats, smooth finish, brushwork		510	.016		.14	.34		.48	.71
	1240	Roller		650	.012		.14	.27		.41	.60
	1280	Spray		1,625	.005		.12	.11		.23	.32
	1300	Sand finish, brushwork		454	.018		.14	.38		.52	.78
	1340	Roller		680	.012		.14	.26		.40	.58
	1380	Spray		1,133	.007		.12	.15		.27	.39
	1600	Glaze coating, 5 coats, spray, clear		900	.009		.60	.19		.79	.98

09910 | Paints		CREW	DAILY OUTPUT	LABOR-HOURS	UNIT	2005 BARE COSTS MAT.	LABOR	EQUIP.	TOTAL	TOTAL INCL O&P	
1640	Multicolor	1 Pord	900	.009	S.F.	.85	.19		1.04	1.26	920
1700	For latex paint, deduct					10%					
1800	For ceiling installations, add						25%				
2000	Masonry or concrete block, oil base, primer or sealer coat										
2100	Smooth finish, brushwork	1 Pord	1,224	.007	S.F.	.05	.14		.19	.28	
2180	Spray		2,400	.003		.07	.07		.14	.20	
2200	Sand finish, brushwork		1,089	.007		.07	.16		.23	.34	
2280	Spray		2,400	.003		.07	.07		.14	.20	
2400	Paint 1 coat, smooth finish, brushwork		1,100	.007		.07	.16		.23	.34	
2480	Spray		2,400	.003		.07	.07		.14	.20	
2500	Sand finish, brushwork		979	.008		.07	.18		.25	.37	
2580	Spray		2,400	.003		.07	.07		.14	.20	
2800	Paint 2 coats, smooth finish, brushwork		756	.011		.15	.23		.38	.54	
2880	Spray		1,360	.006		.14	.13		.27	.36	
2900	Sand finish, brushwork		672	.012		.15	.26		.41	.59	
2980	Spray		1,360	.006		.14	.13		.27	.36	
3200	Paint 3 coats, smooth finish, brushwork		560	.014		.22	.31		.53	.75	
3280	Spray		1,088	.007		.21	.16		.37	.49	
3300	Sand finish, brushwork		498	.016		.22	.35		.57	.81	
3380	Spray		1,088	.007		.21	.16		.37	.49	
3600	Glaze coating, 5 coats, spray, clear		900	.009		.60	.19		.79	.98	
3620	Multicolor		900	.009		.85	.19		1.04	1.26	
4000	Block filler, 1 coat, brushwork		425	.019		.12	.41		.53	.81	
4100	Silicone, water repellent, 2 coats, spray		2,000	.004		.25	.09		.34	.42	
4120	For latex paint, deduct					10%					
8200	For work 8 - 15′ H, add						10%				
8300	For work over 15′ H, add						20%				
0010	**DRY FALL PAINTING**										940
0100	Walls										
0200	Wallboard and smooth plaster, one coat, brush	1 Pord	910	.009	S.F.	.04	.19		.23	.36	
0210	Roll		1,560	.005		.04	.11		.15	.23	
0220	Spray		2,600	.003		.04	.07		.11	.16	
0230	Two coats, brush		520	.015		.09	.33		.42	.65	
0240	Roll		877	.009		.09	.20		.29	.43	
0250	Spray		1,560	.005		.09	.11		.20	.28	
0260	Concrete or textured plaster, one coat, brush		747	.011		.04	.23		.27	.43	
0270	Roll		1,300	.006		.04	.13		.17	.27	
0280	Spray		1,560	.005		.04	.11		.15	.23	
0290	Two coats, brush		422	.019		.09	.41		.50	.78	
0300	Roll		747	.011		.09	.23		.32	.48	
0310	Spray		1,300	.006		.09	.13		.22	.32	
0320	Concrete block, one coat, brush		747	.011		.04	.23		.27	.43	
0330	Roll		1,300	.006		.04	.13		.17	.27	
0340	Spray		1,560	.005		.04	.11		.15	.23	
0350	Two coats, brush		422	.019		.09	.41		.50	.78	
0360	Roll		747	.011		.09	.23		.32	.48	
0370	Spray		1,300	.006		.09	.13		.22	.32	
0380	Wood, one coat, brush		747	.011		.04	.23		.27	.43	
0390	Roll		1,300	.006		.04	.13		.17	.27	
0400	Spray		877	.009		.04	.20		.24	.38	
0410	Two coats, brush		487	.016		.09	.36		.45	.69	
0420	Roll		747	.011		.09	.23		.32	.48	
0430	Spray		650	.012		.09	.27		.36	.54	
0440	Ceilings										
0450	Wallboard and smooth plaster, one coat, brush	1 Pord	600	.013	S.F.	.04	.29		.33	.53	

09900 | Paints & Coatings

	09910 | Paints		CREW	DAILY OUTPUT	LABOR-HOURS	UNIT	2005 BARE COSTS MAT.	LABOR	EQUIP.	TOTAL	TOTAL INCL O&P
940	0460	Roll	1 Pord	1,040	.008	S.F.	.04	.17		.21	.32
	0470	Spray		1,560	.005		.04	.11		.15	.23
	0480	Two coats, brush		341	.023		.09	.51		.60	.94
	0490	Roll		650	.012		.09	.27		.36	.54
	0500	Spray		1,300	.006		.09	.13		.22	.32
	0510	Concrete or textured plaster, one coat, brush		487	.016		.04	.36		.40	.64
	0520	Roll		877	.009		.04	.20		.24	.38
	0530	Spray		1,560	.005		.04	.11		.15	.23
	0540	Two coats, brush		276	.029		.09	.63		.72	1.14
	0550	Roll		520	.015		.09	.33		.42	.65
	0560	Spray		1,300	.006		.09	.13		.22	.32
	0570	Structural steel, bar joists or metal deck, one coat, spray		1,560	.005		.04	.11		.15	.23
	0580	Two coats, spray		1,040	.008		.09	.17		.26	.37
	09930 | Stains/Transp. Finishes										
100	0010	**VARNISH** 1 coat + sealer, on wood trim, no sanding included	1 Pord	400	.020	S.F.	.07	.44		.51	.80
	0100	Hardwood floors, 2 coats, no sanding included, roller	"	1,890	.004	"	.13	.09		.22	.29
	09963 | Glazed Coatings										
200	0010	**WALL COATINGS**									
	0100	Acrylic glazed coatings, minimum	1 Pord	525	.015	S.F.	.26	.33		.59	.83
	0200	Maximum		305	.026		.54	.57		1.11	1.53
	0300	Epoxy coatings, minimum		525	.015		.33	.33		.66	.90
	0400	Maximum		170	.047		1.02	1.02		2.04	2.80
	0600	Exposed aggregate, troweled on, 1/16" to 1/4", minimum		235	.034		.50	.74		1.24	1.77
	0700	Maximum (epoxy or polyacrylate)		130	.062		1.09	1.34		2.43	3.40
	0900	1/2" to 5/8" aggregate, minimum		130	.062		1.01	1.34		2.35	3.31
	1000	Maximum		80	.100		1.72	2.18		3.90	5.45
	1200	1" aggregate size, minimum		90	.089		1.75	1.93		3.68	5.10
	1300	Maximum		55	.145		2.68	3.16		5.84	8.15
	1500	Exposed aggregate, sprayed on, 1/8" aggregate, minimum		295	.027		.47	.59		1.06	1.49
	1600	Maximum		145	.055		.87	1.20		2.07	2.93
	09990 | Paint Restoration										
500	0010	**SCRAPE AFTER FIRE DAMAGE**									
	0050	Boards, 1" x 4"	1 Pord	336	.024	L.F.		.52		.52	.85
	0060	1" x 6"		260	.031			.67		.67	1.10
	0070	1" x 8"		207	.039			.84		.84	1.38
	0080	1" x 10"		174	.046			1		1	1.64
	0500	Framing, 2" x 4"		265	.030			.66		.66	1.08
	0510	2" x 6"		221	.036			.79		.79	1.29
	0520	2" x 8"		190	.042			.92		.92	1.51
	0530	2" x 10"		165	.048			1.05		1.05	1.73
	0540	2" x 12"		144	.056			1.21		1.21	1.99
	1000	Heavy framing, 3" x 4"		226	.035			.77		.77	1.27
	1010	4" x 4"		210	.038			.83		.83	1.36
	1020	4" x 6"		191	.042			.91		.91	1.50
	1030	4" x 8"		165	.048			1.05		1.05	1.73
	1040	4" x 10"		144	.056			1.21		1.21	1.99
	1060	4" x 12"		131	.061			1.33		1.33	2.18
	2900	For sealing, minimum		825	.010	S.F.	.13	.21		.34	.49
	2920	Maximum		460	.017	"	.26	.38		.64	.91
800	0010	**SANDING** and puttying interior trim, compared to									
	0100	Painting 1 coat, on quality work				L.F.		100%			

Important: See the Reference Section for critical supporting data - Reference Nos., Crews, & Location Factors

09990 | Paint Restoration

Line	Description	Crew	Daily Output	Labor-Hours	Unit	2005 Bare Costs Mat.	Labor	Equip.	Total	Total Incl O&P	
0300	Medium work				L.F.		50%				800
0400	Industrial grade						25%				
0500	Surface protection, placement and removal										
0510	Basic drop cloths	1 Pord	6,400	.001	S.F.		.03		.03	.04	
0520	Masking with paper		800	.010		.03	.22		.25	.39	
0530	Volume cover up (using plastic sheathing, or building paper)		16,000	.001			.01		.01	.02	
0010	**SURFACE PREPARATION, EXTERIOR**										900
0015	Doors, per side, not incl. frames or trim										
0020	Scrape & sand										
0030	Wood, flush	1 Pord	616	.013	S.F.		.28		.28	.46	
0040	Wood, detail		496	.016			.35		.35	.58	
0050	Wood, louvered		280	.029			.62		.62	1.02	
0060	Wood, overhead		616	.013			.28		.28	.46	
0070	Wire brush										
0080	Metal, flush	1 Pord	640	.013	S.F.		.27		.27	.45	
0090	Metal, detail		520	.015			.33		.33	.55	
0100	Metal, louvered		360	.022			.48		.48	.79	
0110	Metal or fibr., overhead		640	.013			.27		.27	.45	
0120	Metal, roll up		560	.014			.31		.31	.51	
0130	Metal, bulkhead		640	.013			.27		.27	.45	
0140	Power wash, based on 2500 lb. operating pressure										
0150	Metal, flush	B-9	2,240	.018	S.F.		.32	.06	.38	.61	
0160	Metal, detail		2,120	.019			.33	.07	.40	.64	
0170	Metal, louvered		2,000	.020			.36	.07	.43	.68	
0180	Metal or fibr., overhead		2,400	.017			.30	.06	.36	.56	
0190	Metal, roll up		2,400	.017			.30	.06	.36	.56	
0200	Metal, bulkhead		2,200	.018			.32	.06	.38	.62	
0400	Windows, per side, not incl. trim										
0410	Scrape & sand										
0420	Wood, 1-2 lite	1 Pord	320	.025	S.F.		.54		.54	.89	
0430	Wood, 3-6 lite		280	.029			.62		.62	1.02	
0440	Wood, 7-10 lite		240	.033			.72		.72	1.19	
0450	Wood, 12 lite		200	.040			.87		.87	1.43	
0460	Wood, Bay / Bow		320	.025			.54		.54	.89	
0470	Wire brush										
0480	Metal, 1-2 lite	1 Pord	480	.017	S.F.		.36		.36	.60	
0490	Metal, 3-6 lite		400	.020			.44		.44	.72	
0500	Metal, Bay / Bow		480	.017			.36		.36	.60	
0510	Power wash, based on 2500 lb. operating pressure										
0520	1-2 lite	B-9	4,400	.009	S.F.		.16	.03	.19	.31	
0530	3-6 lite		4,320	.009			.16	.03	.19	.32	
0540	7-10 lite		4,240	.009			.17	.03	.20	.32	
0550	12 lite		4,160	.010			.17	.03	.20	.33	
0560	Bay / Bow		4,400	.009			.16	.03	.19	.31	
0600	Siding, scrape and sand, light=10-30%, med.=30-70%										
0610	Heavy=70-100%, % of surface to sand										
0650	Texture 1-11, light	1 Pord	480	.017	S.F.		.36		.36	.60	
0660	Med.		440	.018			.40		.40	.65	
0670	Heavy		360	.022			.48		.48	.79	
0680	Wood shingles, shakes, light		440	.018			.40		.40	.65	
0690	Med.		360	.022			.48		.48	.79	
0700	Heavy		280	.029			.62		.62	1.02	
0710	Clapboard, light		520	.015			.33		.33	.55	
0720	Med.		480	.017			.36		.36	.60	
0730	Heavy		400	.020			.44		.44	.72	
0740	Wire brush										

	09990 | Paint Restoration		CREW	DAILY OUTPUT	LABOR-HOURS	UNIT	2005 BARE COSTS MAT.	LABOR	EQUIP.	TOTAL	TOTAL INCL O&P
900	0750	Aluminum, light	1 Pord	600	.013	S.F.		.29		.29	.48
	0760	Med.		520	.015			.33		.33	.55
	0770	Heavy		440	.018			.40		.40	.65
	0780	Pressure wash, based on 2500 lb.. operating pressure									
	0790	Stucco	B-9	3,080	.013	S.F.		.23	.05	.28	.44
	0800	Aluminum or vinyl		3,200	.013			.22	.04	.26	.43
	0810	Siding, masonry, brick & block		2,400	.017			.30	.06	.36	.56
	1300	Miscellaneous, wire brush									
	1310	Metal, pedestrian gate	1 Pord	100	.080	S.F.		1.74		1.74	2.86
910	0010	**SURFACE PREPARATION, INTERIOR**									
	0020	Doors									
	0030	Scrape & sand									
	0040	Wood, flush	1 Pord	616	.013	S.F.		.28		.28	.46
	0050	Wood, detail		496	.016			.35		.35	.58
	0060	Wood, louvered		280	.029			.62		.62	1.02
	0070	Wire brush									
	0080	Metal, flush	1 Pord	640	.013	S.F.		.27		.27	.45
	0090	Metal, detail		520	.015			.33		.33	.55
	0100	Metal, louvered		360	.022			.48		.48	.79
	0110	Hand wash									
	0120	Wood, flush	1 Pord	2,160	.004	S.F.		.08		.08	.13
	0130	Wood, detailed		2,000	.004			.09		.09	.14
	0140	Wood, louvered		1,360	.006			.13		.13	.21
	0150	Metal, flush		2,160	.004			.08		.08	.13
	0160	Metal, detail		2,000	.004			.09		.09	.14
	0170	Metal, louvered		1,360	.006			.13		.13	.21
	0400	Windows, per side, not incl. trim									
	0410	Scrape & sand									
	0420	Wood, 1-2 lite	1 Pord	360	.022	S.F.		.48		.48	.79
	0430	Wood, 3-6 lite		320	.025			.54		.54	.89
	0440	Wood, 7-10 lite		280	.029			.62		.62	1.02
	0450	Wood, 12 lite		240	.033			.72		.72	1.19
	0460	Wood, Bay / Bow		360	.022			.48		.48	.79
	0470	Wire brush									
	0480	Metal, 1-2 lite	1 Pord	520	.015	S.F.		.33		.33	.55
	0490	Metal, 3-6 lite		440	.018			.40		.40	.65
	0500	Metal, Bay / Bow		520	.015			.33		.33	.55
	0600	Walls, sanding, light=10-30%									
	0610	Med.=30-70%, heavy=70-100%, % of surface to sand									
	0650	Walls, sand									
	0660	Drywall, gypsum, plaster, light	1 Pord	3,077	.003	S.F.		.06		.06	.09
	0670	Drywall, gypsum, plaster, med.		2,160	.004			.08		.08	.13
	0680	Drywall, gypsum, plaster, heavy		923	.009			.19		.19	.31
	0690	Wood, T&G, light		2,400	.003			.07		.07	.12
	0700	Wood, T&G, med.		1,600	.005			.11		.11	.18
	0710	Wood, T&G, heavy		800	.010			.22		.22	.36
	0720	Walls, wash									
	0730	Drywall, gypsum, plaster	1 Pord	3,200	.002	S.F.		.05		.05	.09
	0740	Wood, T&G		3,200	.002			.05		.05	.09
	0750	Masonry, brick & block, smooth		2,800	.003			.06		.06	.10
	0760	Masonry, brick & block, coarse		2,000	.004			.09		.09	.14
	8000	For Chemical Washing, see Division 04930									

For information about Means Estimating Seminars, see yellow pages 12 and 13 in back of book

Division 10 Specialties

stimating Tips

eneral

The items in this division are usually priced per square foot or each.

Many items in Division 10 require some type of support system or special anchors that are not usually furnished with the item. The required anchors must be added to the estimate in the appropriate division.

Some items in Division 10, such as lockers, may require assembly before installation. Verify the amount of assembly required. Assembly can often exceed installation time.

10150 Compartments & Cubicles

- Support angles and blocking are not included in the installation of toilet compartments, shower/dressing compartments or cubicles. Appropriate line items from Divisions 5 or 6 may need to be added to support the installations.
- Toilet partitions are priced by the stall. A stall consists of a side wall, pilaster and door with hardware. Toilet tissue holders and grab bars are extra.

10600 Partitions

- The required acoustical rating of a folding partition can have a significant impact on costs. Verify the sound transmission coefficient rating of the panel priced to the specification requirements.

10800 Toilet/Bath/Laundry Accessories

- Grab bar installation does not include supplemental blocking or backing to support the required load. When grab bars are installed at an existing facility provisions must be made to attach the grab bars to solid structure.

Reference Numbers

Reference numbers are shown in bold squares at the beginning of some major classifications. These numbers refer to related items in the Reference Section. The reference information may be an estimating procedure, an alternate pricing method or technical information.

Note: Not all subdivisions listed here necessarily appear in this publication.

part of this publication may be reproduced, stored in a retrieval system, or transmitted in any form by any means without prior written permission of Reed Construction Data.

10150 | Compartments & Cubicles

	10185 | Shower/Dressing Compartments		CREW	DAILY OUTPUT	LABOR-HOURS	UNIT	2005 BARE COSTS MAT.	LABOR	EQUIP.	TOTAL	TOTAL INCL O&P
100	0010	**PARTITIONS, SHOWER** Floor mounted, no plumbing									
	0100	Cabinet, incl. base, no door, painted steel, 1″ thick walls	2 Shee	5	3.200	Ea.	730	85		815	945
	0300	With door, fiberglass		4.50	3.556		590	94.50		684.50	810
	0600	Galvanized and painted steel, 1″ thick walls		5	3.200		770	85		855	990
	0800	Stall, 1″ thick wall, no base, enameled steel		5	3.200		830	85		915	1,050
	1500	Circular fiberglass, cabinet 36″ diameter,		4	4		605	106		711	845
	1700	One piece, 36″ diameter, less door		4	4		510	106		616	740
	1800	With door		3.50	4.571		840	121		961	1,125
	2400	Glass stalls, with doors, no receptors, chrome on brass		3	5.333		1,200	142		1,342	1,575
	2700	Anodized aluminum		4	4		830	106		936	1,100
	3200	Receptors, precast terrazzo, 32″ x 32″	2 Marb	14	1.143		225	26.50		251.50	292
	3300	48″ x 34″		9.50	1.684		385	39		424	490
	3500	Plastic, simulated terrazzo receptor, 32″ x 32″		14	1.143		95	26.50		121.50	149
	3600	32″ x 48″		12	1.333		140	31		171	206
	3800	Precast concrete, colors, 32″ x 32″		14	1.143		188	26.50		214.50	251
	3900	48″ x 48″		8	2		200	46.50		246.50	298
	4100	Shower doors, economy plastic, 24″ wide	1 Shee	9	.889		104	23.50		127.50	154
	4200	Tempered glass door, economy		8	1		182	26.50		208.50	245
	4400	Folding, tempered glass, aluminum frame		6	1.333		315	35.50		350.50	405
	4700	Deluxe, tempered glass, chrome on brass frame, minimum		8	1		257	26.50		283.50	325
	4800	Maximum		1	8		700	212		912	1,125
	4850	On anodized aluminum frame, minimum		2	4		122	106		228	310
	4900	Maximum		1	8		410	212		622	810
	5100	Shower enclosure, tempered glass, anodized alum. frame									
	5120	2 panel & door, corner unit, 32″ x 32″	1 Shee	2	4	Ea.	410	106		516	635
	5140	Neo-angle corner unit, 16″ x 24″ x 16″	″	2	4		755	106		861	1,000
	5200	Shower surround, 3 wall, polypropylene, 32″ x 32″	1 Carp	4	2		276	48		324	385
	5220	PVC, 32″ x 32″		4	2		259	48		307	365
	5240	Fiberglass		4	2		300	48		348	410
	5250	2 wall, polypropylene, 32″ x 32″		4	2		215	48		263	320
	5270	PVC		4	2		267	48		315	375
	5290	Fiberglass		4	2		297	48		345	405
	5300	Tub doors, tempered glass & frame, minimum	1 Shee	8	1		175	26.50		201.50	238
	5400	Maximum		6	1.333		405	35.50		440.50	505
	5600	Chrome plated, brass frame, minimum		8	1		229	26.50		255.50	296
	5700	Maximum		6	1.333		455	35.50		490.50	560
	5900	Tub/shower enclosure, temp. glass, alum. frame, minimum		2	4		310	106		416	520
	6200	Maximum		1.50	5.333		605	142		747	905
	6500	On chrome-plated brass frame, minimum		2	4		430	106		536	655
	6600	Maximum		1.50	5.333		885	142		1,027	1,200
	6800	Tub surround, 3 wall, polypropylene	1 Carp	4	2		174	48		222	273
	6900	PVC		4	2		264	48		312	370
	7000	Fiberglass, minimum		4	2		294	48		342	405
	7100	Maximum		3	2.667		505	64		569	665

10200 | Louvers & Vents

	10210 | Wall Louvers		CREW	DAILY OUTPUT	LABOR-HOURS	UNIT	2005 BARE COSTS MAT.	LABOR	EQUIP.	TOTAL	TOTAL INCL O&P
800	0010	**LOUVERS** Aluminum with screen, residential, 8″ x 8″	1 Carp	38	.211	Ea.	9.35	5.05		14.40	18.90
	0100	12″ x 12″		38	.211		10.20	5.05		15.25	19.80

10200 | Louvers & Vents

10210 | Wall Louvers

		CREW	DAILY OUTPUT	LABOR-HOURS	UNIT	2005 BARE COSTS MAT.	LABOR	EQUIP.	TOTAL	TOTAL INCL O&P	
0200	12″ x 18″	1 Carp	35	.229	Ea.	13.30	5.50		18.80	24	800
0250	14″ x 24″		30	.267		17.50	6.40		23.90	30	
0300	18″ x 24″		27	.296		21	7.10		28.10	35.50	
0500	24″ x 30″		24	.333		25.50	8		33.50	41.50	
0700	Triangle, adjustable, small		20	.400		25.50	9.60		35.10	44.50	
0800	Large		15	.533		42	12.80		54.80	67.50	
2100	Midget, aluminum, 3/4″ deep, 1″ diameter		85	.094		.87	2.26		3.13	4.80	
2150	3″ diameter		60	.133		1.81	3.20		5.01	7.45	
2200	4″ diameter		50	.160		2.81	3.84		6.65	9.60	
2250	6″ diameter		30	.267		3.35	6.40		9.75	14.55	
2300	Ridge vent strip, mill finish	1 Shee	155	.052	L.F.	2.53	1.37		3.90	5.10	
2400	Under eaves vent, aluminum, mill finish, 16″ x 4″	1 Carp	48	.167	Ea.	1.78	4		5.78	8.75	
2500	16″ x 8″		48	.167		1.97	4		5.97	8.95	
7000	Vinyl gable vent, 8″ x 8″		38	.211		10.50	5.05		15.55	20	
7020	12″ x 12″		38	.211		21.50	5.05		26.55	32	
7080	12″ x 18″		35	.229		26	5.50		31.50	38	
7200	18″ x 24″		30	.267		29.50	6.40		35.90	43.50	

10300 | Fireplaces & Stoves

10305 | Manufactured Fireplaces

		CREW	DAILY OUTPUT	LABOR-HOURS	UNIT	2005 BARE COSTS MAT.	LABOR	EQUIP.	TOTAL	TOTAL INCL O&P	
0010	**FIREPLACE, PREFABRICATED** Free standing or wall hung										100
0100	with hood & screen, minimum	1 Carp	1.30	6.154	Ea.	1,150	148		1,298	1,500	
0150	Average		1	8		1,375	192		1,567	1,825	
0200	Maximum		.90	8.889		3,350	213		3,563	4,050	
0500	Chimney dbl. wall, all stainless, over 8′-6″, 7″ diam., add		33	.242	V.L.F.	52	5.80		57.80	67	
0600	10″ diameter, add		32	.250		55	6		61	70.50	
0700	12″ diameter, add		31	.258		72	6.20		78.20	89.50	
0800	14″ diameter, add		30	.267		91	6.40		97.40	111	
1000	Simulated brick chimney top, 4′ high, 16″ x 16″		10	.800	Ea.	196	19.20		215.20	249	
1100	24″ x 24″		7	1.143	″	365	27.50		392.50	445	
1500	Simulated logs, gas fired, 40,000 BTU, 2′ long, minimum		7	1.143	Set	480	27.50		507.50	570	
1600	Maximum		6	1.333		670	32		702	790	
1700	Electric, 1,500 BTU, 1′-6″ long, minimum		7	1.143		136	27.50		163.50	196	
1800	11,500 BTU, maximum		6	1.333		293	32		325	375	
2000	Fireplace, built-in, 36″ hearth, radiant		1.30	6.154	Ea.	585	148		733	895	
2100	Recirculating, small fan		1	8		840	192		1,032	1,250	
2150	Large fan		.90	8.889		1,550	213		1,763	2,075	
2200	42″ hearth, radiant		1.20	6.667		750	160		910	1,100	
2300	Recirculating, small fan		.90	8.889		980	213		1,193	1,425	
2350	Large fan		.80	10		1,875	240		2,115	2,450	
2400	48″ hearth, radiant		1.10	7.273		1,400	175		1,575	1,825	
2500	Recirculating, small fan		.80	10		1,725	240		1,965	2,300	
2550	Large fan		.70	11.429		2,675	274		2,949	3,425	
3000	See through, including doors		.80	10		2,200	240		2,440	2,825	
3200	Corner (2 wall)		1	8		1,100	192		1,292	1,525	

10300 | Fireplaces & Stoves

		10310 | Fireplace Specialties & Accessories	CREW	DAILY OUTPUT	LABOR-HOURS	UNIT	2005 BARE COSTS MAT.	LABOR	EQUIP.	TOTAL	TOTAL INCL O&P
100	0010	**FIREPLACE ACCESSORIES** Chimney screens, galv., 13" x 13" flue	1 Bric	8	1	Ea.	33.50	24.50		58	78
	0050	Galv., 24" x 24" flue		5	1.600		100	39.50		139.50	176
	0200	Stainless steel, 13" x 13" flue		8	1		264	24.50		288.50	330
	0250	20" x 20" flue		5	1.600		360	39.50		399.50	460
	0400	Cleanout doors and frames, cast iron, 8" x 8"		12	.667		29	16.45		45.45	59.50
	0450	12" x 12"		10	.800		33.50	19.75		53.25	69.50
	0500	18" x 24"		8	1		105	24.50		129.50	156
	0550	Cast iron frame, steel door, 24" x 30"		5	1.600		227	39.50		266.50	315
	0800	Damper, rotary control, steel, 30" opening		6	1.333		64	33		97	125
	0850	Cast iron, 30" opening		6	1.333		71	33		104	133
	1200	Steel plate, poker control, 60" opening		8	1		225	24.50		249.50	289
	1250	84" opening, special opening		5	1.600		410	39.50		449.50	515
	1400	"Universal" type, chain operated, 32" x 20" opening		8	1		156	24.50		180.50	212
	1450	48" x 24" opening		5	1.600		261	39.50		300.50	355
	1600	Dutch Oven door and frame, cast iron, 12" x 15" opening		13	.615		92	15.20		107.20	127
	1650	Copper plated, 12" x 15" opening		13	.615		177	15.20		192.20	220
	1800	Fireplace forms, no accessories, 32" opening		3	2.667		525	66		591	690
	1900	36" opening		2.50	3.200		635	79		714	830
	2000	40" opening		2	4		765	99		864	1,000
	2100	78" opening		1.50	5.333		1,100	132		1,232	1,425
	2400	Squirrel and bird screens, galvanized, 8" x 8" flue		16	.500		36	12.35		48.35	60.50
	2450	13" x 13" flue	↓	12	.667	↓	41.50	16.45		57.95	73

		10320 | Stoves	CREW	DAILY OUTPUT	LABOR-HOURS	UNIT	MAT.	LABOR	EQUIP.	TOTAL	TOTAL INCL O&P
100	0010	**WOODBURNING STOVES** Cast iron, minimum	2 Carp	1.30	12.308	Ea.	705	295		1,000	1,275
	0020	Average		1	16		1,300	385		1,685	2,075
	0030	Maximum	↓	.80	20		2,175	480		2,655	3,200
	0050	For gas log lighter, add				↓	39			39	43

10340 | Manufactured Exterior Specialties

		10342 | Cupolas	CREW	DAILY OUTPUT	LABOR-HOURS	UNIT	2005 BARE COSTS MAT.	LABOR	EQUIP.	TOTAL	TOTAL INCL O&P
100	0010	**CUPOLA** Stock units, pine, painted, 18" sq., 28" high, alum. roof	1 Carp	4.10	1.951	Ea.	141	47		188	235
	0100	Copper roof		3.80	2.105		150	50.50		200.50	251
	0300	23" square, 33" high, aluminum roof		3.70	2.162		236	52		288	350
	0400	Copper roof		3.30	2.424		252	58		310	375
	0600	30" square, 37" high, aluminum roof		3.70	2.162		360	52		412	485
	0700	Copper roof		3.30	2.424		380	58		438	520
	0900	Hexagonal, 31" wide, 46" high, copper roof		4	2		540	48		588	675
	1000	36" wide, 50" high, copper roof	↓	3.50	2.286		605	55		660	760
	1200	For deluxe stock units, add to above					25%				
	1400	For custom built units, add to above				↓	50%	50%			

		10344 | Weathervanes	CREW	DAILY OUTPUT	LABOR-HOURS	UNIT	MAT.	LABOR	EQUIP.	TOTAL	TOTAL INCL O&P
800	0010	**WEATHERVANES**									
	0020	Residential types, minimum	1 Carp	8	1	Ea.	44	24		68	89.50
	0100	Maximum	"	2	4	"	880	96		976	1,125

10350 | Flagpoles

10355 | Flagpoles		CREW	DAILY OUTPUT	LABOR-HOURS	UNIT	2005 BARE COSTS MAT.	LABOR	EQUIP.	TOTAL	TOTAL INCL O&P	
0010	**FLAGPOLE**, Ground set										400
0050	Not including base or foundation										
0100	Aluminum, tapered, ground set 20′ high	K-1	2	8	Ea.	820	172	88	1,080	1,300	
0200	25′ high	↓	1.70	9.412	↓	965	203	104	1,272	1,500	
0300	30′ high	↓	1.50	10.667	↓	975	230	117	1,322	1,600	
0500	40′ high	↓	1.20	13.333	↓	2,000	287	147	2,434	2,875	

10520 | Fire Protection Specialties

10525 | Fire Prot. Specialties		CREW	DAILY OUTPUT	LABOR-HOURS	UNIT	2005 BARE COSTS MAT.	LABOR	EQUIP.	TOTAL	TOTAL INCL O&P	
0010	**FIRE EXTINGUISHERS**										300
0120	CO2, portable with swivel horn, 5 lb.				Ea.	130			130	143	
0140	With hose and "H" horn, 10 lb.				"	160			160	176	
1000	Dry chemical, pressurized										
1040	Standard type, portable, painted, 2-1/2 lb.				Ea.	27.50			27.50	30.50	
1080	10 lb.				↓	80			80	88	
1100	20 lb.				↓	110			110	121	
1120	30 lb.				↓	157			157	173	
2000	ABC all purpose type, portable, 2-1/2 lb.				↓	30			30	33	
2080	9-1/2 lb.				↓	68			68	75	

10530 | Protective Covers

10535 | Awnings & Canopies		CREW	DAILY OUTPUT	LABOR-HOURS	UNIT	2005 BARE COSTS MAT.	LABOR	EQUIP.	TOTAL	TOTAL INCL O&P	
0010	**CANOPIES, RESIDENTIAL** Prefabricated										100
0500	Carport, free standing, baked enamel, alum., .032″, 40 psf										
0520	16′ x 8′, 4 posts	2 Carp	3	5.333	Ea.	3,100	128		3,228	3,650	
0600	20′ x 10′, 6 posts	↓	2	8	↓	3,250	192		3,442	3,900	
0605	30′ x 10′, 8 posts	↓	2	8	↓	4,875	192		5,067	5,675	
1000	Door canopies, extruded alum., .032″, 42″ projection, 4′ wide	1 Carp	8	1	↓	365	24		389	445	
1020	6′ wide	"	6	1.333	↓	445	32		477	545	
1040	8′ wide	2 Carp	9	1.778	↓	575	42.50		617.50	705	
1060	10′ wide	↓	7	2.286	↓	670	55		725	835	
1080	12′ wide	↓	5	3.200	↓	795	77		872	1,000	
1200	54″ projection, 4′ wide	1 Carp	8	1	↓	470	24		494	560	
1220	6′ wide	"	6	1.333	↓	600	32		632	715	
1240	8′ wide	2 Carp	9	1.778	↓	790	42.50		832.50	945	
1260	10′ wide	↓	7	2.286	↓	895	55		950	1,075	
1280	12′ wide	↓	5	3.200	↓	1,025	77		1,102	1,250	
1300	Painted, add				↓	20%					
1310	Bronze anodized, add				↓	50%					
3000	Window awnings, aluminum, window 3′ high, 4′ wide	1 Carp	10	.800	↓	224	19.20		243.20	279	
3020	6′ wide	"	8	1	↓	261	24		285	330	
3040	9′ wide	2 Carp	9	1.778	↓	425	42.50		467.50	540	

10530 | Protective Covers

10535 | Awnings & Canopies

		10535 Awnings & Canopies	CREW	DAILY OUTPUT	LABOR-HOURS	UNIT	2005 BARE COSTS MAT.	LABOR	EQUIP.	TOTAL	TOTAL INCL O&P
100	3060	12′ wide	2 Carp	5	3.200	Ea.	580	77		657	765
	3100	Window, 4′ high, 4′ wide	1 Carp	10	.800		274	19.20		293.20	335
	3120	6′ wide	″	8	1		370	24		394	445
	3140	9′ wide	2 Carp	9	1.778		500	42.50		542.50	625
	3160	12′ wide	″	5	3.200		645	77		722	840
	3200	Window, 6′ high, 4′ wide	1 Carp	10	.800		415	19.20		434.20	490
	3220	6′ wide	″	8	1		570	24		594	670
	3240	9′ wide	2 Carp	9	1.778		785	42.50		827.50	935
	3260	12′ wide	″	5	3.200		1,075	77		1,152	1,300
	3400	Roll-up aluminum, 2′-6″ wide	1 Carp	14	.571		96	13.70		109.70	130
	3420	3′ wide		12	.667		115	16		131	154
	3440	4′ wide		10	.800		148	19.20		167.20	196
	3460	6′ wide	↓	8	1		183	24		207	242
	3480	9′ wide	2 Carp	9	1.778		266	42.50		308.50	365
	3500	12′ wide	″	5	3.200	↓	325	77		402	490
	3600	Window awnings, canvas, 24″ drop, 3′ wide	1 Carp	30	.267	L.F.	36.50	6.40		42.90	51.50
	3620	4′ wide		40	.200		33	4.80		37.80	44.50
	3700	30″ drop, 3′ wide		30	.267		51.50	6.40		57.90	67.50
	3720	4′ wide		40	.200		44.50	4.80		49.30	56.50
	3740	5′ wide		45	.178		40	4.27		44.27	51.50
	3760	6′ wide		48	.167		37	4		41	47.50
	3780	8′ wide		48	.167		30.50	4		34.50	40.50
	3800	10′ wide	↓	50	.160	↓	28	3.84		31.84	37.50

10550 | Postal Specialties

10555 | Mail Delivery Systems

		10555 Mail Delivery Systems	CREW	DAILY OUTPUT	LABOR-HOURS	UNIT	2005 BARE COSTS MAT.	LABOR	EQUIP.	TOTAL	TOTAL INCL O&P
600	0011	MAIL BOXES									
	1900	Letter slot, residential	1 Carp	20	.400	Ea.	60	9.60		69.60	82.50
	2400	Residential, galv. steel, small 20″ x 7″ x 9″	1 Clab	16	.500		110	8.70		118.70	136
	2410	With galv. steel post, 54″ long		6	1.333		195	23		218	255
	2420	Large, 24″ x 12″ x 15″		16	.500		125	8.70		133.70	153
	2430	With galv. steel post, 54″ long		6	1.333		225	23		248	288
	2440	Decorative, polyethylene, 22″ x 10″ x 10″		16	.500		40	8.70		48.70	59
	2450	With alum. post, decorative, 54″ long	↓	6	1.333	↓	70	23		93	117

10670 | Storage Shelving

10674 | Storage Shelving

		10674 Storage Shelving	CREW	DAILY OUTPUT	LABOR-HOURS	UNIT	2005 BARE COSTS MAT.	LABOR	EQUIP.	TOTAL	TOTAL INCL O&P
500	0010	**SHELVING** Metal, industrial, cross-braced, 3′ wide, 12″ deep	1 Sswk	175	.046	SF Shlf	6.50	1.17		7.67	9.40
	0100	24″ deep		330	.024		4.95	.62		5.57	6.65
	2200	Wide span, 1600 lb. capacity per shelf, 6′ wide, 24″ deep		380	.021		8.10	.54		8.64	9.95
	2400	36″ deep	↓	440	.018	↓	7.10	.46		7.56	8.70

10670 | Storage Shelving

10674	Storage Shelving	Crew	Daily Output	Labor-Hours	Unit	2005 Bare Costs Mat.	Labor	Equip.	Total	Total Incl O&P	
3000	Residential, vinyl covered wire, wardrobe, 12" deep	1 Carp	195	.041	L.F.	3.15	.98		4.13	5.15	500
3100	16" deep		195	.041		3.29	.98		4.27	5.30	
3200	Standard, 6" deep		195	.041		2.92	.98		3.90	4.88	
3300	9" deep		195	.041		2.92	.98		3.90	4.88	
3400	12" deep		195	.041		2.92	.98		3.90	4.88	
3500	16" deep		195	.041		2.98	.98		3.96	4.95	
3600	20" deep		195	.041		3.04	.98		4.02	5	
3700	Support bracket		80	.100	Ea.	1.53	2.40		3.93	5.75	

10800 | Toilet/Bath/Laundry Accessories

10810	Toilet Accessories	Crew	Daily Output	Labor-Hours	Unit	2005 Bare Costs Mat.	Labor	Equip.	Total	Total Incl O&P	
0010	**COMMERCIAL TOILET ACCESSORIES**										100
0200	Curtain rod, stainless steel, 5' long, 1" diameter	1 Carp	13	.615	Ea.	30.50	14.75		45.25	58.50	
0300	1-1/4" diameter		13	.615		29	14.75		43.75	56.50	
0800	Grab bar, straight, 1-1/4" diameter, stainless steel, 18" long		24	.333		18.75	8		26.75	34	
1100	36" long		20	.400		19.75	9.60		29.35	38	
1105	42" long		20	.400		20.50	9.60		30.10	39	
3000	Mirror, with stainless steel 3/4" square frame, 18" x 24"		20	.400		65.50	9.60		75.10	88.50	
3100	36" x 24"		15	.533		104	12.80		116.80	136	
3300	72" x 24"		6	1.333		196	32		228	271	
4300	Robe hook, single, regular		36	.222		4.95	5.35		10.30	14.50	
4400	Heavy duty, concealed mounting		36	.222		11.10	5.35		16.45	21.50	
6400	Towel bar, stainless steel, 18" long		23	.348		35	8.35		43.35	52.50	
6500	30" long		21	.381		56.50	9.15		65.65	78	
7400	Tumbler holder, tumbler only		30	.267		25	6.40		31.40	38.50	
7500	Soap, tumbler & toothbrush		30	.267		22.50	6.40		28.90	36	

10820	Bath Accessories	Crew	Daily Output	Labor-Hours	Unit	Mat.	Labor	Equip.	Total	Total Incl O&P	
0010	**MEDICINE CABINETS** with mirror, st. st. frame, 16" x 22", unlighted	1 Carp	14	.571	Ea.	69.50	13.70		83.20	100	400
0100	Wood frame		14	.571		96.50	13.70		110.20	130	
0300	Sliding mirror doors, 20" x 16" x 4-3/4", unlighted		7	1.143		86.50	27.50		114	142	
0400	24" x 19" x 8-1/2", lighted		5	1.600		136	38.50		174.50	215	
0600	Triple door, 30" x 32", unlighted, plywood body		7	1.143		214	27.50		241.50	283	
0700	Steel body		7	1.143		282	27.50		309.50	355	
0900	Oak door, wood body, beveled mirror, single door		7	1.143		125	27.50		152.50	185	
1000	Double door		6	1.333		320	32		352	405	

For information about Means Estimating Seminars, see yellow pages 12 and 13 in back of book

Division Notes

	CREW	DAILY OUTPUT	LABOR-HOURS	UNIT	2005 BARE COSTS MAT.	LABOR	EQUIP.	TOTAL	TOTAL INCL O&P

Division 11 Equipment

*stimating Tips

eneral

The items in this division are usually priced per square foot or each. Many of these items are purchased by the owner for installation by the contractor. Check the specifications for responsibilities, and include time for receiving, storage, installation and mechanical and electrical hook-ups in the appropriate divisions.

- Many items in Division 11 require some type of support system that is not usually furnished with the item. Examples of these systems include blocking for the attachment of casework and support angles for ceiling hung projection screens. The required blocking or supports must be added to the estimate in the appropriate division.
- Some items in Division 11 may require assembly or electrical hook-ups. Verify the amount of assembly required or the need for a hard electrical connection and add the appropriate costs.

Reference Numbers

Reference numbers are shown in bold squares at the beginning of some major classifications. These numbers refer to related items in the Reference Section. The reference information may be an estimating procedure, an alternate pricing method or technical information.

Note: Not all subdivisions listed here necessarily appear in this publication.

lo part of this publication may be reproduced, stored in a retrieval system, or transmitted in any form r by any means without prior written permission of Reed Construction Data.

11010 | Maintenance Equipment

	11013 | Floor/Wall Cleaning Equipment		CREW	DAILY OUTPUT	LABOR-HOURS	UNIT	2005 BARE COSTS MAT.	LABOR	EQUIP.	TOTAL	TOTAL INCL O&P
800	0010	VACUUM CLEANING									
	0020	Central, 3 inlet, residential	1 Skwk	.90	8.889	Total	610	212		822	1,025
	0400	5 inlet system, residential		.50	16		925	380		1,305	1,675
	0600	7 inlet system, commercial		.40	20		1,025	480		1,505	1,950
	0800	9 inlet system, residential		.30	26.667		1,300	635		1,935	2,500
	4010	Rule of thumb: First 1200 S.F., installed									1,125
	4020	For each additional S.F., add				S.F.					.18

11400 | Food Service Equipment

	11405 | Food Storage Equipment		CREW	DAILY OUTPUT	LABOR-HOURS	UNIT	2005 BARE COSTS MAT.	LABOR	EQUIP.	TOTAL	TOTAL INCL O&P
800	0010	WINE CELLAR, refrigerated, Redwood interior, carpeted, walk-in type									
	0020	6'-8" high, including racks									
	0200	80" W x 48" D for 900 bottles	2 Carp	1.50	10.667	Ea.	2,925	256		3,181	3,625
	0250	80" W x 72" D for 1300 bottles		1.33	12.030		3,850	289		4,139	4,750
	0300	80" W x 94" D for 1900 bottles		1.17	13.675		5,000	330		5,330	6,050

11450 | Residential Equipment

	11454 | Residential Appliances		CREW	DAILY OUTPUT	LABOR-HOURS	UNIT	2005 BARE COSTS MAT.	LABOR	EQUIP.	TOTAL	TOTAL INCL O&P
500	0010	RESIDENTIAL APPLIANCES									
	0020	Cooking range, 30" free standing, 1 oven, minimum	2 Clab	10	1.600	Ea.	241	28		269	310
	0050	Maximum		4	4		1,475	69.50		1,544.50	1,750
	0150	2 oven, minimum		10	1.600		1,475	28		1,503	1,675
	0200	Maximum		10	1.600		1,550	28		1,578	1,750
	0350	Built-in, 30" wide, 1 oven, minimum	1 Elec	6	1.333		445	36.50		481.50	550
	0400	Maximum	2 Carp	2	8		1,325	192		1,517	1,775
	0500	2 oven, conventional, minimum		4	4		930	96		1,026	1,200
	0550	1 conventional, 1 microwave, maximum		2	8		1,475	192		1,667	1,950
	0700	Free-standing, 1 oven, 21" wide range, minimum	2 Clab	10	1.600		247	28		275	320
	0750	21" wide, maximum	"	4	4		273	69.50		342.50	420
	0900	Counter top cook tops, 4 burner, standard, minimum	1 Elec	6	1.333		186	36.50		222.50	264
	0950	Maximum		3	2.667		520	73		593	690
	1050	As above, but with grille and griddle attachment, minimum		6	1.333		465	36.50		501.50	570
	1100	Maximum		3	2.667		640	73		713	825
	1250	Microwave oven, minimum		4	2		82.50	54.50		137	180
	1300	Maximum		2	4		390	109		499	610
	1750	Compactor, residential size, 4 to 1 compaction, minimum	1 Carp	5	1.600		415	38.50		453.50	520
	1800	Maximum	"	3	2.667		470	64		534	625
	2000	Deep freeze, 15 to 23 C.F., minimum	2 Clab	10	1.600		390	28		418	475
	2050	Maximum		5	3.200		515	55.50		570.50	660
	2200	30 C.F., minimum		8	2		750	34.50		784.50	885

11454 | Residential Appliances

		CREW	DAILY OUTPUT	LABOR-HOURS	UNIT	2005 BARE COSTS MAT.	LABOR	EQUIP.	TOTAL	TOTAL INCL O&P	
2250	Maximum	2 Clab	3	5.333	Ea.	850	92.50		942.50	1,100	500
2450	Dehumidifier, portable, automatic, 15 pint					149			149	164	
2550	40 pint					166			166	183	
2750	Dishwasher, built-in, 2 cycles, minimum	L-1	4	2.500		248	67.50		315.50	385	
2800	Maximum		2	5		289	135		424	540	
2950	4 or more cycles, minimum		4	2.500		262	67.50		329.50	400	
2960	Average		4	2.500		350	67.50		417.50	495	
3000	Maximum		2	5		540	135		675	815	
3200	Dryer, automatic, minimum	L-2	3	5.333		282	111		393	500	
3250	Maximum	"	2	8		790	166		956	1,150	
3300	Garbage disposal, sink type, minimum	L-1	10	1		42.50	27		69.50	91	
3350	Maximum	"	10	1		145	27		172	205	
3550	Heater, electric, built-in, 1250 watt, ceiling type, minimum	1 Elec	4	2		72	54.50		126.50	169	
3600	Maximum		3	2.667		118	73		191	249	
3700	Wall type, minimum		4	2		102	54.50		156.50	202	
3750	Maximum		3	2.667		136	73		209	268	
3900	1500 watt wall type, with blower		4	2		127	54.50		181.50	228	
3950	3000 watt		3	2.667		258	73		331	405	
4150	Hood for range, 2 speed, vented, 30" wide, minimum	L-3	5	2		37.50	49.50		87	124	
4200	Maximum		3	3.333		595	82		677	795	
4300	42" wide, minimum		5	2		225	49.50		274.50	330	
4330	Custom		5	2		625	49.50		674.50	770	
4350	Maximum		3	3.333		760	82		842	975	
4500	For ventless hood, 2 speed, add					15			15	16.50	
4650	For vented 1 speed, deduct from maximum					39			39	43	
4850	Humidifier, portable, 8 gallons per day					149			149	164	
5000	15 gallons per day					179			179	197	
5200	Icemaker, automatic, 20 lb. per day	1 Plum	7	1.143		360	31		391	445	
5350	51 lb. per day	"	2	4		1,000	108		1,108	1,275	
5380	Oven, built in, standard	1 Elec	4	2		385	54.50		439.50	510	
5390	Deluxe	"	2	4		1,925	109		2,034	2,275	
5450	Refrigerator, no frost, 6 C.F.	2 Clab	15	1.067		285	18.50		303.50	345	
5500	Refrigerator, no frost, 10 C.F. to 12 C.F. minimum		10	1.600		440	28		468	530	
5600	Maximum		6	2.667		680	46.50		726.50	830	
5750	14 C.F. to 16 C.F., minimum		9	1.778		445	31		476	540	
5800	Maximum		5	3.200		480	55.50		535.50	625	
5950	18 C.F. to 20 C.F., minimum		8	2		510	34.50		544.50	620	
6000	Maximum		4	4		840	69.50		909.50	1,050	
6150	21 C.F. to 29 C.F., minimum		7	2.286		645	39.50		684.50	780	
6200	Maximum		3	5.333		2,075	92.50		2,167.50	2,425	
6400	Sump pump cellar drainer, pedestal, 1/3 H.P., molded PVC base	1 Plum	3	2.667		87	72		159	214	
6450	Solid brass	"	2	4		179	108		287	375	
6460	Sump pump, see also division 15440-940										
6650	Washing machine, automatic, minimum	1 Plum	3	2.667	Ea.	279	72		351	425	
6700	Maximum	"	1	8		1,025	216		1,241	1,475	
6900	Water heater, electric, glass lined, 30 gallon, minimum	L-1	5	2		252	54		306	365	
6950	Maximum		3	3.333		350	90		440	535	
7100	80 gallon, minimum		2	5		485	135		620	750	
7150	Maximum		1	10		670	270		940	1,175	
7180	Water heater, gas, glass lined, 30 gallon, minimum	2 Plum	5	3.200		345	86		431	520	
7220	Maximum		3	5.333		480	144		624	765	
7260	50 gallon, minimum		2.50	6.400		480	172		652	815	
7300	Maximum		1.50	10.667		665	287		952	1,200	
7310	Water heater, see also division 15480-200										
7350	Water softener, automatic, to 30 grains per gallon	2 Plum	5	3.200	Ea.	430	86		516	615	
7400	To 100 grains per gallon	"	4	4		600	108		708	835	

	11454 | Residential Appliances		CREW	DAILY OUTPUT	LABOR-HOURS	UNIT	2005 BARE COSTS MAT.	LABOR	EQUIP.	TOTAL	TOTAL INCL O&P
500	7450	Vent kits for dryers	1 Carp	10	.800	Ea.	12.65	19.20		31.85	46.50
550	0010	**DISAPPEARING STAIRWAY** No trim included									
	0020	One piece, yellow pine, 8′-0″ ceiling	2 Carp	4	4	Ea.	980	96		1,076	1,250
	0030	9′-0″ ceiling		4	4		990	96		1,086	1,275
	0040	10′-0″ ceiling		3	5.333		1,050	128		1,178	1,375
	0050	11′-0″ ceiling		3	5.333		1,225	128		1,353	1,575
	0060	12′-0″ ceiling	↓	3	5.333		1,275	128		1,403	1,625
	0100	Custom grade, pine, 8′-6″ ceiling, minimum	1 Carp	4	2		104	48		152	197
	0150	Average		3.50	2.286		105	55		160	209
	0200	Maximum		3	2.667		174	64		238	300
	0500	Heavy duty, pivoted, from 7′-7″ to 12′-10″ floor to floor		3	2.667		340	64		404	485
	0600	16′-0″ ceiling		2	4		1,150	96		1,246	1,425
	0800	Economy folding, pine, 8′-6″ ceiling		4	2		95.50	48		143.50	187
	0900	9′-6″ ceiling	↓	4	2		103	48		151	196
	1000	Fire escape, galvanized steel, 8′-0″ to 10′-4″ ceiling	2 Carp	1	16		1,200	385		1,585	1,975
	1010	10′-6″ to 13′-6″ ceiling		1	16		1,525	385		1,910	2,325
	1100	Automatic electric, aluminum, floor to floor height, 8′ to 9′	↓	1	16	↓	6,250	385		6,635	7,525

	11460 | Unit Kitchens		CREW	DAILY OUTPUT	LABOR-HOURS	UNIT	MAT.	LABOR	EQUIP.	TOTAL	TOTAL INCL O&P
100	0010	**UNIT KITCHENS**									
	1500	Combination range, refrigerator and sink, 30″ wide, minimum	L-1	2	5	Ea.	730	135		865	1,025
	1550	Maximum		1	10		1,450	270		1,720	2,050
	1570	60″ wide, average		1.40	7.143		2,400	193		2,593	2,975
	1590	72″ wide, average	↓	1.20	8.333	↓	2,725	225		2,950	3,375

For information about Means Estimating Seminars, see yellow pages 12 and 13 in back of book

Division 12 Furnishings

Estimating Tips

General

- The items in this division are usually priced per square foot or each. Most of these items are purchased by the owner and placed by the supplier. Do not assume the items in Division 12 will be purchased and installed by the supplier. Check the specifications for responsibilities and include receiving, storage, installation and mechanical and electrical hook-ups in the appropriate divisions.
- Some items in this division require some type of support system that is not usually furnished with the item. Examples of these systems include blocking for the attachment of casework and heavy drapery rods. The required blocking must be added to the estimate in the appropriate division.

Reference Numbers

Reference numbers are shown in bold squares at the beginning of some major classifications. These numbers refer to related items in the Reference Section. The reference information may be an estimating procedure, an alternate pricing method or technical information.

Note: Not all subdivisions listed here necessarily appear in this publication.

No part of this publication may be reproduced, stored in a retrieval system, or transmitted in any form or by any means without prior written permission of Reed Construction Data.

12300 | Manufactured Casework

		12310 | Metal Casework	CREW	DAILY OUTPUT	LABOR-HOURS	UNIT	2005 BARE COSTS MAT.	LABOR	EQUIP.	TOTAL	TOTAL INCL O&P
560	0010	IRONING CENTER									
	0020	Including cabinet, board & light, minimum	1 Carp	2	4	Ea.	310	96		406	510

12400 | Furnishings & Accessories

		12492 | Blinds and Shades	CREW	DAILY OUTPUT	LABOR-HOURS	UNIT	2005 BARE COSTS MAT.	LABOR	EQUIP.	TOTAL	TOTAL INCL O&P
100	0010	BLINDS, INTERIOR									
	0020	Horizontal, 1″ aluminum slats, solid color, stock	1 Carp	590	.014	S.F.	2.89	.33		3.22	3.73
	0090	Custom, minimum		590	.014		2.28	.33		2.61	3.06
	0100	Maximum		440	.018		6.50	.44		6.94	7.90
	0450	Stock, minimum		590	.014		4.21	.33		4.54	5.20
	0500	Maximum		440	.018		6.85	.44		7.29	8.25
	3000	Wood folding panels with movable louvers, 7″ x 20″ each		17	.471	Pr.	42	11.30		53.30	65.50
	3300	8″ x 28″ each		17	.471		61	11.30		72.30	86
	3450	9″ x 36″ each		17	.471		72.50	11.30		83.80	98.50
	3600	10″ x 40″ each		17	.471		82	11.30		93.30	109
	4000	Fixed louver type, stock units, 8″ x 20″ each		17	.471		63	11.30		74.30	88.50
	4150	10″ x 28″ each		17	.471		86.50	11.30		97.80	114
	4300	12″ x 36″ each		17	.471		110	11.30		121.30	140
	4450	18″ x 40″ each		17	.471		131	11.30		142.30	163
	5000	Insert panel type, stock, 7″ x 20″ each		17	.471		15.60	11.30		26.90	36.50
	5150	8″ x 28″ each		17	.471		28.50	11.30		39.80	50.50
	5300	9″ x 36″ each		17	.471		36.50	11.30		47.80	59
	5450	10″ x 40″ each		17	.471		39	11.30		50.30	62
	5600	Raised panel type, stock, 10″ x 24″ each		17	.471		109	11.30		120.30	139
	5650	12″ x 26″ each		17	.471		126	11.30		137.30	158
	5700	14″ x 30″ each		17	.471		143	11.30		154.30	176
	5750	16″ x 36″ each		17	.471		160	11.30		171.30	195
	6000	For custom built pine, add					22%				
	6500	For custom built hardwood blinds, add					42%				
600	0011	**SHADES** Basswood roll-up, stain finish, 3/8″ slats	1 Carp	300	.027	S.F.	9.85	.64		10.49	11.90
	5011	Insulative shades		125	.064		8.50	1.54		10.04	11.95
	6011	Solar screening, fiberglass		85	.094		3.83	2.26		6.09	8.05
	8011	Interior insulative shutter									
	8111	Stock unit, 15″ x 60″	1 Carp	17	.471	Pr.	8.50	11.30		19.80	28.50

		12493 | Curtains and Drapes	CREW	DAILY OUTPUT	LABOR-HOURS	UNIT	MAT.	LABOR	EQUIP.	TOTAL	TOTAL INCL O&P
200	0010	DRAPERY HARDWARE									
	0030	Standard traverse, per foot, minimum	1 Carp	59	.136	L.F.	1.93	3.25		5.18	7.65
	0100	Maximum		51	.157	″	10.40	3.76		14.16	17.85
	0200	Decorative traverse, 28″-48″, minimum		22	.364	Ea.	14.65	8.75		23.40	31
	0220	Maximum		21	.381		34	9.15		43.15	53
	0300	48″-84″, minimum		20	.400		19.55	9.60		29.15	38
	0320	Maximum		19	.421		55.50	10.10		65.60	78
	0400	66″-120″, minimum		18	.444		22	10.65		32.65	42
	0420	Maximum		17	.471		85	11.30		96.30	113
	0500	84″-156″, minimum		16	.500		24.50	12		36.50	47.50
	0520	Maximum		15	.533		97.50	12.80		110.30	129
	0600	130″-240″, minimum		14	.571		29.50	13.70		43.20	56

12493 | Curtains and Drapes		CREW	DAILY OUTPUT	LABOR-HOURS	UNIT	2005 BARE COSTS MAT.	LABOR	EQUIP.	TOTAL	TOTAL INCL O&P	
0620	Maximum	1 Carp	13	.615	Ea.	148	14.75		162.75	187	200
0700	Slide rings, each, minimum					.66			.66	.73	
0720	Maximum					2.13			2.13	2.34	
3000	Ripplefold, snap-a-pleat system, 3′ or less, minimum	1 Carp	15	.533		45.50	12.80		58.30	71.50	
3020	Maximum	″	14	.571		66	13.70		79.70	96.50	
3200	Each additional foot, add, minimum				L.F.	2.09			2.09	2.30	
3220	Maximum				″	6.35			6.35	6.95	
4000	Traverse rods, adjustable, 28″ to 48″	1 Carp	22	.364	Ea.	17.05	8.75		25.80	33.50	
4020	48″ to 84″		20	.400		24.50	9.60		34.10	43.50	
4040	66″ to 120″		18	.444		28.50	10.65		39.15	49.50	
4060	84″ to 156″		16	.500		31	12		43	54.50	
4080	100″ to 180″		14	.571		35.50	13.70		49.20	62.50	
4100	228″ to 312″		13	.615		55.50	14.75		70.25	86	
4500	Curtain rod, 28″ to 48″, single		22	.364		4.78	8.75		13.53	20	
4510	Double		22	.364		8.15	8.75		16.90	24	
4520	48″ to 86″, single		20	.400		8.20	9.60		17.80	25.50	
4530	Double		20	.400		13.65	9.60		23.25	31.50	
4540	66″ to 120″, single		18	.444		13.70	10.65		24.35	33	
4550	Double		18	.444		21.50	10.65		32.15	41.50	
4600	Valance, pinch pleated fabric, 12″ deep, up to 54″ long, minimum					34			34	37.50	
4610	Maximum					85.50			85.50	94	
4620	Up to 77″ long, minimum					52.50			52.50	58	
4630	Maximum					138			138	152	
5000	Stationary rods, first 2 feet					8.65			8.65	9.50	

or information about Means Estimating Seminars, see yellow pages 12 and 13 in back of book

Division Notes

		CREW	DAILY OUTPUT	LABOR-HOURS	UNIT	2005 BARE COSTS				TOTAL INCL O&P
						MAT.	LABOR	EQUIP.	TOTAL	

Division 13 Special Construction

Estimating Tips

General

- The items and systems in this division are usually estimated, purchased, supplied and installed as a unit by one or more subcontractors. The estimator must ensure that all parties are operating from the same set of specifications and assumptions and that all necessary items are estimated and will be provided. Many times the complex items and systems are covered but the more common ones such as excavation or a crane are overlooked for the very reason that everyone assumes nobody could miss them. The estimator should be the central focus and be able to ensure that all systems are complete.
- Another area where problems can develop in this division is at the interface between systems. The estimator must ensure, for instance, that anchor bolts, nuts and washers are estimated and included for the air-supported structures and pre-engineered buildings to be bolted to their foundations. Utility supply is a common area where essential items or pieces of equipment can be missed or overlooked due to the fact that each subcontractor may feel it is the others' responsibility. The estimator should also be aware of certain items which may be supplied as part of a package but installed by others, and ensure that the installing contractor's estimate includes the cost of installation. Conversely, the estimator must also ensure that items are not costed by two different subcontractors, resulting in an inflated overall estimate.

13120 Pre-Engineered Structures

- The foundations and floor slab, as well as rough mechanical and electrical, should be estimated, as this work is required for the assembly and erection of the structure. Generally, as noted in the book, the pre-engineered building comes as a shell and additional features, such as windows and doors, must be included by the estimator. Here again, the estimator must have a clear understanding of the scope of each portion of the work and all the necessary interfaces.

13200 Storage Tanks

- The prices in this subdivision for above and below ground storage tanks do not include foundations or hold-down slabs. The estimator should refer to Divisions 2 and 3 for foundation system pricing. In addition to the foundations, required tank accessories such as tank gauges, leak detection devices, and additional manholes and piping must be added to the tank prices.

Reference Numbers

Reference numbers are shown in bold squares at the beginning of some major classifications. These numbers refer to related items in the Reference Section. The reference information may be an estimating procedure, an alternate pricing method or technical information.

Note: Not all subdivisions listed here necessarily appear in this publication.

No part of this publication may be reproduced, stored in a retrieval system, or transmitted in any form or by any means without prior written permission of Reed Construction Data.

13030 | Special Purpose Rooms

	13035	Special Purpose Rooms	CREW	DAILY OUTPUT	LABOR-HOURS	UNIT	2005 BARE COSTS MAT.	LABOR	EQUIP.	TOTAL	TOTAL INCL O&P
800	0010	**SAUNA** Prefabricated, incl. heater & controls, 7′ high, 6′ x 4′, C/C	L-7	2.20	11.818	Ea.	3,525	240		3,765	4,275
	0050	6′ x 4′, C/P		2	13		3,275	264		3,539	4,050
	0400	6′ x 5′, C/C		2	13		3,950	264		4,214	4,800
	0450	6′ x 5′, C/P		2	13		3,700	264		3,964	4,525
	0600	6′ x 6′, C/C		1.80	14.444		4,200	293		4,493	5,125
	0650	6′ x 6′, C/P		1.80	14.444		3,925	293		4,218	4,825
	0800	6′ x 9′, C/C		1.60	16.250		5,275	330		5,605	6,350
	0850	6′ x 9′, C/P		1.60	16.250		5,000	330		5,330	6,050
	1000	8′ x 12′, C/C		1.10	23.636		8,175	480		8,655	9,800
	1050	8′ x 12′, C/P		1.10	23.636		7,500	480		7,980	9,050
	1400	8′ x 10′, C/C		1.20	21.667		6,925	440		7,365	8,350
	1450	8′ x 10′, C/P		1.20	21.667		6,425	440		6,865	7,825
	1600	10′ x 12′, C/C		1	26		8,650	525		9,175	10,400
	1650	10′ x 12′, C/P		1	26		7,825	525		8,350	9,525
	1700	Door only, cedar, 2′x6′, with tempered insulated glass window	2 Carp	3.40	4.706		465	113		578	700
	1800	Prehung, incl. jambs, pulls & hardware	"	12	1.333		475	32		507	580
	2500	Heaters only (incl. above), wall mounted, to 200 C.F.					460			460	505
	2750	To 300 C.F.					555			555	615
	3000	Floor standing, to 720 C.F., 10,000 watts, w/controls	1 Elec	3	2.667		1,450	73		1,523	1,700
	3250	To 1,000 C.F., 16,000 watts	"	3	2.667		1,500	73		1,573	1,775
940	0010	**STEAM BATH** Heater, timer & head, single, to 140 C.F.	1 Plum	1.20	6.667	Ea.	975	180		1,155	1,375
	0500	To 300 C.F.	"	1.10	7.273		1,100	196		1,296	1,525
	2700	Conversion unit for residential tub, including door					3,075			3,075	3,375

13100 | Lightning Protection

	13101	Lightning Protection	CREW	DAILY OUTPUT	LABOR-HOURS	UNIT	2005 BARE COSTS MAT.	LABOR	EQUIP.	TOTAL	TOTAL INCL O&P
055	0010	**LIGHTNING PROTECTION**									
	0200	Air terminals & base, copper									
	0400	3/8″ diameter x 10″ (to 75′ high)	1 Elec	8	1	Ea.	26	27.50		53.50	73.50
	1000	Aluminum, 1/2″ diameter x 12″ (to 75′ high)		8	1	"	20	27.50		47.50	66.50
	2000	Cable, copper, 220 lb. per thousand ft. (to 75′ high)		320	.025	L.F.	1.20	.68		1.88	2.43
	2500	Aluminum, 101 lb. per thousand ft. (to 75′ high)		280	.029	"	.65	.78		1.43	1.99
	3000	Arrester, 175 volt AC to ground		8	1	Ea.	40	27.50		67.50	88.50

13120 | Pre-Engineered Structures

	13128	Pre-Engineered Structures	CREW	DAILY OUTPUT	LABOR-HOURS	UNIT	2005 BARE COSTS MAT.	LABOR	EQUIP.	TOTAL	TOTAL INCL O&P
540	0010	**GREENHOUSE** Shell only, stock units, not incl. 2′ stub walls,									
	0020	foundation, floors, heat or compartments									
	0300	Residential type, free standing, 8′-6″ long x 7′-6″ wide	2 Carp	59	.271	SF Flr.	39.50	6.50		46	54.50
	0400	10′-6″ wide		85	.188		30.50	4.52		35.02	41

13120 | Pre-Engineered Structures

13128	Pre-Engineered Structures	CREW	DAILY OUTPUT	LABOR-HOURS	UNIT	2005 BARE COSTS MAT.	LABOR	EQUIP.	TOTAL	TOTAL INCL O&P	
0600	13′-6″ wide	2 Carp	108	.148	SF Flr.	27	3.56		30.56	36	540
0700	17′-0″ wide		160	.100		30.50	2.40		32.90	37.50	
0900	Lean-to type, 3′-10″ wide		34	.471		35	11.30		46.30	57.50	
1000	6′-10″ wide		58	.276		27	6.60		33.60	41.50	
1100	Wall mounted, to existing window, 3′ x 3′	1 Carp	4	2	Ea.	380	48		428	500	
1120	4′ x 5′	″	3	2.667	″	565	64		629	735	
1200	Deluxe quality, free standing, 7′-6″ wide	2 Carp	55	.291	SF Flr.	78.50	7		85.50	98	
1220	10′-6″ wide		81	.198		72.50	4.74		77.24	88	
1240	13′-6″ wide		104	.154		68	3.69		71.69	81.50	
1260	17′-0″ wide		150	.107		58	2.56		60.56	68	
1400	Lean-to type, 3′-10″ wide		31	.516		90.50	12.40		102.90	121	
1420	6′-10″ wide		55	.291		85	7		92	105	
1440	8′-0″ wide		97	.165		79.50	3.96		83.46	94	
0010	**SWIMMING POOL ENCLOSURE** Translucent, free standing,										880
0020	not including foundations, heat or light										
0200	Economy, minimum	2 Carp	200	.080	SF Hor.	11.90	1.92		13.82	16.35	
0300	Maximum		100	.160		32.50	3.84		36.34	42	
0400	Deluxe, minimum		100	.160		34.50	3.84		38.34	44.50	
0600	Maximum		70	.229		175	5.50		180.50	201	

13150 | Swimming Pools

13151	Swimming Pools	CREW	DAILY OUTPUT	LABOR-HOURS	UNIT	2005 BARE COSTS MAT.	LABOR	EQUIP.	TOTAL	TOTAL INCL O&P	
0010	**SWIMMING POOLS** Residential in-ground, vinyl lined, concrete sides										200
0020	Sides including equipment, sand bottom	B-52	300	.187	SF Surf	10.60	3.66	1.31	15.57	19.35	
0100	Metal or polystyrene sides (R13128 -520)	B-14	410	.117		8.85	2.18	.53	11.56	14	
0200	Add for vermiculite bottom					.68			.68	.75	
0500	Gunite bottom and sides, white plaster finish										
0600	12′ x 30′ pool	B-52	145	.386	SF Surf	17.55	7.60	2.72	27.87	35	
0720	16′ x 32′ pool		155	.361		15.85	7.10	2.54	25.49	32.50	
0750	20′ x 40′ pool		250	.224		14.15	4.40	1.58	20.13	25	
0810	Concrete bottom and sides, tile finish										
0820	12′ x 30′ pool	B-52	80	.700	SF Surf	17.75	13.75	4.93	36.43	48.50	
0830	16′ x 32′ pool		95	.589		14.65	11.55	4.15	30.35	40.50	
0840	20′ x 40′ pool		130	.431		11.65	8.45	3.03	23.13	30.50	
1600	For water heating system, see division 15510-880										
1700	Filtration and deck equipment only, as % of total				Total				20%	20%	
1800	Deck equipment, rule of thumb, 20′ x 40′ pool				SF Pool					1.30	
3000	Painting pools, preparation + 3 coats, 20′ x 40′ pool, epoxy	2 Pord	.33	48.485	Total	605	1,050		1,655	2,400	
3100	Rubber base paint, 18 gallons	″	.33	48.485	″	460	1,050		1,510	2,225	
0010	**SWIMMING POOL EQUIPMENT** Diving stand, stainless steel, 3 meter	2 Carp	.40	40	Ea.	4,925	960		5,885	7,050	700
0600	Diving boards, 16′ long, aluminum		2.70	5.926		2,525	142		2,667	3,025	
0700	Fiberglass		2.70	5.926		2,025	142		2,167	2,475	
0900	Filter system, sand or diatomite type, incl. pump, 6,000 gal./hr.	2 Plum	1.80	8.889	Total	1,125	240		1,365	1,625	
1020	Add for chlorination system, 800 S.F. pool	″	3	5.333	Ea.	203	144		347	460	
1200	Ladders, heavy duty, stainless steel, 2 tread	2 Carp	7	2.286		490	55		545	635	
1500	4 tread	″	6	2.667		610	64		674	785	
2100	Lights, underwater, 12 volt, with transformer, 300 watt	1 Elec	.40	20		145	545		690	1,050	

13150 | Swimming Pools

	13151	Swimming Pools	CREW	DAILY OUTPUT	LABOR-HOURS	UNIT	2005 BARE COSTS MAT.	LABOR	EQUIP.	TOTAL	TOTAL INCL O&P
700	2200	110 volt, 500 watt, standard	1 Elec	.40	20	Ea.	130	545		675	1,025
	3000	Pool covers, reinforced vinyl	3 Clab	1,800	.013	S.F.	.30	.23		.53	.72
	3100	Vinyl water tube		3,200	.007		.23	.13		.36	.47
	3200	Maximum		3,000	.008		.47	.14		.61	.76
	3300	Slides, tubular, fiberglass, aluminum handrails & ladder, 5'-0", straight	2 Carp	1.60	10	Ea.	2,200	240		2,440	2,825
	3320	8'-0", curved	"	3	5.333	"	6,075	128		6,203	6,925

13200 | Storage Tanks

	13201	Storage Tanks	CREW	DAILY OUTPUT	LABOR-HOURS	UNIT	2005 BARE COSTS MAT.	LABOR	EQUIP.	TOTAL	TOTAL INCL O&P
300	3001	**STEEL,** storage, above ground, including supports, coating									
	3020	fittings, not including fdn, pumps or piping									
	3040	Single wall, interior, 275 gallon	Q-5	5	3.200	Ea.	305	78		383	470
	3060	550 gallon	"	2.70	5.926		1,400	144		1,544	1,750
	3080	1,000 gallon	Q-7	5	6.400		2,200	156		2,356	2,675
	3320	Double wall, 500 gallon capacity	Q-5	2.40	6.667		2,425	162		2,587	2,950
	3330	2000 gallon capacity	Q-7	4.15	7.711		5,575	188		5,763	6,425
	3340	4000 gallon capacity		3.60	8.889		9,925	216		10,141	11,300
	3350	6000 gallon capacity		2.40	13.333		11,700	325		12,025	13,400
	3360	8000 gallon capacity		2	16		15,000	390		15,390	17,100
	3370	10000 gallon capacity		1.80	17.778		16,600	435		17,035	18,900
	3380	15000 gallon capacity		1.50	21.333		25,200	520		25,720	28,600
	3390	20000 gallon capacity		1.30	24.615		28,600	600		29,200	32,500
	3400	25000 gallon capacity		1.15	27.826		34,800	680		35,480	39,400
	3410	30000 gallon capacity		1	32		38,200	780		38,980	43,300
800	0010	**UNDERGROUND STORAGE TANKS**									
	0210	Fiberglass, underground, single wall, U.L. listed, not including									
	0220	manway or hold-down strap									
	0230	1,000 gallon capacity	Q-5	2.46	6.504	Ea.	1,875	158		2,033	2,325
	0240	2,000 gallon capacity	Q-7	4.57	7.002		2,525	171		2,696	3,050
	0500	For manway, fittings and hold-downs, add					20%	15%			
	2210	Fiberglass, underground, single wall, U.L. listed, including									
	2220	hold-down straps, no manways									
	2230	1,000 gallon capacity	Q-5	1.88	8.511	Ea.	2,125	207		2,332	2,675
	2240	2,000 gallon capacity	Q-7	3.55	9.014	"	2,750	219		2,969	3,375
	5000	Steel underground, sti-P3, set in place, not incl. hold-down bars.									
	5500	Excavation, pad, pumps and piping not included									
	5510	Single wall, 500 gallon capacity, 7 gauge shell	Q-5	2.70	5.926	Ea.	1,000	144		1,144	1,325
	5520	1,000 gallon capacity, 7 gauge shell	"	2.50	6.400		1,650	156		1,806	2,050
	5530	2,000 gallon capacity, 1/4" thick shell	Q-7	4.60	6.957		3,025	169		3,194	3,600
	5535	2,500 gallon capacity, 7 gauge shell	Q-5	3	5.333		3,675	130		3,805	4,275
	5610	25,000 gallon capacity, 3/8" thick shell	Q-7	1.30	24.615		21,500	600		22,100	24,700
	5630	40,000 gallon capacity, 3/8" thick shell		.90	35.556		37,400	865		38,265	42,500
	5640	50,000 gallon capacity, 3/8" thick shell		.80	40		46,800	975		47,775	53,000

	13281 | Hazardous Material Remediation		CREW	DAILY OUTPUT	LABOR-HOURS	UNIT	2005 BARE COSTS MAT.	LABOR	EQUIP.	TOTAL	TOTAL INCL O&P	
440	0010	**REMOVAL** Existing lead paint, by chemicals, per application										440
	0020	See also, Div. 13280, Haz. Mat'l. Abatement										
	0050	Baseboard, to 6" wide	1 Pord	64	.125	L.F.	1.53	2.72		4.25	6.15	
	0070	To 12" wide		32	.250	"	3.01	5.45		8.46	12.25	
	0200	Balustrades, one side		28	.286	S.F.	3.41	6.20		9.61	13.95	
	1400	Cabinets, simple design		32	.250		2.99	5.45		8.44	12.25	
	1420	Ornate design		25	.320		3.84	6.95		10.79	15.65	
	1600	Cornice, simple design		60	.133		1.61	2.90		4.51	6.55	
	1620	Ornate design		20	.400		4.74	8.70		13.44	19.50	
	2800	Doors, one side, flush		84	.095		1.16	2.07		3.23	4.68	
	2820	Two panel		80	.100		1.20	2.18		3.38	4.90	
	2840	Four panel		45	.178		2.12	3.87		5.99	8.70	
	2880	For trim, one side, add		64	.125	L.F.	1.53	2.72		4.25	6.15	
	3000	Fence, picket, one side		30	.267	S.F.	3.21	5.80		9.01	13.10	
	3200	Grilles, one side, simple design		30	.267		3.21	5.80		9.01	13.10	
	3220	Ornate design		25	.320		3.84	6.95		10.79	15.65	
	4400	Pipes, to 4" diameter		90	.089	L.F.	1.10	1.93		3.03	4.39	
	4420	To 8" diameter		50	.160		1.91	3.48		5.39	7.80	
	4440	To 12" diameter		36	.222		2.68	4.83		7.51	10.90	
	4460	To 16" diameter		20	.400		4.77	8.70		13.47	19.55	
	4500	For hangers, add		40	.200	Ea.	2.39	4.35		6.74	9.80	
	4800	Siding		90	.089	S.F.	1.10	1.93		3.03	4.39	
	5000	Trusses, open		55	.145	SF Face	1.75	3.16		4.91	7.15	
	6200	Windows, one side only, double hung, 1/1 light, 24" x 48" high		4	2	Ea.	24	43.50		67.50	98	
	6220	30" x 60" high		3	2.667		32	58		90	131	
	6240	36" x 72" high		2.50	3.200		38.50	69.50		108	157	
	6280	40" x 80" high		2	4		48	87		135	196	
	6400	Colonial window, 6/6 light, 24" x 48" high		2	4		48	87		135	196	
	6420	30" x 60" high		1.50	5.333		64	116		180	262	
	6440	36" x 72" high		1	8		96	174		270	390	
	6480	40" x 80" high		1	8		96	174		270	390	
	6600	8/8 light, 24" x 48" high		2	4		48	87		135	196	
	6620	40" x 80" high		1	8		96	174		270	390	
	6800	12/12 light, 24" x 48" high		1	8		96	174		270	390	
	6820	40" x 80" high		.75	10.667		128	232		360	520	
	6840	Window frame & trim items, included in pricing above										
460	0010	**LEAD PAINT ENCAPSULATION**, water based polymer coating,14 mil DFT										460
	0020	Interior, brushwork, trim, under 6"	1 Pord	240	.033	L.F.	2.26	.72		2.98	3.68	
	0030	6" to 12" wide		180	.044		3	.97		3.97	4.89	
	0040	Balustrades		300	.027		1.81	.58		2.39	2.94	
	0050	Pipe to 4" diameter		500	.016		1.09	.35		1.44	1.77	
	0060	To 8" diameter		375	.021		1.44	.46		1.90	2.34	
	0070	To 12" diameter		250	.032		2.16	.70		2.86	3.52	
	0080	To 16" diameter		170	.047		3.18	1.02		4.20	5.20	
	0090	Cabinets, ornate design		200	.040	S.F.	2.72	.87		3.59	4.42	
	0100	Simple design		250	.032	"	2.16	.70		2.86	3.52	
	0110	Doors, 3'x 7', both sides, incl. frame & trim										
	0120	Flush	1 Pord	6	1.333	Ea.	27.50	29		56.50	78	
	0130	French, 10-15 lite		3	2.667		5.55	58		63.55	102	
	0140	Panel		4	2		33.50	43.50		77	108	
	0150	Louvered		2.75	2.909		30.50	63.50		94	138	
	0160	Windows, per interior side, per 15 S.F.										
	0170	1 to 6 lite	1 Pord	14	.571	Ea.	19.15	12.45		31.60	41.50	
	0180	7 to 10 lite		7.50	1.067		21	23		44	61	
	0190	12 lite		5.75	1.391		28.50	30.50		59	80.50	
	0200	Radiators		8	1		67.50	22		89.50	110	

	13281 | Hazardous Material Remediation		CREW	DAILY OUTPUT	LABOR-HOURS	UNIT	2005 BARE COSTS MAT.	LABOR	EQUIP.	TOTAL	TOTAL INCL O&P
460	0210	Grilles, vents	1 Pord	275	.029	S.F.	1.97	.63		2.60	3.21
	0220	Walls, roller, drywall or plaster		1,000	.008		.54	.17		.71	.88
	0230	With spunbonded reinforcing fabric		720	.011		.62	.24		.86	1.08
	0240	Wood		800	.010		.68	.22		.90	1.11
	0250	Ceilings, roller, drywall or plaster		900	.009		.62	.19		.81	1
	0260	Wood		700	.011		.77	.25		1.02	1.26
	0270	Exterior, brushwork, gutters and downspouts		300	.027	L.F.	1.81	.58		2.39	2.94
	0280	Columns		400	.020	S.F.	1.34	.44		1.78	2.19
	0290	Spray, siding		600	.013	″	.90	.29		1.19	1.47
	0300	Miscellaneous									
	0310	Electrical conduit, brushwork, to 2″ diameter	1 Pord	500	.016	L.F.	1.09	.35		1.44	1.77
	0320	Brick, block or concrete, spray		500	.016	S.F.	1.09	.35		1.44	1.77
	0330	Steel, flat surfaces and tanks to 12″		500	.016		1.09	.35		1.44	1.77
	0340	Beams, brushwork		400	.020		1.34	.44		1.78	2.19
	0350	Trusses		400	.020		1.34	.44		1.78	2.19

	13720 | Detection & Alarm		CREW	DAILY OUTPUT	LABOR-HOURS	UNIT	MAT.	LABOR	EQUIP.	TOTAL	TOTAL INCL O&P
065	0010	**DETECTION SYSTEMS**, not including wires & conduits									
	0100	Burglar alarm, battery operated, mechanical trigger	1 Elec	4	2	Ea.	254	54.50		308.50	370
	0200	Electrical trigger		4	2		305	54.50		359.50	425
	0400	For outside key control, add		8	1		72	27.50		99.50	124
	0600	For remote signaling circuitry, add		8	1		114	27.50		141.50	171
	0800	Card reader, flush type, standard		2.70	2.963		850	81		931	1,075
	1000	Multi-code		2.70	2.963		1,100	81		1,181	1,325
	1200	Door switches, hinge switch		5.30	1.509		53.50	41		94.50	126
	1400	Magnetic switch		5.30	1.509		63	41		104	137
	2800	Ultrasonic motion detector, 12 volt		2.30	3.478		210	95		305	385
	3000	Infrared photoelectric detector		2.30	3.478		173	95		268	345
	3200	Passive infrared detector		2.30	3.478		259	95		354	440
	3420	Switchmats, 30″ x 5′		5.30	1.509		77.50	41		118.50	153
	3440	30″ x 25′		4	2		186	54.50		240.50	293
	3460	Police connect panel		4	2		223	54.50		277.50	335
	3480	Telephone dialer		5.30	1.509		350	41		391	450
	3500	Alarm bell		4	2		71	54.50		125.50	167
	3520	Siren		4	2		134	54.50		188.50	236
	5200	Smoke detector, ceiling type		6.20	1.290		75	35		110	140
	5600	Strobe and horn		5.30	1.509		95	41		136	172
	5800	Fire alarm horn		6.70	1.194		36.50	32.50		69	93
	6600	Drill switch		8	1		86.50	27.50		114	140
	6800	Master box		2.70	2.963		3,100	81		3,181	3,525
	7800	Remote annunciator, 8 zone lamp		1.80	4.444		175	121		296	390
	8000	12 zone lamp	2 Elec	2.60	6.154		300	168		468	605
	8200	16 zone lamp	″	2.20	7.273		300	199		499	655
	8400	Standpipe or sprinkler alarm, alarm device	1 Elec	8	1		125	27.50		152.50	183
	8600	Actuating device	″	8	1		290	27.50		317.50	365

Important: See the Reference Section for critical supporting data - Reference Nos., Crews, & Location Factors

	13838 | Pneumatic/Electric Controls		CREW	DAILY OUTPUT	LABOR-HOURS	UNIT	2005 BARE COSTS MAT.	LABOR	EQUIP.	TOTAL	TOTAL INCL O&P	
0	0010	CONTROL COMPONENTS										200
	5000	Thermostats										
	5030	Manual	1 Shee	8	1	Ea.	27	26.50		53.50	74.50	
	5040	1 set back, electric, timed	↓	8	1	↓	81	26.50		107.50	134	
	5050	2 set back, electric, timed	↓	8	1	↓	178	26.50		204.50	241	

or information about Means Estimating Seminars, see yellow pages 12 and 13 in back of book

Division Notes

		CREW	DAILY OUTPUT	LABOR-HOURS	UNIT	2005 BARE COSTS MAT.	LABOR	EQUIP.	TOTAL	TOTAL INCL O&P

Division 14 Conveying Systems

Estimating Tips

General

- Many products in Division 14 will require some type of support or blocking for installation not included with the item itself. Examples are supports for conveyors or tube systems, attachment points for lifts, and footings for hoists or cranes. Add these supports in the appropriate division.

14100 Dumbwaiters
14200 Elevators

- Dumbwaiters and elevators are estimated and purchased in a method similar to buying a car. The manufacturer has a base unit with standard features. Added to this base unit price will be whatever options the owner or specifications require. Increased load capacity, additional vertical travel, additional stops, higher speed, and cab finish options are items to be considered. When developing an estimate for dumbwaiters and elevators, remember that some items needed by the installers may have to be included as part of the general contract.

  Examples are:
  - shaftway
  - rail support brackets
  - machine room
  - electrical supply
  - sill angles
  - electrical connections
  - pits
  - roof penthouses
  - pit ladders

  Check the job specifications and drawings before pricing.
- Installation of elevators and handicapped lifts in historic structures can require significant additional costs. The associated structural requirements may involve cutting into and repairing finishes, mouldings, flooring, etc. The estimator must account for these special conditions.

14300 Escalators & Moving Walks

- Escalators and moving walks are specialty items installed by specialty contractors. There are numerous options associated with these items. For specific options contact a manufacturer or contractor. In a method similar to estimating dumbwaiters and elevators, you should verify the extent of general contract work and add items as necessary.

14400 Lifts
14500 Material Handling
14600 Hoists & Cranes

- Products such as correspondence lifts, conveyors, chutes, pneumatic tube systems, material handling cranes and hoists as well as other items specified in this subdivision may require trained installers. The general contractor might not have any choice as to who will perform the installation or when it will be performed. Long lead times are often required for these products, making early decisions in scheduling necessary.

Reference Numbers

Reference numbers are shown in bold squares at the beginning of some major classifications. These numbers refer to related items in the Reference Section. The reference information may be an estimating procedure, an alternate pricing method or technical information.

Note: Not all subdivisions listed here necessarily appear in this publication.

No part of this publication may be reproduced, stored in a retrieval system, or transmitted in any form or by any means without prior written permission of Reed Construction Data.

14200 | Elevators

		14210 Electric Traction Elevators	CREW	DAILY OUTPUT	LABOR-HOURS	UNIT	2005 BARE COSTS MAT.	LABOR	EQUIP.	TOTAL	TOTAL INCL O&P
100	0010	ELEVATOR SYSTEMS									
	7000	Residential, cab type, 1 floor, 2 stop, minimum	2 Elev	.20	80	Ea.	8,500	2,425		10,925	13,300
	7100	Maximum		.10	160		14,300	4,825		19,125	23,700
	7200	2 floor, 3 stop, minimum		.12	133		12,600	4,025		16,625	20,500
	7300	Maximum	↓	.06	266	↓	20,600	8,050		28,650	35,800

		14420 Wheelchair Lifts	CREW	DAILY OUTPUT	LABOR-HOURS	UNIT	MAT.	LABOR	EQUIP.	TOTAL	TOTAL INCL O&P
100	0010	CHAIR / WHEELCHAIR LIFT									
	7700	Stair climber (chair lift), single seat, minimum	2 Elev	1	16	Ea.	4,100	485		4,585	5,325
	7800	Maximum	"	.20	80	"	5,650	2,425		8,075	10,200

For information about Means Estimating Seminars, see yellow pages 12 and 13 in back of book

Division 15 Mechanical

'stimating Tips

5100 Building Services Piping

ıis subdivision is primarily basic pipe and ɩated materials. The pipe may be used by any ' the mechanical disciplines, i.e., plumbing, ɛ protection, heating, and air conditioning.

- The piping section lists the add to labor for elevated pipe installation. These adds apply to all elevated pipe, fittings, valves, insulation, etc., that are placed above 10′ high. CAUTION: the correct percentage may vary for the same pipe. For example, the percentage add for the basic pipe installation should be based on the maximum height that the craftsman must install for that particular section. If the pipe is to be located 14′ above the floor but it is suspended on threaded rod from beams, the bottom flange of which is 18′ high (4′ rods), then the height is actually 18′ and the add is 20%. The pipe coverer, however, does not have to go above the 14′ and so his add should be 10%.
- Most pipe is priced first as straight pipe with a joint (coupling, weld, etc.) every 10′ and a hanger usually every 10′. There are exceptions with hanger spacing such as: for cast iron pipe (5′) and plastic pipe (3 per 10′). Following each type of pipe there are several lines listing sizes and the amount to be subtracted to delete couplings and hangers. This is for pipe that is to be buried or supported together on trapeze hangers. The reason that the couplings are deleted is that these runs are usually long and frequently longer lengths of pipe are used. By deleting the couplings the estimator is expected to look up and add back the correct reduced number of couplings.
- When preparing an estimate it may be necessary to approximate the fittings. Fittings usually run between 25% and 50% of the cost of the pipe. The lower percentage is for simpler runs, and the higher number is for complex areas like mechanical rooms.
- For historic restoration projects, the systems must be as invisible as possible, and pathways must be sought for pipes, conduits, and ductwork. While installations in accessible spaces (such as basements and attics) are relatively straightforward to estimate, labor costs may be more difficult to determine when delivery systems must be concealed.

15400 Plumbing Fixtures & Equipment

- Plumbing fixture costs usually require two lines, the fixture itself and its "rough-in, supply and waste".
- In the Assemblies Section (Plumbing D2010) for the desired fixture, the System Components Group in the center of the page shows the fixture itself on the first line while the rest of the list (fittings, pipe, tubing, etc.) will total up to what we refer to in the Unit Price section as "Rough-in, supply, waste and vent". Note that for most fixtures we allow a nominal 5′ of tubing to reach from the fixture to a main or riser.
- Remember that gas and oil fired units need venting.

15500 Heat Generation Equipment

- When estimating the cost of an HVAC system, check to see who is responsible for providing and installing the temperature control system. It is possible to overlook controls, assuming that they would be included in the electrical estimate.
- When looking up a boiler be careful on specified capacity. Some manufacturers rate their products on output while others use input.
- Include HVAC insulation for pipe, boiler and duct (wrap and liner).
- Be careful when looking up mechanical items to get the correct pressure rating and connection type (thread, weld, flange).

15700 Heating/Ventilation/ Air Conditioning Equipment

- Combination heating and cooling units are sized by the air conditioning requirements. (See Reference No. R15710-020 for preliminary sizing guide.)
- A ton of air conditioning is nominally 400 CFM.
- Rectangular duct is taken off by the linear foot for each size, but its cost is usually estimated by the pound. Remember that SMACNA standards now base duct on internal pressure.
- Prefabricated duct is estimated and purchased like pipe: straight sections and fittings.
- Note that cranes or other lifting equipment are not included on any lines in Division 15. For example, if a crane is required to lift a heavy piece of pipe into place high above a gym floor, or to put a rooftop unit on the roof of a four-story building, etc., it must be added. Due to the potential for extreme variation—from nothing additional required, to a major crane or helicopter—we feel that including a nominal amount for "lifting contingency" would be useless and detract from the accuracy of the estimate. When using equipment rental from Means do not forget to include the cost of the operator(s).

Reference Numbers

Reference numbers are shown in bold squares at the beginning of some major classifications. These numbers refer to related items in the Reference Section. The reference information may be an estimating procedure, an alternate pricing method or technical information.

Note: Not all subdivisions listed here necessarily appear in this publication.

Vote: **i2 Trade Service,** *in part, has been used as a reference source for some ɔf the material prices used in Division 15.*

'o part of this publication may be reproduced, stored in a retrieval system, or transmitted in any form r by any means without prior written permission of Reed Construction Data.

15055 | Selective Mech Demolition

		15055 Selective Mech Demolition	CREW	DAILY OUTPUT	LABOR-HOURS	UNIT	2005 BARE COSTS MAT.	LABOR	EQUIP.	TOTAL	TOTAL INCL O&P	
300	0010	**HVAC DEMOLITION**										300
	0100	Air conditioner, split unit, 3 ton	Q-5	2	8	Ea.		195		195	320	
	0150	Package unit, 3 ton	Q-6	3	8	″		188		188	310	
	0260	Baseboard, hydronic fin tube, 1/2″	Q-5	117	.137	L.F.		3.33		3.33	5.45	
	0298	Boilers										
	0300	Electric, up thru 148 kW	Q-19	2	12	Ea.		305		305	500	
	0310	150 thru 518 kW	″	1	24			610		610	995	
	0320	550 thru 2000 kW	Q-21	.40	80			2,050		2,050	3,375	
	0330	2070 kW and up	″	.30	106			2,750		2,750	4,500	
	0340	Gas and/or oil, up thru 150 MBH	Q-7	2.20	14.545			355		355	580	
	0350	160 thru 2000 MBH		.80	40			975		975	1,600	
	0360	2100 thru 4500 MBH		.50	64			1,550		1,550	2,550	
	0370	4600 thru 7000 MBH		.30	106			2,600		2,600	4,275	
	0390	12,200 thru 25,000 MBH		.12	266			6,500		6,500	10,700	
	1000	Ductwork, 4″ high, 8″ wide	1 Clab	200	.040	L.F.		.69		.69	1.18	
	1100	6″ high, 8″ wide		165	.048			.84		.84	1.43	
	1200	10″ high, 12″ wide		125	.064			1.11		1.11	1.88	
	1300	12″-14″ high, 16″-18″ wide		85	.094			1.63		1.63	2.77	
	1500	30″ high, 36″ wide		56	.143			2.48		2.48	4.21	
	2200	Furnace, electric	Q-20	2	10	Ea.		246		246	410	
	2300	Gas or oil, under 120 MBH	Q-9	4	4			95.50		95.50	160	
	2340	Over 120 MBH	″	3	5.333			127		127	214	
	2800	Heat pump, package unit, 3 ton	Q-5	2.40	6.667			162		162	267	
	2840	Split unit, 3 ton		2	8			195		195	320	
	2950	Tank, steel, oil, 275 gal., above ground		10	1.600			39		39	64	
	2960	Remove and reset		3	5.333			130		130	213	
	9000	Minimum labor/equipment charge	Q-6	3	8	Job		188		188	310	
600	0010	**PLUMBING DEMOLITION**										600
	1020	Fixtures, including 10′ piping										
	1100	Bath tubs, cast iron	1 Plum	4	2	Ea.		54		54	88.50	
	1120	Fiberglass		6	1.333			36		36	59	
	1140	Steel		5	1.600			43		43	71	
	1200	Lavatory, wall hung		10	.800			21.50		21.50	35.50	
	1220	Counter top		8	1			27		27	44.50	
	1300	Sink, steel or cast iron, single		8	1			27		27	44.50	
	1320	Double		7	1.143			31		31	50.50	
	1400	Water closet, floor mounted		8	1			27		27	44.50	
	1420	Wall mounted		7	1.143			31		31	50.50	
	2000	Piping, metal, up thru 1-1/2″ diameter		200	.040	L.F.		1.08		1.08	1.77	
	2050	2″ thru 3-1/2″ diameter		150	.053			1.44		1.44	2.36	
	2100	4″ thru 6″ diameter	2 Plum	100	.160			4.31		4.31	7.10	
	2250	Water heater, 40 gal.	1 Plum	6	1.333	Ea.		36		36	59	
	3000	Submersible sump pump		24	.333			9		9	14.75	
	6000	Remove and reset fixtures, minimum		6	1.333			36		36	59	
	6100	Maximum		4	2			54		54	88.50	
	9000	Minimum labor/equipment charge		2	4	Job		108		108	177	

15080 | Mechanical Insulation

		15080 Mechanical Insulation	CREW	DAILY OUTPUT	LABOR-HOURS	UNIT	MAT.	LABOR	EQUIP.	TOTAL	TOTAL INCL O&P	
200	0010	**DUCT INSULATION**										200
	3000	Ductwork										
	3020	Blanket type, fiberglass, flexible										
	3030	Fire resistant liner, black coating one side										
	3050	1/2″ thick, 2 lb. density	Q-14	380	.042	S.F.	.38	.93		1.31	2.02	
	3060	1″ thick, 1-1/2 lb. density	″	350	.046	″	.50	1.01		1.51	2.28	

Important: See the Reference Section for critical supporting data - Reference Nos., Crews, & Location Factors

15080 | Mechanical Insulation		CREW	DAILY OUTPUT	LABOR-HOURS	UNIT	2005 BARE COSTS MAT.	LABOR	EQUIP.	TOTAL	TOTAL INCL O&P	
3140	FRK vapor barrier wrap, .75 lb. density										200
3160	1" thick	Q-14	350	.046	S.F.	.25	1.01		1.26	2.01	
3170	1-1/2" thick	"	320	.050	"	.27	1.10		1.37	2.19	
3490	Board type, fiberglass liner, 3 lb. density										
3500	Fire resistant, black pigmented, 1 side										
3520	1" thick	Q-14	150	.107	S.F.	1.44	2.35		3.79	5.60	
3540	1-1/2" thick	"	130	.123	"	1.76	2.71		4.47	6.60	
9600	Minimum labor/equipment charge	1 Stpi	4	2	Job		54		54	89	
0010	**EQUIPMENT INSULATION**										400
2900	Domestic water heater wrap kit										
2920	1-1/2" with vinyl jacket, 20-60 gal.	1 Plum	8	1	Ea.	16.60	27		43.60	63	
0010	**PIPING INSULATION**										600
2930	Insulated protectors, (ADA)										
2935	For exposed piping under sinks or lavatories.										
2940	Vinyl coated foam, velcro tabs										
2945	P Trap, 1-1/4" or 1-1/2"	1 Plum	32	.250	Ea.	17.15	6.75		23.90	30	
2960	Valve and supply cover										
2965	1/2", 3/8", and 7/16" pipe size	1 Plum	32	.250	Ea.	17.15	6.75		23.90	30	
2970	Extension drain cover										
2975	1-1/4", or 1-1/2" pipe size	1 Plum	32	.250	Ea.	17.60	6.75		24.35	30.50	
2985	1-1/4" pipe size	"	32	.250	"	19.60	6.75		26.35	32.50	
4000	Pipe covering (price copper tube one size less than IPS)										
6600	Fiberglass, with all service jacket										
6840	1" wall, 1/2" iron pipe size	Q-14	240	.067	L.F.	.71	1.47		2.18	3.31	
6860	3/4" iron pipe size		230	.070		.83	1.53		2.36	3.55	
6870	1" iron pipe size		220	.073		.83	1.60		2.43	3.67	
6900	2" iron pipe size	↓	200	.080	↓	1.12	1.76		2.88	4.26	
7879	Rubber tubing, flexible closed cell foam										
8100	1/2" wall, 1/4" iron pipe size	1 Asbe	90	.089	L.F.	.37	2.18		2.55	4.15	
8130	1/2" iron pipe size		89	.090		.57	2.20		2.77	4.41	
8140	3/4" iron pipe size		89	.090		.63	2.20		2.83	4.47	
8150	1" iron pipe size		88	.091		.68	2.23		2.91	4.58	
8170	1-1/2" iron pipe size		87	.092		1.03	2.25		3.28	5	
8180	2" iron pipe size		86	.093		1.20	2.28		3.48	5.25	
8300	3/4" wall, 1/4" iron pipe size		90	.089		.67	2.18		2.85	4.48	
8330	1/2" iron pipe size		89	.090		.81	2.20		3.01	4.67	
8340	3/4" iron pipe size		89	.090		.97	2.20		3.17	4.85	
8350	1" iron pipe size		88	.091		1.18	2.23		3.41	5.15	
8380	2" iron pipe size		86	.093		2.43	2.28		4.71	6.60	
8444	1" wall, 1/2" iron pipe size		86	.093		1.53	2.28		3.81	5.60	
8445	3/4" iron pipe size		84	.095		1.90	2.33		4.23	6.10	
8446	1" iron pipe size		84	.095		2.30	2.33		4.63	6.55	
8447	1-1/4" iron pipe size		82	.098		2.68	2.39		5.07	7.05	
8448	1-1/2" iron pipe size		82	.098		3.02	2.39		5.41	7.45	
8449	2" iron pipe size		80	.100		4.70	2.45		7.15	9.35	
8450	2-1/2" iron pipe size	↓	80	.100	↓	6.15	2.45		8.60	10.95	
8456	Rubber insulation tape, 1/8" x 2" x 30'				Ea.	10.80			10.80	11.90	

15107 | Metal Pipe & Fittings

			CREW	DAILY OUTPUT	LABOR-HOURS	UNIT	2005 BARE COSTS MAT.	LABOR	EQUIP.	TOTAL	TOTAL INCL O&P
320	0010	**PIPE, CAST IRON** Soil, on hangers 5' O.C. R15100-050									
	0020	Single hub, service wt., lead & oakum joints 10' O.C.									
	2120	2" diameter	Q-1	63	.254	L.F.	4.63	6.15		10.78	15.20
	2140	3" diameter		60	.267		6.35	6.45		12.80	17.60
	2160	4" diameter		55	.291		8.10	7.05		15.15	20.50
	4000	No hub, couplings 10' O.C.									
	4100	1-1/2" diameter	Q-1	71	.225	L.F.	5.30	5.45		10.75	14.85
	4120	2" diameter		67	.239		5.45	5.80		11.25	15.50
	4140	3" diameter		64	.250		7.15	6.05		13.20	17.85
	4160	4" diameter		58	.276		9.10	6.70		15.80	21
360	0010	**PIPE, CAST IRON, FITTINGS** Soil									
	0040	Hub and spigot, service weight, lead & oakum joints									
	0080	1/4 bend, 2"	Q-1	16	1	Ea.	9.75	24.50		34.25	51
	0120	3"		14	1.143		13	27.50		40.50	60
	0140	4"		13	1.231		20.50	30		50.50	71.50
	0340	1/8 bend, 2"		16	1		6.95	24.50		31.45	47.50
	0350	3"		14	1.143		10.90	27.50		38.40	57.50
	0360	4"		13	1.231		15.85	30		45.85	66.50
	0500	Sanitary tee, 2"		10	1.600		12.60	39		51.60	78
	0540	3"		9	1.778		22	43		65	95
	0620	4"		8	2		27	48.50		75.50	109
	5990	No hub									
	6000	Cplg. & labor required at joints not incl. in fitting									
	6010	price. Add 1 coupling per joint for installed price									
	6020	1/4 Bend, 1-1/2"				Ea.	5.50			5.50	6.05
	6060	2"					5.95			5.95	6.55
	6080	3"					8.30			8.30	9.10
	6120	4"					11.90			11.90	13.10
	6184	1/4 Bend, long sweep, 1-1/2"					12.85			12.85	14.10
	6186	2"					12.85			12.85	14.10
	6188	3"					15.30			15.30	16.85
	6189	4"					24.50			24.50	27
	6190	5"					45			45	49.50
	6191	6"					25			25	27.50
	6192	8"					133			133	147
	6193	10"					239			239	263
	6200	1/8 Bend, 1-1/2"					4.55			4.55	5
	6210	2"					5.10			5.10	5.60
	6212	3"					6.85			6.85	7.55
	6214	4"					8.70			8.70	9.60
	6380	Sanitary Tee, tapped, 1-1/2"					10.05			10.05	11.05
	6382	2" x 1-1/2"					9.15			9.15	10.05
	6384	2"					10.10			10.10	11.10
	6386	3" x 2"					14.15			14.15	15.60
	6388	3"					26			26	29
	6390	4" x 1-1/2"					12.65			12.65	13.90
	6392	4" x 2"					14.15			14.15	15.60
	6394	6" x 1-1/2"					29.50			29.50	32.50
	6396	6" x 2"					30			30	33
	6459	Sanitary Tee, 1-1/2"					7.45			7.45	8.15
	6460	2"					8.30			8.30	9.10
	6470	3"					10.55			10.55	11.60
	6472	4"					15.60			15.60	17.15
	8000	Coupling, standard (by CISPI Mfrs.)									
	8020	1-1/2"	Q-1	48	.333	Ea.	5.25	8.10		13.35	19.05
	8040	2"		44	.364		5.25	8.80		14.05	20.50

15107 | Metal Pipe & Fittings

Line	Description	Crew	Daily Output	Labor-Hours	Unit	Mat.	Labor	Equip.	Total	Total Incl O&P	
						2005 Bare Costs					
8080	3″	Q-1	38	.421	Ea.	6.20	10.20		16.40	23.50	360
8120	4″		33	.485		7.35	11.75		19.10	27.50	
0010	**PIPE, COPPER** Solder joints										420
1000	Type K tubing, couplings & clevis hangers 10′ O.C.										
1180	3/4″ diameter	1 Plum	74	.108	L.F.	3.02	2.91		5.93	8.10	
1200	1″ diameter	″	66	.121	″	3.97	3.27		7.24	9.70	
2000	Type L tubing, couplings & hangers 10′ O.C.										
2140	1/2″ diameter	1 Plum	81	.099	L.F.	1.50	2.66		4.16	6.05	
2160	5/8″ diameter		79	.101		2.23	2.73		4.96	6.95	
2180	3/4″ diameter		76	.105		2.20	2.84		5.04	7.10	
2200	1″ diameter		68	.118		3.10	3.17		6.27	8.60	
2220	1-1/4″ diameter		58	.138		4.25	3.72		7.97	10.80	
3000	Type M tubing, couplings & hangers 10′ O.C.										
3140	1/2″ diameter	1 Plum	84	.095	L.F.	1.18	2.57		3.75	5.50	
3180	3/4″ diameter		78	.103		1.73	2.76		4.49	6.45	
3200	1″ diameter		70	.114		2.38	3.08		5.46	7.65	
3220	1-1/4″ diameter		60	.133		3.48	3.59		7.07	9.75	
3240	1-1/2″ diameter		54	.148		4.68	3.99		8.67	11.70	
3260	2″ diameter		44	.182		7.30	4.90		12.20	16.10	
4000	Type DWV tubing, couplings & hangers 10′ O.C.										
4100	1-1/4″ diameter	1 Plum	60	.133	L.F.	3.67	3.59		7.26	9.95	
4120	1-1/2″ diameter		54	.148		4.56	3.99		8.55	11.55	
4140	2″ diameter		44	.182		6.10	4.90		11	14.75	
4160	3″ diameter	Q-1	58	.276		10.90	6.70		17.60	23	
4180	4″ diameter	″	40	.400		19.15	9.70		28.85	37	
0010	**PIPE, COPPER, FITTINGS** Wrought unless otherwise noted										460
0040	Solder joints, copper x copper										
0100	1/2″	1 Plum	20	.400	Ea.	.48	10.80		11.28	18.25	
0120	3/4″		19	.421		1.06	11.35		12.41	19.80	
0250	45° elbow, 1/4″		22	.364		2.65	9.80		12.45	19	
0280	1/2″		20	.400		.86	10.80		11.66	18.65	
0290	5/8″		19	.421		4.56	11.35		15.91	23.50	
0300	3/4″		19	.421		1.52	11.35		12.87	20.50	
0310	1″		16	.500		3.82	13.50		17.32	26	
0320	1-1/4″		15	.533		5.45	14.35		19.80	29.50	
0450	Tee, 1/4″		14	.571		3.30	15.40		18.70	29	
0480	1/2″		13	.615		.81	16.60		17.41	28.50	
0490	5/8″		12	.667		5.50	17.95		23.45	35.50	
0500	3/4″		12	.667		1.95	17.95		19.90	31.50	
0510	1″		10	.800		6	21.50		27.50	42	
0520	1-1/4″		9	.889		8.65	24		32.65	49	
0612	Tee, reducing on the outlet, 1/4″		15	.533		5.70	14.35		20.05	30	
0613	3/8″		15	.533		4.99	14.35		19.34	29	
0614	1/2″		14	.571		4.38	15.40		19.78	30.50	
0615	5/8″		13	.615		8.85	16.60		25.45	37	
0616	3/4″		12	.667		2.82	17.95		20.77	32.50	
0617	1″		11	.727		5.75	19.60		25.35	38.50	
0618	1-1/4″		10	.800		9.35	21.50		30.85	46	
0619	1-1/2″		9	.889		9.95	24		33.95	50.50	
0620	2″		8	1		16.25	27		43.25	62.50	
0621	2-1/2″	Q-1	9	1.778		44	43		87	120	
0622	3″		8	2		54	48.50		102.50	139	
0623	4″		6	2.667		110	64.50		174.50	227	
0624	5″		5	3.200		510	77.50		587.50	690	
0625	6″	Q-2	7	3.429		695	80		775	895	

	15107 | Metal Pipe & Fittings		CREW	DAILY OUTPUT	LABOR-HOURS	UNIT	2005 BARE COSTS MAT.	LABOR	EQUIP.	TOTAL	TOTAL INCL O&P
460	0626	8″	Q-2	6	4	Ea.	2,675	93.50		2,768.50	3,100
	0630	Tee, reducing on the run, 1/4″	1 Plum	15	.533		6.65	14.35		21	31
	0631	3/8″		15	.533		7.55	14.35		21.90	32
	0632	1/2″		14	.571		8.90	15.40		24.30	35.50
	0633	5/8″		13	.615		9.40	16.60		26	38
	0634	3/4″		12	.667		2.30	17.95		20.25	32
	0635	1″		11	.727		7.55	19.60		27.15	40.50
	0636	1-1/4″		10	.800		12	21.50		33.50	48.50
	0637	1-1/2″		9	.889		21	24		45	63
	0638	2″		8	1		27	27		54	74.50
	0639	2-1/2″	Q-1	9	1.778		66	43		109	144
	0640	3″		8	2		93.50	48.50		142	183
	0641	4″		6	2.667		207	64.50		271.50	335
	0642	5″		5	3.200		480	77.50		557.50	660
	0643	6″	Q-2	7	3.429		735	80		815	935
	0644	8″	″	6	4		2,675	93.50		2,768.50	3,100
	0650	Coupling, 1/4″	1 Plum	24	.333		.37	9		9.37	15.15
	0680	1/2″		22	.364		.33	9.80		10.13	16.45
	0690	5/8″		21	.381		1.02	10.25		11.27	18
	0700	3/4″		21	.381		.71	10.25		10.96	17.70
	0710	1″		18	.444		1.44	12		13.44	21.50
	0715	1-1/4″		17	.471		2.69	12.70		15.39	24
	2000	DWV, solder joints, copper x copper									
	2030	90° Elbow, 1-1/4″	1 Plum	13	.615	Ea.	4.82	16.60		21.42	33
	2050	1-1/2″		12	.667		7.20	17.95		25.15	37.50
	2070	2″		10	.800		9.40	21.50		30.90	46
	2090	3″	Q-1	10	1.600		23.50	39		62.50	89.50
	2100	4″	″	9	1.778		100	43		143	181
	2250	Tee, Sanitary, 1-1/4″	1 Plum	9	.889		8.55	24		32.55	49
	2270	1-1/2″		8	1		10.65	27		37.65	56.50
	2290	2″		7	1.143		12.45	31		43.45	64
	2310	3″	Q-1	7	2.286		50	55.50		105.50	146
	2330	4″	″	6	2.667		127	64.50		191.50	246
	2400	Coupling, 1-1/4″	1 Plum	14	.571		2.03	15.40		17.43	27.50
	2420	1-1/2″		13	.615		2.53	16.60		19.13	30.50
	2440	2″		11	.727		3.50	19.60		23.10	36
	2460	3″	Q-1	11	1.455		6.80	35.50		42.30	65.50
	2480	4″	″	10	1.600		21.50	39		60.50	88
620	0010	**PIPE, STEEL**									
	0050	Schedule 40, threaded, with couplings, and clevis type									
	0060	hangers sized for covering, 10′ O.C.									
	0540	Black, 1/4″ diameter	1 Plum	66	.121	L.F.	2.09	3.27		5.36	7.65
	0570	3/4″ diameter		61	.131		1.54	3.53		5.07	7.50
	0580	1″ diameter		53	.151		2.16	4.07		6.23	9.10
	0590	1-1/4″ diameter	Q-1	89	.180		2.75	4.36		7.11	10.15
	0600	1-1/2″ diameter		80	.200		3.16	4.85		8.01	11.45
	0610	2″ diameter		64	.250		4.13	6.05		10.18	14.50
640	0010	**PIPE, STEEL, FITTINGS** Threaded									
	5000	Malleable iron, 150 lb.									
	5020	Black									
	5040	90° elbow, straight									
	5090	3/4″	1 Plum	14	.571	Ea.	1.67	15.40		17.07	27.50
	5100	1″	″	13	.615		2.90	16.60		19.50	30.50
	5120	1-1/2″	Q-1	20	.800		6.25	19.40		25.65	39
	5130	2″	″	18	.889		10.80	21.50		32.30	47.50

15107 | Metal Pipe & Fittings		CREW	DAILY OUTPUT	LABOR-HOURS	UNIT	2005 BARE COSTS MAT.	LABOR	EQUIP.	TOTAL	TOTAL INCL O&P	
5450	Tee, straight										640
5500	3/4″	1 Plum	9	.889	Ea.	2.65	24		26.65	42.50	
5510	1″	″	8	1		4.52	27		31.52	49.50	
5520	1-1/4″	Q-1	14	1.143		7.35	27.50		34.85	53.50	
5530	1-1/2″		13	1.231		9.15	30		39.15	59	
5540	2″		11	1.455		15.55	35.50		51.05	75	
5650	Coupling										
5700	3/4″	1 Plum	18	.444	Ea.	2.23	12		14.23	22	
5710	1″	″	15	.533		3.34	14.35		17.69	27	
5730	1-1/2″	Q-1	24	.667		5.85	16.15		22	33	
5740	2″	″	21	.762		8.65	18.50		27.15	40	

15108 | Plastic Pipe & Fittings		CREW	DAILY OUTPUT	LABOR-HOURS	UNIT	MAT.	LABOR	EQUIP.	TOTAL	TOTAL INCL O&P	
0010	**PIPE, PLASTIC**										520
1800	PVC, couplings 10′ O.C., hangers 3 per 10′										
1820	Schedule 40										
1860	1/2″ diameter	1 Plum	54	.148	L.F.	.98	3.99		4.97	7.65	
1870	3/4″ diameter		51	.157		1.06	4.23		5.29	8.10	
1880	1″ diameter		46	.174		1.17	4.69		5.86	9	
1890	1-1/4″ diameter		42	.190		1.34	5.15		6.49	9.95	
1900	1-1/2″ diameter		36	.222		1.44	6		7.44	11.45	
1910	2″ diameter	Q-1	59	.271		1.63	6.60		8.23	12.60	
1920	2-1/2″ diameter		56	.286		2.20	6.95		9.15	13.80	
1930	3″ diameter		53	.302		2.87	7.30		10.17	15.20	
1940	4″ diameter		48	.333		3.67	8.10		11.77	17.35	
4100	DWV type, schedule 40, couplings 10′ O.C., hangers 3 per 10′										
4120	ABS										
4140	1-1/4″ diameter	1 Plum	42	.190	L.F.	1.08	5.15		6.23	9.65	
4150	1-1/2″ diameter	″	36	.222		1.09	6		7.09	11.05	
4160	2″ diameter	Q-1	59	.271		1.19	6.60		7.79	12.10	
4400	PVC										
4410	1-1/4″ diameter	1 Plum	42	.190	L.F.	1.19	5.15		6.34	9.75	
4420	1-1/2″ diameter	″	36	.222		1.20	6		7.20	11.15	
4460	2″ diameter	Q-1	59	.271		1.35	6.60		7.95	12.30	
4470	3″ diameter		53	.302		2.40	7.30		9.70	14.70	
4480	4″ diameter		48	.333		3.13	8.10		11.23	16.75	
5360	CPVC, couplings 10′ O.C., hangers 3 per 10′										
5380	Schedule 40										
5460	1/2″ diameter	1 Plum	54	.148	L.F.	2.31	3.99		6.30	9.10	
5470	3/4″ diameter		51	.157		3.05	4.23		7.28	10.30	
5480	1″ diameter		46	.174		3.74	4.69		8.43	11.80	
5490	1-1/4″ diameter		42	.190		4.35	5.15		9.50	13.25	
5500	1-1/2″ diameter		36	.222		4.86	6		10.86	15.20	
5510	2″ diameter	Q-1	59	.271		6	6.60		12.60	17.40	
6500	Residential installation, plastic pipe										
6510	Couplings 10′ O.C., strap hangers 3 per 10′										
6520	PVC, Schedule 40										
6530	1/2″ diameter	1 Plum	138	.058	L.F.	.29	1.56		1.85	2.89	
6540	3/4″ diameter		128	.063		.35	1.68		2.03	3.15	
6550	1″ diameter		119	.067		.45	1.81		2.26	3.48	
6560	1-1/4″ diameter		111	.072		.57	1.94		2.51	3.82	
6570	1-1/2″ diameter		104	.077		.67	2.07		2.74	4.14	
6580	2″ diameter	Q-1	197	.081		.85	1.97		2.82	4.18	
6590	2-1/2″ diameter		162	.099		1.40	2.40		3.80	5.50	
6600	4″ diameter		123	.130		6.85	3.15		10	12.75	

	15108 | Plastic Pipe & Fittings		CREW	DAILY OUTPUT	LABOR-HOURS	UNIT	2005 BARE COSTS MAT.	LABOR	EQUIP.	TOTAL	TOTAL INCL O&P
520	6700	PVC, DWV, Schedule 40									
	6720	1-1/4" diameter	1 Plum	100	.080	L.F.	.76	2.16		2.92	4.38
	6730	1-1/2" diameter	"	94	.085		.66	2.29		2.95	4.50
	6740	2" diameter	Q-1	178	.090		.84	2.18		3.02	4.50
	6760	4" diameter	"	110	.145		6.75	3.53		10.28	13.25
560	0010	**PIPE, PLASTIC, FITTINGS**									
	2700	PVC (white), schedule 40, socket joints									
	2760	90° elbow, 1/2"	1 Plum	33.30	.240	Ea.	.29	6.45		6.74	10.95
	2770	3/4"		28.60	.280		.32	7.55		7.87	12.75
	2780	1"		25	.320		.57	8.60		9.17	14.85
	2790	1-1/4"		22.20	.360		1	9.70		10.70	17.05
	2800	1-1/2"		20	.400		1.07	10.80		11.87	18.90
	2810	2"	Q-1	36.40	.440		1.68	10.65		12.33	19.35
	2820	2-1/2"		26.70	.599		5.10	14.55		19.65	29.50
	2830	3"		22.90	.699		6.10	16.95		23.05	34.50
	2840	4"		18.20	.879		10.95	21.50		32.45	47
	3180	Tee, 1/2"	1 Plum	22.20	.360		.35	9.70		10.05	16.35
	3190	3/4"		19	.421		.40	11.35		11.75	19.10
	3200	1"		16.70	.479		.75	12.90		13.65	22
	3210	1-1/4"		14.80	.541		1.18	14.55		15.73	25.50
	3220	1-1/2"		13.30	.601		1.43	16.20		17.63	28
	3230	2"	Q-1	24.20	.661		2.07	16.05		18.12	29
	3240	2-1/2"		17.80	.899		6.85	22		28.85	43.50
	3250	3"		15.20	1.053		13.55	25.50		39.05	57
	3260	4"		12.10	1.322		16.25	32		48.25	70.50
	3380	Coupling, 1/2"	1 Plum	33.30	.240		.19	6.45		6.64	10.85
	3390	3/4"		28.60	.280		.26	7.55		7.81	12.70
	3400	1"		25	.320		.44	8.60		9.04	14.70
	3410	1-1/4"		22.20	.360		.61	9.70		10.31	16.60
	3420	1-1/2"		20	.400		.65	10.80		11.45	18.40
	3430	2"	Q-1	36.40	.440		1.01	10.65		11.66	18.60
	3440	2-1/2"		26.70	.599		2.22	14.55		16.77	26.50
	3450	3"		22.90	.699		3.49	16.95		20.44	32
	3460	4"		18.20	.879		5	21.50		26.50	40.50
	4500	DWV, ABS, non pressure, socket joints									
	4540	1/4 Bend, 1-1/4"	1 Plum	20.20	.396	Ea.	2.57	10.65		13.22	20.50
	4560	1-1/2"	"	18.20	.440		1.97	11.85		13.82	21.50
	4570	2"	Q-1	33.10	.483		3.04	11.70		14.74	22.50
	4800	Tee, sanitary									
	4820	1-1/4"	1 Plum	13.50	.593	Ea.	2.91	15.95		18.86	29.50
	4830	1-1/2"	"	12.10	.661		2.68	17.80		20.48	32.50
	4840	2"	Q-1	20	.800		3.92	19.40		23.32	36.50
	5000	DWV, PVC, schedule 40, socket joints									
	5040	1/4 bend, 1-1/4"	1 Plum	20.20	.396	Ea.	3.25	10.65		13.90	21
	5060	1-1/2"	"	18.20	.440		1.24	11.85		13.09	21
	5070	2"	Q-1	33.10	.483		1.93	11.70		13.63	21.50
	5080	3"		20.80	.769		5.55	18.65		24.20	36.50
	5090	4"		16.50	.970		9.05	23.50		32.55	48.50
	5110	1/4 bend, long sweep, 1-1/2"	1 Plum	18.20	.440		3.05	11.85		14.90	23
	5112	2"	Q-1	33.10	.483		2.93	11.70		14.63	22.50
	5114	3"		20.80	.769		6.85	18.65		25.50	38
	5116	4"		16.50	.970		12.80	23.50		36.30	52.50
	5250	Tee, sanitary 1-1/4"	1 Plum	13.50	.593		3.36	15.95		19.31	30
	5254	1-1/2"	"	12.10	.661		2.13	17.80		19.93	32
	5255	2"	Q-1	20	.800		3.18	19.40		22.58	35.50

	15108 Plastic Pipe & Fittings		CREW	DAILY OUTPUT	LABOR-HOURS	UNIT	2005 BARE COSTS MAT.	LABOR	EQUIP.	TOTAL	TOTAL INCL O&P
560	5256	3″	Q-1	13.90	1.151	Ea.	7.05	28		35.05	54
	5257	4″		11	1.455		12.30	35.50		47.80	71.50
	5259	6″		6.70	2.388		60.50	58		118.50	162
	5261	8″	Q-2	6.20	3.871		181	90.50		271.50	350
	5264	2″ x 1-1/2″	Q-1	22	.727		2.94	17.65		20.59	32
	5266	3″ x 1-1/2″		15.50	1.032		4.47	25		29.47	46
	5268	4″ x 3″		12.10	1.322		19.20	32		51.20	73.50
	5271	6″ x 4″		6.90	2.319		65	56		121	164
	5314	Combination Y & 1/8 bend, 1-1/2″	1 Plum	12.10	.661		4.59	17.80		22.39	34.50
	5315	2″	Q-1	20	.800		5.95	19.40		25.35	38.50
	5317	3″		13.90	1.151		11.70	28		39.70	59
	5318	4″		11	1.455		21	35.50		56.50	81
	5324	Combination Y & 1/8 bend, reducing									
	5325	2″ x 2″ x 1-1/2″	Q-1	22	.727	Ea.	6.55	17.65		24.20	36.50
	5327	3″ x 3″ x 1-1/2″		15.50	1.032		11.40	25		36.40	53.50
	5328	3″ x 3″ x 2″		15.30	1.046		8.45	25.50		33.95	51
	5329	4″ x 4″ x 2″		12.20	1.311		15.85	32		47.85	70
	5331	Wye, 1-1/4″	1 Plum	13.50	.593		4.29	15.95		20.24	31
	5332	1-1/2″	″	12.10	.661		2.86	17.80		20.66	32.50
	5333	2″	Q-1	20	.800		3.65	19.40		23.05	36
	5334	3″		13.90	1.151		9.90	28		37.90	57
	5335	4″		11	1.455		16	35.50		51.50	75.50
	5336	6″		6.70	2.388		62	58		120	163
	5337	8″	Q-2	6.20	3.871		73.50	90.50		164	230
	5341	2″ x 1-1/2″	Q-1	22	.727		5.05	17.65		22.70	34.50
	5342	3″ x 1-1/2″		15.50	1.032		6.70	25		31.70	48.50
	5343	4″ x 3″		12.10	1.322		12.75	32		44.75	66.50
	5344	6″ x 4″		6.90	2.319		44.50	56		100.50	142
	5345	8″ x 6″	Q-2	6.40	3.750		124	87.50		211.50	280
	5347	Double wye, 1-1/2″	1 Plum	9.10	.879		6.10	23.50		29.60	45.50
	5348	2″	Q-1	16.60	.964		7.85	23.50		31.35	47
	5349	3″		10.40	1.538		20	37.50		57.50	83.50
	5350	4″		8.25	1.939		41	47		88	123
	5354	2″ x 1-1/2″		16.80	.952		7.15	23		30.15	46
	5355	3″ x 2″		10.60	1.509		15.05	36.50		51.55	76.50
	5356	4″ x 3″		8.45	1.893		32.50	46		78.50	111
	5357	6″ x 4″		7.25	2.207		67.50	53.50		121	163
	5410	Reducer bushing, 2″ x 1-1/4″		36.50	.438		.97	10.65		11.62	18.50
	5412	3″ x 1-1/2″		27.30	.586		4.75	14.20		18.95	29
	5414	4″ x 2″		18.20	.879		9.25	21.50		30.75	45
	5416	6″ x 4″		11.10	1.441		26.50	35		61.50	86.50
	5418	8″ x 6″	Q-2	10.20	2.353		53	55		108	149
	5500	CPVC, Schedule 80, threaded joints									
	5540	90° Elbow, 1/4″	1 Plum	32	.250	Ea.	7.85	6.75		14.60	19.75
	5560	1/2″		30.30	.264		4.55	7.10		11.65	16.70
	5570	3/4″		26	.308		6.80	8.30		15.10	21
	5580	1″		22.70	.352		9.55	9.50		19.05	26
	5590	1-1/4″		20.20	.396		18.40	10.65		29.05	38
	5600	1-1/2″		18.20	.440		19.80	11.85		31.65	41.50
	5610	2″	Q-1	33.10	.483		26.50	11.70		38.20	48.50
	6000	Coupling, 1/4″	1 Plum	32	.250		10	6.75		16.75	22
	6020	1/2″		30.30	.264		8.20	7.10		15.30	21
	6030	3/4″		26	.308		13.30	8.30		21.60	28.50
	6040	1″		22.70	.352		15.10	9.50		24.60	32
	6050	1-1/4″		20.20	.396		16	10.65		26.65	35
	6060	1-1/2″		18.20	.440		17.20	11.85		29.05	38.50

15108 | Plastic Pipe & Fittings

		15108 Plastic Pipe & Fittings	CREW	DAILY OUTPUT	LABOR-HOURS	UNIT	2005 BARE COSTS MAT.	LABOR	EQUIP.	TOTAL	TOTAL INCL O&P
560	6070	2″	Q-1	33.10	.483	Ea.	20.50	11.70		32.20	42

15110 | Valves

		15110 Valves	CREW	DAILY OUTPUT	LABOR-HOURS	UNIT	MAT.	LABOR	EQUIP.	TOTAL	TOTAL INCL O&P
160	0010	**VALVES, BRONZE**									
	1750	Check, swing, class 150, regrinding disc, threaded									
	1860	3/4″	1 Plum	20	.400	Ea.	35	10.80		45.80	56
	1870	1″	″	19	.421	″	52	11.35		63.35	75.50
	2850	Gate, N.R.S., soldered, 125 psi									
	2940	3/4″	1 Plum	20	.400	Ea.	25	10.80		35.80	45
	2950	1″	″	19	.421	″	34.50	11.35		45.85	56.50
	5600	Relief, pressure & temperature, self-closing, ASME, threaded									
	5650	1″	1 Plum	24	.333	Ea.	118	9		127	145
	5660	1-1/4″	″	20	.400	″	237	10.80		247.80	279
	6400	Pressure, water, ASME, threaded									
	6440	3/4″	1 Plum	28	.286	Ea.	47	7.70		54.70	64.50
	6450	1″	″	24	.333	″	93.50	9		102.50	118
	6900	Reducing, water pressure									
	6940	1/2″	1 Plum	24	.333	Ea.	147	9		156	176
	6960	1″	″	19	.421	″	227	11.35		238.35	269
	8350	Tempering, water, sweat connections									
	8400	1/2″	1 Plum	24	.333	Ea.	52.50	9		61.50	72.50
	8440	3/4″	″	20	.400	″	64	10.80		74.80	88
	8650	Threaded connections									
	8700	1/2″	1 Plum	24	.333	Ea.	79	9		88	101
	8740	3/4″	″	20	.400	″	244	10.80		254.80	287

15120 | Piping Specialties

		15120 Piping Specialties	CREW	DAILY OUTPUT	LABOR-HOURS	UNIT	MAT.	LABOR	EQUIP.	TOTAL	TOTAL INCL O&P
320	0010	**EXPANSION TANKS**									
	1505	Fiberglass and steel single / double wall storage, see Div 13201									
	2000	Steel, liquid expansion, ASME, painted, 15 gallon capacity	Q-5	17	.941	Ea.	365	23		388	440
	2040	30 gallon capacity		12	1.333		405	32.50		437.50	500
	3000	Steel ASME expansion, rubber diaphragm, 19 gal. cap. accept.		12	1.333		1,425	32.50		1,457.50	1,625
	3020	31 gallon capacity	↓	8	2	↓	1,600	48.50		1,648.50	1,825
940	0010	**WATER SUPPLY METERS**									
	2000	Domestic/commercial, bronze									
	2020	Threaded									
	2060	5/8″ diameter, to 20 GPM	1 Plum	16	.500	Ea.	40	13.50		53.50	66
	2080	3/4″ diameter, to 30 GPM		14	.571		67.50	15.40		82.90	100
	2100	1″ diameter, to 50 GPM	↓	12	.667	↓	94	17.95		111.95	133

15140 | Domestic Water Piping

		15140 Domestic Water Piping	CREW	DAILY OUTPUT	LABOR-HOURS	UNIT	MAT.	LABOR	EQUIP.	TOTAL	TOTAL INCL O&P
100	0010	**BACKFLOW PREVENTER** Includes valves									
	0020	and four test cocks, corrosion resistant, automatic operation									
	4100	Threaded, bronze, valves are ball									
	4120	3/4″ pipe size	1 Plum	16	.500	Ea.	188	13.50		201.50	229
600	0010	**VACUUM BREAKERS** Hot or cold water									
	1030	Anti-siphon, brass									
	1060	1/2″ size	1 Plum	24	.333	Ea.	16.35	9		25.35	33
	1080	3/4″ size		20	.400		28	10.80		38.80	48
	1100	1″ size	↓	19	.421	↓	43.50	11.35		54.85	66.50
800	0010	**WATER HAMMER ARRESTORS / SHOCK ABSORBERS**									
	0490	Copper									
	0500	3/4″ male I.P.S. For 1 to 11 fixtures	1 Plum	12	.667	Ea.	14.50	17.95		32.45	45.50

15150 | Sanitary Waste and Vent Piping

Section	Line	Description	Crew	Daily Output	Labor-Hours	Unit	2005 Bare Costs Mat.	2005 Bare Costs Labor	2005 Bare Costs Equip.	2005 Bare Costs Total	Total Incl O&P
200	0010	**CLEANOUTS**									
	0080	Round or square, scoriated nickel bronze top									
	0100	2″ pipe size	1 Plum	10	.800	Ea.	77	21.50		98.50	120
	0140	4″ pipe size	″	6	1.333	″	116	36		152	186
250	0010	**CLEANOUT TEE**									
	0100	Cast iron, B&S, with countersunk plug									
	0220	3″ pipe size	1 Plum	3.60	2.222	Ea.	115	60		175	225
	0240	4″ pipe size	″	3.30	2.424		143	65.50		208.50	264
	0500	For round smooth access cover, same price									
	4000	Plastic, tees and adapters. Add plugs									
	4010	ABS, DWV									
	4020	Cleanout tee, 1-1/2″ pipe size	1 Plum	15	.533	Ea.	3.41	14.35		17.76	27.50
300	0010	**FLOOR AND AREA DRAINS**									
	2000	Floor, medium duty, C.I., deep flange, 7″ dia top									
	2040	2″ and 3″ pipe size	Q-1	12	1.333	Ea.	79	32.50		111.50	140
	2080	For galvanized body, add					33.50			33.50	37
	2120	For polished bronze top, add					37			37	41
800	0010	**TRAPS**									
	0030	Cast iron, service weight									
	0050	Running P trap, without vent									
	1100	2″	Q-1	16	1	Ea.	25	24.50		49.50	67
	1150	4″	″	13	1.231		70	30		100	126
	1160	6″	Q-2	17	1.412		325	33		358	410
	3000	P trap, B&S, 2″ pipe size	Q-1	16	1		17.60	24.50		42.10	59.50
	3040	3″ pipe size	″	14	1.143		26.50	27.50		54	74.50
	4700	Copper, drainage, drum trap									
	4840	3″ x 6″ swivel, 1-1/2″ pipe size	1 Plum	16	.500	Ea.	60.50	13.50		74	88.50
	5100	P trap, standard pattern									
	5200	1-1/4″ pipe size	1 Plum	18	.444	Ea.	28	12		40	50.50
	5240	1-1/2″ pipe size		17	.471		27	12.70		39.70	51
	5260	2″ pipe size		15	.533		42	14.35		56.35	69.50
	5280	3″ pipe size		11	.727		101	19.60		120.60	143
	6710	ABS DWV P trap, solvent weld joint									
	6720	1-1/2″ pipe size	1 Plum	18	.444	Ea.	4.31	12		16.31	24.50
	6722	2″ pipe size		17	.471		6.15	12.70		18.85	28
	6724	3″ pipe size		15	.533		22.50	14.35		36.85	48
	6726	4″ pipe size		14	.571		20	15.40		35.40	47.50
	6860	PVC DWV hub x hub, basin trap, 1-1/4″ pipe size		18	.444		6.10	12		18.10	26.50
	6870	Sink P trap, 1-1/2″ pipe size		18	.444		6.10	12		18.10	26.50
	6880	Tubular S trap, 1-1/2″ pipe size		17	.471		11.25	12.70		23.95	33.50
	6890	PVC sch. 40 DWV, drum trap									
	6900	1-1/2″ pipe size	1 Plum	16	.500	Ea.	16.15	13.50		29.65	40
	6910	P trap, 1-1/2″ pipe size		18	.444		4.34	12		16.34	24.50
	6920	2″ pipe size		17	.471		6.45	12.70		19.15	28
	6930	3″ pipe size		15	.533		22.50	14.35		36.85	48
	6940	4″ pipe size		14	.571		51	15.40		66.40	81.50
	6950	P trap w/clean out, 1-1/2″ pipe size		18	.444		7.40	12		19.40	28
	6960	2″ pipe size		17	.471		12.55	12.70		25.25	35
900	0010	**VENT FLASHING, CAPS**									
	0120	Vent caps									
	0140	Cast iron									
	0160	1-1/4″ - 1-1/2″ pipe	1 Plum	23	.348	Ea.	26.50	9.35		35.85	44.50
	0170	2″ - 2-1/8″ pipe		22	.364		30.50	9.80		40.30	49.50
	0180	2-1/2″ - 3-5/8″ pipe		21	.381		35	10.25		45.25	55

MECHANICAL 15

15100 | Building Services Piping

15150 | Sanitary Waste and Vent Piping

			CREW	DAILY OUTPUT	LABOR-HOURS	UNIT	2005 BARE COSTS MAT.	LABOR	EQUIP.	TOTAL	TOTAL INCL O&P
900	0190	4" - 4-1/8" pipe	1 Plum	19	.421	Ea.	41.50	11.35		52.85	64.50
	0200	5" - 6" pipe		17	.471		62	12.70		74.70	89
	0300	PVC									
	0320	1-1/4" - 1-1/2" pipe	1 Plum	24	.333	Ea.	19.80	9		28.80	37
	0330	2" - 2-1/8" pipe		23	.348		23	9.35		32.35	40.50
	0340	2-1/2" - 3-5/8" pipe		22	.364		26	9.80		35.80	44.50
	0350	4" - 4-1/8" pipe		20	.400		31.50	10.80		42.30	52
	0360	5" - 6" pipe		18	.444		46.50	12		58.50	70.50

15160 | Storm Drainage Piping

			CREW	DAILY OUTPUT	LABOR-HOURS	UNIT	MAT.	LABOR	EQUIP.	TOTAL	TOTAL INCL O&P
500	0010	**STORM AREA DRAINS**									
	3860	Roof, flat metal deck, C.I. body, 12" C.I. dome									
	3890	3" pipe size	Q-1	14	1.143	Ea.	155	27.50		182.50	216

15180 | Heating and Cooling Piping

			CREW	DAILY OUTPUT	LABOR-HOURS	UNIT	MAT.	LABOR	EQUIP.	TOTAL	TOTAL INCL O&P
200	0010	**PUMPS, CIRCULATING** Heated or chilled water application									
	0600	Bronze, sweat connections, 1/40 HP, in line									
	0640	3/4" size	Q-1	16	1	Ea.	120	24.50		144.50	172
	1000	Flange connection, 3/4" to 1-1/2" size									
	1040	1/12 HP	Q-1	6	2.667	Ea.	325	64.50		389.50	465
	1060	1/8 HP	"	6	2.667	"	560	64.50		624.50	720

15 MECHANICAL

15400 | Plumbing Fixtures & Equipment

15410 | Plumbing Fixtures

			CREW	DAILY OUTPUT	LABOR-HOURS	UNIT	2005 BARE COSTS MAT.	LABOR	EQUIP.	TOTAL	TOTAL INCL O&P
200	0010	**CARRIERS/SUPPORTS** For plumbing fixtures									
	0600	Plate type with studs, top back plate	1 Plum	7	1.143	Ea.	35	31		66	89
	3000	Lavatory, concealed arm									
	3050	Floor mounted, single									
	3100	High back fixture	1 Plum	6	1.333	Ea.	258	36		294	345
	3200	Flat slab fixture	"	6	1.333	"	300	36		336	390
	8200	Water closet, residential									
	8220	Vertical centerline, floor mount									
	8240	Single, 3" caulk, 2" or 3" vent	1 Plum	6	1.333	Ea.	297	36		333	385
	8260	4" caulk, 2" or 4" vent	"	6	1.333	"	385	36		421	480
300	0010	**FAUCETS/FITTINGS**									
	0150	Bath, faucets, diverter spout combination, sweat	1 Plum	8	1	Ea.	69.50	27		96.50	121
	0200	For integral stops, IPS unions, add					73			73	80.50
	0420	Bath, press-bal mix valve w/diverter, spout, shower hd, arm/flange	1 Plum	8	1		114	27		141	170
	0500	Drain, central lift, 1-1/2" IPS male		20	.400		38	10.80		48.80	59.50
	0600	Trip lever, 1-1/2" IPS male		20	.400		38.50	10.80		49.30	60
	1000	Kitchen sink faucets, top mount, cast spout		10	.800		51.50	21.50		73	92
	1100	For spray, add		24	.333		11.05	9		20.05	27
	2000	Laundry faucets, shelf type, IPS or copper unions		12	.667		42.50	17.95		60.45	76
	2020										
	2100	Lavatory faucet, centerset, without drain	1 Plum	10	.800	Ea.	37.50	21.50		59	76.50
	2200	With pop-up drain		6.66	1.201		52	32.50		84.50	111
	2800	Self-closing, center set		10	.800		112	21.50		133.50	159
	3000	Service sink faucet, cast spout, pail hook, hose end		14	.571		72	15.40		87.40	105

15400 | Plumbing Fixtures & Equipment

15410 | Plumbing Fixtures		Crew	Daily Output	Labor-Hours	Unit	2005 Bare Costs Mat.	Labor	Equip.	Total	Total Incl O&P	
4000	Shower by-pass valve with union	1 Plum	18	.444	Ea.	50.50	12		62.50	75	300
4200	Shower thermostatic mixing valve, concealed		8	1		242	27		269	310	
4300	For inlet strainer, check, and stops, add					32			32	35.50	
5000	Sillcock, compact, brass, IPS or copper to hose	1 Plum	24	.333		4.74	9		13.74	19.95	

15418 | Resi/Comm/Industrial Fixtures		Crew	Daily Output	Labor-Hours	Unit	Mat.	Labor	Equip.	Total	Total Incl O&P	
0010	**BATHS** R15100-420										100
0100	Tubs, recessed porcelain enamel on cast iron, with trim										
0180	48" x 42"	Q-1	4	4	Ea.	1,500	97		1,597	1,800	
0220	72" x 36"		3	5.333		1,525	129		1,654	1,925	
0300	Mat bottom, 4' long		5.50	2.909		1,025	70.50		1,095.50	1,250	
0380	5' long		4.40	3.636		420	88		508	605	
0480	Above floor drain, 5' long		4	4		655	97		752	885	
0560	Corner 48" x 44"		4.40	3.636		1,500	88		1,588	1,800	
2000	Enameled formed steel, 4'-6" long		5.80	2.759		330	67		397	475	
2200	5' long		5.50	2.909		290	70.50		360.50	435	
4600	Module tub & showerwall surround, molded fiberglass										
4610	5' long x 34" wide x 76" high	Q-1	4	4	Ea.	495	97		592	705	
6000	Whirlpool, bath with vented overflow, molded fiberglass										
6100	66" x 48" x 24"	Q-1	1	16	Ea.	2,350	390		2,740	3,250	
6400	72" x 36" x 24"		1	16		2,300	390		2,690	3,200	
6500	60" x 30" x 21"		1	16		2,000	390		2,390	2,850	
6600	72" x 42" x 22"		1	16		3,275	390		3,665	4,250	
6700	83" x 65"		.30	53.333		4,400	1,300		5,700	6,975	
7000	Redwood hot tub system										
7050	4' diameter x 4' deep	Q-1	1	16	Ea.	1,575	390		1,965	2,375	
7150	6' diameter x 4' deep		.80	20		2,575	485		3,060	3,625	
7200	8' diameter x 4' deep		.80	20		3,950	485		4,435	5,150	
9600	Rough-in, supply, waste and vent, for all above tubs, add		2.07	7.729		142	187		329	465	
0010	**LAUNDRY SINKS** With trim										400
0020	Porcelain enamel on cast iron, black iron frame										
0050	24" x 21", single compartment	Q-1	6	2.667	Ea.	345	64.50		409.50	485	
0100	26" x 21", single compartment	"	6	2.667	"	269	64.50		333.50	400	
3000	Plastic, on wall hanger or legs										
3020	18" x 23", single compartment	Q-1	6.50	2.462	Ea.	92	59.50		151.50	199	
3100	20" x 24", single compartment		6.50	2.462		122	59.50		181.50	232	
3200	36" x 23", double compartment		5.50	2.909		147	70.50		217.50	278	
3300	40" x 24", double compartment		5.50	2.909		216	70.50		286.50	355	
5000	Stainless steel, counter top, 22" x 17" single compartment		6	2.667		345	64.50		409.50	485	
5100	22" x 22", single compartment		6	2.667		440	64.50		504.50	590	
5200	33" x 22", double compartment		5	3.200		435	77.50		512.50	605	
9600	Rough-in, supply, waste and vent, for all laundry sinks		2.14	7.477		104	181		285	415	
0010	**LAVATORIES** With trim, white unless noted otherwise										450
0500	Vanity top, porcelain enamel on cast iron										
0600	20" x 18"	Q-1	6.40	2.500	Ea.	206	60.50		266.50	325	
0640	33" x 19" oval		6.40	2.500		415	60.50		475.50	555	
0720	19" round		6.40	2.500		221	60.50		281.50	345	
0860	For color, add					25%					
1000	Cultured marble, 19" x 17", single bowl	Q-1	6.40	2.500	Ea.	153	60.50		213.50	269	
1120	25" x 22", single bowl		6.40	2.500		153	60.50		213.50	269	
1160	37" x 22", single bowl		6.40	2.500		178	60.50		238.50	296	
1560											
1900	Stainless steel, self-rimming, 25" x 22", single bowl, ledge	Q-1	6.40	2.500	Ea.	235	60.50		295.50	360	
1960	17" x 22", single bowl		6.40	2.500		229	60.50		289.50	350	

MECHANICAL 15

	15418 Resi/Comm/Industrial Fixtures		CREW	DAILY OUTPUT	LABOR-HOURS	UNIT	2005 BARE COSTS MAT.	LABOR	EQUIP.	TOTAL	TOTAL INCL O&P
450	2600	Steel, enameled, 20″ x 17″, single bowl	Q-1	5.80	2.759	Ea.	129	67		196	252
	2900	Vitreous china, 20″ x 16″, single bowl		5.40	2.963		222	72		294	360
	3200	22″ x 13″, single bowl		5.40	2.963		231	72		303	370
	3580	Rough-in, supply, waste and vent for all above lavatories		2.30	6.957		90.50	169		259.50	375
	4000	Wall hung									
	4040	Porcelain enamel on cast iron, 16″ x 14″, single bowl	Q-1	8	2	Ea.	320	48.50		368.50	430
	4180	20″ x 18″, single bowl	"	8	2	"	237	48.50		285.50	340
	4580	For color, add					30%				
	6000	Vitreous china, 18″ x 15″, single bowl with backsplash	Q-1	7	2.286	Ea.	198	55.50		253.50	310
	6060	19″ x 17″, single bowl		7	2.286		186	55.50		241.50	295
	6960	Rough-in, supply, waste and vent for above lavatories		1.66	9.639		285	234		519	700
500	0010	**SHOWERS**									
	1500	Stall, with drain only. Add for valve and door/curtain									
	1520	32″ square	Q-1	2	8	Ea.	325	194		519	680
	1530	36″ square		2	8		410	194		604	770
	1540	Terrazzo receptor, 32″ square		2	8		715	194		909	1,100
	1560	36″ square		1.80	8.889		875	216		1,091	1,325
	1580	36″ corner angle		1.80	8.889		800	216		1,016	1,225
	3000	Fiberglass, one piece, with 3 walls, 32″ x 32″ square		2.40	6.667		415	162		577	725
	3100	36″ x 36″ square		2.40	6.667		460	162		622	770
	4200	Rough-in, supply, waste and vent for above showers		2.05	7.805		97	189		286	415
600	0010	**SINKS** With faucets and drain									
	2000	Kitchen, counter top style, P.E. on C.I., 24″ x 21″ single bowl	Q-1	5.60	2.857	Ea.	206	69.50		275.50	340
	2100	31″ x 22″ single bowl		5.60	2.857		252	69.50		321.50	390
	2200	32″ x 21″ double bowl		4.80	3.333		288	81		369	450
	3000	Stainless steel, self rimming, 19″ x 18″ single bowl		5.60	2.857		335	69.50		404.50	480
	3100	25″ x 22″ single bowl		5.60	2.857		370	69.50		439.50	520
	3200	33″ x 22″ double bowl		4.80	3.333		530	81		611	720
	3300	43″ x 22″ double bowl		4.80	3.333		615	81		696	810
	4000	Steel, enameled, with ledge, 24″ x 21″ single bowl		5.60	2.857		110	69.50		179.50	235
	4100	32″ x 21″ double bowl		4.80	3.333		147	81		228	294
	4960	For color sinks except stainless steel, add					10%				
	4980	For rough-in, supply, waste and vent, counter top sinks	Q-1	2.14	7.477		104	181		285	415
	5000	Kitchen, raised deck, P.E. on C.I.									
	5100	32″ x 21″, dual level, double bowl	Q-1	2.60	6.154	Ea.	259	149		408	530
	5790	For rough-in, supply, waste & vent, sinks		1.85	8.649		104	210		314	460
	6650	Service, floor, corner, P.E. on C.I., 28″ x 28″		4.40	3.636		530	88		618	730
	6750	Vinyl coated rim guard, add					54			54	59.50
	6760	Mop sink, molded stone, 24″ x 36″	1 Plum	3.33	2.402		194	64.50		258.50	320
	6770	Mop sink, molded stone, 24″ x 36″, w/rim 3 sides	"	3.33	2.402		218	64.50		282.50	345
	6790	For rough-in, supply, waste & vent, floor service sinks	Q-1	1.64	9.756		263	237		500	680
900	0010	**WATER CLOSETS**									
	0150	Tank type, vitreous china, incl. seat, supply pipe w/stop									
	0200	Wall hung, one piece	Q-1	5.30	3.019	Ea.	390	73		463	550
	0400	Two piece, close coupled		5.30	3.019		480	73		553	645
	0960	For rough-in, supply, waste, vent and carrier		2.73	5.861		335	142		477	600
	1000	Floor mounted, one piece		5.30	3.019		495	73		568	665
	1020	One piece, low profile		5.30	3.019		395	73		468	555
	1100	Two piece, close coupled		5.30	3.019		163	73		236	299
	1960	For color, add					30%				
	1980	For rough-in, supply, waste and vent	Q-1	3.05	5.246	Ea.	152	127		279	375
	3000	Bowl only, with flush valve, seat									
	3100	Wall hung	Q-1	5.80	2.759	Ea.	345	67		412	490
	3200	For rough-in, supply, waste and vent, single WC		2.56	6.250		350	152		502	635
	3300	Floor mounted		5.80	2.759		310	67		377	450

15400 | Plumbing Fixtures & Equipment

15418 | Resi/Comm/Industrial Fixtures		CREW	DAILY OUTPUT	LABOR-HOURS	UNIT	2005 BARE COSTS MAT.	LABOR	EQUIP.	TOTAL	TOTAL INCL O&P	
3400	For rough-in, supply, waste and vent, single WC	Q-1	2.84	5.634	Ea.	168	137		305	410	900

15440 | Plumbing Pumps		CREW	DAILY OUTPUT	LABOR-HOURS	UNIT	MAT.	LABOR	EQUIP.	TOTAL	TOTAL INCL O&P	
0010	**PUMPS, SUBMERSIBLE** Sump										940
7000	Sump pump, automatic										
7100	Plastic, 1-1/4" discharge, 1/4 HP	1 Plum	6	1.333	Ea.	105	36		141	175	
7500	Cast iron, 1-1/4" discharge, 1/4 HP	"	6	1.333	"	122	36		158	193	

15480 | Domestic Water Heaters		CREW	DAILY OUTPUT	LABOR-HOURS	UNIT	MAT.	LABOR	EQUIP.	TOTAL	TOTAL INCL O&P	
0010	**WATER HEATERS**										200
1000	Residential, electric, glass lined tank, 5 yr, 10 gal., single element	1 Plum	2.30	3.478	Ea.	210	93.50		303.50	385	
1060	30 gallon, double element		2.20	3.636		280	98		378	470	
1080	40 gallon, double element		2	4		300	108		408	505	
1100	52 gallon, double element		2	4		355	108		463	565	
1120	66 gallon, double element		1.80	4.444		475	120		595	720	
1140	80 gallon, double element		1.60	5		535	135		670	810	
2000	Gas fired, foam lined tank, 10 yr, vent not incl.,										
2040	30 gallon	1 Plum	2	4	Ea.	385	108		493	600	
2100	75 gallon		1.50	5.333		745	144		889	1,050	
3000	Oil fired, glass lined tank, 5 yr, vent not included, 30 gallon		2	4		745	108		853	995	
3040	50 gallon		1.80	4.444		1,375	120		1,495	1,700	

15500 | Heat Generation Equipment

15510 | Heating Boilers and Accessories		CREW	DAILY OUTPUT	LABOR-HOURS	UNIT	2005 BARE COSTS MAT.	LABOR	EQUIP.	TOTAL	TOTAL INCL O&P	
0010	**BURNERS**										120
0990	Residential, conversion, gas fired, LP or natural										
1000	Gun type, atmospheric input 72 to 200 MBH	Q-1	2.50	6.400	Ea.	655	155		810	975	
1020	120 to 360 MBH		2	8		725	194		919	1,125	
1040	280 to 800 MBH		1.70	9.412		1,400	228		1,628	1,925	
0010	**BOILERS, ELECTRIC, ASME** Standard controls and trim										300
1000	Steam, 6 KW, 20.5 MBH	Q-19	1.20	20	Ea.	2,725	505		3,230	3,825	
1160	60 KW, 205 MBH		1	24		4,600	610		5,210	6,050	
2000	Hot water, 7.5 KW, 25.6 MBH		1.30	18.462		2,900	470		3,370	3,950	
2040	30 KW, 102 MBH		1.20	20		3,050	505		3,555	4,200	
2060	45 KW, 164 MBH		1.20	20		3,450	505		3,955	4,600	
0010	**BOILERS, GAS FIRED** Natural or propane, standard controls										400
1000	Cast iron, with insulated jacket										
3000	Hot water, gross output, 80 MBH	Q-7	1.46	21.918	Ea.	1,550	535		2,085	2,575	
3020	100 MBH	"	1.35	23.704	"	1,750	575		2,325	2,875	
4000	Steel, insulating jacket										
6000	Hot water, including burner & one zone valve, gross output										
6010	51.2 MBH	Q-6	2	12	Ea.	1,650	281		1,931	2,275	
6020	72 MBH		2	12		2,025	281		2,306	2,675	
6040	89 MBH		1.90	12.632		2,075	296		2,371	2,750	
6060	105 MBH		1.80	13.333		2,325	315		2,640	3,075	
6080	132 MBH		1.70	14.118		2,650	330		2,980	3,475	
6100	155 MBH		1.50	16		3,075	375		3,450	4,000	

	15510 | Heating Boilers and Accessories		CREW	DAILY OUTPUT	LABOR-HOURS	UNIT	2005 BARE COSTS MAT.	LABOR	EQUIP.	TOTAL	TOTAL INCL O&P
400	7000	For tankless water heater on smaller gas units, add					10%				
	7050	For additional zone valves up to 312 MBH add				Ea.	117			117	129
460	0010	**BOILERS, GAS/OIL** Combination with burners and controls									
	1000	Cast iron with insulated jacket									
	2000	Steam, gross output, 720 MBH	Q-7	.43	74.074	Ea.	6,950	1,800		8,750	10,600
	2900	Hot water, gross output									
	2910	200 MBH	Q-6	.61	39.024	Ea.	5,700	915		6,615	7,775
	2920	300 MBH		.49	49.080		5,700	1,150		6,850	8,175
	2930	400 MBH		.41	57.971		6,675	1,350		8,025	9,550
	2940	500 MBH	↓	.36	67.039		7,200	1,575		8,775	10,500
	3000	584 MBH	Q-7	.44	72.072	↓	8,800	1,750		10,550	12,600
	4000	Steel, insulated jacket, skid base, tubeless									
	4500	Steam, 150 psi gross output, 335 MBH, 10 BHP	Q-6	.54	44.037	Ea.	12,000	1,025		13,025	14,900
500	0010	**BOILERS, OIL FIRED** Standard controls, flame retention burner									
	1000	Cast iron, with insulated flush jacket									
	2000	Steam, gross output, 109 MBH	Q-7	1.20	26.667	Ea.	1,925	650		2,575	3,200
	2060	207 MBH	"	.90	35.556	"	2,775	865		3,640	4,475
	3000	Hot water, same price as steam									
	7000	Hot water, gross output, 103 MBH	Q-6	1.60	15	Ea.	1,250	350		1,600	1,975
	7020	122 MBH		1.45	16.506		2,475	385		2,860	3,350
	7060	168 MBH		1.30	18.405		3,125	430		3,555	4,125
	7080	225 MBH	↓	1.22	19.704	↓	3,225	460		3,685	4,275
880	0010	**SWIMMING POOL HEATERS** Not including wiring, external									
	0020	piping, base or pad,									
	0060	Gas fired, input, 115 MBH	Q-6	3	8	Ea.	1,850	188		2,038	2,350
	0100	135 MBH		2	12		2,100	281		2,381	2,750
	0160	155 MBH		1.50	16		2,200	375		2,575	3,050
	0200	190 MBH		1	24		2,875	565		3,440	4,075
	0280	500 MBH	↓	.40	60		6,950	1,400		8,350	9,950
	2000	Electric, 12 KW, 4,800 gallon pool	Q-19	3	8		1,725	203		1,928	2,225
	2020	15 KW, 7,200 gallon pool		2.80	8.571		1,725	217		1,942	2,250
	2040	24 KW, 9,600 gallon pool		2.40	10		2,325	253		2,578	3,000
	2100	55 KW, 24,000 gallon pool	↓	1.20	20	↓	3,325	505		3,830	4,475
	15530 | Furnaces										
200	0010	**FURNACE COMPONENTS AND COMBINATIONS**									
	0080	Coils, A/C evaporator, for gas or oil furnaces									
	0090	Add-on, with holding charge									
	0100	Upflow									
	0120	1-1/2 ton cooling	Q-5	4	4	Ea.	127	97.50		224.50	300
	0130	2 ton cooling		3.70	4.324		153	105		258	340
	0140	3 ton cooling		3.30	4.848		191	118		309	405
	0150	4 ton cooling		3	5.333		270	130		400	510
	0160	5 ton cooling	↓	2.70	5.926	↓	345	144		489	615
	0300	Downflow									
	0330	2-1/2 ton cooling	Q-5	3	5.333	Ea.	178	130		308	410
	0340	3-1/2 ton cooling		2.60	6.154		239	150		389	510
	0350	5 ton cooling	↓	2.20	7.273	↓	345	177		522	670
	0600	Horizontal									
	0630	2 ton cooling	Q-5	3.90	4.103	Ea.	180	100		280	360
	0640	3 ton cooling		3.50	4.571		206	111		317	410
	0650	4 ton cooling		3.20	5		259	122		381	485
	0660	5 ton cooling	↓	2.90	5.517	↓	345	134		479	600

15 MECHANICAL

15530 | Furnaces		CREW	DAILY OUTPUT	LABOR-HOURS	UNIT	2005 BARE COSTS MAT.	LABOR	EQUIP.	TOTAL	TOTAL INCL O&P	
2000	Cased evaporator coils for air handlers										200
2100	1-1/2 ton cooling	Q-5	4.40	3.636	Ea.	191	88.50		279.50	355	
2110	2 ton cooling		4.10	3.902		194	95		289	370	
2120	2-1/2 ton cooling		3.90	4.103		218	100		318	405	
2130	3 ton cooling		3.70	4.324		256	105		361	455	
2140	3-1/2 ton cooling		3.50	4.571		240	111		351	445	
2150	4 ton cooling		3.20	5		288	122		410	515	
2160	5 ton cooling		2.90	5.517		330	134		464	585	
3010	Air handler, modular										
3100	With cased evaporator cooling coil										
3120	1-1/2 ton cooling	Q-5	3.80	4.211	Ea.	495	103		598	715	
3130	2 ton cooling		3.50	4.571		520	111		631	760	
3140	2-1/2 ton cooling		3.30	4.848		570	118		688	820	
3150	3 ton cooling		3.10	5.161		620	126		746	885	
3160	3-1/2 ton cooling		2.90	5.517		765	134		899	1,050	
3170	4 ton cooling		2.50	6.400		940	156		1,096	1,275	
3180	5 ton cooling		2.10	7.619		1,025	186		1,211	1,425	
3500	With no cooling coil										
3520	1-1/2 ton coil size	Q-5	12	1.333	Ea.	350	32.50		382.50	435	
3530	2 ton coil size		10	1.600		375	39		414	475	
3540	2-1/2 ton coil size		10	1.600		415	39		454	520	
3554	3 ton coil size		9	1.778		460	43.50		503.50	575	
3560	3-1/2 ton coil size		9	1.778		515	43.50		558.50	635	
3570	4 ton coil size		8.50	1.882		660	46		706	800	
3580	5 ton coil size		8	2		710	48.50		758.50	860	
4000	With heater										
4120	5 kW, 17.1 MBH	Q-5	16	1	Ea.	315	24.50		339.50	385	
4130	7.5 kW, 25.6 MBH		15.60	1.026		335	25		360	410	
4140	10 kW, 34.2 MBH		15.20	1.053		390	25.50		415.50	470	
4150	12.5 KW, 42.7 MBH		14.80	1.081		440	26.50		466.50	525	
4160	15 KW, 51.2 MBH		14.40	1.111		525	27		552	625	
4170	25 KW, 85.4 MBH		14	1.143		595	28		623	695	
4180	30 KW, 102 MBH		13	1.231		695	30		725	815	
0010	**FURNACES** Hot air heating, blowers, standard controls										400
0020	not including gas, oil or flue piping										
1000	Electric, UL listed										
1020	10.2 MBH	Q-20	5	4	Ea.	315	98.50		413.50	510	
1100	34.1 MBH	"	4.40	4.545	"	405	112		517	630	
3000	Gas, AGA certified, upflow, direct drive models										
3020	45 MBH input	Q-9	4	4	Ea.	435	95.50		530.50	640	
3040	60 MBH input		3.80	4.211		580	101		681	810	
3060	75 MBH input		3.60	4.444		615	106		721	860	
3100	100 MBH input		3.20	5		655	120		775	920	
3120	125 MBH input		3	5.333		755	127		882	1,050	
3130	150 MBH input		2.80	5.714		875	137		1,012	1,200	
3140	200 MBH input		2.60	6.154		2,000	147		2,147	2,450	
4000	For starter plenum, add		16	1		61.50	24		85.50	108	
6000	Oil, UL listed, atomizing gun type burner										
6020	56 MBH output	Q-9	3.60	4.444	Ea.	745	106		851	1,000	
6030	84 MBH output		3.50	4.571		775	109		884	1,025	
6040	95 MBH output		3.40	4.706		790	112		902	1,050	
6060	134 MBH output		3.20	5		1,100	120		1,220	1,400	
6080	151 MBH output		3	5.333		1,225	127		1,352	1,550	
6100	200 MBH input		2.60	6.154		2,125	147		2,272	2,575	

15500 | Heat Generation Equipment

		15530 Furnaces	CREW	DAILY OUTPUT	LABOR-HOURS	UNIT	2005 BARE COSTS MAT.	LABOR	EQUIP.	TOTAL	TOTAL INCL O&P
440	3280	184.8 MBH heat, 60 MBH cooling	Q-10	1	24	Ea.	5,725	595		6,320	7,300

		15550 Breechings, Chimneys & Stacks	CREW	DAILY OUTPUT	LABOR-HOURS	UNIT	2005 BARE COSTS MAT.	LABOR	EQUIP.	TOTAL	TOTAL INCL O&P
440	0010	**VENT CHIMNEY** Prefab metal, U.L. listed									
	0020	Gas, double wall, galvanized steel									
	0080	3" diameter	Q-9	72	.222	V.L.F.	3.77	5.30		9.07	13.05
	0100	4" diameter		68	.235	"	4.71	5.60		10.31	14.65
	5000	Vent damper bi-metal 6" flue		16	1	Ea.	105	24		129	156
	5100	Gas, auto., electric	↓	8	2	"	169	48		217	266

15700 | Heating/Ventilating/Air Conditioning Equipment

		15730 Unitary Air Conditioning Equip	CREW	DAILY OUTPUT	LABOR-HOURS	UNIT	2005 BARE COSTS MAT.	LABOR	EQUIP.	TOTAL	TOTAL INCL O&P
500	0010	**PACKAGED TERMINAL AIR CONDITIONER** Cabinet, wall sleeve,									
	0100	louver, electric heat, thermostat, manual changeover, 208 V									
	0200	6,000 BTUH cooling, 8800 BTU heat	Q-5	6	2.667	Ea.	955	65		1,020	1,150
	0220	9,000 BTUH cooling, 13,900 BTU heat		5	3.200		1,000	78		1,078	1,225
	0240	12,000 BTUH cooling, 13,900 BTU heat		4	4		1,100	97.50		1,197.50	1,375
	0260	15,000 BTUH cooling, 13,900 BTU heat	↓	3	5.333	↓	1,300	130		1,430	1,675
600	0010	**ROOF TOP AIR CONDITIONERS** Standard controls, curb, economizer									
	1000	Single zone, electric cool, gas heat									
	1140	5 ton cooling, 112 MBH heating	Q-5	.56	28.520	Ea.	4,050	695		4,745	5,600
	1160	10 ton cooling, 200 MBH heating	Q-6	.67	35.982	"	7,600	845		8,445	9,725
800	0010	**WINDOW UNIT AIR CONDITIONERS**									
	4000	Portable/window, 15 amp 125V grounded receptacle required									
	4060	5000 BTUH	1 Carp	8	1	Ea.	194	24		218	254
	4340	6000 BTUH		8	1		204	24		228	265
	4480	8000 BTUH		6	1.333		290	32		322	375
	4500	10,000 BTUH	↓	6	1.333		390	32		422	480
	4520	12,000 BTUH	L-2	8	2	↓	475	41.50		516.50	590
	4600	Window/thru-the-wall, 15 amp 230V grounded receptacle required									
	4780	17,000 BTUH	L-2	6	2.667	Ea.	615	55.50		670.50	770
	4940	25,000 BTUH		4	4		780	83		863	1,000
	4960	29,000 BTUH	↓	4	4	↓	840	83		923	1,075
840	0010	**SELF-CONTAINED SINGLE PACKAGE**									
	0100	Air cooled, for free blow or duct, not incl. remote condenser									
	0200	3 ton cooling	Q-5	1	16	Ea.	2,575	390		2,965	3,500
	0210	4 ton cooling	"	.80	20	"	2,800	485		3,285	3,900
	1000	Water cooled for free blow or duct, not including tower									
	1100	3 ton cooling	Q-6	1	24	Ea.	2,450	565		3,015	3,625
900	0010	**SPLIT DUCTLESS SYSTEM**									
	0100	Cooling only, single zone									
	0110	Wall mount									
	0120	3/4 ton cooling	Q-5	2	8	Ea.	1,125	195		1,320	1,550
	0130	1 ton cooling		1.80	8.889		1,225	216		1,441	1,700
	0140	1-1/2 ton cooling		1.60	10		1,575	244		1,819	2,150
	0150	2 ton cooling	↓	1.40	11.429	↓	2,275	278		2,553	2,950
	1000	Ceiling mount									

15730 | Unitary Air Conditioning Equip

			CREW	DAILY OUTPUT	LABOR-HOURS	UNIT	2005 BARE COSTS MAT.	LABOR	EQUIP.	TOTAL	TOTAL INCL O&P	
900	1020	2 ton cooling	Q-5	1.40	11.429	Ea.	1,100	278		1,378	1,675	900
	1030	3 ton cooling	"	1.20	13.333	"	3,300	325		3,625	4,150	
	2000	T-Bar mount										
	2010	2 ton cooling	Q-5	1.40	11.429	Ea.	2,450	278		2,728	3,150	
	2020	3 ton cooling		1.20	13.333		2,950	325		3,275	3,775	
	2030	3-1/2 ton cooling	↓	1.10	14.545	↓	3,550	355		3,905	4,475	
	3000	Multizone										
	3010	Wall mount										
	3020	2 @ 3/4 ton cooling	Q-5	1.80	8.889	Ea.	1,075	216		1,291	1,550	
	5000	Cooling / Heating										
	5110	1 ton cooling	Q-5	1.70	9.412	Ea.	795	229		1,024	1,250	
	5120	1-1/2 ton cooling	"	1.50	10.667	"	1,275	260		1,535	1,825	
	5300	Ceiling mount										
	5310	3 ton cooling	Q-5	1	16	Ea.	3,825	390		4,215	4,850	
	7000	Accessories for all split ductless systems										
	7010	Add for ambient frost control	Q-5	8	2	Ea.	120	48.50		168.50	212	
	7020	Add for tube / wiring kit										
	7030	15′ kit	Q-5	32	.500	Ea.	28.50	12.20		40.70	51	
	7040	35′ kit	"	24	.667	"	91.50	16.25		107.75	127	

15740 | Heat Pumps

			CREW	DAILY OUTPUT	LABOR-HOURS	UNIT	MAT.	LABOR	EQUIP.	TOTAL	TOTAL INCL O&P	
100	0010	**AIR-SOURCE HEAT PUMPS** (Not including interconnecting tubing)										100
	1000	Air to air, split system, not including curbs, pads, or ductwork										
	1020	2 ton cooling, 8.5 MBH heat @ 0°F	Q-5	1.20	13.333	Ea.	1,425	325		1,750	2,075	
	1054	4 ton cooling, 24 MBH heat @ 0°F	"	.60	26.667	"	2,150	650		2,800	3,450	
	1500	Single package, not including curbs, pads, or plenums										
	1520	2 ton cooling, 6.5 MBH heat @ 0°F	Q-5	1.50	10.667	Ea.	2,200	260		2,460	2,825	
	1580	4 ton cooling, 13 MBH heat @ 0°F	"	.96	16.667	"	3,075	405		3,480	4,075	
800	0010	**WATER-SOURCE HEAT PUMPS** (Not including interconnecting tubing)										800
	2000	Water source to air, single package										
	2100	1 ton cooling, 13 MBH heat @ 75°F	Q-5	2	8	Ea.	990	195		1,185	1,425	
	2200	4 ton cooling, 31 MBH heat @ 75°F	"	1.20	13.333	"	1,475	325		1,800	2,150	

15760 | Terminal Heating & Cooling Units

			CREW	DAILY OUTPUT	LABOR-HOURS	UNIT	MAT.	LABOR	EQUIP.	TOTAL	TOTAL INCL O&P	
250	0010	**ELECTRIC HEATING**, not incl. conduit or feed wiring										250
	1100	Rule of thumb: Baseboard units, including control	1 Elec	4.40	1.818	kW	77	49.50		126.50	166	
	1300	Baseboard heaters, 2′ long, 375 watt		8	1	Ea.	29	27.50		56.50	76.50	
	1400	3′ long, 500 watt		8	1		34.50	27.50		62	82.50	
	1600	4′ long, 750 watt		6.70	1.194		41	32.50		73.50	98	
	1800	5′ long, 935 watt		5.70	1.404		48.50	38.50		87	116	
	2000	6′ long, 1125 watt		5	1.600		54	43.50		97.50	131	
	2400	8′ long, 1500 watt		4	2		68	54.50		122.50	164	
	2800	10′ long, 1875 watt	↓	3.30	2.424	↓	84.50	66		150.50	201	
	2950	Wall heaters with fan, 120 to 277 volt										
	3600	Thermostats, integral	1 Elec	16	.500	Ea.	18	13.65		31.65	42	
	3800	Line voltage, 1 pole		8	1		22.50	27.50		50	69	
	5000	Radiant heating ceiling panels, 2′ x 4′, 500 watt		16	.500		203	13.65		216.65	245	
	5050	750 watt		16	.500		223	13.65		236.65	268	
	5300	Infrared quartz heaters, 120 volts, 1000 watts		6.70	1.194		125	32.50		157.50	191	
	5350	1500 watt		5	1.600		125	43.50		168.50	209	
	5400	240 volts, 1500 watt		5	1.600		125	43.50		168.50	209	
	5450	2000 watt		4	2		125	54.50		179.50	227	
	5500	3000 watt	↓	3	2.667	↓	145	73		218	279	

15700 | Heating/Ventilating/Air Conditioning Equipment

		15760 | Terminal Heating & Cooling Units	CREW	DAILY OUTPUT	LABOR-HOURS	UNIT	2005 BARE COSTS MAT.	LABOR	EQUIP.	TOTAL	TOTAL INCL O&P
600	0010	**HYDRONIC HEATING** Terminal units, not incl. main supply pipe									
	1000	Radiation									
	1310	Baseboard, pkgd, 1/2" copper tube, alum. fin, 7" high	Q-5	60	.267	L.F.	5.35	6.50		11.85	16.55
	1320	3/4" copper tube, alum. fin, 7" high		58	.276		5.65	6.70		12.35	17.30
	1340	1" copper tube, alum. fin, 8-7/8" high		56	.286		11.70	6.95		18.65	24.50
	1360	1-1/4" copper tube, alum. fin, 8-7/8" high		54	.296		17.30	7.20		24.50	31
	3000	Radiators, cast iron									
	3100	Free standing or wall hung, 6 tube, 25" high	Q-5	96	.167	Section	24	4.06		28.06	33

15800 | Air Distribution

		15810 | Ducts	CREW	DAILY OUTPUT	LABOR-HOURS	UNIT	2005 BARE COSTS MAT.	LABOR	EQUIP.	TOTAL	TOTAL INCL O&P
100	0010	**METAL DUCTWORK**									
	0020	Fabricated rectangular, includes fittings, joints, supports,									
	0030	allowance for flexible connections, no insulation									
	0031	NOTE: Fabrication and installation are combined									
	0040	as LABOR cost. Approx. 25% fittings assumed.									
	0100	Aluminum, alloy 3003-H14, under 100 lb.	Q-10	75	.320	Lb.	2.50	7.95		10.45	16.05
	0110	100 to 500 lb.		80	.300		2	7.45		9.45	14.65
	0120	500 to 1,000 lb.		95	.253		1.70	6.25		7.95	12.35
	0140	1,000 to 2,000 lb.		120	.200		1.60	4.96		6.56	10.05
	0500	Galvanized steel, under 200 lb.		235	.102		1.04	2.53		3.57	5.40
	0520	200 to 500 lb.		245	.098		.84	2.43		3.27	4.99
	0540	500 to 1,000 lb.		255	.094		.74	2.33		3.07	4.72
500	0010	**FLEXIBLE DUCTS**									
	1300	Flexible, coated fiberglass fabric on corr. resist. metal helix									
	1400	pressure to 12" (WG) UL-181									
	1500	Non-insulated, 3" diameter	Q-9	400	.040	L.F.	1.03	.96		1.99	2.73
	1540	5" diameter		320	.050		1.29	1.20		2.49	3.43
	1560	6" diameter		280	.057		1.52	1.37		2.89	3.96
	1580	7" diameter		240	.067		1.81	1.59		3.40	4.66
	1900	Insulated, 1" thick, PE jacket, 3" diameter		380	.042		1.14	1.01		2.15	2.94
	1910	4" diameter		340	.047		1.23	1.12		2.35	3.24
	1920	5" diameter		300	.053		1.35	1.27		2.62	3.63
	1940	6" diameter		260	.062		1.56	1.47		3.03	4.19
	1960	7" diameter		220	.073		1.76	1.74		3.50	4.86
	1980	8" diameter		180	.089		1.96	2.12		4.08	5.70
	2040	12" diameter		100	.160		2.95	3.82		6.77	9.65
		15820 | Duct Accessories									
300	0010	**DUCT ACCESSORIES**									
	0050	Air extractors, 12" x 4"	1 Shee	24	.333	Ea.	15.55	8.85		24.40	32
	0100	8" x 6"		22	.364		15.55	9.65		25.20	33.50
	3000	Fire damper, curtain type, 1-1/2 hr rated, vertical, 6" x 6"		24	.333		13.40	8.85		22.25	29.50
	3020	8" x 6"		22	.364		13.40	9.65		23.05	31
	6000	12" x 12"		21	.381		27.50	10.10		37.60	47
	8000	Multi-blade dampers, parallel blade									
	8100	8" x 8"	1 Shee	24	.333	Ea.	53.50	8.85		62.35	74

Important: See the Reference Section for critical supporting data - Reference Nos., Crews, & Location Factors

15830 | Fans

			CREW	DAILY OUTPUT	LABOR-HOURS	UNIT	2005 BARE COSTS MAT.	LABOR	EQUIP.	TOTAL	TOTAL INCL O&P
100	0010	**FANS**									
	8000	Ventilation, residential									
	8020	Attic, roof type									
	8030	Aluminum dome, damper & curb									
	8040	6" diameter, 300 CFM	1 Elec	16	.500	Ea.	238	13.65		251.65	284
	8050	7" diameter, 450 CFM		15	.533		260	14.55		274.55	310
	8060	9" diameter, 900 CFM		14	.571		415	15.60		430.60	485
	8080	12" diameter, 1000 CFM (gravity)		10	.800		295	22		317	360
	8090	16" diameter, 1500 CFM (gravity)		9	.889		355	24.50		379.50	430
	8100	20" diameter, 2500 CFM (gravity)		8	1		435	27.50		462.50	525
	8160	Plastic, ABS dome									
	8180	1050 CFM	1 Elec	14	.571	Ea.	86.50	15.60		102.10	121
	8200	1600 CFM	"	12	.667	"	130	18.20		148.20	173
	8240	Attic, wall type, with shutter, one speed									
	8250	12" diameter, 1000 CFM	1 Elec	14	.571	Ea.	187	15.60		202.60	232
	8260	14" diameter, 1500 CFM		12	.667		203	18.20		221.20	253
	8270	16" diameter, 2000 CFM		9	.889		230	24.50		254.50	293
	8290	Whole house, wall type, with shutter, one speed									
	8300	30" diameter, 4800 CFM	1 Elec	7	1.143	Ea.	490	31		521	590
	8310	36" diameter, 7000 CFM		6	1.333		535	36.50		571.50	650
	8320	42" diameter, 10,000 CFM		5	1.600		600	43.50		643.50	730
	8330	48" diameter, 16,000 CFM		4	2		745	54.50		799.50	910
	8340	For two speed, add					45			45	49.50
	8350	Whole house, lay-down type, with shutter, one speed									
	8360	30" diameter, 4500 CFM	1 Elec	8	1	Ea.	525	27.50		552.50	620
	8370	36" diameter, 6500 CFM		7	1.143		560	31		591	670
	8380	42" diameter, 9000 CFM		6	1.333		620	36.50		656.50	740
	8390	48" diameter, 12,000 CFM		5	1.600		700	43.50		743.50	840
	8440	For two speed, add					33.50			33.50	37
	8450	For 12 hour timer switch, add	1 Elec	32	.250		33.50	6.85		40.35	48

15850 | Air Outlets & Inlets

			CREW	DAILY OUTPUT	LABOR-HOURS	UNIT	MAT.	LABOR	EQUIP.	TOTAL	TOTAL INCL O&P
300	0010	**DIFFUSERS** Aluminum, opposed blade damper unless noted									
	0100	Ceiling, linear, also for sidewall									
	0120	2" wide	1 Shee	32	.250	L.F.	29	6.65		35.65	42.50
	0160	4" wide		26	.308	"	37.50	8.15		45.65	55
	0500	Perforated, 24" x 24" lay-in panel size, 6" x 6"		16	.500	Ea.	75.50	13.30		88.80	106
	0520	8" x 8"		15	.533		77.50	14.15		91.65	109
	0530	9" x 9"		14	.571		79	15.15		94.15	113
	0590	16" x 16"		11	.727		96	19.30		115.30	138
	1000	Rectangular, 1 to 4 way blow, 6" x 6"		16	.500		40.50	13.30		53.80	67.50
	1010	8" x 8"		15	.533		48.50	14.15		62.65	77
	1014	9" x 9"		15	.533		52	14.15		66.15	81
	1016	10" x 10"		15	.533		61	14.15		75.15	91
	1020	12" x 6"		15	.533		67	14.15		81.15	97.50
	1040	12" x 9"		14	.571		73	15.15		88.15	106
	1060	12" x 12"		12	.667		71.50	17.70		89.20	108
	1070	14" x 6"		13	.615		72	16.35		88.35	107
	1074	14" x 14"		12	.667		93.50	17.70		111.20	133
	1150	18" x 18"		9	.889		137	23.50		160.50	191
	1170	24" x 12"		10	.800		124	21		145	172
	1180	24" x 24"		7	1.143		281	30.50		311.50	360
	1500	Round, butterfly damper, 6" diameter		18	.444		16.85	11.80		28.65	38.50
	1520	8" diameter		16	.500		18.20	13.30		31.50	42.50
	2000	T bar mounting, 24" x 24" lay-in frame, 6" x 6"		16	.500		82	13.30		95.30	113
	2020	9" x 9"		14	.571		90.50	15.15		105.65	125

		15850 | Air Outlets & Inlets	CREW	DAILY OUTPUT	LABOR-HOURS	UNIT	2005 BARE COSTS MAT.	LABOR	EQUIP.	TOTAL	TOTAL INCL O&P
300	2040	12" x 12"	1 Shee	12	.667	Ea.	117	17.70		134.70	159
	2060	15" x 15"		11	.727		150	19.30		169.30	198
	2080	18" x 18"		10	.800		157	21		178	208
	6000	For steel diffusers instead of aluminum, deduct					10%				
500	0010	**GRILLES**									
	0020	Aluminum									
	1000	Air return, 6" x 6"	1 Shee	26	.308	Ea.	13	8.15		21.15	28
	1020	10" x 6"		24	.333		15.60	8.85		24.45	32
	1080	16" x 8"		22	.364		23.50	9.65		33.15	41.50
	1100	12" x 12"		22	.364		23.50	9.65		33.15	41.50
	1120	24" x 12"		18	.444		41.50	11.80		53.30	66
	1180	16" x 16"		22	.364		34	9.65		43.65	53
700	0010	**REGISTERS**									
	0980	Air supply									
	3000	Baseboard, hand adj. damper, enameled steel									
	3012	8" x 6"	1 Shee	26	.308	Ea.	12.45	8.15		20.60	27.50
	3020	10" x 6"		24	.333		14	8.85		22.85	30.50
	3040	12" x 5"		23	.348		16	9.25		25.25	33
	3060	12" x 6"		23	.348		14.75	9.25		24	32
	4000	Floor, toe operated damper, enameled steel									
	4020	4" x 8"	1 Shee	32	.250	Ea.	17.75	6.65		24.40	30.50
	4040	4" x 12"	"	26	.308	"	21	8.15		29.15	36.50
	4300	Spiral pipe supply register									
	4310	Aluminum, double deflection, w/damper extractor									
	4320	4" x 12", for 6" thru 10" diameter duct	1 Shee	25	.320	Ea.	62	8.50		70.50	82.50
	4330	4" x 18", for 6" thru 10" diameter duct		18	.444		76.50	11.80		88.30	104
	4340	6" x 12", for 8" thru 12" diameter duct		19	.421		67.50	11.20		78.70	93.50
	4350	6" x 18", for 8" thru 12" diameter duct		18	.444		88	11.80		99.80	116
	4360	6" x 24", for 8" thru 12" diameter duct		16	.500		109	13.30		122.30	142
	4370	6" x 30", for 8" thru 12" diameter duct		15	.533		138	14.15		152.15	176
	4380	8" x 18", for 10" thru 14" diameter duct		18	.444		93.50	11.80		105.30	123
	4390	8" x 24", for 10" thru 14" diameter duct		15	.533		117	14.15		131.15	153
	4400	8" x 30", for 10" thru 14" diameter duct		14	.571		155	15.15		170.15	197
	4410	10" x 24", for 12" thru 18" diameter duct		13	.615		129	16.35		145.35	170
	4420	10" x 30", for 12" thru 18" diameter duct		12	.667		170	17.70		187.70	217
	4430	10" x 36", for 12" thru 18" diameter duct		11	.727		211	19.30		230.30	265
		15860 | Air Cleaning Devices									
100	0010	**AIR FILTERS**									
	0050	Activated charcoal type, full flow				MCFM	600			600	660
	0060	Activated charcoal type, full flow, impregnated media 12" deep					175			175	193
	0070	Activated charcoal type, HEPA filter & frame for field erection					175			175	193
	0080	Activated charcoal type, HEPA filter-diffuser, ceiling install.					250			250	275
	2000	Electronic air cleaner, duct mounted									
	2150	400 - 1000 CFM	1 Shee	2.30	3.478	Ea.	695	92.50		787.50	920
	2200	1000 - 1400 CFM		2.20	3.636		725	96.50		821.50	955
	2250	1400 - 2000 CFM		2.10	3.810		800	101		901	1,050
	2950	Mechanical media filtration units									
	3000	High efficiency type, with frame, non-supported				MCFM	45			45	49.50
	3100	Supported type					55			55	60.50
	4000	Medium efficiency, extended surface					5			5	5.50
	4500	Permanent washable					20			20	22
	5000	Renewable disposable roll					120			120	132
	5500	Throwaway glass or paper media type				Ea.	4.60			4.60	5.05

For information about Means Estimating Seminars, see yellow pages 12 and 13 in back of book

Division 16 Electrical

Estimating Tips

16060 Grounding & Bonding

When taking off grounding system, identify separately the type and size of wire and list each unique type of ground connection.

16100 Wiring Methods

Conduit should be taken off in three main categories: power distribution, branch power, and branch lighting, so the estimator can concentrate on systems and components, therefore making it easier to ensure all items have been accounted for.

For cost modifications for elevated conduit installation, add the percentages to labor according to the height of installation and only the quantities exceeding the different height levels, not to the total conduit quantities.

Remember that aluminum wiring of equal ampacity is larger in diameter than copper and may require larger conduit.

If more than three wires at a time are being pulled, deduct percentages from the labor hours of that grouping of wires.

- The estimator should take the weights of materials into consideration when completing a takeoff. Topics to consider include: How will the materials be supported? What methods of support are available? How high will the support structure have to reach? Will the final support structure be able to withstand the total burden? Is the support material included or separate from the fixture, equipment and material specified?

16200 Electrical Power

- Do not overlook the costs for equipment used in the installation. If scaffolding or highlifts are available in the field, contractors may use them in lieu of the proposed ladders and rolling staging.

16400 Low-Voltage Distribution

- Supports and concrete pads may be shown on drawings for the larger equipment, or the support system may be just a piece of plywood for the back of a panelboard. In either case, it must be included in the costs.

16500 Lighting

- Fixtures should be taken off room by room, using the fixture schedule, specifications, and the ceiling plan. For large concentrations of lighting fixtures in the same area deduct the percentages from labor hours.

16700 Communications
16800 Sound & Video

- When estimating material costs for special systems, it is always prudent to obtain manufacturers' quotations for equipment prices and special installation requirements which will affect the total costs.

Reference Numbers

Reference numbers are shown in bold squares at the beginning of some major classifications. These numbers refer to related items in the Reference Section. The reference information may be an estimating procedure, an alternate pricing method or technical information.

Note: Not all subdivisions listed here necessarily appear in this publication.

Note: **i2 Trade Service,** *in part, has been used as a reference source for some of the material prices used in Division 16.*

No part of this publication may be reproduced, stored in a retrieval system, or transmitted in any form or by any means without prior written permission of Reed Construction Data.

16055 | Selective Demolition

		Description	CREW	DAILY OUTPUT	LABOR-HOURS	UNIT	2005 BARE COSTS MAT.	LABOR	EQUIP.	TOTAL	TOTAL INCL O&P
300	0010	ELECTRICAL DEMOLITION									
	0020	Conduit to 15′ high, including fittings & hangers									
	0100	Rigid galvanized steel, 1/2″ to 1″ diameter	1 Elec	242	.033	L.F.		.90		.90	1.47
	0120	1-1/4″ to 2″	″	200	.040	″		1.09		1.09	1.78
	0270	Armored cable, (BX) avg. 50′ runs									
	0290	#14, 3 wire	1 Elec	571	.014	L.F.		.38		.38	.62
	0300	#12, 2 wire		605	.013			.36		.36	.59
	0310	#12, 3 wire		514	.016			.42		.42	.69
	0320	#10, 2 wire		514	.016			.42		.42	.69
	0330	#10, 3 wire		425	.019			.51		.51	.84
	0340	#8, 3 wire		342	.023			.64		.64	1.04
	0350	Non metallic sheathed cable (Romex)									
	0360	#14, 2 wire	1 Elec	720	.011	L.F.		.30		.30	.49
	0370	#14, 3 wire		657	.012			.33		.33	.54
	0380	#12, 2 wire		629	.013			.35		.35	.57
	0390	#10, 3 wire		450	.018			.49		.49	.79
	0400	Wiremold raceway, including fittings & hangers									
	0420	No. 3000	1 Elec	250	.032	L.F.		.87		.87	1.42
	0440	No. 4000		217	.037			1.01		1.01	1.64
	0460	No. 6000		166	.048			1.32		1.32	2.14
	0465	Telephone/power pole		12	.667	Ea.		18.20		18.20	29.50
	0470	Non-metallic, straight section		480	.017	L.F.		.46		.46	.74
	0500	Channels, steel, including fittings & hangers									
	0520	3/4″ x 1-1/2″	1 Elec	308	.026	L.F.		.71		.71	1.15
	0540	1-1/2″ x 1-1/2″		269	.030			.81		.81	1.32
	0560	1-1/2″ x 1-7/8″		229	.035			.95		.95	1.55
	1180	400 amp	2 Elec	6.80	2.353	Ea.		64		64	105
	1210	Panel boards, incl. removal of all breakers,									
	1220	conduit terminations & wire connections									
	1230	3 wire, 120/240V, 100A, to 20 circuits	1 Elec	2.60	3.077	Ea.		84		84	137
	1240	200 amps, to 42 circuits	2 Elec	2.60	6.154			168		168	274
	1260	4 wire, 120/208V, 125A, to 20 circuits	1 Elec	2.40	3.333			91		91	148
	1270	200 amps, to 42 circuits	2 Elec	2.40	6.667			182		182	296
	1720	Junction boxes, 4″ sq. & oct.	1 Elec	80	.100			2.73		2.73	4.45
	1760	Switch box		107	.075			2.04		2.04	3.32
	1780	Receptacle & switch plates		257	.031			.85		.85	1.38
	1800	Wire, THW-THWN-THHN, removed from									
	1810	in place conduit, to 15′ high									
	1830	#14	1 Elec	65	.123	C.L.F.		3.36		3.36	5.45
	1840	#12		55	.145			3.97		3.97	6.45
	1850	#10		45.50	.176			4.80		4.80	7.80
	2000	Interior fluorescent fixtures, incl. supports									
	2010	& whips, to 15′ high									
	2100	Recessed drop-in 2′ x 2′, 2 lamp	2 Elec	35	.457	Ea.		12.50		12.50	20.50
	2140	2′ x 4′, 4 lamp	″	30	.533	″		14.55		14.55	23.50
	2180	Surface mount, acrylic lens & hinged frame									
	2220	2′ x 2′, 2 lamp	2 Elec	44	.364	Ea.		9.95		9.95	16.15
	2260	2′ x 4′, 4 lamp	″	33	.485	″		13.25		13.25	21.50
	2300	Strip fixtures, surface mount									
	2320	4′ long, 1 lamp	2 Elec	53	.302	Ea.		8.25		8.25	13.40
	2380	8′ long, 2 lamp	″	40	.400	″		10.90		10.90	17.80
	2460	Interior incandescent, surface, ceiling									
	2470	or wall mount, to 12′ high									
	2480	Metal cylinder type, 75 Watt	2 Elec	62	.258	Ea.		7.05		7.05	11.45
	2600	Exterior fixtures, incandescent, wall mount									
	2620	100 Watt	2 Elec	50	.320	Ea.		8.75		8.75	14.20

16050 | Basic Electrical Materials & Methods

16055 | Selective Demolition

				DAILY	LABOR-		2005 BARE COSTS				TOTAL
			CREW	OUTPUT	HOURS	UNIT	MAT.	LABOR	EQUIP.	TOTAL	INCL O&P
300	3000	Ceiling fan, tear out and remove	1 Elec	18	.444	Ea.		12.15		12.15	19.75
	9000	Minimum labor/equipment charge	″	4	2	Job		54.50		54.50	89

16060 | Grounding & Bonding

			CREW	DAILY OUTPUT	LABOR-HOURS	UNIT	MAT.	LABOR	EQUIP.	TOTAL	TOTAL INCL O&P
800	0010	**GROUNDING**									
	0030	Rod, copper clad, 8′ long, 1/2″ diameter	1 Elec	5.50	1.455	Ea.	13.50	39.50		53	79.50
	0050	3/4″ diameter		5.30	1.509		28	41		69	98
	0080	10′ long, 1/2″ diameter		4.80	1.667		17.55	45.50		63.05	93.50
	0100	3/4″ diameter		4.40	1.818		31	49.50		80.50	116
	0261	Wire, ground bare armored, #8-1 conductor		200	.040	L.F.	.81	1.09		1.90	2.67
	0271	#6-1 conductor		180	.044	″	1.02	1.21		2.23	3.10
	0390	Bare copper wire, #8 stranded		11	.727	C.L.F.	11.95	19.85		31.80	45.50
	0401	Bare copper, #6 wire		1,000	.008	L.F.	.21	.22		.43	.59
	0601	#2 stranded	2 Elec	1,000	.016	″	.48	.44		.92	1.23
	1800	Water pipe ground clamps, heavy duty									
	2000	Bronze, 1/2″ to 1″ diameter	1 Elec	8	1	Ea.	14.40	27.50		41.90	60.50

16100 | Wiring Methods

16120 | Conductors & Cables

				DAILY	LABOR-		2005 BARE COSTS				TOTAL
			CREW	OUTPUT	HOURS	UNIT	MAT.	LABOR	EQUIP.	TOTAL	INCL O&P
120	0010	**ARMORED CABLE**									
	0051	600 volt, copper (BX), #14, 2 conductor, solid	1 Elec	240	.033	L.F.	.59	.91		1.50	2.12
	0101	3 conductor, solid		200	.040		.93	1.09		2.02	2.80
	0151	#12, 2 conductor, solid		210	.038		.60	1.04		1.64	2.35
	0201	3 conductor, solid		180	.044		.95	1.21		2.16	3.02
	0251	#10, 2 conductor, solid		180	.044		1.08	1.21		2.29	3.17
	0301	3 conductor, solid		150	.053		1.49	1.46		2.95	4.01
	0351	#8, 3 conductor, solid		120	.067		2.68	1.82		4.50	5.90
550	0010	**NON-METALLIC SHEATHED CABLE** 600 volt									
	0100	Copper with ground wire, (Romex)									
	0151	#14, 2 wire	1 Elec	250	.032	L.F.	.17	.87		1.04	1.61
	0201	3 wire		230	.035		.29	.95		1.24	1.87
	0251	#12, 2 wire		220	.036		.25	.99		1.24	1.90
	0301	3 wire		200	.040		.42	1.09		1.51	2.24
	0351	#10, 2 wire		200	.040		.51	1.09		1.60	2.34
	0401	3 wire		140	.057		.66	1.56		2.22	3.26
	0451	#8, 3 conductor		130	.062		1.24	1.68		2.92	4.10
	0501	#6, 3 wire		120	.067		1.92	1.82		3.74	5.05
	0550	SE type SER aluminum cable, 3 RHW and									
	0601	1 bare neutral, 3 #8 & 1 #8	1 Elec	150	.053	L.F.	1.12	1.46		2.58	3.60
	0651	3 #6 & 1 #6	″	130	.062		1.27	1.68		2.95	4.13
	0701	3 #4 & 1 #6	2 Elec	220	.073		1.42	1.99		3.41	4.79
	0751	3 #2 & 1 #4		200	.080		2.09	2.18		4.27	5.85
	0801	3 #1/0 & 1 #2		180	.089		3.16	2.43		5.59	7.45
	0851	3 #2/0 & 1 #1		160	.100		3.72	2.73		6.45	8.55
	0901	3 #4/0 & 1 #2/0		140	.114		5.30	3.12		8.42	10.95
	2401	SEU service entrance cable, copper 2 conductors, #8 + #8 neut.	1 Elec	150	.053		1.04	1.46		2.50	3.52
	2601	#6 + #8 neutral		130	.062		1.56	1.68		3.24	4.46
	2801	#6 + #6 neutral		130	.062		1.74	1.68		3.42	4.65
	3001	#4 + #6 neutral	2 Elec	220	.073		2.40	1.99		4.39	5.85

16120 | Conductors & Cables

			CREW	DAILY OUTPUT	LABOR-HOURS	UNIT	2005 BARE COSTS MAT.	LABOR	EQUIP.	TOTAL	TOTAL INCL O&P
550	3201	#4 + #4 neutral	2 Elec	220	.073	L.F.	2.64	1.99		4.63	6.15
	3401	#3 + #5 neutral		210	.076		2.81	2.08		4.89	6.50
	6500	Service entrance cap for copper SEU									
	6600	100 amp	1 Elec	12	.667	Ea.	8.65	18.20		26.85	39
	6700	150 amp		10	.800		12.65	22		34.65	49.50
	6800	200 amp		8	1		26.50	27.50		54	73.50
900	0010	**WIRE**									
	0021	600 volt type THW, copper solid, #14	1 Elec	1,300	.006	L.F.	.04	.17		.21	.31
	0031	#12		1,100	.007		.06	.20		.26	.38
	0041	#10		1,000	.008		.09	.22		.31	.46
	0161	#6		650	.012		.26	.34		.60	.84
	0181	#4	2 Elec	1,060	.015		.40	.41		.81	1.11
	0201	#3		1,000	.016		.50	.44		.94	1.26
	0221	#2		900	.018		.63	.49		1.12	1.48
	0241	#1		800	.020		.80	.55		1.35	1.77
	0261	1/0		660	.024		.94	.66		1.60	2.11
	0281	2/0		580	.028		1.18	.75		1.93	2.52
	0301	3/0		500	.032		1.46	.87		2.33	3.03
	0351	4/0		440	.036		1.81	.99		2.80	3.61

16132 | Conduit & Tubing

			CREW	DAILY OUTPUT	LABOR-HOURS	UNIT	MAT.	LABOR	EQUIP.	TOTAL	TOTAL INCL O&P
205	0010	**CONDUIT** To 15' high, includes 2 terminations, 2 elbows and									
	0020	11 beam clamps per 100 L.F.									
	1750	Rigid galvanized steel, 1/2" diameter	1 Elec	90	.089	L.F.	2.60	2.43		5.03	6.80
	1770	3/4" diameter		80	.100		2.96	2.73		5.69	7.70
	1800	1" diameter		65	.123		4.40	3.36		7.76	10.30
	1830	1-1/4" diameter		60	.133		6	3.64		9.64	12.55
	1850	1-1/2" diameter		55	.145		6.90	3.97		10.87	14.05
	1870	2" diameter		45	.178		8.85	4.85		13.70	17.65
	5000	Electric metallic tubing (EMT), 1/2" diameter		170	.047		.63	1.28		1.91	2.78
	5020	3/4" diameter		130	.062		1.06	1.68		2.74	3.90
	5040	1" diameter		115	.070		1.86	1.90		3.76	5.15
	5060	1-1/4" diameter		100	.080		2.89	2.18		5.07	6.75
	5080	1-1/2" diameter		90	.089		3.66	2.43		6.09	7.95
	9100	PVC, #40, 1/2" diameter		190	.042		.74	1.15		1.89	2.68
	9110	3/4" diameter		145	.055		.87	1.51		2.38	3.41
	9120	1" diameter		125	.064		1.40	1.75		3.15	4.38
	9130	1-1/4" diameter		110	.073		1.85	1.99		3.84	5.25
	9140	1-1/2" diameter		100	.080		2.12	2.18		4.30	5.90
	9150	2" diameter		90	.089		2.67	2.43		5.10	6.90
230	0010	**CONDUIT IN CONCRETE SLAB** Including terminations,									
	0020	fittings and supports									
	3230	PVC, schedule 40, 1/2" diameter	1 Elec	270	.030	L.F.	.49	.81		1.30	1.86
	3250	3/4" diameter		230	.035		.58	.95		1.53	2.19
	3270	1" diameter		200	.040		.77	1.09		1.86	2.63
	3300	1-1/4" diameter		170	.047		1.09	1.28		2.37	3.29
	3330	1-1/2" diameter		140	.057		1.34	1.56		2.90	4.02
	3350	2" diameter		120	.067		1.70	1.82		3.52	4.83
	4350	Rigid galvanized steel, 1/2" diameter		200	.040		2.36	1.09		3.45	4.37
	4400	3/4" diameter		170	.047		2.72	1.28		4	5.10
	4450	1" diameter		130	.062		4.15	1.68		5.83	7.30
	4500	1-1/4" diameter		110	.073		5.55	1.99		7.54	9.35
	4600	1-1/2" diameter		100	.080		6.45	2.18		8.63	10.65
	4800	2" diameter		90	.089		8.25	2.43		10.68	13

Important: See the Reference Section for critical supporting data - Reference Nos., Crews, & Location Factors

16132 | Conduit & Tubing

	Line	Description	Crew	Daily Output	Labor-Hours	Unit	2005 Bare Costs Mat.	Labor	Equip.	Total	Total Incl O&P
240	0010	**CONDUIT IN TRENCH** Includes terminations and fittings									
	0200	Rigid galvanized steel, 2″ diameter	1 Elec	150	.053	L.F.	8	1.46		9.46	11.15
	0400	2-1/2″ diameter	″	100	.080		15.45	2.18		17.63	20.50
	0600	3″ diameter	2 Elec	160	.100		18.90	2.73		21.63	25.50
	0800	3-1/2″ diameter	″	140	.114		23.50	3.12		26.62	31
250	0010	**CONDUIT FITTINGS FOR RIGID GALVANIZED STEEL**									
	2280	LB, LR or LL fittings & covers, 1/2″ diameter	1 Elec	16	.500	Ea.	9.20	13.65		22.85	32
	2290	3/4″ diameter		13	.615		10.50	16.80		27.30	39
	2300	1″ diameter		11	.727		16.45	19.85		36.30	50.50
	2330	1-1/4″ diameter		8	1		24.50	27.50		52	71.50
	2350	1-1/2″ diameter		6	1.333		30.50	36.50		67	93
	2370	2″ diameter		5	1.600		51	43.50		94.50	127
	5280	Service entrance cap, 1/2″ diameter		16	.500		6.80	13.65		20.45	29.50
	5300	3/4″ diameter		13	.615		7.85	16.80		24.65	36
	5320	1″ diameter		10	.800		8.35	22		30.35	44.50
	5340	1-1/4″ diameter		8	1		13.70	27.50		41.20	59.50
	5360	1-1/2″ diameter		6.50	1.231		25	33.50		58.50	82
	5380	2″ diameter		5.50	1.455		39	39.50		78.50	108
320	0010	**FLEXIBLE METALLIC CONDUIT**									
	0050	Steel, 3/8″ diameter	1 Elec	200	.040	L.F.	.34	1.09		1.43	2.15
	0100	1/2″ diameter		200	.040		.37	1.09		1.46	2.19
	0200	3/4″ diameter		160	.050		.51	1.37		1.88	2.78
	0250	1″ diameter		100	.080		.94	2.18		3.12	4.59
	0300	1-1/4″ diameter		70	.114		1.23	3.12		4.35	6.45
	0350	1-1/2″ diameter		50	.160		2.16	4.37		6.53	9.50
	0370	2″ diameter		40	.200		2.65	5.45		8.10	11.80

16133 | Multi-outlet Assemblies

	Line	Description	Crew	Daily Output	Labor-Hours	Unit	Mat.	Labor	Equip.	Total	Total Incl O&P
800	0010	**SURFACE RACEWAY**									
	0100	No. 500	1 Elec	100	.080	L.F.	.81	2.18		2.99	4.45
	0110	No. 700		100	.080		.91	2.18		3.09	4.56
	0200	No. 1000		90	.089		1.55	2.43		3.98	5.65
	0400	No. 1500, small pancake		90	.089		1.65	2.43		4.08	5.75
	0600	No. 2000, base & cover, blank		90	.089		1.61	2.43		4.04	5.70
	0800	No. 3000, base & cover, blank		75	.107		3.25	2.91		6.16	8.30
	2400	Fittings, elbows, No. 500		40	.200	Ea.	1.47	5.45		6.92	10.50
	2800	Elbow cover, No. 2000		40	.200		2.80	5.45		8.25	12
	2880	Tee, No. 500		42	.190		2.82	5.20		8.02	11.55
	2900	No. 2000		27	.296		8.80	8.10		16.90	23
	3000	Switch box, No. 500		16	.500		9.55	13.65		23.20	32.50
	3400	Telephone outlet, No. 1500		16	.500		10.60	13.65		24.25	33.50
	3600	Junction box, No. 1500		16	.500		7.40	13.65		21.05	30
	3800	Plugmold wired sections, No. 2000									
	4000	1 circuit, 6 outlets, 3 ft. long	1 Elec	8	1	Ea.	27	27.50		54.50	74
	4100	2 circuits, 8 outlets, 6 ft. long	″	5.30	1.509	″	45	41		86	117

16136 | Boxes

	Line	Description	Crew	Daily Output	Labor-Hours	Unit	Mat.	Labor	Equip.	Total	Total Incl O&P
600	0010	**OUTLET BOXES**									
	0021	Pressed steel, octagon, 4″	1 Elec	18	.444	Ea.	1.82	12.15		13.97	22
	0060	Covers, blank		64	.125		.76	3.41		4.17	6.40
	0100	Extension rings		40	.200		3.01	5.45		8.46	12.20
	0151	Square 4″		18	.444		2.14	12.15		14.29	22
	0200	Extension rings		40	.200		3.05	5.45		8.50	12.25

		16136 Boxes	CREW	DAILY OUTPUT	LABOR-HOURS	UNIT	2005 BARE COSTS MAT.	LABOR	EQUIP.	TOTAL	TOTAL INCL O&P
600	0250	Covers, blank	1 Elec	64	.125	Ea.	.86	3.41		4.27	6.50
	0300	Plaster rings		64	.125		1.68	3.41		5.09	7.40
	0651	Switchbox		24	.333		2.94	9.10		12.04	18.05
	1101	Concrete, floor, 1 gang		4.80	1.667		67.50	45.50		113	148
620	0010	**OUTLET BOXES, PLASTIC**									
	0051	4" diameter, round, with 2 mounting nails	1 Elec	23	.348	Ea.	2.28	9.50		11.78	17.95
	0101	Bar hanger mounted		23	.348		4.16	9.50		13.66	20
	0201	Square with 2 mounting nails		23	.348		3.44	9.50		12.94	19.25
	0300	Plaster ring		64	.125		1.44	3.41		4.85	7.15
	0401	Switch box with 2 mounting nails, 1 gang		27	.296		1.56	8.10		9.66	14.85
	0501	2 gang		23	.348		3.20	9.50		12.70	18.95
	0601	3 gang		18	.444		5.05	12.15		17.20	25.50
700	0010	**PULL BOXES & CABINETS**									
	0100	Sheet metal, pull box, NEMA 1, type SC, 6" W x 6" H x 4" D	1 Elec	8	1	Ea.	9.75	27.50		37.25	55.50
	0200	8" W x 8" H x 4" D		8	1		13.35	27.50		40.85	59
	0300	10" W x 12" H x 6" D		5.30	1.509		23.50	41		64.50	93
		16139 Residential Wiring									
700	0010	**RESIDENTIAL WIRING**									
	0020	20' avg. runs and #14/2 wiring incl. unless otherwise noted									
	1000	Service & panel, includes 24' SE-AL cable, service eye, meter,									
	1010	Socket, panel board, main bkr., ground rod, 15 or 20 amp									
	1020	1-pole circuit breakers, and misc. hardware									
	1100	100 amp, with 10 branch breakers	1 Elec	1.19	6.723	Ea.	445	184		629	785
	1110	With PVC conduit and wire		.92	8.696		485	237		722	915
	1120	With RGS conduit and wire		.73	10.959		675	299		974	1,225
	1150	150 amp, with 14 branch breakers		1.03	7.767		690	212		902	1,100
	1170	With PVC conduit and wire		.82	9.756		775	266		1,041	1,275
	1180	With RGS conduit and wire		.67	11.940		1,125	325		1,450	1,775
	1200	200 amp, with 18 branch breakers	2 Elec	1.80	8.889		910	243		1,153	1,400
	1220	With PVC conduit and wire		1.46	10.959		990	299		1,289	1,575
	1230	With RGS conduit and wire		1.24	12.903		1,450	350		1,800	2,175
	1800	Lightning surge suppressor for above services, add	1 Elec	32	.250		41.50	6.85		48.35	56.50
	2000	Switch devices									
	2100	Single pole, 15 amp, Ivory, with a 1-gang box, cover plate,									
	2110	Type NM (Romex) cable	1 Elec	17.10	.468	Ea.	8.10	12.75		20.85	30
	2120	Type MC (BX) cable		14.30	.559		20.50	15.25		35.75	47.50
	2130	EMT & wire		5.71	1.401		29	38.50		67.50	94.50
	2150	3-way, #14/3, type NM cable		14.55	.550		12	15		27	37.50
	2170	Type MC cable		12.31	.650		29	17.75		46.75	61
	2180	EMT & wire		5	1.600		31	43.50		74.50	106
	2200	4-way, #14/3, type NM cable		14.55	.550		25.50	15		40.50	52.50
	2220	Type MC cable		12.31	.650		42.50	17.75		60.25	76
	2230	EMT & wire		5	1.600		45	43.50		88.50	120
	2250	S.P., 20 amp, #12/2, type NM cable		13.33	.600		14.05	16.40		30.45	42
	2270	Type MC cable		11.43	.700		25	19.10		44.10	58.50
	2280	EMT & wire		4.85	1.649		35	45		80	112
	2290	S.P. rotary dimmer, 600W, no wiring		17	.471		16.20	12.85		29.05	39
	2300	S.P. rotary dimmer, 600W, type NM cable		14.55	.550		19.65	15		34.65	46
	2320	Type MC cable		12.31	.650		32	17.75		49.75	64.50
	2330	EMT & wire		5	1.600		41	43.50		84.50	117
	2350	3-way rotary dimmer, type NM cable		13.33	.600		17.35	16.40		33.75	45.50
	2370	Type MC cable		11.43	.700		30	19.10		49.10	64
	2380	EMT & wire		4.85	1.649		39	45		84	117
	2400	Interval timer wall switch, 20 amp, 1-30 min., #12/2									
	2410	Type NM cable	1 Elec	14.55	.550	Ea.	31	15		46	58.50

16139 | Residential Wiring

			CREW	DAILY OUTPUT	LABOR-HOURS	UNIT	2005 BARE COSTS MAT.	LABOR	EQUIP.	TOTAL	TOTAL INCL O&P
700	2420	Type MC cable	1 Elec	12.31	.650	Ea.	39	17.75		56.75	71.50
	2430	EMT & wire	↓	5	1.600	↓	52	43.50		95.50	128
	2500	Decorator style									
	2510	S.P., 15 amp, type NM cable	1 Elec	17.10	.468	Ea.	11.65	12.75		24.40	34
	2520	Type MC cable		14.30	.559		24	15.25		39.25	51.50
	2530	EMT & wire		5.71	1.401		32.50	38.50		71	98
	2550	3-way, #14/3, type NM cable		14.55	.550		15.55	15		30.55	41.50
	2570	Type MC cable		12.31	.650		32.50	17.75		50.25	65
	2580	EMT & wire		5	1.600		34.50	43.50		78	109
	2600	4-way, #14/3, type NM cable		14.55	.550		29	15		44	56.50
	2620	Type MC cable		12.31	.650		46	17.75		63.75	80
	2630	EMT & wire		5	1.600		48.50	43.50		92	124
	2650	S.P., 20 amp, #12/2, type NM cable		13.33	.600		17.60	16.40		34	46
	2670	Type MC cable		11.43	.700		29	19.10		48.10	62.50
	2680	EMT & wire		4.85	1.649		38.50	45		83.50	116
	2700	S.P., slide dimmer, type NM cable		17.10	.468		27	12.75		39.75	50.50
	2720	Type MC cable		14.30	.559		39.50	15.25		54.75	68.50
	2730	EMT & wire		5.71	1.401		48.50	38.50		87	116
	2750	S.P., touch dimmer, type NM cable		17.10	.468		23	12.75		35.75	46.50
	2770	Type MC cable		14.30	.559		35.50	15.25		50.75	64
	2780	EMT & wire		5.71	1.401		44.50	38.50		83	112
	2800	3-way touch dimmer, type NM cable		13.33	.600		41	16.40		57.40	71.50
	2820	Type MC cable		11.43	.700		53.50	19.10		72.60	90
	2830	EMT & wire	↓	4.85	1.649	↓	62.50	45		107.50	143
	3000	Combination devices									
	3100	S.P. switch/15 amp recpt., Ivory, 1-gang box, plate									
	3110	Type NM cable	1 Elec	11.43	.700	Ea.	16.35	19.10		35.45	49
	3120	Type MC cable		10	.800		29	22		51	67.50
	3130	EMT & wire		4.40	1.818		38	49.50		87.50	123
	3150	S.P. switch/pilot light, type NM cable		11.43	.700		17	19.10		36.10	49.50
	3170	Type MC cable		10	.800		29.50	22		51.50	68
	3180	EMT & wire		4.43	1.806		38.50	49.50		88	123
	3190	2-S.P. switches, 2-#14/2, no wiring		14	.571		6.75	15.60		22.35	33
	3200	2-S.P. switches, 2-#14/2, type NM cables		10	.800		19.05	22		41.05	56.50
	3220	Type MC cable		8.89	.900		40	24.50		64.50	84
	3230	EMT & wire		4.10	1.951		39.50	53.50		93	130
	3250	3-way switch/15 amp recpt., #14/3, type NM cable		10	.800		23.50	22		45.50	61
	3270	Type MC cable		8.89	.900		40.50	24.50		65	84.50
	3280	EMT & wire		4.10	1.951		42.50	53.50		96	133
	3300	2-3 way switches, 2-#14/3, type NM cables		8.89	.900		31.50	24.50		56	74.50
	3320	Type MC cable		8	1		61	27.50		88.50	112
	3330	EMT & wire		4	2		47	54.50		101.50	141
	3350	S.P. switch/20 amp recpt., #12/2, type NM cable		10	.800		27.50	22		49.50	65.50
	3370	Type MC cable		8.89	.900		35.50	24.50		60	79
	3380	EMT & wire	↓	4.10	1.951	↓	48.50	53.50		102	140
	3400	Decorator style									
	3410	S.P. switch/15 amp recpt., type NM cable	1 Elec	11.43	.700	Ea.	19.90	19.10		39	53
	3420	Type MC cable		10	.800		32.50	22		54.50	71
	3430	EMT & wire		4.40	1.818		41.50	49.50		91	127
	3450	S.P. switch/pilot light, type NM cable		11.43	.700		20.50	19.10		39.60	53.50
	3470	Type MC cable		10	.800		33	22		55	72
	3480	EMT & wire		4.40	1.818		42	49.50		91.50	128
	3500	2-S.P. switches, 2-#14/2, type NM cables		10	.800		22.50	22		44.50	60.50
	3520	Type MC cable		8.89	.900		43.50	24.50		68	88
	3530	EMT & wire		4.10	1.951		43	53.50		96.50	134
	3550	3-way/15 amp recpt., #14/3, type NM cable	↓	10	.800	↓	27	22		49	65

16139 | Residential Wiring

	Line	Description	Crew	Daily Output	Labor-Hours	Unit	2005 Bare Costs Mat.	Labor	Equip.	Total	Total Incl O&P
700	3570	Type MC cable	1 Elec	8.89	.900	Ea.	44	24.50		68.50	88.50
	3580	EMT & wire		4.10	1.951		46	53.50		99.50	137
	3650	2-3 way switches, 2-#14/3, type NM cables		8.89	.900		35	24.50		59.50	78.50
	3670	Type MC cable		8	1		64.50	27.50		92	116
	3680	EMT & wire		4	2		50.50	54.50		105	145
	3700	S.P. switch/20 amp recpt., #12/2, type NM cable		10	.800		31	22		53	69.50
	3720	Type MC cable		8.89	.900		39	24.50		63.50	82.50
	3730	EMT & wire		4.10	1.951		52	53.50		105.50	144
	4000	Receptacle devices									
	4010	Duplex outlet, 15 amp recpt., Ivory, 1-gang box, plate									
	4015	Type NM cable	1 Elec	14.55	.550	Ea.	6.45	15		21.45	31.50
	4020	Type MC cable		12.31	.650		19	17.75		36.75	50
	4030	EMT & wire		5.33	1.501		27	41		68	96.50
	4050	With #12/2, type NM cable		12.31	.650		8.05	17.75		25.80	38
	4070	Type MC cable		10.67	.750		19.20	20.50		39.70	54.50
	4080	EMT & wire		4.71	1.699		29	46.50		75.50	108
	4100	20 amp recpt., #12/2, type NM cable		12.31	.650		15.45	17.75		33.20	46
	4120	Type MC cable		10.67	.750		26.50	20.50		47	63
	4130	EMT & wire		4.71	1.699		36.50	46.50		83	116
	4140	For GFI see line 4300 below									
	4150	Decorator style, 15 amp recpt., type NM cable	1 Elec	14.55	.550	Ea.	10	15		25	35.50
	4170	Type MC cable		12.31	.650		22.50	17.75		40.25	54
	4180	EMT & wire		5.33	1.501		31	41		72	101
	4200	With #12/2, type NM cable		12.31	.650		11.60	17.75		29.35	42
	4220	Type MC cable		10.67	.750		23	20.50		43.50	58.50
	4230	EMT & wire		4.71	1.699		32.50	46.50		79	112
	4250	20 amp recpt. #12/2, type NM cable		12.31	.650		19	17.75		36.75	50
	4270	Type MC cable		10.67	.750		30	20.50		50.50	66.50
	4280	EMT & wire		4.71	1.699		40	46.50		86.50	120
	4300	GFI, 15 amp recpt., type NM cable		12.31	.650		35	17.75		52.75	67.50
	4320	Type MC cable		10.67	.750		47.50	20.50		68	86
	4330	EMT & wire		4.71	1.699		56	46.50		102.50	137
	4350	GFI with #12/2, type NM cable		10.67	.750		36.50	20.50		57	73.50
	4370	Type MC cable		9.20	.870		47.50	23.50		71	91
	4380	EMT & wire		4.21	1.900		57.50	52		109.50	148
	4400	20 amp recpt., #12/2 type NM cable		10.67	.750		38	20.50		58.50	75.50
	4420	Type MC cable		9.20	.870		49.50	23.50		73	93
	4430	EMT & wire		4.21	1.900		59.50	52		111.50	150
	4500	Weather-proof cover for above receptacles, add		32	.250		4.40	6.85		11.25	15.95
	4550	Air conditioner outlet, 20 amp-240 volt recpt.									
	4560	30' of #12/2, 2 pole circuit breaker									
	4570	Type NM cable	1 Elec	10	.800	Ea.	44.50	22		66.50	84
	4580	Type MC cable		9	.889		59	24.50		83.50	105
	4590	EMT & wire		4	2		64.50	54.50		119	160
	4600	Decorator style, type NM cable		10	.800		48	22		70	88.50
	4620	Type MC cable		9	.889		63	24.50		87.50	109
	4630	EMT & wire		4	2		68.50	54.50		123	164
	4650	Dryer outlet, 30 amp-240 volt recpt., 20' of #10/3									
	4660	2 pole circuit breaker									
	4670	Type NM cable	1 Elec	6.41	1.248	Ea.	54	34		88	115
	4680	Type MC cable		5.71	1.401		65.50	38.50		104	135
	4690	EMT & wire		3.48	2.299		69	63		132	178
	4700	Range outlet, 50 amp-240 volt recpt., 30' of #8/3									
	4710	Type NM cable	1 Elec	4.21	1.900	Ea.	80.50	52		132.50	173
	4720	Type MC cable		4	2		128	54.50		182.50	230
	4730	EMT & wire		2.96	2.703		86	74		160	215

16 ELECTRICAL

16139 | Residential Wiring		CREW	DAILY OUTPUT	LABOR-HOURS	UNIT	2005 BARE COSTS MAT.	LABOR	EQUIP.	TOTAL	TOTAL INCL O&P	
4750	Central vacuum outlet, Type NM cable	1 Elec	6.40	1.250	Ea.	48	34		82	109	700
4770	Type MC cable		5.71	1.401		70	38.50		108.50	140	
4780	EMT & wire		3.48	2.299		75	63		138	185	
4800	30 amp-110 volt locking recpt., #10/2 circ. bkr.										
4810	Type NM cable	1 Elec	6.20	1.290	Ea.	58.50	35		93.50	122	
4820	Type MC cable		5.40	1.481		82	40.50		122.50	157	
4830	EMT & wire		3.20	2.500		82.50	68.50		151	202	
4900	Low voltage outlets										
4910	Telephone recpt., 20′ of 4/C phone wire	1 Elec	26	.308	Ea.	7.40	8.40		15.80	22	
4920	TV recpt., 20′ of RG59U coax wire, F type connector	"	16	.500	"	12.65	13.65		26.30	36	
4950	Door bell chime, transformer, 2 buttons, 60′ of bellwire										
4970	Economy model	1 Elec	11.50	.696	Ea.	54.50	19		73.50	91	
4980	Custom model		11.50	.696		85.50	19		104.50	125	
4990	Luxury model, 3 buttons		9.50	.842		231	23		254	292	
6000	Lighting outlets										
6050	Wire only (for fixture), type NM cable	1 Elec	32	.250	Ea.	4.42	6.85		11.27	15.95	
6070	Type MC cable		24	.333		14.10	9.10		23.20	30.50	
6080	EMT & wire		10	.800		21	22		43	59	
6100	Box (4″), and wire (for fixture), type NM cable		25	.320		9.40	8.75		18.15	24.50	
6120	Type MC cable		20	.400		19.10	10.90		30	39	
6130	EMT & wire		11	.727		26	19.85		45.85	61.50	
6200	Fixtures (use with lines 6050 or 6100 above)										
6210	Canopy style, economy grade	1 Elec	40	.200	Ea.	26	5.45		31.45	37.50	
6220	Custom grade		40	.200		48	5.45		53.45	62	
6250	Dining room chandelier, economy grade		19	.421		78	11.50		89.50	105	
6260	Custom grade		19	.421		230	11.50		241.50	272	
6270	Luxury grade		15	.533		510	14.55		524.55	585	
6310	Kitchen fixture (fluorescent), economy grade		30	.267		52	7.30		59.30	69	
6320	Custom grade		25	.320		165	8.75		173.75	196	
6350	Outdoor, wall mounted, economy grade		30	.267		27.50	7.30		34.80	42.50	
6360	Custom grade		30	.267		104	7.30		111.30	126	
6370	Luxury grade		25	.320		235	8.75		243.75	273	
6410	Outdoor PAR floodlights, 1 lamp, 150 watt		20	.400		21.50	10.90		32.40	41.50	
6420	2 lamp, 150 watt each		20	.400		36	10.90		46.90	57.50	
6430	For infrared security sensor, add		32	.250		90	6.85		96.85	110	
6450	Outdoor, quartz-halogen, 300 watt flood		20	.400		39	10.90		49.90	61	
6600	Recessed downlight, round, pre-wired, 50 or 75 watt trim		30	.267		36	7.30		43.30	51.50	
6610	With shower light trim		30	.267		45	7.30		52.30	61.50	
6620	With wall washer trim		28	.286		54	7.80		61.80	72	
6630	With eye-ball trim		28	.286		54	7.80		61.80	72	
6640	For direct contact with insulation, add					1.65			1.65	1.82	
6700	Porcelain lamp holder	1 Elec	40	.200		3.60	5.45		9.05	12.85	
6710	With pull switch		40	.200		3.86	5.45		9.31	13.15	
6750	Fluorescent strip, 1-20 watt tube, wrap around diffuser, 24″		24	.333		53	9.10		62.10	73.50	
6760	1-40 watt tube, 48″		24	.333		65	9.10		74.10	86.50	
6770	2-40 watt tubes, 48″		20	.400		79	10.90		89.90	105	
6780	With residential ballast		20	.400		89.50	10.90		100.40	116	
6800	Bathroom heat lamp, 1-250 watt		28	.286		35	7.80		42.80	51	
6810	2-250 watt lamps		28	.286		56	7.80		63.80	74	
6820	For timer switch, see line 2400										
6900	Outdoor post lamp, incl. post, fixture, 35′ of #14/2										
6910	Type NMC cable	1 Elec	3.50	2.286	Ea.	184	62.50		246.50	305	
6920	Photo-eye, add		27	.296		30	8.10		38.10	46	
6950	Clock dial time switch, 24 hr., w/enclosure, type NM cable		11.43	.700		53.50	19.10		72.60	90	
6970	Type MC cable		11	.727		66	19.85		85.85	105	
6980	EMT & wire		4.85	1.649		74.50	45		119.50	155	

ELECTRICAL 16

16139 | Residential Wiring

	Line	Description	CREW	DAILY OUTPUT	LABOR-HOURS	UNIT	2005 BARE COSTS MAT.	LABOR	EQUIP.	TOTAL	TOTAL INCL O&P
700	7000	Alarm systems									
	7050	Smoke detectors, box, #14/3, type NM cable	1 Elec	14.55	.550	Ea.	29.50	15		44.50	57
	7070	Type MC cable	↓	12.31	.650	↓	43.50	17.75		61.25	77
	7080	EMT & wire	↓	5	1.600	↓	45.50	43.50		89	121
	7090	For relay output to security system, add				↓	12.10			12.10	13.30
	8000	Residential equipment									
	8050	Disposal hook-up, incl. switch, outlet box, 3′ of flex									
	8060	20 amp-1 pole circ. bkr., and 25′ of #12/2									
	8070	Type NM cable	1 Elec	10	.800	Ea.	22.50	22		44.50	60
	8080	Type MC cable	↓	8	1	↓	35.50	27.50		63	83.50
	8090	EMT & wire	↓	5	1.600	↓	46.50	43.50		90	123
	8100	Trash compactor or dishwasher hook-up, incl. outlet box,									
	8110	3′ of flex, 15 amp-1 pole circ. bkr., and 25′ of #14/2									
	8130	Type MC cable	1 Elec	8	1	Ea.	31	27.50		58.50	78.50
	8140	EMT & wire	"	5	1.600	"	40.50	43.50		84	116
	8150	Hot water sink dispensor hook-up, use line 8100									
	8200	Vent/exhaust fan hook-up, type NM cable	1 Elec	32	.250	Ea.	4.42	6.85		11.27	15.95
	8220	Type MC cable	↓	24	.333	↓	14.10	9.10		23.20	30.50
	8230	EMT & wire	↓	10	.800	↓	21	22		43	59
	8250	Bathroom vent fan, 50 CFM (use with above hook-up)									
	8260	Economy model	1 Elec	15	.533	Ea.	22	14.55		36.55	47.50
	8270	Low noise model	↓	15	.533	↓	31	14.55		45.55	57.50
	8280	Custom model	↓	12	.667	↓	114	18.20		132.20	155
	8300	Bathroom or kitchen vent fan, 110 CFM									
	8310	Economy model	1 Elec	15	.533	Ea.	58	14.55		72.55	87.50
	8320	Low noise model	"	15	.533	"	77	14.55		91.55	108
	8350	Paddle fan, variable speed (w/o lights)									
	8360	Economy model (AC motor)	1 Elec	10	.800	Ea.	103	22		125	149
	8370	Custom model (AC motor)	↓	10	.800	↓	178	22		200	232
	8380	Luxury model (DC motor)	↓	8	1	↓	350	27.50		377.50	430
	8390	Remote speed switch for above, add	↓	12	.667	↓	25	18.20		43.20	57
	8500	Whole house exhaust fan, ceiling mount, 36″, variable speed									
	8510	Remote switch, incl. shutters, 20 amp-1 pole circ. bkr.									
	8520	30′ of #12/2, type NM cable	1 Elec	4	2	Ea.	615	54.50		669.50	765
	8530	Type MC cable	↓	3.50	2.286	↓	630	62.50		692.50	790
	8540	EMT & wire	↓	3	2.667	↓	645	73		718	825
	8600	Whirlpool tub hook-up, incl. timer switch, outlet box									
	8610	3′ of flex, 20 amp-1 pole GFI circ. bkr.									
	8620	30′ of #12/2, type NM cable	1 Elec	5	1.600	Ea.	85	43.50		128.50	165
	8630	Type MC cable	↓	4.20	1.905	↓	93.50	52		145.50	188
	8640	EMT & wire	↓	3.40	2.353	↓	103	64		167	219
	8650	Hot water heater hook-up, incl. 1-2 pole circ. bkr., box;									
	8660	3′ of flex, 20′ of #10/2, type NM cable	1 Elec	5	1.600	Ea.	25	43.50		68.50	98.50
	8670	Type MC cable	↓	4.20	1.905	↓	40.50	52		92.50	129
	8680	EMT & wire	↓	3.40	2.353	↓	40.50	64		104.50	150
	9000	Heating/air conditioning									
	9050	Furnace/boiler hook-up, incl. firestat, local on-off switch									
	9060	Emergency switch, and 40′ of type NM cable	1 Elec	4	2	Ea.	45	54.50		99.50	139
	9070	Type MC cable	↓	3.50	2.286	↓	66	62.50		128.50	175
	9080	EMT & wire	↓	1.50	5.333	↓	82	146		228	325
	9100	Air conditioner hook-up, incl. local 60 amp disc. switch									
	9110	3′ sealtite, 40 amp, 2 pole circuit breaker									
	9130	40′ of #8/2, type NM cable	1 Elec	3.50	2.286	Ea.	158	62.50		220.50	276
	9140	Type MC cable	↓	3	2.667	↓	233	73		306	375
	9150	EMT & wire	↓	1.30	6.154	↓	183	168		351	475
	9200	Heat pump hook-up, 1-40 & 1-100 amp 2 pole circ. bkr.									

16100 | Wiring Methods

	16139 | Residential Wiring	CREW	DAILY OUTPUT	LABOR-HOURS	UNIT	2005 BARE COSTS MAT.	LABOR	EQUIP.	TOTAL	TOTAL INCL O&P
700	9210 Local disconnect switch, 3′ sealtite									
	9220 40′ of #8/2 & 30′ of #3/2									
	9230 Type NM cable	1 Elec	1.30	6.154	Ea.	385	168		553	695
	9240 Type MC cable		1.08	7.407		560	202		762	950
	9250 EMT & wire		.94	8.511		425	232		657	850
	9500 Thermostat hook-up, using low voltage wire									
	9520 Heating only	1 Elec	24	.333	Ea.	5.25	9.10		14.35	20.50
	9530 Heating/cooling	″	20	.400	″	6.35	10.90		17.25	25

	16140 | Wiring Devices	CREW	DAILY OUTPUT	LABOR-HOURS	UNIT	MAT.	LABOR	EQUIP.	TOTAL	TOTAL INCL O&P
500	0010 LOW VOLTAGE SWITCHING									
	3600 Relays, 120 V or 277 V standard	1 Elec	12	.667	Ea.	26	18.20		44.20	58
	3800 Flush switch, standard		40	.200		9.05	5.45		14.50	18.85
	4000 Interchangeable		40	.200		11.80	5.45		17.25	22
	4100 Surface switch, standard		40	.200		6.60	5.45		12.05	16.20
	4200 Transformer 115 V to 25 V		12	.667		93	18.20		111.20	132
	4400 Master control, 12 circuit, manual		4	2		94	54.50		148.50	192
	4500 25 circuit, motorized		4	2		102	54.50		156.50	201
	4600 Rectifier, silicon		12	.667		30.50	18.20		48.70	63
	4800 Switchplates, 1 gang, 1, 2 or 3 switch, plastic		80	.100		3	2.73		5.73	7.75
	5000 Stainless steel		80	.100		8.10	2.73		10.83	13.35
	5400 2 gang, 3 switch, stainless steel		53	.151		15.65	4.12		19.77	24
	5500 4 switch, plastic		53	.151		6.70	4.12		10.82	14.05
	5600 2 gang, 4 switch, stainless steel		53	.151		16.35	4.12		20.47	24.50
	5700 6 switch, stainless steel		53	.151		36	4.12		40.12	46
	5800 3 gang, 9 switch, stainless steel		32	.250		50	6.85		56.85	66
910	0010 WIRING DEVICES									
	0200 Toggle switch, quiet type, single pole, 15 amp	1 Elec	40	.200	Ea.	4.84	5.45		10.29	14.20
	0600 3 way, 15 amp		23	.348		7	9.50		16.50	23
	0900 4 way, 15 amp		15	.533		21	14.55		35.55	46.50
	1650 Dimmer switch, 120 volt, incandescent, 600 watt, 1 pole		16	.500		10.80	13.65		24.45	34
	2460 Receptacle, duplex, 120 volt, grounded, 15 amp		40	.200		1.14	5.45		6.59	10.15
	2470 20 amp		27	.296		8.55	8.10		16.65	22.50
	2490 Dryer, 30 amp		15	.533		10.35	14.55		24.90	35
	2500 Range, 50 amp		11	.727		11.40	19.85		31.25	45
	2600 Wall plates, stainless steel, 1 gang		80	.100		1.80	2.73		4.53	6.45
	2800 2 gang		53	.151		4.10	4.12		8.22	11.20
	3200 Lampholder, keyless		26	.308		9.70	8.40		18.10	24.50
	3400 Pullchain with receptacle		22	.364		9.20	9.95		19.15	26.50

	16150 | Wiring Connections	CREW	DAILY OUTPUT	LABOR-HOURS	UNIT	MAT.	LABOR	EQUIP.	TOTAL	TOTAL INCL O&P
275	0010 MOTOR CONNECTIONS									
	0020 Flexible conduit and fittings, 115 volt, 1 phase, up to 1 HP motor	1 Elec	8	1	Ea.	4.96	27.50		32.46	50

16200 | Electrical Power

	16210 | Electrical Utility Services	CREW	DAILY OUTPUT	LABOR-HOURS	UNIT	2005 BARE COSTS MAT.	LABOR	EQUIP.	TOTAL	TOTAL INCL O&P
600	0010 METER CENTERS AND SOCKETS									
	0100 Sockets, single position, 4 terminal, 100 amp	1 Elec	3.20	2.500	Ea.	34.50	68.50		103	149

16200 | Electrical Power

	16210 | Electrical Utility Services		CREW	DAILY OUTPUT	LABOR-HOURS	UNIT	2005 BARE COSTS MAT.	LABOR	EQUIP.	TOTAL	TOTAL INCL O&P
600	0200	150 amp	1 Elec	2.30	3.478	Ea.	38.50	95		133.50	197
	0300	200 amp		1.90	4.211		51	115		166	244
	0500	Double position, 4 terminal, 100 amp		2.80	2.857		135	78		213	275
	0600	150 amp		2.10	3.810		153	104		257	335
	0700	200 amp		1.70	4.706		330	128		458	570
	2590	Basic meter device									
	2600	1P 3W 120/240V 4 jaw 125A sockets, 3 meter	2 Elec	1	16	Ea.	490	435		925	1,250
	2620	5 meter		.80	20		735	545		1,280	1,700
	2640	7 meter		.56	28.571		1,075	780		1,855	2,450
	2660	10 meter		.48	33.333		1,475	910		2,385	3,075
	2680	Rainproof 1P 3W 120/240V 4 jaw 125A sockets									
	2690	3 meter	2 Elec	1	16	Ea.	490	435		925	1,250
	2710	6 meter		.60	26.667		840	730		1,570	2,100
	2730	8 meter		.52	30.769		1,175	840		2,015	2,650
	2750	1P 3W 120/240V 4 jaw sockets									
	2760	with 125A circuit breaker, 3 meter	2 Elec	1	16	Ea.	915	435		1,350	1,700
	2780	5 meter		.80	20		1,450	545		1,995	2,475
	2800	7 meter		.56	28.571		2,075	780		2,855	3,550
	2820	10 meter		.48	33.333		2,875	910		3,785	4,650
	2830	Rainproof 1P 3W 120/240V 4 jaw sockets									
	2840	with 125A circuit breaker, 3 meter	2 Elec	1	16	Ea.	915	435		1,350	1,700
	2870	6 meter		.60	26.667		1,700	730		2,430	3,050
	2890	8 meter		.52	30.769		2,300	840		3,140	3,900
	3250	1P 3W 120/240V 4 jaw sockets									
	3260	with 200A circuit breaker, 3 meter	2 Elec	1	16	Ea.	1,375	435		1,810	2,200
	3290	6 meter		.60	26.667		2,750	730		3,480	4,200
	3310	8 meter		.56	28.571		3,700	780		4,480	5,350
	3330	Rainproof 1P 3W 120/240V 4 jaw sockets									
	3350	with 200A circuit breaker, 3 meter	2 Elec	1	16	Ea.	1,375	435		1,810	2,200
	3380	6 meter		.60	26.667		2,750	730		3,480	4,200
	3400	8 meter		.52	30.769		3,700	840		4,540	5,450

	16230 | Generator Assemblies		CREW	DAILY OUTPUT	LABOR-HOURS	UNIT	MAT.	LABOR	EQUIP.	TOTAL	TOTAL INCL O&P
450	0010	GENERATOR SET									
	0020	Gas or gasoline operated, includes battery,									
	0050	charger, muffler & transfer switch									
	0200	3 phase 4 wire, 277/480 volt, 7.5 kW	R-3	.83	24.096	Ea.	6,000	650	191	6,841	7,850
	0300	11.5 kW		.71	28.169		8,500	760	223	9,483	10,800
	0400	20 kW		.63	31.746		10,000	855	252	11,107	12,700

16400 | Low-Voltage Distribution

	16410 | Encl Switches & Circuit Breakers		CREW	DAILY OUTPUT	LABOR-HOURS	UNIT	2005 BARE COSTS MAT.	LABOR	EQUIP.	TOTAL	TOTAL INCL O&P
200	0010	CIRCUIT BREAKERS (in enclosure)									
	0100	Enclosed (NEMA 1), 600 volt, 3 pole, 30 amp	1 Elec	3.20	2.500	Ea.	430	68.50		498.50	585
	0200	60 amp		2.80	2.857		430	78		508	600
	0400	100 amp		2.30	3.478		495	95		590	700
800	0010	SAFETY SWITCHES									
	0100	General duty 240 volt, 3 pole NEMA 1, fusible, 30 amp	1 Elec	3.20	2.500	Ea.	78	68.50		146.50	197

16400 | Low-Voltage Distribution

16410 | Encl Switches & Circuit Breakers

			CREW	DAILY OUTPUT	LABOR-HOURS	UNIT	2005 BARE COSTS MAT.	LABOR	EQUIP.	TOTAL	TOTAL INCL O&P	
0	0200	60 amp	1 Elec	2.30	3.478	Ea.	132	95		227	300	800
	0300	100 amp		1.90	4.211		227	115		342	435	
	0400	200 amp		1.30	6.154		490	168		658	810	
	0500	400 amp	2 Elec	1.80	8.889		1,225	243		1,468	1,750	
	9010	Disc. switch, 600V 3 pole fusible, 30 amp, to 10 HP motor	1 Elec	3.20	2.500		228	68.50		296.50	360	
	9050	60 amp, to 30 HP motor		2.30	3.478		525	95		620	730	
	9070	100 amp, to 60 HP motor		1.90	4.211		525	115		640	760	
0	0010	**TIME SWITCHES**										840
	0100	Single pole, single throw, 24 hour dial	1 Elec	4	2	Ea.	82	54.50		136.50	179	
	0200	24 hour dial with reserve power		3.60	2.222		350	60.50		410.50	485	
	0300	Astronomic dial		3.60	2.222		141	60.50		201.50	254	
	0400	Astronomic dial with reserve power		3.30	2.424		455	66		521	610	
	0500	7 day calendar dial		3.30	2.424		126	66		192	247	
	0600	7 day calendar dial with reserve power		3.20	2.500		385	68.50		453.50	530	
	0700	Photo cell 2000 watt		8	1		15.15	27.50		42.65	61	

16415 | Transfer Switches

			CREW	DAILY OUTPUT	LABOR-HOURS	UNIT	MAT.	LABOR	EQUIP.	TOTAL	TOTAL INCL O&P	
00	0010	**AUTOMATIC TRANSFER SWITCHES**										600
	0100	Switches, enclosed 480 volt, 3 pole, 30 amp	1 Elec	2.30	3.478	Ea.	2,900	95		2,995	3,350	
	0200	60 amp	″	1.90	4.211	″	2,900	115		3,015	3,375	

16440 | Swbds, Panels & Control Centers

			CREW	DAILY OUTPUT	LABOR-HOURS	UNIT	MAT.	LABOR	EQUIP.	TOTAL	TOTAL INCL O&P	
00	0010	**LOAD CENTERS** (residential type)										500
	0100	3 wire, 120/240V, 1 phase, including 1 pole plug-in breakers										
	0200	100 amp main lugs, indoor, 8 circuits	1 Elec	1.40	5.714	Ea.	127	156		283	395	
	0300	12 circuits		1.20	6.667		177	182		359	490	
	0400	Rainproof, 8 circuits		1.40	5.714		152	156		308	420	
	0500	12 circuits		1.20	6.667		214	182		396	530	
	0600	200 amp main lugs, indoor, 16 circuits	R-1A	1.80	8.889		288	199		487	645	
	0700	20 circuits		1.50	10.667		360	239		599	790	
	0800	24 circuits		1.30	12.308		475	276		751	975	
	1200	Rainproof, 16 circuits		1.80	8.889		345	199		544	710	
	1300	20 circuits		1.50	10.667		410	239		649	845	
	1400	24 circuits		1.30	12.308		615	276		891	1,125	

16500 | Lighting

16510 | Interior Luminaires

			CREW	DAILY OUTPUT	LABOR-HOURS	UNIT	2005 BARE COSTS MAT.	LABOR	EQUIP.	TOTAL	TOTAL INCL O&P	
440	0010	**INTERIOR LIGHTING FIXTURES** Including lamps, mounting										440
	0030	hardware and connections										
	0100	Fluorescent, C.W. lamps, troffer, recess mounted in grid, RS										
	0130	grid ceiling mount										
	0200	Acrylic lens, 1′W x 4′L, two 40 watt	1 Elec	5.70	1.404	Ea.	48	38.50		86.50	116	
	0300	2′W x 2′L, two U40 watt		5.70	1.404		51.50	38.50		90	119	
	0600	2′W x 4′L, four 40 watt		4.70	1.702		58.50	46.50		105	140	
	1000	Surface mounted, RS										
	1030	Acrylic lens with hinged & latched door frame										
	1100	1′W x 4′L, two 40 watt	1 Elec	7	1.143	Ea.	73.50	31		104.50	132	

ELECTRICAL 16

16510 | Interior Luminaires

				DAILY	LABOR-		2005 BARE COSTS				TOTAL
			CREW	OUTPUT	HOURS	UNIT	MAT.	LABOR	EQUIP.	TOTAL	INCL O&P
440	1200	2′W x 2′L, two U40 watt	1 Elec	7	1.143	Ea.	79	31		110	138
	1500	2′W x 4′L, four 40 watt	↓	5.30	1.509	↓	93.50	41		134.50	170
	2100	Strip fixture									
	2200	4′ long, one 40 watt RS	1 Elec	8.50	.941	Ea.	28	25.50		53.50	73
	2300	4′ long, two 40 watt RS	″	8	1		30.50	27.50		58	78
	2600	8′ long, one 75 watt, SL	2 Elec	13.40	1.194		42.50	32.50		75	99.50
	2700	8′ long, two 75 watt, SL	″	12.40	1.290	↓	51	35		86	114
	4450	Incandescent, high hat can, round alzak reflector, prewired									
	4470	100 watt	1 Elec	8	1	Ea.	58	27.50		85.50	109
	4480	150 watt	″	8	1	″	85	27.50		112.50	138
	5200	Ceiling, surface mounted, opal glass drum									
	5300	8″, one 60 watt lamp	1 Elec	10	.800	Ea.	36	22		58	75
	5400	10″, two 60 watt lamps		8	1		40.50	27.50		68	89
	5500	12″, four 60 watt lamps		6.70	1.194		55	32.50		87.50	114
	6900	Mirror light, fluorescent, RS, acrylic enclosure, two 40 watt		8	1		84.50	27.50		112	138
	6910	One 40 watt		8	1		66	27.50		93.50	117
	6920	One 20 watt	↓	12	.667	↓	52	18.20		70.20	87
800	0010	**RESIDENTIAL FIXTURES**									
	0400	Fluorescent, interior, surface, circline, 32 watt & 40 watt	1 Elec	20	.400	Ea.	76.50	10.90		87.40	102
	0500	2′ x 2′, two U 40 watt		8	1		93	27.50		120.50	147
	0700	Shallow under cabinet, two 20 watt		16	.500		40.50	13.65		54.15	67
	0900	Wall mounted, 4′L, one 40 watt, with baffle		10	.800		99	22		121	145
	2000	Incandescent, exterior lantern, wall mounted, 60 watt		16	.500		31	13.65		44.65	56
	2100	Post light, 150W, with 7′ post		4	2		110	54.50		164.50	210
	2500	Lamp holder, weatherproof with 150W PAR		16	.500		16.90	13.65		30.55	40.50
	2550	With reflector and guard		12	.667		52	18.20		70.20	86.50
	2600	Interior pendent, globe with shade, 150 watt	↓	20	.400	↓	117	10.90		127.90	147

16520 | Exterior Luminaires

			CREW	DAILY OUTPUT	LABOR-HOURS	UNIT	MAT.	LABOR	EQUIP.	TOTAL	TOTAL INCL O&P
300	0010	**EXTERIOR FIXTURES** With lamps									
	0400	Quartz, 500 watt	1 Elec	5.30	1.509	Ea.	54	41		95	127
	1100	Wall pack, low pressure sodium, 35 watt		4	2		214	54.50		268.50	325
	1150	55 watt		4	2		255	54.50		309.50	370
	6420	Wood pole, 4-1/2″ x 5-1/8″, 8′ high		6	1.333		265	36.50		301.50	350
	6440	12′ high		5.70	1.404		380	38.50		418.50	480
	6460	20′ high	↓	4	2	↓	535	54.50		589.50	680
	6500	Bollard light, lamp & ballast, 42″ high with polycarbonate lens									
	7200	Incandescent, 150 watt	1 Elec	3	2.667	Ea.	465	73		538	630
	7380	Landscape recessed uplight, incl. housing, ballast, transformer									
	7390	& reflector									
	7420	Incandescent, 250 watt	1 Elec	5	1.600	Ea.	445	43.50		488.50	560
	7440	Quartz, 250 watt	″	5	1.600	″	425	43.50		468.50	535

16550 | Special Purpose Lighting

			CREW	DAILY OUTPUT	LABOR-HOURS	UNIT	MAT.	LABOR	EQUIP.	TOTAL	TOTAL INCL O&P
820	0010	**TRACK LIGHTING**									
	0100	8′ section	1 Elec	5.30	1.509	Ea.	62	41		103	135
	0300	3 circuits, 4′ section		6.70	1.194		48	32.50		80.50	106
	0400	8′ section		5.30	1.509		74	41		115	149
	0500	12′ section		4.40	1.818		148	49.50		197.50	244
	1000	Feed kit, surface mounting		16	.500		9.10	13.65		22.75	32
	1100	End cover		24	.333		3.20	9.10		12.30	18.30
	1200	Feed kit, stem mounting, 1 circuit		16	.500		25	13.65		38.65	49.50
	1300	3 circuit		16	.500		25	13.65		38.65	49.50
	2000	Electrical joiner, for continuous runs, 1 circuit	↓	32	.250	↓	11.90	6.85		18.75	24

16500 | Lighting

	16550 Special Purpose Lighting		Crew	Daily Output	Labor-Hours	Unit	2005 Bare Costs Mat.	Labor	Equip.	Total	Total Incl O&P	
0	2100	3 circuit	1 Elec	32	.250	Ea.	28	6.85		34.85	41.50	820
	2200	Fixtures, spotlight, 75W PAR halogen		16	.500		87	13.65		100.65	118	
	2210	50W MR16 halogen		16	.500		106	13.65		119.65	139	
	3000	Wall washer, 250 watt tungsten halogen		16	.500		101	13.65		114.65	133	
	3100	Low voltage, 25/50 watt, 1 circuit		16	.500		102	13.65		115.65	134	
	3120	3 circuit		16	.500		105	13.65		118.65	138	

	16585 Lamps		Crew	Daily Output	Labor-Hours	Unit	Mat.	Labor	Equip.	Total	Total Incl O&P	
0	0010	**LAMPS**										600
	0081	Fluorescent, rapid start, cool white, 2′ long, 20 watt	1 Elec	100	.080	Ea.	3.10	2.18		5.28	6.95	
	0101	4′ long, 40 watt		90	.089		2.81	2.43		5.24	7.05	
	1351	High pressure sodium, 70 watt		30	.267		37.50	7.30		44.80	53.50	
	1371	150 watt		30	.267		40.50	7.30		47.80	56.50	

16800 | Sound & Video

	16820 Sound Reinforcement		Crew	Daily Output	Labor-Hours	Unit	2005 Bare Costs Mat.	Labor	Equip.	Total	Total Incl O&P	
00	0010	**DOORBELL SYSTEM** Incl. transformer, button & signal										300
	1000	Door chimes, 2 notes, minimum	1 Elec	16	.500	Ea.	22	13.65		35.65	46.50	
	1020	Maximum		12	.667		118	18.20		136.20	159	
	1100	Tube type, 3 tube system		12	.667		167	18.20		185.20	214	
	1180	4 tube system		10	.800		268	22		290	330	
	1900	For transformer & button, minimum add		5	1.600		12.70	43.50		56.20	85	
	1960	Maximum, add		4.50	1.778		38	48.50		86.50	121	
	3000	For push button only, minimum		24	.333		2.48	9.10		11.58	17.55	
	3100	Maximum		20	.400		20	10.90		30.90	40	

	16850 Television Equipment		Crew	Daily Output	Labor-Hours	Unit	Mat.	Labor	Equip.	Total	Total Incl O&P	
00	0010	**T.V. SYSTEMS** not including rough-in wires, cables & conduits										600
	0100	Master TV antenna system										
	0200	VHF reception & distribution, 12 outlets	1 Elec	6	1.333	Outlet	164	36.50		200.50	241	
	0800	VHF & UHF reception & distribution, 12 outlets		6	1.333	″	163	36.50		199.50	240	
	5000	T.V. Antenna only, minimum		6	1.333	Ea.	36	36.50		72.50	99	
	5100	Maximum		4	2	″	152	54.50		206.50	256	

For information about Means Estimating Seminars, see yellow pages 12 and 13 in back of book

Division Notes

		CREW	DAILY OUTPUT	LABOR-HOURS	UNIT	2005 BARE COSTS MAT.	LABOR	EQUIP.	TOTAL	TOTAL INCL O&P

Reference Section

ll the reference information is in one section making it easy to find what you need to know . . . and easy to use the book on a daily basis. This section is visually identified by a vertical gray bar on the edge of pages.

In the reference number information that follows, you'll see the background that relates to the "reference numbers" that appeared in the Unit Price Sections. You'll find reference tables, explanations and estimating information that support how we arrived at the unit price data. Also included are alternate pricing methods, technical data and estimating procedures along with information on design and economy in construction.

Also in this Reference Section, we've included Change Orders, information on pricing changes to contract documents; Crew Listings, a full listing of all the crews, equipment and their costs; Historical Cost Indexes for cost comparisons over time; City Cost Indexes and Location Factors for adjusting costs to the region you are in; and an explanation of all abbreviations used in the book.

Table of Contents

Reference Numbers

R011 Overhead & Misc. Data 534
R012 Contract Modifications Procedures 540
R015 Construction Aids 541
R020 Paving & Surfacing 542
R022 Site Prep. & Excav. Support 543
R025 Sewerage & Drainage 543
R029 Landscaping 544
R040 Mortar & Masonry Accessories 545
R042 Unit Masonry 546
R049 Restoration & Cleaning 550
R053 Metal Deck 550
R071 Waterproofing & Dampproofing 554
R073 Shingles & Roofing Tiles 554

Reference Numbers (cont.)

R075 Membrane Roofing 555
R081 Metal Doors 555
R085 Windows 555
R087 Hardware 556
R092 Plaster & Gypsum Board 557
R097 Wall Finishes 559
R099 Paints & Coatings 559
R131 Pre-Engineered Structures & Aquatic Facilities 560
R136 Solar & Wind Energy Equip. 560
R151 Plumbing 561

Crew Listings 562

Location Factors 588

Abbreviations 594

R01100-005 Tips for Accurate Estimating

1. Use pre-printed or columnar forms for orderly sequence of dimensions and locations and for recording telephone quotations.
2. Use only the front side of each paper or form except for certain pre-printed summary forms.
3. Be consistent in listing dimensions: For example, length x width x height. This helps in rechecking to ensure that, the total length of partitions is appropriate for the building area.
4. Use printed (rather than measured) dimensions where given.
5. Add up multiple printed dimensions for a single entry where possible.
6. Measure all other dimensions carefully.
7. Use each set of dimensions to calculate multiple related quantities.
8. Convert foot and inch measurements to decimal feet when listing. Memorize decimal equivalents to .01 parts of a foot (1/8″ equals approximately .01′).
9. Do not "round off" quantities until the final summary.
10. Mark drawings with different colors as items are taken off.
11. Keep similar items together, different items separate.
12. Identify location and drawing numbers to aid in future checking for completeness.
13. Measure or list everything on the drawings or mentioned in the specifications.
14. It may be necessary to list items not called for to make the job complete.
15. Be alert for: Notes on plans such as N.T.S. (not to scale); changes in scale throughout the drawings; reduced size drawings; discrepancies between the specifications and the drawings.
16. Develop a consistent pattern of performing an estimate. For example:
    a. Start the quantity takeoff at the lower floor and move to the next higher floor.
    b. Proceed from the main section of the building to the wings.
    c. Proceed from south to north or vice versa, clockwise or counterclockwise.
    d. Take off floor plan quantities first, elevations next, then detail drawings.
17. List all gross dimensions that can be either used again for different quantities, or used as a rough check of other quantities for verification (exterior perimeter, gross floor area, individual floor areas, etc.).
18. Utilize design symmetry or repetition (repetitive floors, repetitive wings symmetrical design around a center line, similar room layouts, etc.). Note: Extreme caution is needed here so as not to omit or duplicate an area.
19. Do not convert units until the final total is obtained. For instance, when estimating concrete work, keep all units to the nearest cubic foot then summarize and convert to cubic yards.
20. When figuring alternatives, it is best to total all items involved in the basic system, then total all items involved in the alternates. Therefore you work with positive numbers in all cases. When adds and deducts are used, it is often confusing whether to add or subtract a portion of an item; especially on a complicated or involved alternate.

R01100-040 Builder's Risk Insurance

Builder's Risk Insurance is insurance on a building during construction. Premiums are paid by the owner or the contractor. Blasting, collapse and underground insurance would raise total insurance costs above those listed. Floater policy for materials delivered to the job runs $.75 to $1.25 per $100 value. Contractor equipment insurance runs $.50 to $1.50 per $100 value. Insurance for miscellaneous tools to $1,500 value runs from $3.00 to $7.50 per $100 value.

Tabulated below are New England Builder's Risk insurance rates in dollars per $100 value for $1,000 deductible. For $25,000 deductible, rates can be reduced 13% to 34%. On contracts over $1,000,000, rates may be lower than those tabulated. Policies are written annually for the total completed value in place. For "all risk" insurance (excluding flood, earthquake and certain other perils) add $.025 to total rates below.

Coverage	Frame Construction (Class 1)				Brick Construction (Class 4)				Fire Resistive (Class 6)			
	Range			Average	Range			Average	Range			Average
Fire Insurance	$.350	to	$ .850	$.600	$.158	to	$.189	$.174	$.052	to	$.080	$.070
Extended Coverage	.115	to	.200	.158	.080	to	.105	.101	.081	to	.105	.100
Vandalism	.012	to	.016	.014	.008	to	.011	.011	.008	to	.011	.010
Total Annual Rate	$.477	to	$1.066	$.772	$.246	to	$.305	$.286	$.141	to	$.196	$.180

R01100-050 General Contractor's Overhead

There are two distinct types of overhead on a construction project: Project Overhead and Main Office Overhead. Project Overhead includes those costs at a construction site not directly associated with the installation of construction materials. Examples of Project Overhead costs include the following:

1. Superintendent
2. Construction office and storage trailers
3. Temporary sanitary facilities
4. Temporary utilities
5. Security fencing
6. Photographs
7. Clean up
8. Performance and payment bonds

The above Project Overhead items are also referred to as General Requirements and therefore are estimated in Division 1. Division 1 is the first division listed in the CSI MasterFormat but it is usually the last division estimated. The sum of the costs in Divisions 1 through 16 is referred to as the sum of the direct costs.

All construction projects also include indirect costs. The primary components of indirect costs are the contractor's Main Office Overhead and profit. The amount of the Main Office Overhead expense varies depending on the the following:

1. Owner's compensation
2. Project managers and estimator's wages
3. Clerical support wages
4. Office rent and utilities
5. Corporate legal and accounting costs
6. Advertising
7. Automobile expenses
8. Association dues
9. Travel and entertainment expenses

These costs are usually calculated as a percentage of annual sales volume. This percentage can range from 35% for a small contractor doing less than $500,000 to 5% for a large contractor with sales in excess of $100 million.

R01100-060 Workers' Compensation Insurance Rates by Trade

The table below tabulates the national averages for Workers' Compensation insurance rates by trade and type of building. The average "Insurance Rate" is multiplied by the "% of Building Cost" for each trade. This produces the "Workers' Compensation Cost" by % of total labor cost, to be added for each trade by building type to determine the weighted average Workers' Compensation rate for the building types analyzed.

Trade	Insurance Rate (% Labor Cost)		% of Building Cost			Workers' Compensation		
	Range	Average	Office Bldgs.	Schools & Apts.	Mfg.	Office Bldgs.	Schools & Apts.	Mfg.
Excavation, Grading, etc.	4.2 % to 19.5%	10.4%	4.8%	4.9%	4.5%	.50%	.51%	.47%
Piles & Foundations	7.7 to 76.7	22.5	7.1	5.2	8.7	1.60	1.17	1.96
Concrete	5.2 to 35.8	16.1	5.0	14.8	3.7	.81	2.38	.60
Masonry	5.1 to 28.3	15.1	6.9	7.5	1.9	1.04	1.13	.29
Structural Steel	7.8 to 103.8	38.6	10.7	3.9	17.6	4.13	1.51	6.79
Miscellaneous & Ornamental Metals	4.9 to 24.8	12.6	2.8	4.0	3.6	.35	.50	.45
Carpentry & Millwork	7.0 to 53.2	18.4	3.7	4.0	0.5	.68	.74	.09
Metal or Composition Siding	5.3 to 32.1	17.0	2.3	0.3	4.3	.39	.05	.73
Roofing	7.8 to 78.9	33.1	2.3	2.6	3.1	.76	.86	1.03
Doors & Hardware	4.5 to 24.9	11.3	0.9	1.4	0.4	.10	.16	.05
Sash & Glazing	4.6 to 33.7	13.9	3.5	4.0	1.0	.49	.56	.14
Lath & Plaster	4.3 to 40.2	14.9	3.3	6.9	0.8	.49	1.03	.12
Tile, Marble & Floors	2.4 to 31.8	9.7	2.6	3.0	0.5	.25	.29	.05
Acoustical Ceilings	2.4 to 29.7	10.9	2.4	0.2	0.3	.26	.02	.03
Painting	4.5 to 33.7	13.1	1.5	1.6	1.6	.20	.21	.21
Interior Partitions	7.0 to 53.2	18.4	3.9	4.3	4.4	.72	.79	.81
Miscellaneous Items	2.6 to 137.6	16.9	5.2	3.7	9.7	.88	.63	1.64
Elevators	2.8 to 42.8	7.6	2.1	1.1	2.2	.16	.08	.17
Sprinklers	3.2 to 22.1	8.8	0.5	—	2.0	.04	—	.18
Plumbing	2.7 to 12.4	8.0	4.9	7.2	5.2	.39	.58	.42
Heat., Vent., Air Conditioning	4.0 to 27.2	11.5	13.5	11.0	12.9	1.55	1.27	1.48
Electrical	2.6 to 12.7	6.5	10.1	8.4	11.1	.66	.55	.72
Total	2.4 % to 137.6%	—	100.0%	100.0%	100.0%	16.45%	15.02%	18.43%
		Overall Weighted Average	16.63%					

Workers' Compensation Insurance Rates by States

The table below lists the weighted average Workers' Compensation base rate for each state with a factor comparing this with the national average of 16.2%.

State	Weighted Average	Factor	State	Weighted Average	Factor	State	Weighted Average	Factor
Alabama	25.2%	156	Kentucky	17.4%	107	North Dakota	11.4%	70
Alaska	19.7	122	Louisiana	26.2	162	Ohio	12.8	79
Arizona	7.4	46	Maine	19.3	119	Oklahoma	13.7	85
Arkansas	13.7	85	Maryland	17.7	109	Oregon	12.6	78
California	17.4	107	Massachusetts	12.4	77	Pennsylvania	13.4	83
Colorado	12.3	76	Michigan	17.6	109	Rhode Island	19.6	121
Connecticut	23.1	143	Minnesota	25.1	155	South Carolina	12.9	80
Delaware	14.2	88	Mississippi	16.1	99	South Dakota	14.2	88
District of Columbia	17.3	107	Missouri	17.6	109	Tennessee	17.9	110
Florida	25.1	155	Montana	18.9	117	Texas	13.5	83
Georgia	21.3	131	Nebraska	18.8	116	Utah	11.4	70
Hawaii	15.7	97	Nevada	13.9	86	Vermont	18.6	115
Idaho	10.4	64	New Hampshire	21.3	131	Virginia	10.8	67
Illinois	17.2	106	New Jersey	10.5	65	Washington	10.4	64
Indiana	5.4	33	New Mexico	16.3	101	West Virginia	12.4	77
Iowa	11.8	73	New York	13.4	83	Wisconsin	15.2	94
Kansas	8.5	52	North Carolina	13.2	81	Wyoming	7.9	49
			Weighted Average for U.S. is	15.5% of payroll = 100%				

Rates in the following table are the base or manual costs per $100 of payroll for Workers' Compensation in each state. Rates are usually applied to straight time wages only and not to premium time wages and bonuses.

The weighted average skilled worker rate for 35 trades is 16.2%. For bidding purposes, apply the full value of Workers' Compensation directly to total labor costs, or if labor is 38%, materials 42% and overhead and profit 20% of total cost, carry 38/80 x 16.2% =7.7% of cost (before overhead and profit) into overhead. Rates vary not only from state to state but also with the experience rating of the contractor.

Rates are the most current available at the time of publication.

R01100-060 Workers' Compensation Insurance Rates by Trade and State (cont.)

State	Carpentry — 3 stories or less 5651	Carpentry — interior cab. work 5437	Carpentry — general 5403	Concrete Work — NOC 5213	Concrete Work — flat (flr., sdwk.) 5221	Electrical Wiring — inside 5190	Excavation — earth NOC 6217	Excavation — rock 6217	Glaziers 5462	Insulation Work 5479	Lathing 5443	Masonry 5022	Painting & Decorating 5474	Pile Driving 6003	Plastering 5480	Plumbing 5183	Roofing 5551	Sheet Metal Work (HVAC) 5538	Steel Erection — door & sash 5102	Steel Erection — inter., ornam. 5102	Steel Erection — structure 5040	Steel Erection — NOC 5057	Tile Work — (interior ceramic) 5348	Waterproofing 9014	Wrecking 5701
AL	26.93	15.28	34.89	12.39	9.90	8.95	13.69	13.69	33.66	17.12	12.78	27.57	33.72	34.63	40.23	11.91	60.85	27.15	23.11	23.11	51.16	39.09	13.16	6.65	51.16
AK	17.02	14.74	13.84	12.07	11.18	12.70	19.54	19.54	19.86	25.41	9.78	24.13	16.28	59.80	19.42	11.79	38.40	12.13	12.24	12.24	42.33	18.60	8.48	10.79	42.33
AZ	7.71	5.37	12.63	6.33	3.44	4.18	4.83	4.83	7.59	9.87	4.70	6.40	5.23	9.47	15.76	3.77	11.96	4.71	7.92	7.92	16.58	7.75	2.37	2.55	50.31
AR	13.04	11.01	16.99	13.77	6.64	6.11	9.42	9.42	15.49	33.95	8.65	11.34	10.50	12.82	12.93	5.69	21.95	10.28	7.70	7.70	37.63	28.19	7.10	3.75	37.63
CA	28.28	9.07	28.28	13.18	13.18	9.93	7.88	7.88	16.37	23.34	10.90	14.55	19.45	21.18	18.12	11.59	39.79	15.33	13.55	13.55	23.93	21.91	7.74	19.45	21.91
CO	17.36	9.17	12.24	14.36	7.69	5.32	10.44	10.44	9.28	13.68	5.57	13.41	10.77	17.70	10.81	8.51	24.47	12.11	7.39	7.39	31.30	15.53	7.99	5.78	31.30
CT	20.96	18.42	31.08	29.77	13.28	8.63	11.18	11.18	15.10	24.22	17.24	25.81	19.56	27.54	27.61	10.37	51.40	14.91	18.42	18.42	70.81	35.84	12.65	6.69	50.86
DE	15.39	15.39	12.92	13.03	7.13	6.79	10.03	10.03	11.96	12.92	13.83	12.33	16.85	21.38	13.83	7.94	27.84	10.69	13.02	13.02	29.92	13.02	10.51	12.33	29.92
DC	13.05	10.16	15.52	17.41	20.87	6.52	10.32	10.32	27.59	11.07	8.44	22.34	10.52	18.70	13.70	12.43	21.23	9.29	17.97	17.97	54.79	20.38	31.80	4.05	54.79
FL	30.55	21.73	31.10	30.26	15.05	10.74	13.79	13.79	23.24	22.05	13.08	23.21	21.38	58.93	36.28	10.71	46.17	18.10	14.78	14.78	58.02	39.86	11.26	9.25	58.02
GA	32.05	17.69	24.16	15.16	10.44	8.97	16.86	16.86	15.69	22.02	17.16	20.30	17.13	31.16	19.94	10.77	46.74	17.52	13.97	13.97	44.46	48.01	9.64	8.55	44.46
HI	17.38	11.62	28.20	13.74	12.27	7.10	7.73	7.73	20.55	21.19	10.94	17.87	11.33	20.15	15.61	6.06	33.24	8.02	11.11	11.11	31.71	21.70	9.78	11.54	31.71
ID	11.48	6.50	13.42	8.40	5.87	5.01	6.24	6.24	9.48	7.60	5.27	9.01	7.40	13.14	9.74	4.67	29.72	9.37	9.80	9.80	30.78	13.76	6.47	4.09	30.78
IL	16.38	13.48	17.63	29.67	11.05	8.51	10.49	10.49	15.50	11.58	9.74	16.53	11.34	28.71	11.92	9.79	28.32	14.36	14.87	14.87	52.98	26.69	14.11	4.36	52.98
IN	5.31	4.50	7.02	5.18	3.17	2.64	4.17	4.17	4.59	4.15	2.43	5.07	4.48	7.71	4.30	2.70	10.93	4.03	4.94	4.94	16.12	9.00	3.08	2.55	16.12
IA	11.67	5.80	10.47	12.55	6.82	4.18	5.98	5.98	13.66	9.22	5.45	8.79	7.73	10.70	9.26	5.92	18.71	7.17	9.92	9.92	51.91	36.28	5.29	4.20	30.67
KS	10.58	8.42	10.47	7.12	6.36	3.72	4.73	4.73	6.49	7.80	4.37	7.45	6.06	9.36	8.73	5.40	20.70	6.76	7.83	7.83	23.76	12.28	4.75	3.48	23.76
KY	15.79	12.91	19.21	20.50	6.16	7.64	17.08	17.08	21.66	24.14	11.20	10.81	11.40	18.49	14.90	6.85	48.37	13.64	11.61	11.61	40.21	30.64	12.56	4.72	40.21
LA	23.14	24.93	53.17	26.43	15.61	9.88	17.49	17.49	20.14	21.43	24.51	28.33	29.58	31.25	22.37	8.64	77.12	21.20	18.98	18.98	51.76	24.21	13.77	13.78	66.41
ME	13.43	10.42	43.59	24.29	11.12	5.03	12.19	12.19	13.35	15.39	14.03	17.26	16.07	31.44	17.95	7.32	32.07	8.82	15.08	15.08	37.84	61.81	11.66	6.06	37.84
MD	13.71	5.95	10.55	18.82	8.70	5.15	12.91	12.91	25.52	20.07	11.37	13.31	9.37	26.22	14.21	8.58	48.72	12.74	13.66	13.66	56.25	37.81	8.03	7.14	26.80
MA	9.93	6.07	16.09	17.95	8.10	3.69	6.49	6.49	7.55	13.48	5.76	13.04	7.54	14.31	5.34	4.55	38.25	7.47	12.35	12.35	35.13	27.39	9.24	3.32	28.82
MI	21.68	13.59	19.80	22.12	9.35	5.71	11.69	11.69	13.87	12.38	13.59	19.33	14.93	39.06	15.59	8.24	41.33	10.41	11.36	11.36	39.06	29.15	11.31	6.40	39.06
MN	21.84	21.96	43.39	16.70	14.64	6.94	16.68	16.68	17.42	11.95	21.02	20.28	17.69	26.83	21.02	10.70	78.89	12.72	14.64	14.64	103.82	38.11	14.59	6.96	132.92
MS	16.92	15.26	19.71	11.11	8.98	6.80	12.22	12.22	12.02	14.11	8.02	13.18	12.86	19.38	22.03	8.15	37.48	21.88	11.67	11.67	33.42	32.27	9.89	6.93	33.42
MO	21.84	12.24	16.70	15.66	11.37	7.09	10.05	10.05	11.75	18.49	12.47	16.14	13.64	21.93	16.25	9.37	33.06	14.20	12.51	12.51	59.20	40.85	9.59	7.02	59.20
MT	22.03	10.79	17.11	12.89	9.98	5.54	17.20	17.20	11.51	16.19	19.02	14.58	12.03	76.71	13.82	8.93	57.20	9.44	9.73	9.73	41.23	17.93	6.54	5.33	41.23
NE	23.15	10.97	20.95	28.60	9.65	8.32	12.20	12.20	15.50	24.47	11.30	18.50	12.32	23.35	14.75	11.17	36.72	13.05	13.15	13.15	52.27	38.35	9.97	6.15	50.40
NV	19.78	7.76	13.18	10.57	11.57	6.18	11.21	11.21	12.58	13.22	7.97	11.38	10.62	14.55	12.30	8.93	21.55	18.17	12.87	12.87	31.33	33.90	6.64	6.15	44.29
NH	21.98	14.43	24.90	35.84	11.10	6.02	17.16	17.16	14.70	33.70	8.60	24.98	15.72	17.80	18.65	10.22	60.89	14.43	14.01	14.01	54.76	25.50	17.84	6.36	54.76
NJ	11.76	8.69	11.76	9.82	7.77	4.22	7.74	7.74	6.91	11.95	10.39	12.07	9.66	13.10	10.39	6.14	29.59	6.69	9.94	9.94	19.67	10.61	4.42	4.65	23.42
NM	28.63	8.03	18.67	17.07	9.38	6.85	9.33	9.33	14.38	12.83	7.20	13.69	12.57	18.59	11.22	8.64	32.00	11.70	24.78	24.78	43.20	25.65	7.98	6.81	43.20
NY	17.66	6.98	15.91	17.16	13.57	6.32	10.02	10.02	12.02	9.09	5.11	19.20	13.75	16.81	8.43	8.53	33.44	17.47	9.63	9.63	18.80	19.14	9.43	6.09	11.29
NC	16.34	11.24	15.16	11.77	7.00	8.63	8.65	8.65	10.30	11.88	8.28	10.32	9.75	13.68	15.22	7.78	26.83	11.26	7.71	7.71	52.96	19.19	5.79	4.27	52.96
ND	9.60	9.60	9.60	5.66	5.66	4.00	5.39	5.39	9.60	9.60	9.04	8.10	6.27	20.12	9.04	5.35	20.74	5.35	20.12	20.12	20.12	20.12	9.60	20.24	14.48
OH	6.32	11.48	9.82	11.66	10.37	5.54	8.42	8.42	9.22	14.76	29.68	12.20	15.20	25.34	4.41	7.14	22.43	8.85	10.53	10.53	25.27	18.05	9.73	5.02	25.27
OK	13.44	9.11	12.70	12.65	6.63	5.82	10.25	10.25	16.10	19.50	8.46	10.98	9.09	20.29	12.21	7.24	22.32	9.49	13.78	13.78	39.78	24.01	6.95	5.88	39.78
OR	19.10	9.61	15.51	13.55	8.43	4.65	9.70	9.70	15.02	8.08	7.48	15.36	12.79	15.81	10.75	5.89	23.26	10.49	9.53	9.53	32.00	13.92	11.23	4.97	32.00
PA	13.85	13.85	12.80	14.90	9.67	6.74	8.44	8.44	9.60	12.80	11.67	12.18	14.33	17.72	11.67	7.51	27.83	8.19	15.17	15.17	26.43	15.17	8.22	12.18	26.43
RI	19.53	11.65	18.07	18.23	16.24	4.43	10.38	10.38	12.85	22.78	11.97	25.11	24.13	37.66	17.25	8.52	33.92	10.29	14.07	14.07	59.49	37.50	14.36	7.70	78.79
SC	19.10	13.03	19.24	12.88	6.19	7.02	8.32	8.32	12.22	10.69	7.27	9.55	11.19	16.64	18.89	6.88	28.98	11.42	9.49	9.49	21.80	23.66	6.23	4.65	21.80
SD	17.83	7.27	14.55	18.00	6.34	5.44	13.24	13.24	10.58	13.24	8.39	8.01	12.63	24.87	11.78	8.53	21.26	9.64	11.53	11.53	53.25	20.15	6.68	4.20	53.25
TN	31.79	14.64	24.26	19.40	9.95	8.72	14.06	14.06	12.09	14.84	13.05	17.02	14.59	17.43	17.47	11.33	38.38	14.51	13.38	13.38	45.26	25.19	9.73	6.57	45.26
TX	16.89	11.33	13.25	12.10	8.85	7.36	10.49	10.49	11.61	15.68	9.53	13.68	9.89	13.95	20.99	8.06	22.80	14.58	11.23	11.23	33.39	14.78	7.29	8.09	17.86
UT	9.45	6.23	10.52	7.68	7.99	7.05	6.38	6.38	9.55	10.95	12.32	13.75	15.29	17.24	10.60	6.41	26.27	6.30	8.99	8.99	23.51	23.51	6.06	5.83	26.53
VT	19.90	11.34	19.46	34.13	12.01	6.10	10.61	10.61	21.25	25.54	10.85	19.53	10.41	23.91	18.23	10.10	28.73	12.62	13.94	13.94	47.05	37.23	8.52	9.59	47.05
VA	11.97	7.81	9.72	10.47	5.32	4.43	6.53	6.53	7.60	7.96	13.79	8.17	9.95	13.55	7.16	5.52	21.62	8.30	11.34	11.34	39.01	16.86	6.32	3.09	39.01
WA	9.13	9.13	9.13	8.17	8.17	3.12	8.54	8.54	13.74	8.92	9.13	13.27	9.39	20.04	11.57	5.04	19.54	4.60	12.96	12.96	9.39	9.59	10.76	10.23	9.59
WV	12.30	12.30	12.30	27.37	27.37	5.86	8.90	8.90	6.99	13.15	13.15	11.94	12.35	8.90	12.35	5.48	14.63	6.99	10.89	10.89	10.89	10.89	13.15	13.15	48.62
WI	12.40	10.86	19.86	13.12	11.66	5.32	6.74	6.74	16.53	14.04	10.60	18.99	14.82	17.24	13.95	6.31	43.57	8.01	14.30	14.30	34.84	21.34	14.42	5.80	34.84
WY	7.77	7.77	7.77	7.77	7.77	7.77	7.77	7.77	7.77	7.77	7.77	7.77	7.77	7.77	7.77	7.77	7.77	7.77	7.77	7.77	7.77	7.77	7.77	7.77	7.77
AVG.	16.96	11.33	18.42	16.07	9.94	6.46	10.43	10.43	13.91	15.53	10.87	15.06	13.12	22.45	14.88	7.96	33.14	11.46	12.57	12.57	38.60	24.71	9.66	7.12	39.48

R01100-060 Workers' Compensation (cont.) (Canada in Canadian dollars)

Province		Alberta	British Columbia	Manitoba	Ontario	New Brunswick	Newfndld. & Labrador	Northwest Territories	Nova Scotia	Prince Edward Island	Quebec	Saskat-chewan	Yukon
Carpentry—3 stories or less	Rate	10.80	6.66	4.57	4.83	4.63	10.05	4.25	7.67	6.15	14.69	7.01	4.36
	Code	42143	721028	40102	723	422	4226	4-41	4226	401	80110	B1317	202
Carpentry—interior cab. work	Rate	2.59	5.94	4.57	4.83	4.21	6.35	4.25	5.74	3.41	14.69	3.65	4.36
	Code	42133	721021	40102	723	427	4270	4-41	4274	402	80110	B11-27	202
CARPENTRY—general	Rate	10.80	6.66	4.57	4.83	4.63	6.35	4.25	7.67	6.15	14.69	7.01	4.36
	Code	42143	721028	40102	723	422	4299	4-41	4226	401	80110	B1317	202
CONCRETE WORK—NOC	Rate	7.26	8.48	7.60	16.47	4.63	10.05	4.25	4.83	6.15	15.84	7.01	4.36
	Code	42104	721010	40110	748	422	4224	4-41	4224	401	80100	B13-14	203
CONCRETE WORK—flat (flr. sidewalk)	Rate	7.26	8.48	7.60	16.47	4.63	10.05	4.25	4.83	6.15	15.84	7.01	4.36
	Code	42104	721010	40110	748	422	4224	4-41	4224	401	80100	B13-14	203
ELECTRICAL Wiring—inside	Rate	2.86	2.67	2.52	3.03	2.43	3.19	3.00	2.23	3.41	7.29	3.65	4.36
	Code	42124	721019	40203	704	426	4261	4-46	4261	402	80170	B11-05	206
EXCAVATION—earth NOC	Rate	3.59	4.29	3.84	4.20	3.69	5.55	3.50	4.11	3.58	8.80	4.02	4.36
	Code	40604	721031	40706	711	421	4214	4-43	4214	404	80030	R11-06	207
EXCAVATION—rock	Rate	3.59	4.29	3.84	4.20	3.69	5.55	3.50	4.11	3.58	8.80	4.02	4.36
	Code	40604	721031	40706	711	421	4214	4-43	4214	404	80030	R11-06	207
GLAZIERS	Rate	4.84	3.00	4.57	8.12	6.12	5.97	4.25	7.67	3.41	14.59	7.01	4.36
	Code	42121	715020	40109	751	423	4233	4-41	4233	402	80150	B13-04	212
INSULATION WORK	Rate	3.67	6.94	4.57	8.12	6.12	5.97	4.25	7.67	6.15	14.69	6.20	4.36
	Code	42184	721029	40102	751	423	4234	4-41	4234	401	80110	B12-07	202
LATHING	Rate	8.30	11.20	4.57	4.83	4.21	6.35	4.25	5.74	3.41	14.69	7.01	4.36
	Code	42135	721033	40102	723	427	4279	4-41	4271	402	80110	B13-16	202
MASONRY	Rate	7.26	11.20	4.57	12.21	6.12	5.97	4.25	7.67	6.15	15.84	7.01	4.36
	Code	42102	721037	40102	741	423	4231	4-41	4231	401	80100	B13-18	202
PAINTING & DECORATING	Rate	5.64	6.94	3.45	6.83	4.21	6.35	4.25	5.74	3.41	14.69	6.20	4.36
	Code	42111	721041	40105	719	427	4275	4-41	4275	402	80110	B12-01	202
PILE DRIVING	Rate	7.26	13.07	3.84	5.84	4.63	10.05	3.50	4.83	6.15	8.80	7.01	4.36
	Code	42159	722004	40706	732	422	4221	4-43	4221	401	80030	B13-10	202
PLASTERING	Rate	8.30	11.20	5.13	6.83	4.21	6.35	4.25	5.74	3.41	14.69	6.20	4.36
	Code	42135	721042	40108	719	427	4271	4-41	4271	402	80110	B12-21	202
PLUMBING	Rate	2.86	5.48	2.88	3.83	3.74	3.89	3.00	2.23	3.41	7.75	3.65	4.36
	Code	42122	721043	40204	707	424	4241	4-46	4241	402	80160	B11-01	214
ROOFING	Rate	11.09	10.45	6.52	12.34	8.00	10.05	4.25	9.49	6.15	23.29	7.01	4.36
	Code	42118	721036	40403	728	430	4236	4-41	4236	401	80130	B13-20	202
SHEET METAL WORK (HVAC)	Rate	2.86	5.48	6.52	3.83	3.74	3.89	3.50	3.23	3.41	7.75	3.65	6.60
	Code	42117	721043	40402	707	424	4244	4-46	4244	402	80160	B11-07	208
STEEL ERECTION—door & sash	Rate	3.67	13.07	9.89	16.47	4.63	10.05	4.25	7.67	6.15	29.23	7.01	4.36
	Code	42106	722005	40502	748	422	4227	4-41	4227	401	80080	B13-22	202
STEEL ERECTION—inter., ornam.	Rate	3.67	13.07	9.89	16.47	4.63	10.05	4.25	7.67	6.15	29.23	7.01	4.36
	Code	42106	722005	40502	748	422	4227	4-41	4227	401	80080	B13-22	202
STEEL ERECTION—structure	Rate	3.67	13.07	9.89	16.47	4.63	10.05	4.25	7.67	6.15	29.23	7.01	4.36
	Code	42106	722005	40502	748	422	4227	4-41	4227	401	80080	B13-22	202
STEEL ERECTION—NOC	Rate	3.67	13.07	9.89	16.47	4.63	10.05	4.25	7.67	6.15	29.23	7.01	4.36
	Code	42106	722005	40502	748	422	4227	4-41	4227	401	80080	B13-22	202
TILE WORK—inter. (ceramic)	Rate	4.86	4.70	2.17	6.83	4.21	6.35	4.25	5.74	3.41	14.69	7.01	4.36
	Code	42113	721054	40103	719	427	4276	4-41	4276	402	80110	B13-01	202
WATERPROOFING	Rate	5.64	6.94	4.57	4.83	6.12	6.35	4.25	7.67	3.41	23.29	6.20	4.36
	Code	42139	721016	40102	723	423	4299	4-41	4239	402	80130	B12-17	202
WRECKING	Rate	3.59	6.53	6.46	16.47	3.69	5.55	3.50	4.11	6.15	14.69	7.01	4.36
	Code	40604	721005	40106	748	421	4211	4-43	4211	401	80110	B13-09	202

R01100-070 Contractor's Overhead & Profit

Below are the **average** installing contractor's percentage mark-ups applied to base labor rates to arrive at typical billing rates.

Column A: Labor rates are based on average open shop wages for 7 major U.S. regions. Base rates including fringe benefits are listed hourly and daily. These figures are the sum of the wage rate and employer-paid fringe benefits such as vacation pay, and employer-paid health costs.

Column B: Workers' Compensation rates are the national average of state rates established for each trade.

Column C: Column C lists average fixed overhead figures for all trades. Included are Federal and State Unemployment costs set at 6.5%; Social Security Taxes (FICA) set at 7.65%; Builder's Risk Insurance costs set at 0.44%; and Public Liability costs set at 2.02%. All the percentages except those for Social Security Taxes vary from state to state as well as from company to company.

Columns D and E: Percentages in Columns D and E are based on the presumption that the installing contractor has annual billing of $2,000,000 and up. Overhead percentages may increase with smaller annual billing. Th overhead percentages for any given contractor may vary greatly and depend on a number of factors, such as the contractor's annual volume, engineering and logistical support costs, and staff requirements. The figures for overhead and profit will also vary depending on the type of job, the job location, and the prevailing economic conditions. All factors should be examined very carefully for each job.

Column F: Column F lists the total of Columns B, C, D, and E.

Column G: Column G is Column A (hourly base labor rate) multiplied by the percentage in Column F (O&P percentage).

Column H: Column H is the total of Column A (hourly base labor rate) pl Column G (Total O&P).

Column I: Column I is Column H multiplied by eight hours.

		A		B	C	D	E	F	G	H	I
		Base Rate Incl. Fringes		Workers' Comp. Ins.	Average Fixed Overhead	Overhead		Total Overhead & Profit		Rate with O & P	
Abbr.	Trade	Hourly	Daily				Profit	%	Amount	Hourly	Daily
Skwk	Skilled Workers Average (35 trades)	$23.90	$191.20	16.2%	16.3%	27.0%	10.0%	69.5%	$16.60	$40.50	$324.00
	Helpers Average (5 trades)	17.55	140.40	18.2		25.0		69.5	$12.20	29.75	238.00
	Foreman Average, Inside ($.50 over trade)	24.40	195.20	16.2		27.0		69.5	16.95	41.35	330.80
	Foreman Average, Outside ($2.00 over trade)	25.90	207.20	16.2		27.0		69.5	18.00	43.90	351.20
Clab	Common Building Laborers	17.35	138.80	18.4		25.0		69.7	12.10	29.45	235.60
Asbe	Asbestos Workers	24.50	196.00	15.5		30.0		71.8	17.60	42.10	336.80
Boil	Boilermakers	28.30	226.40	13.7		30.0		70.0	19.80	48.10	384.80
Bric	Bricklayers	24.70	197.60	15.1		25.0		66.4	16.40	41.10	328.80
Brhe	Bricklayer Helpers	18.85	150.80	15.1		25.0		66.4	12.50	31.35	250.80
Carp	Carpenters	24.00	192.00	18.4		25.0		69.7	16.75	40.75	326.00
Cefi	Cement Finishers	23.00	184.00	9.9		25.0		61.2	14.10	37.10	296.80
Elec	Electricians	27.30	218.40	6.5		30.0		62.8	17.15	44.45	355.60
Elev	Elevator Constructors	30.20	241.60	7.6		30.0		63.9	19.30	49.50	396.00
Eqhv	Equipment Operators, Crane or Shovel	24.75	198.00	10.4		28.0		64.7	16.00	40.75	326.00
Eqmd	Equipment Operators, Medium Equipment	23.90	191.20	10.4		28.0		64.7	15.45	39.35	314.80
Eqlt	Equipment Operators, Light Equipment	22.80	182.40	10.4		28.0		64.7	14.75	37.55	300.40
Eqol	Equipment Operators, Oilers	20.75	166.00	10.4		28.0		64.7	13.45	34.20	273.60
Eqmm	Equipment Operators, Master Mechanics	25.00	200.00	10.4		28.0		64.7	16.20	41.20	329.60
Glaz	Glaziers	23.45	187.60	13.9		25.0		65.2	15.30	38.75	310.00
Lath	Lathers	22.00	176.00	10.9		25.0		62.2	13.70	35.70	285.60
Marb	Marble Setters	23.25	186.00	15.1		25.0		66.4	15.45	38.70	309.60
Mill	Millwrights	24.85	198.80	10.4		25.0		61.7	15.35	40.20	321.60
Mstz	Mosaic and Terrazzo Workers	22.85	182.80	9.7		25.0		61.0	13.95	36.80	294.40
Pord	Painters, Ordinary	21.75	174.00	13.1		25.0		64.4	14.00	35.75	286.00
Psst	Painters, Structural Steel	22.00	176.00	46.8		25.0		98.1	21.60	43.60	348.80
Pape	Paper Hangers	21.50	172.00	13.1		25.0		64.4	13.85	35.35	282.80
Pile	Pile Drivers	23.30	186.40	22.5		30.0		78.8	18.35	41.65	333.20
Plas	Plasterers	21.95	175.60	14.9		25.0		66.2	14.55	36.50	292.00
Plah	Plasterer Helpers	18.95	151.60	14.9		25.0		66.2	12.55	31.50	252.00
Plum	Plumbers	26.95	215.60	8.0		30.0		64.3	17.35	44.30	354.40
Rodm	Rodmen (Reinforcing)	25.45	203.60	24.7		28.0		79.0	20.10	45.55	364.40
Rofc	Roofers, Composition	20.30	162.40	33.1		25.0		84.4	17.15	37.45	299.60
Rots	Roofers, Tile and Slate	20.50	164.00	33.1		25.0		84.4	17.30	37.80	302.40
Rohe	Roofer Helpers (Composition)	14.90	119.20	33.1		25.0		84.4	12.60	27.50	220.00
Shee	Sheet Metal Workers	26.55	212.40	11.5		30.0		67.8	18.00	44.55	356.40
Spri	Sprinkler Installers	26.90	215.20	8.8		30.0		65.1	17.50	44.40	355.20
Stpi	Steamfitters or Pipefitters	27.05	216.40	8.0		30.0		64.3	17.40	44.45	355.60
Ston	Stone Masons	24.00	192.00	15.1		25.0		66.4	15.95	39.95	319.60
Sswk	Structural Steel Workers	25.55	204.40	38.6		28.0		92.9	23.75	49.30	394.40
Tilf	Tile Layers (Floor)	22.90	183.20	9.7		25.0		61.0	13.95	36.85	294.80
Tilh	Tile Layer Helpers	17.75	142.00	9.7		25.0		61.0	10.85	28.60	228.80
Trlt	Truck Drivers, Light	19.05	152.40	15.2		25.0		66.5	12.65	31.70	253.60
Trhv	Truck Drivers, Heavy	19.55	156.40	15.2		25.0		66.5	13.00	32.55	260.40
Sswl	Welders, Structural Steel	25.55	204.40	38.6		28.0		92.9	23.75	49.30	394.40
Wrck	*Wrecking	17.90	143.20	39.5		25.0		90.8	16.25	34.15	273.20

*Not included in Averages.

R01100-090 Sales Tax by State

State sales tax on materials is tabulated below (5 states have no sales tax). Many states allow local jurisdictions, such as a county or city, to levy additional sales tax.

Some projects may be sales tax exempt, particularly those constructed with public funds.

State	Tax (%)	State	Tax (%)	State	Tax (%)	State	Tax (%)
Alabama	4	Illinois	6.25	Montana	0	Rhode Island	7
Alaska	0	Indiana	6	Nebraska	5.5	South Carolina	5
Arizona	5.6	Iowa	5	Nevada	6.5	South Dakota	4
Arkansas	5.125	Kansas	5.3	New Hampshire	0	Tennessee	7
California	7.25	Kentucky	6	New Jersey	6	Texas	6.25
Colorado	2.9	Louisiana	4	New Mexico	5	Utah	4.75
Connecticut	6	Maine	5	New York	4.25	Vermont	6
Delaware	0	Maryland	5	North Carolina	4.5	Virginia	5
District of Columbia	5.75	Massachusetts	5	North Dakota	5	Washington	6.5
Florida	6	Michigan	6	Ohio	6	West Virginia	6
Georgia	4	Minnesota	6.5	Oklahoma	4.5	Wisconsin	5
Hawaii	4	Mississippi	7	Oregon	0	Wyoming	4
Idaho	6	Missouri	4.225	Pennsylvania	6	Average	4.86 %

Sales Tax by Province (Canada)

GST - a value-added tax, which the government imposes on most goods and services provided in or imported into Canada. PST - a retail sales tax, which five of the provinces impose on the price of most goods and some services. QST - a value-added tax, similar to the federal GST, which Quebec imposes. HST - Three provinces have combined their retail sales tax with the federal GST into one harmonized tax.

Province	PST (%)	QST (%)	GST(%)	HST(%)
Alberta	0	0	7	0
British Columbia	7.5	0	7	0
Manitoba	7	0	7	0
New Brunswick	0	0	0	15
Newfoundland	0	0	0	15
Northwest Territories	0	0	7	0
Nova Scotia	0	0	0	15
Ontario	8	0	7	0
Prince Edward Island	10	0	7	0
Quebec	0	7.5	7	0
Saskatchewan	6	0	7	0
Yukon	0	0	7	0

R01100-100 Unemployment Taxes and Social Security Taxes

Mass. State Unemployment tax ranges from 1.325% to 7.225% plus an experience rating assessment the following year, on the first $10,800 of wages. Federal Unemployment tax is 6.2% of the first $7,000 of wages. This is reduced by a credit for payment to the state. The minimum Federal Unemployment tax is 0.8% after all credits.

Combined rates in Mass. thus vary from 2.125% to 8.025% of the first $10,800 of wages. Combined average U.S. rate is about 6.2% of the first $7,000. Contractors with permanent workers will pay less since the average annual wages for skilled workers is $23.90 x 2,000 hours or about $47,800 per year. The average combined rate for U.S. would thus be 6.2% x $7,000 ÷ $47,800 = 0.9% of total wages for permanent employees.

Rates vary not only from state to state but also with the experience rating of the contractor.

Social Security (FICA) for 2005 is estimated at time of publication to be 7.65% of wages up to $87,900.

R01107-010 Architectural Fees

Tabulated below are typical percentage fees by project size, for good professional architectural service. Fees may vary from those listed depending upon degree of design difficulty and economic conditions in any particular area.

Rates can be interpolated horizontally and vertically. Various portions of the same project requiring different rates should be adjusted proportionately. For alterations, add 50% to the fee for the first $500,000 of project cost and add 25% to the fee for project cost over $500,000.

Architectural fees tabulated below include Structural, Mechanical and Electrical Engineering Fees. They do not include the fees for special consultants such as kitchen planning, security, acoustical, interior design, etc.

Civil Engineering fees are included in the Architectural fee for project site requiring minimal design such as city sites. However, separate Civil Engineering fees must be added when utility connections require design, drainage calculations are needed, stepped foundations are required, or provisions are required to protect adjacent wetlands.

Building Types	Total Project Size in Thousands of Dollars						
	100	250	500	1,000	5,000	10,000	50,000
Factories, garages, warehouses, repetitive housing	9.0%	8.0%	7.0%	6.2%	5.3%	4.9%	4.5%
Apartments, banks, schools, libraries, offices, municipal buildings	12.2	12.3	9.2	8.0	7.0	6.6	6.2
Churches, hospitals, homes, laboratories, museums, research	15.0	13.6	12.7	11.9	9.5	8.8	8.0
Memorials, monumental work, decorative furnishings	—	16.0	14.5	13.1	10.0	9.0	8.3

General Requirements R012 Special Project Procedures

R01250-010 Repair and Remodeling

Cost figures are based on new construction utilizing the most cost-effective combination of labor, equipment and material with the work scheduled in proper sequence to allow the various trades to accomplish their work in an efficient manner.

The costs for repair and remodeling work must be modified due to the following factors that may be present in any given repair and remodeling project.

1. Equipment usage curtailment due to the physical limitations of the project, with only hand-operated equipment being used.
2. Increased requirement for shoring and bracing to hold up the building while structural changes are being made and to allow for temporary storage of construction materials on above-grade floors.
3. Material handling becomes more costly due to having to move within the confines of an enclosed building. For multi-story construction, low capacity elevators and stairwells may be the only access to the upper floors.
4. Large amount of cutting and patching and attempting to match the existing construction is required. It is often more economical to remove entire walls rather than create many new door and window openings. This sort of trade-off has to be carefully analyzed.
5. Cost of protection of completed work is increased since the usual sequence of construction usually cannot be accomplished.
6. Economies of scale usually associated with new construction may not be present. If small quantities of components must be custom fabricated due to job requirements, unit costs will naturally increase. Also, if only small work areas are available at a given time, job scheduling between trades becomes difficult and subcontractor quotations may reflect the excessive start-up and shut-down phases of the job.
7. Work may have to be done on other than normal shifts and may have to be done around an existing production facility which has to stay in production during the course of the repair and remodeling.
8. Dust and noise protection of adjoining non-construction areas can involve substantial special protection and alter usual construction methods.
9. Job may be delayed due to unexpected conditions discovered during demolition or removal. These delays ultimately increase construction costs.
10. Piping and ductwork runs may not be as simple as for new construction. Wiring may have to be snaked through walls and floors.
11. Matching "existing construction" may be impossible because materials may no longer be manufactured. Substitutions may be expensive.
12. Weather protection of existing structure requires additional temporary structures to protect building at openings.
13. On small projects, because of local conditions, it may be necessary to pay a tradesman for a minimum of four hours for a task that is completed in one hour.

All of the above areas can contribute to increased costs for a repair and remodeling project. Each of the above factors should be considered in the planning, bidding and construction stage in order to minimize the increased costs associated with repair and remodeling jobs.

R01540-100 Steel Tubular Scaffolding

On new construction, tubular scaffolding is efficient up to 60′ high or five stories. Above this it is usually better to use a hung scaffolding if construction permits. Swing scaffolding operations may interfere with tenants. In this case, the tubular is more practical at all heights.

In repairing or cleaning the front of an existing building the cost of tubular scaffolding per S.F. of building front increases as the height increases above the first tier. The first tier cost is relatively high due to leveling and alignment.

The minimum efficient crew for erection is three workers. For heights over 50′, a crew of four is more efficient. Use two or more on top and two at the bottom for handing up or hoisting. Four workers can erect and dismantle about nine frames per hour up to five stories. From five to eight stories they will average six frames per hour. With 7′ horizontal spacing this will run about 400 S.F. and 265 S.F. of wall surface, respectively. Time for placing planks must be added to the above. On heights above 50′, five planks can be placed per labor-hour.

The table below shows the number of pieces required to erect tubular steel scaffolding for 1000 S.F. of building frontage. This area is made up of a scaffolding system that is 12 frames (11 bays) long by 2 frames high.

For jobs under twenty-five frames, add 50% to rental cost. Rental rates will be lower for jobs over three months duration. Large quantities for long periods can reduce rental rates by 20%.

Description of Component	CSI Line Item	Number of Pieces for 1000 S.F. of Building Front	Unit
5′ Wide Standard Frame, 6′-4″ High	01540-750-2200	24	Ea.
Leveling Jack & Plate	01540-750-2650	24	
Cross Brace	01540-750-2500	44	
Side Arm Bracket, 21″	01540-750-2700	12	
Guardrail Post	01540-750-2550	12	
Guardrail, 7′ section	01540-750-2600	22	
Stairway Section	01540-750-2900	2	
Stairway Starter Bar	01540-750-2910	1	
Stairway Inside Handrail	01540-750-2920	2	
Stairway Outside Handrail	01540-750-2930	2	
Walk-Thru Frame Guardrail	01540-750-2940	2	

Scaffolding is often used as falsework over 15′ high during construction of cast-in-place concrete beams and slabs. Two foot wide scaffolding is generally used for heavy beam construction. The span between frames depends upon the load to be carried with a maximum span of 5′.

Heavy duty shoring frames with a capacity of 10,000#/leg can be spaced up to 10′ O.C. depending upon form support design and loading.

Scaffolding used as horizontal shoring requires less than half the material required with conventional shoring.

On new construction, erection is done by carpenters.

Rolling towers supporting horizontal shores can reduce labor and speed the job. For maintenance work, catwalks with spans up to 70′ can be supported by the rolling towers.

R01540-200 Pump Staging

Pump staging is generally not available for rent. The table below shows the number of pieces required to erect pump staging for 2400 S.F. of building frontage. This area is made up of a pump jack system that is 3 poles (2 bays) wide by 2 poles high.

Item	CSI Line Item	Number of Pieces for 2400 S.F. of Building Front	Unit
Aluminum pole section, 24′ long	01540-550-0200	6	Ea.
Aluminum splice joint, 6′ long	01540-550-0600	3	
Aluminum foldable brace	01540-550-0900	3	
Aluminum pump jack	01540-550-0700	3	
Aluminum support for workbench/back safety rail	01540-550-1000	3	
Aluminum scaffold plank/workbench, 14″ wide x 24′ long	01540-550-1100	4	
Safety net, 22′ long	01540-550-1250	2	
Aluminum plank end safety rail	01540-550-1200	2	

The cost in place for this 2400 S.F. will depend on how many uses are realized during the life of the equipment. Several options are given in Division 01540-550.

R02065-300 Bituminous Paving

City	Sidewalk Mix Bituminous Asphalt per Ton*	Sidewalks (2") 9.2 S.Y./ton Cost per S.Y. Material*	Installation	Total	Per Ton Total	Pavement (3") 6.13 S.Y./ton Cost per S.Y. Material*	Installation	Total	Per Ton Total
Atlanta	$30.00	$3.26	$1.33	$4.59	$42.23	$4.89	$.80	$5.69	$34.88
Baltimore	34.73	3.78	1.30	5.08	46.74	5.67	.78	6.45	39.54
Boston	39.50	4.29	1.38	5.67	52.16	6.44	.83	7.27	44.57
Buffalo	34.60	3.76	1.30	5.06	46.55	5.64	.78	6.42	39.35
Chicago	31.75	3.45	1.43	4.88	44.90	5.18	.86	6.04	37.03
Cincinnati	41.00	4.46	1.51	5.97	54.92	6.69	.91	7.60	46.59
Cleveland	29.50	3.21	1.49	4.70	43.24	4.81	.89	5.70	34.94
Columbus	30.50	3.32	1.42	4.74	43.61	4.98	.85	5.83	35.74
Dallas	29.25	3.18	1.20	4.38	40.30	4.77	.72	5.49	33.65
Denver	30.00	3.26	1.48	4.74	43.61	4.89	.89	5.78	35.43
Detroit	32.00	3.48	1.36	4.84	44.53	5.22	.81	6.03	36.96
Houston	34.50	3.75	1.23	4.98	45.82	5.63	.74	6.37	39.05
Indianapolis	29.75	3.23	1.40	4.63	42.60	4.85	.84	5.69	34.88
Kansas City	29.50	3.21	1.36	4.57	42.04	4.81	.81	5.62	34.45
Los Angeles	36.00	3.91	1.50	5.41	49.77	5.87	.90	6.77	41.50
Memphis	38.75	4.21	1.28	5.49	50.51	6.32	.77	7.09	43.46
Milwaukee	32.50	3.53	1.30	4.83	44.44	5.30	.78	6.08	37.27
Minneapolis	31.38	3.41	1.52	4.93	45.36	5.12	.91	6.03	36.96
Nashville	27.75	3.02	1.38	4.40	40.48	4.53	.83	5.36	32.86
New Orleans	33.75	3.67	1.23	4.90	45.08	5.51	.73	6.24	38.25
New York City	49.50	5.38	1.60	6.98	64.22	8.08	.96	9.04	55.42
Philadelphia	32.50	3.53	1.32	4.85	44.62	5.30	.79	6.09	37.33
Phoenix	30.00	3.26	1.46	4.72	43.42	4.89	.88	5.77	35.37
Pittsburgh	35.00	3.80	1.52	5.32	48.94	5.71	.91	6.62	40.58
St. Louis	32.00	3.48	1.34	4.82	44.34	5.22	.81	6.03	36.96
San Antonio	33.25	3.61	1.24	4.85	44.62	5.42	.74	6.16	37.76
San Diego	37.00	4.02	1.40	5.42	49.86	6.04	.84	6.88	42.17
San Francisco	40.25	4.38	1.55	5.93	54.56	6.57	.93	7.50	45.98
Seattle	39.50	4.29	1.58	5.87	54.00	6.44	.95	7.39	45.30
Washington, D.C.	35.41	3.85	1.25	5.10	46.92	5.78	.75	6.53	40.03
Average	$34.00	$3.70	$1.39	$5.09	$46.83	$5.55	$.83	$6.38	$39.11

Assumed density is 145 lb. per C.F.
*Includes delivery within 20 miles for quantities over 300 tons only

Table below shows quantities and bare costs for 1000 S.Y. of Bituminous Paving.

Item	Sidewalks, 2" Thick Quantities	Cost	Roads and Parking Areas, 3" Thick Quantities	Cost
Bituminous asphalt	108.7 tons @ $34.00 per ton	$3,696.00	163.1 tons@ $34.00 per ton	$5,546.00
Installation using	Crew B-37 @ $999.80 /720 SY/day x 1000	1,388.61	Crew B-25B @ $4,081.00 /4900SY/ day x 1000	832.86
Total per 1000 S.Y.		$5,084.61		$6,378.86
Total per S.Y.		$ 5.09		$ 6.38
Total per Ton		$ 46.83		$ 39.11

R02220-510 Demolition Defined

Whole Building Demolition (Division 02220-110) - Demolition of the whole building with no concern for any particular building element, component, or material type being demolished. This type of demolition is accomplished with large pieces of construction equipment that break up the structure, load it into trucks and haul it to a disposal site, but disposal or dump fees are not included (Divisions 02220-320 and 02220-330). Demolition of below-grade foundation elements, such as footings, foundation walls, grade beams, slabs on grade, etc. (Division 02220-130), is not included. Certain mechanical equipment containing flammable liquids or ozone-depleting refrigerants, electric lighting elements, communication equipment components, and other building elements may contain hazardous waste, and must be removed, either selectively or carefully, as hazardous waste before the building can be demolished (Division 13281).

Foundation Demolition (Division 02220-130) - Demolition of below-grade foundation footings, foundation walls, grade beams, and slabs on grade. This type of demolition is accomplished by hand or pneumatic hand tools, and does not include saw cutting (Division 02220-360), or handling, loading, hauling, or disposal of the debris (Divisions 02220-320, 02220-330 and 02220-350).

Gutting (Division 02220-340) - Removal of building interior finishes and electrical/mechanical systems down to the load-bearing and sub-floor elements of the rough building frame, with no concern for any particular building element, component, or material type being demolished. This type of demolition is accomplished by hand or pneumatic hand tools, and includes loading into trucks, but not hauling, disposal or dump fees (Divisions 02220-320 and 02220-330); scaffolding (01540-750); or shoring (03150-600). Certain mechanical equipment containing flammable liquids or ozone-depleting refrigerants, electric lighting elements, communication equipment components, and other building elements may contain hazardous waste, and must be removed, either selectively or carefully, as hazardous waste, before the building is gutted (Division 13281).

Selective Demolition - Demolition of a selected building element, component, or finish, with some concern for surrounding or adjacent elements, components, or finishes (see the first Subdivision (s) at the beginning of Divisions 3 through 16). This type of demolition is accomplished by hand or pneumatic hand tools, and does not include handling, loading, storing, hauling, or disposal of the debris (Divisions 02220-320, 02220-330 and 02220-350); scaffolding (Division 01540-750); or shoring (Division 03150-600). "Gutting" methods may be used in order to save time, but damage that was caused to surrounding or adjacent elements, components, or finishes may have to be repaired at a later time.

Careful Removal - Removal of a piece of service equipment, building element or component, or material type, with great concern for both the removed item and surrounding or adjacent elements, components or finishes. The purpose of careful removal may be to protect the removed item for later re-use, preserve a higher salvage value of the removed item, or replace an item while taking care to protect surrounding or adjacent elements, components, connections, or finishes from cosmetic and/or structural damage. An approximation of the time required to perform this type of removal is 1/3 to 1/2 the time it would take to install a new item of like kind (see Reference Numbers R15050-720 and R16055-300). This type of removal is accomplished by hand or pneumatic hand tools, and does not include loading, hauling, or storing the removed item (Divisions 01840-100 and 02220-350); scaffolding (Division 01540-750); shoring (Division 03150-600); or lifting equipment (Division 01590-600).

Cutout Demolition (Division 02220-310) - Demolition of a small quantity of floor, wall, roof, or other assembly, with concern for the appearance and structural integrity of the surrounding materials. This type of demolition is accomplished by hand or pneumatic hand tools, and does not include saw cutting (Division 02220-360); handling, loading, hauling, or disposal of debris (Divisions 02220-320, 02220-330 and 02220-350); scaffolding (Division 01540-750); or shoring (Division 03150-600).

Rubbish Handling (Division 02220-350) - Work activities that involve handling, loading or hauling of debris. Generally, the cost of rubbish handling must be added to the cost of all types of demolition, with the exception of whole building demolition.

Minor Site Demolition (Division 02220-240) - Demolition of site elements outside the footprint of a building. This type of demolition is accomplished by hand or pneumatic hand tools, or with larger pieces of construction equipment, and may include loading a removed item onto a truck (check the Crew for equipment used). It does not include saw cutting (Division 02220-360), hauling or disposal of debris (Divisions 02220-320 and 02220-330), and, sometimes, handling or loading (Division 02220-350).

R02510-800 Piping Designations

There are several systems currently in use to describe pipe and fittings. The following paragraphs will help to identify and clarify classifications of piping systems used for water distribution.

Piping may be classified by schedule. Piping schedules include 5S, 10S, 10, 20, 30, Standard, 40, 60, Extra Strong, 80, 100, 120, 140, 160 and Double Extra Strong. These schedules are dependent upon the pipe wall thickness. The wall thickness of a particular schedule may vary with pipe size.

Ductile iron pipe for water distribution is classified by Pressure Classes such as Class 150, 200, 250, 300 and 350. These classes are actually the rated water working pressure of the pipe in pounds per square inch (psi). The pipe in these pressure classes is designed to withstand the rated water working pressure plus a surge allowance of 100 psi.

The American Water Works Association (AWWA) provides standards for various types of **plastic pipe.** C-900 is the specification for polyvinyl chloride (PVC) piping used for water distribution in sizes ranging from 4″ through 12″. C-901 is the specification for polyethylene (PE) pressure pipe, tubing and fittings used for water distribution in sizes ranging from 1/2″ through 3″. C-905 is the specification for PVC piping sizes 14″ and greater.

PVC pressure-rated pipe is identified using the standard dimensional ratio (SDR) method. This method is defined by the American Society for Testing and Materials (ASTM) Standard D 2241. This pipe is available in SDR numbers 64, 41, 32.5, 26, 21, 17, and 13.5. Pipe with an SDR of 64 will have the thinnest wall while pipe with an SDR of 13.5 will have the thickest wall. When the pressure rating (PR) of a pipe is given in psi, it is based on a line supplying water at 73 degrees F.

The National Sanitation Foundation (NSF) seal of approval is applied to products that can be used with potable water. These products have been tested to ANSI/NSF Standard 14.

Valves and strainers are classified by American National Standards Institute (ANSI) Classes. These Classes are 125, 150, 200, 250, 300, 400, 600, 900, 1500 and 2500. Within each class there is an operating pressure range dependent upon temperature. Design parameters should be compared to the appropriate material dependent, pressure-temperature rating chart for accurate valve selection.

R02920-500 Seeding

The type of grass is determined by light, shade and moisture content of soil plus intended use. Fertilizer should be disked 4″ before seeding. For steep slopes disk five tons of mulch and lay two tons of hay or straw on surface per acre after seeding. Surface mulch can be staked, lightly disked or tar emulsion sprayed. Material for mulch can be wood chips, peat moss, partially rotted hay or straw, wood fibers and sprayed emulsions. Hemp seed blanke with fertilizer are also available. For spring seeding, watering is necessary. Late fall seeding may have to be reseeded in the spring. Hydraulic seeding power mulching, and aerial seeding can be used on large areas.

R02930-900 Cost of Trees: Based on Pin Oak (Quercus palustris)

Tree Diameter	Normal Height	Catalog List Price of Tree	Guying Material	Bare Equipment Charge	Bare Installation Labor	Bare Total
2 to 3 inch	14 feet	$ 146	$15.35	$ 54.24	$ 61.64	$ 277.23
3 to 4 inch	16 feet	323	18.35	$ 90.40	$102.73	534.48
4 to 5 inch	18 feet	373	75.50	$108.48	$123.28	680.26
6 to 7 inch	22 feet	813	80.00	$135.60	$123.28	1,151.88
8 to 9 inch	26 feet	1,023	90.00	$180.80	$205.47	1,499.27

Installation Time & Cost for Planting Trees, Bare Costs														
Ball Size Diam. X Depth	Soil in Ball	Weight of Ball	Hole Diam. Req'd	Hole Excavation	Amount of Soil Displ.	Topsoil Handled	Time Required in Labor-Hours						Cost	
							Dig & Lace	Handle Ball	Dig Hole	Plant & Prune	Water & Guy	Total L.H.	Crew	Bare Total per Tree
Inches	C.F.	Lbs.	Feet	C.F.	C.F.	C.F.								
12 x 12	0.70	56.00	2.00	4.00	3.00	11.00	0.25	0.17	0.33	0.25	0.07	1.10	1 Clab	$ 19.09
18 x 16	2.00	160.00	2.50	8.00	6.00	21.00	0.50	0.33	0.47	0.35	0.08	1.70	2 Clab	29.50
24 x 18	4.00	320.00	3.00	13.00	9.00	38.00	1.00	0.67	1.08	0.82	0.20	3.80	3 Clab	65.93
30 x 21	7.50	600.00	4.00	27.00	19.50	76.00	0.82	0.71	0.79	1.22	0.26	3.80		107.01
36 x 24	12.50	980.00	4.50	38.00	25.50	114.00	1.08	0.95	1.11	1.32	0.30	4.76		134.04
42 x 27	19.00	1,520.00	5.50	64.00	45.00	185.00	1.90	1.27	1.87	1.43	0.34	6.80		191.49
48 x 30	28.00	2,040.00	6.00	85.00	57.00	254.00	2.41	1.60	2.06	1.55	0.39	8.00		225.28
54 x 33	38.50	3,060.00	7.00	127.00	88.50	370.00	2.86	1.90	2.39	1.76	0.45	9.40	B-6	264.70
60 x 36	52.00	4,160.00	7.50	159.00	107.00	474.00	3.26	2.17	2.73	2.00	0.51	10.70		301.31
66 x 39	68.00	5,440.00	8.00	196.00	128.00	596.00	3.61	2.41	3.07	2.26	0.58	11.90		335.10
72 x 42	87.00	7,160.00	9.00	267.00	180.00	785.00	3.90	2.60	3.71	2.78	0.70	13.70		385.79

For information about Means Estimating Seminars, see yellow pages 12 and 13 in back of boo

R04060-100 Cement Mortar (material only)

Type N - 1:1:6 mix by volume. Use everywhere above grade except as noted below.
- 1:3 mix using conventional masonry cement which saves handling two separate bagged materials.

Type M - 1:1/4:3 mix by volume, or 1 part cement, 1/4 (10% by wt.) lime, 3 parts sand. Use for heavy loads and where earthquakes or hurricanes may occur. Also for reinforced brick, sewers, manholes and everywhere below grade.

Cost and Mix Proportions of Various Types of Mortar

Components	Type Mortar and Mix Proportions by Volume										
	M		S		N		O		K	PM	PL
	1:1:6	1:1/4:3	1/2:1:4	1:1/2:4	1:3	1:1:6	1:3	1:2:9	1:3:12	1:1:6	1:1/2:4
Portland cement	$ 8.25	$ 8.25	$ 4.13	$ 8.25	—	$ 8.25	—	$ 8.25	$ 8.25	$ 8.25	$ 8.25
Masonry cement	6.15	—	6.15	—	$ 6.15	—	$ 6.15	—	—	6.15	—
Lime	—	1.45	—	2.89	—	5.78	—	11.56	17.34	—	2.89
Masonry sand*	4.28	2.14	2.85	2.85	2.14	4.28	2.14	6.42	8.56	4.28	2.85
Mixing machine incl. fuel**	3.70	1.85	2.46	2.46	1.85	3.70	1.85	5.54	7.39	3.70	2.46
Total for Materials	$22.38	$13.69	$15.59	$16.45	$10.14	$22.01	$10.14	$31.77	$41.54	$22.38	$16.45
Total C.F.	6	3	4	4	3	6	3	9	12	6	4
Approximate Cost per C.F.	$ 3.73	$ 4.56	$ 3.90	$ 4.11	$ 3.38	$ 3.67	$ 3.38	$ 3.53	$ 3.46	$ 3.73	$ 4.11

*Includes 10 mile haul
**Based on a daily rental, 10 C.F., 25 H.P. mixer, mix 200 C.F./Day

Mix Proportions by Volume, Compressive Strength and Cost of Mortar

Where Used	Allowable Proportions by Volume					Compressive Strength @ 28 days	Cost per Cubic Foot
	Mortar Type	Portland Cement	Masonry Cement	Hydrated Lime	Masonry Sand		
Plain Masonry	M	1	1	—	6	2500 psi	$3.73
		1	—	1/4	3		4.56
	S	1/2	1	—	4	1800 psi	3.90
		1	—	1/4 to 1/2	4		4.11
	N	—	1	—	3	750 psi	3.38
		1	—	1/2 to 1 1/4	6		3.67
	O	—	1	—	3	350 psi	3.38
		1	—	1 1/4 to 2 1/2	9		3.53
	K	1	—	2 1/2 to 4	12	75 psi	3.46
Reinforced Masonry	PM	1	1	—	6	2500 psi	3.73
	PL	1	—	1/4 to 1/2	4	2500 psi	4.11

Note: The total aggregate should be between 2.25 to 3 times the sum of the cement and lime used.

The labor cost to mix the mortar is included in the labor cost on brickwork.

Machine mixing is usually specified on jobs of any size. There is a large price saving over hand mixing and mortar is more uniform.

There are two types of mortar color used. Prices in Section 04060-540 are for the inert additive type with about 100 lbs. per M brick as the typical quantity required. These colors are also available in smaller batch size bags (1 lb. to 15 lb.) which can be placed directly into the mixer without measuring. The other type is premixed and replaces the masonry cement. Dark green color has the highest cost.

R04060-200 Miscellaneous Mortar (material only)

Quantities	Glass Block Mortar		Gypsum Cement Mortar	
White Portland cement, 94 Lb bag	7 bags	$136.15		
Gypsum cement, 80 Lb bag			11.25 bags	$258.75
Lime, 50 Lb bag	280 lbs.	32.37		
Sand*	1 C.Y.	19.25	1 C.Y.	19.25
Mixing machine and fuel**		16.65		16.65
Total per C.Y.		$204.42		$294.65
Approximate Total per C.F.		$ 7.57		$ 10.91

* Includes 10 mile haul

** Based on a daily rental, 10 C.F., 25 HP mixer, mix 200 C.F./Day = 7.4 C.Y./Day

R04080-500 Masonry Reinforcing

Horizontal joint reinforcing helps prevent wall cracks where wall movement may occur and in many locations is required by code. Horizontal joint reinforcing is generally not considered to be structural reinforcing and an unreinforced wall may still contain joint reinforcing.

Reinforcing strips come in 10′ and 12′ lengths and in truss and ladder shapes, with and without drips. Field labor runs between 2.7 to 5.3 hours per 1000 L.F. for wall thicknesses up to 12″.

The wire meets ASTM A82 for cold drawn steel wire and the typical size is 9 ga. sides and ties with 3/16″ diameter also available. Typical finish is mill galvanized with zinc coating at .10 oz. per S.F. Class I (.40 oz. per S.F.) and Class III (.80 oz per S.F.) are also available, as is hot dipped galvanizing at 1.50 oz. per S.F.

R04210-100 Economy in Bricklaying

Have adequate supervision. Be sure bricklayers are always supplied with materials so there is no waiting. Place best bricklayers at corners and openings.

Use only screened sand for mortar. Otherwise, labor time will be wasted picking out pebbles. Use seamless metal tubs for mortar as they do not leak or catch the trowel. Locate stack and mortar for easy wheeling.

Have brick delivered for stacking. This makes for faster handling, reduces chipping and breakage, and requires less storage space. Many dealers will deliver select common in 2′ x 3′ x 4′ pallets or face brick packaged. This affords quick handling with a crane or forklift and easy tonging in units of ten, which reduces waste.

Use wider bricks for one wythe wall construction. Keep scaffolding away from wall to allow mortar to fall clear and not stain wall.

On large jobs develop specialized crews for each type of masonry unit.

Consider designing for prefabricated panel construction on high rise projects.

Avoid excessive corners or openings. Each opening adds about 50% to labor cost for area of opening.

Bolting stone panels and using window frames as stops reduces labor costs and speeds up erection.

R04210-120 Common and Face Brick Prices

Prices are based on truckload lot purchases for Common Building Brick and Facing Brick. Prices are per M, (thousand), brick.

City	Material			Installation				Total			
	Brick per M Delivered		Mortar	Common in 8″ Wall		Face Brick, 4″ Veneer		Common in 8″ Wall		Face Brick, 4″ Veneer	
	Common	Face	3/8″ Joint	Bare Costs	Incl. O & P	Bare Costs	Incl. O & P	Bare Costs	Incl. O & P	Bare Costs	Incl. O & P
Atlanta	$315	$438	$46.00	$330	$ 549	$396	$ 659	$ 700	$ 919	$ 885	$1,148
Baltimore	230	325	for 8″ Wall	400	666	480	799	683	949	853	1,172
Boston	395	550	and	719	1,197	863	1,436	1,172	1,650	1,468	2,041
Buffalo	410	575	$38.00	592	985	711	1,182	1,060	1,453	1,341	1,812
Chicago	275	388	for 4″ Wall	636	1,058	763	1,270	965	1,387	1,201	1,708
Cincinnati	260	367		477	794	573	953	791	1,108	989	1,369
Cleveland	270	375		540	899	648	1,079	864	1,223	1,072	1,503
Columbus	250	350		462	768	554	922	766	1,072	953	1,321
Dallas	305	425		322	536	386	643	682	896	862	1,119
Denver	275	388		389	647	466	776	718	976	904	1,214
Detroit	330	465		606	1,008	727	1,209	992	1,394	1,244	1,726
Houston	270	375		331	551	398	662	655	875	822	1,086
Indianapolis	305	425		477	794	572	952	837	1,154	1,048	1,428
Kansas City	245	344		542	902	651	1,083	840	1,200	1,043	1,475
Los Angeles	255	355		564	938	676	1,126	873	1,247	1,080	1,530
Memphis	225	313		407	677	488	813	685	955	848	1,173
Milwaukee	290	405		557	926	668	1,111	902	1,271	1,123	1,566
Minneapolis	320	450		607	1,010	729	1,212	983	1,386	1,231	1,714
Nashville	195	275		341	568	410	682	588	815	731	1,003
New Orleans	395	555		308	513	370	615	761	966	980	1,225
New York City	285	400		788	1,312	946	1,574	1,128	1,652	1,396	2,024
Philadelphia	440	615		626	1,041	751	1,250	1,125	1,540	1,422	1,921
Phoenix	350	490		355	591	427	710	762	998	970	1,253
Pittsburgh	205	290		482	802	579	963	739	1,059	916	1,300
St. Louis	250	350		569	947	683	1,136	873	1,251	1,082	1,535
San Antonio	215	300		333	555	400	665	600	822	747	1,012
San Diego	270	380		527	877	633	1,053	851	1,201	1,062	1,482
San Francisco	470	655		636	1,059	763	1,270	1,166	1,589	1,476	1,983
Seattle	430	600		554	922	665	1,107	1,043	1,411	1,321	1,763
Washington, D.C.	220	305		414	689	497	826	687	962	849	1,178
Average	$300	$420	▼	$495	$ 825	$595	$ 990	$ 850	$1,180	$1,065	$1,460

Common building brick manufactured according to ASTM C62 and facing brick manufactured according to ASTM C216 are the two standard bricks available for general building use.

Building brick is made in three grades; SW, where high resistance to damage caused by cyclic freezing is required; MW, where moderate resistance to cyclic freezing is needed; and NW, where little resistance to cyclic freezing is needed. Facing brick is made in only the two grades SW and MW. Additionally, facing brick is available in three types; FBS, for general use; FBX, for general use where a higher degree of precision and lower permissible variation in size than FBS is needed; and FBA, for general use to produce characteristic architectural effects resulting from non-uniformity in size and texture of the units.

In figuring above installation costs, a D-8 Crew (with a daily output of 1.5 M) was used for the 4″ veneer. A D-8 Crew (with a daily output of 1.8 M) was used for the 8″ solid wall.

In figuring the total cost including overhead and profit, an allowance of 10% was added to the sum of the cost of the brick and mortar. Also, 3% breakage was included for both the bare costs and the costs with overhead and profit. If bricks are delivered palletized with 280 to 300 per pallet, or packaged, allow only 1-1/2% for breakage. Then add $10 per M to the cost of brick and deduct two hours helper time. The net result is a savings of $30 to $40 per M in place. Packaged or palletized delivery is practical when a job is big enough to have a crane or other equipment available to handle a package of brick. This is so on all industrial work but not always true on small commercial buildings.

The prices above are for bricks used in commercial, apartment house or industial construction. If it is possible to obtain the price of the actual brick to be used, it should be done and substituted in the table. The use of buff and gray face is increasing, and there is a continuing trend to the Norman, Roman, Jumbo and SCR brick.

See R04210-500 for brick quantities per S.F. and mortar quantities per M brick. (Average prices for the various sizes are listed in Division 4.)

Common red clay brick for backup is not used that often. Concrete block is the most usual backup material with occasional use of sand lime or cement brick. Sand lime cost about $15 per M less than red clay and cement brick are about $5 per M less than red clay. These figures may be substituted in the common brick breakdown for the cost of these items in place, as labor is about the same. Building brick is commonly used in solid walls for strength and as a fire stop.

Brick panels built on the ground and then crane erected to the upper floors have proven to be economical. This allows the work to be done under cover and without scaffolding.

R04210-180 Brick in Place

Table below is for common bond with 3/8″ concave joints and includes 3% waste for brick and 25% waste for mortar.
Crew costs are bare costs.

Item	8″ Common Brick Wall 8″ x 2-2/3″ x 4″		Select Common Face 8″ x 2-2/3″ x 4″		Red Face Brick 8″ x 2-2/3″ x 4″	
1030 brick delivered		$306.94		$ 365.65		$ 384.19
Type N mortar	12.5 C.F.	45.88	10.3 C.F.	37.80	10.3 C.F.	37.80
Installation using indicated crew	Crew D 8 @ 0.556 Days	497.29	Crew D 8 @ 0.667 Days	596.56	Crew D 8 @ 0.667 Days	596.56
Total per M in place		$850.11		$1,000.01		$1,018.55
Total per S.F. of wall	13.5 bricks/S.F.	$ 11.48	6.75 bricks/S.F.	$ 6.75	6.75 bricks/S.F.	$ 6.88

R04210-500 Brick, Block & Mortar Quantities

Running Bond						For Other Bonds Standard Size Add to S.F. Quantities in Table to Left		
Number of Brick per S.F. of Wall - Single Wythe with 3/8″ Joints				C.F. of Mortar per M Bricks, Waste Included				
Type Brick	Nominal Size (incl. mortar) L H W	Modular Coursing	Number of Brick per S.F.	3/8″ Joint	1/2″ Joint	Bond Type	Description	Factor
Standard	8 x 2 2/3 x 4	3C=8″	6.75	10.3	12.9	Common	full header every fifth course	+20%
Economy	8 x 4 x 4	1C=4″	4.50	11.4	14.6		full header every sixth course	+16.7%
Engineer	8 x 3 1/5 x 4	5C=16″	5.63	10.6	13.6	English	full header every second course	+50%
Fire	9 x 2 1/2 x 4 1/2	2C=5″	6.40	550 # Fireclay	—	Flemish	alternate headers every course	+33.3%
Jumbo	12 x 4 x 6 or 8	1C=4″	3.00	23.8	30.8		every sixth course	+5.6%
Norman	12 x 2 2/3 x 4	3C=8″	4.50	14.0	17.9	Header = W x H exposed		+100%
Norwegian	12 x 3 1/5 x 4	5C=16″	3.75	14.6	18.6	Rowlock = H x W exposed		+100%
Roman	12 x 2 x 4	2C=4″	6.00	13.4	17.0	Rowlock stretcher = L x W exposed		+33.3%
SCR	12 x 2 2/3 x 6	3C=8″	4.50	21.8	28.0	Soldier = H x L exposed		—
Utility	12 x 4 x 4	1C=4″	3.00	15.4	19.6	Sailor = W x L exposed		33.3%

Concrete Blocks Nominal Size	Approximate Weight per S.F.		Blocks per 100 S.F.	Mortar per M block, waste included	
	Standard	Lightweight		Partitions	Back up
2″ x 8″ x 16″	20 PSF	15 PSF	113	27 C.F.	36 C.F.
4″	30	20	↓	41	51
6″	42	30	↓	56	66
8″	55	38	↓	72	82
10″	70	47	↓	87	97
12″ ↓	85 ↓	55 ↓	↓	102 ↓	112 ↓

R04220-200 Concrete Block

Concrete masonry units, 8" high x 16" long, sand aggregate, 3/8 joints for partitions, tooled joints two sides, 113 blocks per 100 S.F., bare costs.

City	Material				Mortar 3/8" Joint	Bare Installation		Bare Total Per 100 S.F.	
	Per Block, Delivered		113 Block, Delivered						
	4" Thick	8" Thick	4" Thick	8" Thick		4" Thick	8" Thick	4" Thick	8" Thick
Atlanta	$ .60	$1.00	$ 67.80	$113.00	$16.88 for 4"	$138.44	$158.64	$223.12	$301.37
Baltimore	.54	.89	61.02	100.57	$29.73 for 8"	167.78	192.27	245.68	322.57
Boston	.68	1.13	76.84	127.69		302.44	346.57	396.16	503.99
Buffalo	.68	1.12	76.84	126.56		248.34	284.58	342.06	440.87
Chicago	.59	.98	66.67	110.74		266.65	305.56	350.20	446.03
Cincinnati	.66	1.10	74.58	124.30		200.09	229.29	291.55	383.32
Cleveland	.65	1.07	73.45	120.91		226.62	259.69	316.95	410.33
Columbus	.57	.94	64.41	106.22		193.67	221.93	274.96	357.88
Dallas	.71	1.17	80.23	132.21		135.07	154.78	232.18	316.72
Denver	1.10	1.83	124.30	206.79		163.00	186.78	304.18	423.30
Detroit	.83	1.37	93.79	154.81		254.02	291.09	364.69	475.63
Houston	.66	1.10	74.58	124.30		138.96	159.24	230.42	313.27
Indianapolis	.77	1.27	87.01	143.51		200.04	229.23	303.93	402.47
Kansas City	.74	1.22	83.62	137.86		227.37	260.55	327.87	428.14
Los Angeles	.57	.95	64.41	107.35		236.42	270.92	317.71	408.00
Memphis	.65	1.08	73.45	122.04		170.67	195.57	261.00	347.34
Milwaukee	.65	1.07	73.45	120.91		233.43	267.49	323.76	418.13
Minneapolis	.65	1.08	73.45	122.04		254.63	291.79	344.96	443.56
Nashville	.67	1.11	75.71	125.43		143.15	164.04	235.74	319.20
New Orleans	.74	1.23	83.62	138.99		129.24	148.10	229.74	316.82
New York City	.65	1.08	73.45	122.04		330.60	378.84	420.93	530.61
Philadelphia	.55	.91	62.15	102.83		262.47	300.77	341.50	433.33
Phoenix	.49	.81	55.37	91.53		149.07	170.83	221.32	292.09
Pittsburgh	.60	1.00	67.80	113.00		202.19	231.70	286.87	374.43
St. Louis	.80	1.33	90.40	150.29		238.57	273.38	345.85	453.40
San Antonio	.72	1.20	81.36	135.60		139.77	160.17	238.01	325.50
San Diego	.53	.88	59.89	99.44		221.07	253.33	297.84	382.50
San Francisco	1.15	1.90	129.95	214.70		266.81	305.74	413.64	550.17
Seattle	1.02	1.69	115.26	190.97		232.50	266.43	364.64	487.13
Washington D.C.	.75	1.24	84.75	140.12		173.58	198.91	275.21	368.76
Average	$ .70	$1.16	$ 79.10	$131.08	▼	$208.22	$238.61	$304.20	$399.42

Cost for 100 S.F. of 8" x 16" Concrete Block Partitions to Four Stories High, Tooled Joints Two Sides						
8" x 16" Sand Aggregate	4" Thick Block		8" Thick Block		12" Thick Block	
113 block delivered		$ 79.10		$131.08		$184.19
Mortar Type N, 1:3	4.6 C.F.	16.88	8.1 C.F.	29.73	11.5 C.F.	42.21
Installation Crew	D 8 @ 0.233 days	208.22	D 8 @ .267 days	238.61	D 9 @ .294 days	307.01
Total bare cost per 100 S.F.		$304.20		$399.42		$533.41
Add for filling cores solid	6.7 C.F.	$103.45	25.8 C.F.	$221.10	42.2 C.F.	$283.80

Cost for 100 S.F. of 8" x 16" Concrete Block Backup, Tooled Joints One Side						
8" x 16" Sand Aggregate	4" Thick Block		8" Thick Block		12" Thick Block	
113 block delivered		$ 79.10		$131.08		$184.19
Mortar Type N, 1:3	5.8 C.F.	21.29	9.3 C.F.	34.13	12.7 C.F.	46.61
Installation crew	D 8 @ .217 days	193.92	D 8 @ .250 days	223.41	D 9 @ .323 days	337.30
Total bare cost per 100 S.F.		$294.31		$388.62		$568.10

Special block: corner, jamb and head block are same price as ordinary block of same size. Tabulated on the next page are national average prices per block. Labor on specials is about the same as equal sized regular block. Bond beam and 16" high lintel blocks cost 30% more than regular units of equal size. Lintel blocks are 8" long and 8" or 16" high. Costs in individual cities may be factored from the table above.

Use of motorized mortar spreader box will speed construction of continuous walls. Hollow non-load bearing units are made according to ASTM C129 and hollow load bearing units according to ASTM C90.

Masonry | R049 | Restoration & Cleaning

R04930-100 Cleaning Face Brick

On smooth brick a person can clean 70 S.F. an hour; on rough brick 50 S.F. per hour. Use one gallon muriatic acid to 20 gallons of water for 1000 S.F. Do not use acid solution until wall is at least seven days old, but a mild soap solution may be used after two days. Commercial cleaners cost from $9 to $12 per gallon.

Time has been allowed for clean-up in brick prices.

Metals | R053 | Metal Deck

R05310-100 Decking Descriptions

General - All Deck Products

Steel deck is made by cold forming structural grade sheet steel into a repeating pattern of parallel ribs. The strength and stiffness of the panels are the result of the ribs and the material properties of the steel. Deck lengths can be varied to suit job conditions, but because of shipping considerations, are usually less than 40 feet. Standard deck width varies with the product used but full sheets are usually 12″, 18″, 24″, 30″, or 36″. Deck is typically furnished in a standard width with the ends cut square. Any cutting for width, such as at openings or for angular fit, is done at the job site.

Deck is typically attached to the building frame with arc puddle welds, self-drilling screws, or powder or pneumatically driven pins. Sheet to sheet fastening is done with screws, button punching (crimping), or welds.

Composite Floor Deck

After installation and adequate fastening, floor deck serves several purposes. It (a) acts as a working platform, (b) stabilizes the frame, (c) serves as a concrete form for the slab, and (d) reinforces the slab to carry the design loads applied during the life of the building. Composite decks are distinguished by the presence of shear connector devices as part of the deck. These devices are designed to mechanically lock the concrete and deck together so that the concrete and the deck work together to carry subsequent floor loads. These shear connector devices can be rolled-in embossments, lugs, holes, or wires welded to the panels. The deck profile can also be used to interlock concrete and steel.

Composite deck finishes are either galvanized (zinc coated) or phosphatized/painted. Galvanized deck has a zinc coating on both the top and bottom surfaces. The phosphatized/painted deck has a bare (phosphatized) top surface that will come into contact with the concrete. This bare top surface can be expected to develop rust before the concrete is placed. The bottom side of the deck has a primer coat of paint.

Composite floor deck is normally installed so the panel ends do not overlap on the supporting beams. Shear lugs or panel profile shape often prevent a tight metal to metal fit if the panel ends overlap; the air gap caused by overlapping will prevent proper fusion with the structural steel supports when the panel end laps are shear stud welded.

Adequate end bearing of the deck must be obtained as shown on the drawings. If bearing is actually less in the field than shown on the drawings, further investigation is required.

Roof Deck

Roof deck is not designed to act compositely with other materials. Roof deck acts alone in transferring horizontal and vertical loads into the building frame. Roof deck rib openings are usually narrower than floor deck rib openings. This provides adequate support of rigid thermal insulation board.

Roof deck is typically installed to endlap approximately 2″ over supports. However, it can be butted (or lapped more than 2″) to solve field fit problems. Since designers frequently use the installed deck system as part of the horizontal bracing system (the deck as a diaphragm), any fastening substitution or change should be approved by the designer. Continuous perimeter support of the deck is necessary to limit edge deflection in the finished roof and may be required for diaphragm shear transfer.

Standard roof deck finishes are galvanized or primer painted. The standard factory applied paint for roof deck is a primer paint and is not intended to weather for extended periods of time. Field painting or touching up of abrasions and deterioration of the primer coat or other protective finishes is the responsibility of the contractor.

Cellular Deck

Cellular deck is made by attaching a bottom steel sheet to a roof deck or composite floor deck panel. Cellular deck can be used in the same manner as floor deck. Electrical, telephone, and data wires are easily run through the chase created between the deck panel and the bottom sheet.

When used as part of the electrical distribution system, the cellular deck must be installed so that the ribs line up and create a smooth cell transition at abutting ends. The joint that occurs at butting cell ends must be taped or otherwise sealed to prevent wet concrete from seeping into the cell. Cell interiors must be free of welding burrs, or other sharp intrusions, to prevent damage to wires.

When used as a roof deck, the bottom flat plate is usually left exposed to view. Care must be maintained during erection to keep good alignment and prevent damage.

Cellular deck is sometimes used with the flat plate on the top side to provide a flat working surface. Installation of the deck for this purpose requires special methods for attachment to the frame because the flat plate, now on the top, can prevent direct access to the deck material that is bearing on the structural steel. It may be advisable to treat the flat top surface to prevent slipping.

Cellular deck is always furnished galvanized or painted over galvanized.

Form Deck

Form deck can be any floor or roof deck product used as a concrete form. Connections to the frame are by the same methods used to anchor floor and roof deck. Welding washers are recommended when welding deck that is less than 20 gauge thickness.

Form deck is furnished galvanized, prime painted, or uncoated. Galvanized deck must be used for those roof deck systems where form deck is used to carry a lightweight insulating concrete fill.

For information about Means Estimating Seminars, see yellow pages 12 and 13 in back of book

R06100-010 Thirty City Lumber Prices

rices for boards are for #2 or better or sterling, whichever is in best supply. imension lumber is "Standard or Better" either Southern Yellow Pine S.Y.P.), Spruce-Pine-Fir (S.P.F.), Hem-Fir (H.F.) or Douglas Fir (D.F.). The pecies of lumber used in a geographic area is listed by city. Plyform is 3/4" B oil sealed fir or S.Y.P. whichever prevails locally, 3/4" CDX is S.Y.P. r Fir.

These are prices at the time of publication and should be checked against the current market price. Relative differences between cities will stay approximately constant.

City	Species	Contractor Purchases per M.B.F. S4S: Dimensions						Boards		Contractor Purchases per M.S.F.	
		2"x4"	2"x6"	2"x8"	2"x10"	2"x12"	4"x4"	1"x6"	1"x12"	3/4" Ext. Plyform	3/4" Thick CDX T&G
Atlanta	S.Y.P.	$428	$453	$523	$595	$675	$ 793	$ 970	$1,316	$1,223	$1,067
Baltimore	S.P.F.	536	566	654	743	843	991	1212	1645	1529	1334
Boston	S.P.F.	517	547	631	718	814	957	1171	1588	1476	1288
Buffalo	S.P.F.	519	549	634	721	818	961	1175	1594	1482	1293
Chicago	H.F.	519	549	634	721	818	961	1175	1594	1482	1293
Cincinnati	S.P.F.	536	566	654	743	843	991	1212	1645	1529	1334
Cleveland	S.P.F.	520	550	635	722	819	963	1178	1598	1485	1295
Columbus	S.P.F.	530	561	647	736	835	982	1201	1629	1514	1321
Dallas	S.Y.P.	431	456	527	599	680	798	977	1325	1232	1074
Denver	H.F.	517	547	631	718	814	957	1171	1588	1476	1288
Detroit	H.F.	531	562	649	738	837	983	1203	1632	1517	1323
Houston	S.Y.P.	427	452	521	593	673	790	967	1312	1219	1064
Indianapolis	S.P.F.	524	554	640	728	826	970	1187	1610	1497	1306
Kansas City	D.F.	517	547	631	718	814	957	1171	1588	1476	1288
Los Angeles	D.F.	527	557	644	732	831	976	1194	1620	1505	1313
Memphis	S.Y.P.	508	537	620	705	800	940	1150	1560	1450	1265
Milwaukee	H.F.	519	549	634	721	818	961	1175	1594	1482	1293
Minneapolis	H.F.	517	547	631	718	814	957	1171	1588	1476	1288
Nashville	S.Y.P.	508	537	620	705	800	940	1150	1560	1450	1265
New Orleans	S.Y.P.	440	465	537	611	693	814	996	1352	1256	1096
New York City	S.P.F.	532	563	650	739	839	985	1205	1635	1520	1326
Philadelphia	S.P.F.	532	563	650	739	839	985	1205	1635	1520	1326
Phoenix	D.F.	536	566	654	743	843	991	1212	1645	1529	1334
Pittsburgh	S.P.F.	530	561	647	736	835	982	1201	1629	1514	1321
St. Louis	H.F.	524	554	640	728	826	970	1187	1610	1497	1306
San Antonio	S.Y.P.	436	461	532	605	686	806	987	1338	1244	1085
San Diego	D.F.	533	563	650	739	839	986	1206	1636	1520	1326
San Francisco	D.F.	508	537	620	705	800	940	1150	1560	1450	1265
Seattle	D.F.	503	532	614	698	792	931	1139	1544	1436	1252
Washington, DC	S.P.F.	541	572	660	751	852	1001	1224	1661	1544	1347
Average		$508	$537	$620	$705	$800	$ 940	$1,150	$1,560	$1,450	$1,265

To convert square feet of surface to board feet, 4% waste included.

S4S Size	Multiply S.F. by	T & G Size	Multiply S.F. by	Flooring Size	Multiply S.F. by
				25/32" x 2 1/4"	1.37
1 x 4	1.18	1 x 4	1.27	25/32" x 3 1/4"	1.29
1 x 6	1.13	1 x 6	1.18	15/32" x 1 1/2"	1.54
1 x 8	1.11	1 x 8	1.14	1" x 3"	1.28
1 x 10	1.09	2 x 6	2.36	1" x 4"	1.24

R06110-030 Lumber Product Material Prices

The price of forest products fluctuates widely from location to location and from season to season depending upon economic conditions. The table below indicates National Average material prices in effect Jan. 1 of this book year. The table shows relative differences between various sizes, grades and species. These percentage differentials remain fairly constant even though lumber prices in general may change significantly during the year.

Availability of certain items depends upon geographic location and must be checked prior to firm price bidding.

National Average Contractor Price, Quantity Purchase

	Dimension Lumber, S4S, #2 & Better, KD					
	Species	2″x4″	2″x6″	2″x8″	2″x10″	2″x12″
Framing Lumber per MBF	Douglas Fir	$ 608	$ 540	$ 567	$ 527	$ 594
	Spruce	543	473	527	533	578
	Southern Yellow Pine	540	358	459	493	709
	Hem-Fir	486	473	493	520	473

Heavy Timbers, Fir	
3″ x 4″ thru 3″ x 12″	$872
4″ x 4″ thru 4″ x 12″	872
6″ x 6″ thru 6″ x 12″	1,060
8″ x 8″ thru 8″ x 12″	1,222
10″ x 10″ and 10″ x 12″	1,285

	S4S "D" Quality or Clear, KD					
	Species	1″x4″	1″x6″	1″x8″	1″x10″	1″x12″
Boards per MBF *See also Cedar Siding	Sugar Pine	$1,485	$1,667	$1,688	$2,329	$3,071
	Idaho Pine	979	986	1,033	1,013	1,087
	Engleman Spruce	905	1,168	1,127	1,202	1,715
	So. Yellow Pine	662	878	965	898	1,121
	Ponderosa Pine	1,033	1,175	1,046	1,249	2,025

S4S # 2 & Better or Sterling, KD					
Species	1″x4″	1″x6″	1″x8″	1″x10″	1″x12″
Sugar Pine	$648	$770	$ 790	$ 797	$1,073
Idaho Pine	628	986	1,033	1,013	1,019
Engleman Spruce	702	736	682	675	992
So. Yellow Pine	513	608	689	554	1,121
Ponderosa Pine	689	648	655	662	945

R06160-020 Plywood

There are two types of plywood used in construction: interior, which is moisture resistant but not waterproofed, and exterior, which is waterproofed.

The grade of the exterior surface of the plywood sheets is designated by the first letter: A, for smooth surface with patches allowed; B, for solid surface with patches and plugs allowed; C, which may be surface plugged or may have knot holes up to 1″ wide; and D, which is used only for interior type plywood and may have knot holes up to 2-1/2″ wide. "Structural Grade" is specifically designed for engineered applications such as box beams. All CC & DD grades have roof and floor spans marked on them.

Underlayment grade plywood runs from 1/4″ to 1-1/4″ thick. Thicknesses 5/8″ and over have optional tongue and groove joints which eliminates the need for blocking the edges. Underlayment 19/32″ and over may be referred to as Sturd-i-Floor.

The price of plywood can fluctuate widely due to geographic and economic conditions. When one or two local prices are known, the relative prices for other types and sizes may be found by direct factoring of the prices in the table below.

Typical uses for various plywood grades are as follows:

AA-AD Interior — cupboards, shelving, paneling, furniture
BB Plyform — concrete form plywood
CDX — wall and roof sheathing
Structural — box beams, girders, stressed skin panels
AA-AC Exterior — fences, signs, siding, soffits, etc.
Underlayment — base for resilient floor coverings
Overlaid HDO — high density for concrete forms & highway signs
Overlaid MDO — medium density for painting, siding, soffits & signs
303 Siding — exterior siding, textured, striated, embossed, etc.

Grade	National Average Price in Lots of 10 MSF, per MSF-January				
	Type	4′x8′	Type	4′x8′	4′x10′
Sanded Grade	1/4″ Interior AD	$ 779	1/4″ Exterior AC	$ 780	$ 779
	3/8″	869	3/8″	870	869
	1/2″	1,059	1/2″	1,060	1,059
	5/8″	1,215	5/8″	1,215	1,215
	3/4″	1,399	3/4″	1,399	1,339
	1″	2,014	1″	2,014	2,014
	1-1/4″	2,196	Exterior AA, add	225	230
	Interior AA, add	225	Exterior AB, add	180	195
			CD Structural 1	Underlayment	
Unsanded Grade 4′ x 8′ Sheets	5/16″ CDX	$ 630			
	3/8″	652		3/8″, 4′x8′ sheets	$ 869
	1/2″	619	3/8″ 4′x8′ sheets 340	1/2″	1,059
	5/8″	780	1/2″ 592	5/8″T&G	800
	3/4″	940	3/4″ 840	3/4″T&G	1,050
	3/4″ T&G	1,050		1-1/8″ 2-4-1T&G	1,185
Form Plywood	5/8″ Exterior, oiled BB, plyform	$1,364	5/8″ HDO (overlay 2 sides)		$2,340
	3/4″ Exterior, oiled BB, plyform	1,450	3/4″ HDO (overlay 2 sides)		2,410
Overlaid 4′x8′ Sheets	Overlay 2 Sides MDO		Overlay 1 Side MDO		
	3/8″ thick	$1,345	3/8″ thick		$1,250
	1/2″	1,576	1/2″		1,316
	5/8″	1,776	5/8″		1,360
	3/4″	2,034	3/4″		1,400
303 Siding	Fir, rough sawn, natural finish, 3/8″ thick	$ 730	Texture 1-11	5/8″ thick, Fir	$1,320
	Redwood	2,514		Redwood	2,650
	Cedar	2,380		Cedar	2,150
	Southern Yellow Pine	585		Southern Yellow	1,070
Waferboard/O.S.B.	1/4″ sheathing	$ 335	19/32″ T&G		$ 780
	7/16″ sheathing	624	23/32″ T&G		937

For 2 MSF to 10 MSF, add 10%. For less than 2 MSF, add 15%.

R06170-100 Wood Roof Trusses

Loading figures represent live load. An additional load of 10 psf on the top chord and 10 psf on the bottom chord is included in the truss design. Spacing is 24″ O.C.

Span in Feet	Cost per Truss for Different Live Loads and Roof Pitches					
	Flat	4 in 12 Pitch		5 in 12 Pitch		8 in 12 Pitch
	40 psf	30 psf	40 psf	30 psf	40 psf	30 psf
20	$ 64	$ 54	$ 51	$ 59	$ 54	$ 60
22	70	57	55	62	57	65
24	77	62	47	64	55	70
26	80	72	52	73	60	72
28	88	73	60	75	65	78
30	111	81	67	92	72	82
32	119	99	72	93	130	115
34	126	99	106	113	113	124
36	133	105	112	115	115	130
38	141	120	120	121	122	140
40	148	124	122	121	122	146

Thermal & Moist. Protection | R071 | Waterproofing & Dampproofing

R07110-010 1/2″ Pargetting (rough dampproofing plaster)

1:2-1/2 Mix, 4.5 C.F. Covers 100 S.F., 2 Coats, Waste Included	Regular Portland Cement		Waterproofed Portland Cement	
1.7 Lbs. integral waterproofing admixture			$.85 per lb.	$ 1.45
1.7 Bags Portland cement	$8.25 per bag	$ 14.03	8.25 per bag	14.03
4.25 C.F. sand	15.80 per C.Y.	2.49	15.80 per C.Y.	2.49
.40 Days of labor to mix and apply (crew D-1)	348.40 per day	139.36	348.40 per day	139.36
Total Bare Cost per 100 S.F.		$155.88		$157.33

Thermal & Moist. Protection | R073 | Shingles & Roofing Tiles

R07310-020 Roof Slate

16″, 18″ and 20″ are standard lengths, and slate usually comes in random widths. For standard 3/16″ thickness use 1-1/2″ copper nails. Allow for 3% breakage.

Quantities per Square	Unfading Vermont Colored	Weathering Sea Green	Buckingham, Virginia Black
Slate delivered (incl. punching)	$346.00	$330.00	$550.00
# 30 Felt, 2.5 lbs. copper nails	19.45	19.45	19.45
Slate roofer 4.6 hrs. @ $20.50 per hr.	94.30	94.30	94.30
Total Bare Cost per Square	$459.75	$443.75	$663.75

For information about Means Estimating Seminars, see yellow pages 12 and 13 in back of book

R07550-030 Modified Bitumen Roofing

The cost of modified bitumen roofing is highly dependent on the type of installation that is planned. Installation is based on the type of modifier used in the bitumen. The two most popular modifiers are atactic polypropylene (APP) and styrene butadiene styrene (SBS). The modifiers are added to heated bitumen during the manufacturing process to change its characteristics. A polyethylene, polyester or fiberglass reinforcing sheet is then sandwiched between layers of this bitumen. When completed, the result is a pre-assembled, built-up roof that has increased elasticity and weatherablility. Some manufacturers include a surfacing material such as ceramic or mineral granules, metal particles or sand.

The preferred method of adhering SBS-modified bitumen roofing to the substrate is with hot-mopped asphalt (much the same as built-up roofing). This installation method requires a tar kettle/pot to heat the asphalt, as well as the labor, tools and equipment necessary to distribute and spread the hot asphalt.

The alternative method for applying APP and SBS modified bitumen is as follows. A skilled installer uses a torch to melt a small pool of bitumen off the membrane. This pool must form across the entire roll for proper adhesion. The installer must unroll the roofing at a pace slow enough to melt the bitumen, but fast enough to prevent damage to the rest of the membrane.

Modified bitumen roofing provides the advantages of both built-up and single-ply roofing. Labor costs are reduced over those of built-up roofing because only a single ply is necessary. The elasticity of single-ply roofing is attained with the reinforcing sheet and polymer modifiers. Modifieds have some self-healing characteristics and because of their multi-layer construction, they offer the reliability and safety of built-up roofing.

Doors & Windows | R081 | Metal Doors

R08110-020 Steel Door Selection Guide

Standard steel doors are classified into four levels, as recommended by the Steel Door Institute in the chart below. Each of the four levels offers a range of construction models and designs, to meet architectural requirements for preference and appearance, including full flush, seamless, and stile & rail. Recommended minimum gauge requirements are also included.

For complete standard steel door construction specifications and available sizes, refer to the Steel Door Institute Technical Data Series, ANSI A250.8-98 (SDI-100), and ANSI A250.4-94 Test Procedure and Acceptance Criteria for Physical Endurance of Steel Door and Hardware Reinforcements.

Level		Model	Construction	For Full Flush or Seamless		
				Min. Gauge	Thickness (in)	Thickness (mm)
I	Standard Duty	1 2	Full Flush Seamless	20	0.032	0.8
II	Heavy Duty	1 2	FullFlush Seamless	18	0.042	1.0
III	Extra Heavy Duty	1 2 3	FullFlush Seamless *Stile&Rail	16	0.053	1.3
IV	Maximum Duty	1 2	FullFlush Seamless	14	0.067	1.6

*Stiles & rails are 16 gauge; flush panels, when specified, are 18 gauge

Doors & Windows | R085 | Windows

R08550-010 Double Hung Windows - Tilt Wash

Ponderosa pine and vinyl clad sash, exterior primed with double insulated, low E glass.

Description	2'-0" x 3'-0"				3'-0" x 4'-0"			
	Wood		Vinyl Clad		Wood		Vinyl Clad	
Window, 2 lights w/ screens & grilles		$239.50		$278.90		$313.50		$372.00
Interior trim set		$ 19.90		$ 19.90		$ 23.50		$ 23.50
Carpenter @ $24.00 per hr.	1.7 hr.	40.80	1.7 hr.	40.80	2 hr.	48.00	2 hr.	48.00
Complete bare cost in place		$300.20		$339.60		$385.00		$443.50

DOORS & WINDOWS 8

REFERENCE NOS.

R08550-200 Replacement Windows

Replacement windows are typically measured per United Inch.

United Inches are calculated by rounding the width and height of the window opening up to the nearest inch, then adding the two figures.

The labor cost for replacement windows includes removal of sash, existing sash balance or weights, parting bead where necessary and installation of new window.

Debris hauling and dump fees are not included.

R08700-100 Hinges

All closer equipped doors should have ball bearing hinges. Lead lined or extremely heavy doors require special strength hinges. Usually 1-1/2 pair of hinges are used per door up to 7′-6″ high openings. Table below shows typical hinge requirements.

Use Frequency	Type Hinge Required	Type of Opening	Type of Structure
High	Heavy weight ball bearing	Entrances Toilet Rooms	Banks, Office buildings, Schools, Stores & Theaters Office buildings and Schools
Average	Standard weight ball bearing	Entrances Corridors Toilet Rooms	Dwellings Office buildings and Schools Stores
Low	Plain bearing	Interior	Dwellings

Door Thickness	Weight of Doors in Pounds per Square Foot				
	White Pine	Oak	Hollow Core	Solid Core	Hollow Metal
1-3/8″	3 psf	6 psf	1-1/2 psf	3-1/2 — 4 psf	6-1/2 psf
1-3/4″	3-1/2	7	2-1/2	4-1/2 — 5-1/4	6-1/2
2-1/4″	4-1/2	9	—	5-1/2 — 6-3/4	6-1/2

For information about Means Estimating Seminars, see yellow pages 12 and 13 in back of book

R09250-050 Lath, Plaster and Gypsum Board

Gypsum board lath is available in 3/8″ thick x 16″ wide x 4′ long sheets as a base material for multi-layer plaster applications. It is also available as a base for either multi-layer or veneer plaster applications in 1/2″ and 5/8″ thick–4′ wide x 8′, 10′ or 12′ long sheets. Fasteners are screws or blued ring shank nails for wood framing and screws for metal framing.

Metal lath is available in diamond mesh pattern with flat or self-furring profiles. Paper backing is available for applications where excessive plaster waste needs to be avoided. A slotted mesh ribbed lath should be used in areas where the span between structural supports is greater than normal. Most metal lath comes in 27″ x 96″ sheets. Diamond mesh weighs 1.75, 2.5 or 3.4 pounds per square yard, slotted mesh lath weighs 2.75 or 3.4 pounds per square yard. Metal lath can be nailed, screwed or tied in place.

Many **accessories** are available. Corner beads, flat reinforcing strips, casing beads, control and expansion joints, furring brackets and channels are some examples. Note that accessories are not included in plaster or stucco line items.

Plaster is defined as a material or combination of materials that when mixed with a suitable amount of water, forms a plastic mass or paste. When applied to a surface, the paste adheres to it and subsequently hardens, preserving in a rigid state the form or texture imposed during the period of elasticity.

Gypsum plaster is made from ground calcined gypsum. It is mixed with aggregates and water for use as a base coat plaster.

Vermiculite plaster is a fire-retardant plaster covering used on steel beams, concrete slabs and other heavy construction materials. Vermiculite is a group name for certain clay minerals, hydrous silicates or aluminum, magnesium and iron that have been expanded by heat.

Perlite plaster is a plaster using perlite as an aggregate instead of sand. Perlite is a volcanic glass that has been expanded by heat.

Gauging plaster is a mix of gypsum plaster and lime putty that when applied produces a quick drying finish coat.

Veneer plaster is a one or two component gypsum plaster used as a thin finish coat over special gypsum board.

Keenes cement is a white cementitious material manufactured from gypsum that has been burned at a high temperature and ground to a fine powder. Alum is added to accelerate the set. The resulting plaster is hard and strong and accepts and maintains a high polish, hence it is used as a finishing plaster.

Stucco is a Portland cement based plaster used primarily as an exterior finish.

Plaster is used on both interior and exterior surfaces. Generally it is applied in multiple-coat systems. A three-coat system uses the terms scratch, brown and finish to identify each coat. A two-coat system uses base and finish to describe each coat. Each type of plaster and application system has attributes that are chosen by the designer to best fit the intended use.

Gypsum Plaster Quantities for 100 S.Y.	2 Coat, 5/8″ Thick		3 Coat, 3/4″ Thick		
	Base	Finish	Scratch	Brown	Finish
	1:3 Mix	2:1 Mix	1:2 Mix	1:3 Mix	2:1 Mix
Gypsum plaster	1,300 lb.		1,350 lb.	650 lb.	
Sand	1.75 C.Y.		1.85 C.Y.	1.35 C.Y.	
Finish hydrated lime		340 lb.			340 lb.
Gauging plaster		170 lb.			170 lb.

Vermiculite or Perlite Plaster Quantities for 100 S.Y.	2 Coat, 5/8″ Thick		3 Coat, 3/4″ Thick		
	Base	Finish	Scratch	Brown	Finish
Gypsum plaster	1,250 lb.		1,450 lb.	800 lb.	
Vermiculite or perlite	7.8 bags		8.0 bags	3.3 bags	
Finish hydrated lime		340 lb.			340 lb.
Gauging plaster		170 lb.			170 lb.

Stucco–Three Coat System Quantities for 100 S.Y.	On Wood Frame	On Masonry
Portland cement	29 bags	21 bags
Sand	2.6 C.Y.	2.0 C.Y.
Hydrated lime	180 lb.	120 lb.

R09250-100 Levels of Gypsum Drywall Finish

In the past, contract documents often used phrases such as "industry standard" and "workmanlike finish" to specify the expected quality of gypsum board wall and ceiling installations. The vagueness of these descriptions led to unacceptable work and disputes.

In order to resolve this problem, four major trade associations concerned with the manufacture, erection, finish and decoration of gypsum board wall and ceiling systems have developed an industry-wide *Recommended Levels of Gypsum Board Finish.*

The finish of gypsum board walls and ceilings for specific final decoration is dependent on a number of factors. A primary consideration is the location of the surface and the degree of decorative treatment desired. Painted and unpainted surfaces in warehouses and other areas where appearance is normally not critical may simply require the taping of wallboard joints and 'spotting' of fastener heads. Blemish-free, smooth, monolithic surfaces often intended for painted and decorated walls and ceilings in habitated structures, ranging from single-family dwellings through monumental buildings, require additional finishing prior to the application of the final decoration.

Other factors to be considered in determining the level of finish of the gypsum board surface are (1) the type of angle of surface illumination (both natural and artificial lighting), and (2) the paint and method of application or the type and finish of wallcovering specified as the final decoration. Critical lighting conditions, gloss paints, and thin wallcoverings require a higher level of gypsum board finish than do heavily textured surfaces which are subsequently painted or surfaces which are to be decorated with heavy grade wallcoverings.

The following descriptions were developed jointly by the Association of the Wall and Ceiling Industries-International (AWCI), Ceiling & Interior Systems Construction Association (CISCA), Gypsum Association (GA), and Painting and Decorating Contractors of America (PDCA) as a guide.

Level 0: No taping, finishing, or accessories required. This level of finish may be useful in temporary construction or whenever the final decoration has not been determined.

Level 1: All joints and interior angles shall have tape set in joint compound. Surface shall be free of excess joint compound. Tool marks and ridges are acceptable. Frequently specified in plenum areas above ceilings, in attics, in areas where the assembly would generally be concealed or in building service corridors, and other areas not normally open to public view.

Level 2: All joints and interior angles shall have tape embedded in joint compound and wiped with a joint knife leaving a thin coating of joint compound over all joints and interior angles. Fastener heads and accessories shall be covered with a coat of joint compound. Surface shall be free of excess joint compound. Tool marks and ridges are acceptable. Joint compound applied over the body of the tape at the time of tape embedment shall be considered a separate coat of joint compound and shall satisfy the conditions of this level. Specified where water-resistant gypsum backing board is used as a substrate for tile; may be specified in garages, warehouse storage, or other similar areas where surface appearance is not of primary concern.

Level 3: All joints and interior angles shall have tape embedded in joint compound and one additional coat of joint compound applied over all joints and interior angles. Fastener heads and accessories shall be covered with two separate coats of joint compound. All joint compound shall be smooth and free of tool marks and ridges. Typically specified in appearance areas which are to receive heavy- or medium-texture (spray or hand applied) finishes before final painting, or where heavy-grade wallcoverings are to be applied as the final decoration. This level of finish is not recommended where smooth painted surfaces or light to medium wallcoverings are specified.

Level 4: All joints and interior angles shall have tape embedded in joint compound and two separate coats of joint compound applied over all flat joints and one separate coat of joint compound applied over interior angles. Fastener heads and accessories shall be covered with three separate coats of joint compound. All joint compound shall be smooth and free of tool marks and ridges. This level should be specified where flat paints, light textures, or wallcoverings are to be applied. In critical lighting areas, flat paints applied over light textures tend to reduce joint photographing. Gloss, semi-gloss, and enamel paints are not recommended over this level of finish. The weight, texture, and sheen level of wallcoverings applied over this level of finish should be carefully evaluated. Joints and fasteners must be adequately concealed if the wallcovering material is lightweight, contains limited pattern, has a gloss finish, or any combination of these finishes is present. Unbacked vinyl wallcoverings are not recommended over this level of finish.

Level 5: All joints and interior angles shall have tape embedded in joint compound and two separate coats of joint compound applied over all flat joints and one separate coat of joint compound applied over interior angles. Fastener heads and accessories shall be covered with three separate coats of joint compound. A thin skim coat of joint compound or a material manufactured especially for this purpose, shall be applied to the entire surface. The surface shall be smooth and free of tool marks and ridges. This level of finish is highly recommended where gloss, semi-gloss, enamel, or nontextured flat paints are specified or where severe lighting conditions occur. This highest quality finish is the most effective method to provide a uniform surface and minimize the possibility of joint photographing and of fasteners showing through the final decoration.

R09700-700 Wall Covering

Quantities for 100 S.F.		Medium Price Paper			Expensive Paper		
		Quantities	Bare Cost	Incl. O & P	Quantities	Bare Cost	Incl. O & P
Paper @	$30.00 and $57.00 per double roll	1.6 dbl. rolls	$48.00	$ 52.80	1.6 dbl. rolls	$ 91.20	$100.32
Wall sizing @	$16.50 per gallon	.25 gallon	4.13	4.54	.25 gallon	4.13	4.54
Vinyl wall paste @	$ 8.60 per gallon	.6 gallon	5.16	5.68	.6 gallon	5.16	5.68
Apply sizing @	$21.50 and $35.35 per hour	0.3 hour	6.45	10.61	0.3 hour	6.45	10.61
Apply paper @	$21.50 and $35.35 per hour	1.2 hours	25.80	42.42	1.5 hours	32.25	53.03
Total cost in place per 100 S.F.			$89.54	$116.05		$139.19	$174.18
Total cost in place per double roll			$55.96	$ 72.53		$ 86.99	$108.86

Most wallpapers now come in double rolls only.
To remove old paper, allow 1.3 hours per 100 S.F.

R09910-220 Painting

Item	Coat	One Gallon Covers			In 8 Hours a Laborer Covers			Labor Hours per 100 S.F.		
		Brush	Roller	Spray	Brush	Roller	Spray	Brush	Roller	Spray
Paint wood siding	prime	250 S.F.	225 S.F.	290 S.F.	1150 S.F.	1300 S.F.	2275 S.F.	.695	.615	.351
	others	270	250	290	1300	1625	2600	.615	.492	.307
Paint exterior trim	prime	400	—	—	650	—	—	1.230	—	—
	1st	475	—	—	800	—	—	1.000	—	—
	2nd	520	—	—	975	—	—	.820	—	—
Paint shingle siding	prime	270	255	300	650	975	1950	1.230	.820	.410
	others	360	340	380	800	1150	2275	1.000	.695	.351
Stain shingle siding	1st	180	170	200	750	1125	2250	1.068	.711	.355
	2nd	270	250	290	900	1325	2600	.888	.603	.307
Paint brick masonry	prime	180	135	160	750	800	1800	1.066	1.000	.444
	1st	270	225	290	815	975	2275	.981	.820	.351
	2nd	340	305	360	815	1150	2925	.981	.695	.273
Paint interior plaster or drywall	prime	400	380	495	1150	2000	3250	.695	.400	.246
	others	450	425	495	1300	2300	4000	.615	.347	.200
Paint interior doors and windows	prime	400	—	—	650	—	—	1.230	—	—
	1st	425	—	—	800	—	—	1.000	—	—
	2nd	450	—	—	975	—	—	.820	—	—

Special Construction | R131 | Pre-Engin. Struct. & Aquatic Facil.

R13128-520 Swimming Pools

Pool prices given per square foot of surface area include pool structure, filter and chlorination equipment, pumps, related piping, ladders/steps, maintenance kit, skimmer and vacuum system. Decks and electrical service to equipment are not included.

Residential in-ground pool construction can be divided into two categories: vinyl lined and gunite. Vinyl lined pool walls are constructed of different materials including wood, concrete, plastic or metal. The bottom is often graded with sand over which the vinyl liner is installed. Vermiculite or soil cement bottoms may be substituted for an added cost.

Gunite pool construction is used both in residential and municipal installations. These structures are steel reinforced for strength and finished with a white cement limestone plaster.

Municipal pools will have a higher cost because plumbing codes require more expensive materials, chlorination equipment and higher filtration rates.

Municipal pools greater than 1,800 S.F. require gutter systems to control waves. This gutter may be formed into the concrete wall. Often a vinyl/stainless steel gutter or gutter/wall system is specified, which will raise the pool cost.

Competition pools usually require tile bottoms and sides with contrasting lane striping, which will also raise the pool cost.

Special Construction | R136 | Solar & Wind Energy Equip.

R13600-610 Solar Heating (Space and Hot Water)

Collectors should face as close to due South as possible, however, variations of up to 20 degrees on either side of true South are acceptable. Local climate and collector type may influence the choice between east or west deviations. Obviously they should be located so they are not shaded from the sun's rays. Incline collectors at a slope of latitude minus 5 degrees for domestic hot water and latitude plus 15 degrees for space heating.

Flat plate collectors consist of a number of components as follows: Insulation to reduce heat loss through the bottom and sides of the collector. The enclosure which contains all the components in this assembly is usually weatherproof and prevents dust, wind and water from coming in contact with the absorber plate. The cover plate usually consists of one or more layers of a variety of glass or plastic and reduces the reradiation by creating an air space which traps the heat between the cover and the absorber plates.

The absorber plate must have a good thermal bond with the fluid passages. The absorber plate is usually metallic and treated with a surface coating which improves absorptivity. Black or dark paints or selective coatings are used for this purpose, and the design of this passage and plate combination helps determine a solar system's effectiveness.

Heat transfer fluid passage tubes are attached above and below or integral with an absorber plate for the purpose of transferring thermal energy from the absorber plate to a heat transfer medium. The heat exchanger is a device for transferring thermal energy from one fluid to another.

Piping and storage tanks should be well insulated to minimize heat losses.

Size domestic water heating storage tanks to hold 20 gallons of water per user, minimum, plus 10 gallons per dishwasher or washing machine. For domestic water heating an optimum collector size is approximately 3/4 square foot of area per gallon of water storage. For space heating of residences and small commercial applications the collector is commonly sized between 30% and 50% of the internal floor area. For space heating of large commercial applications, collector areas less than 30% of the internal floor area can still provide significant heat reductions.

A supplementary heat source is recommended for Northern states for December through February.

The solar energy transmission per square foot of collector surface varies greatly with the material used. Initial cost, heat transmittance and useful life are obviously interrelated.

For information about Means Estimating Seminars, see yellow pages 12 and 13 in back of book

R15100-050 Pipe Material Considerations

1. Malleable fittings should be used for gas service.
2. Malleable fittings are used where there are stresses/strains due to expansion and vibration.
3. Cast fittings may be broken as an aid to disassembling of heating lines frozen by long use, temperature and minerals.
4. Cast iron pipe is extensively used for underground and submerged service.
5. Type M (light wall) copper tubing is available in hard temper only and is used for nonpressure and less severe applications than K and L.
6. Type L (medium wall) copper tubing, available hard or soft for interior service.
7. Type K (heavy wall) copper tubing, available in hard or soft temper for use where conditions are severe. For underground and interior service.
8. Hard drawn tubing requires fewer hangers or supports but should not be bent. Silver brazed fittings are recommended, however soft solder is normally used.
9. Type DMV (very light wall) copper tubing designed for drainage, waste and vent plus other non-critical pressure services.

Domestic/Imported Pipe and Fittings Cost

The prices shown in this publication for steel/cast iron pipe and steel, cast iron, malleable iron fittings are based on domestic production sold at the normal trade discounts. The above listed items of foreign manufacture may be available at prices of 1/3 to 1/2 those shown. Some imported items after minor machining or finishing operations are being sold as domestic to further complicate the system.

Caution: Most pipe prices in this book also include a coupling and pipe hangers which for the larger sizes can add significantly to the per foot cost and should be taken into account when comparing "book cost" with quoted supplier's cost.

R15100-420 Plumbing Fixture Installation Time

Item	Rough In	Set	Total Hours	Item	Rough In	Set	Total Hours
Bathtub	5	5	10	Shower head only	2	1	3
Bathtub and shower, cast iron	6	6	12	Shower drain	3	1	4
Fire hose reel and cabinet	4	2	6	Shower stall, slate		15	15
Floor drain to 4 inch diameter	3	1	4	Slop sink	5	3	8
Grease trap, single, cast iron	5	3	8	Test 6 fixtures			14
Kitchen gas range		4	4	Urinal, wall	6	2	8
Kitchen sink, single	4	4	8	Urinal, pedestal or floor	6	4	10
Kitchen sink, double	6	6	12	Water closet and tank	4	3	7
Laundry tubs	4	2	6	Water closet and tank, wall hung	5	3	8
Lavatory wall hung	5	3	8	Water heater, 45 gals. gas, automatic	5	2	7
Lavatory pedestal	5	3	8	Water heaters, 65 gals. gas, automatic	5	2	7
Shower and stall	6	4	10	Water heaters, electric, plumbing only	4	2	6

Fixture prices in front of book are based on the cost per fixture set in place. The rough-in cost, which must be added for each fixture, includes carrier, if required, some supply, waste and vent pipe connecting fittings and stops. The lengths of rough-in pipe are nominal runs which would connect to the larger runs and stacks. The supply runs and DWV runs and stacks must be accounted for in separate entries. In the eastern half of the United States it is common for the plumber to carry these to a point 5′ outside the building.

Crew No.	Bare Costs Hr.	Bare Costs Daily	Incl. Subs O & P Hr.	Incl. Subs O & P Daily	Cost Per Labor-Hour Bare Costs	Cost Per Labor-Hour Incl. O&P
Crew A-1	**Hr.**	**Daily**	**Hr.**	**Daily**	**Bare Costs**	**Incl. O&P**
1 Building Laborer	$17.35	$138.80	$29.45	$235.60	$17.35	$29.45
1 Concrete saw, gas manual		49.80		54.80	6.23	6.85
8 L.H., Daily Totals		$188.60		$290.40	$23.58	$36.30

Crew A-1A	Hr.	Daily	Hr.	Daily	Bare Costs	Incl. O&P
1 Skilled Worker	$23.90	$191.20	$40.50	$324.00	$23.90	$40.50
1 Shot Blaster, 20″		338.35		372.20	42.29	46.52
8 L.H., Daily Totals		$529.55		$696.20	$66.19	$87.02

Crew A-1B	Hr.	Daily	Hr.	Daily	Bare Costs	Incl. O&P
1 Building Laborer	$17.35	$138.80	$29.45	$235.60	$17.35	$29.45
1 Concr. saw, gas, self prop.		114.40		125.85	14.30	15.73
8 L.H., Daily Totals		$253.20		$361.45	$31.65	$45.18

Crew A-1C	Hr.	Daily	Hr.	Daily	Bare Costs	Incl. O&P
1 Building Laborer	$17.35	$138.80	$29.45	$235.60	$17.35	$29.45
1 Brush saw		19.20		21.10	2.40	2.64
8 L.H., Daily Totals		$158.00		$256.70	$19.75	$32.09

Crew A-1D	Hr.	Daily	Hr.	Daily	Bare Costs	Incl. O&P
1 Building Laborer	$17.35	$138.80	$29.45	$235.60	$17.35	$29.45
1 Vibrating plate, gas, 18″		25.20		27.70	3.15	3.47
8 L.H., Daily Totals		$164.00		$263.30	$20.50	$32.92

Crew A-1E	Hr.	Daily	Hr.	Daily	Bare Costs	Incl. O&P
1 Building Laborer	$17.35	$138.80	$29.45	$235.60	$17.35	$29.45
1 Vibrating plate, gas, 21″		36.80		40.50	4.60	5.06
8 L.H., Daily Totals		$175.60		$276.10	$21.95	$34.51

Crew A-1F	Hr.	Daily	Hr.	Daily	Bare Costs	Incl. O&P
1 Building Laborer	$17.35	$138.80	$29.45	$235.60	$17.35	$29.45
1 Rammer/tamper, gas, 8″		34.80		38.30	4.35	4.79
8 L.H., Daily Totals		$173.60		$273.90	$21.70	$34.24

Crew A-1G	Hr.	Daily	Hr.	Daily	Bare Costs	Incl. O&P
1 Building Laborer	$17.35	$138.80	$29.45	$235.60	$17.35	$29.45
1 Rammer/tamper, gas, 8″		34.80		38.30	4.35	4.79
8 L.H., Daily Totals		$173.60		$273.90	$21.70	$34.24

Crew A-1H	Hr.	Daily	Hr.	Daily	Bare Costs	Incl. O&P
1 Building Laborer	$17.35	$138.80	$29.45	$235.60	$17.35	$29.45
1 Pressure washer		55.00		60.50	6.88	7.56
8 L.H., Daily Totals		$193.80		$296.10	$24.23	$37.01

Crew A-1J	Hr.	Daily	Hr.	Daily	Bare Costs	Incl. O&P
1 Building Laborer	$17.35	$138.80	$29.45	$235.60	$17.35	$29.45
1 Rototiller		84.50		92.95	10.56	11.61
8 L.H., Daily Totals		$223.30		$328.55	$27.91	$41.06

Crew A-1K	Hr.	Daily	Hr.	Daily	Bare Costs	Incl. O&P
1 Building Laborer	$17.35	$138.80	$29.45	$235.60	$17.35	$29.45
1 Lawn aerator		23.95		26.35	2.99	3.29
8 L.H., Daily Totals		$162.75		$261.95	$20.34	$32.74

Crew A-1L	Hr.	Daily	Hr.	Daily	Bare Costs	Incl. O&P
1 Building Laborer	$17.35	$138.80	$29.45	$235.60	$17.35	$29.45
1 Power blower/vacuum		23.95		26.35	2.99	3.29
8 L.H., Daily Totals		$162.75		$261.95	$20.34	$32.74

Crew A-1M	Hr.	Daily	Hr.	Daily	Bare Costs	Incl. O&P
1 Building Laborer	$17.35	$138.80	$29.45	$235.60	$17.35	$29.45
1 Snow blower		84.50		92.95	10.56	11.61
8 L.H., Daily Totals		$223.30		$328.55	$27.91	$41.06

Crew A-2	Hr.	Daily	Hr.	Daily	Bare Costs	Incl. O&P
2 Laborers	$17.35	$277.60	$29.45	$471.20	$17.92	$30.20
1 Truck Driver (light)	19.05	152.40	31.70	253.60		
1 Light Truck, 1.5 Ton		128.80		141.70	5.37	5.90
24 L.H., Daily Totals		$558.80		$866.50	$23.29	$36.10

Crew A-2A	Hr.	Daily	Hr.	Daily	Bare Costs	Incl. O&P
2 Laborers	$17.35	$277.60	$29.45	$471.20	$17.92	$30.20
1 Truck Driver (light)	19.05	152.40	31.70	253.60		
1 Light Truck, 1.5 Ton		128.80		141.70		
1 Concrete Saw		114.40		125.85	10.13	11.15
24 L.H., Daily Totals		$673.20		$992.35	$28.05	$41.35

Crew A-3	Hr.	Daily	Hr.	Daily	Bare Costs	Incl. O&P
1 Truck Driver (heavy)	$19.55	$156.40	$32.55	$260.40	$19.55	$32.55
1 Dump Truck, 12 Ton		326.60		359.25	40.83	44.91
8 L.H., Daily Totals		$483.00		$619.65	$60.38	$77.46

Crew A-3A	Hr.	Daily	Hr.	Daily	Bare Costs	Incl. O&P
1 Truck Driver (light)	$19.05	$152.40	$31.70	$253.60	$19.05	$31.70
1 Pickup Truck (4x4)		85.20		93.70	10.65	11.72
8 L.H., Daily Totals		$237.60		$347.30	$29.70	$43.42

Crew A-3B	Hr.	Daily	Hr.	Daily	Bare Costs	Incl. O&P
1 Equip. Oper. (medium)	$23.90	$191.20	$39.35	$314.80	$21.73	$35.95
1 Truck Driver (heavy)	19.55	156.40	32.55	260.40		
1 Dump Truck, 16 Ton		476.60		524.25		
1 F.E. Loader, 3 C.Y.		305.00		335.50	48.85	53.74
16 L.H., Daily Totals		$1129.20		$1434.95	$70.58	$89.69

Crew A-3C	Hr.	Daily	Hr.	Daily	Bare Costs	Incl. O&P
1 Equip. Oper. (light)	$22.80	$182.40	$37.55	$300.40	$22.80	$37.55
1 Wheeled Skid Steer Loader		197.80		217.60	24.73	27.20
8 L.H., Daily Totals		$380.20		$518.00	$47.53	$64.75

Crew A-3D	Hr.	Daily	Hr.	Daily	Bare Costs	Incl. O&P
1 Truck Driver, Light	$19.05	$152.40	$31.70	$253.60	$19.05	$31.70
1 Pickup Truck (4x4)		85.20		93.70		
1 Flatbed Trailer, 25 Ton		88.00		96.80	21.65	23.82
8 L.H., Daily Totals		$325.60		$444.10	$40.70	$55.52

Crew A-3E	Hr.	Daily	Hr.	Daily	Bare Costs	Incl. O&P
1 Equip. Oper. (crane)	$24.75	$198.00	$40.75	$326.00	$22.15	$36.65
1 Truck Driver (heavy)	19.55	156.40	32.55	260.40		
1 Pickup Truck (4x4)		85.20		93.70	5.33	5.86
16 L.H., Daily Totals		$439.60		$680.10	$27.48	$42.51

Crew A-3F	Hr.	Daily	Hr.	Daily	Bare Costs	Incl. O&P
1 Equip. Oper. (crane)	$24.75	$198.00	$40.75	$326.00	$22.15	$36.65
1 Truck Driver (heavy)	19.55	156.40	32.55	260.40		
1 Pickup Truck (4x4)		85.20		93.70		
1 Tractor, 6x2, 40 Ton Cap.		301.60		331.75		
1 Lowbed Trailer, 75 Ton		170.20		187.20	34.81	38.29
16 L.H., Daily Totals		$911.40		$1199.05	$56.96	$74.94

Crew No.	Bare Costs		Incl. Subs O & P		Cost Per Labor-Hour	

Crew A-3G	Hr.	Daily	Hr.	Daily	Bare Costs	Incl. O&P
Equip. Oper. (crane)	$24.75	$198.00	$40.75	$326.00	$22.15	$36.65
Truck Driver (heavy)	19.55	156.40	32.55	260.40		
Pickup Truck (4x4)		85.20		93.70		
Tractor, 6x4, 45 Ton Cap.		353.60		388.95		
Lowbed Trailer, 75 Ton		170.20		187.20	38.06	41.87
6 L.H., Daily Totals		$963.40		$1256.25	$60.21	$78.52

Crew A-4	Hr.	Daily	Hr.	Daily	Bare Costs	Incl. O&P
Carpenters	$24.00	$384.00	$40.75	$652.00	$23.25	$39.08
Painter, Ordinary	21.75	174.00	35.75	286.00		
24 L.H., Daily Totals		$558.00		$938.00	$23.25	$39.08

Crew A-5	Hr.	Daily	Hr.	Daily	Bare Costs	Incl. O&P
2 Laborers	$17.35	$277.60	$29.45	$471.20	$17.54	$29.70
.25 Truck Driver (light)	19.05	38.10	31.70	63.40		
.25 Light Truck, 1.5 Ton		32.20		35.40	1.79	1.97
18 L.H., Daily Totals		$347.90		$570.00	$19.33	$31.67

Crew A-6	Hr.	Daily	Hr.	Daily	Bare Costs	Incl. O&P
1 Instrument Man	$23.90	$191.20	$40.50	$324.00	$23.68	$39.63
1 Rodman/Chainman	23.45	187.60	38.75	310.00		
1 Laser Transit/Level		60.75		66.85	3.79	4.17
16 L.H., Daily Totals		$439.55		$700.85	$27.47	$43.80

Crew A-7	Hr.	Daily	Hr.	Daily	Bare Costs	Incl. O&P
1 Chief Of Party	$28.30	$226.40	$48.10	$384.80	$25.22	$42.45
1 Instrument Man	23.90	191.20	40.50	324.00		
1 Rodman/Chainman	23.45	187.60	38.75	310.00		
1 Laser Transit/Level		60.75		66.85	2.53	2.78
24 L.H., Daily Totals		$665.95		$1085.65	$27.75	$45.23

Crew A-8	Hr.	Daily	Hr.	Daily	Bare Costs	Incl. O&P
1 Chief Of Party	$28.30	$226.40	$48.10	$384.80	$24.78	$41.53
1 Instrument Man	23.90	191.20	40.50	324.00		
2 Rodmen/Chainmen	23.45	375.20	38.75	620.00		
1 Laser Transit/Level		60.75		66.85	1.90	2.09
32 L.H., Daily Totals		$853.55		$1395.65	$26.68	$43.62

Crew A-9	Hr.	Daily	Hr.	Daily	Bare Costs	Incl. O&P
1 Asbestos Foreman	$25.00	$200.00	$42.95	$343.60	$24.56	$42.21
7 Asbestos Workers	24.50	1372.00	42.10	2357.60		
64 L.H., Daily Totals		$1572.00		$2701.20	$24.56	$42.21

Crew A-10	Hr.	Daily	Hr.	Daily	Bare Costs	Incl. O&P
1 Asbestos Foreman	$25.00	$200.00	$42.95	$343.60	$24.56	$42.21
7 Asbestos Workers	24.50	1372.00	42.10	2357.60		
64 L.H., Daily Totals		$1572.00		$2701.20	$24.56	$42.21

Crew A-10A	Hr.	Daily	Hr.	Daily	Bare Costs	Incl. O&P
1 Asbestos Foreman	$25.00	$200.00	$42.95	$343.60	$24.67	$42.38
2 Asbestos Workers	24.50	392.00	42.10	673.60		
24 L.H., Daily Totals		$592.00		$1017.20	$24.67	$42.38

Crew A-10B	Hr.	Daily	Hr.	Daily	Bare Costs	Incl. O&P
1 Asbestos Foreman	$25.00	$200.00	$42.95	$343.60	$24.63	$42.31
3 Asbestos Workers	24.50	588.00	42.10	1010.40		
32 L.H., Daily Totals		$788.00		$1354.00	$24.63	$42.31

Crew No.	Bare Costs		Incl. Subs O & P		Cost Per Labor-Hour	

Crew A-10C	Hr.	Daily	Hr.	Daily	Bare Costs	Incl. O&P
3 Asbestos Workers	$24.50	$588.00	$42.10	$1010.40	$24.50	$42.10
1 Flatbed Truck		128.80		141.70	5.37	5.90
24 L.H., Daily Totals		$716.80		$1152.10	$29.87	$48.00

Crew A-10D	Hr.	Daily	Hr.	Daily	Bare Costs	Incl. O&P
2 Asbestos Workers	$24.50	$392.00	$42.10	$673.60	$23.63	$39.79
1 Equip. Oper. (crane)	24.75	198.00	40.75	326.00		
1 Equip. Oper. Oiler	20.75	166.00	34.20	273.60		
1 Hydraulic Crane, 33 Ton		641.60		705.75	20.05	22.06
32 L.H., Daily Totals		$1397.60		$1978.95	$43.68	$61.85

Crew A-11	Hr.	Daily	Hr.	Daily	Bare Costs	Incl. O&P
1 Asbestos Foreman	$25.00	$200.00	$42.95	$343.60	$24.56	$42.21
7 Asbestos Workers	24.50	1372.00	42.10	2357.60		
2 Chipping Hammers		32.00		35.20	.50	.55
64 L.H., Daily Totals		$1604.00		$2736.40	$25.06	$42.76

Crew A-12	Hr.	Daily	Hr.	Daily	Bare Costs	Incl. O&P
1 Asbestos Foreman	$25.00	$200.00	$42.95	$343.60	$24.56	$42.21
7 Asbestos Workers	24.50	1372.00	42.10	2357.60		
1 Large Prod. Vac. Loader		480.00		528.00	7.50	8.25
64 L.H., Daily Totals		$2052.00		$3229.20	$32.06	$50.46

Crew A-13	Hr.	Daily	Hr.	Daily	Bare Costs	Incl. O&P
1 Equip. Oper. (light)	$22.80	$182.40	$37.55	$300.40	$22.80	$37.55
1 Large Prod. Vac. Loader		480.00		528.00	60.00	66.00
8 L.H., Daily Totals		$662.40		$828.40	$82.80	$103.55

Crew B-1	Hr.	Daily	Hr.	Daily	Bare Costs	Incl. O&P
1 Labor Foreman (outside)	$19.35	$154.80	$32.85	$262.80	$18.02	$30.58
2 Laborers	17.35	277.60	29.45	471.20		
24 L.H., Daily Totals		$432.40		$734.00	$18.02	$30.58

Crew B-1A	Hr.	Daily	Hr.	Daily	Bare Costs	Incl. O&P
1 Laborer Foreman	$19.35	$154.80	$32.85	$262.80	$18.02	$30.58
2 Laborers	17.35	277.60	29.45	471.20		
2 Cutting Torches		36.00		39.60		
2 Gases		129.60		142.55	6.90	7.59
24 L.H., Daily Totals		$598.00		$916.15	$24.92	$38.17

Crew B-1B	Hr.	Daily	Hr.	Daily	Bare Costs	Incl. O&P
1 Laborer Foreman	$19.35	$154.80	$32.85	$262.80	$19.70	$33.13
2 Laborers	17.35	277.60	29.45	471.20		
1 Equip. Oper. (crane)	24.75	198.00	40.75	326.00		
2 Cutting Torches		36.00		39.60		
2 Gases		129.60		142.55		
1 Hyd. Crane, 12 Ton		602.60		662.85	24.01	26.41
32 L.H., Daily Totals		$1398.60		$1905.00	$43.71	$59.54

Crew B-2	Hr.	Daily	Hr.	Daily	Bare Costs	Incl. O&P
1 Labor Foreman (outside)	$19.35	$154.80	$32.85	$262.80	$17.75	$30.13
4 Laborers	17.35	555.20	29.45	942.40		
40 L.H., Daily Totals		$710.00		$1205.20	$17.75	$30.13

Crew No.	Bare Costs		Incl. Subs O & P		Cost Per Labor-Hour	
Crew B-3	**Hr.**	**Daily**	**Hr.**	**Daily**	**Bare Costs**	**Incl. O&P**
1 Labor Foreman (outside)	$19.35	$154.80	$32.85	$262.80	$19.51	$32.70
2 Laborers	17.35	277.60	29.45	471.20		
1 Equip. Oper. (med.)	23.90	191.20	39.35	314.80		
2 Truck Drivers (heavy)	19.55	312.80	32.55	520.80		
1 F.E. Loader, T.M., 2.5 C.Y.		788.00		866.80		
2 Dump Trucks, 16 Ton		953.20		1048.50	36.28	39.90
48 L.H., Daily Totals		$2677.60		$3484.90	$55.79	$72.60

Crew B-3A	**Hr.**	**Daily**	**Hr.**	**Daily**	**Bare Costs**	**Incl. O&P**
4 Laborers	$17.35	$555.20	$29.45	$942.40	$18.66	$31.43
1 Equip. Oper. (med.)	23.90	191.20	39.35	314.80		
1 Hyd. Excavator, 1.5 C.Y.		720.40		792.45	18.01	19.81
40 L.H., Daily Totals		$1466.80		$2049.65	$36.67	$51.24

Crew B-3B	**Hr.**	**Daily**	**Hr.**	**Daily**	**Bare Costs**	**Incl. O&P**
2 Laborers	$17.35	$277.60	$29.45	$471.20	$19.54	$32.70
1 Equip. Oper. (med.)	23.90	191.20	39.35	314.80		
1 Truck Driver (heavy)	19.55	156.40	32.55	260.40		
1 Backhoe Loader, 80 H.P.		256.40		282.05		
1 Dump Truck, 16 Ton		476.60		524.25	22.91	25.20
32 L.H., Daily Totals		$1358.20		$1852.70	$42.45	$57.90

Crew B-3C	**Hr.**	**Daily**	**Hr.**	**Daily**	**Bare Costs**	**Incl. O&P**
3 Laborers	$17.35	$416.40	$29.45	$706.80	$18.99	$31.93
1 Equip. Oper. (med.)	23.90	191.20	39.35	314.80		
1 F.E. Crawler Ldr, 4 C.Y.		1094.00		1203.40	34.19	37.61
32 L.H., Daily Totals		$1701.60		$2225.00	$53.18	$69.54

Crew B-4	**Hr.**	**Daily**	**Hr.**	**Daily**	**Bare Costs**	**Incl. O&P**
1 Labor Foreman (outside)	$19.35	$154.80	$32.85	$262.80	$18.05	$30.53
4 Laborers	17.35	555.20	29.45	942.40		
1 Truck Driver (heavy)	19.55	156.40	32.55	260.40		
1 Tractor, 4 x 2, 195 H.P.		208.20		229.00		
1 Platform Trailer		119.40		131.35	6.83	7.51
48 L.H., Daily Totals		$1194.00		$1825.95	$24.88	$38.04

Crew B-5	**Hr.**	**Daily**	**Hr.**	**Daily**	**Bare Costs**	**Incl. O&P**
1 Labor Foreman (outside)	$19.35	$154.80	$32.85	$262.80	$19.06	$32.11
3 Laborers	17.35	416.40	29.45	706.80		
1 Equip. Oper. (med.)	23.90	191.20	39.35	314.80		
1 Air Compr., 250 C.F.M.		112.40		123.65		
2 Air Tools & Accessories		19.60		21.55		
2 50 Ft. Air Hoses, 1.5" Dia.		9.40		10.35		
1 F.E. Loader, T.M., 2.5 C.Y.		788.00		866.80	23.24	25.56
40 L.H., Daily Totals		$1691.80		$2306.75	$42.30	$57.67

Crew B-5A	**Hr.**	**Daily**	**Hr.**	**Daily**	**Bare Costs**	**Incl. O&P**
1 Foreman	$19.35	$154.80	$32.85	$262.80	$19.43	$32.58
6 Laborers	17.35	832.80	29.45	1413.60		
2 Equip. Oper. (med.)	23.90	382.40	39.35	629.60		
1 Equip. Oper. (light)	22.80	182.40	37.55	300.40		
2 Truck Drivers (heavy)	19.55	312.80	32.55	520.80		
1 Air Compr. 365 C.F.M.		146.80		161.50		
2 Pavement Breakers		19.60		21.55		
8 Air Hoses w/Coup.,1"		29.20		32.10		
2 Dump Trucks, 12 Ton		653.20		718.50	8.84	9.73
96 L.H., Daily Totals		$2714.00		$4060.85	$28.27	$42.31

Crew No.	Bare Costs		Incl. Subs O & P		Cost Per Labor-Hou	
Crew B-5B	**Hr.**	**Daily**	**Hr.**	**Daily**	**Bare Costs**	**Incl. O&P**
1 Powderman	$23.90	$191.20	$40.50	$324.00	$21.73	$36.14
2 Equip. Oper. (med.)	23.90	382.40	39.35	629.60		
3 Truck Drivers (heavy)	19.55	469.20	32.55	781.20		
1 F.E. Ldr. 2 1/2 CY		305.00		335.50		
3 Dump Trucks, 16 Ton		1429.80		1572.80		
1 Air Compr. 365 C.F.M.		146.80		161.50	39.20	43.12
48 L.H., Daily Totals		$2924.40		$3804.60	$60.93	$79.26

Crew B-5C	**Hr.**	**Daily**	**Hr.**	**Daily**	**Bare Costs**	**Incl. O&P**
3 Laborers	$17.35	$416.40	$29.45	$706.80	$20.07	$33.47
1 Equip. Oper. (medium)	23.90	191.20	39.35	314.80		
2 Truck Drivers (heavy)	19.55	312.80	32.55	520.80		
1 Equip. Oper. (crane)	24.75	198.00	40.75	326.00		
1 Equip. Oper. Oiler	20.75	166.00	34.20	273.60		
2 Dump Trucks, 16 Ton		953.20		1048.50		
1 F.E. Crawler Ldr, 4 C.Y.		1094.00		1203.40		
1 Hyd. Crane, 25 Ton		575.00		632.50	40.97	45.07
64 L.H., Daily Totals		$3906.60		$5026.40	$61.04	$78.54

Crew B-6	**Hr.**	**Daily**	**Hr.**	**Daily**	**Bare Costs**	**Incl. O&P**
2 Laborers	$17.35	$277.60	$29.45	$471.20	$19.17	$32.15
1 Equip. Oper. (light)	22.80	182.40	37.55	300.40		
1 Backhoe Loader, 48 H.P.		215.80		237.40	8.99	9.89
24 L.H., Daily Totals		$675.80		$1009.00	$28.16	$42.04

Crew B-6B	**Hr.**	**Daily**	**Hr.**	**Daily**	**Bare Costs**	**Incl. O&P**
2 Labor Foreman (out)	$19.35	$309.60	$32.85	$525.60	$18.02	$30.58
4 Laborers	17.35	555.20	29.45	942.40		
1 Winch Truck		317.00		348.70		
1 Flatbed Truck		128.80		141.70		
1 Butt Fusion Machine		432.80		476.10	18.30	20.13
48 L.H., Daily Totals		$1743.40		$2434.50	$36.32	$50.71

Crew B-7	**Hr.**	**Daily**	**Hr.**	**Daily**	**Bare Costs**	**Incl. O&P**
1 Labor Foreman (outside)	$19.35	$154.80	$32.85	$262.80	$18.78	$31.67
4 Laborers	17.35	555.20	29.45	942.40		
1 Equip. Oper. (med.)	23.90	191.20	39.35	314.80		
1 Chipping Machine		164.80		181.30		
1 F.E. Loader, T.M., 2.5 C.Y.		788.00		866.80		
2 Chain Saws, 36"		66.80		73.50	21.24	23.37
48 L.H., Daily Totals		$1920.80		$2641.60	$40.02	$55.04

Crew B-7A	**Hr.**	**Daily**	**Hr.**	**Daily**	**Bare Costs**	**Incl. O&P**
2 Laborers	$17.35	$277.60	$29.45	$471.20	$19.17	$32.15
1 Equip. Oper. (light)	22.80	182.40	37.55	300.40		
1 Rake w/Tractor		191.50		210.65		
2 Chain Saws, 18"		38.40		42.25	9.58	10.54
24 L.H., Daily Totals		$689.90		$1024.50	$28.75	$42.69

Crew B-8	**Hr.**	**Daily**	**Hr.**	**Daily**	**Bare Costs**	**Incl. O&P**
1 Labor Foreman (outside)	$19.35	$154.80	$32.85	$262.80	$20.14	$33.65
2 Laborers	17.35	277.60	29.45	471.20		
2 Equip. Oper. (med.)	23.90	382.40	39.35	629.60		
2 Truck Drivers (heavy)	19.55	312.80	32.55	520.80		
1 Hyd. Crane, 25 Ton		616.80		678.50		
1 F.E. Loader, T.M., 2.5 C.Y.		788.00		866.80		
2 Dump Trucks, 16 Ton		953.20		1048.50	42.11	46.32
56 L.H., Daily Totals		$3485.60		$4478.20	$62.25	$79.97

Crew No.	Bare Costs		Incl. Subs O & P		Cost Per Labor-Hour	
Crew B-9	**Hr.**	**Daily**	**Hr.**	**Daily**	**Bare Costs**	**Incl. O&P**
Labor Foreman (outside)	$19.35	$154.80	$32.85	$262.80	$17.75	$30.13
Laborers	17.35	555.20	29.45	942.40		
Air Compr., 250 C.F.M.		112.40		123.65		
Air Tools & Accessories		19.60		21.55		
50 Ft. Air Hoses, 1.5″ Dia.		9.40		10.35	3.54	3.89
0 L.H., Daily Totals		$851.40		$1360.75	$21.29	$34.02
Crew B-9A	**Hr.**	**Daily**	**Hr.**	**Daily**	**Bare Costs**	**Incl. O&P**
Laborers	$17.35	$277.60	$29.45	$471.20	$18.08	$30.48
Truck Driver (heavy)	19.55	156.40	32.55	260.40		
Water Tanker		116.80		128.50		
Tractor		208.20		229.00		
50 Ft. Disch. Hoses		5.00		5.50	13.75	15.13
4 L.H., Daily Totals		$764.00		$1094.60	$31.83	$45.61
Crew B-9B	**Hr.**	**Daily**	**Hr.**	**Daily**	**Bare Costs**	**Incl. O&P**
Laborers	$17.35	$277.60	$29.45	$471.20	$18.08	$30.48
Truck Driver (heavy)	19.55	156.40	32.55	260.40		
50 Ft. Disch. Hoses		5.00		5.50		
Water Tanker		116.80		128.50		
Tractor		208.20		229.00		
Pressure Washer		46.20		50.80	15.68	17.24
4 L.H., Daily Totals		$810.20		$1145.40	$33.76	$47.72
Crew B-9C	**Hr.**	**Daily**	**Hr.**	**Daily**	**Bare Costs**	**Incl. O&P**
1 Labor Foreman (outside)	$19.35	$154.80	$32.85	$262.80	$17.75	$30.13
4 Laborers	17.35	555.20	29.45	942.40		
1 Air Compr., 250 C.F.M.		112.40		123.65		
2 50 Ft. Air Hoses, 1.5″ Dia.		9.40		10.35		
2 Breaker, Pavement, 60 lb.		19.60		21.55	3.54	3.89
40 L.H., Daily Totals		$851.40		$1360.75	$21.29	$34.02
Crew B-9D	**Hr.**	**Daily**	**Hr.**	**Daily**	**Bare Costs**	**Incl. O&P**
1 Labor Foreman (Outside)	$19.35	$154.80	$32.85	$262.80	$17.75	$30.13
4 Common Laborers	17.35	555.20	29.45	942.40		
1 Air Compressor, 250 CFM		112.40		123.65		
2 Air hoses, 1.5″ x 50′		9.40		10.35		
2 Air tamper		55.30		60.85	4.43	4.87
40 L.H., Daily Totals		$887.10		$1400.05	$22.18	$35.00
Crew B-10	**Hr.**	**Daily**	**Hr.**	**Daily**	**Bare Costs**	**Incl. O&P**
1 Equip. Oper. (med.)	$23.90	$191.20	$39.35	$314.80	$23.90	$39.35
8 L.H., Daily Totals		$191.20		$314.80	$23.90	$39.35
Crew B-10A	**Hr.**	**Daily**	**Hr.**	**Daily**	**Bare Costs**	**Incl. O&P**
1 Equip. Oper. (med.)	$23.90	$191.20	$39.35	$314.80	$23.90	$39.35
1 Walk behind compactor, 7.5 HP		122.80		135.10	15.35	16.89
8 L.H., Daily Totals		$314.00		$449.90	$39.25	$56.24
Crew B-10B	**Hr.**	**Daily**	**Hr.**	**Daily**	**Bare Costs**	**Incl. O&P**
1 Equip. Oper. (med.)	$23.90	$191.20	$39.35	$314.80	$23.90	$39.35
1 Dozer, 200 H.P.		919.60		1011.55	114.95	126.45
8 L.H., Daily Totals		$1110.80		$1326.35	$138.85	$165.80
Crew B-10C	**Hr.**	**Daily**	**Hr.**	**Daily**	**Bare Costs**	**Incl. O&P**
1 Equip. Oper. (med.)	$23.90	$191.20	$39.35	$314.80	$23.90	$39.35
1 Dozer, 200 H.P.		919.60		1011.55		
1 Vibratory Roller, Towed		591.60		650.75	188.90	207.79
8 L.H., Daily Totals		$1702.40		$1977.10	$212.80	$247.14

Crew No.	Bare Costs		Incl. Subs O & P		Cost Per Labor-Hour	
Crew B-10D	**Hr.**	**Daily**	**Hr.**	**Daily**	**Bare Costs**	**Incl. O&P**
1 Equip. Oper. (med.)	$23.90	$191.20	$39.35	$314.80	$23.90	$39.35
1 Dozer, 200 H.P.		919.60		1011.55		
1 Sheepsft. Roller, Towed		610.20		671.20	191.23	210.35
8 L.H., Daily Totals		$1721.00		$1997.55	$215.13	$249.70
Crew B-10E	**Hr.**	**Daily**	**Hr.**	**Daily**	**Bare Costs**	**Incl. O&P**
1 Equip. Oper. (med.)	$23.90	$191.20	$39.35	$314.80	$23.90	$39.35
1 Tandem Roller, 5 Ton		107.40		118.15	13.43	14.77
8 L.H., Daily Totals		$298.60		$432.95	$37.33	$54.12
Crew B-10F	**Hr.**	**Daily**	**Hr.**	**Daily**	**Bare Costs**	**Incl. O&P**
1 Equip. Oper. (med.)	$23.90	$191.20	$39.35	$314.80	$23.90	$39.35
1 Tandem Roller, 10 Ton		179.20		197.10	22.40	24.64
8 L.H., Daily Totals		$370.40		$511.90	$46.30	$63.99
Crew B-10G	**Hr.**	**Daily**	**Hr.**	**Daily**	**Bare Costs**	**Incl. O&P**
1 Equip. Oper. (med.)	$23.90	$191.20	$39.35	$314.80	$23.90	$39.35
1 Sheepsft. Roll., 130 H.P.		794.80		874.30	99.35	109.29
8 L.H., Daily Totals		$986.00		$1189.10	$123.25	$148.64
Crew B-10H	**Hr.**	**Daily**	**Hr.**	**Daily**	**Bare Costs**	**Incl. O&P**
1 Equip. Oper. (med.)	$23.90	$191.20	$39.35	$314.80	$23.90	$39.35
1 Diaphr. Water Pump, 2″		50.60		55.65		
1 20 Ft. Suction Hose, 2″		2.55		2.80		
2 50 Ft. Disch. Hoses, 2″		3.80		4.20	7.12	7.83
8 L.H., Daily Totals		$248.15		$377.45	$31.02	$47.18
Crew B-10I	**Hr.**	**Daily**	**Hr.**	**Daily**	**Bare Costs**	**Incl. O&P**
1 Equip. Oper. (med.)	$23.90	$191.20	$39.35	$314.80	$23.90	$39.35
1 Diaphr. Water Pump, 4″		74.40		81.85		
1 20 Ft. Suction Hose, 4″		5.05		5.55		
2 50 Ft. Disch. Hoses, 4″		7.10		7.80	10.82	11.90
8 L.H., Daily Totals		$277.75		$410.00	$34.72	$51.25
Crew B-10J	**Hr.**	**Daily**	**Hr.**	**Daily**	**Bare Costs**	**Incl. O&P**
1 Equip. Oper. (med.)	$23.90	$191.20	$39.35	$314.80	$23.90	$39.35
1 Centr. Water Pump, 3″		53.60		58.95		
1 20 Ft. Suction Hose, 3″		4.25		4.70		
2 50 Ft. Disch. Hoses, 3″		5.00		5.50	7.86	8.64
8 L.H., Daily Totals		$254.05		$383.95	$31.76	$47.99
Crew B-10K	**Hr.**	**Daily**	**Hr.**	**Daily**	**Bare Costs**	**Incl. O&P**
1 Equip. Oper. (med.)	$23.90	$191.20	$39.35	$314.80	$23.90	$39.35
1 Centr. Water Pump, 6″		217.00		238.70		
1 20 Ft. Suction Hose, 6″		12.80		14.10		
2 50 Ft. Disch. Hoses, 6″		17.80		19.60	30.95	34.05
8 L.H., Daily Totals		$438.80		$587.20	$54.85	$73.40
Crew B-10L	**Hr.**	**Daily**	**Hr.**	**Daily**	**Bare Costs**	**Incl. O&P**
1 Equip. Oper. (med.)	$23.90	$191.20	$39.35	$314.80	$23.90	$39.35
1 Dozer, 80 H.P.		314.80		346.30	39.35	43.29
8 L.H., Daily Totals		$506.00		$661.10	$63.25	$82.64
Crew B-10M	**Hr.**	**Daily**	**Hr.**	**Daily**	**Bare Costs**	**Incl. O&P**
1 Equip. Oper. (med.)	$23.90	$191.20	$39.35	$314.80	$23.90	$39.35
1 Dozer, 300 H.P.		1195.00		1314.50	149.38	164.31
8 L.H., Daily Totals		$1386.20		$1629.30	$173.28	$203.66

Crew No.	Bare Costs		Incl. Subs O & P		Cost Per Labor-Hour	
Crew B-10N	**Hr.**	**Daily**	**Hr.**	**Daily**	**Bare Costs**	**Incl. O&P**
1 Equip. Oper. (med.)	$23.90	$191.20	$39.35	$314.80	$23.90	$39.35
1 F.E. Loader, T.M., 1.5 C.Y		312.60		343.85	39.08	42.98
8 L.H., Daily Totals		$503.80		$658.65	$62.98	$82.33
Crew B-10O	**Hr.**	**Daily**	**Hr.**	**Daily**	**Bare Costs**	**Incl. O&P**
1 Equip. Oper. (med.)	$23.90	$191.20	$39.35	$314.80	$23.90	$39.35
1 F.E. Loader, T.M., 2.25 C.Y.		551.80		607.00	68.98	75.87
8 L.H., Daily Totals		$743.00		$921.80	$92.88	$115.22
Crew B-10P	**Hr.**	**Daily**	**Hr.**	**Daily**	**Bare Costs**	**Incl. O&P**
1 Equip. Oper. (med.)	$23.90	$191.20	$39.35	$314.80	$23.90	$39.35
1 F.E. Loader, T.M., 2.5 C.Y.		788.00		866.80	98.50	108.35
8 L.H., Daily Totals		$979.20		$1181.60	$122.40	$147.70
Crew B-10Q	**Hr.**	**Daily**	**Hr.**	**Daily**	**Bare Costs**	**Incl. O&P**
1 Equip. Oper. (med.)	$23.90	$191.20	$39.35	$314.80	$23.90	$39.35
1 F.E. Loader, T.M., 5 C.Y.		1094.00		1203.40	136.75	150.43
8 L.H., Daily Totals		$1285.20		$1518.20	$160.65	$189.78
Crew B-10R	**Hr.**	**Daily**	**Hr.**	**Daily**	**Bare Costs**	**Incl. O&P**
1 Equip. Oper. (med.)	$23.90	$191.20	$39.35	$314.80	$23.90	$39.35
1 F.E. Loader, W.M., 1 C.Y.		193.80		213.20	24.23	26.65
8 L.H., Daily Totals		$385.00		$528.00	$48.13	$66.00
Crew B-10S	**Hr.**	**Daily**	**Hr.**	**Daily**	**Bare Costs**	**Incl. O&P**
1 Equip. Oper. (med.)	$23.90	$191.20	$39.35	$314.80	$23.90	$39.35
1 F.E. Loader, W.M., 1.5 C.Y.		241.00		265.10	30.13	33.14
8 L.H., Daily Totals		$432.20		$579.90	$54.03	$72.49
Crew B-10T	**Hr.**	**Daily**	**Hr.**	**Daily**	**Bare Costs**	**Incl. O&P**
1 Equip. Oper. (med.)	$23.90	$191.20	$39.35	$314.80	$23.90	$39.35
1 F.E. Loader, W.M.,2.5 C.Y.		305.00		335.50	38.13	41.94
8 L.H., Daily Totals		$496.20		$650.30	$62.03	$81.29
Crew B-10U	**Hr.**	**Daily**	**Hr.**	**Daily**	**Bare Costs**	**Incl. O&P**
1 Equip. Oper. (med.)	$23.90	$191.20	$39.35	$314.80	$23.90	$39.35
1 F.E. Loader, W.M., 5.5 C.Y.		700.40		770.45	87.55	96.31
8 L.H., Daily Totals		$891.60		$1085.25	$111.45	$135.66
Crew B-10V	**Hr.**	**Daily**	**Hr.**	**Daily**	**Bare Costs**	**Incl. O&P**
1 Equip. Oper. (med.)	$23.90	$191.20	$39.35	$314.80	$23.90	$39.35
1 Dozer, 700 H.P.		3139.00		3452.90	392.38	431.61
8 L.H., Daily Totals		$3330.20		$3767.70	$416.28	$470.96
Crew B-10W	**Hr.**	**Daily**	**Hr.**	**Daily**	**Bare Costs**	**Incl. O&P**
1 Equip. Oper. (med.)	$23.90	$191.20	$39.35	$314.80	$23.90	$39.35
1 Dozer, 105 H.P.		453.80		499.20	56.73	62.40
8 L.H., Daily Totals		$645.00		$814.00	$80.63	$101.75
Crew B-10X	**Hr.**	**Daily**	**Hr.**	**Daily**	**Bare Costs**	**Incl. O&P**
1 Equip. Oper. (med.)	$23.90	$191.20	$39.35	$314.80	$23.90	$39.35
1 Dozer, 410 H.P.		1524.00		1676.40	190.50	209.55
8 L.H., Daily Totals		$1715.20		$1991.20	$214.40	$248.90
Crew B-10Y	**Hr.**	**Daily**	**Hr.**	**Daily**	**Bare Costs**	**Incl. O&P**
1 Equip. Oper. (med.)	$23.90	$191.20	$39.35	$314.80	$23.90	$39.35
1 Vibratory Drum Roller		344.80		379.30	43.10	47.41
8 L.H., Daily Totals		$536.00		$694.10	$67.00	$86.76

Crew No.	Bare Costs		Incl. Subs O & P		Cost Per Labor-Hour	
Crew B-11A	**Hr.**	**Daily**	**Hr.**	**Daily**	**Bare Costs**	**Incl. O&P**
1 Equipment Oper. (med.)	$23.90	$191.20	$39.35	$314.80	$20.63	$34.40
1 Laborer	17.35	138.80	29.45	235.60		
1 Dozer, 200 H.P.		919.60		1011.55	57.48	63.22
16 L.H., Daily Totals		$1249.60		$1561.95	$78.11	$97.62
Crew B-11B	**Hr.**	**Daily**	**Hr.**	**Daily**	**Bare Costs**	**Incl. O&P**
1 Equipment Oper. (light)	$22.80	$182.40	$37.55	$300.40	$20.08	$33.50
1 Laborer	17.35	138.80	29.45	235.60		
1 Air Powered Tamper		27.65		30.40		
1 Air Compr. 365 C.F.M.		146.80		161.50		
2 50 Ft. Air Hoses, 1.5″ Dia.		9.40		10.35	11.49	12.64
16 L.H., Daily Totals		$505.05		$738.25	$31.57	$46.14
Crew B-11C	**Hr.**	**Daily**	**Hr.**	**Daily**	**Bare Costs**	**Incl. O&P**
1 Equipment Oper. (med.)	$23.90	$191.20	$39.35	$314.80	$20.63	$34.40
1 Laborer	17.35	138.80	29.45	235.60		
1 Backhoe Loader, 48 H.P.		215.80		237.40	13.49	14.84
16 L.H., Daily Totals		$545.80		$787.80	$34.12	$49.24
Crew B-11K	**Hr.**	**Daily**	**Hr.**	**Daily**	**Bare Costs**	**Incl. O&P**
1 Equipment Oper. (med.)	$23.90	$191.20	$39.35	$314.80	$20.63	$34.40
1 Laborer	17.35	138.80	29.45	235.60		
1 Trencher, 8′ D., 16″ W.		1420.00		1562.00	88.75	97.63
16 L.H., Daily Totals		$1750.00		$2112.40	$109.38	$132.03
Crew B-11L	**Hr.**	**Daily**	**Hr.**	**Daily**	**Bare Costs**	**Incl. O&P**
1 Equipment Oper. (med.)	$23.90	$191.20	$39.35	$314.80	$20.63	$34.40
1 Laborer	17.35	138.80	29.45	235.60		
1 Grader, 30,000 Lbs.		456.60		502.25	28.54	31.39
16 L.H., Daily Totals		$786.60		$1052.65	$49.17	$65.79
Crew B-11M	**Hr.**	**Daily**	**Hr.**	**Daily**	**Bare Costs**	**Incl. O&P**
1 Equipment Oper. (med.)	$23.90	$191.20	$39.35	$314.80	$20.63	$34.40
1 Laborer	17.35	138.80	29.45	235.60		
1 Backhoe Loader, 80 H.P.		256.40		282.05	16.03	17.63
16 L.H., Daily Totals		$586.40		$832.45	$36.66	$52.03
Crew B-11W	**Hr.**	**Daily**	**Hr.**	**Daily**	**Bare Costs**	**Incl. O&P**
1 Equipment Operator (med.)	$23.90	$191.20	$39.35	$314.80	$19.73	$32.86
1 Common Laborer	17.35	138.80	29.45	235.60		
10 Truck Drivers, Heavy	19.55	1564.00	32.55	2604.00		
1 Dozer, 200 H.P.		919.60		1011.55		
1 Vib. roller, smth, towed, 23 Ton		591.60		650.75		
10 Dump Truck, 10 Ton		3266.00		3592.60	49.76	54.74
96 L.H., Daily Totals		$6671.20		$8409.30	$69.49	$87.60
Crew B-11Y	**Hr.**	**Daily**	**Hr.**	**Daily**	**Bare Costs**	**Incl. O&P**
1 Labor Foreman (Outside)	$19.35	$154.80	$32.85	$262.80	$19.76	$33.13
5 Common Laborers	17.35	694.00	29.45	1178.00		
3 Equipment Operator (med.)	23.90	573.60	39.35	944.40		
1 Dozer, 80 H.P.		314.80		346.30		
2 Walk behind compactor, 7.5 HP		245.60		270.15		
4 Vibratory plate, gas, 21″		147.20		161.90	9.83	10.81
72 L.H., Daily Totals		$2130.00		$3163.55	$29.59	$43.94
Crew B-12A	**Hr.**	**Daily**	**Hr.**	**Daily**	**Bare Costs**	**Incl. O&P**
1 Equip. Oper. (crane)	$24.75	$198.00	$40.75	$326.00	$21.05	$35.10
1 Laborer	17.35	138.80	29.45	235.60		
1 Hyd. Excavator, 1 C.Y.		557.80		613.60	34.86	38.35
16 L.H., Daily Totals		$894.60		$1175.20	$55.91	$73.45

Crew No.	Bare Costs		Incl. Subs O & P		Cost Per Labor-Hour	
Crew B-12B	**Hr.**	**Daily**	**Hr.**	**Daily**	**Bare Costs**	**Incl. O&P**
Equip. Oper. (crane)	$24.75	$198.00	$40.75	$326.00	$21.05	$35.10
Laborer	17.35	138.80	29.45	235.60		
Hyd. Excavator, 1.5 C.Y.		720.40		792.45	45.03	49.53
6 L.H., Daily Totals		$1057.20		$1354.05	$66.08	$84.63

Crew B-12C	Hr.	Daily	Hr.	Daily	Bare Costs	Incl. O&P
Equip. Oper. (crane)	$24.75	$198.00	$40.75	$326.00	$21.05	$35.10
Laborer	17.35	138.80	29.45	235.60		
Hyd. Excavator, 2 C.Y.		909.80		1000.80	56.86	62.55
6 L.H., Daily Totals		$1246.60		$1562.40	$77.91	$97.65

Crew B-12D	Hr.	Daily	Hr.	Daily	Bare Costs	Incl. O&P
Equip. Oper. (crane)	$24.75	$198.00	$40.75	$326.00	$21.05	$35.10
Laborer	17.35	138.80	29.45	235.60		
Hyd. Excavator, 3.5 C.Y.		2020.00		2222.00	126.25	138.88
6 L.H., Daily Totals		$2356.80		$2783.60	$147.30	$173.98

Crew B-12E	Hr.	Daily	Hr.	Daily	Bare Costs	Incl. O&P
Equip. Oper. (crane)	$24.75	$198.00	$40.75	$326.00	$21.05	$35.10
Laborer	17.35	138.80	29.45	235.60		
Hyd. Excavator, .5 C.Y.		332.40		365.65	20.78	22.85
6 L.H., Daily Totals		$669.20		$927.25	$41.83	$57.95

Crew B-12F	Hr.	Daily	Hr.	Daily	Bare Costs	Incl. O&P
Equip. Oper. (crane)	$24.75	$198.00	$40.75	$326.00	$21.05	$35.10
1 Laborer	17.35	138.80	29.45	235.60		
1 Hyd. Excavator, .75 C.Y.		474.00		521.40	29.63	32.59
16 L.H., Daily Totals		$810.80		$1083.00	$50.68	$67.69

Crew B-12G	Hr.	Daily	Hr.	Daily	Bare Costs	Incl. O&P
1 Equip. Oper. (crane)	$24.75	$198.00	$40.75	$326.00	$21.05	$35.10
1 Laborer	17.35	138.80	29.45	235.60		
1 Power Shovel, .5 C.Y.		508.10		558.90		
1 Clamshell Bucket, .5 C.Y.		33.80		37.20	33.87	37.25
16 L.H., Daily Totals		$878.70		$1157.70	$54.92	$72.35

Crew B-12H	Hr.	Daily	Hr.	Daily	Bare Costs	Incl. O&P
1 Equip. Oper. (crane)	$24.75	$198.00	$40.75	$326.00	$21.05	$35.10
1 Laborer	17.35	138.80	29.45	235.60		
1 Power Shovel, 1 C.Y.		833.20		916.50		
1 Clamshell Bucket, 1 C.Y.		43.40		47.75	54.79	60.27
16 L.H., Daily Totals		$1213.40		$1525.85	$75.84	$95.37

Crew B-12I	Hr.	Daily	Hr.	Daily	Bare Costs	Incl. O&P
1 Equip. Oper. (crane)	$24.75	$198.00	$40.75	$326.00	$21.05	$35.10
1 Laborer	17.35	138.80	29.45	235.60		
1 Power Shovel, .75 C.Y.		630.45		693.50		
1 Dragline Bucket, .75 C.Y.		18.80		20.70	40.58	44.63
16 L.H., Daily Totals		$986.05		$1275.80	$61.63	$79.73

Crew B-12J	Hr.	Daily	Hr.	Daily	Bare Costs	Incl. O&P
1 Equip. Oper. (crane)	$24.75	$198.00	$40.75	$326.00	$21.05	$35.10
1 Laborer	17.35	138.80	29.45	235.60		
1 Gradall, 3 Ton, .5 C.Y.		837.80		921.60	52.36	57.60
16 L.H., Daily Totals		$1174.60		$1483.20	$73.41	$92.70

Crew No.	Bare Costs		Incl. Subs O & P		Cost Per Labor-Hour	
Crew B-12K	**Hr.**	**Daily**	**Hr.**	**Daily**	**Bare Costs**	**Incl. O&P**
1 Equip. Oper. (crane)	$24.75	$198.00	$40.75	$326.00	$21.05	$35.10
1 Laborer	17.35	138.80	29.45	235.60		
1 Gradall, 3 Ton, 1 C.Y.		971.00		1068.10	60.69	66.76
16 L.H., Daily Totals		$1307.80		$1629.70	$81.74	$101.86

Crew B-12L	Hr.	Daily	Hr.	Daily	Bare Costs	Incl. O&P
1 Equip. Oper. (crane)	$24.75	$198.00	$40.75	$326.00	$21.05	$35.10
1 Laborer	17.35	138.80	29.45	235.60		
1 Power Shovel, .5 C.Y.		508.10		558.90		
1 F.E. Attachment, .5 C.Y.		47.40		52.15	34.72	38.19
16 L.H., Daily Totals		$892.30		$1172.65	$55.77	$73.29

Crew B-12M	Hr.	Daily	Hr.	Daily	Bare Costs	Incl. O&P
1 Equip. Oper. (crane)	$24.75	$198.00	$40.75	$326.00	$21.05	$35.10
1 Laborer	17.35	138.80	29.45	235.60		
1 Power Shovel, .75 C.Y.		630.45		693.50		
1 F.E. Attachment, .75 C.Y.		51.80		57.00	42.64	46.90
16 L.H., Daily Totals		$1019.05		$1312.10	$63.69	$82.00

Crew B-12N	Hr.	Daily	Hr.	Daily	Bare Costs	Incl. O&P
1 Equip. Oper. (crane)	$24.75	$198.00	$40.75	$326.00	$21.05	$35.10
1 Laborer	17.35	138.80	29.45	235.60		
1 Power Shovel, 1 C.Y.		833.20		916.50		
1 F.E. Attachment, 1 C.Y.		58.60		64.45	55.74	61.31
16 L.H., Daily Totals		$1228.60		$1542.55	$76.79	$96.41

Crew B-12O	Hr.	Daily	Hr.	Daily	Bare Costs	Incl. O&P
1 Equip. Oper. (crane)	$24.75	$198.00	$40.75	$326.00	$21.05	$35.10
1 Laborer	17.35	138.80	29.45	235.60		
1 Power Shovel, 1.5 C.Y.		888.20		977.00		
1 F.E. Attachment, 1.5 C.Y.		67.80		74.60	59.75	65.73
16 L.H., Daily Totals		$1292.80		$1613.20	$80.80	$100.83

Crew B-12P	Hr.	Daily	Hr.	Daily	Bare Costs	Incl. O&P
1 Equip. Oper. (crane)	$24.75	$198.00	$40.75	$326.00	$21.05	$35.10
1 Laborer	17.35	138.80	29.45	235.60		
1 Crawler Crane, 40 Ton		888.20		977.00		
1 Dragline Bucket, 1.5 C.Y.		30.60		33.65	57.43	63.17
16 L.H., Daily Totals		$1255.60		$1572.25	$78.48	$98.27

Crew B-12Q	Hr.	Daily	Hr.	Daily	Bare Costs	Incl. O&P
1 Equip. Oper. (crane)	$24.75	$198.00	$40.75	$326.00	$21.05	$35.10
1 Laborer	17.35	138.80	29.45	235.60		
1 Hyd. Excavator, 5/8 C.Y.		432.40		475.65	27.03	29.73
16 L.H., Daily Totals		$769.20		$1037.25	$48.08	$64.83

Crew B-12R	Hr.	Daily	Hr.	Daily	Bare Costs	Incl. O&P
1 Equip. Oper. (crane)	$24.75	$198.00	$40.75	$326.00	$21.05	$35.10
1 Laborer	17.35	138.80	29.45	235.60		
1 Hyd. Excavator, 1.5 C.Y.		720.40		792.45	45.03	49.53
16 L.H., Daily Totals		$1057.20		$1354.05	$66.08	$84.63

Crew B-12S	Hr.	Daily	Hr.	Daily	Bare Costs	Incl. O&P
1 Equip. Oper. (crane)	$24.75	$198.00	$40.75	$326.00	$21.05	$35.10
1 Laborer	17.35	138.80	29.45	235.60		
1 Hyd. Excavator, 2.5 C.Y.		1215.00		1336.50	75.94	83.53
16 L.H., Daily Totals		$1551.80		$1898.10	$96.99	$118.63

Crew No.	Bare Costs		Incl. Subs O & P		Cost Per Labor-Hour	
Crew B-12T	**Hr.**	**Daily**	**Hr.**	**Daily**	**Bare Costs**	**Incl. O&P**
1 Equip. Oper. (crane)	$24.75	$198.00	$40.75	$326.00	$21.05	$35.10
1 Laborer	17.35	138.80	29.45	235.60		
1 Crawler Crane, 75 Ton		1170.00		1287.00		
1 F.E. Attachment, 3 C.Y.		90.20		99.20	78.76	86.64
16 L.H., Daily Totals		$1597.00		$1947.80	$99.81	$121.74

Crew B-12V	Hr.	Daily	Hr.	Daily	Bare Costs	Incl. O&P
1 Equip. Oper. (crane)	$24.75	$198.00	$40.75	$326.00	$21.05	$35.10
1 Laborer	17.35	138.80	29.45	235.60		
1 Crawler Crane, 75 Ton		1170.00		1287.00		
1 Dragline Bucket, 3 C.Y.		47.40		52.15	76.09	83.70
16 L.H., Daily Totals		$1554.20		$1900.75	$97.14	$118.80

Crew B-13	Hr.	Daily	Hr.	Daily	Bare Costs	Incl. O&P
1 Labor Foreman (outside)	$19.35	$154.80	$32.85	$262.80	$18.92	$31.90
4 Laborers	17.35	555.20	29.45	942.40		
1 Equip. Oper. (crane)	24.75	198.00	40.75	326.00		
1 Hyd. Crane, 25 Ton		616.80		678.50	12.85	14.14
48 L.H., Daily Totals		$1524.80		$2209.70	$31.77	$46.04

Crew B-13A	Hr.	Daily	Hr.	Daily	Bare Costs	Incl. O&P
1 Foreman	$19.35	$154.80	$32.85	$262.80	$20.14	$33.65
2 Laborers	17.35	277.60	29.45	471.20		
2 Equipment Operators	23.90	382.40	39.35	629.60		
2 Truck Drivers (heavy)	19.55	312.80	32.55	520.80		
1 Crane, 75 Ton		1170.00		1287.00		
1 F.E. Lder, 3.75 C.Y.		1094.00		1203.40		
2 Dump Trucks, 12 Ton		653.20		718.50	52.09	57.30
56 L.H., Daily Totals		$4044.80		$5093.30	$72.23	$90.95

Crew B-13B	Hr.	Daily	Hr.	Daily	Bare Costs	Incl. O&P
1 Labor Foreman (outside)	$19.35	$154.80	$32.85	$262.80	$19.18	$32.23
4 Laborers	17.35	555.20	29.45	942.40		
1 Equip. Oper. (crane)	24.75	198.00	40.75	326.00		
1 Equip. Oper. Oiler	20.75	166.00	34.20	273.60		
1 Hyd. Crane, 55 Ton		895.40		984.95	15.99	17.59
56 L.H., Daily Totals		$1969.40		$2789.75	$35.17	$49.82

Crew B-13C	Hr.	Daily	Hr.	Daily	Bare Costs	Incl. O&P
1 Labor Foreman (outside)	$19.35	$154.80	$32.85	$262.80	$19.18	$32.23
4 Laborers	17.35	555.20	29.45	942.40		
1 Equip. Oper. (crane)	24.75	198.00	40.75	326.00		
1 Equip. Oper. Oiler	20.75	166.00	34.20	273.60		
1 Crawler Crane, 100 Ton		1519.00		1670.90	27.13	29.84
56 L.H., Daily Totals		$2593.00		$3475.70	$46.31	$62.07

Crew B-14	Hr.	Daily	Hr.	Daily	Bare Costs	Incl. O&P
1 Labor Foreman (outside)	$19.35	$154.80	$32.85	$262.80	$18.59	$31.37
4 Laborers	17.35	555.20	29.45	942.40		
1 Equip. Oper. (light)	22.80	182.40	37.55	300.40		
1 Backhoe Loader, 48 H.P.		215.80		237.40	4.50	4.95
48 L.H., Daily Totals		$1108.20		$1743.00	$23.09	$36.32

Crew B-15	Hr.	Daily	Hr.	Daily	Bare Costs	Incl. O&P
1 Equipment Oper. (med)	$23.90	$191.20	$39.35	$314.80	$20.48	$34.05
.5 Laborer	17.35	69.40	29.45	117.80		
2 Truck Drivers (heavy)	19.55	312.80	32.55	520.80		
2 Dump Trucks, 16 Ton		953.20		1048.50		
1 Dozer, 200 H.P.		919.60		1011.55	66.89	73.57
28 L.H., Daily Totals		$2446.20		$3013.45	$87.37	$107.62

Crew B-16	Hr.	Daily	Hr.	Daily	Bare Costs	Incl. O&P
1 Labor Foreman (outside)	$19.35	$154.80	$32.85	$262.80	$18.40	$31.08
2 Laborers	17.35	277.60	29.45	471.20		
1 Truck Driver (heavy)	19.55	156.40	32.55	260.40		
1 Dump Truck, 16 Ton		476.60		524.25	14.89	16.38
32 L.H., Daily Totals		$1065.40		$1518.65	$33.29	$47.46

Crew B-17	Hr.	Daily	Hr.	Daily	Bare Costs	Incl. O&P
2 Laborers	$17.35	$277.60	$29.45	$471.20	$19.26	$32.25
1 Equip. Oper. (light)	22.80	182.40	37.55	300.40		
1 Truck Driver (heavy)	19.55	156.40	32.55	260.40		
1 Backhoe Loader, 48 H.P.		215.80		237.40		
1 Dump Truck, 12 Ton		326.60		359.25	16.95	18.65
32 L.H., Daily Totals		$1158.80		$1628.65	$36.21	$50.90

Crew B-17A	Hr.	Daily	Hr.	Daily	Bare Costs	Incl. O&P
2 Laborer Foreman	$19.35	$309.60	$32.85	$525.60	$19.26	$32.68
6 Laborers	17.35	832.80	29.45	1413.60		
1 Skilled Worker Foreman	25.90	207.20	43.90	351.20		
1 Skilled Worker	23.90	191.20	40.50	324.00		
80 L.H., Daily Totals		$1540.80		$2614.40	$19.26	$32.68

Crew B-18	Hr.	Daily	Hr.	Daily	Bare Costs	Incl. O&P
1 Labor Foreman (outside)	$19.35	$154.80	$32.85	$262.80	$18.02	$30.58
2 Laborers	17.35	277.60	29.45	471.20		
1 Vibrating Compactor		36.80		40.50	1.53	1.69
24 L.H., Daily Totals		$469.20		$774.50	$19.55	$32.27

Crew B-19	Hr.	Daily	Hr.	Daily	Bare Costs	Incl. O&P
1 Pile Driver Foreman	$25.30	$202.40	$45.25	$362.00	$22.94	$40.29
4 Pile Drivers	23.30	745.60	41.65	1332.80		
1 Equip. Oper. (crane)	24.75	198.00	40.75	326.00		
1 Building Laborer	17.35	138.80	29.45	235.60		
1 Crane, 40 Ton & Access.		888.20		977.00		
60 L.F. Pile Leads		75.00		82.50		
1 Hammer, Diesel, 22k Ft Lb		564.00		620.40	27.33	30.06
56 L.H., Daily Totals		$2812.00		$3936.30	$50.27	$70.35

Crew B-19A	Hr.	Daily	Hr.	Daily	Bare Costs	Incl. O&P
1 Pile Driver Foreman	$25.30	$202.40	$45.25	$362.00	$23.59	$40.94
4 Pile Drivers	23.30	745.60	41.65	1332.80		
2 Equip. Oper. (crane)	24.75	396.00	40.75	652.00		
1 Equip. Oper. Oiler	20.75	166.00	34.20	273.60		
1 Crawler Crane, 75 Ton		1170.00		1287.00		
60 L.F. Leads, 25K Ft. Lbs.		120.00		132.00		
1 Hammer, Diesel, 41k Ft Lb		662.60		728.85	30.51	33.56
64 L.H., Daily Totals		$3462.60		$4768.25	$54.10	$74.50

Crew B-20	Hr.	Daily	Hr.	Daily	Bare Costs	Incl. O&P
1 Labor Foreman (out)	$19.35	$154.80	$32.85	$262.80	$18.02	$30.58
2 Laborers	17.35	277.60	29.45	471.20		
24 L.H., Daily Totals		$432.40		$734.00	$18.02	$30.58

Crew B-20A	Hr.	Daily	Hr.	Daily	Bare Costs	Incl. O&P
1 Labor Foreman	$19.35	$154.80	$32.85	$262.80	$21.30	$35.50
1 Laborer	17.35	138.80	29.45	235.60		
1 Plumber	26.95	215.60	44.30	354.40		
1 Plumber Apprentice	21.55	172.40	35.40	283.20		
32 L.H., Daily Totals		$681.60		$1136.00	$21.30	$35.50

Crew B-21	Bare Costs Hr.	Bare Costs Daily	Incl. Subs O & P Hr.	Incl. Subs O & P Daily	Cost Per Labor-Hour Bare Costs	Cost Per Labor-Hour Incl. O&P
Labor Foreman (out)	$19.35	$154.80	$32.85	$262.80	$18.98	$32.04
2 Laborers	17.35	277.60	29.45	471.20		
.5 Equip. Oper. (crane)	24.75	99.00	40.75	163.00		
.5 S.P. Crane, 5 Ton		158.50		174.35	5.66	6.23
28 L.H., Daily Totals		$689.90		$1071.35	$24.64	$38.27

Crew B-21A	Hr.	Daily	Hr.	Daily	Bare Costs	Incl. O&P
1 Labor Foreman	$19.35	$154.80	$32.85	$262.80	$21.99	$36.55
1 Laborer	17.35	138.80	29.45	235.60		
1 Plumber	26.95	215.60	44.30	354.40		
1 Plumber Apprentice	21.55	172.40	35.40	283.20		
1 Equip. Oper. (crane)	24.75	198.00	40.75	326.00		
1 S.P. Crane, 12 Ton		505.80		556.40	12.65	13.91
40 L.H., Daily Totals		$1385.40		$2018.40	$34.64	$50.46

Crew B-21B	Hr.	Daily	Hr.	Daily	Bare Costs	Incl. O&P
1 Laborer Foreman	$19.35	$154.80	$32.85	$262.80	$19.23	$32.39
3 Laborers	17.35	416.40	29.45	706.80		
1 Equip. Oper. (crane)	24.75	198.00	40.75	326.00		
1 Hyd. Crane, 12 Ton		602.60		662.85	15.07	16.57
40 L.H., Daily Totals		$1371.80		$1958.45	$34.30	$48.96

Crew B-21C	Hr.	Daily	Hr.	Daily	Bare Costs	Incl. O&P
1 Laborer Foreman	$19.35	$154.80	$32.85	$262.80	$19.18	$32.23
4 Laborers	17.35	555.20	29.45	942.40		
1 Equip. Oper. (crane)	24.75	198.00	40.75	326.00		
1 Equip. Oper. Oiler	20.75	166.00	34.20	273.60		
2 Cutting Torches		36.00		39.60		
2 Gases		129.60		142.55		
1 Crane, 90 Ton		1325.00		1457.50	26.62	29.28
56 L.H., Daily Totals		$2564.60		$3444.45	$45.80	$61.51

Crew B-22	Hr.	Daily	Hr.	Daily	Bare Costs	Incl. O&P
1 Labor Foreman (out)	$19.35	$154.80	$32.85	$262.80	$19.36	$32.62
2 Laborers	17.35	277.60	29.45	471.20		
.75 Equip. Oper. (crane)	24.75	148.50	40.75	244.50		
.75 S.P. Crane, 5 Ton		237.75		261.50	7.93	8.72
30 L.H., Daily Totals		$818.65		$1240.00	$27.29	$41.34

Crew B-22A	Hr.	Daily	Hr.	Daily	Bare Costs	Incl. O&P
1 Labor Foreman (out)	$19.35	$154.80	$32.85	$262.80	$20.32	$34.28
1 Skilled Worker	23.90	191.20	40.50	324.00		
2 Laborers	17.35	277.60	29.45	471.20		
.75 Equipment Oper. (crane)	24.75	148.50	40.75	244.50		
.75 Crane, 5 Ton		237.75		261.50		
1 Generator, 5 KW		32.20		35.40		
1 Butt Fusion Machine		432.80		476.10	18.49	20.34
38 L.H., Daily Totals		$1474.85		$2075.50	$38.81	$54.62

Crew B-22B	Hr.	Daily	Hr.	Daily	Bare Costs	Incl. O&P
1 Skilled Worker	$23.90	$191.20	$40.50	$324.00	$20.63	$34.98
1 Laborer	17.35	138.80	29.45	235.60		
1 Electro Fusion Machine		171.80		189.00	10.74	11.81
16 L.H., Daily Totals		$501.80		$748.60	$31.37	$46.79

Crew B-23	Hr.	Daily	Hr.	Daily	Bare Costs	Incl. O&P
1 Labor Foreman (outside)	$19.35	$154.80	$32.85	$262.80	$17.75	$30.13
4 Laborers	17.35	555.20	29.45	942.40		
1 Drill Rig, Wells		3179.00		3496.90		
1 Light Truck, 3 Ton		176.20		193.80	83.88	92.27
40 L.H., Daily Totals		$4065.20		$4895.90	$101.63	$122.40

Crew B-23A	Hr.	Daily	Hr.	Daily	Bare Costs	Incl. O&P
1 Labor Foreman (outside)	$19.35	$154.80	$32.85	$262.80	$20.20	$33.88
1 Laborer	17.35	138.80	29.45	235.60		
1 Equip. Operator (medium)	23.90	191.20	39.35	314.80		
1 Drill Rig, Wells		3179.00		3496.90		
1 Pickup Truck, 3/4 Ton		78.00		85.80	135.71	149.28
24 L.H., Daily Totals		$3741.80		$4395.90	$155.91	$183.16

Crew B-23B	Hr.	Daily	Hr.	Daily	Bare Costs	Incl. O&P
1 Labor Foreman (outside)	$19.35	$154.80	$32.85	$262.80	$20.20	$33.88
1 Laborer	17.35	138.80	29.45	235.60		
1 Equip. Operator (medium)	23.90	191.20	39.35	314.80		
1 Drill Rig, Wells		3179.00		3496.90		
1 Pickup Truck, 3/4 Ton		78.00		85.80		
1 Pump, Cntfgl, 6"		217.00		238.70	144.75	159.23
24 L.H., Daily Totals		$3958.80		$4634.60	$164.95	$193.11

Crew B-24	Hr.	Daily	Hr.	Daily	Bare Costs	Incl. O&P
1 Cement Finisher	$23.00	$184.00	$37.10	$296.80	$21.45	$35.77
1 Laborer	17.35	138.80	29.45	235.60		
1 Carpenter	24.00	192.00	40.75	326.00		
24 L.H., Daily Totals		$514.80		$858.40	$21.45	$35.77

Crew B-25	Hr.	Daily	Hr.	Daily	Bare Costs	Incl. O&P
1 Labor Foreman	$19.35	$154.80	$32.85	$262.80	$19.32	$32.46
7 Laborers	17.35	971.60	29.45	1649.20		
3 Equip. Oper. (med.)	23.90	573.60	39.35	944.40		
1 Asphalt Paver, 130 H.P		1590.00		1749.00		
1 Tandem Roller, 10 Ton		179.20		197.10		
1 Roller, Pneumatic Wheel		241.80		266.00	22.85	25.14
88 L.H., Daily Totals		$3711.00		$5068.50	$42.17	$57.60

Crew B-25B	Hr.	Daily	Hr.	Daily	Bare Costs	Incl. O&P
1 Labor Foreman	$19.35	$154.80	$32.85	$262.80	$19.70	$33.03
7 Laborers	17.35	971.60	29.45	1649.20		
4 Equip. Oper. (medium)	23.90	764.80	39.35	1259.20		
1 Asphalt Paver, 130 H.P.		1590.00		1749.00		
2 Rollers, Steel Wheel		358.40		394.25		
1 Roller, Pneumatic Wheel		241.80		266.00	22.81	25.10
96 L.H., Daily Totals		$4081.40		$5580.45	$42.51	$58.13

Crew B-26	Hr.	Daily	Hr.	Daily	Bare Costs	Incl. O&P
1 Labor Foreman (outside)	$19.35	$154.80	$32.85	$262.80	$19.97	$33.72
6 Laborers	17.35	832.80	29.45	1413.60		
2 Equip. Oper. (med.)	23.90	382.40	39.35	629.60		
1 Rodman (reinf.)	25.45	203.60	45.55	364.40		
1 Cement Finisher	23.00	184.00	37.10	296.80		
1 Grader, 30,000 Lbs.		456.60		502.25		
1 Paving Mach. & Equip.		1910.00		2101.00	26.89	29.58
88 L.H., Daily Totals		$4124.20		$5570.45	$46.86	$63.30

Crew B-27	Hr.	Daily	Hr.	Daily	Bare Costs	Incl. O&P
1 Labor Foreman (outside)	$19.35	$154.80	$32.85	$262.80	$17.85	$30.30
3 Laborers	17.35	416.40	29.45	706.80		
1 Berm Machine		216.20		237.80	6.76	7.43
32 L.H., Daily Totals		$787.40		$1207.40	$24.61	$37.73

Crew B-28	Hr.	Daily	Hr.	Daily	Bare Costs	Incl. O&P
2 Carpenters	$24.00	$384.00	$40.75	$652.00	$21.78	$36.98
1 Laborer	17.35	138.80	29.45	235.60		
24 L.H., Daily Totals		$522.80		$887.60	$21.78	$36.98

Crews

Crew No.	Bare Costs		Incl. Subs O & P		Cost Per Labor-Hour	
Crew B-29	**Hr.**	**Daily**	**Hr.**	**Daily**	**Bare Costs**	**Incl. O&P**
1 Labor Foreman (outside)	$19.35	$154.80	$32.85	$262.80	$18.92	$31.90
4 Laborers	17.35	555.20	29.45	942.40		
1 Equip. Oper. (crane)	24.75	198.00	40.75	326.00		
1 Gradall, 3 Ton, 1/2 C.Y.		837.80		921.60	17.45	19.20
48 L.H., Daily Totals		$1745.80		$2452.80	$36.37	$51.10

Crew B-30	**Hr.**	**Daily**	**Hr.**	**Daily**	**Bare Costs**	**Incl. O&P**
1 Equip. Oper. (med.)	$23.90	$191.20	$39.35	$314.80	$21.00	$34.82
2 Truck Drivers (heavy)	19.55	312.80	32.55	520.80		
1 Hyd. Excavator, 1.5 C.Y.		720.40		792.45		
2 Dump Trucks, 16 Ton		953.20		1048.50	69.73	76.71
24 L.H., Daily Totals		$2177.60		$2676.55	$90.73	$111.53

Crew B-31	**Hr.**	**Daily**	**Hr.**	**Daily**	**Bare Costs**	**Incl. O&P**
1 Labor Foreman (outside)	$19.35	$154.80	$32.85	$262.80	$17.75	$30.13
4 Laborers	17.35	555.20	29.45	942.40		
1 Air Compr., 250 C.F.M.		112.40		123.65		
1 Sheeting Driver		7.20		7.90		
2 50 Ft. Air Hoses, 1.5" Dia.		9.40		10.35	3.23	3.55
40 L.H., Daily Totals		$839.00		$1347.10	$20.98	$33.68

Crew B-32	**Hr.**	**Daily**	**Hr.**	**Daily**	**Bare Costs**	**Incl. O&P**
1 Laborer	$17.35	$138.80	$29.45	$235.60	$22.26	$36.88
3 Equip. Oper. (med.)	23.90	573.60	39.35	944.40		
1 Grader, 30,000 Lbs.		456.60		502.25		
1 Tandem Roller, 10 Ton		179.20		197.10		
1 Dozer, 200 H.P.		919.60		1011.55	48.61	53.47
32 L.H., Daily Totals		$2267.80		$2890.90	$70.87	$90.35

Crew B-32A	**Hr.**	**Daily**	**Hr.**	**Daily**	**Bare Costs**	**Incl. O&P**
1 Laborer	$17.35	$138.80	$29.45	$235.60	$21.72	$36.05
2 Equip. Oper. (medium)	23.90	382.40	39.35	629.60		
1 Grader, 30,000 Lbs.		456.60		502.25		
1 Roller, Vibratory, 29,000 Lbs.		433.20		476.50	37.08	40.78
24 L.H., Daily Totals		$1411.00		$1843.95	$58.80	$76.83

Crew B-32B	**Hr.**	**Daily**	**Hr.**	**Daily**	**Bare Costs**	**Incl. O&P**
1 Laborer	$17.35	$138.80	$29.45	$235.60	$21.72	$36.05
2 Equip. Oper. (medium)	23.90	382.40	39.35	629.60		
1 Dozer, 200 H.P.		919.60		1011.55		
1 Roller, Vibratory, 29,000 Lbs.		433.20		476.50	56.37	62.00
24 L.H., Daily Totals		$1874.00		$2353.25	$78.09	$98.05

Crew B-32C	**Hr.**	**Daily**	**Hr.**	**Daily**	**Bare Costs**	**Incl. O&P**
1 Labor Foreman	$19.35	$154.80	$32.85	$262.80	$20.96	$34.97
2 Laborers	17.35	277.60	29.45	471.20		
3 Equip. Oper. (medium)	23.90	573.60	39.35	944.40		
1 Grader, 30,000 Lbs.		456.60		502.25		
1 Roller, Steel Wheel		179.20		197.10		
1 Dozer, 200 H.P.		919.60		1011.55	32.40	35.64
48 L.H., Daily Totals		$2561.40		$3389.30	$53.36	$70.61

Crew B-33A	**Hr.**	**Daily**	**Hr.**	**Daily**	**Bare Costs**	**Incl. O&P**
1 Equip. Oper. (med.)	$23.90	$191.20	$39.35	$314.80	$23.90	$39.35
.25 Equip. Oper. (med.)	23.90	47.80	39.35	78.70		
1 Scraper, Towed, 7 C.Y.		150.75		165.80		
1.25 Dozer, 300 H.P.		1493.75		1643.15	164.46	180.90
10 L.H., Daily Totals		$1883.50		$2202.45	$188.36	$220.25

Crew B-33B	**Hr.**	**Daily**	**Hr.**	**Daily**	**Bare Costs**	**Incl. O&P**
1 Equip. Oper. (med.)	$23.90	$191.20	$39.35	$314.80	$23.90	$39.35
.25 Equip. Oper. (med.)	23.90	47.80	39.35	78.70		
1 Scraper, Towed, 10 C.Y.		169.20		186.10		
1.25 Dozer, 300 H.P.		1493.75		1643.15	166.30	182.92
10 L.H., Daily Totals		$1901.95		$2222.75	$190.20	$222.27

Crew B-33C	**Hr.**	**Daily**	**Hr.**	**Daily**	**Bare Costs**	**Incl. O&P**
1 Equip. Oper. (med.)	$23.90	$191.20	$39.35	$314.80	$23.90	$39.35
.25 Equip. Oper. (med.)	23.90	47.80	39.35	78.70		
1 Scraper, Towed, 12 C.Y.		169.20		186.10		
1.25 Dozer, 300 H.P.		1493.75		1643.15	166.30	182.92
10 L.H., Daily Totals		$1901.95		$2222.75	$190.20	$222.27

Crew B-33D	**Hr.**	**Daily**	**Hr.**	**Daily**	**Bare Costs**	**Incl. O&P**
1 Equip. Oper. (med.)	$23.90	$191.20	$39.35	$314.80	$23.90	$39.35
.25 Equip. Oper. (med.)	23.90	47.80	39.35	78.70		
1 S.P. Scraper, 14 C.Y.		1552.00		1707.20		
.25 Dozer, 300 H.P.		298.75		328.65	185.08	203.58
10 L.H., Daily Totals		$2089.75		$2429.35	$208.98	$242.93

Crew B-33E	**Hr.**	**Daily**	**Hr.**	**Daily**	**Bare Costs**	**Incl. O&P**
1 Equip. Oper. (med.)	$23.90	$191.20	$39.35	$314.80	$23.90	$39.35
.25 Equip. Oper. (med.)	23.90	47.80	39.35	78.70		
1 S.P. Scraper, 24 C.Y.		2317.00		2548.70		
.25 Dozer, 300 H.P.		298.75		328.65	261.58	287.73
10 L.H., Daily Totals		$2854.75		$3270.85	$285.48	$327.08

Crew B-33F	**Hr.**	**Daily**	**Hr.**	**Daily**	**Bare Costs**	**Incl. O&P**
1 Equip. Oper. (med.)	$23.90	$191.20	$39.35	$314.80	$23.90	$39.35
.25 Equip. Oper. (med.)	23.90	47.80	39.35	78.70		
1 Elev. Scraper, 11 C.Y.		819.40		901.35		
.25 Dozer, 300 H.P.		298.75		328.65	111.82	123.00
10 L.H., Daily Totals		$1357.15		$1623.50	$135.72	$162.35

Crew B-33G	**Hr.**	**Daily**	**Hr.**	**Daily**	**Bare Costs**	**Incl. O&P**
1 Equip. Oper. (med.)	$23.90	$191.20	$39.35	$314.80	$23.90	$39.35
.25 Equip. Oper. (med.)	23.90	47.80	39.35	78.70		
1 Elev. Scraper, 20 C.Y.		1648.00		1812.80		
.25 Dozer, 300 H.P.		298.75		328.65	194.68	214.14
10 L.H., Daily Totals		$2185.75		$2534.95	$218.58	$253.49

Crew B-34A	**Hr.**	**Daily**	**Hr.**	**Daily**	**Bare Costs**	**Incl. O&P**
1 Truck Driver (heavy)	$19.55	$156.40	$32.55	$260.40	$19.55	$32.55
1 Dump Truck, 12 Ton		326.60		359.25	40.83	44.91
8 L.H., Daily Totals		$483.00		$619.65	$60.38	$77.46

Crew B-34B	**Hr.**	**Daily**	**Hr.**	**Daily**	**Bare Costs**	**Incl. O&P**
1 Truck Driver (heavy)	$19.55	$156.40	$32.55	$260.40	$19.55	$32.55
1 Dump Truck, 16 Ton		476.60		524.25	59.58	65.53
8 L.H., Daily Totals		$633.00		$784.65	$79.13	$98.08

Crew B-34C	**Hr.**	**Daily**	**Hr.**	**Daily**	**Bare Costs**	**Incl. O&P**
1 Truck Driver (heavy)	$19.55	$156.40	$32.55	$260.40	$19.55	$32.55
1 Truck Tractor, 40 Ton		301.60		331.75		
1 Dump Trailer, 16.5 C.Y.		103.00		113.30	50.58	55.63
8 L.H., Daily Totals		$561.00		$705.45	$70.13	$88.18

Crews

Crew No.	Bare Costs		Incl. Subs O & P		Cost Per Labor-Hour	
Crew B-34D	**Hr.**	**Daily**	**Hr.**	**Daily**	**Bare Costs**	**Incl. O&P**
Truck Driver (heavy)	$19.55	$156.40	$32.55	$260.40	$19.55	$32.55
Truck Tractor, 40 Ton		301.60		331.75		
Dump Trailer, 20 C.Y.		115.80		127.40	52.18	57.39
L.H., Daily Totals		$573.80		$719.55	$71.73	$89.94
Crew B-34E	**Hr.**	**Daily**	**Hr.**	**Daily**	**Bare Costs**	**Incl. O&P**
Truck Driver (heavy)	$19.55	$156.40	$32.55	$260.40	$19.55	$32.55
Truck, Off Hwy., 25 Ton		912.20		1003.40	114.03	125.43
L.H., Daily Totals		$1068.60		$1263.80	$133.58	$157.98
Crew B-34F	**Hr.**	**Daily**	**Hr.**	**Daily**	**Bare Costs**	**Incl. O&P**
Truck Driver (heavy)	$19.55	$156.40	$32.55	$260.40	$19.55	$32.55
Truck, Off Hwy., 22 C.Y.		933.40		1026.75	116.68	128.34
L.H., Daily Totals		$1089.80		$1287.15	$136.23	$160.89
Crew B-34G	**Hr.**	**Daily**	**Hr.**	**Daily**	**Bare Costs**	**Incl. O&P**
Truck Driver (heavy)	$19.55	$156.40	$32.55	$260.40	$19.55	$32.55
Truck, Off Hwy., 34 C.Y.		1181.00		1299.10	147.63	162.39
8 L.H., Daily Totals		$1337.40		$1559.50	$167.18	$194.94
Crew B-34H	**Hr.**	**Daily**	**Hr.**	**Daily**	**Bare Costs**	**Incl. O&P**
1 Truck Driver (heavy)	$19.55	$156.40	$32.55	$260.40	$19.55	$32.55
1 Truck, Off Hwy., 42 C.Y.		1276.00		1403.60	159.50	175.45
8 L.H., Daily Totals		$1432.40		$1664.00	$179.05	$208.00
Crew B-34J	**Hr.**	**Daily**	**Hr.**	**Daily**	**Bare Costs**	**Incl. O&P**
1 Truck Driver (heavy)	$19.55	$156.40	$32.55	$260.40	$19.55	$32.55
1 Truck, Off Hwy., 60 C.Y.		1649.00		1813.90	206.13	226.74
8 L.H., Daily Totals		$1805.40		$2074.30	$225.68	$259.29
Crew B-34K	**Hr.**	**Daily**	**Hr.**	**Daily**	**Bare Costs**	**Incl. O&P**
1 Truck Driver (heavy)	$19.55	$156.40	$32.55	$260.40	$19.55	$32.55
1 Truck Tractor, 240 H.P.		353.60		388.95		
1 Low Bed Trailer		170.20		187.20	65.48	72.02
8 L.H., Daily Totals		$680.20		$836.55	$85.03	$104.57
Crew B-34N	**Hr.**	**Daily**	**Hr.**	**Daily**	**Bare Costs**	**Incl. O&P**
1 Truck Driver (heavy)	$19.55	$156.40	$32.55	$260.40	$19.55	$32.55
1 Dump Truck, 12 Ton		326.60		359.25		
1 Flatbed Trailer, 40 Ton		119.40		131.35	55.75	61.33
8 L.H., Daily Totals		$602.40		$751.00	$75.30	$93.88
Crew B-34P	**Hr.**	**Daily**	**Hr.**	**Daily**	**Bare Costs**	**Incl. O&P**
1 Pipe Fitter	$27.05	$216.40	$44.45	$355.60	$23.33	$38.50
1 Truck Driver (light)	19.05	152.40	31.70	253.60		
1 Equip. Oper. (medium)	23.90	191.20	39.35	314.80		
1 Flatbed Truck, 3 Ton		176.20		193.80		
1 Backhoe Loader, 48 H.P.		215.80		237.40	16.33	17.97
24 L.H., Daily Totals		$952.00		$1355.20	$39.66	$56.47
Crew B-34Q	**Hr.**	**Daily**	**Hr.**	**Daily**	**Bare Costs**	**Incl. O&P**
1 Pipe Fitter	$27.05	$216.40	$44.45	$355.60	$23.62	$38.97
1 Truck Driver (light)	19.05	152.40	31.70	253.60		
1 Eqip. Oper. (crane)	24.75	198.00	40.75	326.00		
1 Flatbed Trailer, 25 Ton		88.00		96.80		
1 Dump Truck, 12 Ton		326.60		359.25		
1 Hyd. Crane, 25 Ton		616.80		678.50	42.98	47.27
24 L.H., Daily Totals		$1598.20		$2069.75	$66.60	$86.24

Crew No.	Bare Costs		Incl. Subs O & P		Cost Per Labor-Hour	
Crew B-34R	**Hr.**	**Daily**	**Hr.**	**Daily**	**Bare Costs**	**Incl. O&P**
1 Pipe Fitter	$27.05	$216.40	$44.45	$355.60	$23.62	$38.97
1 Truck Driver (light)	19.05	152.40	31.70	253.60		
1 Eqip. Oper. (crane)	24.75	198.00	40.75	326.00		
1 Flatbed Trailer, 25 Ton		88.00		96.80		
1 Dump Truck, 12 Ton		326.60		359.25		
1 Hyd. Crane, 25 Ton		616.80		678.50		
1 Hyd. Excavator, 1 C.Y.		557.80		613.60	66.22	72.84
24 L.H., Daily Totals		$2156.00		$2683.35	$89.84	$111.81
Crew B-34S	**Hr.**	**Daily**	**Hr.**	**Daily**	**Bare Costs**	**Incl. O&P**
2 Pipe Fitter	$27.05	$432.80	$44.45	$711.20	$24.60	$40.55
1 Truck Driver (heavy)	19.55	156.40	32.55	260.40		
1 Eqip. Oper. (crane)	24.75	198.00	40.75	326.00		
1 Flatbed Trailer, 40 Ton		119.40		131.35		
1 Truck Tractor, 40 Ton		301.60		331.75		
1 Truck Crane, 80 Ton		995.00		1094.50		
1 Hyd. Excavator, 2 C.Y.		909.80		1000.80	72.68	79.95
32 L.H., Daily Totals		$3113.00		$3856.00	$97.28	$120.50
Crew B-34T	**Hr.**	**Daily**	**Hr.**	**Daily**	**Bare Costs**	**Incl. O&P**
2 Pipe Fitter	$27.05	$432.80	$44.45	$711.20	$24.60	$40.55
1 Truck Driver (heavy)	19.55	156.40	32.55	260.40		
1 Eqip. Oper. (crane)	24.75	198.00	40.75	326.00		
1 Flatbed Trailer, 40 Ton		119.40		131.35		
1 Truck Tractor, 40 Ton		301.60		331.75		
1 Truck Crane, 80 Ton		995.00		1094.50	44.25	48.68
32 L.H., Daily Totals		$2203.20		$2855.20	$68.85	$89.23
Crew B-35	**Hr.**	**Daily**	**Hr.**	**Daily**	**Bare Costs**	**Incl. O&P**
1 Laborer Foreman (out)	$19.35	$154.80	$32.85	$262.80	$22.46	$37.57
1 Skilled Worker	23.90	191.20	40.50	324.00		
1 Welder (plumber)	26.95	215.60	44.30	354.40		
1 Laborer	17.35	138.80	29.45	235.60		
1 Equip. Oper. (crane)	24.75	198.00	40.75	326.00		
1 Electric Welding Mach.		80.65		88.70		
1 Hyd. Excavator, .75 C.Y.		474.00		521.40	13.87	15.25
40 L.H., Daily Totals		$1453.05		$2112.90	$36.33	$52.82
Crew B-35A	**Hr.**	**Daily**	**Hr.**	**Daily**	**Bare Costs**	**Incl. O&P**
1 Laborer Foreman (out)	$19.35	$154.80	$32.85	$262.80	$21.49	$35.93
2 Laborers	17.35	277.60	29.45	471.20		
1 Skilled Worker	23.90	191.20	40.50	324.00		
1 Welder (plumber)	26.95	215.60	44.30	354.40		
1 Equip. Oper. (crane)	24.75	198.00	40.75	326.00		
1 Equip. Oper. Oiler	20.75	166.00	34.20	273.60		
1 Welder, 300 amp		81.20		89.30		
1 Crane, 75 Ton		1170.00		1287.00	22.34	24.58
56 L.H., Daily Totals		$2454.40		$3388.30	$43.83	$60.51
Crew B-36	**Hr.**	**Daily**	**Hr.**	**Daily**	**Bare Costs**	**Incl. O&P**
1 Labor Foreman (outside)	$19.35	$154.80	$32.85	$262.80	$20.37	$34.09
2 Laborers	17.35	277.60	29.45	471.20		
2 Equip. Oper. (med.)	23.90	382.40	39.35	629.60		
1 Dozer, 200 H.P.		919.60		1011.55		
1 Aggregate Spreader		35.00		38.50		
1 Tandem Roller, 10 Ton		179.20		197.10	28.35	31.18
40 L.H., Daily Totals		$1948.60		$2610.75	$48.72	$65.27

Crew B-36A	Bare Costs Hr.	Bare Costs Daily	Incl. Subs O & P Hr.	Incl. Subs O & P Daily	Cost Per Labor-Hour Bare Costs	Cost Per Labor-Hour Incl. O&P
1 Labor Foreman (outside)	$19.35	$154.80	$32.85	$262.80	$21.38	$35.59
2 Laborers	17.35	277.60	29.45	471.20		
4 Equip. Oper. (med.)	23.90	764.80	39.35	1259.20		
1 Dozer, 200 H.P.		919.60		1011.55		
1 Aggregate Spreader		35.00		38.50		
1 Roller, Steel Wheel		179.20		197.10		
1 Roller, Pneumatic Wheel		241.80		266.00	24.56	27.02
56 L.H., Daily Totals		$2572.80		$3506.35	$45.94	$62.61

Crew B-36B	Hr.	Daily	Hr.	Daily	Bare Costs	Incl. O&P
1 Labor Foreman (outside)	$19.35	$154.80	$32.85	$262.80	$21.15	$35.21
2 Laborers	17.35	277.60	29.45	471.20		
4 Equip. Oper. (medium)	23.90	764.80	39.35	1259.20		
1 Truck Driver, Heavy	19.55	156.40	32.55	260.40		
1 Grader, 30,000 Lbs.		456.60		502.25		
1 F.E. Loader, crl, 1.5 C.Y.		368.80		405.70		
1 Dozer, 300 H.P.		1195.00		1314.50		
1 Roller, Vibratory		433.20		476.50		
1 Truck, Tractor, 240 H.P.		353.60		388.95		
1 Water Tanker, 5000 Gal.		116.80		128.50	45.69	50.26
64 L.H., Daily Totals		$4277.60		$5470.00	$66.84	$85.47

Crew B-36C	Hr.	Daily	Hr.	Daily	Bare Costs	Incl. O&P
1 Labor Foreman (outside)	$19.35	$154.80	$32.85	$262.80	$22.12	$36.69
3 Equip. Oper. (medium)	23.90	573.60	39.35	944.40		
1 Truck Driver, Heavy	19.55	156.40	32.55	260.40		
1 Grader, 30,000 Lbs.		456.60		502.25		
1 Dozer, 300 H.P.		1195.00		1314.50		
1 Roller, Vibratory		433.20		476.50		
1 Truck, Tractor, 240 H.P.		353.60		388.95		
1 Water Tanker, 5000 Gal.		116.80		128.50	63.88	70.27
40 L.H., Daily Totals		$3440.00		$4278.30	$86.00	$106.96

Crew B-36E	Hr.	Daily	Hr.	Daily	Bare Costs	Incl. O&P
1 Labor Foreman (outside)	$19.35	$154.80	$32.85	$262.80	$22.42	$37.13
4 Equip. Oper. (medium)	23.90	764.80	39.35	1259.20		
1 Truck Driver, Heavy	19.55	156.40	32.55	260.40		
1 Grader, 30,000 Lbs.		456.60		502.25		
1 Dozer, 300 H.P.		1195.00		1314.50		
1 Roller, Vibratory		433.20		476.50		
1 Truck, Tractor, 240 H.P.		353.60		388.95		
1 Dist. Tanker, 3000 G		233.80		257.20	55.67	61.24
48 L.H., Daily Totals		$3748.20		$4721.80	$78.09	$98.37

Crew B-37	Hr.	Daily	Hr.	Daily	Bare Costs	Incl. O&P
1 Labor Foreman (outside)	$19.35	$154.80	$32.85	$262.80	$18.59	$31.37
4 Laborers	17.35	555.20	29.45	942.40		
1 Equip. Oper. (light)	22.80	182.40	37.55	300.40		
1 Tandem Roller, 5 Ton		107.40		118.15	2.24	2.46
48 L.H., Daily Totals		$999.80		$1623.75	$20.83	$33.83

Crew B-38	Hr.	Daily	Hr.	Daily	Bare Costs	Incl. O&P
2 Laborers	$17.35	$277.60	$29.45	$471.20	$19.17	$32.15
1 Equip. Oper. (light)	22.80	182.40	37.55	300.40		
1 Backhoe Loader, 48 H.P.		215.80		237.40		
1 Hyd.Hammer, (1200 lb)		112.00		123.20	13.66	15.02
24 L.H., Daily Totals		$787.80		$1132.20	$32.83	$47.17

Crew B-39	Bare Costs Hr.	Bare Costs Daily	Incl. Subs O & P Hr.	Incl. Subs O & P Daily	Cost Per Labor-Hour Bare Costs	Cost Per Labor-Hour Incl. O&P
1 Labor Foreman (outside)	$19.35	$154.80	$32.85	$262.80	$17.68	$30.02
5 Laborers	17.35	694.00	29.45	1178.00		
1 Air Compr., 250 C.F.M.		112.40		123.65		
2 Air Tools & Accessories		19.60		21.55		
2 50 Ft. Air Hoses, 1.5" Dia.		9.40		10.35	2.95	3.24
48 L.H., Daily Totals		$990.20		$1596.35	$20.63	$33.26

Crew B-40	Hr.	Daily	Hr.	Daily	Bare Costs	Incl. O&P
1 Pile Driver Foreman (out)	$25.30	$202.40	$45.25	$362.00	$22.94	$40.29
4 Pile Drivers	23.30	745.60	41.65	1332.80		
1 Building Laborer	17.35	138.80	29.45	235.60		
1 Equip. Oper. (crane)	24.75	198.00	40.75	326.00		
1 Crane, 40 Ton		888.20		977.00		
1 Vibratory Hammer & Gen.		1403.00		1543.30	40.91	45.01
56 L.H., Daily Totals		$3576.00		$4776.70	$63.85	$85.30

Crew B-40B	Hr.	Daily	Hr.	Daily	Bare Costs	Incl. O&P
1 Laborer Foreman	$19.35	$154.80	$32.85	$262.80	$19.48	$32.69
3 Laborers	17.35	416.40	29.45	706.80		
1 Equip. Oper. (crane)	24.75	198.00	40.75	326.00		
1 Equip. Oper. Oiler	20.75	166.00	34.20	273.60		
1 Crane, 40 Ton		951.90		1047.10	19.83	21.81
48 L.H., Daily Totals		$1887.10		$2616.30	$39.31	$54.50

Crew B-41	Hr.	Daily	Hr.	Daily	Bare Costs	Incl. O&P
1 Labor Foreman (outside)	$19.35	$154.80	$32.85	$262.80	$18.20	$30.80
4 Laborers	17.35	555.20	29.45	942.40		
.25 Equip. Oper. (crane)	24.75	49.50	40.75	81.50		
.25 Equip. Oper. Oiler	20.75	41.50	34.20	68.40		
.25 Crawler Crane, 40 Ton		222.05		244.25	5.05	5.55
44 L.H., Daily Totals		$1023.05		$1599.35	$23.25	$36.35

Crew B-42	Hr.	Daily	Hr.	Daily	Bare Costs	Incl. O&P
1 Labor Foreman (outside)	$19.35	$154.80	$32.85	$262.80	$20.06	$33.67
4 Laborers	17.35	555.20	29.45	942.40		
1 Equip. Oper. (crane)	24.75	198.00	40.75	326.00		
1 Welder	26.95	215.60	44.30	354.40		
1 Hyd. Crane, 25 Ton		616.80		678.50		
1 Gas Welding Machine		81.20		89.30		
1 Horz. Boring Csg. Mch.		372.40		409.65	19.11	21.03
56 L.H., Daily Totals		$2194.00		$3063.05	$39.17	$54.70

Crew B-43	Hr.	Daily	Hr.	Daily	Bare Costs	Incl. O&P
1 Labor Foreman (outside)	$19.35	$154.80	$32.85	$262.80	$17.75	$30.13
4 Laborers	17.35	555.20	29.45	942.40		
1 Drill Rig & Augers		3179.00		3496.90	79.48	87.42
40 L.H., Daily Totals		$3889.00		$4702.10	$97.23	$117.55

Crew B-44	Hr.	Daily	Hr.	Daily	Bare Costs	Incl. O&P
1 Pile Driver Foreman	$25.30	$202.40	$45.25	$362.00	$22.24	$38.94
4 Pile Drivers	23.30	745.60	41.65	1332.80		
1 Equip. Oper. (crane)	24.75	198.00	40.75	326.00		
2 Laborers	17.35	277.60	29.45	471.20		
1 Crane, 40 Ton, & Access.		888.20		977.00		
45 L.F. Leads, 15K Ft. Lbs.		56.25		61.90	14.79	16.27
64 L.H., Daily Totals		$2368.05		$3530.90	$37.03	$55.21

Crew No.	Bare Costs		Incl. Subs O & P		Cost Per Labor-Hour	
Crew B-45	**Hr.**	**Daily**	**Hr.**	**Daily**	**Bare Costs**	**Incl. O&P**
1 Building Laborer	$17.35	$138.80	$29.45	$235.60	$18.45	$31.00
1 Truck Driver (heavy)	19.55	156.40	32.55	260.40		
1 Dist. Tank Truck, 3K Gal.		233.80		257.20	14.61	16.07
16 L.H., Daily Totals		$529.00		$753.20	$33.06	$47.07

Crew B-46	Hr.	Daily	Hr.	Daily	Bare Costs	Incl. O&P
1 Pile Driver Foreman	$25.30	$202.40	$45.25	$362.00	$20.66	$36.15
2 Pile Drivers	23.30	372.80	41.65	666.40		
3 Laborers	17.35	416.40	29.45	706.80		
1 Chain Saw, 36″ Long		33.40		36.75	.70	.77
48 L.H., Daily Totals		$1025.00		$1771.95	$21.36	$36.92

Crew B-47	Hr.	Daily	Hr.	Daily	Bare Costs	Incl. O&P
1 Blast Foreman	$19.35	$154.80	$32.85	$262.80	$18.35	$31.15
1 Driller	17.35	138.80	29.45	235.60		
1 Crawler Type Drill, 4″		664.40		730.85		
1 Air Compr., 600 C.F.M.		297.80		327.60		
2 50 Ft. Air Hoses, 3″ Dia.		35.40		38.95	62.35	68.59
16 L.H., Daily Totals		$1291.20		$1595.80	$80.70	$99.74

Crew B-47A	Hr.	Daily	Hr.	Daily	Bare Costs	Incl. O&P
1 Drilling Foreman	$19.35	$154.80	$32.85	$262.80	$21.62	$35.93
1 Equip. Oper. (heavy)	24.75	198.00	40.75	326.00		
1 Oiler	20.75	166.00	34.20	273.60		
1 Quarry Drill		822.60		904.85	34.28	37.70
24 L.H., Daily Totals		$1341.40		$1767.25	$55.90	$73.63

Crew B-47C	Hr.	Daily	Hr.	Daily	Bare Costs	Incl. O&P
1 Laborer	$17.35	$138.80	$29.45	$235.60	$20.08	$33.50
1 Equip. Oper. (light)	22.80	182.40	37.55	300.40		
1 Air Compressor, 750 CFM		317.40		349.15		
2 50′ Air Hose, 3″		35.40		38.95		
1 Air Track Drill, 4″		664.40		730.85	63.58	69.93
16 L.H., Daily Totals		$1338.40		$1654.95	$83.66	$103.43

Crew B-47E	Hr.	Daily	Hr.	Daily	Bare Costs	Incl. O&P
1 Laborer Foreman	$19.35	$154.80	$32.85	$262.80	$17.85	$30.30
3 Laborers	17.35	416.40	29.45	706.80		
1 Truck, Flatbed, 3 Ton		176.20		193.80	5.51	6.06
32 L.H., Daily Totals		$747.40		$1163.40	$23.36	$36.36

Crew B-47G	Hr.	Daily	Hr.	Daily	Bare Costs	Incl. O&P
1 Laborer Foreman	$19.35	$154.80	$32.85	$262.80	$18.02	$30.58
2 Laborers	17.35	277.60	29.45	471.20		
1 Air Track Drill, 4″		664.40		730.85		
1 Air Compr., 600 C.F.M.		297.80		327.60		
2 50 Ft. Air Hoses, 3″ Dia.		35.40		38.95		
1 Grout Pump		101.40		111.55	45.79	50.37
24 L.H., Daily Totals		$1531.40		$1942.95	$63.81	$80.95

Crew B-48	Hr.	Daily	Hr.	Daily	Bare Costs	Incl. O&P
1 Labor Foreman (outside)	$19.35	$154.80	$32.85	$262.80	$18.92	$31.90
4 Laborers	17.35	555.20	29.45	942.40		
1 Equip. Oper. (crane)	24.75	198.00	40.75	326.00		
1 Centr. Water Pump, 6″		217.00		238.70		
1 20 Ft. Suction Hose, 6″		12.80		14.10		
1 50 Ft. Disch. Hose, 6″		8.90		9.80		
1 Drill Rig & Augers		3179.00		3496.90	71.20	78.32
48 L.H., Daily Totals		$4325.70		$5290.70	$90.12	$110.22

Crew No. | Bare Costs | Incl. Subs O & P | Cost Per Labor-Hour

Crew B-49	Hr.	Daily	Hr.	Daily	Bare Costs	Incl. O&P
1 Labor Foreman (outside)	$19.35	$154.80	$32.85	$262.80	$19.72	$33.79
5 Laborers	17.35	694.00	29.45	1178.00		
1 Equip. Oper. (crane)	24.75	198.00	40.75	326.00		
2 Pile Drivers	23.30	372.80	41.65	666.40		
1 Hyd. Crane, 25 Ton		616.80		678.50		
1 Centr. Water Pump, 6″		217.00		238.70		
1 20 Ft. Suction Hose, 6″		12.80		14.10		
1 50 Ft. Disch. Hose, 6″		8.90		9.80		
1 Drill Rig & Augers		3179.00		3496.90	56.03	61.64
72 L.H., Daily Totals		$5454.10		$6871.20	$75.75	$95.43

Crew B-50	Hr.	Daily	Hr.	Daily	Bare Costs	Incl. O&P
1 Pile Driver Foremen	$25.30	$202.40	$45.25	$362.00	$21.28	$37.17
6 Pile Drivers	23.30	1118.40	41.65	1999.20		
1 Equip. Oper. (crane)	24.75	198.00	40.75	326.00		
5 Laborers	17.35	694.00	29.45	1178.00		
1 Crane, 40 Ton		888.20		977.00		
60 L.F. Leads, 15K Ft. Lbs.		75.00		82.50		
1 Hammer, 15K Ft. Lbs.		359.80		395.80		
1 Air Compr., 600 C.F.M.		297.80		327.60		
2 50 Ft. Air Hoses, 3″ Dia.		35.40		38.95		
1 Chain Saw, 36″ Long		33.40		36.75	16.28	17.90
104 L.H., Daily Totals		$3902.40		$5723.80	$37.56	$55.07

Crew B-51	Hr.	Daily	Hr.	Daily	Bare Costs	Incl. O&P
1 Labor Foreman (outside)	$19.35	$154.80	$32.85	$262.80	$17.97	$30.39
4 Laborers	17.35	555.20	29.45	942.40		
1 Truck Driver (light)	19.05	152.40	31.70	253.60		
1 Light Truck, 1.5 Ton		128.80		141.70	2.68	2.95
48 L.H., Daily Totals		$991.20		$1600.50	$20.65	$33.34

Crew B-52	Hr.	Daily	Hr.	Daily	Bare Costs	Incl. O&P
1 Labor Foreman	$19.35	$154.80	$32.85	$262.80	$19.63	$33.41
1 Carpenter	24.00	192.00	40.75	326.00		
4 Laborers	17.35	555.20	29.45	942.40		
.5 Rodman (reinf.)	25.45	101.80	45.55	182.20		
.5 Equip. Oper. (med.)	23.90	95.60	39.35	157.40		
.5 F.E. Ldr., T.M., 2.5 C.Y.		394.00		433.40	7.04	7.74
56 L.H., Daily Totals		$1493.40		$2304.20	$26.67	$41.15

Crew B-53	Hr.	Daily	Hr.	Daily	Bare Costs	Incl. O&P
1 Building Laborer	$17.35	$138.80	$29.45	$235.60	$17.35	$29.45
1 Trencher, Chain, 12 H.P.		48.00		52.80	6.00	6.60
8 L.H., Daily Totals		$186.80		$288.40	$23.35	$36.05

Crew B-54	Hr.	Daily	Hr.	Daily	Bare Costs	Incl. O&P
1 Equip. Oper. (light)	$22.80	$182.40	$37.55	$300.40	$22.80	$37.55
1 Trencher, Chain, 40 H.P.		213.40		234.75	26.68	29.34
8 L.H., Daily Totals		$395.80		$535.15	$49.48	$66.89

Crew B-54A	Hr.	Daily	Hr.	Daily	Bare Costs	Incl. O&P
.17 Labor Foreman (outside)	$19.35	$26.32	$32.85	$44.68	$23.24	$38.41
1 Equipment Operator (med.)	23.90	191.20	39.35	314.80		
1 Wheel Trencher, 67 H.P.		816.20		897.80	87.20	95.92
9.36 L.H., Daily Totals		$1033.72		$1257.28	$110.44	$134.33

Crew B-54B	Hr.	Daily	Hr.	Daily	Bare Costs	Incl. O&P
.25 Labor Foreman (outside)	$19.35	$38.70	$32.85	$65.70	$22.99	$38.05
1 Equipment Operator (med.)	23.90	191.20	39.35	314.80		
1 Wheel Trencher, 150 H.P.		1460.00		1606.00	146.00	160.60
10 L.H., Daily Totals		$1689.90		$1986.50	$168.99	$198.65

Crew No.	Bare Costs		Incl. Subs O & P		Cost Per Labor-Hour	
Crew B-55	**Hr.**	**Daily**	**Hr.**	**Daily**	**Bare Costs**	**Incl. O&P**
1 Laborers	$17.35	$138.80	$29.45	$235.60	$18.20	$30.58
1 Truck Driver (light)	19.05	152.40	31.70	253.60		
1 Auger, 4" to 36" Dia		604.60		665.05		
1 Flatbed 3 Ton Truck		176.20		193.80	48.80	53.68
16 L.H., Daily Totals		$1072.00		$1348.05	$67.00	$84.26
Crew B-56	**Hr.**	**Daily**	**Hr.**	**Daily**	**Bare Costs**	**Incl. O&P**
2 Laborers	$17.35	$277.60	$29.45	$471.20	$17.35	$29.45
1 Crawler Type Drill, 4"		664.40		730.85		
1 Air Compr., 600 C.F.M.		297.80		327.60		
1 50 Ft. Air Hose, 3" Dia.		17.70		19.45	61.24	67.37
16 L.H., Daily Totals		$1257.50		$1549.10	$78.59	$96.82
Crew B-57	**Hr.**	**Daily**	**Hr.**	**Daily**	**Bare Costs**	**Incl. O&P**
1 Labor Foreman (outside)	$19.35	$154.80	$32.85	$262.80	$19.23	$32.39
3 Laborers	17.35	416.40	29.45	706.80		
1 Equip. Oper. (crane)	24.75	198.00	40.75	326.00		
1 Barge, 400 Ton		273.20		300.50		
1 Power Shovel, 1 C.Y.		833.20		916.50		
1 Clamshell Bucket, 1 C.Y.		43.40		47.75		
1 Centr. Water Pump, 6"		217.00		238.70		
1 20 Ft. Suction Hose, 6"		12.80		14.10		
20 50 Ft. Disch. Hoses, 6"		178.00		195.80	38.94	42.83
40 L.H., Daily Totals		$2326.80		$3008.95	$58.17	$75.22
Crew B-58	**Hr.**	**Daily**	**Hr.**	**Daily**	**Bare Costs**	**Incl. O&P**
2 Laborers	$17.35	$277.60	$29.45	$471.20	$19.17	$32.15
1 Equip. Oper. (light)	22.80	182.40	37.55	300.40		
1 Backhoe Loader, 48 H.P.		215.80		237.40		
1 Small Helicopter, w/pilot		2153.00		2368.30	98.70	108.57
24 L.H., Daily Totals		$2828.80		$3377.30	$117.87	$140.72
Crew B-59	**Hr.**	**Daily**	**Hr.**	**Daily**	**Bare Costs**	**Incl. O&P**
1 Truck Driver (heavy)	$19.55	$156.40	$32.55	$260.40	$19.55	$32.55
1 Truck, 30 Ton		208.20		229.00		
1 Water tank, 6000 Gal.		116.80		128.50	40.63	44.69
8 L.H., Daily Totals		$481.40		$617.90	$60.18	$77.24
Crew B-60	**Hr.**	**Daily**	**Hr.**	**Daily**	**Bare Costs**	**Incl. O&P**
1 Labor Foreman (outside)	$19.35	$154.80	$32.85	$262.80	$19.83	$33.25
3 Laborers	17.35	416.40	29.45	706.80		
1 Equip. Oper. (crane)	24.75	198.00	40.75	326.00		
1 Equip. Oper. (light)	22.80	182.40	37.55	300.40		
1 Crawler Crane, 40 Ton		888.20		977.00		
45 L.F. Leads, 15K Ft. Lbs.		56.25		61.90		
1 Backhoe Loader, 48 H.P.		215.80		237.40	24.22	26.64
48 L.H., Daily Totals		$2111.85		$2872.30	$44.05	$59.89
Crew B-61	**Hr.**	**Daily**	**Hr.**	**Daily**	**Bare Costs**	**Incl. O&P**
1 Labor Foreman (outside)	$19.35	$154.80	$32.85	$262.80	$17.75	$30.13
4 Laborers	17.35	555.20	29.45	942.40		
1 Cement Mixer, 2 C.Y.		162.00		178.20		
1 Air Compr., 160 C.F.M.		87.40		96.15	6.24	6.86
40 L.H., Daily Totals		$959.40		$1479.55	$23.99	$36.99
Crew B-62	**Hr.**	**Daily**	**Hr.**	**Daily**	**Bare Costs**	**Incl. O&P**
2 Laborers	$17.35	$277.60	$29.45	$471.20	$19.17	$32.15
1 Equip. Oper. (light)	22.80	182.40	37.55	300.40		
1 Loader, Skid Steer		153.40		168.75	6.39	7.03
24 L.H., Daily Totals		$613.40		$940.35	$25.56	$39.18

Crew No.	Bare Costs		Incl. Subs O & P		Cost Per Labor-Hour	
Crew B-63	**Hr.**	**Daily**	**Hr.**	**Daily**	**Bare Costs**	**Incl. O&P**
5 Laborers	$17.35	$694.00	$29.45	$1178.00	$17.35	$29.45
1 Loader, Skid Steer		153.40		168.75	3.84	4.22
40 L.H., Daily Totals		$847.40		$1346.75	$21.19	$33.67
Crew B-64	**Hr.**	**Daily**	**Hr.**	**Daily**	**Bare Costs**	**Incl. O&P**
1 Laborer	$17.35	$138.80	$29.45	$235.60	$18.20	$30.58
1 Truck Driver (light)	19.05	152.40	31.70	253.60		
1 Power Mulcher (small)		103.20		113.50		
1 Light Truck, 1.5 Ton		128.80		141.70	14.50	15.95
16 L.H., Daily Totals		$523.20		$744.40	$32.70	$46.53
Crew B-65	**Hr.**	**Daily**	**Hr.**	**Daily**	**Bare Costs**	**Incl. O&P**
1 Laborer	$17.35	$138.80	$29.45	$235.60	$18.20	$30.58
1 Truck Driver (light)	19.05	152.40	31.70	253.60		
1 Power Mulcher (large)		184.00		202.40		
1 Light Truck, 1.5 Ton		128.80		141.70	19.55	21.51
16 L.H., Daily Totals		$604.00		$833.30	$37.75	$52.09
Crew B-66	**Hr.**	**Daily**	**Hr.**	**Daily**	**Bare Costs**	**Incl. O&P**
1 Equip. Oper. (light)	$22.80	$182.40	$37.55	$300.40	$22.80	$37.55
1 Backhoe Ldr. w/Attchmt.		173.20		190.50	21.65	23.82
8 L.H., Daily Totals		$355.60		$490.90	$44.45	$61.37
Crew B-67	**Hr.**	**Daily**	**Hr.**	**Daily**	**Bare Costs**	**Incl. O&P**
1 Millwright	$24.85	$198.80	$40.20	$321.60	$23.83	$38.88
1 Equip. Oper. (light)	22.80	182.40	37.55	300.40		
1 Forklift		233.00		256.30	14.56	16.02
16 L.H., Daily Totals		$614.20		$878.30	$38.39	$54.90
Crew B-68	**Hr.**	**Daily**	**Hr.**	**Daily**	**Bare Costs**	**Incl. O&P**
2 Millwrights	$24.85	$397.60	$40.20	$643.20	$24.17	$39.32
1 Equip. Oper. (light)	22.80	182.40	37.55	300.40		
1 Forklift		233.00		256.30	9.71	10.68
24 L.H., Daily Totals		$813.00		$1199.90	$33.88	$50.00
Crew B-69	**Hr.**	**Daily**	**Hr.**	**Daily**	**Bare Costs**	**Incl. O&P**
1 Labor Foreman (outside)	$19.35	$154.80	$32.85	$262.80	$19.48	$32.69
3 Laborers	17.35	416.40	29.45	706.80		
1 Equip Oper. (crane)	24.75	198.00	40.75	326.00		
1 Equip Oper. Oiler	20.75	166.00	34.20	273.60		
1 Truck Crane, 80 Ton		995.00		1094.50	20.73	22.80
48 L.H., Daily Totals		$1930.20		$2663.70	$40.21	$55.49
Crew B-69A	**Hr.**	**Daily**	**Hr.**	**Daily**	**Bare Costs**	**Incl. O&P**
1 Labor Foreman	$19.35	$154.80	$32.85	$262.80	$19.72	$32.94
3 Laborers	17.35	416.40	29.45	706.80		
1 Equip. Oper. (medium)	23.90	191.20	39.35	314.80		
1 Concrete Finisher	23.00	184.00	37.10	296.80		
1 Curb Paver		582.80		641.10	12.14	13.36
48 L.H., Daily Totals		$1529.20		$2222.30	$31.86	$46.30
Crew B-69B	**Hr.**	**Daily**	**Hr.**	**Daily**	**Bare Costs**	**Incl. O&P**
1 Labor Foreman	$19.35	$154.80	$32.85	$262.80	$19.72	$32.94
3 Laborers	17.35	416.40	29.45	706.80		
1 Equip. Oper. (medium)	23.90	191.20	39.35	314.80		
1 Cement Finisher	23.00	184.00	37.10	296.80		
1 Curb/Gutter Paver		725.00		797.50	15.10	16.61
48 L.H., Daily Totals		$1671.40		$2378.70	$34.82	$49.55

Crew No.	Bare Costs		Incl. Subs O & P		Cost Per Labor-Hour	
Crew B-70	**Hr.**	**Daily**	**Hr.**	**Daily**	**Bare Costs**	**Incl. O&P**
Labor Foreman (outside)	$19.35	$154.80	$32.85	$262.80	$20.44	$34.18
Laborers	17.35	416.40	29.45	706.80		
Equip. Oper. (med.)	23.90	573.60	39.35	944.40		
Motor Grader, 30,000 Lb.		456.60		502.25		
Grader Attach., Ripper		71.40		78.55		
Road Sweeper, S.P.		445.00		489.50		
F.E. Loader, 1 3/4 C.Y.		241.00		265.10	21.68	23.85
56 L.H., Daily Totals		$2358.80		$3249.40	$42.12	$58.03

Crew B-71	**Hr.**	**Daily**	**Hr.**	**Daily**	**Bare Costs**	**Incl. O&P**
1 Labor Foreman (outside)	$19.35	$154.80	$32.85	$262.80	$20.44	$34.18
3 Laborers	17.35	416.40	29.45	706.80		
3 Equip. Oper. (med.)	23.90	573.60	39.35	944.40		
1 Pvmt. Profiler, 750 H.P.		4266.00		4692.60		
1 Road Sweeper, S.P.		445.00		489.50		
1 F.E. Loader, 1 3/4 C.Y.		241.00		265.10	88.43	97.27
56 L.H., Daily Totals		$6096.80		$7361.20	$108.87	$131.45

Crew B-72	**Hr.**	**Daily**	**Hr.**	**Daily**	**Bare Costs**	**Incl. O&P**
1 Labor Foreman (outside)	$19.35	$154.80	$32.85	$262.80	$20.88	$34.83
3 Laborers	17.35	416.40	29.45	706.80		
4 Equip. Oper. (med.)	23.90	764.80	39.35	1259.20		
1 Pvmt. Profiler, 750 H.P.		4266.00		4692.60		
1 Hammermill, 250 H.P.		1344.00		1478.40		
1 Windrow Loader		791.20		870.30		
1 Mix Paver 165 H.P.		1656.00		1821.60		
1 Roller, Pneu. Tire, 12 T.		241.80		266.00	129.67	142.64
64 L.H., Daily Totals		$9635.00		$11357.70	$150.55	$177.47

Crew B-73	**Hr.**	**Daily**	**Hr.**	**Daily**	**Bare Costs**	**Incl. O&P**
1 Labor Foreman (outside)	$19.35	$154.80	$32.85	$262.80	$21.69	$36.06
2 Laborers	17.35	277.60	29.45	471.20		
5 Equip. Oper. (med.)	23.90	956.00	39.35	1574.00		
1 Road Mixer, 310 H.P.		1546.00		1700.60		
1 Roller, Tandem, 12 Ton		179.20		197.10		
1 Hammermill, 250 H.P.		1344.00		1478.40		
1 Motor Grader, 30,000 Lb.		456.60		502.25		
.5 F.E. Loader, 1 3/4 C.Y.		120.50		132.55		
.5 Truck, 30 Ton		104.10		114.50		
.5 Water Tank 5000 Gal.		58.40		64.25	59.51	65.46
64 L.H., Daily Totals		$5197.20		$6497.65	$81.20	$101.52

Crew B-74	**Hr.**	**Daily**	**Hr.**	**Daily**	**Bare Costs**	**Incl. O&P**
1 Labor Foreman (outside)	$19.35	$154.80	$32.85	$262.80	$21.43	$35.60
1 Laborer	17.35	138.80	29.45	235.60		
4 Equip. Oper. (med.)	23.90	764.80	39.35	1259.20		
2 Truck Drivers (heavy)	19.55	312.80	32.55	520.80		
1 Motor Grader, 30,000 Lb.		456.60		502.25		
1 Grader Attach., Ripper		71.40		78.55		
2 Stabilizers, 310 H.P.		2090.00		2299.00		
1 Flatbed Truck, 3 Ton		176.20		193.80		
1 Chem. Spreader, Towed		54.60		60.05		
1 Vibr. Roller, 29,000 Lb.		433.20		476.50		
1 Water Tank 5000 Gal.		116.80		128.50		
1 Truck, 30 Ton		208.20		229.00	56.36	62.00
64 L.H., Daily Totals		$4978.20		$6246.05	$77.79	$97.60

Crew B-75	**Hr.**	**Daily**	**Hr.**	**Daily**	**Bare Costs**	**Incl. O&P**
1 Labor Foreman (outside)	$19.35	$154.80	$32.85	$262.80	$21.69	$36.04
1 Laborer	17.35	138.80	29.45	235.60		
4 Equip. Oper. (med.)	23.90	764.80	39.35	1259.20		
1 Truck Driver (heavy)	19.55	156.40	32.55	260.40		
1 Motor Grader, 30,000 Lb.		456.60		502.25		
1 Grader Attach., Ripper		71.40		78.55		
2 Stabilizers, 310 H.P.		2090.00		2299.00		
1 Dist. Truck, 3000 Gal.		233.80		257.20		
1 Vibr. Roller, 29,000 Lb.		433.20		476.50	58.66	64.53
56 L.H., Daily Totals		$4499.80		$5631.50	$80.35	$100.57

Crew B-76	**Hr.**	**Daily**	**Hr.**	**Daily**	**Bare Costs**	**Incl. O&P**
1 Dock Builder Foreman	$25.30	$202.40	$45.25	$362.00	$23.56	$41.02
5 Dock Builders	23.30	932.00	41.65	1666.00		
2 Equip. Oper. (crane)	24.75	396.00	40.75	652.00		
1 Equip. Oper. Oiler	20.75	166.00	34.20	273.60		
1 Crawler Crane, 50 Ton		1206.00		1326.60		
1 Barge, 400 Ton		273.20		300.50		
1 Hammer, 15K Ft. Lbs.		359.80		395.80		
60 L.F. Leads, 15K Ft. Lbs.		75.00		82.50		
1 Air Compr., 600 C.F.M.		297.80		327.60		
2 50 Ft. Air Hoses, 3" Dia.		35.40		38.95	31.25	34.38
72 L.H., Daily Totals		$3943.60		$5425.55	$54.81	$75.40

Crew B-76A	**Hr.**	**Daily**	**Hr.**	**Daily**	**Bare Costs**	**Incl. O&P**
1 Laborer Foreman	$19.35	$154.80	$32.85	$262.80	$18.95	$31.88
5 Laborers	17.35	694.00	29.45	1178.00		
1 Equip. Oper. (crane)	24.75	198.00	40.75	326.00		
1 Equip. Oper. Oiler	20.75	166.00	34.20	273.60		
1 Crawler Crane, 50 Ton		1206.00		1326.60		
1 Barge, 400 Ton		273.20		300.50	23.11	25.42
64 L.H., Daily Totals		$2692.00		$3667.50	$42.06	$57.30

Crew B-77	**Hr.**	**Daily**	**Hr.**	**Daily**	**Bare Costs**	**Incl. O&P**
1 Labor Foreman	$19.35	$154.80	$32.85	$262.80	$18.09	$30.58
3 Laborers	17.35	416.40	29.45	706.80		
1 Truck Driver (light)	19.05	152.40	31.70	253.60		
1 Crack Cleaner, 25 H.P.		44.40		48.85		
1 Crack Filler, Trailer Mtd.		157.00		172.70		
1 Flatbed Truck, 3 Ton		176.20		193.80	9.44	10.38
40 L.H., Daily Totals		$1101.20		$1638.55	$27.53	$40.96

Crew B-78	**Hr.**	**Daily**	**Hr.**	**Daily**	**Bare Costs**	**Incl. O&P**
1 Labor Foreman	$19.35	$154.80	$32.85	$262.80	$17.75	$30.13
4 Laborers	17.35	555.20	29.45	942.40		
1 Paint Striper, S.P.		134.20		147.60		
1 Flatbed Truck, 3 Ton		176.20		193.80		
1 Pickup Truck, 3/4 Ton		78.00		85.80	9.71	10.68
40 L.H., Daily Totals		$1098.40		$1632.40	$27.46	$40.81

Crew B-79	**Hr.**	**Daily**	**Hr.**	**Daily**	**Bare Costs**	**Incl. O&P**
1 Labor Foreman	$19.35	$154.80	$32.85	$262.80	$17.85	$30.30
3 Laborers	17.35	416.40	29.45	706.80		
1 Thermo. Striper, T.M.		549.20		604.10		
1 Flatbed Truck, 3 Ton		176.20		193.80		
2 Pickup Truck, 3/4 Ton		156.00		171.60	27.54	30.30
32 L.H., Daily Totals		$1452.60		$1939.10	$45.39	$60.60

Crews

Crew B-80	Bare Costs Hr.	Bare Costs Daily	Incl. Subs O & P Hr.	Incl. Subs O & P Daily	Cost Per Labor-Hour Bare Costs	Cost Per Labor-Hour Incl. O&P
1 Labor Foreman	$19.35	$154.80	$32.85	$262.80	$18.02	$30.58
2 Laborers	17.35	277.60	29.45	471.20		
1 Flatbed Truck, 3 Ton		176.20		193.80		
1 Fence Post Auger, T.M.		357.40		393.15	22.23	24.46
24 L.H., Daily Totals		$966.00		$1320.95	$40.25	$55.04

Crew B-80A	Hr.	Daily	Hr.	Daily	Bare Costs	Incl. O&P
3 Laborers	$17.35	$416.40	$29.45	$706.80	$17.35	$29.45
1 Flatbed Truck, 3 Ton		176.20		193.80	7.34	8.08
24 L.H., Daily Totals		$592.60		$900.60	$24.69	$37.53

Crew B-80B	Hr.	Daily	Hr.	Daily	Bare Costs	Incl. O&P
3 Laborers	$17.35	$416.40	$29.45	$706.80	$18.71	$31.48
1 Equip. Oper. (light)	22.80	182.40	37.55	300.40		
1 Crane, Flatbed Mnt.		209.60		230.55	6.55	7.21
32 L.H., Daily Totals		$808.40		$1237.75	$25.26	$38.69

Crew B-80C	Hr.	Daily	Hr.	Daily	Bare Costs	Incl. O&P
2 Laborers	$17.35	$277.60	$29.45	$471.20	$17.92	$30.20
1 Truck Driver (light)	19.05	152.40	31.70	253.60		
1 Light Truck, 1.5 Ton		128.80		141.70		
1 Manual fence post auger, gas		5.00		5.50	5.58	6.13
24 L.H., Daily Totals		$563.80		$872.00	$23.50	$36.33

Crew B-81	Hr.	Daily	Hr.	Daily	Bare Costs	Incl. O&P
1 Laborer	$17.35	$138.80	$29.45	$235.60	$18.45	$31.00
1 Truck Driver (heavy)	19.55	156.40	32.55	260.40		
1 Hydromulcher, T.M.		196.00		215.60		
1 Tractor Truck, 4x2		208.20		229.00	25.26	27.79
16 L.H., Daily Totals		$699.40		$940.60	$43.71	$58.79

Crew B-82	Hr.	Daily	Hr.	Daily	Bare Costs	Incl. O&P
1 Laborer	$17.35	$138.80	$29.45	$235.60	$20.08	$33.50
1 Equip. Oper. (light)	22.80	182.40	37.55	300.40		
1 Horiz. Borer, 6 H.P.		62.60		68.85	3.91	4.30
16 L.H., Daily Totals		$383.80		$604.85	$23.99	$37.80

Crew B-83	Hr.	Daily	Hr.	Daily	Bare Costs	Incl. O&P
1 Tugboat Captain	$23.90	$191.20	$39.35	$314.80	$20.63	$34.40
1 Tugboat Hand	17.35	138.80	29.45	235.60		
1 Tugboat, 250 H.P.		440.60		484.65	27.54	30.29
16 L.H., Daily Totals		$770.60		$1035.05	$48.17	$64.69

Crew B-84	Hr.	Daily	Hr.	Daily	Bare Costs	Incl. O&P
1 Equip. Oper. (med.)	$23.90	$191.20	$39.35	$314.80	$23.90	$39.35
1 Rotary Mower/Tractor		228.80		251.70	28.60	31.46
8 L.H., Daily Totals		$420.00		$566.50	$52.50	$70.81

Crew B-85	Hr.	Daily	Hr.	Daily	Bare Costs	Incl. O&P
3 Laborers	$17.35	$416.40	$29.45	$706.80	$19.10	$32.05
1 Equip. Oper. (med.)	23.90	191.20	39.35	314.80		
1 Truck Driver (heavy)	19.55	156.40	32.55	260.40		
1 Aerial Lift Truck, 80'		496.80		546.50		
1 Brush Chipper, 130 H.P.		164.80		181.30		
1 Pruning Saw, Rotary		8.75		9.65	16.76	18.43
40 L.H., Daily Totals		$1434.35		$2019.45	$35.86	$50.48

Crew B-86	Hr.	Daily	Hr.	Daily	Bare Costs	Incl. O&P
1 Equip. Oper. (med.)	$23.90	$191.20	$39.35	$314.80	$23.90	$39.35
1 Stump Chipper, S.P.		70.70		77.75	8.84	9.72
8 L.H., Daily Totals		$261.90		$392.55	$32.74	$49.07

Crew B-86A	Hr.	Daily	Hr.	Daily	Bare Costs	Incl. O&P
1 Equip. Oper. (medium)	$23.90	$191.20	$39.35	$314.80	$23.90	$39.35
1 Grader, 30,000 Lbs.		456.60		502.25	57.08	62.78
8 L.H., Daily Totals		$647.80		$817.05	$80.98	$102.13

Crew B-86B	Hr.	Daily	Hr.	Daily	Bare Costs	Incl. O&P
1 Equip. Oper. (medium)	$23.90	$191.20	$39.35	$314.80	$23.90	$39.35
1 Dozer, 200 H.P.		919.60		1011.55	114.95	126.45
8 L.H., Daily Totals		$1110.80		$1326.35	$138.85	$165.80

Crew B-87	Hr.	Daily	Hr.	Daily	Bare Costs	Incl. O&P
1 Laborer	$17.35	$138.80	$29.45	$235.60	$22.59	$37.37
4 Equip. Oper. (med.)	23.90	764.80	39.35	1259.20		
2 Feller Bunchers, 50 H.P.		864.80		951.30		
1 Log Chipper, 22" Tree		1062.00		1168.20		
1 Dozer, 105 H.P.		453.80		499.20		
1 Chainsaw, Gas, 36" Long		33.40		36.75	60.35	66.39
40 L.H., Daily Totals		$3317.60		$4150.25	$82.94	$103.76

Crew B-88	Hr.	Daily	Hr.	Daily	Bare Costs	Incl. O&P
1 Laborer	$17.35	$138.80	$29.45	$235.60	$22.96	$37.94
6 Equip. Oper. (med.)	23.90	1147.20	39.35	1888.80		
2 Feller Bunchers, 50 H.P.		864.80		951.30		
1 Log Chipper, 22" Tree		1062.00		1168.20		
2 Log Skidders, 50 H.P.		1490.80		1639.90		
1 Dozer, 105 H.P.		453.80		499.20		
1 Chainsaw, Gas, 36" Long		33.40		36.75	69.73	76.70
56 L.H., Daily Totals		$5190.80		$6419.75	$92.69	$114.64

Crew B-89	Hr.	Daily	Hr.	Daily	Bare Costs	Incl. O&P
1 Skilled Worker	$23.90	$191.20	$40.50	$324.00	$20.63	$34.98
1 Building Laborer	17.35	138.80	29.45	235.60		
1 Cutting Machine		50.30		55.35	3.14	3.45
16 L.H., Daily Totals		$380.30		$614.95	$23.77	$38.43

Crew B-89A	Hr.	Daily	Hr.	Daily	Bare Costs	Incl. O&P
1 Skilled Worker	$23.90	$191.20	$40.50	$324.00	$20.63	$34.98
1 Laborer	17.35	138.80	29.45	235.60		
1 Core Drill (large)		103.05		113.35	6.44	7.08
16 L.H., Daily Totals		$433.05		$672.95	$27.07	$42.06

Crew B-89B	Hr.	Daily	Hr.	Daily	Bare Costs	Incl. O&P
1 Equip. Oper. (light)	$22.80	$182.40	$37.55	$300.40	$20.93	$34.63
1 Truck Driver, Light	19.05	152.40	31.70	253.60		
1 Wall Saw, Hydraulic, 10 H.P.		73.30		80.65		
1 Generator, Diesel, 100 KW		176.60		194.25		
1 Water Tank, 65 Gal.		13.60		14.95		
1 Flatbed Truck, 3 Ton		176.20		193.80	27.48	30.23
16 L.H., Daily Totals		$774.50		$1037.65	$48.41	$64.86

Crew No.	Bare Costs		Incl. Subs O & P		Cost Per Labor-Hour	
Crew B-90	**Hr.**	**Daily**	**Hr.**	**Daily**	**Bare Costs**	**Incl. O&P**
Labor Foreman (outside)	$19.35	$154.80	$32.85	$262.80	$19.51	$32.68
Laborers	17.35	416.40	29.45	706.80		
Equip. Oper. (light)	22.80	364.80	37.55	600.80		
Truck Drivers (heavy)	19.55	312.80	32.55	520.80		
Road Mixer, 310 H.P.		1546.00		1700.60		
Dist. Truck, 2000 Gal.		204.80		225.30	27.36	30.09
4 L.H., Daily Totals		$2999.60		$4017.10	$46.87	$62.77
Crew B-90A	**Hr.**	**Daily**	**Hr.**	**Daily**	**Bare Costs**	**Incl. O&P**
Labor Foreman	$19.35	$154.80	$32.85	$262.80	$21.38	$35.59
Laborers	17.35	277.60	29.45	471.20		
Equip. Oper. (medium)	23.90	764.80	39.35	1259.20		
Graders, 30,000 Lbs.		913.20		1004.50		
Roller, Steel Wheel		179.20		197.10		
Roller, Pneumatic Wheel		241.80		266.00	23.83	26.21
56 L.H., Daily Totals		$2531.40		$3460.80	$45.21	$61.80
Crew B-90B	**Hr.**	**Daily**	**Hr.**	**Daily**	**Bare Costs**	**Incl. O&P**
1 Labor Foreman	$19.35	$154.80	$32.85	$262.80	$20.96	$34.97
2 Laborers	17.35	277.60	29.45	471.20		
3 Equip. Oper. (medium)	23.90	573.60	39.35	944.40		
1 Roller, Steel Wheel		179.20		197.10		
1 Roller, Pneumatic Wheel		241.80		266.00		
1 Road Mixer, 310 H.P.		1546.00		1700.60	40.98	45.08
48 L.H., Daily Totals		$2973.00		$3842.10	$61.94	$80.05
Crew B-91	**Hr.**	**Daily**	**Hr.**	**Daily**	**Bare Costs**	**Incl. O&P**
1 Labor Foreman (outside)	$19.35	$154.80	$32.85	$262.80	$21.15	$35.21
2 Laborers	17.35	277.60	29.45	471.20		
4 Equip. Oper. (med.)	23.90	764.80	39.35	1259.20		
1 Truck Driver (heavy)	19.55	156.40	32.55	260.40		
1 Dist. Truck, 3000 Gal.		233.80		257.20		
1 Aggreg. Spreader, S.P.		771.20		848.30		
1 Roller, Pneu. Tire, 12 Ton		241.80		266.00		
1 Roller, Steel, 10 Ton		179.20		197.10	22.28	24.51
64 L.H., Daily Totals		$2779.60		$3822.20	$43.43	$59.72
Crew B-92	**Hr.**	**Daily**	**Hr.**	**Daily**	**Bare Costs**	**Incl. O&P**
1 Labor Foreman (outside)	$19.35	$154.80	$32.85	$262.80	$17.85	$30.30
3 Laborers	17.35	416.40	29.45	706.80		
1 Crack Cleaner, 25 H.P.		44.40		48.85		
1 Air Compressor		55.40		60.95		
1 Tar Kettle, T.M.		40.95		45.05		
1 Flatbed Truck, 3 Ton		176.20		193.80	9.90	10.90
32 L.H., Daily Totals		$888.15		$1318.25	$27.75	$41.20
Crew B-93	**Hr.**	**Daily**	**Hr.**	**Daily**	**Bare Costs**	**Incl. O&P**
1 Equip. Oper. (med.)	$23.90	$191.20	$39.35	$314.80	$23.90	$39.35
1 Feller Buncher, 50 H.P.		432.40		475.65	54.05	59.46
8 L.H., Daily Totals		$623.60		$790.45	$77.95	$98.81
Crew B-94A	**Hr.**	**Daily**	**Hr.**	**Daily**	**Bare Costs**	**Incl. O&P**
1 Laborer	$17.35	$138.80	$29.45	$235.60	$17.35	$29.45
1 Diaph. Water Pump, 2"		50.60		55.65		
1 20 Ft. Suction Hose, 2"		2.55		2.80		
2 50 Ft. Disch. Hoses, 2"		3.80		4.20	7.12	7.83
8 L.H., Daily Totals		$195.75		$298.25	$24.47	$37.28

Crew No.	Bare Costs		Incl. Subs O & P		Cost Per Labor-Hour	
Crew B-94B	**Hr.**	**Daily**	**Hr.**	**Daily**	**Bare Costs**	**Incl. O&P**
1 Laborer	$17.35	$138.80	$29.45	$235.60	$17.35	$29.45
1 Diaph. Water Pump, 4"		74.40		81.85		
1 20 Ft. Suction Hose, 4"		5.05		5.55		
2 50 Ft. Disch. Hoses, 4"		7.10		7.80	10.82	11.90
8 L.H., Daily Totals		$225.35		$330.80	$28.17	$41.35
Crew B-94C	**Hr.**	**Daily**	**Hr.**	**Daily**	**Bare Costs**	**Incl. O&P**
1 Laborer	$17.35	$138.80	$29.45	$235.60	$17.35	$29.45
1 Centr. Water Pump, 3"		53.60		58.95		
1 20 Ft. Suction Hose, 3"		4.25		4.70		
2 50 Ft. Disch. Hoses, 3"		5.00		5.50	7.86	8.64
8 L.H., Daily Totals		$201.65		$304.75	$25.21	$38.09
Crew B-94D	**Hr.**	**Daily**	**Hr.**	**Daily**	**Bare Costs**	**Incl. O&P**
1 Laborer	$17.35	$138.80	$29.45	$235.60	$17.35	$29.45
1 Centr. Water Pump, 6"		217.00		238.70		
1 20 Ft. Suction Hose, 6"		12.80		14.10		
2 50 Ft. Disch. Hoses, 6"		17.80		19.60	30.95	34.05
8 L.H., Daily Totals		$386.40		$508.00	$48.30	$63.50
Crew B-95A	**Hr.**	**Daily**	**Hr.**	**Daily**	**Bare Costs**	**Incl. O&P**
1 Equip. Oper. (crane)	$24.75	$198.00	$40.75	$326.00	$21.05	$35.10
1 Laborer	17.35	138.80	29.45	235.60		
1 Hyd. Excavator, 5/8 C.Y.		432.40		475.65	27.03	29.73
16 L.H., Daily Totals		$769.20		$1037.25	$48.08	$64.83
Crew B-95B	**Hr.**	**Daily**	**Hr.**	**Daily**	**Bare Costs**	**Incl. O&P**
1 Equip. Oper. (crane)	$24.75	$198.00	$40.75	$326.00	$21.05	$35.10
1 Laborer	17.35	138.80	29.45	235.60		
1 Hyd. Excavator, 1.5 C.Y.		720.40		792.45	45.03	49.53
16 L.H., Daily Totals		$1057.20		$1354.05	$66.08	$84.63
Crew B-95C	**Hr.**	**Daily**	**Hr.**	**Daily**	**Bare Costs**	**Incl. O&P**
1 Equip. Oper. (crane)	$24.75	$198.00	$40.75	$326.00	$21.05	$35.10
1 Laborer	17.35	138.80	29.45	235.60		
1 Hyd. Excavator, 2.5 C.Y.		1215.00		1336.50	75.94	83.53
16 L.H., Daily Totals		$1551.80		$1898.10	$96.99	$118.63
Crew C-1	**Hr.**	**Daily**	**Hr.**	**Daily**	**Bare Costs**	**Incl. O&P**
2 Carpenters	$24.00	$384.00	$40.75	$652.00	$20.73	$35.18
1 Carpenter Helper	17.55	140.40	29.75	238.00		
1 Laborer	17.35	138.80	29.45	235.60		
32 L.H., Daily Totals		$663.20		$1125.60	$20.73	$35.18
Crew C-2	**Hr.**	**Daily**	**Hr.**	**Daily**	**Bare Costs**	**Incl. O&P**
1 Carpenter Foreman (out)	$26.00	$208.00	$44.10	$352.80	$21.08	$35.76
2 Carpenters	24.00	384.00	40.75	652.00		
2 Carpenter Helpers	17.55	280.80	29.75	476.00		
1 Laborer	17.35	138.80	29.45	235.60		
48 L.H., Daily Totals		$1011.60		$1716.40	$21.08	$35.76
Crew C-2A	**Hr.**	**Daily**	**Hr.**	**Daily**	**Bare Costs**	**Incl. O&P**
1 Carpenter Foreman (out)	$26.00	$208.00	$44.10	$352.80	$23.06	$38.82
3 Carpenters	24.00	576.00	40.75	978.00		
1 Cement Finisher	23.00	184.00	37.10	296.80		
1 Laborer	17.35	138.80	29.45	235.60		
48 L.H., Daily Totals		$1106.80		$1863.20	$23.06	$38.82

Crew No.	Bare Costs		Incl. Subs O & P		Cost Per Labor-Hour	
Crew C-3	**Hr.**	**Daily**	**Hr.**	**Daily**	**Bare Costs**	**Incl. O&P**
1 Rodman Foreman	$27.45	$219.60	$49.15	$393.20	$22.33	$38.96
3 Rodmen (reinf.)	25.45	610.80	45.55	1093.20		
1 Equip. Oper. (light)	22.80	182.40	37.55	300.40		
3 Laborers	17.35	416.40	29.45	706.80		
3 Stressing Equipment		40.80		44.90		
.5 Grouting Equipment		76.40		84.05	1.83	2.01
64 L.H., Daily Totals		$1546.40		$2622.55	$24.16	$40.97
Crew C-4	**Hr.**	**Daily**	**Hr.**	**Daily**	**Bare Costs**	**Incl. O&P**
1 Rodman Foreman	$27.45	$219.60	$49.15	$393.20	$23.93	$42.43
2 Rodmen (reinf.)	25.45	407.20	45.55	728.80		
1 Building Laborer	17.35	138.80	29.45	235.60		
3 Stressing Equipment		40.80		44.90	1.28	1.40
32 L.H., Daily Totals		$806.40		$1402.50	$25.21	$43.83
Crew C-5	**Hr.**	**Daily**	**Hr.**	**Daily**	**Bare Costs**	**Incl. O&P**
1 Rodman Foreman	$27.45	$219.60	$49.15	$393.20	$22.97	$39.98
2 Rodmen (reinf.)	25.45	407.20	45.55	728.80		
1 Equip. Oper. (crane)	24.75	198.00	40.75	326.00		
2 Building Laborers	17.35	277.60	29.45	471.20		
1 Hyd. Crane, 25 Ton		616.80		678.50	12.85	14.14
48 L.H., Daily Totals		$1719.20		$2597.70	$35.82	$54.12
Crew C-6	**Hr.**	**Daily**	**Hr.**	**Daily**	**Bare Costs**	**Incl. O&P**
1 Labor Foreman (outside)	$19.35	$154.80	$32.85	$262.80	$18.63	$31.29
4 Laborers	17.35	555.20	29.45	942.40		
1 Cement Finisher	23.00	184.00	37.10	296.80		
2 Gas Engine Vibrators		48.00		52.80	1.00	1.10
48 L.H., Daily Totals		$942.00		$1554.80	$19.63	$32.39
Crew C-7	**Hr.**	**Daily**	**Hr.**	**Daily**	**Bare Costs**	**Incl. O&P**
1 Labor Foreman (outside)	$19.35	$154.80	$32.85	$262.80	$19.31	$32.31
5 Laborers	17.35	694.00	29.45	1178.00		
1 Cement Finisher	23.00	184.00	37.10	296.80		
1 Equip. Oper. (med.)	23.90	191.20	39.35	314.80		
1 Equip. Oper. (oiler)	20.75	166.00	34.20	273.60		
2 Gas Engine Vibrators		48.00		52.80		
1 Concrete Bucket, 1 C.Y.		16.00		17.60		
1 Hyd. Crane, 55 Ton		895.40		984.95	13.33	14.66
72 L.H., Daily Totals		$2349.40		$3381.35	$32.64	$46.97
Crew C-8	**Hr.**	**Daily**	**Hr.**	**Daily**	**Bare Costs**	**Incl. O&P**
1 Labor Foreman (outside)	$19.35	$154.80	$32.85	$262.80	$20.19	$33.54
3 Laborers	17.35	416.40	29.45	706.80		
2 Cement Finishers	23.00	368.00	37.10	593.60		
1 Equip. Oper. (med.)	23.90	191.20	39.35	314.80		
1 Concrete Pump (small)		704.20		774.60	12.58	13.83
56 L.H., Daily Totals		$1834.60		$2652.60	$32.77	$47.37
Crew C-8A	**Hr.**	**Daily**	**Hr.**	**Daily**	**Bare Costs**	**Incl. O&P**
1 Labor Foreman (outside)	$19.35	$154.80	$32.85	$262.80	$19.57	$32.57
3 Laborers	17.35	416.40	29.45	706.80		
2 Cement Finishers	23.00	368.00	37.10	593.60		
48 L.H., Daily Totals		$939.20		$1563.20	$19.57	$32.57

Crew No.	Bare Costs		Incl. Subs O & P		Cost Per Labor-Hour	
Crew C-8B	**Hr.**	**Daily**	**Hr.**	**Daily**	**Bare Costs**	**Incl. O&P**
1 Labor Foreman (outside)	$19.35	$154.80	$32.85	$262.80	$19.06	$32.11
3 Laborers	17.35	416.40	29.45	706.80		
1 Equipment Operator	23.90	191.20	39.35	314.80		
1 Vibrating Screed		43.05		47.35		
1 Vibratory Roller		433.20		476.50		
1 Dozer, 200 H.P.		919.60		1011.55	34.90	38.38
40 L.H., Daily Totals		$2158.25		$2819.80	$53.96	$70.49
Crew C-8C	**Hr.**	**Daily**	**Hr.**	**Daily**	**Bare Costs**	**Incl. O&P**
1 Labor Foreman (outside)	$19.35	$154.80	$32.85	$262.80	$19.72	$32.94
3 Laborers	17.35	416.40	29.45	706.80		
1 Cement Finisher	23.00	184.00	37.10	296.80		
1 Equipment Operator (med.)	23.90	191.20	39.35	314.80		
1 Shotcrete Rig, 12 CY/hr		223.80		246.20	4.66	5.13
48 L.H., Daily Totals		$1170.20		$1827.40	$24.38	$38.07
Crew C-8D	**Hr.**	**Daily**	**Hr.**	**Daily**	**Bare Costs**	**Incl. O&P**
1 Labor Foreman (outside)	$19.35	$154.80	$32.85	$262.80	$20.63	$34.24
1 Laborer	17.35	138.80	29.45	235.60		
1 Cement Finisher	23.00	184.00	37.10	296.80		
1 Equipment Operator (light)	22.80	182.40	37.55	300.40		
1 Compressor, 250 CFM		112.40		123.65		
2 Hoses, 1", 50'		7.30		8.05	3.74	4.11
32 L.H., Daily Totals		$779.70		$1227.30	$24.37	$38.35
Crew C-8E	**Hr.**	**Daily**	**Hr.**	**Daily**	**Bare Costs**	**Incl. O&P**
1 Labor Foreman (outside)	$19.35	$154.80	$32.85	$262.80	$20.63	$34.24
1 Laborer	17.35	138.80	29.45	235.60		
1 Cement Finisher	23.00	184.00	37.10	296.80		
1 Equipment Operator (light)	22.80	182.40	37.55	300.40		
1 Compressor, 250 CFM		112.40		123.65		
2 Hoses, 1", 50'		7.30		8.05		
1 Concrete Pump (small)		704.20		774.60	25.75	28.32
32 L.H., Daily Totals		$1483.90		$2001.90	$46.38	$62.56
Crew C-10	**Hr.**	**Daily**	**Hr.**	**Daily**	**Bare Costs**	**Incl. O&P**
1 Laborer	$17.35	$138.80	$29.45	$235.60	$21.12	$34.55
2 Cement Finishers	23.00	368.00	37.10	593.60		
24 L.H., Daily Totals		$506.80		$829.20	$21.12	$34.55
Crew C-10B	**Hr.**	**Daily**	**Hr.**	**Daily**	**Bare Costs**	**Incl. O&P**
3 Laborers	$17.35	$416.40	$29.45	$706.80	$19.61	$32.51
2 Cement Finishers	23.00	368.00	37.10	593.60		
1 Concrete mixer, 10 CF		123.20		135.50		
2 Concrete finisher, 48" dia		49.20		54.10	4.31	4.74
40 L.H., Daily Totals		$956.80		$1490.00	$23.92	$37.25
Crew C-11	**Hr.**	**Daily**	**Hr.**	**Daily**	**Bare Costs**	**Incl. O&P**
1 Skilled Worker Foreman	$25.90	$207.20	$43.90	$351.20	$24.31	$41.02
5 Skilled Worker	23.90	956.00	40.50	1620.00		
1 Equip. Oper. (crane)	24.75	198.00	40.75	326.00		
1 Truck Crane, 150 Ton		1445.00		1589.50	25.80	28.38
56 L.H., Daily Totals		$2806.20		$3886.70	$50.11	$69.40
Crew C-12	**Hr.**	**Daily**	**Hr.**	**Daily**	**Bare Costs**	**Incl. O&P**
1 Carpenter Foreman (out)	$26.00	$208.00	$44.10	$352.80	$23.35	$39.43
3 Carpenters	24.00	576.00	40.75	978.00		
1 Laborer	17.35	138.80	29.45	235.60		
1 Equip. Oper. (crane)	24.75	198.00	40.75	326.00		
1 Hyd. Crane, 12 Ton		602.60		662.85	12.55	13.81
48 L.H., Daily Totals		$1723.40		$2555.25	$35.90	$53.24

Crew No.	Bare Costs		Incl. Subs O & P		Cost Per Labor-Hour	
Crew C-13	**Hr.**	**Daily**	**Hr.**	**Daily**	**Bare Costs**	**Incl. O&P**
Struc. Steel Worker	$25.55	$408.80	$49.30	$788.80	$25.03	$46.45
Carpenter	24.00	192.00	40.75	326.00		
Gas Welding Machine		81.20		89.30	3.38	3.72
4 L.H., Daily Totals		$682.00		$1204.10	$28.41	$50.17
Crew C-14	**Hr.**	**Daily**	**Hr.**	**Daily**	**Bare Costs**	**Incl. O&P**
Carpenter Foreman (out)	$26.00	$208.00	$44.10	$352.80	$21.13	$35.84
Carpenters	24.00	576.00	40.75	978.00		
Carpenter Helpers	17.55	280.80	29.75	476.00		
Laborers	17.35	555.20	29.45	942.40		
Rodmen (reinf.)	25.45	407.20	45.55	728.80		
Rodman Helpers	17.55	280.80	29.75	476.00		
Cement Finishers	23.00	368.00	37.10	593.60		
Equip. Oper. (crane)	24.75	198.00	40.75	326.00		
Crane, 80 Ton, & Tools		995.00		1094.50	7.32	8.05
136 L.H., Daily Totals		$3869.00		$5968.10	$28.45	$43.89
Crew C-14A	**Hr.**	**Daily**	**Hr.**	**Daily**	**Bare Costs**	**Incl. O&P**
Carpenter Foreman (out)	$26.00	$208.00	$44.10	$352.80	$23.74	$40.55
16 Carpenters	24.00	3072.00	40.75	5216.00		
4 Rodmen (reinf.)	25.45	814.40	45.55	1457.60		
2 Laborers	17.35	277.60	29.45	471.20		
1 Cement Finisher	23.00	184.00	37.10	296.80		
1 Equip. Oper. (med.)	23.90	191.20	39.35	314.80		
1 Gas Engine Vibrator		24.00		26.40		
1 Concrete Pump (small)		704.20		774.60	3.64	4.01
200 L.H., Daily Totals		$5475.40		$8910.20	$27.38	$44.56
Crew C-14B	**Hr.**	**Daily**	**Hr.**	**Daily**	**Bare Costs**	**Incl. O&P**
1 Carpenter Foreman (out)	$26.00	$208.00	$44.10	$352.80	$23.71	$40.41
16 Carpenters	24.00	3072.00	40.75	5216.00		
4 Rodmen (reinf.)	25.45	814.40	45.55	1457.60		
2 Laborers	17.35	277.60	29.45	471.20		
2 Cement Finishers	23.00	368.00	37.10	593.60		
1 Equip. Oper. (med.)	23.90	191.20	39.35	314.80		
1 Gas Engine Vibrator		24.00		26.40		
1 Concrete Pump (small)		704.20		774.60	3.50	3.85
208 L.H., Daily Totals		$5659.40		$9207.00	$27.21	$44.26
Crew C-14C	**Hr.**	**Daily**	**Hr.**	**Daily**	**Bare Costs**	**Incl. O&P**
1 Carpenter Foreman (out)	$26.00	$208.00	$44.10	$352.80	$22.38	$38.19
6 Carpenters	24.00	1152.00	40.75	1956.00		
2 Rodmen (reinf.)	25.45	407.20	45.55	728.80		
4 Laborers	17.35	555.20	29.45	942.40		
1 Cement Finisher	23.00	184.00	37.10	296.80		
1 Gas Engine Vibrator		24.00		26.40	.21	.24
112 L.H., Daily Totals		$2530.40		$4303.20	$22.59	$38.43
Crew C-14D	**Hr.**	**Daily**	**Hr.**	**Daily**	**Bare Costs**	**Incl. O&P**
1 Carpenter Foreman (out)	$26.00	$208.00	$44.10	$352.80	$23.62	$40.16
18 Carpenters	24.00	3456.00	40.75	5868.00		
2 Rodmen (reinf.)	25.45	407.20	45.55	728.80		
2 Laborers	17.35	277.60	29.45	471.20		
1 Cement Finisher	23.00	184.00	37.10	296.80		
1 Equip. Oper. (med.)	23.90	191.20	39.35	314.80		
1 Gas Engine Vibrator		24.00		26.40		
1 Concrete Pump (small)		704.20		774.60	3.64	4.01
200 L.H., Daily Totals		$5452.20		$8833.40	$27.26	$44.17

Crew No.	Bare Costs		Incl. Subs O & P		Cost Per Labor-Hour	
Crew C-14E	**Hr.**	**Daily**	**Hr.**	**Daily**	**Bare Costs**	**Incl. O&P**
1 Carpenter Foreman (out)	$26.00	$208.00	$44.10	$352.80	$22.80	$39.39
2 Carpenters	24.00	384.00	40.75	652.00		
4 Rodmen (reinf.)	25.45	814.40	45.55	1457.60		
3 Laborers	17.35	416.40	29.45	706.80		
1 Cement Finisher	23.00	184.00	37.10	296.80		
1 Gas Engine Vibrator		24.00		26.40	.27	.30
88 L.H., Daily Totals		$2030.80		$3492.40	$23.07	$39.69
Crew C-14F	**Hr.**	**Daily**	**Hr.**	**Daily**	**Bare Costs**	**Incl. O&P**
1 Laborer Foreman (out)	$19.35	$154.80	$32.85	$262.80	$21.34	$34.93
2 Laborers	17.35	277.60	29.45	471.20		
6 Cement Finishers	23.00	1104.00	37.10	1780.80		
1 Gas Engine Vibrator		24.00		26.40	.33	.37
72 L.H., Daily Totals		$1560.40		$2541.20	$21.67	$35.30
Crew C-14G	**Hr.**	**Daily**	**Hr.**	**Daily**	**Bare Costs**	**Incl. O&P**
1 Laborer Foreman (out)	$19.35	$154.80	$32.85	$262.80	$20.86	$34.31
2 Laborers	17.35	277.60	29.45	471.20		
4 Cement Finishers	23.00	736.00	37.10	1187.20		
1 Gas Engine Vibrator		24.00		26.40	.43	.47
56 L.H., Daily Totals		$1192.40		$1947.60	$21.29	$34.78
Crew C-14H	**Hr.**	**Daily**	**Hr.**	**Daily**	**Bare Costs**	**Incl. O&P**
1 Carpenter Foreman (out)	$26.00	$208.00	$44.10	$352.80	$23.30	$39.62
2 Carpenters	24.00	384.00	40.75	652.00		
1 Rodman (reinf.)	25.45	203.60	45.55	364.40		
1 Laborer	17.35	138.80	29.45	235.60		
1 Cement Finisher	23.00	184.00	37.10	296.80		
1 Gas Engine Vibrator		24.00		26.40	.50	.55
48 L.H., Daily Totals		$1142.40		$1928.00	$23.80	$40.17
Crew C-15	**Hr.**	**Daily**	**Hr.**	**Daily**	**Bare Costs**	**Incl. O&P**
1 Carpenter Foreman (out)	$26.00	$208.00	$44.10	$352.80	$21.94	$37.08
2 Carpenters	24.00	384.00	40.75	652.00		
3 Laborers	17.35	416.40	29.45	706.80		
2 Cement Finishers	23.00	368.00	37.10	593.60		
1 Rodman (reinf.)	25.45	203.60	45.55	364.40		
72 L.H., Daily Totals		$1580.00		$2669.60	$21.94	$37.08
Crew C-16	**Hr.**	**Daily**	**Hr.**	**Daily**	**Bare Costs**	**Incl. O&P**
1 Labor Foreman (outside)	$19.35	$154.80	$32.85	$262.80	$21.36	$36.21
3 Laborers	17.35	416.40	29.45	706.80		
2 Cement Finishers	23.00	368.00	37.10	593.60		
1 Equip. Oper. (med.)	23.90	191.20	39.35	314.80		
2 Rodmen (reinf.)	25.45	407.20	45.55	728.80		
1 Concrete Pump (small)		704.20		774.60	9.78	10.76
72 L.H., Daily Totals		$2241.80		$3381.40	$31.14	$46.97
Crew C-17	**Hr.**	**Daily**	**Hr.**	**Daily**	**Bare Costs**	**Incl. O&P**
2 Skilled Worker Foremen	$25.90	$414.40	$43.90	$702.40	$24.30	$41.18
8 Skilled Workers	23.90	1529.60	40.50	2592.00		
80 L.H., Daily Totals		$1944.00		$3294.40	$24.30	$41.18
Crew C-17A	**Hr.**	**Daily**	**Hr.**	**Daily**	**Bare Costs**	**Incl. O&P**
2 Skilled Worker Foremen	$25.90	$414.40	$43.90	$702.40	$24.31	$41.17
8 Skilled Workers	23.90	1529.60	40.50	2592.00		
.125 Equip. Oper. (crane)	24.75	24.75	40.75	40.75		
.125 Crane, 80 Ton, & Tools		124.38		136.80	1.54	1.69
81 L.H., Daily Totals		$2093.13		$3471.95	$25.85	$42.86

Crew No.	Bare Costs		Incl. Subs O & P		Cost Per Labor-Hour	
Crew C-17B	**Hr.**	**Daily**	**Hr.**	**Daily**	**Bare Costs**	**Incl. O&P**
2 Skilled Worker Foremen	$25.90	$414.40	$43.90	$702.40	$24.31	$41.17
8 Skilled Workers	23.90	1529.60	40.50	2592.00		
.25 Equip. Oper. (crane)	24.75	49.50	40.75	81.50		
.25 Crane, 80 Ton, & Tools		248.75		273.65		
.25 Walk Behind Power Tools		6.15		6.75	3.11	3.42
82 L.H., Daily Totals		$2248.40		$3656.30	$27.42	$44.59

Crew C-17C	Hr.	Daily	Hr.	Daily	Bare Costs	Incl. O&P
2 Skilled Worker Foremen	$25.90	$414.40	$43.90	$702.40	$24.32	$41.16
8 Skilled Workers	23.90	1529.60	40.50	2592.00		
.375 Equip. Oper. (crane)	24.75	74.25	40.75	122.25		
.375 Crane, 80 Ton & Tools		373.13		410.45	4.50	4.95
83 L.H., Daily Totals		$2391.38		$3827.10	$28.82	$46.11

Crew C-17D	Hr.	Daily	Hr.	Daily	Bare Costs	Incl. O&P
2 Skilled Worker Foremen	$25.90	$414.40	$43.90	$702.40	$24.32	$41.16
8 Skilled Workers	23.90	1529.60	40.50	2592.00		
.5 Equip. Oper. (crane)	24.75	99.00	40.75	163.00		
.5 Crane, 80 Ton & Tools		497.50		547.25	5.92	6.51
84 L.H., Daily Totals		$2540.50		$4004.65	$30.24	$47.67

Crew C-17E	Hr.	Daily	Hr.	Daily	Bare Costs	Incl. O&P
2 Skilled Worker Foremen	$25.90	$414.40	$43.90	$702.40	$24.30	$41.18
8 Skilled Workers	23.90	1529.60	40.50	2592.00		
1 Hyd. Jack with Rods		79.00		86.90	.99	1.09
80 L.H., Daily Totals		$2023.00		$3381.30	$25.29	$42.27

Crew C-18	Hr.	Daily	Hr.	Daily	Bare Costs	Incl. O&P
.125 Labor Foreman (out)	$19.35	$19.35	$32.85	$32.85	$17.57	$29.83
1 Laborer	17.35	138.80	29.45	235.60		
1 Concrete Cart, 10 C.F.		49.80		54.80	5.53	6.09
9 L.H., Daily Totals		$207.95		$323.25	$23.10	$35.92

Crew C-19	Hr.	Daily	Hr.	Daily	Bare Costs	Incl. O&P
.125 Labor Foreman (out)	$19.35	$19.35	$32.85	$32.85	$17.57	$29.83
1 Laborer	17.35	138.80	29.45	235.60		
1 Concrete Cart, 18 C.F.		76.80		84.50	8.53	9.39
9 L.H., Daily Totals		$234.95		$352.95	$26.10	$39.22

Crew C-20	Hr.	Daily	Hr.	Daily	Bare Costs	Incl. O&P
1 Labor Foreman (outside)	$19.35	$154.80	$32.85	$262.80	$19.13	$32.07
5 Laborers	17.35	694.00	29.45	1178.00		
1 Cement Finisher	23.00	184.00	37.10	296.80		
1 Equip. Oper. (med.)	23.90	191.20	39.35	314.80		
2 Gas Engine Vibrators		48.00		52.80		
1 Concrete Pump (small)		704.20		774.60	11.75	12.93
64 L.H., Daily Totals		$1976.20		$2879.80	$30.88	$45.00

Crew C-21	Hr.	Daily	Hr.	Daily	Bare Costs	Incl. O&P
1 Labor Foreman (outside)	$19.35	$154.80	$32.85	$262.80	$19.13	$32.07
5 Laborers	17.35	694.00	29.45	1178.00		
1 Cement Finisher	23.00	184.00	37.10	296.80		
1 Equip. Oper. (med.)	23.90	191.20	39.35	314.80		
2 Gas Engine Vibrators		48.00		52.80		
1 Concrete Conveyer		152.80		168.10	3.14	3.45
64 L.H., Daily Totals		$1424.80		$2273.30	$22.27	$35.52

Crew No.	Bare Costs		Incl. Subs O & P		Cost Per Labor-Hour	
Crew C-22	**Hr.**	**Daily**	**Hr.**	**Daily**	**Bare Costs**	**Incl. O&P**
1 Rodman Foreman	$27.45	$219.60	$49.15	$393.20	$25.70	$45.85
4 Rodmen (reinf.)	25.45	814.40	45.55	1457.60		
.125 Equip. Oper. (crane)	24.75	24.75	40.75	40.75		
.125 Equip. Oper. Oiler	20.75	20.75	34.20	34.20		
.125 Hyd. Crane, 25 Ton		77.10		84.80	1.84	2.02
42 L.H., Daily Totals		$1156.60		$2010.55	$27.54	$47.87

Crew C-23	Hr.	Daily	Hr.	Daily	Bare Costs	Incl. O&P
2 Skilled Worker Foremen	$25.90	$414.40	$43.90	$702.40	$24.07	$40.58
6 Skilled Workers	23.90	1147.20	40.50	1944.00		
1 Equip. Oper. (crane)	24.75	198.00	40.75	326.00		
1 Equip. Oper. Oiler	20.75	166.00	34.20	273.60		
1 Crane, 90 Ton		1325.00		1457.50	16.56	18.22
80 L.H., Daily Totals		$3250.60		$4703.50	$40.63	$58.80

Crew C-24	Hr.	Daily	Hr.	Daily	Bare Costs	Incl. O&P
2 Skilled Worker Foremen	$25.90	$414.40	$43.90	$702.40	$24.07	$40.58
6 Skilled Workers	23.90	1147.20	40.50	1944.00		
1 Equip. Oper. (crane)	24.75	198.00	40.75	326.00		
1 Equip. Oper. Oiler	20.75	166.00	34.20	273.60		
1 Truck Crane, 150 Ton		1445.00		1589.50	18.06	19.87
80 L.H., Daily Totals		$3370.60		$4835.50	$42.13	$60.45

Crew C-25	Hr.	Daily	Hr.	Daily	Bare Costs	Incl. O&P
2 Rodmen (reinf.)	$25.45	$407.20	$45.55	$728.80	$20.18	$36.53
2 Rodmen Helpers	14.90	238.40	27.50	440.00		
32 L.H., Daily Totals		$645.60		$1168.80	$20.18	$36.53

Crew C-27	Hr.	Daily	Hr.	Daily	Bare Costs	Incl. O&P
2 Cement Finishers	$23.00	$368.00	$37.10	$593.60	$23.00	$37.10
1 Concrete Saw		114.40		125.85	7.15	7.87
16 L.H., Daily Totals		$482.40		$719.45	$30.15	$44.97

Crew C-28	Hr.	Daily	Hr.	Daily	Bare Costs	Incl. O&P
1 Cement Finisher	$23.00	$184.00	$37.10	$296.80	$23.00	$37.10
1 Portable Air Compressor		17.30		19.05	2.16	2.37
8 L.H., Daily Totals		$201.30		$315.85	$25.16	$39.47

Crew D-1	Hr.	Daily	Hr.	Daily	Bare Costs	Incl. O&P
1 Bricklayer	$24.70	$197.60	$41.10	$328.80	$21.78	$36.23
1 Bricklayer Helper	18.85	150.80	31.35	250.80		
16 L.H., Daily Totals		$348.40		$579.60	$21.78	$36.23

Crew D-2	Hr.	Daily	Hr.	Daily	Bare Costs	Incl. O&P
3 Bricklayers	$24.70	$592.80	$41.10	$986.40	$22.36	$37.20
2 Bricklayer Helpers	18.85	301.60	31.35	501.60		
40 L.H., Daily Totals		$894.40		$1488.00	$22.36	$37.20

Crew D-3	Hr.	Daily	Hr.	Daily	Bare Costs	Incl. O&P
3 Bricklayers	$24.70	$592.80	$41.10	$986.40	$22.44	$37.37
2 Bricklayer Helpers	18.85	301.60	31.35	501.60		
.25 Carpenter	24.00	48.00	40.75	81.50		
42 L.H., Daily Totals		$942.40		$1569.50	$22.44	$37.37

Crew No.	Bare Costs		Incl. Subs O & P		Cost Per Labor-Hour	
Crew D-4	**Hr.**	**Daily**	**Hr.**	**Daily**	**Bare Costs**	**Incl. O&P**
Bricklayer	$24.70	$197.60	$41.10	$328.80	$19.72	$32.92
Bricklayer Helpers	18.85	452.40	31.35	752.40		
Building Laborer	17.35	138.80	29.45	235.60		
Grout Pump, 50 C.F./hr		107.75		118.55		
Hoses & Hopper		15.20		16.70		
Accessories		11.90		13.10	3.37	3.71
) L.H., Daily Totals		$923.65		$1465.15	$23.09	$36.63
Crew D-5	**Hr.**	**Daily**	**Hr.**	**Daily**	**Bare Costs**	**Incl. O&P**
Block Mason Helper	$18.85	$150.80	$31.35	$250.80	$18.85	$31.35
L.H., Daily Totals		$150.80		$250.80	$18.85	$31.35
Crew D-6	**Hr.**	**Daily**	**Hr.**	**Daily**	**Bare Costs**	**Incl. O&P**
Bricklayers	$24.70	$592.80	$41.10	$986.40	$21.78	$36.23
Bricklayer Helpers	18.85	452.40	31.35	752.40		
8 L.H., Daily Totals		$1045.20		$1738.80	$21.78	$36.23
Crew D-7	**Hr.**	**Daily**	**Hr.**	**Daily**	**Bare Costs**	**Incl. O&P**
Tile Layer	$22.90	$183.20	$36.85	$294.80	$20.33	$32.73
Tile Layer Helper	17.75	142.00	28.60	228.80		
6 L.H., Daily Totals		$325.20		$523.60	$20.33	$32.73
Crew D-8	**Hr.**	**Daily**	**Hr.**	**Daily**	**Bare Costs**	**Incl. O&P**
3 Bricklayers	$24.70	$592.80	$41.10	$986.40	$22.36	$37.20
2 Bricklayer Helpers	18.85	301.60	31.35	501.60		
40 L.H., Daily Totals		$894.40		$1488.00	$22.36	$37.20
Crew D-9	**Hr.**	**Daily**	**Hr.**	**Daily**	**Bare Costs**	**Incl. O&P**
3 Bricklayers	$24.70	$592.80	$41.10	$986.40	$21.78	$36.23
3 Bricklayer Helpers	18.85	452.40	31.35	752.40		
48 L.H., Daily Totals		$1045.20		$1738.80	$21.78	$36.23
Crew D-10	**Hr.**	**Daily**	**Hr.**	**Daily**	**Bare Costs**	**Incl. O&P**
1 Stone Mason Foreman	$26.00	$208.00	$43.25	$346.00	$22.49	$37.33
1 Stone Mason	24.00	192.00	39.95	319.60		
2 Bricklayer Helpers	18.85	301.60	31.35	501.60		
1 Equip. Oper. (crane)	24.75	198.00	40.75	326.00		
1 Truck Crane, 12.5 Ton		505.80		556.40	12.65	13.91
40 L.H., Daily Totals		$1405.40		$2049.60	$35.14	$51.24
Crew D-11	**Hr.**	**Daily**	**Hr.**	**Daily**	**Bare Costs**	**Incl. O&P**
2 Stone Masons	$24.00	$384.00	$39.95	$639.20	$22.28	$37.08
1 Stone Mason Helper	18.85	150.80	31.35	250.80		
24 L.H., Daily Totals		$534.80		$890.00	$22.28	$37.08
Crew D-12	**Hr.**	**Daily**	**Hr.**	**Daily**	**Bare Costs**	**Incl. O&P**
2 Stone Masons	$24.00	$384.00	$39.95	$639.20	$21.43	$35.65
2 Bricklayer Helpers	18.85	301.60	31.35	501.60		
32 L.H., Daily Totals		$685.60		$1140.80	$21.43	$35.65
Crew D-13	**Hr.**	**Daily**	**Hr.**	**Daily**	**Bare Costs**	**Incl. O&P**
1 Stone Mason Foreman	$26.00	$208.00	$43.25	$346.00	$22.74	$37.77
2 Stone Masons	24.00	384.00	39.95	639.20		
2 Bricklayer Helpers	18.85	301.60	31.35	501.60		
1 Equip. Oper. (crane)	24.75	198.00	40.75	326.00		
1 Truck Crane, 12.5 Ton		505.80		556.40	10.54	11.59
48 L.H., Daily Totals		$1597.40		$2369.20	$33.28	$49.36

Crew No.	Bare Costs		Incl. Subs O & P		Cost Per Labor-Hour	
Crew E-1	**Hr.**	**Daily**	**Hr.**	**Daily**	**Bare Costs**	**Incl. O&P**
2 Struc. Steel Workers	$25.55	$408.80	$49.30	$788.80	$25.55	$49.30
1 Gas Welding Machine		81.20		89.30	5.08	5.58
16 L.H., Daily Totals		$490.00		$878.10	$30.63	$54.88
Crew E-2	**Hr.**	**Daily**	**Hr.**	**Daily**	**Bare Costs**	**Incl. O&P**
1 Struc. Steel Foreman	$27.55	$220.40	$53.15	$425.20	$25.75	$48.52
4 Struc. Steel Workers	25.55	817.60	49.30	1577.60		
1 Equip. Oper. (crane)	24.75	198.00	40.75	326.00		
1 Crane, 90 Ton		1325.00		1457.50	27.60	30.36
48 L.H., Daily Totals		$2561.00		$3786.30	$53.35	$78.88
Crew E-3	**Hr.**	**Daily**	**Hr.**	**Daily**	**Bare Costs**	**Incl. O&P**
1 Struc. Steel Foreman	$27.55	$220.40	$53.15	$425.20	$26.22	$50.58
2 Struc. Steel Worker	25.55	408.80	49.30	788.80		
1 Gas Welding Machine		81.20		89.30	3.38	3.72
24 L.H., Daily Totals		$710.40		$1303.30	$29.60	$54.30
Crew E-4	**Hr.**	**Daily**	**Hr.**	**Daily**	**Bare Costs**	**Incl. O&P**
1 Struc. Steel Foreman	$27.55	$220.40	$53.15	$425.20	$26.05	$50.26
3 Struc. Steel Workers	25.55	613.20	49.30	1183.20		
1 Gas Welding Machine		81.20		89.30	2.54	2.79
32 L.H., Daily Totals		$914.80		$1697.70	$28.59	$53.05
Crew E-5	**Hr.**	**Daily**	**Hr.**	**Daily**	**Bare Costs**	**Incl. O&P**
1 Struc. Steel Foremen	$27.55	$220.40	$53.15	$425.20	$25.68	$48.78
7 Struc. Steel Workers	25.55	1430.80	49.30	2760.80		
1 Equip. Oper. (crane)	24.75	198.00	40.75	326.00		
1 Crane, 90 Ton		1325.00		1457.50		
1 Gas Welding Machine		81.20		89.30	19.53	21.48
72 L.H., Daily Totals		$3255.40		$5058.80	$45.21	$70.26
Crew E-6	**Hr.**	**Daily**	**Hr.**	**Daily**	**Bare Costs**	**Incl. O&P**
1 Struc. Steel Foreman	$27.55	$220.40	$53.15	$425.20	$25.45	$48.20
12 Struc. Steel Workers	25.55	2452.80	49.30	4732.80		
1 Equip. Oper. (crane)	24.75	198.00	40.75	326.00		
1 Equip. Oper. (light)	22.80	182.40	37.55	300.40		
1 Crane, 90 Ton		1325.00		1457.50		
1 Gas Welding Machine		81.20		89.30		
1 Air Compr., 160 C.F.M.		87.40		96.15		
2 Impact Wrenches		24.00		26.40	12.65	13.91
120 L.H., Daily Totals		$4571.20		$7453.75	$38.10	$62.11
Crew E-7	**Hr.**	**Daily**	**Hr.**	**Daily**	**Bare Costs**	**Incl. O&P**
1 Struc. Steel Foreman	$27.55	$220.40	$53.15	$425.20	$25.68	$48.78
7 Struc. Steel Workers	25.55	1430.80	49.30	2760.80		
1 Equip. Oper. (crane)	24.75	198.00	40.75	326.00		
1 Crane, 90 Ton		1325.00		1457.50		
2 Gas Welding Machines		162.40		178.65	20.66	22.72
72 L.H., Daily Totals		$3336.60		$5148.15	$46.34	$71.50
Crew E-8	**Hr.**	**Daily**	**Hr.**	**Daily**	**Bare Costs**	**Incl. O&P**
1 Struc. Steel Foreman	$27.55	$220.40	$53.15	$425.20	$25.66	$48.87
9 Struc. Steel Workers	25.55	1839.60	49.30	3549.60		
1 Equip. Oper. (crane)	24.75	198.00	40.75	326.00		
1 Crane, 90 Ton		1325.00		1457.50		
4 Gas Welding Machines		324.80		357.30	18.75	20.62
88 L.H., Daily Totals		$3907.80		$6115.60	$44.41	$69.49

Crew No.	Bare Costs		Incl. Subs O & P		Cost Per Labor-Hour	
Crew E-9	**Hr.**	**Daily**	**Hr.**	**Daily**	**Bare Costs**	**Incl. O&P**
2 Struc. Steel Foremen	$27.55	$440.80	$53.15	$850.40	$25.40	$47.81
5 Struc. Steel Workers	25.55	1022.00	49.30	1972.00		
1 Welder Foreman	27.55	220.40	53.15	425.20		
5 Welders	25.55	1022.00	49.30	1972.00		
1 Equip. Oper. (crane)	24.75	198.00	40.75	326.00		
1 Equip. Oper. Oiler	20.75	166.00	34.20	273.60		
1 Equip. Oper. (light)	22.80	182.40	37.55	300.40		
1 Crane, 90 Ton		1325.00		1457.50		
5 Gas Welding Machines		406.00		446.60	13.52	14.88
128 L.H., Daily Totals		$4982.60		$8023.70	$38.92	$62.69
Crew E-10	**Hr.**	**Daily**	**Hr.**	**Daily**	**Bare Costs**	**Incl. O&P**
1 Struc. Steel Foreman	$27.55	$220.40	$53.15	$425.20	$26.22	$50.58
2 Struc. Steel Workers	25.55	408.80	49.30	788.80		
1 Gas Welding Machines		81.20		89.30		
1 Truck, 3 Ton		176.20		193.80	10.73	11.80
24 L.H., Daily Totals		$886.60		$1497.10	$36.95	$62.38
Crew E-11	**Hr.**	**Daily**	**Hr.**	**Daily**	**Bare Costs**	**Incl. O&P**
2 Painters, Struc. Steel	$22.00	$352.00	$43.60	$697.60	$20.45	$38.88
1 Building Laborer	17.35	138.80	29.45	235.60		
1 Air Compressor 250 C.F.M.		112.40		123.65		
1 Sand Blaster		15.20		16.70		
1 Sand Blasting Accessories		11.90		13.10	5.81	6.39
24 L.H., Daily Totals		$630.30		$1086.65	$26.26	$45.27
Crew E-12	**Hr.**	**Daily**	**Hr.**	**Daily**	**Bare Costs**	**Incl. O&P**
1 Welder Foreman	$27.55	$220.40	$53.15	$425.20	$25.18	$45.35
1 Equip. Oper. (light)	22.80	182.40	37.55	300.40		
1 Gas Welding Machine		81.20		89.30	5.08	5.58
16 L.H., Daily Totals		$484.00		$814.90	$30.26	$50.93
Crew E-13	**Hr.**	**Daily**	**Hr.**	**Daily**	**Bare Costs**	**Incl. O&P**
1 Welder Foreman	$27.55	$220.40	$53.15	$425.20	$25.97	$47.95
.5 Equip. Oper. (light)	22.80	91.20	37.55	150.20		
1 Gas Welding Machine		81.20		89.30	6.77	7.44
12 L.H., Daily Totals		$392.80		$664.70	$32.74	$55.39
Crew E-14	**Hr.**	**Daily**	**Hr.**	**Daily**	**Bare Costs**	**Incl. O&P**
1 Struc. Steel Worker	$25.55	$204.40	$49.30	$394.40	$25.55	$49.30
1 Gas Welding Machine		81.20		89.30	10.15	11.17
8 L.H., Daily Totals		$285.60		$483.70	$35.70	$60.47
Crew E-16	**Hr.**	**Daily**	**Hr.**	**Daily**	**Bare Costs**	**Incl. O&P**
1 Welder Foreman	$27.55	$220.40	$53.15	$425.20	$26.55	$51.23
1 Welder	25.55	204.40	49.30	394.40		
1 Gas Welding Machine		81.20		89.30	5.08	5.58
16 L.H., Daily Totals		$506.00		$908.90	$31.63	$56.81
Crew E-17	**Hr.**	**Daily**	**Hr.**	**Daily**	**Bare Costs**	**Incl. O&P**
1 Structural Steel Foreman	$27.55	$220.40	$53.15	$425.20	$26.55	$51.23
1 Structural Steel Worker	25.55	204.40	49.30	394.40		
1 Power Tool		5.20		5.70	.33	.36
16 L.H., Daily Totals		$430.00		$825.30	$26.88	$51.59

Crew No.	Bare Costs		Incl. Subs O & P		Cost Per Labor-Hour	
Crew E-18	**Hr.**	**Daily**	**Hr.**	**Daily**	**Bare Costs**	**Incl. O&P**
1 Structural Steel Foreman	$27.55	$220.40	$53.15	$425.20	$25.62	$48.08
3 Structural Steel Workers	25.55	613.20	49.30	1183.20		
1 Equipment Operator (med.)	23.90	191.20	39.35	314.80		
1 Crane, 20 Ton		797.10		876.80	19.93	21.92
40 L.H., Daily Totals		$1821.90		$2800.00	$45.55	$70.00
Crew E-19	**Hr.**	**Daily**	**Hr.**	**Daily**	**Bare Costs**	**Incl. O&P**
1 Structural Steel Worker	$25.55	$204.40	$49.30	$394.40	$25.30	$46.67
1 Structural Steel Foreman	27.55	220.40	53.15	425.20		
1 Equip. Oper. (light)	22.80	182.40	37.55	300.40		
1 Power Tool		5.20		5.70		
1 Crane, 20 Ton		797.10		876.80	33.43	36.77
24 L.H., Daily Totals		$1409.50		$2002.50	$58.73	$83.44
Crew E-20	**Hr.**	**Daily**	**Hr.**	**Daily**	**Bare Costs**	**Incl. O&P**
1 Structural Steel Foreman	$27.55	$220.40	$53.15	$425.20	$25.10	$46.83
5 Structural Steel Workers	25.55	1022.00	49.30	1972.00		
1 Equip. Oper. (crane)	24.75	198.00	40.75	326.00		
1 Oiler	20.75	166.00	34.20	273.60		
1 Power Tool		5.20		5.70		
1 Crane, 40 Ton		951.90		1047.10	14.95	16.45
64 L.H., Daily Totals		$2563.50		$4049.60	$40.05	$63.28
Crew E-22	**Hr.**	**Daily**	**Hr.**	**Daily**	**Bare Costs**	**Incl. O&P**
1 Skilled Worker Foreman	$25.90	$207.20	$43.90	$351.20	$24.57	$41.63
2 Skilled Workers	23.90	382.40	40.50	648.00		
24 L.H., Daily Totals		$589.60		$999.20	$24.57	$41.63
Crew E-24	**Hr.**	**Daily**	**Hr.**	**Daily**	**Bare Costs**	**Incl. O&P**
3 Structural Steel Workers	$25.55	$613.20	$49.30	$1183.20	$25.14	$46.81
1 Equipment Operator (medium)	23.90	191.20	39.35	314.80		
1 25 Ton Crane		616.80		678.50	19.28	21.20
32 L.H., Daily Totals		$1421.20		$2176.50	$44.42	$68.01
Crew E-25	**Hr.**	**Daily**	**Hr.**	**Daily**	**Bare Costs**	**Incl. O&P**
1 Welder	$25.55	$204.40	$49.30	$394.40	$25.55	$49.30
1 Cutting Torch		18.00		19.80		
1 Gases		64.80		71.30	10.35	11.39
8 L.H., Daily Totals		$287.20		$485.50	$35.90	$60.69
Crew F-3	**Hr.**	**Daily**	**Hr.**	**Daily**	**Bare Costs**	**Incl. O&P**
2 Carpenters	$24.00	$384.00	$40.75	$652.00	$21.57	$36.35
2 Carpenter Helpers	17.55	280.80	29.75	476.00		
1 Equip. Oper. (crane)	24.75	198.00	40.75	326.00		
1 Hyd. Crane, 12 Ton		602.60		662.85	15.07	16.57
40 L.H., Daily Totals		$1465.40		$2116.85	$36.64	$52.92
Crew F-4	**Hr.**	**Daily**	**Hr.**	**Daily**	**Bare Costs**	**Incl. O&P**
2 Carpenters	$24.00	$384.00	$40.75	$652.00	$21.57	$36.35
2 Carpenter Helpers	17.55	280.80	29.75	476.00		
1 Equip. Oper. (crane)	24.75	198.00	40.75	326.00		
1 Hyd. Crane, 55 Ton		895.40		984.95	22.39	24.62
40 L.H., Daily Totals		$1758.20		$2438.95	$43.96	$60.97
Crew F-5	**Hr.**	**Daily**	**Hr.**	**Daily**	**Bare Costs**	**Incl. O&P**
2 Carpenters	$24.00	$384.00	$40.75	$652.00	$20.78	$35.25
2 Carpenter Helpers	17.55	280.80	29.75	476.00		
32 L.H., Daily Totals		$664.80		$1128.00	$20.78	$35.25

Crew F-6	Bare Costs Hr.	Bare Costs Daily	Incl. Subs O & P Hr.	Incl. Subs O & P Daily	Cost Per Labor-Hour Bare Costs	Cost Per Labor-Hour Incl. O&P
Carpenters	$24.00	$384.00	$40.75	$652.00	$21.49	$36.23
Building Laborers	17.35	277.60	29.45	471.20		
Equip. Oper. (crane)	24.75	198.00	40.75	326.00		
Hyd. Crane, 12 Ton		602.60		662.85	15.07	16.57
L.H., Daily Totals		$1462.20		$2112.05	$36.56	$52.80

Crew F-7	Hr.	Daily	Hr.	Daily	Bare Costs	Incl. O&P
Carpenters	$24.00	$384.00	$40.75	$652.00	$20.68	$35.10
Building Laborers	17.35	277.60	29.45	471.20		
2 L.H., Daily Totals		$661.60		$1123.20	$20.68	$35.10

Crew G-1	Hr.	Daily	Hr.	Daily	Bare Costs	Incl. O&P
Roofer Foreman	$22.30	$178.40	$41.10	$328.80	$19.04	$35.13
Roofers, Composition	20.30	649.60	37.45	1198.40		
Roofer Helpers	14.90	238.40	27.50	440.00		
Application Equipment		146.60		161.25		
Tar Kettle/Pot		52.50		57.75		
Crew Truck		98.80		108.70	5.32	5.85
6 L.H., Daily Totals		$1364.30		$2294.90	$24.36	$40.98

Crew G-2	Hr.	Daily	Hr.	Daily	Bare Costs	Incl. O&P
Plasterer	$21.95	$175.60	$36.50	$292.00	$19.42	$32.48
Plasterer Helper	18.95	151.60	31.50	252.00		
Building Laborer	17.35	138.80	29.45	235.60		
Grouting Equipment		107.75		118.55	4.49	4.94
4 L.H., Daily Totals		$573.75		$898.15	$23.91	$37.42

Crew G-3	Hr.	Daily	Hr.	Daily	Bare Costs	Incl. O&P
2 Sheet Metal Workers	$26.55	$424.80	$44.55	$712.80	$21.95	$37.00
2 Building Laborers	17.35	277.60	29.45	471.20		
32 L.H., Daily Totals		$702.40		$1184.00	$21.95	$37.00

Crew G-4	Hr.	Daily	Hr.	Daily	Bare Costs	Incl. O&P
1 Labor Foreman (outside)	$19.35	$154.80	$32.85	$262.80	$18.02	$30.58
2 Building Laborers	17.35	277.60	29.45	471.20		
1 Light Truck, 1.5 Ton		128.80		141.70		
1 Air Compr., 160 C.F.M.		87.40		96.15	9.01	9.91
24 L.H., Daily Totals		$648.60		$971.85	$27.03	$40.49

Crew G-5	Hr.	Daily	Hr.	Daily	Bare Costs	Incl. O&P
1 Roofer Foreman	$22.30	$178.40	$41.10	$328.80	$18.54	$34.20
2 Roofers, Composition	20.30	324.80	37.45	599.20		
2 Roofer Helpers	14.90	238.40	27.50	440.00		
1 Application Equipment		146.60		161.25	3.67	4.03
40 L.H., Daily Totals		$888.20		$1529.25	$22.21	$38.23

Crew G-6A	Hr.	Daily	Hr.	Daily	Bare Costs	Incl. O&P
2 Roofers Composition	$20.30	$324.80	$37.45	$599.20	$20.30	$37.45
1 Small Compressor		12.15		13.35		
2 Pneumatic Nailers		39.00		42.90	3.19	3.51
16 L.H., Daily Totals		$375.95		$655.45	$23.49	$40.96

Crew G-7	Hr.	Daily	Hr.	Daily	Bare Costs	Incl. O&P
1 Carpenter	$24.00	$192.00	$40.75	$326.00	$24.00	$40.75
1 Small Compressor		12.15		13.35		
1 Pneumatic Nailer		19.50		21.45	3.94	4.34
8 L.H., Daily Totals		$223.65		$360.80	$27.94	$45.09

Crew H-1	Bare Costs Hr.	Bare Costs Daily	Incl. Subs O & P Hr.	Incl. Subs O & P Daily	Cost Per Labor-Hour Bare Costs	Cost Per Labor-Hour Incl. O&P
2 Glaziers	$23.45	$375.20	$38.75	$620.00	$24.50	$44.03
2 Struc. Steel Workers	25.55	408.80	49.30	788.80		
32 L.H., Daily Totals		$784.00		$1408.80	$24.50	$44.03

Crew H-2	Hr.	Daily	Hr.	Daily	Bare Costs	Incl. O&P
2 Glaziers	$23.45	$375.20	$38.75	$620.00	$21.42	$35.65
1 Building Laborer	17.35	138.80	29.45	235.60		
24 L.H., Daily Totals		$514.00		$855.60	$21.42	$35.65

Crew H-3	Hr.	Daily	Hr.	Daily	Bare Costs	Incl. O&P
1 Glazier	$23.45	$187.60	$38.75	$310.00	$20.50	$34.25
1 Helper	17.55	140.40	29.75	238.00		
16 L.H., Daily Totals		$328.00		$548.00	$20.50	$34.25

Crew J-1	Hr.	Daily	Hr.	Daily	Bare Costs	Incl. O&P
3 Plasterers	$21.95	$526.80	$36.50	$876.00	$20.75	$34.50
2 Plasterer Helpers	18.95	303.20	31.50	504.00		
1 Mixing Machine, 6 C.F.		103.00		113.30	2.58	2.83
40 L.H., Daily Totals		$933.00		$1493.30	$23.33	$37.33

Crew J-2	Hr.	Daily	Hr.	Daily	Bare Costs	Incl. O&P
3 Plasterers	$21.95	$526.80	$36.50	$876.00	$20.96	$34.70
2 Plasterer Helpers	18.95	303.20	31.50	504.00		
1 Lather	22.00	176.00	35.70	285.60		
1 Mixing Machine, 6 C.F.		103.00		113.30	2.15	2.36
48 L.H., Daily Totals		$1109.00		$1778.90	$23.11	$37.06

Crew J-3	Hr.	Daily	Hr.	Daily	Bare Costs	Incl. O&P
1 Terrazzo Worker	$22.85	$182.80	$36.80	$294.40	$20.55	$33.10
1 Terrazzo Helper	18.25	146.00	29.40	235.20		
1 Terrazzo Grinder, Electric		70.00		77.00		
1 Terrazzo Mixer		139.60		153.55	13.10	14.41
16 L.H., Daily Totals		$538.40		$760.15	$33.65	$47.51

Crew J-4	Hr.	Daily	Hr.	Daily	Bare Costs	Incl. O&P
1 Tile Layer	$22.90	$183.20	$36.85	$294.80	$20.33	$32.73
1 Tile Layer Helper	17.75	142.00	28.60	228.80		
16 L.H., Daily Totals		$325.20		$523.60	$20.33	$32.73

Crew K-1	Hr.	Daily	Hr.	Daily	Bare Costs	Incl. O&P
1 Carpenter	$24.00	$192.00	$40.75	$326.00	$21.53	$36.23
1 Truck Driver (light)	19.05	152.40	31.70	253.60		
1 Truck w/Power Equip.		176.20		193.80	11.01	12.11
16 L.H., Daily Totals		$520.60		$773.40	$32.54	$48.34

Crew K-2	Hr.	Daily	Hr.	Daily	Bare Costs	Incl. O&P
1 Struc. Steel Foreman	$27.55	$220.40	$53.15	$425.20	$24.05	$44.72
1 Struc. Steel Worker	25.55	204.40	49.30	394.40		
1 Truck Driver (light)	19.05	152.40	31.70	253.60		
1 Truck w/Power Equip.		176.20		193.80	7.34	8.08
24 L.H., Daily Totals		$753.40		$1267.00	$31.39	$52.80

Crew L-1	Hr.	Daily	Hr.	Daily	Bare Costs	Incl. O&P
.25 Electrician	$27.30	$54.60	$44.45	$88.90	$27.02	$44.33
1 Plumber	26.95	215.60	44.30	354.40		
10 L.H., Daily Totals		$270.20		$443.30	$27.02	$44.33

Crew L-2	Hr.	Daily	Hr.	Daily	Bare Costs	Incl. O&P
1 Carpenter	$24.00	$192.00	$40.75	$326.00	$20.78	$35.25
1 Carpenter Helper	17.55	140.40	29.75	238.00		
16 L.H., Daily Totals		$332.40		$564.00	$20.78	$35.25

Crew L-3	Hr.	Daily	Hr.	Daily	Bare Costs	Incl. O&P
1 Carpenter	$24.00	$192.00	$40.75	$326.00	$24.66	$41.49
.25 Electrician	27.30	54.60	44.45	88.90		
10 L.H., Daily Totals		$246.60		$414.90	$24.66	$41.49

Crew L-3A	Hr.	Daily	Hr.	Daily	Bare Costs	Incl. O&P
1 Carpenter Foreman (outside)	$26.00	$208.00	$44.10	$352.80	$26.18	$44.25
.5 Sheet Metal Worker	26.55	106.20	44.55	178.20		
12 L.H., Daily Totals		$314.20		$531.00	$26.18	$44.25

Crew L-4	Hr.	Daily	Hr.	Daily	Bare Costs	Incl. O&P
1 Skilled Workers	$23.90	$191.20	$40.50	$324.00	$20.73	$35.13
1 Helper	17.55	140.40	29.75	238.00		
16 L.H., Daily Totals		$331.60		$562.00	$20.73	$35.13

Crew L-5	Hr.	Daily	Hr.	Daily	Bare Costs	Incl. O&P
1 Struc. Steel Foreman	$27.55	$220.40	$53.15	$425.20	$25.72	$48.63
5 Struc. Steel Workers	25.55	1022.00	49.30	1972.00		
1 Equip. Oper. (crane)	24.75	198.00	40.75	326.00		
1 Hyd. Crane, 25 Ton		616.80		678.50	11.01	12.12
56 L.H., Daily Totals		$2057.20		$3401.70	$36.73	$60.75

Crew L-5A	Hr.	Daily	Hr.	Daily	Bare Costs	Incl. O&P
1 Structural Steel Foreman	$27.55	$220.40	$53.15	$425.20	$25.85	$48.13
2 Structural Steel Workers	25.55	408.80	49.30	788.80		
1 Equip. Oper. (crane)	24.75	198.00	40.75	326.00		
1 Crane,SP, 25 Ton		575.00		632.50	17.97	19.77
32 L.H., Daily Totals		$1402.20		$2172.50	$43.82	$67.90

Crew L-6	Hr.	Daily	Hr.	Daily	Bare Costs	Incl. O&P
1 Plumber	$26.95	$215.60	$44.30	$354.40	$27.07	$44.35
.5 Electrician	27.30	109.20	44.45	177.80		
12 L.H., Daily Totals		$324.80		$532.20	$27.07	$44.35

Crew L-7	Hr.	Daily	Hr.	Daily	Bare Costs	Incl. O&P
1 Carpenters	$24.00	$192.00	$40.75	$326.00	$20.28	$34.27
2 Carpenter Helpers	17.55	280.80	29.75	476.00		
.25 Electrician	27.30	54.60	44.45	88.90		
26 L.H., Daily Totals		$527.40		$890.90	$20.28	$34.27

Crew L-8	Hr.	Daily	Hr.	Daily	Bare Costs	Incl. O&P
1 Carpenters	$24.00	$192.00	$40.75	$326.00	$22.01	$37.06
1 Carpenter Helper	17.55	140.40	29.75	238.00		
.5 Plumber	26.95	107.80	44.30	177.20		
20 L.H., Daily Totals		$440.20		$741.20	$22.01	$37.06

Crew L-9	Hr.	Daily	Hr.	Daily	Bare Costs	Incl. O&P
1 Skilled Worker Foreman	$25.90	$207.20	$43.90	$351.20	$21.90	$36.92
1 Skilled Worker	23.90	191.20	40.50	324.00		
2 Helpers	17.55	280.80	29.75	476.00		
.5 Electrician	27.30	109.20	44.45	177.80		
36 L.H., Daily Totals		$788.40		$1329.00	$21.90	$36.92

Crew L-10	Hr.	Daily	Hr.	Daily	Bare Costs	Incl. O&P
1 Structural Steel Foreman	$27.55	$220.40	$53.15	$425.20	$25.95	$47.73
1 Structural Steel Worker	25.55	204.40	49.30	394.40		
1 Equip. Oper. (crane)	24.75	198.00	40.75	326.00		
1 Hyd. Crane, 12 Ton		602.60		662.85	25.11	27.62
24 L.H., Daily Totals		$1225.40		$1808.45	$51.06	$75.35

Crew L-11	Hr.	Daily	Hr.	Daily	Bare Costs	Incl. O&P
2 Wrecker	$17.90	$286.40	$34.15	$546.40	$20.84	$36.65
1 Equip. Oper. (crane)	24.75	198.00	40.75	326.00		
1 Equip. Oper. (light)	22.80	182.40	37.55	300.40		
1 Hyd. Excavator, 2.5 C.Y.		1215.00		1336.50		
1 Skid steer loader		197.80		217.60	44.15	48.57
32 L.H., Daily Totals		$2079.60		$2726.90	$64.99	$85.22

Crew M-1	Hr.	Daily	Hr.	Daily	Bare Costs	Incl. O&P
3 Elevator Constructors	$30.20	$724.80	$49.50	$1188.00	$28.69	$47.03
1 Elevator Apprentice	24.15	193.20	39.60	316.80		
5 Hand Tools		63.00		69.30	1.97	2.17
32 L.H., Daily Totals		$981.00		$1574.10	$30.66	$49.20

Crew M-3	Hr.	Daily	Hr.	Daily	Bare Costs	Incl. O&P
1 Electrician Foreman (out)	$29.30	$234.40	$47.70	$381.60	$25.17	$41.43
1 Common Laborer	17.35	138.80	29.45	235.60		
.25 Equipment Operator, Medium	23.90	47.80	39.35	78.70		
1 Elevator Constructor	30.20	241.60	49.50	396.00		
1 Elevator Apprentice	24.15	193.20	39.60	316.80		
.25 Crane, SP, 4 x 4, 20 ton		145.35		159.90	4.28	4.70
34 L.H., Daily Totals		$1001.15		$1568.60	$29.45	$46.13

Crew M-4	Hr.	Daily	Hr.	Daily	Bare Costs	Incl. O&P
1 Electrician Foreman (out)	$29.30	$234.40	$47.70	$381.60	$24.97	$41.11
1 Common Laborer	17.35	138.80	29.45	235.60		
.25 Equipment Operator, Crane	24.75	49.50	40.75	81.50		
.25 Equipment Operator, Oiler	20.75	41.50	34.20	68.40		
1 Elevator Constructor	30.20	241.60	49.50	396.00		
1 Elevator Apprentice	24.15	193.20	39.60	316.80		
.25 Crane, Hyd, SP, 4WD, 40 Ton		217.40		239.15	6.04	6.64
36 L.H., Daily Totals		$1116.40		$1719.05	$31.01	$47.75

Crew Q-1	Hr.	Daily	Hr.	Daily	Bare Costs	Incl. O&P
1 Plumber	$26.95	$215.60	$44.30	$354.40	$24.25	$39.85
1 Plumber Apprentice	21.55	172.40	35.40	283.20		
16 L.H., Daily Totals		$388.00		$637.60	$24.25	$39.85

Crew Q-1C	Hr.	Daily	Hr.	Daily	Bare Costs	Incl. O&P
1 Plumber	$26.95	$215.60	$44.30	$354.40	$24.13	$39.68
1 Plumber Apprentice	21.55	172.40	35.40	283.20		
1 Equip. Oper. (medium)	23.90	191.20	39.35	314.80		
1 Trencher, Chain		1420.00		1562.00	59.17	65.08
24 L.H., Daily Totals		$1999.20		$2514.40	$83.30	$104.76

Crew Q-2	Hr.	Daily	Hr.	Daily	Bare Costs	Incl. O&P
1 Plumber	$26.95	$215.60	$44.30	$354.40	$23.35	$38.37
2 Plumber Apprentices	21.55	344.80	35.40	566.40		
24 L.H., Daily Totals		$560.40		$920.80	$23.35	$38.37

Crew Q-3	Hr.	Daily	Hr.	Daily	Bare Costs	Incl. O&P
2 Plumbers	$26.95	$431.20	$44.30	$708.80	$24.25	$39.85
2 Plumber Apprentices	21.55	344.80	35.40	566.40		
32 L.H., Daily Totals		$776.00		$1275.20	$24.25	$39.85

Crew No.	Bare Costs		Incl. Subs O & P		Cost Per Labor-Hour	
Crew Q-4	**Hr.**	**Daily**	**Hr.**	**Daily**	**Bare Costs**	**Incl. O&P**
Plumbers	$26.95	$431.20	$44.30	$708.80	$25.60	$42.08
Welder (plumber)	26.95	215.60	44.30	354.40		
Plumber Apprentice	21.55	172.40	35.40	283.20		
Electric Welding Mach.		80.65		88.70	2.52	2.77
2 L.H., Daily Totals		$899.85		$1435.10	$28.12	$44.85
Crew Q-5	**Hr.**	**Daily**	**Hr.**	**Daily**	**Bare Costs**	**Incl. O&P**
Steamfitter	$27.05	$216.40	$44.45	$355.60	$24.35	$40.00
Steamfitter Apprentice	21.65	173.20	35.55	284.40		
5 L.H., Daily Totals		$389.60		$640.00	$24.35	$40.00
Crew Q-6	**Hr.**	**Daily**	**Hr.**	**Daily**	**Bare Costs**	**Incl. O&P**
Steamfitters	$27.05	$216.40	$44.45	$355.60	$23.45	$38.52
Steamfitter Apprentices	21.65	346.40	35.55	568.80		
4 L.H., Daily Totals		$562.80		$924.40	$23.45	$38.52
Crew Q-7	**Hr.**	**Daily**	**Hr.**	**Daily**	**Bare Costs**	**Incl. O&P**
Steamfitters	$27.05	$432.80	$44.45	$711.20	$24.35	$40.00
Steamfitter Apprentices	21.65	346.40	35.55	568.80		
2 L.H., Daily Totals		$779.20		$1280.00	$24.35	$40.00
Crew Q-8	**Hr.**	**Daily**	**Hr.**	**Daily**	**Bare Costs**	**Incl. O&P**
Steamfitters	$27.05	$432.80	$44.45	$711.20	$25.70	$42.23
Welder (steamfitter)	27.05	216.40	44.45	355.60		
Steamfitter Apprentice	21.65	173.20	35.55	284.40		
Electric Welding Mach.		80.65		88.70	2.52	2.77
32 L.H., Daily Totals		$903.05		$1439.90	$28.22	$45.00
Crew Q-9	**Hr.**	**Daily**	**Hr.**	**Daily**	**Bare Costs**	**Incl. O&P**
1 Sheet Metal Worker	$26.55	$212.40	$44.55	$356.40	$23.90	$40.10
1 Sheet Metal Apprentice	21.25	170.00	35.65	285.20		
16 L.H., Daily Totals		$382.40		$641.60	$23.90	$40.10
Crew Q-10	**Hr.**	**Daily**	**Hr.**	**Daily**	**Bare Costs**	**Incl. O&P**
2 Sheet Metal Workers	$26.55	$424.80	$44.55	$712.80	$24.78	$41.58
1 Sheet Metal Apprentice	21.25	170.00	35.65	285.20		
24 L.H., Daily Totals		$594.80		$998.00	$24.78	$41.58
Crew Q-11	**Hr.**	**Daily**	**Hr.**	**Daily**	**Bare Costs**	**Incl. O&P**
2 Sheet Metal Workers	$26.55	$424.80	$44.55	$712.80	$23.90	$40.10
2 Sheet Metal Apprentices	21.25	340.00	35.65	570.40		
32 L.H., Daily Totals		$764.80		$1283.20	$23.90	$40.10
Crew Q-12	**Hr.**	**Daily**	**Hr.**	**Daily**	**Bare Costs**	**Incl. O&P**
1 Sprinkler Installer	$26.90	$215.20	$44.40	$355.20	$24.20	$39.95
1 Sprinkler Apprentice	21.50	172.00	35.50	284.00		
16 L.H., Daily Totals		$387.20		$639.20	$24.20	$39.95
Crew Q-13	**Hr.**	**Daily**	**Hr.**	**Daily**	**Bare Costs**	**Incl. O&P**
2 Sprinkler Installers	$26.90	$430.40	$44.40	$710.40	$24.20	$39.95
2 Sprinkler Apprentices	21.50	344.00	35.50	568.00		
32 L.H., Daily Totals		$774.40		$1278.40	$24.20	$39.95
Crew Q-14	**Hr.**	**Daily**	**Hr.**	**Daily**	**Bare Costs**	**Incl. O&P**
1 Asbestos Worker	$24.50	$196.00	$42.10	$336.80	$22.05	$37.88
1 Asbestos Apprentice	19.60	156.80	33.65	269.20		
16 L.H., Daily Totals		$352.80		$606.00	$22.05	$37.88

Crew No.	Bare Costs		Incl. Subs O & P		Cost Per Labor-Hour	
Crew Q-15	**Hr.**	**Daily**	**Hr.**	**Daily**	**Bare Costs**	**Incl. O&P**
1 Plumber	$26.95	$215.60	$44.30	$354.40	$24.25	$39.85
1 Plumber Apprentice	21.55	172.40	35.40	283.20		
1 Electric Welding Mach.		80.65		88.70	5.04	5.54
16 L.H., Daily Totals		$468.65		$726.30	$29.29	$45.39
Crew Q-16	**Hr.**	**Daily**	**Hr.**	**Daily**	**Bare Costs**	**Incl. O&P**
2 Plumbers	$26.95	$431.20	$44.30	$708.80	$25.15	$41.33
1 Plumber Apprentice	21.55	172.40	35.40	283.20		
1 Electric Welding Mach.		80.65		88.70	3.36	3.70
24 L.H., Daily Totals		$684.25		$1080.70	$28.51	$45.03
Crew Q-17	**Hr.**	**Daily**	**Hr.**	**Daily**	**Bare Costs**	**Incl. O&P**
1 Steamfitter	$27.05	$216.40	$44.45	$355.60	$24.35	$40.00
1 Steamfitter Apprentice	21.65	173.20	35.55	284.40		
1 Electric Welding Mach.		80.65		88.70	5.04	5.54
16 L.H., Daily Totals		$470.25		$728.70	$29.39	$45.54
Crew Q-17A	**Hr.**	**Daily**	**Hr.**	**Daily**	**Bare Costs**	**Incl. O&P**
1 Steamfitter	$27.05	$216.40	$44.45	$355.60	$24.48	$40.25
1 Steamfitter Apprentice	21.65	173.20	35.55	284.40		
1 Equip. Oper. (crane)	24.75	198.00	40.75	326.00		
1 Truck Crane, 12 Ton		602.60		662.85		
1 Electric Welding Mach.		80.65		88.70	28.47	31.32
24 L.H., Daily Totals		$1270.85		$1717.55	$52.95	$71.57
Crew Q-18	**Hr.**	**Daily**	**Hr.**	**Daily**	**Bare Costs**	**Incl. O&P**
2 Steamfitters	$27.05	$432.80	$44.45	$711.20	$25.25	$41.48
1 Steamfitter Apprentice	21.65	173.20	35.55	284.40		
1 Electric Welding Mach.		80.65		88.70	3.36	3.70
24 L.H., Daily Totals		$686.65		$1084.30	$28.61	$45.18
Crew Q-19	**Hr.**	**Daily**	**Hr.**	**Daily**	**Bare Costs**	**Incl. O&P**
1 Steamfitter	$27.05	$216.40	$44.45	$355.60	$25.33	$41.48
1 Steamfitter Apprentice	21.65	173.20	35.55	284.40		
1 Electrician	27.30	218.40	44.45	355.60		
24 L.H., Daily Totals		$608.00		$995.60	$25.33	$41.48
Crew Q-20	**Hr.**	**Daily**	**Hr.**	**Daily**	**Bare Costs**	**Incl. O&P**
1 Sheet Metal Worker	$26.55	$212.40	$44.55	$356.40	$24.58	$40.97
1 Sheet Metal Apprentice	21.25	170.00	35.65	285.20		
.5 Electrician	27.30	109.20	44.45	177.80		
20 L.H., Daily Totals		$491.60		$819.40	$24.58	$40.97
Crew Q-21	**Hr.**	**Daily**	**Hr.**	**Daily**	**Bare Costs**	**Incl. O&P**
2 Steamfitters	$27.05	$432.80	$44.45	$711.20	$25.76	$42.23
1 Steamfitter Apprentice	21.65	173.20	35.55	284.40		
1 Electrician	27.30	218.40	44.45	355.60		
32 L.H., Daily Totals		$824.40		$1351.20	$25.76	$42.23
Crew Q-22	**Hr.**	**Daily**	**Hr.**	**Daily**	**Bare Costs**	**Incl. O&P**
1 Plumber	$26.95	$215.60	$44.30	$354.40	$24.25	$39.85
1 Plumber Apprentice	21.55	172.40	35.40	283.20		
1 Truck Crane, 12 Ton		602.60		662.85	37.66	41.43
16 L.H., Daily Totals		$990.60		$1300.45	$61.91	$81.28

Crew No.	Bare Costs		Incl. Subs O & P		Cost Per Labor-Hour	
Crew Q-22A	**Hr.**	**Daily**	**Hr.**	**Daily**	**Bare Costs**	**Incl. O&P**
1 Plumber	$26.95	$215.60	$44.30	$354.40	$22.65	$37.48
1 Plumber Apprentice	21.55	172.40	35.40	283.20		
1 Laborer	17.35	138.80	29.45	235.60		
1 Equip. Oper. (crane)	24.75	198.00	40.75	326.00		
1 Truck Crane, 12 Ton		602.60		662.85	18.83	20.71
32 L.H., Daily Totals		$1327.40		$1862.05	$41.48	$58.19
Crew Q-23	**Hr.**	**Daily**	**Hr.**	**Daily**	**Bare Costs**	**Incl. O&P**
1 Plumber Foreman	$28.95	$231.60	$47.55	$380.40	$26.60	$43.73
1 Plumber	26.95	215.60	44.30	354.40		
1 Equip. Oper. (medium)	23.90	191.20	39.35	314.80		
1 Power Tools		5.20		5.70		
1 Crane, 20 Ton		797.10		876.80	33.43	36.77
24 L.H., Daily Totals		$1440.70		$1932.10	$60.03	$80.50
Crew R-1	**Hr.**	**Daily**	**Hr.**	**Daily**	**Bare Costs**	**Incl. O&P**
1 Electrician Foreman	$27.80	$222.40	$45.25	$362.00	$24.13	$39.68
3 Electricians	27.30	655.20	44.45	1066.80		
2 Helpers	17.55	280.80	29.75	476.00		
48 L.H., Daily Totals		$1158.40		$1904.80	$24.13	$39.68
Crew R-1A	**Hr.**	**Daily**	**Hr.**	**Daily**	**Bare Costs**	**Incl. O&P**
1 Electrician	$27.30	$218.40	$44.45	$355.60	$22.43	$37.10
1 Helper	17.55	140.40	29.75	238.00		
16 L.H., Daily Totals		$358.80		$593.60	$22.43	$37.10
Crew R-2	**Hr.**	**Daily**	**Hr.**	**Daily**	**Bare Costs**	**Incl. O&P**
1 Electrician Foreman	$27.80	$222.40	$45.25	$362.00	$24.22	$39.84
3 Electricians	27.30	655.20	44.45	1066.80		
2 Helpers	17.55	280.80	29.75	476.00		
1 Equip. Oper. (crane)	24.75	198.00	40.75	326.00		
1 S.P. Crane, 5 Ton		317.00		348.70	5.66	6.23
56 L.H., Daily Totals		$1673.40		$2579.50	$29.88	$46.07
Crew R-3	**Hr.**	**Daily**	**Hr.**	**Daily**	**Bare Costs**	**Incl. O&P**
1 Electrician Foreman	$27.80	$222.40	$45.25	$362.00	$26.99	$44.03
1 Electrician	27.30	218.40	44.45	355.60		
.5 Equip. Oper. (crane)	24.75	99.00	40.75	163.00		
.5 S.P. Crane, 5 Ton		158.50		174.35	7.93	8.72
20 L.H., Daily Totals		$698.30		$1054.95	$34.92	$52.75
Crew R-4	**Hr.**	**Daily**	**Hr.**	**Daily**	**Bare Costs**	**Incl. O&P**
1 Struc. Steel Foreman	$27.55	$220.40	$53.15	$425.20	$26.30	$49.10
3 Struc. Steel Workers	25.55	613.20	49.30	1183.20		
1 Electrician	27.30	218.40	44.45	355.60		
1 Gas Welding Machine		81.20		89.30	2.03	2.23
40 L.H., Daily Totals		$1133.20		$2053.30	$28.33	$51.33
Crew R-5	**Hr.**	**Daily**	**Hr.**	**Daily**	**Bare Costs**	**Incl. O&P**
1 Electrician Foreman	$27.80	$222.40	$45.25	$362.00	$23.80	$39.18
4 Electrician Linemen	27.30	873.60	44.45	1422.40		
2 Electrician Operators	27.30	436.80	44.45	711.20		
4 Electrician Groundmen	17.55	561.60	29.75	952.00		
1 Crew Truck		98.80		108.70		
1 Tool Van		122.15		134.35		
1 Pickup Truck, 3/4 Ton		78.00		85.80		
.2 Crane, 55 Ton		179.08		197.00		
.2 Crane, 12 Ton		120.52		132.55		
.2 Auger, Truck Mtd.		635.80		699.40		
1 Tractor w/Winch		260.00		286.00	16.98	18.68
88 L.H., Daily Totals		$3588.75		$5091.40	$40.78	$57.86

Crew No.	Bare Costs		Incl. Subs O & P		Cost Per Labor-Hour	
Crew R-6	**Hr.**	**Daily**	**Hr.**	**Daily**	**Bare Costs**	**Incl. O&P**
1 Electrician Foreman	$27.80	$222.40	$45.25	$362.00	$23.80	$39.18
4 Electrician Linemen	27.30	873.60	44.45	1422.40		
2 Electrician Operators	27.30	436.80	44.45	711.20		
4 Electrician Groundmen	17.55	561.60	29.75	952.00		
1 Crew Truck		98.80		108.70		
1 Tool Van		122.15		134.35		
1 Pickup Truck, 3/4 Ton		78.00		85.80		
.2 Crane, 55 Ton		179.08		197.00		
.2 Crane, 12 Ton		120.52		132.55		
.2 Auger, Truck Mtd.		635.80		699.40		
1 Tractor w/Winch		260.00		286.00		
3 Cable Trailers		490.65		539.70		
.5 Tensioning Rig		160.00		176.00		
.5 Cable Pulling Rig		934.00		1027.40	34.99	38.49
88 L.H., Daily Totals		$5173.40		$6834.50	$58.79	$77.67
Crew R-7	**Hr.**	**Daily**	**Hr.**	**Daily**	**Bare Costs**	**Incl. O&P**
1 Electrician Foreman	$27.80	$222.40	$45.25	$362.00	$19.26	$32.33
5 Electrician Groundmen	17.55	702.00	29.75	1190.00		
1 Crew Truck		98.80		108.70	2.06	2.26
48 L.H., Daily Totals		$1023.20		$1660.70	$21.32	$34.59
Crew R-8	**Hr.**	**Daily**	**Hr.**	**Daily**	**Bare Costs**	**Incl. O&P**
1 Electrician Foreman	$27.80	$222.40	$45.25	$362.00	$24.13	$39.68
3 Electrician Linemen	27.30	655.20	44.45	1066.80		
2 Electrician Groundmen	17.55	280.80	29.75	476.00		
1 Pickup Truck, 3/4 Ton		78.00		85.80		
1 Crew Truck		98.80		108.70	3.68	4.05
48 L.H., Daily Totals		$1335.20		$2099.30	$27.81	$43.73
Crew R-9	**Hr.**	**Daily**	**Hr.**	**Daily**	**Bare Costs**	**Incl. O&P**
1 Electrician Foreman	$27.80	$222.40	$45.25	$362.00	$22.49	$37.20
1 Electrician Lineman	27.30	218.40	44.45	355.60		
2 Electrician Operators	27.30	436.80	44.45	711.20		
4 Electrician Groundmen	17.55	561.60	29.75	952.00		
1 Pickup Truck, 3/4 Ton		78.00		85.80		
1 Crew Truck		98.80		108.70	2.76	3.04
64 L.H., Daily Totals		$1616.00		$2575.30	$25.25	$40.24
Crew R-10	**Hr.**	**Daily**	**Hr.**	**Daily**	**Bare Costs**	**Incl. O&P**
1 Electrician Foreman	$27.80	$222.40	$45.25	$362.00	$25.76	$42.13
4 Electrician Linemen	27.30	873.60	44.45	1422.40		
1 Electrician Groundman	17.55	140.40	29.75	238.00		
1 Crew Truck		98.80		108.70		
3 Tram Cars		345.75		380.35	9.26	10.19
48 L.H., Daily Totals		$1680.95		$2511.45	$35.02	$52.32
Crew R-11	**Hr.**	**Daily**	**Hr.**	**Daily**	**Bare Costs**	**Incl. O&P**
1 Electrician Foreman	$27.80	$222.40	$45.25	$362.00	$25.59	$41.89
4 Electricians	27.30	873.60	44.45	1422.40		
1 Equip. Oper. (crane)	24.75	198.00	40.75	326.00		
1 Common Laborer	17.35	138.80	29.45	235.60		
1 Crew Truck		98.80		108.70		
1 Crane, 12 Ton		602.60		662.85	12.53	13.78
56 L.H., Daily Totals		$2134.20		$3117.55	$38.12	$55.67

Crews

Crew No.	Bare Costs		Incl. Sub O & P		Cost Per Labor-Hour	
Crew R-12	**Hr.**	**Daily**	**Hr.**	**Daily**	**Bare Costs**	**Incl. O&P**
Carpenter Foreman	$24.50	$196.00	$41.60	$332.80	$21.76	$37.37
Carpenters	24.00	768.00	40.75	1304.00		
Common Laborers	17.35	555.20	29.45	942.40		
Equip. Oper. (med.)	23.90	191.20	39.35	314.80		
Steel Worker	25.55	204.40	49.30	394.40		
Dozer, 200 H.P.		919.60		1011.55		
Pickup Truck, 3/4 Ton		78.00		85.80	11.34	12.47
8 L.H., Daily Totals		$2912.40		$4385.75	$33.10	$49.84
Crew R-15	**Hr.**	**Daily**	**Hr.**	**Daily**	**Bare Costs**	**Incl. O&P**
Electrician Foreman	$27.80	$222.40	$45.25	$362.00	$26.63	$43.43
Electricians	27.30	873.60	44.45	1422.40		
Equipment Operator	22.80	182.40	37.55	300.40		
Aerial Lift Truck		252.60		277.85	5.26	5.79
48 L.H., Daily Totals		$1531.00		$2362.65	$31.89	$49.22
Crew R-15A	**Hr.**	**Daily**	**Hr.**	**Daily**	**Bare Costs**	**Incl. O&P**
Electrician Foreman	$27.80	$222.40	$45.25	$362.00	$23.32	$38.43
2 Electricians	27.30	436.80	44.45	711.20		
2 Common Laborer	17.35	277.60	29.45	471.20		
1 Equipment Operator	22.80	182.40	37.55	300.40		
1 Aerial Lift Truck		252.60		277.85	5.26	5.79
48 L.H., Daily Totals		$1371.80		$2122.65	$28.58	$44.22
Crew R-18	**Hr.**	**Daily**	**Hr.**	**Daily**	**Bare Costs**	**Incl. O&P**
.25 Electrician Foreman	$27.80	$55.60	$45.25	$90.50	$21.34	$35.47
1 Electrician	27.30	218.40	44.45	355.60		
2 Helpers	17.55	280.80	29.75	476.00		
26 L.H., Daily Totals		$554.80		$922.10	$21.34	$35.47
Crew R-19	**Hr.**	**Daily**	**Hr.**	**Daily**	**Bare Costs**	**Incl. O&P**
.5 Electrician Foreman	$27.80	$111.20	$45.25	$181.00	$27.40	$44.61
2 Electricians	27.30	436.80	44.45	711.20		
20 L.H., Daily Totals		$548.00		$892.20	$27.40	$44.61
Crew R-21	**Hr.**	**Daily**	**Hr.**	**Daily**	**Bare Costs**	**Incl. O&P**
1 Electrician Foreman	$27.80	$222.40	$45.25	$362.00	$27.34	$44.52
3 Electricians	27.30	655.20	44.45	1066.80		
.1 Equip. Oper. (med.)	23.90	19.12	39.35	31.48		
.1 Hyd. Crane 25 Ton		57.50		63.25	1.75	1.93
32. L.H., Daily Totals		$954.22		$1523.53	$29.09	$46.45
Crew R-22	**Hr.**	**Daily**	**Hr.**	**Daily**	**Bare Costs**	**Incl. O&P**
.66 Electrician Foreman	$27.80	$146.78	$45.25	$238.92	$23.19	$38.25
2 Helpers	17.55	280.80	29.75	476.00		
2 Electricians	27.30	436.80	44.45	711.20		
37.28 L.H., Daily Totals		$864.38		$1426.12	$23.19	$38.25
Crew R-30	**Hr.**	**Daily**	**Hr.**	**Daily**	**Bare Costs**	**Incl. O&P**
.25 Electrician	$29.30	$58.60	$47.70	$95.40	$21.33	$35.47
1 Electrician	27.30	218.40	44.45	355.60		
2 Laborers, (Semi Skilled)	17.35	277.60	29.45	471.20		
26 L.H., Daily Totals		$554.60		$922.20	$21.33	$35.47

Location Factors

Costs shown in *Means cost data publications* are based on National Averages for materials and installation. To adjust these costs to a specific location, simply multiply the base cost by the factor for that city. The data is arranged alphabetically by state and postal zip code numbers. For a city not listed, use the factor for a nearby city with similar economic characteristics.

STATE	CITY	Residential
ALABAMA		
350-352	Birmingham	.87
354	Tuscaloosa	.73
355	Jasper	.71
356	Decatur	.77
357-358	Huntsville	.84
359	Gadsden	.73
360-361	Montgomery	.76
362	Anniston	.68
363	Dothan	.75
364	Evergreen	.71
365-366	Mobile	.80
367	Selma	.72
368	Phenix City	.73
369	Butler	.71
ALASKA		
995-996	Anchorage	1.26
997	Fairbanks	1.28
998	Juneau	1.26
999	Ketchikan	1.27
ARIZONA		
850,853	Phoenix	.87
852	Mesa/Tempe	.84
855	Globe	.80
856-857	Tucson	.84
859	Show Low	.83
860	Flagstaff	.85
863	Prescott	.82
864	Kingman	.81
865	Chambers	.81
ARKANSAS		
716	Pine Bluff	.75
717	Camden	.65
718	Texarkana	.70
719	Hot Springs	.64
720-722	Little Rock	.81
723	West Memphis	.74
724	Jonesboro	.73
725	Batesville	.71
726	Harrison	.72
727	Fayetteville	.68
728	Russellville	.70
729	Fort Smith	.76
CALIFORNIA		
900-902	Los Angeles	1.06
903-905	Inglewood	1.04
906-908	Long Beach	1.03
910-912	Pasadena	1.04
913-916	Van Nuys	1.07
917-918	Alhambra	1.08
919-921	San Diego	1.04
922	Palm Springs	1.02
923-924	San Bernardino	1.04
925	Riverside	1.06
926-927	Santa Ana	1.04
928	Anaheim	1.07
930	Oxnard	1.07
931	Santa Barbara	1.07
932-933	Bakersfield	1.04
934	San Luis Obispo	1.07
935	Mojave	1.05
936-938	Fresno	1.10
939	Salinas	1.12
940-941	San Francisco	1.21
942,956-958	Sacramento	1.11
943	Palo Alto	1.16
944	San Mateo	1.19
945	Vallejo	1.13
946	Oakland	1.18
947	Berkeley	1.21
948	Richmond	1.22
949	San Rafael	1.20
950	Santa Cruz	1.14
951	San Jose	1.18

STATE	CITY	Residential
952	Stockton	1.09
953	Modesto	1.08
954	Santa Rosa	1.14
955	Eureka	1.09
959	Marysville	1.10
960	Redding	1.10
961	Susanville	1.11
COLORADO		
800-802	Denver	.96
803	Boulder	.94
804	Golden	.92
805	Fort Collins	.91
806	Greeley	.80
807	Fort Morgan	.94
808-809	Colorado Springs	.92
810	Pueblo	.93
811	Alamosa	.90
812	Salida	.91
813	Durango	.93
814	Montrose	.88
815	Grand Junction	.93
816	Glenwood Springs	.92
CONNECTICUT		
060	New Britain	1.06
061	Hartford	1.05
062	Willimantic	1.06
063	New London	1.05
064	Meriden	1.05
065	New Haven	1.06
066	Bridgeport	1.06
067	Waterbury	1.06
068	Norwalk	1.06
069	Stamford	1.07
D.C.		
200-205	Washington	.93
DELAWARE		
197	Newark	1.00
198	Wilmington	1.01
199	Dover	1.00
FLORIDA		
320,322	Jacksonville	.79
321	Daytona Beach	.86
323	Tallahassee	.72
324	Panama City	.67
325	Pensacola	.75
326,344	Gainesville	.77
327-328,347	Orlando	.85
329	Melbourne	.87
330-332,340	Miami	.83
333	Fort Lauderdale	.84
334,349	West Palm Beach	.84
335-336,346	Tampa	.87
337	St. Petersburg	.77
338	Lakeland	.83
339,341	Fort Myers	.81
342	Sarasota	.85
GEORGIA		
300-303,399	Atlanta	.89
304	Statesboro	.67
305	Gainesville	.74
306	Athens	.74
307	Dalton	.70
308-309	Augusta	.76
310-312	Macon	.78
313-314	Savannah	.79
315	Waycross	.71
316	Valdosta	.71
317	Albany	.75
318-319	Columbus	.80
HAWAII		
967	Hilo	1.22

LOCATION FACTORS

Location Factors

STATE	CITY	Residential
968	Honolulu	1.23
STATES & POSS.		
969	Guam	1.62
IDAHO		
832	Pocatello	.88
833	Twin Falls	.72
834	Idaho Falls	.72
835	Lewiston	.97
836-837	Boise	.89
838	Coeur d'Alene	.84
ILLINOIS		
600-603	North Suburban	1.09
604	Joliet	1.11
605	South Suburban	1.09
606-608	Chicago	1.15
609	Kankakee	1.00
610-611	Rockford	1.03
612	Rock Island	.96
613	La Salle	1.01
614	Galesburg	.98
615-616	Peoria	1.01
617	Bloomington	.96
618-619	Champaign	.98
620-622	East St. Louis	.97
623	Quincy	.97
624	Effingham	.98
625	Decatur	.97
626-627	Springfield	.98
628	Centralia	.96
629	Carbondale	.96
INDIANA		
460	Anderson	.91
461-462	Indianapolis	.96
463-464	Gary	1.00
465-466	South Bend	.91
467-468	Fort Wayne	.91
469	Kokomo	.92
470	Lawrenceburg	.86
471	New Albany	.85
472	Columbus	.92
473	Muncie	.91
474	Bloomington	.94
475	Washington	.90
476-477	Evansville	.90
478	Terre Haute	.90
479	Lafayette	.91
IOWA		
500-503,509	Des Moines	.92
504	Mason City	.77
505	Fort Dodge	.76
506-507	Waterloo	.80
508	Creston	.81
510-511	Sioux City	.86
512	Sibley	.73
513	Spencer	.75
514	Carroll	.75
515	Council Bluffs	.81
516	Shenandoah	.74
520	Dubuque	.84
521	Decorah	.77
522-524	Cedar Rapids	.93
525	Ottumwa	.83
526	Burlington	.87
527-528	Davenport	.98
KANSAS		
660-662	Kansas City	.95
664-666	Topeka	.79
667	Fort Scott	.85
668	Emporia	.72
669	Belleville	.74
670-672	Wichita	.81
673	Independence	.75
674	Salina	.73
675	Hutchinson	.68
676	Hays	.74
677	Colby	.76
678	Dodge City	.73
679	Liberal	.68

STATE	CITY	Residential
KENTUCKY		
400-402	Louisville	.93
403-405	Lexington	.84
406	Frankfort	.82
407-409	Corbin	.67
410	Covington	.93
411-412	Ashland	.94
413-414	Campton	.68
415-416	Pikeville	.77
417-418	Hazard	.67
420	Paducah	.89
421-422	Bowling Green	.89
423	Owensboro	.82
424	Henderson	.91
425-426	Somerset	.67
427	Elizabethtown	.86
LOUISIANA		
700-701	New Orleans	.85
703	Thibodaux	.80
704	Hammond	.78
705	Lafayette	.77
706	Lake Charles	.79
707-708	Baton Rouge	.78
710-711	Shreveport	.78
712	Monroe	.73
713-714	Alexandria	.73
MAINE		
039	Kittery	.80
040-041	Portland	.88
042	Lewiston	.88
043	Augusta	.83
044	Bangor	.86
045	Bath	.81
046	Machias	.82
047	Houlton	.86
048	Rockland	.80
049	Waterville	.75
MARYLAND		
206	Waldorf	.84
207-208	College Park	.84
209	Silver Spring	.85
210-212	Baltimore	.89
214	Annapolis	.85
215	Cumberland	.86
216	Easton	.68
217	Hagerstown	.85
218	Salisbury	.75
219	Elkton	.81
MASSACHUSETTS		
010-011	Springfield	1.05
012	Pittsfield	1.02
013	Greenfield	1.01
014	Fitchburg	1.08
015-016	Worcester	1.11
017	Framingham	1.12
018	Lowell	1.13
019	Lawrence	1.12
020-022, 024	Boston	1.19
023	Brockton	1.11
025	Buzzards Bay	1.10
026	Hyannis	1.09
027	New Bedford	1.11
MICHIGAN		
480,483	Royal Oak	1.03
481	Ann Arbor	1.03
482	Detroit	1.10
484-485	Flint	.97
486	Saginaw	.94
487	Bay City	.95
488-489	Lansing	.96
490	Battle Creek	.93
491	Kalamazoo	.92
492	Jackson	.94
493,495	Grand Rapids	.83
494	Muskegon	.89
496	Traverse City	.81
497	Gaylord	.84
498-499	Iron Mountain	.90

STATE	CITY	Residential
MINNESOTA		
550-551	Saint Paul	1.13
553-555	Minneapolis	1.18
556-558	Duluth	1.10
559	Rochester	1.04
560	Mankato	1.00
561	Windom	.84
562	Willmar	.85
563	St. Cloud	1.07
564	Brainerd	.96
565	Detroit Lakes	.97
566	Bemidji	.94
567	Thief River Falls	.92
MISSISSIPPI		
386	Clarksdale	.61
387	Greenville	.68
388	Tupelo	.64
389	Greenwood	.65
390-392	Jackson	.73
393	Meridian	.66
394	Laurel	.62
395	Biloxi	.75
396	McComb	.74
397	Columbus	.64
MISSOURI		
630-631	St. Louis	1.01
633	Bowling Green	.90
634	Hannibal	.87
635	Kirksville	.80
636	Flat River	.93
637	Cape Girardeau	.86
638	Sikeston	.83
639	Poplar Bluff	.83
640-641	Kansas City	1.02
644-645	St. Joseph	.93
646	Chillicothe	.85
647	Harrisonville	.94
648	Joplin	.82
650-651	Jefferson City	.88
652	Columbia	.87
653	Sedalia	.85
654-655	Rolla	.88
656-658	Springfield	.84
MONTANA		
590-591	Billings	.88
592	Wolf Point	.84
593	Miles City	.86
594	Great Falls	.89
595	Havre	.81
596	Helena	.88
597	Butte	.83
598	Missoula	.83
599	Kalispell	.82
NEBRASKA		
680-681	Omaha	.89
683-685	Lincoln	.78
686	Columbus	.69
687	Norfolk	.77
688	Grand Island	.77
689	Hastings	.76
690	Mccook	.69
691	North Platte	.75
692	Valentine	.66
693	Alliance	.65
NEVADA		
889-891	Las Vegas	1.00
893	Ely	.90
894-895	Reno	.97
897	Carson City	.97
898	Elko	.97
NEW HAMPSHIRE		
030	Nashua	.90
031	Manchester	.90
032-033	Concord	.86
034	Keene	.73
035	Littleton	.81
036	Charleston	.70

STATE	CITY	Residential
037	Claremont	.72
038	Portsmouth	.84
NEW JERSEY		
070-071	Newark	1.12
072	Elizabeth	1.15
073	Jersey City	1.12
074-075	Paterson	1.12
076	Hackensack	1.11
077	Long Branch	1.12
078	Dover	1.12
079	Summit	1.12
080,083	Vineland	1.09
081	Camden	1.10
082,084	Atlantic City	1.13
085-086	Trenton	1.11
087	Point Pleasant	1.10
088-089	New Brunswick	1.12
NEW MEXICO		
870-872	Albuquerque	.85
873	Gallup	.85
874	Farmington	.85
875	Santa Fe	.85
877	Las Vegas	.85
878	Socorro	.85
879	Truth/Consequences	.84
880	Las Cruces	.83
881	Clovis	.85
882	Roswell	.85
883	Carrizozo	.85
884	Tucumcari	.86
NEW YORK		
100-102	New York	1.36
103	Staten Island	1.27
104	Bronx	1.29
105	Mount Vernon	1.15
106	White Plains	1.19
107	Yonkers	1.21
108	New Rochelle	1.21
109	Suffern	1.13
110	Queens	1.27
111	Long Island City	1.30
112	Brooklyn	1.32
113	Flushing	1.29
114	Jamaica	1.29
115,117,118	Hicksville	1.19
116	Far Rockaway	1.28
119	Riverhead	1.20
120-122	Albany	.95
123	Schenectady	.96
124	Kingston	1.02
125-126	Poughkeepsie	1.06
127	Monticello	1.03
128	Glens Falls	.88
129	Plattsburgh	.92
130-132	Syracuse	.96
133-135	Utica	.93
136	Watertown	.91
137-139	Binghamton	.92
140-142	Buffalo	1.06
143	Niagara Falls	1.01
144-146	Rochester	.98
147	Jamestown	.88
148-149	Elmira	.87
NORTH CAROLINA		
270,272-274	Greensboro	.73
271	Winston-Salem	.73
275-276	Raleigh	.74
277	Durham	.73
278	Rocky Mount	.64
279	Elizabeth City	.61
280	Gastonia	.74
281-282	Charlotte	.74
283	Fayetteville	.71
284	Wilmington	.72
285	Kinston	.62
286	Hickory	.62
287-288	Asheville	.72
289	Murphy	.66
NORTH DAKOTA		

STATE	CITY	Residential
580-581	Fargo	.80
582	Grand Forks	.77
583	Devils Lake	.80
584	Jamestown	.75
585	Bismarck	.80
586	Dickinson	.77
587	Minot	.80
588	Williston	.77
OHIO		
430-432	Columbus	.96
433	Marion	.93
434-436	Toledo	1.01
437-438	Zanesville	.91
439	Steubenville	.96
440	Lorain	1.01
441	Cleveland	1.02
442-443	Akron	.99
444-445	Youngstown	.96
446-447	Canton	.94
448-449	Mansfield	.95
450	Hamilton	.96
451-452	Cincinnati	.96
453-454	Dayton	.92
455	Springfield	.94
456	Chillicothe	.96
457	Athens	.90
458	Lima	.92
OKLAHOMA		
730-731	Oklahoma City	.80
734	Ardmore	.78
735	Lawton	.81
736	Clinton	.77
737	Enid	.77
738	Woodward	.76
739	Guymon	.67
740-741	Tulsa	.79
743	Miami	.82
744	Muskogee	.72
745	Mcalester	.74
746	Ponca City	.77
747	Durant	.76
748	Shawnee	.76
749	Poteau	.77
OREGON		
970-972	Portland	1.02
973	Salem	1.01
974	Eugene	1.01
975	Medford	.99
976	Klamath Falls	1.00
977	Bend	1.02
978	Pendleton	.99
979	Vale	.99
PENNSYLVANIA		
150-152	Pittsburgh	.99
153	Washington	.94
154	Uniontown	.91
155	Bedford	.89
156	Greensburg	.95
157	Indiana	.91
158	Dubois	.90
159	Johnstown	.90
160	Butler	.93
161	New Castle	.93
162	Kittanning	.94
163	Oil City	.89
164-165	Erie	.97
166	Altoona	.89
167	Bradford	.89
168	State College	.92
169	Wellsboro	.88
170-171	Harrisburg	.94
172	Chambersburg	.89
173-174	York	.88
175-176	Lancaster	.91
177	Williamsport	.85
178	Sunbury	.89
179	Pottsville	.89
180	Lehigh Valley	.98
181	Allentown	1.01
182	Hazleton	.89

STATE	CITY	Residential
183	Stroudsburg	.92
184-185	Scranton	.95
186-187	Wilkes-Barre	.91
188	Montrose	.89
189	Doylestown	1.04
190-191	Philadelphia	1.14
193	Westchester	1.07
194	Norristown	1.03
195-196	Reading	.96
PUERTO RICO		
009	San Juan	.84
RHODE ISLAND		
028	Newport	1.08
029	Providence	1.08
SOUTH CAROLINA		
290-292	Columbia	.73
293	Spartanburg	.71
294	Charleston	.72
295	Florence	.66
296	Greenville	.70
297	Rock Hill	.65
298	Aiken	.83
299	Beaufort	.67
SOUTH DAKOTA		
570-571	Sioux Falls	.77
572	Watertown	.73
573	Mitchell	.75
574	Aberdeen	.76
575	Pierre	.76
576	Mobridge	.73
577	Rapid City	.76
TENNESSEE		
370-372	Nashville	.85
373-374	Chattanooga	.77
375,380-381	Memphis	.85
376	Johnson City	.72
377-379	Knoxville	.75
382	Mckenzie	.70
383	Jackson	.71
384	Columbia	.72
385	Cookeville	.68
TEXAS		
750	Mckinney	.75
751	Waxahackie	.76
752-753	Dallas	.83
754	Greenville	.69
755	Texarkana	.74
756	Longview	.68
757	Tyler	.74
758	Palestine	.67
759	Lufkin	.72
760-761	Fort Worth	.83
762	Denton	.77
763	Wichita Falls	.79
764	Eastland	.72
765	Temple	.75
766-767	Waco	.77
768	Brownwood	.68
769	San Angelo	.72
770-772	Houston	.85
773	Huntsville	.69
774	Wharton	.70
775	Galveston	.83
776-777	Beaumont	.82
778	Bryan	.74
779	Victoria	.74
780	Laredo	.73
781-782	San Antonio	.80
783-784	Corpus Christi	.77
785	Mc Allen	.75
786-787	Austin	.79
788	Del Rio	.66
789	Giddings	.70
790-791	Amarillo	.78
792	Childress	.76
793-794	Lubbock	.76
795-796	Abilene	.75
797	Midland	.76

Location Factors

STATE	CITY	Residential
798-799,885	El Paso	.75
UTAH		
840-841	Salt Lake City	.83
842,844	Ogden	.81
843	Logan	.82
845	Price	.72
846-847	Provo	.83
VERMONT		
050	White River Jct.	.73
051	Bellows Falls	.75
052	Bennington	.74
053	Brattleboro	.75
054	Burlington	.80
056	Montpelier	.82
057	Rutland	.81
058	St. Johnsbury	.75
059	Guildhall	.74
VIRGINIA		
220-221	Fairfax	.85
222	Arlington	.87
223	Alexandria	.90
224-225	Fredericksburg	.75
226	Winchester	.71
227	Culpeper	.77
228	Harrisonburg	.67
229	Charlottesville	.72
230-232	Richmond	.81
233-235	Norfolk	.81
236	Newport News	.79
237	Portsmouth	.77
238	Petersburg	.78
239	Farmville	.68
240-241	Roanoke	.72
242	Bristol	.67
243	Pulaski	.66
244	Staunton	.68
245	Lynchburg	.69
246	Grundy	.67
WASHINGTON		
980-981,987	Seattle	1.01
982	Everett	1.03
983-984	Tacoma	1.00
985	Olympia	1.00
986	Vancouver	.97
988	Wenatchee	.90
989	Yakima	.95
990-992	Spokane	1.00
993	Richland	.97
994	Clarkston	.96
WEST VIRGINIA		
247-248	Bluefield	.88
249	Lewisburg	.89
250-253	Charleston	.97
254	Martinsburg	.84
255-257	Huntington	.96
258-259	Beckley	.90
260	Wheeling	.92
261	Parkersburg	.91
262	Buckhannon	.91
263-264	Clarksburg	.90
265	Morgantown	.91
266	Gassaway	.92
267	Romney	.86
268	Petersburg	.88
WISCONSIN		
530,532	Milwaukee	1.05
531	Kenosha	1.03
534	Racine	1.01
535	Beloit	.99
537	Madison	.99
538	Lancaster	.97
539	Portage	.96
540	New Richmond	.98
541-543	Green Bay	1.01
544	Wausau	.95
545	Rhinelander	.95
546	La Crosse	.94
547	Eau Claire	.98

STATE	CITY	Residential
548	Superior	.97
549	Oshkosh	.95
WYOMING		
820	Cheyenne	.75
821	Yellowstone Nat. Pk.	.70
822	Wheatland	.71
823	Rawlins	.69
824	Worland	.69
825	Riverton	.70
826	Casper	.76
827	Newcastle	.68
828	Sheridan	.73
829-831	Rock Springs	.73
ALBERTA		
	Calgary	1.06
	Edmonton	1.05
	Fort McMurray	1.03
	Lethbridge	1.04
	Lloydminster	1.03
	Medicine Hat	1.04
	Red Deer	1.04
BRITISH COLUMBIA		
	Kamloops	1.01
	Prince George	1.01
	Vancouver	1.08
	Victoria	1.01
MANITOBA		
	Brandon	1.00
	Portage la Prairie	1.00
	Winnipeg	1.01
NEW BRUNSWICK		
	Bathurst	.91
	Dalhousie	.91
	Fredericton	.99
	Moncton	.91
	Newcastle	.91
	Saint John	.99
NEWFOUNDLAND		
	Corner Brook	.92
	St. John's	.93
NORTHWEST TERRITORIES		
	Yellowknife	.98
NOVA SCOTIA		
	Dartmouth	.93
	Halifax	.93
	New Glasgow	.93
	Sydney	.91
	Yarmouth	.93
ONTARIO		
	Barrie	1.10
	Brantford	1.12
	Cornwall	1.11
	Hamilton	1.12
	Kingston	1.11
	Kitchener	1.06
	London	1.10
	North Bay	1.08
	Oshawa	1.10
	Ottawa	1.12
	Owen Sound	1.09
	Peterborough	1.09
	Sarnia	1.12
	Sudbury	1.03
	Thunder Bay	1.08
	Toronto	1.15
	Windsor	1.09
PRINCE EDWARD ISLAND		
	Charlottetown	.88
	Summerside	.88
QUEBEC		
	Cap-de-la-Madeleine	1.11
	Charlesbourg	1.11
	Chicoutimi	1.11

Location Factors

STATE	CITY	Residential
	Gatineau	1.10
	Laval	1.11
	Montreal	1.11
	Quebec	1.13
	Sherbrooke	1.10
	Trois Rivieres	1.11
SASKATCHEWAN		
	Moose Jaw	.91
	Prince Albert	.90
	Regina	.92
	Saskatoon	.91
YUKON		
	Whitehorse	.90

Abbreviations

A	Area Square Feet; Ampere
ABS	Acrylonitrile Butadiene Stryrene; Asbestos Bonded Steel
A.C.	Alternating Current; Air-Conditioning; Asbestos Cement; Plywood Grade A & C
A.C.I.	American Concrete Institute
AD	Plywood, Grade A & D
Addit.	Additional
Adj.	Adjustable
af	Audio-frequency
A.G.A.	American Gas Association
Agg.	Aggregate
A.H.	Ampere Hours
A hr.	Ampere-hour
A.H.U.	Air Handling Unit
A.I.A.	American Institute of Architects
AIC	Ampere Interrupting Capacity
Allow.	Allowance
alt.	Altitude
Alum.	Aluminum
a.m.	Ante Meridiem
Amp.	Ampere
Anod.	Anodized
Approx.	Approximate
Apt.	Apartment
Asb.	Asbestos
A.S.B.C.	American Standard Building Code
Asbe.	Asbestos Worker
A.S.H.R.A.E.	American Society of Heating, Refrig. & AC Engineers
A.S.M.E.	American Society of Mechanical Engineers
A.S.T.M.	American Society for Testing and Materials
Attchmt.	Attachment
Avg.	Average
A.W.G.	American Wire Gauge
AWWA	American Water Works Assoc.
Bbl.	Barrel
B&B	Grade B and Better; Balled & Burlapped
B.&S.	Bell and Spigot
B.&W.	Black and White
b.c.c.	Body-centered Cubic
B.C.Y.	Bank Cubic Yards
BE	Bevel End
B.F.	Board Feet
Bg. cem.	Bag of Cement
BHP	Boiler Horsepower; Brake Horsepower
B.I.	Black Iron
Bit.; Bitum.	Bituminous
Bk.	Backed
Bkrs.	Breakers
Bldg.	Building
Blk.	Block
Bm.	Beam
Boil.	Boilermaker
B.P.M.	Blows per Minute
BR	Bedroom
Brg.	Bearing
Brhe.	Bricklayer Helper
Bric.	Bricklayer
Brk.	Brick
Brng.	Bearing
Brs.	Brass
Brz.	Bronze
Bsn.	Basin
Btr.	Better
BTU	British Thermal Unit
BTUH	BTU per Hour
B.U.R.	Built-up Roofing
BX	Interlocked Armored Cable
c	Conductivity, Copper Sweat
C	Hundred; Centigrade
C/C	Center to Center, Cedar on Cedar
Cab.	Cabinet
Cair.	Air Tool Laborer
Calc	Calculated
Cap.	Capacity
Carp.	Carpenter
C.B.	Circuit Breaker
C.C.A.	Chromate Copper Arsenate
C.C.F.	Hundred Cubic Feet
cd	Candela
cd/sf	Candela per Square Foot
CD	Grade of Plywood Face & Back
CDX	Plywood, Grade C & D, exterior glue
Cefi.	Cement Finisher
Cem.	Cement
CF	Hundred Feet
C.F.	Cubic Feet
CFM	Cubic Feet per Minute
c.g.	Center of Gravity
CHW	Chilled Water; Commercial Hot Water
C.I.	Cast Iron
C.I.P.	Cast in Place
Circ.	Circuit
C.L.	Carload Lot
Clab.	Common Laborer
Clam	Common maintenance laborer
C.L.F.	Hundred Linear Feet
CLF	Current Limiting Fuse
CLP	Cross Linked Polyethylene
cm	Centimeter
CMP	Corr. Metal Pipe
C.M.U.	Concrete Masonry Unit
CN	Change Notice
Col.	Column
CO_2	Carbon Dioxide
Comb.	Combination
Compr.	Compressor
Conc.	Concrete
Cont.	Continuous; Continued
Corr.	Corrugated
Cos	Cosine
Cot	Cotangent
Cov.	Cover
C/P	Cedar on Paneling
CPA	Control Point Adjustment
Cplg.	Coupling
C.P.M.	Critical Path Method
CPVC	Chlorinated Polyvinyl Chloride
C.Pr.	Hundred Pair
CRC	Cold Rolled Channel
Creos.	Creosote
Crpt.	Carpet & Linoleum Layer
CRT	Cathode-ray Tube
CS	Carbon Steel, Constant Shear Bar Joist
Csc	Cosecant
C.S.F.	Hundred Square Feet
CSI	Construction Specifications Institute
C.T.	Current Transformer
CTS	Copper Tube Size
Cu	Copper, Cubic
Cu. Ft.	Cubic Foot
cw	Continuous Wave
C.W.	Cool White; Cold Water
Cwt.	100 Pounds
C.W.X.	Cool White Deluxe
C.Y.	Cubic Yard (27 cubic feet)
C.Y./Hr.	Cubic Yard per Hour
Cyl.	Cylinder
d	Penny (nail size)
D	Deep; Depth; Discharge
Dis.;Disch.	Discharge
Db.	Decibel
Dbl.	Double
DC	Direct Current
DDC	Direct Digital Control
Demob.	Demobilization
d.f.u.	Drainage Fixture Units
D.H.	Double Hung
DHW	Domestic Hot Water
Diag.	Diagonal
Diam.	Diameter
Distrib.	Distribution
Dk.	Deck
D.L.	Dead Load; Diesel
DLH	Deep Long Span Bar Joist
Do.	Ditto
Dp.	Depth
D.P.S.T.	Double Pole, Single Throw
Dr.	Driver
Drink.	Drinking
D.S.	Double Strength
D.S.A.	Double Strength A Grade
D.S.B.	Double Strength B Grade
Dty.	Duty
DWV	Drain Waste Vent
DX	Deluxe White, Direct Expansion
dyn	Dyne
e	Eccentricity
E	Equipment Only; East
Ea.	Each
E.B.	Encased Burial
Econ.	Economy
E.C.Y	Embankment Cubic Yards
EDP	Electronic Data Processing
EIFS	Exterior Insulation Finish System
E.D.R.	Equiv. Direct Radiation
Eq.	Equation
Elec.	Electrician; Electrical
Elev.	Elevator; Elevating
EMT	Electrical Metallic Conduit; Thin Wall Conduit
Eng.	Engine, Engineered
EPDM	Ethylene Propylene Diene Monomer
EPS	Expanded Polystyrene
Eqhv.	Equip. Oper., Heavy
Eqlt.	Equip. Oper., Light
Eqmd.	Equip. Oper., Medium
Eqmm.	Equip. Oper., Master Mechanic
Eqol.	Equip. Oper., Oilers
Equip.	Equipment
ERW	Electric Resistance Welded
E.S.	Energy Saver
Est.	Estimated
esu	Electrostatic Units
E.W.	Each Way
EWT	Entering Water Temperature
Excav.	Excavation
Exp.	Expansion, Exposure
Ext.	Exterior
Extru.	Extrusion
f.	Fiber stress
F	Fahrenheit; Female; Fill
Fab.	Fabricated
FBGS	Fiberglass
F.C.	Footcandles
f.c.c.	Face-centered Cubic
f'c.	Compressive Stress in Concrete; Extreme Compressive Stress
F.E.	Front End
FEP	Fluorinated Ethylene Propylene (Teflon)
F.G.	Flat Grain
F.H.A.	Federal Housing Administration
Fig.	Figure
Fin.	Finished
Fixt.	Fixture
Fl. Oz.	Fluid Ounces
Flr.	Floor
F.M.	Frequency Modulation; Factory Mutual
Fmg.	Framing
Fndtn.	Foundation

ori.	Foreman, Inside
oro.	Foreman, Outside
ount.	Fountain
PM	Feet per Minute
PT	Female Pipe Thread
:	Frame
R.	Fire Rating
RK	Foil Reinforced Kraft
RP	Fiberglass Reinforced Plastic
S	Forged Steel
SC	Cast Body; Cast Switch Box
:	Foot; Feet
tng.	Fitting
tg.	Footing
t. Lb.	Foot Pound
urn.	Furniture
VNR	Full Voltage Non-Reversing
XM	Female by Male
y.	Minimum Yield Stress of Steel
	Gram
	Gauss
a.	Gauge
al.	Gallon
al./Min.	Gallon per Minute
alv.	Galvanized
en.	General
.F.I.	Ground Fault Interrupter
laz.	Glazier
PD	Gallons per Day
PH	Gallons per Hour
PM	Gallons per Minute
R	Grade
ran.	Granular
rnd.	Ground
I	High; High Strength Bar Joist; Henry
I.C.	High Capacity
I.D.	Heavy Duty; High Density
I.D.O.	High Density Overlaid
Hdr.	Header
Hdwe.	Hardware
Help.	Helper Average
HEPA	High Efficiency Particulate Air Filter
Hg	Mercury
HIC	High Interrupting Capacity
HM	Hollow Metal
H.O.	High Output
Horiz.	Horizontal
H.P.	Horsepower; High Pressure
H.P.F.	High Power Factor
Hr.	Hour
Hrs./Day	Hours per Day
HSC	High Short Circuit
Ht.	Height
Htg.	Heating
Htrs.	Heaters
HVAC	Heating, Ventilation & Air-Conditioning
Hvy.	Heavy
HW	Hot Water
Hyd.;Hydr.	Hydraulic
Hz.	Hertz (cycles)
I.	Moment of Inertia
I.C.	Interrupting Capacity
ID	Inside Diameter
I.D.	Inside Dimension; Identification
I.F.	Inside Frosted
I.M.C.	Intermediate Metal Conduit
In.	Inch
Incan.	Incandescent
Incl.	Included; Including
Int.	Interior
Inst.	Installation
Insul.	Insulation/Insulated
I.P.	Iron Pipe
I.P.S.	Iron Pipe Size
I.P.T.	Iron Pipe Threaded
I.W.	Indirect Waste
J	Joule
J.I.C.	Joint Industrial Council
K	Thousand; Thousand Pounds; Heavy Wall Copper Tubing, Kelvin
K.A.H.	Thousand Amp. Hours
KCMIL	Thousand Circular Mils
KD	Knock Down
K.D.A.T.	Kiln Dried After Treatment
kg	Kilogram
kG	Kilogauss
kgf	Kilogram Force
kHz	Kilohertz
Kip.	1000 Pounds
KJ	Kiljoule
K.L.	Effective Length Factor
K.L.F.	Kips per Linear Foot
Km	Kilometer
K.S.F.	Kips per Square Foot
K.S.I.	Kips per Square Inch
kV	Kilovolt
kVA	Kilovolt Ampere
K.V.A.R.	Kilovar (Reactance)
KW	Kilowatt
KWh	Kilowatt-hour
L	Labor Only; Length; Long; Medium Wall Copper Tubing
Lab.	Labor
lat	Latitude
Lath.	Lather
Lav.	Lavatory
lb.; #	Pound
L.B.	Load Bearing; L Conduit Body
L. & E.	Labor & Equipment
lb./hr.	Pounds per Hour
lb./L.F.	Pounds per Linear Foot
lbf/sq.in.	Pound-force per Square Inch
L.C.L.	Less than Carload Lot
L.C.Y.	Loose Cubic Yard
Ld.	Load
LE	Lead Equivalent
LED	Light Emitting Diode
L.F.	Linear Foot
Lg.	Long; Length; Large
L & H	Light and Heat
LH	Long Span Bar Joist
L.H.	Labor Hours
L.L.	Live Load
L.L.D.	Lamp Lumen Depreciation
lm	Lumen
lm/sf	Lumen per Square Foot
lm/W	Lumen per Watt
L.O.A.	Length Over All
log	Logarithm
L-O-L	Lateralolet
L.P.	Liquefied Petroleum; Low Pressure
L.P.F.	Low Power Factor
LR	Long Radius
L.S.	Lump Sum
Lt.	Light
Lt. Ga.	Light Gauge
L.T.L.	Less than Truckload Lot
Lt. Wt.	Lightweight
L.V.	Low Voltage
M	Thousand; Material; Male; Light Wall Copper Tubing
M^2CA	Meters Squared Contact Area
m/hr; M.H.	Man-hour
mA	Milliampere
Mach.	Machine
Mag. Str.	Magnetic Starter
Maint.	Maintenance
Marb.	Marble Setter
Mat; Mat'l.	Material
Max.	Maximum
MBF	Thousand Board Feet
MBH	Thousand BTU's per hr.
MC	Metal Clad Cable
M.C.F.	Thousand Cubic Feet
M.C.F.M.	Thousand Cubic Feet per Minute
M.C.M.	Thousand Circular Mils
M.C.P.	Motor Circuit Protector
MD	Medium Duty
M.D.O.	Medium Density Overlaid
Med.	Medium
MF	Thousand Feet
M.F.B.M.	Thousand Feet Board Measure
Mfg.	Manufacturing
Mfrs.	Manufacturers
mg	Milligram
MGD	Million Gallons per Day
MGPH	Thousand Gallons per Hour
MH, M.H.	Manhole; Metal Halide; Man-Hour
MHz	Megahertz
Mi.	Mile
MI	Malleable Iron; Mineral Insulated
mm	Millimeter
Mill.	Millwright
Min., min.	Minimum, minute
Misc.	Miscellaneous
ml	Milliliter, Mainline
M.L.F.	Thousand Linear Feet
Mo.	Month
Mobil.	Mobilization
Mog.	Mogul Base
MPH	Miles per Hour
MPT	Male Pipe Thread
MRT	Mile Round Trip
ms	Millisecond
M.S.F.	Thousand Square Feet
Mstz.	Mosaic & Terrazzo Worker
M.S.Y.	Thousand Square Yards
Mtd.	Mounted
Mthe.	Mosaic & Terrazzo Helper
Mtng.	Mounting
Mult.	Multi; Multiply
M.V.A.	Million Volt Amperes
M.V.A.R.	Million Volt Amperes Reactance
MV	Megavolt
MW	Megawatt
MXM	Male by Male
MYD	Thousand Yards
N	Natural; North
nA	Nanoampere
NA	Not Available; Not Applicable
N.B.C.	National Building Code
NC	Normally Closed
N.E.M.A.	National Electrical Manufacturers Assoc.
NEHB	Bolted Circuit Breaker to 600V.
N.L.B.	Non-Load-Bearing
NM	Non-Metallic Cable
nm	Nanometer
No.	Number
NO	Normally Open
N.O.C.	Not Otherwise Classified
Nose.	Nosing
N.P.T.	National Pipe Thread
NQOD	Combination Plug-on/Bolt on Circuit Breaker to 240V.
N.R.C.	Noise Reduction Coefficient
N.R.S.	Non Rising Stem
ns	Nanosecond
nW	Nanowatt
OB	Opposing Blade
OC	On Center
OD	Outside Diameter
O.D.	Outside Dimension
ODS	Overhead Distribution System
O.G.	Ogee
O.H.	Overhead
O&P	Overhead and Profit
Oper.	Operator
Opng.	Opening
Orna.	Ornamental
OSB	Oriented Strand Board

Carlinville Public Library
P.O. Box 17
Carlinville, Illinois 62626

Abbreviations

O.S.&Y.	Outside Screw and Yoke
Ovhd.	Overhead
OWG	Oil, Water or Gas
Oz.	Ounce
P.	Pole; Applied Load; Projection
p.	Page
Pape.	Paperhanger
P.A.P.R.	Powered Air Purifying Respirator
PAR	Parabolic Reflector
Pc., Pcs.	Piece, Pieces
P.C.	Portland Cement; Power Connector
P.C.F.	Pounds per Cubic Foot
P.C.M.	Phase Contract Microscopy
P.E.	Professional Engineer; Porcelain Enamel; Polyethylene; Plain End
Perf.	Perforated
Ph.	Phase
P.I.	Pressure Injected
Pile.	Pile Driver
Pkg.	Package
Pl.	Plate
Plah.	Plasterer Helper
Plas.	Plasterer
Pluh.	Plumbers Helper
Plum.	Plumber
Ply.	Plywood
p.m.	Post Meridiem
Pntd.	Painted
Pord.	Painter, Ordinary
pp	Pages
PP; PPL	Polypropylene
P.P.M.	Parts per Million
Pr.	Pair
P.E.S.B.	Pre-engineered Steel Building
Prefab.	Prefabricated
Prefin.	Prefinished
Prop.	Propelled
PSF; psf	Pounds per Square Foot
PSI; psi	Pounds per Square Inch
PSIG	Pounds per Square Inch Gauge
PSP	Plastic Sewer Pipe
Pspr.	Painter, Spray
Psst.	Painter, Structural Steel
P.T.	Potential Transformer
P. & T.	Pressure & Temperature
Ptd.	Painted
Ptns.	Partitions
Pu	Ultimate Load
PVC	Polyvinyl Chloride
Pvmt.	Pavement
Pwr.	Power
Q	Quantity Heat Flow
Quan.;Qty.	Quantity
Q.C.	Quick Coupling
r	Radius of Gyration
R	Resistance
R.C.P.	Reinforced Concrete Pipe
Rect.	Rectangle
Reg.	Regular
Reinf.	Reinforced
Req'd.	Required
Res.	Resistant
Resi.	Residential
Rgh.	Rough
RGS	Rigid Galvanized Steel
R.H.W.	Rubber, Heat & Water Resistant; Residential Hot Water
rms	Root Mean Square
Rnd.	Round
Rodm.	Rodman
Rofc.	Roofer, Composition
Rofp.	Roofer, Precast
Rohe.	Roofer Helpers (Composition)
Rots.	Roofer, Tile & Slate
R.O.W.	Right of Way
RPM	Revolutions per Minute
R.S.	Rapid Start
Rsr	Riser
RT	Round Trip
S.	Suction; Single Entrance; South
SCFM	Standard Cubic Feet per Minute
Scaf.	Scaffold
Sch.; Sched.	Schedule
S.C.R.	Modular Brick
S.D.	Sound Deadening
S.D.R.	Standard Dimension Ratio
S.E.	Surfaced Edge
Sel.	Select
S.E.R.; S.E.U.	Service Entrance Cable
S.F.	Square Foot
S.F.C.A.	Square Foot Contact Area
S.F. Flr.	Square Foot of Floor
S.F.G.	Square Foot of Ground
S.F. Hor.	Square Foot Horizontal
S.F.R.	Square Feet of Radiation
S.F. Shlf.	Square Foot of Shelf
S4S	Surface 4 Sides
Shee.	Sheet Metal Worker
Sin.	Sine
Skwk.	Skilled Worker
SL	Saran Lined
S.L.	Slimline
Sldr.	Solder
SLH	Super Long Span Bar Joist
S.N.	Solid Neutral
S-O-L	Socketolet
sp	Standpipe
S.P.	Static Pressure; Single Pole; Self-Propelled
Spri.	Sprinkler Installer
spwg	Static Pressure Water Gauge
S.P.D.T.	Single Pole, Double Throw
SPF	Spruce Pine Fir
S.P.S.T.	Single Pole, Single Throw
SPT	Standard Pipe Thread
Sq.	Square; 100 Square Feet
Sq. Hd.	Square Head
Sq. In.	Square Inch
S.S.	Single Strength; Stainless Steel
S.S.B.	Single Strength B Grade
sst	Stainless Steel
Sswk.	Structural Steel Worker
Sswl.	Structural Steel Welder
St.;Stl.	Steel
S.T.C.	Sound Transmission Coefficient
Std.	Standard
STK	Select Tight Knot
STP	Standard Temperature & Pressure
Stpi.	Steamfitter, Pipefitter
Str.	Strength; Starter; Straight
Strd.	Stranded
Struct.	Structural
Sty.	Story
Subj.	Subject
Subs.	Subcontractors
Surf.	Surface
Sw.	Switch
Swbd.	Switchboard
S.Y.	Square Yard
Syn.	Synthetic
S.Y.P.	Southern Yellow Pine
Sys.	System
t.	Thickness
T	Temperature; Ton
Tan	Tangent
T.C.	Terra Cotta
T & C	Threaded and Coupled
T.D.	Temperature Difference
T.E.M.	Transmission Electron Microscopy
TFE	Tetrafluoroethylene (Teflon)
T. & G.	Tongue & Groove; Tar & Gravel
Th.; Thk.	Thick
Thn.	Thin
Thrded	Threaded
Tilf.	Tile Layer, Floor
Tilh.	Tile Layer, Helper
THHN	Nylon Jacketed Wire
THW.	Insulated Strand Wire
THWN;	Nylon Jacketed Wire
T.L.	Truckload
T.M.	Track Mounted
Tot.	Total
T-O-L	Threadolet
T.S.	Trigger Start
Tr.	Trade
Transf.	Transformer
Trhv.	Truck Driver, Heavy
Trlr	Trailer
Trlt.	Truck Driver, Light
TV	Television
T.W.	Thermoplastic Water Resistant Wire
UCI	Uniform Construction Index
UF	Underground Feeder
UGND	Underground Feeder
U.H.F.	Ultra High Frequency
U.L.	Underwriters Laboratory
Unfin.	Unfinished
URD	Underground Residential Distribution
US	United States
USP	United States Primed
UTP	Unshielded Twisted Pair
V	Volt
V.A.	Volt Amperes
V.C.T.	Vinyl Composition Tile
VAV	Variable Air Volume
VC	Veneer Core
Vent.	Ventilation
Vert.	Vertical
V.F.	Vinyl Faced
V.G.	Vertical Grain
V.H.F.	Very High Frequency
VHO	Very High Output
Vib.	Vibrating
V.L.F.	Vertical Linear Foot
Vol.	Volume
VRP	Vinyl Reinforced Polyester
W	Wire; Watt; Wide; West
w/	With
W.C.	Water Column; Water Closet
W.F.	Wide Flange
W.G.	Water Gauge
Wldg.	Welding
W. Mile	Wire Mile
W-O-L	Weldolet
W.R.	Water Resistant
Wrck.	Wrecker
W.S.P.	Water, Steam, Petroleum
WT., Wt.	Weight
WWF	Welded Wire Fabric
XFER	Transfer
XFMR	Transformer
XHD	Extra Heavy Duty
XHHW; XLPE	Cross-Linked Polyethylene Wire Insulation
XLP	Cross-linked Polyethylene
Y	Wye
yd	Yard
yr	Year
Δ	Delta
%	Percent
~	Approximately
Ø	Phase
@	At
#	Pound; Number
<	Less Than
>	Greater Than

ESIDENTIAL OST ESTIMATE

WNER'S NAME: ______ APPRAISER: ______

ESIDENCE ADDRESS: ______ PROJECT: ______

ITY, STATE, ZIP CODE: ______ DATE: ______

LASS OF CONSTRUCTION
- ☐ ECONOMY
- ☐ AVERAGE
- ☐ CUSTOM
- ☐ LUXURY

RESIDENCE TYPE
- ☐ 1 STORY
- ☐ 1-1/2 STORY
- ☐ 2 STORY
- ☐ 2-1/2 STORY
- ☐ 3 STORY
- ☐ BI-LEVEL
- ☐ TRI-LEVEL

CONFIGURATION
- ☐ DETACHED
- ☐ TOWN/ROW HOUSE
- ☐ SEMI-DETACHED

OCCUPANCY
- ☐ ONE STORY
- ☐ TWO FAMILY
- ☐ THREE FAMILY
- ☐ OTHER ______

EXTERIOR WALL SYSTEM
- ☐ WOOD SIDING—WOOD FRAME
- ☐ BRICK VENEER—WOOD FRAME
- ☐ STUCCO ON WOOD FRAME
- ☐ PAINTED CONCRETE BLOCK
- ☐ SOLID MASONRY (AVERAGE & CUSTOM)
- ☐ STONE VENEER—WOOD FRAME
- ☐ SOLID BRICK (LUXURY)
- ☐ SOLID STONE (LUXURY)

LIVING AREA (Main Building)		*LIVING AREA (Wing or Ell) ( )		*LIVING AREA (Wing or Ell) ( )	
irst Level	______ S.F.	First Level	______ S.F.	First Level	______ S.F.
Second Level	______ S.F.	Second Level	______ S.F.	Second Level	______ S.F.
Third Level	______ S.F.	Third Level	______ S.F.	Third Level	______ S.F.
Total	______ S.F.	Total	______ S.F.	Total	______ S.F.

asement Area is not part of living area.

MAIN BUILDING	COSTS PER S.F. LIVING AREA
Cost per Square Foot of Living Area, from Page ______	$
Basement Addition: ______ % Finished, ______ % Unfinished	+
Roof Cover Adjustment: ______ Type, Page ______ (Add or Deduct)	( )
Central Air Conditioning: ☐ Separate Ducts ☐ Heating Ducts, Page ______	+
Heating System Adjustment: ______ Type, Page ______ (Add or Deduct)	( )
Main Building: Adjusted Cost per S.F. of Living Area	$

MAIN BUILDING TOTAL COST	$ /S.F.	x	S.F.	x		=	$
	Cost per S.F. Living Area		Living Area		Town/Row House Multiplier (Use 1 for Detached)		TOTAL COST

WING OR ELL ( ) ______ STORY	COSTS PER S.F. LIVING AREA
Cost per Square Foot of Living Area, from Page ______	$
Basement Addition: ______ % Finished, ______ % Unfinished	+
Roof Cover Adjustment: ______ Type, Page ______ (Add or Deduct)	( )
Central Air Conditioning: ☐ Separate Ducts ☐ Heating Ducts, Page ______	+
Heating System Adjustment: ______ Type, Page ______ (Add or Deduct)	( )
Wing or Ell ( ): Adjusted Cost per S.F. of Living Area	$

WING OR ELL ( ) TOTAL COST	$ /S.F.	x	S.F.	=	$
	Cost per S.F. Living Area		Living Area		TOTAL COST

WING OR ELL ( ) ______ STORY	COSTS PER S.F. LIVING AREA
Cost per Square Foot of Living Area, from Page ______	$
Basement Addition: ______ % Finished, ______ % Unfinished	+
Roof Cover Adjustment: ______ Type, Page ______ (Add or Deduct)	( )
Central Air Conditioning: ☐ Separate Ducts ☐ Heating Ducts, Page ______	+
Heating System Adjustment: ______ Type, Page ______ (Add or Deduct)	( )
Wing or Ell ( ): Adjusted Cost per S.F. of Living Area	$

WING OR ELL ( ) TOTAL COST	$ /S.F.	x	S.F.	=	$
	Cost per S.F. Living Area		Living Area		TOTAL COST

TOTAL THIS PAGE ______

Page 1 of 2

RESIDENTIAL COST ESTIMATE

Total Page 1			$
	QUANTITY	UNIT COST	
Additional Bathrooms: _______ Full, _______ Half			
Finished Attic: _______ Ft. x _______ Ft.	S.F.		+
Breezeway: ☐ Open ☐ Enclosed _______ Ft. x _______ Ft.	S.F.		+
Covered Porch: ☐ Open ☐ Enclosed _______ Ft. x _______ Ft.	S.F.		+
Fireplace: ☐ Interior Chimney ☐ Exterior Chimney ☐ No. of Flues ☐ Additional Fireplaces			+
Appliances:			+
Kitchen Cabinets Adjustment: (+/–)			
☐ Garage ☐ Carport: _______ Car(s) Description _______________ (+/–)			
Miscellaneous:			+

ADJUSTED TOTAL BUILDING COST $

REPLACEMENT COST	
ADJUSTED TOTAL BUILDING COST	$ _______
Site Improvements	
(A) Paving & Sidewalks	$ _______
(B) Landscaping	$ _______
(C) Fences	$ _______
(D) Swimming Pool	$ _______
(E) Miscellaneous	$ _______
TOTAL	$ _______
Location Factor	x _______
Location Replacement Cost	$ _______
Depreciation	–$ _______
LOCAL DEPRECIATED COST	$ _______

INSURANCE COST	
ADJUSTED TOTAL BUILDING COST	$ _______
Insurance Exclusions	
(A) Footings, Site Work, Underground Piping	–$ _______
(B) Architects' Fees	–$ _______
Total Building Cost Less Exclusion	$ _______
Location Factor	x _______
LOCAL INSURABLE REPLACEMENT COST	$ _______

SKETCH AND ADDITIONAL CALCULATIONS

Page 2 of 2

A

ABC extinguisher 473
Abrasive floor tile 450
  tread 450
ABS DWV pipe 501
A/C coils furnace 510
Access door and panels 425
  door basement 332
  door duct 514
Accessories bathroom 449, 475
  door 439
  door and window 439
  drywall 448
  duct 514
  plaster 445
  roof 413
  toilet 475
Accessory drainage 412
  fireplace 472
  formwork 327, 328
  masonry 334
Accordion door 422
Acid proof floor 452
Acoustic ceiling board 451
Acoustical ceiling 451
  sealant 414, 448
  wallboard 448
Acrylic carpet 228
  ceiling 451
  wall coating 466
  wallcovering 456
  wood block 453
Adhesive cement 454
  roof 408
  wallpaper 456
Adobe brick 336
Aerial bucket 300
  lift 295
Aggregate base course 313
  exposed 331
  spreader 299
  stone 321
Aids staging 291
Air cleaner electronic 516
  compressor 295
  conditioner cooling & heating 512
  conditioner packaged terminal 512
  conditioner portable 512
  conditioner receptacle 524
  conditioner removal 496
  conditioner rooftop 512
  conditioner self-contained 512
  conditioner thru-wall 512
  conditioner window 512
  conditioner wiring 526
  conditioning 268, 270
  conditioning ventilating 512, 515
  extractor 514
  filter 516
  filter roll type 516
  handler heater 511
  handler modular 511
  hose 295
  return grille 516
  spade 296
  supply register 516
  tool 295
Air-source heat pumps 513
Alarm burglar 490
  residential 526
Alteration fee 288
Aluminum ceiling tile 451
  chain link 317
  column 394
  coping 339
  diffuser perforated 515
  downspout 411
  drip edge 412
  ductwork 514
  edging 323
  flagpole 473
  flashing 211, 410
  foil 399, 402
  gravel stop 412
  grille 516
  gutter 412, 413
  louver 470
  nail 367
  rivet 346
  roof 405
  service entrance cable 519
  sheet metal 411
  siding 166, 405
  siding accessories 405
  siding paint 462
  sliding door 424
  storm door 419
  tile 450
  window 426
  window demolition 416
Anchor bolt 327, 335
  brick 335
  buck 335
  chemical 344
  epoxy 344
  expansion 345
  framing 368
  hollow wall 345
  jute fiber 346
  lead screw 346
  masonry 335
  metal nailing 346
  nylon nailing 346
  partition 335
  plastic screw 346
  rafter 368
  rigid 335
  sill 368
  steel 335
  stone 335
Anchors drilling 344
Antenna system 531
  T.V. 531
Appliance 478
  residential 478, 526
Apron wood 385
Arch laminated 383
  radial 383
Architectural equipment 478
  fee 288, 540
  precast 331
  woodwork 388
Armored cable 519
Arrester lightning 486
Ashlar stone 340
Asphalt base sheet 408
  block 315
  block floor 315
  coating 398
  curb 314
  distributor 296
  driveway 110
  felt 408
  flood coat 408
  paper 402
  paver 297
  primer 454
  sheathing 381
  shingle 403
  sidewalk 315
  tile 229
Asphaltic emulsion 320
  paving 542
Astragal 439
  molding 386
Attachment grader 299
Attic stair 480
  ventilation fan 515
Auger 292
  hole 304
Autoclaved concrete block 337
Automatic transfer switch 529
  washing machine 479
Awning canvas 474
  window 174, 427

B

Backerboard 446
Backfill 308, 309
  general 308
  planting pit 320
  trench 310
Backflow preventer 504
Backhoe 293
Backhoe 309
  excavation 309
  extension 299
Backsplash countertop 392
Backup block 338
Baked enamel frame 418
Ball wrecking 299
Baluster 231, 393
  birch 231
  pine 231
Balustrade painting 461
Band joist framing 353
  molding 385
Bankrun gravel 304
Bar grab 475
  towel 475
  Zee 448
Barge construction 301
Bark mulch 320
  mulch redwood 320
Barrel 296
Barricade 296
Barrier slipform 299
Base cabinet 388, 389
  carpet 455
  ceramic tile 448
  column 368
  course 313
  course aggregate 313
  cove 453
  gravel 314
  molding 384
  quarry tile 449
  resilient 453
  sheet 408
  sink 389
  stabilizer 299
  stone 314, 340
  terrazzo 450
  vanity 392
  wood 384, 385, 395
Baseboard demolition 366
  heat 514
  heat electric 513
  heating 279
  register 516
Basement stair 231, 332
Basic finish materials 442
  meter device 528
  thermal materials 398
Basket weave fence, vinyl 317
Basketweave fence 318
Bath accessory 237
  communal 507
  steam 486
  whirlpool 507
Bathroom 248, 250, 252, 254, 256, 258, 260, 262, 264, 266, 507
  accessories 449, 475
  faucet 506
  fixture 507, 508
Bathtub 507
  enclosure 470
  removal 496
Batt insulation 168, 400
Bay window 178
Bead casing 448
  corner 445
  parting 386
Beam & girder framing 371
  bondcrete 446
  ceiling 394
  drywall 446, 447
  hanger 368
  laminated 382, 383
  mantel 394
  plaster 445
  steel 348
  wood 371, 372
Bed molding 385
Bedding brick 315
  pipe 310
Beech tread 393
Belgian block 315
Bell & spigot pipe 498
Bench park 319
  players 320
Benches 319
Berm pavement 314
  road 314
Bevel siding 407
Bi-fold door 226, 417, 422
Bin retaining walls metal 319
Bi-passing closet door 422
  door 226
Birch door 419, 422, 423
  molding 386
  paneling 387
  stair 393
  wood frame 423
Bituminous block 315
  coating 398
  concrete curbs 314
  expansion joint 327
  paver 297
  paving 542
Blanket insulation 400, 496
Blaster shot 298
Blind 191
  exterior 395
  venetian 482
  window 482
Block asphalt 315
  backup 338
  belgian 315
  bituminous 315
Block, brick and mortar 548
Block concrete 338, 339
  concrete bond beam 338
  concrete exterior 338
  decorative concrete 338
  filler 465
  floor 453
  glass 339
  insulation 338
  lightweight 339
  lintel 339

partition 339
profile 338
reflective 339
split rib 338
wall 158
wall foundation 120
wall removal 305
Blocking 370
carpentry 370
wood 374
Blockout slab 326
Blower insulation 297
Blown in cellulose 399
in fiberglass 399
in insulation 399
Blueboard 446
Bluegrass sod 321
Bluestone sidewalk 315
sill 341
step 316
Board & batten fence 319
& batten siding 407
ceiling 451
fence 318
gypsum 446
insulation 399
paneling 388
sheathing 381
siding 162
valance 389
verge 385
Boiler 509
demolition 496
electric 509
electric steam 509
gas fired 509
gas/oil combination 510
hot water 509, 510
oil fired 510
removal 496
steam 509, 510
Bollard 347
light 530
Bolt anchor 335
steel 368
toggle 345
Bondcrete 446
Bookcase 388, 389
Boom lift 295
truck 300
Borer horizontal 299
Boring and exploratory drilling 304
cased 304
Borrow 304
Bow window 178, 427
Bowstring truss 383
Box 521
distribution 312
fence shadow 318
out 326
pull 522
stair 393
Boxed beam headers 350, 353
Boxes & wiring device 522
utility 304
Bracing 370
let-in 370
metal joist 352
roof rafter 355
stud wall 349
Bracket sidewall 291
Brass hinge 438
screw 367
Breaker circuit 528
vacuum 504
Brick 548
adobe 336
anchor 335
bedding 315
Brick, block and mortar 548
Brick cart 296
chimney simulated 471
cleaning 342
demolition 334, 442
driveway 110
edging 323
face 336
forklift 296
molding 385
oversized 337
paving 315
removal 305
shelf 326
sidewalk 315
sill 341
step 316
veneer 160, 337
veneer demolition 334
wall 160
wash 342
Bricklaying 546
Bridging 370
metal joist 352
roof rafter 355
stud wall 349
Bronze valve 504
Broom cabinet 389
finish concrete 331
Brownstone 340
Brush clearing 307
cutter 293, 294
Buck anchor 335
rough 373
Bucket aerial 300
concrete 292
crane 293
excavator 299
pavement 299
Buggy concrete 292, 330
Builder's risk insurance 534
Building demolition 304
excavation 102, 104
greenhouse 486
hardware 437
insulation 399
paper 402
permit 288
prefabricated 487
Built-in range 478
Built-up roof 408
roofing 210
Bulk bank measure excavating 309
Bulkhead door 425
formwork 327
Bulldozer 294, 309
Bumper door 437
wall 437
Buncher feller 293
Burglar alarm 490
Burlap curing 331
rubbing 331
Burner gas conversion 509
gun type 509
residential 509
Bush hammer 331
hammer concrete 331
Butt fusion machine 296
Butterfly valve 311
Butyl caulking 414

C

Cabinet 234
base 388, 389
broom 389
casework 389
corner base 389
corner wall 389
demolition 366
door 390, 391
electrical 522
hardboard 388
hardware 391
hinge 391
kitchen 388
medicine 475
oven 389
shower 470
stain 456
varnish 456
wall 389
Cable armored 519
electric 519, 520
jack 301
pulling 298
sheathed nonmetallic 519
sheathed Romex 519
tensioning 298
trailer 298
Cafe door 421
Canopy door 473
framing 377
Cant roof 376
Cantilever retaining wall 319
Canvas awning 474
Cap post 368
service entrance 520
Caps vent 506
Car muck 300
Carbon dioxide extinguisher 473
Carpentry finish 384
rough 370
Carpet 228, 454, 455
base 455
felt pad 454
floor 455
nylon 455
olefin 455
pad 454
padding 454
removal 442
stair 455
urethane pad 454
wool 455
Carrier ceiling 444
channel 444
fixture 506
Cart brick 296
concrete 292, 330
Case work 389
Cased boring 304
evaporator coils 511
Casement window 172, 427, 428
Casework cabinet 389
custom 388
ground 380
metal 482
painting 456
stain 456
varnish 456
Casing bead 448
wood 385
Cast in place concrete 329
in place retaining walls 319
iron bench 320
iron damper 472
iron fitting 498
iron pipe 498
iron pipe fitting 498
iron radiator 514
iron trap 505
iron vent caps 505
Catch basin 313
basin precast 313
basin removal 305
door 391
Caulking 414
polyurethane 414
sealant 414
Cavity wall insulation 399
wall reinforcing 335
Cedar closet 388
fence 318
paneling 388
post 395
roof deck 382
roof plank 381
shake siding 164
shingle 403
siding 407
Ceiling 216, 220, 451
acoustical 451
beam 394
board 451
board acoustic 451
board fiberglass 451
bondcrete 445
carrier 444
demolition 442
diffuser 515
drilling 345
drywall 447
eggcrate 451
framing 371
furring 380, 444
heater 479
insulation 399
lath 445
luminous 451, 531
molding 385
painting 464, 465
plaster 445
stair 480
support 347
suspended 444, 451
tile 451
Cellar door 425
wine 478
Cellulose blown in 399
insulation 399
Cement adhesive 454
concrete curbs 314
flashing 409
gypsum 335
masonry 335
masonry unit 338, 339
mortar 449, 545
parging 398
terrazzo portland 450
Central vacuum 478
Centrifugal pump 297
Ceramic mulch 320
tile 229, 448, 449
tile countertop 393
tile demolition 442
tile floor 448
Chain hoist 301
link fence 114, 316, 317
link fence paint 459
link fence removal 305
saw 298
trencher 295

Chair lifts 494
molding 386
rail demolition 366
Chamfer strip 327
strip wood 327
Channel carrier 444
furring 444, 448
siding 407
steel 445
Charge powder 346
Charges dump 307
Chase formwork 328
Chemical anchor 344
dry extinguisher 473
spreader 299
toilet 298
Chime door 531
Chimney 337
accessories 471
brick 337
demolition 334
flue 336
foundation 329
metal 471
screen 472
simulated brick 471
vent 512
China cabinet 388
Chipper brush 293
log 294
stump 294
Chipping hammer 295
Chips wood 320
Chlorination system 487
Chute rubbish 307
C.I.P. concrete 329
Circline fixture 530
Circuit wiring 280
Circuit-breaker 528
Circular saw 298
Circulating pump 506
Cladding 405
sheet metal 410
Clamp water pipe ground 519
Clamshell bucket 293
Clapboard painting 462
Clay masonry 336
roofing tile 404
tile 404
Cleaner crack 299
steam 298
Cleaning brick 342
face brick 550
masonry 342
up 302
Cleanout door 472
floor 505
pipe 505
tee 505
Clear & grub 307
Clearing brush 307
selective 307
site 307
Climbing crane 300
jack 301
Clip plywood 368
Clock timer 525
Closet cedar 388
door 226, 417, 422
pole 386
rod 388
water 508
Clothes dryer residential 479
Coal tar pitch 408
Coat glaze 464
Coating bituminous 398
roof 409
rubber 398
silicone 398
spray 398
trowel 398
wall 466
water repellent 398
waterproofing 398
Coil tie 328
Cold-formed framing 349
joists 352
roof framing 355
Colonial door 419, 421
wood frame 423
Column 394
base 368
bondcrete 446
brick 337
demolition 334
drywall 447
lally 347
laminated wood 384
lath 445
plaster 445
removal 334
steel 348
wood 371, 372, 394
Columns formwork 326
lightweight 347
structural 347
Combination device 523
storm door 419, 421
Commercial gutting 307
Common brick 336
nail 367
rafter roof framing 138
Communal bath 507
Compact fill 309
Compaction 309
soil 309
structural 309
Compactor earth 293
landfill 294
residential 478
Compartments shower 470
Compensation workers' 289
Components furnace 510
Composite decking 375
insulation 402
rafter 375
Compressor air 295
Concrete block 338, 339, 549
block autoclaved 337
block back-up 338
block bond beam 338
block decorative 338
block demolition 306
block exterior 338
block foundation 338
block grout 335
block insulation 399
block painting 465
block wall 158
broom finish 331
bucket 292
buggy 330
bush hammer 331
cart 292, 330
cast in place 329
C.I.P. 329
conveyer 292
coping 339
curb 314
curing 331
cutout 306
demolition 306
drilling 344
driveway 110
finish 314, 330
float 292
float finish 330
footing 329, 330
formwork 326
foundation 122, 330
furring 380
hand trowel finish 331
lintel 332
mixer 292
monolithic finish 330
paver 297
paving 314
pipe 312, 313
placing 330
precast 331, 332
protection 326
ready mix 329
rehabilitation 332
reinforcement 328
removal 306
restoration 332
retaining wall 319
saw 292
septic tank 312
shingle 404
sidewalk 315
sill 331
slab 126, 330
spreader 297
stair 330
stamping 331
structural 329
tile 404
trowel 292
truck 292
utility vault 304
vibrator 292
wall 330
wheeling 330
Conditioner portable air 512
Conditioning ventilating air 512
Conductive floor 454
Conductor 519
& grounding 519
wire 520
Conduit & fitting flexible 527
electrical 520
fitting 521
flexible metallic 521
greenfield 521
in slab 520
in slab PVC 520
in trench electrical 521
in trench steel 521
PVC 520
rigid in slab 520
rigid steel 520
Connection motor 527
Connector joist 368
timber 368
Construction aids 289
barge 301
management fee 288
time 288
Contractor equipment 292
pump 297
Control component 491
joint 331
Conveyor 292
Cooking equipment 478
range 478
Cooling 268, 270
& heating A/C 512
Coping 339, 340
aluminum 339
concrete 339
removal 334
terra cotta 339
Copper cable 519
downspout 411
drum trap 505
DWV tubing 499
fitting 499
flashing 410
gutter 413
pipe 499
rivet 347
roof 409
wall covering 455
wire 520
Core drill 292
Cork floor 454
tile 454
wall tile 455
Corner base cabinet 389
bead 445
wall cabinet 389
Cornice 339
molding 385
painting 461
Corrugated metal pipe 312
roof tile 404
siding 405, 407
void form 326
Costs of trees 544
Counter top 392, 393
top demolition 366
Countertop backsplash 392
sink 508
Cove base 453
base ceramic tile 448
base terrazzo 450
molding 385
molding scotia 385
Cover ground 321
pool 488
CPVC pipe 501
Crack cleaner 299
filler 299
Crane bucket 293
climbing 300
crawler 300
hydraulic 300
material handling 300
tower 300
Crawler crane 300
Crew survey 288
Crown molding 385
Crushed stone sidewalk 315
Cubicle shower 508
Cupola 472
Curb 314
asphalt 314
concrete 314
edging 314
formwork 326
granite 314
inlet 315
precast 314
removal 306
roof 376
terrazzo 450
Curing concrete 331
paper 402
Curtain damper fire 514
rod 475
Custom casework 388
Cut stone trim 340
Cutout counter 392

demolition 306
slab 306
Cutter brush 293, 294
Cutting block 393
torch 298
Cylinder lockset 438

D

Damper fireplace 472
multi-blade 514
Deadbolt 437
Deciduous shrub 322
tree 322
Deck framing 374, 375
metal 348
metal roof 348
roof 381
slab form 348
steel 348
wood 244, 381, 382
Decking 550
composite 375
Decorator device 523
switch 523
Deep freeze 478
Dehumidifier 296, 479
Delivery charge 308
Demolition 304, 305, 335, 442, 496, 543
baseboard 366
boiler 496
brick 334, 442
building 304
cabinet 366
ceiling 442
ceramic tile 442
chimney 334
column 334
concrete 306
concrete block 306
cutout 306
door 416
ductwork 496
electric 518
fireplace 334
flooring 442
framing 363
furnace 496
glass 416
granite 335
gutter 398
hammer 293
house 305
HVAC 496
joist 363
masonry 306, 334
metal stud 442
millwork 366
paneling 366
partition 442
pavement 306
plaster 442
plenum 442
plumbing 496
plywood 398, 442
post 364
rafter 364
railing 366
roofing 398
selective 362, 416
siding 398
site 304
tile 442
truss 365
walls and partitions 442
window 416
wood 442
Derrick 300
guyed 300
stiffleg 301
Detection system 490
systems 490
Detector infra-red 490
motion 490
smoke 490
Device combination 523
decorator 523
GFI 524
receptacle 524
residential 522
wiring 527
Dewater 301
Diaphragm pump 297
Diesel hammer 293
tugboat 301
Diffuser ceiling 515
linear 515
opposed blade damper 515
perforated aluminum 515
rectangular 515
steel 516
T-bar mount 515
Dimmer switch 522, 527
Disappearing stairs 480
Disc harrow 293
Discharge hose 297
Disconnect safety switch 529
Dishwasher 479
Disposal 305
field 312
garbage 479
Disposer garbage 236
Distribution box 312
pipe water 311
system gas 312
Distributor asphalt 296
Diving board 487
Door accessories 439
accordion 422
and panels access 425
and window matls 416
bell residential 525
bell system 531
bi-fold 417, 422
bi-passing closet 422
birch 423
blind 395
bulkhead 425
bumper 437
cabinet 390, 391
cafe 421
canopy 473
catch 391
chime 531
cleanout 472
closet 417, 422
combination storm 421
demolition 416
double 417
dutch 421
dutch oven 472
entrance 421
exterior 182
fire 417
flush 422
folding accordion 422
frame 418, 423
frame interior 424
french 419
garage 186, 425
glass 424
handle 391
hardware 437
jamb molding 386
kick plate 438
knocker 439
labeled 417
metal 417
mirror 440
molding 386
moulded 421
opener 425
overhead 425
panel 421
paneled 417
passage 420, 422
plastic 419, 424
pre-hung 419, 424
removal 416
residential 417, 421, 423
residential garage 425
rough buck 373
sauna 486
sectional overhead 425
shower 470
side light 424
sidelight 417
sill 386, 423, 439
sliding 424
special 425
stain 458
steel 417, 555
stop 437
storm 190, 419
switch burglar alarm 490
threshold 424
trim vinyl 407
varnish 458
weatherstrip 438
weatherstrip garage 439
wood 224, 226, 419
Doors & windows interior paint 457
Dormer framing 150, 152
gable 380
roofing 204, 206
Double hung window 170, 428, 429
Dowels reinforcing 328
Downspout 411, 412
aluminum 411
copper 411
elbow 412
lead coated copper 412
painting 460
steel 412
strainer 411
Dozer 294
excavation 295
Dragline bucket 293
Drain floor 505
Drainage accessory 412
pipe 311, 313
site 313
trap 505
Drains roof 506
Drapery hardware 482, 483
rings 483
Drawer 392
track 391
wood 392
Drill & tap main 311
core 292
hammer 296
main 311
rig 304
rotary 296
steel 296
track 295
wood 368
Drilling anchors 344
ceiling 345
concrete 344
plaster 345
steel 345
Drinking fountain support 506
Drip edge 412
edge aluminum 412
Driver post 299
sheeting 296
Driveway 110, 315
removal 306
Drop pan ceiling 451
Drum trap copper 505
Dry fall painting 465
Dryer clothes 479
receptacle 524
vent 480
wiring 280
Drywall 214, 216, 446
accessories 448
column 447
cutout 307
demolition 442
finish 558
frame 418
gypsum 446
nail 367
painting 464
prefinished 447
screw 448
Duck tarpaulin 291
Duct access door 514
accessories 514
flexible insulated 514
flexible noninsulated 514
HVAC 514
insulation 496
liner 497
mechanical 514
Ductile iron fitting 311
iron fitting mech joint 311
iron pipe 311
Ductless split system 512
Ducts flexible 514
Ductwork aluminum 514
demolition 496
fabric coated flexible 514
fabricated 514
galvanized 514
metal 514
rectangular 514
rigid 514
Dump charge 307
charges 307
truck 295
Dumpster 307
Duplex receptacle 280, 527
Dust partition 307
Dutch door 182, 421
oven door 472
DWV pipe ABS 501
PVC pipe 501
tubing copper 499

E

Earth compactor 293
vibrator 309
Earthwork 308
equipment 292, 309
Eave vent 413
Edge drip 412

form 326
Edging 323
aluminum 323
curb 314
Eggcrate ceiling 451
EIFS 402
Elbow downspout 412
Electric appliance 478
baseboard heater 513
boiler 509
cable 519, 520
demolition 518
fixture 529, 530
furnace 511
generator 296
generator set 528
heater 479
heating 279, 513
lamp 531
log 471
metallic tubing 520
pool heater 510
service 278, 528
stair 480
switch 527, 529
water heater 509
wire 520
Electrical cabinet 522
conduit 520
conduit greenfield 521
Electronic air cleaner 516
Elevator 494
construction 301
windrow 300
Embossed metal door 420
print door 421
Employer liability 289
EMT 520
Emulsion adhesive 454
asphaltic 320
pavement 316
penetration 313
sprayer 296
Enclosure bathtub 470
shower 470
swimming pool 487
Entrance cable service 519
door 182, 419, 421
frame 423
lock 437
weather cap 521
Epoxy anchor 344
grout 449
wall coating 466
Equipment 292
earthwork 292, 309
hydromulch 299
insulation 497
insurance 289
rental 292, 300
Erosion control synthetic 310
Estimate electrical heating 513
Estimating 534
Evaporator coils cased 511
Evergreen shrub 321, 322
tree 321
Excavating bulk bank measure 309
Excavation 308, 309
backhoe 309
dozer 295
footing 102
foundation 104
hand 309
planting pit 321
septic tank 312
structural 309
trench 309
utility 106
Excavator bucket 299
rental 292
Exhaust hood 479
Expansion anchor 345
joint 327, 445
joint bituminous 327
joint neoprene 327
joint premolded 327
shield 345
tank 504
tank steel 504
Exposed aggregate 331
aggregate coating 466
Extension backhoe 299
Exterior blind 395
concrete block 338
door 182
door frame 423
insulation 402
insulation finish system 402
light 281
lighting fixture 530
molding 385
plaster 445
pre-hung door 419
residential door 421
wall framing 136
wood frame 423
Extinguisher carbon dioxide 473
chemical dry 473
fire 473
standard 473
Extractor 294
air 514

F

Fabric welded wire 329
Fabricated ductwork 514
Fabrication metal 358
Fabrications plastic 396
Face brick 336, 547
brick cleaning 550
Facing stone 340
Fall painting dry 465
Fan 515
attic ventilation 515
house ventilation 515
paddle 526
residential 526
ventilation 526
wiring 280, 526
Fascia aluminum 406
board 375
board demolition 363
metal 406
vinyl 407
wood 385
Fastener timber 367
wood 367
Fastenings metal 344
Faucet & fitting 506
bathroom 506
Fee architectural 288
Feller buncher 293
Felt asphalt 408
carpet pad 454
Fence and gate 316
auger, truck mounted 292
basketweave 318
board 318
board & batten 319
cedar 318
metal 317
open rail 318
picket paint 459
removal 305
residential 316
vinyl 317
wire 317
wood 318
Fences 459
Fiber cement shingle 403
cement siding 406
reinforcing 329
tube formwork 326
Fiberboard insulation 401
Fibergalss tank 488
Fiberglass blown in 399
ceiling board 451
insulation 399, 400
panel 405
planter 323
shower stall 470
tank 488
wool 399
Field disposal 312
septic 112
Fieldstone 340
Fill 309
gravel 304, 309
Filler block 465
crack 299
Film polyethylene 320
Filter air 516
mechanical media 516
swimming pool 487
Filtration equipment 487
Fine grade 308, 321
Finish carpentry 384
concrete 314, 330
floor 453, 466
grading 308
nail 367
wall 331
Finisher floor 292
Finishes wall 455
Fir column 394
floor 453
molding 386
roof deck 382
roof plank 381
Fire call pullbox 490
damage repair 466
damper curtain type 514
door 417
door frame 418
escape disappearing 480
escape stairs 358
extinguisher 473
horn 490
protection 473
resistant drywall 447
retardant lumber 366
retardant plywood 366
Fireplace accessory 472
box 341
built-in 471
damper 472
demolition 334
form 472
free standing 471
mantel 394
masonry 238, 341
prefabricated 240, 471
Firestop gypsum 445
wood 373
Fitting cast iron 498
conduit 521
copper 499
ductile iron 311
malleable iron 500
plastic 502
PVC 502
steel 500
Fixed window 180
Fixture bathroom 507, 508
carrier 506
electric 529
fluorescent 529, 530
incandescent 530
interior light 529
lantern 530
mirror light 530
plumbing 506-508
removal 496
residential 525, 530
Flagging 315, 452
slate 316
Flagpole 473
aluminum 473
Flashing 410
aluminum 410
cement 409
copper 410
lead coated 410
masonry 410
stainless 411
vent 505
Flatbed truck 295
Flexible conduit & fitting 527
ducts 514
ductwork fabric coated 514
insulated duct 514
metallic conduit 521
Float concrete 292
finish concrete 330
glass 440
Floater equipment 289
Floating floor 453
pin 438
Floodlight 296, 297
Floor 452
acid proof 452
asphalt block 315
brick 452
carpet 455
ceramic tile 449
cleanout 505
concrete 126
conductive 454
cork 454
drain 505
finish 466
finisher 292
flagging 316
framing 130, 132, 134
framing removal 307
insulation 399
marble 340
nail 367
oak 453
paint 460
parquet 452
plank 372
plywood 382
polyethylene 452
quarry tile 449
register 516
removal 305
rubber 453
sander 298
stain 460
subfloor 382
tile terrazzo 450

underlayment 382
varnish 460
vinyl 454
wood 453
wood composition 452
Flooring 229, 452
demolition 442
Flue chimney 336
chimney metal 512
liner 336, 337
screen 472
tile 336
Fluorescent fixture 529, 530
Flush door 422
Foam glass insulation 400
insulation 399
Fog seal 316
Foil aluminum 399, 402
back insulation 401
Folding accordion door 422
door shower 470
Footing 118
concrete 329, 330
excavation 102
reinforcing 328
removal 305
spread 326
Forklift brick 296
rigging 296
Form edge 326
fireplace 472
slab deck 348
Formwork 326
accessory 327, 328
bulkhead 327
chase 328
columns 326
concrete 326
curb 326
fiber tube 326
insert 328
plywood 327
sleeve 328
snap tie 328
steel frame 326
structural 326
wall 326
Foundation 118
chimney 329
concrete 122, 330
concrete block 338
excavation 104
masonry 120
mat 330
removal 304
wall 338
wood 124
Fountain outdoor 316
Fountains yard 316
Frame baked enamel 418
door 418, 423
drywall 418
entrance 423
fire door 418
labeled 418
metal 418
metal butt 418
steel 418
welded 418
window 432
wood 389, 423
Framing anchor 368
band joist 353
beam & girder 371
canopy 377
cold-formed 349
deck 374, 375
demolition 363
dormer 150, 152
floor 130, 132, 134
heavy 372
laminated 383
metal 347
metal joist 354
metal partition 350
metal stud 350
partition 154, 442, 443
removal 362
roof 138, 140, 142, 144, 146, 148
roof metal rafter 356
roof metal truss 357
roof rafters 375
roof soffits 357
sleepers 377
timber 372
wall 377
wood 370-372
Freeze deep 478
French door 419
Front end loader 309
shovel 294
Furnace A/C coils 510
components 510
demolition 496
electric 511
gas fired 511
hot air 511
oil fired 511
wiring 280
Furnishings site 319
Furring and lathing 444
ceiling 380, 444
channel 444, 448
metal 444
steel 444
wall 380, 444
Fusion machine butt 296

G

Gable dormer 380
dormer framing 150
dormer roofing 204
roof framing 138
Galvanized ductwork 514
Gambrel roof framing 144
Garage door 186, 425
door weatherstrip 439
Garbage disposal 479
Gas conversion burner 509
distribution system 312
fired boiler 509
fired furnace 511
fired pool heater 510
generator 528
heat air conditioner 512
heating 268, 272
log 471
pipe 312
vent 512
Gasoline generator 528
Gate metal 316, 317
General contractor's overhead 534
Generator construction 296
emergency 528
gas 528
gasoline 528
set 528
GFI receptacle 524
Girder wood 371, 372
Glass 440
block 339
demolition 416
door 424
door shower 470
float 440
heat reflective 440
lined water heater 479
low emissivity 440
mirror 440
shower stall 470
tempered 440
tile 440
tinted 440
window 440
Glaze coat 464
Glazed ceramic tile 448
wall coating 466
Glazing 440
Glued laminated 382, 383
laminated construction 383
Grab bar 475
Grade fine 308, 321
Grader attachment 299
motorized 293
Grading 308
slope 308
Granite building 340
chips 320
curb 314
demolition 335
indian 315
paver 340
sidewalk 316
Grass cloth wallpaper 456
lawn 321
seed 321
sprinkler 316
Gravel bankrun 304
base 314
fill 304, 309
pea 320
stop 412
Gravity retaining wall 319
Greenfield conduit 521
Greenhouse 242, 486
Grid spike 368
Grille air return 516
aluminum 516
decorative wood 394
painting 460
window 431
Grinder terrazzo 292
Ground 380
clamp water pipe 519
cover 321
fault 280
rod 519
wire 519
Grounding 519
& conductor 519
Grout 335
concrete block 335
epoxy 449
tile 448
wall 335
Guard gutter 413
snow 413
Guards snow 413
Gunite pool 243
Gutter 413
aluminum 412, 413
copper 413
demolition 398
guard 413
lead coated copper 413
stainless 413
steel 413
strainer 413
vinyl 413
wood 413
Gutting 307
Guyed derrick 300
Gymnasium floor 452
Gypsum block demolition 306
board 446, 447
board accessories 448
cement 335
drywall 446
fabric wallcovering 456
firestop 445
lath 444
lath nail 367
plaster 445
restoration 448
sheathing 381
wallboard repairs 448
weatherproof 381

H

Half round molding 386
Hammer bush 331
chipping 295
demolition 293
drill 296
hydraulic 296
pile 293
Hammermill 299
Hand carved door 420
clearing 308
excavation 309
hole 304
split shake 404
trowel finish concrete 331
Handicap lever 437
ramp 330
Handle door 391
Handling rubbish 307
Handrail and railing 358
oak 231
wood 386, 393
Hand-split shake 164
Hanger beam 368
joist 368
Hardboard cabinet 388
overhead door 425
paneling 387
siding 407
tempered 387
underlayment 382
Hardware 437
cabinet 391
door 437
drapery 482, 483
finish 437
rough 370
window 437
Hardwood floor 453
grille 394
Harrow disc 293
Hauling 309
truck 309
Hay 320
Header pipe 301
wood 378
Headers boxed beam 350, 353
Hearth 341
Heat baseboard 514
electric baseboard 513
pump 513
pump residential 527

pumps air-source . . . 513
pumps water-source . . . 513
reflective glass . . . 440
Heater air handler . . . 511
contractor . . . 296
electric . . . 479
infrared quartz . . . 513
sauna . . . 486
swimming pool . . . 510
water . . . 479, 509
wiring . . . 280
Heating 268, 270, 272, 488, 496, 497, 509, 510
electric . . . 279, 513
estimate electrical . . . 513
hot air . . . 511
hydronic . . . 510
panel radiant . . . 513
Heavy duty shoring . . . 289, 290
framing . . . 372
Hemlock column . . . 394
Hex bolt steel . . . 344
Highway paver . . . 316
Hinge brass . . . 438
cabinet . . . 391
residential . . . 438
steel . . . 437, 438
Hinges . . . 556
Hip rafter . . . 375
roof framing . . . 142
Hoist and tower . . . 301
contractor . . . 301
lift equipment . . . 300
personnel . . . 301
Holdown . . . 368
Hollow core door . . . 423
metal . . . 418
wall anchor . . . 345
Hood range . . . 479
Hook robe . . . 475
Horizontal aluminum siding . . . 405
auger . . . 292
borer . . . 299
vinyl siding . . . 406
Horn fire . . . 490
Hose air . . . 295
bibb sillcock . . . 507
discharge . . . 297
suction . . . 297
water . . . 297
Hospital tip pin . . . 438
Hot air furnace . . . 511
air heating . . . 511
tub . . . 507
water boiler . . . 509, 510
water heating . . . 509, 514
Hot-air heating . . . 268, 270
Hot-water heating . . . 272
House demolition . . . 305
ventilation fan . . . 515
Housewrap . . . 402
HTRW . . . 489
Humidifier . . . 479
Humus peat . . . 320
HVAC demolition . . . 496
duct . . . 514
Hydraulic crane . . . 300
hammer . . . 296
jack . . . 301
Hydromulch equipment . . . 299
Hydronic heating . . . 510

I

Icemaker . . . 479
Impact wrench . . . 296
In insulation blown . . . 399
place retaining walls cast . . . 319
slab conduit . . . 520
Incandescent fixture . . . 530
light . . . 281
Indian granite . . . 315
Infra-red detector . . . 490
Infrared quartz heater . . . 513
Inlet curb . . . 315
Insecticide . . . 310
Insert formwork . . . 328
Insulated panels . . . 402
protectors ADA . . . 497
Insulation . . . 399, 400
batt . . . 400
blanket . . . 400, 496
blower . . . 297
board . . . 211, 399
building . . . 399
cavity wall . . . 399
cellulose . . . 399
composite . . . 402
duct . . . 496
equipment . . . 497
exterior . . . 402
fiberglass . . . 399, 400
finish system exterior . . . 402
foam . . . 399
foam glass . . . 400
insert . . . 338
isocyanurate . . . 400
masonry . . . 399
mineral fiber . . . 401
pipe . . . 497
polystyrene . . . 399, 400
roof . . . 401
roof deck . . . 401
vapor barrier . . . 402
vermiculite . . . 399
wall . . . 338, 400
water heater . . . 497
Insurance . . . 289, 534
builder risk . . . 289
equipment . . . 289
public liability . . . 289
Interior door . . . 224, 226
door frame . . . 424
light . . . 281
light fixture . . . 529
partition framing . . . 154
pre-hung door . . . 420
residential door . . . 422
shutter . . . 396
wood frame . . . 424
Interlocking retaining walls . . . 319
Interval timer . . . 522
Intrusion system . . . 490
Investigation subsurface . . . 304
Ironing center . . . 482
Ironspot brick . . . 452
Irrigation system . . . 316
Isocyanurate insulation . . . 400

J

Jack cable . . . 301
hydraulic . . . 301
rafter . . . 376
Jackhammer . . . 295
Jalousie . . . 427
Jetting pump . . . 301
Joint control . . . 331
expansion . . . 327, 445
push-on . . . 311
reinforcing . . . 335
sealer . . . 414
tyton . . . 311
Joist connector . . . 368
demolition . . . 363
hanger . . . 368
metal framing . . . 354
removal . . . 363
sister . . . 372
web stiffeners . . . 354
wood . . . 372, 383
Joists cold-formed . . . 352
Jute fiber anchor . . . 346
mesh . . . 310

K

Kennel fence . . . 317
Kettle tar . . . 299
Kettle/pot tar . . . 298
Keyway . . . 326
Kick plate . . . 438
plate door . . . 438
Kiln dried lumber . . . 366
Kitchen . . . 234
appliance . . . 478
cabinet . . . 388
sink . . . 508
sink faucet . . . 506
Knocker door . . . 439

L

Labeled door . . . 417
frame . . . 418
Ladder . . . 297
swimming pool . . . 487
Lag screw . . . 346
screw shield . . . 346
Lally column . . . 347
Laminated beam . . . 382, 383
construction glued . . . 383
countertop . . . 393
countertop plastic . . . 392
framing . . . 383
glued . . . 382
roof deck . . . 382
veneer members . . . 384
wood . . . 382, 383
Lamp post . . . 359
Lampholder . . . 527
Landfill compactor . . . 294
fees . . . 307
Landing newel . . . 393
Landscape light . . . 530
surface . . . 452
Lantern fixture . . . 530
Laser level . . . 297
Latch set . . . 437, 438
Latex caulking . . . 414
underlayment . . . 454
Lath gypsum . . . 444
metal . . . 445
Lath, plaster and gypsum board . . . 557
Lattice molding . . . 386
Lauan door . . . 422
Laundry faucet . . . 506
sink . . . 507
tray . . . 507
Lavatory . . . 248, 507
faucet . . . 506
removal . . . 496
support . . . 506
vanity top . . . 507
wall hung . . . 508
Lawn grass . . . 321
mower, 22" rotary, 5HP . . . 297
Lazy susan . . . 389
Leaching pit . . . 312
Lead coated copper downspout . . . 412
coated copper gutter . . . 413
coated downspout . . . 412
coated flashing . . . 410
flashing . . . 410
paint encapsulation . . . 489
paint removal . . . 489
pile . . . 293
screw anchor . . . 346
Lean-to type greenhouse . . . 487
Leeching field . . . 112
Let-in bracing . . . 370
Letter slot . . . 474
Level laser . . . 297
Lever handicap . . . 437
Liability employer . . . 289
insurance . . . 534
Lift aerial . . . 295
Light bollard . . . 530
fixture interior . . . 529
landscape . . . 530
portable . . . 296
post . . . 530
support . . . 347
tower . . . 297
track . . . 531
Lighting . . . 280, 530, 531
fixture exterior . . . 530
incandescent . . . 530
outdoor . . . 296
outlet . . . 525
residential . . . 525
strip . . . 530
track . . . 530
Lightning arrester . . . 486
protection . . . 486
suppressor . . . 522
Lightweight block . . . 339
columns . . . 347
Limestone . . . 321, 340
coping . . . 339
Linear diffuser . . . 515
Linen wallcovering . . . 456
Liner duct . . . 497
flue . . . 336
Lintel . . . 332, 340
block . . . 339
concrete . . . 332
precast . . . 332
steel . . . 348
Load center indoor . . . 529
center plug-in breaker . . . 529
center rainproof . . . 529
Load-bearing studs . . . 349
Loadcenter residential . . . 522
Loader front end . . . 309
tractor . . . 294
Loam . . . 308
Lock entrance . . . 437
Locking receptacle . . . 525
Lockset cylinder . . . 438
Locomotive . . . 299
Log chipper . . . 294
electric . . . 471
gas . . . 471
skidder . . . 294
Louver . . . 471
aluminum . . . 470
midget . . . 471
redwood . . . 394
ventilation . . . 394

wall ... 471
wood ... 394
Louvered blind ... 395
door ... 224, 226, 420, 423
Low-voltage silicon rectifier ... 527
switching ... 527
switchplate ... 527
transformer ... 527
Lumber ... 371, 551, 552
core paneling ... 388
kiln dried ... 366
treatment ... 366
Luminous ceiling ... 451, 531
panel ... 451

M

Macadam ... 313
penetration ... 313
Machine trowel finish ... 331
welding ... 299
Mahogany door ... 420
Malleable iron fitting ... 500
iron pipe fitting ... 500
Management fee construction ... 288
Manhole precast ... 313
removal ... 305
Mansard roof framing ... 146
Mantel beam ... 394
fireplace ... 394
Maple countertop ... 393
Marble ... 340
chips ... 320
coping ... 339
countertop ... 393
floor ... 340
sill ... 341
synthetic ... 452
tile ... 452
Masonry accessory ... 334
anchor ... 335
brick ... 337, 452
cement ... 335
clay ... 336
cleaning ... 342
cornice ... 339
demolition ... 306, 334
fireplace ... 238, 341
flashing ... 410
foundation ... 120
furring ... 380
insulation ... 399
nail ... 367
painting ... 465
pointing ... 341, 342
reinforcing ... 335, 546
removal ... 305, 334
restoration ... 341
saw ... 298
sill ... 341
step ... 316
toothing ... 306
unit ... 336
wall ... 158, 160, 319, 338
wall tie ... 335
Mat foundation ... 330
Material hoist ... 301
Mechanical duct ... 514
media filter ... 516
Medicine cabinet ... 475
Membrane ... 211
curing ... 331
roofing ... 408
Metal bin retaining walls ... 319
butt frame ... 418
casework ... 482
chimney ... 471
clad window ... 170, 172, 174, 176, 178, 180
deck ... 348
door ... 417
door residential ... 417
ductwork ... 514
fabrication ... 358
faced door ... 419
fascia ... 406
fastenings ... 344
fence ... 317
fireplace ... 240
flue chimney ... 512
frame ... 418
framing ... 347
framing parapet ... 356
furring ... 444
gate ... 316, 317
halide fixture ... 281
hollow ... 418
joist bracing ... 352
joist bridging ... 352
joist framing ... 354
lath ... 445
nailing anchor ... 346
overhead door ... 425
partition framing ... 350
pipe removal ... 496
rafter framing ... 356
roof ... 405
roof deck ... 348
roof parapet framing ... 356
sheet ... 409
shelf ... 474
siding ... 166
soffit ... 407
stair ... 358
stud ... 350, 442, 443
stud demolition ... 442
stud framing ... 350
support assemblies ... 442
threshold ... 439
tile ... 450
truss framing ... 357
window ... 426
Metallic foil ... 399
Meter center rainproof ... 528
device basic ... 528
socket ... 527
water supply ... 504
water supply domestic ... 504
Microtunneling ... 310
Microwave oven ... 478
Mill construction ... 372
Millwork ... 384, 389
demolition ... 366
Mineral fiber ceiling ... 451
fiber insulation ... 401
roof ... 409
wool blown in ... 399
Minor site demolition ... 305
Mirror ... 440
ceiling board ... 451
door ... 440
glass ... 440
light fixture ... 530
wall ... 440
Mix paver ... 299
planting pit ... 321
Mixer concrete ... 292
mortar ... 292, 297
plaster ... 297
road ... 299
transit ... 292
Mixing valve ... 507
Mobilization or demob. ... 308
Modified bitumen roofing ... 555
Modular air handler ... 511
Module tub-shower ... 507
Moil point ... 296
Molding ... 384
base ... 384
brick ... 385
ceiling ... 385
chair ... 386
cornice ... 385
exterior ... 385
hardboard ... 387
pine ... 385
trim ... 386
wood ... 393
wood transition ... 453
Monel rivet ... 347
Monitor support ... 347
Monolithic finish concrete ... 330
Monument survey ... 288
Mortar ... 545
Mortar, brick and block ... 548
Mortar cement ... 449
mixer ... 292, 297
thinset ... 450
Moss peat ... 320
Motion detector ... 490
Motor connection ... 527
generator ... 296
support ... 347
Moulded door ... 421
Mounting board plywood ... 381
Movable louver blind ... 482
Moving shrub ... 320
tree ... 320
Mower, 22" rotary, 5HP lawn ... 297
Muck car ... 300
Mulch bark ... 320
ceramic ... 320
stone ... 320
Mulching ... 320
Multi-blade damper ... 514
Muntin window ... 431

N

Nail ... 366, 367
Nailer ... 297
pneumatic ... 296
steel ... 373
wood ... 373
Neoprene expansion joint ... 327
Newel ... 231, 393
No hub pipe ... 498
Non-removable pin ... 438
Nylon carpet ... 228, 455, 456
nailing anchor ... 346

O

Oak door frame ... 423
floor ... 229, 453
molding ... 386
paneling ... 387
stair tread ... 393
threshold ... 424
Off highway truck ... 295
Oil fired boiler ... 510
fired furnace ... 511
heater temporary ... 297
heating ... 270, 272
water heater ... 509
Olefin carpet ... 455
Onyx ... 450
Open rail fence ... 318
Opener door ... 425
Operating cost ... 292
cost equipment ... 292
Outdoor fountain ... 316
lighting ... 296
Outlet box plastic ... 522
box steel ... 521
lighting ... 525
Oven ... 478
cabinet ... 389
microwave ... 478
Overhaul ... 307
Overhead and profit ... 538
contractor ... 534
door ... 186, 425
Oversized brick ... 337

P

P trap ... 505
trap running ... 505
Padding ... 228
carpet ... 454
Paddle fan ... 526
Paint aluminum siding ... 462
chain link fence ... 459
doors & windows interior ... 457
encapsulation lead ... 489
fence picket ... 459
floor ... 460
floor concrete ... 459, 460
floor wood ... 460
removal ... 489
siding ... 462
striper ... 299
trim exterior ... 463
walls masonry, exterior ... 464
Painting ... 559
balustrade ... 461
casework ... 456
ceiling ... 464, 465
clapboard ... 462
concrete block ... 465
cornice ... 461
decking ... 460
downspout ... 460
drywall ... 464
grille ... 460
masonry ... 465
pipe ... 461
plaster ... 464
siding ... 461
sprayer ... 297
steel siding ... 461
stucco ... 462
swimming pool ... 487
trim ... 461
truss ... 461
wall ... 462, 464, 465
window ... 459
Paints & coatings ... 456
Palladian windows ... 430
Pan shower ... 411
Panel door ... 421
fiberglass ... 405
luminous ... 451
radiant heat ... 513
Paneled door ... 224, 417
pine door ... 423
Paneling ... 387
board ... 388
cutout ... 307

demolition 366
hardboard 387
plywood 387
wood 387
Panels insulated 402
prefabricated 402
Paper building 402
curing 331
sheathing 402
Paperhanging 455
Parapet metal framing 356
Pargeting 554
Parging cement 398
Park bench 319
Parquet floor 229, 452
Particle board siding 408
board underlayment 382
Parting bead 386
Partition 214, 218
anchor 335
block 339
demolition 442
dust 307
framing 154, 442, 443
shower 470
support 347
wood frame 373
Passage door 224, 420, 422
Patch roof 409
Patching concrete floor 332
Patio door 424
Pavement 315
berm 314
bucket 299
demolition 306
emulsion 316
planer 299
profiler 299
widener 299
Paver bituminous 297
concrete 297
floor 452
highway 316
shoulder 299
slipform 299
Paving asphaltic 542
brick 315
concrete 314
Pea gravel 320
Peastone 321
Peat humus 320
moss 320
Pegboard 387
Penetration macadam 313
Perforated aluminum pipe 313
ceiling 451
pipe 313
Perlite insulation 400
plaster 445
Permit building 288
Personnel hoist 301
Picket fence 318
fence, vinyl 317
Pickup truck 298
Picture window 180, 430
Pier brick 337
Pilaster wood column 394
Pile driver 293
hammer 293
sod 321
Pin floating 438
non-removable 438
powder 346
Pine door 421
door frame 423
fireplace mantel 394
floor 453
molding 385
roof deck 382
shelving 388
siding 407
stair 393
stair tread 393
Pipe & fittings 499, 501, 502, 504
and fittings 561
bedding 310
cast iron 498
cleanout 505
concrete 312, 313
copper 499
corrugated metal 312
covering 497
covering fiberglass 497
CPVC 501
drainage 311, 313
ductile iron 311
DWV ABS 501
DWV PVC 501
fitting cast iron 498
fitting copper 499
fitting DWV 502
fitting no hub 498
fitting plastic 502, 503
fitting soil 498
gas 312
header 301
insulation 497
no hub 498
painting 461
perforated aluminum 313
plastic 501, 502
polyethylene 312
PVC 311, 501
quick coupling 301
rail 358
rail aluminum 358
rail galvanized 359
rail stainless 359
rail steel 358
rail wall 359
removal 305
removal metal 496
residential PVC 501
sewage 311
single hub 498
soil 498
steel 312, 313
subdrainage 312
supply register spiral 516
Piping designations 543
Piping subdrainage 312
Pit leaching 312
Pitch coal tar 408
emulsion tar 316
Placing concrete 330
Planer pavement 299
Plank floor 372
roof 381
scaffolding 290
sheathing 381
Plant bed preparation 320
screening 294
Planter 323
fiberglass 323
Planting 320
Plants, planting, transplanting 320
Plaster 218, 220, 444
accessories 445
beam 445
board 214, 216
ceiling 445
column 445
cutout 307
demolition 442
drilling 345
ground 380
gypsum 445
mixer 297
painting 464
perlite 445
thincoat 445
wall 445
Plasterboard 446
Plastic blind 191
clad window 170, 172, 174, 176, 178, 180
door 419, 424
fabrications 396
faced hardboard 387
fitting 502
laminated countertop 392
outlet box 522
pipe 501, 502
pipe fitting 502, 503
screw anchor 346
siding 166
skylight 436
trap 505
vent caps 506
window 432
Plate roadway 299
shear 368
wall switch 527
Platform trailer 298
Plating zinc 367
Players bench 320
Plenum demolition 442
Plow vibrator 295
Plugmold raceway 521
Plumbing 504
demolition 496
fixture 506-508, 561
fixtures removal 496
Plywood 553
clip 368
demolition 398, 442
floor 382
formwork 327
joist 383
mounting board 381
paneling 387
sheathing roof & walls 381
shelving 388
siding 408
soffit 387
subfloor 382
treatment 366
underlayment 382
Pneumatic nailer 296
Point moil 296
Pointing masonry 341, 342
Pole closet 386
Police connect panel 490
Polyethylene film 320
floor 452
pipe 312
pool cover 488
septic tank 312
tarpaulin 291
Polystyrene blind 396
ceiling 451
ceiling panel 451
insulation 399, 400
Polysulfide caulking 414
Polyurethane caulking 414
Polyvinyl soffit 387, 407
Pool accessory 487
cover 488
cover polyethylene 488
heater electric 510
heater gas fired 510
swimming 487
Porcelain tile 449
Porch molding 385
Portable air compressor 295
heater 296
light 296
Portland cement terrazzo 450
Post cap 368
cedar 395
demolition 364
driver 299
lamp 359
light 530
wood 371
Poured insulation 168
Powder actuated tool 346
charge 346
pin 346
Power wiring 528
Precast architectural 331
catch basin 313
concrete 331, 332
coping 339
curb 314
lintel 332
manhole 313
receptor 470
sill 331
stair 332
terrazzo 450
window sills 331
Pre-engineered structure 486
Prefabricated building 487
fireplace 240, 471
panels 402
Prefinished drywall 447
floor 229, 453
shelving 388
Preformed roof panel 405
roofing & siding 405
Pre-hung door 419, 424
Premolded expansion joint 327
Preparation plant bed 320
surface 467, 468
Pressure valve relief 504
wash 468
Pretreatment termite 310
Prime coat 313
Primer asphalt 454
Privacy fence, vinyl 317
Profile block 338
Profiler pavement 299
Projected window 426
Property line survey 288
Protection fire 473
lightning 486
slope 310
termite 310
Protectors ADA insulated 497
P&T relief valve 504
Pull box 522
door 391
Pump 509
circulating 506
diaphragm 297
heat 513
jetting 301
rental 297
staging 289, 541
submersible 297, 509
sump 479
trash 297
water 297, 301

wellpoint 301
Purlin roof 372
Push button lock 438
Push-on joint 311
Puttying 467
PVC conduit 520
conduit in slab 520
DWV pipe 501, 502
fitting 502
gravel stop 412
pipe 311, 501
pipe residential 501
siding 406
waterstop 328

Q

Quarry tile 449
Quarter round molding 386
Quartz 320
heater 513
Quoins 340

R

Raceway plugmold 521
surface 521
wiremold 521
Radial arch 383
Radiant heating panel 513
Radiator cast iron 514
Rafter 375
anchor 368
composite 375
demolition 364
framing metal 356
metal bracing 355
metal bridging 355
roof framing 138
wood 375
Rail aluminum pipe 358
galvanized pipe 359
pipe 358
stainless pipe 359
steel pipe 358
wall pipe 359
Railing demolition 366
wood 386, 393
Railroad tie 316, 323
tie step 316
Rainproof meter center 528
Ramp handicap 330
Ranch plank floor 453
Range circuit 280
cooking 478
hood 236, 479
receptacle 524, 527
Ready mix concrete 329
Receptacle air conditioner 524
device 524
dryer 524
duplex 527
GFI 524
locking 525
range 524, 527
telephone 525
television 525
weatherproof 524
Receptor precast 470
shower 470
terrazzo 470
Rectangular diffuser 515
ductwork 514
Rectifier low-voltage silicon 527
Redwood bark mulch 320
cupola 472
louver 394
paneling 388
siding 162, 407
trim 385
wine cellar 478
Refinish floor 453
Reflective block 339
insulation 399
Refrigerated wine cellar 478
Refrigeration residential 479
Register air supply 516
baseboard 516
Rehabilitation concrete 332
Reinforcement concrete 328
Reinforcing dowels 328
fiber 329
footing 328
joint 335
masonry 335
steel 328
wall 328
Removal air conditioner 496
block wall 305
boiler 496
catch basin 305
chain link fence 305
concrete 306
curb 306
driveway 306
fence 305
floor 305
foundation 304
lavatory 496
masonry 305
paint 489
pipe 305
plumbing fixtures 496
shingle 398
sidewalk 305
sod 321
stone 305
tree 307
water closet 496
window 416
Rental equipment 292, 300
equipment rate 293
generator 296
Repair fire damage 466
Repellent water 465
Replacement sash 432
sliding door 424
windows 435, 556
Residential alarm 526
appliance 478, 526
burner 509
closet door 417
device 522
door 417, 421, 423
door bell 525
elevator 494
fan 526
fence 316
fixture 525, 530
garage door 425
greenhouse 486
gutting 307
heat pump 527
hinge 438
lighting 525
loadcenter 522
lock 438
overhead door 425
PVC pipe 501
refrigeration 479
service 522
smoke detector 526
stair 393
storm door 419
switch 522
ventilation 515
water heater 509, 526
wiring 522, 526
Resilient base 453
floor 453, 454
flooring 229
Resquared shingle 404
Restoration concrete 332
gypsum 448
masonry 341
window 489
Retaining wall 319
wall segmental 319
walls interlocking 319
walls stone 319
Retarder vapor 402
Ribbed waterstop 328
Ridge board 376
shingle 404
shingle clay 404
shingle slate 403
vent 413, 471
Rig drill 304
Rigging forklift 296
Rigid anchor 335
conduit in-trench 521
in slab conduit 520
insulation 168, 399
Ring split 368
toothed 369
Rings drapery 483
Riser beech 231
oak 231
wood stair 393
Rivet 346
aluminum 346
copper 347
monel 347
stainless 347
steel 347
tool 347
Road berm 314
mixer 299
sweeper 299
Roadway plate 299
Robe hook 475
Rod closet 388
curtain 475
ground 519
Roll roof 408
roofing 409
type air filter 516
Roller earth 294
vibratory 294
Romex copper 519
Roof accessories 413
adhesive 408
aluminum 405
beam 383
built-up 408
cant 376, 408
clay tile 404
coating 409
copper 409
deck 381
deck insulation 401
deck laminated 382
deck wood 382
drains 506
fiberglass 405
framing 138, 140, 142, 144, 146, 148
framing cold-formed 355
framing removal 307
insulation 401
metal 405
metal rafter framing 356
metal truss framing 357
mineral 409
modified bitumen 409
nail 367
panel preformed 405
patch 409
purlin 372
rafter 375
rafter bracing 355
rafter bridging 355
rafters framing 375
roll 408
sheathing 381
skylight 436
slate 403, 554
soffits framing 357
specialties, prefab 411
tile 404
truss 372, 383
zinc 410
Roofing 194, 196, 198, 200, 202
& siding preformed 405
demolition 398
gable end 194
gambrel 198
hip roof 196
mansard 200
membrane 408
modified bitumen 409
roll 409
shed 202
sheet metal 409
Rooftop air conditioner 512
Rosewood door 420
Rosin paper 402
Rotary drill 296
hammer drill 296
Rough buck 373
carpentry 370
hardware 370
stone wall 340
Rough-in sink countertop 508
sink raised deck 508
sink service floor 508
tub 507
Round diffuser 515
Rubber base 453
coating 398
floor 453
floor tile 454
sheet 453
threshold 439
tile 454
tired roller 294
Rubbish chute 307
handling 307

S

Safety switch 528
switch disconnect 529
Salamander 297
Sales tax 288, 539
Salt treatment lumber 366
Sand fill 304
Sandblasting equipment 297
Sander floor 298
Sanding 466

floor 453
Sandstone 340
flagging 316
Sanitary base cove 448
Sash replacement 432
wood 431
Sauna 486
door 486
Saw 298
chain 298
circular 298
concrete 292
masonry 298
Scaffolding 540, 541
plank 290
tubular 289
Scrape after damage 466
Scraper 294
Screed, gas engine, vibrating 292
Screen chimney 472
fence 318
molding 385
security 426
squirrel and bird 472
window 426, 432
wood 432
Screening plant 294
Screw brass 367
drywall 448
lag 346
sheet metal 367
steel 367
wood 367
Seal fog 316
pavement 316
Sealant 398
acoustical 448
caulking 414
Sealcoat 316
Sealer joint 414
Sectional door 425
overhead door 425
Security screen 426
Seeding 321, 544
Seeding, general 321
Segmental retaining wall 319
Selective clearing 307
demolition 334, 362, 416
Self propelled crane 300
Self-contained air conditioner 512
Septic system 112, 312
tank 312
tank concrete 312
Service electric 278, 528
entrance cable aluminum 519
entrance cap 520
residential 522
sink 508
sink faucet 506
Sewage pipe 311
Shadow box fence 318
Shake siding 164
wood 404
Shear plate 368
wall 381
Sheathed nonmetallic cable 519
romex cable 519
Sheathing 381
asphalt 381
gypsum 381
paper 402
roof 381
roof & walls plywood 381
Shed dormer framing 152
dormer roofing 206
roof framing 148
Sheet metal 409
metal aluminum 411
metal cladding 410
metal roofing 409
metal screw 367
rock 214, 216
Sheeting driver 296
Shelf brick 326
metal 474
Shellac door 458
Shelving 388
pine 388
plywood 388
prefinished 388
storage 474
wood 388
Shield expansion 345
lag screw 346
Shingle 194, 196, 198, 200, 202
asphalt 403
concrete 404
removal 398
ridge 404
siding 164
stain 462
strip 403
wood 403
Shock absorber 504
Shoring heavy duty 289, 290
Shot blaster 298
Shoulder paver 299
Shovel front 294
Shower by-pass valve 507
compartments 470
cubicle 508
door 470
enclosure 470
glass door 470
pan 411
partition 470
receptor 470
stall 508
surround 470
Shrub broadleaf evergreen 322
deciduous 322
evergreen 321, 322
moving 320
Shutter 191, 482
interior 396
wood 396
Side light door 424
Sidelight 423
door 417
Sidewalk 315, 452
asphalt 315
brick 315
concrete 315
removal 305
Sidewall bracket 291
Siding aluminum 405
bevel 407
cedar 407
demolition 398
fiber cement 406
fiberglass 405
hardboard 407
metal 166
nail 367
paint 462
painting 461
plywood 408
PVC 406
redwood 407
removal 398
shingle 164
stain 462
steel 407
vinyl 406
wood 162, 407
wood product 407
Silicone caulking 414
coating 398
water repellent 399
Sill 340, 376
anchor 368
door 386, 423, 439
masonry 341
precast 331
quarry tile 450
Sillcock hose bibb 507
Silt fence 310
Single hub pipe 498
hung window 426
zone rooftop unit 512
Sink 236, 508
base 389
countertop 508
countertop rough-in 508
kitchen 508
laundry 507
lavatory 507
raised deck rough-in 508
removal 496
service 508
service floor rough-in 508
Siren 490
Sister joist 372
Site clearing 307
demolition 304
demolition minor 305
drainage 313
furnishings 319
improvement 316, 320
preparation 304
Sitework 102
Skidder log 294
Skim coat 214, 216
Skirtboard 393
Skylight 208, 436
removal 398
roof 436
Skywindow 208
Slab blockout 326
concrete 126, 330
cutout 306
deck form 348
on grade 326, 330
on grade removal 305
textured 330
Slate 341
flagging 316
removal 398
roof 403
shingle 403
sidewalk 316
sill 341
stair 341
tile 452
Slatwall 456
Sleeper 377
wood 229
Sleepers framing 377
Sleeve and tap 311
formwork 328
Sliding door 185, 424
glass door 424
mirror 475
window 176, 426, 430
Slipform barrier 299
paver 299
Slop sink 507, 508
Slope grading 308
protection 310
Slot letter 474
Slotted pipe 313
Small tools 291
Smoke detector 490
Snap tie formwork 328
Snow guard 413
guards 413
Soap holder 475
Socket meter 527
Sod 321
Sodding 321
Sodium low pressure fixture 530
Soffit 377
aluminum 406
drywall 447
metal 407
plaster 445
plywood 387
vent 413
vinyl 407
wood 386, 387
Softener water 479
Soil compaction 309
compactor 293
pipe 498
tamping 309
treatment 310
Solar heating 560
Solid surface countertops 396
wood door 421
Spa bath 507
Space heater rental 296
Spade air 296
tree 295
Spanish roof tile 404
Special door 425
purpose room 486
stairway 231
systems 490
Specialties & accessories, roof 411
Spike grid 368
Spiral pipe supply register 516
stair 358, 393
Split rib block 338
ring 368
system ductless 512
Spray coating 398
Sprayer emulsion 296
Spread footing 326, 329
Spreader aggregate 299
chemical 299
concrete 297
Sprinkler grass 316
Stabilizer base 299
Staging aids 291
pump 289
Stain cabinet 456
casework 456
door 458
floor 460
lumber 383
shingle 462
siding 408, 462
truss 461
Stainless flashing 411
gutter 413
rivet 347
steel gravel stop 412
steel hinge 438
Stair 393
basement 332, 393
carpet 455
ceiling 480
climber 494
concrete 330

electric ... 480
metal ... 358
precast ... 332
railroad tie ... 316
removal ... 365
residential ... 393
slate ... 341
spiral ... 358, 393
stringer ... 373
stringer wood ... 373
tread ... 231
tread tile ... 450
tread wood ... 393
wood ... 393
Stairs disappearing ... 480
fire escape ... 358
Stairway ... 230
Stairwork and handrails ... 393
Stall shower ... 508
Stamping concrete ... 331
texture ... 331
Standard extinguisher ... 473
Starting newel ... 393
Steam bath ... 486
bath residential ... 486
boiler ... 509, 510
boiler electric ... 509
cleaner ... 298
Steel anchor ... 335
beam ... 348
bin wall ... 319
bolt ... 368
bridging ... 370
chain link ... 316
channel ... 445
column ... 348
conduit in slab ... 520
conduit in trench ... 521
conduit rigid ... 520
deck ... 348
diffuser ... 516
door ... 417, 555
downspout ... 412
drill ... 296
drilling ... 345
edging ... 316, 323
expansion tank ... 504
fireplace ... 240
fitting ... 500
flashing ... 411
frame ... 418
frame formwork ... 326
furring ... 444
gravel stop ... 412
gutter ... 413
hex bolt ... 344
hinge ... 437, 438
lintel ... 348
members structural ... 348
nailer ... 373
pipe ... 312, 313
reinforcing ... 328
rivet ... 347
screw ... 367
siding ... 407
structural ... 347
stud ... 442, 443
window ... 426
Steeple tip pin ... 438
Step bluestone ... 316
brick ... 316
masonry ... 316
railroad tie ... 316
stone ... 340
Stiffeners joist web ... 354
Stiffleg derrick ... 301
Stockade fence ... 318
Stone aggregate ... 321
anchor ... 335
ashlar ... 340
base ... 314, 340
curbs ... 314
dust ... 315
fill ... 304
floor ... 340
ground cover ... 321
mulch ... 320
paver ... 315, 340
removal ... 305
retaining walls ... 319
step ... 340
threshold ... 331
trim cut ... 340
wall ... 160, 319
Stool cap ... 386
window ... 340, 341
Stop door ... 386
gravel ... 412
Storage shelving ... 474
tank ... 488
tanks underground ... 488
trailer ... 298
Storm door ... 190, 419
drainage manholes, frames ... 313
window ... 190, 434, 435
Stove ... 472
wood burning ... 472
Strainer downspout ... 411
gutter ... 413
roof ... 411
wire ... 412
Strap tie ... 368, 369
Straw ... 320
Stringer ... 231
stair ... 373
Strip chamfer ... 327
floor ... 453
footing ... 330
lighting ... 530
shingle ... 403
soil ... 308
Striper paint ... 299
thermal ... 299
Stripping topsoil ... 308
Structural columns ... 347
concrete ... 329
excavation ... 309
formwork ... 326
steel ... 347
steel members ... 348
Structure pre-engineered ... 486
Stucco ... 220, 445
interior ... 218
painting ... 462
wall ... 158
Stud demolition ... 365
metal ... 442, 443
partition ... 373, 374
steel ... 442, 443
wall ... 136, 373, 442, 443
wall bracing ... 349
wall bridging ... 349
Studs load-bearing ... 349
metal ... 350
Stump chipper ... 294
Subdrainage pipe ... 312
piping ... 312
system ... 312
Subfloor ... 382
plywood ... 229, 382
Submersible pump ... 297, 311, 509
Subsurface investigation ... 304
Suction hose ... 297
Sump pump ... 479
Support ceiling ... 347
drinking fountain ... 506
lavatory ... 506
light ... 347
monitor ... 347
motor ... 347
partition ... 347
X-ray ... 347
Suppressor lightning ... 522
Surface countertops solid ... 396
landscape ... 452
preparation ... 467, 468
raceway ... 521
Surfacing ... 316
Surround shower ... 470
tub ... 470
Survey crew ... 288
monument ... 288
property line ... 288
topographic ... 288
Suspended ceiling ... 222, 444, 451
Suspension system ceiling ... 444
Sweeper road ... 299
Swimming pool ... 243, 487
pool enclosure ... 487
pool heater ... 510
pool ladder ... 487
pools ... 560
Swing check valve ... 504
Swing-up door ... 186
overhead door ... 425
Switch box ... 522
box plastic ... 522
decorator ... 523
dimmer ... 522, 527
electric ... 527, 529
general duty ... 529
residential ... 522
safety ... 528
time ... 529
toggle ... 527
wiring ... 280
Switching low-voltage ... 527
Switchplate low-voltage ... 527
Synthetic erosion control ... 310
marble ... 452
System antenna ... 531
ceiling suspension ... 444
irrigation ... 316
septic ... 312
subdrainage ... 312
T.V. ... 531
VHF ... 531
Systems detection ... 490

T

T & G siding ... 162
Tamper ... 296
Tamping soil ... 309
Tandem roller ... 294
Tank expansion ... 504
fibergalss ... 488
septic ... 112, 312
storage ... 488
Tanks septic ... 312
Tap and sleeve ... 311
Tapping main ... 311
Tar kettle ... 299
kettle/pot ... 298
paper ... 320
pitch emulsion ... 316
Tarpaulin ... 291
duck ... 291
polyethylene ... 291
Tax ... 288
sales ... 288
social security ... 288
unemployment ... 288
Taxes ... 539
T-bar mount diffuser ... 515
Teak floor ... 229, 452
molding ... 385, 386
paneling ... 387
Tee cleanout ... 505
Telephone receptacle ... 525
Television receptacle ... 525
Temperature relief valve ... 504
Tempered glass ... 440
hardboard ... 387
Tempering valve ... 504
Temporary oil heater ... 296
toilet ... 298
Tennis court fence ... 114, 317
Terminal A/C packaged ... 512
Termite pretreatment ... 310
protection ... 310
Terne coated flashing ... 411
Terra cotta coping ... 339
cotta demolition ... 306
Terrazzo ... 450
precast ... 450
receptor ... 470
wainscot ... 450
Texture stamping ... 331
Textured slab ... 330
Thermal striper ... 299
Thermostat ... 491
integral ... 513
wire ... 527
Thincoat ... 214, 216
plaster ... 445
Thinset ceramic tile ... 448
mortar ... 450
Threshold ... 439
door ... 424
stone ... 331, 340
wood ... 386
Thru-wall air conditioner ... 512
Tie formwork snap ... 328
rafter ... 376
railroad ... 316, 323
strap ... 368, 369
wall ... 335
Tile ... 448, 453
ceiling ... 451
ceramic ... 448, 449
clay ... 404
concrete ... 404
cork ... 454
cork wall ... 455
demolition ... 442
flue ... 336
glass ... 440
grout ... 448
marble ... 452
metal ... 450
porcelain ... 449
quarry ... 449
roof ... 404
rubber ... 454
slate ... 452
stainless steel ... 450
stair tread ... 450
vinyl ... 454
wall ... 448
window sill ... 450
Timber connector ... 368
fastener ... 367

framing 372
laminated 383
Time switch 529
Timer clock 525
interval 522
switch ventilator 515
Tinted glass 440
Toggle bolt 345
switch 527
Toilet accessories 475
bowl 508
chemical 298
temporary 298
trailer 298
Tongue and groove siding 162
Tool air 295
powder actuated 346
rivet 347
van 298
Tools small 291
Toothed ring 369
Toothing masonry 306
Top demolition counter 366
dressing 321
Topographic survey 288
Topsoil 308, 321
stripping 308
Torch cutting 298
Towel bar 475
Tower crane 300
hoist 301
light 297
Track drawer 391
drill 295
light 531
lighting 530
traverse 482
Tractor 294, 309
loader 294
truck 298
Trailer platform 298
storage 298
toilet 298
truck 295, 298
Tram car 298
Transfer switch automatic 529
Transformer low-voltage 527
Transit mixer 292
Transition molding wood 453
Transom lite frame 419
windows 431
Trap cast iron 505
drainage 505
P 505
plastic 505
Trapezoid windows 431
Trash pump 297
Traverse 482, 483
track 482
Travertine 340
Tray laundry 507
Tread abrasive 450
beech 231
oak 231
stone 341
wood 393
Treatment lumber 366
plywood 366
wood 366
Tree 322
deciduous 322
evergreen 321
guying 323
moving 320
removal 307
spade 295
Trench backfill 310
box 298
excavation 309
utility 309
Trencher 295
chain 295
Trenching 106
Trim exterior 385
painting 461
redwood 385
tile 449
wood 385
Trowel coating 398
concrete 292
Truck boom 300
concrete 292
dump 295
flatbed 295
hauling 309
loading 309
mounted crane 300
off highway 295
pickup 298
rental 295
tractor 298
trailer 295, 298
vacuum 298
Truss bowstring 383
demolition 365
flat wood 383
framing metal 357
painting 461
plate 369
roof 372, 383
roof framing 140
stain 461
varnish 461
Tub hot 507
redwood 507
rough-in 507
surround 470
Tubing copper 499
electric metallic 520
Tub-shower module 507
Tubular scaffolding 289
steel joist 383
Tugboat diesel 301
Tumbler holder 475
Tunnel ventilator 300
Tunneling 310
Turned column 394
TV antenna 531
system 531
Tyton joint 311

U

Undereave vent 471
Underground storage tanks 488
Underlayment 382
latex 454
Unemployment tax 288
Unit masonry 336
Utility 106
boxes 304
excavation 106
trench 309

V

Vacuum breaker 504
central 478
cleaning 478
truck 298
Valance board 389
Valley rafter 376
Valve 504
bronze 504
mixing 507
relief pressure 504
shower by-pass 507
swing check 504
tempering 504
water pressure 504
Vanity base 392
top lavatory 507
Vapor barrier 402
barrier sheathing 381
retarder 402
Varnish cabinet 456
casework 456
door 458
floor 460, 466
truss 461
VCT removal 442
Veneer brick 160, 337
core paneling 387
members laminated 384
removal 335
Venetian blind 482
Vent caps 506
caps cast iron 505
caps plastic 506
chimney 512
dryer 480
eave 413
flashing 505
ridge 413, 471
ridge strip 471
soffit 413
Ventilating air conditioning 512, 515
Ventilation fan 526
louver 394
residential 515
Ventilator timer switch 515
tunnel 300
Verge board 385
Vermiculite insulation 399
Vertical aluminum siding 405
vinyl siding 406
VHF system 531
Vibrating screed, gas engine 292
Vibrator concrete 292
earth 294, 309
plow 295
Vibratory equipment 293
roller 294
Vinyl blind 191, 396
chain link 317
composition floor 454
door trim 407
downspout 412
faced wallboard 447
fascia 407
fence 317
floor 454
flooring 229
gutter 413
shutter 191
siding 166, 406
siding accessories 406
soffit 407
tile 454
wallpaper 456
window 432, 434
window trim 407
Void form corrugated 326
Volume control damper 514

W

Wainscot ceramic tile 449
molding 386
quarry tile 450
terrazzo 450
Walk 315
Wall bumper 437
cabinet 389
ceramic tile 449
coating 466
concrete 330
covering 559
cutout 306
drywall 446
finish 331
finishes 455
formwork 326
foundation 338
framing 136, 377
framing removal 307
furring 380, 444
grout 335
heater 479
hung lavatory 508
insulation 338, 399, 400
interior 214, 218
lath 444
louver 471
masonry 158, 160, 319, 338
mirror 440
painting 462, 464, 465
paneling 387
plaster 445
reinforcing 328
removal 305
retaining 319
shear 381
sheathing 381
siding 407
steel bin 319
stucco 445
stud 373, 442, 443
switch plate 527
tie 335
tie masonry 335
tile 448
tile cork 455
Wallboard acoustical 448
Wallcovering 455
acrylic 456
gypsum fabric 456
Wallpaper 455, 456
grass cloth 456
vinyl 456
Walls and partitions demolition 442
Walnut door frame 423
floor 452
Wardrobe wood 390
Wash bowl 507
brick 342
Washer 369
residential 479
Washing machine automatic 479
Water closet 508
closet removal 496
closet support 506
distribution pipe 311
heater 479, 509, 526
heater electric 509
heater insulation 497
heater oil 509
heater removal 496
heater residential 509
heater wrap kit 497
heating hot 509, 514

  - heating wiring 280
  - hose 297
  - pipe ground clamp 519
  - pressure relief valve 504
  - pressure valve 504
  - pump 297, 301, 479, 506
  - repellent 465
  - repellent coating 398
  - repellent silicone 399
  - softener 479
  - supply domestic meter 504
  - supply meter 504
  - tank sprayer 298
  - tempering valve 504
  - trailer 298
  - well 311
- Waterproofing 398
  - coating 398
- Water-source heat pumps 513
- Waterstop 328
  - PVC 328
  - ribbed 328
- Weather cap entrance 521
- Weatherproof receptacle 280, 524
- Weatherstrip 438
  - and seals 438
  - door 438
- Weathervanes 472
- Welded frame 418
  - wire fabric 329
- Welding machine 299
- Well 311
  - water 311
- Wellpoint equipment 301
  - header pipe 301
  - pump 301
- Wheelbarrow 299
- Whirlpool bath 507
- Widener pavement 299
- Winch truck 298
- Window 170, 172, 174, 176, 178, 180, 426, 437
  - air conditioner 512
  - aluminum 426
  - awning 427
  - blind 395, 482
  - casement 427, 428
  - casing 385
  - demolition 416
  - double hung 428, 429
  - double-hung 170
  - frame 432
  - glass 440
  - grille 431
  - hardware 437
  - metal 426
  - muntin 431
  - painting 459
  - picture 430
  - plastic 432
  - removal 416
  - restoration 489
  - screen 426, 432
  - sill 341
  - sill marble 340
  - sill tile 450
  - sills precast 331
  - sliding 430
  - steel 426
  - stool 340, 341
  - storm 190, 434, 435
  - trim set 386
  - trim vinyl 407
  - vinyl 432, 434
  - wood 427-430, 432
  - [illegible]s casement 428
  - transom 431
  - trapezoid 431
- Windrow elevator 300
- Wine cellar 478
- Winter protection 326
- Wire copper 520
  - electric 520
  - fence 317
  - fences, misc. metal 317
  - ground 519
  - mesh 446
  - strainer 412
  - thermostat 527
  - THW 520
- Wiremold raceway 521
- Wiring air conditioner 526
  - device 527
  - fan 526
  - methods 519
  - power 528
  - residential 522, 526
- Wood base 384, 395
  - beam 371, 372
  - blind 191, 395, 482
  - block floor demolition 442
  - blocking 211, 374
  - canopy 377
  - casing 385
  - chamfer strip 327
  - chips 320
  - column 371, 372, 394
  - cupola 472
  - deck 244, 381, 382
  - demolition 442
  - door 224, 226, 419, 421
  - drawer 392
  - fascia 385
  - fastener 367
  - fence 318
  - fiber ceiling 451
  - fiber sheathing 381
  - fiber underlayment 382
  - floor 130, 132, 134, 229, 452, 453
  - floor demolition 442
  - foundation 124
  - frame 389, 423
  - framing 370, 371
  - furring 380
  - girder 371
  - gutter 413
  - handrail 386
  - joist 372, 383
  - laminated 382, 383
  - louver 394
  - molding 393
  - nailer 373
  - overhead door 425
  - panel door 421
  - paneling 387
  - parquet 452
  - partition 373
  - planter 323
  - product siding 407
  - rafter 375
  - railing 393
  - roof deck 381
  - roof deck demolition 398
  - roof trusses 554
  - sash 431
  - screen 432
  - screw 367
  - shake 404
  - sheathing 381
  - shelving 388
  - shingle 403
  - shutter 191
  - sidewalk 315
  - siding 162, 407
  - siding demolition 398
  - sill 376
  - soffit 386, 387
  - stair 393
  - stair stringer 373
  - storm door 190, 419
  - storm window 190
  - subfloor 382
  - threshold 386
  - tread 393
  - treatment 366
  - trim 385
  - truss 383
  - veneer wallpaper 456
  - wall framing 136
  - wardrobe 390
  - window 427-430, 432, 555
  - window demolition 416
- Woodwork architectural 388
- Wool fiberglass 399
- Workers' compensation 289, 535-537
- Wrecking ball 299
- Wrench impact 296

X

- X-ray support 347

Y

- Yard fountains 316
- Yellow pine floor 453

Z

- Z bar suspension 444
- Zee bar 448
- Zinc plating 367
  - roof 410
  - weatherstrip 438

Notes

Notes

Reed Construction Data, Inc.

Reed Construction Data, Inc., a leading worldwide provider of total construction information solutions, is comprised of three main product groups designed specifically to help construction professionals advance their businesses with timely, accurate and actionable project, product, and cost data. Reed Construction Data is a division of Reed Business Information, a member of the Reed Elsevier plc group of companies.

The *Project, Product, and Cost & Estimating* divisions offer a variety of innovative products and services designed for the full spectrum of design, construction, and manufacturing professionals. Through it's *International* companies, Reed Construction Data's reputation for quality construction market data is growing worldwide.

Cost Information

RSMeans, the undisputed market leader and authority on construction costs, publishes current cost and estimating information in annual cost books and on the CostWorks CD-ROM. RSMeans furnishes the construction industry with a rich library of complementary reference books and a series of professional seminars that are designed to sharpen professional skills and maximize the effective use of cost estimating and management tools. RSMeans also provides construction cost consulting for Owners, Manufacturers, Designers, and Contractors.

Project Data

Reed Construction Data provides complete, accurate and relevant project information through all stages of construction. Customers are supplied industry data through leads, project reports, contact lists, plans and specifications surveys, market penetration analyses and sales evaluation reports. Any of these products can pinpoint a county, look at a state, or cover the country. Data is delivered via paper, e-mail, CD-ROM or the Internet.

Building Product Information

The First Source suite of products is the only integrated building product information system offered to the commercial construction industry for comparing and specifying building products. These print and online resources include *First Source,* CSI's SPEC-DATA™, CSI's MANU-SPEC™, First Source CAD, and Manufacturer Catalogs. Written by industry professionals and organized using CSI's MasterFormat™, construction professionals use this information to make better design decisions.

FirstSourceONL.com combines Reed Construction Data's project, product and cost data with news and information from Reed Business Information's *Building Design & Construction* and *Consulting-Specifying Engineer,* this industry-focused site offers easy and unlimited access to vital information for all construction professionals.

International

BIMSA/Mexico provides construction project news, product information, cost-data, seminars and consulting services to construction professionals in Mexico. Its subsidiary, PRISMA, provides job costing software.

Byggfakta Scandinavia AB, founded in 1936, is the parent company for the leaders of customized construction market data for Denmark, Estonia, Finland, Norway and Sweden. Each company fully covers the local construction market and provides information across several platforms including subscription, ad-hoc basis, electronically and on paper.

Reed Construction Data Canada serves the Canadian construction market with reliable and comprehensive project and product information services that cover all facets of construction. Core services include: *BuildSource, BuildSpec, BuildSelect,* product selection and specification tools available in print and on the Internet; Building Reports, a national construction project lead service; CanaData, statistical and forecasting information; *Daily Commercial News,* a construction newspaper reporting on news and projects in Ontario; and *Journal of Commerce,* reporting news in British Columbia and Alberta.

Cordell Building Information Services, with its complete range of project and cost and estimating services, is Australia's specialist in the construction information industry. Cordell provides in-depth and historical information on all aspects of construction projects and estimation, including several customized reports, construction and sales leads, and detailed cost information among others.

For more information, please visit our Web site at www.reedconstructiondata.com.

Reed Construction Data, Inc., Corporate Office
30 Technology Parkway South
Norcross, GA 30092-2912
(800) 322-6996
(800) 895-8661 (fax)
info@reedbusiness.com
www.reedconstructiondata.com

Means Project Cost Report

By filling out and returning the Project Description, you can receive a discount of $20.00 off any one of the Means products advertised in the following pages. The cost information required includes all items marked (✔) except those where no costs occurred. The sum of all major items should equal the Total Project Cost.

$20.00 Discount per product for each report you submit.

DISCOUNT PRODUCTS AVAILABLE—FOR U.S. CUSTOMERS ONLY—STRICTLY CONFIDENTIAL

Project Description (No remodeling projects, please.)

✔ Type Building ______
✔ Location ______
Capacity ______
✔ Frame ______
✔ Exterior ______
✔ Basement: full ☐ partial ☐ none ☐ crawl ☐
✔ Height in Stories ______
✔ Total Floor Area ______
Ground Floor Area ______
✔ Volume in C.F. ______
% Air Conditioned ______ Tons ______
Comments ______

	✔	**Total Project Cost**			$
A	✔	**General Conditions**			$
B	✔	**Site Work**			$
BS		**Site Clearing & Improvement**			
BE		*Excavation*	(	C.Y.)	
BF		*Caissons & Piling*	(	L.F.)	
BU		*Site Utilities*			
BP		*Roads & Walks Exterior Paving*	(	S.Y.)	
C	✔	**Concrete**			$
C		*Cast in Place*	(	C.Y.)	
CP		*Precast*	(	S.F.)	
D	✔	**Masonry**			$
DB		*Brick*	(	M)	
DC		*Block*	(	M)	
DT		*Tile*	(	S.F.)	
DS		*Stone*	(	S.F.)	
E	✔	**Metals**			$
ES		*Structural Steel*	(	Tons)	
EM		*Misc. & Ornamental Metals*			
F	✔	**Wood & Plastics**			$
FR		*Rough Carpentry*	(	MBF)	
FF		*Finish Carpentry*			
FM		*Architectural Millwork*			
G	✔	**Thermal & Moisture Protection**			$
GW		*Waterproofing-Dampproofing*	(	S.F.)	
GN		*Insulation*	(	S.F.)	
GR		*Roofing & Flashing*	(	S.F.)	
GM		*Metal Siding/Curtain Wall*	(	S.F.)	
H	✔	**Doors and Windows**			$
HD		*Doors*	(	Ea.)	
HW		*Windows*	(	S.F.)	
HH		*Finish Hardware*			
HG		*Glass & Glazing*	(	S.F.)	
HS		*Storefronts*	(	S.F.)	

Product Name ______
Product Number ______
Your Name ______
Title ______
Company ______
☐ Company
☐ Home Street Address ______
City, State, Zip ______
☐ Please send ______ forms.

Owner ______
Architect ______
General Contractor ______
✔ Bid Date ______
Typical Bay Size ______
✔ Labor Force: ____ % Union ____ % Non-Union
✔ Project Description (Circle one number in each line)
1. Economy 2. Average 3. Custom 4. Luxury
1. Square 2. Rectangular 3. Irregular 4. Very Irregular

J	✔	**Finishes**			$
JL		*Lath & Plaster*	(	S.Y.)	
JD		*Drywall*	(	S.F.)	
JM		*Tile & Marble*	(	S.F.)	
JT		*Terrazzo*	(	S.F.)	
JA		*Acoustical Treatment*	(	S.F.)	
JC		*Carpet*	(	S.Y.)	
JF		*Hard Surface Flooring*	(	S.F.)	
JP		*Painting & Wall Covering*	(	S.F.)	
K	✔	**Specialties**			$
KB		*Bathroom Partitions & Access.*	(	S.F.)	
KF		*Other Partitions*	(	S.F.)	
KL		*Lockers*	(	Ea.)	
L	✔	**Equipment**			$
LK		*Kitchen*			
LS		*School*			
LO		*Other*			
M	✔	**Furnishings**			$
MW		*Window Treatment*			
MS		*Seating*	(	Ea.)	
N	✔	**Special Construction**			$
NA		*Acoustical*	(	S.F.)	
NB		*Prefab. Bldgs.*	(	S.F.)	
NO		*Other*			
P	✔	**Conveying Systems**			$
PE		*Elevators*	(	Ea.)	
PS		*Escalators*	(	Ea.)	
PM		*Material Handling*			
Q	✔	**Mechanical**			$
QP		*Plumbing* (*No. of fixtures*		)	
QS		*Fire Protection (Sprinklers)*			
QF		*Fire Protection (Hose Standpipes)*			
QB		*Heating, Ventilating & A.C.*			
QH		*Heating & Ventilating* (*BTU Output*		)	
QA		*Air Conditioning*	(	Tons)	
R	✔	**Electrical**			$
RL		*Lighting*	(	S.F.)	
RP		*Power Service*			
RD		*Power Distribution*			
RA		*Alarms*			
RG		*Special Systems*			
S	✔	**Mech./Elec. Combined**			$

Please specify the Means product you wish to receive. Complete the address information as requested and return this form with your check (product cost less $20.00) to address below.

RSMeans Company, Inc.,
Square Foot Costs Department
P.O. Box 800
Kingston, MA 02364-9988

For more information
visit Means Web Site
at www.rsmeans.com

Reed Construction Data/RSMeans . . . a tradition of excellence in Construction Cost Information and Services since 1942.

Table of Contents

Annual Cost Guides, Page 2
Reference Books, Page 6
Seminars, Page 12
Electronic Data, Page 14
New Titles, Page 15
Order Form, Page 16

Book Selection Guide

The following table provides definitive information on the content of each cost data publication. The number of lines of data provided in each unit price or assemblies division, as well as the number of reference tables and crews is listed for each book. The presence of other elements such as an historical cost index, city cost indexes, square foot models or cross-referenced index is also indicated. You can use the table to help select the Means' book that has the quantity and type of information you most need in your work.

Unit Cost Divisions	Building Construction Costs	Mechanical	Electrical	Repair & Remodel.	Square Foot	Site Work Landsc.	Assemblies	Interior	Concrete Masonry	Open Shop	Heavy Construc.	Light Commercial	Facil. Construc.	Plumbing	Western Construction Costs	Residential
1	1104	822	907	1013		1045		847	1014	1102	1051	752	1528	874	1102	710
2	2505	1468	460	1450		7794		547	1387	2467	4978	692	4305	1584	2490	772
3	1401	113	100	726		1259		201	1782	1396	1404	229	1312	82	1399	248
4	835	18	0	647		657		574	1056	811	594	407	1058	0	821	334
5	1820	239	193	957		768		927	691	1788	1023	805	1802	316	1802	752
6	1489	82	78	1475		452		1406	318	1481	608	1626	1596	47	1826	1757
7	1264	158	74	1241		468		489	408	1264	349	956	1318	168	1264	752
8	1839	28	0	1907		313		1673	645	1821	51	1227	2034	0	1841	1193
9	1595	47	0	1424		233		1688	360	1547	159	1337	1809	47	1586	1216
10	861	47	25	498		193		700	170	863	0	394	906	233	861	217
11	1018	322	169	502		137		813	44	925	107	218	1173	291	925	104
12	307	0	0	47		210		1416	27	298	0	62	1435	0	298	64
13	1153	1005	375	494		388		884	75	1136	279	484	1783	928	1119	193
14	345	36	0	258		36		292	0	344	30	12	343	35	343	6
15	1987	13138	636	1779		1569		1174	59	1996	1766	1217	10865	9695	2017	826
16	1277	470	10063	1002		739		1108	55	1265	760	1081	9834	415	1226	552
17	427	354	427	0		0		0	0	427	0	0	427	356	427	0
Totals	**21227**	**18347**	**13507**	**15420**		**16261**		**14739**	**8091**	**20961**	**13159**	**11499**	**43528**	**15071**	**21347**	**9696**

Assembly Divisions	Building Construction Costs	Mechanical	Electrical	Repair & Remodel.	Square Foot	Site Work Landsc.	Assemblies	Interior	Concrete Masonry	Open Shop	Heavy Construc.	Light Commercial	Facil. Construc.	Plumbing	Western Construction Costs	Asm Div	Residential
A		19	0	192	150	540	612	0	550		542	149	24	0		1	374
B		0	0	809	2480	0	5590	333	1914		0	2024	144	0		2	217
C		0	0	635	862	0	1220	1568	145		0	767	238	0		3	588
D		1031	780	693	1823	0	2439	753	0		0	1310	1027	896		4	871
E		0	0	85	255	0	292	5	0		0	255	5	0		5	393
F		0	0	0	123	0	126	0	0		0	123	3	0		6	358
G		465	172	332	111	1856	584	0	482		432	110	113	560		7	299
																8	760
																9	80
																10	0
																11	0
																12	0
Totals		**1515**	**952**	**2746**	**5804**	**2396**	**10863**	**2659**	**3091**		**974**	**4738**	**1554**	**1456**			**3940**

Reference Section	Building Construction Costs	Mechanical	Electrical	Repair & Remodel.	Square Foot	Site Work Landsc.	Assemblies	Interior	Concrete Masonry	Open Shop	Heavy Construc.	Light Commercial	Facil. Construc.	Plumbing	Western Construction Costs	Residential
Tables	130	45	85	69	4	80	219	46	71	130	61	57	104	51	131	42
Models					102							43				32
Crews	410	410	410	391		410		410	410	393	410	393	391	410	410	393
City Cost Indexes	yes	yes	yes	yes	yes	yes	yes	yes	yes	yes	yes	yes	yes	yes	yes	yes
Historical Cost Indexes	yes	yes	yes	yes	yes	yes	yes	yes	yes	yes	yes	yes	yes	yes	yes	no

For more informatic
visit Means Web Si
at www.rsmeans.co

Means Building Construction Cost Data 2005

Available in Both Softbound and Looseleaf Editions

The "Bible" of the industry comes in the standard softcover edition or the looseleaf edition.

Many customers enjoy the convenience and flexibility of the looseleaf binder, which increases the usefulness of *Means Building Construction Cost Data 2005* by making it easy to add and remove pages. You can insert your own cost information pages, so everything is in one place. Copying pages for faxing is easier also. Whichever edition you prefer, softbound or the convenient looseleaf edition, you'll be eligible to receive *The Change Notice* FREE. Current subscribers receive *The Change Notice* via e-mail.

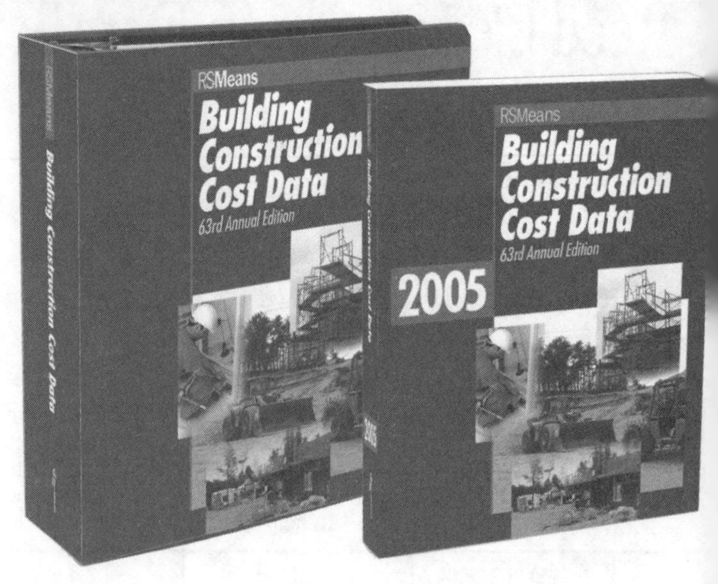

$115.95 per copy, Softbound
Catalog No. 60015

$144.95 per copy, Looseleaf
Catalog No. 61015

Means Building Construction Cost Data 2005

Offers you unchallenged unit price reliability in an easy-to-use arrangement. Whether used for complete, finished estimates or for periodic checks, it supplies more cost facts better and faster than any comparable source. Over 23,000 unit prices for 2005. The City Cost Indexes cover over 930 areas, for indexing to any project location in North America. Order and get *The Change Notice* FREE. You'll have year-long access to the Means Estimating **HOTLINE** FREE with your subscription. Expert assistance when using Means data is just a phone call away.

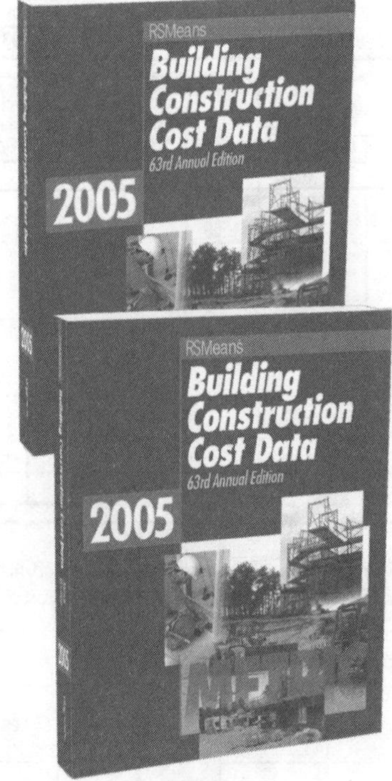

$115.95 per copy
Over 700 pages, illustrated, available Oct. 2004
Catalog No. 60015

Means Building Construction Cost Data 2005

Metric Version

The Federal Government has stated tha all federal construction projects must now use metric documentation. The *Metric Version* of *Means Building Construction Cost Data 2005* is presented in metric measurements covering all construction areas. Don't miss out on these billion dollar opportunities. Make the switch to metric today.

$115.95 per copy
Over 700 pages, illus., available Dec. 2004
Catalog No. 63015

For more information
visit Means Web Site
at www.rsmeans.com

Annual Cost Guides

Means Mechanical Cost Data 2005

• HVAC • Controls

Total unit and systems price guidance for mechanical construction...materials, parts, fittings, and complete labor cost information. Includes prices for piping, heating, air conditioning, ventilation, and all related construction.

Plus new 2005 unit costs for:

- Over 2500 installed HVAC/controls assemblies
- "On Site" Location Factors for over 930 cities and towns in the U.S. and Canada
- Crews, labor and equipment

$115.95 per copy
Over 600 pages, illustrated, available Oct. 2004
Catalog No. 60025

Means Plumbing Cost Data 2005

Comprehensive unit prices and assemblies for plumbing, irrigation systems, commercial and residential fire protection, point-of-use water heaters, and the latest approved materials. This publication and its companion, *Means Mechanical Cost Data*, provide full-range cost estimating coverage for all the mechanical trades.

$115.95 per copy
Over 550 pages, illustrated, available Oct. 2004
Catalog No. 60215

Means Electrical Cost Data 2005

Pricing information for every part of electrical cost planning: More than 15,000 unit and systems costs with design tables; clear specifications and drawings; engineering guides and illustrated estimating procedures; complete labor-hour and materials costs for better scheduling and procurement; the latest electrical products and construction methods.

- A Variety of Special Electrical Systems including Cathodic Protection
- Costs for maintenance, demolition, HVAC/ mechanical, specialties, equipment, and more

$115.95 per copy
Over 450 pages, illustrated, available Oct. 2004
Catalog No. 60035

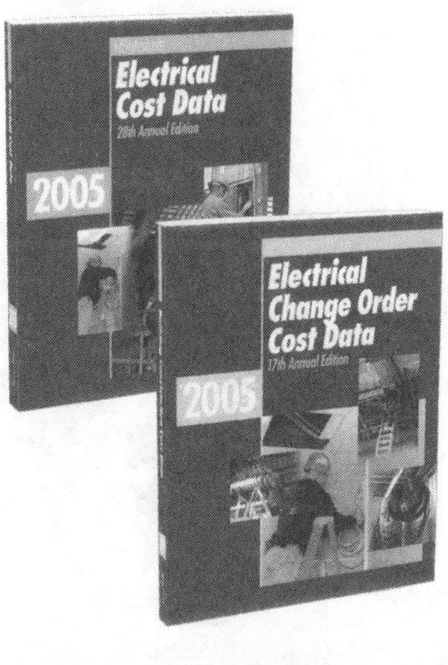

Means Electrical Change Order Cost Data 2005

You are provided with electrical unit prices exclusively for pricing change orders based on the recent, direct experience of contractors and suppliers. Analyze and check your own change order estimates against the experience others have had doing the same work. It also covers productivity analysis and change order cost justifications. With useful information for calculating the effects of change orders and dealing with their administration.

$115.95 per copy
Over 450 pages, available Oct. 2004
Catalog No. 60235

Means Facilities Maintenance & Repair Cost Data 2005

Published in a looseleaf format, *Means Facilities Maintenance & Repair Cost Data* gives you a complete system to manage and plan your facility repair and maintenance costs and budget efficiently. Guidelines for auditing a facility and developing an annual maintenance plan. Budgeting is included, along with reference tables on cost and management and information on frequency and productivity of maintenance operations.

The only nationally recognized source of maintenance and repair costs. Developed in cooperation with the Army Corps of Engineers.

$252.95 per copy
Over 600 pages, illustrated, available Dec. 2004
Catalog No. 60305

Means Square Foot Costs 2005

It's Accurate and Easy To Use!

- **Updated 2005 price information,** based on nationwide figures from suppliers, estimators, labor experts and contractors.
- "How-to-Use" Sections, with **clear examples** of commercial, residential, industrial, and institutional structures.
- Realistic graphics, offering true-to-life illustrations of building projects.
- Extensive information on using square foot cost data, including **sample estimates** and **alternate pricing methods.**

$126.95 per copy
Over 450 pages, illustrated, available Nov. 2004
Catalog No. 60055

Annual Cost Guides

Means Repair & Remodeling Cost Data 2005

Commercial/Residential

You can use this valuable tool to estimate commercial and residential renovation and remodeling.

Includes: New costs for hundreds of unique methods, materials and conditions that only come up in repair and remodeling. PLUS:

- Unit costs for over 16,000 construction components
- Installed costs for over 90 assemblies
- Costs for 300+ construction crews
- Over 930 "On Site" localization factors for the U.S. and Canada.

$99.95 per copy
Over 650 pages, illustrated, available Nov. 2004
Catalog No. 60045

Means Facilities Construction Cost Data 2005

For the maintenance and construction of commercial, industrial, municipal, and institutional properties. Costs are shown for new and remodeling construction and are broken dow into materials, labor, equipment, overhead, and profit. Special emphasis is given to sections on mechanical, electrical, furnishings, site work, building maintenance, finish work, and demolition. More than 45,000 unit costs, plus assemblies and reference sections are included.

$279.95 per copy
Over 1200 pages, illustrated, available Nov. 2004
Catalog No. 60205

Means Residential Cost Data 2005

Contains square foot costs for 30 basic home models with the look of today, plus hundreds of custom additions and modifications you can quote right off the page. With costs for the 100 residential systems you're most likely to use in the year ahead. Complete with blank estimating forms, sample estimates and step-by-step instructions.

$99.95 per copy
Over 600 pages, illustrated, available Oct. 2004
Catalog No. 60175

Means Light Commercial Cost Data 2005

Specifically addresses the light commercial market, which is an increasingly specialized niche in the industry. Aids you, the owner/designer/contractor, in preparing all types of estimates, from budgets to detailed bids. Includes new advances in methods and materials. Assemblies section allows you to evaluate alternatives in the early stages of design/plannin

Over 13,000 unit costs for 2005 ensure you have the prices you need...when you need them.

$99.95 per copy
Over 650 pages, illustrated, available Nov. 2004
Catalog No. 60185

Means Assemblies Cost Data 2005

Means Assemblies Cost Data 2005 takes the guesswork out of preliminary or conceptual estimates. Now you don't have to try to calculate the assembled cost by working up individual components costs. We've done all the work for you.

Presents detailed illustrations, descriptions, specifications and costs for every conceivable building assembly—240 types in all—arranged in the easy-to-use UNIFORMAT II system. Each illustrated "assembled" cost includes a complete grouping of materials and associated installation costs including the installing contractor's overhead and profit.

$189.95 per copy
Over 600 pages, illustrated, available Oct. 2004
Catalog No. 60065

Means Site Work & Landscape Cost Data 2005

Means Site Work & Landscape Cost Data 2005 is organized to assist you in all your estimating needs. Hundreds of fact-filled pages help you mak accurate cost estimates efficiently.

Updated for 2005!

- Demolition features—including ceilings, doors, electrical, flooring, HVAC, millwork, plumbing, roofing, walls and windows
- State-of-the-art segmental retaining walls
- Flywheel trenching costs and details
- Updated Wells section
- Landscape materials, flowers, shrubs and trees

$115.95 per copy
Over 600 pages, illustrated, available Nov. 2004
Catalog No. 60285

For more information
visit Means Web Site
at www.rsmeans.com

Annual Cost Guides

Means Open Shop Building Construction Cost Data 2005

The latest costs for accurate budgeting and estimating of new commercial and residential construction... renovation work... change orders... cost engineering. *Means Open Shop BCCD* will assist you to...

- Develop benchmark prices for change orders
- Plug gaps in preliminary estimates, budgets
- Estimate complex projects
- Substantiate invoices on contracts
- Price ADA-related renovations

$115.95 per copy
Over 700 pages, illustrated, available Dec. 2004
Catalog No. 60155

Means Heavy Construction Cost Data 2005

A comprehensive guide to heavy construction costs. Includes costs for highly specialized projects such as tunnels, dams, highways, airports, and waterways. Information on different labor rates, equipment, and material costs is included. Has unit price costs, systems costs, and numerous reference tables for costs and design. Valuable not only to contractors and civil engineers, but also to government agencies and city/town engineers.

$115.95 per copy
Over 450 pages, illustrated, available Nov. 2004
Catalog No. 60165

Means Building Construction Cost Data 2005

Western Edition

This regional edition provides more precise cost information for western North America. Labor rates are based on union rates from 13 western states and western Canada. Included are western practices and materials not found in our national edition: tilt-up concrete walls, glu-lam structural systems, specialized timber construction, seismic restraints, landscape and irrigation systems.

$115.95 per copy
Over 650 pages, illustrated, available Dec. 2004
Catalog No. 60225

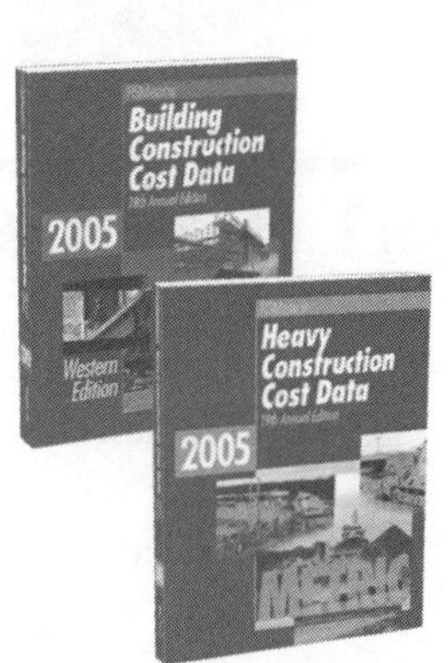

Means Heavy Construction Cost Data 2005

Metric Version

Make sure you have the Means industry standard metric costs for the federal, state, municipal and private marketplace. With thousands of up-to-date metric unit prices in tables by CSI standard divisions. Supplies you with assemblies costs using the metric standard for reliable cost projections in the design stage of your project. Helps you determine sizes, material amounts, and has tips for handling metric estimates.

$115.95 per copy
Over 450 pages, illustrated, available Dec. 2004
Catalog No. 63165

Means Construction Cost Indexes 2005

Who knows what 2005 holds? What materials and labor costs will change unexpectedly? By how much?

- Breakdowns for 316 major cities.
- National averages for 30 key cities.
- Expanded five major city indexes.
- Historical construction cost indexes.

$252.95 per year
$63.00 individual quarters
Catalog No. 60145 A,B,C,D

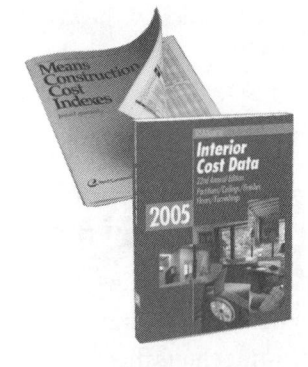

Means Interior Cost Data 2005

Provides you with prices and guidance needed to make accurate interior work estimates. Contains costs on materials, equipment, hardware, custom installations, furnishings, labor costs . . . every cost factor for new and remodel commercial and industrial interior construction, including updated information on office furnishings, plus more than 50 reference tables. For contractors, facility managers, owners.

$115.95 per copy
Over 600 pages, illustrated, available Oct. 2004
Catalog No. 60095

Means Concrete & Masonry Cost Data 2005

Provides you with cost facts for virtually all concrete/masonry estimating needs, from complicated formwork to various sizes and face finishes of brick and block, all in great detail. The comprehensive unit cost section contains more than 8,500 selected entries. Also contains an assemblies cost section, and a detailed reference section which supplements the cost data.

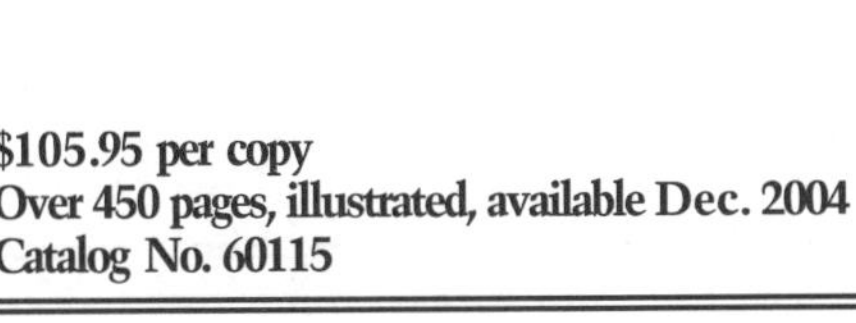

$105.95 per copy
Over 450 pages, illustrated, available Dec. 2004
Catalog No. 60115

Means Labor Rates for the Construction Industry 2005

Complete information for estimating labor costs, making comparisons and negotiating wage rates by trade for over 300 cities (United States and Canada). With 46 construction trades listed by local union number in each city, and historical wage rates included for comparison. No similar book is available through the trade.

Each city chart lists the county and is alphabetically arranged with handy visual flip tabs for quick reference.

$253.95 per copy
Over 300 pages, available Dec. 2004
Catalog No. 60125

For more informatio
visit Means Web Si
at www.rsmeans.co

Building Security: Strategies & Costs

By David Owen

Unauthorized systems access, terrorism, hurricanes and tornadoes, sabotage, vandalism, fire, explosions, and other threats... All are considerations as building owners, facility managers, and design and construction professionals seek ways to meet security needs.

This comprehensive resource will help you evaluate your facility's security needs — and design and budget for the materials and devices needed to fulfill them. The text and cost data will help you to:

- Identify threats, probability of occurrence, and the potential losses— to determine and address your real vulnerabilities.
- Perform a detailed risk assessment of an existing facility, and prioritize and budget for security enhancement.
- Evaluate and price security systems and construction solutions, so you can make cost-effective choices.

Includes over 130 pages of Means Cost Data for installation of security systems and materials, plus a review of more than 50 security devices and construction solutions—how they work, and how they compare.

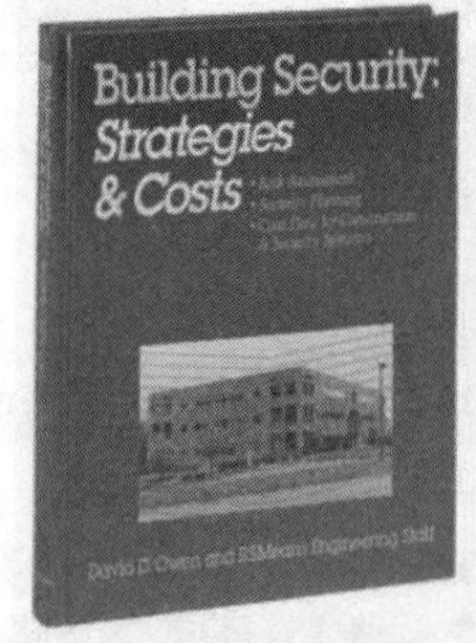

$89.95 per copy
Over 400 pages, Hardcover
Catalog No. 37339

Green Building: Project Planning & Cost Estimating

By RSMeans and Contributing Authors

Written by a team of leading experts in sustainable design, this new book is a complete guide to planning and estimating green building projects, a growing trend in building design and construction – commercial, industrial, institutional and residential. It explains:

- All the different criteria for "green-ness"
- What criteria your building needs to meet to get a LEED, Energy Star, or other recognized rating for green buildings
- How the project team works differently on a green versus a traditional building project
- How to select and specify green products
- How to evaluate the cost and value of green products versus conventional ones — not only for their first (installation) cost, but their cost over time (in maintenance and operation).

Features an extensive Green Building Cost Data section, which details the available products, how they are specified, and how much they cost.

$89.95 per copy
350 pages, illustrated, Hardcover
Catalog No. 67338

Life Cycle Costing for Facilities

By Alphonse Dell'Isola and Dr. Steven Kirk

Guidance for achieving higher quality design and construction projects at lower costs!

Facility designers and owners are frustrated with cost-cutting efforts that yield the cheapest product, but sacrifice quality. Life Cycle Costing, properly done, enables them to achieve both—high quality, incorporating innovative design, and costs that meet their budgets.

The authors show how LCC can work for a broad variety of projects – from several types of buildings, to roads and bridges, to HVAC and electrical upgrades, to materials and equipment procurement. Case studies include:

- Health care and nursing facilities
- College campus and high schools
- Office buildings, courthouses, and banks
- Exterior walls, elevators, lighting, HVAC, and more

The book's extensive cost section provides maintenance and replacement costs for facility elements.

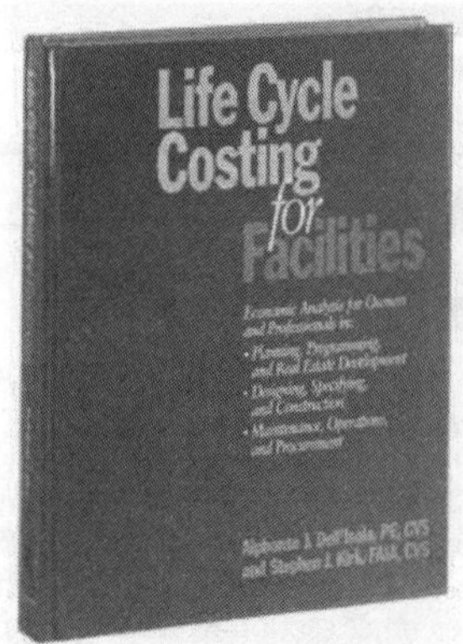

$99.95 per copy
450 pages, Hardcover
Catalog No. 67341

For more information
visit Means Web Site
at www.rsmeans.com

Reference Books

Value Engineering: Practical Applications

. . For Design, Construction, Maintenance & Operations

By Alphonse Dell'Isola, PE

A tool for immediate application—for engineers, architects, facility managers, owners, and contractors. Includes: making the case for VE—the management briefing, integrating VE into planning and budgeting, conducting life cycle costing, integrating VE into the design process, using VE methodology in design review and consultant selection, case studies, VE workbook, and a life cycle costing program on disk.

$79.95 per copy
Over 450 pages, illustrated, Hardcover
Catalog No. 67319

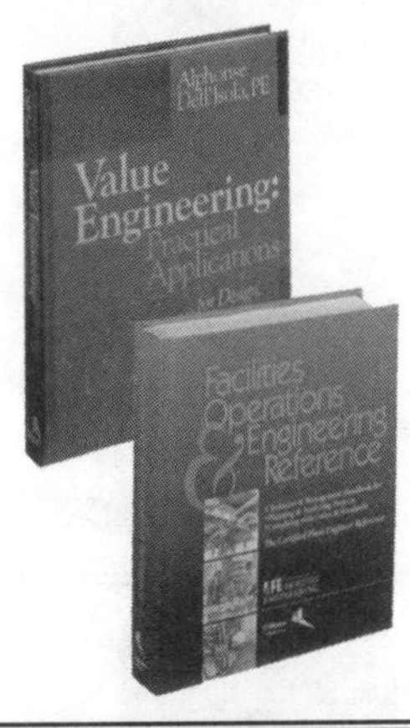

Facilities Operations & Engineering Reference

By the Association for Facilities Engineering and RSMeans

An all-in-one technical referance for planning and managing facility projects and solving day-to-day operations problems. Selected as the official Certified Plant Engineer reference, this handbook covers financial analysis, maintenance, HVAC and energy efficiency, and more.

$109.95 per copy
Over 700 pages, illustrated, Hardcover
Catalog No. 67318

The Building Professional's Guide to Contract Documents

3rd Edition

By Waller S. Poage, AIA, CSI, CVS

This comprehensive treatment of Contract Documents is an important reference for owners, design professionals, contractors, and students.

- Structure your Documents for Maximum Efficiency
- Effectively communicate construction requirements to all concerned
- Understand the Roles and Responsibilities of Construction Professionals
- Improve Methods of Project Delivery

$64.95 per copy, 400 pages
Diagrams and construction forms, Hardcover
Catalog No. 67261A

HVAC: Design Criteria, Options, Selection

Expanded 2nd Edition

By William H. Rowe III, AIA, PE

Includes Indoor Air Quality, CFC Removal, Energy Efficient Systems, and Special Systems by Building Type. Helps you solve a wide range of HVAC system design and selection problems effectively and economically. Gives you clear explanations of the latest ASHRAE standards.

$84.95 per copy
Over 600 pages, illustrated, Hardcover
Catalog No. 67306

Cost Planning & Estimating for Facilities Maintenance

In this unique book, a team of facilities management authorities shares their expertise at:

- Evaluating and budgeting maintenance operations
- Maintaining & repairing key building components
- Applying *Means Facilities Maintenance & Repair Cost Data* to your estimating

Covers special maintenance requirements of the 10 major building types.

$89.95 per copy
Over 475 pages, Hardcover
Catalog No. 67314

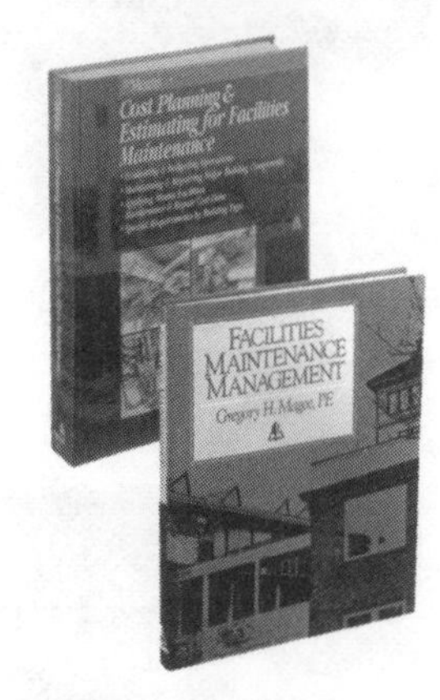

Facilities Maintenance Management

By Gregory H. Magee, PE

Now you can get successful management methods and techniques for all aspects of facilities maintenance. This comprehensive reference explains and demonstrates successful management techniques for all aspects of maintenance, repair, and improvements for buildings, machinery, equipment, and grounds. Plus, guidance for outsourcing and managing internal staffs.

$86.95 per copy
Over 280 pages with illustrations, Hardcover
Catalog No. 67249

Builder's Essentials: Advanced Framing Methods

By Scot Simpson

A highly illustrated, "framer-friendly" approach to advanced framing elements. Provides expert, but easy-to-interpret, instruction for laying out and framing complex walls, roofs, and stairs, and special requirements for earthquake and hurricane protection. Also helps bring framers up to date on the latest building code changes, and provides tips on the lead framer's role and responsibilities, how to prepare for a job, and how to get the crew started.

$24.95 per copy
250 pages, illustrated, Softcover
Catalog No. 67330

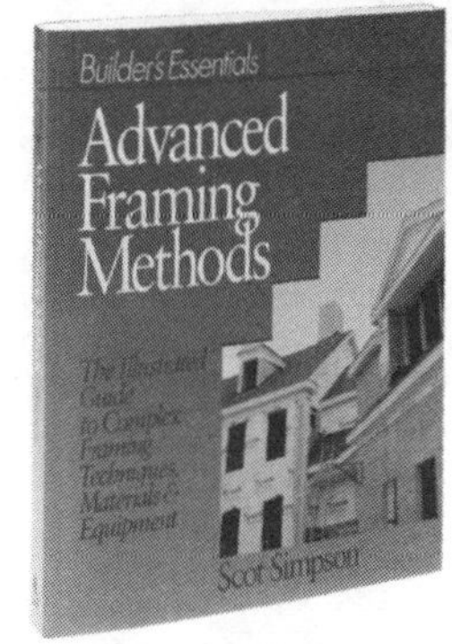

Reference Books

For more informatio
visit Means Web Sit
at www.rsmeans.co

Interior Home Improvement Costs,

New 9th Edition

Estimates for the most popular remodeling and repair projects—from small, do-it-yourself jobs—to major renovations and new construction. Includes: Kitchens & Baths; New Living Space from Your Attic, Basement or Garage; New Floors, Paint & Wallpaper; Tearing Out or Building New Walls; Closets, Stairs & Fireplaces; New Energy-Saving Improvements, Home Theatres, and More!

$24.95 per copy
250 pages, illustrated, Softcover
Catalog No. 67308E

Exterior Home Improvement Costs

New 9th Edition

Estimates for the most popular remodeling and repair projects—from small, do-it-yourself jobs, to major renovatior and new construction. Includes: Curb Appeal Projects—Landscaping, Patios, Porches, Driveways and Walkways; New Windows and Doors; Decks, Greenhouses, and Sunrooms; Roo Additions and Garages; Roofing, Siding, and Painting; "Green Improvements to Save Energy & Water

$24.95 per copy
Over 275 pages, illustrated, softcover
Catalog No. 67309E

Builder's Essentials: Plan Reading & Material Takeoff

By Wayne J. DelPico

For Residential and Light Commercial Construction

A valuable tool for understanding plans and specs, and accurately calculating material quantities.
Step-by-step instructions and takeoff procedures based on a full set of working drawings.

$35.95 per copy
Over 420 pages, Softcover
Catalog No. 67307

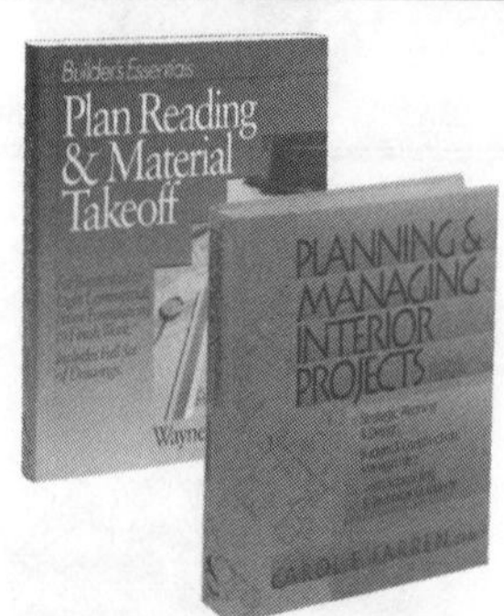

Planning & Managing Interior Projects

2nd Edition

By Carol E. Farren, CFM

Addresses changes in technology and business, guiding you through commercial design and construction from initial clie meetings to post-project administration. Includes: evaluating space requirements, alternative work models, telecommunications and data management, and environment issues.

$69.95 per copy
Over 400 pages, illustrated, Hardcover
Catalog No. 67245A

Builder's Essentials: Best Business Practices for Builders & Remodelers: An Easy-to-Use Checklist System

By Thomas N. Frisby

A comprehensive guide covering all aspects of running a construction business, with more than 40 user–friendly checklists. This book provides expert guidance on: increasing your revenue and keeping more of your profit; planning for long-term growth; keeping good employees and managing subcontractors.

$29.95 per copy
Over 220 pages, Softcover
Catalog No. 67329

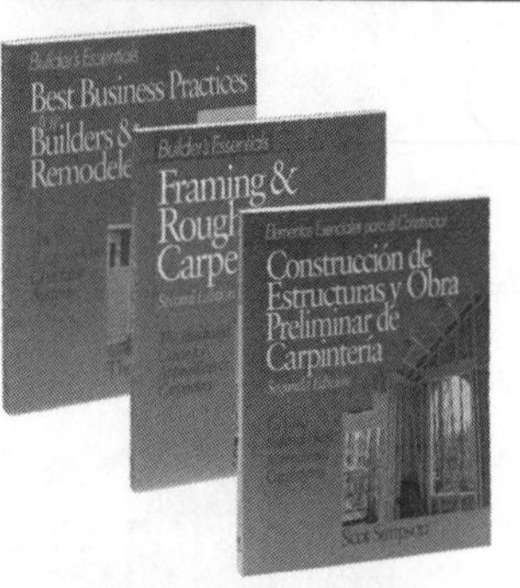

Builder's Essentials: Framing & Rough Carpentry 2nd Edition

By Scot Simpson

A complete training manual for apprentice and experienced carpenters. Develop and improve your skills with "framer-friendly," easy-to-follow instructions, and step-by-step illustrations. Learn proven techniques for framing walls, floors roofs, stairs, doors, and windows. Updated guidance on standards, building codes, safety requirements, and more. Also available in Spanish!

$24.95 per copy
Over 150 pages, Softcover
Catalog No. 67298A   Spanish Catalog No. 67298AS

How to Estimate with Means Data & CostWorks

By RSMeans and Saleh Mubarak

New 2nd Edition!

Learn estimating techniques using Means cost data. Includes an instructional version of Means CostWorks CD-ROM with Sample Building Plans.
The step-by-step guide takes you through all the major construction items. Over 300 sample estimating problems are included.

$59.95 per copy
Over 190 pages, Softcover
Catalog No. 67324A

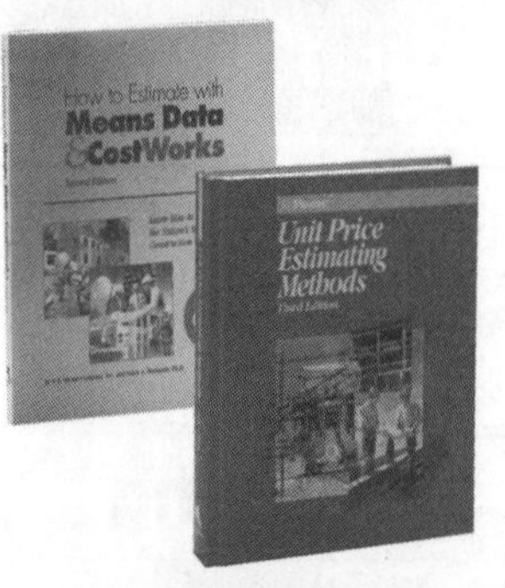

Unit Price Estimating Methods

New 3rd Edition

This new edition includes up-to-date cost data and estimating examples, updated to reflect changes to the CSI numbering system and new features of Means cost data. It describes the most productive, universally accepted ways to estimate, and uses checklists and forms to illustrate shortcuts and timesavers. A model estimate demonstrates procedures. A new chapter explores computer estimating alternatives.

$59.95 per copy
Over 350 pages, illustrated, Hardcover
Catalog No. 67303A

Means Landscape Estimating Methods

4th Edition

By Sylvia H. Fee

This revised edition offers expert guidance for preparing accurate estimates for new landscape construction and grounds maintenance. Includes a complete project estimate featuring the latest equipment and methods, and **two chapters on Life Cycle Costing, and Landscape Maintenance Estimating.**

$62.95 per copy
Over 300 pages, illustrated, Hardcover
Catalog No. 67295B

Means Environmental Remediation Estimating Methods, 2nd Edition

By Richard R. Rast

Guidelines for estimating 50 standard remediation technologies Use it to prepare preliminary budgets, develop estimates, compare costs and solutions, estimate liability, review quotes, negotiate settlements.

A valuable support tool for ***Means Environmental Remediation Unit Price*** and ***Assemblies*** books.

$99.95 per copy
Over 750 pages, illustrated, Hardcover
Catalog No. 64777A

For more information visit Means Web Site at www.rsmeans.com

Reference Books

Means Illustrated Construction Dictionary, Condensed Edition, 2nd Edition

By RSMeans

Recognized in the industry as the best resource of its kind, this has been further enhanced with updates to existing terms and the addition of hundreds of new terms and illustrations . . . in keeping with the most recent developments in the industry.
The best portable dictionary for office or field use—an essential tool for contractors, architects, insurance and real estate personnel, homeowners, and anyone who needs quick, clear definitions for construction terms.

$59.95 per copy
Over 500 pages
Catalog No. 67282A

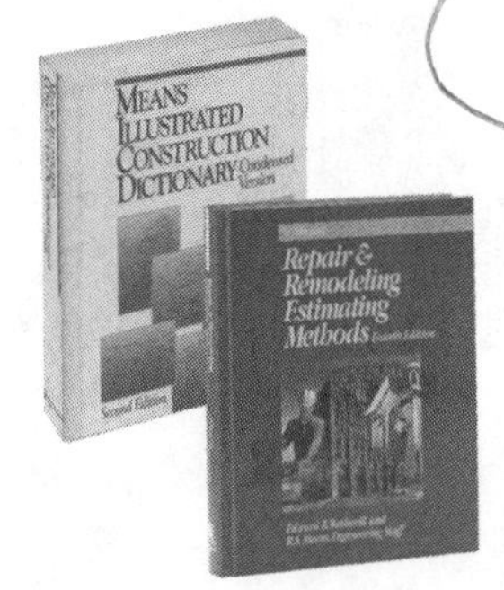

Means Repair and Remodeling Estimating

New 4th Edition
By Edward B. Wetherill & RSMeans

Focuses on the unique problems of estimating renovations of existing structures. It helps you determine the true costs of remodeling through careful evaluation of architectural details and a site visit.
New section on disaster restoration costs.

$69.95 per copy
Over 450 pages, illustrated, Hardcover
Catalog No. 67265B

Facilities Planning & Relocation

New, lower price and user-friendly format.
By David D. Owen

A complete system for planning space needs and managing relocations. Includes step-by-step manual, over 50 forms, and extensive reference section on materials and furnishings.

$89.95 per copy
Over 450 pages, Softcover
Catalog No. 67301

Means Square Foot & Assemblies Estimating Methods

3rd Edition!

Develop realistic Square Foot and Assemblies Costs for budgeting and construction funding. The new edition features updated guidance on square foot and assemblies estimating using UNIFORMAT II. An essential reference for anyone who performs conceptual estimates.

$69.95 per copy
Over 300 pages, illustrated, Hardcover
Catalog No. 67145B

Means Electrical Estimating Methods 3rd Edition

Expanded new edition includes sample estimates and cost information in keeping with the latest version of the CSI MasterFormat and UNIFORMAT II. Complete coverage of Fiber Optic and Uninterruptible Power Supply electrical systems, broken down by components and explained in detail. Includes a new chapter on computerized estimating methods. A practical companion to *Means Electrical Cost Data.*

$64.95 per copy
Over 325 pages, Hardcover
Catalog No. 67230A

Means Mechanical Estimating Methods 3rd Edition

This guide assists you in making a review of plans, specs, and bid packages with suggestions for takeoff procedures, listings, substitutions and pre-bid scheduling. Includes suggestions for budgeting labor and equipment usage. Compares materials and construction methods to allow you to select the best option.

$64.95 per copy
Over 350 pages, illustrated, Hardcover
Catalog No. 67294A

Means Spanish/English Construction Dictionary

By RSMeans, The International Conference of Building Officials (ICBO), and Rolf Jensen & Associates (RJA)

Designed to facilitate communication among Spanish- and English-speaking construction personnel—improving performance and job-site safety. Features the most common words and phrases used in the construction industry, with easy-to-follow pronunciations. Includes extensive building systems and tools illustrations.

$22.95 per copy
250 pages, illustrated, Softcover
Catalog No. 67327

Project Scheduling & Management for Construction

New 3rd Edition
By David R. Pierce, Jr.

A comprehensive yet easy-to-follow guide to construction project scheduling and control—from vital project management principles through the latest scheduling, tracking, and controlling techniques. The author is a leading authority on scheduling with years of field and teaching experience at leading academic institutions. Spend a few hours with this book and come away with a solid understanding of this essential management topic.

$64.95 per copy
Over 300 pages, illustrated, Hardcover
Catalog No. 67247B

Reference Books

For more informatio
visit Means Web Sit
at www.rsmeans.co

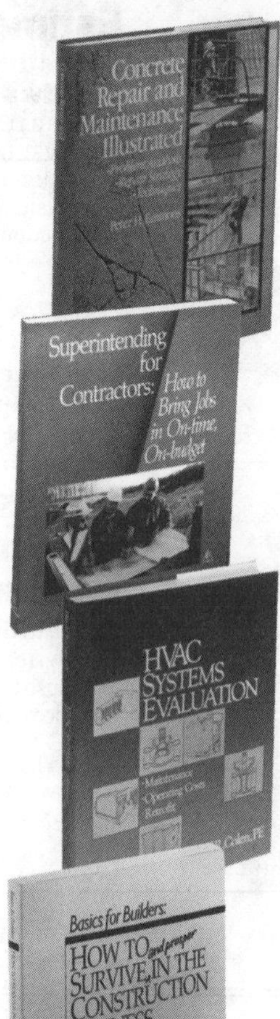

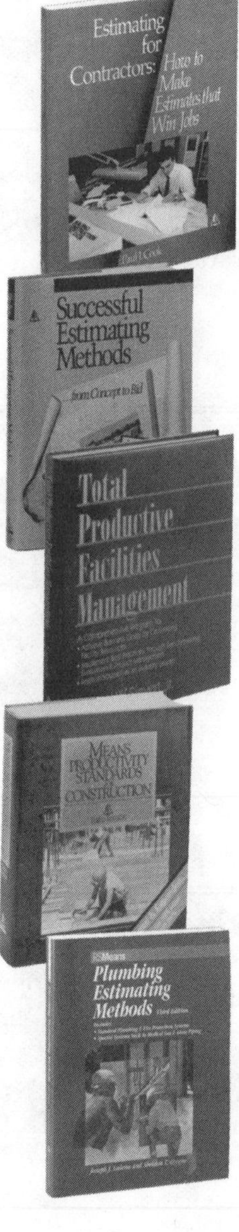

Concrete Repair and Maintenance Illustrated
By Peter H. Emmons
$69.95 per copy
Catalog No. 67146

Superintending for Contractors: How to Bring Jobs in On-Time, On-Budget
By Paul J. Cook
$35.95 per copy
Catalog No. 67233

HVAC Systems Evaluation
By Harold R. Colen, PE
$84.95 per copy
Catalog No. 67281

Basics for Builders: How to Survive and Prosper in Construction
By Thomas N. Frisby
$34.95 per copy
Catalog No. 67273

Successful Interior Projects Through Effective Contract Documents
By Joel Downey & Patricia K. Gilbert
Now $24.98 per copy, limited quantity
Catalog No. 67313

Building Spec Homes Profitably
By Kenneth V. Johnson
$29.95 per copy
Catalog No. 67312

Estimating for Contractors
How to Make Estimates that Win Jobs
By Paul J. Cook
$35.95 per copy
Catalog No. 67160

Successful Estimating Methods:
From Concept to Bid
By John D. Bledsoe, PhD, PE
$32.48 per copy
Catalog No. 67287

Total Productive Facilities Management
By Richard W. Sievert, Jr.
$29.98 per copy
Over 270 pages, illustrated, Hardcover
Catalog No. 67321

Means Productivity Standards for Construction
Expanded Edition *(Formerly Man-Hour Standards)*
$49.98 per copy
Over 800 pages, Hardcover
Catalog No. 67236A

Means Plumbing Estimating Methods, 3rd Edition
By Joseph Galeno and Sheldon Greene
Now $29.98 per copy
Catalog No. 67283B

Reference Books

Preventive Maintenance Guidelines for School Facilities

By John C. Maciha

A complete PM program for K-12 schools that ensures sustained security, safety, property integrity, user satisfaction, and reasonable ongoing expenditures.

Includes schedules for weekly, monthly, semiannual, and annual maintenance with hard copy and electronic forms.

$149.95 per copy
Over 225 pages, Hardcover
Catalog No. 67326

Preventive Maintenance for Higher Education Facilities

By Applied Management Engineering, Inc.

An easy-to-use system to help facilities professionals establish the value of PM, and to develop and budget for an appropriate PM program for their college or university. Features interactive campus building models typical of those found in different-sized higher education facilities, and PM checklists linked to each piece of equipment or system in hard copy and electronic format.

$149.95 per copy
150 pages, Hardcover
Catalog No. 67337

Historic Preservation: Project Planning & Estimating

By Swanke Hayden Connell Architects

Expert guidance on managing historic restoration, rehabilitation, and preservation building projects and determining and controlling their costs. Includes:

- How to determine whether a structure qualifies as historic
- Where to obtain funding and other assistance
- How to evaluate and repair more than 75 historic building materials

$99.95 per copy
Over 675 pages, Hardcover
Catalog No. 67323

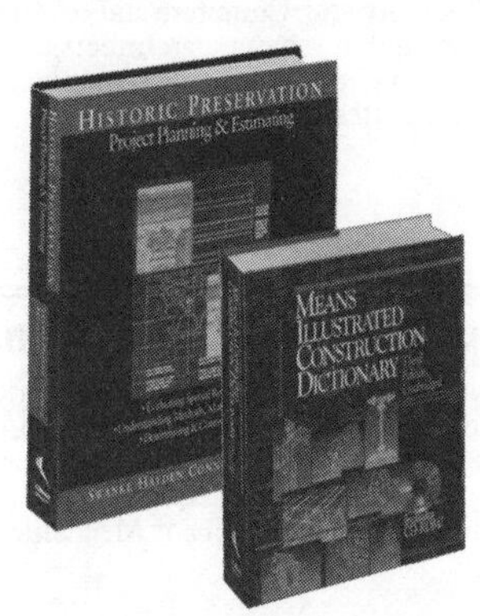

Means Illustrated Construction Dictionary, 3rd Edition

Long regarded as the Industry's finest, the Means *Illustrated Construction Dictionary* is now even better.

With the addition of over 1,000 new terms and hundreds of new illustrations, it is the clear choice for the most comprehensive and current information. **The companion CD-ROM that comes with this new edition adds many extra features: larger graphics, expanded definitions, and links to both CSI MasterFormat numbers and product information.**

$99.95 per copy
Over 790 pages, Illustrated, Hardcover
Catalog No. 67292A

Designing & Building with the IBC, 2nd Edition

By Rolf Jensen & Associates, Inc.

This updated comprehensive guide helps building professionals make the transition to the 2003 *International Building Code®*. Includes a side-by-side code comparison of the IBC 2003 to the IBC 2000 and the three primary model codes, a quick-find index, and professional code commentary. With illustrations, abbreviations key, and an extensive Resource section.

$99.95 per copy
Over 875 pages
Catalog No. 67328A

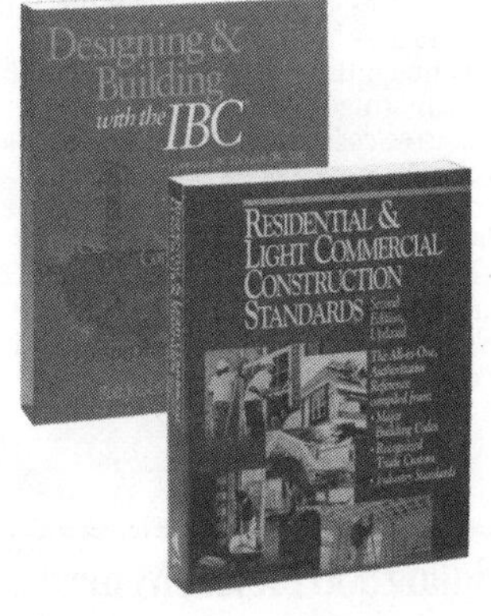

Residential & Light Commercial Construction Standards, 2nd Edition

By RSMeans and Contributing Authors

New, updated second edition of this unique collection of industry standards that define quality construction. For contractors, subcontractors, owners, developers, architects, engineers, attorneys, and insurance personnel, this book provides authoritative requirements and recommendations compiled from the nation's leading professional associations, industry publications, and building code organizations.

$59.95 per copy
600 pages, illustrated, Softcover
Catalog No. 67322A

Means Estimating Handbook, 2nd Edition

By RSMeans

Updated new Second Edition answers virtually any estimating technical question - all organized by CSI MasterFormat. This comprehensive reference covers the full spectrum of technical data required to estimate construction costs. The book includes information on sizing, productivity, equipment requirements, code-mandated specifications, design standards and engineering factors.

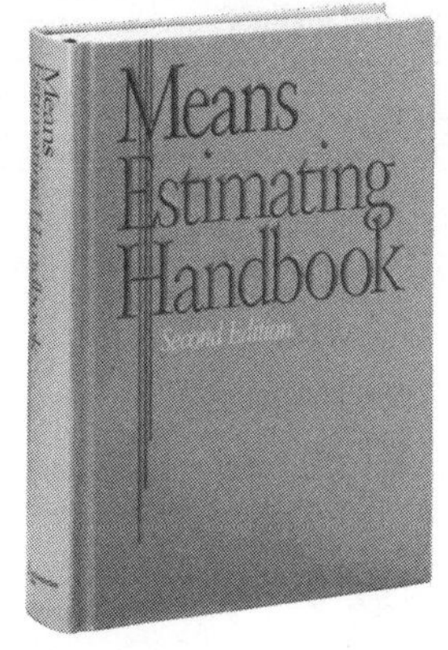

$99.95 per copy
Over 900 pages, Hardcover
Catalog No. 67276A

Seminars

Means CostWorks Training

This one-day seminar course has been designed with the intention of assisting both new and existing users to become more familiar with CostWorks program. The class is broken into two unique sections: (1) A one-half day presentation on the function of each icon; and each student will be shown how to use the software to develop a cost estimate. (2) Hands-on estimating exercises that will ensure that each student thoroughly understands how to use CostWorks. You must bring your own laptop computer to this course.

CostWorks Benefits/Features:
- Estimate in your own spreadsheet format
- Power of Means National Database
- Database automatically regionalized
- Save time with keyword searches
- Save time by establishing common estimate items in "Bookmark" files
- Customize your spreadsheet template
- Hot key to Product Manufacturers' listings and specs
- Merge capability for networking environments
- View crews and assembly components
- AutoSave capability
- Enhanced sorting capability

Unit Price Estimating

This interactive two-day seminar teaches attendees how to interpret project information and process it into final, detailed estimates with the greatest accuracy level.

The single most important credential an estimator can take to the job is the ability to visualize construction in the mind's eye, and thereby estimate accurately.

Some Of What You'll Learn:
- Interpreting the design in terms of cost
- The most detailed, time tested methodology for accurate "pricing"
- Key cost drivers—material, labor, equipment, staging and subcontracts
- Understanding direct and indirect costs for accurate job cost accounting and change order management

Who Should Attend: Corporate and government estimators and purchasers, architects, engineers...and others needing to produce accurate project estimates.

Square Foot and Assemblies Cost Estimating

This two-day course teaches attendees how to quickly deliver accurate square foot estimates using limited budget and design information.

Some Of What You'll Learn:
- How square foot costing gets the estimate done faster
- Taking advantage of a "systems" or "assemblies" format
- The Means "building assemblies/square foot cost approach"
- How to create a very reliable preliminary and systems estimate using bare-bones design information

Who Should Attend: Facilities managers, facilities engineers, estimators, planners, developers, construction finance professionals...and others needing to make quick, accurate construction cost estimates at commercial, government, educational and medical facilities.

Repair and Remodeling Estimating

This two-day seminar emphasizes all the underlying considerations unique to repair/remodeling estimating and presents the correct methods for generating accurate, reliable R&R project costs using the unit price and assemblies methods.

Some Of What You'll Learn:
- Estimating considerations—like labor-hours, building code compliance, working within existing structures, purchasing materials in smaller quantities, unforeseen deficiencies
- Identifies problems and provides solutions to estimating building alterations
- Rules for factoring in minimum labor costs, accurate productivity estimates and allowances for project contingencies
- R&R estimating examples are calculated using unite prices and assemblies data

Who Should Attend: Facilities managers, plant engineers, architects, contractors, estimators, builders...and others who are concerned with the proper preparation and/or evaluation of repair and remodeling estimates.

Mechanical and Electrical Estimating

This two-day course teaches attendees how to prepare more accurate and complete mechanical/electrical estimates, avoiding the pitfalls of omission and double-counting, while understanding the composition and rationale within the Means Mechanical/Electrical database.

Some Of What You'll Learn:
- The unique way mechanical and electrical systems are interrelated
- M&E estimates, conceptual, planning, budgeting and bidding stages
- Order of magnitude, square foot, assemblies and unit price estimating
- Comparative cost analysis of equipment and design alternatives

Who Should Attend: Architects, engineers, facilities managers, mechanical and electrical contractors...and others needing a highly reliable method for developing, understanding and evaluating mechanical and electrical contracts.

Plan Reading and Material Takeoff

This two-day program teaches attendees to read and understand construction documents and to use them in the preparation of material takeoffs.

Some of What You'll Learn:
- Skills necessary to read and understand typical contract documents—blueprints and specifications
- Details and symbols used by architects and engineers
- Construction specifications' importance in conjunction with blueprints
- Accurate takeoff of construction materials and industry-accepted takeoff methods

Who Should Attend: Facilities managers, construction supervisors, office managers...and other responsible for the execution and administration of a construction project including government, medical, commercial, educational or retail facilities.

Facilities Maintenance and Repair Estimating

This two-day course teaches attendees how to plan, budget, and estimate the cost of ongoing and preventive maintenance and repair for existing buildings and grounds.

Some Of What You'll Learn:
- The most financially favorable maintenance, repair and replacement scheduling and estimating
- Auditing and value engineering facilities
- Preventive planning and facilities upgrading
- Determining both in-house and contract-out service costs; annual, asset-protecting M&R plan

Who Should Attend: Facility managers, maintenance supervisors, buildings and grounds superintendents, plant managers, planners, estimators...and others involved in facilities planning and budgeting.

Scheduling and Project Management

This two-day course teaches attendees the most current and proven scheduling and management techniques needed to bring projects in on time and on budget.

Some Of What You'll Learn:
- Crucial phases of planning and scheduling
- How to establish project priorities, develop realistic schedules and management techniques
- Critical Path and Precedence Methods
- Special emphasis on cost control

Who Should Attend: Construction project managers, supervisors, engineers, estimators, contractors...and others who want to improve their project planning, scheduling and management skills.

Advanced Project Management

This two-day seminar will teach you how to effectively manage and control the entire design-build process and allow you to take home tangible skills that will be immediately applicable on existing projects.

Some Of What You'll Learn:
- Value engineering, bonding, fast-tracking and bid package creation
- How estimates and schedules can be integrated to provide advanced project management tools
- Cost engineering, quality control, productivity measurement and improvement
- Front loading a project and predicting its cash flow

Who Should Attend: Owners, project managers, architectural and engineering managers, construction managers, contractors...and anyone else who is responsible for the timely design and completion of construction projects.

Seminars

For more information visit Means Web Site at www.rsmeans.com

2005 Means Seminar Schedule

Location	Dates
Las Vegas, NV	March
Washington, DC	April
Phoenix, AZ	April
Denver, CO	May
San Francisco, CA	June
Philadelphia, PA	June
Washington, DC	September
Dallas, TX	September
Las Vegas, NV	October
Orlando, FL	November
Atlantic City, NJ	November
San Diego, CA	December

Note: Call for exact dates and details.

Registration Information

Register Early... Save up to $100! Register 30 days before the start date of a seminar and save $100 off your total fee. *Note: This discount can be applied only once per order. It cannot be applied to team discount registrations or any other special offer.*

How to Register Register by phone today! Means toll-free number for making reservations is: **1-800-334-3509.**

Individual Seminar Registration Fee $895. Individual CostWorks Training Registration Fee $349. To register by mail, complete the registration form and return with your full fee to: Seminar Division, Reed Construction Data, RSMeans Seminars, 63 Smiths Lane, Kingston, MA 02364.

Federal Government Pricing All Federal Government employees save 25% off regular seminar price. Other promotional discounts cannot be combined with Federal Government discount.

Team Discount Program Two to four seminar registrations: Call for pricing.

Multiple Course Discounts When signing up for two or more courses, call for pricing.

Refund Policy Cancellations will be accepted up to ten days prior to the seminar start. There are no refunds for cancellations received later than ten working days prior to the first day of the seminar. A $150 processing fee will be applied for all cancellations. Written notice of cancellation is required . Substitutions can be made at anytime before the session starts. **No-shows are subject to the full seminar fee.**

AACE Approved Courses The RSMeans Construction Estimating and Management Seminars described and offered to you here have each been approved for 14 hours (1.4 recertification credits) of credit by the AACE International Certification Board toward meeting the continuing education requirements for re-certification as a Certified Cost Engineer/Certified Cost Consultant.

AIA Continuing Education We are registered with the AIA Continuing Education System (AIA/CES) and are committed to developing quality learning activities in accordance with the CES criteria. RSMeans seminars meet the AIA/CES criteria for Quality Level 2. AIA members will receive (14) learning units (LUs) for each two-day RSMeans Course.

NASBA CPE Sponsor Credits We are part of the National Registry of CPE Sponsors. Attendees may be eligable for (16) CPE credits.

Daily Course Schedule The first day of each seminar session begins at 8:30 A.M. and ends at 4:30 P.M. The second day is 8:00 A.M.–4:00 P.M. Participants are urged to bring a hand-held calculator since many actual problems will be worked out in each session.

Continental Breakfast Your registration includes the cost of a continental breakfast, a morning coffee break, and an afternoon break. These informal segments will allow you to discuss topics of mutual interest with other members of the seminar. (You are free to make your own lunch and dinner arrangements.)

Hotel/Transportation Arrangements RSMeans has arranged to hold a block of rooms at each hotel hosting a seminar. To take advantage of special group rates when making your reservation, be sure to mention that you are attending the Means Seminar. You are, of course, free to stay at the lodging place of your choice. **(Hotel reservations and transportation arrangements should be made directly by seminar attendees.)**

Important Class sizes are limited, so please register as soon as possible.

Note: *Pricing subject to change.*

Registration Form

Call 1-800-334-3509 to register or FAX 1-800-632-6732. Visit our Web site www.rsmeans.com

Please register the following people for the Means Construction Seminars as shown here. Full payment or deposit is enclosed, and we understand that we must make our own hotel reservations if overnight stays are necessary.

☐ Full payment of $ ______________________ enclosed.

☐ Bill me

Name of Registrant(s)
(To appear on certificate of completion)

____

____

____

P.O. #: ______________________________

GOVERNMENT AGENCIES MUST SUPPLY PURCHASE ORDER NUMBER

Firm Name ______________________________

Address ______________________________

City/State/Zip ______________________________

Telephone No. fax No. ______________________________

E-mail Address ______________________________

Charge our registration(s) to: ☐ MasterCard ☐ VISA ☐ American Express ☐ Discover

Account No. ______________________ Exp. Date __________

Cardholder's Signature ______________________________

Seminar Name City Dates

____

____

Please mail check to: Seminar Division, Reed Construction Data, RSMeans Seminars, 63 Smiths Lane, P.O.Box 800, Kingston, MA 02364 USA

MeansData™

CONSTRUCTION COSTS FOR SOFTWARE APPLICATIONS

Your construction estimating software is only as good as your cost data.

A proven construction cost database is a mandatory part of any estimating package. The following list of softwa providers can offer you MeansData™ as an added feature for their estimating systems. See the table below for w types of products and services they offer (match their numbers). Visit online at **www.rsmeans.com/demosource/** for more information and free demos. Or call their numbers listed below.

1. **3D International**
   713-871-7000
   venegas@3di.com
2. **4Clicks-Solutions, LLC**
   719-574-7721
   mbrown@4clicks-solutions.com
3. **Aepco, Inc.**
   301-670-4642
   blueworks@aepco.com
4. **Applied Flow Technology**
   800-589-4943
   info@aft.com
5. **ArenaSoft Estimating**
   888-370-8806
   info@arenasoft.com
6. **ARES Corporation**
   925-299-6700
   sales@arescorporation.com
7. **BSD - Building Systems Design, Inc.**
   888-273-7638
   bsd@bsdsoftlink.com
8. **CMS - Computerized Micro Solutions**
   800-255-7407
   cms@proest.com
9. **Corecon Technologies, Inc.**
   714-895-7222
   sales@corecon.com
10. **CorVet Systems**
    301-622-9069
    sales@corvetsys.com
11. **Discover Software**
    727-559-0161
    sales@discoversoftware.com
12. **Estimating Systems, Inc.**
    800-967-8572
    esipulsar@adelphia.net
13. **MAESTRO Estimator Schwaab Technology Solutions, Inc.**
    281-578-3039
    Stefan@schwaabtech.com
14. **Magellan K-12**
    936-447-1744
    sam.wilson@magellan-K12.com
15. **MC2 - Management Computer**
    800-225-5622
    vkeys@mc2-ice.com
16. **Maximus Asset Solutions**
    800-659-9001
    assetsolutions@maximus.com
17. **Prime Time**
    610-964-8200
    dick@primetime.com
18. **Prism Computer Corp.**
    800-774-7622
    famis@prismcc.com
19. **Quest Solutions, Inc.**
    800-452-2342
    info@questsolutions.com
20. **RIB Software (Americas), Inc.**
    800-945-7093
    san@rib-software.com
21. **Shaw Beneco Enterprises, Inc.**
    877-719-4748
    inquire@beneco.com
22. **Timberline Software Corp.**
    800-628-6583
    product.info@timberline.com
23. **TMA SYSTEMS, Inc.**
    800-862-1130
    sales@tmasys.com
24. **US Cost, Inc.**
    800-372-4003
    sales@uscost.com
25. **Vanderweil Facility Advisors**
    617-451-5100
    info@VFA.com
26. **Vertigraph, Inc.**
    800-989-4243
    info-request@vertigraph.com
27. **WinEstimator, Inc.**
    800-950-2374
    sales@winest.com

TYPE	1	2	3	4	5	6	7	8	9	10	11	12	13	14	15	16	17	18	19	20	21	22	23	24	25	26	27
BID					●		●	●		●	●				●		●		●	●		●				●	●
Estimating		●			●	●	●	●	●	●	●	●	●		●		●		●	●	●	●		●		●	●
DOC/JOC/SABER		●			●		●			●		●				●					●	●					●
IDIQ		●								●		●				●					●						●
Asset Mgmt.											●			●		●		●							●		●
Facility Mgmt.	●		●								●		●	●		●	●	●		●			●		●		
Project Mgmt.	●	●				●			●	●	●		●	●			●			●	●	●					
TAKE-OFF					●			●	●	●	●				●		●		●	●		●		●		●	●
EARTHWORK									●		●				●		●		●							●	
Pipe Flow				●							●						●										
HVAC/Plumbing					●			●		●	●						●										
Roofing					●			●		●	●						●										
Design	●				●						●						●				●	●					●
Other Offers/Links:																											
Accounting/HR		●			●						●									●	●	●					
Scheduling					●						●		●							●		●		●		●	●
CAD											●			●								●					●
PDA																						●		●			●
Lt. Versions		●						●			●						●					●		●			●
Consulting	●	●			●			●		●	●		●			●	●	●		●		●		●	●		●
Training		●			●		●	●	●	●	●	●	●			●	●	●	●		●	●	●	●	●	●	●

Reseller applications now being accepted. Call Carol Polio Ext. 5107.

FOR MORE INFORMATION

CALL 1-800-448-8182, EXT. 5107 OR FAX 1-800-632-6732

For more information
visit Means Web Site
at www.rsmeans.com

New Titles

Builder's Essentials: Estimating Building Costs for the Residential & Light Commercial Contractor

By Wayne J. DelPico

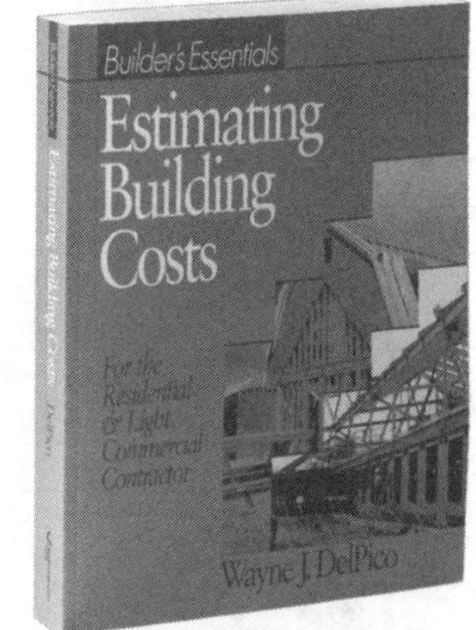

Step-by-step estimating methods for residential and light commercial contractors. Includes a detailed look at every construction specialty—explaining all the components, takeoff units, and labor needed for well-organized, complete estimates covers:

- Correctly interpreting plans and specifications.
- Developing accurate and complete labor and material costs.
- Understanding direct and indirect overhead costs... and accounting for time-sensitive costs.
- Using historical cost data to generate new project budgets.

Plus hard-to-find, professional guidance on what to consider so you can allocate the right amount for profit and contingencies.

$29.95 per copy
Over 400 pages, illustrated, softcover
Catalog No. 67343

Building & Renovating Schools

This all-inclusive guide covers every step of the school construction process—from initial planning, needs assessment, and design, right through moving into the new facility. A must-have resource for anyone concerned with new school construction or renovation, including architects and engineers, contractors and project managers, facility managers, school administrators and school board members, building committees, community leaders, and anyone else who wants to ensure that the project meets the schools' needs in a cost-effective, timely manner. With square foot cost models for elementary, middle, and high school facilities and real-life case studies of recently completed school projects.

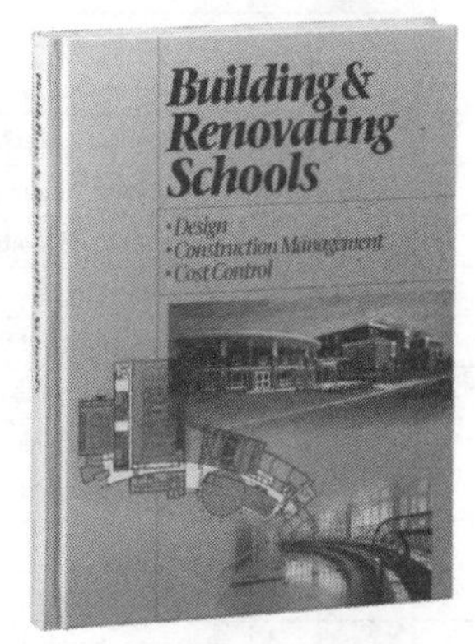

The contributors to this book – architects, construction project managers, contractors, and estimators who specialize in school construction – provide start-to-finish, expert guidance on the process.

$99.95 per copy
Over 425 pages
Catalog No. 67342

Means ADA Compliance Pricing Guide, New Second Edition

By Adaptive Environments and RSMeans

Completely updated and revised to the new 2004 Americans with Disabilities Act Accessible Guidelines, this book features more than 70 of the most commonly needed modifications for ADA compliance—their design requirements, suggestions, and final cost. Projects range from installing ramps and walkways, widening doorways and entryways, and installing and refitting elevators, to relocating light switches and signage, and remodeling bathrooms and kitchens. Also provided are:

- Detailed cost estimates for budgeting modification projects, including estimates for each of 260 alternates.
- An assembly estimate for every project, with detailed cost breakdown including materials, labor hours, and contractor's overhead.
- 3,000 Additional ADA compliance-related unit cost line items.
- Costs that are easily adjusted to over 900 cities and towns.

Over 350 pages
$79.99 per copy
Catalog No. 67310A

ORDER TOLL FREE 1-800-334-3509
OR FAX 1-800-632-6732.

2005 Order Form

Qty.	Book No.	COST ESTIMATING BOOKS	Unit Price	Total
	60065	Assemblies Cost Data 2005	$189.95	
	60015	Building Construction Cost Data 2005	115.95	
	61015	Building Const. Cost Data-Looseleaf Ed. 2005	144.95	
	63015	Building Const. Cost Data-Metric Version 2005	115.95	
	60225	Building Const. Cost Data-Western Ed. 2005	115.95	
	60115	Concrete & Masonry Cost Data 2005	105.95	
	60145	Construction Cost Indexes 2005	251.95	
	60145A	Construction Cost Index-January 2005	63.00	
	60145B	Construction Cost Index-April 2005	63.00	
	60145C	Construction Cost Index-July 2005	63.00	
	60145D	Construction Cost Index-October 2005	63.00	
	60345	Contr. Pricing Guide: Resid. R & R Costs 2005	39.95	
	60335	Contr. Pricing Guide: Resid. Detailed 2005	39.95	
	60325	Contr. Pricing Guide: Resid. Sq. Ft. 2005	39.95	
	64025	ECHOS Assemblies Cost Book 2005	189.95	
	64015	ECHOS Unit Cost Book 2005	126.95	
	54005	ECHOS (Combo set of both books)	251.95	
	60235	Electrical Change Order Cost Data 2005	115.95	
	60035	Electrical Cost Data 2005	115.95	
	60205	Facilities Construction Cost Data 2005	279.95	
	60305	Facilities Maintenance & Repair Cost Data 2005	252.95	
	60165	Heavy Construction Cost Data 2005	115.95	
	63165	Heavy Const. Cost Data-Metric Version 2005	115.95	
	60095	Interior Cost Data 2005	115.95	
	60125	Labor Rates for the Const. Industry 2005	253.95	
	60185	Light Commercial Cost Data 2005	99.95	
	60025	Mechanical Cost Data 2005	115.95	
	60155	Open Shop Building Const. Cost Data 2005	115.95	
	60215	Plumbing Cost Data 2005	115.95	
	60045	Repair and Remodeling Cost Data 2005	99.95	
	60175	Residential Cost Data 2005	99.95	
	60285	Site Work & Landscape Cost Data 2005	115.95	
	60055	Square Foot Costs 2005	126.95	
		REFERENCE BOOKS		
	67147A	ADA in Practice	59.98	
	67310A	ADA Compliance Pricing Guide, 2nd Ed.	79.95	
	67273	Basics for Builders: How to Survive and Prosper	34.95	
	67330	Bldrs Essentials: Adv. Framing Methods	24.95	
	67329	Bldrs Essentials: Best Bus. Practices for Bldrs	29.95	
	67298A	Bldrs Essentials: Framing/Carpentry 2nd Ed.	24.95	
	67298AS	Bldrs Essentials: Framing/Carpentry Spanish	24.95	
	67307	Bldrs Essentials: Plan Reading & Takeoff	35.95	
	67261A	Bldg. Prof. Guide to Contract Documents 3rd Ed.	64.95	
	67342	Building & Renovating Schools	99.95	
	67339	Building Security: Strategies & Costs	89.95	
	67312	Building Spec Homes Profitably	29.95	
	67146	Concrete Repair & Maintenance Illustrated	69.95	
	67314	Cost Planning & Est. for Facil. Maint.	89.95	
	67317A	Cyberplaces: The Internet Guide 2nd Ed.	59.95	
	67328A	Designing & Building with the IBC, 2nd Ed.	99.95	
	67230B	Electrical Estimating Methods 3rd Ed.	64.95	

Qty.	Book No.	REFERENCE BOOKS (Cont.)	Unit Price	Total
	64777A	Environmental Remediation Est. Methods 2nd Ed.	$ 99.95	
	67160	Estimating for Contractors	35.95	
	67276A	Estimating Handbook 2nd Ed.	99.95	
	67249	Facilities Maintenance Management	86.95	
	67246	Facilities Maintenance Standards	79.95	
	67318	Facilities Operations & Engineering Reference	109.95	
	67301	Facilities Planning & Relocation	89.95	
	67231	Forms for Building Const. Professional	47.48	
	67260	Fundamentals of the Construction Process	34.98	
	67323	Historic Preservation: Proj. Planning & Est.	99.95	
	67308E	Home Improvement Costs-Int. Projects 9th Ed.	29.95	
	67309E	Home Improvement Costs-Ext. Projects 9th Ed.	29.95	
	67324A	How to Est. w/Means Data & CostWorks 2nd Ed.	59.95	
	67304	How to Estimate with Metric Units	9.98	
	67306	HVAC: Design Criteria, Options, Select. 2nd Ed.	84.95	
	67281	HVAC Systems Evaluation	84.95	
	67282A	Illustrated Construction Dictionary, Condensed	59.95	
	67292A	Illustrated Construction Dictionary, w/CD-ROM	99.95	
	67295B	Landscape Estimating 4th Ed.	62.95	
	67341	Life Cycle Costing	99.95	
	67302	Managing Construction Purchasing	19.98	
	67294A	Mechanical Estimating 3rd Ed.	64.95	
	67245A	Planning and Managing Interior Projects 2nd Ed.	69.95	
	67283A	Plumbing Estimating Methods 2nd Ed.	59.95	
	67326	Preventive Maint. Guidelines for School Facil.	149.95	
	67337	Preventive Maint. for Higher Education Facilities	149.95	
	67236A	Productivity Standards for Constr.-3rd Ed.	49.98	
	67247B	Project Scheduling & Management for Constr. 3rd Ed.	64.95	
	67265B	Repair & Remodeling Estimating 4th Ed.	69.95	
	67322A	Resi. & Light Commercial Const. Stds. 2nd Ed.	59.95	
	67327	Spanish/English Construction Dictionary	22.95	
	67145A	Sq. Ft. & Assem. Estimating Methods 3rd Ed.	69.95	
	67287	Successful Estimating Methods	32.48	
	67313	Successful Interior Projects	24.98	
	67233	Superintending for Contractors	35.95	
	67321	Total Productive Facilities Management	29.98	
	67284	Understanding Building Automation Systems	29.98	
	67303A	Unit Price Estimating Methods 3rd Ed.	59.95	
	67319	Value Engineering: Practical Applications	79.95	

MA residents add 5% state sales tax	
Shipping & Handling**	
Total (U.S. Funds)*	

Prices are subject to change and are for U.S. delivery only. *Canadian customers may call for current prices. **Shipping & handling charges: Add 7% of total order for check and credit card payments. Add 9% of total order for invoiced orders.

Send Order To: **ADDV-1001**

Name (Please Print) ______________________

Company ______________________

☐ **Company**
☐ **Home** Address ______________________

____

City/State/Zip ______________________

Phone # ______________ P.O. # ______________

(Must accompany all orders being billed)

Mail To: **RSMeans** P.O. Box 800, Kingston, MA 02364-0800